Handbook of Monte Carlo Methods

Published by John Wiley & Sons, Inc., Hoboken, New Jersey
Published simultaneously in Canada

Library of Congress Cataloging-in-Publication Data:

Kroese, Dirk P.
Handbook for Monte Carlo methods / Dirk P. Kroese, Thomas Taimre, Zdravko I. Botev.
p. cm. — (Wiley series in probability and statistics ; 706)
Includes index.
ISBN 978-0-470-17793-8 (hardback)
1. Monte Carlo method. I. Taimre, Thomas, 1983– II. Botev, Zdravko I., 1982– III. Title.
QA298.K76 2011
518'.282—dc22 2010042348

10 9 8 7 6 5 4 3 2 1

To Lesley

— *DPK*

To Aita and Ilmar

— *TT*

To my parents, Maya and Ivan

— *ZIB*

CONTENTS

PREFACE

Many numerical problems in science, engineering, finance, and statistics are solved nowadays through **Monte Carlo methods**; that is, through random experiments on a computer. As the popularity of these methods continues to grow, and new methods are developed in rapid succession, the staggering number of related techniques, ideas, concepts, and algorithms makes it difficult to maintain an overall picture of the Monte Carlo approach. In addition, the study of Monte Carlo techniques requires detailed knowledge in a wide range of fields; for example, *probability* to describe the random experiments and processes, *statistics* to analyze the data, *computational science* to efficiently implement the algorithms, and *mathematical programming* to formulate and solve optimization problems. This knowledge may not always be readily available to the Monte Carlo practitioner or researcher.

The purpose of this Handbook is to provide an accessible and comprehensive compendium of Monte Carlo techniques and related topics. It contains a mix of theory (summarized), algorithms (pseudo + actual), and applications. The book is intended to be an essential guide to Monte Carlo methods, to be used by both advanced undergraduates and graduates/researchers to quickly look up ideas, procedures, formulas, pictures, etc., rather than purely a research monograph or a textbook.

As Monte Carlo methods can be used in many ways and for many different purposes, the Handbook is organized as a collection of independent chapters, each focusing on a separate topic, rather than following a mathematical development. The theory is cross-referenced with other parts of the book where a related topic is discussed — the symbol ☞ in the margin points to the corresponding page number. The theory is illustrated with worked examples and MATLAB code, so that it is easy to implement in practice. The code in this book can also be downloaded from the Handbook's website: `www.montecarlohandbook.org`.

Accessible references to proofs and literature are provided within the text and at the end of each chapter. Extensive appendices on probability, statistics, and optimization have been included to provide the reader with a review of the main ideas and results in these areas relevant to Monte Carlo simulation. A comprehensive index is given at the end of the book.

The Handbook starts with a discussion on uniform (pseudo)random number generators, which are at the heart of any Monte Carlo method. We discuss what constitutes a "good" uniform random number generator, give various approaches for constructing such generators, and provide theoretical and empirical tests for randomness. Chapter 2 discusses methods for generating quasirandom numbers, which exhibit much more regularity than their pseudorandom counterparts, and are well-suited to estimating multidimensional integrals. Chapter 3 discusses general methods for random variable generation from arbitrary distributions, whereas Chapter 4 gives a list of specific generation algorithms for the major univariate and multivariate probability distributions. Chapter 5 lists the main random processes used in Monte Carlo simulation, along with their properties and how to generate them. Various Markov chain Monte Carlo techniques are discussed in Chapter 6, all of which aim to (approximately) generate samples from complicated distributions. Chapter 7 deals with simulation modeling and discrete event simulation, using the fundamental random variables and processes in Chapters 4 and 5 as building blocks. The simulation of such models then allows one to estimate quantities of interest related to the system.

The statistical analysis of simulation data is discussed in Chapter 8, which surveys a number of techniques available to obtain estimates and confidence intervals for quantities of interest, as well as methods to test hypotheses related to the data. Chapter 9 provides a comprehensive overview of variance reduction techniques for use in Monte Carlo simulation. The efficient estimation of rare-event probabilities is discussed in Chapter 10, including specific variance reduction techniques. Chapter 11 details the main methods for estimating derivatives with respect to the parameters of interest.

Monte Carlo is not only used for estimation but also for optimization. Chapter 12 discusses various randomized optimization techniques, including stochastic gradient methods, the simulated annealing technique, and the cross-entropy method. The cross-entropy method, which relates rare-event simulation to randomized optimization, is further explored in Chapter 13, while Chapter 14 focuses on particle splitting methods for rare-event simulation and combinatorial optimization.

Applications of Monte Carlo methods in finance and in network reliability are given in Chapters 15 and 16, respectively. Chapter 17 highlights the use of Monte Carlo to obtain approximate solutions to complex systems of differential equations.

Appendix A provides background material on probability theory and stochastic processes. Fundamental material from mathematical statistics is summarized in Appendix B. Appendix C reviews a number of key optimization concepts and techniques, and presents some common optimization problems. Finally, Appendix D summarizes miscellaneous results on exponential families, tail probabilities, differentiation, and the EM algorithm.

DIRK KROESE, THOMAS TAIMRE, AND ZDRAVKO BOTEV

Brisbane and Montreal
September, 2010

ACKNOWLEDGMENTS

This book has benefited from the input of many people. We thank Tim Brereton, Josh Chan, Nicolas Chopin, Georgina Davies, Adam Grace, Pierre L'Ecuyer, Ben Petschel, Ad Ridder, and Virgil Stokes, for their valuable feedback on the manuscript. Most of all, we would like to thank our families — without their support, love, and patience this book could not have been written.

This work was financially supported by the Australian Research Council under grant number DP0985177 and the Isaac Newton Institute for Mathematical Sciences, Cambridge, U.K.

DPK, TT, ZIB

CHAPTER 1

UNIFORM RANDOM NUMBER GENERATION

This chapter gives an overview of the main techniques and algorithms for generating uniform random numbers, including those based on linear recurrences, modulo 2 arithmetic, and combinations of these. A range of theoretical and empirical tests is provided to assess the quality of a uniform random number generator. We refer to Chapter 3 for a discussion on methods for random variable generation from arbitrary distributions — such methods are invariably based on uniform random number generators. ☞ 43

1.1 RANDOM NUMBERS

At the heart of any Monte Carlo method is a **random number generator**: a procedure that produces an infinite stream

$$U_1, U_2, U_3, \ldots \overset{\text{iid}}{\sim} \mathsf{Dist}$$

of random variables that are independent and identically distributed (iid) according to some probability distribution Dist. When this distribution is the uniform distribution on the interval (0,1) (that is, $\mathsf{Dist} = \mathsf{U}(0,1)$), the generator is said to be a **uniform random number generator**. Most computer languages already contain a built-in uniform random number generator. The user is typically requested only to input an initial number, called the **seed**, and upon invocation the random number generator produces a sequence of independent uniform random variables on the interval $(0, 1)$. In MATLAB, for example, this is provided by the **rand** function.

The concept of an infinite iid sequence of random variables is a mathematical abstraction that may be impossible to implement on a computer. The best one can hope to achieve in practice is to produce a sequence of "random" numbers with statistical properties that are indistinguishable from those of a true sequence of iid random variables. Although physical generation methods based on universal background radiation or quantum mechanics seem to offer a stable source of such true randomness, the vast majority of current random number generators are based on simple algorithms that can be easily implemented on a computer. Following L'Ecuyer [10], such algorithms can be represented as a tuple $(\mathcal{S}, f, \mu, \mathcal{U}, g)$, where

- $\mathcal{S}$ is a finite set of **states**,
- f is a function from $\mathcal{S}$ to $\mathcal{S}$,
- μ is a probability distribution on $\mathcal{S}$,
- $\mathcal{U}$ is the **output space**; for a uniform random number generator $\mathcal{U}$ is the interval $(0, 1)$, *and we will assume so from now on, unless otherwise specified,*
- g is a function from $\mathcal{S}$ to $\mathcal{U}$.

A random number generator then has the following structure:

Algorithm 1.1 (Generic Random Number Generator)

1. **Initialize**: *Draw the seed* S_0 *from the distribution* μ *on* $\mathcal{S}$. *Set* $t = 1$.
2. **Transition**: *Set* $S_t = f(S_{t-1})$.
3. **Output**: *Set* $U_t = g(S_t)$.
4. **Repeat**: *Set* $t = t + 1$ *and return to Step 2.*

The algorithm produces a sequence $U_1, U_2, U_3, \ldots$ of **pseudorandom numbers** — we will refer to them simply as **random numbers**. Starting from a certain seed, the sequence of states (and hence of random numbers) must repeat itself, because the state space is finite. The smallest number of steps taken before entering a previously visited state is called the **period length** of the random number generator.

1.1.1 Properties of a Good Random Number Generator

What constitutes a good random number generator depends on many factors. It is always advisable to have a variety of random number generators available, as different applications may require different properties of the random generator. Below are some desirable, or indeed essential, properties of a good uniform random number generator; see also [39].

1. *Pass statistical tests*: The ultimate goal is that the generator should produce a stream of uniform random numbers that is indistinguishable from a genuine uniform iid sequence. Although from a theoretical point of view this criterion is too imprecise and even infeasible (see Remark 1.1.1), from a practical point

of view this means that the generator should pass a battery of simple statistical tests designed to detect deviations from uniformity and independence. We discuss such tests in Section 1.5.2.

2. *Theoretical support*: A good generator should be based on sound mathematical principles, allowing for a rigorous analysis of essential properties of the generator. Examples are linear congruential generators and multiple-recursive generators discussed in Sections 1.2.1 and 1.2.2.

3. *Reproducible*: An important property is that the stream of random numbers is reproducible without having to store the complete stream in memory. This is essential for testing and variance reduction techniques. Physical generation methods cannot be repeated unless the entire stream is recorded.

4. *Fast and efficient*: The generator should produce random numbers in a fast and efficient manner, and require little storage in computer memory. Many Monte Carlo techniques for optimization and estimation require billions or more random numbers. Current physical generation methods are no match for simple algorithmic generators in terms of speed.

5. *Large period*: The period of a random number generator should be extremely large — on the order of 10^{50} — in order to avoid problems with duplication and dependence. Evidence exists [36] that in order to produce N random numbers, the period length needs to be at least $10N^2$. Most early algorithmic random number generators were fundamentally inadequate in this respect.

6. *Multiple streams*: In many applications it is necessary to run multiple independent random streams in parallel. A good random number generator should have easy provisions for multiple independent streams.

7. *Cheap and easy*: A good random number generator should be cheap and not require expensive external equipment. In addition, it should be easy to install, implement, and run. In general such a random number generator is also more easily portable over different computer platforms and architectures.

8. *Not produce 0 or 1*: A desirable property of a random number generator is that both 0 and 1 are excluded from the sequence of random numbers. This is to avoid division by 0 or other numerical complications.

Remark 1.1.1 (Computational Complexity) From a theoretical point of view, a finite-state random number generator can *always* be distinguished from a true iid sequence, after observing the sequence longer than its period. However, from a practical point of view this may not be feasible within a "reasonable" amount of time. This idea can be formalized through the notion of *computational complexity*; see, for example, [33].

1.1.2 Choosing a Good Random Number Generator

As Pierre L'Ecuyer puts it [12], choosing a good random generator is like choosing a new car: for some people or applications speed is preferred, while for others robustness and reliability are more important. For Monte Carlo simulation the distributional properties of random generators are paramount, whereas in coding and cryptography unpredictability is crucial.

Nevertheless, as with cars, there are many poorly designed and outdated models available that should be avoided. Indeed several of the standard generators that come with popular programming languages and computing packages can be appallingly poor [13].

Two classes of generators that have overall good performance are:

1. *Combined multiple recursive generators*, some of which have excellent statistical properties, are simple, have large period, support multiple streams, and are relatively fast. A popular choice is L'Ecuyer's `MRG32k3a` (see Section 1.3), which has been implemented as one of the core generators in MATLAB (from version 7), VSL, SAS, and the simulation packages SSJ, Arena, and Automod.

2. *Twisted general feedback shift register generators*, some of which have very good equidistributional properties, are among the fastest generators available (due to their essentially binary implementation), and can have extremely long periods. A popular choice is Matsumoto and Nishimura's Mersenne twister `MT19937ar` (see Section 1.2.4), which is currently the default generator in MATLAB.

In general, a good uniform number generator has *overall* good performance, in terms of the criteria mentioned above, but is not usually the top performer over all these criteria. In choosing an appropriate generator it pays to remember the following.

- Faster generators are not necessarily better (indeed, often the contrary is true).
- A small period is in general bad, but a larger period is not necessarily better.
- Good equidistribution is a necessary requirement for a good generator but not a sufficient requirement.

1.2 GENERATORS BASED ON LINEAR RECURRENCES

The most common methods for generating pseudorandom sequences use simple linear recurrence relations.

1.2.1 Linear Congruential Generators

A **linear congruential generator** (LCG) is a random number generator of the form of Algorithm 1.1, with state $S_t = X_t \in \{0, \ldots, m-1\}$ for some strictly positive integer m called the **modulus**, and state transitions

$$X_t = (aX_{t-1} + c) \bmod m \,, \quad t = 1, 2, \ldots \,, \tag{1.1}$$

where the **multiplier** a and the **increment** c are integers. Applying the modulo-m operator in (1.1) means that $aX_{t-1} + c$ is divided by m, and the remainder is taken as the value for X_t. Note that the multiplier and increment may be chosen in the set $\{0, \ldots, m-1\}$. When $c = 0$, the generator is sometimes called a **multiplicative congruential generator**. Most existing implementations of LCGs are of this form

— in general the increment does not have a large impact on the quality of an LCG. The output function for an LCG is simply

$$U_t = \frac{X_t}{m} \; .$$

■ **EXAMPLE 1.1 (Minimal Standard LCG)**

An often-cited LCG is that of Lewis, Goodman, and Miller [24], who proposed the choice $a = 7^5 = 16807$, $c = 0$, and $m = 2^{31} - 1 = 2147483647$. This LCG passes many of the standard statistical tests and has been successfully used in many applications. For this reason it is sometimes viewed as the *minimal standard* LCG, against which other generators should be judged.

Although the generator has good properties, its period $(2^{31} - 2)$ and statistical properties no longer meet the requirements of modern Monte Carlo applications; see, for example, [20].

A comprehensive list of classical LCGs and their properties can be found on Karl Entacher's website:

`http://random.mat.sbg.ac.at/results/karl/server/`

The following recommendations for LCGs are reported in [20]:

- All LCGs with modulus 2^p for some integer p are badly behaved and should not be used.

- All LCGs with modulus up to $2^{61} \approx 2 \times 10^{18}$ fail several tests and should be avoided.

1.2.2 Multiple-Recursive Generators

A **multiple-recursive generator** (MRG) of **order** k is a random number generator of the form of Algorithm 1.1, with state $S_t = \mathbf{X}_t = (X_{t-k+1}, \ldots, X_t)^\top \in \{0, \ldots, m-1\}^k$ for some modulus m and state transitions defined by

$$X_t = (a_1 X_{t-1} + \cdots + a_k X_{t-k}) \bmod m \; , \quad t = k, k+1, \ldots \; , \tag{1.2}$$

where the **multipliers** $\{a_i, i = 1, \ldots, k\}$ lie in the set $\{0, \ldots, m-1\}$. The output function is often taken as

$$U_t = \frac{X_t}{m} \; .$$

The maximum period length for this generator is $m^k - 1$, which is obtained if (a) m is a prime number and (b) the polynomial $p(z) = z^k - \sum_{i=1}^{k-1} a_i z^{k-i}$ is *primitive* using modulo m arithmetic. Methods for testing primitivity can be found in [8, Pages 30 and 439]. To yield fast algorithms, all but a few of the $\{a_i\}$ should be 0.

MRGs with very large periods can be implemented efficiently by combining several smaller-period MRGs (see Section 1.3).

1.2.3 Matrix Congruential Generators

An MRG can be interpreted and implemented as a **matrix multiplicative congruential generator**, which is a random number generator of the form of Algorithm 1.1, with state $S_t = \mathbf{X}_t \in \{0, \ldots, m-1\}^k$ for some modulus m, and state transitions

$$\mathbf{X}_t = (A\mathbf{X}_{t-1}) \bmod m, \quad t = 1, 2, \ldots , \tag{1.3}$$

where A is an invertible $k \times k$ matrix and $\mathbf{X}_t$ is a $k \times 1$ vector. The output function is often taken as

$$\mathbf{U}_t = \frac{\mathbf{X}_t}{m} , \tag{1.4}$$

yielding a vector of uniform numbers in $(0, 1)$. Hence, here the output space $\mathcal{U}$ for the algorithm is $(0, 1)^k$. For fast random number generation, the matrix A should be sparse.

To see that the multiple-recursive generator is a special case, take

$$A = \begin{pmatrix} 0 & 1 & \cdots & 0 \\ \vdots & \vdots & \ddots & \vdots \\ 0 & 0 & \cdots & 1 \\ a_k & a_{k-1} & \cdots & a_1 \end{pmatrix} \quad \text{and} \quad \mathbf{X}_t = \begin{pmatrix} X_t \\ X_{t+1} \\ \vdots \\ X_{t+k-1} \end{pmatrix} . \tag{1.5}$$

Obviously, the matrix multiplicative congruential generator is the k-dimensional generalization of the multiplicative congruential generator. A similar generalization of the multiplicative recursive generator — replacing the multipliers $\{a_i\}$ with matrices, and the scalars $\{X_t\}$ with vectors in (1.2) —, yields the class of **matrix multiplicative recursive generators**; see, for example, [34].

1.2.4 Modulo 2 Linear Generators

Good random generators must have very large state spaces. For an LCG this means that the modulus m must be a large integer. However, for multiple recursive and matrix generators it is not necessary to take a large modulus, as the state space can be as large as m^k. Because binary operations are in general faster than floating point operations (which are in turn faster than integer operations), it makes sense to consider random number generators that are based on linear recurrences modulo 2. A general framework for such random number generators is given in [18], where the state is a k-bit vector $\mathbf{X}_t = (X_{t,1}, \ldots, X_{t,k})^\top$ that is mapped via a linear transformation to a w-bit output vector $\mathbf{Y}_t = (Y_{t,1}, \ldots, Y_{t,w})^\top$, from which the random number $U_t \in (0, 1)$ is obtained by *bitwise decimation* as follows. More precisely, the procedure is as follows.

Algorithm 1.2 (Generic Linear Recurrence Modulo 2 Generator)

1. **Initialize**: *Draw the seed* $\mathbf{X}_0$ *from the distribution* μ *on the state space* $\mathcal{S} = \{0, 1\}^k$. *Set* $t = 1$.

2. **Transition**: *Set* $\mathbf{X}_t = A\mathbf{X}_{t-1}$.

3. **Output**: *Set* $\mathbf{Y}_t = B\mathbf{X}_t$ *and return*

$$U_t = \sum_{\ell=1}^{w} Y_{t,\ell}\, 2^{-\ell} .$$

4. **Repeat**: *Set* $t = t + 1$ *and return to Step 2.*

Here, A and B are $k \times k$ and $w \times k$ binary matrices, respectively, and all operations are performed modulo 2. In algebraic language, the operations are performed over the finite field $\mathbb{F}_2$, where addition corresponds to the bitwise XOR operation (in particular, $1 + 1 = 0$). The integer w can be thought of as the word length of the computer (that is, $w = 32$ or 64). Usually (but there are exceptions, see [18]) k is taken much larger than w.

■ EXAMPLE 1.2 (Linear Feedback Shift Register Generator)

The **Tausworthe** or **linear feedback shift register** (LFSR) generator is an MRG of the form (1.2) with $m = 2$, but with output function

$$U_t = \sum_{\ell=1}^{w} X_{ts+\ell-1}\, 2^{-\ell} ,$$

for some $w \leqslant k$ and $s \geqslant 1$ (often one takes $s = w$). Thus, a binary sequence $X_0, X_1, \ldots$ is generated according to the recurrence (1.2), and the t-th "word" $(X_{ts}, \ldots, X_{ts+w-1})^\top$, $t = 0, 1, \ldots$ is interpreted as the binary representation of the t-th random number.

This generator can be put in the framework of Algorithm 1.2. Namely, the state at iteration t is given by the vector $\mathbf{X}_t = (X_{ts}, \ldots, X_{ts+k-1})^\top$, and the state is updated by advancing the recursion (1.2) over s time steps. As a result, the transition matrix A in Algorithm 1.2 is equal to the s-th power of the "1-step" transition matrix given in (1.5). The output vector $\mathbf{Y}_t$ is obtained by simply taking the first w bits of $\mathbf{X}_t$; hence $B = [I_w \;\; O_{w\times(k-w)}]$, where I_w is the identity matrix of dimension w and $O_{w\times(k-w)}$ the $w \times (k - w)$ matrix of zeros.

For fast generation most of the multipliers $\{a_i\}$ are 0; in many cases there is often only *one* other non-zero multiplier a_r apart from a_k, in which case

$$X_t = X_{t-r} \oplus X_{t-k} , \tag{1.6}$$

where $\oplus$ signifies addition modulo 2. The same recurrence holds for the states (vectors of bits); that is,

$$\mathbf{X}_t = \mathbf{X}_{t-r} \oplus \mathbf{X}_{t-k} ,$$

where addition is defined componentwise.

The LFSR algorithm derives its name from the fact that it can be implemented very efficiently on a computer via **feedback shift registers** — binary arrays that allow fast shifting of bits; see, for example, [18, Algorithm L] and [7, Page 40].

Generalizations of the LFSR generator that all fit the framework of Algorithm 1.2 include the **generalized feedback shift register** generators [25] and the **twisted** versions thereof [30], the most popular of which are the **Mersenne twisters** [31]. A particular instance of the Mersenne twister, MT19937, has become widespread, and has been implemented in software packages such as SPSS and MATLAB. It has a huge period length of $2^{19937} - 1$, is very fast, has good equidistributional properties, and passes most statistical tests. The latest version of the code may be found at

`http://www.math.sci.hiroshima-u.ac.jp/~m-mat/MT/emt.html`

Two drawbacks are that the initialization procedure and indeed the implementation itself is not straightforward. Another potential problem is that the algorithm recovers too slowly from the states near zero. More precisely, after a state with very few 1s is hit, it may take a long time (several hundred thousand steps) before getting back to some state with a more equal division between 0s and 1s. Some other weakness are discussed in [20, Page 23].

The development of good and fast modulo 2 generators is important, both from a practical and theoretical point of view, and is still an active area of research, not in the least because of the close connection to coding and cryptography. Some recent developments include the WELL (well-equidistributed long-period linear) generators by Panneton et al. [35], which correct some weaknesses in `MT19937`, and the SIMD-oriented fast Mersenne twister [38], which is significantly faster than the standard Mersenne twister, has better equidistribution properties, and recovers faster from states with many 0s.

1.3 COMBINED GENERATORS

A significant leap forward in the development of random number generators was made with the introduction of **combined generators**. Here the output of several generators, which individually may be of poor quality, is combined, for example by shuffling, adding, and/or selecting, to make a superior quality generator.

■ **EXAMPLE 1.3 (Wichman–Hill)**

One of the earliest combined generators is the Wichman–Hill generator [41], which combines three LCGs:

$$\begin{aligned} X_t &= (171\, X_{t-1}) \bmod m_1 \qquad (m_1 = 30269)\,, \\ Y_t &= (172\, Y_{t-1}) \bmod m_2 \qquad (m_2 = 30307)\,, \\ Z_t &= (170\, Z_{t-1}) \bmod m_3 \qquad (m_3 = 30323)\,. \end{aligned}$$

These random integers are then combined into a single random number

$$U_t = \frac{X_t}{m_1} + \frac{Y_t}{m_2} + \frac{Z_t}{m_3} \mod 1\,.$$

The period of the sequence of triples (X_t, Y_t, Z_t) is shown [42] to be $(m_1 - 1)(m_2 - 1)(m_3 - 1)/4 \approx 6.95 \times 10^{12}$, which is much larger than the individual periods. Zeisel [43] shows that the generator is in fact equivalent (produces the same output) as a multiplicative congruential generator with modulus $m = 27817185604309$ and multiplier $a = 16555425264690$.

The Wichman–Hill algorithm performs quite well in simple statistical tests, but since its period is not sufficiently large, it fails various of the more sophisticated tests, and is no longer suitable for high-performance Monte Carlo applications.

One class of combined generators that has been extensively studied is that of the **combined multiple-recursive generators**, where a small number of MRGs are combined. This class of generators can be analyzed theoretically in the same way as single MRG: under appropriate initialization the output stream of random numbers of a combined MRG is exactly the same as that of some larger-period

MRG [23]. Hence, to assess the quality of the generator one can employ the same well-understood theoretical analysis of MRGs. As a result, the multipliers and moduli in the combined MRG can be searched and chosen in a systematic and principled manner, leading to random number generators with excellent statistical properties. An important added bonus is that such algorithms lead to easy multi-stream generators [21].

In [12] L'Ecuyer conducts an extensive numerical search and detailed theoretical analysis to find good combined MRGs. One of the combined MRGs that stood out was MRG32k3a, which employs two MRGs of order 3,

$$X_t = (1403580\, X_{t-2} - 810728\, X_{t-3}) \bmod m_1 \quad (m_1 = 2^{32} - 209 = 4294967087)\,,$$

$$Y_t = (527612\, Y_{t-1} - 1370589\, Y_{t-3}) \bmod m_2 \quad (m_2 = 2^{32} - 22853 = 4294944443)\,,$$

and whose output is

$$U_t = \begin{cases} \dfrac{X_t - Y_t + m_1}{m_1 + 1} & \text{if } X_t \leqslant Y_t\,, \\[2ex] \dfrac{X_t - Y_t}{m_1 + 1} & \text{if } X_t > Y_t\,. \end{cases}$$

The period length is approximately 3×10^{57}. The generator MRG32k3a passes all statistical tests in today's most comprehensive test suit *TestU01* [20] (see also Section 1.5) and has been implemented in many software packages, including MATLAB, Mathematica, Intel's MKL Library, SAS, VSL, Arena, and Automod. It is also the core generator in L'Ecuyer's SSJ simulation package, and is easily extendable to generate multiple random streams. An implementation in MATLAB is given below.

```
%MRG32k3a.m
m1=2^32-209; m2=2^32-22853;
ax2p=1403580; ax3n=810728;
ay1p=527612; ay3n=1370589;

X=[12345 12345 12345]; % Initial X
Y=[12345 12345 12345]; % Initial Y

N=100; % Compute the sequence for N steps
U=zeros(1,N);
for t=1:N
   Xt=mod(ax2p*X(2)-ax3n*X(3),m1);
   Yt=mod(ay1p*Y(1)-ay3n*Y(3),m2);
   if Xt <= Yt
       U(t)=(Xt - Yt + m1)/(m1+1);
   else
       U(t)=(Xt - Yt)/(m1+1);
   end
   X(2:3)=X(1:2); X(1)=Xt; Y(2:3)=Y(1:2); Y(1)=Yt;
end
```

Different *types* of generators can also be combined. For example, Marsaglia's KISS99 (keep it simple stupid) generator [26] combines two shift register generators with an LCG. This generator performs very well in *TestU01* [20]. The following MATLAB code implements the KISS99 generator.

```
% KISS99.m
% Seeds: Correct variable types crucial!
A=uint32(12345); B=uint32(65435); Y=12345; Z=uint32(34221);
N=100; % Compute the sequence for N steps
U=zeros(1,N);
for t=1:N
    % Two Multiply with Carry Generators
    A=36969*bitand(A,uint32(65535))+bitshift(A,-16);
    B=18000*bitand(B,uint32(65535))+bitshift(B,-16);
    % MWC: Low and High 16 bits are A and B
    X=bitshift(A,16)+B;
    % CONG: Linear Congruential Generator
    Y = mod(69069*Y+1234567,4294967296);
    % SHR3: 3-Shift Register Generator
    Z=bitxor(Z,bitshift(Z,17));
    Z=bitxor(Z,bitshift(Z,-13));
    Z=bitxor(Z,bitshift(Z,5));
    % Combine them to form the KISS99 generator
    KISS=mod(double(bitxor(X,uint32(Y)))+double(Z),4294967296);
    U(t)=KISS/4294967296; % U[0,1] output
end
```

1.4 OTHER GENERATORS

Many variations on linear congruential methods have been proposed. Of the ones not discussed in the previous section we mention the following:

- *Multiply with carry:* This is a variation of the LCG where the increment c changes per iteration. Specifically, the recurrence is given by

$$X_t = (aX_{t-1} + c_{t-1}) \bmod m ,$$

  where c_t (the **carry**) satisfies, for a given **lag** k,

$$c_t = \lfloor (aX_{t-k} + c_{t-1})/m \rfloor , \quad t \geqslant k .$$

- *XOR shift:* This is a generalization of an LFSR generator, and is a special case of a matrix MRG [34], where the state at iteration t is given by a binary vector $\mathbf{X}_t$ satisfying the linear recursion

$$\mathbf{X}_t = A_1\mathbf{X}_{t-k_1} \oplus \cdots \oplus A_r\mathbf{X}_{t-k_r} ,$$

  where $k_1, \ldots, k_r$ are strictly positive integers and $A_1, \ldots, A_r$ are either identity matrices or the products of XOR shift matrices.

- *Lagged Fibonacci generators:* This is a generalization of the LFSR generator (1.6), where the `XOR` operator $\oplus$ is replaced by a general binary operator, for example, the product.

More details on these generators can be found, for example, in [18, 29]. The above generators in general do not pass all statistical tests for randomness in the test suite *TestU01*, but combining them, as for example in the `KISS99` generator, may produce high-quality generators. The multiply with carry and lagged Fibonacci generators are known to have poor theoretical properties [11, 40].

Congruential generators based on *nonlinear* recurrences,

$$X_t = g(X_{t-1}, \ldots, X_{t-k}) \bmod m ,$$

for some nonlinear function g are currently not in much use in Monte Carlo simulations, since they tend to be slower, are more difficult to analyze theoretically, and often fail empirical tests for uniformity. However, nonlinear generators are important in cryptography, as the output sequence of linear congruential methods is easy to predict — in particular, the parameters of a linear congruential method can be easily estimated from previously generated output; see, for example, [29].

A famous nonlinear method in cryptography is that of Blum, Blum, and Shub [2], who proposed the quadratic recurrence

$$X_t = X_{t-1}^2 \bmod m ,$$

where $m = p\,q$ and p and q are (large) primes that divided by 4 give a remainder of 3 (so-called **Blum primes**; for example, $p =$ 12676506002282294014967039815l9 and $q =$ 1267650600228229401496704318359). Each iteration of the Blum–Blum–Shub generator produces only one bit of output, being either the even or odd bit parity, or the last bit (least significant bit) of X_t. It is shown in [2] that the output sequence of such a generator is not predictable in polynomial time. The generator is not appropriate for Monte Carlo simulation, due to its low speed.

Another example of a nonlinear congruential generator is the **inverse congruential generator** where the recurrence is of the form

$$X_t = (a X_{t-1}^{-} + c) \bmod m ,$$

where X^- is the multiplicative inverse of X modulo m (that is, $XX^- = 1 \bmod m$ if it exists, or 0 otherwise). A survey of nonlinear generators may be found in [4].

1.5 TESTS FOR RANDOM NUMBER GENERATORS

The quality of random number generators can be assessed in two ways. The first is to investigate the theoretical properties of the random number generator. Such properties include the period length of the generator and various measures of uniformity and independence. This type of random number generator testing is called **theoretical**, as it does not require the actual output of the generator but only its algorithmic structure and parameters. Powerful theoretical tests are only feasible if the generators have a sufficiently simple structure, such as those of linear congruential and multiple-recursive methods and combined versions thereof.

A second type of test involves the application of a battery of statistical tests to the output of the generator, with the objective to detect deviations from uniformity and independence. Such tests are said to be **empirical**.

1.5.1 Spectral Test

One of the most useful theoretical tests concerns the structural properties of the generator. Suppose that $U_0, U_1, \ldots,$ is the sequence of numbers produced by a random number generator. It is well known [3, 5, 9, 27] that if the generator is of LCG or MRG type, then vectors of successive values $\mathbf{U}_0 = (U_0, \ldots, U_{d-1})^\top, \mathbf{U}_1 = (U_1, \ldots, U_d)^\top, \ldots,$ lie on a d-dimensional **lattice**; that is, a set $L \subset \mathbb{R}^d$ of the form

$$L = \left\{ \sum_{i=1}^{d} z_i \, \mathbf{b}_i, \quad z_1, \ldots, z_d \in \mathbb{Z} \right\},$$

for some set of linearly independent **basis** vectors $\mathbf{b}_1, \ldots, \mathbf{b}_d$. In other words, the elements of L are simply linear combinations of the basis vectors, using only integer coefficients. The lattice L is said to be **generated** by the basis matrix $B = (\mathbf{b}_1, \ldots, \mathbf{b}_d)$.

For an MRG satisfying the recursion (1.2), the basis vectors can be chosen as [15]

$$\begin{aligned}
\mathbf{b}_1 &= (1, 0, \ldots, 0, X_{1,k}, \ldots, X_{1,d-1})^\top / m \\
&\;\vdots \\
\mathbf{b}_k &= (0, 0, \ldots, 1, X_{k,k}, \ldots, X_{k,d-1})^\top / m \\
\mathbf{b}_{k+1} &= (0, 0, \ldots, 0, 1, \ldots, 0)^\top \\
&\;\vdots \\
\mathbf{b}_d &= (0, 0, \ldots, 0, 0, \ldots, 1)^\top ,
\end{aligned}$$

where $X_{i,0}, X_{i,1}, \ldots$ is the sequence of states produced by the generator when starting with states $X_i = 1, X_t = 0,\ t \neq i,\ t \leqslant k$.

For a good generator the set $L \cap (0,1)^d$ should cover the d-dimensional unit hypercube $(0,1)^d$ in a uniform manner. One way to quantify this is to measure the distance between hyperplanes in the lattice L. The maximal distance between such hyperplanes is called the **spectral gap**, denoted here as g_d. A convenient way to compute the spectral gap is to consider first the **dual lattice** of L, which is the lattice generated by the inverse matrix of B. The dual lattice is denoted by L^*. Each vector $\mathbf{v}$ in L^* defines a family of equidistant hyperplanes in L, at a distance $1/\|\mathbf{v}\|$ apart. Hence, the length of the shortest non-zero vector in L^* corresponds to $1/g_d$.

For any d-dimensional lattice with m points there is a lower bound g_d^* on the spectral gap for dimension d. Specifically, for dimensions less than 8 it can be shown (see, for example, [8, Section 3.3.4]) that $g_d \geqslant g_d^* = \gamma_d^{-1/2} m^{-1/d}$, where $\gamma_1, \ldots, \gamma_8$ take the values

$$1, \quad (4/3)^{1/2}, \quad 2^{1/3}, \quad 2^{1/2}, \quad 2^{3/5}, \quad (64/3)^{1/6}, \quad 4^{3/7}, \quad 2\,.$$

An often-used *figure of merit* for the quality of a random number generator is the quotient

$$S_d = \frac{g_d^*}{g_d} = \frac{1}{g_d \, m^{1/d} \gamma_d^{1/2}} ,$$

or the minimum of K of such values: $S = \min_{d \leqslant K} S_d$, where $K \leqslant 8$. High values of S (close to 1) indicate that the generator has good structural properties.

The following example illustrates the main points; see also [8, Section 3.3.4].

■ **EXAMPLE 1.4 (Lattice Structure and Spectral Gap)**

Consider the LCG (1.1) with $a = 3$, $c = 0$, and $m = 31$. For $d = 2$, the corresponding lattice is generated by the basis matrix

$$B = \begin{pmatrix} 1/m & 0 \\ a/m & 1 \end{pmatrix},$$

since this LCG is an MRG with $k = 1$ and $X_{1,1} = a/m$. The dual lattice, which is depicted in Figure 1.1, is generated by the basis matrix

$$B^{-1} = \begin{pmatrix} m & 0 \\ -a & 1 \end{pmatrix}.$$

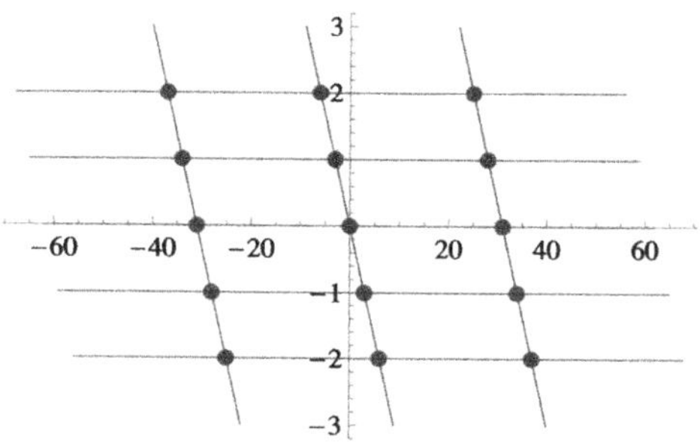

Figure 1.1 The dual lattice L^*.

The shortest non-zero vector in L^* is $(-3,1)^\top$; hence, the spectral gap for dimension 2 is $g_2 = 1/\sqrt{10} \approx 0.316$. Figure 1.2 shows the normalized vector $g_2^2\,(-3,1)^\top$ to be perpendicular to hyperplanes in L that are a distance g_2 apart. The figure of merit S_2 is here $3^{1/4}(5/31)^{1/2} \approx 0.53$.

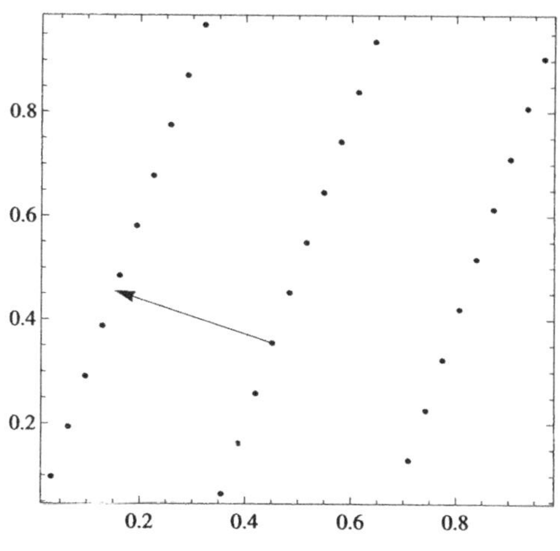

Figure 1.2 The lattice L truncated to the unit square. The length of the arrow corresponds to the spectral gap.

In order to select a good random number generator, it is important that the spectral gap is computed over a range of dimensions d. Some generators may display good structure at lower dimensions and bad structure at higher dimensions (the opposite is also possible). A classical example is IBM's RANDU LCG, with $a = 2^{16} + 3$, $c = 0$, and $m = 2^{31}$, which has reasonable structure for $d = 1$ and 2, but bad structure for $d = 3$; the latter is illustrated in Figure 1.3.

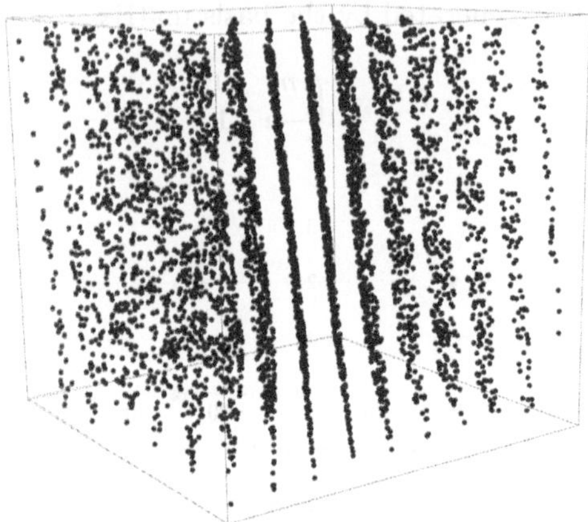

Figure 1.3 Structural deficiency of RANDU.

Structural properties of combined MRGs can be analyzed in the same way, as such generators are equivalent (under appropriate initialization conditions) to a single MRG with large modulus [23].

The computational effort required to compute the spectral gap grows rapidly with the dimension d and becomes impractical for dimensions over about 60. A fast implementation for analyzing the lattice structure of LCGs and MRGs is the LatMRG software package described in [17].

Modulo 2 linear generators do not have a lattice structure in Euclidean space, but they do in the space of formal power series. Much of the theory and algorithms developed for lattices in $\mathbb{R}^d$ carries over to the modulo 2 case [14].

Other theoretical tests of random number generators include discrepancy tests [32] and serial correlation tests [8, Section 3.3.3]. See also [1].

1.5.2 Empirical Tests

While theoretical tests are important for the elimination of bad generators and the search for potentially good generators [6, 12], the ultimate goal remains to find uniform random number generators whose output is statistically indistinguishable (within reasonable computational time) from a sequence of iid uniform random variables. Hence, any candidate generator should pass a wide range of statistical tests that examine uniformity and independence. The general structure of such tests is often of the following form.

Algorithm 1.3 (Two-Stage Empirical Test for Randomness) *Suppose that* $\mathbf{U} = \{U_i\}$ *represents the output stream of the uniform random generator. Let* H_0 *be the hypothesis that the* $\{U_i\}$ *are iid from a* $\mathsf{U}(0,1)$ *distribution. Let* Z *be some deterministic function of* $\mathbf{U}$.

1. *Generate* N *independent copies* $Z_1, \ldots, Z_N$ *of* Z *and evaluate a test statistic* $T = T(Z_1, \ldots, Z_N)$ *for testing* H_0 *versus the alternative that* H_0 *is not true. Suppose that under* H_0 *the test statistic* T *has distribution or asymptotic (for large* N*) distribution* Dist_0.

2. *Generate* K *independent copies* $T_1, \ldots, T_K$ *of* T *and perform a goodness of fit test to test the hypothesis that the* $\{T_i\}$ *are iid from* Dist_0.

Such a test procedure is called a **two-stage** or **second-order** statistical test. The first stage corresponds to an ordinary statistical test, such as a χ^2 goodness of fit test, and the second stage combines K such tests by means of another goodness of fit test, such as the Kolmogorov–Smirnov or Anderson–Darling test; see also Section 8.7.2. The following example demonstrates the procedure. ☞ 336

■ **EXAMPLE 1.5 (Binary Rank Test for the `drand48` Generator)**

The default random number generator in the C library is `drand48`, which implements an LCG with $a = 25214903917$, $m = 2^{48}$, and $c = 11$. We wish to examine if the output stream of this generator passes the *binary rank test* described in Section 1.5.2.11. For this test, the sequence $U_1, U_2, \ldots$ is first transformed to a binary sequence $B_1, B_2, \ldots$, for example, by taking $B_i = \mathrm{I}_{\{U_i \leqslant 1/2\}}$, and then the $\{B_i\}$ are arranged in a binary array, say with 32 rows and 32 columns. The first row of the matrix is $B_1, \ldots, B_{32}$, the second row is $B_{33}, \ldots B_{64}$, etc. Under H_0 the distribution of the rank (in modulo 2 arithmetic) R of this random matrix is given in (1.9). We generate $N = 200$ copies of R, and divide these into three classes: $R \leqslant 30$, $R = 31$, and $R = 32$. The expected number of ranks in these classes is by (1.9) equal to $E_1 = 200 \times 0.1336357$, $E_2 = 200 \times 0.5775762$, and $E_3 = 200 \times 0.2887881$. This is compared with the observed number of ranks O_1, O_2, and O_3, via the χ^2 goodness of fit statistic

$$T = \sum_{i=1}^{3} \frac{(O_i - E_i)^2}{E_i} \,. \tag{1.7}$$

Under H_0, the random variable T approximately has a χ^2_2 distribution (the number of degrees of freedom is the number of classes, 3, minus 1). This completes the first stage of the empirical test. ☞ 341

In the second stage, $K = 20$ replications of T are generated. The test statistics for the χ^2 test were 2.5556, 11.3314, 146.2747, 24.9729, 1.6850, 50.7449, 2.6507, 12.9015, 40.9470, 8.3449, 11.8191, 9.4470, 91.1219, 37.7246, 18.6256, 1.2965, 1.2267, 0.8346, 23.3909, 14.7596.

Notice that the null hypothesis would not be rejected if it were based only on the first outcome, 2.5556, as the p-value, $\mathbb{P}_{H_0}(T \geqslant 2.5556) \approx 0.279$ is quite large (and therefore the observed outcome is not uncommon under the null hypothesis). However, other values, such as 50.7449 are very large and lead to very small p-values (and a rejection of H_0). The second stage combines these findings into a single number, using a Kolmogorov–Smirnov test, to test whether the distribution of T does indeed follow a χ_2^2 distribution. The empirical cdf (of the 20 values for T) and the cdf of the χ_2^2 distribution are depicted in Figure 1.4. The figure shows a clear disagreement between the two cdfs. The maximal gap between the cdfs is 0.6846 in this case, leading to a Kolmogorov–Smirnov test statistic value of $\sqrt{20} \times 0.6846 \approx 3.06$, which gives a p-value of around 3.7272×10^{-9}, giving overwhelming evidence that the output sequence of the **drand48** generator does not behave like an iid $\mathsf{U}(0,1)$ sequence.

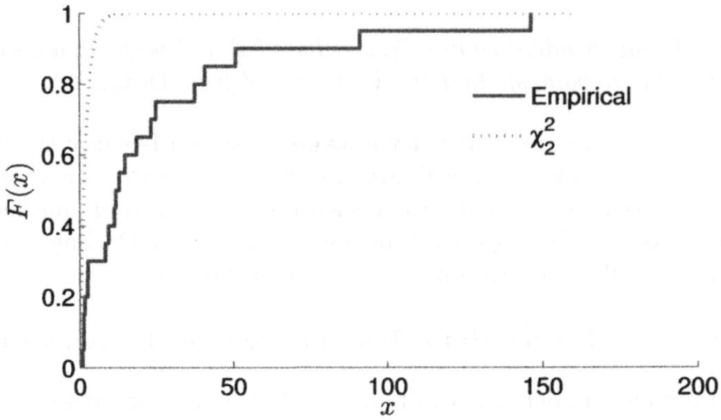

Figure 1.4 Kolmogorov–Smirnov test for the binary rank test using the **drand48** generator.

By comparison, we repeated the same procedure using the default MATLAB generator. The result of the Kolmogorov–Smirnov test is given in Figure 1.5. In this case the empirical and theoretical cdfs have a close match, and the p-value is large, indicating that the default MATLAB generator passes the binary rank test.

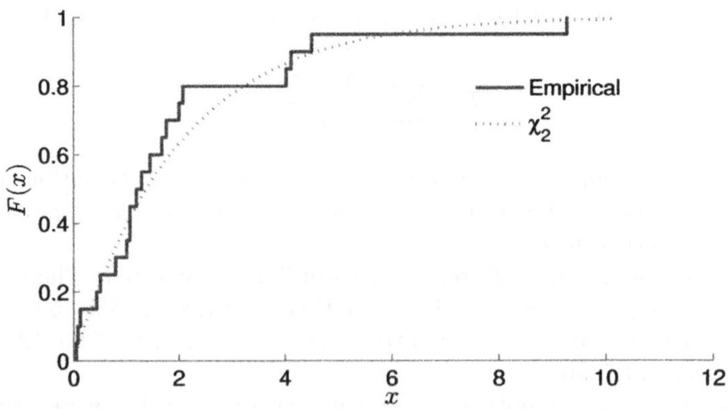

Figure 1.5 Kolmogorov–Smirnov test for the binary rank test using the default MATLAB random number generator (in this case the Mersenne twister).

Today's most complete library for the empirical testing of random number generators is the *TestU01* software library by L'Ecuyer and Simard [20]. The library is comprised of three predefined test suites: *Small Crush, Crush*, and *Big Crush*, in increasing order of complexity. *TestU01* includes the *standard tests* by Knuth [8, Section 3.3.2], and adapted version of the *Diehard* suite of tests by Marsaglia [28], the ones implemented by the National Institute of Standards and Technology (NIST) [37], and various other tests.

We conclude with a selection of empirical tests. Below, $U_0, U_1, \ldots$ is the original test sequence. The null hypothesis H_0 is that $\{U_i\} \sim_{\text{iid}} \mathsf{U}(0,1)$. Other random variables and processes derived from the $\{U_i\}$ are:

- $Y_0, Y_1, \ldots$, with $Y_i = \lfloor m U_i \rfloor$, $i = 0, 1, \ldots$, for some integer (*size*) $m \geqslant 1$. Under H_0 the $\{Y_i\}$ are iid with a discrete uniform distribution on $\{0, 1, \ldots, m-1\}$.
- $\mathbf{U}_0, \mathbf{U}_1, \ldots$, with $\mathbf{U}_i = (U_{id}, \ldots, U_{id+d-1})$, $i = 0, 1, \ldots$ for some dimension $d \geqslant 1$. Under H_0 the $\{\mathbf{U}_i\}$ are independent random vectors, each uniformly distributed on the d-dimensional hypercube $(0,1)^d$.
- $\mathbf{Y}_0, \mathbf{Y}_1, \ldots$, with $\mathbf{Y}_i = (Y_{id}, \ldots, Y_{id+d-1})$, $i = 0, 1, \ldots$ for some dimension $d \geqslant 1$. Under H_0 the $\{\mathbf{Y}_i\}$ are independent random vectors, each from the discrete uniform distribution on the d-dimensional set $\{0, 1, \ldots, m-1\}^d$.

1.5.2.1 Equidistribution (or Frequency) Tests This is to test whether the $\{U_i\}$ have a $\mathsf{U}(0,1)$ distribution. Two possible approaches are:

1. Apply a Kolmogorov–Smirnov test to ascertain whether the empirical cdf of $U_0, \ldots, U_{n-1}$ matches the theoretical cdf of the $\mathsf{U}(0,1)$ distribution; that is, $F(x) = x$, $0 \leqslant x \leqslant 1$.
2. Apply a χ^2 test on $Y_0, \ldots, Y_{n-1}$, comparing for each $k = 0, \ldots, m-1$ the observed number of occurrences in class k, $O_k = \sum_{i=0}^{n-1} \mathrm{I}_{\{Y_i = k\}}$, with the expected number $E_k = n/m$. Under H_0 the χ^2 statistic (1.7) asymptotically has (as $n \to \infty$) a χ^2_{m-1} distribution.

1.5.2.2 Serial Tests This is to test whether successive values of the random number generator are uniformly distributed. More precisely, generate vectors $\mathbf{Y}_0, \ldots, \mathbf{Y}_{n-1}$ for a given dimension d and size m. Count the number of times that the vector $\mathbf{Y}$ satisfies $\mathbf{Y} = \mathbf{y}$, for $\mathbf{y} \in \{0, \ldots, m-1\}^d$, and compare with the expected count n/m^d via a χ^2 goodness of fit test. It is usually recommended that each class should have enough samples, say at least 5 in expectation, so that $n \geqslant 5m^d$; however, see [22] for sparse serial tests. Typically, d is small, say 2 or 3.

1.5.2.3 Nearest Pairs Tests This is to detect spatial clustering (or repulsion) of the $\{\mathbf{U}_i\}$ vectors. Generate points (vectors) $\mathbf{U}_0, \ldots, \mathbf{U}_{n-1}$ in the d-dimensional unit hypercube $(0,1)^d$. For each pair of points $\mathbf{U}_i = (U_{i1}, \ldots, U_{id})^\top$ and $\mathbf{U}_j = (U_{j1}, \ldots, U_{jd})^\top$ let D_{ij} be the distance between them, defined by

$$D_{ij} = \begin{cases} \left[\sum_{k=1}^{d} (\min\{|U_{ik} - U_{jk}|, 1 - |U_{ik} - U_{jk}|\})^p\right]^{1/p} & \text{if } 1 \leqslant p < \infty \\ \max_{k=1}^{d} \min\{|U_{ik} - U_{jk}|, 1 - |U_{ik} - U_{jk}|\} & \text{if } p = \infty\,, \end{cases}$$

for some $1 \leqslant p \leqslant \infty$. This corresponds to the L^p norm on the *torus* $(0,1)^d$, whereby opposite sides of the unit hypercube are identified.

For $t \geqslant 0$, let N_t be the number of pairs (i,j) with $i < j$ such that $D_{ij} \leqslant (t/\lambda)^{1/d}$, where $\lambda = n(n-1)V_d/2$ and $V_d = [2\Gamma(1+1/p)]^d/\Gamma(1+d/p)$ (corresponding to the volume of the unit d-ball in L^p norm). It can be shown [16] that under H_0 the stochastic process $\{N_t, 0 \leqslant t \leqslant t_1\}$ converges in distribution (as $n \to \infty$) to a Poisson process with rate 1, for any fixed choice of t_1. It follows that if $T_1, T_2, \ldots$ are the jump times of $\{N_t\}$, then the spacings $A_i = T_i - T_{i-1}$, $i = 1, 2 \ldots$ are approximately iid $\mathsf{Exp}(1)$ distributed and the transformed spacings $Z_i = 1 - \exp(-A_i)$, $i = 1, 2 \ldots$ are approximately iid $\mathsf{U}(0,1)$ distributed.

The **q-nearest pair** test assesses the hypothesis that the first q transformed spacings, $Z_1, \ldots, Z_q$, are iid from $\mathsf{U}(0,1)$, by using a Kolmogorov–Smirnov or Anderson–Darling test statistic. By creating N copies of the test statistic, a two-stage test can be obtained.

Typically, ranges for the testing parameters are $1 \leqslant q \leqslant 8$, $1 \leqslant N \leqslant 30$, $2 \leqslant d \leqslant 8$, and $10^3 \leqslant n \leqslant 10^5$. The choice $p = \infty$ is often convenient in terms of computational speed. It is recommended [16] that $n \geqslant 4q^2\sqrt{N}$.

1.5.2.4 Gap Tests Let $T_1, T_2, \ldots$ denote the times when the output process $U_0, U_1, \ldots,$ visits a specified interval $(\alpha, \beta) \subset (0,1)$, and let $Z_1, Z_2, \ldots$ denote the **gap** lengths between subsequent visits; that is, $Z_i = T_i - T_{i-1} - 1$, $i = 1, 2, \ldots,$ with $T_0 = 0$. Under H_0, the $\{Z_i\}$ are iid with a $\mathsf{Geom}_0(p)$ distribution, with $p = \beta - \alpha$; that is,

$$\mathbb{P}(Z = z) = p\,(1-p)^z, \quad z = 0, 1, 2, \ldots .$$

The gap test assesses this hypothesis by tallying the number of gaps that fall in certain classes. In particular, a χ^2 test is performed with classes $Z = 0, Z = 1, \ldots, Z = r - 1$, and $Z \geqslant r$, with probabilities $p(1-p)^z$, $z = 0, \ldots, r-1$ for the first r classes and $(1-p)^r$ for the last class. The integers n and r should be chosen so that the expected number per class is $\geqslant 5$.

When $\alpha = 0$ and $\beta = 1/2$, this is sometimes called **runs above the mean**, and when $\alpha = 1/2$ and $\beta = 1$ this is sometimes called **runs below the mean**.

1.5.2.5 Poker or Partition Tests Consider the sequence of d-dimensional vectors $\mathbf{Y}_1, \ldots, \mathbf{Y}_n$, each taking values in $\{0, \ldots, m-1\}^d$. For such a vector $\mathbf{Y}$, let Z be the number of distinct components; for example if $\mathbf{Y} = (4, 2, 6, 4, 2, 5, 1, 4)$, then $Z = 5$. Under H_0, Z has probability distribution

$$\mathbb{P}(Z = z) = \frac{m(m-1)\cdots(m-z+1)\left\{ {d \atop z} \right\}}{m^d}, \quad z = 1, \ldots, \min\{d, m\} . \tag{1.8}$$

Here, $\left\{ {d \atop z} \right\}$ represents the **Stirling number of the second kind**, which gives the number of ways a set of size d can be partitioned into z non-empty subsets. For example, $\left\{ {4 \atop 2} \right\} = 7$. Such Stirling numbers can be expressed in terms of binomial coefficients as

$$\left\{ {d \atop z} \right\} = \frac{1}{z!} \sum_{k=0}^{z} (-1)^{z-k} \binom{z}{k} k^d .$$

Using the above probabilities, the validity of H_0 can now be tested via a χ^2 test.

1.5.2.6 Coupon Collector's Tests Consider the sequence $Y_1, Y_2, \ldots$, each Y_i taking values in $\{0, \ldots, m-1\}$. Let T be the first time that a "complete" set $\{0, \ldots, m-1\}$

is obtained among $Y_1, \ldots, Y_T$. The probability that $(Y_1, \ldots, Y_t)$ is incomplete is, by (1.8), equal to $\mathbb{P}(T > t) = 1 - m!\left\{{t \atop m}\right\}/m^t$, so that

$$\mathbb{P}(T = t) = \frac{m!}{m^t}\left\{{t-1 \atop m-1}\right\}, \quad t = m, m+1, \ldots .$$

The coupon collector's test proceeds by generating successive times $T_1, \ldots, T_n$ and applying a χ^2 goodness of fit test using classes $T = t$, $t = m, \ldots, r-1$ and $T > r - 1$, with probabilities given above.

1.5.2.7 Permutation Tests Consider the d-dimensional random vector $\mathbf{U} = (U_1, \ldots, U_d)^\top$. Order the components from smallest to largest and let $\mathbf{\Pi}$ be the corresponding ordering of indices. Under H_0,

$$\mathbb{P}(\mathbf{\Pi} = \boldsymbol{\pi}) = \frac{1}{d!} \quad \text{for all permutations } \boldsymbol{\pi} .$$

The permutation test assesses this uniformity of the permutations via a χ^2 goodness of fit test with $d!$ permutation classes, each with class probability $1/d!$.

1.5.2.8 Run Tests Consider the sequence $U_1, U_2, \ldots$. Let Z be the **run-up length**; that is, $Z = \min\{k : U_{k+1} < U_k\}$. Under H_0, $\mathbb{P}(Z \geqslant z) = 1/z!$, so that

$$\mathbb{P}(Z = z) = \frac{1}{z!} - \frac{1}{(z+1)!}, \quad z = 1, 2, \ldots .$$

In the run test, n of such run lengths $Z_1, \ldots, Z_n$ are obtained, and a χ^2 test is performed on the counts, using the above probabilities. It is important to start from fresh after each run. In practice this is done by throwing away the number immediately after a run. For example the second run is started with U_{Z_1+2} rather than U_{Z_1+1}, since the latter is not $\mathsf{U}(0,1)$ distributed, as it is by definition smaller than U_{Z_1}.

1.5.2.9 Maximum-of-d Tests Generate $\mathbf{U}_1, \ldots, \mathbf{U}_n$ for some dimension d. For each $\mathbf{U} = (U_1, \ldots, U_d)^\top$ let $Z = \max\{U_1, \ldots, U_d\}$ be the maximum. Under H_0, Z has cdf

$$F(z) = \mathbb{P}(Z \leqslant z) = z^d, \quad 0 \leqslant z \leqslant 1 .$$

Apply the Kolmogorov–Smirnov test to $Z_1, \ldots, Z_n$ with distribution function $F(z)$. Another option is to define $W_k = Z_k^d$ and apply the equidistribution test to $W_1, \ldots, W_n$.

1.5.2.10 Collision Tests Consider a sequence of d-dimensional vectors $\mathbf{Y}_1, \ldots, \mathbf{Y}_b$, each taking values in $\{0, \ldots, m-1\}^d$. There are $r = m^d$ possible values for each $\mathbf{Y}$. Typically, $r \gg b$. Think of throwing b balls into r urns. As there are many more urns than balls, most balls will land in an empty urn, but sometimes a "collision" occurs. Let C be the number of such collisions. Under H_0 the probability of c collisions (that is, the probability that exactly $b - c$ urns are occupied) is given, as in (1.8), by

$$\mathbb{P}(C = c) = \frac{r(r-1)\cdots(r-(b-c)+1)\left\{{b \atop b-c}\right\}}{r^b}, \quad c = 0, \ldots, b-1 .$$

A χ^2 goodness of fit test can be applied to compare the empirical distribution of n such collision values, $C_1, \ldots, C_n$, with the above distribution under H_0. One may need to group various of the classes $C = c$ in order to obtain a sufficient number of observations in each class.

1.5.2.11 Rank of Binary Matrix Tests Transform the sequence $U_1, U_2, \ldots$ to a binary sequence $B_1, B_2, \ldots$ and arrange these in a binary array of dimension $r \times c$ (assume $r \leqslant c$). Under H_0 the distribution of the rank (in modulo 2 arithmetic) Z of this matrix is given by

$$\mathbb{P}(Z = z) = 2^{(c-z)(z-r)} \prod_{i=0}^{z-1} \frac{(1 - 2^{i-c})(1 - 2^{i-r})}{1 - 2^{i-z}}, \quad z = 0, 1, \ldots, r\,. \tag{1.9}$$

☞ 632 This can be seen, for example, by defining a Markov chain $\{Z_t, t = 0, 1, 2, \ldots\}$, starting at 0 and with transition probabilities $p_{i,i} = 2^{-c+i}$ and $p_{i,i+1} = 1 - 2^{-c+i}$, $i = 0, \ldots, r$. The interpretation is that Z_t is the rank of a $t \times c$ matrix which is constructed from a $(t-1) \times c$ matrix by adding a $1 \times c$ random binary row; this row is either dependent on the $t - 1$ previous rows (rank stays the same) or not (rank is increased by 1). The distribution of Z_r corresponds to (1.9).

For $c = r = 32$ we have

$$\begin{aligned} \mathbb{P}(Z \leqslant 30) &\approx 0.1336357 \\ \mathbb{P}(Z = 31) &\approx 0.5775762 \\ \mathbb{P}(Z = 32) &\approx 0.2887881\,. \end{aligned}$$

These probabilities can be compared with the observed frequencies, via a χ^2 goodness of fit test.

1.5.2.12 Birthday Spacings Tests Consider the sequence $Y_1, \ldots, Y_n$ taking values in $\{0, \ldots, m-1\}$. Sort the sequence as $Y_{(1)} \leqslant \cdots \leqslant Y_{(n)}$ and define spacings $S_1 = Y_{(2)} - Y_{(1)}, \ldots, S_{n-1} = Y_{(n)} - Y_{(n-1)}$, and $S_n = Y_{(1)} + m - Y_{(n)}$. Sort the spacings and denote them as $S_{(1)} \leqslant \cdots \leqslant S_{(n)}$.

Let R be the number of times that we have $S_{(j)} = S_{(j-1)}$ for $j = 1, \ldots, n$. The distribution of R depends on m and n, but for example when $m = 2^{25}$ and $n = 512$, we have [8, Page 71]:

$$\begin{aligned} \mathbb{P}(R = 0) &\approx 0.368801577 \\ \mathbb{P}(R = 1) &\approx 0.369035243 \\ \mathbb{P}(R = 2) &\approx 0.183471182 \\ \mathbb{P}(R \geqslant 3) &\approx 0.078691997\,. \end{aligned}$$

The idea is to repeat the test many times, say $N = 1000$, and perform a χ^2 test on the collected data. Asymptotically, for large n, R has a $\mathsf{Poi}(\lambda)$ distribution, with $\lambda = n^3/(4m)$, where λ should not be large; see [8, Page 570]. An alternative is to use $N = 1$ and base the decision whether to reject H_0 or not on the approximate p-value $\mathbb{P}(R \geqslant r) \approx 1 - \sum_{k=0}^{r-1} \mathrm{e}^{-\lambda} \lambda^k / k!$ (reject H_0 for small values). As a rule of thumb [19] the Poisson approximation is accurate when $m \geqslant (4N\lambda)^4$; that is, $Nn^3 \leqslant m^{5/4}$.

Further Reading

The problem of producing a collection of random numbers has been extensively studied, though as von Neumann said: "Any one who considers arithmetical methods of producing random digits is, of course, in a state of sin." Nevertheless, we can produce numbers that are "sufficiently random" for much of the Monte Carlo simulation that occurs today. A comprehensive overview of random number generation can be found in [15]. The poor lattice structure of certain linear congruential generators was pointed out in [36], adding the concept of "good lattice structure" to the list of qualities a generator ought to have. Afflerbach [1] discusses a number of theoretical criteria for the assessment of random number generators. The celebrated Mersenne twister was introduced in [31], paving the way for generators with massive periods, which have become a necessity in the random number hungry world of Monte Carlo. A discussion of good multiple-recursive generators can be found in [12]. Niederreiter [33] covers many theoretical aspects of random number sequences, and Knuth [8] gives a classic treatment, discussing both the generation of random numbers and evaluation of the quality of same through the use of theoretical and empirical tests. The book by Tezuka [39] is exclusively on random numbers and proves a handy aid when implementing generators and tests. Books by Fishman [5] and Gentle [7] discuss the generation of random numbers for use in Monte Carlo applications. Our treatment of the spectral test draws from [5].

REFERENCES

1. L. Afflerbach. Criteria for the assessment of random number generators. *Journal of Computational and Applied Mathematics*, 31(1):3–10, 1990.
2. L. Blum, M. Blum, and M. Shub. A simple unpredictable pseudo-random number generator. *SIAM Journal on Computing*, 15(2):364–383, 1986.
3. R. R. Coveyou and R. D. MacPherson. Fourier analysis of uniform random number generators. *Journal of the ACM*, 14(1):100–119, 1967.
4. J. Eichenauer-Herrmann. Pseudorandom number generation by nonlinear methods. *International Statistics Review*, 63(2):247–255, 1985.
5. G. S. Fishman. *Monte Carlo: Concepts, Algorithms and Applications.* Springer-Verlag, New York, 1996.
6. G. S. Fishman and L. R. Moore III. An exhaustive analysis of multiplicative congruential random number generators with modulus $2^{31} - 1$. *SIAM Journal on Scientific and Statistical Computing*, 7(1):24–45, 1986.
7. J. E. Gentle. *Random Number Generation and Monte Carlo Methods.* Springer-Verlag, New York, second edition, 2003.
8. D. E. Knuth. *The Art of Computer Programming*, volume 2: Seminumerical Algorithms. Addison-Wesley, Reading, MA, third edition, 1997.
9. P. L'Ecuyer. Random numbers for simulation. *Communications of the ACM*, 33(10):85–97, 1990.
10. P. L'Ecuyer. Uniform random number generation. *Annals of Operations Research*, 53(1):77–120, 1994.
11. P. L'Ecuyer. Bad lattice structure for vectors of non-successive values produced by linear recurrences. *INFORMS Journal of computing*, 9(1):57–60, 1997.

12. P. L'Ecuyer. Good parameters and implementations for combined multiple recursive random number generators. *Operations Research*, 47(1):159 – 164, 1999.

13. P. L'Ecuyer. Software for uniform random number generation: distinguishing the good and the bad. In B. A. Peters, J. S. Smith, D. J. Medeiros, and M. W. Rohrer, editors, *Proceedings of the 2001 Winter Simulation Conference*, pages 95–105, Arlington, VA, December 2001.

14. P. L'Ecuyer. Polynomial integration lattices. In H. Niederreiter, editor, *Monte Carlo and Quasi-Monte Carlo methods*, pages 73–98, Berlin, 2002. Springer-Verlag.

15. P. L'Ecuyer. *Handbooks in Operations Research and Management Science: Simulation. S. G. Henderson and B. L. Nelson, eds.*, chapter 3: Random Number Generation. Elsevier, Amsterdam, 2006.

16. P. L'Ecuyer, J.-F. Cordeau, and R. Simard. Close-point spatial tests and their application to random number generators. *Operations Research*, 48(2):308–317, 2000.

17. P. L'Ecuyer and R. Couture. An implementation of the lattice and spectral tests for multiple recursive linear random number generators. *INFORMS Journal on Computing*, 9(2):206–217, 1997.

18. P. L'Ecuyer and F. Panneton. $\mathbb{F}_2$-linear random number generators. In C. Alexopoulos, D. Goldsman, and J. R. Wilson, editors, *Advancing the Frontiers of Simulation: A Festschrift in Honor of George Samuel Fishman*, pages 175–200, New York, 2009. Springer-Verlag.

19. P. L'Ecuyer and R. Simard. On the performance of birthday spacings tests with certain families of random number generators. *Mathematics and Computers in Simulation*, 55(1–3):131–137, 2001.

20. P. L'Ecuyer and R. Simard. TestU01: A C library for empirical testing of random number generators. *ACM Transactions on Mathematical Software*, 33(4), 2007. Article 22.

21. P. L'Ecuyer, R. Simard, E. J. Chen, and W. W. Kelton. An object-oriented random-number package with many long streams and substreams. *Operations Research*, 50(6):1073–1075, 2002.

22. P. L'Ecuyer, R. Simard, and S. Wegenkittl. Sparese serial tests of uniformity for random number generators. *SIAM Journal of Scientific Computing*, 24(2):652–668, 2002.

23. P. L'Ecuyer and S. Tezuka. Structural properties for two classes of combined random number generators. *Mathematics of Computation*, 57(196):735–746, 1991.

24. P. A. Lewis, A. S. Goodman, and J. M. Miller. A pseudo-random number generator for the system/360. *IBM Systems Journal*, 8(2):136–146, 1969.

25. T. G. Lewis and W. H. Payne. Generalized feedback shift register pseudorandom number algorithm. *Journal of the ACM*, 20(3):456–468, 1973.

26. G. Marsaglia. KISS99. `http://groups.google.com/group/sci.stat.math/msg/b555f463a2959bb7/`.

27. G. Marsaglia. Random numbers fall mainly in the planes. *Proceedings of the National Academy of Sciences of the United States of America*, 61(1):25–28, 1968.

28. G. Marsaglia. DIEHARD: A battery of tests of randomness, 1996. `http://www.stat.fsu.edu/pub/diehard/`.

29. G. Marsaglia. Random number generators. *Journal of Modern Applied Statistical Methods*, 2(1):2–13, 2003.

30. M. Matsumoto and Y. Kurita. Twisted GFSR generators. *ACM Transactions on Modeling and Computer Simulation*, 2(3):179–194, 1992.

31. M. Matsumoto and T. Nishimura. Mersenne twister: A 623-dimensionally equidistributed uniform pseudo-random number generator. *ACM Transactions on Modeling and Computer Simulation*, 8(1):3–30, 1998.

32. H. Niederreiter. Recent trends in random number and random vector generation. *Annals of Operations Research*, 31(1):323–345, 1991.

33. H. Niederreiter. *Random Number Generation and Quasi-Monte Carlo Methods.* SIAM, Philadelphia, 1992.

34. H. Niederreiter. New developments in uniform pseudorandom number and vector generation. In H. Niederreiter and P. J.-S. Shiue, editors, *Monte Carlo and Quasi-Monte Carlo Methods in Scientific Computing*, pages 87–120, New York, 1995. Springer-Verlag.

35. F. Panneton, P. L'Ecuyer, and M. Matsumoto. Improved long-period generators based on linear reccurences modulo 2. *ACM Transactions on Mathematical Software*, 32(1):1–16, 2006.

36. B. D. Ripley. The lattice structure of pseudo-random number generators. *Proceedings of the Royal Society, Series A*, 389(1796):197–204, 1983.

37. A. Rukhin, J. Soto, J. Nechvatal, M. Smid, E. Barker, S. Leigh, M. Levenson, M. Vangel, D. Banks, A. Heckert, J. Dray, and S. Vo. A statistical test suite for random and pseudorandom number generators for cryptographic applications. NIST special publication 800-22, National Institute of Standards and Technology, Gaithersburg, Maryland, USA, 2001. `http://csrc.nist.gov/rng/`.

38. M. Saito and M. Matsumoto. SIMD-oriented fast Mersenne twister: a 128-bit pseudorandom number generator. In *Monte Carlo and Quasi-Monte Carlo Methods 2006*, pages 607 – 622, Berlin, 2008. Springer-Verlag.

39. S. Tezuka. *Uniform Random Numbers: Theory and Practice.* Springer-Verlag, New York, 1995.

40. S. Tezuka, P. L'Ecuyer, and R. Couture. On the add-with-carry and subtract-with-borrow random number generators. *ACM Transactions on Modeling and Computer Simulation*, 3(4):315–331, 1994.

41. B. A. Wichmann and I. D. Hill. Algorithm AS 183: An efficient and portable pseudorandom number generator. *Applied Statistics*, 31(2):188–190, 1982.

42. B. A. Wichmann and I. D. Hill. Correction to algorithm 183. *Applied Statistics*, 33(123), 1984.

43. H. Zeisel. Remark ASR 61: A remark on algorithm AS 183. an efficient and portable pseudo-random number generator. *Applied Statistics*, 35(1):89, 1986.

CHAPTER 2

QUASIRANDOM NUMBER GENERATION

Quasirandom numbers are akin to random numbers but exhibit much more regularity. This makes them well-suited for numerical evaluation of multidimensional integrals. This chapter discusses the main types of quasirandom sequences, including Halton, Faure, Sobol', and Korobov sequences. ☞ 1

2.1 MULTIDIMENSIONAL INTEGRATION

Recall that the purpose of a uniform random number generator is to produce an unlimited stream of numbers $U_1, U_2, \ldots$ that behave statistically as independent and uniformly distributed random variables on $(0, 1)$. From such a stream it is easy to construct an infinite sequence of independent and uniformly distributed random *vectors* (points) in $(0, 1)^d$, by defining $\mathbf{U}_1 = (U_1, \ldots, U_d)$, $\mathbf{U}_2 = (U_{d+1}, \ldots, U_{2d}), \ldots$. For any real-valued function h on $(0, 1)^d$ these random vectors can then be used to approximate the d-dimensional integral

$$\ell = \int_{(0,1)^d} h(\mathbf{u}) \, \mathrm{d}\mathbf{u} \tag{2.1}$$

via the sample average

$$\widehat{\ell} = \frac{1}{N} \sum_{i=1}^{N} h(\mathbf{U}_i) \,. \tag{2.2}$$

Handbook of Monte Carlo Methods. By D. P. Kroese, T. Taimre, Z. I. Botev

Precise error bounds on the approximation error can be found through simple sta- ☞ 306 tistical procedures; see, for example, Algorithm 8.2. In particular, the standard error $\{\mathbb{E}(\widehat{\ell} - \ell)^2\}^{1/2}$ decreases at a rate $\mathcal{O}(N^{-1/2})$. Hence, asymptotically, to decrease the error by a factor 2, one needs 4 times as many samples. This convergence rate can often be improved by constructing **quasirandom** points $\mathbf{u}_1, \mathbf{u}_2, \ldots, \mathbf{u}_N$ that fill the unit cube in a much more regular way than is achieved via iid random points. In general, the components of such points can be zero, so we assume from now on that quasirandom points lie in the unit cube $[0,1)^d$ rather than $(0,1)^d$. **Quasi Monte Carlo** methods are Monte Carlo methods in which the ordinary uniform random points are replaced by quasirandom points. Quasirandom points are no longer independent, but do have a high degree of uniformity, which is often expressed in terms of their discrepancy (first introduced by Roth [27]). Specifically, let $\mathscr{C}$ be a collection of subsets of $[0,1)^d$ and $\mathcal{P}_N = \{\mathbf{u}_1, \ldots, \mathbf{u}_N\}$ a set of points in $[0,1)^d$. The **discrepancy** of $\mathcal{P}_N$ relative to $\mathscr{C}$ is defined as

$$D_{\mathscr{C}}(\mathcal{P}_N) = \sup_{C \in \mathscr{C}} \left| \frac{1}{N} \sum_{i=1}^{N} \mathrm{I}_{\{\mathbf{u}_i \in C\}} - \int \mathrm{I}_{\{\mathbf{u} \in C\}} \, \mathrm{d}\mathbf{u} \right| . \tag{2.3}$$

Special cases are the **ordinary discrepancy**, where $\mathscr{C}$ is the collection of rectangles $[a_1, b_1) \times \cdots \times [a_d, b_d)$, and the **star discrepancy**, where $\mathscr{C}$ is the collection of rectangles $[0, b_1) \times \cdots \times [0, b_d)$.

The sum in (2.3) is simply the number of points in C, whereas the integral is the d-dimensional volume of C. The integration error for all indicator functions $\mathrm{I}_{\{\mathbf{u} \in C\}}$, $C \in \mathscr{C}$ is thus bounded by the discrepancy of the point set. Similarly, the **Koksma–Hlawka inequality** provides, for a suitable class of functions h in (2.1) and (2.2), a bound $|\widehat{\ell} - \ell| \leqslant D^* \, k_h$ on the integration error, where D^* is the star discrepancy and k_h is a constant that depends only on the function h; see [24, Page 19]. Discrepancy measures are therefore useful tools for studying convergence rates for multidimensional integration. Note that the star discrepancy may be viewed as ☞ 336 the d-dimensional generalization of the Kolmogorov–Smirnov test statistic.

■ **EXAMPLE 2.1 (Regular Grid)**

☞ 12 Consider the d-dimensional lattice $\mathbb{Z}^d N^{-1/d}$, where we assume that $N = m^d$ for some strictly positive integer m. By intersecting the lattice with the hypercube $[0,1)^d$ we obtain a regular grid of N points on $[0,1)^d$. The ordinary and star discrepancy for this point set are both $1-(1-m^{-1})^d$, which is of the order $\mathcal{O}(m^{-1}) = \mathcal{O}(N^{-1/d})$.

To see this, take in (2.3) the "worst-case" set $C = [0, 1-m^{-1}]^d = [0, 1-m^{-1}+\varepsilon)^d$ for some infinitesimally small $\varepsilon > 0$. The number of grid points in C is N, while its volume is $(1-m^{-1})^d$. It follows that the integration error for the indicator $\mathrm{I}_{\{\mathbf{u} \in C\}}$ is $1-(1-m^{-1})^d$.

The above example indicates that for $d > 2$ integration with a regular grid is inferior to ordinary Monte Carlo integration. However, it is possible to construct infinite sequences $\mathbf{u}_1, \mathbf{u}_2, \ldots$ of points in $[0,1)^d$ so that any point set $\{\mathbf{u}_1, \mathbf{u}_2, \ldots, \mathbf{u}_N\}$ has star discrepancy

$$D^*(\{\mathbf{u}_1, \ldots, \mathbf{u}_N\}) \approx c_d \, \frac{(\ln N)^d}{N} . \tag{2.4}$$

Note that this is close to $\mathcal{O}(N^{-1})$ for fixed d. Using such **low-discrepancy** sequences instead of ordinary random numbers therefore has the potential of significantly improving the accuracy of the integration.

There are two main classes of low-discrepancy sequences: those based on *van der Corput sequences*, such as the Halton, Faure, and Sobol' point sets; and those based on *lattice methods*, such as the Korobov lattice rule. These are discussed in the sections that follow.

2.2 VAN DER CORPUT AND DIGITAL SEQUENCES

Let $b \geqslant 2$ be an integer. Any number $k \in \mathbb{N}$ admits a ***b*-ary expansion** of the form

$$k = \sum_{i=1}^{r} a_i \, b^{i-1} = a_1 + a_2 b + \cdots + a_r b^{r-1} ,$$

for some finite r and **digits** $a_1, \ldots, a_r \in \{0, \ldots, b-1\}$. The corresponding b-ary representation of k is written as $(a_r \ldots a_1)_b$, or simply $a_r \ldots a_1$ if the **base** or **radix** b is implicitly understood. For example, the number 12345 in decimal representation ($b = 10$) has binary representation 11000000111001_2 and ternary representation 121221020_3.

Many low-discrepancy sequences are based on the following transformation of natural numbers. Let $k \in \mathbb{N}$ have b-ary representation $a_r \ldots a_2 a_1$. The base-b **radical inverse** of k is the number (in [0,1))

$$\sum_{i=1}^{r} a_i \, b^{-i} = a_1 b^{-1} + a_2 b^{-2} + \cdots + a_r b^{-r} ,$$

written in b-ary form as $0.a_1 a_2 \ldots a_r$. The radical inverse transformation thus simply reverses the order of the digits and puts a b-ary point in front to obtain a number in the interval $[0, 1)$. As an example, for $b = 2$ the radical inverse of $880 = 1101110000_2$ is $0.0000111011_2 = \frac{59}{2^{10}}$.

The **base-*b* van der Corput sequence** is the sequence obtained by applying the base-b radical inverse to the numbers $0, 1, 2, \ldots$. The main significance of van der Corput sequences for quasi Monte Carlo is that for each base b the sequence constitutes a one-dimensional low-discrepancy sequence. Another important property of a base-b van der Corput sequence $x_1, x_2, \ldots$ is that it is linearly increasing in subsequent segments of length b; that is, $x_1 < x_2 < \cdots < x_b, \quad x_{b+1} < x_{b+2} < \cdots < x_{2b}$ and so on, with the increments within each segment equal to $1/b$.

■ EXAMPLE 2.2 (Van der Corput Sequence Generation)

The following MATLAB function `vdc.m` calculates the first N elements of the base-b van der Corput sequence. At each iteration the next b-ary representation is determined by the function `nbe.m`. For example, for the binary ($b = 2$) case and starting with 0, subsequent calls to `nbe.m` give the sequence of binary numbers $0, 1, 10, 11, 100, 101, \ldots$ (stored as row vectors). To each of these numbers a reversal of the digits is applied. Finally, each of these "flipped" b-ary numbers is translated back to an ordinary (that is, decimal) representation. The first ten elements of the base-2 van der Corput sequence are thus: $0, 0.5, 0.25, 0.75, 0.125, 0.625, 0.375, 0.875, 0.0625, 0.5625$.

Note that this sequence can be divided into segments of length 2 in which the points are increasing with increment 1/2.

```
function out = vdc(b,N)
out = zeros(N,1);
numd = ceil(log(N)/log(b)); %maximal number of digits required
bb = 1./b.^(1:numd);
a = [];
out(1)=0;
for i=2:N
    a = nbe(a,b); %next b-ary expansion
    fa = fliplr(a); %flip the digits
    out(i) = sum(fa.*bb(1:numel(a)));
end
```

```
function na = nbe(a,b)
numd = numel(a);
na = a;
carry = true;
for i=numd:-1:1
    if carry
        if a(i) == b-1
            na(i) = 0;
        else
            na(i) = a(i) + 1;
            carry = false;
        end
    end
end
if carry
    na = [1,na];
end
```

A base-b van der Corput sequence can thus be thought of in the following way:

1. *Construct a sequence of vectors representing the reverse b-ary representation of the numbers 0,1,2,...:*

$$\begin{pmatrix}0\\0\\0\\\vdots\end{pmatrix},\begin{pmatrix}1\\0\\0\\\vdots\end{pmatrix},\ldots,\begin{pmatrix}b-1\\0\\0\\\vdots\end{pmatrix},\begin{pmatrix}0\\1\\0\\\vdots\end{pmatrix},\begin{pmatrix}1\\1\\0\\\vdots\end{pmatrix},\ldots,\begin{pmatrix}b-1\\1\\0\\\vdots\end{pmatrix},\begin{pmatrix}0\\2\\0\\\vdots\end{pmatrix},\ldots \tag{2.5}$$

2. *Premultiply each vector with the row vector $(b^{-1}, b^{-2}, \ldots)$ to obtain a number in $[0, 1)$.*

Digital sequences are low-discrepancy sequences in d dimensions constructed from van der Corput sequences with bases $b_1, \ldots, b_d$ for each of the components. Their general construction is as follows.

Algorithm 2.1 (Digital Sequence) *For each component $k = 1, \ldots, d$, execute the following steps.*

1. *Construct the reverse b_k-ary representation vectors of the numbers 0,1,2,...,* *as in* (2.5).
2. *Premultiply each such vector with some fixed* **generator matrix** G_k.
3. *Premultiply the resulting vectors with the row vector $(b^{-1}, b^{-2}, \ldots)$ to obtain the k-th coordinate of the digital sequence.*

The generator matrices are usually taken to be right-triangular and such that each $r \times r$ submatrix is non-singular. Notice that a van der Corput sequence is a one-dimensional digital sequence where the generator matrix is the identity matrix.

2.3 HALTON SEQUENCES

The simplest way to construct low-discrepancy sequences in d dimensions is by taking van der Corput sequences with different bases $b_1, \ldots, b_d$ for each of the components. The bases must be pairwise relatively prime; that is, each pair does not share a common factor greater than 1. It is customary to take $b_1, \ldots, b_d$ equal to the first b prime numbers greater than 1. In the framework of Algorithm 2.1, the Halton sequence is a digital sequence in which the generator matrices for the components are simply identity matrices. More precisely, let $b_1, \ldots, b_d \geqslant 2$ be relative prime integers. Let $u_{1j}, u_{2j}, \ldots$ be the base-b_j van der Corput sequence, $j = 1, \ldots, d$. The sequence of d-dimensional points $\mathbf{u}_1 = (u_{11}, \ldots, u_{1d}), \mathbf{u}_2 = (u_{21}, \ldots, u_{2d}), \ldots$ is called the **Halton sequence** corresponding to $b_1, \ldots, b_d \geqslant 2$.

Let $\mathbf{u}_1, \mathbf{u}_2, \ldots$ be a Halton sequence. The set $\{\mathbf{u}_1, \ldots, \mathbf{u}_N\}$ formed by the first N elements of the Halton sequence is the corresponding **Halton point set.** Halton [10] showed that the star discrepancy of such sets is of the order $\mathcal{O}((\ln N)^d/N)$. Hammersley [11] modified the Halton set by defining points $\mathbf{u}_1^* = (u_{11}^*, \ldots, u_{1d}^*), \ldots, \mathbf{u}_N^* = (u_{N1}^*, \ldots, u_{Nd}^*)$ with $u_{i1}^* = (i-1)/N$ and $u_{ij}^* = u_{i,(j-1)}, j = 2, \ldots, d, \ i = 1, \ldots, N$. The point set thus obtained, called a **Hammersley set**, has a discrepancy of $\mathcal{O}((\ln N)^{d-1}/N)$.

■ EXAMPLE 2.3 (Halton and Hammersley Points)

The following two MATLAB programs produce Halton and Hammersley point sets of size N for a given vector of bases $\mathbf{b} = (b_1, \ldots, b_d)$. The difference in uniformity between the two point sets is illustrated in Figure 2.1 for the two-dimensional case with bases $b_1 = 2$ and $b_2 = 3$.

```
%halton.m
function out = halton(b,N);
```

```
dim = numel(b);
out = zeros(N,dim);
for i=1:dim
    out(:,i) = vdc(b(i),N);
end
```

```
function out = hammersley(b,N)
dim = numel(b);
out = zeros(N,dim);
out(2:N,2:dim) = halton(b(1:dim-1),N-1);
out(:,1) = [0:N-1]/N;
```

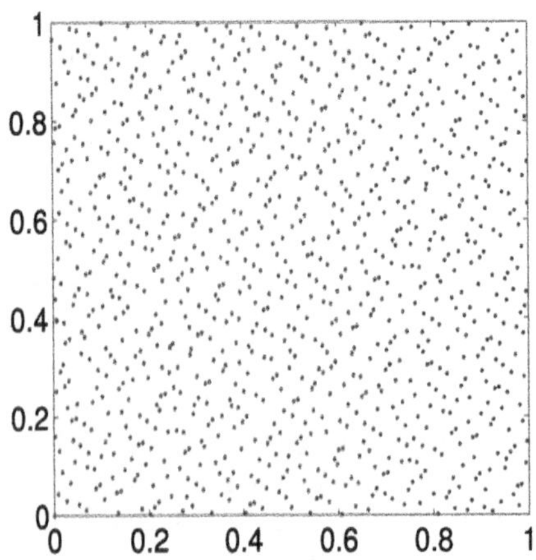

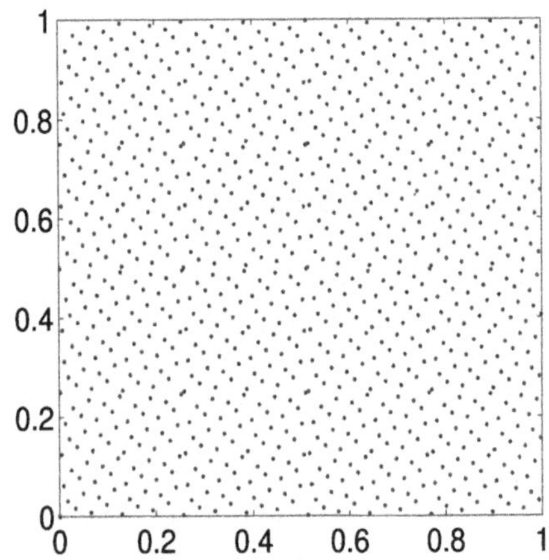

Figure 2.1 The Halton (left) and Hammersley (right) point sets of size $N = 1000$, with bases 2 and 3.

Halton sequences are particularly useful for quasi Monte Carlo integration when d is relatively small. For large d, say $d > 20$, Halton sequences are less effective, because they do not fill the unit hypercube in a uniform way. The reason is that the corresponding van der Corput sequences with large bases produce long linearly increasing segments, as noted in Section 2.2. The situation is illustrated in the following example.

■ EXAMPLE 2.4 (Poor Space-Filling of a Halton Sequence)

Consider a Halton sequence in dimension $d = 40$, where the bases of the van der Corput sequences are chosen to be the first 40 primes. The van der Corput sequences corresponding to the last two coordinates thus form a two-dimensional Halton sequence $(x_1, y_1), (x_2, y_2), \ldots$ with bases 167 and 173. But this sequence does not fill the unit square in an even fashion, as shown in Figure 2.2, where the first 1000 and 6000 points are plotted. The van der Corput sequences $x_1, x_2, \ldots$, and $y_1, y_2, \ldots$ follow the recursion $x_n = x_{n-1} + 1/167$ and $y_n = y_{n-1} + 1/173$, in each subsequent segment of 167 and 173 points. This results in the unit square being filled in a highly linear fashion.

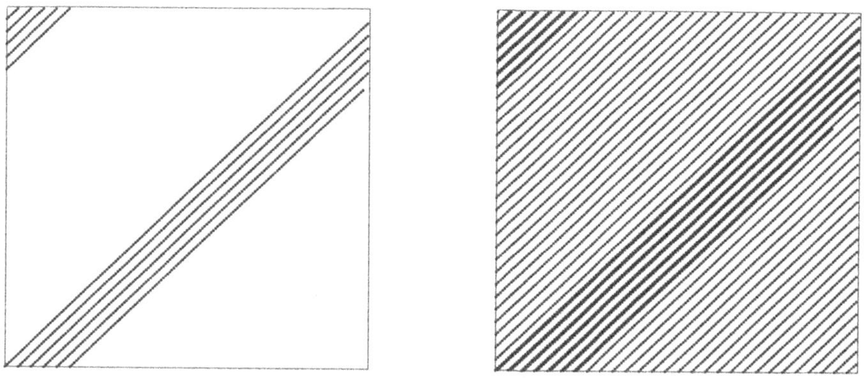

Figure 2.2 Point sets of a two-dimensional Halton sequence with bases 167 and 171 with $N = 1000$ (left) and $N = 6000$ (right) points.

2.4 FAURE SEQUENCES

Faure [5] developed low-discrepancy sequences that, like the sequences of Halton, are based on van der Corput sequences. But, unlike Halton sequences, Faure sequences have a *common* base. This base needs to be a prime number at least as large as the dimension d.

The construction of a d-dimensional Faure sequence is based on d permutations (one for each component) of a base-b van der Corput sequence. More precisely, the Faure sequence is a digital sequence in which the generator matrix for component $k = 1, \ldots, d$ is given by G^{k-1}, where G is the upper-triangular **Pascal matrix** in base b; that is, the matrix with (i, j)-th element

$$G_{ij} = \binom{j-1}{i-1} \bmod b, \quad i = 1, \ldots, j, \quad j = 1, 2, \ldots .$$

For each $r = 0, 1, 2, \ldots$ and $k = 1, \ldots, d$ the submatrix formed by the first b^r rows and columns of G^{k-1} is invertible. As a consequence, premultiplication by G^{k-1} in Step 2 of Algorithm 2.1 leads to a permutation of the first b^r vectors obtained in Step 1, and results (in Step 3) in a permutation of the first b^r numbers in the original van der Corput sequence. To obtain a sequence of length N it suffices to truncate G to the r-th row and column with $r = \lfloor \ln N / \ln b \rfloor + 1$, as no more than r digits will be needed. This leads to the following algorithm.

Algorithm 2.2 (Faure Sequence With Dimension d and Base b)

1. *Set* $r = \lfloor \ln N / \ln b \rfloor + 1$. *Define* $\mathbf{b} = (b^{-1}, b^{-2}, \ldots, b^{-r})$ *and let* $n = 0$.
2. *Calculate the* b*-ary representation* $(a_r \ldots a_1)_b$ *of* n*. Store it in the vector* $\mathbf{a} = (a_1, a_2, \ldots, a_r)^\top$.
3. *For* $k = 1, \ldots, d$ *compute* $\widetilde{\mathbf{a}}_k = G^{k-1}\mathbf{a}$.
4. *For* $k = 1, \ldots, d$ *compute* $x_{nk} = \mathbf{b}\,\widetilde{\mathbf{a}}_k$ *and let* $\mathbf{x}_n = (x_{n1}, \ldots, x_{nd})^\top$ *be the* n*-th Faure point.*
5. *If* $n = N$ *stop; otherwise, set* $n = n + 1$ *and go to Step 2.*

In contrast to Halton sequences, the discrepancy of a Faure sequence does not deteriorate when d grows large. Although both sequences satisfy the low discrepancy growth (2.4), the constant of proportionality c_d grows rapidly in d for the Halton sequence, whereas it decreases to 0 for the Faure sequence; see [5]. Even though the Faure sequence has better discrepancy and uniformity properties than the Halton sequence, for small N and large d (and hence large b) it can still exhibit poor space-filling behavior similar to the Halton sequence illustrated in Figure 2.2.

The amount of uniformity in a Faure sequence is further demonstrated by the fact that certain subsets of points form a $(0, m, d)$-net. More precisely, a set of b^m points in $[0,1)^d$ is said to be **(t, m, d)-net** in base b if every elementary rectangle of volume b^{t-m} contains exactly b^t points. Here an **elementary rectangle** means a d-dimensional rectangle of the form

$$\prod_{i=1}^{d} \left[\frac{a_i}{b^{k_i}}, \frac{a_i + 1}{b^{k_i}} \right) ,$$

with $a_i \in \{0, \ldots, b-1\}$ and $k_i \in \{1, 2, \ldots\}$ for $i = 1, \ldots, n$. The volume of this elementary rectangle is $1/\prod_{i=1}^{d} b^{k_i}$. In terms of (2.3), the discrepancy of a (t, m, d)-net relative to each elementary rectangle of volume b^{t-m} is 0. Note that a (t, m, d)-net is automatically a $(t+1, m, d)$-net.

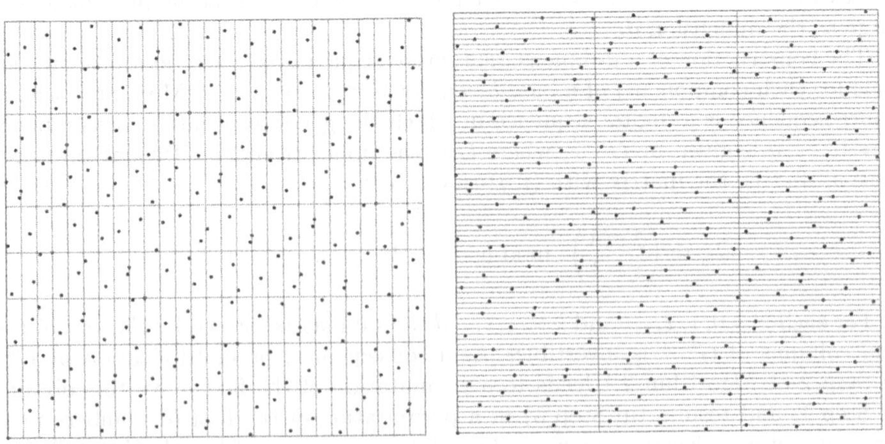

Figure 2.3 The first 3^5 points of a two-dimensional Faure sequence with base $b = 3$ are plotted in each unit square. Both squares are divided into elementary boxes of volume $1/b^5$. Each such box has exactly one point, because the Faure points form a $(0, 5, 2)$-net.

The first b^m points of a d-dimensional Faure sequence form a $(0, m, d)$-net. As an illustration, Figure 2.3 shows various elementary boxes of volume $1/b^m$, for the case where $b = 3$, $m = 5$, and $d = 2$. Each elementary box of volume $1/b^m$ contains exactly one of the $b^m = 243$ points.

■ **EXAMPLE 2.5 (Faure Sequence Generation)**

The following MATLAB function produces a d-dimensional Faure point set of size N, using a base b. The latter must be chosen to be prime and greater than or equal to d. Note that for convenience we work below with the transpose of the generator matrix G and the vector **a**.

```
function p = faure(b,d,N)
r = floor(log(N)/log(b))+1; %largest number of digits
bb=repmat(1./b.^(1:r),N+1,1); %rows (1/b, 1/b^2,...)
p = zeros(N+1,d);
G = zeros(r,r);
for j=1:r
    for i=1:j
        G(i,j) = mod(nchoosek(j-1,i-1),b);
    end
end
G=G';
a=repmat((0:N)',1,r);
for i=1:r-1
    a(:,i)=mod(a(:,i),b);
    a(:,(i+1):r)=floor(a(:,(i+1):r)/b);
    % a now contains the b-ary expansion of 0:N
end
p(:,1)=sum(bb.*a,2);
for k=2:d
    a=mod(a*G,b); %permuted b-ary expansion of 0:N
    p(:,k)=sum(bb.*a,2);
end
```

2.5 SOBOL' SEQUENCES

A d-dimensional Sobol' sequence [29] is a digital sequence (see Algorithm 2.1) where each component has the *same* base 2, and where the $r \times r$ generator matrices (r is the maximal number of digits, that is, $r = \lfloor \ln N / \ln b \rfloor + 1$ if N is the total number of points required) are chosen in the following way. Each generator matrix G (a different one for each component) is defined by r **direction numbers**, $g_1, \ldots, g_r$, whose binary representations form the columns of the generator matrix. The j-th direction number is of the form

$$g_j = \frac{m_j}{2^j}, \quad j = 1, \ldots, r ,$$

in which $m_1, m_2, \ldots, m_r$ satisfy the recursion ($\oplus$ denotes binary addition via the XOR operation)

$$m_i = \left(\bigoplus_{j=1}^{q} 2^j \, c_j \, m_{i-j} \right) \oplus m_{i-q} \,, \tag{2.6}$$

where the $\{c_i\}$ are the (binary) coefficients and q is the degree of a primitive polynomial in binary arithmetic

$$x^q + c_1 \, x^{q-1} + \cdots + c_{q-1} \, x + 1 \,.$$

Each such polynomial can be represented by a **polynomial number**, whose binary representation corresponds to the coefficients. For example, the polynomial number $37 = 100101_2$ corresponds to the primitive polynomial $x^5 + x^2 + 1$. Each generator matrix is thus completely specified by

1. a primitive polynomial, or the corresponding polynomial number, and

2. the initial values $m_1, \ldots, m_q$ for the recursion (2.6).

Lists of primitive polynomial numbers and starting values may be found in [15]; see also the MATLAB implementation in Example 2.6 below. It is standard to take the d lowest polynomial numbers for a d-dimensional problem, starting with the 0-degree polynomial 1, so that the first component is generated according to the base-2 van der Corput sequence. Because Sobol' sequences use binary arithmetic they can be implemented very efficiently on a computer. A benchmark implementation may be found in [1]; see also [16].

Similar to the Faure sequence, the first 2^m points of a d-dimensional Sobol' sequence form a (t, m, d)-net in base 2, for some $t \leqslant 1 - d + \sum_{i=1}^{d} q_i$, where the $\{q_i\}$ are the degrees of the primitive polynomials for the first d components [29]. For example, for a two-dimensional Sobol' sequence $q_1 = 0$ and $q_2 = 1$, so that $t = 0$. Figure 2.4 illustrates that the corresponding Sobol' point set of size $N = 2^{10} = 1024$ forms a $(0, 10, 2)$-net.

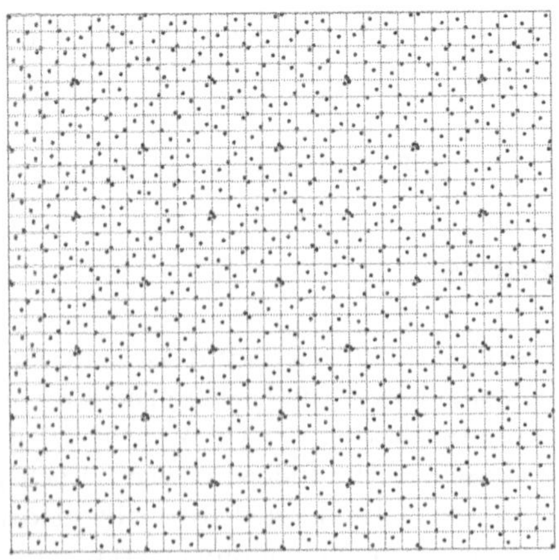

Figure 2.4 The first 2^{10} points of a two-dimensional Sobol' sequence form a $(0, 10, 2)$-net: each elementary box of volume $1/2^{10}$ has exactly one point.

■ **EXAMPLE 2.6 (Sobol' Generator)**

The following MATLAB function `sobol.m` produces d-dimensional Sobol' point sets of size N. The generator matrices are computed via the function `sobmat.m`. This function in turn uses the function `cbe.m` which returns the b-ary representation of a number.

```
function p = sobol(d,N)
b=2; %always base 2
r =  floor(log(N)/log(b))+1;
bb = 1./b.^(1:r);
bbb = repmat(bb,N+1,1);
p = zeros(N+1,d);
G=zeros(r,r);
GG=sobmat(d,r);
a=repmat((0:N)',1,r);
for i=1:r
    a(:,i) = mod(a(:,i),b);
    a(:,(i+1):r) = floor(a(:,(i+1):end)./b);
    % a contains now the b-ary expansion of 0:N
end
for i=1:d
    G(:,:)=GG(:,:,i);
    p(:,i) = sum(bbb.*mod(a*G',b),2);
end
```

```
function G = sobmat(d,r)
polys=[1,3,7,11,13,19,25,37,59,47,61,55,41,67,97,91,109,103,...
      115,131,193,137,145,143,241,157,185,167,229,171,213,191,...
      253,203,211,239,247,285,369,299,425,301,361,333,357,351,...
      501,355,397,391,451,463,487];
ivals=[1 1 1 1 1 1 1 1;          %1
       1 3 5 15 17 51 85 255;  %2
       1 1 7 11 13 61 67 79;   %3
       1 3 7 5 7 43 49 147;    %4
       1 1 5 3 15 51 125 141;  %5
       1 3 1 1 9 59 25 89;     %6
       1 1 3 7 31 47 109 173;  %7
       1 3 3 9 9 57 43 43;     %8
       1 3 7 13 3 35 89 9;     %9
       1 1 5 11 27 53 69 25;  %10
       1 3 5 1 15 19 113 115; %11
       1 1 7 3 29 51 47 97;   %12
       1 3 7 7 21 61 55 19;   %13
       1 1 1 9 23 39 97 97;   %14
       1 3 3 5 19 33 3 197;   %15
       1 1 3 13 11 7 37 101;  %16
```

```
        1 1 7 13 25 5 83 255;   %17
        1 3 5 11 7 11 103 29;   %18
        1 1 1 3 13 39 27 203;   %19
        1 3 1 15 17 63 13 65]; %20

m = zeros(1,r);
G = zeros(r,r,d);
G(:,:,1) = eye(r);

for k=2:d
    ppn=polys(k); %polynomial number
    m=ivals(k,:); %initial values
    c= cbe(ppn,2);
    c = c(2:end); % coefficients of the primitive polynomial
    deg = numel(c); %degree of the polynomial
    for i=9:r %first 8 values already given
        s = 0;
        for j = 1:deg
            s = bitxor(s,2^j*c(j)*m(i-j));
        end
    m(i)= bitxor(s,m(i-deg));
    end
    for j=1:r
        h = cbe(m(j),2); % binary representation
        numdigs = numel(h);
        G(j- numdigs+1:j,j,k)= h';
    end
end
```

```
function a = cbe(k,b) % coefficients of base-b expansion of k
numd = max(0,floor(log(k)/log(b))) + 1; %number of digits
a = zeros(1,numd);
q = b^(numd-1);
for i = 1:numd
    a(i) = floor(k/q);
    k = k - q*a(i);
    q = q/b;
end
```

2.6 LATTICE METHODS

Not all quasirandom point sets are constructed via van der Corput sequences. An important alternative employs the theory of lattices. The most common construction is Korobov's **method of good lattice points** [17], which defines a set of N points in $[0,1)^d$ of the form

$$\left\{ \frac{i}{N}(1, a, a^2, \ldots, a^{d-1}) \bmod 1, \quad i = 0, \ldots, N-1 \right\}, \tag{2.7}$$

depending on the choice of an integer a. The points can be related to a *linear congruential generator* (LCG) with modulus N, multiplier a, and increment $c = 0$: ☞ 4

$$x_t = a\,x_{t-1} \bmod N, \quad t = 1, 2\ldots, N-1 \,. \tag{2.8}$$

Namely, let $\mathcal{P}_{x_0} = \{(x_t, x_{t+1}, \ldots, x_{t+d-1})/N,\, t = 0, 1, \ldots\}$ be the point set of d successive values of the LCG (divided by N), starting from some x_0. Then, the union $\cup_{x_0 \in \{0,1,\ldots,N-1\}} \mathcal{P}_{x_0}$ coincides with the Korobov point set (2.7). A consequence of this viewpoint is that the coordinates of the Korobov point set can also be generated "on the fly", without specifying the dimension in advance; see also [22, Page 205].

By replacing the vector $(1, a, \ldots, a^{d-1})$ in (2.7) with a general integer vector $\mathbf{v}$ or with linear combinations of r such vectors, one obtains **rank-1** and **rank-r lattice rules**, respectively [28]. Another class of lattice rules is defined through more algebraic means, using polynomial rings and formal Laurent series. Details and references on such **polynomial lattices** may be found in [3] and [22].

The selection of good multipliers and generating vectors is discussed, for example, in [4, 12, 20]. To obtain a full period of length $N - 1$ in each coordinate for any a with the Korobov lattice, N should be chosen to be a prime number. If N is a power of 2 instead, a should be an odd number to ensure a full period.

One drawback of the standard Korobov lattice is that the point set is fixed. The introduction of **extensible** lattice rules [13] alleviates this difficulty. The idea, for the Korobov lattice, is to replace (2.7) with the infinite set

$$\{\psi_b(i)(1, a, a^2, \ldots, a^{d-1}) \mod 1, \quad i = 0, 1, 2, \ldots\} \,, \tag{2.9}$$

where $\{\psi_b(k)\}$ is the van der Corput sequence with base b. Hickernell et al. [13] recommend the choice $b = 2$ and $a = 17797$ or $a = 1267$; see also [8].

■ EXAMPLE 2.7 (Korobov Lattice Generation)

The following MATLAB functions implement the Korobov and extensible Korobov lattice rules, using (2.7) rather than the recursion (2.8). The last function uses the van der Corput function vdc in Example 2.2 and assumes by default a base $b = 2$. Figure 2.5 shows the 10-th and 21-st coordinates of a 30-dimensional Korobov point set of size $N = 2^{10}$ with $a = 17797$.

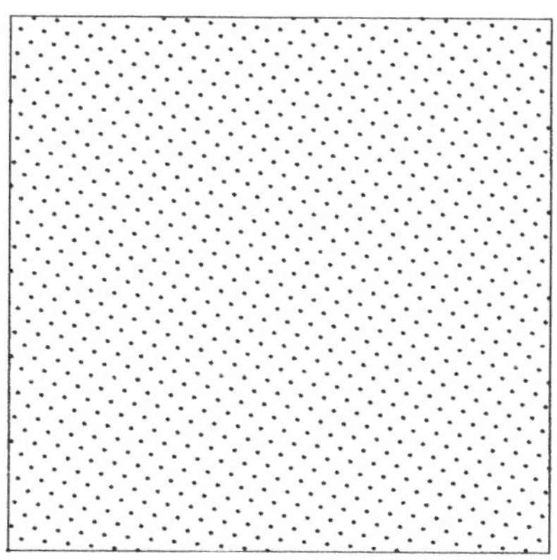

Figure 2.5 The projections onto coordinates 10 and 21 of a 30-dimensional Korobov lattice.

```
function P = korobov(a,d,N)
z(1) = 1;
for i=2:d
    z(i) = mod(z(i-1)*a,N);
end
Z = repmat(z,N,1);
B = repmat((1:N)',1,d);
P = mod(B.*Z/N, 1);
```

```
function P = extkorobov(a,d,N)
b = 2;
z(1) = 1;
for i=2:d
    z(i) = mod(z(i-1)*a,N);
end
Z = repmat(z,N,1);
v = vdc(b,N);
B = repmat(v,1,d);
P = mod(B.*Z, 1);
```

2.7 RANDOMIZATION AND SCRAMBLING

One of the appealing features of ordinary Monte Carlo integration is that an assessment of the error in the sample average approximation (2.2) of the integral (2.1) ☞ 301 is readily available in the form of standard errors and confidence intervals. For quasi Monte Carlo integration this is no longer the case, as the points, $\{\mathbf{u}_i\}$ say, are deterministic and not $\mathsf{U}[0,1)^d$ distributed.

However, the situation can be remedied by simply adding a fixed random vector $\mathbf{Z} \sim \mathsf{U}[0,1)^d$ to each point and then taking the fractional part of the resulting point. It is easy to see that each point $\widetilde{\mathbf{U}}_i = (\mathbf{u}_i + \mathbf{Z}) \mod 1$ is $\mathsf{U}[0,1)^d$ distributed. This procedure is called **random shifting** and was first proposed by Cranley and Patterson [2]. Using a random shift renders the quasi Monte Carlo approximation

$$\widetilde{\ell} = \frac{1}{N} \sum_{i=1}^{N} h(\widetilde{\mathbf{U}}_i) \tag{2.10}$$

a random variable with expectation ℓ in (2.1). By repeating the quasi Monte Carlo procedure independently with K different shift vectors one obtains K independent copies of $\widetilde{\ell}$, to which one can apply the standard statistical techniques for evalu- ☞ 306 ating confidence intervals and standard error, as described in Algorithm 8.2. See ☞ 376 Algorithm 9.11 for a more detailed description.

For digital sequences a random shift can also be applied directly to the digits. Specifically, suppose $\mathbf{a}_k = (a_{k1}, a_{k2}, \ldots)^\top$ is the infinite-dimensional vector that corresponds to the b-ary expansion of the k-th coordinate of a point $\mathbf{u}$; thus, the k-th coordinate of $\mathbf{u}$ is given by

$$u_k = \sum_{i=1}^{\infty} a_{ki}\, b^{-i} \, .$$

Let $\mathbf{W} = (W_1, W_2, \ldots)^\top$ be an infinite-dimensional random vector in which the $\{W_i\}$ are independent and discrete uniformly distributed on $\{0, 1, \ldots, b-1\}$. In other words, $\mathbf{W}$ is the vector representing the b-ary expansion of a $\mathsf{U}[0,1)$-distributed random number. Next, let $\mathbf{W}_1, \ldots, \mathbf{W}_d$ be independent copies of $\mathbf{W}$. By adding $\mathbf{W}_k$ to $\mathbf{a}_k$ modulo b, for $k = 1, \ldots, d$, one obtains vectors $\widetilde{\mathbf{a}}_1, \ldots, \widetilde{\mathbf{a}}_d$ representing the b-ary expansion of $(\mathbf{u} + \mathbf{Z}) \bmod 1$, where $\mathbf{Z} \sim \mathsf{U}[0,1)^d$. By adding the same $\{\mathbf{W}_k\}$ to all the points in the quasi Monte Carlo point set, this **digital shift** procedure yields a point set that has exactly the same distribution as one obtained using the original random shift method.

Digital sequences such as the Halton, Faure, and Sobol' sequences are sometimes "shuffled" with the aim of improving their uniformity and convergence properties. A general procedure, called **nested permutation scrambling**, introduced by Owen (see, for example, [25] and [26]) is to permute the digits in the b_k-ary expansion between Steps 2 and 3 of Algorithm 2.1 for each component $k = 1, \ldots, d$. This can be done in a deterministic or random way. A convenient subset of such procedures is obtained by premultiplying the digital vectors obtained after Step 2 in Algorithm 2.1 with a random lower-triangular matrix L_k with elements in $\{0, 1, \ldots, b_k\}$, for each dimension $k = 1, \ldots, d$. There exist several variants of this procedure (see [23] and [22, Page 207]), but the most common approach is to choose the lower off-diagonal elements of L_k independently and uniformly from $\{0, 1, \ldots, b_k\}$ and the diagonal elements independently and uniformly from $\{1, \ldots, b_k\}$. This leads to the following modification of Algorithm 2.1. We assume for simplicity that the bases for the d components are all equal to b. All matrix operations are carried out modulo b.

Algorithm 2.3 (Uniform Linear Scrambling) *Let* $r = \lfloor \ln N / \ln b \rfloor + 1$. *For each component* $k = 1, \ldots, d$, *execute the following steps.*

1. *Create a random* $r \times r$ *matrix* L_k *as follows:*
   (a) *For* $i = 1, \ldots, r$ *and* $j = 1, \ldots, i-1$ *draw the* (i,j)*-th element of* L_k *uniformly and independently from* $\{0, 1, \ldots, b\}$.
   (b) *For* $i = 1, \ldots, r$ *draw the* (i,i)*-th element of* L_k *uniformly and independently from* $\{1, \ldots, b\}$.
   (c) *Set the remaining elements to 0.*
2. *Construct the reverse* b_k*-ary representation vectors of the numbers* $0, 1, \ldots, N$ *as in* (2.5).
3. *Premultiply each such vector with some fixed* generator matrix G_k.
4. *Premultiply the resulting vectors with* L_k.
5. *Premultiply the resulting vectors with the row vector* $(b^{-1}, b^{-2}, \ldots)$ *to obtain the* k*-th coordinates of the digital sequence* $\mathbf{u}_1, \ldots, \mathbf{u}_N$.

Finally, generate $\mathbf{Z} \sim \mathsf{U}[0,1)^d$ *and return* $\{(\mathbf{u}_i + \mathbf{Z}) \bmod 1, i = 1, \ldots, N\}$ *as the scrambled set.*

Further Reading

A concise introduction to quasi Monte Carlo methods is given in Glasserman [9, Chapter 5], whereas more comprehensive treatments may be found in Niederreiter [24] and Lemieux [22]. An extensive state-of-the-art survey on the theory of quasi Monte Carlo and randomized quasi Monte Carlo is given in [18]. Fox [7] focuses on using quasi Monte Carlo techniques in practice, and Jäckel [15] provides a useful resource for applications in finance. Randomized quasi Monte Carlo is surveyed in [21], and various useful scrambling procedures are discussed in [6, 14, 20, 23]. Efficient methods to implement scrambled digital sequences are discussed in Hong and Hickernell [14]. Related scrambling procedures can be found in [6, 21, 23]. For a collection of research papers on quasi Monte Carlo see [19].

REFERENCES

1. P. Bratley and B. L. Fox. Algorithm 659: Implementing Sobol's quasirandom sequence generator. *ACM Transactions on Mathematical Software*, 14(1):88–100, 1988.
2. R. Cranley and T. N. L. Patterson. Randomisation of number theoretic methods for multiple integration. *SIAM Journal on Numerical Analalysis*, 13(6):904–914, 1976.
3. J. Dick, F. Y. Kuo, F. Pillichshammer, and I. H. Sloan. Construction algorithms for polynomial lattice rules for multivariate integration. *Mathematics of Computation*, 74(252):1895–1921, 2005.
4. K.-T. Fang and Y. Wang. *Number-Theoretic Methods in Statistics*. Chapman & Hall, London, 1994.
5. H. Faure. Discrépance de suites associées à un système de numération. *Comptes Rendus Mathematique*, 286-A:293–296, 1978.
6. H. Faure and S. Tezuka. Another random scrambing of digital (t, s)-sequences. In K. T. Fang, F. J. Hickernell, and H. Niederreiter, editors, *Monte Carlo and Quasi-Monte Carlo Methods 2000*, pages 242–256. Springer-Verlag, Berlin, 2002.
7. B. L. Fox. *Strategies for Quasi-Monte Carlo*. Kluwer Academic Publishers, Norwell, MA, 1999.
8. H. S. Gill and C. Lemieux. Searching for extensible Korobov rules. *Journal of Complexity*, 23(4-6):603–613, 2007.
9. P. Glasserman. *Monte Carlo Methods in Financial Engineering*. Springer-Verlag, New York, 2004.
10. J. H. Halton. On the efficiency of certain quasi-random sequences of points in evaluating multi-dimensional integral. *Numerische Mathematik*, 2(1):84–90, 1960.
11. J. M. Hammersley. Monte Carlo methods for solving multivariable problems. *Annals of the New York Academy of Sciences*, 86(3):844–874, 1960.
12. P. Hellekalek. On the assessment of random and quasi-random point sets. In *Random and Quasi-Random Point Sets*, volume 1:38 of *Lecture Notes in Statistics*, pages 49–108. Springer-Verlag, New York, 1998.
13. F. J. Hickernell, H. S. Hong, P. L'Ecuyer, and C. Lemieux. Extensible lattice sequences for quasi-Monte Carlo quadrature. *SIAM Journal on Scientific Computing*, 22(3):1117–1138, 2000.
14. H. S. Hong and F. J. Hickernell. Algorithm 823: Implementing scrambled digital sequences. *ACM Transactions on Mathematical Software*, 29(2):95–109, 2003.

15. P. Jäckel. *Monte Carlo Methods in Finance.* John Wiley & Sons, New York, 2002.

16. S. Joe and F. Y. Kuo. Remark on algorithm 659: Implementing Sobol's quasirandom sequence generator. *ACM Transactions on Mathematical Software*, 29(1):49–57, 2003.

17. N. M. Korobov. The approximate calculation of multiple integrals using number theoretic methods. *Doklady Academii Nauk SSSR*, 115:1062–1065, 1957.

18. P. L'Ecuyer. Quasi-Monte Carlo methods with applications in finance. *Finance and Stochastics*, 13(3):307–349, 2009.

19. P. L'Ecuyer and A. B. Owen (editors). *Monte Carlo and Quasi-Monte Carlo Methods.* Springer-Verlag, New York, 2010.

20. P. L'Ecuyer and C. Lemieux. Variance reduction via lattice rules. *Management Science*, 46(9):1214–1235, 2000.

21. P. L'Ecuyer and C. Lemieux. *Recent Advances in Randomized Quasi-Monte Carlo Methods*, pages 419–474. Kluwer Academic Publishers, Boston, 2002.

22. C. Lemieux. *Monte Carlo and Quasi-Monte Carlo Sampling.* Springer-Verlag, New York, 2009.

23. J. Matoušek. On the $\mathcal{L}_2$-discrepancy for anchored boxes. *Journal of Complexity*, 14(4):527–556, 1998.

24. H. Niederreiter. *Random Number Generation and Quasi-Monte Carlo Methods.* SIAM, Philadelphia, PA, 1992.

25. A. B. Owen. Randomly permuted (t, m, s)-nets and (t, s)-sequences. In H. Niederreiter and J.-S. Shiue, editors, *Monte Carlo and Quasi-Monte Carlo Methods in Scientific Computing*, pages 299–317. Springer-Verlag, New York, 1995.

26. A. B. Owen. Scrambled net variance for integrals of smooth functions. *Annals of Statistics*, 25(4):1541–1562, 1997.

27. K. F. Roth. On irregularities of distribution. *Mathematika*, 1(2):73–79, 1954.

28. I. H. Sloan and S. Joe. *Lattice Methods for Multiple Integration.* Clarendon Press, Oxford, 1994.

29. I. M. Sobol'. On the distribution of points in a cube and the approximate evaluation of integrals. *USSR Computational Mathematics and Mathematical Physics*, 7(4):86–112, 1967.

CHAPTER 3

RANDOM VARIABLE GENERATION

Generating a random vector $\mathbf{X}$ from an arbitrary distribution in some Euclidean space $\mathbb{R}^d$ invariably involves the following two steps:

1. Draw uniform random numbers $U_1, \ldots, U_k$, for some $k = 1, 2, \ldots$.

2. Return $\mathbf{X} = g(U_1, \ldots, U_k)$, where g is some function from $(0, 1)^k$ to $\mathbb{R}^d$.

The generation of uniform random numbers in the first step is discussed in Chapter 1. The present chapter considers how the second step is implemented. In Section 3.1 we consider various general methods for generating one-dimensional random variables and in Section 3.2 we consider methods for generation of multivariate random variables. Section 3.3 discusses generation methods for miscellaneous random objects, such as random vectors that are uniformly distributed over hyperspheres, ellipsoids, and simplexes. Specific generation algorithms for common discrete and continuous distributions are given in Chapter 4.

☞ 1

☞ 85

All generation methods in this chapter are *exact*, in the sense that each generated random variable has exactly the required distribution (assuming the uniform number generation and computer arithmetic are exact). For an increasing number of Monte Carlo applications exact random variable generation is difficult or impossible to achieve, and *approximate* generation methods are called for, the most prominent being Markov chain Monte Carlo methods; see Chapter 6.

☞ 225

3.1 GENERIC ALGORITHMS BASED ON COMMON TRANSFORMATIONS

Many common distributions and families of distributions are related to each other via simple transformations. Such relations lead to general rules for generating random variables. For example, generating random variables from any location–scale family of distributions can be carried out by generating random variables from the base distribution of the family, followed by an affine transformation. A selection of common transformations is discussed in Section 3.1.2. Universal procedures for generating random variables include the inverse-transform method (Section 3.1.1), the alias method (Section 3.1.4), the composition method (Section 3.1.2.6), and the acceptance–rejection method (Section 3.1.5).

Remark 3.1.1 (Computing With Finite Precision) In our generation algorithms we will assume that we have a source for generating perfect iid $\mathsf{U}(0,1)$ random variables, and that our computing device can manipulate real numbers to infinite precision. However, usually neither of these assumptions is satisfied in practice due to the limitations of finite precision computation.

In computer implementations it is common to use floating-point arithmetic. Floating point numbers (in the IEEE 754-2008 standard) are represented by binary vectors of length d, where $d = 32$ is called **single precision**, $d = 64$ is called **double precision**, and $d = 128$ is called **quadruple precision**. These binary vectors are ordered as

$$(s, e_1, \ldots, e_p, m_1, \ldots, m_q)\,,$$

where s is the **sign**, $(e_1, \ldots, e_p)$ is the **biased exponent**, and $(m_1, \ldots, m_q)$ is the **mantissa** (or **trailing significant field**), and represent the numbers given by

$$\begin{cases} (-1)^s \times \infty & e_1 = \cdots = e_p = 1, m_1 = \cdots = m_q = 0 \\ \texttt{qNaN} & e_1 = \cdots = e_p = 1, m_1 = 1 \\ \texttt{sNaN} & e_1 = \cdots = e_p = 1, m_1 = 0, m_k = 1 \text{ for some } k \\ (-1)^s\, 2^{-b}(\sum_{j=1}^{q} 2^{1-j} m_j) & e_1 = \cdots = e_p = 0 \ \text{(subnormal number)} \\ (-1)^s\, 2^{-b+\sum_{i=1}^{p} 2^{p-i} e_i}(1 + \sum_{j=1}^{q} 2^{-j}\, m_j) & \text{otherwise (normal number)}, \end{cases}$$

where $b = 2^{p-1} - 1$ is called the **bias**. sNaN and qNaN represent two fictitious "numbers" different from $\pm\infty$. The values of p and q for the different precisions are given in the following table.

d	32	64	128
p	8	11	15
q	23	52	112

A practical consequence of this lack of arbitrary precision is that one can inadvertently map a large proportion of uniform random numbers to a single point mass using an absolutely continuous distribution. As an example [11], consider sampling $X \sim \mathsf{Beta}(1, 0.01)$ via the inverse-transform method (Section 3.1.1) — which yields $X = 1 - (1-U)^{100}$. Suppose we use the IEEE 754 standard for double precision floating point numbers and round to the nearest floating point number. In this case, all numbers $X \in (1 - 2^{-52}, 1]$ will be mapped to the floating point 1. Thus, all $U \in (1 - 2^{-13/25}, 1]$ will be mapped to 1. Thus a proportion $2^{-13/25} \approx 0.69737$ of all uniform random variables converted by this inversion are mapped to the single floating point value of 1. But the beta distribution does not have any point masses. Hence, caution should be exercised when dealing with distributions that have significant mass in ranges that cannot be represented using the computer number format in which one is computing.

3.1.1 Inverse-Transform Method

Let X be a random variable with cdf F. Since F is a nondecreasing function, the inverse function F^{-1} may be defined as

$$F^{-1}(y) = \inf\{x : F(x) \geqslant y\} \,, \quad 0 \leqslant y \leqslant 1 \,. \tag{3.1}$$

Let $U \sim \mathsf{U}(0,1)$. The cdf of the inverse transform $F^{-1}(U)$ is given by

$$\mathbb{P}(F^{-1}(U) \leqslant x) = \mathbb{P}(U \leqslant F(x)) = F(x) \,. \tag{3.2}$$

Thus, to generate a random variable X with cdf F, draw $U \sim \mathsf{U}(0,1)$ and set $X = F^{-1}(U)$. This leads to the following general method, illustrated in Figure 3.1, for generating from an arbitrary cdf F.

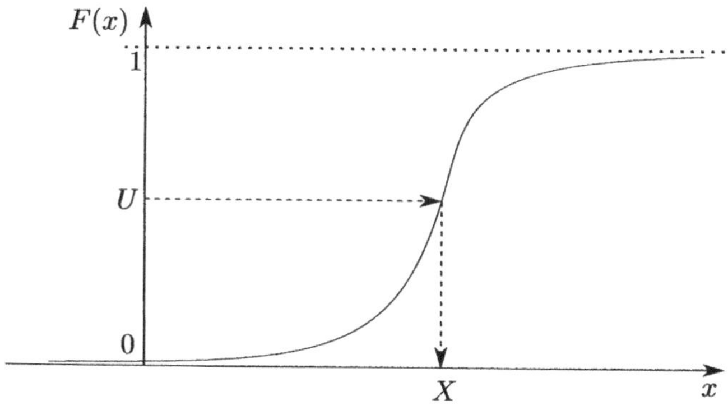

Figure 3.1 Inverse-transform method.

Algorithm 3.1 (Inverse-Transform Method)

1. *Generate* $U \sim \mathsf{U}(0,1)$.
2. *Return* $X = F^{-1}(U)$.

■ **EXAMPLE 3.1 (Illustration of the Inverse-Transform Method)**

Generate a random variable from the pdf

$$f(x) = \begin{cases} 2x, & 0 \leqslant x \leqslant 1 \\ 0, & \text{otherwise.} \end{cases} \tag{3.3}$$

The cdf F is defined by $F(x) = \int_0^x 2y \, \mathrm{d}y = x^2$, $0 \leqslant x \leqslant 1$, the inverse function of which is given by $F^{-1}(u) = \sqrt{u}$ for $0 \leqslant u \leqslant 1$. Therefore, to generate a random variable X from the pdf (3.3), first generate a random variable U from $\mathsf{U}(0,1)$, and then take its square root.

In general, the inverse-transform method requires that the underlying cdf, F, exists in a form for which the corresponding inverse function F^{-1} can be found analytically or algorithmically. Applicable distributions are, for example, the exponential, uniform, Weibull, logistic, and Cauchy distributions. Unfortunately, for many other probability distributions, it is either impossible or difficult to find the inverse transform, that is, to solve

$$F(x) = \int_{-\infty}^{x} f(t)\,dt = u\,,$$

with respect to x. Even in the case where F^{-1} exists in an explicit form, the inverse-transform method may not necessarily be the most efficient random variable generation method (see [5]).

The inverse-transform method applies to both absolutely continuous and discrete distributions. For a discrete random variable X taking values $x_1 < x_2 < \ldots$ with probabilities $p_1, p_2, \ldots$, where $\sum_i p_i = 1$, the cdf is a step function, as illustrated in Figure 3.2.

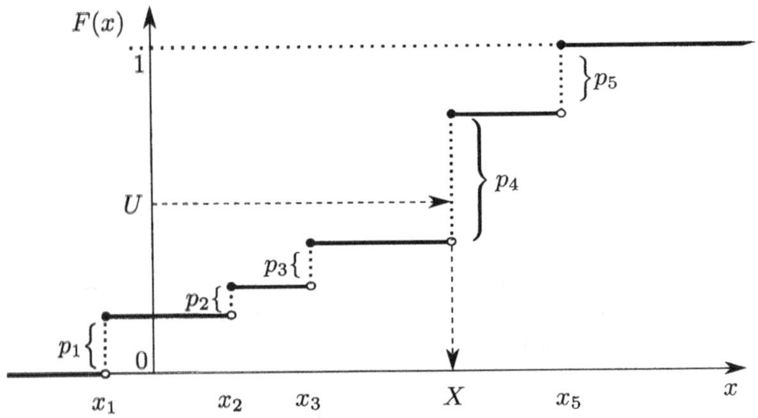

Figure 3.2 Inverse-transform method for a discrete random variable.

For the discrete case the inverse-transform method can be written as follows.

Algorithm 3.2 (Discrete Inverse-Transform Method)

1. *Generate* $U \sim \mathsf{U}(0,1)$.
2. *Find the smallest positive integer* k *such that* $F(x_k) \geqslant U$, *and return* $X = x_k$.

■ **EXAMPLE 3.2 (Discrete Inverse-Transform Implementation)**

Suppose we wish to draw $N = 10^5$ independent copies of a discrete random variable taking values $1, \ldots, 5$ with probabilities $0.2, 0.3, 0.1, 0.05, 0.35$, respectively. The following MATLAB program implements the inverse transform method to achieve this, and records the frequencies of occurrences of $1, \ldots, 5$.

```
%discIT.m
p = [0.2,0.3,0.1,0.05,0.35];
N = 10^5;
x = zeros(N,1);
for i=1:N
    x(i) = min(find(rand<cumsum(p))); %draws from p
end
freq = hist(x,1:5)/N
```

Note that `cumsum(p)` corresponds to the vector of cdf values $(F(1), \ldots, F(5))$. By applying the function `find` first and then `min`, one finds the smallest index k such that $F(k) \geqslant$ `rand`, where `rand` presents a uniform random number. A faster generation program, which uses the function `histc(x,e)` to efficiently count the number of values in a vector `x` that fall between the elements of a vector `e`, is given next.

```
%discinvtrans.m
p = [0.2,0.3,0.1,0.05,0.35];
N = 10^5;
[dummy,x]=histc(rand(1,N),[0,cumsum(p)]);
freq = hist(x,1:5)/N
```

3.1.2 Other Transformation Methods

Many distributions used in Monte Carlo simulation are the result of simple operations on random variables. We list some of the main examples.

3.1.2.1 Affine Transformation Let $\mathbf{X} = (X_1, \ldots, X_n)^\top$ be a random vector, A an $m \times n$ matrix, and $\mathbf{b}$ an $m \times 1$ vector. The $m \times 1$ random vector

$$\mathbf{Z} = A\mathbf{X} + \mathbf{b}$$

is said to be an **affine transformation** of $\mathbf{X}$. If $\mathbf{X}$ has an expectation vector $\boldsymbol{\mu}_\mathbf{X}$, then the expectation vector of $\mathbf{Z}$ is $\boldsymbol{\mu}_\mathbf{Z} = A\boldsymbol{\mu}_\mathbf{X} + \mathbf{b}$. If $\mathbf{X}$ has a covariance matrix $\Sigma_\mathbf{X}$, then the covariance matrix of $\mathbf{Z}$ is $\Sigma_\mathbf{Z} = A\,\Sigma_\mathbf{X}\,A^\top$. Finally, if A is an invertible $n \times n$ matrix and $\mathbf{X}$ has a pdf $f_\mathbf{X}$, then the pdf of $\mathbf{Z}$ is given by

$$f_\mathbf{Z}(\mathbf{z}) = \frac{f_\mathbf{X}(A^{-1}(\mathbf{z} - \mathbf{b}))}{|\det(A)|}, \quad \mathbf{z} \in \mathbb{R}^n ,$$

where $|\det(A)|$ denotes the absolute value of the determinant of A (see Section A.6.1). ☞ 620

3.1.2.2 Location–Scale Family A family of continuous distributions with pdfs $\{f(x; \mu, \sigma), \mu \in \mathbb{R}, \sigma > 0\}$ of the form

$$f(x; \mu, \sigma) = \frac{1}{\sigma} \mathring{f}\left(\frac{x - \mu}{\sigma}\right), \quad x \in \mathbb{R} \tag{3.4}$$

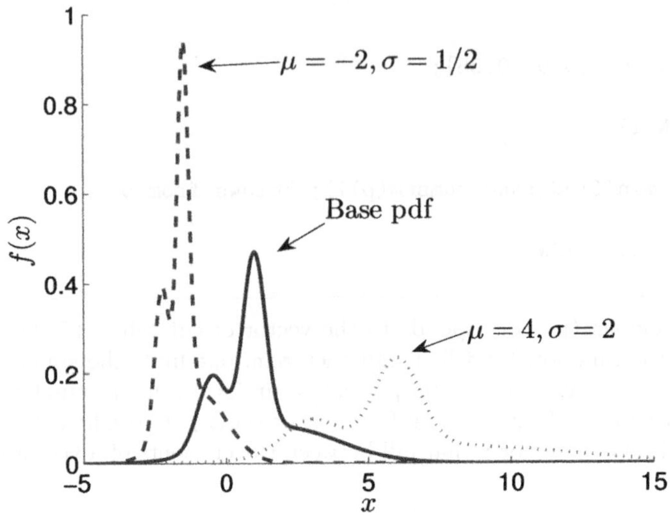

Figure 3.3 A location–scale family of pdfs.

is called a **location–scale family** with **base** (or standard) pdf $\mathring{f}(x)$. Parameter μ is called the **location** and σ is called the **scale**. Families for which (3.4) holds with $\mu = 0$ are called **scale families**. Families for which (3.4) holds with $\sigma = 1$ are called **location families**.

In a location–scale family the graph of the pdf $f(\cdot; \mu, \sigma)$ has the same shape as that of $\mathring{f}(\cdot)$ but is shifted over a distance μ and scaled by a factor σ, as illustrated in Figure 3.3.

☞ 85 Location–scale families include the following distributions (note the anomalous notation for the normal, Student's t, and uniform distributions):

Cauchy(μ, σ)	Fréchet(α, μ, σ)	Gumbel(μ, σ)	Laplace(μ, σ)
Logistic(μ, σ)	N(μ, σ^2)	U$[a, b]$	t$_\nu(\mu, \sigma^2)$

Scale families are often parameterized via $\lambda = 1/\sigma$, where λ is again called the scale parameter. Examples include:

Exp(λ)	Gamma(α, λ)	Pareto(α, λ)	Weib(α, λ)

Location–scale families of distributions arise from the affine transformation

$$Z = \mu + \sigma X ,$$

where X is distributed according to the base or "standard" pdf of the family. In particular, if $X \sim \mathring{f} \equiv f(\cdot; 0, 1)$, then

$$\mu + \sigma X \sim f(\cdot;\mu,\sigma)\,.$$

Thus, to generate a random variable from a location–scale family of pdfs, first generate a random variable from the base pdf and then apply an affine transformation to that random variable.

■ **EXAMPLE 3.3 (Normal Distribution and Location–Scale)**

A typical example of a location–scale family is the normal family of distributions $\{\mathsf{N}(\mu,\sigma^2)\}$ with location parameter μ and scale parameter σ. Here ☞ 122

$$f(x;\mu,\sigma) = \frac{1}{\sigma}\mathring{f}\left(\frac{x-\mu}{\sigma}\right) = \frac{1}{\sqrt{2\pi\sigma^2}}\,\mathrm{e}^{-\frac{1}{2}\frac{(x-\mu)^2}{\sigma^2}}\,,$$

and $\mathring{f}(x) = (2\pi)^{-1/2}\,\mathrm{e}^{-x^2/2}$ is the base pdf. Hence, to draw $Z \sim \mathsf{N}(\mu,\sigma^2)$, first draw $X \sim \mathsf{N}(0,1)$ and then return $Z = \mu + \sigma X$. In MATLAB, drawing from the standard normal distribution is implemented via the function **randn**. For example, the following MATLAB program draws 10^5 samples from $\mathsf{N}(4,9)$ and plots the corresponding histogram.

```
X = randn(1,10^5);  Z = 4 + 3*X;   hist(Z,100)
```

3.1.2.3 Reciprocation Another common transformation is inversion or reciprocation. Specifically, if X is a univariate random variable, then the **inverse** or **reciprocal** of X is

$$Z = \frac{1}{X}\,.$$

If X has pdf f_X, then (see Section A.6.2) Z has pdf ☞ 620

$$f_Z(z) = \frac{f_X(z^{-1})}{z^2}\,, \quad z \in \mathbb{R}\,. \tag{3.5}$$

Distributions obtained in this way are called **inverted** or **inverse distributions**.

■ **EXAMPLE 3.4 (Inverse-Gamma Distribution via Reciprocation)**

The **inverse-gamma** distribution, denoted by $\mathsf{InvGamma}(\alpha,\lambda)$, has pdf

$$f_Z(z;\alpha,\lambda) = \frac{\lambda^\alpha z^{-\alpha-1}\mathrm{e}^{-\lambda z^{-1}}}{\Gamma(\alpha)}\,, \quad z > 0\,,$$

which is of the form (3.5), with f_X the pdf of the $\mathsf{Gamma}(\alpha,\lambda)$ distribution. To generate a random variable $Z \sim \mathsf{InvGamma}(\alpha,\lambda)$, draw $X \sim \mathsf{Gamma}(\alpha,\lambda)$ and return $Z = 1/X$. ☞ 112

Similarly, if $\mathbf{X}$ is an $n \times n$ invertible random matrix with pdf $f_{\mathbf{X}}$, then the inverse matrix $\mathbf{Z} = \mathbf{X}^{-1}$ has pdf

$$f_{\mathbf{Z}}(\mathbf{z}) = \frac{f_{\mathbf{X}}(\mathbf{z}^{-1})}{|\det(J(\mathbf{z}))|}\,, \quad \mathbf{z} \in \mathbb{R}^{n\times n}\,,$$

where $|\det(J(\mathbf{z}))|$ is the absolute value of the determinant of the matrix of Jacobi corresponding to the transformation $\mathbf{x} \mapsto \mathbf{z} = \mathbf{x}^{-1}$. For example, if $\mathbf{x}$ is a general invertible $n \times n$ matrix, then $|\det(J(\mathbf{z}))| = |\det(\mathbf{z})|^{2n}$, and if $\mathbf{x}$ is an $n \times n$ positive definite random matrix, then $|\det(J(\mathbf{z}))| = |\det(\mathbf{z})|^{n+1}$. An example is the distri-
☞ 148 bution of $\mathbf{Z} = \mathbf{X}^{-1}$, where $\mathbf{X} \sim \mathsf{Wishart}(\nu, \Sigma)$. In this case, $\mathbf{X}$ is a positive definite random matrix, and $\mathbf{Z}$ is said to have an **inverse Wishart** distribution.

3.1.2.4 Truncation Let $\mathsf{Dist}_{\mathscr{A}}$ and $\mathsf{Dist}_{\mathscr{B}}$ be two distributions on sets $\mathscr{A}$ and $\mathscr{B} \subset \mathscr{A}$, respectively. Let $\mathbf{X} \sim \mathsf{Dist}_{\mathscr{A}}$ and $\mathbf{Z} \sim \mathsf{Dist}_{\mathscr{B}}$. If the conditional distribution of $\mathbf{X}$ given $\mathbf{X} \in \mathscr{B}$ coincides with the distribution of $\mathbf{Z}$ (that is, $\mathsf{Dist}_{\mathscr{B}}$), then the latter distribution is said to be the **truncation** of $\mathsf{Dist}_{\mathscr{A}}$ to $\mathscr{B}$. In particular, if $f_{\mathbf{X}}$ is the pdf of $\mathbf{X}$, then the pdf of $\mathbf{Z}$ is (in the continuous case)

$$f_{\mathbf{Z}}(\mathbf{z}) = \frac{f_{\mathbf{X}}(\mathbf{z})}{\int_{\mathscr{B}} f_{\mathbf{X}}(\mathbf{x})\, \mathrm{d}\mathbf{x}}, \quad \mathbf{z} \in \mathscr{B}.$$

In the continuous univariate case, the truncation of a pdf $f(x)$ to an interval $[a, b]$ gives the pdf

$$f_Z(z) = \frac{f(z)}{\int_a^b f(x)\, \mathrm{d}x}, \quad a \leqslant z \leqslant b,$$

and in the discrete case we replace the integral with a sum. In terms of cdfs we have:

$$F_Z(z) = \frac{F(z) - F(a-)}{F(b) - F(a-)}, \quad a \leqslant z \leqslant b, \tag{3.6}$$

where $F(a-) = \lim_{x \uparrow a} F(x)$. To generate random variables from a truncated distribution on $[a, b]$ one can simply use the acceptance–rejection method (see Section 3.1.5) by generating $X \sim F$ until $X \in [a, b]$. When the generation of X can be readily performed via the inverse-transform method, a more direct approach can be taken. In particular, the inverse of (3.6) yields the following inverse-transform method.

Algorithm 3.3 (Truncation via the Inverse-Transform Method)

1. *Generate* $U \sim \mathsf{U}(0, 1)$.
2. *Return* $Z = F^{-1}(F(a-) + U(F(b) - F(a-)))$.

Note that the only difference with the inverse-transform method is that in Step 2 the argument of F^{-1} is uniformly distributed on the interval $(F(a-), F(b))$ rather than on $(0, 1)$.

■ **EXAMPLE 3.5 (Truncated Exponential Generator)**

☞ 108 Consider the pdf of the $\mathsf{Exp}(1)$ distribution truncated to the interval $[0, 2]$:

$$f_Z(z) = \frac{\mathrm{e}^{-z}}{1 - \mathrm{e}^{-2}}, \quad 0 \leqslant z \leqslant 2. \tag{3.7}$$

The inverse of the cdf of the $\mathsf{Exp}(1)$ distribution is $F^{-1}(u) = -\ln(1 - u)$, so that

$$Z = -\ln(1 + U(\mathrm{e}^{-2} - 1)) \sim f_Z.$$

The following MATLAB program provides an implementation for generating 10^5 samples from this truncated distribution and plotting the corresponding histogram.

```
%truncexp.m
U= rand(1,10^5); Z = -log(1 + U *(exp(-2) - 1)); hist(Z,100)
```

■ **EXAMPLE 3.6 (Truncated Normal Generator)**

Consider the $\mathsf{N}(\mu, \sigma^2)$ pdf truncated to the interval $[a, b]$:

$$f_Z(z) = \frac{1}{\sigma C}\, \varphi\left(\frac{z-\mu}{\sigma}\right), \quad a \leqslant z \leqslant b\,,$$

where $C = \Phi(\frac{b-\mu}{\sigma}) - \Phi(\frac{a-\mu}{\sigma})$, and φ and Φ are the pdf and cdf of the $\mathsf{N}(0,1)$ distribution, respectively. The following MATLAB function implements the inverse-transform method.

```
function out=normt(mu,sig,a,b)
pb=normcdf((b-mu)./sig);
pa=normcdf((a-mu)./sig);
C=pb-pa;
out=mu+sig.*norminv(C.*rand(size(mu))+pa);
```

■ **EXAMPLE 3.7 (Sampling from the Tail of a Normal Distribution)**

Consider the problem of sampling from the truncated normal pdf

$$f_Z(z) = \frac{\varphi(z)\, \mathrm{I}_{\{z \geqslant a\}}}{\Phi(-a)}\,,$$

where the truncation point $a > 0$ is large, say $a > 10$. A straightforward implementation of the inverse-transform method gives:

$$Z = \Phi^{-1}(\Phi(a) + U\,(1 - \Phi(a))), \quad U \sim \mathsf{U}[0,1]\,.$$

However, this approach is numerically unstable, and in most computer implementations one obtains infinity for the value of Z or an error message when $a > 6.4$. A theoretically equivalent but more numerically stable generator is:

$$Z = -\Phi^{-1}(U\,\Phi(-a)), \quad U \sim \mathsf{U}[0,1]\,.$$

This generator works well for values of a up to $a = 37$. However, it still breaks down in MATLAB for values of $a > 37$. The improved reliability is due to the fact that it is easier to approximate Φ^{-1} in the left tail than in the right tail. This example shows that Algorithm 3.3 should be used with caution and is not prescriptive for all problems.

3.1.2.5 Wrapping A continuous random variable Y is said to be **wrapped** onto the interval $[0, p)$, if

$$Y = X \bmod p$$

for some continuous random variable X; that is, X is the remainder of Y after dividing by $p > 0$. In the univariate case with support on all of $\mathbb{R}$, we have the following result.

Proposition 3.1.1 (Wrapped Random Variables) *Let X be a continuous random variable with pdf f_X on $\mathbb{R}$. Suppose that $\sum_{k=-\infty}^{\infty} f_X(x + k\,p) < \infty$ converges uniformly for $x \in [0, p]$. Then, $Y = (X \bmod p)$ has pdf*

$$f_Y(y) = \sum_{k=-\infty}^{\infty} f_X(y + k\,p), \quad y \in [0, p). \tag{3.8}$$

Generating a random variable Y from the pdf (3.8) can thus be accomplished as follows.

Algorithm 3.4 (Wrapped Random Variable Generator)

1. *Draw $X \sim f_X$.*
2. *Output $Y = X \bmod p$.*

■ **EXAMPLE 3.8 (Wrapped Cauchy Distribution)**

Suppose that X has a $\mathsf{Cauchy}(\mu, \sigma)$ distribution. Then, using (3.8) and the *Poisson summation formula*, the pdf of $Y = (X \bmod p)$ is given by: ☞ 714

$$f_Y(y) = \sum_{k=-\infty}^{\infty} \frac{\sigma/\pi}{\sigma^2 + (y - \mu + k\,p)^2} = \frac{(1 - r^2)/p}{1 - 2\,r\cos(2\pi(y-\mu)/p) + r^2}, \quad y \in [0, p),$$

where $r = \mathrm{e}^{-2\pi\sigma/p}$. The corresponding distribution is known as the **wrapped Cauchy** distribution.

■ **EXAMPLE 3.9 (Wrapped Normal Distribution)**

Suppose that X has a $\mathsf{N}(\mu, \sigma^2)$ distribution. Then, the pdf of $Y = (X \bmod p)$ is given by:

$$\begin{aligned} f_Y(y) &= \sum_{k=-\infty}^{\infty} \frac{1}{\sqrt{2\pi\sigma^2}}\, \mathrm{e}^{-\frac{(y-\mu+kp)^2}{2\sigma^2}} \\ &= \frac{1}{p} \sum_{k=-\infty}^{\infty} \mathrm{e}^{-2k^2\pi^2\sigma^2/p^2} \cos(2\pi k(y-\mu)/p), \quad y \in [0, p), \end{aligned}$$

where the second equality follows from the Poisson summation formula. The distribution of Y is known as the **wrapped normal** distribution.

3.1.2.6 Composition Method Of great practical importance are distributions that are probabilistic mixtures of other distributions. Let $\mathscr{T}$ be an index set and $\{H_t, t \in \mathscr{T}\}$ be a collection of cdfs (possibly multidimensional). Suppose that G is the cdf of a distribution on $\mathscr{T}$. Then

$$F(\mathbf{x}) = \int_{\mathscr{T}} H_t(\mathbf{x})\, \mathrm{d}G(t)\,,$$

is again a cdf and the corresponding distribution is called a **mixture distribution** or simply **mixture**, with **mixing components** $\{H_t, t \in \mathscr{T}\}$. It is useful to think of G as the cdf of a random variable T and H_t as the conditional cdf of a random variable $\mathbf{X}_t$ given $T = t$. Then, F is cdf of the random variable $\mathbf{X}_T$. In other words, if $T \sim G$ and $X_t \sim H_t$, then $X = X_T$ has cdf F. This yields the following generator.

Algorithm 3.5 (Composition Method Generator)

1. *Generate the random variable T according to the cdf G.*
2. *Given $T = t$, generate X from the cdf H_t.*

In many applications G is a distribution on $\{1, \ldots, n\}$ for some strictly positive integer n, in which case the mixture cdf is of the form $F(x) = \sum_{t=1}^{n} p_t\, F_t(x)$ for some collection of cdfs $\{F_t\}$ and probabilities $\{p_t\}$ summing to 1. Denoting the corresponding pdfs by $\{f_t\}$, the pdf f of the finite mixture is given by

$$f(x) = \sum_{t=1}^{n} p_t\, f_t(x)\,. \tag{3.9}$$

■ **EXAMPLE 3.10 (Mixture of Normals)**

We wish to draw samples from a mixture of normal pdfs. Specifically, suppose that the pdf from which to draw has the form (3.9) with $n = 3$ and $(p_1, p_2, p_3) = (0.2, 0.4, 0.4)$, and suppose that the means and standard deviations of the normal pdfs are given by $\boldsymbol{\mu} = (-0.5, 1, 2)$ and $\boldsymbol{\sigma} = (0.5, 0.4, 2)$. A useful shorthand notation for this distribution is

$$0.2\,\mathsf{N}(-0.5, 0.5^2) + 0.4\,\mathsf{N}(1, 0.4^2) + 0.4\,\mathsf{N}(2, 2^2)\,. \tag{3.10}$$

A graph of the corresponding pdf is given as the base pdf in Figure 3.3. The following MATLAB code implements the composition method and plots a histogram of the generated data.

```
%mixturefin.m
p = [0.2, 0.4, 0.4];
mu = [-0.5, 1, 2];
sigma = [0.5, 0.4, 2];
N = 10^5;
[dummy,t]=histc(rand(1,N),[0,cumsum(p)]); % draw from p
x = randn(1,N).*sigma(t) + mu(t);      % draw a normal r.v.
hist(x,200)                     % make a histogram of the data
```

■ **EXAMPLE 3.11 (Composition Method in Bayesian Inference)**

☞ 672 Composition methods appear often in Bayesian analysis. As an example, consider the following Bayesian model for a coin toss experiment. Let θ (random) denote the probability of success (heads) and let X be the number of successes in n tosses. Define the joint distribution of X and θ via the hierarchical model

$$\begin{aligned} \theta &\sim \mathsf{Beta}(\alpha,\beta) && \text{prior distribution,} \\ (X \mid \theta) &\sim \mathsf{Bin}(n,\theta) && \text{likelihood distribution} \end{aligned}$$

for some given $\alpha > 0$ and $\beta > 0$. Using Bayesian notation, we can write for the pdf of X:

$$f(x) = \int f(x \mid \theta)\, f(\theta)\, \mathrm{d}\theta, \quad x = 0, \ldots, n\,,$$

where $f(\theta)$ is the pdf of the $\mathsf{Beta}(\alpha,\beta)$ distribution and $f(x \mid \theta)$ is the pdf of the $\mathsf{Bin}(n,\theta)$ distribution. Note that the distribution of X is a continuous mixture. The mechanism for simulating samples from this distribution using the composition method is given precisely in the Bayesian hierarchical model: first draw θ from $\mathsf{Beta}(\alpha,\beta)$, and then, given θ, draw X from $\mathsf{Bin}(n,\theta)$.

3.1.2.7 Polar Transformation The **polar method** is based on the polar coordinate transformation $X = R\cos\Theta$, $Y = R\sin\Theta$, where $\Theta \sim \mathsf{U}(0,2\pi)$ and $R \sim f_R$ are independent. By the transformation rule (A.33) it follows that the joint pdf of X and Y satisfies ☞ 620

$$f_{X,Y}(x,y) = \frac{f_R(r)}{2\pi r}\,,$$

with $r = \sqrt{x^2+y^2}$, so that

$$f_X(x) = \int_0^\infty \frac{f_R(\sqrt{x^2+y^2})}{\pi\sqrt{x^2+y^2}}\, \mathrm{d}y\,.$$

For example, if $f_R(r) = r\,\mathrm{e}^{-r^2/2}$, then $f_X(x) = \mathrm{e}^{-x^2/2}/\sqrt{2\pi}$. Note that in this case the pdf of R is the same as that of $\sqrt{2E}$ with $E \sim \mathsf{Exp}(1)$. Equivalently, R has the same distribution as $\sqrt{-2\ln U}$ with $U \sim \mathsf{U}(0,1)$. These observations lead to the *Box–Muller* method for generating standard normal random variables. ☞ 123

Interesting relationships between distributions can be obtained from a slight modification of the polar method. Specifically, suppose $R \in [0,\infty)$ and $Z_1, Z_2 \sim_{\text{iid}} \mathsf{N}(0,1)$ are independent random variables. Then, $(X_1, X_2) = R\,(Z_1, Z_2) = (RZ_1, RZ_2)$ has a radially symmetric pdf with radius distributed according to the distribution of $R\sqrt{Z_1^2+Z_2^2}$, or, equivalently, according to the distribution of $R\sqrt{2E}$, where $E \sim \mathsf{Exp}(1)$ is independent of R. For some choices of R the pdf of $R\sqrt{2E}$ is easy, leading to simple generation algorithms for X_1; see, for example, [6].

3.1.2.8 Order Statistics Let $X_1, \ldots, X_n \overset{\text{iid}}{\sim} f$, with cdf F. In many applications one is interested in the distribution of the **order statistics** $X_{(1)}, X_{(2)}, \ldots, X_{(n)}$, where $X_{(1)}$ is the smallest of the $\{X_i\}$, $X_{(2)}$ is the second smallest, and so on. The random variable $R = X_{(n)} - X_{(1)}$ is called the **range** of the data or **sample range** and is a measure for the spread of the data. Some well-known facts about order statistics are given next; see, for example, [2]. Generation of order statistic is discussed in Section 3.3.1.

1. *Cdf of maximum*: $\mathbb{P}(X_{(n)} \leqslant x) = (F(x))^n$.

2. *Cdf of minimum*: $\mathbb{P}(X_{(1)} \leqslant x) = 1 - (1 - F(x))^n$.

3. *Joint pdf*: The joint pdf of the order statistics is given by

$$f_{X_{(1)},\ldots,X_{(n)}}(x_1, \ldots, x_n) = n! \prod_{i=1}^{n} f(x_i) \quad \text{for } x_1 \leqslant x_2 \leqslant \cdots \leqslant x_n \, . \tag{3.11}$$

4. *Marginal pdf*: $f_{X_{(i)}}(x) = n!\, f(x) \, \dfrac{F(x)^{i-1} \, (1 - F(x))^{n-i}}{(i-1)! \, (n-i)!}$.

5. *Subvectors*: The joint pdf of $X_{(i)}$ and $X_{(j)}$ (with $i < j$) is given by

$$\begin{aligned} f_{X_{(i)},X_{(j)}}(x,y) = & \frac{n!}{(i-1)!\,(j-1-i)!\,(n-j)!} f(x) f(y) \\ & \times F(x)^{i-1}\,(F(y) - F(x))^{j-1-i}\,(1 - F(y))^{n-j}, \quad x < y \, . \end{aligned}$$

3.1.2.9 Products Taking products of independent random variables provides a convenient way to generate random variables from many distributions. The distribution of the product of two independent random variables is sometimes called a **scale mixture** [6].

■ **EXAMPLE 3.12 (Normal Scale Mixture)**

Consider the random variable $Z = XY$ with $X \sim \mathsf{N}(0,1)$ and $Y = \sqrt{2V}$, where $V \geqslant 0$ is independent of X. Then, the characteristic function of Z is given by

$$\phi_Z(t) = \mathbb{E}\mathrm{e}^{\mathrm{i}\,tZ} = \mathbb{E}\,\mathbb{E}[\mathrm{e}^{\mathrm{i}(t\sqrt{2V})X} \mid V] = \mathbb{E}\mathrm{e}^{-t^2 V} = \phi_V(\mathrm{i}\,t^2) \, ,$$

where ϕ_V is the characteristic function of V. As a particular case, if $V \sim \mathsf{Exp}(1)$, then $\phi_V(t) = 1/(1 - \mathrm{i}\,t)$, and hence $\phi_Z(t) = 1/(1 + t^2)$, which is the characteristic function of the $\mathsf{Laplace}(0,1)$ distribution. ☞ 118

3.1.3 Table Lookup Method

One of the easiest and fastest general methods for generating discrete random variables is Marsaglia's **table lookup method** [15].

Algorithm 3.6 (Table Lookup Method)

1. *Draw* $U \sim \mathsf{U}(0,1)$.

2. *Set* $I = \lceil Un \rceil$.

3. *Return* $\mathbf{X} = a_I$.

Here $(a_1, \ldots, a_n)$ is a predetermined table of numbers or, more generally, objects such as vectors, trees, etc. Duplication among the $\{a_i\}$ is allowed. If the set of *distinct* objects is $\{b_1, \ldots, b_k\}$, then the algorithm generates random variables $\mathbf{X}$ that satisfy

$$\mathbb{P}(\mathbf{X} = b_i) = \frac{\sum_{j=1}^{n} \mathrm{I}_{\{a_j = b_i\}}}{n} = \frac{\#\{j : a_j = b_i\}}{n}, \quad i = 1, \ldots, k .$$

■ **EXAMPLE 3.13 (Random Variable Generation via Table Lookup)**

Suppose we wish to generate from the discrete pdf f with

$$f(x) = \frac{x}{55}, \quad x = 1, \ldots, 10 .$$

This can be done via table lookup using a table of size $n = 55$ with elements $1, 2, 2, 3, 3, 3, \ldots, 10, \ldots, 10$. The following MATLAB program creates the lookup table, generates 10^5 random variables from f via the lookup method, and plots the histogram of the generated data.

```
%tablook.m
r = 10;
a = zeros(1,(r+1)*r/2);
n=0;
for i=1:r
    for j=1:i
        n = n+1;
        a(n) = i;
    end
end
I = ceil(rand(1,10^5)*n);
X = a(I);
hist(X,1:r)
```

The table lookup method is a *resampling* technique: given data $\{a_i\}$ the algorithm resamples the data by selecting one of the a_i uniformly and independently each time. In other words, Algorithm 3.6 generates samples from the empirical distribution of the data $\{a_i\}$. This is of particular importance in the *bootstrap method*; see Section 8.6.

☞ 331

3.1.4 Alias Method

The **alias method** [22] is an alternative to the inverse-transform method for generating discrete random variables, which does not require time-consuming search techniques as in Step 2 of Algorithm 3.2. It is based on the fact that an arbitrary n-point distribution can be represented as an equally weighted mixture of n two-point distributions. The idea is to redistribute the probability mass into n bins of equal weight $1/n$, as illustrated in Figure 3.4.

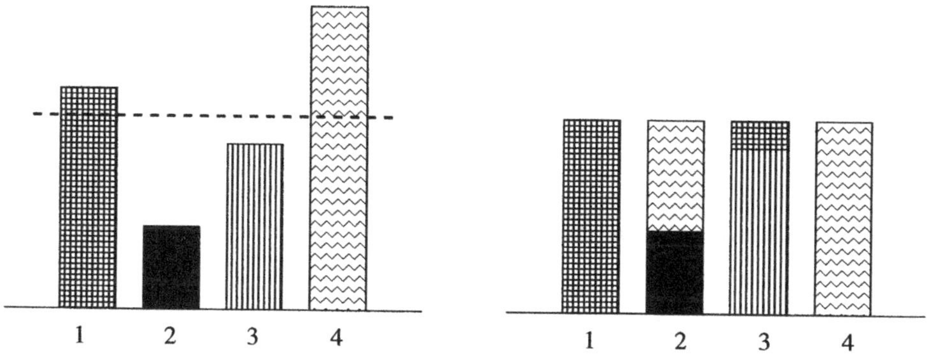

Figure 3.4 Redistribution of probability mass.

Here, a probability distribution on $\{1, 2, 3, 4\}$ is depicted on the left side, with probability masses $8/28, 3/28, 6/28$, and $11/28$. These masses are redistributed over four bins such that (1) the total capacity of each bin is $1/4$, (2) each bin has masses corresponding to at most two variables, (3) bin i contains mass corresponding to variable i, $i = 1, 2, 3, 4$.

To see that such a redistribution can be done generally, consider a probability distribution on $\{1, \ldots, n\}$ with probability mass $p_i > 0$ assigned to i, $i = 1, \ldots, n$. If $p_1 = \ldots = p_n$, then, trivially, the original distribution is an equal mixture of 1-point (and hence 2-point) distributions. If not all $\{p_k\}$ are equal, then there must exist indices i and j such that $p_i < 1/n$ and $p_j \geqslant 1/n$. Now fill bin i by first adding p_i and then transferring an amount $1/n - p_i$ from p_j. This leaves $n - 1$ bins to be filled with $n - 1$ probabilities that sum up to $(n-1)/n$, which can be done in exactly the same way by choosing i' and j' from the remaining indices such that $p_{i'} < 1/n$ and $p_{j'} \geqslant 1/n$, and redistributing their weights, and so on. At the end, each bin $k = 1, \ldots, n$ corresponds to a 2-point distribution at the points k and another point a_k, with probabilities q_k and $1 - q_k$, respectively. For example, in Figure 3.4, $a_2 = 4$ and $q_2 = 3/28 \times 4 = 3/7$. The $\{a_k\}$ are called the **alias** values and the $\{q_k\}$ the **cut-off** values. These can be determined by the following algorithm, which formalizes the bin-filling procedure described above.

Algorithm 3.7 (Set-up for the Alias Method) *Let $\{p_k, k = 1, \ldots, n\}$ be a distribution on $\{1, \ldots, n\}$.*

1. *Let $q_k = n\, p_k, k = 1, \ldots, n$. Let $\mathscr{S} = \{k : q_k < 1\}$ and $\mathscr{G} = \{k : q_k \geqslant 1\}$.*
2. *While $\mathscr{S}$ and $\mathscr{G}$ are not empty,*
   (a) *Choose some $i \in \mathscr{S}$ and $j \in \mathscr{G}$.*
   (b) *Set $a_i = j$ and $q_j = q_j - (1 - q_i)$.*
   (c) *If $q_j < 1$, remove j from $\mathscr{G}$ and add to $\mathscr{S}$.*
   (d) *Remove i from $\mathscr{S}$.*

The set-up algorithm can be implemented to run in $\mathcal{O}(n)$ time [5, 19]. Once the alias and cut-off values have been established, generation of a random variable X from the distribution $\{p_k\}$ is simple and can be written as follows.

Algorithm 3.8 (Alias Method)

1. *Generate* $U \sim \mathsf{U}(0,1)$ *and set* $K = \lceil nU \rceil$.
2. *Draw* $V \sim \mathsf{U}(0,1)$. *If* $V \leqslant q_K$, *return* $X = K$; *otherwise, return* $X = a_K$.

■ **EXAMPLE 3.14 (Alias Method)**

The following MATLAB program shows how the alias method works in practice. The objective is to generate 10^6 samples from a fixed 400-point pdf that is itself randomly generated. In the first part of the program the alias and cut-off values are calculated. The second part checks that the original probabilities are faithfully reconstructed. In the last part the data are generated.

```
%aliasfin.m
p =rand(1,400);p = p/sum(p); %the distribution from which to sample
n = size(p,2);
a = 1:n; %alias values
q = zeros(1,n); % cut-off values
q = n*p;
greater = find(q >= 1);
smaller = find(q < 1);
while (~isempty(smaller) && ~isempty(greater))
    i = smaller(1);
    j = greater(1);
    a(i) = j;
    q(j) = q(j) -(1- q(i));
    if (q(j) < 1)
        greater = setdiff(greater,j);
        smaller = union(smaller,j);
    end
    smaller = setdiff(smaller,i);
end
pp = q/n;
for i = 1:n
    ind = find(a == i);
    pp(i) = pp(i) + sum((1 - q(ind)))/n;
end
max(abs(pp - p))
N = 10^6; % generate sample of  size N
X = zeros(1,N);
for i = 1:N
    K = ceil(rand*n);
    if (rand > q(K));
        X(i) = a(K);
    else
        X(i) = K;
    end
end
```

3.1.5 Acceptance–Rejection Method

The acceptance–rejection method is one of the most useful general methods for sampling from general distributions. It can be applied to both discrete and continuous distributions, and even to multidimensional distributions — although its efficiency rapidly decreases with the number of dimensions (see Section 3.3.3). The method is based on the following observation.

Theorem 3.1.1 (Acceptance–Rejection) *Let $f(\mathbf{x})$ and $g(\mathbf{x})$ be two pdfs such that for some $C \geqslant 1$, $C\, g(\mathbf{x}) \geqslant f(\mathbf{x})$ for all $\mathbf{x}$. Let $\mathbf{X} \sim g(\mathbf{x})$ and $U \sim \mathsf{U}(0,1)$ be independent. Then, the conditional pdf of $\mathbf{X}$ given $U \leqslant f(\mathbf{X})/(Cg(\mathbf{X}))$ is $f(\mathbf{x})$.*

Proof: Consider the joint pdf of $\mathbf{X}$ and U, which is

$$\begin{aligned} f_{\mathbf{X},U}(\mathbf{x},u) &= \frac{g(\mathbf{x})\,\mathrm{I}_{\left\{u \leqslant \frac{f(\mathbf{x})}{Cg(\mathbf{x})}\right\}}}{\int \int_0^1 g(\mathbf{x})\,\mathrm{I}_{\left\{u \leqslant \frac{f(\mathbf{x})}{Cg(\mathbf{x})}\right\}}\,\mathrm{d}u\,\mathrm{d}\mathbf{x}} = \frac{g(\mathbf{x})\,\mathrm{I}_{\left\{u \leqslant \frac{f(\mathbf{x})}{Cg(\mathbf{x})}\right\}}}{\int g(\mathbf{x})\Big(\int_0^{\frac{f(\mathbf{x})}{Cg(\mathbf{x})}} 1\,\mathrm{d}u\Big)\mathrm{d}\mathbf{x}} \\ &= C\, g(\mathbf{x})\,\mathrm{I}_{\left\{u \leqslant \frac{f(\mathbf{x})}{Cg(\mathbf{x})}\right\}}\,. \end{aligned}$$

The (marginal) pdf of $\mathbf{X}$ is therefore

$$f_{\mathbf{X}}(\mathbf{x}) = \int_0^1 f_{\mathbf{X},U}(\mathbf{x},u)\,\mathrm{d}u = C\, g(\mathbf{x})\frac{f(\mathbf{x})}{Cg(\mathbf{x})} = f(\mathbf{x})\,,$$

as required.

We call $g(\mathbf{x})$ the **proposal** pdf and assume that it is easy to generate random variables from it. The acceptance–rejection method can be formulated as follows (see, for example, [21, Page 55]).

Algorithm 3.9 (Acceptance–Rejection)

1. *Generate $\mathbf{X}$ from $g(\mathbf{x})$.*

2. *Generate U from $\mathsf{U}(0,1)$, independently of $\mathbf{X}$.*

3. *If $U \leqslant f(\mathbf{X})/(C\, g(\mathbf{X}))$ output $\mathbf{X}$; otherwise, return to Step 1.*

In other words, generate $\mathbf{X} \sim g$ and accept it with probability $f(\mathbf{X})/(C\, g(\mathbf{X}))$; otherwise, reject $\mathbf{X}$ and try again.

The **efficiency** of the acceptance–rejection method is defined as the probability of acceptance, which is,

$$\mathbb{P}\left(U \leqslant \frac{f(\mathbf{X})}{Cg(\mathbf{X})}\right) = \int g(\mathbf{x}) \int_0^1 \mathrm{I}_{\left\{u \leqslant \frac{f(\mathbf{x})}{Cg(\mathbf{x})}\right\}}\mathrm{d}u\,\mathrm{d}\mathbf{x} = \int \frac{f(\mathbf{x})}{C}\,\mathrm{d}\mathbf{x} = \frac{1}{C}\,.$$

Since the trials are independent, the number of trials required to obtain a successful pair $(\mathbf{X}, U)$ has a $\mathsf{Geom}(1/C)$ distribution, so that the expected number of trials is equal to C.

■ **EXAMPLE 3.15 (Generating from the Positive Normal Distribution)**

Suppose we wish to generate random variables from the **positive normal** pdf

$$f(x) = \sqrt{\frac{2}{\pi}}\, \mathrm{e}^{-x^2/2}, \qquad x \geqslant 0\,, \tag{3.12}$$

using acceptance–rejection. We can bound $f(x)$ by $C\,g(x)$, where $g(x) = \mathrm{e}^{-x}$ is the pdf of the $\mathsf{Exp}(1)$ distribution. The smallest constant C such that $f(x) \leqslant Cg(x)$ is $\sqrt{2\mathrm{e}/\pi}$. The pdf $f(x)$ and the dominating function $Cg(x)$ are depicted in Figure 3.5. The efficiency of this method is $\sqrt{\pi/2\mathrm{e}} \approx 0.76$.

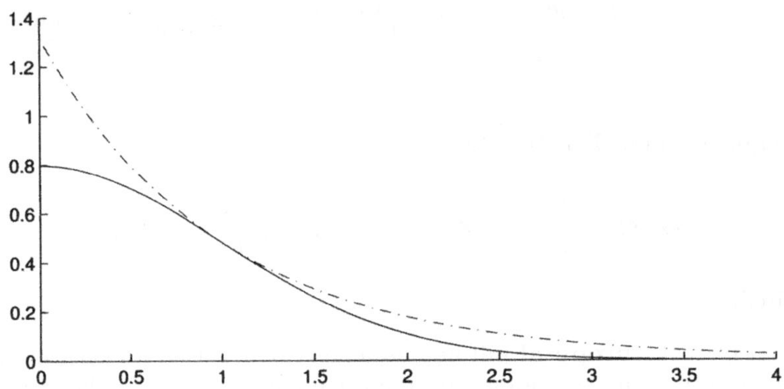

Figure 3.5 Bounding the positive normal density (solid curve).

Since $f(x)$ is the pdf of the absolute value of a standard normal random variable, we can generate $Z \sim \mathsf{N}(0,1)$ by first generating $X \sim f$ as above and then returning $Z = XS$, where S is a random sign; for example, $S = 1 - 2\mathrm{I}_{\{U \leqslant 1/2\}}$ with $U \sim \mathsf{U}(0,1)$. This procedure for generating $\mathsf{N}(0,1)$ random variables is summarized in
☞ 124 Algorithm 4.50.

An acceptance–rejection algorithm can often be made more efficient and faster by employing a so-called **squeeze**. The idea is to "squeeze" the target function $f(\mathbf{x})$ between the dominating function $C\,g(\mathbf{x})$ and a simple (for example, piecewise linear) function $s(\mathbf{x})$, such that

$$s(\mathbf{x}) \leqslant f(\mathbf{x}) \leqslant C\,g(\mathbf{x}) \quad \text{for all } \mathbf{x}\,,$$

as is illustrated in Figure 3.6 for a one-dimensional pdf.

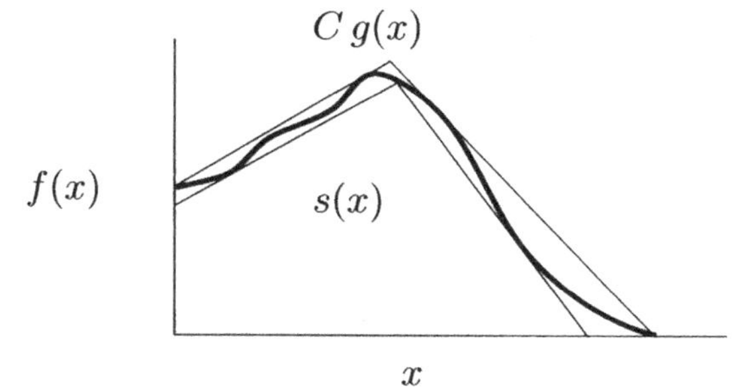

Figure 3.6 Squeeze function.

The advantage of using a squeeze function is that the acceptance–rejection condition (Step 3 of Algorithm 3.9) can be carried out more efficiently: to check if $U \leqslant f(\mathbf{X})/(C\,g(\mathbf{X}))$, first check if $U \leqslant s(\mathbf{X})/(C\,g(\mathbf{X}))$. The latter step is usually much less time-consuming.

EXAMPLE 3.16 (Nearly Linear Densities)

To illustrate the squeeze principle, consider the generation of random variables from "nearly linear" densities. These are one-dimensional densities that can be bounded as

$$a - b\,x/h \leqslant f(x) \leqslant b - b\,x/h\,, \quad 0 \leqslant x \leqslant h\,,$$

where a is less than but close to b, as illustrated on the left-hand side of Figure 3.7. By shifting the pdf over a distance d one obtains nearly linear densities on $[d, d+h]$ rather than $[0, h]$. Often a complicated pdf such as that of the standard normal distribution can be decomposed as a mixture of nearly linear densities; see for example [14, Page 124].

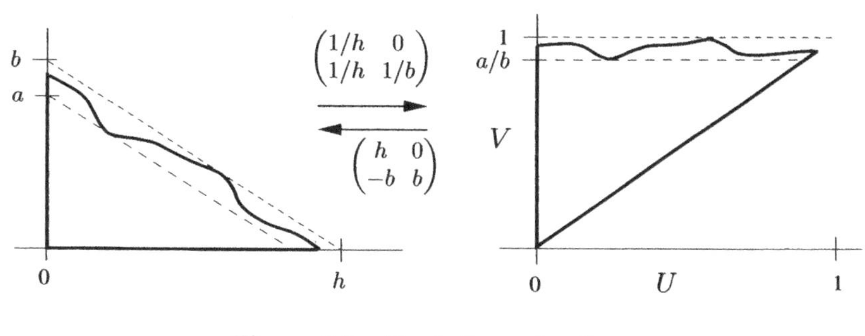

Figure 3.7 Nearly linear density.

To sample from the above almost linear density using acceptance–rejection, it is convenient to first linearly transform (x, y) to $(u, v) = (x/h, x/h + y/b)$. This transforms the triangle $(0,0) - (h,0) - (0,b)$ into the triangle $(0,0) - (1,1) - (0,1)$, and a point $(x, f(x))$ is mapped to $(x/h, x/h + f(x)/b) = (u, u + f(hu)/b)$ — see Figure 3.7. The acceptance–rejection procedure on the original pdf is now equivalent to the following algorithm, which uses the constant a/b as a squeeze.

Algorithm 3.10 (Sampling From a Nearly Linear Density)

1. *Draw* $U \sim \mathsf{U}(0,1)$ *and* $V \sim \mathsf{U}(0,1)$ *independently. If* $U > V$, *exchange* U *and* V.

2. *If* $V \leqslant a/b$ *go to Step 4.*

3. *If* $V > U + f(h\,U)/b$, *go back to Step 1; otherwise, continue with Step 4.*

4. *Return* $X = h\,U$.

3.1.5.1 Transformed Acceptance–Rejection The **transformed acceptance–rejection** method [23] combines ideas from the acceptance–rejection and the inverse-transform methods. The method uses a transformation in order to increase the acceptance probability in an acceptance–rejection step and improve overall efficiency. The transformation has to be simple and yet it must be a good approximation to the inverse cdf.

Theorem 3.1.2 (Transformation of a Random Variable) *Let* f *be the pdf of an absolutely continuous distribution. Let* G *be a nondecreasing and almost everywhere differentiable function, with derivative* G' *and inverse* G^{-1}. *If* $U \sim h(u) = f(G(u))\,G'(u)$, *then* $X = G(U) \sim f$.

☞ 620 *Proof:* This is an immediate consequence of the transformation rule (A.33). Namely, with $x = G(u)$, we have

$$f_X(x) = \frac{f_U(u)}{G'(u)} = \frac{f(G(G^{-1}(x)))\,G'(u)}{G'(u)} = f(x)\,,$$

as had to be shown.

Typically, h is a pdf on either the interval $(0,1)$ or on $(-1/2,1/2)$. Note that $h(u) = (F(G(u)))'$, where F is the cdf of f. Hence, the choice $G = F^{-1}$ simply recovers the inverse-transform method for generating $X \sim f$. However, choosing $G = F^{-1}$ is not always computationally efficient. Instead one can choose G to be a simple function such that h is as close as possible to some uniform pdf, and use this uniform pdf as a proposal density in an acceptance–rejection procedure. Assuming that h is a pdf on $(-1/2,1/2)$, the general algorithm is as follows.

Algorithm 3.11 (Transformed Acceptance–Rejection Method) *Let* $C \geqslant \max_u h(u)$.

1. *Generate independently* $U \sim \mathsf{U}(-1/2,1/2)$ *and* $V \sim \mathsf{U}(0,1)$.

2. *If* $V \leqslant h(U)/C$ *output* $G(U)$; *otherwise, return to Step 1.*

■ EXAMPLE 3.17 (Normal Distribution)

Hörmann and Derflinger [10] describe a transformed acceptance–rejection method for the $\mathsf{N}(0,1)$ distribution that uses the simple function

$$G(u) = \frac{2\,a\,u}{1/2 - |u|} + b\,u, \quad -1/2 < u < 1/2\,,$$

so that

$$G'(u) = \frac{a}{(1/2 - |u|)^2} + b\,.$$

By numerical search the following optimal parameters were found:

$$a = 0.062794, \quad b = 2.530885\,,$$

which gives an acceptance probability of $1/C \approx 0.8904302215$. The function $h(u)/C$ is plotted in Figure 3.8, along with a squeeze function $s(u) = v_r\,\mathrm{I}_{\{-u_r \leqslant u \leqslant u_r\}}$, with $u_r = 0.4359971734$ and $v_r = 0.9296123611$.

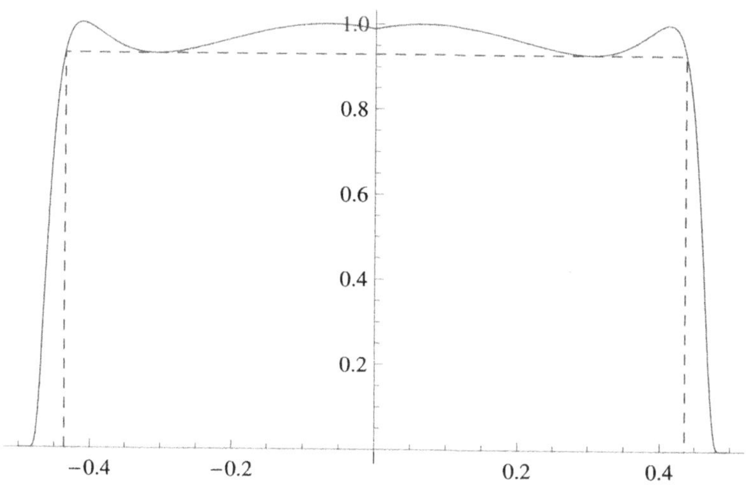

Figure 3.8 Transformed acceptance–rejection for the $\mathsf{N}(0,1)$ distribution. The squeeze function is indicated by dashed lines.

3.1.5.2 Generation From a Log-Concave Density We say that a density f on $\mathscr{X} \subseteq \mathbb{R}$ is **log-concave** if the logarithm of f is a *concave* function. In particular, if f is differentiable, log-concavity implies that

$$\frac{\mathrm{d}^2}{\mathrm{d}x^2} \ln f(x) \leqslant 0 \quad \text{for all } x \in \mathscr{X}\,,$$

and if f is discrete, log-concavity implies that

$$\frac{\ln f(a) + \ln f(c)}{2} \leqslant 2 \ln f(b) \quad \text{for all } a \leqslant b \leqslant c\,.$$

Examples of log-concave distributions include the following:

$\mathsf{N}(\mu, \sigma^2)$	$\mathsf{Exp}(\lambda)$	$\mathsf{Logistic}(\mu, \sigma)$	$\chi^2_n,\ n \geqslant 2$
$\mathsf{Beta}(\alpha, \beta),\ \alpha, \beta \geqslant 1$	$\mathsf{Weib}(\alpha, \lambda),\ \alpha \geqslant 1$	$\mathsf{Gamma}(\alpha, \lambda),\ \alpha \geqslant 1$	

The **adaptive rejection sampling** method by Gilks et al. [8] is an acceptance–rejection algorithm for log-concave densities in which the proposal density and the (normalized) squeeze function are both finite mixtures of truncated exponential pdfs. Note that generation from a truncated exponential distribution is straightforward and computationally fast, and that the exponential pdf is frequently a suitable proposal density in the acceptance–rejection algorithm (see Example 3.15). The idea is as follows.

Suppose that $f(x) = p(x)/\mathcal{Z}$ is a log-concave density on $\mathscr{X}$, where $p(x)$ is a known function and $\mathcal{Z}$ is a known or unknown constant. Let $\mathcal{X}_n$ be a collection of points $x_1 < \cdots < x_n$ on $\mathscr{X}$. For $i = 1, \ldots, n$, let $\ell_i(x) = \alpha_i\, x + \beta_i$ denote the line through the points $(x_i, \ln p(x_i))$ and $(x_{i+1}, \ln p(x_{i+1}))$; see the upper panel of Figure 3.9. In other words, the line ℓ_i has slope and intercept

$$\alpha_i = \frac{\ln p(x_{i+1}) - \ln p(x_i)}{x_{i+1} - x_i} \quad \text{and} \quad \beta_i = \frac{x_{i+1} \ln p(x_i) - x_i \ln p(x_{i+1})}{x_{i+1} - x_i}\,,$$

respectively. From the concavity of $\ln p(x)$ we have for $i = 1, \ldots, n$,

$$\ell_i(x) \leqslant \ln p(x) \leqslant \min\{\ell_{i-1}(x),\, \ell_{i+1}(x)\}, \quad x \in [x_i, x_{i+1}]\,,$$

where $\ell_0(x) = \ell_{n+1}(x) \stackrel{\text{def}}{=} \infty$. Define the piecewise exponential functions

$$\overline{p}_n(x) = \begin{cases} \mathrm{e}^{\ell_1(x)} & x \leqslant x_1 \\ \mathrm{e}^{\min\{\ell_{i-1}(x),\, \ell_{i+1}(x)\}} & x \in [x_i, x_{i+1}],\ i = 1, \ldots, n \\ \mathrm{e}^{\ell_n(x)} & x \geqslant x_{n+1}\,, \end{cases}$$

$$\underline{p}_n(x) = \begin{cases} 0 & x \leqslant x_1 \\ \mathrm{e}^{\ell_i(x)} & x \in [x_i, x_{i+1}],\ i = 1, \ldots, n \\ 0 & x \geqslant x_{n+1}\,. \end{cases}$$

Then, we have the enveloping property that $\underline{p}_n(x) \leqslant p(x) \leqslant \overline{p}_n(x)$ for all x. Let $\overline{\mathcal{Z}}_n$ be the normalization constant of $\overline{p}_n(x)$ so that $\overline{f}_n(x) = \overline{p}_n(x)/\overline{\mathcal{Z}}_n$ is a pdf. The idea of the adaptive rejection sampling is to use $\overline{f}_n(x)$ as a proposal in an acceptance–rejection algorithm and the lower bound $\underline{p}_n(x)$ as a squeezing function (see Figure 3.6), while gradually increasing the size of the point set $\mathcal{X}_n$ to obtain tighter enveloping functions $\underline{p}_n$ and $\overline{p}_n$. More precisely, we have the following algorithm for sampling N iid random variables from f.

Algorithm 3.12 (Adaptive Rejection Sampling) *For $t = 1, 2, \ldots, N$ iterate the following steps.*

1. *Given the point set $\mathcal{X}_n$, construct the enveloping functions $\underline{p}_n(x)$ and $\overline{p}_n(x)$.*

2. *Generate $Y \sim \overline{f}_n(y)$ and $U \sim \mathsf{U}(0, 1)$, independently.*

3. *If $U \leqslant \underline{p}_n(Y)/\overline{p}_n(Y)$, go to Step 5; otherwise, go to Step 4.*

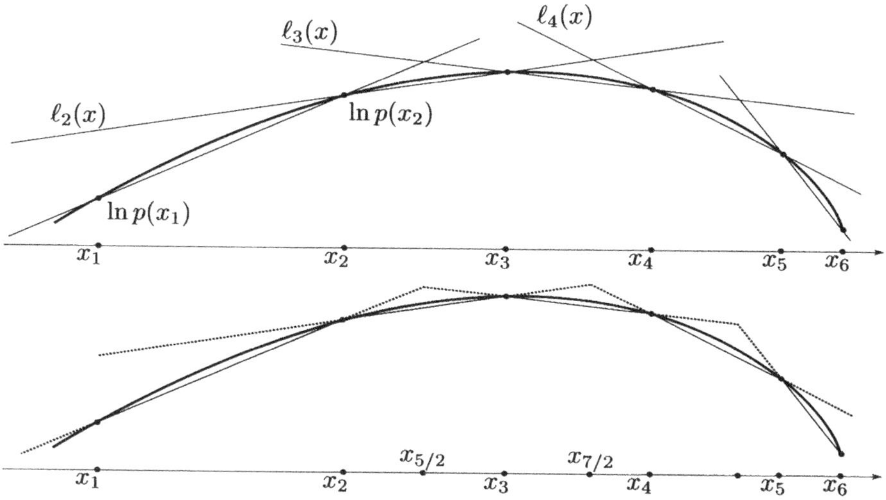

Figure 3.9 Construction of the dominating and squeeze functions for a log-concave density $f(x)$. The upper panel shows how the lines $\{\ell_i\}$ are constructed using points on the concave curve $\ln p(x)$. The lower panel shows how the upper (dotted line segments) and lower (solid line segments) bounds are constructed from the lines $\{\ell_i\}$.

4. *If* $U \leqslant p(Y)/\overline{p}_n(Y)$, *go to Step 5; otherwise, repeat from Step 2.*

5. *Output* $X_t = Y$ *as a random variable from* f. *Let* $\mathcal{X}_{n+1} = \mathcal{X}_n \cup \{Y\}$ *be the point set* $\mathcal{X}_n$ *with the addition of point* Y *for a total of* $n + 2$ *sorted points. Increment* $n = n + 1$.

We now explain how to generate random variables from $\overline{f}_n(x)$ efficiently. Let $x_{i+1/2}$ be the x-coordinate of the intersection point of the lines ℓ_{i-1} and ℓ_{i+1}; see the lower panel of Figure 3.9. In other words,

$$x_{i+1/2} \stackrel{\text{def}}{=} -\frac{\beta_{i+1} - \beta_{i-1}}{\alpha_{i+1} - \alpha_{i-1}}, \quad i = 2, \dots, n-1 .$$

It follows that for all $i = 1, \dots, n$ we have

$$\overline{p}_n(x) = \mathrm{e}^{\alpha_i x + \beta_i}, \quad x \in [x_{i-1/2}, x_i] \cup [x_{i+1}, x_{i+3/2}] ,$$

where $x_{3/2} \stackrel{\text{def}}{=} x_1$ and $x_{n+1/2} \stackrel{\text{def}}{=} x_{n+1}$, and $x_{1/2}$ is the lower bound of $\mathscr{X}$ (possibly $-\infty$) and $x_{n+3/2}$ is the upper bound of $\mathscr{X}$ (possibly ∞). Hence, we can write $\overline{f}_n(x)$ as a mixture of truncated exponential pdfs:

$$\overline{f}_n(x) = \frac{1}{\overline{Z}_n} \sum_{i=1}^{n} \mathrm{e}^{\alpha_i x + \beta_i} \left(\mathrm{I}_{\{x_{i-1/2} \leqslant x \leqslant x_i\}} + \mathrm{I}_{\{x_{i+1} \leqslant x \leqslant x_{i+3/2}\}} \right) ,$$

and we can use the composition method (see Section 3.1.2.6) to generate from $\overline{f}_n$ efficiently.

We now discuss the choice of the point set $\mathcal{X}_n$ used to initialize the algorithm. The algorithm fails if one of the following conditions holds: (1) the slope of ℓ_1 is negative and $\mathscr{X}$ is not bounded from below; (2) the slope of ℓ_n is positive and $\mathscr{X}$ is not bounded from above. As a consequence of this observation, the points $x_1 < x_2 < \cdots < x_{n+1}$ should be selected such that the interval $[x_1, x_{n+1}]$ contains most of the probability mass of f and the mode of f is bracketed by $[x_1, x_{n+1}]$. More specifically, if $\mathscr{X}$ is unbounded, then x_1 and x_{n+1} should be placed far into the tails of f, and if $\mathscr{X}$ is bounded, then x_1 and x_{n+1} should be placed close to the boundaries of $\mathscr{X}$. The choice of the interior points $x_2, \ldots, x_n$ is less important, because of the progressive addition of newly sampled points to the set $\mathcal{X}_n$. As a rule of thumb one can start with $n = 6$ points.

For a version of the adaptive rejection sampling that uses the first derivative of f, see [9]. For alternatives to the linear interpolation used here in the construction of the enveloping functions, see [16].

3.1.6 Ratio of Uniforms Method

The **ratio of uniforms method**, first proposed by Kinderman and Monahan [12], is closely related to the acceptance–rejection method. An advantage of the method is that the form of the density need only be known up to a normalizing constant. Thus, the density from which to sample has the form $f(z) = c\,h(z)$, where $h(z)$ is known, but the constant $c > 0$ could be unknown or difficult to compute. The method is summarized as follows.

Algorithm 3.13 (Ratio of Uniforms Method)

1. *Generate (X, Y) uniformly over the set*

$$\mathscr{R} = \{(x, y) : 0 \leqslant x \leqslant \sqrt{h(y/x)}\} ,$$

2. *Return $Z = Y/X$.*

To see that indeed $Z \sim f(z)$, consider the coordinate transformation $x = w$, $y = w\,z$, so that $w = x$ and $z = y/x$. The determinant of the corresponding matrix of
☞ 620 Jacobi is w, so that the joint pdf of W and Z, by the transformation rule (A.33), is given by

$$f_{W,Z}(w, z) = \widetilde{c}\,w, \quad 0 \leqslant w \leqslant \sqrt{h(z)} ,$$

for some constant $\widetilde{c}$. Consequently, the (marginal) pdf of Z is

$$f_Z(z) = \int_0^{\sqrt{h(z)}} \widetilde{c}\,w\,\mathrm{d}w = \frac{\widetilde{c}\,h(z)}{2} = f(z) .$$

The last equality follows from the fact that if f_Z is a pdf that is proportional to f, then it must be identical to f.

■ **EXAMPLE 3.18 (Sampling from the Positive Normal Distribution)**

As in Example 3.15 we wish to sample from the positive normal distribution, but now using the ratio of uniforms method. The region $\mathscr{R}$ on which we need to draw uniform samples is enclosed by the curve $x = \sqrt{f(y/x)}$ in Figure 3.10.

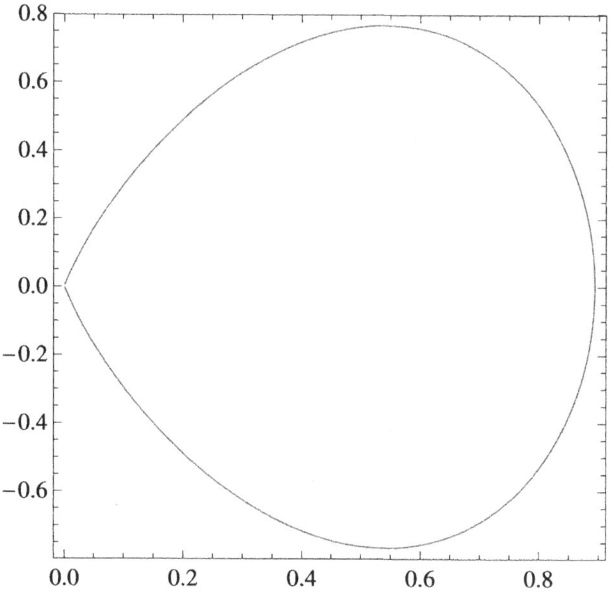

Figure 3.10 The curve encloses the sampling region $\mathscr{R}$ for the ratio of uniforms method.

Direct computation shows that $\mathscr{R}$ can be enclosed by the box $[a, b] \times [c, d]$, with

$$a = 0, \quad b = \sqrt[4]{\frac{2}{\pi}} \approx 0.893, \quad c = -\frac{2^{3/4}}{\sqrt{\mathrm{e}}\sqrt[4]{\pi}} \approx -0.766 \quad \text{and} \quad d = -c\,.$$

The total volume of the region $\mathscr{R}$ is 1 (by definition), and the bounding box has area $2\,b\,d = \frac{4}{\sqrt{\mathrm{e}\,\pi}} \approx 1.369$. Hence, if acceptance–rejection is used to sample uniformly on $\mathscr{R}$, the efficiency of the procedure is $\sqrt{\mathrm{e}\,\pi}/4 \approx 0.73$, which is comparable to the acceptance–rejection method in Example 3.15.

3.2 GENERATION METHODS FOR MULTIVARIATE RANDOM VARIABLES

In this section we consider some general procedures for generating a random vector $\mathbf{X} = (X_1, \ldots, X_n)^\top$ from a given n-dimensional distribution with pdf $f(\mathbf{x})$. Algorithms for generating from specific multivariate distributions are given in Section 4.3. ☞ 138

When the components $X_1, \ldots, X_n$ are *independent* the situation is easy. Suppose that the component pdfs are f_i, $i = 1, \ldots, n$, so that $f(\mathbf{x}) = f_1(x_1) \cdots f_n(x_n)$. To generate $\mathbf{X}$, simply generate each component $X_i \sim f_i$ individually — for example, via the inverse-transform method or acceptance–rejection.

Algorithm 3.14 (Independent Components Generation)

1. *Independently generate* $X_i \sim f_i$, $i = 1, \ldots, n$.

2. *Return* $\mathbf{X} = (X_1, \ldots, X_n)^\top$.

For *dependent* components $X_1, \ldots, X_n$, we can represent the joint pdf $f(\mathbf{x})$, using
☞ 616 the product rule (A.21), as

$$f(\mathbf{x}) = f(x_1, \ldots, x_n) = f_1(x_1)\, f_2(x_2 \,|\, x_1) \cdots f_n(x_n \,|\, x_1, \ldots, x_{n-1}) \;, \tag{3.13}$$

where $f_1(x_1)$ is the marginal pdf of X_1 and $f_k(x_k \,|\, x_1, \ldots, x_{k-1})$ is the conditional pdf of X_k given $X_1 = x_1, X_2 = x_2, \ldots, X_{k-1} = x_{k-1}$. This observation leads to the following procedure.

Algorithm 3.15 (Dependent Components Generation)

1. *Generate* $X_1 \sim f_1$. *Set* $t = 1$.
2. *While* $t < n$, *given* $X_1 = x_1, \ldots, X_t = x_t$, *generate* $X_{t+1} \sim f_{t+1}(x_{t+1} \,|\, x_1, \ldots, x_t)$ *and set* $t = t + 1$.
3. *Return* $\mathbf{X} = (X_1, \ldots, X_n)^\top$.

The applicability of this approach depends, of course, on knowledge of the conditional distributions. In certain models, for example Markov models, this knowledge
☞ 162 is easily obtainable.

Another, usually simpler, approach is to generate the random vector $\mathbf{X}$ by multidimensional acceptance–rejection; for instance, when generating a random vector uniformly over an n-dimensional region. In Example 3.18 such an approach is employed.

For high-dimensional distributions, efficient *exact* random variable generation is often difficult to achieve, and *approximate* generation methods are used instead. Such methods are discussed in Chapter 6.

Copulas, described next, provide an alternative framework for random vector generation.

3.2.1 Copulas

Suppose we wish to generate a random vector $\mathbf{X} = (X_1, \ldots, X_n)^\top$ of *dependent* components $X_i \sim f_i(x_i)$, $i = 1, \ldots, n$, where $\{f_i\}$ are *known* univariate densities with corresponding cdfs $\{F_i\}$. Copulas provide a convenient method for imposing dependency structure among the components of $\mathbf{X}$ while keeping the marginal distributions fixed. A **copula** is a cdf $C : [0,1]^n \to [0,1]$ of n *dependent* uniform random variables $U_1, \ldots, U_n \sim \mathsf{U}(0,1)$:

$$C(u_1, \ldots, u_n) = \mathbb{P}(U_1 \leqslant u_1, \ldots, U_n \leqslant u_n) \;.$$

For a given copula C and marginal cdfs $\{F_i\}$, define

$$F(x_1, \ldots, x_n) = C(F_1(x_1), \ldots, F_n(x_n)) \;.$$

If $\mathbf{U} = (U_1, \ldots, U_n)^\top$ has cdf $C(u_1, \ldots, u_n)$, then the random vector $\mathbf{X} = (X_1, \ldots, X_n)^\top = (F_1^{-1}(U_1), \ldots, F_n^{-1}(U_n))^\top$ has joint cdf F and marginal cdfs $F_1(x_1), \ldots, F_n(x_n)$. Thus, given a copula C and marginal cdfs $\{F_i\}$ we can simulate a vector $\mathbf{X}$ with cdf F as follows.

Algorithm 3.16 (Dependent Components Generation Using a Copula)

1. *Generate* $\mathbf{U} \sim C(u_1, \ldots, u_n)$.
2. *Output* $\mathbf{X} = (X_1, \ldots, X_n)^\top = (F_1^{-1}(U_1), \ldots, F_n^{-1}(U_n))^\top$.

A commonly used copula is the **Student's *t* copula**:

$$C(u_1, \ldots, u_n) = T_{\nu,\Sigma}\left(T_\nu^{-1}(u_1), \ldots, T_\nu^{-1}(u_n)\right),$$

where $T_{\nu,\Sigma}$ is the cdf of the multivariate $\mathsf{t}_\nu(\mathbf{0}, \Sigma)$ distribution, with ν degrees of freedom, mean vector $\mathbf{0}$, and *correlation matrix* Σ (that is, Σ is a covariance matrix with ones down the main diagonal), and T_ν^{-1} is the inverse cdf of the univariate t_ν distribution. This includes the special case ($\nu = \infty$) of the **Gaussian copula** model. In this setting the dependency structure in the random vector $\mathbf{X}$ is determined by the correlation matrix Σ. Another way to specify the dependency structure is to use *rank correlation* measures such as *Spearman's* or *Kendall's rank correlation*, see [18] for more details.

☞ 147

☞ 131

■ **EXAMPLE 3.19 (Student's *t* Copula)**

Suppose we wish to generate $N = 10^4$ random vectors $\mathbf{X} = (X_1, X_2)^\top$ with $X_1 \sim \mathsf{Gamma}(2, 1)$ and $X_2 \sim \mathsf{N}(0, 1)$. We use a t copula model with $\nu = 10$ and correlation coefficient 0.7, that is, the off-diagonal entries of Σ are 0.7 and the diagonal entries are 1.

Although the cdf $T_{\nu,\Sigma}$ of the multivariate $\mathsf{t}_\nu(\mathbf{0}, \Sigma)$ distribution is not available in closed form, the generation of $\mathbf{U} = (U_1, U_2)^\top$ with cdf $C(u_1, u_2)$ is straightforward. Namely, we first generate $\mathbf{Y} \sim \mathsf{t}_\nu(\mathbf{0}, \Sigma)$ via Algorithm 4.72 and then compute $\mathbf{U} = (T_\nu(Y_1), T_\nu(Y_2))^\top$. Finally, we apply Step 2 of Algorithm 3.16 to obtain the desired dependent vector $\mathbf{X}$. Figure 3.11 shows an outcome of this experiment. In addition to the sampled points, we show a kernel density estimate for X_1 and X_2 (constructed from the sampled points). The kernel density estimates suggest that the marginals of $\mathbf{X}$ follow the desired distributions. The following code is used to generate the data.

☞ 148

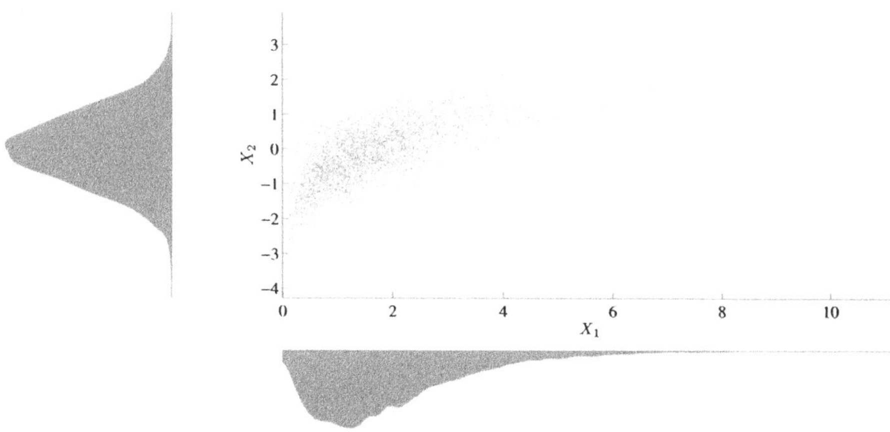

Figure 3.11 An outcome from the t copula model. Plotted along the axes are kernel density estimates of the marginal distributions.

```
% copula.m
clear all, N = 10^4; nu=10;
Sig=[1,.7;.7,1]; % correlation matrix
A=chol(Sig);
% generate multivariate Student dist.
Y=repmat(sqrt(nu./gamrnd(nu/2,2,[N,1])),1,2).*randn(N,2);
Y=Y*A;
U = tcdf(Y,nu); % a sample from C(u_1,...,u_n)
X=[gaminv(U(:,1),2,1), norminv(U(:,2))];
plot(X(:,1),X(:,2),'r.','MarkerSize',1)
```

☞ 567 For another example of using a copula model see Section 16.5. Numerous other copula models have been proposed with applications in finance and risk assessment. The interested reader is referred to [7, 13].

3.3 GENERATION METHODS FOR VARIOUS RANDOM OBJECTS

3.3.1 Generating Order Statistics

☞ 54 Generating the order statistics $X_{(1)} \leqslant \cdots \leqslant X_{(n)}$ from a collection of independent and identically distributed random variables $X_1, \ldots, X_n \sim_{\text{iid}} \mathsf{Dist}$ is established most easily by sorting.

Algorithm 3.17 (Generating Order Statistics)

1. *Generate* $X_1, \ldots, X_n \sim_{\text{iid}} \mathsf{Dist}$.
2. *Sort the* $\{X_i\}$ *in ascending order and return* $\min_i X_i = X_{(1)} \leqslant \cdots \leqslant X_{(n)} = \max_i X_i$.

For the uniform distribution an alternative algorithm is based on the fact that the spacings $Y_1 = X_{(1)}, Y_2 = X_{(2)} - X_{(1)}, \ldots, Y_n = X_{(n)} - X_{(n-1)}$ are uniformly distributed in the unit simplex $\{\mathbf{y} : y_i \geqslant 0,\ \ i = 1, \ldots, n,\ \sum_{i=1}^{n} y_i \leqslant 1\}$. Moreover, by Property 5 on Page 140, $\mathbf{Y} = (Y_1, \ldots, Y_n) \sim \mathsf{Dirichlet}(1, \ldots, 1)$. Property 1 of the Dirichlet distribution (see Page 140) leads now to the following algorithm.

Algorithm 3.18 (Generating $\mathsf{U}(0,1)$ Order Statistics (I))

1. *Generate* $U_1, \ldots, U_{n+1} \overset{\text{iid}}{\sim} \mathsf{U}(0,1)$.
2. *Set* $E_i = -\ln U_i, i = 1, \ldots, n+1$, $S = \sum_{i=1}^{n+1} E_i$, $U_{(0)} = 0$, *and* $t = 1$.
3. *While* $t \leqslant n$ *set* $U_{(t)} = U_{(t-1)} + \frac{E_t}{S}$ *and* $t = t + 1$.
4. *Return* $U_{(1)}, \ldots, U_{(n)}$.

Another algorithm for generating order statistics for the $\mathsf{U}(0,1)$ distribution is based on the following two facts, which hold for any $k = 1, \ldots, n$:

- $\max\{U_1, \ldots, U_k\}$ is distributed as $U^{1/k}$, with $U \sim \mathsf{U}(0,1)$.

- Conditional upon $\max\{U_1, \ldots, U_k\} = u_{(k)}$ the random variables $U_1, \ldots, U_{k-1}$ are independent and $\mathsf{U}(0, u_{(k)})$ distributed.

Algorithm 3.19 (Generating U(0, 1) Order Statistics (II))

1. *Set* $U_{(n+1)} = 1$ *and* $t = n$.

2. *While* $t \geqslant 1$, *generate* $U \sim \mathsf{U}(0,1)$, *set* $U_{(t)} = U^{\frac{1}{t}} U_{(t+1)}$ *and* $t = t - 1$.

3. *Return* $U_{(1)}, \ldots, U_{(n)}$.

To generate an ordered sample of n exponentials, we have, in addition to Algorithm 3.17, the following algorithm, which is based on the fact that the order statistics can be viewed as the jump times of a pure birth process on $\{0, 1, \ldots, n\}$ ☞ 637 with birth rates $\lambda_i = n - i, i = 0, \ldots, n$.

Algorithm 3.20 (Generating Exp(1) Order Statistics)

1. *Set* $X_{(0)} = 0$ *and* $t = 1$.

2. *While* $t \leqslant n$, *generate* $U \sim \mathsf{U}(0,1)$ *and set* $X_{(t)} = X_{(t-1)} - \frac{\ln U}{n-t+1}$ *and* $t = t+1$.

3. *Return* $X_{(1)}, \ldots, X_{(n)}$.

3.3.2 Generating Uniform Random Vectors in a Simplex

A simplex is the multidimensional analogue of a triangle in $\mathbb{R}^n$. More precisely, a **simplex** is a set of vectors in $\mathbb{R}^n$ of the form $\mathbf{z}_0 + C\mathbf{y}$, where $\mathbf{y} = (y_1, \ldots, y_n)^T$ is a vector of positive components with a sum less than or equal to 1, and C is an invertible matrix with columns $\mathbf{z}_1 - \mathbf{z}_0, \ldots, \mathbf{z}_n - \mathbf{z}_0$. The simplex is thus the point set within the convex hull spanned by the vectors $\mathbf{z}_0, \ldots, \mathbf{z}_n$.

As an example, consider the n-dimensional set

$$\mathscr{X} = \{\mathbf{x} \in \mathbb{R}^n : x_i \geqslant 0, \quad i = 1, \ldots, n, \; x_1 \leqslant x_2 \leqslant \cdots \leqslant x_n \leqslant 1\} .$$

$\mathscr{X}$ is a simplex on the points $\mathbf{0}, \mathbf{e}_n, \mathbf{e}_n + \mathbf{e}_{n-1}, \ldots, \mathbf{1}$, where $\mathbf{e}_i$ is the i-th unit vector in $\mathbb{R}^n$, $i = 1, \ldots, n$, and $\mathbf{0}$ and $\mathbf{1}$ are vectors of all 0s and 1s, respectively. Figure 3.12 gives an illustration of the three-dimensional case.

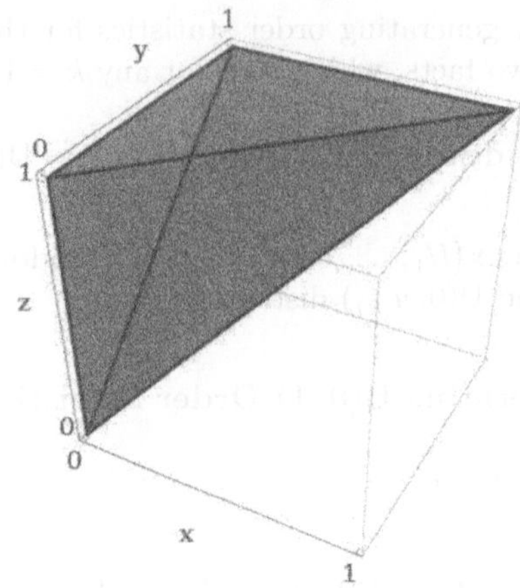

Figure 3.12 Simplex $\mathscr{X}$ for $n = 3$.

Generating random vectors $\mathbf{X}$ uniformly in the "wedge" $\mathscr{X}$ is easy, as the coordinates of $\mathbf{X}$ are distributed according to the order statistics of an iid sample from the uniform distribution on $(0, 1)$.

Algorithm 3.21 (Generating Uniformly in the Simplex $\mathscr{X}$)

1. *Generate* $U_1, \ldots, U_n \overset{\text{iid}}{\sim} \mathsf{U}(0,1)$.

2. *Sort the* $\{U_i\}$ *to give the order statistics* $U_{(1)}, \ldots, U_{(n)}$.

3. *Return* $\mathbf{X} = (U_{(1)}, \ldots, U_{(n)})^\top$.

Using the linear transformation $\mathbf{y} = A\mathbf{x}$ with

$$A = \begin{pmatrix} 1 & 0 & \cdots & 0 \\ -1 & 1 & \cdots & 0 \\ \vdots & \ddots & \ddots & \vdots \\ 0 & \cdots & -1 & 1 \end{pmatrix},$$

the simplex $\mathscr{X}$ is transformed into the n-dimensional **unit simplex** (see Figure 3.13), that is, the simplex on the points $\mathbf{0}, \mathbf{e}_1, \ldots, \mathbf{e}_n$:

$$\mathscr{Y} = \left\{ \mathbf{y} : y_i \geqslant 0, \ \ i = 1, \ldots, n, \ \sum_{i=1}^{n} y_i \leqslant 1 \right\} . \tag{3.14}$$

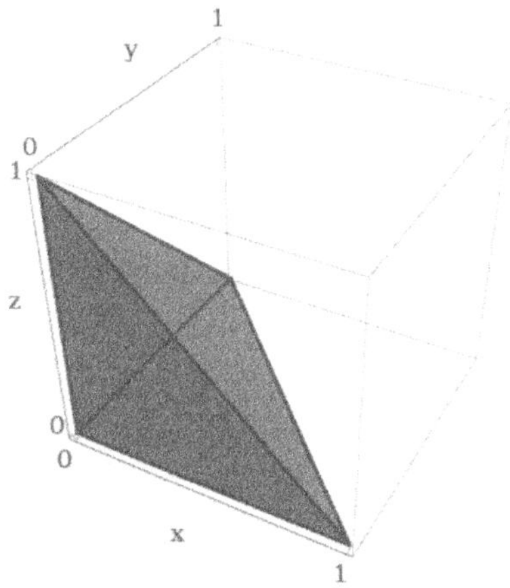

Figure 3.13 The unit simplex $\mathscr{Y}$ for $n = 3$.

This results in the following algorithm.

Algorithm 3.22 (Generating Uniformly in the Unit Simplex $\mathscr{Y}$)

1. *Generate* $U_1, \ldots, U_n \overset{\text{iid}}{\sim} \mathsf{U}(0, 1)$.
2. *Sort* $U_1, \ldots, U_n$ *into the order statistics* $U_{(1)}, \ldots, U_{(n)}$.
3. *Define*

$$\begin{aligned} Y_1 &= U_{(1)} \ , \\ Y_2 &= U_{(2)} - U_{(1)} \ , \\ &\vdots \\ Y_n &= U_{(n)} - U_{(n-1)} \ , \end{aligned} \tag{3.15}$$

and return the vector $\mathbf{Y} = (Y_1, \ldots, Y_n)^\top$.

An alternative method is based on the fact (see Property 5 on Page 140) that $\mathbf{Y} \sim \mathsf{Dirichlet}(\underbrace{1, \ldots, 1}_{n+1})$.

Algorithm 3.23 (Dirichlet Sampling for the Unit Simplex $\mathscr{Y}$)

1. *Generate* $X_1, \ldots, X_{n+1} \overset{\text{iid}}{\sim} \mathsf{Exp}(1)$.
2. *Output* $\mathbf{Y} = (Y_1, \ldots, Y_n)^\top$, *where*

$$Y_i = \frac{X_i}{\sum_{k=1}^{n+1} X_k}, \quad i = 1, \ldots, n \ .$$

Finally, in order to generate random vectors uniformly distributed in an n-dimensional simplex defined by arbitrary vertices, say, $\mathbf{z}_0, \mathbf{z}_1, \ldots, \mathbf{z}_n$, we simply generate $\mathbf{Y}$ uniformly in $\mathscr{Y}$ and apply an affine transformation.

Algorithm 3.24 (Generating Uniformly in a General Simplex)

1. *Generate* $\mathbf{Y} \in \mathscr{Y}$ *as in Algorithm 3.22 or 3.23.*
2. *Return* $\mathbf{Z} = C\mathbf{Y} + \mathbf{z}_0$, *where* C *is the matrix whose columns are* $\mathbf{z}_1 - \mathbf{z}_0, \ldots, \mathbf{z}_n - \mathbf{z}_0$.

3.3.3 Generating Random Vectors Uniformly Distributed in a Unit Hyperball and Hypersphere

Consider the n-dimensional unit hyperball, $\mathscr{B}_n = \{\mathbf{x} \in \mathbb{R}^n : \|\mathbf{x}\| \leqslant 1\}$. Generating uniform random vectors in $\mathscr{B}_n$ is straightforward via the acceptance–rejection method.

Algorithm 3.25 (Generating Uniformly in $\mathscr{B}_n$ (I))

1. *Generate* $U_1, \ldots, U_n \overset{\text{iid}}{\sim} \mathsf{U}(0,1)$.
2. *Set* $X_1 = 1 - 2U_1, \ldots, X_n = 1 - 2U_n$, *and* $R = \sum_{i=1}^n X_i^2$.
3. *If* $R \leqslant 1$, *accept* $\mathbf{X} = (X_1, \ldots, X_n)^\top$ *as the desired vector; otherwise, go to Step 1.*

The efficiency of this n-dimensional acceptance–rejection method is equal to the ratio

$$\frac{1}{C} = \frac{\text{volume of the hyperball}}{\text{volume of the hypercube}} = \frac{\frac{\pi^{n/2}}{(n/2)\Gamma(n/2)}}{2^n} = \frac{1}{n\,2^{n-1}}\frac{\pi^{n/2}}{\Gamma(n/2)},$$

which rapidly decreases to 0 as $n \to \infty$; for example, for $n = 8$ the efficiency is approximately 0.016. The next algorithm is more efficient for higher dimensions, and utilizes the following facts.

- If $X_1, \ldots, X_n \overset{\text{iid}}{\sim} \mathsf{N}(0,1)$, then the normalized vector

$$\mathbf{Y} = \left(\frac{X_1}{\|\mathbf{X}\|}, \ldots, \frac{X_n}{\|\mathbf{X}\|}\right), \tag{3.16}$$

  where $\|\mathbf{X}\| = (\sum_{i=1}^n X_i^2)^{\frac{1}{2}}$, is distributed uniformly on the n-dimensional hypersphere $\mathscr{S}_n = \{\mathbf{y} : \|\mathbf{y}\| = 1\}$.
- The radius R of a uniformly chosen point in $\mathscr{B}_n$ has cdf $F_R(r) = r^n, 0 \leqslant r \leqslant 1$.

Algorithm 3.26 (Generating Uniformly in $\mathscr{B}_n$ (II))

1. *Generate a random vector* $\mathbf{X} = (X_1, \ldots, X_n)^\top$ *with iid* $\mathsf{N}(0,1)$ *components.*
2. *Generate* $U \sim \mathsf{U}(0,1)$ *and set* $R = U^{1/n}$.
3. *Return* $\mathbf{Z} = R\mathbf{X}/\|\mathbf{X}\|$.

To generate a random vector that is uniformly distributed over the *surface* of an n-dimensional unit ball — in other words, uniformly on the unit hypersphere $\mathscr{S}_n$, we simplify the previous algorithm and arrive at the following one.

Algorithm 3.27 (Generating Uniform Random Vectors on $\mathscr{S}_n$)

1. *Generate a random vector* $\mathbf{X} = (X_1, \ldots, X_n)^\top$ *with iid* $\mathsf{N}(0,1)$ *components.*
2. *Return* $\mathbf{Y} = \mathbf{X}/\|\mathbf{X}\|$.

3.3.4 Generating Random Vectors Uniformly Distributed in a Hyperellipsoid

Consider the interior (including the boundary) of an n-dimensional hyperellipsoid $\mathscr{X}$ centered at the origin, written in the form

$$\mathscr{X} = \{\mathbf{x} \in \mathbb{R}^n : \mathbf{x}^\top \Sigma \mathbf{x} \leqslant r^2\} , \tag{3.17}$$

where Σ is a positive definite $n \times n$ matrix ($\mathbf{x}$ is interpreted as a column vector). Since Σ is positive definite, there exists a Cholesky matrix B such that $\Sigma = BB^\top$. We may therefore view the set $\mathscr{X}$ as a linear transformation $\mathbf{y} = B^\top \mathbf{x}$ of the n-dimensional ball $\mathscr{Y} = \{\mathbf{y} : \mathbf{y}^\top \mathbf{y} \leqslant r^2\}$. Since linear transformations preserve uniformity, if the vector $\mathbf{Y}$ is uniformly distributed in $\mathscr{Y}$, then $\mathbf{X} = (B^\top)^{-1}\mathbf{Y}$ is uniformly distributed in $\mathscr{X}$. The corresponding generation algorithm is given below.

Algorithm 3.28 (Generating Random Vectors in a Hyperellipsoid)

1. *Calculate the Cholesky matrix B of Σ.*
2. *Generate $\mathbf{Z} = (Z_1, \ldots, Z_n)^\top$ uniformly distributed in the unit hyperball $\mathscr{B}_n$. Set $\mathbf{Y} = r\mathbf{Z}$.*
3. *Solve the matrix equation $B^\top \mathbf{X} = \mathbf{Y}$ for $\mathbf{X}$ using backward substitution and return $\mathbf{X}$.*

3.3.5 Uniform Sampling on a Curve

The Cartesian coordinates $\mathbf{x} = (x_1, \ldots, x_n)^\top$ of any curve $\mathcal{C}$ in $\mathbb{R}^n$ can be parameterized as $\mathbf{x}(t) = (x_1(t), \ldots, x_n(t))^\top$ for some parameter $t \in [t_0, t_1]$ such that as t varies in its domain, $\mathbf{x}(t)$ traces out the curve $\mathcal{C}$ only once. Assuming that the curve is *rectifiable*, that is, has finite arc length, the length of the curve traced out until some $t < t_1$ is given by $s(t) = \int_{t_0}^{t} \|\dot{\mathbf{x}}(u)\| \, du$, where $\dot{\mathbf{x}}(t) = (\frac{dx_1}{dt}, \ldots, \frac{dx_n}{dt})$, and $\|\mathbf{x}\| = \sqrt{x_1^2 + \cdots + x_n^2}$. This motivates the following algorithm for generating points uniformly on the curve $\mathcal{C}$ traced out by $\mathbf{x}(t)$ for $t \in [t_0, t_1]$.

Algorithm 3.29 (Uniform Sampling on a Curve $\mathcal{C}$)

1. *Generate a point T from the cdf $s(t)/(s(t_1) - s(t_0))$, or, equivalently, from the pdf*
$$d(t) = \frac{\|\dot{\mathbf{x}}(t)\|}{s(t_1) - s(t_0)}, \quad t_0 \leqslant t \leqslant t_1 .$$
2. *Output the point on the curve $\mathbf{X} = (x_1(T), \ldots, x_n(T))^\top$.*

■ **EXAMPLE 3.20 (Generating Uniformly on an Ellipse)**

An occasionally suggested but incorrect approach of generating points uniformly distributed on an ellipse is to generate points uniformly on a circle and then linearly transform the points. To illustrate the difference between this approach and the correct one of Algorithm 3.29, consider the ellipse

$$\frac{x^2}{9} + y^2 = 1 ,$$

which can be parameterized as

$$x(t) = 3\sin t, \quad y(t) = \cos t, \quad t \in [0, 2\pi]\,.$$

A straightforward calculation shows that density $d(t)$ is given by

$$d(t) = \sqrt{4\cos(2t) + 5}/c\,,$$

where c is the circumference of the ellipse. We can draw from this density simply by acceptance–rejection. Figure 3.14 displays the subtle difference between 150 points generated using Algorithm 3.29 (left) and the linear transformation method (right). On the right the points are more "bunched up" near the left and right ends of the ellipse.

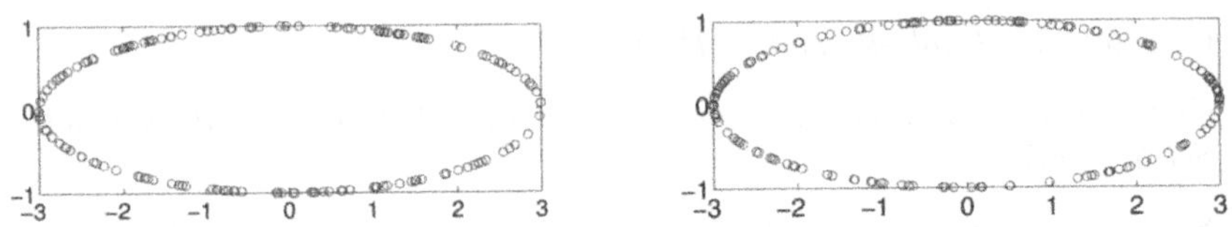

Figure 3.14 The points on the left ellipse are uniformly distributed, but not so for the right ellipse.

3.3.6 Uniform Sampling on a Surface

The method for generating random points on a curve can be extended to surfaces. To illustrate the general procedure, we consider here only two-dimensional surfaces. The set of all points $(x_1, x_2, x_3) \in \mathbb{R}^3$ on any two-dimensional surface $\mathscr{S}$ can be described by a set of parametric equations

$$x_1 = x_1(v_1, v_2), \quad x_2 = x_2(v_1, v_2), \quad x_3 = x_3(v_1, v_2), \quad (v_1, v_2) \in \mathscr{V}, \tag{3.18}$$

where the surface $\mathscr{S}$ is traced out only once as (v_1, v_2) moves throughout the region $\mathscr{V}$. Then, a small rectangular region in $\mathscr{V}$ enclosing the point (v_1, v_2) and having area $\mathrm{d}v_1\,\mathrm{d}v_2$ is mapped into a patch on the surface $\mathscr{S}$ with approximate area $\|\mathbf{r}_1 \times \mathbf{r}_2\|\,\mathrm{d}v_1\,\mathrm{d}v_2$, where $\mathbf{r}_1 = (\frac{\partial x_1}{\partial v_1}, \frac{\partial x_2}{\partial v_1}, \frac{\partial x_3}{\partial v_1})^\top$ and $\mathbf{r}_2 = (\frac{\partial x_1}{\partial v_2}, \frac{\partial x_2}{\partial v_2}, \frac{\partial x_3}{\partial v_2})^\top$, and $\mathbf{r}_1 \times \mathbf{r}_2$ denotes the vector cross product. It follows that the surface area of $\mathscr{S}$ is given by the integral $\iint_{\mathscr{V}} \|\mathbf{r}_1 \times \mathbf{r}_2\|\,\mathrm{d}v_1\,\mathrm{d}v_2$. This motivates the following algorithm.

Algorithm 3.30 (Uniform Sampling on a Surface $\mathscr{S}$)

1. *Given the surface parameterized by* (3.18), *generate a random pair* (V_1, V_2) *with pdf proportional to*

$$\|\mathbf{r}_1 \times \mathbf{r}_2\|, \quad (v_1, v_2) \in \mathscr{V}\,.$$

2. *Output* $(X_1, X_2, X_3) = (x_1(V_1, V_2), x_2(V_1, V_2), x_3(V_1, V_2))$.

■ **EXAMPLE 3.21 (Sampling Uniformly on the Surface of an Ellipsoid)**

Consider the ellipsoid parameterized using spherical coordinates:

$$x_1 = a\sin(v_2)\cos(v_1), \quad x_2 = b\sin(v_2)\sin(v_1), \quad x_3 = c\cos(v_2)\,, \tag{3.19}$$

where $0 \leqslant v_1 \leqslant 2\pi$ and $0 \leqslant v_2 \leqslant \pi$. Then,

$$\mathbf{r}_1 \times \mathbf{r}_2 = -\begin{pmatrix} b\,c\,\sin^2(v_2)\cos(v_1) \\ a\,c\,\sin^2(v_2)\sin(v_1) \\ a\,b\,\sin(v_2)\cos(v_2) \end{pmatrix}.$$

Hence, $\|\mathbf{r}_1 \times \mathbf{r}_2\|$ is given by

$$|\sin(v_2)|\sqrt{(bc)^2\sin^2(v_2)\cos^2(v_1) + (ac)^2\sin^2(v_2)\sin^2(v_1) + (ab)^2\cos^2(v_2)}\,. \tag{3.20}$$

To sample from the surface density $\|\mathbf{r}_1 \times \mathbf{r}_2\|/S(a,b,c)$, where $S(a,b,c)$ is the surface area of the ellipsoid, note that $\|\mathbf{r}_1 \times \mathbf{r}_2\| \leqslant r^2\sin(v_2)$, where $r = \max\{a,b,c\}$. Thus, we can use acceptance–rejection with proposal pdf $g(v_1,v_2) = \sin(v_2)/(4\pi)$. Then, the probability of acceptance in the acceptance–rejection algorithm is given by $S(a,b,c)/(4\pi r^2)$. Thus, from the bound $S(a,b,c) \geqslant 4\pi(ab+ac+bc)/3$ we conclude that the efficiency is bounded from below by $(ab+ac+bc)/(3r^2)$.

As a specific case, let $a = 2, b = 4$, and $c = 1$. The (unnormalized) surface density is depicted in Figure 3.15. Figure 3.16 shows a uniform sample.

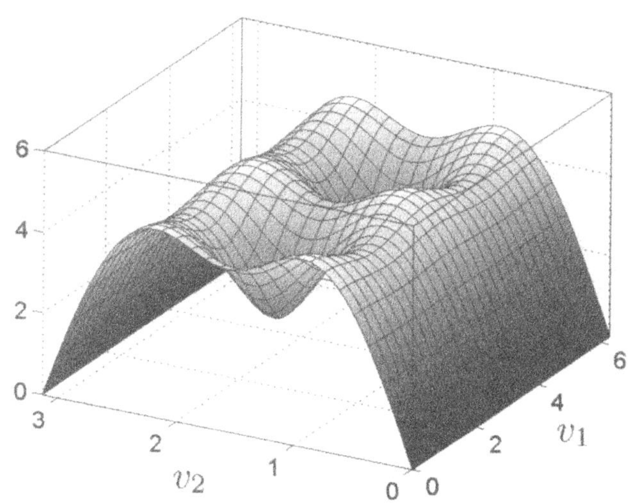

Figure 3.15 Unnormalized surface density of the ellipsoid in parameter space.

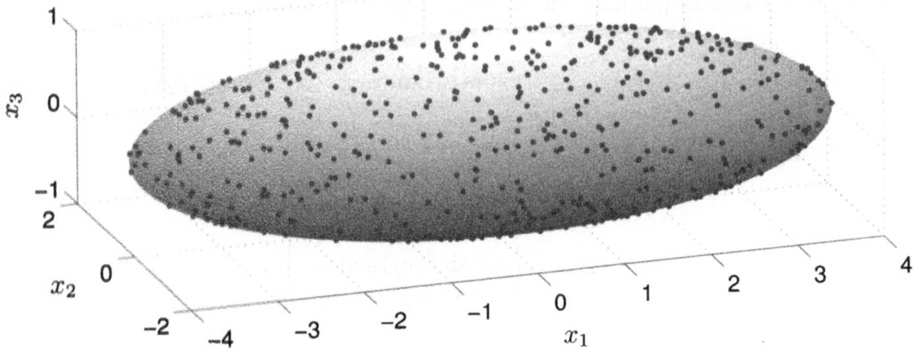

Figure 3.16 Uniform sample on the surface of the ellipsoid $\frac{x_1^2}{a^2} + \frac{x_2^2}{b^2} + \frac{x_3^2}{c^2} = 1$ with $(a, b, c) = (2, 4, 1)$.

```
%ellipsoid_sample.m
clear all,clf
global a b c
a=4;b=2;c=1;h=2*pi/100;
[v1,v2]=meshgrid([0:h:2*pi],[0:h:pi]);
X1=@(v1,v2)(a*sin(v2).*cos(v1));
X2=@(v1,v2)(b*sin(v2).*sin(v1));
X3=@(v1,v2)(c*cos(v2));
for i=1:10^3
    [V1,V2]=rand_ellipsoid;
    data(i,:)=[V1,V2];
end
hold on
surf(X1(v1,v2),X2(v1,v2),X3(v1,v2),'LineStyle','none'), axis equal,
colormap(gray)
plot3(X1(data(:,1),data(:,2)),...
X2(data(:,1),data(:,2)),X3(data(:,1),data(:,2)),...
'.','MarkerSize',10), axis equal
alpha(0.8)
```

```
function L=ellipsoid(v1,v2)
global a b c
A1=b*c*sin(v2).^2.*cos(v1);
B1=a*c*sin(v2).^2.*sin(v1);
C1=a*b*cos(v2).*sin(v2);
L=sqrt(A1.^2+B1.^2+C1.^2);
```

```
function [V1,V2]=rand_ellipsoid
global a b c
V1=rand*2*pi;
V2=acos(1-2*rand);
C=4*pi*max([a,b,c])^2;
while rand>ellipsoid(V1,V2)/(C*sin(V2)/4*pi)
    V1=rand*2*pi;
    V2=acos(1-2*rand);
end
```

3.3.7 Generating Random Permutations

Suppose we have a collection of n objects, labeled $1, 2, \ldots, n$, and we wish to generate a random permutation of these labels such that each of the $n!$ possible orderings occurs with equal probability. A simple algorithm for generating such uniform permutations is based on the ordering of uniform random numbers.

Algorithm 3.31 (Generating Random Permutations by Sorting)

1. *Generate* $U_1, \ldots, U_n \overset{\text{iid}}{\sim} \mathsf{U}(0,1)$.
2. *Sort these in increasing order:* $U_{X_1} \leqslant U_{X_2} \leqslant \cdots \leqslant U_{X_n}$.
3. *Return* $\mathbf{X} = (X_1, \ldots, X_n)$.

■ **EXAMPLE 3.22 (Drawing Without Replacement)**

Suppose we wish to select 30 numbers out of 100 uniformly without replacement. This can be accomplished by generating a uniform random permutation of $1, \ldots, 100$ and selecting the first 30 components thereof. The following MATLAB program achieves this via an implementation of Algorithm 3.31. This procedure is most efficient when the number of draws k is close to n. For $k \ll n$, a more efficient approach for sampling without replacement is given in the function `resample.m` on Page 484.

```
%unifperm.m
n = 100;
k = 30;
[s,ix] = sort(rand(1,n));
x = ix(1:k)
```

The next algorithm for drawing uniform random permutation is faster than Algorithm 3.31 and builds the permutation component by component, requiring only n uniform random numbers and no sorting; see [20].

Algorithm 3.32 (Generating Uniform Random Permutations)

1. *Set* $\mathbf{a} = (1, \ldots, n)$ *and* $i = 1$.
2. *Generate an index* I *uniformly from* $\{1, \ldots, n - i + 1\}$.
3. *Set* $X_i = a_I$ *followed by setting* $a_I = a_{n-i+1}$.
4. *Set* $i = i + 1$. *If* $i \leqslant n$ *go to Step 2.*
5. *Return* $\mathbf{X} = (X_1, \ldots, X_n)$.

3.3.8 Exact Sampling From a Conditional Bernoulli Distribution

Suppose the vector $\mathbf{X} = (X_1, \ldots, X_n)$ has independent components, with $X_i \sim \mathsf{Ber}(p_i)$, $i = 1, \ldots, n$. The conditional distribution of $\mathbf{X}$ given $\sum_i X_i = k$ is given by

$$\mathbb{P}\left(X_1 = x_1, \ldots, X_n = x_n \,\middle|\, \sum_{i=1}^{n} X_i = k\right) = \frac{\prod_{i=1}^{n} w_i^{x_i}}{R_k}, \tag{3.21}$$

where

$$R_k = \mathbb{P}\left(\sum_{i=1}^{n} X_i = k\right) \prod_{i=1}^{n} (1 + w_i)$$

is a normalization constant and $w_i = p_i/(1 - p_i)$, $i = 1, \ldots, n$. Similarly, let $R_{k-1,j}$ denote the normalization constant for the conditional distribution of $\{X_i, i \neq j\}$ given $\sum_{i \neq j} X_i = k - 1$; that is,

$$R_{k-1,j} = \mathbb{P}\left(\sum_{i \neq j} X_i = k - 1\right) \prod_{i \neq j} (1 + w_i) \,.$$

The normalization constants can be computed efficiently via the following result (see [4, Theorem 3]).

Theorem 3.3.1 (Computation of the Normalization Constants) *Define* $T_i = \sum_{j=1}^{n} w_j^i$, $i = 1, \ldots, k$ *and* $T_{i,j} = T_i - w_j^i$, $i = 1, \ldots, k$, $j = 1, \ldots, n$, *and put* $R_0 = 1$ *and* $R_{0,j} = 1$, $j = 1, \ldots, n$. *Then,*

$$R_k = \frac{1}{k} \sum_{i=1}^{k} (-1)^{i+1}\, T_i\, R_{k-i} \,, \tag{3.22}$$

$$R_{k-1,j} = \frac{1}{k-1} \sum_{i=1}^{k-1} (-1)^{i+1}\, T_{i,j}\, R_{k-1-i,j}, \quad j = 1, \ldots, n \,. \tag{3.23}$$

The above normalization constants play a prominent role in the following sampling procedure, called **drafting**. The idea is to draft the positions $J_1, \ldots, J_k$ of the unities (so $X_{J_i} = 1, i = 1, \ldots, k$) one by one. The first position, J_1 is selected from the probabilities proportional to the conditional probabilities $\mathbb{P}(X_j = 1 \,|\, \sum_i X_i = k)$, $j = 1, \ldots, n$. It is not difficult to see that the corresponding normalization constant is k. More precisely, the **drafting vector** $\mathbf{a} = (a_1, \ldots, a_n)$ with $a_j = \mathbb{P}(X_j = 1 \,|\, \sum_i X_i = k)/k$, $j = 1, \ldots, n$ forms a probability distribution vector, and $J_1 \sim \mathbf{a}$. The probabilities $\{a_j\}$, the normalization constants R_k, and $\{R_{k,j}\}$ are related via

$$\begin{aligned} a_j &= \frac{\mathbb{P}(X_j = 1, \sum_{i \neq j} X_i = k-1)}{k\,\mathbb{P}(\sum_{i=1}^n X_i = k)} = \frac{p_j\, R_{k-1,j} \prod_{i\neq j}(1+w_i)^{-1}}{k\, R_k \prod_{i=1}^n (1+w_i)^{-1}} \\ &= \frac{w_j\, R_{k-1,j}}{k\, R_k}\ , j = 1,\ldots,n\ . \end{aligned}$$

Once J_1 is chosen, J_2 can be selected in a similar way, by first calculating R_{k-1} and $\{R_{k-2,j}\}$ for the conditional Bernoulli distribution defined by $\{w_i, i \neq J\}$, and so on. This gives the following procedure (several other generation algorithms may be found in [3]).

Algorithm 3.33 (Sampling From a Conditional Bernoulli Distribution)

1. *Set $\mathscr{C} = \{1,\ldots,n\}$ (current indices) and $\mathscr{S} = \emptyset$ (selected indices). Set $i = 1$.*
2. *While $i \leqslant k$, calculate R_{k-i+1}, $R_{k-i,j}$, $j \in \mathscr{C}$ based on $\{w_i, i \in \mathscr{C}\}$, and compute the corresponding drafting vector* $\mathbf{a}$.
3. *Draw $J \sim \mathbf{a}$.*
4. *Set $\mathscr{S} = \mathscr{S} \cup \{J\}$, $\mathscr{C} = \mathscr{C} \setminus \{J\}$, and $i = i + 1$. Return to Step 2.*
5. *Set $X_i = 1$, $i \in \mathscr{S}$ and $X_i = 0$, $i \in \mathscr{C}$. Output $\mathbf{X} = (X_1,\ldots,X_n)$.*

■ **EXAMPLE 3.23 (Exact Conditional Bernoulli Sampling)**

Suppose $\mathbf{p} = (1/2, 1/3, 1/4, 1/5)$ and $k = 2$. Then $\mathbf{w} = (w_1,\ldots,w_4) = (1, 1/2, 1/3, 1/4)$. We have $R_2 = 35/24 \approx 1.45833$. Thus, for example,

$$\mathbb{P}\left(X_1 = 0, X_2 = 1, X_3 = 0, X_4 = 1 \,\middle|\, \sum_{i=1}^4 X_i = 2\right) = \frac{w_2\, w_4}{35/24} = \frac{3}{35} \approx 0.08571\ .$$

The following MATLAB program calculates the normalization constants. Note that the first element of `Rvals` corresponds to R_k and element `j+1` to $R_{k,j}$.

```
function Rvals = Rgens(k,W)
N=length(W);
T=zeros(k,N+1);
R=zeros(k+1,N+1);
for i=1:k
    for j=1:N, T(i,1)=T(i,1)+W(j)^i; end
    for j=1:N, T(i,j+1)=T(i,1)-W(j)^i; end
end
R(1,:)=ones(1,N+1);
for j=1:k
    for l=1:N+1
        for i=1:j
```

```
            R(j+1,1)=R(j+1,1)+(-1)^(i+1)*T(i,1)*R(j-i+1,1);
        end
    end
    R(j+1,:)=R(j+1,:)/j;
end
Rvals=[R(k+1,1),R(k,2:N+1)];
```

To generate random vectors according to this conditional Bernoulli distribution call the following MATLAB function `condbern(p,k)`, where `k` is the number of unities (here 2) and `p` is the vector of probabilities **p**. This function returns the positions of the unities, such as $(1,2)$ or $(2,4)$.

```
function sample = condbern(k,p)
w=zeros(1,length(p));
sample=zeros(1,k);
ind1=find(p==1);
sample(1:length(ind1))=ind1;
k=k-length(ind1);
ind=find(p<1 & p>0);
w(ind)=p(ind)./(1-p(ind));
for i=1:k
    a=zeros(1,length(ind));
    Rvals=Rgens(k-i+1,w(ind));
    for j=1:length(ind)
        a(j)=w(ind(j))*Rvals(j+1)/((k-i+1)*Rvals(1));
    end
    a=cumsum(a);
    entry=ind(min(find(a>rand)));
    ind=ind(find(ind~=entry));
    sample(length(ind1)+i)=entry;
end
sample=sort(sample);
```

Further Reading

For an overview of the different approaches to generating random variables, see the classic book of Devroye [5]. (See the author's home page `http://cg.scs.carleton.ca/~luc/` for a list of errata, and an electronic copy.)

Hörmann et al. [11] examine the problem of automatically constructing generation algorithms. The same group is also responsible for the UNU.RAN library: a software implementation in C for automatic random number generation (available at `http://statmath.wu.ac.at/unuran/`).

For a survey of random variable generation algorithms with emphasis on efficient software implementation, see [17].

The incorrectness of the usual method for generating variates uniformly on the surface of an ellipsoid is pointed out in [1].

REFERENCES

1. K. A. Borovkov. On simulation of random vectors with given densities in regions and on their boundaries. *Journal of Applied Probability*, 31(1):205–220, 1994.
2. G. Casella and R. L. Berger. *Statistical Inference*. Duxbury Press, Pacific Grove, CA, second edition, 2001.
3. S. X. Chen and J. S. Liu. Statistical applications of the Poisson-binomial and conditional Bernoulli distributions. *Statistica Sinica*, 7:875–892, 1997.
4. Y. Chen, A. P. Dempster, and J. S. Liu. Weighted finite population sampling to maximize entropy. *Biometrika*, 81(3):457–469, 1997.
5. L. Devroye. *Non-Uniform Random Variate Generation*. Springer-Verlag, New York, 1986.
6. L. Devroye. Random variate generation in one line of code. In J. M. Charnes, D. J. Morrice, D. T. Brunner, and J. J. Swain, editors, *Proceedings of the 1996 Winter Simulation Conference*, pages 265–272, Coronado, CA, December 1996.
7. G. Fusai and A. Roncoroni. *Implementing Models in Quantitative Finance: Methods and Cases*. Springer-Verlag, Berlin, second edition, 2008.
8. W. R. Gilks, N. G. Best, and K. K. C. Tan. Adaptive rejection Metropolis sampling within Gibbs sampling. *Journal of the Royal Statistical Society, Series C*, 44(4):455–472, 1995.
9. W. R. Gilks and P. Wild. Adaptive rejection sampling for Gibbs sampling. *Journal of the Royal Statistical Society, Series C*, 41(2):337–348, 1992.
10. W. Hörmann and G. Derflinger. The transformed rejection method for generation random variables, an alternative to the ratio of uniforms method. *Communications in Statistics: Simulation and Computation*, 23(3):847–860, 1994.
11. W. Hörmann, J. Leydold, and G. Derflinger. *Automatic Nonuniform Random Variate Generation*. Springer-Verlag, Berlin, 2004.
12. A. J. Kinderman and J. F. Monahan. Computer generation of random variables using the ratio of uniform deviates. *ACM Transactions on Mathematical Software*, 3(3):257–260, 1977.
13. C. Klüppelberg and S. I. Resnick. The Pareto copula, aggregation of risks, and the emperor's socks. *Journal of Applied Probability*, 45(1):67–84, 2008.
14. D. E. Knuth. *The Art of Computer Programming*, volume 2: *Seminumerical Algorithms*. Addison-Wesley, Reading, MA, third edition, 1997.
15. R. A. Kronmal and A. V. Peterson. *Marsaglia's Table Method*, volume 5 of *Encyclopedia of Statistical Sciences*, pages 275–276. John Wiley & Sons, New York, 1985.
16. R. Meyer, B. Cai, and F. Perron. Adaptive rejection Metropolis sampling using Lagrange interpolation polynomials of degree 2. *Computational Statistics and Data Analysis*, 52(7):3408–3423, 2008.
17. J. F. Monahan. *Numerical Methods of Statistics*. Cambridge Series in Statistical and Probabilistic Mathematics. Cambridge University Press, London, 2010.
18. R. B. Nelsen. *An Introduction to Copulas*. Springer-Verlag, New York, second edition, 2006.
19. A. V. Peterson Jr. and R. A. Kronmal. A representation for discrete distributions by equiprobable mixture. *Journal of Applied Probability*, 17(1):102–111, 1980.
20. S. M. Ross. *Simulation*. Academic Press, New York, third edition, 2002.

21. R. Y. Rubinstein and D. P. Kroese. *Simulation and the Monte Carlo Method.* John Wiley & Sons, New York, second edition, 2007.

22. A. J. Walker. An efficient method for generating discrete random variables with general distributions. *ACM Transactions on Mathematical Software*, 3(3):253–256, 1977.

23. C. S. Wallace. Transformed rejection generators for gamma and normal pseudo-random variables. *Australian Computer Journal*, 8(1):103–105, 1976.

CHAPTER 4

PROBABILITY DISTRIBUTIONS

This chapter lists the major discrete and continuous probability distributions used in Monte Carlo simulation, along with their main properties and specific algorithms for random variable generation. For general random variable generation procedures, see Chapter 3. Further information on families of distributions and their properties can be found in Section D.1 (exponential families), Section D.2 (tail and stability properties), and Section 3.1.2 (transformations and location–scale families).

☞ 43
☞ 701
☞ 47

4.1 DISCRETE DISTRIBUTIONS

We list various discrete distributions in alphabetical order. Recall that a discrete distribution is completely specified by its discrete pdf. ☞ 610

4.1.1 Bernoulli Distribution

The pdf of the **Bernoulli** distribution is given by

$$f(x;p) = p^x (1-p)^{1-x}, \quad x \in \{0,1\} ,$$

where $p \in [0,1]$ is the **success parameter**. We write the distribution as $\mathsf{Ber}(p)$.

The Bernoulli distribution is used to describe experiments with only two outcomes: 1 (success) or 0 (failure). Such an experiment is called a **Bernoulli trial**. A sequence of iid Bernoulli random variables, $X_1, X_2, \ldots \sim_{\text{iid}} \mathsf{Ber}(p)$, is called a

Bernoulli process. Such a process is a model for the random experiment where a biased coin is tossed repeatedly; see also Example A.5. ☞ 626

Table 4.1 Moment properties of the $\text{Ber}(p)$ distribution.

Property		Condition
Expectation	p	
Variance	$p(1-p)$	
Probability generating function	$1-p+zp$	$\|z\| \leqslant 1$

The inverse-transform method leads to the following generation algorithm.

Algorithm 4.1 (Ber(p) Generator)

1. *Generate* $U \sim \text{U}(0,1)$.
2. *If* $U \leqslant p$, *return* $X = 1$; *otherwise, return* $X = 0$.

■ **EXAMPLE 4.1 (Bernoulli Generation)**

The following MATLAB code generates one hundred $\text{Ber}(0.25)$ random variables, and plots a bar graph of the binary data.

```
X = (rand(1,100) <= 0.25);    bar(X)
```

4.1.2 Binomial Distribution

The pdf of the **binomial** distribution is given by

$$f(x;n,p) = \mathbb{P}(X = x) = \binom{n}{x} p^x (1-p)^{n-x}, \quad x = 0, 1, \ldots, n\,,$$

where $0 \leqslant p \leqslant 1$. We write the distribution as $\text{Bin}(n,p)$. The binomial distribution is used to describe the total number of successes in a sequence of n independent Bernoulli trials. That is, a $\text{Bin}(n,p)$-distributed random variable X can be written as the sum $X = B_1 + \cdots + B_n$ of independent $\text{Ber}(p)$ random variables $\{B_i\}$. Examples of the graph of the pdf are given in Figure 4.1.

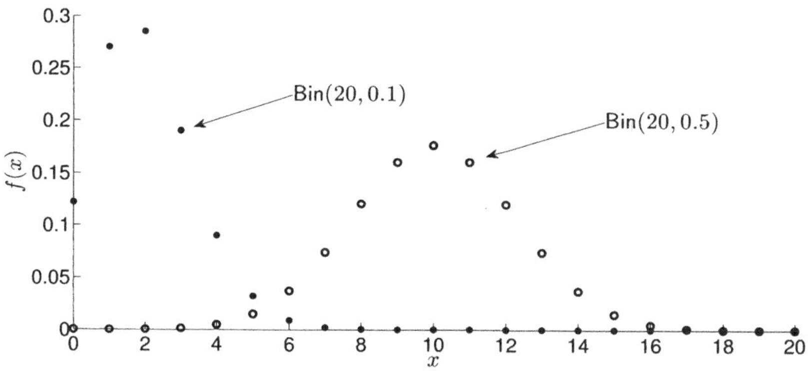

Figure 4.1 The pdfs of the Bin(20, 0.1) (solid dot) and Bin(20, 0.5) (circle) distributions.

Table 4.2 Moment and tail properties of the $\mathsf{Bin}(n, p)$ distribution.

Property		Condition
Expectation	np	
Variance	$np(1-p)$	
Probability generating function	$(1-p+zp)^n$	$\lvert z\rvert \leqslant 1$
Tail probability $\mathbb{P}(X > x)$	$I_p(x+1, n-x)$	$x = 0, 1, \ldots$

Here, $I_x(\alpha, \beta)$ denotes the *incomplete beta function*. Other properties and relations are: ☞ 715

1. *Order statistics*: Let $U_1, \ldots, U_n \overset{\text{iid}}{\sim} \mathsf{U}(0,1)$, and denote by $U_{(1)} < \cdots < U_{(n)}$ the order statistics. Let $X = \max\{i : U_{(i)} < p\}$. Then, $X \sim \mathsf{Bin}(n, p)$.

2. *Number of failures*: Denote $X \sim \mathsf{Bin}(n, p)$ the number of successes in n Bernoulli experiments and $Y = n - X$ the number of failures. Then, $Y \sim \mathsf{Bin}(n, 1-p)$.

3. *Sum of binomial random variables*: Let $X_k \sim \mathsf{Bin}(n_k, p)$, $k = 1, \ldots, K$, independently. Then,
$$\sum_{k=1}^{K} X_k \sim \mathsf{Bin}\left(\sum_{k=1}^{K} n_k, p\right).$$

4. *Convergence to the normal distribution*: Let $X_n \sim \mathsf{Bin}(n, p)$. A direct consequence of the central limit theorem is
$$\frac{X_n - np}{\sqrt{np(1-p)}} \xrightarrow{d} Y \sim \mathsf{N}(0,1) \quad \text{as} \quad n \to \infty\,.$$

An often used rule of thumb is that for finite n the corresponding approximation is accurate if both np and $n(1-p)$ are greater than 5. For p close to $1/2$ the cdf of $\mathsf{N}(np-1/2, np(1-p))$ approximates the cdf of X_n even better. This is called the **continuity correction.**

5. *Convergence to the Poisson distribution:* Let $X_n \sim \mathsf{Bin}(n, \lambda/n)$. Then,

$$X_n \xrightarrow{d} Y \sim \mathsf{Poi}(\lambda) \quad \text{as} \quad n \to \infty\,.$$

6. *Geometric distribution:* Let $Y_0, Y_1, \ldots \stackrel{\text{iid}}{\sim} \mathsf{Geom}(p)$. Then,

$$X = \min\left\{k : \sum_{i=0}^{k} Y_i > n\right\} \sim \mathsf{Bin}(n,p)\,.$$

7. *Exponential distribution:* Let $Y_1, \ldots, Y_n \stackrel{\text{iid}}{\sim} \mathsf{Exp}(1)$. Then,

$$X = \min\left\{k : \sum_{i=0}^{k} \frac{Y_i}{n-i} > -\ln(1-p)\right\} \sim \mathsf{Bin}(n,p)\,.$$

The fact that binomial random variables can be viewed as sums of Bernoulli random variables leads to the following generation algorithm.

Algorithm 4.2 ($\mathsf{Bin}(n,p)$ Generator)

1. *Generate* $X_1, \ldots, X_n \stackrel{\text{iid}}{\sim} \mathsf{Ber}(p)$.
2. *Return* $X = \sum_{i=1}^{n} X_i$.

Since the execution time of Algorithm 4.2 is proportional to n, one is motivated to use alternative methods for large n. When n is large but p small such that np is moderate (say $\leqslant 10$), the following method, which is based on Property 6 above, provides fast random variable generation. The algorithm can also be used when p is close to 1, by generating $Y = n - X$.

Algorithm 4.3 ($\mathsf{Bin}(n,p)$ Generator via the Geometric Method)

1. *Set* $X = 0$, $c = \ln(1-p)$, *and* $S = \lceil \ln(U)/c \rceil$, *where* $U \sim \mathsf{U}(0,1)$.
2. *While* $S < n+1$, *generate* $U \sim \mathsf{U}(0,1)$, *and set* $X = X+1$ *and* $S = S + \lceil \ln(U)/c \rceil$.
3. *Return* X.

■ **EXAMPLE 4.2 (Binomial Generation)**

A MATLAB implementation of Algorithm 4.3 is given next. In the last part of the code the observed counts are compared with the expected counts.

```
%bingen.m
n = 100; p = 0.1; mu = n*p; N = 10^5;
x = zeros(1,N); c = log(1-p);
for i=1:N
    s = ceil(log(rand)/c);
    while s < n + 1
       x(i) = x(i)+ 1;
       s = s + ceil(log(rand)/c);
    end
end

xx = [floor(mu - 4*sqrt(mu)):1:ceil(mu + 4*sqrt(mu))];
count = hist(x,xx);
ex = binopdf(xx,n,p)*N;
hold on
plot(xx,count,'or')
plot(xx,ex,'.b')
hold off
```

Finally, as a result of the central limit theorem for the binomial distribution (see Property 4 above), the distribution of $X \sim \mathsf{Bin}(n,p)$ is close to that of $Y \sim \mathsf{N}(np - 1/2, np(1-p))$ as n becomes large. This leads to the following *approximate* generation algorithm.

Algorithm 4.4 ($\mathsf{Bin}(n,p)$ Generator via the Normal Approximation)

1. *Generate* $Y \sim \mathsf{N}(0,1)$.
2. *Return* $X = \max\left\{0, \left\lfloor np + \frac{1}{2} + Z\sqrt{np(1-p)}\right\rfloor\right\}$.

As the approximation can be quite inaccurate for certain choices of n and p, we recommend the following exact recursive methods instead for $n > 10$.

The first method relies on the fast $\mathsf{NegBin}(r,p)$ generator given in Algorithm 4.10 and Property 2 on Page 95.

Algorithm 4.5 (Recursive $\mathsf{Bin}(n,p)$ Generator (I))

1. *If* $n \leqslant 10$, *output* $X \sim \mathsf{Bin}(n,p)$ *using Algorithm 4.2; otherwise, proceed with the next step.*
2. *Set* $k = \lceil np \rceil$. *Generate* $Y \sim \mathsf{NegBin}(k,p)$ *via Algorithm 4.10 and set* $T = Y + k$.
3. *If* $T \leqslant n$, *generate* $Z \sim \mathsf{Bin}(n-T,p)$ *and output:*
$$X = k + Z \;;$$
otherwise (that is, if $T > n$*), generate* $Z \sim \mathsf{Bin}(T-n,p)$ *and output:*
$$X = k - Z \;.$$

The following MATLAB code implements the algorithm. With $n = 10^{10}$ and $p = 1/2$, the number of recursive calls to the function binomialrnd.m is typically 4 or 5.

```
function x=binomialrnd(n,p)
% recursive binomial generator
if n<=10
      x=sum(rand(1,n)<p);
else
    k=ceil(n*p);Y=nbinrnd(k,p);% generate NegBin(k,p)
    T=k+Y;
    if T<=n
        x=k+binomialrnd(n-T,p);
    else
        x=k-binomialrnd(T-n,p);
    end
end
```

The second recursive method relies on a fast $\mathsf{Beta}(\alpha, \beta)$ generator (for example, Algorithm 4.25) and is based on Property 1 on Page 87 and Property 6 on Page 104.

Algorithm 4.6 (Recursive $\mathsf{Bin}(n, p)$ Generator (II))

1. *If* $n \leqslant 10$, *output* $X \sim \mathsf{Bin}(n, p)$ *using Algorithm 4.2; otherwise, proceed with the next step.*
2. *Set* $k = \lceil np \rceil$. *Generate* $U_{(k)} \sim \mathsf{Beta}(k, n + 1 - k)$ *via Algorithm 4.25.*
3. *If* $U_{(k)} \leqslant p$, *generate* $Z \sim \mathsf{Bin}\left(n - k, \frac{p - U_{(k)}}{1 - U_{(k)}}\right)$ *and output:*

$$X = k + Z\ ;$$

otherwise (that is, if $U_{(k)} > p$*), generate* $Z \sim \mathsf{Bin}\left(k - 1, \frac{U_{(k)} - p}{U_{(k)}}\right)$ *and output:*

$$X = k - Z\ .$$

The following MATLAB code implements the algorithm. With $n = 10^{10}$ and $p = 1/2$, the number of recursive calls to the function binomrnd_beta.m is typically 6 or 7, but despite this the recursive algorithm based on the $\mathsf{Beta}(\alpha, \beta)$ generator is faster than the one based on the $\mathsf{NegBin}(r, p)$ generator.

```
function x=binomrnd_beta(n,p)
% recursive binomial generator based on Beta dist.
if n<=10
      x=sum(rand(1,n)<p);
else
    k=ceil(n*p);Uk=betarnd(k,n+1-k);% generate beta r.v.
    if Uk<p
```

```
        x=k+binomrnd_beta(n-k,(p-Uk)/(1-Uk));
    else
        x=k-binomrnd_beta(k-1,(Uk-p)/Uk);
    end
end
```

4.1.3 Geometric Distribution

The pdf of the **geometric** distribution is given by

$$f(x;p) = (1-p)^{x-1}p, \quad x = 1,2,3,\ldots \tag{4.1}$$

where $0 \leqslant p \leqslant 1$. We write the distribution as $\mathsf{Geom}(p)$. The geometric distribution is used to describe the time of first success in an infinite sequence of independent Bernoulli trials with success probability p. Examples of the graph of the pdf are given in Figure 4.2.

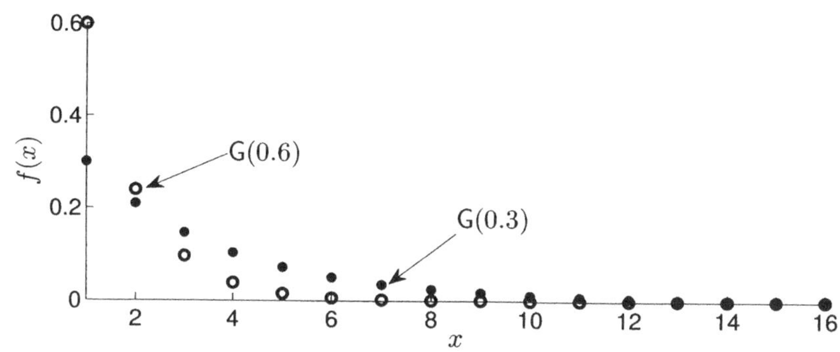

Figure 4.2 The pdfs of the $\mathsf{Geom}(0.3)$ (solid dot) and $\mathsf{Geom}(0.6)$ (circle) distributions.

Remark 4.1.1 (Alternative Definition) There is another, nonequivalent, definition of the geometric distribution, where the pdf is given by

$$f(x;p) = (1-p)^{x}p, \; x = 0,1,2,\ldots \;,$$

that is, a shifted version of (4.1). This alternate definition describes the number of trials *before* the first success in an infinite sequence of independent Bernoulli trials. We will always assume the definition in (4.1), unless otherwise specified. To distinguish the two cases, we denote the geometric distribution starting at 0 with $\mathsf{Geom}_0(p)$. If $X \sim \mathsf{Geom}(p)$, then $(X-1) \sim \mathsf{Geom}_0(p)$.

Table 4.3 Moment and tail properties of the Geom(p) distribution.

Property		Condition
Expectation	$\frac{1}{p}$	
Variance	$\frac{1-p}{p^2}$	
Probability generating function	$\frac{z\,p}{1-z\,(1-p)}$	$\|z\| \leqslant 1$
Tail probability $\mathbb{P}(X > x)$	$(1-p)^x$	$x = 1, 2, \ldots$

Other properties and relations are:

1. *Memoryless property:* $\mathbb{P}(X > x + y \mid X > x) = \mathbb{P}(X > y)$ for $x, y = 1, 2, \ldots$.

2. *Sum:* If $X_1, \ldots, X_n \overset{\text{iid}}{\sim} \mathsf{Geom}_0(p)$, then

$$\sum_{k=1}^{n} X_k \sim \mathsf{NegBin}(n, p) .$$

3. *Exponential distribution:* Let $Y \sim \mathsf{Exp}(\lambda)$, with $\lambda = -\ln(1-p)$. Then, $\lceil Y \rceil \sim \mathsf{Geom}(p)$.

4. *Convergence to exponential distribution:* Let $X_n \sim \mathsf{Geom}(p_n)$, with $p_n \to 0$ as $n \to \infty$. Then,

$$p_n\, X_n \xrightarrow{d} Y \sim \mathsf{Exp}(1) \quad \text{as} \quad n \to \infty .$$

The relationship between the exponential and geometric distributions described in Property 3 yields the following generator.

Algorithm 4.7 (Geom(p) Generator (I))

1. *Generate* $Y \sim \mathsf{Exp}(-\ln(1-p))$.

2. *Output* $X = \lceil Y \rceil$.

Writing Algorithm 4.7 directly in terms of uniform random variables gives:

Algorithm 4.8 (Geom(p) Generator (II))

1. *Generate* $U \sim \mathsf{U}(0,1)$.

2. *Output* $X = \left\lceil \frac{\ln(U)}{\ln(1-p)} \right\rceil$.

4.1.4 Hypergeometric Distribution

The **hypergeometric** distribution has pdf

$$f(x; n, r, N) = \frac{\binom{r}{x}\binom{N-r}{n-x}}{\binom{N}{n}}, \quad \max\{0, r+n-N\} \leqslant x \leqslant \min\{n, r\},$$

where N, $n \leqslant N$, and $r \leqslant N$ are positive integers. We write the distribution as $\mathsf{Hyp}(n, r, N)$. The hypergeometric distribution is used in the following situation. Consider an urn with N balls, r of which are red. Draw n balls from the urn at random *without* replacement. Then, the number of red balls among the n chosen balls has a $\mathsf{Hyp}(n, r, N)$ distribution. A consequence is that the sequence $X_1, X_2, \ldots$ with $X_N \sim \mathsf{Hyp}(n, pN, N)$ converges in distribution to $\mathsf{Bin}(n, p)$ as $N \to \infty$.

Examples of the graph of the pdf are given in Figure 4.3. Here, the same values for p are used as in Figure 4.1, namely, $p = 0.1$ (solid dot) and $p = 0.5$ (circle).

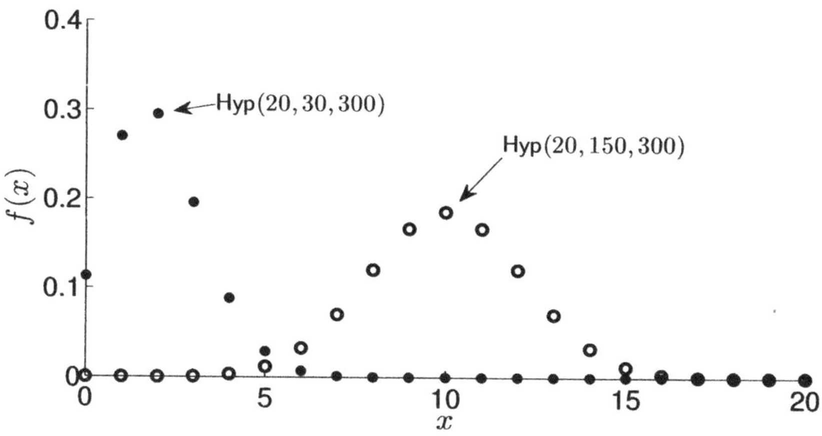

Figure 4.3 The pdfs of the $\mathsf{Hyp}(20, 30, 300)$ (solid dot) and $\mathsf{Hyp}(20, 150, 300)$ (circle) distributions.

Table 4.4 Moment properties of the $\mathsf{Hyp}(n, r, N)$ distribution.

Property		Condition
Expectation	$n\,\frac{r}{N}$	
Variance	$n\,\frac{r}{N}\left(1-\frac{r}{N}\right)\frac{N-n}{N-1}$	
Prob. gen. function	$\frac{(N-n)!\,(N-r)!}{N!\,(N-n-r)!}\,{}_2F_1(-n, -r; N-n-r+1; z)$	$\lvert z\rvert \leqslant 1$

Here, ${}_2F_1$ is the *hypergeometric function* (see Section D.9.7). The urn description of the $\mathsf{Hyp}(n, r, N)$ distribution motivates the following generation algorithm. ☞ 716

Algorithm 4.9 (Hyp(n, r, N) Generation) *Consider an urn with N balls, numbered $1, \ldots, N$. The first r of these are red.*

1. *Draw n balls uniformly without replacement from the N numbered balls.*

2. *Let X be the number of red balls.*

■ **EXAMPLE 4.3 (Hypergeometric Generation)**

The following MATLAB code gives an implementation of the above algorithm.

```
%hyperg.m
N = 100; %total number of balls
n = 20; % take n balls
r = 30; % number of red balls
w = zeros(1,N);
w(1:r) = 1;
K = 10^5; %sample size
x = zeros(1,K);
for i=1:K
    [s,ix] = sort(rand(1,N));
    x(i) = sum(w(ix(1:n)));
end
```

Algorithm 4.9 is not efficient for large N. A universally fast, but much more complicated generation algorithm can be found in [8, Page 545].

4.1.5 Negative Binomial Distribution

The **negative binomial distribution** has pdf

$$f(x; r, p) = \frac{\Gamma(r + x)}{\Gamma(r)\, x!}\, p^r (1 - p)^x, \quad x = 0, 1, 2, \ldots\,,$$

where $r \geqslant 0$ and $0 \leqslant p \leqslant 1$. We write the distribution as $\mathsf{NegBin}(r, p)$. In many applications r is a positive integer, $r = n$. The distribution is then also known as the **Pascal distribution**, in which case the pdf can be written as

$$f(x; n, p) = \binom{n + x - 1}{n - 1} p^n (1 - p)^x, \quad x = 0, 1, 2, \ldots\,.$$

The Pascal distribution describes the number of failures before the n-th success in a Bernoulli process. The dependence of the distribution on r, for a fixed mean, is illustrated in Figure 4.4.

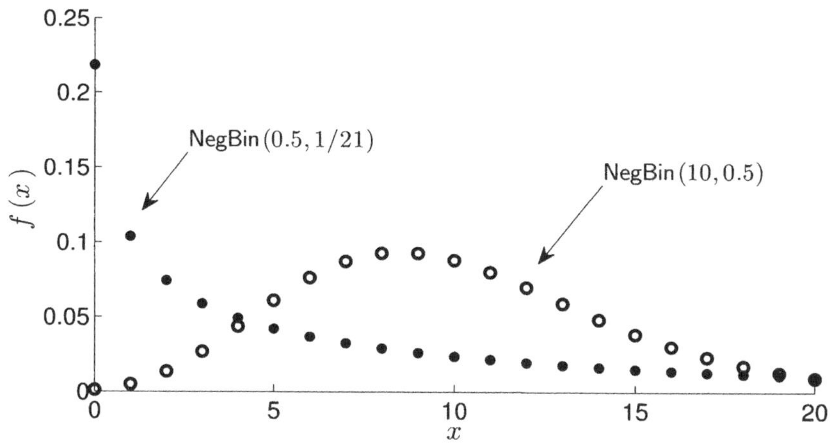

Figure 4.4 The pdfs of the NegBin(0.5, 1/21) (solid dot) and NegBin(10, 0.5) (circle) distributions.

Table 4.5 Moment and tail properties of the $\mathsf{NegBin}(r,p)$ distribution.

Property		Condition
Expectation	$\dfrac{r(1-p)}{p}$	
Variance	$\dfrac{r(1-p)}{p^2}$	
Probability generating function	$\left(\dfrac{p}{1-z(1-p)}\right)^r$	$\lvert z\rvert \leqslant 1$
Tail probability $\mathbb{P}(X > x)$	$I_{1-p}(x+1,r)$	$x = 0,1,2,\ldots$

Here, $I_x(\alpha,\beta)$ denotes the *incomplete beta function*. Other properties and relations are: ☞ 715

1. *Bernoulli process*: Consider an infinite sequence of independent Bernoulli trials with success parameter p. Let Y_n denote the number of failures before the n-th success and let N_t denote the number of successes in the first t trials. Then,

$$\{Y_n \leqslant x\} \Leftrightarrow \{N_{x+n} \geqslant n\} \quad \text{for all } n \text{ and } x \in \{0,1,\ldots\} .$$

Moreover, $N_{x+n} \sim \mathsf{Bin}(x+n,p)$ and $Y_n \sim \mathsf{NegBin}(n,p)$.

2. *Geometric distribution*: If $X_1,\ldots,X_n \overset{\text{iid}}{\sim} \mathsf{Geom}_0(p)$, then

$$\sum_{k=1}^{n} X_k \sim \mathsf{NegBin}(n,p) .$$

3. *Negative binomial distribution:* Let $X_k \sim \mathsf{NegBin}(r_k, p)$, $k = 1, \ldots, n$, independently. Then,
$$\sum_{k=1}^{n} X_k \sim \mathsf{NegBin}\left(\sum_{k=1}^{n} r_k,\, p\right).$$

4. *Convergence to the Poisson distribution:* Let $X_n \sim \mathsf{NegBin}\left(n, 1 - \frac{\lambda}{n}\right)$. Then,
$$X_n \xrightarrow{d} Y \sim \mathsf{Poi}(\lambda) \quad \text{as} \quad n \to \infty .$$

5. *Gamma–Poisson mixture:* If $\Lambda \sim \mathsf{Gamma}(n, \frac{p}{1-p})$ and $(X \mid \Lambda = \lambda) \sim \mathsf{Poi}(\lambda)$, then $X \sim \mathsf{NegBin}(n, p)$.

For small integer-valued r ($r = n$), we can generate a $\mathsf{NegBin}(n,p)$ random variable via the sum of n $\mathsf{Geom}_0(p)$ random variables (see Property 2 above), leading to the following algorithm.

Algorithm 4.10 (NegBin(n, p) Generator via Geometric Distribution)

1. *Generate* $U_1, \ldots, U_n \overset{\text{iid}}{\sim} \mathsf{U}(0,1)$.
2. *Set* $c = \ln(1-p)$ *and* $Y_i = \lfloor \ln(U_i)/c \rfloor, i = 1, \ldots, n$.
3. *Return* $X = \sum_{i=1}^{n} Y_i$.

This direct approach becomes inefficient for large r and does not work for non-integer values of r. We recommend the following uniformly fast generator (valid for all values of r and p), based on Property 5 above.

Algorithm 4.11 (Uniformly Fast NegBin(r, p) Generator)

1. *Generate* $\Lambda \sim \mathsf{Gamma}\left(r, \frac{p}{1-p}\right)$.
2. *Conditional on* Λ, *generate and output* $X \sim \mathsf{Poi}(\Lambda)$.

4.1.6 Phase-Type Distribution (Discrete Case)

The **discrete phase-type** distribution has pdf
$$f(x; \boldsymbol{\alpha}, A) = \boldsymbol{\alpha} A^{x-1}(I - A)\mathbf{1} \quad x = 1, 2, \ldots,$$
where $\boldsymbol{\alpha}$ is a $1 \times m$ probability vector, $\mathbf{1}$ is a $m \times 1$ vector of 1s, A is an $m \times m$ matrix such that $I - A$ has an inverse and such that
$$P = \begin{pmatrix} A & \mathbf{A}_0 \\ \mathbf{0}^\top & 1 \end{pmatrix}$$
☞ 632 is the transition matrix of a Markov chain, and $\mathbf{A}_0 = (I - A)\mathbf{1}$. We write the distribution as $\mathsf{DPH}(\boldsymbol{\alpha}, A)$. A random variable $X \sim \mathsf{DPH}(\boldsymbol{\alpha}, A)$ can be thought of as the time of absorption in state $m+1$ of a discrete-time Markov chain on $\{1, \ldots, m, m+1\}$ with transition matrix P and initial distribution $(\boldsymbol{\alpha}, 0)$.

Table 4.6 Moment and tail properties of the $\mathsf{DPH}(\boldsymbol{\alpha}, A)$ distribution.

Property		Condition
Expectation (μ)	$\boldsymbol{\alpha}(I - A)^{-1}\mathbf{1}$	
Variance	$\boldsymbol{\alpha}A(I - A)^2\mathbf{1} + \mu - \mu^2$	
Probability generating function	$z\boldsymbol{\alpha}(I - zA)^{-1}\mathbf{A}_0$	$\lvert z\rvert \leqslant 1$
Tail probability $\mathbb{P}(X > x)$	$\boldsymbol{\alpha}A^x\mathbf{1}$	$x = 1, 2, \ldots$

Other properties and relations (see, for example, [26]) are:

1. *Geometric distribution*: The geometric distribution $\mathsf{Geom}(p)$ is the simplest discrete phase-type distribution, with $m = 1$, $A = 1 - p$ and $\alpha = 1$.

2. *Negative binomial*: If $X \sim \mathsf{NegBin}(n, p)$, then $X + n \sim \mathsf{DPH}(\boldsymbol{\alpha}, A)$, with $\boldsymbol{\alpha}$ the $1 \times n$ vector $(1, 0, \ldots, 0)$ and A the $n \times n$ matrix

$$A = \begin{pmatrix} 1-p & p & 0 & \ldots & 0 \\ 0 & 1-p & p & \ldots & 0 \\ \vdots & \vdots & \ddots & \ddots & \vdots \\ 0 & \ldots & 0 & 1-p & p \\ 0 & \ldots & 0 & 0 & 1-p \end{pmatrix}.$$

3. *Sum*: The sum of a finite number of independent phase-type random variables has again a phase-type distribution.

4. *Mixture*: The mixture of a finite number of phase-type distributions is again a phase-type distribution.

The fact that $X \sim \mathsf{DPH}(\boldsymbol{\alpha}, A)$ can be viewed as the time of absorption of a Markov chain leads to the following algorithm, where $P(y, \cdot)$ denotes the y-th row of the transition matrix P and $\boldsymbol{\alpha}$ denotes the initial distribution.

Algorithm 4.12 ($\mathsf{DPH}(\alpha, A)$ Generator)

1. *Draw* $Y_0 \sim \boldsymbol{\alpha}$. *Set* $t = 0$.

2. *While* $Y_t \neq m + 1$, *draw* $Y_{t+1} \sim P(Y_t, \cdot)$ *and set* $t = t + 1$.

3. *Return* $X = t$.

■ **EXAMPLE 4.4** (NegBin as a DPH **Distribution**)

The negative binomial distribution can be seen as a phase-type distribution — see Property 2 above. The following MATLAB program implements Algorithm 4.12 for the corresponding phase-type distribution with $m = 4$ and $p = 0.2$ and compares the samples (with m subtracted) with the true, $\mathsf{NegBin}(m, p)$, distribution.

```
%negbin.m
alpha = [1 0 0 0];
p = 0.2;
m = 4;
A = [1-p, p, 0, 0; 0, 1-p ,p ,0 ; 0 ,0, 1-p, p; 0, 0, 0 ,1-p] ;
A0 = abs((eye(m) - A)*ones(m,1));
P = [A A0; zeros(1,m) 1]
N = 10^5;
x = zeros(N,1);
for i=1:N;
    X = 0;
    Y = min(find(cumsum(alpha)> rand));
    while Y ~= m+1
        Y = min(find(cumsum(P(Y,:))> rand));
        X = X+1;
    end
    x(i) = X;
end
x = x - m; % shifted samples come from a NegBin(n,p) distribution.

xx = [0: 2*m/p];
count = hist(x,xx);
ex = nbinpdf(xx,m,p)*N;
hold on
plot(xx,count,'.r')
plot(xx,ex,'ob')
hold off
```

4.1.7 Poisson Distribution

The pdf of the **Poisson** distribution is given by

$$f(x;\lambda) = \frac{\lambda^x}{x!}\,\mathrm{e}^{-\lambda}, \quad x = 0, 1, 2, \ldots,$$

where $\lambda > 0$ is the **rate** parameter. We write the distribution as $\mathsf{Poi}(\lambda)$. Examples of the graph of the pdf are given in Figure 4.5.

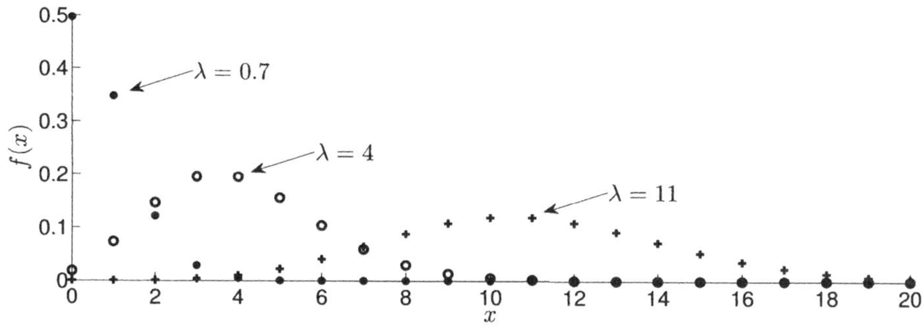

Figure 4.5 The pdfs of the Poi(0.7) (solid dot), Poi(4) (circle), and Poi(11) (plus) distributions.

The Poisson distribution is often used to model the number of arrivals of some sort during a fixed period of time. The Poisson distribution is closely related to the exponential distribution via the **Poisson process**; see Section 5.4. ☞ 170

Table 4.7 Moment and tail properties of the $\mathsf{Poi}(\lambda)$ distribution.

Property		Condition
Expectation	λ	
Variance	λ	
Probability generating function	$e^{-\lambda(1-z)}$	$\lvert z\rvert \leqslant 1$
Tail probability $\mathbb{P}(X > x)$	$P(\lfloor x+1 \rfloor, \lambda)$	$x = 0, 1, 2, \ldots$

Here, $P(\alpha, x)$ is the *incomplete gamma function*. Other properties and relations are: ☞ 716

1. *Binomial limit*: Let $X_n \sim \mathsf{Bin}(n, \lambda/n)$. Then,
$$X_n \xrightarrow{d} Y \sim \mathsf{Poi}(\lambda) \quad \text{as} \quad n \to \infty\,.$$
2. *Sum of Poisson random variables*: Let $X \sim \mathsf{Poi}(\lambda)$ and $Y \sim \mathsf{Poi}(\nu)$ be independent. Then, $Z = X + Y \sim \mathsf{Poi}(\lambda + \nu)$.
3. *Multinomial distribution*: Let $0 \leqslant p_i \leqslant 1$, $i = 1, \ldots, m$ and $\mathbf{p} = (p_1, \ldots, p_m)$. If $N \sim \mathsf{Poi}(\lambda)$ and $((X_1, \ldots, X_m) \mid N) \sim \mathsf{Mnom}(N, \mathbf{p})$, then
$$X_i \sim \mathsf{Poi}(\lambda\, p_i),\ i = 1, \ldots, m, \quad \text{independently}.$$
4. *Normal approximation*: $X \overset{\text{approx.}}{\sim} \mathsf{N}(\lambda, \lambda)$ for large λ.
5. *Binomial conditional distribution*: Let $X \sim \mathsf{Poi}(\lambda)$ and $Y \sim \mathsf{Poi}(\mu)$ be independent, and let $Z = X + Y$. Then,
$$(X \mid Z = z) \sim \mathsf{Bin}\left(z, \frac{\lambda}{\lambda + \mu}\right)\,.$$

6. *Exponential distribution*: Let $\{Y_i\} \stackrel{\text{iid}}{\sim} \mathsf{Exp}(\lambda)$. Then,

$$X = \max\left\{n : \sum_{j=1}^{n} Y_j \leqslant 1\right\} \sim \mathsf{Poi}(\lambda) . \tag{4.2}$$

That is, the Poisson random variable X can be interpreted as the maximal number of iid exponential variables whose sum does not exceed 1.

Let $\{U_i\} \stackrel{\text{iid}}{\sim} \mathsf{U}(0,1)$. Rewriting (4.2), we see that

$$\begin{aligned} X &= \max\left\{n : \sum_{j=1}^{n} -\ln U_j \leqslant \lambda\right\} \\ &= \max\left\{n : \ln\left(\prod_{j=1}^{n} U_j\right) \geqslant -\lambda\right\} \\ &= \max\left\{n : \prod_{j=1}^{n} U_j \geqslant \mathrm{e}^{-\lambda}\right\} \end{aligned} \tag{4.3}$$

has a $\mathsf{Poi}(\lambda)$ distribution. This leads to the following algorithm.

Algorithm 4.13 ($\mathsf{Poi}(\lambda)$ Generator)

1. *Set* $n = 1$ *and* $a = 1$.
2. *Generate* $U_n \sim \mathsf{U}(0,1)$ *and set* $a = a\,U_n$.
3. *If* $a \geqslant \mathrm{e}^{-\lambda}$, *set* $n = n + 1$ *and go to Step 2.*
4. *Otherwise, return* $X = n - 1$ *as a random variable from* $\mathsf{Poi}(\lambda)$.

As $\mathrm{e}^{-\lambda}$ is small for large λ, for large λ this algorithm becomes slow and more random numbers U_j are required to satisfy $\prod_{j=1}^{n} U_j < \mathrm{e}^{-\lambda}$. However, for large n we can skip a considerable number of steps in Algorithm 4.13 (say $m = \alpha n$ steps for some α less than but close to 1) by drawing $\sum_{j=1}^{m} -\ln U_j$ directly from a $\mathsf{Gamma}(m,1)$ distribution. This observation gives rise to the following recursive algorithm, due to Ahrens and Dieter [2], for generating from a Poisson distribution with large λ, say $\lambda > 100$.

Algorithm 4.14 ($\mathsf{Poi}(\lambda)$ Generator for Large λ)

1. *Set* $m = \lfloor (7/8)\,\lambda \rfloor$.
2. *Generate* $Y \sim \mathsf{Gamma}(m,1)$.
3. *If* $Y \leqslant \lambda$, *generate* $Z \sim \mathsf{Poi}(\lambda - Y)$ *and set* $X = m + Z$; *otherwise, generate* $X \sim \mathsf{Bin}(m-1, \lambda/Y)$.

Algorithm 4.14 relies on fast generation of $\mathsf{Gamma}(n,1)$ and $\mathsf{Bin}(n,p)$ random variables for large n, and p close to 1. Atkinson [3] proposes an acceptance–rejection algorithm for generating Poisson random variables using the $\mathsf{Logistic}(\mu,\sigma)$ distribution, with $\sigma = \frac{\pi}{\sqrt{3\lambda}}$ and $\mu = \lambda\,\sigma$, as the proposal distribution. Although the logistic distribution is continuous, it can be related to a discrete distribution via the floor function $\lfloor \cdot \rfloor$. The details of the algorithm are as follows.

Algorithm 4.15 (Poi(λ) Generator via Logistic(μ, σ))

1. *Set* $\sigma = \pi/\sqrt{3\lambda}$, $\mu = \lambda\,\sigma$, $c = 0.767 - 3.36/\lambda$, *and* $k = \ln c - \lambda - \ln \sigma$.
2. *Until* $X > -0.5$, *keep generating:*

$$X = \frac{\mu - \ln\left(\frac{1-U}{U}\right)}{\sigma}, \quad U \sim \mathsf{U}(0,1)\,.$$

3. *Let* $N = \lfloor X + 0.5 \rfloor$ *and* $V \sim \mathsf{U}(0,1)$. *If*

$$\mu - \sigma X + \ln\left(\frac{V}{(1 + \mathrm{e}^{\mu - \sigma X})^2}\right) \leqslant k + N \ln \lambda - \ln(N!)\,,$$

accept N *as a random variable from* Poi(λ); *otherwise, repeat from Step 2.*

In cases where a large number of independent Poisson random variables are desired, one can generate them in batches as in the following algorithm, based on Property 3 above.

Algorithm 4.16 (Generating Poisson Random Variables in Batches)

1. *Draw* $N \sim \mathsf{Poi}(\lambda)$.
2. *Conditional on* N, *draw* $\mathbf{X} = (X_1, \ldots, X_m) \sim \mathsf{Mnom}(N, \mathbf{p})$. *Output* $\mathbf{X}$, *where* $X_i \sim \mathsf{Poi}(\lambda\, p_i)$, $i = 1, \ldots, m$, *independently.*

4.1.8 Uniform Distribution (Discrete Case)

The **discrete uniform** distribution has pdf

$$f(x; a, b) = \frac{1}{b - a + 1}, \quad x \in \{a, \ldots, b\}\,,$$

where $a, b \in \mathbb{Z}$, $b \geqslant a$ are parameters. The discrete uniform distribution is used as a model for choosing a random element from $\{a, \ldots, b\}$ such that each element is equally likely to be drawn. We denote this distribution by $\mathsf{DU}(a, b)$. The uniform distribution can also be defined on any finite set $\mathscr{X}$, in which case

$$f(x; |\mathscr{X}|) = \frac{1}{|\mathscr{X}|}, \quad x \in \mathscr{X}\,,$$

where $|\mathscr{X}|$ denotes the number of elements in $\mathscr{X}$. We write the distribution as $\mathsf{DU}(\mathscr{X})$ or simply $\mathsf{U}(\mathscr{X})$.

As with its continuous counterpart, generating uniform draws from a discrete uniform distribution is central in Monte Carlo simulation. There are numerous specialized algorithms for doing so when the set $\mathscr{X}$ is nontrivial. Such nontrivial sets include the set of all permutations of the integers $\{1, 2, \ldots, n\}$, the set of all (free) trees, and the set of k-regular bipartite graphs with m and n vertices for the two sets in the partition. See also Section 3.3. ☞ 70

Table 4.8 Moment properties of the $\mathsf{DU}(a, b)$ distribution.

Property		Condition
Expectation	$\dfrac{a+b}{2}$	
Variance	$\dfrac{(b-a)(b-a+2)}{12}$	
Probability generating function	$\dfrac{z^a - z^{b+1}}{(b-a+1)(1-z)}$	$\lvert z\rvert \leqslant 1$

Drawing from a discrete uniform distribution on $\{a, \ldots, b\}$, where a and b are integers, is carried out via a simple table lookup method.

Algorithm 4.17 ($\mathsf{DU}(a, b)$ Generator)

Draw $U \sim \mathsf{U}(0, 1)$ *and output* $X = \lfloor a + (b + 1 - a)U \rfloor$.

4.2 CONTINUOUS DISTRIBUTIONS

We list various continuous distributions in alphabetical order. Recall that an absolutely continuous distribution is completely specified by its pdf. ☞ 611

4.2.1 Beta Distribution

The **beta** distribution has pdf

$$f(x; \alpha, \beta) = \frac{x^{\alpha-1}(1-x)^{\beta-1}}{B(\alpha, \beta)}, \quad x \in [0, 1],$$

☞ 715 where $\alpha > 0$ and $\beta > 0$ are called **shape** parameters and B is the *beta function*:

$$B(\alpha, \beta) = \frac{\Gamma(\alpha)\,\Gamma(\beta)}{\Gamma(\alpha + \beta)}.$$

We write the distribution as $\mathsf{Beta}(\alpha, \beta)$. The dependence of the beta distribution on its shape parameters is illustrated in Figure 4.6.

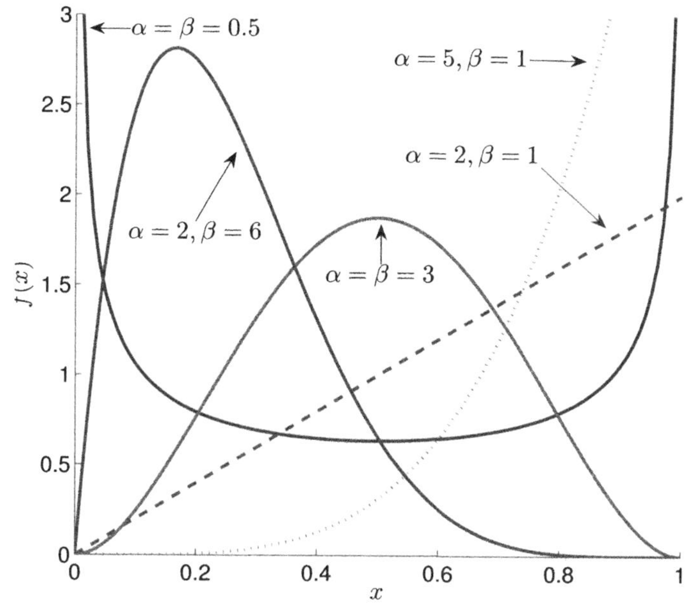

Figure 4.6 Various pdfs of the beta distribution.

The $\mathsf{Beta}(1/2, 1/2)$ distribution is known as the **arcsine** distribution, which we will denote as Arcsine. A related distribution is the **beta-prime** or **inverted beta** distribution, obtained from the standard beta distribution via $Y = X/(1 - X)$.

For a multivariate generalization of the beta distribution, see the Dirichlet distribution in Section 4.3.1.

Table 4.9 Moment and tail properties of the $\mathsf{Beta}(\alpha, \beta)$ distribution.

Property		Condition
Expectation	$\dfrac{\alpha}{\alpha+\beta}$	
Variance	$\dfrac{\alpha\beta}{(\alpha+\beta)^2(\alpha+\beta+1)}$	
Moment $\mathbb{E}X^k$	$\dfrac{B(\alpha+k,\beta)}{B(\alpha,\beta)}$	
Characteristic function	${}_1F_1(\alpha; \alpha+\beta; \mathrm{i}t)$	
Tail probability $\mathbb{P}(X > x)$	$I_{1-x}(\beta, \alpha)$	$x \in [0, 1]$

Here, $I_x(\alpha, \beta)$ denotes the *incomplete beta function* and ${}_1F_1(\alpha; \gamma; x)$ the *confluent*
☞ 715 *hypergeometric function.* Other properties and relations are:

1. *Uniform distribution:* $\mathsf{Beta}(1,1) \equiv \mathsf{U}(0,1)$.

2. *Arcsine distribution:* If $U \sim \mathsf{U}(0,1)$, then

$$X = 1/2 + \cos(\pi U)/2 = \cos^2(\pi U/2) \sim \mathsf{Beta}(1/2, 1/2) \equiv \mathsf{Arcsine}\ .$$

3. *Student's t distribution:* If $X \sim \mathsf{Beta}(\alpha, \alpha)$, then

$$T = \frac{(X - \frac{1}{2})\sqrt{2\alpha}}{\sqrt{X(1-X)}} \sim \mathsf{t}_{2\alpha}\ .$$

   In addition, if $X \sim \mathsf{Beta}\left(\frac{1}{2}, \frac{\alpha}{2}\right)$ and $B \sim \mathsf{Ber}(1/2)$ independently, then

$$(2B - 1)\sqrt{\frac{\nu X}{1 - X}} \sim \mathsf{t}_\nu\ .$$

4. *Gamma distribution:* Let $X \sim \mathsf{Gamma}(\alpha, \theta)$ be independent of $Y \sim \mathsf{Gamma}(\beta, \theta)$. Then,

$$\frac{X}{X + Y} \sim \mathsf{Beta}(\alpha, \beta)\ .$$

   More generally, suppose $X_k \sim \mathsf{Gamma}(\alpha_k, 1)$, $k = 1, \ldots, n$, independently. Then, the random variables

$$Y_k = \frac{X_1 + \cdots + X_k}{X_1 + \cdots + X_{k+1}}, \quad k = 1, \ldots, n-1\ ,$$

   and $S_n = X_1 + \cdots + X_n$ are independent. Moreover,

$$Y_k \sim \mathsf{Beta}(\alpha_1 + \cdots + \alpha_k, \alpha_{k+1}) \quad \text{and} \quad S_n \sim \mathsf{Gamma}(\alpha_1 + \cdots + \alpha_n, 1)\ .$$

5. *Uniform distribution:* Let $U, V \overset{\text{iid}}{\sim} \mathsf{U}(0,1)$. Then, conditional on $U^{1/\alpha} + V^{1/\beta} \leqslant 1$, we have

$$\frac{U^{1/\alpha}}{U^{1/\alpha} + V^{1/\beta}} \sim \mathsf{Beta}(\alpha, \beta) \quad \text{and} \quad U^{1/\alpha} \sim \mathsf{Beta}(\alpha, \beta + 1)\ .$$

6. *Order statistics:* Let $X_1, \ldots, X_n \overset{\text{iid}}{\sim} \mathsf{U}(0,1)$ and let $X_{(1)} \leqslant X_{(2)} \leqslant \cdots \leqslant X_{(n)}$ be the order statistics. Then,

$$X_{(i)} \sim \mathsf{Beta}(i, n + 1 - i)\ .$$

7. *Moments:* Let $X_1, \ldots, X_n \overset{\text{iid}}{\sim} \mathsf{Beta}(\alpha, \beta)$. If $\gamma > \min\left\{\frac{1}{n}, \frac{\alpha}{n-1}, \frac{\beta}{n-1}\right\}$, then (see [10])

$$\mathbb{E} \prod_{1 \leqslant i < j \leqslant n} |X_i - X_j|^{2\gamma} = \prod_{j=0}^{n-1} \frac{\Gamma(\alpha + j\gamma)\,\Gamma(\beta + j\gamma)\,\Gamma(1 + (j+1)\gamma)}{\Gamma(\alpha + \beta + (n + j - 1)\gamma)\,\Gamma(1 + \gamma)}\ .$$

Depending on the values of the parameters of the beta distribution, we have the following algorithms, the last being the most generally applicable.

If α or β equals 1, the inverse-transform method yields:

Algorithm 4.18 (Beta(α, 1) Generator)

Draw $U \sim \mathsf{U}(0,1)$ *and output* $X = U^{1/\alpha}$.

Algorithm 4.19 (Beta(1, β) Generator)

Draw $U \sim \mathsf{U}(0,1)$ *and output* $X = 1 - U^{1/\beta}$.

From Property 2 above we have:

Algorithm 4.20 (Beta(1/2, 1/2) $\equiv$ Arcsine Generator)

Draw $U \sim \mathsf{U}(0,1)$ *and output* $X = \cos^2(\pi U/2)$.

For generating from a symmetric beta distribution $X \sim \mathsf{Beta}(\alpha, \alpha)$, with $\alpha > 1/2$, the *polar method* (see Section 3.1.2.7) can be employed. In particular, using $f_R(r) = 2\,c\,r(1-r^2)^{c-1}$ with $c = \alpha - 1/2$, we find that the pdf of $X = R\cos\Theta$ is that of $2B - 1$ with $B \sim \mathsf{Beta}(\alpha, \alpha)$. Since R can be written as $\sqrt{1 - U^{1/c}}$, with $U \sim \mathsf{U}(0,1)$, we obtain the following algorithm. ☞ 54

Algorithm 4.21 (Beta(α, α) Generator (I), $\alpha > 1/2$)

Draw $U_1, U_2 \stackrel{\text{iid}}{\sim} \mathsf{U}(0,1)$ *and output* $B = \frac{1}{2}\left(1 + \sqrt{1 - U_1^{\frac{2}{2\alpha-1}}}\cos(2\pi U_2)\right)$.

The evaluation of the cosine function can be avoided by a rejection step as in the following algorithm.

Algorithm 4.22 (Beta(α, α) Generator (II), $\alpha > 1/2$)

1. *Draw* $U \sim \mathsf{U}(0,1)$ *and* $V \sim \mathsf{U}(-1,1)$, *independently. Set* $S = X^2 + Y^2$.
2. *If* $S > 1$, *repeat from Step 1; otherwise, output*

$$B = \frac{1}{2} + \frac{U\,V}{S}\sqrt{1 - S^{\frac{2}{2\alpha-1}}}\ .$$

Another generation method for symmetric beta distributions, which holds for general $\alpha > 0$ follows from Property 3 above:

Algorithm 4.23 (Beta(α, α) Generator)

Draw $T \sim \mathsf{t}_{2a}$ *and return* $X = \frac{1}{2}\left(1 + \frac{T}{\sqrt{2\alpha + T^2}}\right)$.

An alternative technique for generating from the symmetric $\mathsf{Beta}(\alpha, \alpha)$ distribution that uses approximate inversion can be found in [22]. Unlike Algorithms 4.21 and 4.22 it applies when $\alpha \leqslant 1/2$, and can be more efficient for small α.

For integer $\alpha = m$ and $\beta = n$, Property 6 yields the following algorithm.

Algorithm 4.24 (Beta(α, β) Generator with Integer $\alpha = m$ and $\beta = n$)

1. *Generate* $U_1, \ldots, U_{m+n-1} \stackrel{\text{iid}}{\sim} \mathsf{U}(0,1)$.
2. *Return the* m*-th order statistic* $U_{(m)}$ *as a random variable from* $\mathsf{Beta}(m, n)$.

It can be shown that the total number of comparisons needed to find $U_{(m)}$ is $(m/2)(m + 2n - 1)$, so that this procedure loses efficiency for large m and n.

Property 4, however, yields the most generally applicable algorithm.

Algorithm 4.25 (Beta(α, β) Generator)

1. *Generate independently* $Y_1 \sim \mathsf{Gamma}(\alpha, 1)$ *and* $Y_2 \sim \mathsf{Gamma}(\beta, 1)$.
2. *Return* $X = Y_1/(Y_1 + Y_2)$ *as a random variable from* $\mathsf{Beta}(\alpha, \beta)$.

4.2.2 Cauchy Distribution

The **Cauchy** distribution has pdf

$$f(x) = \frac{1}{\pi} \frac{1}{1 + x^2}, \quad x \in \mathbb{R}.$$

A location–scale version of this distribution is given by the pdf $f(x; \mu, \sigma) = f((x - \mu)/\sigma)/\sigma$; we write $\mathsf{Cauchy}(\mu, \sigma)$ for the corresponding distribution. The graph of the pdf of the $\mathsf{Cauchy}(0, 1)$ distribution is given in Figure 4.7.

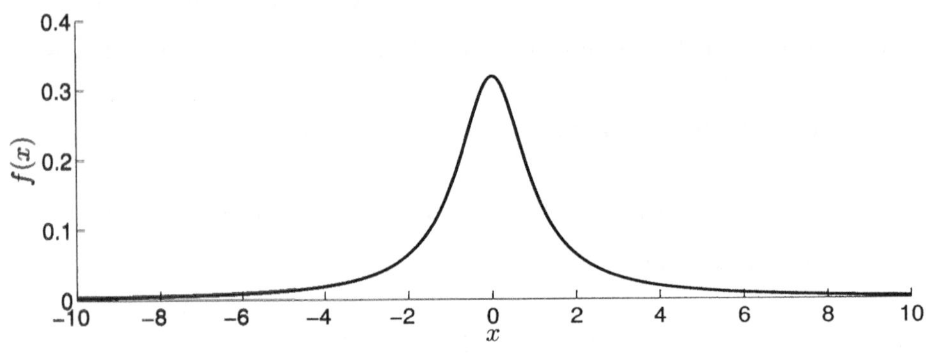

Figure 4.7 The pdf of the standard Cauchy distribution.

Table 4.10 Moment and tail properties of the $\mathsf{Cauchy}(0, 1)$ distribution.

Property	
Expectation	undefined ($\infty - \infty$)
Characteristic function	$e^{-\lvert t \rvert}$
Tail probability $\mathbb{P}(X > x)$	$\frac{1}{2} - \frac{1}{\pi} \arctan(x)$

Other properties and relations are:

1. *Student's t distribution*: $\mathsf{Cauchy}(0, 1) \equiv \mathsf{t}_1$.

2. *Normal distribution*: Let $Y_1, Y_2 \overset{\text{iid}}{\sim} \mathsf{N}(0, 1)$. Then,

$$\frac{Y_1}{Y_2} \sim \mathsf{Cauchy}(0, 1) .$$

In addition,

$$\frac{Y_1 - Y_2}{2\sqrt{Y_1 Y_2}} \sim \mathsf{Cauchy}(0, 1) .$$

3. *Reciprocal*: If $X \sim \mathsf{Cauchy}(\mu, \sigma)$, then $\frac{1}{X} \sim \mathsf{Cauchy}\left(\frac{\mu}{\mu^2+\sigma^2}, \frac{\sigma}{\mu^2+\sigma^2}\right)$.

4. *Sum*: Let $X_i \sim \mathsf{Cauchy}(\mu_i, \sigma_i)$, $i = 1, 2, \ldots, n$ be independent. Then,

$$\sum_{i=1}^{n} X_i \sim \mathsf{Cauchy}\left(\sum_{i=1}^{n} \mu_i, \sum_{i=1}^{n} \sigma_i\right) .$$

Since $\mathsf{Cauchy}(\mu, \sigma)$ forms a location–scale family, we only consider generation from $\mathsf{Cauchy}(0, 1)$. The following algorithm is a direct consequence of the inverse-transform method and the fact that $\cot(\pi x) = \tan\left(\pi x - \frac{\pi}{2}\right)$.

Algorithm 4.26 (Cauchy(0, 1) Generator)

Draw $U \sim \mathsf{U}(0, 1)$ and output $X = \cot(\pi U)$ (or $X = \tan(\pi U - \pi/2)$).

The *ratio of uniforms* method yields the following algorithm. ☞ 66

Algorithm 4.27 (Cauchy(0, 1) Generator via Ratio of Uniforms)

1. *Generate $U, V \overset{\text{iid}}{\sim} \mathsf{U}(0, 1)$. Set $V = V - 1/2$.*

2. *If $U^2 + V^2 \leqslant 1$, set $X = V/U$ and return X. Otherwise, repeat from Step 1.*

Finally, Property 2 above leads to the following algorithm.

Algorithm 4.28 (Cauchy(0, 1) Generator via Ratio of Normals)

1. *Generate $Y_1, Y_2 \overset{\text{iid}}{\sim} \mathsf{N}(0, 1)$.*

2. *Return $X = Y_1/Y_2$.*

4.2.3 Exponential Distribution

The **exponential** distribution has pdf

$$f(x;\lambda) = \lambda \mathrm{e}^{-\lambda x}, \quad x \geqslant 0\,,$$

where $\lambda > 0$ is the **rate** parameter. We write the distribution as $\mathsf{Exp}(\lambda)$. The exponential distribution can be viewed as a continuous version of the geometric distribution. It plays a central role in the theory and application of Markov jump processes, and in stochastic modeling in general, due to its **memoryless property** (see Property 1 below). Graphs of the pdf for various values of λ are given in Figure 4.8.

☞ 166

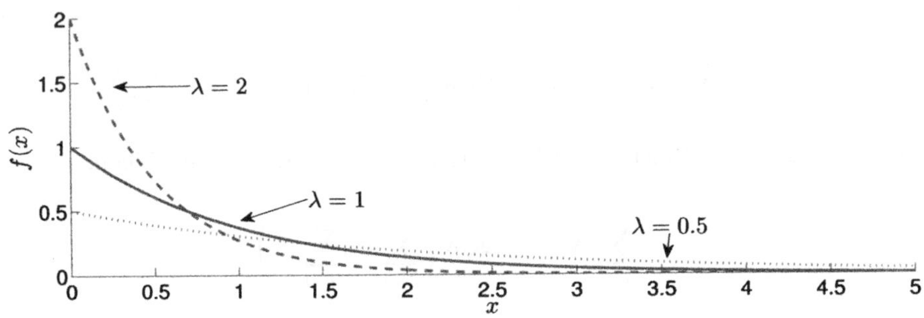

Figure 4.8 Pdfs of the $\mathsf{Exp}(\lambda)$ distribution for various values of λ.

Table 4.11 Moment and tail properties of the $\mathsf{Exp}(\lambda)$ distribution.

Property		Condition
Expectation	$\dfrac{1}{\lambda}$	
Variance	$\dfrac{1}{\lambda^2}$	
Moment generating function	$\dfrac{\lambda}{\lambda - t}$	$t < \lambda$
Tail probability $\mathbb{P}(X > x)$	$\mathrm{e}^{-\lambda x}$	$x \geqslant 0$

Other properties and relations are:

1. *Memoryless property*: If $X \sim \mathsf{Exp}(\lambda)$, then

$$\mathbb{P}(X > s + t \,|\, X > s) = \mathbb{P}(X > t), \quad s, t \geqslant 0\,.$$

2. *Sum*: If $X_1, \ldots, X_n \overset{\text{iid}}{\sim} \mathsf{Exp}(\lambda)$, then $\displaystyle\sum_{k=1}^{n} X_k \sim \mathsf{Gamma}(n, \lambda)$.

3. *Minimum*: Let $X_i \sim \mathsf{Exp}(\lambda_i)$, $i = 1, \ldots, n$, independently. Then, $M \stackrel{\text{def}}{=} \min\{X_1, \ldots, X_n\} \sim \mathsf{Exp}(\lambda_1 + \cdots + \lambda_n)$. In addition, $\mathbb{P}(M = X_i) = \lambda_i/(\lambda_1 + \cdots + \lambda_n)$.

4. *Poisson and geometric distributions*: Let $\mu > 0$ be an arbitrary constant and $Z \geqslant 1$ be a truncated $\mathsf{Poi}(\mu)$ random variable with pdf $\mathbb{P}(Z = z) = \frac{\mu^z}{z!\,(\mathrm{e}^{\mu}-1)}$. Let $G \sim \mathsf{Geom}_0(1 - \mathrm{e}^{-\mu})$ and $U_1, U_2, \ldots \sim_{\text{iid}} \mathsf{U}(0,1)$, then

$$X = \mu\,(G + \min\{U_1, \ldots, U_Z\}) \sim \mathsf{Exp}(1)\,.$$

Noting that $U \sim \mathsf{U}(0,1)$ implies $1 - U \sim \mathsf{U}(0,1)$, we obtain the following inverse-transform algorithm.

Algorithm 4.29 (Exp(λ) Generator)

Draw $U \sim \mathsf{U}(0,1)$ *and output* $X = -\frac{1}{\lambda} \ln U$.

There are many alternative procedures for generating variables from the exponential distribution. The interested reader is referred to [8]. Note that $\mathsf{Exp}(\lambda)$ forms a scale family. Hence, it suffices to specify how to generate $Y \sim \mathsf{Exp}(1)$ and then return Y/λ as a random variable from $\mathsf{Exp}(\lambda)$. ☞ 47

4.2.4 *F* Distribution

The F or **Fisher–Snedecor** distribution has pdf

$$f(x; m, n) = \frac{\Gamma(\frac{m+n}{2})\,(m/n)^{m/2} x^{(m-2)/2}}{\Gamma(\frac{m}{2})\,\Gamma(\frac{n}{2})\,[1 + (m/n)x]^{(m+n)/2}}, \quad x \geqslant 0\,,$$

where m and n are positive integer parameters known as the **degrees of freedom**. We write this distribution as $\mathsf{F}(m,n)$. Graphs of the pdf for various values of the parameters are given in Figure 4.9. The F distribution arises frequently in statistics in the context of hypothesis testing and analysis of variance; see also Section B.1.4. ☞ 660

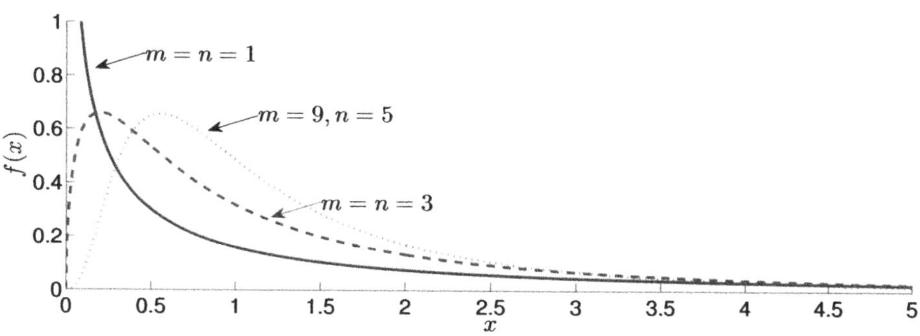

Figure 4.9 Pdfs of the F distribution for various values of the parameters.

Table 4.12 Moment and tail properties of the $\mathsf{F}(m,n)$ distribution.

Property		Condition
Expectation	$\frac{n}{n-2}$	$n > 2$
Variance	$\frac{2n^2(m+n-2)}{m(n-2)^2(n-4)}$	$n > 5$
Tail probability $\mathbb{P}(X > x)$	$I_{n/(n+m\,x)}\left(\frac{n}{2}, \frac{m}{2}\right)$	$x \geqslant 0$

☞ 715 Here, $I_x(\alpha, \beta)$ denotes the *incomplete beta function.* Other properties and relations are:

1. *Expectation and variance*: In addition to the cases listed above, for $n = 1, 2$ the expectation is ∞ and the variance does not exist. For $n = 3, 4$ the variance is ∞.

2. *Reciprocal*: If $X \sim \mathsf{F}(m,n)$, then $\frac{1}{X} \sim \mathsf{F}(n,m)$.

3. *Chi-square distribution*: Let $X \sim \chi^2_m$ and $Y \sim \chi^2_n$ be independent. Then,

$$\frac{X/m}{Y/n} \sim \mathsf{F}(m,n) .$$

4. *Exponential distribution*: If $X_1, X_2 \stackrel{\text{iid}}{\sim} \mathsf{Exp}(1)$, then $Z = X/Y \sim \mathsf{F}(2,2)$. That is, $f_Z(z) = 1/(1+z)^2, z \geqslant 0$.

A simple generator is based on the defining Property 3 above.

Algorithm 4.30 ($\mathsf{F}(m,n)$ Generator)

1. *Draw* $X \sim \chi^2_m$ *and* $Y \sim \chi^2_n$ *independently.*

2. *Output*

$$Z = \frac{X/m}{Y/n} .$$

Let X and Y be as in Algorithm 4.30. Using the fact that $\chi^2_d \equiv \mathsf{Gamma}(d/2, 1/2)$ and Property 4 of the beta distribution on Page 104, we can see that $B = X/(X + Y) \sim \mathsf{Beta}(m/2, n/2)$, which leads to the following algorithm.

Algorithm 4.31 ($\mathsf{F}(m,n)$ Generator via Beta Distribution)

1. *Draw* $B \sim \mathsf{Beta}(m/2, n/2)$.

2. *Output*

$$X = \frac{n\,B}{m\,(1-B)} .$$

4.2.5 Fréchet Distribution

The **Fréchet** or **type II extreme value** distribution has pdf

$$f(x;\alpha) = \alpha\, x^{-\alpha-1} \mathrm{e}^{-x^{-\alpha}}, \quad x > 0 ,$$

where $\alpha > 0$ is the **shape** parameter. Graphs of the pdf for various values of α are given in Figure 4.10. A location–scale version of this distribution is given by the pdf $f(x;\alpha,\mu,\sigma) = f((x-\mu)/\sigma;\alpha)/\sigma$, $x \geqslant \mu$; we write this distribution as $\mathsf{Fréchet}(\alpha,\mu,\sigma)$.

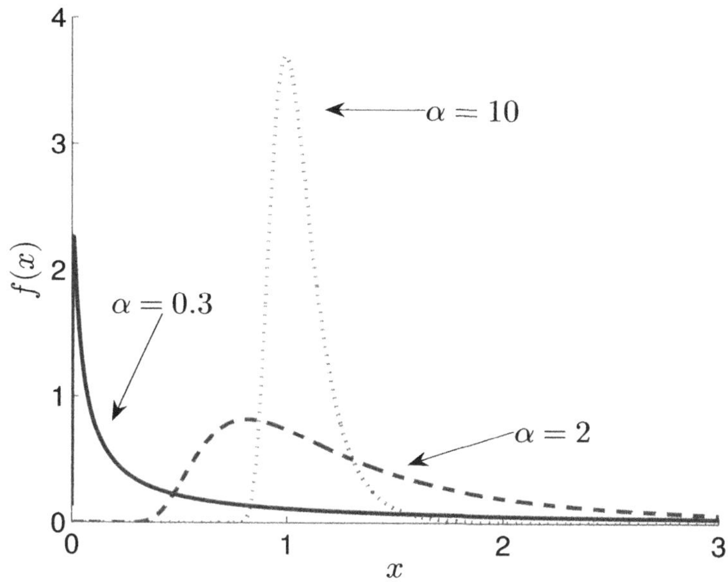

Figure 4.10 Pdfs of the standard Fréchet distribution for various values of the shape parameter α.

The Fréchet distribution is one of three possible limiting distributions for the *maximum* of iid random variables. Similarly, the "reflected" distribution, with pdf $f(-x), x < 0$, is one of the three possible limiting distributions for the *minimum* of iid random variables. ☞ 705

Table 4.13 Moment and tail properties of the $\mathsf{Fréchet}(\alpha, 0, 1)$ distribution.

Property		Condition
Expectation	$\Gamma(1-\alpha^{-1})$	$\alpha > 1$
Variance	$\Gamma(1-2\alpha^{-1}) - \Gamma(1-\alpha^{-1})^2$	$\alpha > 2$
Tail probability $\mathbb{P}(X > x)$	$1 - \exp(-x^{-\alpha})$	$x > 0$

Since $\mathsf{Fréchet}(\alpha, \mu, \sigma)$ forms a location–scale family, it suffices to describe how to generate from the standard Fréchet distribution. This can be done directly via the inverse-transform method.

Algorithm 4.32 ($\mathsf{Fréchet}(\alpha, 0, 1)$ Generator)

Draw $U \sim \mathsf{U}(0,1)$ *and output* $X = (-\ln U)^{-1/\alpha}$.

4.2.6 Gamma Distribution

The **gamma** distribution has pdf

$$f(x; \alpha, \lambda) = \frac{\lambda^\alpha x^{\alpha-1} \mathrm{e}^{-\lambda x}}{\Gamma(\alpha)}, \quad x \geqslant 0, \tag{4.4}$$

where $\alpha > 0$ is called the **shape** parameter and $\lambda > 0$ the **scale** parameter. In the formula for the pdf, Γ is the *gamma function* (see Section D.9.5). We write the distribution as $\mathsf{Gamma}(\alpha, \lambda)$. ☞ 716

An important special case is the $\mathsf{Gamma}(n/2, 1/2)$ distribution with $n \in \{1, 2, \ldots\}$, which is called a **chi-square** distribution; the parameter n is then referred to as the **number of degrees of freedom**. The distribution is written as χ^2_n. A graph of the pdf of the χ^2_n distribution, for various n, is given in Figure 4.11.

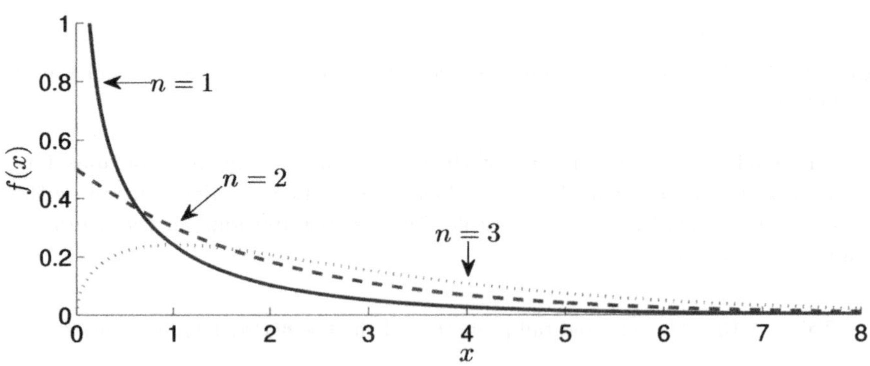

Figure 4.11 Pdfs of the χ^2_n distribution for various degrees of freedom n.

Another well-known special case is the $\mathsf{Gamma}(n, \lambda)$ distribution, where n is a strictly positive integer. In this case, the distribution is known as an **Erlang** distribution with shape parameter n and scale parameter λ. It is denoted by $\mathsf{Erl}(n, \lambda)$.

Table 4.14 Moment and tail properties of the $\mathsf{Gamma}(\alpha, \lambda)$ distribution.

Property		Condition
Expectation	$\dfrac{\alpha}{\lambda}$	
Variance	$\dfrac{\alpha}{\lambda^2}$	
Moment generating function	$\left(\dfrac{\lambda}{\lambda - t}\right)^{\alpha}$	$t < \lambda$
Tail probability $\mathbb{P}(X > x)$	$1 - P(\alpha, \lambda x)$	

Here, $P(\alpha, x)$ is the *incomplete gamma function*. Other properties and relations are: ☞ 716

1. *Exponential distribution*: $\mathsf{Gamma}(1, \lambda) \equiv \mathsf{Exp}(\lambda)$.

2. *Sums*: If $X_1, \ldots, X_n \overset{\text{iid}}{\sim} \mathsf{Gamma}(\alpha, \lambda)$, then
$$\sum_{k=1}^{n} X_k \sim \mathsf{Gamma}(n\alpha, \lambda) .$$

3. *Beta distribution (I)*: Let $Z \sim \mathsf{Exp}(1)$ and $Y \sim \mathsf{Beta}(\alpha, 1 - \alpha)$ for $0 < \alpha < 1$ be independent. Then,
$$Y Z \sim \mathsf{Gamma}(\alpha, 1) .$$

4. *Beta distribution (II)*: Let $Y_k \sim \mathsf{Beta}(\alpha_1 + \cdots + \alpha_k, \alpha_{k+1})$, $\alpha_k > 0$, $k = 1, \ldots, n-1$, independently. Let $Y_n \sim \mathsf{Gamma}(\alpha_1 + \cdots + \alpha_n, 1)$ be independent of the $\{Y_k\}$. Then, the random variables
$$X_k = (1 - Y_{k-1}) \prod_{j=k}^{n} Y_j, \quad k = 1, \ldots, n ,$$
where $Y_0 = 0$, are independent and $X_k \sim \mathsf{Gamma}(\alpha_k, 1)$, $k = 1, \ldots, n$.

5. *F distribution*: Let $U \sim \chi^2_m$ and $V \sim \chi^2_n$ be independent. Then
$$\frac{U/m}{V/n} \sim \mathsf{F}(m, n) .$$

6. *Dirichlet distribution*: Let $X_1, \ldots, X_{n+1}$ be independent random variables with $X_k \sim \mathsf{Gamma}(\alpha_k, \lambda)$, $k = 1, \ldots, n+1$. Then, with $\boldsymbol{\alpha} = (\alpha_1, \ldots, \alpha_{n+1})$ and $\mathbf{X} = (X_1, \ldots, X_n)^{\top}$,
$$\frac{\mathbf{X}}{\sum_{k=1}^{n+1} X_k} \sim \mathsf{Dirichlet}(\boldsymbol{\alpha}) .$$

Since $\mathsf{Gamma}(\alpha, \lambda)$ is a scale family, it suffices to only give algorithms for generating random variables $X \sim \mathsf{Gamma}(\alpha, 1)$, because $X/\lambda \sim \mathsf{Gamma}(\alpha, \lambda)$. Since the cdf of the gamma distribution does not generally exist in explicit form, the inverse-transform method cannot always be applied to generate random variables from this distribution. Thus, alternative methods are called for. The following algorithm, by Marsaglia and Tsang [25], provides a highly efficient acceptance–rejection method for generating $\mathsf{Gamma}(\alpha, 1)$ random variables with $\alpha \geqslant 1$; see also [28, Pages 60–61] for a detailed explanation.

Algorithm 4.33 ($\mathsf{Gamma}(\alpha, 1)$ Generator for $\alpha \geqslant 1$)

1. *Set* $d = \alpha - 1/3$ *and* $c = 1/\sqrt{9\,d}$.
2. *Generate* $Z \sim \mathsf{N}(0,1)$ *and* $U \sim \mathsf{U}(0,1)$ *independently.*
3. *If* $Z > -1/c$ *and* $\ln U < \frac{1}{2}Z^2 + d - d\,V + d\ln V$, *where* $V = (1 + c\,Z)^3$, *return* $X = d\,V$; *otherwise, go back to Step 2.*

Ahrens and Dieter [1] propose the following acceptance–rejection procedure that does not rely on a normal generator for $\alpha > 1$.

Algorithm 4.34 ($\mathsf{Gamma}(\alpha, 1)$ Generator for $\alpha > 1$)

1. *Set* $b = \alpha - 1$ *and* $c = \alpha + b$.
2. *Generate* $U \sim \mathsf{U}(0,1)$. *Set* $Y = \sqrt{c}\,\tan(\pi(U - 1/2))$ *and* $X = b + Y$.
3. *If* $X < 0$, *repeat from Step 2; otherwise, proceed to the next step.*
4. *Generate* $V \sim \mathsf{U}(0,1)$. *If*
$$V > \exp\left(b\ln\left(\frac{X}{b}\right) - Y + \ln\left(1 + \frac{Y^2}{c}\right)\right),$$
repeat from Step 2; otherwise, output X.

Ahrens and Dieter [1] show that the expected number of trials before a sample is generated from $\mathsf{Gamma}(\alpha, 1)$ decreases from π (when α is close to 1) to $\sqrt{\pi}$ (when $\alpha \to \infty$).

Cheng and Feast [7] propose a slightly longer algorithm based on the acceptance–rejection method with proposal given by the **Burr distribution**, with pdf:

$$f(x; \mu, \lambda) = \lambda\mu\frac{x^{\lambda-1}}{(\mu + x^{\lambda})^2}, \quad x > 0,\ \mu > 0,\ \lambda > 0\,.$$

The algorithm does not involve evaluation of the $\tan(\cdot)$ function.

Algorithm 4.35 ($\mathsf{Gamma}(\alpha, 1)$ Generator for $\alpha > 1$)

1. *Set* $c_1 = \alpha - 1$, $c_2 = \frac{\alpha - 1/(6\alpha)}{c_1}$, $c_3 = 2/c_1$, $c_4 = 1 + c_3$, *and* $c_5 = \alpha^{-1/2}$.
2. *Generate* $U, U_2 \sim \mathsf{U}(0,1)$, *independently. If* $\alpha > 2.5$, *set*
$$U_1 = U_2 + c_5(1 - 1.86\,U)\,;$$
otherwise, set $U_1 = U$.

3. *If* $0 < U_1 < 1$, *set* $W = c_2 U_2/U_1$ *and proceed to Step 4; otherwise, repeat from Step 2.*

4. *If*

$$c_3 U_1 + W + W^{-1} < c_4 \quad or \quad c_3 \ln U_1 - \ln W + W < 1\,,$$

output $X = c_1 W$*; otherwise, repeat from Step 2.*

For the case $\alpha < 1$ one can use the fact that if $X \sim \mathsf{Gamma}(1+\alpha, 1)$, and $U \sim \mathsf{U}(0,1)$ are independent, then $XU^{1/a} \sim \mathsf{Gamma}(\alpha, 1)$. Alternatively, one can use the following algorithm based on Property 3.

Algorithm 4.36 (Gamma(α, 1) Generator for $\alpha < 1$)

1. *Generate* $Y \sim \mathsf{Beta}(\alpha, 1-\alpha)$ *and* $Z \sim \mathsf{Exp}(1)$ *independently.*

2. *Output* $X = Y\,Z$ *as a random variable from* $\mathsf{Gamma}(\alpha, 1)$.

If a fast beta generator is not available, then one could use the following acceptance–rejection algorithm by Best [5], based on [1].

Algorithm 4.37 (Gamma(α, 1) Generator for $\alpha < 1$)

1. *Set* $d = 0.07 + 0.75\sqrt{1-\alpha}$ *and* $b = 1 + \mathrm{e}^{-d}\alpha/d$.

2. *Generate* $U_1, U_2 \overset{\text{iid}}{\sim} \mathsf{U}(0,1)$ *and set* $V = b\,U_1$.

3. *If* $V \leqslant 1$, *then set* $X = d\,V^{1/\alpha}$. *Check whether* $U_2 \leqslant (2-X)/(2+X)$. *If true, return* X*; otherwise, check whether* $U_2 \leqslant \mathrm{e}^{-X}$. *If true, return* X*; otherwise, go back to Step 2.*

   If $V > 1$, *then set* $X = -\ln(d(b-V)/\alpha)$ *and* $Y = X/d$. *Check whether* $U_2(\alpha + y(1-\alpha)) \leqslant 1$. *If true, return* X*; otherwise, check if* $U_2 < Y^{\alpha-1}$. *If true, return* X*; otherwise, go back to Step 2.*

■ EXAMPLE 4.5 (Best's Gamma(α, 1) Generator for $\alpha < 1$)

The following MATLAB code is an implementation of Algorithm 4.37.

```
%gamma_best.m
N = 10^5; alpha = 0.3;
d= 0.07 + 0.75*sqrt(1-alpha); b = 1 + exp(-d)*alpha/d;
x = zeros(N,1);
for i = 1:N
    cont = true;
    while cont
        U1 = rand;
        U2 = rand;
        V = b*U1;
        if V <= 1
            X = d*V^(1/alpha);
            if U2 <= (2-X)/(2+X)
```

```
                    cont = false; break;
                else
                    if U2 <= exp(-X)
                        cont = false; break;
                    end
                end
            else
                X = -log(d*(b-V)/alpha);
                y = X/d;
                if U2*(alpha + y*(1-alpha)) < 1
                    cont= false; break;
                else
                    if U2 <= y^(alpha - 1)
                        cont= false;break;
                    end
                end
            end
        end
        x(i) = X;
end
```

A random variable $X \sim \mathsf{Gamma}(n, 1)$ for integer n can be viewed as the sum of n independent exponential random variables (see Property 2), resulting in the following algorithm.

Algorithm 4.38 (Gamma($n, 1$) Generator With Integer n)

1. *Generate* $U_1, \ldots, U_n \overset{\text{iid}}{\sim} \mathsf{U}(0, 1)$.

2. *Return* $X = -\ln \prod_{k=1}^{n} U_k$.

A random variable $X \sim \chi_1^2 \equiv \mathsf{Gamma}(1/2, 1/2)$ can be simply generated as the square of a standard normal random variable (Property 3 on Page 123).

Algorithm 4.39 (Gamma($1/2, 1/2$) Generator)

Draw $Z \sim \mathsf{N}(0, 1)$ *and output* $X = Z^2$.

4.2.7 Gumbel Distribution

The **Gumbel distribution** or **type I extreme value** distribution has pdf

$$f(x) = \mathrm{e}^{-x - \mathrm{e}^{-x}}, \quad x \in \mathbb{R} .$$

A location–scale version of this distribution is given by the pdf $f(x; \mu, \sigma) = f((x - \mu)/\sigma)/\sigma$; we write $\mathsf{Gumbel}(\mu, \sigma)$ for the corresponding distribution. A graph of the standard Gumbel pdf is given in Figure 4.12.

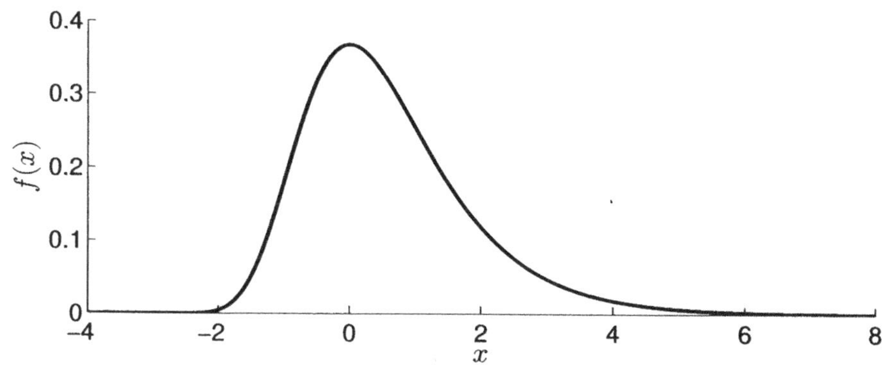

Figure 4.12 The pdf of the standard Gumbel distribution.

The Gumbel distribution is one of three possible limiting distributions for the *maximum* of iid random variables. For example, the maximum of iid standard normal random variables, when properly scaled, converges in distribution to the Gumbel distribution [16, Page 275]. Similarly, the Gumbel distribution reflected around 0, that is, with pdf $f(-x)$, is one of the three possible limiting distributions for the *minimum* of iid random variables. Note also that if $X \sim \mathsf{Exp}(1)$, then $-\ln X \sim \mathsf{Gumbel}(0,1)$. ☞ 705

Table 4.15 Moment and tail properties of the $\mathsf{Gumbel}(0,1)$ distribution.

Property		Condition
Expectation	$-\Gamma'(1) = 0.577216\cdots$	
Variance	$\frac{\pi^2}{6}$	
Moment generating function	$\Gamma(1-t)$	$t < 1$
Tail probability $\mathbb{P}(X > x)$	$1 - \mathrm{e}^{-\mathrm{e}^{-x}}$	

As the class of $\mathsf{Gumbel}(\mu, \sigma)$ forms a location–scale family, it suffices to only consider generating from the $\mathsf{Gumbel}(0,1)$ distribution. The following algorithm is a direct consequence of the inverse-transform method, as the cdf of the $\mathsf{Gumbel}(0,1)$ distribution is $F(x) = \exp(-\exp(-x))$.

Algorithm 4.40 (Gumbel(0, 1) Generator)

Draw $U \sim \mathsf{U}(0,1)$ *and output* $X = -\ln(-\ln U)$.

4.2.8 Laplace Distribution

The **Laplace** distribution, also called the **double-exponential** distribution, is defined via the pdf

$$f(x) = \frac{1}{2}\mathrm{e}^{-|x|}, \quad x \in \mathbb{R} .$$

A location–scale version of this distribution is given by the pdf $f(x; \mu, \sigma) = f((x - \mu)/\sigma)/\sigma$; we write $\mathsf{Laplace}(\mu, \sigma)$ for the corresponding distribution. The pdf of the standard Laplace distribution is given in Figure 4.13.

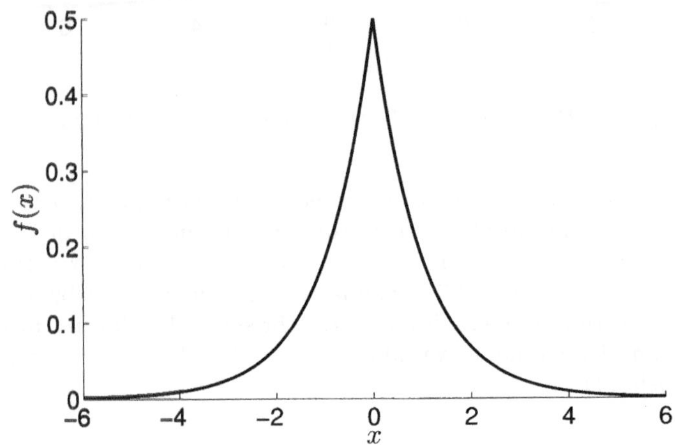

Figure 4.13 The pdf of the standard Laplace distribution.

Table 4.16 Moment and tail properties of the $\mathsf{Laplace}(0,1)$ distribution.

Property		Condition
Expectation	0	
Variance	2	
Moment generating function	$\dfrac{1}{1-t^2}$	$\lvert t\rvert < 1$
Tail probability $\mathbb{P}(X > x)$	$\dfrac{\mathrm{sgn}(x)(\mathrm{e}^{-\lvert x\rvert} - 1) + 1}{2}$	

Since $\mathsf{Laplace}(\mu, \sigma)$ forms a location–scale family, we only consider generation from $\mathsf{Laplace}(0, 1)$. The latter is obtained by assigning a random sign to an $\mathsf{Exp}(1)$ random variable, leading to the following algorithm.

Algorithm 4.41 ($\mathsf{Laplace}(0, 1)$ Generator (I))

1. *Generate* $Y \sim \mathsf{Exp}(1)$ *and* $B \sim \mathsf{Ber}(\frac{1}{2})$ *independently.*
2. *Output* $X = (2B - 1)Y$.

The above method requires the generation of two independent random variables (one exponential and one Bernoulli random variable). However, as the next algorithm illustrates, it is sufficient to use only one random variable.

Algorithm 4.42 (Laplace(0, 1) Generator (II))

1. *Generate* $U \sim \mathsf{U}(-1/2, 1/2)$.
2. *Output* $X = \text{sgn}(U)\ln(1 - 2|U|)$.

An alternative is based on the fact that if V and W are independent $\mathsf{Exp}(1)$ random variables, then $V - W \sim \mathsf{Laplace}(0, 1)$.

Algorithm 4.43 (Laplace(0, 1) Generator (III))

1. *Generate* $V, W \overset{\text{iid}}{\sim} \mathsf{Exp}(1)$.
2. *Output* $X = V - W$.

Example 3.12 provides yet another generation method. ☞ 55

Algorithm 4.44 (Laplace(0, 1) Generator (IV))

1. *Generate* $E \sim \mathsf{Exp}(1)$ *and* $Y \sim \mathsf{N}(0, 1)$ *independently.*
2. *Output* $X = Y\sqrt{2E}$.

4.2.9 Logistic Distribution

The **logistic** distribution has pdf

$$f(x) = \frac{\mathrm{e}^{-x}}{(1 + \mathrm{e}^{-x})^2}, \qquad x \in \mathbb{R}.$$

A location–scale version of this distribution is given by the pdf $f(x; \mu, \sigma) = f((x - \mu)/\sigma)/\sigma$. We write this distribution as $\mathsf{Logistic}(\mu, \sigma)$. The distribution derives its name from the fact that the cdf $F(x) = 1/(1 + \mathrm{e}^{-x})$ satisfies the logistic differential equation $f(x) = F'(x) = F(x)(1 - F(x))$, so that $x = \ln[F(x)/(1 - F(x))]$. The graph of the pdf of the standard logistic distribution is given in Figure 4.14. Note that the pdf is symmetric around 0.

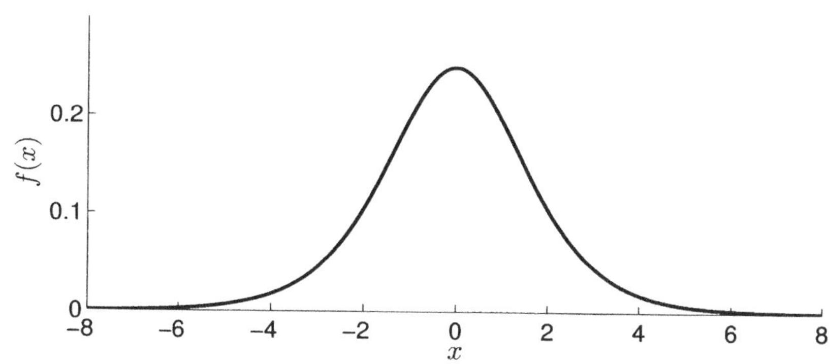

Figure 4.14 The pdf of the standard logistic distribution.

Table 4.17 Moment and tail properties of the Logistic$(0, 1)$ distribution.

Property		Condition
Expectation	0	
Variance	$\pi^2/3$	
Moment generating function	$\Gamma(1-t)\,\Gamma(1+t) = \frac{\pi t}{\sin(\pi t)}$	$\lvert t\rvert < 1$
Tail probability $\mathbb{P}(X > x)$	$1/(1+e^x)$	

Other properties and relations are:

1. *Uniform:* If $U \sim \mathsf{U}(0,1)$, then

$$\ln\left(\frac{U}{1-U}\right) \sim \mathsf{Logistic}(0,1)\,.$$

2. *Exponential quotient:* Let $X, Y \overset{\text{iid}}{\sim} \mathsf{Exp}(1)$. Then,

$$\ln\left(\frac{X}{Y}\right) \sim \mathsf{Logistic}(0,1)\,.$$

Since Logistic(μ, σ) forms a location–scale family, we only consider generation from Logistic$(0, 1)$. The following algorithm follows directly from the inverse-transform method.

Algorithm 4.45 (Logistic(0, 1) Generator)

Generate $U \sim \mathsf{U}(0,1)$ *and output* $X = \ln\left(\frac{U}{1-U}\right)$.

4.2.10 Log-Normal Distribution

The **log-normal** distribution with scale parameter $\sigma > 0$ and location parameter $\mu \in \mathbb{R}$ is defined via the pdf

$$f(x;\mu,\sigma) = \frac{1}{x\,\sigma\sqrt{2\pi}}\,\exp\left(-\frac{(\ln(x)-\mu)^2}{2\,\sigma^2}\right), \quad x > 0\,.$$

We write LogN(μ, σ^2) for the distribution. The characterizing property of the distribution is that if $X \sim \mathsf{LogN}(\mu,\sigma^2)$, then $\ln X \sim \mathsf{N}(\mu,\sigma^2)$. As a consequence, it arises as the scaled limit of products of iid random variables, in the same way that the normal distribution arises for sums via the central limit theorem. Note that the log-normal distribution is not determined uniquely by its moments, see [13]. The dependence of the log-normal distribution on the scale parameter for $\mu = 0$ is illustrated in Figure 4.15.

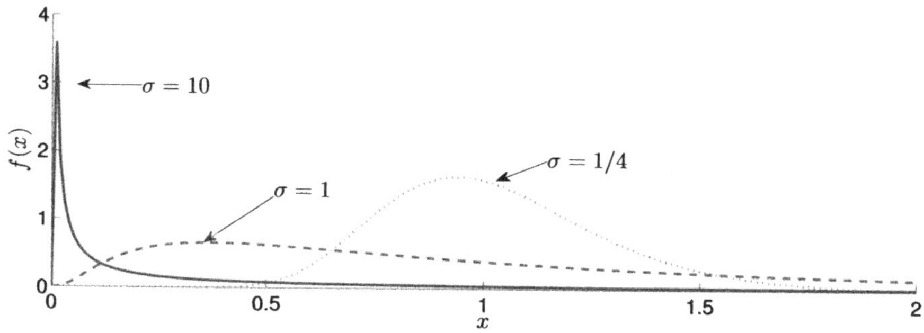

Figure 4.15 Pdfs of the log-normal distribution for $\mu = 0$ and various values of the scale parameter σ.

Table 4.18 Moment and location properties of the $\mathsf{LogN}(\mu, \sigma^2)$ distribution.

Property	
Expectation	$e^{\mu+\sigma^2/2}$
Variance	$e^{2\mu+\sigma^2}\left(e^{\sigma^2} - 1\right)$
Moment $\mathbb{E}X^k$	$e^{k\mu+k^2\sigma^2/2}$
Mode	$e^{\mu-\sigma^2}$
Median	e^{μ}

Other properties and relations are:

1. *Normal distribution*: If $X \sim \mathsf{N}(\mu, \sigma^2)$, then $e^X \sim \mathsf{LogN}(\mu, \sigma^2)$.
2. *Power transformation*: If $Y \sim \mathsf{LogN}(\mu, \sigma^2)$, then $e^a Y^b \sim \mathsf{LogN}(a + b\mu, (b\sigma)^2)$.
3. *Product*: If $X_i \sim \mathsf{LogN}(\mu_i, \sigma_i^2)$ independently, $i = 1, \ldots, n$, then

$$\prod_{i=1}^{n} X_i \sim \mathsf{LogN}\left(\sum_{i=1}^{n} \mu_i, \sum_{i=1}^{n} \sigma_i^2\right).$$

4. *Reciprocal*: If $X \sim \mathsf{LogN}(\mu, \sigma^2)$, then $1/X \sim \mathsf{LogN}(-\mu, \sigma^2)$.

The generation of a $\mathsf{LogN}(\mu, \sigma^2)$ distributed random variable follows immediately from its relation to the normal distribution as given in Property 1 above.

Algorithm 4.46 ($\mathsf{LogN}(\mu, \sigma^2)$ Generator)

1. *Generate* $Y \sim \mathsf{N}(\mu, \sigma^2)$.
2. *Output* $X = e^Y$.

4.2.11 Normal Distribution

The standard **normal** or standard **Gaussian** distribution has pdf

$$f(x) = \frac{1}{\sqrt{2\pi}} \mathrm{e}^{-x^2/2}, \quad x \in \mathbb{R}.$$

The corresponding location–scale family of pdfs is therefore

$$f(x; \mu, \sigma^2) = \frac{1}{\sigma\sqrt{2\pi}} \mathrm{e}^{-\frac{1}{2}\left(\frac{x-\mu}{\sigma}\right)^2}, \; x \in \mathbb{R}. \tag{4.5}$$

We write the distribution as $\mathsf{N}(\mu, \sigma^2)$. We denote the pdf and cdf of the $\mathsf{N}(0,1)$ distribution as φ and Φ, respectively. Here, $\Phi(x) = \int_{-\infty}^{x} \varphi(t)\,\mathrm{d}t = \frac{1}{2} + \frac{1}{2}\mathrm{erf}\left(\frac{x}{\sqrt{2}}\right)$,
☞ 715 where $\mathrm{erf}(x)$ is the *error function*.

The normal distribution plays a central role in statistics and arises naturally as the limit of the sum of iid random variables via the central limit theorem (see
☞ 625 Section A.8.3). Its crucial property is that any affine combination of independent normal random variables is again normal. In Figure 4.16 the probability densities for three different normal distributions are depicted. More information on the Gaussian distribution, especially regarding the multidimensional case, can be found in Section 4.3.3.

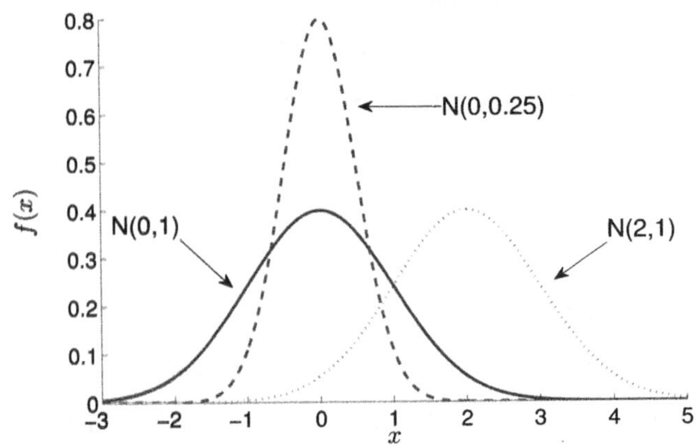

Figure 4.16 Pdfs of the normal distribution for various values of the parameters.

Table 4.19 Moment and tail properties of the $\mathsf{N}(\mu, \sigma^2)$ distribution.

Property	
Expectation	μ
Variance	σ^2
Moment generating function	$\mathrm{e}^{t\mu + t^2\sigma^2/2}$
Tail probability $\mathbb{P}(X > x)$	$\frac{1}{2} - \frac{1}{2}\mathrm{erf}\left(\frac{x-\mu}{\sigma\sqrt{2}}\right)$

Other properties and relations are:

1. *Affine combinations*: Let $X_1, X_2, \ldots, X_n$ be independent with $X_i \sim \mathsf{N}(\mu_i, \sigma_i^2)$, $i = 1, \ldots, n$. Then,

$$a + \sum_{i=1}^{n} b_i X_i \sim \mathsf{N}\left(a + \sum_{i=1}^{n} b_i \mu_i,\ \sum_{i=1}^{n} b_i^2 \sigma_i^2\right).$$

2. *Central limit theorem*: Let $X_1, \ldots, X_n$ be independent and identically distributed with expectation μ and variance $\sigma^2 < \infty$. Let $S_n = X_1 + \cdots + X_n$. Then

$$\frac{S_n - n\mu}{\sigma\sqrt{n}} \xrightarrow{d} Z \sim \mathsf{N}(0,1) \quad \text{as} \quad n \to \infty .$$

3. *Square*: If $X \sim \mathsf{N}(0,1)$, then $X^2 \sim \mathsf{Gamma}(1/2, 1/2)$.

4. *Quotient*: Let $X_1, X_2 \overset{\text{iid}}{\sim} \mathsf{N}(0,1)$. Then, $\frac{X_1}{X_2} \sim \mathsf{Cauchy}(0,1)$.

5. *Exponential*: If $X \sim \mathsf{N}(\mu, \sigma^2)$, then $\mathrm{e}^X \sim \mathsf{LogN}(\mu, \sigma^2)$.

6. *Stable distribution*: $\mathsf{N}(0,2) \equiv \mathsf{Stable}(2, \beta)$. In addition, if $X \sim \mathsf{N}(0,1)$, then $X^{-2} \sim \mathsf{Stable}(\frac{1}{2}, 1) \equiv$ Lévy.

7. *Fractional moments*: If $X \sim \mathsf{N}(0,1)$, then

$$\mathbb{E}|X|^{\alpha} = \frac{\Gamma\left(\frac{\alpha+1}{2}\right) 2^{\alpha/2}}{\sqrt{\pi}}, \quad \alpha \geqslant 0 .$$

In addition, if $X_1, \ldots, X_n \overset{\text{iid}}{\sim} \mathsf{N}(0,1)$, then (see [10])

$$\mathbb{E} \prod_{1 \leqslant i < j \leqslant n} |X_i - X_j|^{2\alpha} = \prod_{j=1}^{n} \frac{\Gamma(1 + j\alpha)}{\Gamma(1 + \alpha)},$$

and for $c \geqslant 0$,

$$\mathbb{E} \prod_{i=1}^{n} (2X_i^2)^c \prod_{1 \leqslant i < j \leqslant n} |X_i^2 - X_j^2|^{2\alpha} = \prod_{j=0}^{n-1} \frac{\Gamma(1 + 2c + 2j\alpha)\, \Gamma(1 + (j+1)\alpha)}{\Gamma(1 + c + j\alpha)\, \Gamma(1 + \alpha)} .$$

Since $\mathsf{N}(\mu, \sigma)$ forms a location–scale family, we only consider generation from $\mathsf{N}(0,1)$. The most prominent application of the polar method (see Section 3.1.2.7) ☞ 54 lies in the generation of standard normal random variables, leading to the celebrated Box–Muller method.

Algorithm 4.47 (N(0, 1) Generator, Box–Muller Approach)

1. *Generate* $U_1, U_2 \overset{\text{iid}}{\sim} \mathsf{U}(0,1)$.

2. *Return two independent standard normal variables,* X *and* Y*, via*

$$\begin{aligned} X &= \sqrt{-2 \ln U_1}\, \cos(2\pi U_2)\, , \\ Y &= \sqrt{-2 \ln U_1}\, \sin(2\pi U_2)\, . \end{aligned} \tag{4.6}$$

Marsaglia [23] (see also [24]) developed an alternative generation method for $\mathsf{N}(0,1)$ random variable generation which is also based on the polar method, but which avoids the evaluation of the relatively costly cos and sin functions. The method uses a simple acceptance–rejection approach to generate random vectors within the unit circle: draw $V_1, V_2 \sim_{\text{iid}} \mathsf{U}(-1,1)$ and accept $(V_1, V_2)/\sqrt{S}$ if $S = V_1^2 + V_2^2 \leqslant 1$. As in the Box–Muller method, a pair of independent $\mathsf{N}(0,1)$-distributed random variables is obtained by multiplying the coordinates of the generated point with the radius $R = \sqrt{-2\ln U}$, where $U \sim \mathsf{U}(0,1)$. This leads to the following algorithm.

Algorithm 4.48 ($\mathsf{N}(0,1)$ Generator, Polar/Acceptance–Rejection)

1. *Generate* $U_1, U_2 \overset{\text{iid}}{\sim} \mathsf{U}(0,1)$. *Set* $V_i = 2U_i - 1, i = 1,2$ *and* $S = V_1^2 + V_2^2$.

2. *If* $S \leqslant 1$. *Draw* $U \sim \mathsf{U}(0,1)$ *and return*

$$X = V_1 \sqrt{-2\ln U/S}\,,$$
$$Y = V_2 \sqrt{-2\ln U/S}\,.$$

Otherwise, go back to Step 1.

The following method [8, Pages 197–199] is based on the ratio of uniforms method
☞ 66 as illustrated in Example 3.18 and uses a squeeze function to further speed up the calculation.

Algorithm 4.49 ($\mathsf{N}(0,1)$ Generator, Ratio of Uniforms with Squeezing)

1. *Set* $a_+ = \sqrt{2/\mathrm{e}}$ *and* $a_- = -\sqrt{2/\mathrm{e}}$.

2. *Generate* $U \sim \mathsf{U}(0,1)$ *and* $V \sim \mathsf{U}(a_-, a_+)$ *independently, and set* $X = V/U$.

3. *If* $X^2 \leqslant 6 - 8U + 2U^2$, *return* X.

4. *Else if* $X^2 \geqslant \frac{2}{U} - 2U$, *repeat from Step 2.*

5. *Else if* $X^2 \leqslant -4\ln U$, *return* X. *Otherwise, restart from Step 2.*

Finally, the following algorithm uses acceptance–rejection with an exponential proposal distribution. This gives a probability of acceptance of $\sqrt{\pi/(2\mathrm{e})} \approx 0.76$.
☞ 60 The theory behind it is given in Example 3.15.

Algorithm 4.50 ($\mathsf{N}(0,1)$ Generator, Acceptance–Rejection from $\mathsf{Exp}(1)$)

1. *Generate* $X \sim \mathsf{Exp}(1)$ *and* $U' \sim \mathsf{U}(0,1)$, *independently.*

2. *If* $U' \leqslant \mathrm{e}^{-(X-1)^2/2}$, *generate* $U \sim \mathsf{U}(0,1)$ *and output* $Z = (1 - 2\,\mathrm{I}_{\{U \leqslant 1/2\}})\,X$; *otherwise, repeat from Step 1.*

4.2.12 Pareto Distribution

The **Pareto (type II)** distribution, sometimes known as the **Lomax** distribution [20], with **scale** parameter $\lambda > 0$ and **shape** parameter $\alpha > 0$ is defined via the pdf

$$f(x;\alpha,\lambda) = \alpha\,\lambda(1+\lambda\,x)^{-(\alpha+1)}, \quad x \geqslant 0\,.$$

We write the distribution as $\mathsf{Pareto}(\alpha,\lambda)$. The effect of the parameter α on the shape of the distribution is illustrated in Figure 4.17.

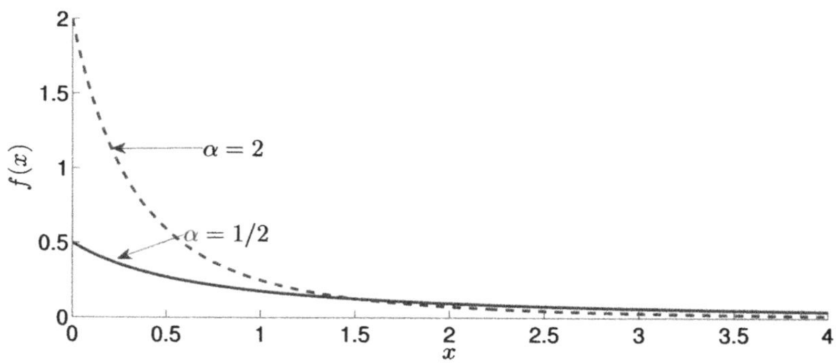

Figure 4.17 Pdfs of the Pareto distribution for two values of the shape parameter α and fixed scale parameter $\lambda = 1$.

Remark 4.2.1 (Alternative Definition) Also commonly used is the Pareto (type I) distribution, which has pdf

$$f(x;\alpha,x_0) = \frac{\alpha}{x_0}\left(\frac{x}{x_0}\right)^{-(\alpha+1)}, \quad x \geqslant x_0\,,$$

with shape parameter $\alpha > 0$ as before, and scale parameter $x_0 > 0$. We will denote this distribution by $\mathsf{ParetoI}(\alpha, x_0)$. It is related to the type II distribution for any $\lambda > 0$ via $X \sim \mathsf{Pareto}(\alpha,\lambda) \Leftrightarrow (X+\lambda^{-1}) \sim \mathsf{ParetoI}(\alpha,\lambda^{-1})$.

Table 4.20 Moment and tail properties of the $\mathsf{Pareto}(\alpha,\lambda)$ distribution.

Property		Condition
Expectation	$\frac{1}{\lambda(\alpha-1)}$	$\alpha > 1$
Variance	$\frac{\alpha}{\lambda^2(\alpha-1)^2(\alpha-2)}$	$\alpha > 2$
Moment $\mathbb{E}X^k$	$\frac{k!}{\lambda^k}\frac{\Gamma(\alpha-k)}{\Gamma(\alpha)}$	$\alpha > k$
Tail probability $\mathbb{P}(X > x)$	$(1+\lambda\,x)^{-\alpha}$	$x \geqslant 0$

Other properties and relations are:

1. *Expectation and variance*: In addition to the cases listed above, for $0 < \alpha \leqslant 1$ the expectation is ∞ and the variance does not exist. For $1 < \alpha \leqslant 2$ the variance is ∞.

2. *Exponential*: If $Y \sim \mathsf{Exp}(1)$, then
$$e^{Y/\alpha} - 1 \sim \mathsf{Pareto}(\alpha, 1) .$$

3. *Gamma mixture*: If $Y \sim \mathsf{Gamma}(\alpha, 1)$ and $(X \mid Y) \sim \mathsf{Exp}(Y)$, then $X \sim \mathsf{Pareto}(\alpha, 1)$.

The $\mathsf{Pareto}(\alpha, \lambda)$ family of distribution is a scale family. As a consequence, if $X \sim \mathsf{Pareto}(\alpha, 1)$, then $X/\lambda \sim \mathsf{Pareto}(\alpha, \lambda)$. Generating from $\mathsf{Pareto}(\alpha, 1)$ is easily established via the inverse-transform method, using the fact that the cdf is given by $F(x) = 1 - (1 + x)^{-\alpha}$, $x \geqslant 0$.

Algorithm 4.51 ($\mathsf{Pareto}(\alpha, 1)$ Generator via Inverse-Transform)

Generate $U \sim \mathsf{U}(0, 1)$ *and return* $X = U^{-1/\alpha} - 1$.

An equivalent algorithm (see also Property 2 above) is the following.

Algorithm 4.52 ($\mathsf{Pareto}(\alpha, 1)$ Generator)

Generate $Y \sim \mathsf{Exp}(1)$ *and return* $X = e^{Y/\alpha} - 1$.

Property 3 results in the next algorithm.

Algorithm 4.53 ($\mathsf{Pareto}(\alpha, 1)$ Generator)

1. *Generate* $Y \sim \mathsf{Gamma}(\alpha, 1)$.
2. *Given* Y, *return* $X \sim \mathsf{Exp}(Y)$.

4.2.13 Phase-Type Distribution (Continuous Case)

The **continuous phase-type** distribution has pdf
$$f(x; \boldsymbol{\alpha}, A) = -\boldsymbol{\alpha}\, e^{Ax} A\mathbf{1}, \quad x \geqslant 0 ,$$
where $\boldsymbol{\alpha}$ is a $1 \times m$ probability vector and A is an $m \times m$ invertible matrix such that
$$Q = \begin{pmatrix} A & -A\mathbf{1} \\ \mathbf{0}^\top & 0 \end{pmatrix} \tag{4.7}$$
☞ 166 is the Q-matrix (infinitesimal generator) of a Markov jump process. Above, $\mathbf{0}$ and $\mathbf{1}$ are the $m \times 1$ vectors of zeros and ones, respectively, and e^{Ax} is a matrix exponential, defined as
$$e^{Ax} = \sum_{n=0}^{\infty} \frac{A^n x^n}{n!} .$$

We write the distribution as $\mathsf{PH}(\boldsymbol{\alpha}, A)$. In the table below, $\varrho(A) < 0$ denotes the largest eigenvalue of A.

Table 4.21 Moment and tail properties of the $\mathsf{PH}(\boldsymbol{\alpha}, A)$ distribution.

Property		Condition
Expectation	$-\boldsymbol{\alpha}A^{-1}\mathbf{1}$	
Variance	$2\boldsymbol{\alpha}A^{-2}\mathbf{1} - (\boldsymbol{\alpha}A^{-1}\mathbf{1})^2$	
Moment $\mathbb{E}X^k$	$(-1)^k\, k!\, \boldsymbol{\alpha}A^{-k}\mathbf{1}$	
Moment generating function	$\boldsymbol{\alpha}(tI + A)^{-1}A\mathbf{1}$	$t \leqslant -\varrho(A)$
Tail probability $\mathbb{P}(X > x)$	$\boldsymbol{\alpha}\, \mathrm{e}^{Ax}\mathbf{1}$	

Other properties and relations (see, for example, [26]) are:

1. *Absorption time*: A random variable $X \sim \mathsf{PH}(\boldsymbol{\alpha}, A)$ can be thought of as the time to absorption in state $m+1$ of a Markov jump process $\{Y_t, t \geqslant 0\}$ taking values in $\{1, \ldots, m, m+1\}$. The Markov jump process has Q-matrix Q and initial distribution $(\boldsymbol{\alpha}, 0)$. ☞ 635

2. *Exponential distribution*: The exponential distribution $\mathsf{Exp}(\lambda)$ is the simplest continuous phase-type distribution, with $m = 1$, $A = -\lambda$, and $\alpha = 1$.

3. *Generalized Erlang distribution*: A random variable with a **generalized Erlang** distribution can be thought of as the sum of n independent exponential random variables with rates $\lambda_1, \ldots, \lambda_n$. The distribution is of phase-type with $m = n$, $\boldsymbol{\alpha} = (1, 0, \ldots, 0)$ and

$$A = \begin{pmatrix} -\lambda_1 & \lambda_1 & 0 & \cdots & 0 \\ 0 & -\lambda_2 & \lambda_2 & \cdots & 0 \\ \vdots & \vdots & \ddots & \ddots & \vdots \\ 0 & \cdots & 0 & -\lambda_{n-1} & \lambda_{n-1} \\ 0 & \cdots & 0 & 0 & -\lambda_n \end{pmatrix}.$$

   If all $\lambda_i = \lambda$ for all i, then we have the ordinary Erlang distribution, $\mathsf{Erl}(n, \lambda) \equiv \mathsf{Gamma}(n, \lambda)$.

   Algorithm 16.2 describes the computation of the cdf of the generalized Erlang distribution for the case where the $\{\lambda_k\}$ are distinct and sorted in descending order: $\lambda_1 > \lambda_2 > \cdots > \lambda_n$. ☞ 556

4. *Hyperexponential distribution*: The **hyperexponential** distribution is the mixture of n exponential distributions, with parameters $\lambda_i > 0$ and mixing probabilities $\alpha_i > 0$, $i = 1, 2, \ldots, n$. It can be represented as the phase-type distribution with $\boldsymbol{\alpha} = (\alpha_1, \ldots, \alpha_n)$ and $A = \operatorname{diag}(-\lambda_1, \ldots, -\lambda_n)$.

5. *Coxian distribution*: The **Coxian** distribution is a phase-type distribution with $\boldsymbol{\alpha} = (1, 0, \ldots, 0)$ and

$$A = \begin{pmatrix} -\lambda_1 & p_1\lambda_1 & 0 & \cdots & 0 \\ 0 & -\lambda_2 & p_2\lambda_2 & \cdots & 0 \\ \vdots & \vdots & \ddots & \ddots & \vdots \\ 0 & \cdots & 0 & -\lambda_{n-1} & p_{n-1}\lambda_{n-1} \\ 0 & \cdots & 0 & 0 & -\lambda_n \end{pmatrix} .$$

6. *Sum*: The sum of a finite number of independent phase-type distributed random variables has again a phase-type distribution.

7. *Mixture*: The mixture of a finite number of phase-type distributions is again a phase-type distribution.

Let Q be the Q-matrix as in (4.7). Denote the transition matrix of the corresponding jump chain (see Property 1) by K and the exponential holding time
☞ 635 parameters by $\{q_i, i = 1, \ldots, m\}$. Let $K(y, \cdot)$ denote the y-th row of K and $\boldsymbol{\alpha}$ the initial distribution.

Algorithm 4.54 (PH(α, A) Generator)

1. *Draw* $Y \sim \boldsymbol{\alpha}$. *Set* $T = 0$.

2. *Draw* $S \sim \mathsf{Exp}(q_Y)$. *Set* $T = T + S$.

3. *Draw* $Y \sim K(Y, \cdot)$.

4. *If* $Y = m + 1$, *return* $X = T$; *otherwise, go back to Step 2.*

■ EXAMPLE 4.6 (Coxian Distribution Generation)

Consider the generation from the Coxian distribution given in Property 5 above. The following MATLAB program implements Algorithm 4.54 for the case $m = 4$, $\lambda_i = 3, p_i = 0.9, i = 1, \ldots, m$, and $\boldsymbol{\alpha} = (1, 0, 0, 0)$. The sample mean and variance are compared with the true expectation of 1.1463 and true variance of 0.4961 (rounded).

```
%coxian_ex.m
alpha = [1 0 0 0];
p = 0.9;
m = 4;
lam = 3;
A = [-lam , lam*p, 0, 0; 0, -lam, lam*p ,0 ; ...
    0 ,0, -lam, lam*p; 0, 0, 0 ,-lam] ;
A1 = - A*ones(m,1);
q = - diag(A);
Q = [A A1; zeros(1,m) 0];
K = diag([1./q; 1])*(Q + diag([q;0]));
K(m+1,m+1) = 1;
N = 10^5;
```

```
x = zeros(N,1);
for i=1:N;
    T = 0;
    Y = min(find(cumsum(alpha)> rand));
    while Y ~= m+1
        T= T -log(rand)/q(Y);
        Y = min(find(cumsum(K(Y,:))> rand));
    end
    x(i) = T;
end

truemean = -alpha*inv(A)*ones(m,1)
mx = mean(x)
truevar = alpha*inv(A)*(2*eye(m,m)-ones(m,1)*alpha)*inv(A)*ones(m,1)
vx = var(x)
hist(x,100)
```

4.2.14 Stable Distribution

The **stable** (also called **α-stable**, **Lévy α-stable** or **Lévy skew α-stable**) distribution has no closed-form pdf in general. Instead, it is defined in terms of its characteristic function [12, 27, 32], with

$$\phi(t) = \mathbb{E}\,\mathrm{e}^{\mathrm{i}tX} = \begin{cases} \exp\left(-|t|^\alpha \left(1 - \mathrm{i}\beta\,\mathrm{sgn}(t)\tan\left(\frac{\pi}{2}\alpha\right)\right)\right) & \alpha \neq 1 \\ \exp\left(-|t|\left(1 + \mathrm{i}\frac{2}{\pi}\beta\,\mathrm{sgn}(t)\ln|t|\right)\right) & \alpha = 1\,, \end{cases} \tag{4.8}$$

where $0 < \alpha \leqslant 2$ is the **characteristic exponent** and $-1 \leqslant \beta \leqslant 1$ the **skewness parameter**. Here $\mathrm{sgn}(t)$ denotes the sign of t, that is, $\mathrm{sgn}(t) = -1, 0$, and 1, for $t < 0, t = 0$, and $t > 0$, respectively. Stable distributions arise naturally in the study of *Lévy processes*; see Section 5.13. ☞ 208

The pdf can be obtained via numerical inverse–Fourier transform techniques. In particular (note that $\Re[\phi(-t)] = \Re[\phi(t)]$ and $\Im[\phi(-t)] = -\Im[\phi(t)]$),

$$f(x) = \frac{1}{\pi}\int_0^\infty \mathrm{e}^{-\mathrm{i}tx}\,\phi(t)\,\mathrm{d}t\,. \tag{4.9}$$

The support of the distribution is $\mathbb{R}$, except in the cases $\beta = 1, \alpha < 1$ (where the support is $[0,\infty)$) and $\beta = -1, \alpha < 1$ (where the support is $(-\infty, 0]$); see, for example, [32]. When $|\beta| = 1$ the stable law is said to be **extremal**. When $\beta = 0$ the distribution is symmetric around 0.

A location–scale version of this distribution is given by the transformation $Y = \mu + \sigma X$, where X has the characteristic function above. The defining property of a stable distribution is that any affine combination of independent stable random variables is again a stable random variable. We write the distribution as $\mathsf{Stable}(\alpha, \beta, \mu, \sigma)$. The standard stable distributions $\mathsf{Stable}(\alpha, \beta, 0, 1)$ are denoted by $\mathsf{Stable}(\alpha, \beta)$. Graphs of the standard stable pdf for various values of the parameters are given in Figure 4.18.

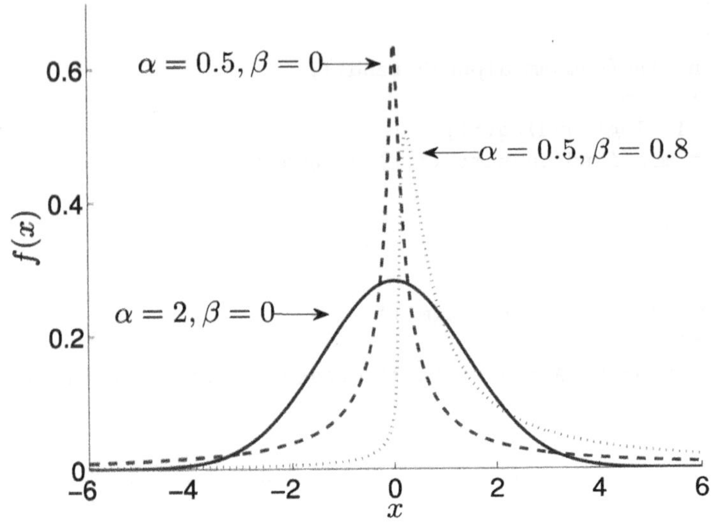

Figure 4.18 Pdfs of the standard stable distribution for various values of α and β.

Three special cases of stable distributions for which the pdf has an easy form are (1) the $\mathsf{Stable}(2, \beta)$ distribution, which is equal to the $\mathsf{N}(0, 2)$ distribution; (2) the $\mathsf{Stable}(1, 0)$ distribution, which is equal to the standard Cauchy distribution; (3) the $\mathsf{Stable}(\frac{1}{2}, 1)$, which is known as the **Lévy** distribution (denoted here as Lévy), with pdf

$$f(x) = \frac{1}{\sqrt{2\pi}} x^{-3/2} e^{-\frac{1}{2x}}, \quad x > 0 .$$

Other properties and relations include:

1. *Expectation and variance*: The expectation of the $\mathsf{Stable}(\alpha, \beta)$ distribution is 0 for $1 < \alpha \leqslant 2$ and ∞ otherwise. The second moment is ∞ except when $\alpha = 2$ where it is 2.

2. *Sum*: If $X_i \sim \mathsf{Stable}(\alpha, \beta_i)$, $i = 1, 2$ are independent, then

$$\frac{X_1 + X_2}{2^{1/\alpha}} \sim \mathsf{Stable}\left(\alpha, \frac{\beta_1 + \beta_2}{2}\right) .$$

3. *Extremals*: If $X_1 \sim \mathsf{Stable}(\alpha, 1)$ and $X_2 \sim \mathsf{Stable}(\alpha, -1)$ are independent, then

$$\left(\frac{1+\beta}{2}\right)^{\frac{1}{\alpha}} X_1 + \left(\frac{1-\beta}{2}\right)^{\frac{1}{\alpha}} X_2 \sim \mathsf{Stable}(\alpha, \beta) .$$

4. *Normal distribution*: If $X \sim \mathsf{N}(0, 1)$, then $(1/X^2) \sim$ Lévy.

5. *Reflection*: If $X \sim \mathsf{Stable}(\alpha, \beta)$, then $(-X) \sim \mathsf{Stable}(\alpha, -\beta)$.

Zolotarev [32, Page 78] derives integral expressions for the pdf and the cdf of the $\mathsf{Stable}(\alpha, \beta)$ distribution from (4.9) containing only real-valued (trigonometric) functions. Based on this, Chambers et al. [6] develop the following generation method for general $\mathsf{Stable}(\alpha, \beta)$ distributions. For details and proofs we refer to [32] and [31] (the last is accompanied by a small correction).

Algorithm 4.55 (Stable(α, β) Generator, $\alpha \neq 1$)

1. *Set* $B = \arctan(\beta \tan(\pi\alpha/2))/\alpha$.
2. *Generate* $V \sim \mathsf{U}(-\pi/2, \pi/2)$ *and* $W \sim \mathsf{Exp}(1)$ *independently.*
3. *Output*

$$X = \frac{\sin(\alpha(V+B))}{[\cos(\alpha B)\cos(V)]^{1/\alpha}} \left(\frac{\cos[V - \alpha(V+B)]}{W} \right)^{(1-\alpha)/\alpha} .$$

When $\alpha = 1$ one can instead use the following algorithm.

Algorithm 4.56 (Stable($1, \beta$) Generator)

1. *Generate* $V \sim \mathsf{U}(-\pi/2, \pi/2)$ *and* $W \sim \mathsf{Exp}(1)$ *independently.*
2. *Output*

$$X = \frac{2}{\pi} \left[\left(\frac{\pi}{2} + \beta V \right) \tan(V) - \beta \ln \left(\frac{W \cos(V)}{1 + 2\beta V/\pi} \right) \right] .$$

For $\mathsf{Stable}(2, \beta) \equiv \mathsf{N}(0, 2)$ and $\mathsf{Stable}(1, 0) \equiv \mathsf{Cauchy}$, we refer to the corresponding random variable generation sections. For the Lévy distribution, Property 4 above yields the following algorithm.

Algorithm 4.57 (Stable($1/2, 1$) $\equiv$ Lévy Generator)

Generate $Y \sim \mathsf{N}(0, 1)$ *and output* $X = 1/Y^2$.

4.2.15 Student's t Distribution

Student's t distribution or simply the **t** distribution has pdf

$$f(x; \nu) = \frac{\Gamma(\frac{\nu+1}{2})}{\sqrt{\nu\pi}\,\Gamma(\frac{\nu}{2})} \left(1 + \frac{x^2}{\nu} \right)^{-(\nu+1)/2} , \quad x \in \mathbb{R} , \tag{4.10}$$

where $\nu > 0$. We write the distribution as t_ν. If the parameter ν takes integer values, then it is referred to as the **degrees of freedom** of the t distribution. A location–scale version of this distribution is given by the pdf $f(x; \mu, \sigma) = f((x - \mu)/\sigma)/\sigma$; we write $\mathsf{t}_\nu(\mu, \sigma^2)$ for the corresponding distribution.

The distribution arises in statistics in the problem of estimating the mean of a normally distributed population when the population variance needs to be estimated and the sample size is small. In particular, let $X_1, \ldots, X_n \sim_{\text{iid}} \mathsf{N}(\mu, \sigma^2)$, with sample mean $\bar{X}_n = (X_1 + \cdots + X_n)/n$ and sample variance $S_n^2 = \sum_{i=1}^n (X_i -$ ☞ 662

$\bar{X}_n)^2/(n-1)$. Then,

$$T = \frac{\bar{X}_n - \mu}{S_n/\sqrt{n}} \sim \mathsf{t}_{n-1}\,.$$

The effect of the parameter ν on the shape of the distribution is illustrated in Figure 4.19.

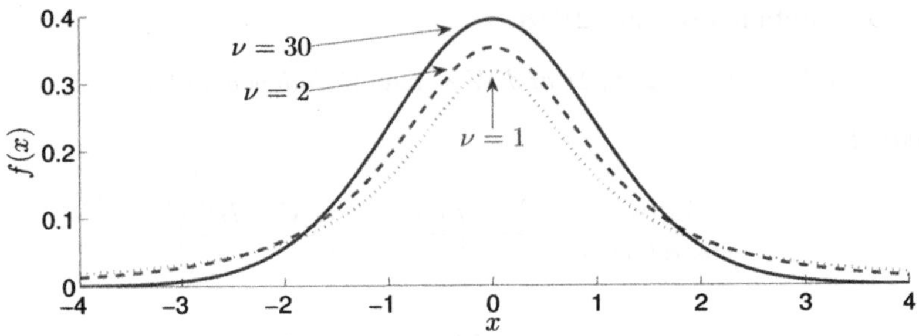

Figure 4.19 Pdfs of the t_ν distribution for various values of ν.

Table 4.22 Moment properties of the t_ν distribution.

Property		Condition
Expectation	0	$\nu > 1$
Variance	$\dfrac{\nu}{\nu-2}$	$\nu > 2$
Characteristic function	$\dfrac{2^{1-\frac{\nu}{2}} \lvert\sqrt{\nu}t\rvert^{\nu/2} K_{\frac{\nu}{2}}\left(\sqrt{\nu}\lvert t\rvert\right)}{\Gamma(\frac{\nu}{2})}$	$\nu > 0$
Tail probability $\mathbb{P}(X > x)$	$\dfrac{1}{2}\, I_{\nu/(\nu+x^2)}\left(\dfrac{\nu}{2}, \dfrac{1}{2}\right)$	

Here, $K_\nu(x)$ denotes the *modified Bessel function of the third kind* and $I_x(\alpha, \beta)$ denotes the *incomplete beta function*. Other properties and relations are:

☞ 715

1. *Cauchy distribution*: $\mathsf{t}_1 \equiv \mathsf{Cauchy}(0,1)$.

2. *Normal distribution*: If $X_\nu \sim \mathsf{t}_\nu$, then $X_\nu \xrightarrow{d} Z \sim \mathsf{N}(0,1)$ as $\nu \to \infty$.

3. *Normal and gamma mixture*: If $X \sim \mathsf{N}(0,1)$ and $Y \sim \chi^2_\nu \equiv \mathsf{Gamma}(\nu/2, 1/2)$ are independent, then

$$\frac{X}{\sqrt{Y/\nu}} \sim \mathsf{t}_\nu\,.$$

4. *Beta distribution*: If $T \sim \mathsf{t}_{2\alpha}$, then (equivalent to Property 3 of the beta distribution on Page 104)

$$B = \frac{1}{2}\left(1 + \frac{T}{\sqrt{2\alpha + T^2}}\right) \sim \mathsf{Beta}(\alpha, \alpha) .$$

5. *Gamma distribution*: Let $X \sim \mathsf{Gamma}(1/2, 1)$ and $Y \sim \mathsf{Gamma}(\nu/2, 1)$, independently. If $B \sim \mathsf{Ber}(1/2)$ is independent of X and Y, then

$$(2B - 1)\sqrt{\frac{X}{Y/\nu}} \sim \mathsf{t}_\nu .$$

In addition, if $Z \sim \mathsf{Gamma}(\nu/2, 1)$, independently of Y, then

$$\frac{\sqrt{\nu}}{2}\frac{Y - Z}{\sqrt{Y\,Z}} \sim \mathsf{t}_\nu .$$

6. *Square*: If $X \sim \mathsf{t}_n$, then $X^2 \sim \mathsf{F}(1, n)$.

7. *F distribution*: If $X \sim \mathsf{F}(n, n)$, then $\frac{\sqrt{n}}{2}\left(\sqrt{X} - 1/\sqrt{X}\right) \sim \mathsf{t}_n$.

Since $\mathsf{t}_\nu(\mu, \sigma^2)$ forms a location–scale family, we only consider generation from t_ν. There are many simple and effective generation algorithms for the t distribution. Property 3 above, which is fundamental in classical statistics, yields the following algorithm.

Algorithm 4.58 (t_ν Generator via Gamma)

1. *Generate* $Z \sim \mathsf{N}(0, 1)$ *and* $Y \sim \mathsf{Gamma}(\nu/2, 1/2)$ *independently.*
2. *Output* $X = Z/\sqrt{Y/\nu}$.

An alternative approach is to use the relation between the Student and beta distribution as in Property 4 above, for example.

Algorithm 4.59 (t_ν Generator via Beta)

1. *Generate* $Y \sim \mathsf{Beta}\left(\frac{\nu}{2}, \frac{\nu}{2}\right)$.
2. *Output*

$$X = \sqrt{\nu}\frac{Y - \frac{1}{2}}{\sqrt{Y(1 - Y)}} .$$

The following simple algorithm, from [14], is based on the ratio of uniforms method. ☞ 66

Algorithm 4.60 (t_ν Generator via Ratio of Uniforms)

1. *Generate* $U \sim \mathsf{U}(-\sqrt{\nu}, \sqrt{\nu})$ *and* $V \sim \mathsf{U}(0, 1)$.
2. *Set* $W = V^{1/\nu}$.
3. *If* $W^2 + \frac{U^2}{\nu} \leqslant 1$, *return* $X = U/W$. *Otherwise, restart from Step 1.*

The polar method (see Section 3.1.2.7), with radial pdf ☞ 54

$$f_R(r) = r\left(1 + r^2/\nu\right)^{-1-\nu/2}, \quad r \geqslant 0 ,$$

results in the following algorithm.

Algorithm 4.61 (t_ν Generation via Polar Method (I))

1. *Generate* $U, V \overset{\text{iid}}{\sim} \mathsf{U}[0,1]$.
2. *Set* $\Theta = 2\pi U$ *and* $R = \sqrt{\nu\left(V^{-2/\nu} - 1\right)}$.
3. *Return* $X = R\cos(\Theta)$ *and* $Y = R\sin(\Theta)$. *Both* X *and* Y *are* t_ν *random variables, but are* not *independent.*

The following algorithm, from [4], is again based on the polar method, but replaces trigonometric function calculation with (faster) acceptance–rejection.

Algorithm 4.62 (t_ν Generator via Polar Method (II))

1. *Generate* $U, V \overset{\text{iid}}{\sim} \mathsf{U}[-1,1]$.
2. *Set* $W = U^2 + V^2$. *If* $W > 1$, *return to Step 1.*
3. *Set* $C = U^2/W$ *and* $R = \nu\left(W^{-2/\nu} - 1\right)$.
4. *Return* $X = \operatorname{sgn}(U)\sqrt{RC}$.

Finally, Kinderman et al. [19] proposed the following acceptance–rejection algorithm.

Algorithm 4.63 (t_ν Generator via Acceptance–Rejection)

1. *Generate* $U_1, U_2 \overset{\text{iid}}{\sim} \mathsf{U}(0,1)$.
2. *If* $U_1 < 0.5$, *set* $X = \frac{1}{4U_1 - 1}$ *and* $V = \frac{U_2}{X^2}$; *otherwise, set* $X = 4U_1 - 3$ *and* $V = U_2$.
3. *If* $V < 1 - \frac{|X|}{2}$ *or* $V < \left(1 + \frac{X^2}{\nu}\right)^{-\frac{\nu+1}{2}}$, *accept* X *as a random variable from* t_ν; *otherwise, repeat from Step 1.*

4.2.16 Uniform Distribution (Continuous Case)

The **uniform** distribution on the interval $[a,b]$ has pdf

$$f(x; a, b) = \frac{1}{b-a}, \qquad a \leqslant x \leqslant b\,.$$

We write the distribution as $\mathsf{U}[a,b]$. A graph of the pdf is given in Figure 4.20.

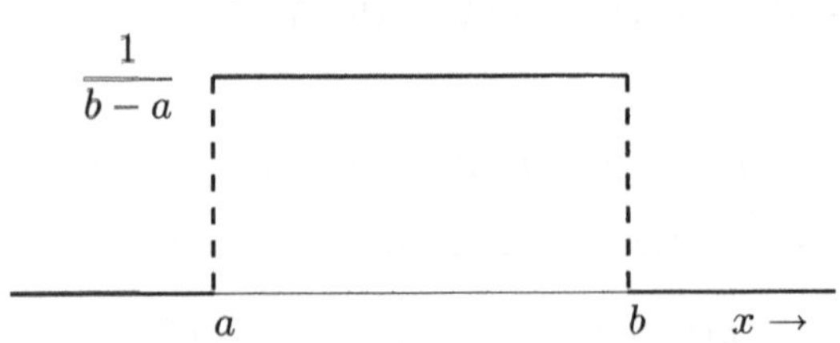

Figure 4.20 The pdf of the uniform distribution on $[a,b]$.

The uniform distribution is used as a model for choosing a point randomly from the interval $[a, b]$, such that each point is equally likely to be drawn. The uniform distribution on an arbitrary Borel set B in $\mathbb{R}^n$ with non-zero Lebesgue measure (for example, area, volume) $|B|$ is defined similarly: its pdf is constant, taking value $1/|B|$ on B and 0 otherwise. We write $\mathsf{U}(B)$ or simply $\mathsf{U}B$. The $\mathsf{U}[a, b]$ distribution is a location–scale family, as $Z \sim \mathsf{U}[a, b]$ has the same distribution as $a + (b - a)X$, with $X \sim \mathsf{U}[0, 1]$.

Table 4.23 Moment and tail properties of the $\mathsf{U}(a, b)$ distribution.

Property		Condition
Expectation	$\dfrac{a+b}{2}$	
Variance	$\dfrac{(a-b)^2}{12}$	
Moment generating function	$\dfrac{\mathrm{e}^{bt} - \mathrm{e}^{at}}{(b-a)t}$	
Tail probability $\mathbb{P}(X > x)$	$\dfrac{b-x}{b-a}$	$x \in (a, b)$

The generation of $\mathsf{U}(0, 1)$ random variables, crucial for any Monte Carlo method, is discussed in detail in Chapter 1. $\mathsf{U}(a, b)$ random variable generation follows immediately from the inverse-transform method. ☞ 1

Algorithm 4.64 ($\mathsf{U}(a, b)$ Generation)

Generate $U \sim \mathsf{U}(0, 1)$ *and return* $X = a + (b - a)U$.

4.2.17 Wald Distribution

The **Wald** or **inverse Gaussian** distribution has pdf

$$f(x; \mu, \lambda) = \sqrt{\frac{\lambda}{2\pi x^3}}\, \mathrm{e}^{-\frac{\lambda(x-\mu)^2}{2\mu^2 x}}, \quad x > 0,$$

where $\mu > 0$ is a **location** parameter and $\lambda > 0$ is a **scale** parameter. We write this distribution as $\mathsf{Wald}(\mu, \lambda)$. The Wald distribution arises as the first time that a Brownian motion with positive drift hits a positive level. More specifically, if $\{B_t, t \geqslant 0\}$ is a Brownian motion with drift $\mu > 0$ and diffusion coefficient $\sigma^2 > 0$ and if $\tau_a = \inf\{t \geqslant 0 : B_t \geqslant a\}$, then $\tau_a \sim \mathsf{Wald}(a/\mu, a^2/\sigma^2)$. Graphs of the pdf for various values of the parameters are given in Figure 4.21. ☞ 182

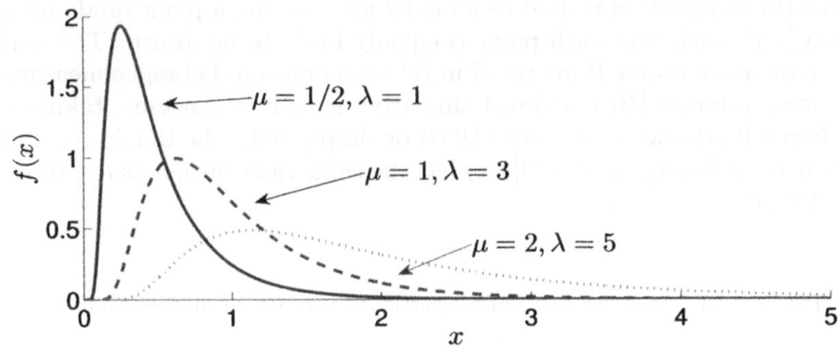

Figure 4.21 Pdfs of the Wald distribution for various values of the parameters.

Table 4.24 Moment and tail properties of the $\mathsf{Wald}(\mu, \lambda)$ distribution.

Property		Condition
Expectation	μ	
Variance	$\frac{\mu^3}{\lambda}$	
Moment gen. function	$\mathrm{e}^{\frac{\lambda}{\mu}\left(1-\sqrt{1-\frac{2\mu^2 t}{\lambda}}\right)}$	$t \leqslant \frac{\lambda}{2\mu^2}$
Tail probability $\mathbb{P}(X > x)$	$\Phi\left(\sqrt{\frac{\lambda}{x}}\left(1-\frac{x}{\mu}\right)\right) - \Phi\left(-\sqrt{\frac{\lambda}{x}}\left(1+\frac{x}{\mu}\right)\right)\mathrm{e}^{\frac{2\lambda}{\mu}}$	$x > 0$

Here, $\Phi(x)$ is the cdf of the standard normal distribution. Other properties and relations are:

1. *Chi-square distribution*: If $X \sim \mathsf{Wald}(\mu, \lambda)$, then

$$\lambda \frac{(X-\mu)^2}{\mu^2 X} \sim \chi_1^2 \, .$$

☞ 49

2. *Inverse gamma distribution*: Let $X_\mu \sim \mathsf{Wald}(\mu, \lambda)$. Then,

$$X_\mu \xrightarrow{d} Y \sim \mathsf{InvGamma}(1/2, \lambda/2) \quad \text{as } \mu \to \infty \, .$$

3. *Sum*: If $X_i \sim \mathsf{Wald}(\mu, \lambda\, a_i)$, $i = 1, \ldots, n$, independently, then

$$\sum_{i=1}^{n} X_i \sim \mathsf{Wald}(\mu\, a,\ \lambda\, a^2), \quad \text{with} \quad a = \sum_{i=1}^{n} a_i \, .$$

A simple generator can be derived from the following transformation. Let $X \sim \mathsf{Wald}(\mu, \lambda)$ and $g(x) = \frac{\lambda(x-\mu)^2}{\mu^2 x}$. For each $y > 0$ there are *two* solutions, say x_1 and

x_2, to $g(x) = y$. It is easy to check that $x_1 \geqslant \mu$ and $x_2 \leqslant \mu$ are the roots of a quadratic equation, and that their product is simply μ^2. So we can set $x_1 = x$ and $x_2 = \mu^2/x$, with $x \geqslant \mu$ satisfying $g(x) = y$. A simple modification of the standard transformation rule (A.33) yields that the pdf of Y satisfies

$$f_Y(y) = \frac{f_X(\mu^2/x)}{-g'(\mu^2/x)} + \frac{f_X(x)}{g'(x)} ,$$

where f_X is the pdf of X, $g'(x) = \lambda(\mu^{-2} - x^{-2})$, and $x = \mu + \frac{\mu^2 y}{2\lambda} + \frac{\mu}{2\lambda}\sqrt{4\mu\lambda y + \mu^2 y^2}$. Evaluation of the above expression yields $f_Y(y) = \mathrm{e}^{-y/2}(2\pi y)^{-1/2}$ for all $y > 0$, which is the pdf of the χ^2_1 distribution, so that we have derived Property 1 above.

In order to generate the appropriate X-value for a given Y-value we have to choose the appropriate root (μ^2/X or X) with probabilities in the proportion $\frac{f_X(\mu^2/x)}{f_X(x)} \times \frac{g'(x)}{-g'(\mu^2/x)} = x^3/\mu^3 \times \mu^2/x^2 = x/\mu$ to 1. This leads to the following algorithm.

Algorithm 4.65 (Wald(μ, λ) Generator)

1. *Generate* $W \sim \mathsf{N}(0,1)$ *and set* $Y = W^2$.

2. *Set*

$$Z = \mu + \frac{\mu^2\, Y}{2\lambda} + \frac{\mu}{2\lambda}\sqrt{4\,\mu\,\lambda\, Y + \mu^2\, Y^2} .$$

3. *Generate* $B \sim \mathsf{Ber}(\frac{\mu}{\mu+Z})$. *If* $B = 1$, *output* $X = Z$; *otherwise, output* $X = \frac{\mu^2}{Z}$.

4.2.18 Weibull Distribution

The **Weibull** distribution has pdf

$$f(x; \alpha, \lambda) = \alpha\, \lambda\, (\lambda x)^{\alpha - 1}\, \mathrm{e}^{-(\lambda x)^{\alpha}}, \quad x \geqslant 0 ,$$

where $\lambda > 0$ is called the **scale parameter** and $\alpha > 0$ the **shape parameter**. We write the distribution as $\mathsf{Weib}(\alpha, \lambda)$. A location–scale version of this distribution is given by the pdf $f(x; \alpha, \mu, \sigma) = f((x - \mu)/\sigma; \alpha, 1)/\sigma$, $x \geqslant \mu$; we write this distribution as $\mathsf{Weib}(\alpha, \mu, \sigma)$. Note that $\mathsf{Weib}(\alpha, \lambda) \equiv \mathsf{Weib}(\alpha, 0, \lambda^{-1})$.

In reliability theory the Weibull distribution is often used as the distribution of a lifetime of a component. The reflection of this distribution, with pdf $f(-x; \alpha, \mu, \sigma)$, which we will call the **reversed Weibull** distribution, is one of the three extreme value type distributions (type III), denoted here by $-\mathsf{Weib}(\alpha, \mu, \sigma)$.

A special case is the $\mathsf{Weib}(2, 1/(\sigma\sqrt{2}))$ distribution, known as the **Rayleigh** distribution, denoted by $\mathsf{Rayleigh}(\sigma)$. The dependence of the Weibull distribution on its parameters is illustrated in Figure 4.22.

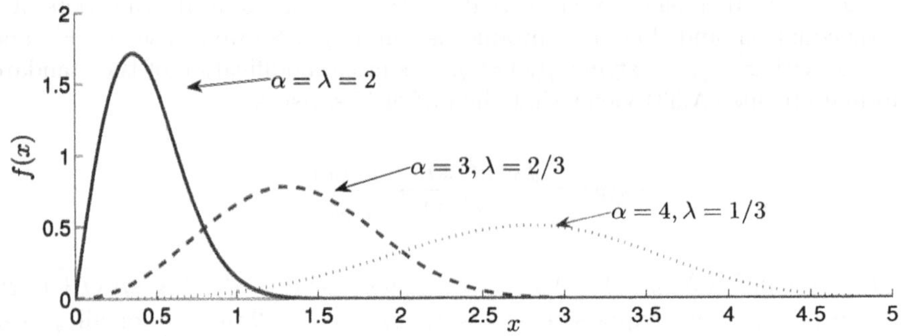

Figure 4.22 Pdfs of the Weib(α, λ) distribution for various values of its parameters.

Table 4.25 Moment and tail properties of the Weib(α, λ) distribution.

Property		Condition
Expectation	$\lambda^{-1}\Gamma(1+\alpha^{-1})$	
Variance	$\lambda^{-2}(\Gamma(1+2\alpha^{-1}) - \Gamma(1+\alpha^{-1})^2)$	
Moment $\mathbb{E}X^k$	$\lambda^{-k}\Gamma(1+k\alpha^{-1})$	
Tail probability $\mathbb{P}(X > x)$	$\mathrm{e}^{-(\lambda x)^\alpha}$	$x \geqslant 0$

Other properties and relations are:

1. *Exponential distribution*: If $Y \sim \mathsf{Exp}(\lambda)$ and $\beta > 0$, then $Y^\beta \sim \mathsf{Weib}(\beta^{-1}, \lambda^\beta)$.
2. *Normal distribution*: If $X, Y \overset{\text{iid}}{\sim} \mathsf{N}(0, \sigma^2)$, then $\sqrt{X^2 + Y^2} \sim \mathsf{Rayleigh}(\sigma)$.
3. *Gumbel distribution*: If $X \sim \mathsf{Weib}(\alpha, \lambda)$, then $-\ln X \sim \mathsf{Gumbel}(\ln \lambda, 1/\alpha)$.

Since Weib(α, μ, σ) defines a location–scale family, it suffices to consider generating Weib(α, λ) random variables only. This is easily established via the inverse-transform method, as the cdf satisfies $F(x) = 1 - \exp(-(\lambda x)^\alpha)$, $x \geqslant 0$.

Algorithm 4.66 (Weib(α, λ) Generator)

Generate $U \sim \mathsf{U}(0,1)$ *and output* $X = \frac{1}{\lambda}(-\ln U)^{\frac{1}{\alpha}}$.

4.3 MULTIVARIATE DISTRIBUTIONS

We list various multivariate distributions in alphabetical order. Recall that an absolutely continuous or discrete distribution is completely specified by its pdf. ☞ 611

Remark 4.3.1 (Singular Continuous Distributions) Many practically encountered multivariate singular continuous distributions involve absolutely continuous random vectors that are mapped to a lower-dimensional manifold. For example, the uniform distribution on the perimeter of a circle is singular with respect to the Lebesgue measure on $\mathbb{R}^2$ and so its two-dimensional pdf is zero. However, it is readily thought of as the distribution of a one-dimensional uniform random variable bent into a circle.

4.3.1 Dirichlet Distribution

The **standard Dirichlet** (or **type I Dirichlet**) distribution has pdf

$$f(\mathbf{x};\boldsymbol{\alpha}) = \frac{\Gamma\left(\sum_{i=1}^{n+1}\alpha_i\right)}{\prod_{i=1}^{n+1}\Gamma(\alpha_i)}\prod_{i=1}^{n} x_i^{\alpha_i-1}\left(1-\sum_{i=1}^{n}x_i\right)^{\alpha_{n+1}-1}, \quad x_i \geqslant 0, i=1,\ldots,n, \sum_{i=1}^{n} x_i \leqslant 1 ,$$

where $\alpha_i > 0$, $i = 1,\ldots,n+1$ are **shape** parameters. We write this distribution as $\mathsf{Dirichlet}(\alpha_1,\ldots,\alpha_{n+1})$ or $\mathsf{Dirichlet}(\boldsymbol{\alpha})$, with $\boldsymbol{\alpha} = (\alpha_1,\ldots,\alpha_{n+1})^\top$. The Dirichlet distribution can be regarded as a multivariate generalization of the beta distribution (see Section 4.2.1), in the sense that each marginal X_k has a beta distribution (a more general statement is made in Property 6 below).

In Bayesian statistics, the Dirichlet distribution is the conjugate prior for the success probabilities of the multinomial distribution (see Section 4.3.2). A graph of the pdf for the two-dimensional case is given in Figure 4.23. ☞ 675

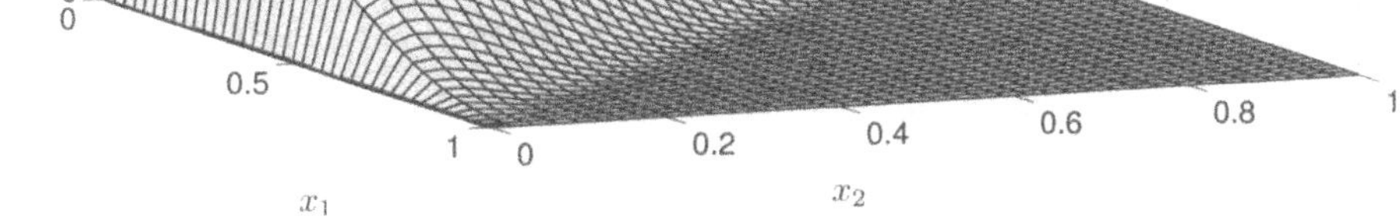

Figure 4.23 The Dirichlet pdf in two dimensions, with parameter vector $\boldsymbol{\alpha} = (1.5, 1.5, 3)^\top$.

Table 4.26 Moment properties of the Dirichlet($\boldsymbol{\alpha}$) distribution.

Property	
Expectation vector	$\dfrac{\widetilde{\boldsymbol{\alpha}}}{c}$
Covariance matrix	$\dfrac{c\,\mathrm{diag}(\widetilde{\boldsymbol{\alpha}}) - \widetilde{\boldsymbol{\alpha}}\widetilde{\boldsymbol{\alpha}}^{\top}}{c^2(c+1)}$

Here, $\widetilde{\boldsymbol{\alpha}} = (\alpha_1, \ldots, \alpha_n)^{\top}$ and $c = \sum_{i=1}^{n+1} \alpha_i$. Other properties and relations are:

1. *Gamma to Dirichlet:* Let $Y_1, \ldots, Y_{n+1}$ be independent random variables with $Y_k \sim \mathsf{Gamma}(\alpha_k, 1)$, $k = 1, \ldots, n+1$, and define

$$X_k = \frac{Y_k}{\sum_{i=1}^{n+1} Y_i}, \qquad k = 1, \ldots, n \,.$$

Then, $\mathbf{X} = (X_1, \ldots, X_n) \sim \mathsf{Dirichlet}(\boldsymbol{\alpha})$.

2. *Dirichlet to gamma:* Let $Y \sim \mathsf{Gamma}(\sum_{i=1}^{n+1} \alpha_i, 1)$ be a random variable independent of $\mathbf{X} = (X_1, \ldots, X_n) \sim \mathsf{Dirichlet}(\boldsymbol{\alpha})$. Define

$$Y_i = Y X_i, \quad i = 1, \ldots, n \quad \text{and} \quad Y_{n+1} = Y\Big(1 - \sum_{i=1}^{n} X_i\Big) \,.$$

Then, $Y_1, \ldots, Y_{n+1}$ are independent with $Y_i \sim \mathsf{Gamma}(\alpha_i, 1)$, $i = 1, \ldots, n+1$.

3. *Beta to Dirichlet:* Let $Y_1, \ldots, Y_n$ be independent random variables with $Y_i \sim \mathsf{Beta}(\alpha_i, \sum_{j=i+1}^{n+1} \alpha_j)$, $i = 1, \ldots, n$. Define

$$X_i = Y_i \prod_{j=1}^{i-1} (1 - Y_j) \,, \quad i = 1, \ldots, n \,.$$

Then, $\mathbf{X} = (X_1, \ldots, X_n) \sim \mathsf{Dirichlet}(\boldsymbol{\alpha})$.

4. *Dirichlet to beta:* If $\mathbf{X} = (X_1, \ldots, X_n) \sim \mathsf{Dirichlet}(\boldsymbol{\alpha})$ and

$$Y_i = \frac{X_i}{1 - X_1 - \cdots - X_{i-1}}, \qquad i = 1, \ldots, n \,,$$

then $Y_1, \ldots, Y_n$ are independent with $Y_i \sim \mathsf{Beta}(\alpha_i, \sum_{j=i+1}^{n+1} \alpha_j)$, $i = 1, \ldots, n$.

5. *Uniform:* The n-dimensional $\mathsf{Dirichlet}(1, \ldots, 1)$ distribution has a constant density on the unit simplex $\{\mathbf{x} \in \mathbb{R}^n : x_i \geqslant 0, i = 1, \ldots, n, \sum_{i=1}^{n} x_i \leqslant 1\}$ and thus corresponds to the uniform distribution on that set. ☞ 71

6. *Beta distribution of marginals:* As a consequence of Property 4, $X_1 \sim \mathsf{Beta}(\alpha_1, \sum_{i \neq 1} \alpha_i)$, and (by relabeling) $X_k \sim \mathsf{Beta}(\alpha_k, \sum_{i \neq k} \alpha_i)$ in general.

7. *Subvectors*: From Property 1 it follows immediately that for any choice $\{i_1, \ldots, i_k\}$ of distinct indices from $\{1, \ldots, n\}$,

$$(X_{i_1}, \ldots, X_{i_k}) \sim \mathsf{Dirichlet}(\alpha_{i_1}, \ldots, \alpha_{i_k}, \beta) ,$$

with $\beta = \sum_{i=1}^{n+1} \alpha_i - \sum_{j=1}^{k} \alpha_{i_j}$.

8. *Moments*: Let $\mathbf{X} \sim \mathsf{Dirichlet}(\boldsymbol{\alpha})$, with $\boldsymbol{\alpha} = (\alpha, \alpha, \ldots, \alpha, \beta)^\top$. Then, for $\gamma \geqslant 0$, we have (see [10])

$$\mathbb{E} \prod_{1 \leqslant i < j \leqslant n} |X_i - X_j|^{2\gamma} = \frac{\Gamma(\alpha n + \beta)}{\Gamma^n(\alpha)\, \Gamma(\alpha n + \beta + n(n-1)\gamma)} \prod_{j=0}^{n-1} \frac{\Gamma(\alpha + j\gamma)\, \Gamma(1 + (j+1)\gamma)}{\Gamma(1+\gamma)} .$$

The fundamental relation between the Dirichlet and the gamma distribution given in Property 1 provides the following generation method.

Algorithm 4.67 (Dirichlet(α) Generator)

1. *Generate* $Y_k \sim \mathsf{Gamma}(\alpha_k, 1)$, $k = 1, \ldots, n+1$ *independently.*
2. *Output* $\mathbf{X} = (X_1, \ldots, X_n)$, *where*

$$X_k = \frac{Y_k}{\sum_{i=1}^{n+1} Y_i}, \quad k = 1, \ldots, n .$$

An alternative generation mechanism is based on the relation between the Dirichlet distribution and the beta distribution, see Property 3 above. In general, beta generators require a gamma generator, unless the parameter values are integers, in which case one can use order statistics (see Algorithm 4.24 on Page 106). The following algorithm can be useful when the parameters are small and integer-valued.

Algorithm 4.68 (Dirichlet(α) Generator via Beta Random Variables)

1. *Generate*

$$Y_i \sim \mathsf{Beta}\left(\alpha_i, \sum_{j=i+1}^{n+1} \alpha_j\right), \quad i = 1, \ldots, n .$$

2. *Output* $\mathbf{X} = (X_1, \ldots, X_n)$, *where*

$$X_i = Y_i \prod_{j=1}^{i-1} (1 - Y_j), \quad i = 1, \ldots, n .$$

4.3.2 Multinomial Distribution

The **multinomial** distribution has joint pdf given by

$$f(\mathbf{x}; n, \mathbf{p}) = n! \prod_{i=1}^{k} \frac{p_i^{x_i}}{x_i!} \quad \text{for all } \mathbf{x} \in \{0, \ldots, n\}^k \text{ for which } \sum_{i=1}^{k} x_i = n ,$$

where k and n are strictly positive integers, each $p_i > 0$, and $\sum_{i=1}^{k} p_i = 1$. Summarizing the $\{p_i\}$ in a k-dimensional (column) vector $\mathbf{p}$, we write the distribution as $\mathsf{Mnom}(n, \mathbf{p})$.

The multinomial distribution arises in the experiment in which n balls are thrown into k bins, where the probability of a ball falling in the i-th bin is p_i. This situation arises, for example, when a histogram is constructed for an iid sample of size n from some distribution. The multinomial distribution can be seen as a generalization of the binomial distribution. In Bayesian statistics, the multinomial distribution is the conjugate prior for the shape parameters of the Dirichlet distribution. A graph of the pdf for a three-dimensional case is given in Figure 4.24.

☞ 675

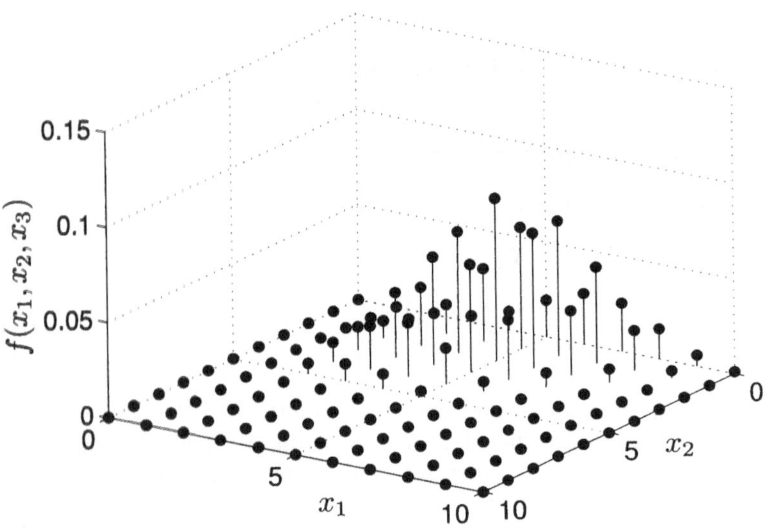

Figure 4.24 The multinomial pdf with parameters $n = 10$ and $\mathbf{p} = (0.2, 0.5, 0.3)^\top$. Note that $x_3 = n - x_1 - x_2$ is not shown.

Table 4.27 Moment properties of the $\mathsf{Mnom}(n, \mathbf{p})$ distribution.

Property		Condition
Expectation vector	$n\mathbf{p}$	
Covariance matrix	$n(\text{diag}(\mathbf{p}) - \mathbf{p}\mathbf{p}^\top)$	
Probability generating function	$\mathbb{E} z_1^{X_1} \cdots z_k^{X_k} = \left(\sum_{i=1}^{k} p_i z_i\right)^n$	$\lvert z_i \rvert \leqslant 1, i = 1, \ldots, k$

Other properties are:

1. *Binomial*: If $(X_1, \ldots, X_k) \sim \mathsf{Mnom}(n, \mathbf{p})$, then $X_i \sim \mathsf{Bin}(n, p_i)$, $i = 1, \ldots, k$.

2. *Conditionals*: $(X_{t+1} \mid X_1, \ldots, X_t) \sim \mathsf{Bin}\Big(n - \sum_{i=1}^{t} X_i,\ \frac{p_{t+1}}{\sum_{i=t+1}^{k} p_i}\Big)$.

3. *Subvectors*: For any choice $\{i_1, \ldots, i_m\}$ of distinct indices from $\{1, \ldots, k\}$,

$$\left(X_{i_1}, \ldots, X_{i_m}, n - \sum_{j=1}^{m} X_{i_j}\right) \sim \mathsf{Mnom}(n, \widetilde{\mathbf{p}})\ ,$$

where $\widetilde{\mathbf{p}} = (p_{i_1}, \ldots, p_{i_m}, 1 - \sum_{j=1}^{m} p_{i_j})^\top$.

The following algorithm for generating multinomial vectors mimics throwing n balls into k bins with probabilities $p_1, \ldots, p_k$.

Algorithm 4.69 (Direct $\mathsf{Mnom}(n, \mathbf{p})$ Generation)

1. *Set* $\mathbf{X} = (X_1, \ldots, X_k) = (0, \ldots, 0)$.
2. *For* $i = 1, \ldots, n$ *draw* $Z_i \sim \mathbf{p}$ *and set* $X_{Z_i} = X_{Z_i} + 1$.
3. *Return* $\mathbf{X}$.

From Property 2, the conditional densities of a multinomially distributed vector are binomial. Hence, the following algorithm can be used to generate a random vector $\mathbf{X} = (X_1, \ldots, X_k) \sim \mathsf{Mnom}(n, \mathbf{p})$.

Algorithm 4.70 ($\mathsf{Mnom}(n, \mathbf{p})$ Generator)

1. *Set* $s = 0$, $q = 1$, *and* $t = 1$.
2. *While* $t \leqslant k$ *generate*

$$X_t \sim \mathsf{Bin}\Big(n - s,\ \frac{p_t}{q}\Big),$$

and set $s = s + X_t$, $q = q - p_t$, *and* $t = t + 1$.

3. *Output* $\mathbf{X} = (X_1, \ldots, X_k)$.

4.3.3 Multivariate Normal Distribution

The **standard multivariate normal** or **standard multivariate Gaussian** distribution in n dimensions has pdf

$$f(\mathbf{x}) = \frac{1}{\sqrt{(2\pi)^n}}\, \mathrm{e}^{-\frac{1}{2}\,\mathbf{x}^\top \mathbf{x}}, \quad \mathbf{x} \in \mathbb{R}^n\ . \tag{4.11}$$

All marginals of $\mathbf{X} = (X_1, \ldots, X_n)^\top$ are iid standard normal random variables.

Suppose that $\mathbf{Z}$ has an m-dimensional standard normal distribution. If A is an $n \times m$ matrix and $\boldsymbol{\mu}$ is an $n \times 1$ vector, then the affine transformation

$$\mathbf{X} = \boldsymbol{\mu} + A\mathbf{Z}$$

is said to have a **multivariate normal** or **multivariate Gaussian** distribution with **mean vector** $\boldsymbol{\mu}$ and **covariance matrix** $\Sigma = AA^\top$. We write the distribution as $\mathsf{N}(\boldsymbol{\mu}, \Sigma)$.

The covariance matrix Σ is always symmetric and positive semidefinite. When the matrix Σ is singular (that is, $\det(\Sigma) = 0$), the distribution of $\mathbf{X}$ is singular with respect to the Lebesgue measure on $\mathbb{R}^n$. When A is of full rank (that is, $\text{rank}(A) = \min\{m, n\}$) the covariance matrix Σ is positive definite. In this case Σ has an inverse and the distribution of $\mathbf{X}$ has pdf

$$f(\mathbf{x}; \boldsymbol{\mu}, \Sigma) = \frac{1}{\sqrt{(2\pi)^n \det(\Sigma)}} \, \mathrm{e}^{-\frac{1}{2}(\mathbf{x}-\boldsymbol{\mu})^\top \Sigma^{-1}(\mathbf{x}-\boldsymbol{\mu})}, \quad \mathbf{x} \in \mathbb{R}^n . \tag{4.12}$$

The matrix $\Lambda = \Sigma^{-1}$ is called the **precision matrix**. A random vector $\mathbf{X}$ is thus multivariate normal with precision matrix Λ and mean vector $\boldsymbol{\mu}$ if and only if the logarithm of its pdf $\widetilde{f}(\mathbf{x}; \boldsymbol{\mu}, \Lambda) = f(\mathbf{x}; \boldsymbol{\mu}, \Sigma)$ satisfies

$$\ln \widetilde{f}(\mathbf{x}; \boldsymbol{\mu}, \Lambda) = -\frac{1}{2}\left(\mathbf{x}^\top \Lambda \mathbf{x} - 2\mathbf{x}^\top \Lambda \boldsymbol{\mu}\right) + \text{constant}, \quad \mathbf{x} \in \mathbb{R}^n . \tag{4.13}$$

The multivariate normal distribution is a natural extension of the normal distribution (see Section 4.2.11) and plays a correspondingly important role in multivariate statistics. This multidimensional counterpart also has the property that any affine combination of independent multivariate normal random variables is again multivariate normal. A graph of the standard normal pdf for the two-dimensional case is given in Figure 4.25.

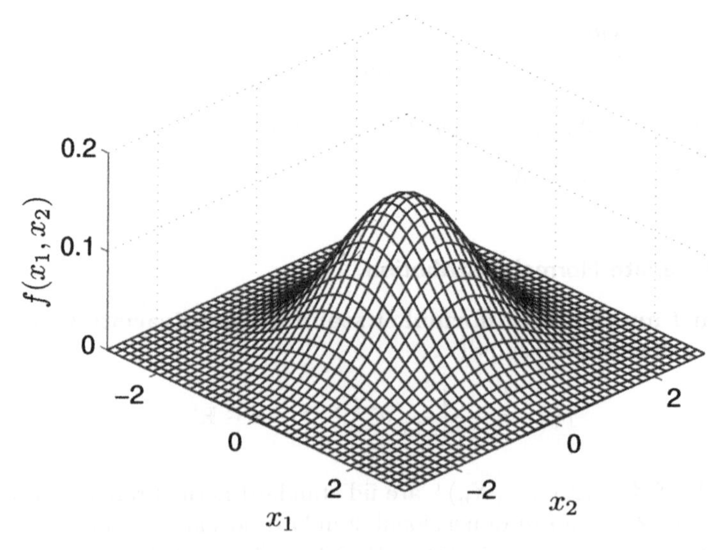

Figure 4.25 The standard multivariate normal pdf in two dimensions.

Table 4.28 Moment properties of the $\mathsf{N}(\boldsymbol{\mu}, \Sigma)$ distribution.

Property		Condition
Expectation vector	$\boldsymbol{\mu}$	
Covariance matrix	Σ	
Moment generating function	$\mathbb{E}\, \mathrm{e}^{\mathbf{t}^\top \mathbf{X}} = \exp\left(\mathbf{t}^\top \boldsymbol{\mu} + \frac{1}{2}\mathbf{t}^\top \Sigma\, \mathbf{t}\right)$	$\mathbf{t} \in \mathbb{R}^n$

Other properties and relations are:

1. *Affine combinations*: Let $\mathbf{X}_1, \mathbf{X}_2, \ldots, \mathbf{X}_r$ be independent m_i-dimensional normal variables, with $\mathbf{X}_i \sim \mathsf{N}(\boldsymbol{\mu}_i, \Sigma_i)$, $i = 1, \ldots, r$. Let $\mathbf{a}$ be an $n \times 1$ vector and let each A_i be an $n \times m_i$ matrix for $i = 1, \ldots, r$. Then,

$$\mathbf{a} + \sum_{i=1}^{r} A_i\, \mathbf{X}_i \sim \mathsf{N}\left(\mathbf{a} + \sum_{i=1}^{r} A_i\, \boldsymbol{\mu}_i,\ \sum_{i=1}^{r} A_i\, \Sigma_i\, A_i^\top\right).$$

   In other words, any affine combination of independent multivariate normal random variables is again multivariate normal.

2. *Standardization (whitening)*: A particular case of the affine combinations property is the following. Suppose $\mathbf{X} \sim \mathsf{N}(\boldsymbol{\mu}, \Sigma)$ is an n-dimensional normal random variable with $\det(\Sigma) > 0$. Let A be the Cholesky factor of the matrix Σ. That is, A is an $n \times n$ lower triangular matrix of the form

$$A = \begin{pmatrix} a_{11} & 0 & \cdots & 0 \\ a_{21} & a_{22} & \cdots & 0 \\ \vdots & \vdots & & \vdots \\ a_{n1} & a_{n2} & \cdots & a_{nn} \end{pmatrix}, \tag{4.14}$$

   and such that $\Sigma = AA^\top$. It follows that

$$A^{-1}(\mathbf{X} - \boldsymbol{\mu}) \sim \mathsf{N}(\mathbf{0}, I).$$

   Note that the Cholesky factor can be obtained efficiently via the *Cholesky square root method* (see Section D.3). ☞ 706

3. *Marginal distributions*: Let $\mathbf{X}$ be an n-dimensional normal variable, with $\mathbf{X} \sim \mathsf{N}(\boldsymbol{\mu}, \Sigma)$. Separate the vector $\mathbf{X}$ into a part of size p and one of size $q = n - p$ and, similarly, partition the mean vector and covariance matrix:

$$\mathbf{X} = \begin{pmatrix} \mathbf{X}_p \\ \mathbf{X}_q \end{pmatrix}, \quad \boldsymbol{\mu} = \begin{pmatrix} \boldsymbol{\mu}_p \\ \boldsymbol{\mu}_q \end{pmatrix}, \quad \Sigma = \begin{pmatrix} \Sigma_p & \Sigma_r \\ \Sigma_r^\top & \Sigma_q \end{pmatrix}, \tag{4.15}$$

   where Σ_p is the upper left $p \times p$ corner of Σ, Σ_q is the lower right $q \times q$ corner of Σ, and Σ_r is the $p \times q$ upper right block of Σ. Then, the distributions of the marginal vectors $\mathbf{X}_p$ and $\mathbf{X}_q$ are also multivariate normal, with $\mathbf{X}_p \sim \mathsf{N}(\boldsymbol{\mu}_p, \Sigma_p)$ and $\mathbf{X}_q \sim \mathsf{N}(\boldsymbol{\mu}_q, \Sigma_q)$.

Note that an arbitrary selection of p and q elements can be achieved by first performing a linear transformation $\mathbf{Z} = A\mathbf{X}$, where A is the $n \times n$ *permutation matrix* that appropriately rearranges the elements in $\mathbf{X}$.

4. *Conditional distributions*: Suppose we again have an n-dimensional vector $\mathbf{X} \sim \mathsf{N}(\boldsymbol{\mu}, \Sigma)$, partitioned as for the marginal distribution property above, but with $\det(\Sigma) > 0$. Then, we have the conditional distributions

$$(\mathbf{X}_p \,|\, \mathbf{X}_q = \mathbf{x}_q) \sim \mathsf{N}(\boldsymbol{\mu}_p + \Sigma_r \Sigma_q^{-1}(\mathbf{x}_q - \boldsymbol{\mu}_q),\; \Sigma_p - \Sigma_r \Sigma_q^{-1} \Sigma_r^\top),$$

and

$$(\mathbf{X}_q \,|\, \mathbf{X}_p = \mathbf{x}_p) \sim \mathsf{N}(\boldsymbol{\mu}_q + \Sigma_r^\top \Sigma_p^{-1}(\mathbf{x}_p - \boldsymbol{\mu}_p),\; \Sigma_q - \Sigma_r^\top \Sigma_p^{-1} \Sigma_r).$$

As with the marginals property, arbitrary conditioning can be achieved by first permuting the elements of $\mathbf{X}$ by way of an affine transformation using a permutation matrix.

The conditional distributions can also be conveniently characterized via the precision matrix. In particular, if the latter is decomposed as

$$\Lambda = \begin{pmatrix} \Lambda_p & \Lambda_r \\ \Lambda_r^\top & \Lambda_q \end{pmatrix}, \tag{4.16}$$

then the logarithm of the conditional pdf of $\mathbf{X}_p$ given $\mathbf{X}_q = \mathbf{x}_q$ is given by

$$\begin{aligned} &-\frac{1}{2}\left[(\mathbf{x}_p^\top \mathbf{x}_q^\top)\begin{pmatrix} \Lambda_p & \Lambda_r \\ \Lambda_r^\top & \Lambda_q \end{pmatrix}\begin{pmatrix} \mathbf{x}_p - 2\boldsymbol{\mu}_p \\ \mathbf{x}_q - 2\boldsymbol{\mu}_q \end{pmatrix}\right] + \text{constant} \\ &= -\frac{1}{2}\left[\mathbf{x}_p^\top \Lambda_p \mathbf{x}_p - 2\mathbf{x}_p^\top(\Lambda_p \boldsymbol{\mu}_p + \Lambda_r \boldsymbol{\mu}_q - \Lambda_r \mathbf{x}_q)\right] + \text{constant} \\ &= -\frac{1}{2}\left[\mathbf{x}_p^\top \Lambda_p \mathbf{x}_p - 2\mathbf{x}_p^\top \Lambda_p \boldsymbol{\mu}_{p\,|\,q}\right] + \text{constant} \end{aligned}$$

for some p-dimensional vector $\boldsymbol{\mu}_{p\,|\,q}$. This shows that conditional on $\mathbf{X}_q = \mathbf{x}_q$ the random vector $\mathbf{X}_p$ has a multivariate Gaussian distribution with precision matrix Λ_p and mean vector $\boldsymbol{\mu}_{p\,|\,q}$. Moreover, this mean vector can be found by solving the linear equation

$$\Lambda_p(\boldsymbol{\mu}_{p\,|\,q} - \boldsymbol{\mu}_p) = \Lambda_r(\boldsymbol{\mu}_q - \mathbf{x}_q). \tag{4.17}$$

5. *Square*: If $\mathbf{X} = (X_1, \ldots, X_n)^\top \sim \mathsf{N}(\mathbf{0}, I)$, then $\mathbf{X}^\top\mathbf{X} \sim \mathsf{Gamma}(n/2, 1/2) \equiv \chi^2_n$. More generally, combining with the standardization property, if $\mathbf{X} \sim \mathsf{N}(\boldsymbol{\mu}, \Sigma)$ is n-dimensional normal with $\det(\Sigma) > 0$, then $(\mathbf{X}-\boldsymbol{\mu})^\top \Sigma^{-1}(\mathbf{X}-\boldsymbol{\mu}) \sim \chi^2_n$.

Property 2 is the key to generating a multivariate normal random vector $\mathbf{X} \sim \mathsf{N}(\boldsymbol{\mu}, \Sigma)$, and leads to the following algorithm.

Algorithm 4.71 ($\mathsf{N}(\boldsymbol{\mu}, \Sigma)$ Generator)

1. *Derive the Cholesky decomposition* $\Sigma = AA^\top$.
2. *Generate* $Z_1, \ldots, Z_n \overset{\text{iid}}{\sim} \mathsf{N}(0,1)$ *and let* $\mathbf{Z} = (Z_1, \ldots, Z_n)^\top$.
3. *Output* $\mathbf{X} = \boldsymbol{\mu} + A\mathbf{Z}$.

4.3.4 Multivariate Student's t Distribution

The **multivariate Student's t** distribution in n dimensions has pdf

$$f(\mathbf{x};\nu) = \frac{\Gamma\left(\frac{\nu+n}{2}\right)}{(\pi\nu)^{n/2}\,\Gamma\left(\frac{\nu}{2}\right)}\left(1+\frac{1}{\nu}\mathbf{x}^\top\mathbf{x}\right)^{-\frac{\nu+n}{2}}, \quad \mathbf{x}\in\mathbb{R}^n, \tag{4.18}$$

where $\nu \geqslant 0$ is the **degrees of freedom** or **shape** parameter. We write the distribution as $\mathbf{t}_\nu$. Suppose $\mathbf{Y} = (Y_1,\ldots,Y_m)^\top$ has an m-dimensional $\mathbf{t}_\nu$ distribution. If A is an $n\times m$ matrix and $\boldsymbol{\mu}$ is an $n\times 1$ vector, then the affine transformation

$$\mathbf{X} = \boldsymbol{\mu} + A\mathbf{Y}$$

is said to have a **multivariate Student's** or **multivariate t** distribution with **mean vector $\boldsymbol{\mu}$** and **scale matrix Σ** $= AA^\top$. We write the distribution as $\mathbf{t}_\nu(\boldsymbol{\mu},\Sigma)$. If Σ is positive definite (see Section 4.3.3), then $\mathbf{X}$ has pdf

$$f(\mathbf{x};\nu,\boldsymbol{\mu},\Sigma) = \frac{\Gamma\left(\frac{\nu+n}{2}\right)}{(\pi\nu)^{n/2}\,\Gamma\left(\frac{\nu}{2}\right)\sqrt{\det(\Sigma)}}\left(1+\frac{1}{\nu}(\mathbf{x}-\boldsymbol{\mu})^\top\Sigma^{-1}(\mathbf{x}-\boldsymbol{\mu})\right)^{-\frac{\nu+n}{2}}. \tag{4.19}$$

The multivariate t distribution is a *radially symmetric* extension of the univariate t distribution and plays an important role as a proposal distribution in Markov chain Monte Carlo algorithms and Bayesian statistical modeling; see Example 6.2. ☞ 231
In particular, it arises as the posterior distribution of the mean of a multivariate normal distribution [21]. Note that an alternative *product form* extension of the multivariate t distribution is also used in, for example, kernel density estimation [30, Page 103]. In this alternative extension, the pdf of the multivariate t distribution in n dimensions is defined as the product of n univariate pdfs: $\prod_{i=1}^n f(x_i;\nu)$, where $f(x_i;\nu)$ is the Student's t pdf in (4.10).

Table 4.29 Moment properties of the $\mathbf{t}_\nu(\boldsymbol{\mu},\Sigma)$ distribution.

Property		Condition
Expectation vector	$\boldsymbol{\mu}$	
Covariance matrix	$\frac{\nu}{\nu-2}\Sigma$	$\nu > 2$
Characteristic Func.	$e^{\mathrm{i}\,\mathbf{t}^\top\boldsymbol{\mu}}\,\frac{\lVert\sqrt{\nu}\,\Sigma^{1/2}\mathbf{t}\rVert^{\nu/2}}{2^{\nu/2-1}\Gamma(\nu/2)}\,K_{\nu/2}(\lVert\sqrt{\nu}\,\Sigma^{1/2}\mathbf{t}\rVert)$	$\mathbf{t}\in\mathbb{R}^n$

Here, $K_\nu(x)$ denotes the *modified Bessel function of the second kind.* Other properties and relations are: ☞ 717

1. *Special cases*: Let $\mathbf{X}\sim\mathbf{t}_\nu(\boldsymbol{\mu},\Sigma)$, then

$$\mathbf{X} \xrightarrow{d} \mathbf{Z} \sim \mathsf{N}(\boldsymbol{\mu},\Sigma) \quad \text{as } \nu\to\infty\,.$$

   If $\nu = 1$, then (4.18) gives the pdf of the multivariate **radially symmetric Cauchy** distribution.

2. *Normal distribution:* Let $\mathbf{Z} \sim \mathsf{N}(\mathbf{0}, I)$ and $S \sim \mathsf{Gamma}(\nu/2, 1/2)$ be independent. Then,
$$\mathbf{X} = \sqrt{\frac{\nu}{S}}\, \mathbf{Z} \sim \mathbf{t}_\nu \,.$$

3. *Wishart distribution:* Let V be an $n \times n$ random matrix with a $\mathsf{Wishart}(\nu + n - 1, I)$ distribution. If $\mathbf{Y} \sim \mathsf{N}(\mathbf{0}, I)$, then
$$\mathbf{X} = \sqrt{\nu}\left(V^{1/2}\right)^{-1} \mathbf{Y} \sim \mathbf{t}_\nu \,,$$
where $V^{1/2}$ is a symmetric matrix such that $V^{1/2}V^{1/2} = V$.

4. *Marginal distributions:* If $\mathbf{X} \sim \mathbf{t}_\nu(\boldsymbol{\mu}, \Sigma)$ and the vector $\mathbf{X}$ is decomposed as in (4.15), then
$$\mathbf{X}_p \sim \mathbf{t}_\nu(\boldsymbol{\mu}_p, \Sigma_p) \quad \text{and} \quad \mathbf{X}_q \sim \mathbf{t}_\nu(\boldsymbol{\mu}_q, \Sigma_q)\,.$$

5. *Product moments:* If $\mathbf{X} \sim \mathbf{t}_\nu$, then [21, Page 11]
$$\mathbb{E}\prod_{i=1}^n |X_i|^{r_i} = \frac{\nu^{\frac{r}{2}}\Gamma\left(\frac{\nu - r}{2}\right)\prod_{i=1}^n \Gamma\left(\frac{r_i+1}{2}\right)}{\pi^{\frac{n}{2}}\,\Gamma\left(\frac{\nu}{2}\right)}, \qquad r = \sum_{j=1}^n r_j < \nu,\ r_j \geqslant 0\,.$$

Generation from the multivariate t distribution follows from its relation to the multivariate normal distribution in Property 2 above.

Algorithm 4.72 ($\mathbf{t}_\nu(\boldsymbol{\mu}, \Sigma)$ Generator)

1. *Draw the random column vector* $\mathbf{Z} \sim \mathsf{N}(\mathbf{0}, I)$.
2. *Draw* $S \sim \mathsf{Gamma}\left(\frac{\nu}{2}, \frac{1}{2}\right) \equiv \chi^2_\nu$.
3. *Compute* $\mathbf{Y} = \sqrt{\frac{\nu}{S}}\,\mathbf{Z}$.

☞ 706
4. *Return* $\mathbf{X} = \boldsymbol{\mu} + A\mathbf{Y}$, *where A is the Cholesky factor of Σ, so that* $AA^\top = \Sigma$.

4.3.5 Wishart Distribution

Let $\mathbf{x} = (x_{ij}, i, j = 1, \ldots, n)$ denote an $n \times n$ positive definite matrix; the latter property is written as $\mathbf{x} \succ 0$. Note that $\mathbf{x}$ is necessarily symmetric. The **Wishart** distribution on the space of such matrices has pdf
$$f(\mathbf{x}) = c^{-1}\, \det(\mathbf{x})^{(\nu - n - 1)/2} \exp\left(-\frac{1}{2}\mathrm{tr}\left(\Sigma^{-1}\mathbf{x}\right)\right), \quad \mathbf{x} \succ 0\,,$$
where $\nu \geqslant n$ is the **number of degrees of freedom**, $\Sigma \succ 0$ is an $n \times n$ covariance matrix, and the normalization constant c is
$$c = \det(\Sigma)^{\nu/2}\, 2^{\nu n/2} \pi^{n(n-1)/4} \prod_{j=1}^n \Gamma\left(\frac{\nu - j + 1}{2}\right).$$

We denote the distribution by $\mathsf{Wishart}(\nu, \Sigma)$.

The Wishart distribution is closely related to the multivariate normal distribution. In particular, if $\mathbf{Y}_1, \ldots, \mathbf{Y}_r \sim_{\text{iid}} \mathsf{N}(\mathbf{0}, \Sigma)$ are n-dimensional normal variables with $\det(\Sigma) > 0$, then

$$\sum_{k=1}^{r} \mathbf{Y}_k \mathbf{Y}_k^\top \sim \mathsf{Wishart}(r, \Sigma) .$$

In Bayesian statistics, the Wishart distribution is the conjugate prior for Σ^{-1} of a $\mathsf{N}(\boldsymbol{\mu}, \Sigma)$ random variable, when $\boldsymbol{\mu}$ is known. ☞ 675

Table 4.30 Moment properties of the $\mathsf{Wishart}(\nu, \Sigma)$ distribution.

Property		Condition
Expectation matrix	$\nu\Sigma$	
Moment gen. function	$\mathbb{E}\, e^{\text{tr}(T\mathbf{X})} = \det(I - 2\,\Sigma\, T)^{-\nu/2}$	$(\Sigma^{-1} - 2T) \succ 0$

Other properties and relations are:

1. *Sums*: If $\mathbf{X}_1 \sim \mathsf{Wishart}(\nu_1, \Sigma)$ and $\mathbf{X}_2 \sim \mathsf{Wishart}(\nu_2, \Sigma)$ are independent, then

$$\mathbf{X}_1 + \mathbf{X}_2 \sim \mathsf{Wishart}(\nu_1 + \nu_2, \Sigma) .$$

2. χ^2_r *distribution*: $\mathsf{Wishart}(r, 1) \equiv \chi^2_r$, if r is a positive integer. Thus, the Wishart distribution is the multivariate analogue of the χ^2_r distribution.

3. *Scaling*: If $\mathbf{X} \sim \mathsf{Wishart}(\nu, \Sigma)$ is an $n \times n$ random matrix and A is an $n \times m$ constant matrix, then $A^\top \mathbf{X} A$ is an $m \times m$ random matrix with

$$A^\top \mathbf{X} A \sim \mathsf{Wishart}(\nu, A^\top \Sigma A) .$$

The special case where $A = \mathbf{a}$ is an $n \times 1$ vector ($m = 1$) and $\nu = r$ is a positive integer gives

$$\frac{\mathbf{a}^\top \mathbf{X}\, \mathbf{a}}{\mathbf{a}^\top \Sigma\, \mathbf{a}} \sim \chi^2_r .$$

4. *Bartlett decomposition*: Let $\Sigma = CC^\top$ be the Cholesky factorization of the $n \times n$ matrix Σ, and let A be an $n \times n$ lower diagonal matrix of the form given in (4.14), such that $a_{ij} \sim_{\text{iid}} \mathsf{N}(0, 1)$ for all $j < i$ and $a_{ii} = \sqrt{Y_i}$, where $Y_i \sim \chi^2_{r-i+1}$, $i = 1, \ldots, n$ are independent. Then, we have

$$\mathbf{X} = CAA^\top C^\top \sim \mathsf{Wishart}(r, \Sigma) .$$

Generation from the $\mathsf{Wishart}(r, \Sigma)$ distribution for small integer values of $r \geqslant n$ follows directly from its defining property.

Algorithm 4.73 (Wishart(r, Σ) Generator)

1. *Draw the random column vectors* $\mathbf{Y}_1, \ldots, \mathbf{Y}_r \stackrel{\text{iid}}{\sim} \mathsf{N}(\mathbf{0}, \Sigma)$.

2. *Return the matrix* $\mathbf{X} = \sum_{k=1}^{r} \mathbf{Y}_k \mathbf{Y}_k^\top$.

The above algorithm becomes inefficient for large values of r. An alternative and more efficient algorithm is based on the Bartlett decomposition; see Property 4 and [29].

Algorithm 4.74 (Wishart(r, Σ) Generator via Bartlett Decomposition)

☞ 706

1. *Compute C in the Cholesky factorization $\Sigma = CC^\top$ of the $n \times n$ matrix Σ.*

2. *Generate the lower diagonal matrix A in (4.14), with $a_{ij} \sim_{\text{iid}} \mathsf{N}(0,1)$ for all $j < i$ and $a_{ii} = \sqrt{Y_i}$, where $Y_i \sim \chi^2_{r-i+1}$, $i = 1, \ldots, n$ are independent.*

3. *Output the matrix $\mathbf{X} = CAA^\top C^\top$.*

Further Reading

Further information on discrete and continuous distributions may be found in [9, 15] and [9, 16, 17, 18, 20], respectively. Phase-type distributions and their applications are discussed in [26]. Details on stable distributions are given in [11, 12, 27, 32]. For a comprehensive reference on random variable generation algorithms, see [8].

REFERENCES

1. J. H. Ahrens and U. Dieter. Computer methods for sampling from gamma, beta, Poisson, and binomial distributions. *Computing*, 12(3):223–246, 1974.
2. J. H. Ahrens and U. Dieter. Sampling from binomial and Poisson distributions: A method with bounded computation times. *Computing*, 25(3):193–208, 1980.
3. A. C. Atkinson. The computer generation of Poisson random variables. *Journal of the Royal Statistical Society, Series C*, 28(1):29–35, 1979.
4. R. W. Bailey. Polar generation of random variates with the t-distribution. *Mathematics of Computation*, 62(206):779–781, 1994.
5. D. J. Best. A note on gamma variate generators with shape parameters less than unity. *Computing*, 30(2):185–188, 1983.
6. J. M. Chambers, C. L. Mallows, and B. W. Stuck. A method for simulating stable random variables. *Journal of the American Statistics Association*, 71(354):340–344, 1976.
7. R. C. H. Cheng and G. M. Feast. Some simple gamma variate generators. *Computing*, 28(3):290–295, 1979.
8. L. Devroye. *Non-Uniform Random Variate Generation.* Springer-Verlag, New York, 1986.
9. M. Evans, N. Hastings, and B. Peacock. *Statistical Distributions.* John Wiley & Sons, New York, second edition, 1993.

10. P. J. Forrester and S. O. Warnaar. The importance of the Selberg integral. *Bulletin of the American Mathematical Society*, 45(4):489–534, 2008.

11. B. V. Gnedenko and A. N. Kolmogorov. *Limit Distributions for Sums of Independent Random Variables.* Addison-Wesley, Reading, Massachusetts, 1954.

12. P. Hall. A comedy of errors: The canonical form for a stable characteristic function. *The Bulletin of the London Mathematical Society*, 13(1):23–27, 1981.

13. C. C. Heyde. On a property of the lognormal distribution. *Journal of the Royal Statistical Society, Series B*, 25(2):392–393, 1963.

14. W. Hörmann. A simple generator for the t distribution. *Computing*, 81(4):317–322, 2007.

15. N. L. Johnson and S. Kotz. *Distributions in Statistics: Discrete Distributions.* Houghton Mifflin Company, New York, 1969.

16. N. L. Johnson and S. Kotz. *Distributions in Statistics: Continuous Univariate Distributions, Volume 1.* Houghton Mifflin Company, New York, 1970.

17. N. L. Johnson and S. Kotz. *Distributions in Statistics: Continuous Univariate Distributions, Volume 2.* Houghton Mifflin Company, New York, 1970.

18. N. L. Johnson and S. Kotz. *Distributions in Statistics: Continuous Multivariate Distributions.* John Wiley & Sons, New York, 1972.

19. A. J. Kinderman, J. F. Monahan, and J. G. Ramage. Computer methods for sampling from Student's t distribution. *Mathematics of Computation*, 31(140):1009–1018, 1977.

20. C. Kleiber and S. Kotz. *Statistical Size Distributions in Economics and Actuarial Sciences.* John Wiley & Sons, New York, 2003.

21. S. Kotz and S. Nadarajah. *Multivariate t Distributions and Their Applications.* Cambridge University Press, Cambridge, 2004.

22. P. L'Ecuyer and R. Simard. Inverting the symmetrical beta distribution. *ACM Transactions on Mathematical Software*, 32(4):509–520, 2006.

23. G. Marsaglia. Improving the polar method for generating a pair of normal random variables. Technical Report D1-82-0203, Boeing Scientific Research Laboratories, September 1962.

24. G. Marsaglia and T. A. Bray. A convenient method for generating normal variables. *SIAM Review*, 6(3):260–264, 1964.

25. G. Marsaglia and W. Tsang. A simple method for generating gamma variables. *ACM Transactions on Mathematical Software*, 26(3):363–372, 2000.

26. M. F. Neuts. *Matrix-Geometric Solutions in Stochastic Models: An Algorithmic Approach.* Dover Publications, New York, 1981. Unabridged and corrected edition from 1994.

27. J. P. Nolan. *Stable Distributions - Models for Heavy Tailed Data.* Birkhäuser, Boston, 2009. In progress, Chapter 1 online at `http://academic2.american.edu/~jpnolan`.

28. R. Y. Rubinstein and D. P. Kroese. *Simulation and the Monte Carlo Method.* John Wiley & Sons, New York, second edition, 2007.

29. W. B. Smith and R. R. Hocking. Algorithm AS 53: Wishart variate generator. *Journal of the Royal Statistical Society, Series C*, 21(3):341–345, 1972.

30. M. P. Wand and M. C. Jones. *Kernel Smoothing.* Chapman & Hall, London, 1995.

31. R. Weron. On the Chambers–Mallows–Stuck method for simulating skewed stable random variables. *Statistics and Probability Letters*, 28(2):165–171, 1996.

32. V. M. Zolotarev. *One-Dimensional Stable Distributions.* American Mathematical Society, Providence, Rhode Island, 1986.

CHAPTER 5

RANDOM PROCESS GENERATION

This chapter lists the major random processes used in Monte Carlo simulation, along with their main properties and how to generate them. Further background on stochastic processes is given in Sections A.9–A.13 of the appendix. The following processes are discussed. ☞ 626

Handbook of Monte Carlo Methods. By D. P. Kroese, T. Taimre, Z. I. Botev

5.1 GAUSSIAN PROCESSES

A real-valued stochastic process $\{\widetilde{X}_t, t \in \mathscr{T}\}$ is said to be a **Gaussian process** if all its finite-dimensional distributions are Gaussian (normal); that is, if $\mathbf{X} = (X_1, \ldots, X_n) = (\widetilde{X}_{t_1}, \ldots, \widetilde{X}_{t_n})^\top$ has a multivariate Gaussian distribution for any choice of n and $t_1, \ldots, t_n \in \mathscr{T}$, or equivalently, if any linear combination $\sum_{i=1}^n b_i \widetilde{X}_{t_i}$ has a normal distribution. ☞ 143

The probability distribution of a Gaussian process is determined completely by its **expectation function**

$$\widetilde{\mu}_t = \mathbb{E}\widetilde{X}_t, \quad t \in \mathscr{T}$$

and **covariance function**

$$\widetilde{\Sigma}_{s,t} = \mathrm{Cov}(\widetilde{X}_s, \widetilde{X}_t), \quad s, t \in \mathscr{T} .$$

A **zero-mean** Gaussian process is one for which $\widetilde{\mu}_t = 0$ for all t.

Gaussian processes can be thought of as generalizations of Gaussian random vectors. To generate a realization of a Gaussian process with expectation function $(\widetilde{\mu}_t)$ and covariance function $(\widetilde{\Sigma}_{s,t})$ at times $t_1, \ldots, t_n$ we can simply sample a multivariate normal random vector $\mathbf{X} = (X_1, \ldots, X_n)^\top = (\widetilde{X}_{t_1}, \ldots, \widetilde{X}_{t_n})^\top$. As such, the fundamental generation method is the same as given in Algorithm 4.71. ☞ 146

Algorithm 5.1 (Gaussian Process Generator)

1. *Construct the mean vector* $\boldsymbol{\mu} = (\mu_1, \ldots, \mu_n)^\top$ *and covariance matrix* $\Sigma = (\Sigma_{ij})$ *by setting* $\mu_i = \widetilde{\mu}_{t_i}, i = 1, \ldots, n$ *and* $\Sigma_{ij} = \widetilde{\Sigma}_{t_i, t_j}$, $i, j = 1, \ldots, n$.
2. *Derive the Cholesky decomposition* $\Sigma = AA^\top$. ☞ 706
3. *Generate* $Z_1, \ldots, Z_n \overset{\text{iid}}{\sim} \mathsf{N}(0, 1)$. *Let* $\mathbf{Z} = (Z_1, \ldots, Z_n)^\top$.
4. *Output* $\mathbf{X} = \boldsymbol{\mu} + A\mathbf{Z}$.

It is sometimes useful to employ an "online" version of the above algorithm, especially when the length of the Gaussian vector $\mathbf{X}$ is not known in advance, for example when the length is a stopping time. Specifically, given a fixed sequence of times $t_1, t_2, \ldots$, denote $\mathbf{X}_n = (X_1, \ldots, X_n)^\top = (\widetilde{X}_{t_1}, \ldots, \widetilde{X}_{t_n})^\top$. Write $\Sigma_n = \mathrm{Cov}(\mathbf{X}_n, \mathbf{X}_n)$ and let A_n be the corresponding lower-triangular Cholesky matrix, $n = 1, 2, \ldots$. We can partition Σ_{n+1} and A_{n+1} as

$$\Sigma_{n+1} = \begin{pmatrix} \Sigma_n & \mathbf{b}_n \\ \mathbf{b}_n^\top & b_{n+1} \end{pmatrix} \quad \text{and} \quad A_{n+1} = \begin{pmatrix} A_n & \mathbf{0} \\ \mathbf{a}_n^\top & a_{n+1} \end{pmatrix} ,$$

for some column vectors $\mathbf{b}_n, \mathbf{a}_n$, and constants b_{n+1} and a_{n+1}. It follows that the Cholesky matrix can be updated row by row, where the $(n+1)$-st row is obtained by forward substitution from the linear equation

$$A_n \mathbf{a}_n = \mathbf{b}_n \tag{5.1}$$

and

$$a_{n+1} = \sqrt{b_{n+1} - \mathbf{a}_n^\top \mathbf{a}_n} . \tag{5.2}$$

A Gaussian vector can also be generated from its *precision matrix* $\Lambda = \Sigma^{-1}$ instead of its covariance matrix Σ. Namely, if $DD^\top$ is the Cholesky factorization of Λ, and if $\mathbf{Y}$ satisfies $\mathbf{Z} = D\mathbf{Y}$, where $\mathbf{Z}$ is a vector of iid $\mathsf{N}(0,1)$ random variables, then $\mathbf{Y}$ is a zero-mean multivariate normal vector with covariance matrix ☞ 144

$$\mathbb{E}\mathbf{Y}\mathbf{Y}^\top = D^{-1}\mathbb{E}\mathbf{Z}\mathbf{Z}^\top(D^{-1})^\top = (D^\top D)^{-1} = (\Lambda^\top)^{-1} = (\Lambda^{-1})^\top = \Sigma \,.$$

This leads to the following algorithm.

Algorithm 5.2 (Gaussian Process Generator Using a Precision Matrix)

1. *Derive the Cholesky decomposition* $\Lambda = DD^\top$ *of the precision matrix.*
2. *Generate* $Z_1, \ldots, Z_n \overset{\text{iid}}{\sim} \mathsf{N}(0,1)$. *Let* $\mathbf{Z} = (Z_1, \ldots, Z_n)^\top$.
3. *Solve* $\mathbf{Y}$ *from* $\mathbf{Z} = D\mathbf{Y}$, *using forward substitution.*
4. *Output* $\mathbf{X} = \boldsymbol{\mu} + \mathbf{Y}$.

As the Cholesky decomposition of a general n-dimensional positive definite matrix takes $\mathcal{O}(n^3)$ floating point operations, the generation of large-dimensional Gaussian vectors becomes cumbersome for large n, unless some extra structure is introduced that allows efficient square root factorization of the covariance or precision matrix. This is the case, for example, when the covariance matrix or the precision matrix is *sparse*. The Cholesky matrix can then be evaluated efficiently, for example via the band-Cholesky method; see Algorithm 5.33.

■ **EXAMPLE 5.1 (Gaussian Markov Random Field Generation)**

A **Gaussian Markov random field** can be viewed as Gaussian vector $\mathbf{X} = (X_1, \ldots, X_n)$ that is defined by means of an undirected graph $\mathcal{G} = (V, E)$, where the vertex set V corresponds to the indices $\{1, \ldots, n\}$ and the edge set E specifies the dependencies between the variables. Specifically, the (i,j)-th element of the precision matrix $\Lambda = (\lambda_{ij})$ is 0 if and only if $(i,j) \notin E$ (see also Page 208). A consequence of this construction is that the conditional distribution of each random variable X_i given all other random variables is equal to the conditional distribution of X_i given only its neighbors; more precisely,

$$(X_i \,|\, X_1, \ldots, X_n) \sim (X_i \,|\, X_j, j \in \mathcal{N}_i) \,,$$

where $\mathcal{N}_i = \{j : (i,j) \in E\}$ is the set of indices to which i is adjacent. Thus, if each vertex i has only a small number of adjacent vertices, then the precision matrix is typically sparse and the Gaussian vector can be generated efficiently.

An important application is found in image analysis, where the vertices correspond to positions on a grid $\{1, \ldots, m\} \times \{1, \ldots, m\}$ of $n = m^2$ points, as in Figure 5.1. In this particular case internal vertices have four neighbors, vertices on the edge have three neighbors and the corner vertices each have two neighbors.

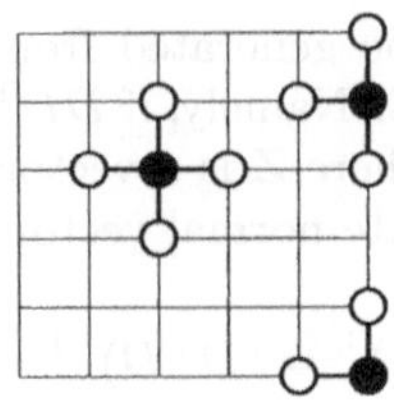

Figure 5.1 Adjacency graph for a Gaussian Markov random field.

Figure 5.2 depicts an outcome of a zero-mean Gaussian Markov random field for a 200×200 grid. Note that each of the 200^2 rows of the precision matrix has at most 5 elements; so that the matrix is very sparse. For each row/vertex i the diagonal element of the precision matrix is taken as $\lambda_{ii} = 1$; and $\lambda_{ij} = -0.25$ for each neighboring element j of i.

Figure 5.2 A realization of a Gaussian Markov random field.

The following MATLAB code is used.

```
%gp_sparsechol.m
m = 200; d1 = 1; d2 = -0.25; %at most d1/4 in absolute value
nels = m*(5*m-4);
%preallocate memory to form sparse precision matrix
a = zeros(1, nels); b = zeros(1,nels); c = zeros(1,nels);
%compute the links and weights for the precision matrix
k=0;
for i=1:m
```

```
    for j=1:m
        A = findneigh(i,j,m);
        nnb = size(A,1);
        for h=1:nnb
            a(k+h)= ij2k(i,j,m);
            b(k+h)= ij2k(A(h,1),A(h,2),m);
            if h==1
                c(k+h) = d1;
            else
                c(k+h) = d2;
            end
        end
        k = k+nnb;
    end
end
Lambda = sparse(a,b,c,m^2,m^2); %Construct the precision matrix
D = chol(Lambda,'lower'); %calculate the cholesky matrix
Z = randn(m^2,1);
x = D\Z;  % generate the Gaussian process
colormap gray, brighten(-0.2)
imagesc(reshape(x,m,m)) % plot the result
```

The function `findneigh.m` returns the set of neighboring sites to the (i, j)-th site on an $m \times m$ lattice.

```
%findneigh.m
function A = findneigh(i,j,m)
% find neighbors of the (i,j)-th site of an m by m grid
if i==1
    if j==1
        A = [1,1;1,2;2,1];
    elseif j==m
        A = [1,m;1,m-1;2,m];
    else
        A = [1,j;1,j-1;1,j+1;2,j];
    end
elseif i==m
    if j==1
        A = [m,1;m,2;m-1,1];
    elseif j==m
        A = [m,m;m,m-1;m-1,m];
    else
        A = [m,j;m,j-1;m,j+1;m-1,j];
    end
else
   if j==1
        A = [i,1;i,2;i-1,1;i+1,1];
    elseif j==m
```

```
        A = [i,m;i,m-1;i+1,m;i-1,m];
    else
        A = [i,j;i,j-1;i,j+1;i+1,j;i-1,j];
    end
end
```

```
%ij2k.m
function k = ij2k(i,j,m)
k = (i-1)*m + j;
```

The *conditional distributions* property of Gaussian vectors (Property 4 on Page 146) makes it easy to generate Gaussian Markov random fields in which some of the values are specified in advance. For example, if $\mathbf{X}$ is decomposed as

$$\mathbf{X} = \begin{pmatrix} \mathbf{X}_p \\ \mathbf{X}_q \end{pmatrix},$$

leading to a decomposition of the precision matrix as in (4.16), then, conditional on $\mathbf{X}_q = \mathbf{x}_q$, the vector $\mathbf{X}_p$ is again Gaussian with (sparse) precision matrix Λ_p and mean vector $\boldsymbol{\mu}_{p\,|\,q}$ that is solved from (4.17). Since this linear equation involves sparse matrices, it can be solved efficiently. The left pane of Figure 5.3 depicts the same Gaussian Markov random field as in Figure 5.2 but now conditioned upon the values at $(1, i), i = 1, \ldots, m$ being equal to 20. The right pane depicts the mean vector $\boldsymbol{\mu}_{p\,|\,q}$. The corresponding MATLAB program `gp_sparsechol_cond.m` can be found on the Handbook website.

Figure 5.3 Conditional Gaussian Markov random field and its conditional expectation.

5.1.1 Markovian Gaussian Processes

Let $\{\widetilde{X}_t, t \geqslant 0\}$ be a real-valued *Markovian* Gaussian process. Thus, in addition to being Gaussian, the process also satisfies the Markov property ☞ 628

$$(\widetilde{X}_{t+s} \,|\, \widetilde{X}_u, u \leqslant t) \sim (\widetilde{X}_{t+s} \,|\, \widetilde{X}_t) \quad \text{for all } s, t \geqslant 0 \,.$$

If the mean $(\widetilde{\mu}_t)$ and covariance function $(\widetilde{\Sigma}_{s,t})$ are known, then it is straightforward to generate realizations $(X_1, \ldots, X_n) = (\widetilde{X}_{t_1}, \ldots, \widetilde{X}_{t_n})$ of the process at any selection of times $0 \leqslant t_1 < \ldots < t_n$ by using the *conditional distributions* property of the multivariate normal distribution; see Property 4 on Page 146. Denote the expectation and variance of $X_i = \widetilde{X}_{t_i}$ by μ_i and $\sigma_{i,i}$, respectively, and let $\sigma_{i,i+1} = \text{Cov}(X_i, X_{i+1})$, $i = 1, \ldots, n-1$. Then, by the marginal distributions property (Property 3 on Page 145),

$$\begin{pmatrix} X_i \\ X_{i+1} \end{pmatrix} \sim \mathsf{N}\left(\begin{pmatrix} \mu_i \\ \mu_{i+1} \end{pmatrix}, \begin{pmatrix} \sigma_{i,i} & \sigma_{i,i+1} \\ \sigma_{i,i+1} & \sigma_{i+1,i+1} \end{pmatrix} \right) .$$

Hence, by the conditional distributions property we have

$$(X_{i+1} \,|\, X_i = x) \sim \mathsf{N}\left(\mu_{i+1} + \frac{\sigma_{i,i+1}}{\sigma_{i,i}}(x - \mu_i), \quad \sigma_{i+1,i+1} - \frac{\sigma_{i,i+1}^2}{\sigma_{i,i}} \right) .$$

This leads to the following algorithm.

Algorithm 5.3 (Generating a Markovian Gaussian Process)

1. *Draw* $Z \sim \mathsf{N}(0,1)$ *and set* $X_1 = \mu_1 + \sqrt{\sigma_{1,1}}\, Z$.
2. *For* $i = 1, \ldots, n-1$, *draw* $Z \sim \mathsf{N}(0,1)$ *and set*

$$X_{i+1} = \mu_{i+1} + \frac{\sigma_{i+1,i}}{\sigma_{i,i}}\,(X_i - \mu_i) + \sqrt{\sigma_{i+1,i+1} - \frac{\sigma_{i,i+1}^2}{\sigma_{i,i}}}\; Z \,.$$

The algorithm can be easily generalized to generate multidimensional Markovian Gaussian processes. In particular, let $\{\widetilde{\mathbf{X}}_t, t \geqslant 0\}$ be a d-dimensional Markovian Gaussian process with expectation function $\widetilde{\boldsymbol{\mu}}_t = \mathbb{E}\widetilde{\mathbf{X}}_t$, $t \geqslant 0$, and covariance function $\widetilde{\Sigma}_{s,t} = \text{Cov}(\widetilde{\mathbf{X}}_s, \widetilde{\mathbf{X}}_t)$, $s, t \geqslant 0$. The following algorithm generates realizations of the process at times $0 \leqslant t_1 < \ldots < t_n$.

Algorithm 5.4 (Generating a Multidimensional Markovian Gaussian Process)

1. *Draw* $\mathbf{Z} \sim \mathsf{N}(\mathbf{0}, I)$ *and set* $\widetilde{\mathbf{X}}_{t_1} = \widetilde{\boldsymbol{\mu}}_{t_1} + B\mathbf{Z}$, *where* B *is the (lower-triangular) Cholesky square root matrix of* $\widetilde{\Sigma}_{t_1,t_1}$.
2. *For* $k = 1, \ldots, n-1$,

   (a) *Compute the Cholesky square-root of*

   $$\widetilde{\Sigma}_{t_{k+1},t_{k+1}} - \widetilde{\Sigma}_{t_k,t_{k+1}} \widetilde{\Sigma}_{t_k,t_k}^{-1} \widetilde{\Sigma}_{t_k,t_{k+1}} \,,$$

   and denote it C.

   (b) *Draw* $\mathbf{Z} \sim \mathsf{N}(\mathbf{0}, I)$ *and set*

   $$\widetilde{\mathbf{X}}_{t_{k+1}} = \widetilde{\boldsymbol{\mu}}_{t_{k+1}} + \widetilde{\Sigma}_{t_{k+1},t_k} \widetilde{\Sigma}_{t_k,t_k}^{-1} \left(\mathbf{X}_{t_k} - \widetilde{\boldsymbol{\mu}}_{t_k}\right) + C\,\mathbf{Z} \,.$$

5.1.2 Stationary Gaussian Processes and the FFT

☞ 631 Let $\{\widetilde{X}_t, t \in \mathbb{R}\}$ be a real-valued *stationary* Gaussian process. Thus, in addition to having the Gaussian property, the expectation $\mathbb{E}\widetilde{X}_t$ and covariance $\mathrm{Cov}(\widetilde{X}_t, \widetilde{X}_{t+s})$ do not depend on t. Without loss of generality we will assume that $\mathbb{E}\widetilde{X}_t = 0$ for all t. We wish to generate realizations of the process at equidistant times $\delta k, k = 0, \ldots, n$ for some $\delta > 0$. Let $X_k = \widetilde{X}_{\delta k}$, $k = 0, \ldots, n$. Then $\{X_k\}$ is a discrete-time zero-mean stationary Gaussian process. Its covariance matrix, Σ say, takes the form of a *symmetric Toeplitz matrix*:

$$\Sigma = \begin{pmatrix} \sigma_0 & \sigma_1 & \sigma_2 & \cdots & \sigma_{n-2} & \sigma_{n-1} & \sigma_n \\ \sigma_1 & \sigma_0 & \sigma_1 & \cdots & \sigma_{n-3} & \sigma_{n-2} & \sigma_{n-1} \\ \sigma_2 & \sigma_1 & \sigma_0 & \ddots & \sigma_{n-4} & \sigma_{n-3} & \sigma_{n-2} \\ \vdots & \vdots & \ddots & \ddots & \ddots & \vdots & \vdots \\ \sigma_{n-2} & \sigma_{n-3} & \sigma_{n-4} & \ddots & \sigma_0 & \sigma_1 & \sigma_2 \\ \sigma_{n-1} & \sigma_{n-2} & \sigma_{n-3} & \cdots & \sigma_1 & \sigma_0 & \sigma_1 \\ \sigma_n & \sigma_{n-1} & \sigma_{n-2} & \cdots & \sigma_2 & \sigma_1 & \sigma_0 \end{pmatrix}.$$

Such a matrix can be uniquely embedded in a $2n \times 2n$ *symmetric circulant matrix* C, given by

$$C = \left(\begin{array}{ccccc|ccccc} \sigma_0 & \sigma_1 & \cdots & \sigma_{n-1} & \sigma_n & \sigma_{n-1} & \sigma_{n-2} & \cdots & \sigma_2 & \sigma_1 \\ \sigma_1 & \sigma_0 & \cdots & \sigma_{n-2} & \sigma_{n-1} & \sigma_n & \sigma_{n-1} & \cdots & \sigma_3 & \sigma_2 \\ \vdots & \ddots & \ddots & \ddots & \ddots & \ddots & \ddots & \ddots & \ddots & \vdots \\ \sigma_{n-1} & \sigma_{n-2} & \cdots & \sigma_0 & \sigma_1 & \sigma_2 & \sigma_3 & \cdots & \sigma_{n-1} & \sigma_n \\ \sigma_n & \sigma_{n-1} & \cdots & \sigma_1 & \sigma_0 & \sigma_1 & \sigma_2 & \cdots & \sigma_{n-2} & \sigma_{n-1} \\ \hline \sigma_{n-1} & \sigma_n & \cdots & \sigma_2 & \sigma_1 & \sigma_0 & \sigma_1 & \cdots & \sigma_{n-3} & \sigma_{n-2} \\ \sigma_{n-2} & \sigma_{n-1} & \cdots & \sigma_3 & \sigma_2 & \sigma_1 & \sigma_0 & \cdots & \sigma_{n-4} & \sigma_{n-3} \\ \vdots & \ddots & \ddots & \ddots & \ddots & \ddots & \ddots & \ddots & \ddots & \vdots \\ \sigma_2 & \sigma_3 & \cdots & \sigma_{n-1} & \sigma_{n-2} & \sigma_{n-3} & \sigma_{n-4} & \cdots & \sigma_0 & \sigma_1 \\ \sigma_1 & \sigma_2 & \cdots & \sigma_n & \sigma_{n-1} & \sigma_{n-2} & \sigma_{n-3} & \cdots & \sigma_1 & \sigma_0 \end{array}\right).$$

Note that the upper-left $(n+1) \times (n+1)$ quadrant of C is Σ.

☞ 706 The fundamental relation between circulant matrices and the discrete Fourier transform (see Section D.4) provides the opportunity to use the fast Fourier transform (FFT) to generate realizations of the process $\{X_k\}$. The main condition that has to be satisfied is that the matrix C has nonnegative eigenvalues — so that C itself is a covariance matrix. A sufficient condition [11] is

- $\sigma_0 \geqslant \sigma_1 \geqslant \ldots \geqslant \sigma_n \geqslant 0$,
- $2\sigma_k \leqslant \sigma_{k-1} + \sigma_{k+1}$ for $k = 1, 2, \ldots, n-1$.

This implies that all variables are nonnegatively correlated. Another sufficient condition is $\sigma_k \leqslant 0$ for $k \neq 0$; see [10].

Let $\mathbf{c}$ denote the first row of C and let F be the $2n \times 2n$ discrete Fourier transform matrix, with entries $F_{pq} = \exp(-2\pi \mathrm{i}\, p q/(2n))$, $p, q = 0, 1, \ldots, 2n-1$. As is explained in Section D.4, the vector of eigenvalues $\boldsymbol{\lambda}$ of C is given by $\boldsymbol{\lambda} = \overline{F}\mathbf{c}$

($= F\mathbf{c}$ since the eigenvalues of a symmetric matrix are real), and the matrix $D = F\sqrt{\text{diag}(\boldsymbol{\lambda}/2n)}$ is a complex square root matrix of C, in the sense that $D\overline{D}^{\top} = C$. Moreover, the real and imaginary parts of the random vector $\mathbf{X} = D\mathbf{Z}$, where $\mathbf{Z}$ is complex-valued normal, are dependent $2n$-dimensional Gaussian vectors with covariance matrix C. The first n components of these vectors therefore are Gaussian with covariance matrix Σ. This leads to the following algorithm, which employs the FFT to perform fast $\mathcal{O}(n \ln n)$ evaluations of linear transformations of the form $\mathbf{b} = F\mathbf{a}$.

Algorithm 5.5 (Generating a Zero-Mean Stationary Gaussian Process)

1. *Compute the vector* $\boldsymbol{\lambda} = F\mathbf{c}$ *via the FFT.*

2. *Compute the vector* $\boldsymbol{\eta}$ *with entries* $\eta_k = \sqrt{\lambda_k/(2n)}$.

3. *Generate* $\mathbf{Z} = \mathbf{Y}_1 + \mathrm{i}\,\mathbf{Y}_2$, *where* $\mathbf{Y}_1, \mathbf{Y}_2 \overset{\text{iid}}{\sim} \mathsf{N}(\mathbf{0}, I)$ *and* $\dim(I) = 2n$.

4. *Compute the vector* $\boldsymbol{\zeta}$ *with entries* $\zeta_k = Z_k\,\eta_k$.

5. *Compute* $\mathbf{V} = F\boldsymbol{\zeta}$ *via the FFT.*

6. *Let* $\mathbf{A}$ *be the vector of the first* $n+1$ *components of* $\mathbf{V}$.

7. *Output* $\mathbf{X} = \Re(\mathbf{A})$.

■ **EXAMPLE 5.2 (Stationary Simulation via Circulant Embedding)**

The following MATLAB program provides an implementation for simulating a stationary, zero-mean Gaussian process on an equally spaced mesh of $n+1 = 10^4+1$ points on $[a,b] = [0,5]$, with $\sigma_k = \exp(-(b-a)k/N)$, $k = 0, 1, \ldots, N$. A typical realization is given in Figure 5.4.

```
%statgaus.m
n=10^4; a=0; b=5;
t=linspace(a,b,n+1); sigma=exp(-(t-t(1)));
c=[sigma sigma((end-1):-1:2)]';
lambda=fft(c); %eigenvalues
eta=sqrt(lambda./(2*n));
Z=randn(2*n,1)+sqrt(-1).*randn(2*n,1); %complex normal vectors
Zeta= Z.*eta;
X2n=fft(Zeta);
A=X2n(1:(n+1));
X=real(A);
plot(t,X)
```

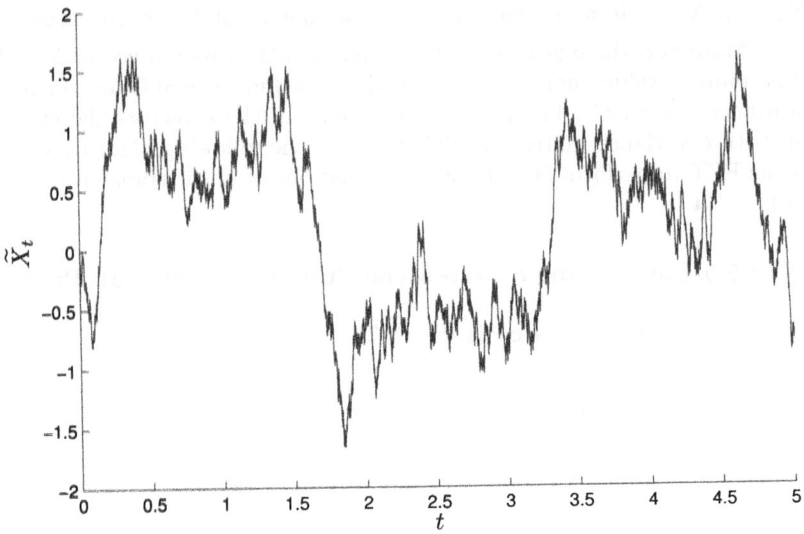

Figure 5.4 Realization of a stationary Gaussian process.

5.2 MARKOV CHAINS

A **Markov chain** is a stochastic process $\{X_t, t \in \mathscr{T}\}$ with a countable index set $\mathscr{T} \subset \mathbb{R}$ which satisfies the **Markov property**

$$(X_{t+s} \mid X_u, u \leqslant t) \sim (X_{t+s} \mid X_t) .$$

☞ 628
☞ 632
Markov chains are discussed in more detail in Sections A.9.2 and A.10. We discuss here only the main points pertinent to the simulation of such processes. We assume throughout that the index set is $\mathscr{T} = \{0, 1, 2, \ldots\}$.

A direct consequence of the Markov property is that Markov chains can be generated *sequentially*: $X_0, X_1, \ldots$, as expressed in the following generic recipe.

Algorithm 5.6 (Generating a Markov Chain)

1. *Draw* X_0 *from its distribution. Set* $t = 0$.
2. *Draw* X_{t+1} *from the conditional distribution of* X_{t+1} *given* X_t.
3. *Set* $t = t + 1$ *and repeat from Step 2.*

The conditional distribution of X_{t+1} given X_t can be specified in two common ways as follows.

- The process $\{X_t, t = 0, 1, 2, \ldots\}$ satisfies a recurrence relation

$$X_{t+1} = g(t, X_t, U_t) , \quad t = 0, 1, 2 \ldots , \tag{5.3}$$

where g is an easily evaluated function and U_t is an easily generated random variable whose distribution may depend on X_t and t.

- The conditional distribution of X_{t+1} given X_t is known and is easy to sample from.

An important instance of the second case occurs when the Markov chain $\{X_0, X_1, \ldots\}$ has a discrete state space E and is time-homogeneous. Its distribution is then completely specified by the distribution of X_0 (the initial distribution) and the matrix of one-step transition probabilities $P = (p_{ij})$, where

$$p_{ij} = \mathbb{P}(X_{t+1} = j \mid X_t = i), \quad i, j \in E .$$

The conditional distribution of X_{n+1} given $X_n = i$ is therefore a discrete distribution given by the i-th row of P. This leads to the following specification of Algorithm 5.6.

Algorithm 5.7 (Generating a Time-Homogeneous Markov Chain on a Discrete State Space)

1. *Draw X_0 from the initial distribution. Set $t = 0$.*
2. *Draw X_{t+1} from the discrete distribution corresponding to the X_t-th row of P.*
3. *Set $t = t + 1$ and go to Step 2.*

■ EXAMPLE 5.3 (A Markov Chain Maze)

At time $t = 0$ a robot is placed in compartment 3 of the maze in Figure 5.5. At each time $t = 1, 2, \ldots$ the robot chooses one of the adjacent compartments with equal probability. Let X_t be the robot's compartment at time t. Then $\{X_t\}$ is a time-homogeneous Markov chain with transition matrix P given below.

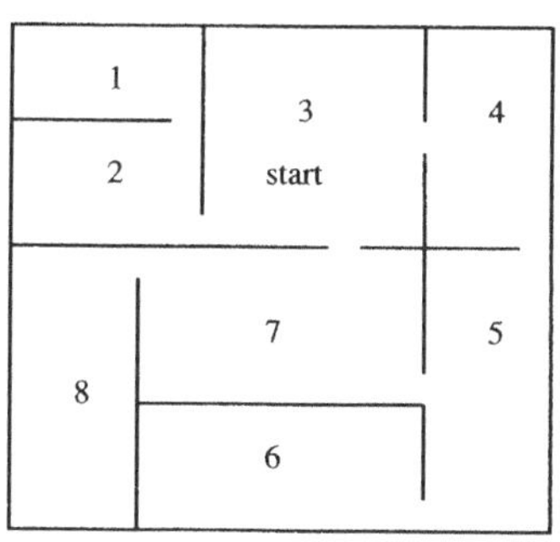

$$P = \begin{pmatrix} 0 & 1 & 0 & 0 & 0 & 0 & 0 & 0 \\ \frac{1}{2} & 0 & \frac{1}{2} & 0 & 0 & 0 & 0 & 0 \\ 0 & \frac{1}{3} & 0 & \frac{1}{3} & 0 & 0 & \frac{1}{3} & 0 \\ 0 & 0 & \frac{1}{2} & 0 & \frac{1}{2} & 0 & 0 & 0 \\ 0 & 0 & 0 & \frac{1}{3} & 0 & \frac{1}{3} & \frac{1}{3} & 0 \\ 0 & 0 & 0 & 0 & 1 & 0 & 0 & 0 \\ 0 & 0 & \frac{1}{3} & 0 & \frac{1}{3} & 0 & 0 & \frac{1}{3} \\ 0 & 0 & 0 & 0 & 0 & 0 & 1 & 0 \end{pmatrix} .$$

Figure 5.5 A maze and the corresponding transition matrix.

The following MATLAB program implements Algorithm 5.7. The first 100 values of the process are given in Figure 5.6.

```
%maze.m
n = 101
a = 0.5; b = 1/3;
P = [0, 1, 0, 0, 0, 0, 0, 0; a, 0, a, 0, 0, 0, 0, 0;
    0, b, 0, b, 0, 0, b, 0; 0, 0, a, 0, a, 0, 0, 0;
    0, 0, 0, b, 0, b, b, 0; 0, 0, 0, 0, 1, 0, 0, 0;
    0, 0, b, 0, b, 0, 0, b; 0, 0, 0, 0, 0, 0, 1, 0 ]
x = zeros(1,n);
x(1)= 3;
for t=1:n-1
    x(t+1) = min(find(cumsum(P(x(t),:))> rand));
end
hold on
plot(0:n-1,x,'.')
plot(0:n-1,x)
hold off
```

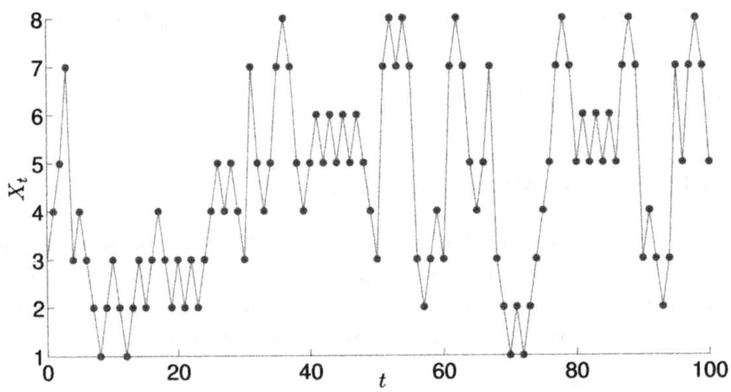

Figure 5.6 Realization of the maze process.

■ EXAMPLE 5.4 (Random Walk on an n-Dimensional Hypercube)

A typical example of a Markov chain that is specified by a recurrence relation such as (5.3) is the **random walk** process. Here the recurrence is simply

$$X_{t+1} = X_t + U_t \ , \quad t = 1, 2, \dots \ ,$$

where $U_1, U_2, \dots$ is an iid sequence of random variables from some discrete or continuous distribution.

A similar recurrence can be used to generate a random walk on the set of vertices of the unit n-dimensional hypercube — that is, the set of binary vectors of length n. Denote by $\mathbf{e}_1, \dots, \mathbf{e}_n$ the unit vectors in $\mathbb{R}^n$. Starting with $\mathbf{X}_0$ somewhere on the unit hypercube, define

$$\mathbf{X}_{t+1} = \mathbf{X}_t + \mathbf{e}_{I_t} \mod 2 ,$$

where $I_1, I_2, \ldots \overset{\text{iid}}{\sim} \mathsf{DU}(1, \ldots, n)$. Note that $\mathbf{e}_I$ is simply a uniformly chosen unit vector. Thus, the process performs a random walk on the unit hypercube, where at each step the process jumps from one vertex to one of the n adjacent vertices with equal probability. Since $\mathbf{X}_{t+1}$ and $\mathbf{X}_t$ only differ in position I_t, each state transition can be generated by simply flipping the component of $\mathbf{X}_t$ at a randomly chosen index I_t. The following MATLAB program gives an implementation of the generation algorithm for a 20-dimensional hypercube. In Figure 5.7 the progress of the Markov chain $\mathbf{X}_t = (X_{t1}, \ldots, X_{tn})^\top$, $t = 0, 1, 2, \ldots, 200$ can be observed through its representation $Y_t = \sum_{i=1}^n 2^{-i} X_{ti}$, $t = 0, 1, 2, \ldots, 200$ on the interval $[0, 1]$.

```
%hypercube.m
N = 200; % number of samples
n = 20; %dimension
x = zeros(N,n);
for t=1:N
    I = ceil(rand*n); %choose random position
    x(t+1,:) = x(t,:); %copy
    x(t+1,I) = ~x(t+1,I); %flip bit at position I
end
b = 0.5.^[1:n];
y = x*b';
hold on
plot(0:N,y,'.'), plot(0:N,y)
hold off
```

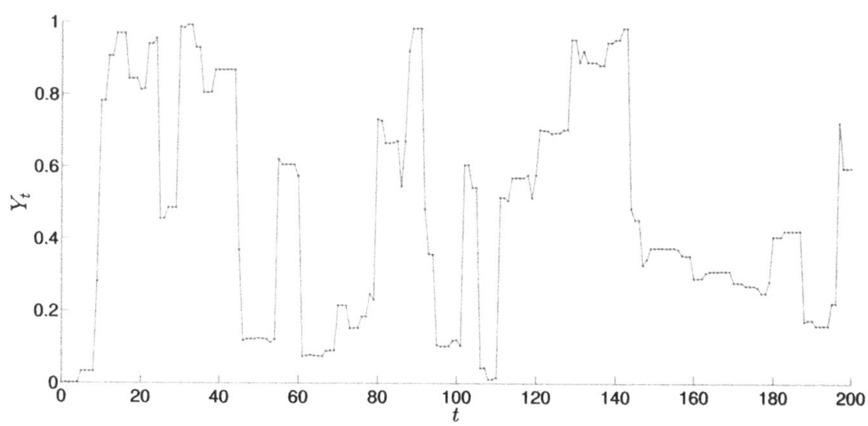

Figure 5.7 Realization of the process $\{Y_t\}$.

5.3 MARKOV JUMP PROCESSES

A **Markov jump process** is a stochastic process $\{X_t, t \in \mathscr{T}\}$ with a continuous index set $\mathscr{T} \subseteq \mathbb{R}$ and a discrete state space E, which satisfies the **Markov property**

$$(X_{t+s} \mid X_u, u \leqslant t) \sim (X_{t+s} \mid X_t) .$$

☞ 628 Markov jump processes are discussed in more detail in Section A.11. We discuss here only the main points pertinent to the simulation of such processes. We assume throughout that the index set is $\mathscr{T} = [0, \infty)$ and that the state space is $E = \{1, 2, \ldots\}$.

A time-homogeneous Markov jump process is often defined via its Q-matrix,

$$Q = \begin{pmatrix} -q_1 & q_{12} & q_{13} & \cdots \\ q_{21} & -q_2 & q_{23} & \cdots \\ q_{31} & q_{32} & -q_3 & \cdots \\ \vdots & \vdots & \vdots & \ddots \end{pmatrix},$$

where q_{ij} is the **transition rate** from i to j:

$$q_{ij} = \lim_{h \downarrow 0} \frac{\mathbb{P}(X_{t+h} = j \mid X_t = i)}{h}, \quad i \neq j, \quad i, j \in E \tag{5.4}$$

and q_i is the **holding rate** in i:

$$q_i = \lim_{h \downarrow 0} \frac{1 - \mathbb{P}(X_{t+h} = i \mid X_t = i)}{h}, \quad i \in E .$$

A typical assumption is that $0 \leqslant q_{ij} < \infty$ and that $q_i = \sum_{j \neq i} q_{ij}$, so that all row sums of Q are 0. Theorem A.11.1 specifies the behavior of such a Markov jump process: if the process is in some state i at time t, it will remain there for an additional $\mathsf{Exp}(q_i)$-distributed amount of time. When the process leaves a state i, it will jump to a state j with probability $K_{ij} = q_{ij}/q_i$, independent of the history of the process. In particular, the jump states $Y_0, Y_1, \ldots$ form a Markov chain with transition matrix $K = (K_{ij})$. Defining the holding times as $A_1, A_2, \ldots$ and the jump times as $T_1, T_2, \ldots$, the generation algorithm is as follows.

Algorithm 5.8 (Generating a Time-Homogeneous Markov Jump Process)

1. *Set* $T_0 = 0$. *Draw* Y_0 *from its distribution. Set* $X_0 = Y_0$ *and* $n = 0$.
2. *Draw* $A_{n+1} \sim \mathsf{Exp}(q_{Y_n})$.
3. *Set* $T_{n+1} = T_n + A_{n+1}$.
4. *Set* $X_t = Y_n$ *for* $T_n \leqslant t < T_{n+1}$.
5. *Draw* Y_{n+1} *from the distribution corresponding to the* Y_n*-th row of* K. *Set* $n = n + 1$ *and go to Step 2.*

■ **EXAMPLE 5.5 (Repairable System)**

Consider a reliability system with two unreliable machines and one repairman. Both machines have exponentially distributed life and repair times. The failure and repair rates are λ_1, μ_1 and λ_2, μ_2 for machine 1 and machine 2, respectively. The repairman can only work on one machine at a time, and if both have failed the repair man keeps working on the machine that has failed first, while the other machine remains idle. All life and repair times are independent of each other.

Because of the exponentiality and independence assumptions, the system can be described via a Markov jump process with 5 states: 1 (both machines working), 2 (machine 2 working, machine 1 failed), 3 (machine 1 working, machine 2 failed), 4 (both failed, machine 1 failed first), 5 (both failed, machine 2 failed first). The transition rate graph and Q-matrix are given in Figure 5.8.

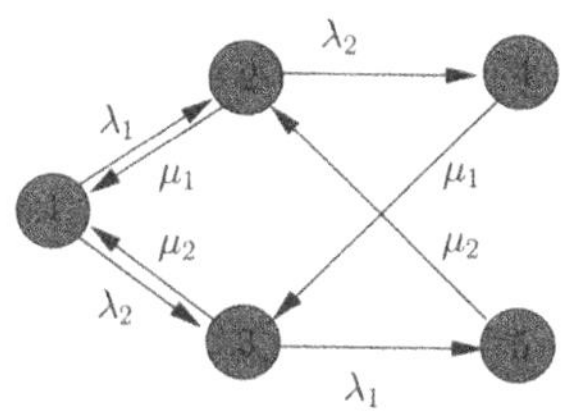

$$Q = \begin{pmatrix} -(\lambda_1+\lambda_2) & \lambda_1 & \lambda_2 & 0 & 0 \\ \mu_1 & -(\mu_1+\lambda_2) & 0 & \lambda_2 & 0 \\ \mu_2 & 0 & -(\mu_2+\lambda_1) & 0 & \lambda_1 \\ 0 & 0 & \mu_1 & -\mu_1 & 0 \\ 0 & \mu_2 & 0 & 0 & -\mu_2 \end{pmatrix}.$$

Figure 5.8 Transition rate graph and Q-matrix of the repairable system.

The following MATLAB program implements Algorithm 5.8 for the case where $\lambda_1 = 1, \lambda_2 = 2$, $\mu_1 = 3$, and $\mu_4 = 4$. A realization of the process on the interval $[0, 5]$, starting in state 1, is given in Figure 5.9.

```
%mjprep.m
clear all, clf
lam1= 1; lam2 = 2; mu1= 3; mu2 = 4;
Q = [-(lam1 + lam2), lam1, lam2, 0, 0;
     mu1, -(mu1+ lam2), 0, lam2, 0;
     mu2, 0, -(mu2 + lam1), 0, lam1;
     0, 0, mu1, -mu1, 0;
     0, mu2, 0, 0, -mu2];

q = -diag(Q);
K = diag(1./q)*Q + eye(5);
T = 5;
n=0;
t = 0; y = 1;
yy = [y]; tt = [t];
while t < T
    A = -log(rand)/q(y);
    y =  min(find(cumsum(K(y,:))> rand));
    t = t + A;
    tt = [tt,t];
    yy= [yy,y];
```

```
    n= n+1;
end
for i=1:n
    line([tt(i),tt(i+1)],[yy(i),yy(i)],'Linewidth',3);
    line([tt(i+1),tt(i+1)],[yy(i),yy(i+1)],'LineStyle',':');
end
axis([0,T,1,5.1])
```

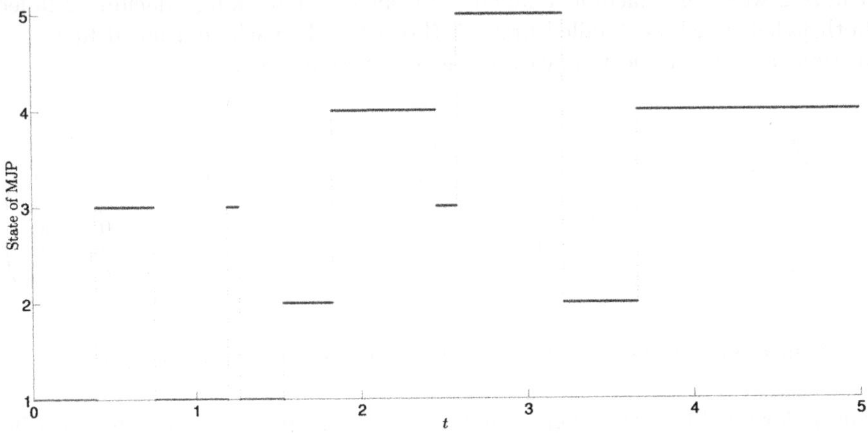

Figure 5.9 Realization of the reliability Markov jump process.

The above algorithm is easily extended to the nonhomogeneous case, where the rates in (5.4) depend on t. Let $0 \leqslant q_{ij}(t) < \infty$ denote the right-hand side of (5.4) and let $q_i(t) = \sum_{j \neq i} q_{ij}(t)$. The process jumps from state to state according to a time-*non*homogeneous Markov chain, while staying a certain amount of time in each state. Suppose at some time T_n the process jumps to state $Y_n = i$. Let A_{n+1} denote the holding time in state i. We have

$$
\begin{aligned}
q_i(t) &= \lim_{h \downarrow 0} \frac{\mathbb{P}(t - T_n < A_{n+1} < t + h - T_n \,|\, A_{n+1} > t - T_n)}{h} \\
&= \lim_{h \downarrow 0} \frac{F(t + h - T_n) - F(t - T_n)}{(1 - F(t - T_n))h} = \frac{f(t - T_n)}{1 - F(t - T_n)} \\
&= -\frac{\mathrm{d}}{\mathrm{d}t} \ln(1 - F(t - T_n)) \,,
\end{aligned}
$$

where $F(t)$ denotes the cdf of A_{n+1} and $f(t)$ its pdf. It follows that

$$
F(t) = \mathbb{P}(A_{n+1} \leqslant t) = 1 - \mathrm{e}^{-\int_{T_n}^{T_n + t} q_i(s)\,\mathrm{d}s} \,, \quad t \geqslant 0 \,. \tag{5.5}
$$

At time $T_{n+1} = T_n + A_{n+1}$ the process jumps to state j with probability $q_{ij}(T_{n+1})/q_i(T_{n+1})$, $j \in E$. We thus have the following algorithm.

Algorithm 5.9 (Generating a Nonhomogeneous Markov Jump Process)

1. *Set $T_0 = 0$. Draw Y_0 from its distribution. Set $X_0 = Y_0$ and set $n = 0$.*
2. *Draw A_{n+1} from the cdf given in* (5.5).
3. *Set $T_{n+1} = T_n + A_{n+1}$.*
4. *Set $X_t = Y_n$ for $T_n \leqslant t < T_{n+1}$.*
5. *Draw Y_{n+1} from the distribution $\{q_{Y_n,y}(T_{n+1})/q_{Y_n}(T_{n+1}),\ y \in E\}$. Set $n = n + 1$ and repeat from Step 2.*

■ EXAMPLE 5.6 (A Nonhomogeneous Markov Jump Process)

Consider a nonhomogeneous Markov jump process with three states, 1, 2, and 3, with transition rates $q_{12}(t) = \sin^2(t)$, $q_{21}(t) = 1 + \sin(t)$, $q_{23}(t) = 1 - \cos(t)$, $t \geqslant 0$, and the other rates are 0. Note that state 3 is an absorbing state. From (5.5) the holding time cdf for state 1 is given by

$$F_1(t) = 1 - \mathrm{e}^{\frac{1}{4}(-2t - \sin(2T_n) + \sin(2(t+T_n)))}, \quad t \geqslant 0\,,$$

with pdf

$$f_1(t) = \mathrm{e}^{\frac{1}{4}(-2t - \sin(2T_n) + \sin(2(t+T_n)))} \sin^2(t + T_n) \leqslant \frac{1}{2}\mathrm{e}^{-t/2}\, 2\mathrm{e}^{1/2}\,, \quad t \geqslant 0\,. \quad (5.6)$$

Similarly, for state 2 the holding time cdf is

$$F_2(t) = 1 - \mathrm{e}^{-2t - \cos(T_n) + \cos(t+T_n) - \sin(T_n) + \sin(t+T_n)}, \quad t \geqslant 0\,,$$

and the pdf satisfies

$$\begin{aligned} f_2(t) &= \mathrm{e}^{-2t - \cos(T_n) + \cos(t+T_n) - \sin(T_n) + \sin(t+T_n)}(-\cos(t + T_n) + \sin(t + T_n) + 2) \\ &\leqslant 2\mathrm{e}^{-2t}\mathrm{e}^{2\sqrt{2}}(1 + \sqrt{2}/2), \quad t \geqslant 0\,. \end{aligned}$$

Holding times for states 1 and 2 can thus be generated via acceptance–rejection using an exponential proposal. The following MATLAB program implements Algorithm 5.9 and simulates the nonhomogeneous Markov jump process until the absorbing state is reached. See also Example 5.8 for an alternative holding time generation method. ☞ 59

```
%nonhommjp.m
q12 = @(t) sin(t)^2;
q21 = @(t) 1 + sin(t);
q23 = @(t) 1 - cos(t);
q2 = @(t) q21(t) + q23(t);
f1 = @(t,Tn) exp(-t/2 + (-sin(2*Tn) + sin(2*(t+Tn)))/4)*sin(t+Tn)^2;
f2 = @(t,Tn) exp(-2*t-cos(Tn)+cos(t+Tn)-sin(Tn)+sin(t+Tn))*...
    (2-cos(t+Tn)+sin(t+Tn));
```

```
y=1; tn=0; n=0; yy=[y]; tt=[tn];
while y ~= 3
    if y==1
        accept=false;
        while ~accept
            A = -log(rand)*2;
            accept = rand < f1(A,tn)/exp(-(A-1)/2);
        end
        tn = tn + A; y = 2;
    else %y==2
        accept =false;
        while ~accept
            A = -log(rand)/2;
            accept=rand<f2(A,tn)/(exp(-2*(A-sqrt(2)))*(2+sqrt(2)));
        end
        tn = tn + A;
        if rand < q21(tn)/q2(tn)
            y =1;
        else
            y=3;
        end
    end
    yy = [yy,y]; tt = [tt,tn]; n = n+1;
end
% for plotting
for i=1:n
    line([tt(i),tt(i+1)],[yy(i),yy(i)],'Linewidth',3);
    line([tt(i+1),tt(i+1)],[yy(i),yy(i+1)],'LineStyle',':');
end
```

5.4 POISSON PROCESSES

Poisson processes are used to model random configurations of points in space and time. Specifically, let E be some subset of $\mathbb{R}^d$ and let $\mathcal{E}$ be the collection of Borel sets on E. To any collection of random points $\{T_n\}$ in E corresponds a **random counting measure** N defined by

$$N(A) = \sum_k \mathrm{I}_{\{T_k \in A\}}, \quad A \in \mathcal{E} ,$$

counting the random number of points in A. Such a random counting measure is said to be a **Poisson random measure** with **mean measure** μ if the following properties hold:

1. $N(A) \sim \mathsf{Poi}(\mu(A))$ for any set $A \in \mathcal{E}$, where $\mu(A)$ denotes the mean measure of A.
2. For any disjoint sets $A_1, \ldots, A_n \in \mathcal{E}$, the random variables $N(A_1), \ldots, N(A_n)$ are independent.

In most practical cases the mean measure has a density, called the **intensity** or **rate** function, $\lambda(\mathbf{x})$; so that

$$\mu(A) = \int_A \lambda(\mathbf{x})\,\mathrm{d}\mathbf{x}\,.$$

We will assume from now on that such a rate function exists.

Informally, both the collection $\{T_k\}$ and the random measure N are referred to as a **Poisson process** on E. The Poisson process is said to be **homogeneous** if the rate function is constant. An important corollary of Properties 1 and 2 above is:

3. Conditional upon $N(A) = n$, the n points in A are independent of each other and have pdf $f(\mathbf{x}) = \lambda(\mathbf{x})/\mu(A)$.

This leads immediately to the following generic algorithm for generating a Poisson process on E, assuming that $\mu(E) = \int_E \lambda(\mathbf{x})\,\mathrm{d}\mathbf{x} < \infty$.

Algorithm 5.10 (Generating a General Poisson Random Measure)

1. *Generate a Poisson random variable* $N \sim \mathsf{Poi}(\mu(E))$.
2. *Given* $N = n$, *draw* $\mathbf{X}_1, \ldots, \mathbf{X}_n \overset{\text{iid}}{\sim} f$, *where* $f(\mathbf{x})$ *is the mean density* $\lambda(\mathbf{x})/\mu(E)$, *and return these as the points of the Poisson process.*

■ **EXAMPLE 5.7 (Convex Hull of a Poisson Process)**

Figure 5.10 shows six realizations of the point sets and their *convex hulls* of a homogeneous Poisson process on the unit square with rate 20. The MATLAB code is given below. A particular object of interest could be the random volume of the convex hull formed in this way.

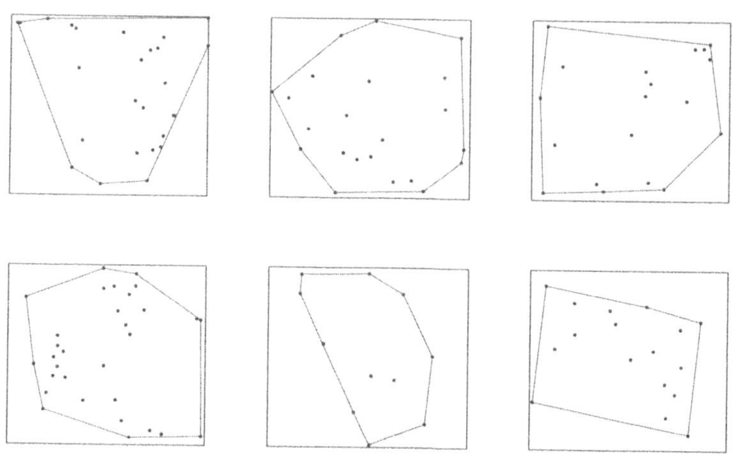

Figure 5.10 Realizations of a homogeneous Poisson process with rate 20. For each case the convex hull is also plotted.

```
%hompoich.m
for i=1:6
   N = poissrnd(20);
   x = rand(N,2);
   k = convhull(x(:,1),x(:,2));
   %[K,v] =  convhulln(x); %v is the area
   subplot(2,3,i);
   plot(x(k,1),x(k,2),'r-',x(:,1),x(:,2),'.')
end
```

For one-dimensional Poisson processes more direct generation algorithms can be formulated, using the additional properties of such processes. Consider first a homogeneous Poisson process with rate λ on $\mathbb{R}_+$. Denote the points of the process by $0 < T_1, T_2, \ldots$, interpreted as *arrival* points of some sort, and let $A_i = T_i - T_{i-1}$ be the i-th *interarrival* time, $i = 1, 2, \ldots$, setting $T_0 = 0$. The interarrival times $\{A_i\}$ are iid and $\mathsf{Exp}(\lambda)$ distributed; see, for example, [8, Pages 79–80]. We can thus generate the points of the Poisson process on some interval $[0, T]$ as follows.

Algorithm 5.11 (One-Dimensional Homogeneous Poisson Process)

1. *Set* $T_0 = 0$ *and* $n = 1$.

2. *Generate* $U \sim \mathsf{U}(0, 1)$.

3. *Set* $T_n = T_{n-1} - \frac{1}{\lambda} \ln U$.

4. *If* $T_n > T$, *stop; otherwise, set* $n = n + 1$ *and go to Step 2.*

Notice that the corresponding **Poisson counting process** $\{N_t, t \geqslant 0\}$, defined by $N_t = N([0, t])$, is a Markov jump process on $\{0, 1, 2, \ldots\}$ with $N_0 = 0$ and transition rates $q_{i,i+1} = \lambda$, $i = 0, 1, 2, \ldots$ and $q_{i,j} = 0$ otherwise. The process jumps at times $T_1, T_2, \ldots$ to states $1, 2, \ldots$, staying an $\mathsf{Exp}(\lambda)$-distributed amount of time in each state (including in state 0). In a similar way, the counting process corresponding to a *nonhomogeneous* one-dimensional Poisson process on $\mathbb{R}_+$ with rate function $\lambda(t), t \geqslant 0$ is a nonhomogeneous Markov jump process with transition rates $q_{i,i+1}(t) = \lambda(t)$, $i = 0, 1, 2, \ldots$. Similar to (5.5), the tail probabilities of the interarrival times are

$$\mathbb{P}(A_{n+1} > t) = \exp\left(- \int_{T_n}^{T_n + t} \lambda(s)\, \mathrm{d}s \right) , \quad t \geqslant 0 .$$

Therefore, a variety of generation methods are available to generate the interarrival times directly. However, it is often easier to construct the points indirectly, as illustrated in Figure 5.11: First select a constant $\lambda \geqslant \sup_{s \leqslant t} \lambda(s)$, assuming it exists. Then, generate the points of a two-dimensional homogeneous Poisson process, M say, on $[0, t] \times [0, \lambda]$ with rate 1. Finally, project all points of M that lie below the graph of $\lambda(s)$, $s \leqslant t$ onto the t-axis.

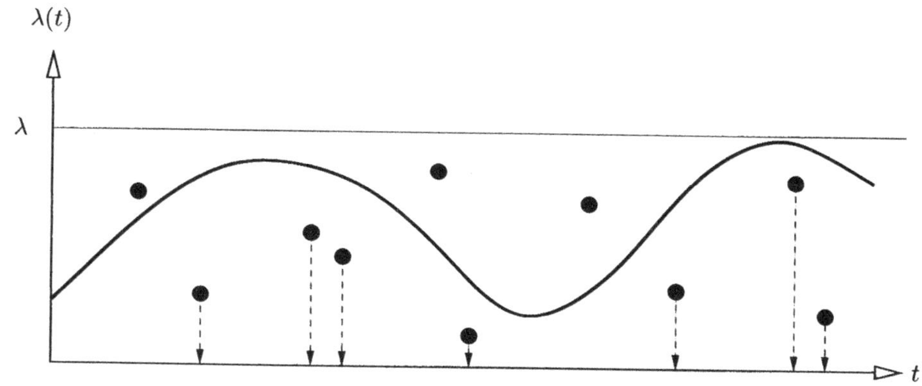

Figure 5.11 Constructing a nonhomogeneous Poisson process.

Let $\mathscr{R}_t = \{(s,y), 0 \leqslant s \leqslant t, y \leqslant \lambda(s)\}$. For each $t \geqslant 0$, we have

$$\mathbb{P}(N_t = 0) = \mathbb{P}(M(\mathscr{R}_t) = 0) = \exp\left(-\int_0^t \lambda(s)\,\mathrm{d}s\right),$$

which shows that the stochastic process $\{N_t, t \geqslant 0\}$ constructed in this way is a nonhomogeneous Poisson counting process with rate function $\lambda(t), t \geqslant 0$. If instead *all* points of M are projected onto the t-axis, we obtain a homogeneous Poisson counting process with rate λ. To obtain the nonhomogeneous process we accept each point τ with probability $\frac{\lambda(\tau)}{\lambda}$. This leads to the following algorithm.

Algorithm 5.12 (One-Dimensional Nonhomogeneous Poisson Process)

1. *Set* $t = 0$ *and* $n = 0$.
2. *Generate* $U \sim \mathsf{U}(0,1)$.
3. *Set* $t = t - \frac{1}{\lambda} \ln U$.
4. *If* $t > T$, *stop; otherwise, continue.*
5. *Generate* $V \sim \mathsf{U}(0,1)$.
6. *If* $V \leqslant \frac{\lambda(t)}{\lambda}$, *increase* n *by 1 and set* $T_n = t$. *Repeat from Step 2.*

■ EXAMPLE 5.8 (A Nonhomogeneous Poisson Counting Process)

Figure 5.12 gives a typical realization of a nonhomogeneous Poisson counting process $\{N_t, t \geqslant 0\}$ with rate function $\lambda(t) = \sin^2(t)$ on the interval $[0, 50]$. The realization is obtained using the following MATLAB code, which implements Algorithm 5.12. An alternative is to generate the interarrival times via acceptance–rejection, as in Example 5.6, with an $\mathsf{Exp}(1/2)$ proposal distribution and target density f_1 in (5.6).

```
%pois.m
T = 50;
t = 0; n = 0;
tt = [t];
while t < T
    t = t - log(rand);
      if (rand < sin(t)^2)
        tt = [tt,t];
        n = n+1;
    end
end
nn = 0:n;
for i =1:n
    line([tt(i),tt(i+1)],[nn(i),nn(i)],'Linewidth',2);
end
```

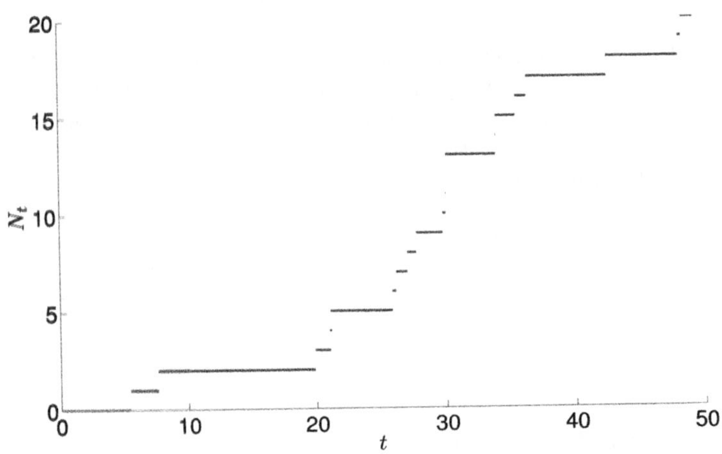

Figure 5.12 A typical realization of a nonhomogeneous Poisson counting process with rate function $\lambda(t) = \sin^2(t)$.

5.4.1 Compound Poisson Process

Let N be a Poisson random measure on $\mathbb{R}_+ \times \mathbb{R}^d$ with mean measure $\mathrm{d}t\, \nu(\mathrm{d}\mathbf{y})$. We assume that $\lambda = \nu(\mathbb{R}^d) < \infty$. The process $\{K_t\}$ with $K_t = N([0,t] \times \mathbb{R}^d)$ is a homogeneous Poisson process with rate λ. The process $\{\mathbf{X}_t, t \geqslant 0\}$ defined as

$$\mathbf{X}_t = \int_0^t \int_{\mathbb{R}^d} \mathbf{y}\, N(\mathrm{d}u, \mathrm{d}\mathbf{y}), \quad t \geqslant 0\,,$$

is the **compound Poisson process** corresponding to the measure ν. The process can be thought of as a "batch" Poisson process, where arrivals occur according to a Poisson process with rate λ, and each arrival adds a batch of size $\mathbf{Y} \sim \nu(\mathrm{d}\mathbf{y})/\lambda$ to the total, so that we may also write

$$\mathbf{X}_t = \sum_{i=1}^{K_t} \mathbf{Y}_i \,,$$

where $\mathbf{Y}_1, \mathbf{Y}_2, \ldots \overset{\text{iid}}{\sim} \nu(\mathrm{d}\mathbf{y})/\lambda$ are independent of K_t. The compound Poisson process is an important example of a **Lévy process**: a stochastic process with independent and stationary increments; see Section 5.13. In this context the measure ν is called the **Lévy measure**.

The characteristic function of $\mathbf{X}_t$ can be found by conditioning on K_t:

$$\begin{aligned}\mathbb{E}\, \mathrm{e}^{\mathrm{i}\,\mathbf{s}^\top \mathbf{X}_t} &= \mathbb{E}\,\mathbb{E}\left[\mathrm{e}^{\mathrm{i}\,\mathbf{s}^\top \sum_{i=1}^{K_t} \mathbf{Y}_i} \,\middle|\, K_t\right] = \mathbb{E}\left(\mathbb{E}\, \mathrm{e}^{\mathrm{i}\,\mathbf{s}^\top \mathbf{Y}}\right)^{K_t} = \exp(-\lambda t\,(1 - \mathbb{E}\, \mathrm{e}^{\mathrm{i}\,\mathbf{s}^\top \mathbf{Y}})) \\ &= \exp\left(\lambda t \int (\mathrm{e}^{\mathrm{i}\,\mathbf{s}^\top \mathbf{y}} - 1)\, \nu(\mathrm{d}\mathbf{y})\right) . \end{aligned} \tag{5.7}$$

Denoting the jump times of the compound Poisson process by $\{T_k\}$ and the jump sizes by $\{\mathbf{Y}_k\}$, we have the following generation algorithm.

Algorithm 5.13 (Generating a Compound Poisson Process (I))

1. *Set* $k = 0$, $T_k = 0$, *and* $\mathbf{X}_{T_k} = 0$.
2. *Generate* $A_k \sim \mathsf{Exp}(\lambda)$.
3. *Generate* $\mathbf{Y}_k \sim \nu(\mathrm{d}\mathbf{y})/\lambda$.
4. *Set* $T_{k+1} = T_k + A_k$ *and* $\mathbf{X}_{T_{k+1}} = \mathbf{X}_{T_k} + \mathbf{Y}_k$.
5. *Set* $k = k + 1$ *and repeat from Step 2.*

■ EXAMPLE 5.9 (Compound Poisson Process)

Consider a compound Poisson process with Lévy measure

$$\nu(\mathrm{d}y) = |y|^{-3/2}\, \mathrm{I}_{\{\delta < |y| < \varepsilon\}}\, \mathrm{d}y \,, \quad y \in \mathbb{R} \tag{5.8}$$

for some $0 < \delta < \varepsilon \leqslant \infty$. We have $\lambda = 4(\delta^{-1/2} - \varepsilon^{-1/2})$. The reader may verify that if $U \sim \mathsf{U}(0,1)$ and $R \sim \mathsf{Ber}(1/2)$ are independent, then

$$\frac{(2R-1)\delta}{(1 - U + U\sqrt{\delta/\varepsilon})^2}$$

has distribution ν/λ, so generating random variables from this distribution is easy.

The MATLAB program below implements Algorithm 5.13. A typical realization for $\delta = 10^{-6}$ and $\varepsilon = \infty$ is given in the left-hand pane of Figure 5.13. Note that this process can be seen as the superposition of independent compound Poisson processes, having jump sizes $|y|$ restricted to disjoint intervals (δ, δ_1), $(\delta_1, \delta_2), \ldots, (\delta_n, \varepsilon)$. Note also that for small (δ_1, δ_2) — as on the right pane of Figure 5.13 — the process has sample paths that start to resemble those of a Brownian motion process. The path on the left-hand pane is thus not as smooth as it seems.

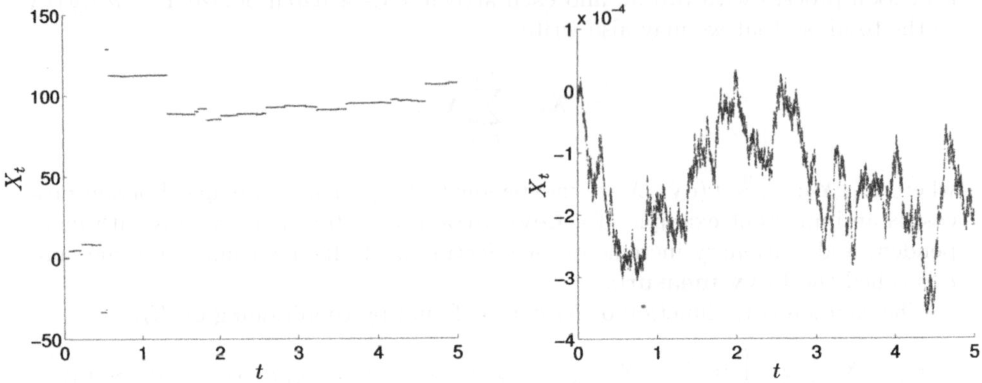

Figure 5.13 Typical realizations of the compound Poisson process with Lévy measure (5.8). On the left pane $\delta = 10^{-6}$ and $\varepsilon = \infty$; on the right pane $\delta = 10^{-6}$ and $\varepsilon = 10^{-5}$.

```
%compp.m
T = 5; delta = 10^-6; epsilon = inf;
lambda = 4*(1/sqrt(delta) - 1/sqrt(epsilon));
X = []; tt  = []; t=0; x= 0;
while t < T
    a = -log(rand)/lambda;
    t = t + a;
    R = (rand < 0.5);
    U = rand;
    y = (2*R-1)*delta/(1-U + sqrt(delta/epsilon)*U)^2;
    x = x + y;
    X = [X,x];
    tt = [tt,t];
end
N = numel(tt);
hold on
for i=1:N-1
    line([tt(i),tt(i+1)],[X(i),X(i)],'Linewidth',1);
end
```

An alternative procedure to generate compound Poisson processes is based on Algorithm 5.10. It generates a compound Poisson process on a fixed interval $[0, T]$.

Algorithm 5.14 (Generating a Compound Poisson Process (II))

1. *Generate* $N \sim \mathsf{Poi}(\lambda T)$.
2. *Generate* $U_1, \ldots, U_N \overset{\text{iid}}{\sim} \mathsf{U}(0, T)$.
3. *Generate* $Y_1, , \ldots, Y_N \overset{\text{iid}}{\sim} \nu(\mathrm{d}y)/\lambda$.
4. *Return* $X_t = \sum_{i:U_i \leqslant t} Y_i$, $t \in [0, T]$.

5.5 WIENER PROCESS AND BROWNIAN MOTION

A **Wiener process** is a stochastic process $W = \{W_t, t \geqslant 0\}$ characterized by the following properties.

1. *Independent increments*: W has **independent increments**; that is, for any $t_1 < t_2 \leqslant t_3 < t_4$
$$W_{t_4} - W_{t_3} \quad \text{and} \quad W_{t_2} - W_{t_1}$$
are independent random variables. In other words, $W_t - W_s,\ t > s$ is independent of the past history of $\{W_u,\ 0 \leqslant u \leqslant s\}$.

2. *Gaussian stationarity*: For all $t \geqslant s \geqslant 0$,
$$W_t - W_s \sim \mathsf{N}(0, t - s) .$$

3. *Continuity of paths*: $\{W_t\}$ has continuous paths, with $W_0 = 0$.

The Wiener process plays a central role in probability and forms the basis of many other stochastic processes. It can be viewed as a continuous version of a random walk process. Two typical sample paths are depicted in Figure 5.14. ☞ 164

Remark 5.5.1 (Starting Position) Although by definition the Wiener process starts at position 0, it is useful to consider Wiener processes starting from some arbitrary state x under a probability measure $\mathbb{P}^x$.

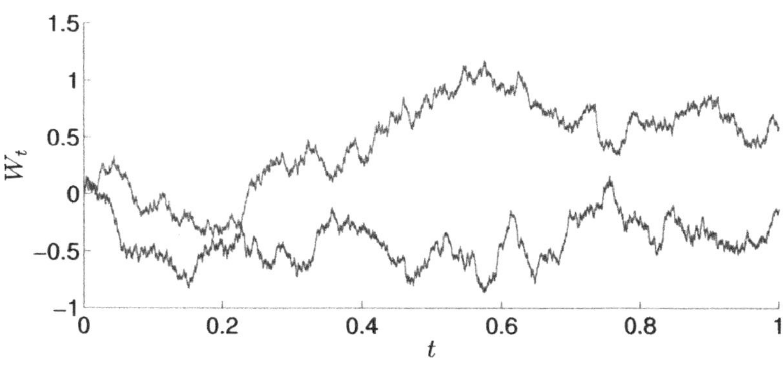

Figure 5.14 Two realizations of the Wiener process on the interval [0,1].

We list some of the many properties of the Wiener process $W = \{W_t, t \geqslant 0\}$. For more details see [19, 25].

1. *Gaussian process*: W is a Gaussian process with $\mathbb{E}W_t = 0$ and $\text{Cov}(W_s, W_t) = \min\{s, t\}$. It is the only Gaussian process with continuous sample paths that has these properties.

2. *Infinite variation sample path*: The sample paths of W on $[0, T]$ do not have bounded variation; that is,
$$\lim_{n \to \infty} \sum_{i=1}^{n} |W_{t_i} - W_{t_{i-1}}| = \infty \quad \text{a.s.} ,$$

where $0 = t_0 < t_1 < \cdots < t_n = t$ and $\lim_{n\to\infty} \max_i\{t_{i+1} - t_i\} = 0$. In particular, almost surely each sample path has infinite "length" and is not differentiable.

3. *Quadratic variation*: The quadratic variation of W on $[0, t]$ is given by

$$\lim_{n\to\infty} \sum_{i=1}^{n} (W_{t_{i+1}} - W_{t_i})^2 = t \quad \text{a.s.} ,$$

where $0 = t_0 < t_1 < \cdots < t_n = t$ and $\lim_{n\to\infty} \max_i\{t_{t+1} - t_i\} = 0$. Indeed, a Wiener process is the unique continuous martingale with $W_0 = 0$ and quadratic variation process t.

4. *Martingales*: Both $\{W_t, t \geqslant 0\}$ and $\{W_t^2 - t, t \geqslant 0\}$ are martingales with respect to their own filtration. Indeed, any real-valued continuous stochastic process for which this holds must be a Wiener process. In addition, for any θ the process $\{e^{\theta W_t - \theta^2 t/2}, t \geqslant 0\}$ is a martingale.

5. *Donsker's invariance principle*: Let $X_1, X_2, \ldots$ be iid random variables with mean 0 and variance 1. Define $X_0 = 0$ and

$$S_t = \sum_{i=0}^{\lfloor t \rfloor} X_i + (t - \lfloor t \rfloor)\, X_{\lfloor t \rfloor + 1}, \quad t \geqslant 0 ,$$

where $\lfloor t \rfloor$ is the largest integer smaller than or equal to t. Denote by $S^{(n)}$ the scaled process

$$S_t^{(n)} = \frac{S_{tn}}{\sqrt{n}}, \quad t \in [0, 1] .$$

Then, as $n \to \infty$

$$S^{(n)} \xrightarrow{d} W = \{W_t, \ t \in [0, 1]\} ,$$

where convergence in distribution is with respect to the space $C[0, 1]$ of continuous functions on $[0, 1]$ with supremum norm $\|s\| = \sup_{0 \leqslant t \leqslant 1} s(t)$. In particular, for any continuous functional h on $C[0, 1]$ we have

$$\lim_{n\to\infty} \mathbb{E} h(S^{(n)}) = \mathbb{E} h(W) .$$

☞ 619

6. *Orthogonal series expansions*: Let $E = \mathbb{R}_+$ or $[0, b]$ for some $b > 0$. Viewed as a random element of $L^2(E)$, $\{W_t, t \in E\}$ has the series expansion

$$W_t = \sum_{n=0}^{\infty} Z_n \int_0^t h_n(x)\, dx ,$$

where $Z_0, Z_1, \ldots \stackrel{\text{iid}}{\sim} \mathsf{N}(0, 1)$ and $\{h_n(x)\}_{n=0}^{\infty}$ is any complete orthonormal basis of $L^2(E)$ for which the above random series converges in L^2-norm. Typical examples on $E = [0, 1]$ are the *Haar* functions [13] and the basis of cosine functions $h_0(x) \equiv 1, h_n(x) = \sqrt{2}\cos(n\pi x), n = 1, 2, \ldots$. The latter gives the **sine series expansion**:

$$W_t = Z_0\, t + \frac{\sqrt{2}}{\pi} \sum_{n=1}^{\infty} Z_n \frac{\sin(n\pi t)}{n} , \quad t \in [0, 1] .$$

Similarly, the orthonormal basis $\sqrt{2/b}\cos((1+2n)\pi x/(2b)), x \in [0,b], n = 0,1,2,\ldots$ gives the **Karhunen–Loève expansion**:

$$W_t = \sum_{n=0}^{\infty} Z_n \frac{2\sqrt{2b}}{(2n+1)\pi} \sin\left(\frac{(2n+1)\pi t}{2b}\right), \quad t \in [0,b]. \tag{5.9}$$

7. *Markov property*: W is a time-homogeneous strong Markov process. In particular, for any finite stopping time τ and for all $t \geqslant 0$,

$$(W_{\tau+t} \mid W_u, u \leqslant \tau) \sim (W_{\tau+t} \mid W_\tau).$$

The transition density $p_t(x,y)$ of W is given by the Gaussian kernel

$$p_t(x,y) = \frac{1}{\sqrt{2\pi t}} \mathrm{e}^{-\frac{1}{2}\frac{(y-x)^2}{t}}, \quad t \geqslant 0, \quad x,y \in \mathbb{R}.$$

8. *Time-reversal*: The time-reversed process $\{\widehat{W}_s, s \in [0,t]\}$ defined by $\widehat{W}_s = W_{t-s} - W_t$ is a Wiener process on $[0,t]$.

9. *Reflection principle*: Let τ_x be the first time that $\{W_t\}$ hits x. Then $\{\widetilde{W}_t, t \geqslant 0\}$ defined by

$$\widetilde{W}_t = W_t \, \mathrm{I}_{\{t \leqslant \tau_x\}} + (2x - W_t) \, \mathrm{I}_{\{t > \tau_x\}}, \quad t \geqslant 0$$

is a Wiener process.

10. *Reciprocal time:* If $\{W_t\}$ is a Wiener process, then so is $\{X_t\}$, with $X_t = t\,W_{1/t}$, $t > 0$, $X_0 = 0$.

11. *Invariance under scaling:* If $\{W_t\}$ is a Wiener process, then so is $\{X_t\}$, with $X_t = W_{a\,t}/\sqrt{a}$, $t \geqslant 0$ for any $a > 0$.

12. *Maximum and hitting time*: Let $M_t = \max_{0 \leqslant s \leqslant t} W_s$ be the maximum of the Wiener process, and let $\tau_x = \inf\{t \geqslant 0 : W_t = x\}$ be the hitting time of (or first passage time to) x. For $x \geqslant 0$ the two are related via

$$\{M_t \geqslant x\} = \{\tau_x \leqslant t\}. \tag{5.10}$$

Consequently,

$$\begin{aligned}
\mathbb{P}(M_t \geqslant x) = \mathbb{P}(\tau_x \leqslant t) &= \mathbb{P}(\tau_x \leqslant t, W_t < x) + \mathbb{P}(\tau_x \leqslant t, W_t > x) \\
&= 2\,\mathbb{P}(\tau_x \leqslant t, W_t > x) = 2\,\mathbb{P}(W_t > x) \qquad (5.11)\\
&= 2 - 2\,\Phi\left(\frac{x}{\sqrt{t}}\right), \quad x \geqslant 0. \qquad (5.12)
\end{aligned}$$

The second equality in (5.11) is due to the reflection principle, as from τ_x onwards the Wiener process and its reflection around x have the same distribution. By differentiating (5.12) with respect to x it follows that

$$f_{M_t}(x) = \sqrt{\frac{2}{\pi t}}\,\mathrm{e}^{-\frac{1}{2}\frac{x^2}{t}}, \quad x \geqslant 0.$$

That is, M_t has a $\mathsf{N}(0,t)$ distribution truncated to $[0,\infty)$. Similarly, by differentiating (5.12) with respect to t we find

$$f_{\tau_x}(t) = \frac{1}{\sqrt{2\pi}}\, x\, t^{-3/2}\, \mathrm{e}^{-\frac{x^2}{2t}}, \quad t, x \geqslant 0\,.$$

Thus, $\tau_x \sim \mathsf{Stable}(\frac{1}{2}, 1, 0, x^2) \equiv \mathsf{InvGamma}(1/2, x^2/2)$. Finally, M_t and W_t have joint cdf

$$\mathbb{P}(M_t \leqslant x, W_t \leqslant y) = \Phi\left(\frac{2x-y}{\sqrt{t}}\right) - \Phi\left(\frac{-y}{\sqrt{t}}\right), \quad x \geqslant 0,\ x \geqslant y\,, \qquad (5.13)$$

which can again be derived from the reflection principle, as

$$\begin{aligned}\mathbb{P}(M_t \geqslant x, W_t \leqslant y) &= \mathbb{P}(M_t \geqslant x, W_t \geqslant 2x - y)\\ &= \mathbb{P}(W_t \geqslant 2x - y) = 1 - \Phi\left(\frac{2x-y}{\sqrt{t}}\right)\,.\end{aligned}$$

13. *Exit time:* Starting from $x \in (a,b)$, the expected time to exit the interval $[a,b]$ is $(x-a)(b-x)$. Moreover, W exits through b rather than a with probability $(x-a)/(b-a)$.

14. *The arcsine law of zeros*: The probability that $\{W_t\}$ has no zeros in the time interval (t_1, t_2) is given by

$$\mathbb{P}(W_t \neq 0 \text{ for all } t \in (t_1, t_2)) = \frac{2}{\pi} \arcsin \sqrt{\frac{t_1}{t_2}}\,.$$

15. *Law of iterated logarithm*: W satisfies almost surely,

$$\limsup_{t\to\infty} \frac{W_t}{\sqrt{2t \ln\ln t}} = 1\,,$$

$$\liminf_{t\to\infty} \frac{W_t}{\sqrt{2t \ln\ln t}} = -1\,.$$

The basic generation algorithm below uses the Markovian and Gaussian properties of the Wiener process.

Algorithm 5.15 (Generating the Wiener Process)

1. *Let* $0 = t_0 < t_1 < t_2 < \cdots < t_n$ *be the set of distinct times for which simulation of the process is desired.*

2. *Generate* $Z_1, \ldots, Z_n \stackrel{\text{iid}}{\sim} \mathsf{N}(0,1)$ *and output*

$$W_{t_k} = \sum_{i=1}^{k} \sqrt{t_k - t_{k-1}}\, Z_i, \quad k = 1, \ldots, n\,.$$

The algorithm is *exact*, in that the $\{W_{t_k}\}$ are drawn exactly according to their respective distributions. Nevertheless, the algorithm returns only a discrete skeleton of the true continuous process. To obtain a continuous path approximation to the exact path of the Wiener process, one could use linear interpolation on the points $W_{t_1}, \dots, W_{t_n}$. In other words, within each interval $[t_{k-1}, t_k]$, $k = 1, \dots, n$ approximate the continuous process $\{W_s, s \in [t_{k-1}, t_k]\}$ via:

$$\widehat{W}_s = \frac{W_{t_k}(s - t_{k-1}) + W_{t_{k-1}}(t_k - s)}{(t_k - t_{k-1})}, \quad s \in [t_{k-1}, t_k] \, .$$

It is possible to adaptively refine the path by using a Brownian bridge process, see Section 5.7. An alternative to interpolation is to exploit the Karhunen–Loève expansion (5.9) to approximately generate a Wiener process on the interval $[0, b]$.

Algorithm 5.16 (Wiener Process Generation via Karhunen–Loève)

1. *Generate* $Z_1, \dots, Z_n \stackrel{\text{iid}}{\sim} \mathsf{N}(0,1)$ *for a sufficiently large* n.
2. *Output the approximation to the Wiener process at time* t*:*

$$\widetilde{W}(t) = \sum_{k=1}^{n} Z_k \frac{2\sqrt{2b}}{(2k-1)\pi} \sin\left(\frac{(2k-1)\pi t}{2b}\right) .$$

A process $\{B_t, t \geqslant 0\}$ satisfying

$$B_t = \mu\, t + \sigma\, W_t, \quad t \geqslant 0 \, ,$$

where $\{W_t\}$ is a Wiener process, is called a **Brownian motion** with **drift** μ and **diffusion coefficient** σ^2. It is called a **standard Brownian motion** if $\mu = 0$ and $\sigma^2 = 1$ (Wiener process). Many properties of the Brownian motion follow directly from those of the Wiener process. Some additional properties are given next.

1. *Stochastic differential equation:* The Brownian motion process is the solution to the stochastic differential equation (see Section 5.6)

$$\mathrm{d}B_t = \mu\,\mathrm{d}t + \sigma\,\mathrm{d}W_t \, , \quad t \geqslant 0, \quad B_0 = 0 \, .$$

Since $B_t - B_0 \sim \mathsf{N}(\mu\, t, \sigma^2 t)$, the transition density $p_t(x, y)$ is given by the Gaussian kernel

$$p_t(x,y) = \frac{1}{\sqrt{2\pi\sigma^2 t}}\, \mathrm{e}^{-\frac{1}{2}\frac{(y-x-\mu t)^2}{\sigma^2 t}}, \quad t \geqslant 0, \quad x, y \in \mathbb{R} \, ,$$

which satisfies the Kolmogorov backward equations (A.81) ☞ 647

$$\frac{\partial}{\partial t} p_t(x,y) = \mu \frac{\partial}{\partial x} p_t(x,y) + \frac{\sigma^2}{2} \frac{\partial^2}{\partial x^2} p_t(x,y)$$

and Kolmogorov forward equations (A.82)

$$\frac{\partial}{\partial t} p_t(x,y) = -\mu \frac{\partial}{\partial y} p_t(x,y) + \frac{\sigma^2}{2} \frac{\partial^2}{\partial y^2} p_t(x,y) \, .$$

Note that for a standard Brownian motion (Wiener process) this reduces to Laplace's **heat equation**.

2. *Distribution of hitting time*: For $\mu > 0$ and $\sigma > 0$, let τ_x be the first time that $\{B_t\}$ hits $x \geqslant 0$. Then $\tau_x \sim \mathsf{Wald}(x/\mu, x^2/\sigma^2)$. ☞ 135

3. *Exit probabilities:* By Theorem A.13.5, the probability that the Brownian motion process exits the interval $[a, b]$ through b, starting from $x \in [a, b]$, is

$$\mathbb{P}^x(\tau_b < \tau_a) = \begin{cases} \dfrac{e^{-\frac{2\mu a}{\sigma^2}} - e^{-\frac{2\mu x}{\sigma^2}}}{e^{-\frac{2\mu a}{\sigma^2}} - e^{-\frac{2\mu b}{\sigma^2}}}, & \mu \neq 0 \\[2ex] \dfrac{x-a}{b-a}, & \mu = 0\,, \end{cases} \tag{5.14}$$

and the corresponding exit time $\tau = \min\{\tau_a, \tau_b\}$ has expectation

$$\mathbb{E}^x\tau = \begin{cases} \dfrac{b-a}{\mu}\,\dfrac{e^{-\frac{2\mu a}{\sigma^2}} - e^{-\frac{2\mu x}{\sigma^2}}}{e^{-\frac{2\mu a}{\sigma^2}} - e^{-\frac{2\mu b}{\sigma^2}}} - \dfrac{x-a}{\mu}, & \mu \neq 0 \\[2ex] \dfrac{(b-x)(x-a)}{\sigma^2}, & \mu = 0\,. \end{cases}$$

4. *Maximum:* A consequence of (5.14) is that for $\mu < 0$ and $x \geqslant 0$,

$$\mathbb{P}^0\Big(\max_t B_t > x\Big) = \mathbb{P}^0(\tau_x < \infty) = e^{2\mu x/\sigma^2},$$

which shows that, for a Brownian motion starting at 0 and with negative drift μ, the maximum has an $\mathsf{Exp}(-2\mu/\sigma^2)$ distribution.

The generation of a Brownian motion at times $t_1, \ldots, t_n$ follows directly from its definition.

Algorithm 5.17 (Generating Brownian Motion)

1. *Generate outcomes $W_{t_1}, \ldots, W_{t_n}$ of a Wiener process at times $t_1, \ldots, t_n$.*

2. *Return $B_{t_i} = \mu\, t_i + \sigma\, W_{t_i}$, $i = 1, \ldots, n$ as the outcomes of the Brownian motion at times $t_1, \ldots, t_n$.*

Let $\{W_{t,i}, t \geqslant 0\}$, $i = 1, \ldots, n$ be independent Wiener processes and let $\mathbf{W}_t = (W_{t,1}, \ldots, W_{t,n})$. The process $\{\mathbf{W}_t, t \geqslant 0\}$ is called an **n-dimensional Wiener process.**

■ EXAMPLE 5.10 (Three-Dimensional Wiener Process)

The following MATLAB program generates a realization of the three-dimensional Wiener process at times $0, 1/N, 2/N, \ldots, 1$, for $N = 10^4$. Figure 5.15 shows a typical realization.

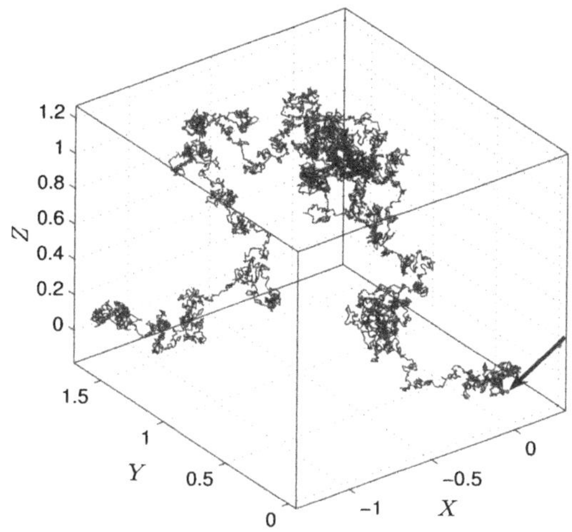

Figure 5.15 Three-dimensional Wiener process $\{\mathbf{W}_t, 0 \leqslant t \leqslant 1\}$. The arrow points at the origin.

```
%wp3d.m
N=10^4; T=1; dt=T/N; %step size
X=cumsum([0,0,0;randn(N,3)*sqrt(dt)],1);
plot3(X(:,1),X(:,2),X(:,3))
```

5.6 STOCHASTIC DIFFERENTIAL EQUATIONS AND DIFFUSION PROCESSES

A **stochastic differential equation** (SDE) for a stochastic process $\{X_t, t \geqslant 0\}$ is an expression of the form

$$\mathrm{d}X_t = a(X_t, t)\,\mathrm{d}t + b(X_t, t)\,\mathrm{d}W_t\,, \tag{5.15}$$

where $\{W_t, t \geqslant 0\}$ is a Wiener process and $a(x, t)$ and $b(x, t)$ are deterministic functions. The coefficient (function) a is called the **drift** and b^2 is called the **diffusion** coefficient — some authors call b the diffusion coefficient. The resulting process $\{X_t, t \geqslant 0\}$ is referred to as an **(Itô) diffusion**.

Stochastic differential equations are based on the same principle as ordinary differential equations, relating an unknown function to its derivatives, but with the additional feature that part of the unknown function is driven by randomness. Intuitively, (5.15) expresses that the infinitesimal change in $\mathrm{d}X_t$ at time t is the sum of an infinitesimal displacement $a(X_t, t)\,\mathrm{d}t$ and an infinitesimal noise term $b(X_t, t)\,\mathrm{d}W_t$. The precise mathematical meaning of (5.15) is that the stochastic process $\{X_t, t \geqslant 0\}$ satisfies the integral equation

$$X_t = X_0 + \int_0^t a(X_s, s)\,\mathrm{d}s + \int_0^t b(X_s, s)\,\mathrm{d}W_s\,, \tag{5.16}$$

where the last integral is an **Itô integral**. The definition of such integrals is ☞ 639 discussed in Section A.12, and the theory behind SDEs is discussed in more detail ☞ 643 in Section A.13. In this section we focus on the computer generation of SDEs, after recalling some of their main properties:

1. *Markov process*: The diffusion process $\{X_t\}$ solving (5.15) is a *Markov process* with continuous sample paths.

2. *Time-homogeneous*: When a and b do not depend on t explicitly (that is, $a(x,t) = \widetilde{a}(x)$, and $b(x,t) = \widetilde{b}(x)$), the diffusion process is a time-homogeneous Markov process. The corresponding SDE is then referred to as being **homogeneous** or **autonomous**.

3. ☞ 644 *Existence and uniqueness*: Theorem A.13.1 gives conditions on functions a and b to guarantee the existence and uniqueness of strong solutions to (5.16) on an interval $[0, T]$. Such solutions express X_t as a functional of the underlying Wiener process $\{W_s, s \leqslant t\}$ and t.

4. *Linear SDEs*: When a and b are linear in x, the SDE is said to be **linear**. For ☞ 645 such SDEs the strong solution X_t can be given explicitly; see Example A.12.

5. *Kolmogorov equations*: If $(p_t(x,y))$ is the transition density of a diffusion that satisfies a homogeneous SDE

$$\mathrm{d}X_t = a(X_t)\,\mathrm{d}t + b(X_t)\,\mathrm{d}W_t\,,$$

☞ 646 then under mild conditions (see Section A.13.1) the transition density satisfies the **Kolmogorov backward equations**

$$\frac{\partial}{\partial t} p_t(x,y) = a(x)\frac{\partial}{\partial x} p_t(x,y) + \frac{1}{2} b^2(x) \frac{\partial^2}{\partial x^2} p_t(x,y) \tag{5.17}$$

and the **Kolmogorov forward equations** (or Fokker–Planck equations)

$$\frac{\partial}{\partial t} p_t(x,y) = -\frac{\partial}{\partial y}\left(a(y)\, p_t(x,y)\right) + \frac{1}{2}\frac{\partial^2}{\partial y^2}\left(b^2(y)\, p_t(x,y)\right)\,. \tag{5.18}$$

Multidimensional SDEs can be defined in a similar way as in (5.15). A stochastic differential equation in $\mathbb{R}^m$ is an expression of the form

$$\mathrm{d}\mathbf{X}_t = \mathbf{a}(\mathbf{X}_t, t)\,\mathrm{d}t + B(\mathbf{X}_t, t)\,\mathrm{d}\mathbf{W}_t\,, \tag{5.19}$$

where $\{\mathbf{W}_t\}$ is an n-dimensional Wiener process, $\mathbf{a}(\mathbf{x}, t)$ is an m-dimensional vector (the drift) and $B(\mathbf{x}, t)$ an $m \times n$ matrix, for each $\mathbf{x} \in \mathbb{R}^m$ and $t \geqslant 0$. The $m \times m$ matrix $C = BB^\top$ is called the **diffusion matrix**. The resulting diffusion process is Markov, and if $\mathbf{a}$ and B do not depend explicitly on t then the diffusion process is time-homogeneous. Existence and uniqueness follow from exactly the same ☞ 646 conditions as in Theorem A.13.2.

We next discuss three general techniques for approximately simulating diffusion processes, and one less general technique for exactly simulating certain diffusion processes. The approximate methods discussed are the *direct Euler* method, *Milstein's* method, and the *implicit Euler* method. The exact method is due to Beskos and Roberts [6].

5.6.1 Euler's Method

Let $\{X_t, t \geqslant 0\}$ be a diffusion process defined by the SDE

$$\mathrm{d}X_t = a(X_t, t)\,\mathrm{d}t + b(X_t, t)\,\mathrm{d}W_t\,, \quad t \geqslant 0\,, \tag{5.20}$$

where X_0 has a known distribution.

The **Euler** or **Euler–Maruyama** method for solving SDEs is a simple generalization of Euler's method for solving ordinary differential equations. The idea is to replace the SDE with the stochastic difference equation

$$Y_{k+1} = Y_k + a(Y_k, kh)\,h + b(Y_k, kh)\sqrt{h}\,Z_k\,, \tag{5.21}$$

where $Z_1, Z_2, \ldots \sim_{\text{iid}} \mathsf{N}(0,1)$. For a small step size h the time series $\{Y_k, k = 0, 1, 2, \ldots\}$ approximates the process $\{X_t, t \geqslant 0\}$; that is, $Y_k \approx X_{kh}$, $k = 0, 1, 2, \ldots$.

Algorithm 5.18 (Euler's Method)

1. *Generate Y_0 from the distribution of X_0. Set $k = 0$.*
2. *Draw $Z_k \sim \mathsf{N}(0,1)$.*
3. *Evaluate Y_{k+1} from (5.21) as an approximation to X_{kh}.*
4. *Set $k = k + 1$ and go to Step 2.*

Remark 5.6.1 (Interpolation) The Euler procedure only returns approximations to $\{X_t\}$ at times that are multiples of the step size h. To obtain approximations for times $s \neq kh$ one could simply approximate X_t by Y_k for $t \in [kh, (k+1)h)$, or use the linear interpolation

$$\left(k + 1 - \frac{t}{h}\right) Y_k + \left(\frac{t}{h} - k\right) Y_{k+1}, \quad t \in [kh, (k+1)h]\,.$$

Remark 5.6.2 (Non-Gaussian Noise) Instead of using a Gaussian noise term $\xi = \sqrt{h}\,Z \sim \mathsf{N}(0,h)$, one could use any random variable ξ for which

$$\mathbb{E}\,\xi = 0, \qquad \mathbb{E}\,\xi^2 = h, \qquad \mathbb{E}\,\xi^3 = \mathcal{O}(h^2), \qquad \mathbb{E}\,\xi^4 = \mathcal{O}(h^2)\,.$$

A convenient choice is to take $\xi = (2B-1)\sqrt{h}$, where $B \sim \mathsf{Ber}(1/2)$; that is, $\mathbb{P}(\xi = \sqrt{h}) = \mathbb{P}(\xi = -\sqrt{h}) = \frac{1}{2}$. See [2, Page 296] and [18, Page 464] for implementations of this idea.

For a multidimensional SDE of the form (5.19) the Euler method has the following simple generalization.

Algorithm 5.19 (Multidimensional Euler Method)

1. *Generate* $\mathbf{Y}_0$ *from the distribution of* $\mathbf{X}_0$. *Set* $k = 0$.

2. *Draw* $\mathbf{Z}_k \sim \mathsf{N}(\mathbf{0}, I)$.

3. *Evaluate*
$$\mathbf{Y}_{k+1} = \mathbf{Y}_k + \mathbf{a}(\mathbf{Y}_k, kh)\, h + B(\mathbf{Y}_k, kh)\, \sqrt{h}\, \mathbf{Z}_k$$
as an approximation to $\mathbf{X}_{kh}$.

4. *Set* $k = k + 1$ *and go to Step 2.*

■ **EXAMPLE 5.11 (Simplified Duffing–Van der Pol Oscillator)**

Consider the following two-dimensional SDE [18, Page 427]:

$$\begin{aligned} \mathrm{d}X_t &= Y_t\, \mathrm{d}t\,, \\ \mathrm{d}Y_t &= \left(X_t\left(\alpha - X_t^2\right) - Y_t\right)\, \mathrm{d}t + \sigma X_t\, \mathrm{d}W_t\,. \end{aligned}$$

The following MATLAB code generates the process with parameters $\alpha = 1$ and $\sigma = 1/2$ for $t \in [0, 1000]$, using a step size $h = 10^{-3}$ and starting at $(-2, 0)$; see Figure 5.16. Note that the process oscillates between two modes.

```
%vdpol.m
alpha = 1; sigma = 0.5;
a1 = @(x1,x2,t) x2;
a2 = @(x1,x2,t) x1*(alpha-x1^2)-x2;
b1 = @(x1,x2,t) 0 ;
b2 = @(x1,x2,t) sigma*x1;
n=10^6; h=10^(-3); t=h.*(0:1:n); x1=zeros(1,n+1); x2=x1;
x1(1)=-2;
x2(1)=0;
for k=1:n
    x1(k+1)=x1(k)+a1(x1(k),x2(k),t(k))*h+ ...
               b1(x1(k),x2(k),t(k))*sqrt(h)*randn;
    x2(k+1)=x2(k)+a2(x1(k),x2(k),t(k))*h+ ...
               b2(x1(k),x2(k),t(k))*sqrt(h)*randn;
end
step = 100; %plot each 100th value
figure(1),plot(t(1:step:n),x1(1:step:n),'k-')
figure(2), plot(x1(1:step:n),x2(1:step:n),'k-');
```

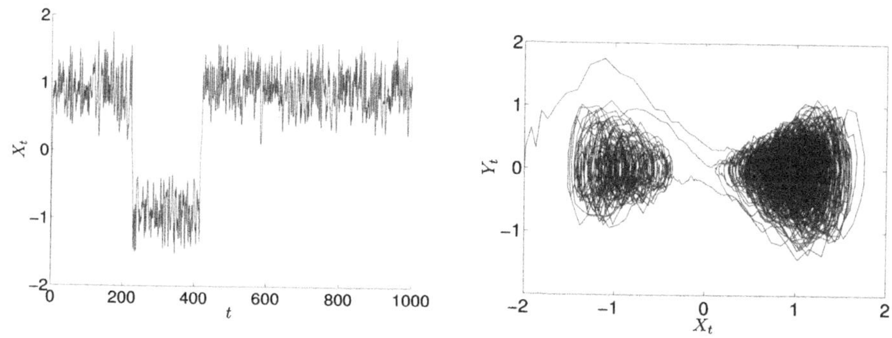

Figure 5.16 A typical realization with parameters $\alpha = 1$ and $\sigma = 1/2$, starting at $(-2, 0)$.

5.6.2 Milstein's Method

By Itô's lemma in $\mathbb{R}^m$, ☞ 642

$$\begin{aligned}
\mathrm{d}b(X_s, s) &= b_x(X_s, s)\,\mathrm{d}X_s + b_s(X_s, s)\,\mathrm{d}s + \frac{1}{2} b_{xx}(X_s, s)\,\mathrm{d}[X_s, X_s] \\
&= b_x(X_s, s)\,\{a(X_s, s)\mathrm{d}s + b(X_s, s)\,\mathrm{d}W_s\} \\
&\quad + b_s(X_s, s)\mathrm{d}s + \frac{1}{2} b_{xx}(X_s, s)\, b(X_s, s)^2\,\mathrm{d}s\;,
\end{aligned}$$

where b_x, b_t, and b_{xx} are the corresponding partial derivatives of $b(x, t)$. It follows that

$$\begin{aligned}
X_{t+h} &= X_t + \int_t^{t+h} a(X_u, u)\,\mathrm{d}u + \int_t^{t+h} b(X_u, u)\,\mathrm{d}W_u \\
&= X_t + h\,a(X_t, t) + b(X_t, t)(W_{t+h} - W_t) + \mathcal{O}(h\sqrt{h}) + \\
&\quad + \int_t^{t+h} \int_t^u b_x(X_s, s)\, b(X_s, s)\,\mathrm{d}W_s\,\mathrm{d}W_u\;,
\end{aligned}$$

where the last term can be written as

$$b_x(X_t, t)\, b(X_t, t) \frac{1}{2}((W_{t+h} - W_t)^2 - h) + \mathcal{O}(h^2)\;.$$

This suggests that we can replace the SDE (5.20) with the difference equation

$$Y_{k+1} = Y_k + a(Y_k, kh)\,h + b(Y_k, kh)\sqrt{h}\,Z_k + \underbrace{b_x(Y_k, kh)\, b(Y_k, kh)(Z_k^2 - 1)\frac{h}{2}}_{\text{additional term}}\;, \qquad (5.22)$$

where $Z_1, Z_2, \ldots \stackrel{\text{iid}}{\sim} \mathsf{N}(0, 1)$. This is **Milstein's method**. The only difference with the Euler method is the additional term involving the partial derivative of $b(x, t)$ with respect to x. The approximation algorithm is now as follows.

Algorithm 5.20 (Milstein's Method)

1. *Generate* Y_0 *from the distribution of* X_0. *Set* $k = 0$.
2. *Draw* $Z_k \sim \mathsf{N}(0, 1)$.
3. *Evaluate* Y_{k+1} *from* (5.22) *as an approximation to* X_{kh}.
4. *Set* $k = k + 1$ *and go to Step 2.*

5.6.3 Implicit Euler

In the **implicit Euler** method the difference equation (5.21) is replaced with

$$Y_{k+1} = Y_k + a(\underbrace{Y_{k+1}}_{\text{differs from Euler}}, kh)\, h + b(Y_k, kh)\, \sqrt{h}\, Z_k\,. \tag{5.23}$$

Note that Y_k now has to be solved from (5.23).

■ **EXAMPLE 5.12 (Geometric Brownian Motion)**

We illustrate the different SDE generation methods via the geometric Brownian motion SDE (see Section 5.8):

$$\mathrm{d}X_t = \mu\, X_t\, \mathrm{d}t + \sigma X_t\, \mathrm{d}W_t$$

for $t \in [0, T]$ and initial value X_0. The (strong) solution is given by

$$X_t = X_0 \exp\left(\left(\mu - \frac{\sigma^2}{2}\right)t + \sigma\, W_t\right),$$

for $t \in [0, T]$ and the Wiener process $\{W_t, t \geqslant 0\}$.

In order to compare the Euler and Milstein schemes to the above exact solution, we use the *same* random variables $V_1, V_2, \ldots \sim_{\text{iid}} \mathsf{N}(0, h)$ for a given step size h.

The exact solution at time nh can be written as

$$X_{nh} = X_0 \exp\left(\left(\mu - \frac{\sigma^2}{2}\right)nh + \sigma \sum_{k=1}^{n} V_k\right).$$

Euler's method gives the approximation

$$Y_n^{(e)} = Y_{n-1}^{(e)}\,(1 + \mu\, h + \sigma V_n)\,, \quad n = 1, 2, \ldots\,.$$

Milstein's method gives the approximation

$$Y_n^{(m)} = Y_{n-1}^{(m)}\left(1 + \mu\, h + \sigma V_n + \frac{\sigma^2}{2}\,(V_n^2 - h)\right), \quad n = 1, 2, \ldots\,.$$

For values not at the approximation points, we use linear interpolation.

To compare the different approximations we use several different step sizes, $h, 2h, 4h, \ldots,$ while ensuring that the appropriate common random numbers are used. In particular, if we use a step size $h = mh$, then for $k = 1, 2, \ldots$ set

$$\widetilde{V}_k = \sum_{j=1}^{m} V_{(k-1)m+j}\,.$$

The exact solution at time $n\widetilde{h}$ on the $\widetilde{h}$ time scale is generated as

$$X_{n\widetilde{h}} = X_0 \exp\left(\left(\mu - \frac{\sigma^2}{2}\right)n\widetilde{h} + \sigma \sum_{k=1}^{n} \widetilde{V}_k\right) = X_0 \exp\left(\left(\mu - \frac{\sigma^2}{2}\right)n\widetilde{h} + \sigma \sum_{j=1}^{mn} V_j\right).$$

Euler's approximation on this time scale is computed as

$$\widetilde{Y}_n^{(e)} = \widetilde{Y}_{n-1}^{(e)}\left(1 + \mu\widetilde{h} + \sigma\widetilde{V}_n\right) = \widetilde{Y}_{n-1}^{(e)}\left(1 + \mu\widetilde{h} + \sigma\sum_{j=1}^{m} V_{(n-1)m+j}\right),$$

and similarly for the Milstein scheme.

Figure 5.17 depicts the output of the exact, Euler, and Milstein schemes for the case $\mu = 2$, $\sigma = 1$, and initial value $X_0 = 1$. The different time steps used were $2^{-2}, 2^{-4}, 2^{-6}$, and 2^{-8}. The corresponding MATLAB program `gbm_comp.m` can be found on the Handbook website.

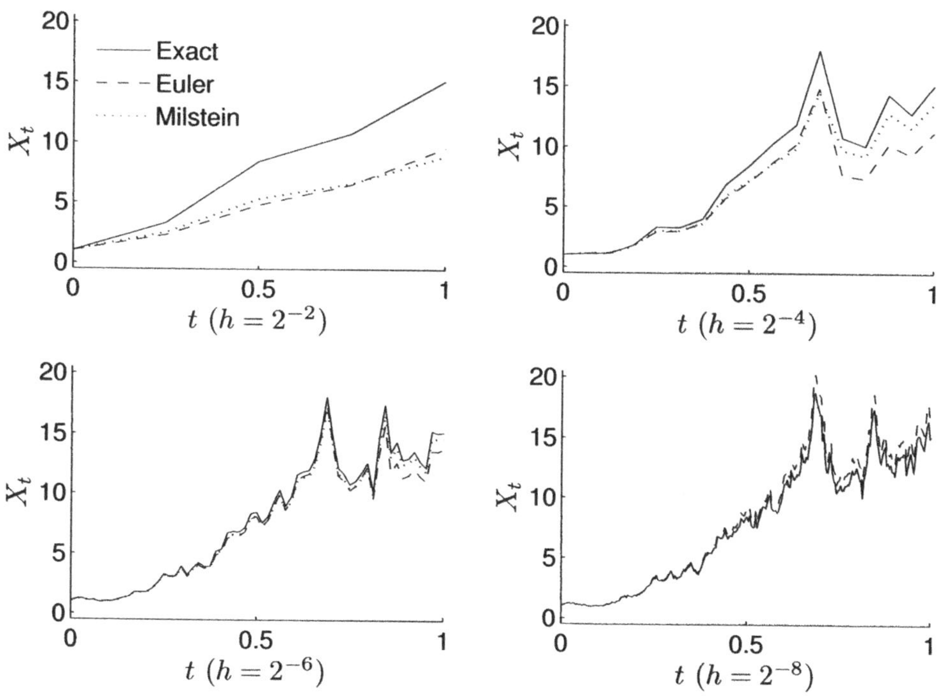

Figure 5.17 Approximation schemes and exact solution for a geometric Brownian motion trajectory.

5.6.4 Exact Methods

In contrast to approximate simulation schemes, for certain classes of SDEs *exact* simulation algorithms are known [6]. Suppose that we have a general one-dimensional autonomous SDE, given by

$$\mathrm{d}Y_t = a(Y_t)\,\mathrm{d}t + b(Y_t)\,\mathrm{d}W_t\,, \quad 0 \leqslant t \leqslant T\,,\, Y_0 = y_0\,.$$

Such an SDE can be transformed into one with unit diffusion coefficient via a **Lamperti transform** (see, for example, [15, Page 40]):

$$X_t = F(Y_t) - F(y_0)\,,$$

where

$$F(y) = \int_z^y \frac{1}{b(u)}\,\mathrm{d}u\,,$$

and z is an arbitrary point in the state space of $\{Y_t\}$. The transformed process $\{X_t\}$ satisfies the unit variance autonomous SDE

$$\mathrm{d}X_t = \alpha(X_t)\,\mathrm{d}t + \mathrm{d}W_t\,, \quad 0 \leqslant t \leqslant T\,,\, X_0 = 0\,, \tag{5.24}$$

where

$$\alpha(x) = \frac{a(F^{-1}(x + F[y_0]))}{b(F^{-1}(x + F[y_0]))} - \frac{1}{2}\,b'(F^{-1}(x + F[y_0]))\,.$$

We now give conditions on the SDE (5.24) and its drift coefficient α under which exact simulation can be performed [6].

Assumptions 5.6.1 (Exact Simulation)

☞ 643

1. *There is a unique weak solution to* (5.24) *(see Section A.13).*

2. *The drift coefficient α is everywhere differentiable, with derivative α'.*

3. *There exist constants $k_1, k_2 \in \mathbb{R}$ such that $k_1 \leqslant \frac{1}{2}\left(\alpha^2(u) + \alpha'(u)\right) \leqslant k_2$ for any $u \in \mathbb{R}$.*

4. *The following condition is satisfied:*

$$c = \int_{-\infty}^{\infty} \exp\left(A(u) - \frac{u^2}{2T}\right)\,\mathrm{d}u < \infty\,,$$

where $A(u) = \int_0^u \alpha(y)\,\mathrm{d}y$ for all $y \in \mathbb{R}$.

In particular, Points 3 and 4 of Assumptions 5.6.1 hold if there exist constants $k_1', k_2' \in \mathbb{R}$ such that $k_1' \leqslant \alpha(u)$ and $\alpha'(u) \leqslant k_2'$ for any $u \in \mathbb{R}$.

Before giving the exact algorithm, we briefly provide some intuition. The exact algorithm is of acceptance–rejection type, where candidate paths over $[0, T]$ are drawn from a *Brownian motion* process, conditional on an exactly sampled endpoint (that is, a *Brownian bridge* process, see Section 5.7). Unlike typical acceptance–rejection algorithms, a complete realization of the candidate path is not needed before deciding whether to accept or reject the sample path. Instead, if a particular random skeleton of a candidate path is accepted or rejected, then further refinement of the path does not change the decision. As a consequence, valid paths on any time discretization can be drawn by first simulating an accepted random skeleton, and subsequently filling in the gaps using Brownian bridge processes. The exact algorithm that returns such a random skeleton is as follows [6].

Algorithm 5.21 (Exact Acceptance–Rejection Algorithm for SDEs) *Given a fixed terminal time T satisfying $0 < T < 1/(k_2 - k_1)$, execute the following steps.*

1. *Fix the endpoints of the candidate Brownian bridge $\{Y_t, 0 \leqslant t \leqslant T\}$ by setting $Y_0 = 0$ and $Y_T = Z$, where $Z \sim h$, with $h(y) = \exp\left(A(y) - \frac{y^2}{2T}\right)/c$.*
2. *Draw $U \sim \mathsf{U}(0,1)$ and set $k = 0$.*
3. *Draw $V \sim \mathsf{U}(0,T)$ and $W \sim \mathsf{U}(0,1/T)$ independently, and set $k = k+1$.*
4. *Sample Y_V from a Brownian bridge (see Algorithm 5.23), given the current random skeleton.*
5. *Let $\phi(u) = \frac{1}{2}\left(\alpha^2(u) + \alpha'(u)\right) - k_1$.*
   (a) *If $\phi(Y_V) < W$ or $U > 1/k!$, then:*
      i. *If k is even, the random skeleton is rejected. Return to Step 1.*
      ii. *If k is odd, the random skeleton is accepted. Proceed to Step 6.*
   (b) *Otherwise, return to step 3.*
6. *Output the current random skeleton as a realization from* (5.24).

If further refinement at some time s is required (for example, if we wish to simulate on a regular time-mesh), simply proceed as in Step 4 and sample Y_s from a Brownian bridge given the current skeleton. To draw from the univariate distribution h, apply the techniques discussed in Chapter 3. ☞ 43

5.6.5 Error and Accuracy

Most approaches for solving SDEs, such as Euler's or Milstein's methods, are not exact (Algorithm 5.21 is an exception). For example, Y_k in (5.21) is not exactly distributed as X_{kh}. Moreover, even if the process $\{X_t\}$ could be simulated exactly at discrete time steps $t_1, \ldots, t_n$, the path between those time steps would still need to be approximated — for example, via linear interpolation.

The accuracy of approximating an m-dimensional stochastic process $\{\mathbf{X}_t, t \geqslant 0\}$ with a simulated process $\{\widetilde{\mathbf{X}}_t, t \geqslant 0\}$ can be measured using various error criteria. If one is interested solely in generating from the distribution of $\mathbf{X}_T$ accurately, then one criterion is

$$\sup_{g \in \mathscr{C}} \left| \mathbb{E} g(\mathbf{X}_T) - \mathbb{E} g(\widetilde{\mathbf{X}}_T) \right| , \tag{5.25}$$

where $\mathscr{C}$ is a suitable class of smooth functions $g : \mathbb{R}^m \to \mathbb{R}$. If instead one is interested in the accuracy of the entire path and $\mathbf{X}$ and $\widetilde{\mathbf{X}}$ can be generated on the same probability space, then natural error criteria are

$$\mathbb{E} \sup_{t \in [0,T]} \left\| \mathbf{X}_t - \widetilde{\mathbf{X}}_t \right\| \quad \text{and} \quad \mathbb{E} \int_0^T \left\| \mathbf{X}_t - \widetilde{\mathbf{X}}_t \right\|^2 \mathrm{d}t . \tag{5.26}$$

We assume from now on that the process of interest, $\{\mathbf{X}_t, t \geqslant 0\}$, is approximated by a discrete-time process $\mathbf{X}^h = \{\mathbf{X}_t^h, t = 0, h, 2h, \ldots\}$, as is typically the case in the SDE setting. Denote by $\mathscr{C}_P^r$ the space of r times continuously differentiable functions $g : \mathbb{R}^m \to \mathbb{R}$ that have polynomial growth as well as polynomial growth of their partial derivatives of order up to and including r. The following definitions are motivated by the criteria (5.25) and (5.26).

The process $\mathbf{X}^h$ is said to **converge weakly with order $\beta > 0$** to $\mathbf{X}$ at time T as $h \downarrow 0$ if for each $g \in \mathscr{C}_P^{2(\beta+1)}$ there exists a positive constant C, independent of h, as well as a finite $h_0 > 0$, such that

$$\left|\mathbb{E}g(\mathbf{X}_T) - \mathbb{E}g(\mathbf{X}_T^h)\right| \leqslant C h^\beta \quad \text{for all } h \in (0, h_0)\,.$$

The process $\mathbf{X}^h$ is said to **converge strongly with order $\gamma > 0$** to $\mathbf{X}$ at time T if there exists a positive constant C, independent of h, as well as some $h_0 > 0$, such that

$$\mathbb{E}\left\|\mathbf{X}_T - \mathbf{X}_T^h\right\| \leqslant C h^\gamma \quad \text{for all } h \in (0, h_0)\,.$$

We now discuss the convergence order of Euler's and Milstein's method. Kloeden and Platen [18] show that Euler's method converges strongly with order $\gamma = 1/2$ under assumptions slightly stronger than those given in Theorem A.13.2. In particular, we have the following result from [18, Page 342]. ☞ 646

Theorem 5.6.1 (Weak Order Convergence of Euler's method) *Suppose the coefficients of the diffusion SDE* (5.19) *satisfy the assumptions of Theorem A.13.2, and additionally:*

$$\mathbb{E}\|\mathbf{X}_0 - \mathbf{X}_0^h\| \leqslant C\, h^{1/2}$$

and (for all $\mathbf{x} \in \mathbb{R}^m$ and $s, t \in [0, T]$)

$$\|\mathbf{a}(\mathbf{x}, s) - \mathbf{a}(\mathbf{x}, t)\| + \|B(\mathbf{x}, s) - B(\mathbf{x}, t)\| \leqslant C(1 + \|\mathbf{x}\|)|s - t|^{1/2} \qquad (5.27)$$

for some constant C. Then, Euler's method converges strongly with order $\gamma = 1/2$.

Kloeden and Platen [18, Page 473] show that if $\mathbf{a}$ and B are four times continuously differentiable in all components (spatial and temporal) and a number of its derivatives satisfy the linear growth condition (A.73), then Euler's scheme converges with weak order $\beta = 1$. They note that if the coefficients are only Lipschitz continuous (see (A.74)), then Euler's method still converges weakly, but with weak order $\beta < 1$. ☞ 644

Kloeden and Platen [18, Page 350] also show that if $\mathbf{a}$ and B, and a number of its derivatives satisfy the linear growth condition (A.73), Lipschitz continuity (A.74), and (5.27), then Milstein's method converges strongly with order $\gamma = 1$. However, the weak order convergence is still $\beta = 1$ — the same as for Euler's method.

Finally, if we use interpolation between the points of the approximation, then under some technical conditions (see [18, Pages 361–362]) a strongly convergent scheme of order γ at time T is also strongly convergent uniformly over the whole path in the sense that

$$\mathbb{E} \sup_{0 \leqslant t \leqslant T} \left\|\mathbf{X}_t - \mathbf{X}_t^h\right\| \leqslant C h^\gamma\,.$$

5.7 BROWNIAN BRIDGE

The **standard Brownian bridge** process $\{X_t, t \in [0,1]\}$ is a stochastic process whose distribution is that of the Wiener process on $[0,1]$ *conditioned* upon $X_1 = 0$. In other words, the Wiener process is "tied-down" to be 0 at both time 0 and time 1. It plays a crucial role in the Kolmogorov–Smirnov test. The standard Brownian bridge process can be viewed as a nonhomogeneous diffusion process satisfying the linear SDE ☞ 336

$$\mathrm{d}X_t = -\frac{X_t}{1-t}\,\mathrm{d}t + \mathrm{d}W_t, \quad 0 \leqslant t < 1, \quad X_0 = 0\,.$$

Using the general solution of linear SDEs in Example A.12 and the fact that $X_1 = 0$, we find that the strong solution is given by ☞ 645

$$X_t = \int_0^t \frac{1-t}{1-s}\,\mathrm{d}W_s\,, \quad 0 \leqslant t \leqslant 1\,.$$

A realization of the process is given in Figure 5.18.

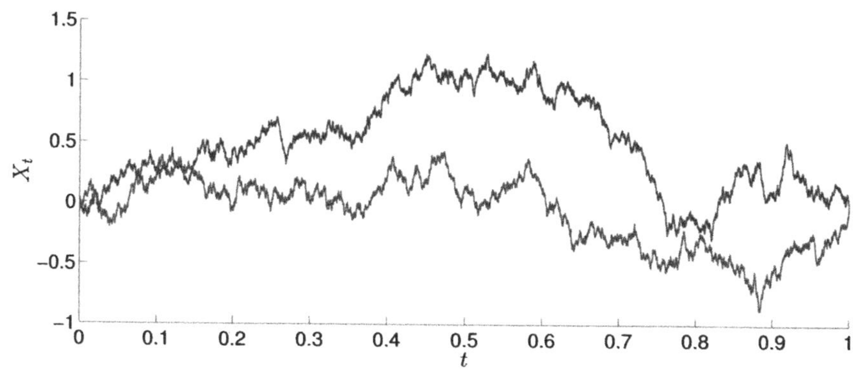

Figure 5.18 Two realizations of the standard Brownian bridge process on the interval $[0,1]$.

Some properties of the standard Brownian bridge process include:

1. If $\{W_t\}$ is a Wiener process, then,

$$X_t = (1-t)\,W_{\frac{t}{1-t}}, \quad 0 \leqslant t \leqslant 1\,,$$

defines a standard Brownian bridge (see the time-change property, Property 8 on Page 642), where $X_1 = \lim_{t\uparrow 1}(1-t)\,W_{\frac{t}{1-t}} = 0$, almost surely. ☞ 642

2. If $\{W_t\}$ is a Wiener process, then,

$$X_t = W_t - t\,W_1, \quad 0 \leqslant t \leqslant 1\,,$$

defines a standard Brownian bridge.

3. A standard Brownian bridge is a Gaussian process with mean 0 and covariance function $\mathrm{Cov}(X_s, X_t) = \min\{s,t\} - s\,t$.

4. The random variable $K = \sup_{t\in[0,1]} |X_t|$, where $\{X_t\}$ is the standard Brownian bridge process, has the *Kolmogorov distribution*

$$\mathbb{P}(K \leqslant x) = 1 - 2\sum_{n=1}^{\infty}(-1)^{n-1}\mathrm{e}^{-2n^2x^2} = \frac{\sqrt{2\pi}}{x}\sum_{n=1}^{\infty}\mathrm{e}^{-(2n-1)^2\pi^2/(8x^2)}.$$

Generation of a sample path of the standard Brownian bridge is most easily accomplished using Property 2 above.

Algorithm 5.22 (Generating a Standard Brownian Bridge)

1. *Let $0 = t_0 < t_1 < t_2 < \cdots < t_{n+1} = 1$ be the set of distinct times for which simulation of the process is desired.*

2. *Generate a Wiener process on $t_1, \ldots, t_n$ using, for example, Algorithm 5.15.*

3. *Set $X_0 = 0$ and $X_1 = 0$, and output for each $k = 1, \ldots, n$:*

$$X_{t_k} = W_{t_k} - t_k\, W_{t_n}.$$

A general (nonstandard) **Brownian bridge** is a stochastic process $\{X_t, t \in [t_0, t_{n+1}]\}$ whose distribution in that of the Wiener process on $[t_0, t_{n+1}]$ conditioned upon $X_{t_0} = a$ and $X_{t_{n+1}} = b$. Generalizing Property 3, the Brownian bridge is a Gaussian process with mean function

$$\mu_t = a + \frac{(b-a)(t-t_0)}{(t_{n+1}-t_0)}$$

and covariance function

$$\mathrm{Cov}(X_s, X_t) = \min\{s - t_0,\, t - t_0\} - \frac{(s-t_0)(t-t_0)}{(t_{n+1}-t_0)}\,.$$

A frequent use of the Brownian bridge is for adaptive refinement of the discrete-time approximation of the Wiener process. Suppose that we have generated the Wiener process at certain time instants and wish to generate the process for additional times. Specifically, suppose $W_{t_0} = a$ and $W_{t_{n+1}} = b$ and we wish to generate the Wiener process for additional times $t_1, \ldots, t_n$ in the interval $[t_0, t_{n+1}]$.

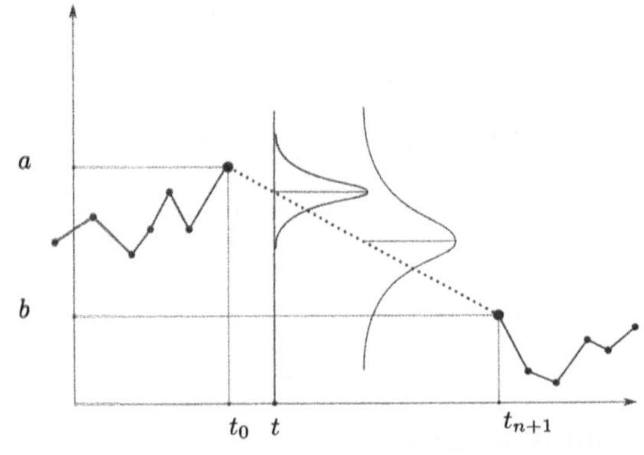

Figure 5.19 Conditional on $W_{t_0} = a$ and $W_{t_{n+1}} = b$, the point $W_t, t \in [t_0, t_{n+1}]$ has a normal distribution.

Conditional on $W_{t_0} = a$ and $W_{t_{n+1}} = b$, the process $\{W_t, t \in [t_0, t_{n+1}]\}$ is a Brownian bridge; see Figure 5.19. In particular, the distribution of W_t with $t \in [t_0, t_{n+1}]$ is normal with mean $a + (b - a)(t - t_0)/(t_{n+1} - t_0)$ and variance $(t_{n+1} - t)(t - t_0)/(t_{n+1} - t_0)$. We can thus generate the process at any number of additional points $t_1 < \cdots < t_n$ within the interval $[t_0, t_{n+1}]$ using the following algorithm.

Algorithm 5.23 (Brownian Bridge Sampling Between W_{t_0} and $W_{t_{n+1}}$)

1. *Generate* $Z_1, \ldots, Z_n \overset{\text{iid}}{\sim} \mathsf{N}(0, 1)$.
2. *For each* $k = 1, \ldots, n$, *output:*

$$W_{t_k} = W_{t_{k-1}} + (b - W_{t_{k-1}})\frac{t_k - t_{k-1}}{t_{n+1} - t_{k-1}} + \sqrt{\frac{(t_{n+1} - t_k)(t_k - t_{k-1})}{t_{n+1} - t_{k-1}}}\, Z_k\,.$$

In principle we need not generate the intermediate points $W_{t_1}, \ldots, W_{t_n}$ in any particular order as long as at each step we condition on the two closest time points already sampled. The following code implements the algorithm given above.

```
function X=brownian_bridge(t,x_r,x_s,Z)
n=length(t)-2;X=nan(1,n+2);
X(1)=x_r; X(n+2)=x_s;s=t(n+2);
for k=2:n+1
    mu=X(k-1)+(x_s-X(k-1))*(t(k)-t(k-1))/(s-t(k-1));
    sig2=(s-t(k))*(t(k)-t(k-1))/(s-t(k-1));
    X(k)=mu+sqrt(sig2)*Z(k-1);
end
```

Sampling using a Brownian bridge is useful in combination with *stratification.* ☞ 356
For example, in option pricing via simulation (see Chapter 15) one frequently has to ☞ 521
compute the expectation of payoffs that depend on the terminal value of the underlying asset price process. The variance in the estimate can be reduced by stratifying the asset price process along the terminal value. In the following example we generate Brownian motion that is stratified along its terminal value. Stratification for geometric Brownian motion — the most widely used asset price model — is essentially the same, because it can be represented as a monotone transformation of Brownian motion.

■ EXAMPLE 5.13 (Stratified Brownian Motion)

Suppose we wish to construct a Brownian motion path starting at 0 and finishing at $t = 1$ in K distinct strata, as in Figure 5.20. For simplicity we assume $\mu = 0$ and $\sigma = 1$, so that the Brownian motion reduces to a Wiener process. The following code generates K stratified values at $t = 1$ and then conditional on these values samples Brownian paths in the interval $[0, 1]$ using Algorithm 5.23. Figure 5.20 depicts an outcome involving $K = 7$ strata.

```
%brownian_bridge_stratification.m
K=7;    %number of strata
n=10^3; %number of points for path construction
T=1; % terminal time at which we stratify
t=[0:T/(n+1):T];
for i=1:K
   U=(i-1+rand)/K;   % stratified uniforms
   x_s=sqrt(T)*norminv(U); % stratified terminal value
   X=brownian_bridge(t,0,x_s,randn(n,1));
   plot(t,X), hold all
end
```

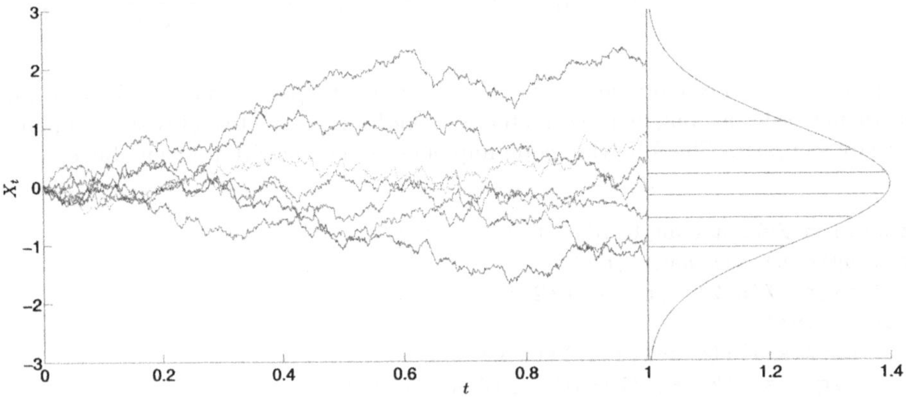

Figure 5.20 Terminal stratification of the Wiener process at $t = 1$. There are seven different paths which finish in the $K = 7$ strata depicted at $t = 1$.

5.8 GEOMETRIC BROWNIAN MOTION

The **geometric Brownian motion** process satisfies the homogeneous linear SDE

$$\mathrm{d}X_t = \mu\, X_t\, \mathrm{d}t + \sigma\, X_t\, \mathrm{d}W_t\ ,$$

which has strong solution

$$X_t = X_0\, \mathrm{e}^{(\mu - \frac{\sigma^2}{2})\, t + \sigma W_t}, \quad t \geqslant 0\ .$$

The special case where $\mu = 0$ and $\sigma = 1$ is called the **stochastic exponential** of the Wiener process $\{W_t\}$.

☞ 524 An important application is the Black–Scholes equity model (see Example 15.1), where X_t is the price of the equity at time t, and the interest rate at time t is modeled as the sum of a fixed rate μ and an uncertain rate $\sigma\eta_t$, with $\eta_t = \mathrm{d}W_t/\mathrm{d}t$ denoting Gaussian "white noise" and σ a volatility factor.

One usually takes $X_0 = 1$. The expectation and variance of X_t are then given by

$$\mathbb{E}X_t = e^{\mu t} \qquad \text{and} \qquad \text{Var}(X_t) = e^{2\mu t}\left(e^{\sigma^2 t} - 1\right) .$$

The strong solution of the geometric Brownian motion suggests the following exact simulation algorithm.

Algorithm 5.24 (Geometric Brownian Motion) *Let $0 = t_0 < t_1 < t_2 < \cdots < t_n$ be the set of distinct times for which simulation of the process is desired.*

1. *Generate $Z_1, \ldots, Z_n \stackrel{\text{iid}}{\sim} \mathsf{N}(0,1)$.*
2. *Output*

$$X_{t_k} = X_0 \exp\left(\left(\mu - \frac{\sigma^2}{2}\right)t_k + \sigma \sum_{i=1}^{k} \sqrt{t_i - t_{i-1}}\, Z_i\right), \quad k = 1, \ldots, n .$$

The following code implements the algorithm.

```
%geometricbm.m
T=1; % final time
n=10000; h=T/(n-1); t= 0:h:T;
mu = 1; sigma = 0.2; xo=1; % parameters
W = sqrt(h)*[0, cumsum(randn(1,n-1))];
x = xo*exp((mu - sigma^2/2)*t + sigma*W);
plot(t,x)
hold on
plot(t, exp(mu*t),'r'); %plot exact mean function
```

Figure 5.21 depicts two realizations of a geometric Brownian motion on the interval $[0, 1]$, with parameters $\mu = 1$ and $\sigma = 0.2$.

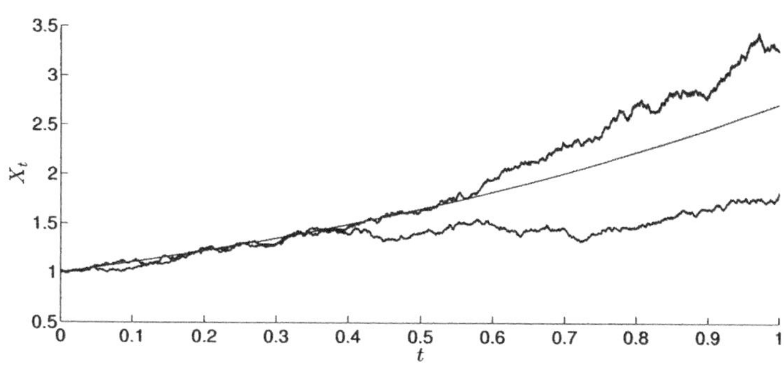

Figure 5.21 Two realizations of a geometric Brownian motion on the interval $[0, 1]$, with parameters $\mu = 1$ and $\sigma = 0.2$. The smooth line depicts the expectation function.

5.9 ORNSTEIN–UHLENBECK PROCESS

The **Ornstein–Uhlenbeck** process satisfies the SDE

$$\mathrm{d}X_t = \theta(\nu - X_t)\,\mathrm{d}t + \sigma\,\mathrm{d}W_t\,, \tag{5.28}$$

with $\sigma > 0$, $\theta > 0$, and $\nu \in \mathbb{R}$. The (strong) solution of this linear SDE is given by

$$X_t = \mathrm{e}^{-\theta t} X_0 + \nu(1 - \mathrm{e}^{-\theta t}) + \sigma \mathrm{e}^{-\theta t} \int_0^t \mathrm{e}^{\theta s}\,\mathrm{d}W_s\,.$$

It follows that $\{X_t\}$ is a Gaussian process whenever X_0 is Gaussian, with mean function

$$\mathbb{E}X_t = \mathrm{e}^{-\theta t}\,\mathbb{E}X_0 + \nu(1 - \mathrm{e}^{-\theta t})$$

and covariance function

$$\mathrm{Cov}(X_s, X_t) = \frac{\sigma^2}{2\theta}\,\mathrm{e}^{-\theta(s+t)}\left(\mathrm{e}^{2\theta \min\{s,t\}} - 1\right).$$

In particular,

$$\mathrm{Var}(X_t) = \sigma^2 \frac{1 - \mathrm{e}^{-2\theta t}}{2\theta}\,.$$

This shows that X_t converges in distribution to a $\mathsf{N}(\nu, \sigma^2/(2\theta))$ random variable as $t \to \infty$. Moreover, when X_0 has this limiting distribution, the Markov process
☞ 631 $\{X_t\}$ is stationary and time-reversible. Another way to see this is to consider the forward equation for the Ornstein–Uhlenbeck SDE. The forward equation gives the following ODE for the stationary distribution ($\theta > 0$):

$$\frac{\sigma^2}{2}\frac{\mathrm{d}^2}{\mathrm{d}y^2}\left(\pi(y)\right) - \frac{\mathrm{d}}{\mathrm{d}y}\left(\theta(\nu - y)\pi(y)\right) = 0\,.$$

The solution is (see Section A.13.2 on Page 648)

$$\pi(x) = \frac{c}{\sigma^2}\exp\left(\int_{x_0}^{x} \frac{2\theta(\nu - y)\,\mathrm{d}y}{\sigma^2}\right) \propto \exp\left(-\frac{(x-\nu)^2}{2\,\sigma^2/(2\theta)}\right),$$

which we recognize as the density of the $\mathsf{N}(\nu, \sigma^2/(2\theta))$ distribution. Also, by the time-change property (Property 8 on Page 642) we have that

$$X_t = \mathrm{e}^{-\theta t} X_0 + \nu(1 - \mathrm{e}^{-\theta t}) + \sigma \mathrm{e}^{-\theta t}\, W\left(\frac{\mathrm{e}^{2t\theta} - 1}{2\theta}\right), \quad t \geqslant 0$$

defines an Ornstein–Uhlenbeck process, where $\{W(t) = W_t\}$ is a Wiener process. In particular, if $Y_t = X_t - \nu(1 - \mathrm{e}^{-\theta t})$, $t \geqslant 0$, then $\{Y_t\}$ is an Ornstein–Uhlenbeck process with $\nu = 0$, $Y_0 = X_0$ and exact solution:

$$Y_t = \mathrm{e}^{-\theta t} Y_0 + \sigma W\left(\frac{1 - \mathrm{e}^{-2t\theta}}{2\theta}\right), \quad t \geqslant 0\,. \tag{5.29}$$

It follows that if we can simulate from an Ornstein–Uhlenbeck process with $\nu = 0$, say $\{Y_t\}$, then we can also simulate from an Ornstein–Uhlenbeck process, say $\{X_t\}$, with $\nu \neq 0$. The following exact algorithm simulates $\{Y_t\}$ exactly and then uses the relation $X_t = Y_t + \nu(1 - \mathrm{e}^{-\theta t})$ to construct a sample path for $\{X_t\}$. The algorithm simulates $\{X_t\}$ exactly for any discretization, see [12].

Algorithm 5.25 (Generating an Ornstein–Uhlenbeck Process) *Let $0 = t_0 < t_1 < t_2 < \cdots < t_n$ be the set of distinct times for which simulation of the process is desired.*

1. *If X_0 is random, draw X_0 from its distribution. Set $Y_0 = X_0$.*

2. *For $k = 1, 2, \ldots, n$ compute:*

$$Y_k = \mathrm{e}^{-\theta(t_k - t_{k-1})} Y_{k-1} + \sigma \sqrt{\frac{1 - \mathrm{e}^{-2\theta(t_k - t_{k-1})}}{2\theta}}\, Z_k\,,$$

where $Z_1, \ldots, Z_n \overset{\text{iid}}{\sim} \mathsf{N}(0,1)$.

3. *Output $(X_0, X_{t_1}, \ldots, X_{t_n})$, where*

$$X_{t_k} = Y_k + \nu(1 - \mathrm{e}^{-\theta t_k}), \quad k = 1, \ldots, n\,.$$

A realization of the process is given in Figure 5.22 using three different starting conditions. The following code implements the algorithm.

```
%ou_timechange_ex.m
T=4; % final time
N=10^4; % number of steps
theta=2;nu=1;sig=0.2; % parameters
x=nan(N,1); x(1)=0;     % initial point
h=T/(N-1); % step size
t=0:h:T;   % time
% code the right-hand side of the updating formula
f=@(z,x)(exp(-theta*h)*x+sig*sqrt((1-exp(-2*h*theta))/(2*theta))*z);
for i=2:N
    x(i)=f(randn,x(i-1));
end
x=x+nu*(1-exp(-theta*t'));
plot(t,x)
```

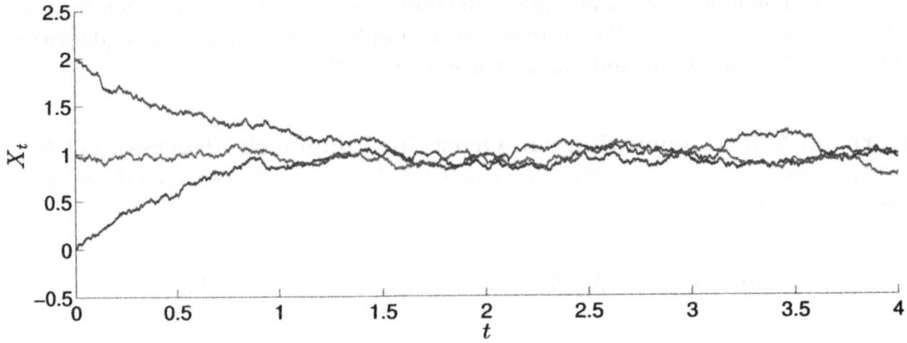

Figure 5.22 Three realizations of an Ornstein–Uhlenbeck process with parameters $\nu = 1, \theta = 2$, and $\sigma = 0.2$, starting from 0, 1, and 2. After about 2 time units all paths have reached "stationarity" and fluctuate around the long-term mean ν.

The Ornstein–Uhlenbeck process has applications in physics and finance. For example, X_t is used to describe the *velocity* of a Brownian particle. The term ν is then usually taken to be 0, and the resulting SDE is said to be of *Langevin* type. In finance, the process is used to describe price fluctuations around a mean price ν such that the process "reverts to the mean" — that is, when $X_t > \nu$ the drift is negative, and when $X_t < \nu$ the drift is positive, so that at all times the process tends to drift toward ν.

5.10 REFLECTED BROWNIAN MOTION

Consider a Brownian motion process $\{B_t, t \geqslant 0\}$ starting at $b_0 \geqslant 0$ with drift $\mu \leqslant 0$ and diffusion coefficient σ^2. Thus,

$$B_t = b_0 + \mu\, t + \sigma\, W_t \,,$$

where $\{W_t, t \geqslant 0\}$ is a Wiener process (starting at 0). The process $X = \{X_t, t \geqslant 0\}$ defined by

$$X_t = B_t + \max\left\{0, -\inf_{s \leqslant t} B_s\right\}$$

is called a **Brownian motion reflected at 0**. We write $X \sim \mathsf{RBM}(b_0, \mu, \sigma^2)$.

Some properties of the process include:

1. *Scaling:* If $X \sim \mathsf{RBM}(b_0, \mu, \sigma^2)$, then $X = \sigma \widetilde{X}$, with $\widetilde{X} \sim \mathsf{RBM}(b_0/\sigma, \mu/\sigma, 1)$.

2. *Markov process:* X is a time-homogeneous strong Markov process whose sample paths are almost surely continuous.

3. *Marginal distribution:* When $b_0 = 0$ the distribution of the random variable $X_t = B_t - \inf_{s \leqslant t} B_s = \inf_{s \leqslant t}(B_t - B_s)$ is the same as that of $\sup_{s \leqslant t} B_s$.

4. *Stationarity:* If $\mu < 0$ then X_t converges in distribution to an $\mathsf{Exp}(-2\mu/\sigma^2)$ distributed random variable as $t \to \infty$. Moreover, if b_0 is chosen according to this distribution, then the process is stationary.

5. *Absolute value*: The reflected Wiener process $W = \mathsf{RBM}(0,0,1)$ has the same distribution as the absolute value process $\{|W_t|, t \geqslant 0\}$ of the Wiener process.

For the purpose of generating a reflected Brownian motion we may assume by Property 1 that $\sigma = 1$. We make this assumption from now on.

Since $\{X_t, t \geqslant 0\}$ is a Markov process, we may generate it consecutively at times $t_1 < t_2 < \ldots$, starting from $X_0 = b_0$. In particular, to generate X_h, given $X_0 = b_0$, we have

$$X_h = B_h + \max\Big\{ -\inf_{s \leqslant h} B_s,\ 0\Big\} = \max\{T_1 + T_2,\ T_1\}\ ,$$

where $T_1 = b_0 + \mu\, h - (-W_h)$ and $T_2 = -b_0 - \mu\, h - \inf_{s \leqslant h} W_s$. The process $\{(-W_t, -\inf_{s\leqslant t} W_s), t \geqslant 0\}$ has the same distribution as $\{(Y_t, M_t), t \geqslant 0\}$, where $Y_t = -W_t$ and $M_t = \sup_{s \leqslant t} W_s$. The joint distribution of (Y_h, M_h) is given in (5.13). In particular, $Y_h \sim \mathsf{N}(0,h)$ and the conditional cdf of M_h given $Y_h = y$ is

$$F(m \,|\, y) = \mathbb{P}(M_h \leqslant m \,|\, W_h = y) = 1 - \mathrm{e}^{\frac{-2m(m-y)}{h}}, \quad m \geqslant y\ ,$$

with inverse

$$F^{-1}(u \,|\, y) = \frac{1}{2}\left(y + \sqrt{y^2 - 2\,h\,\ln(1-u)}\right)\ .$$

Thus, the random variable X_h can be generated in three steps:

1. Draw $Y_h = \mathsf{N}(0,h)$.

2. Draw $U \sim \mathsf{U}(0,1)$ and set $M_h = \frac{1}{2}\left(Y_h + \sqrt{Y_h^2 - 2h\ln U}\right)$.

3. Return $X_h = \max\{M_h - Y_h,\ x + \mu h - Y_h\}$.

To generate the process at time $2h$, we can use the same steps, where we first generate a realization at time h of a reflected Brownian motion $\{\widetilde{X}_t\}$ with the same drift, but starting at X_h, and then set $X_{2h} = \widetilde{X}_h$, and so on. This leads to the following algorithm.

Algorithm 5.26 (Generating a Reflected Brownian Motion)

1. *Set $k = 0$ and generate X_0.*

2. *Draw $Z \sim \mathsf{N}(0,1)$ and set $Y = Z\sqrt{h}$.*

3. *Draw $U \sim \mathsf{U}(0,1)$ and set*

$$M = \frac{Y + \sqrt{Y^2 - 2h\ln U}}{2}\ .$$

4. *Return $X_{(k+1)h} = \max\{M - Y, X_{kh} + \mu h - Y\}$. Set $k = k+1$ and go to Step 2.*

It is interesting to note that when $\mu = 0$, the random variable $\max\{M - Y, x - Y\}$ in Step 4 of the algorithm has the same distribution as $|x + Y|$. Namely, for any

$a > 0$ we have $\mathbb{P}(\max\{M-Y, x-Y\} > a) = \mathbb{P}(M-Y > a, x-Y < a) + \mathbb{P}(x-Y > a)$. The first term satisfies

$$\begin{aligned}\mathbb{P}(M - Y > a,\ x - Y < a) &= \mathbb{P}(M - Y > a,\ x - Y < a) \\ &= \mathbb{P}(\sup_{s \leqslant t}(W_s - W_t) > a,\ x - W_t < a) \\ &= \mathbb{P}(\sup_{s \leqslant t} \widehat{W}_s > a,\ x + \widehat{W}_t < a) \\ &= \mathbb{P}(\widehat{W}_t > a + x) = \mathbb{P}(Y + x < -a)\,, \end{aligned} \tag{5.30}$$

where $\widehat{W}_u = W_{t-u} - W_t$, and where the fourth equation follows from the reflection principle. Adding (5.30) to $\mathbb{P}(x - Y > a) = \mathbb{P}(Y + x > a)$ gives $\mathbb{P}(|Y + x| > a)$ as had to be shown.

It follows that for a Wiener process $\{W_t, t \geqslant 0\}$, its reflected process, $\{W_t - \inf_{s \leqslant t} W_s, t \geqslant 0\}$, has the *same* distribution as its absolute value process $\{|W_s|, s \geqslant t\}$. This is Property 5 above and suggests the following simpler algorithms for generating the reflected Wiener process at times $0, h, 2h, \ldots$.

Algorithm 5.27 (Generating the Reflected Wiener Process (I))

1. *Set* $k = 0$ *and* $X_0 = 0$.
2. *Draw* $Z \sim \mathsf{N}(0, 1)$ *and set* $Y = Z\sqrt{h}$.
3. *Return* $X_{(k+1)h} = |X_{kh} + Y|$. *Set* $k = k + 1$ *and go to Step 2.*

Algorithm 5.28 (Generating the Reflected Wiener Process (II))

1. *Set* $k = 0$, $X_0 = 0$, *and* $W_0 = 0$.
2. *Draw* $Z \sim \mathsf{N}(0, 1)$ *and set* $Y = Z\sqrt{h}$.
3. *Let* $W_{(k+1)h} = W_{kh} + Y$.
4. *Return* $X_{(k+1)h} = |W_{(k+1)h}|$. *Set* $k = k + 1$ *and go to Step 2.*

■ EXAMPLE 5.14 (Simulating Reflected Brownian Motion)

The following MATLAB program provides an implementation for simulating a sample path of a reflected Brownian motion $\{X_t, t \geqslant 0\}$ at equally spaced times $t_1 = 0, t_2 = h, t_3 = 2h, \ldots$, starting at $X_0 = 3$ and with drift $\mu = -1$. A typical sample path is depicted in Figure 5.23. In addition, a realization of the process $\{X_t - B_t, t \geqslant 0\}$ is plotted, where $\{B_t, t \geqslant 0\}$ is the Brownian motion process. Notice that the
☞ 609 path of this process resembles the Cantor function: the path is continuous and nondecreasing (it increases only at times when the reflected process hits 0) and its derivative is almost surely 0.

```
%rbm.m
n=10^4; h=10^(-3); t=h.*(0:n); mu=-1;
X=zeros(1,n+1); M= X; B=X;
B(1) =3; X(1) = 3;
```

```
for k= 2:n+1
  Y= sqrt(h)*randn; U=rand(1);
  B(k) = B(k-1) + mu*h - Y;
  M=(Y + sqrt(Y^2-2*h*log(U)))/2;
  X(k)=max(M-Y, X(k-1)+h*mu-Y);
end
subplot(2,1,1)
plot(t,X,'k-');
subplot(2,1,2)
plot(t,X-B,'k-');
```

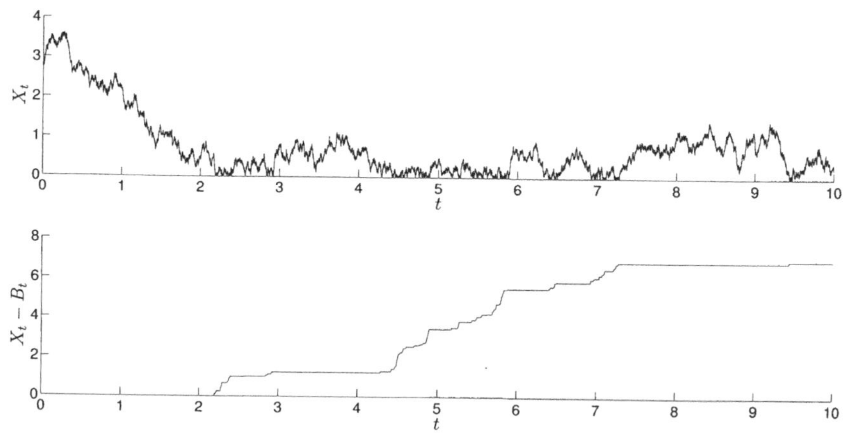

Figure 5.23 A realization of a reflected Brownian motion and the process $X_t - B_t = \max\{0, -\inf_{s\leqslant t} B_s\}$.

5.11 FRACTIONAL BROWNIAN MOTION

An important property of the Wiener process (standard Brownian motion) is its invariance under scaling; that is, $\{W_t, t \geqslant 0\}$ has the same distribution as the scaled process $\{W_{ct}/\sqrt{c}, t \geqslant 0\}$. More generally, a stochastic process $Z = \{Z_t, t \geqslant 0\}$ is said to be **self-similar** with **self-similarity** or **Hurst** parameter $H > 0$, if for any $c > 0$ the rescaled process $\{c^{-H} Z_{ct}, t \geqslant 0\}$ has the same distribution as Z. ☞ 179

Self-similar processes with **stationary increments** (that is, for any $t > s$ the distribution of $Z_t - Z_s$ only depends on $t - s$) and $Z_0 = 0$ have a specific covariance structure. First, note that $\mathbb{E}Z_t = 0$ for all t, because $\mathbb{E}Z_m = m\,\mathbb{E}Z_1$ and, by self-similarity, also $\mathbb{E}m^{-H} Z_m = \mathbb{E}Z_1$. Second, if we define $\sigma^2 = \mathbb{E}\,Z_1^2$, then

$$\mathbb{E}\,(Z_s - Z_t)^2 = \mathbb{E}\,(Z_{s-t} - Z_0)^2 = \sigma^2\,(s-t)^{2H}.$$

On the other hand,

$$\mathbb{E}\,(Z_s - Z_t)^2 = \mathbb{E}\,Z_s^2 + \mathbb{E}\,Z_t^2 - 2\mathbb{E}\,Z_s\,Z_t = \sigma^2\,s^{2H} + \sigma^2\,t^{2H} - 2\,\mathrm{Cov}(Z_s, Z_t)\,.$$

This leads to the following result.

Theorem 5.11.1 (Self-Similar Process) *Let $\{Z_t, t \geqslant 0\}$ be a self-similar process with stationary increments, Hurst parameter $H \in (0,1)$, and $Z_0 = 0$. Define $\sigma^2 = \mathbb{E}Z_1^2$. Then $\mathbb{E}Z_t = 0$ for all t and*

$$\mathrm{Cov}(Z_s, Z_t) = \frac{1}{2}\sigma^2\left(s^{2H} - (t-s)^{2H} + t^{2H}\right), \quad 0 \leqslant s \leqslant t\,. \tag{5.31}$$

A continuous zero-mean Gaussian process $\{Z_t\}$ with $Z_0 = 0$ and covariance function (5.31) — which is therefore a self-similar process with Hurst parameter $H \in (0,1)$ — is called a **fractional Brownian motion**. The time series $\{X_i\}$ of increments $X_i = Z_i - Z_{i-1}$, $i = 1, 2, \ldots$ is called **fractional Gaussian noise**.

The fractional Gaussian noise time series is a discrete-time zero-mean Gaussian process with variance $\mathrm{Var}(X_i) = \sigma^2, i = 1, 2, \ldots$ and covariance function

$$\mathrm{Cov}(X_i, X_{i+k}) = \frac{1}{2}\sigma^2\left((k+1)^{2H} - 2k^{2H} + (k-1)^{2H}\right), \quad k = 1, 2, \ldots$$

☞ 631 and is therefore weakly stationary. It follows that for large k the autocorrelations $\varrho_k = \mathrm{Cov}(X_i, X_{i+k})/\sigma^2$ has the asymptotic behavior

$$\varrho_k \approx \frac{H(2H-1)}{k^{2(1-H)}}\,. \tag{5.32}$$

In particular, for $H \in (1/2, 1)$ the autocorrelations are positive and can decay arbitrarily slowly for H close to 1. Such a process is said to be **long-range dependent**. For $H = 1/2$ the autocorrelations are zero — in fact the $\{X_i\}$ are independent and the process $\{Z_t\}$ is a Wiener process. For $H \in (0, 1/2)$ the autocorrelations are negative and increase to 0 at a rate faster than $1/k$. Figure 5.24 depicts the realizations of a fractional Brownian motion for $H = 0.9$ and $H = 0.3$. Note that the first path is much smoother than a Wiener path, whereas the second path is rougher than a Wiener path.

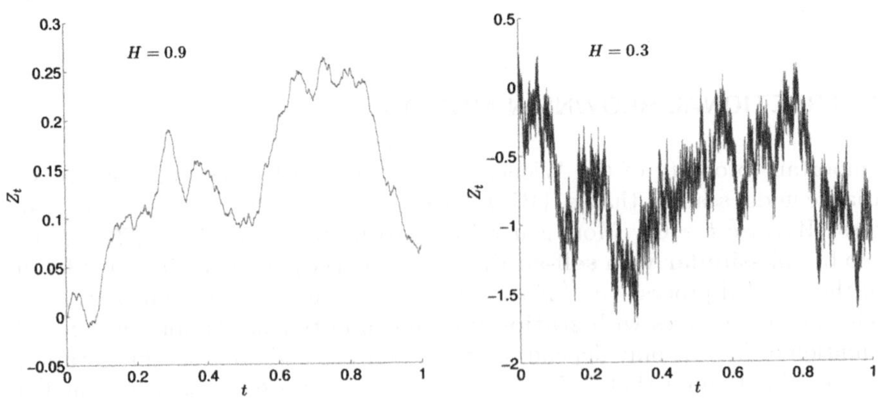

Figure 5.24 Fractional Brownian motion realizations.

Fractional Gaussian noise is used in telecommunications, for example, to model the number of arriving bytes during consecutive time intervals. Specifically, let

$\{X_1, X_2, \ldots\}$ be fractional Gaussian noise with variance σ^2 and Hurst parameter $H \in (0,1)$. The stochastic process $\{Y_1, Y_2, \ldots\}$ with $Y_i = X_i + \mu$, where μ is a fixed constant, is a 3-parameter model for the arrival process of traffic. The corresponding cumulative process approximates the sum of a fractional Brownian motion and a linear drift term (μt).

Fractional Brownian motion with $\sigma = 1$ can be represented in several ways as the deterministic integral with respect to a Wiener process. For example, Norros et al. [22] give the following representation:

$$Z_t = \int_0^t z(t,s)\,\mathrm{d}W_s\,,$$

with

$$z(t,s) = c_H \left[\left(\frac{t(t-s)}{s} \right)^{\alpha} - \alpha\, s^{\alpha} \int_s^t u^{\alpha-1}(u-s)^{\alpha} \mathrm{d}u \right],$$

where

$$\alpha = H - \frac{1}{2}, \quad c_H^2 = \frac{2H\,\Gamma(\frac{3}{2} - H)}{\Gamma(H + \frac{1}{2})\,\Gamma(2 - 2H)}.$$

For $H \in (1/2, 1)$ (or $\alpha > 0$) another expression for $z(t,s)$ is

$$z(t,s) = c_H\, \alpha\, s^{\alpha} \int_s^t u^{\alpha}(u-s)^{\alpha-1} \mathrm{d}u = c_H\,(t-s)^{\alpha}\; {}_2F_1\Big(-\alpha,\ \alpha;\ \alpha+1;\ 1 - \frac{t}{s} \Big),$$

where ${}_2F_1$ denotes the *hypergeometric* function. Other properties and references may be found in [20]. ☞ 716

Since the fractional Gaussian noise process is a stationary Gaussian process, one can use Algorithm 5.5 to generate sample paths. ☞ 161

■ EXAMPLE 5.15 (Generating Fractional Gaussian Noise)

The following MATLAB code implements Algorithm 5.5 to generate a fractional Gaussian noise process with Hurst parameter 0.9. The fractional Gaussian noise process is scaled and then summed to obtain a realization of the fractional Brownian motion on $[0,1]$ at the grid points $k/N, k = 0, \ldots, N$. A typical realization is given in Figure 5.24.

```
%fbm.m
N=2^15;
sig = 1;
t= 0:N;
H = 0.9; %Hurst parameter
sigma(1) = sig;
for k=1:N
    sigma(k+1) = 0.5*sig*((k+1)^(2*H) - 2*k^(2*H) + (k-1)^(2*H));
end
c=[sigma sigma((end-1):-1:2)]';
lambda=fft(c); %eigenvalues
eta=sqrt(lambda./(2*N));
Z=randn(2*N,1)+sqrt(-1).*randn(2*N,1); %complex normal vectors
```

```
Zeta= Z.*eta;
X2n=fft(Zeta);
A=X2n(1:(N+1));
X=real(A);
c = 1/N;
X1 = c^H*cumsum(X); %scaled process
plot(t*c,X1);
```

5.12 RANDOM FIELDS

A **random field** is a real-valued stochastic process $\mathbf{Z} = \{Z_{\mathbf{t}}, \mathbf{t} \in \mathscr{T}\}$ with a discrete or continuous index set $\mathscr{T} \subseteq \mathbb{R}^n, n \geqslant 2$, typically $\mathbb{R}^2$ or $\mathbb{R}^3$. A random field is said to
☞ 627 be **Gaussian** if it is a Gaussian process. A random field is said to be a **Markovian** for a neighborhood structure $\mathscr{N} = \{\mathscr{N}_{\mathbf{t}}\}$ if

$$(Z_{\mathbf{t}} \,|\, Z_{\mathbf{s}},\ \mathbf{s} \in \mathscr{T} \setminus \{\mathbf{t}\}) \quad \sim \quad (Z_{\mathbf{t}} \,|\, Z_{\mathbf{s}},\ \mathbf{s} \in \mathscr{N}_{\mathbf{t}}) \,, \tag{5.33}$$

where $\mathscr{N}_{\mathbf{t}}$ is the set of neighbors of $\mathbf{t}$. A random field $\mathbf{Z}$ is called a **Gibbs** random field if its pdf is of the form

$$f(\mathbf{z}) = \frac{\mathrm{e}^{-E(\mathbf{z})}}{\mathcal{Z}} \,.$$

The corresponding distribution is called a **Gibbs** or **Boltzmann** distribution with **energy function** E. The constant $\mathcal{Z} = \sum_{\mathbf{z}} \mathrm{e}^{-E(\mathbf{z})}$ is called the **partition function**.

■ **EXAMPLE 5.16 (Brownian Sheet)**

The **Brownian sheet** or **Wiener sheet** process on the unit square is the continuous Gaussian random field $\{W_{x,y},\ (x,y) \in [0,1]^2\}$ with zero mean and covariance function

$$\mathrm{Cov}(W_{x_1,y_1}, W_{x_2,y_2}) = \min\{x_1, x_2\}\ \min\{y_1, y_2\} \,.$$

It can be viewed as a spatial version of the Wiener process and as a limiting version (as $n \to \infty$) of the lattice process $\{W_{i,j}^{(n)}, i = 0, 1, \ldots, n, j = 0, 1, \ldots, n\}$, defined by $W_{0,j}^{(n)} = W_{i,0}^{(n)} = 0$ and the partial sums

$$W_{i,j}^{(n)} = \sum_{k=1}^{i} \sum_{l=1}^{j} Z_{k,l} \,,$$

where $\{Z_{k,l}\}$ are independent $\mathsf{N}(0, 1/n^2)$ random variables. The process $\{W_{i,j}^{(n)}\}$ is a Gaussian Markov random field with the natural (nearest neighbor) neighborhood structure on the lattice. Other properties of the Brownian sheet include:

1. *Self-similarity*: $W_{ax,by} \sim \sqrt{ab}\ W_{x,y}$.

2. *Marginal process*: For each fixed y, $\{W_{x,y}/\sqrt{y}, x \geqslant 0\}$ is a Wiener process.

3. *Symmetry*: $W_{x,y}$ has the same distribution as $W_{y,x}$.

4. *Stationary Gaussian increments*: For each rectangle $\mathbf{a} \times \mathbf{b} = [a_1, a_2] \times [b_1, b_2]$ the increment $W_{b_1,b_2} - W_{a_1,a_2}$ has a $\mathsf{N}(0, (b_2 - b_1)(a_2 - a_1))$ distribution.

5. *Independent increments*: For each pair of nonoverlapping rectangles $\mathbf{a} \times \mathbf{b}$ and $\mathbf{c} \times \mathbf{d}$ the increments $W_{b_1,b_2} - W_{a_1,a_2}$ and $W_{d_1,d_2} - W_{c_1,c_2}$ are independent.

A realization of the process is given in Figure 5.25.

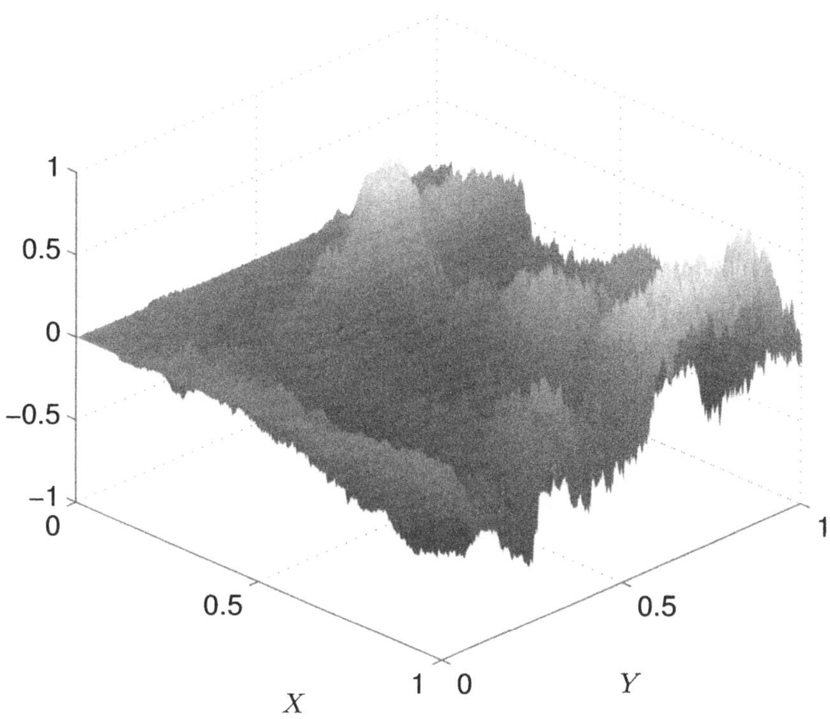

Figure 5.25 Brownian sheet.

Markov random fields are usually defined on a graph $\mathcal{G} = (V, E)$, where V is the vertex set and E the edge set. The Markovian property (5.33) extends that of the ordinary Markov chain (A.34). The **Hammersley–Clifford** theorem [5] links the "local" behavior of Markov random fields to the "global" behavior of Gibbs random fields. ☞ 233

Theorem 5.12.1 (Hammersley–Clifford) *A Markov random field satisfying the positivity condition $f(\mathbf{x}) > 0$ for all $\mathbf{x}$ is a Gibbs random field, and any Gibbs random field is a Markov random field.*

The Gibbs distribution for a Gaussian Markov random field on a graph $\mathcal{G}$ with $|V| = n$ vertices, labeled $1, \ldots, n$, is simply the Gaussian pdf

$$f(\mathbf{z}) = (2\pi)^{-n/2} \sqrt{\det(\Lambda)}\, \mathrm{e}^{-\frac{1}{2}(\mathbf{z}-\boldsymbol{\mu})^\top \Lambda (\mathbf{z}-\boldsymbol{\mu})} ,$$

where $\Lambda = (\lambda_{ij}, i, j \in \{1, \ldots, n\})$ is the *precision matrix* (the inverse of the covariance matrix) and $\boldsymbol{\mu}$ the mean vector. The elements of Λ can be interpreted in the following way, see, for example, [23, Page 22]:

1. $\mathbb{E}[Z_i \mid Z_k, k \neq i] = \mu_i - \dfrac{1}{\lambda_{ii}} \sum_{k \in \mathcal{N}_i} \lambda_{ik}(x_k - \mu_k)$,

2. $\mathrm{Var}(Z_i \mid Z_k, k \neq i) = \dfrac{1}{\lambda_{ii}}$,

3. $\mathrm{Corr}(Z_i, Z_j \mid Z_k, j \neq i, j) = -\dfrac{\lambda_{ij}}{\sqrt{\lambda_{ii}\lambda_{jj}}}$.

Gaussian Markov fields are widely used in image analysis. In practice, the matrix Λ is a sparse matrix, which greatly speeds up the simulation and analysis of Gaussian Markov fields. The simulation of Gaussian Markov random fields is illustrated in Example 5.1.

5.13 LÉVY PROCESSES

A d-dimensional **Lévy process** is a stochastic process $\{\mathbf{X}_t, t \geqslant 0\}$ taking values in $\mathbb{R}^d$ with the following properties:

1. *Independent increments*: For any $t_1 < t_2 \leqslant t_3 < t_4$ the random variables $\mathbf{X}_{t_4} - \mathbf{X}_{t_3}$ and $\mathbf{X}_{t_2} - \mathbf{X}_{t_1}$ are independent.

2. *Stationarity*: The law of $\mathbf{X}_{t+h} - \mathbf{X}_t$ does not depend on t.

3. *Stochastic continuity*: For all $\varepsilon > 0$, $\lim_{h \to 0} \mathbb{P}(\|\mathbf{X}_{t+h} - \mathbf{X}_t\| \geqslant \varepsilon) = 0$.

4. *Zero initial value*: $\mathbf{X}_0 = \mathbf{0}$ almost surely.

A Lévy process can be viewed as a continuous-time generalization of a random walk process. Indeed the process observed at times $0 = t_0 < t_1 < t_2 < \ldots$ forms a random walk,

$$\mathbf{X}_{t_n} = \sum_{i=1}^{n} (\mathbf{X}_{t_i} - \mathbf{X}_{t_{i-1}}) , \tag{5.34}$$

whose increments $\{\mathbf{X}_{t_i} - \mathbf{X}_{t_{i-1}}\}$ are independent. Moreover, if the times are chosen at an equal distance from each other, $t_i - t_{i-1} = h$, then the increments are iid, and so the distribution of the Lévy process is completely specified by its increment distribution in any time interval of length $h > 0$, for example by the distribution of $\mathbf{X}_1$. Moreover, the increment distributions are *infinitely divisible*. ☞ 705

Let $\{\mathbf{X}_t, t \geqslant 0\}$ be a Lévy process on $\mathbb{R}^d$ and let $N([0, t] \times A)$ denote the number of jumps of $\mathbf{X}$ during the interval $[0, t]$ whose size lies in $A \in \mathcal{B}^d$, excluding $\mathbf{0}$.

Let $\Delta \mathbf{X}_t$ denote the size of the jump of the process at time t. The measure ν defined by

$$\nu(A) = \mathbb{E}N([0,1] \times A) = \mathbb{E}\left[\#\{t \in [0,1] : \Delta\mathbf{X}_t \neq \mathbf{0}, \Delta\mathbf{X}_t \in A\}\right], \quad A \in \mathcal{B}^d$$

is called the **Lévy measure** of $\mathbf{X}$. The random measure $N(\mathrm{d}t, \mathrm{d}\mathbf{x})$ is called the **jump measure**.

Lévy processes can be thought of as a combination of a diffusion process and a jump process. Both Brownian motion (a pure diffusion process) and Poisson processes (pure jump processes) are Lévy processes. Lévy processes represent a tractable extension of Brownian motion to processes based on a far richer family of distributions. All infinitely divisible distributions, including the gamma, Cauchy, and stable distributions can serve as a basis for Lévy processes. In addition, Lévy processes allow the modeling of discontinuous sample paths, whose properties match those of empirical phenomena such as financial time series.

A main theorem on Lévy processes is the following. A complete proof can be found in [1, Pages 96–111] using a martingale-based approach.

Theorem 5.13.1 (Lévy–Itô Decomposition) *Let $\{\mathbf{X}_t, t \geqslant 0\}$ be a Lévy process with Lévy measure ν. Then, the following properties hold.*

1. $\int_{\mathbb{R}^d} \min\{\|\mathbf{x}\|^2, 1\}\, \nu(\mathrm{d}\mathbf{x}) < \infty$.

2. *The jump measure $N(\mathrm{d}t, \mathrm{d}\mathbf{x})$ is a Poisson random measure on $\mathbb{R}_+ \times \mathbb{R}^d$ with mean measure $\mathbb{E}N(\mathrm{d}t, \mathrm{d}\mathbf{x}) = \mathrm{d}t\, \nu(\mathrm{d}\mathbf{x})$.* ☞ 170

3. *$\mathbf{X}_t$ can be decomposed as $\mathbf{X}_t = \mathbf{B}_t + \mathbf{Y}_t + \mathbf{Z}_t$, where $\{\mathbf{B}_t, t \geqslant 0\}$ is a Brownian motion process, $\{\mathbf{Y}_t, t \geqslant 0\}$ is the compound Poisson process* ☞ 174

$$\mathbf{Y}_t = \int_0^t \int_{\|\mathbf{x}\|>1} \mathbf{x}\, N(\mathrm{d}s, \mathrm{d}\mathbf{x}), \quad t \geqslant 0,$$

and $\{\mathbf{Z}_t, t \geqslant 0\}$ is the limit of a compensated compound Poisson process:

$$\mathbf{Z}_t = \lim_{\delta \downarrow 0} \mathbf{Z}_t^\delta \quad \textit{with} \quad \mathbf{Z}_t^\delta = \int_0^t \int_{\delta \leqslant \|\mathbf{x}\| \leqslant 1} \mathbf{x}\, [N(\mathrm{d}s, \mathrm{d}\mathbf{x}) - \mathrm{d}s\, \nu(\mathrm{d}\mathbf{x})] . \quad (5.35)$$

4. *The processes $\{\mathbf{B}_t\}$, $\{\mathbf{Y}_t\}$, and $\{\mathbf{Z}_t^\delta\}$ are independent of each other, and so are $\{\mathbf{B}_t\}$, $\{\mathbf{Y}_t\}$, and $\{\mathbf{Z}_t\}$.*

5. *Convergence of $\mathbf{Z}_t^\delta$ to $\mathbf{Z}_t$ is with probability 1 and uniform in t on any interval $[0, T]$.*

Remark 5.13.1 (Truncation Level) The choice of 1 for the truncation level in the above theorem is arbitrary. It may be changed to any positive number. However, this will change the drift of $\{\mathbf{B}_t\}$.

A second main result for Lévy processes is a direct result of the Lévy–Itô decomposition, the characteristic functions of a compound Poisson random vector (5.7), and the moment generating function of a normal random vector (see Table 4.28). ☞ 145

Theorem 5.13.2 (Lévy–Khinchin Representation) *The characteristic function of* $\mathbf{X}_t$ *satisfies* $\mathbb{E}\,\mathrm{e}^{\mathrm{i}\mathbf{s}^\top \mathbf{X}_t} = \mathrm{e}^{t\psi(\mathbf{s})}$, *where the* **characteristic exponent** $\psi(\mathbf{s})$ *is of the form*

$$\underbrace{-\frac{1}{2}\,\mathbf{s}^\top \Sigma \mathbf{s} + \mathrm{i}\,\mathbf{s}^\top \gamma}_{\textit{from } \mathbf{B}_t} + \underbrace{\int_{\|\mathbf{x}\|>1} \left(\mathrm{e}^{\mathrm{i}\mathbf{s}^\top \mathbf{x}} - 1\right) \nu(\mathrm{d}\mathbf{x})}_{\textit{from } \mathbf{Y}_t} + \underbrace{\int_{\|\mathbf{x}\|\leqslant 1} \left(\mathrm{e}^{\mathrm{i}\mathbf{s}^\top \mathbf{x}} - 1 - \mathrm{i}\,\mathbf{s}^\top \mathbf{x}\right) \nu(\mathrm{d}\mathbf{x})}_{\textit{from } \mathbf{Z}_t}$$

$$= -\frac{1}{2}\,\mathbf{s}^\top \Sigma \mathbf{s} + \mathrm{i}\,\mathbf{s}^\top \gamma + \int \left(\mathrm{e}^{\mathrm{i}\mathbf{s}^\top \mathbf{x}} - 1 - \mathrm{i}\,\mathbf{s}^\top \mathbf{x}\,\mathrm{I}_{\{\|\mathbf{x}\|\leqslant 1\}}\right) \nu(\mathrm{d}\mathbf{x}) \tag{5.36}$$

for some vector γ *and covariance matrix* Σ. *It follows that each Lévy process is characterized by a* **characteristic triplet** (γ, Σ, ν).

Let $\{\mathbf{X}_t, t \geqslant 0\}$ be a Lévy process with characteristic triplet (γ, Σ, ν). Properties of the process include:

☞ 705

1. *Infinite divisibility*: For each t the random variable $\mathbf{X}_t$ has an infinitely divisible distribution with characteristic function given in Theorem 5.13.2. Conversely, for any infinitely divisible distribution Dist there is a Lévy process $\{X_t, t \geqslant 0\}$ for which $X_1 \sim$ Dist.

2. *Finite versus infinite activity*: If the Lévy measure $\nu(\mathbb{R}^d) < \infty$ then almost all paths of $\{\mathbf{X}_t, t \geqslant 0\}$ have a finite number of jumps on every compact interval. Such a process is said to be a **finite activity** process. If $\nu(\mathbb{R}^d) = \infty$ then almost all paths display infinite jumps on every compact interval. Such a process is said to be an **infinite activity** Lévy process.

3. *Expectation and variance*: For a Lévy process on $\mathbb{R}$ with characteristic triplet (γ, σ^2, ν), if $\int_{|x|>1} |x|^2 \nu(\mathrm{d}x) < \infty$ (in particular, when the jump sizes are bounded), then

$$\mathbb{E}X_t = t\left(\gamma + \int_{|x|>1} x\,\nu(\mathrm{d}x)\right)$$

   and

$$\mathrm{Var}(X_t) = t\left(\sigma^2 + \int x^2 \nu(\mathrm{d}x)\right) .$$

4. *Finite variation*: A Lévy process has finite variation if and only if its characteristic triplet (γ, Σ, ν) satisfies

$$\Sigma = 0 \quad \text{and} \quad \int_{\|\mathbf{x}\|\leqslant 1} \|\mathbf{x}\|\,\nu(\mathrm{d}\mathbf{x}) < \infty .$$

   For a finite variation process the Lévy–Itô decomposition takes the form

$$\mathbf{X}_t = \mathbf{b}\,t + \int_0^t \int_{\mathbb{R}^d} \mathbf{x}\,N(\mathrm{d}s, \mathrm{d}\mathbf{x}) ,$$

   where $\mathbf{b} = \gamma - \int_{\|\mathbf{x}\|\leqslant 1} \mathbf{x}\,\nu(\mathrm{d}\mathbf{x})$.

5. *Linear transformation*: Let $\{\mathbf{X}_t\}$ be a Lévy process on $\mathbb{R}^n$ with characteristic triplet (γ, Σ, ν). If A is a $m \times n$ matrix, then $\mathbf{Y}_t = A\mathbf{X}_t$ defines a Lévy process on $\mathbb{R}^m$ with characteristic triplet $(\widetilde{\gamma}, \widetilde{\Sigma}, \widetilde{\nu})$, where (see [9])

$$\widetilde{\gamma} = A\gamma + \int_{\mathbb{R}^m} \mathbf{y} \left(\mathrm{I}_{\{\|\mathbf{y}\| \leqslant 1\}} - \mathrm{I}_{\{\mathbf{y} \in S_1\}} \right) \widetilde{\nu}(\mathrm{d}\mathbf{y}) \ , \ S_1 = \{A\mathbf{x} : \|\mathbf{x}\| \leqslant 1\} \ ,$$
$$\widetilde{\Sigma} = A\Sigma A^\top,$$
$$\widetilde{\nu}(B) = \nu\left(\{\mathbf{x} \in \mathbb{R}^n : A\mathbf{x} \in B\}\right) \quad \text{for all } B \in \mathcal{B}^m \ .$$

6. *Markov property*: A Lévy process is a strong Markov process; that is, for each finite stopping time τ and for all $t \geqslant 0$

$$(\mathbf{X}_{\tau+t} \,|\, \mathbf{X}_u \,, u \leqslant \tau) \sim (\mathbf{X}_{\tau+t} \,|\, \mathbf{X}_\tau) \,.$$

Moreover, an m-dimensional Lévy process with characteristic triplet (γ, Σ, ν) has infinitesimal generator ☞ 646

$$Lf(\mathbf{x}) = \sum_{i=1}^m \gamma_i \frac{\partial}{\partial x_i} f(\mathbf{x}) + \frac{1}{2} \sum_{i=1}^m \sum_{j=1}^m \Sigma_{ij} \frac{\partial^2}{\partial x_i \partial x_j} f(\mathbf{x})$$
$$+ \int_{\mathbb{R}^m} \left(f(\mathbf{x} + \mathbf{y}) - f(\mathbf{x}) - \sum_{i=1}^m y_i \frac{\partial}{\partial x_i} f(\mathbf{x}) \, \mathrm{I}_{\{\|\mathbf{y}\| \leqslant 1\}} \right) \nu(\mathrm{d}\mathbf{y}) \,.$$

In the one-dimensional case with triple (γ, σ^2, ν), this reduces to

$$Lf(x) = \gamma f'(x) + \frac{1}{2}\sigma^2 f''(x) + \int_{\mathbb{R}} \left(f(x+y) - f(x) - y f'(x) \, \mathrm{I}_{\{|y| \leqslant 1\}} \right) \nu(\mathrm{d}y) \,.$$

5.13.1 Increasing Lévy Processes

A special class of one-dimensional Lévy processes are *increasing* Lévy processes. They play an important role in practice, because they can be used to time-change simpler Lévy processes such as Brownian motion. Specifically, a **Lévy subordinator** $\{X_t, t \geqslant 0\}$ is an almost surely increasing Lévy process on $\mathbb{R}$. For a Lévy subordinator, the following are equivalent (see, for example, [9, Page 88]):

1. $X_t \geqslant 0$ for some $t \geqslant 0$.

2. $X_t \geqslant 0$ for all $t \geqslant 0$.

3. $X_t \geqslant X_s$ for all $t \geqslant s$.

4. The characteristic triplet (γ, σ^2, ν) of $\{X_t, t \geqslant 0\}$ satisfies

   (a) *Positive jumps:* $\nu((-\infty, 0]) = 0$.

   (b) *Positive drift:* $\gamma - \int_0^1 x \, \nu(\mathrm{d}x) \geqslant 0$.

   (c) *Finite variation:* $\sigma^2 = 0$ and $\int_0^1 x \, \nu(\mathrm{d}x) < \infty$.

■ **EXAMPLE 5.17 (Gamma Process)**

Let N be a Poisson random measure on $\mathbb{R}_+ \times \mathbb{R}_+$ with mean measure $\mathrm{d}t\,\nu(\mathrm{d}x) = \mathrm{d}t\,g(x)\mathrm{d}x$, where

$$g(x) = \frac{\alpha\,\mathrm{e}^{-\lambda x}}{x}, \quad x \geqslant 0\,.$$

Define

$$X_t = \int_0^t \int_0^\infty x N(\mathrm{d}s, \mathrm{d}x)\,, \quad t \geqslant 0\,.$$

Then $\{X_t, t \geqslant 0\}$ is, by construction, an increasing Lévy process. The characteristic exponent is

$$\psi(s) = \int (\mathrm{e}^{\mathrm{i}\,sx} - 1)\,g(x)\,\mathrm{d}x = \alpha(\ln\lambda - \ln(\lambda - \mathrm{i}\,s))\,,$$

☞ 113 which shows that $X_1 \sim \mathsf{Gamma}(\alpha, \lambda)$, hence the name **gamma process**. The characteristic triplet is thus $(\gamma, 0, \nu)$, with $\gamma = \int_0^1 g(x)\,\mathrm{d}x = \frac{\alpha}{\lambda}(1 - \mathrm{e}^{-\lambda})$. This is an example of an infinite activity subordinator. Note that an increment $X_{t+s} - X_t$ has a $\mathsf{Gamma}(\alpha s, \lambda)$ distribution. A typical realization on $[0, 1]$ with $\alpha = 10$ and $\lambda = 1$ is given in Figure 5.26.

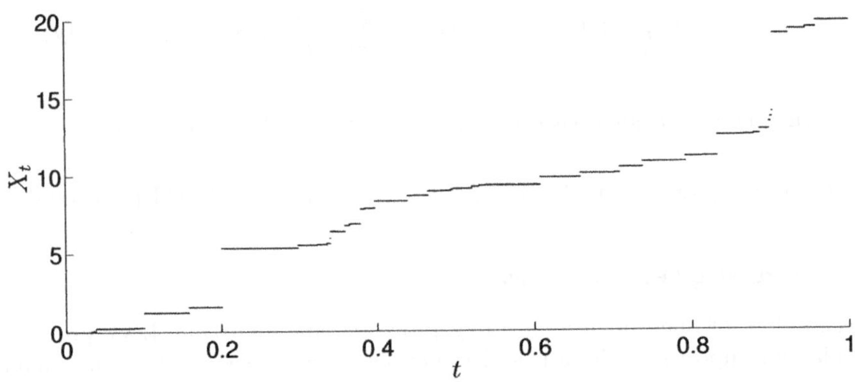

Figure 5.26 Gamma process realization for $\alpha = 3$ and $\lambda = 1$.

Let $\{S_t, t \geqslant 0\}$ be a subordinator of the form

$$S_t = \beta\,t + \int_0^\infty N(\mathrm{d}s, \mathrm{d}x)\,,$$

where $\beta \geqslant 0$ and N is a Poisson random measure on $\mathbb{R}_+ \times \mathbb{R}_+$ with $\mathbb{E}N(\mathrm{d}t, \mathrm{d}x) = \mathrm{d}t\varrho(\mathrm{d}x)$ — thus, $\varrho(\mathrm{d}x)$ is the Lévy measure. Let $\{\mathbf{X}_t, t \geqslant 0\}$ be a Lévy process on $\mathbb{R}^d$ with characteristic triplet (γ, Σ, ν) independent of $\{S_t, t \geqslant 0\}$. The process $\{\mathbf{Y}_t, t \geqslant 0\}$ defined by

$$\mathbf{Y}_t = \mathbf{X}_{S_t}, \quad t \geqslant 0$$

is called **subordinate** to the process $\{\mathbf{X}_t, t \geqslant 0\}$. The following theorem summarizes the main results for subordinated processes.

Theorem 5.13.3 (Subordination of a Lévy Process)

1. *The subordinate process* $\{\mathbf{Y}_t, t \geqslant 0\}$ *is again a Lévy process.*

2. *Let* ζ *and* ψ *be the characteristic exponents of* $\{S_t\}$ *and* $\{\mathbf{X}_t\}$*, respectively. The characteristic exponent of* $\{\mathbf{Y}_t, t \geqslant 0\}$ *is given by*

$$\mathbb{E}\,\mathrm{e}^{\mathrm{i}\mathbf{r}^\top \mathbf{Y}_t} = \mathbb{E}\,\mathbb{E}\left[\mathrm{e}^{\mathrm{i}\mathbf{r}^\top \mathbf{X}_{S_t}} \mid S_t\right] = \mathbb{E}\,\mathrm{e}^{S_t \psi(\mathbf{r})} = \mathrm{e}^{t\zeta(-\mathrm{i}\,\psi(\mathbf{r}))}\,.$$

3. *The characteristic triplet* $(\widetilde{\gamma}, \widetilde{\Sigma}, \widetilde{\nu})$ *of* $\{\mathbf{Y}_t\}$ *is given by*

$$\widetilde{\gamma} = \beta\,\gamma + \int_0^\infty \left(\int_{\|\mathbf{x}\| \leqslant 1} \mathbf{x}\, p_s(\mathrm{d}\mathbf{x})\right) \varrho(\mathrm{d}s)\,,$$

$$\widetilde{\Sigma} = \beta\,\Sigma\,,$$

$$\widetilde{\nu}(B) = \beta\,\nu(B) + \int_0^\infty p_s(B)\varrho(\mathrm{d}s) \qquad \text{for all } B \in \mathcal{B}^d\,,$$

where p_s *is the probability distribution of* $\mathbf{X}_s$*.*

■ EXAMPLE 5.18 (Variance Gamma Process)

Let $\{B_t\}$ be a one-dimensional Brownian motion with drift μ and diffusion coefficient σ^2. Let $\{S_t\}$ be a gamma process with $\alpha = \lambda$ and independent of $\{B_t\}$. Consider the subordinate process $\{Y_t, t \geqslant 0\}$, where $Y_t = B_{S_t}$. From Theorem 5.13.3 we see that $\{Y_t, t \geqslant 0\}$ has Lévy measure

$$\begin{aligned}
\widetilde{\nu}(\mathrm{d}x) &= \mathrm{d}x \int_0^\infty \frac{1}{\sqrt{2\pi\sigma^2 s}} \exp\left(-\frac{1}{2}\frac{(x-\mu s)^2}{\sigma^2 s}\right) \frac{\alpha\,\mathrm{e}^{-\alpha s}}{s}\,\mathrm{d}s \\
&= \mathrm{d}x\;\alpha\,\exp\left(\frac{x(\mu - \sqrt{\mu^2 + 2\,\alpha\,\sigma^2}\,\mathrm{sgn}(x))}{\sigma^2}\right)\Big/|x| \\
&= \mathrm{d}x\,\frac{\alpha\,\mathrm{e}^{-\lambda_1 x}}{x}\,\mathrm{I}_{\{x>0\}} - \mathrm{d}x\,\frac{\alpha\,\mathrm{e}^{\lambda_2 x}}{x}\,\mathrm{I}_{\{x<0\}}\,,
\end{aligned} \tag{5.37}$$

where

$$\lambda_1 = \frac{\sqrt{\mu^2 + 2\alpha\sigma^2} - \mu}{\sigma^2} \quad \text{and} \quad \lambda_2 = \frac{\sqrt{\mu^2 + 2\alpha\sigma^2} + \mu}{\sigma^2}\,.$$

We recognize (5.37) as the Lévy measure belonging to the difference of two gamma processes, with parameters (α, λ_i), $i = 1, 2$. Since both the linear drift term of $\{Y_t, t \geqslant 0\}$ and the diffusion coefficient are 0, the process is completely determined by its Lévy measure. Therefore, we may write

$$Y_t = X_t^{(1)} - X_t^{(2)}, \quad t \geqslant 0\,,$$

where $\{X_t^{(i)}, t \geqslant 0\}$, $i = 1, 2$ are gamma processes with parameters $(\alpha, \lambda_i), i = 1, 2$. The subordinated process $\{Y_t, t \geqslant 0\}$ is thus a **variance gamma process**: a Lévy process that is the difference of two independent gamma processes.

5.13.2 Generating Lévy Processes

We describe various approaches to simulating one-dimensional Lévy processes $\{X_t, t \geqslant 0\}$.

5.13.2.1 Random Walk The simplest method to generate certain Lévy processes is based on the random walk property (5.34). For this approach to work, the distribution of X_t needs to be known for all t.

Algorithm 5.29 (Known Marginal Distributions) *Suppose X_t has a known distribution* $\mathsf{Dist}(t)$, $t \geqslant 0$. *Generate a realization of the Lévy process at times* $0 = t_0 < t_1 < \ldots < t_n$ *as follows.*

1. *Set $X_0 = 0$ and $k = 1$.*

2. *Draw $A \sim \mathsf{Dist}(t_k - t_{k-1})$.*

3. *Set $X_{t_k} = X_{t_{k-1}} + A$.*

4. *If $k = n$ then stop; otherwise, set $k = k + 1$ and return to Step 2.*

An important class of Lévy processes with known increment distributions is the class of **stable processes.** Here the increments have a *stable* distribution. In particular, if $X_1 \sim \mathsf{Stable}(\alpha, \beta, \mu, \sigma)$, then by (4.8) and the Lévy–Khinchin Theorem 5.13.2 we have $X_t \sim \mathsf{Stable}(\alpha, \beta, \mu\, t,\, t^{1/\alpha}\sigma)$. Hence,

$$X_t = \mu t + t^{1/\alpha}\sigma Z, \quad \text{where} \quad Z \sim \mathsf{Stable}(\alpha, \beta, 0, 1) \equiv \mathsf{Stable}(\alpha, \beta) \, .$$

☞ 129 The generation of $\mathsf{Stable}(\alpha, \beta)$ random variables is discussed in Section 4.2.14.

■ **EXAMPLE 5.19 (Cauchy Process)**

Let $\{X_t\}$ be a Lévy process such that $X_1 \sim \mathsf{Stable}(1, 0) \equiv \mathsf{Cauchy}(0, 1) \equiv \mathsf{t}_1$. We use Algorithm 5.29 and the ratio-of-normals method for Cauchy random variables
☞ 107 (see Section 4.2.2) to simulate this process at times $t_k = k\Delta$, for $k = 0, 1, \ldots$, and $\Delta = 10^{-5}$. Sample MATLAB code is given below, and a typical realization on $[0, 1]$ is given in Figure 5.27. Note that the process is a pure jump process with occasional very large increments.

```
%rpcauchy.m
Delta=10^(-5); N=10^5; times=(0:1:N).*Delta;
Z=randn(1,N+1)./randn(1,N+1);
Z=Delta.*Z; Z(1)=0;
X=cumsum(Z);
plot(times,X)
```

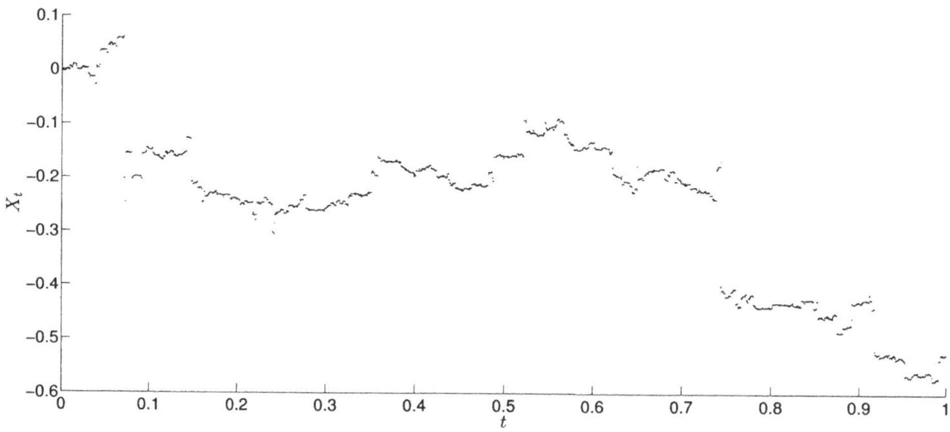

Figure 5.27 Cauchy process realization.

Another example where the increment distributions are known is the gamma process introduced in Example 5.17, where $X_t \sim \mathsf{Gamma}(\alpha t, \lambda), t \geqslant 0$. The generation of $\mathsf{Gamma}(\alpha, \lambda)$ random variables is discussed in Section 4.2.6. The *normal inverse Gaussian process* provides another example; see Example 5.22. ☞ 112

5.13.2.2 Compound Poisson Process and Brownian Motion A second general approach to generate Lévy processes is based on the Lévy–Itô decomposition in Theorem 5.13.1. It is assumed that the characteristic triplet (γ, σ^2, ν) is known.

For simulation purposes, there are two broad cases, namely finite activity ($\nu(\mathbb{R}) < \infty$) versus infinite activity ($\nu(\mathbb{R}) = \infty$). In the *finite activity* case, we have the decomposition

$$X_t = \gamma t + \sigma W_t + Y_t \; , \tag{5.38}$$

where $\{W_t\}$ is a Wiener process and $\{Y_t\}$ is a compound Poisson process, with Lévy measure ν; that is, with rate $\lambda = \nu(\mathbb{R})$ and jump-size distribution ν/λ. The process is called a **jump diffusion**. It was introduced as a model for stock prices by Merton [21]. Since $\{W_t\}$ and $\{Y_t\}$ are independent of each other, simulation in this case is achieved by melding appropriate algorithms for the Wiener process and for compound Poisson processes; for example, Algorithms 5.15 and 5.14.

■ EXAMPLE 5.20 (Jump Diffusion)

The following MATLAB program generates a jump diffusion of the form (5.38) on the interval $[0, 1]$. The jump size distribution is $\mathsf{N}(a, b^2)$. The specific parameters are $\gamma = 5$, $\sigma = 2$, $\lambda = 7$, $a = 0$, and $b = 1$. A typical realization is given in Figure 5.28.

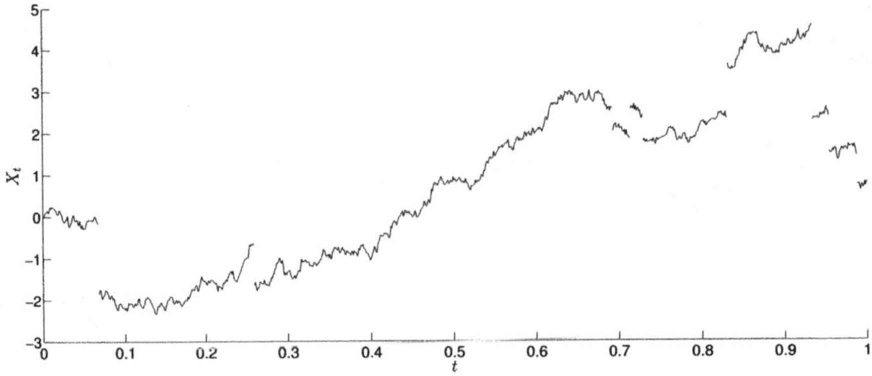

Figure 5.28 A jump-diffusion process.

```
%jumpdiff.m
T = 1; nsteps = 10^3; h = T/nsteps;
lambda = 7; a = 0; b = 1; %Jump Parameters
gamma = 5;  sigma = 2; %BM Parameters
%Generate BM increments
dX = gamma*h + sigma*sqrt(h)*randn(nsteps,1);
dX(1) = 0;
%Generate Jump Process Part
N = poissrnd(lambda);
jumpidx = zeros(N,1); jumpsize = zeros(N,1);
for i = 1:N
    jumpidx(i) = ceil(rand*T*nsteps);
    jumpsize(i) = a + b*randn;
    dX(jumpidx(i)) = dX(jumpidx(i)) + jumpsize(i);
end
t = h:h:T;
X = cumsum(dX);
if N==0
    plot(t,X,'k-')
else
    jidx=sort(jumpidx);
    plot(t(1:jidx(1)-1),X(1:jidx(1)-1),'k-'),hold on
    for k=2:N
        plot(t(jidx(k-1):jidx(k)-1),X(jidx(k-1):jidx(k)-1),'k-')
    end
    plot(t(jidx(N):nsteps),X(jidx(N):nsteps),'k-'),hold off
end
```

Next, consider an *infinite* activity Lévy process $\{X_t\}$, with characteristic triplet (γ, σ^2, ν). Write the Lévy–Itô decomposition as

$$X_t = \gamma\, t + \sigma\, W_t + Y_t + Z_t\ ,$$

where $\{Y_t\}$ is a compound Poisson process with Lévy measure $\nu(\mathrm{d}x)\, \mathrm{I}_{\{|x|>1\}}$, $\{W_t\}$ is a Wiener process, and $\{Z_t\}$ is the limit of compensated compound Poisson processes, as in (5.35). This suggests the following approximation to X_t:

$$X_t^\delta = \gamma\, t + \sigma\, W_t + Y_t + Z_t^\delta \tag{5.39}$$

$$= t\left(\gamma - \int_{\delta \leqslant |x| \leqslant 1} x\, \nu(\mathrm{d}x)\right) + \sigma\, W_t + Y_t^\delta\ , \tag{5.40}$$

where $\{Y_t^\delta\}$ is a compound Poisson process with Lévy measure $\nu(\mathrm{d}x)\, \mathrm{I}_{\{|x|>\delta\}}$. The process $\{X_t^\delta\}$ can thus be generated by separately generating the Wiener and the compound Poisson parts.

The stochastic process $\{R_t, t \geqslant 0\}$, where

$$R_t = X_t - X_t^\delta = \lim_{\varepsilon \downarrow 0} Z_t^\varepsilon - Z_t^\delta = \lim_{\varepsilon \downarrow 0} (Z_t^\varepsilon - Z_t^\delta)$$

is the error of the approximation at time t, is again a Lévy process, with characteristic triplet $(0, 0, \nu(\mathrm{d}x)\, \mathrm{I}_{\{|x| \leqslant \delta\}})$. Since the jumps are bounded, the expectation and variance of R_t are given by (see Property 3 on Page 210) $\mathbb{E}R_t = 0$ and

$$\mathrm{Var}(R_t) = t \int_{|x|<\delta} x^2\, \nu(\mathrm{d}x) \stackrel{\text{def}}{=} t\, \sigma_\delta^2\ .$$

This suggests that the error process $\{R_t\}$ could be approximated by a zero-mean Brownian motion process with diffusion coefficient σ_δ^2. The following theorem (see [2, Page 334] and [3]) specifies a practical condition under which such an approximation is justified. A sufficient condition is that $\delta/\sigma_\delta \to 0$ as $\delta \to 0$.

Theorem 5.13.4 (Convergence of Error Process) *Assume ν has a density of the form $L(x)/|x|^{\alpha+1}$ for all small x, where* ☞ 704

$$\lim_{x \to 0} \frac{L(tx)}{L(x)} = 1 \quad \textit{for all } t > 0\ ,$$

and $0 < \alpha < 2$. Then, $\{R_t/\sigma_\delta, t \geqslant 0\}$ converges in distribution a Wiener process $\{W_t, t \geqslant 0\}$ as $\delta \to 0$.

5.13.2.3 Subordination A third general approach to generate Lévy processes is to use subordination (see Section 5.13.1). In particular, suppose we wish to generate a subordinated Lévy process $\{Y_t\}$, with $Y_t = X_{S_t}$, where $\{X_t\}$ is a Lévy process and $\{S_t\}$ the subordinator. If both $\{X_t\}$ and $\{S_t\}$ are easy to generate at times $t_1, t_2, \ldots, t_n$ (for example, via Algorithm 5.29), then generating $\{Y_t\}$ at those times is straightforward.

Algorithm 5.30 (Generation via Subordination) *Suppose that realizations of $\{X_t\}$ and $\{S_t\}$ at arbitrary times are easily obtained. Generate a realization of $\{Y_t = X_{S_t}\}$ at times $t_1 < \ldots < t_n$ as follows:*

1. *Draw $S_{t_i}, i = 1, \ldots, n$.*

2. *Draw $Y_{t_i} = X_{S_{t_i}}, i = 1, \ldots, n$.*

■ **EXAMPLE 5.21 (Variance Gamma Process Generation)**

Let $\{Y_t\}$ be the variance gamma process in Example 5.18 defined by subordinating a Brownian motion $\{B_t\}$ (with drift μ and diffusion coefficient σ^2) with a Gamma process $\{S_t\}$ (with parameters α and $\lambda = \alpha$).

Such a process is easily simulated at time points $t_1, t_2, \ldots, t_n$ (with the convention that $t_0 = 0$ and $X_0 = 0$), using the facts that $S_t \sim \mathsf{Gamma}(\alpha t, \alpha)$ and $B_t = \mu t + \sigma\sqrt{t}\, Z$, $Z \sim \mathsf{N}(0,1)$. This gives the following algorithm.

Algorithm 5.31 (Variance Gamma Process Simulation)

1. *Set $k = 1$.*

2. *Generate $S \sim \mathsf{Gamma}(\alpha(t_k - t_{k-1}), \alpha)$.*

3. *Generate $Z \sim \mathsf{N}(0,1)$.*

4. *Set $X_{t_k} = X_{t_{k-1}} + \sigma Z \sqrt{S} + \mu\, S$.*

5. *Set $k = k + 1$, and repeat from Step 2.*

■ **EXAMPLE 5.22 (Normal Inverse Gaussian Process)**

Let S_t be the first time a Brownian motion with drift $\gamma > 0$ and diffusion coefficient κ^2 hits $t \geqslant 0$. The process $\{S_t, t \geqslant 0\}$ is a subordinator, as it has independent and stationary increments and is increasing. Moreover, S_t has a $\mathsf{Wald}(t/\gamma, t^2/\kappa^2)$
☞ 135 (inverse Gaussian) distribution; see Section 4.2.17. The process is called an **inverse Gaussian process.**

Let $\{B_t\}$ be Brownian motion with drift μ and diffusion coefficient σ^2, independent of $\{S_t, t \geqslant 0\}$. The subordinated process $\{B_{S_t}, t \geqslant 0\}$ is called a **normal inverse Gaussian process** (the qualifier "normal" indicates that the subordination is with respect to the Brownian motion). The procedure for generating such processes is exactly the same as in Algorithm 5.31 except that Step 2 is replaced by

2'. *Generate $S \sim \mathsf{Wald}((t_k - t_{k-1})/\gamma,\ (t_k - t_{k-1})^2/\kappa^2)$.*

☞ 137 A simple method for generating Wald random variables is given in Algorithm 4.65.

5.14 TIME SERIES

A **time series** is an ordered sequence of random variables $X_{t_1}, X_{t_2}, \ldots$ measured at successive times $t_1 < t_2 < \ldots$. In other words, a time series is a stochastic process with a discrete index set corresponding to an ordered set of time epochs. In most cases the index set is chosen to be either $\mathbb{N}$ or $\mathbb{Z}$, for example, by defining $\widetilde{X}_i = X_{t_i}$. Moreover, for many applications it is assumed that the process is (weakly) *stationary* (see Section A.9.5), in which case it is convenient to take the index set as $\mathbb{Z}$. In that case the autocovariance function $R(s) = \mathrm{Cov}(X_t, X_{t-s})$ is symmetric: $R(s) = R(-s)$. ☞ 631

From now on we consider time series $\{X_t, t \in \mathbb{Z}\}$, unless otherwise specified. Below, $\{\varepsilon_t\}$ is a **white noise** process; that is, the $\{\varepsilon_t\}$ satisfy:

1. *Zero mean:* $\mathbb{E}\,\varepsilon_t = 0$
2. *Constant variance:* $\mathrm{Var}(\varepsilon_t) = \sigma^2$,
3. *Uncorrelated:* $\mathrm{Cov}(\varepsilon_t, \varepsilon_{t-k}) = 0,\ k \neq 0$.

If, in addition, $\varepsilon_t \sim \mathsf{N}(0, \sigma^2)$ for all t, the process $\{\varepsilon_t\}$ is said to be **Gaussian white noise**; in this case the $\{\varepsilon_t\}$ being uncorrelated is equivalent to them being *independent*.

A time series $\{X_t, t \in \mathbb{Z}\}$ is said to be a ***p*-th order autoregressive** process, denoted $\mathsf{AR}(p)$, if

$$X_t = \sum_{k=1}^{p} a_k X_{t-k} + \varepsilon_t\,, \tag{5.41}$$

where $a_1, \ldots, a_p$ are constant coefficients. The current observation in this time series is thus a linear combination of the p previous values in the series, perturbed by a zero-mean, constant variance error term that is independent of past errors and the past terms in the time series.

If, further, the time series is assumed to be weakly stationary, then $\mathbb{E}X_t = 0$ for all t, and the autocovariance function $R(s) = \mathrm{Cov}(X_t, X_{t-s})$ satisfies the **Yule–Walker equations**:

$$R(s) = \sum_{k=1}^{p} a_k\, R(s-k) \quad \text{for all } s = 1, 2, \ldots\,.$$

This is obtained from (5.41) by multiplying both sides by X_{t-s} and then taking expectations. Similarly, by multiplying both sides by X_t and then taking expectations one finds

$$R(0) = \sum_{k=1}^{p} a_k R(k) + \sigma^2\,,$$

from which, in combination with the Yule–Walker equations, $R(s)$ can be found. A necessary and sufficient condition for such a stationary version to exist is that the roots of the polynomial $q(z) = 1 - \sum_{k=1}^{p} a_k z^k$ lie outside the unit circle [17, Page 63]. In particular, for the $\mathsf{AR}(1)$ process

$$R(s) = \frac{\sigma^2}{1 - a_1^2}\, a_1^{|s|}\,,$$

and the stationarity condition is equivalent to $|a_1| < 1$.

If we can sample from the white-noise process $\{\varepsilon_t\}$, and are given the initial p values $X_{-p+1}, \ldots, X_0$, a straightforward generation algorithm is the following.

Algorithm 5.32 (Generating an Autoregressive Process)

1. *Set* $t = 1$.
2. *Generate* ε_t.
3. *Set* $X_t = \varepsilon_t + \sum_{k=1}^{p} a_k X_{t-k}$.
4. *Set* $t = t + 1$ *and repeat from Step 2.*

If $\{\varepsilon_t\}$ is a Gaussian white noise process with $\text{Var}(\varepsilon_t) = \sigma^2$, and we wish to generate a number of autoregressive time series of length n, we can do so using a band-Cholesky version of the multivariate normal sampling algorithm given in Section 4.3.3. The main work is in computing the Cholesky factorization of the precision matrix $\Lambda = \Sigma^{-1} = DD^{\top}$, where $\Sigma = \text{Cov}(\mathbf{X}, \mathbf{X})$, and $\mathbf{X} = (X_1, X_2, \ldots, X_n)^{\top}$.

☞ 146

Let p be the bandwidth of the matrix Λ; that is $p = \max_{i,j}\{|i - j| \,:\, \lambda_{ij} \neq 0\}$. In this case, the bandwidth is the order of our autoregressive process p.

Algorithm 5.33 (Band-Cholesky Factorization)

For $j = 1$ *to* n:

1. *Set* $r = \min\{j + p, n\}$
2. *For* $l = j$ *to* r, *set* $v_l = \lambda_{l,j}$.
3. *For* $k = \max\{1, j - p\}$ *to* $j - 1$,
   (a) *Set* $i = \min\{k + p, n\}$.
   (b) *For* $l = j$ *to* i, *set* $v_l = v_l - D_{l,k} D_{j,k}$.
4. *For* $l = j$ *to* r, *set* $D_{l,j} = v_l / \sqrt{v_j}$.

This algorithm returns the Cholesky factorization of any positive definite matrix Λ, and takes $n(p^2 + 3p)$ flops or in other words $\mathcal{O}(n)$ when $p \ll n$. Once the matrix D is determined, a realization of $\mathbf{X}$ can be generated using the standard generation algorithm for multivariate normal random variables (see Section 4.3.3).

☞ 146

A time series $\{X_t, t \in \mathbb{Z}\}$ is said to be a ***q*-th order moving average** process, denoted $\mathsf{MA}(q)$, if

$$X_t = \sum_{k=1}^{q} b_k \, \varepsilon_{t-k} + \varepsilon_t \,, \tag{5.42}$$

where $b_1, \ldots, b_q$ are constant coefficients. The $\mathsf{MA}(q)$ process is weakly stationary, with $\mathbb{E}X_t = 0$ and autocovariance function $R(s) = \text{Cov}(X_t, X_{t-s})$ given by

$$R(s) = \sigma^2 \sum_{k=1}^{q} b_k \, b_{t-s} \,,$$

$s = 1, 2, \ldots, q$, and $R(0) = \text{Var}(X_t) = \sigma^2(1 + \sum_{k=1}^{q} b_k^2)$. For $|s| > q$ the autocovariance $R(s)$ is zero. An algorithm for generating $X_0, X_1, \ldots$ follows directly from the definition.

Algorithm 5.34 (Generating a Moving Average Time Series)

1. *Define* $\mathbf{b} = (b_q, b_{q-1}, \ldots, b_1, 1)^\top$. *Set* $t = 0$ *and generate* $\varepsilon_{-1}, \ldots, \varepsilon_{-q}$.

2. *Generate* ε_t *and set* $\boldsymbol{\varepsilon}_t = (\varepsilon_{t-q}, \ldots, \varepsilon_t)^\top$.

3. *Set* $X_t = \mathbf{b}^\top \boldsymbol{\varepsilon}_t$.

4. *Increase* t *by 1 and return to Step 2.*

■ **EXAMPLE 5.23 (Generating a Moving Average Time Series)**

The following MATLAB program generates realizations of a moving average time series of order $q = 20$, with $b_i = i, i = 1, \ldots, 20$, and where the $\{\varepsilon_t\}$ are iid and standard normal. A typical realization is given in Figure 5.29.

```
%movav.m
pars = [1:20]';
b = [pars;1];
q = numel(pars);
N = 10^3;
eps = randn(q+1,1);
for i=1:N
    X(i) = b'*eps;
    eps = [eps(2:q+1);randn];
end
plot(X)
```

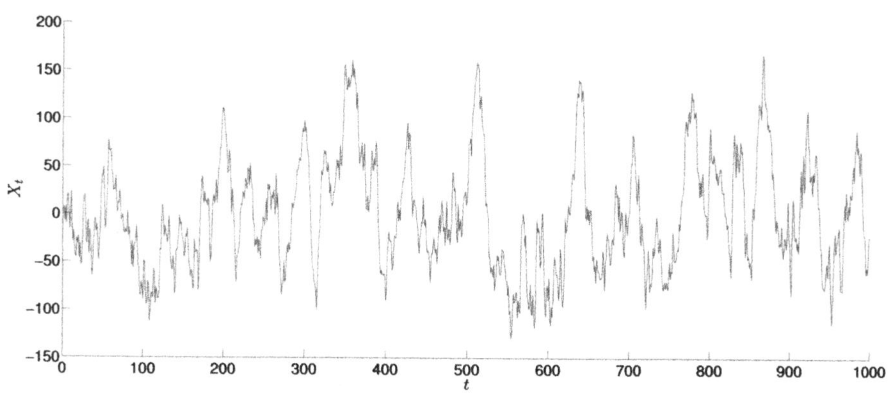

Figure 5.29 A moving average time series.

A time series $\{X_t, t \in \mathbb{Z}\}$ is said to be an **autoregressive moving average process of order (p, q)**, denoted $\mathsf{ARMA}(p, q)$ if

$$X_t = \sum_{i=1}^{p} a_i \, X_{t-i} - \sum_{k=1}^{q} b_k \, \varepsilon_{t-k} + \varepsilon_t \,,$$

where $a_1, \ldots, a_p$ and $b_1, \ldots, b_q$ are constant coefficients.

A time series $\{X_t, t \in \mathbb{Z}\}$ is said to be **autoregressive integrated moving-average process of order (p, q, d)**, denoted $\mathsf{ARIMA}(p, q, d)$ if the d-th backward difference process $\{Y_t\}$, defined by

$$Y_t = \nabla^d X_t = \sum_{k=0}^{d} (-1)^k \binom{d}{k} X_{t-k}$$

is an $\mathsf{ARMA}(p, q)$ process.

For example, the first backward difference is $\nabla X_t = X_t - X_{t-1}$, the second backward difference is $\nabla^2 X_t = \nabla(\nabla X_t) = \nabla(X_t - X_{t-1}) = X_t - 2X_{t-1} + X_{t-2}$, and so on.

Further Reading

A general reference on stochastic process generation is Asmussen and Glynn [2]. Kloeden and Platen [18] is the standard reference on the numerical simulation of stochastic differential equations, giving convergence properties of the schemes, as well as instructive numerical exercises. Burrage et al. [7] covers stability issues for numerical schemes with variable step sizes. Cont and Tankov [9] provide an accessible treatment of Lévy processes, with a particular view to simulation and applications in finance. Various sampling techniques for generating variance gamma processes are proposed in [4]. Karatzas and Shreve [16] provide a comprehensive account of Brownian motion, and Sato [24] and Applebaum [1] give encompassing treatments on Lévy processes. Çinlar [8] is a resource for Markov chains and jump processes. A reference for Gaussian Markov random fields and their generation is the monograph of Rue and Held [23]. A useful reference for time series analysis is [17], and a reference for time series models is [14].

REFERENCES

1. D. Applebaum. *Lévy Processes and Stochastic Calculus.* Cambridge University Press, Cambridge, 2004.
2. S. Asmussen and P. W. Glynn. *Stochastic Simulation: Algorithms and Analysis.* Springer-Verlag, New York, 2007.
3. S. Asmussen and J. Rosiński. Approximations of small jumps of Lévy processes with a view towards simulation. *Journal of Applied Probability*, 38(2):482–493, 2001.
4. T. Avramidis and P. L'Ecuyer. Efficient Monte Carlo and quasi-Monte Carlo option pricing under the variance-gamma model. *Management Science*, 52(12):1930–1944, 2006.

5. J. Besag. Spatial interaction and the statistical analysis of lattice systems. *Journal of the Royal Statistical Society, Series B*, 36(2):192–236, 1974.

6. A. Beskos and G. O. Roberts. Exact simulation of diffusions. *The Annals of Applied Probability*, 15(4):2422–2444, 2005.

7. K. Burrage, P. Burrage, and T. Mitsui. Numerical solutions of stochastic differential equations implementation and stability issues. *Journal of Computational and Applied Mathematics*, 125(1&2):171–182, 2005.

8. E. Çinlar. *Introduction to Stochastic Processes.* Prentice Hall, Englewood Cliffs, 1975.

9. R. Cont and P. Tankov. *Financial Modelling with Jump Processes.* Chapman & Hall/CRC, Boca Raton, FL, 2004.

10. P. F. Craigmile. Simulating a class of stationary Gaussian processes using the Davies-Harte algorithm, with application to long memory processes. *Journal of Time Series Analysis*, 24(5):505–511, 2003.

11. C. R. Dietrich and G. N. Newsam. Fast and exact simulation of stationary Gaussian processes through circulant embedding of the covariance matrix. *SIAM Journal on Scientific Computing*, 18(4):1088–1107, 1997.

12. D. T. Gillespie. Exact numerical simulation of the Ornstein-Uhlenbeck process and its integral. *Physical Review E*, 54(2):2084–2091, 1996.

13. A. Haar. Zur Theorie der orthogonalen Funktionensysteme. *Mathematische Annalen*, 69(3):331–371, 1910.

14. A. C. Harvey. *Time Series Models.* Harvester Wheatsheaf, New York, second edition, 1993.

15. S. M. Iacus. *Simulation and Inference for Stochastic Differential Equations: With R Examples.* Springer-Verlag, New York, 2008.

16. I. Karatzas and S. E. Shreve. *Brownian Motion and Stochastic Calculus.* Springer-Verlag, Berlin, second edition, 2000.

17. M. Kendall and J. K. Ord. *Time Series.* Oxford University Press, Oxford, third edition, 1990.

18. P. E. Kloeden and E. Platen. *Numerical Solution of Stochastic Differential Equations.* Springer-Verlag, Berlin, 1999.

19. N. V. Krylov. *Introduction to the Theory of Random Processes*, volume 43 of *Graduate Studies in Mathematics.* American Mathematical Society, Providence, RI, 2002.

20. B. B. Mandelbrot and J. W. van Ness. Fractional Brownian motions, fractional noises and applications. *SIAM Review*, 10(4):422–437, 1968.

21. R. C. Merton. Option pricing when underlying stock returns are discontinuous. *Journal of Financial Economics*, 3(1–2):125–144, 1976.

22. I. Norros, E. Valkeila, and J. Virtamo. An elementary approach to a Girsanov formula and other analytical results on fractional brownian motions. *Bernoulli*, 5(4):571–587, 1999.

23. H. Rue and L. Held. *Gaussian Markov Random Fields: Theory and Applications.* Chapman & Hall, London, 2005.

24. K. Sato. *Lévy Processes and Infinitely Divisible Distributions.* Cambridge University Press, Cambridge, 1999.

25. D. W. Stroock and S. R. S. Varadhan. *Multidimensional Diffusion Processes.* Springer-Verlag, New York, 2005.

CHAPTER 6

MARKOV CHAIN MONTE CARLO

Markov chain Monte Carlo (MCMC) is a generic method for *approximate* sampling from an arbitrary distribution. The main idea is to generate a Markov chain whose limiting distribution is equal to the desired distribution. In this chapter we describe the most prominent MCMC algorithms:

1. The *Metropolis–Hastings* algorithm and in particular the *independence sampler* and *random walk sampler*;
2. The *Gibbs sampler*, which is particularly useful in Bayesian analysis;
3. The *hit-and-run sampler* — commonly used in Bayesian settings with a highly constrained parameter space and for generic rare-event simulation problems;
4. The *shake-and-bake* algorithm — a practical approach for generating points uniformly distributed on the surface of a polytope;
5. *Metropolis–Gibbs hybrids* and the *multiple-try Metropolis–Hastings* method, in which ideas from different MCMC algorithms are combined;
6. *Auxiliary variable* samplers such as the *slice sampler* and the *Swendsen–Wang* algorithm;
7. The *reversible-jump sampler*, which has applications in Bayesian model selection.

For *exact* methods for random variable generation from commonly used distributions, see Chapter 3. Applications of MCMC to optimization can be found in Chapter 12, in particular Section 12.3, which discusses *simulated annealing*. We refer to Appendix A.10 for more details on Markov chains. MCMC algorithms

☞ 43
☞ 449
☞ 632

are frequently used in statistical data analysis, in particular in Bayesian statistics.
☞ 653 Elements of mathematical and Bayesian statistics are discussed in Appendix B. Analysis of statistical data, for example, generated from an MCMC algorithm, is
☞ 301 discussed in Chapter 8.

6.1 METROPOLIS–HASTINGS ALGORITHM

The MCMC method originates from Metropolis et al. [54] and applies to the following setting. Suppose that we wish to generate samples from an arbitrary multidimensional pdf

$$f(\mathbf{x}) = \frac{p(\mathbf{x})}{\mathcal{Z}}, \quad \mathbf{x} \in \mathscr{X},$$

where $p(\mathbf{x})$ is a known positive function and $\mathcal{Z}$ is a known or unknown normalizing constant. Let $q(\mathbf{y} \,|\, \mathbf{x})$ be a **proposal** or **instrumental** density: a Markov transition density describing how to go from state $\mathbf{x}$ to $\mathbf{y}$. Similar to the acceptance–rejection
☞ 59 method, the Metropolis–Hastings algorithm is based on the following "trial-and-error" strategy.

Algorithm 6.1 (Metropolis–Hastings Algorithm) *To sample from a density $f(\mathbf{x})$ known up to a normalizing constant, initialize with some $\mathbf{X}_0$ for which $f(\mathbf{X}_0) > 0$. Then, for each $t = 0, 1, 2, \ldots, T-1$ execute the following steps:*

1. *Given the current state $\mathbf{X}_t$, generate $\mathbf{Y} \sim q(\mathbf{y} \,|\, \mathbf{X}_t)$.*

2. *Generate $U \sim \mathsf{U}(0,1)$ and deliver*

$$\mathbf{X}_{t+1} = \begin{cases} \mathbf{Y} & \text{if } U \leqslant \alpha(\mathbf{X}_t, \mathbf{Y}) \\ \mathbf{X}_t & \text{otherwise}, \end{cases} \tag{6.1}$$

where

$$\alpha(\mathbf{x}, \mathbf{y}) = \min\left\{ \frac{f(\mathbf{y})\, q(\mathbf{x} \,|\, \mathbf{y})}{f(\mathbf{x})\, q(\mathbf{y} \,|\, \mathbf{x})}, 1 \right\}. \tag{6.2}$$

The probability $\alpha(\mathbf{x}, \mathbf{y})$ is called the **acceptance probability**. Note that in (6.2) we may replace f by p.

We thus obtain the so-called **Metropolis–Hastings Markov chain**, $\mathbf{X}_0, \mathbf{X}_1, \ldots, \mathbf{X}_T$, with $\mathbf{X}_T$ approximately distributed according to $f(\mathbf{x})$ for large T. A single Metropolis–Hastings iteration is equivalent to generating a point from the transition density $\kappa(\mathbf{x}_{t+1} \,|\, \mathbf{x}_t)$, where

$$\kappa(\mathbf{y} \,|\, \mathbf{x}) = \alpha(\mathbf{x}, \mathbf{y})\, q(\mathbf{y} \,|\, \mathbf{x}) + (1 - \alpha^*(\mathbf{x}))\, \delta_{\mathbf{x}}(\mathbf{y}), \tag{6.3}$$

with $\alpha^*(\mathbf{x}) = \int \alpha(\mathbf{x}, \mathbf{y})\, q(\mathbf{y} \,|\, \mathbf{x})\, \mathrm{d}\mathbf{y}$ and $\delta_{\mathbf{x}}(\mathbf{y})$ denoting the Dirac delta function. Since

$$f(\mathbf{x})\, \alpha(\mathbf{x}, \mathbf{y})\, q(\mathbf{y} \,|\, \mathbf{x}) = f(\mathbf{y})\, \alpha(\mathbf{y}, \mathbf{x})\, q(\mathbf{x} \,|\, \mathbf{y})$$

and

$$(1 - \alpha^*(\mathbf{x}))\, \delta_{\mathbf{x}}(\mathbf{y})\, f(\mathbf{x}) = (1 - \alpha^*(\mathbf{y}))\, \delta_{\mathbf{y}}(\mathbf{x})\, f(\mathbf{y}),$$

☞ 635 the transition density satisfies the *detailed balance equation* (A.48):

$$f(\mathbf{x})\,\kappa(\mathbf{y}\,|\,\mathbf{x}) = f(\mathbf{y})\,\kappa(\mathbf{x}\,|\,\mathbf{y}),$$

from which it follows that f is the stationary pdf of the chain. In addition, if the transition density q satisfies the conditions

$$\mathbb{P}(\alpha(\mathbf{X}_t, \mathbf{Y}) < 1\,|\,\mathbf{X}_t) > 0,$$

that is, the event $\{\mathbf{X}_{t+1} = \mathbf{X}_t\}$ has positive probability, and

$$q(\mathbf{y}\,|\,\mathbf{x}) > 0 \quad \text{for all } \mathbf{x}, \mathbf{y} \in \mathscr{X},$$

then f is the limiting pdf of the chain. As a consequence, to estimate an expectation $\mathbb{E}H(\mathbf{X})$, with $\mathbf{X} \sim f$, one can use the following **ergodic** estimator (see also Section 8.3): ☞ 309

$$\frac{1}{T+1}\sum_{t=0}^{T} H(\mathbf{X}_t)\,. \tag{6.4}$$

The original Metropolis algorithm [54] is suggested for symmetric proposal functions; that is, for $q(\mathbf{y}\,|\,\mathbf{x}) = q(\mathbf{x}\,|\,\mathbf{y})$. Hastings [37] modified the original MCMC algorithm to allow nonsymmetric proposal functions, hence the name Metropolis–Hastings algorithm.

6.1.1 Independence Sampler

If the proposal function $q(\mathbf{y}\,|\,\mathbf{x})$ does not depend on $\mathbf{x}$, that is, $q(\mathbf{y}\,|\,\mathbf{x}) = g(\mathbf{y})$ for some pdf $g(\mathbf{y})$, then the acceptance probability is

$$\alpha(\mathbf{x}, \mathbf{y}) = \min\left\{\frac{f(\mathbf{y})\,g(\mathbf{x})}{f(\mathbf{x})\,g(\mathbf{y})}, 1\right\},$$

and Algorithm 6.1 is referred to as the **independence sampler**. The independence sampler is very similar to the acceptance–rejection method in Chapter 3. Just as in that method, it is important that the proposal density g is close to the target f. Note, however, that in contrast to the acceptance–rejection method the independence sampler produces *dependent* samples. In addition, if there is a constant C such that

$$f(\mathbf{x}) = \frac{p(\mathbf{x})}{\int p(\mathbf{x})\,\mathrm{d}\mathbf{x}} \leqslant C g(\mathbf{x})$$

for all $\mathbf{x}$, then the acceptance rate in (6.1) is at least $1/C$ whenever the chain is in stationarity; namely,

$$\begin{aligned}
\mathbb{P}(U \leqslant \alpha(\mathbf{X}, \mathbf{Y})) &= \iint \min\left\{\frac{f(\mathbf{y})\,g(\mathbf{x})}{f(\mathbf{x})\,g(\mathbf{y})}, 1\right\} f(\mathbf{x})\,g(\mathbf{y})\,\mathrm{d}\mathbf{x}\,\mathrm{d}\mathbf{y} \\
&= 2\iint \mathrm{I}\left\{\frac{f(\mathbf{y})\,g(\mathbf{x})}{f(\mathbf{x})\,g(\mathbf{y})} \geqslant 1\right\} f(\mathbf{x})\,g(\mathbf{y})\,\mathrm{d}\mathbf{x}\,\mathrm{d}\mathbf{y} \\
&\geqslant \frac{2}{C}\iint \mathrm{I}\left\{\frac{f(\mathbf{y})\,g(\mathbf{x})}{f(\mathbf{x})\,g(\mathbf{y})} \geqslant 1\right\} f(\mathbf{x})\,f(\mathbf{y})\,\mathrm{d}\mathbf{x}\,\mathrm{d}\mathbf{y} \\
&\geqslant \frac{2}{C}\,\mathbb{P}\left(\frac{f(\mathbf{Y})}{g(\mathbf{Y})} \geqslant \frac{f(\mathbf{X})}{g(\mathbf{X})}\right) = \frac{1}{C}\,.
\end{aligned}$$

In contrast, the acceptance rate in Algorithm 3.9 (acceptance–rejection) using g as a proposal is always equal to $1/C$. ☞ 59

■ **EXAMPLE 6.1 (Sampling on the Surface of an Ellipsoid)**

Consider the problem of sampling uniformly on the surface of an ellipsoid defined by
☞ 77 the parametric equations (3.19); see Example 3.21. This problem requires that we sample from the surface density $\|\mathbf{r}_1 \times \mathbf{r}_2\|/S(a,b,c)$ of the ellipsoid, where $S(a,b,c)$ is the surface area of the ellipsoid and $\|\mathbf{r}_1 \times \mathbf{r}_2\|$ is defined in (3.20). In Example 3.21 we show that the efficiency (probability of acceptance) of the acceptance–rejection approach is given by $S(a,b,c)/(4\pi r^2)$, where $r = \max\{a,b,c\}$. It can be shown [61] that

$$\frac{ab + ac + bc}{3r^2} \leqslant \frac{S(a,b,c)}{4\pi r^2} \leqslant \frac{\sqrt{a^2b^2 + a^2c^2 + b^2c^2}}{r^2\sqrt{3}} .$$

Thus, the acceptance–rejection approach in Example 3.21 becomes inefficient when one of the parameters a, b, c is significantly larger than the other two. For example, for $(a,b,c) = (400, 2, 1)$ the acceptance probability lies in the interval [0.0050, 0.0065]. An alternative is to use the independence sampler with $\mathbf{x} = (v_1, v_2)^\top$ and proposal pdf $q(\mathbf{y}\,|\,\mathbf{x}) = g(\mathbf{y}) = g(v_1, v_2) = \sin(v_2)/(4\pi)$. This gives the following MATLAB implementation for generating an array `data` with (approximate) surface density corresponding to $(a,b,c) = (400, 2, 1)$. An estimate of the acceptance probability $\mathbb{P}(U \leqslant \alpha(\mathbf{X}, \mathbf{Y}))$ of the independence sampler is 0.88, which is significantly larger than the acceptance probability of the acceptance–rejection Algorithm 3.9.

```
%independence_sampler.m
clear all, a=400; b=2; c=1;
p=@(x)sqrt((b*c)^2*sin(x(2)).^2.*cos(x(1)).^2+...
    (a*c)^2*sin(x(2)).^2.*sin(x(1)).^2+...
    (a*b)^2*cos(x(2)).^2);
%define alpha(x,y) in the Metropolis algorithm
alpha=@(x,y)min(1,sqrt  ( p(y) ./ p(x) ) );
X=ones(2,1)*pi/2; % initial starting point
T=10^4; data=nan(T,2); % preallocate memory
accept_prob=0;
for t=1:T
    Y=[rand*2*pi;acos(1-2*rand)]; % make proposal
    if rand<alpha(X,Y) % Metropolis criterion
        X=Y; accept_prob=accept_prob+1;
    end
    data(t,:)=X';
end
accept_prob=accept_prob/T
K=20; x=data(:,1); ell=mean(x);
for k=0:K
    R(k+1)= (x(1:end-k)-ell)'*(x(k+1:end)-ell);
    R(k+1)=R(k+1)/(length(x)-k-1);
end
plot([0:K],R)
```

☞ 309 Figure 6.1 shows the estimated autocovariance function ($k = 0, \ldots, K,\ K < T$)

$$\widehat{R}(k) = \frac{1}{T-k-1}\sum_{t=1}^{T-k}(X_{t,1} - \bar{X}_{\bullet,1})(X_{t+k,1} - \bar{X}_{\bullet,1}), \quad \bar{X}_{\bullet,1} = \frac{1}{T}\sum_{t=1}^{T} X_{t,1} ,$$

based on the simulated values $\{X_{t,1}\} = \{V_{t,1}\}$ in the code above. The covariance function decays fast and is negligible after the first 10 lags. Failure to detect such fast decay may indicate that the independence sampler converges slowly to its stationary distribution (that is, it is not mixing well, see Section 6.4). In practice, we can consider the **thinned** chain $\mathbf{X}_0, \mathbf{X}_{t^*}, \mathbf{X}_{2t^*}, \ldots$, constructed from the output $\mathbf{X}_0, \mathbf{X}_1, \mathbf{X}_2, \ldots$, of the independence sampler so that $\mathbf{X}_0, \mathbf{X}_{t^*}, \mathbf{X}_{2t^*}, \ldots$, are approximately iid random variables. A possible choice for t^* is 10, because $\widehat{R}(k)$ is negligible after $k \geqslant 10$.

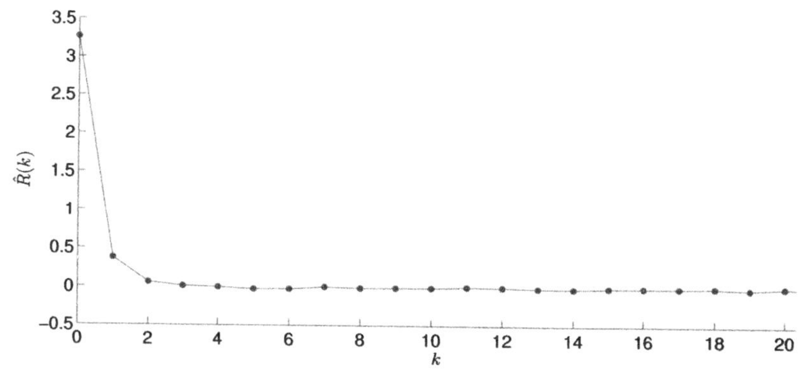

Figure 6.1 The estimated autocovariance function for the $\{X_{t,1}\}$ for lags k up to 20.

As an additional assessment of the performance of the sampler Figure 6.2 shows the first 10^4 points generated by the sampler on the surface of the ellipsoid. A visual inspection suggests that the points cover the whole surface uniformly.

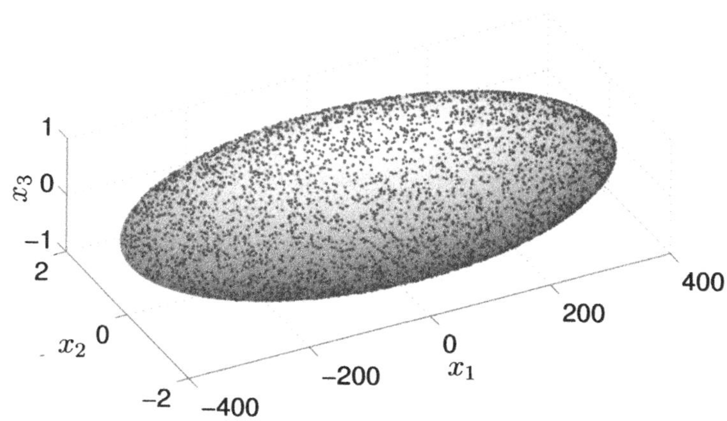

Figure 6.2 The first 10^4 points (generated by the independence sampler) on the surface of the ellipsoid. The estimated acceptance probability of the independence sampler is 0.88.

An assessment of the quality of the approximation obtained from any MCMC ☞ 509 sampler is in general a difficult task, see Section 6.4 and Section 14.7. For the independence sampler it is possible to compute an upper bound on the total variation distance between the sampling pdf and the target pdf. The **total variation distance** between two probability measures μ and ν is defined as

$$\sup_A |\mu(A) - \nu(A)|$$

for all sets A in the corresponding σ-algebra. Mengersen and Tweedie [52] show that the convergence of the independence sampler is geometric, that is,

$$\sup_A \left| \int_A (\kappa_t(\mathbf{y} \,|\, \mathbf{x}) - f(\mathbf{y}))\, \mathrm{d}\mathbf{y} \right| \leqslant 2\left(1 - \frac{1}{C}\right)^t ,$$

where κ_t is the t-step transition density of the independence sampler, when there exists an enveloping constant C such that $f(\mathbf{x}) = \mathcal{Z}^{-1} p(\mathbf{x}) \leqslant C\, g(\mathbf{x})$ for all $\mathbf{x}$.

While this bound rigorously quantifies the convergence of the independence sampler, the bound is typically loose and cannot be used to adequately assess the convergence of the chain. For instance, for the surface density of the ellipsoid in Example 6.1 we have

$$\frac{\|\mathbf{r}_1 \times \mathbf{r}_2\|}{S(a,b,c)} \leqslant \frac{r^2 \sin(v_2)}{S(a,b,c)} = \frac{4\pi r^2}{S(a,b,c)}\, g(v_1, v_2), \qquad r = \max\{a, b, c\} .$$

Hence, the constant $C = \frac{3r^2}{ab+ac+bc} \geqslant \frac{4\pi r^2}{S(a,b,c)}$ is such that the total variation distance between the transition density of the independence sampler and the target pdf is less than $2\left(1 - \frac{1}{C}\right)^t$. Therefore, to guarantee a discrepancy smaller than 10%, we need to run the independence sampler for $t \geqslant \lceil -\ln(10)/\ln(1 - 1/C) \rceil$ steps, which is larger than the expected number of trials before a success in the acceptance–rejection method. Figure 6.2 and the estimated autocovariance plot on Figure 6.1, however, suggest that it is not necessary to run the chain for so long to achieve satisfactory performance.

6.1.2 Random Walk Sampler

If the proposal is symmetric, that is, $q(\mathbf{y} \,|\, \mathbf{x}) = q(\mathbf{x} \,|\, \mathbf{y})$, then the acceptance probability (6.2) is

$$\alpha(\mathbf{x}, \mathbf{y}) = \min\left\{\frac{f(\mathbf{y})}{f(\mathbf{x})}, 1\right\}, \tag{6.5}$$

and Algorithm 6.1 is referred to as the **random walk sampler**. An example of a random walk sampler is when $\mathbf{Y} = \mathbf{X}_t + \sigma \mathbf{Z}$ in Step 1 of Algorithm 6.1, where $\mathbf{Z}$ is typically generated from some spherically symmetrical distribution (in the continuous case), such as $\mathsf{N}(\mathbf{0}, I)$.

☞ 643 Consider the **Langevin diffusion** defined by the SDE

$$\mathrm{d}\mathbf{X}_t = \frac{1}{2} \nabla \ln f(\mathbf{X}_t)\, \mathrm{d}t + \mathrm{d}\mathbf{W}_t ,$$

where $\nabla \ln f(\mathbf{X}_t)$ denotes the gradient of $\ln f(\mathbf{x})$ evaluated at $\mathbf{X}_t$. The Langevin diffusion has stationary pdf f, and is nonexplosive and reversible. Suppose the proposal state $\mathbf{Y}$ in Step 1 of Algorithm 6.1 corresponds to the Euler discretization of the Langevin SDE for some step size h: ☞ 185

$$\mathbf{Y} = \mathbf{X}_t + \frac{h}{2}\nabla \ln f(\mathbf{X}_t) + \sqrt{h}\,\mathbf{Z}, \quad \mathbf{Z} \sim \mathsf{N}(\mathbf{0}, I)\,.$$

This gives a more sophisticated random walk sampler with a "drift" term $\nabla \ln f(\mathbf{x}_t)$. Such random walk samplers are collectively known as **Langevin Metropolis–Hastings** algorithms [70]. Note that the gradient can be approximated numerically via finite differences and does not require knowledge of the normalizing constant of $f(\mathbf{x})$. In some cases the Langevin Metropolis–Hastings algorithms are more efficient than the simple random walk algorithms [66, 70]. For a discussion of the optimal tuning of Langevin Metropolis–Hastings algorithms see [58]. ☞ 423

■ EXAMPLE 6.2 (Bayesian Analysis of the Logit Model)

Consider the Bayesian analysis of the logistic regression model or **logit model**. This is a commonly used generalized linear model [25], where binary data $y_1, \ldots, y_n$ (the responses) are assumed to be conditionally independent realizations from $\mathsf{Ber}(p_i)$ given $p_1, \ldots, p_n$ (that is, $y_i \,|\, p_i \sim \mathsf{Ber}(p_i)$, $i = 1, \ldots, n$, independently), with

$$p_i = \frac{1}{1 + \mathrm{e}^{-\mathbf{x}_i^\top \boldsymbol{\beta}}}, \quad i = 1, \ldots, n\,.$$

Here, $\mathbf{x}_i = (x_{i1}, x_{i2}, \ldots, x_{ik})^\top$ are the explanatory variables or covariates for the i-th response and $\boldsymbol{\beta} = (\beta_1, \ldots, \beta_k)^\top$ are the parameters of the model with multivariate normal prior: $\mathsf{N}(\boldsymbol{\beta}_0, V_0)$. Thus, the Bayesian logit model can be summarized as:

- Prior: $f(\boldsymbol{\beta}) \propto \exp\left(-\frac{1}{2}(\boldsymbol{\beta} - \boldsymbol{\beta}_0)^\top V_0^{-1}(\boldsymbol{\beta} - \boldsymbol{\beta}_0)\right)$, $\boldsymbol{\beta} \in \mathbb{R}^k$.
- Likelihood: $f(\mathbf{y} \,|\, \boldsymbol{\beta}) = \prod_{i=1}^n p_i^{y_i}(1 - p_i)^{1-y_i}$, $p_i^{-1} = 1 + \exp(-\mathbf{x}_i^\top \boldsymbol{\beta})$.

Since the posterior pdf $f(\boldsymbol{\beta} \,|\, \mathbf{y}) \propto f(\boldsymbol{\beta}, \mathbf{y}) = f(\boldsymbol{\beta}) f(\mathbf{y} \,|\, \boldsymbol{\beta})$ cannot be written in a simple analytical form, the Bayesian analysis proceeds by (approximately) drawing a sample from the posterior $f(\boldsymbol{\beta} \,|\, \mathbf{y})$ in order to obtain estimates of various quantities of interest such as $\mathbb{E}[\boldsymbol{\beta} \,|\, \mathbf{y}]$ and $\mathrm{Cov}(\boldsymbol{\beta} \,|\, \mathbf{y})$. In addition, simulation allows a convenient way to explore the marginal posterior densities of each model parameter.

To approximately draw from the posterior we use the random walk sampler with a multivariate $\mathbf{t}_\nu(\boldsymbol{\mu}, \Sigma)$ proposal tailored to match the overall shape of the posterior around its mode. The vector $\boldsymbol{\mu}$ is taken as the mode of the posterior; that is, as $\mathrm{argmax}_{\boldsymbol{\beta}} \ln f(\boldsymbol{\beta} \,|\, \mathbf{y})$, which can be obtained approximately via a Newton–Raphson procedure (see Section C.2.2.1) with gradient ☞ 147 ☞ 688

$$\nabla \ln f(\boldsymbol{\beta} \,|\, \mathbf{y}) = \sum_{i=1}^n \left(y_i - \frac{1}{1 + \mathrm{e}^{-\mathbf{x}_i^\top \boldsymbol{\beta}}}\right) \mathbf{x}_i - V_0^{-1}(\boldsymbol{\beta} - \boldsymbol{\beta}_0)\,,$$

and Hessian

$$H = -\sum_{i=1}^n \frac{\mathrm{e}^{-\mathbf{x}_i^\top \boldsymbol{\beta}}}{(1 + \mathrm{e}^{-\mathbf{x}_i^\top \boldsymbol{\beta}})^2}\, \mathbf{x}_i \mathbf{x}_i^\top - V_0^{-1}\,,$$

where we have used the fact that the logarithm of the posterior (ignoring constant terms) is:

$$-\frac{1}{2}(\boldsymbol{\beta}-\boldsymbol{\beta}_0)^\top V_0^{-1}(\boldsymbol{\beta}-\boldsymbol{\beta}_0) - \sum_{i=1}^n y_i \ln\left(1 + \mathrm{e}^{-\mathbf{x}_i^\top \boldsymbol{\beta}}\right) + (1-y_i)\left(\mathbf{x}_i^\top \boldsymbol{\beta} + \ln\left(1 + \mathrm{e}^{-\mathbf{x}_i^\top \boldsymbol{\beta}}\right)\right)\,.$$

☞ 147 The scale matrix Σ of the proposal distribution $\mathbf{t}_\nu(\boldsymbol{\mu}, \Sigma)$ is chosen as the inverse of the observed Fisher information matrix: $\Sigma = -H^{-1}$. Finally, the shape ν (degrees of freedom) is arbitrarily set to 10. The random walk sampler is initialized at the mode $\boldsymbol{\mu}$ of the posterior. If $\boldsymbol{\beta}^*$ is a newly generated proposal and $\boldsymbol{\beta}$ is the current value, the acceptance criterion (6.2) in the Metropolis–Hastings algorithm can be written as:

$$\alpha(\boldsymbol{\beta}, \boldsymbol{\beta}^*) = \min\left\{\frac{f(\boldsymbol{\beta}^*, \mathbf{y})}{f(\boldsymbol{\beta}, \mathbf{y})}, 1\right\}.$$

The following MATLAB code implements this procedure for the logit model using an artificial data set.

```
%logit_model.m
clear all,clc
n=5000; % number of data points (y_1,...,y_n)
k=3;   % number of explanatory variables
% generate artificial dataset
randn('seed', 12345);  rand('seed', 67890);
truebeta = [1 -5.5 1]';
X = [ ones(n,1) randn(n,k-1)*0.1 ];  % design matrix
Y = binornd(1,1./(1+exp(-X*truebeta)));
bo=zeros(k,1); % we set Vo=100*eye(k);
% determine the mode using Newton Raphson
err=inf; b=bo; % initial guess
while norm(err)>10^(-3)
    p=1./(1+exp(-X*b));
    g=X'*(Y-p)-(b-bo)/100;
    H=-X'*diag(p.^2.*(1./p-1))*X-eye(k)/100;
    err=H\g; % compute Newton-Raphson correction
    b=b-err; % update Newton guess
end
% scale parameter for proposal
Sigma=-H\eye(k); B=chol(Sigma);
% logarithm of joint density (up to a constant)
logf=@(b)(-.5*(b-bo)'*(b-bo)/100-Y'*log(1+exp(-X*b))...
    -(1-Y)'*(X*b+log(1+exp(-X*b))));
alpha=@(x,y)min(1,exp(logf(y)-logf(x)));
df=10; T=10^4; data=nan(T,k); %allocate memory
for t=1:T
    % make proposal from multivariate t
    b_star= b + B*(sqrt(df/gamrnd(df/2,2))*randn(k,1));
    if rand<alpha(b,b_star)
        b=b_star;
    end
    data(t,:)=b';
end
b_hat=mean(data)
Cov_hat=cov(data)
```

Typical estimates for the posterior mean $\mathbb{E}[\boldsymbol{\beta}\,|\,\mathbf{y}]$ and covariance $\mathrm{Cov}(\boldsymbol{\beta}\,|\,\mathbf{y})$ are

$$\widehat{\mathbb{E}[\boldsymbol{\beta}\,|\,\mathbf{y}]} = \begin{pmatrix} 0.980 \\ -5.313 \\ 1.136 \end{pmatrix} \text{ and } \widehat{\mathrm{Cov}(\boldsymbol{\beta}\,|\,\mathbf{y})} = \begin{pmatrix} 0.0011 & -0.0025 & 0.0005 \\ -0.0025 & 0.1116 & -0.0095 \\ 0.0005 & -0.0095 & 0.1061 \end{pmatrix}.$$

Since the prior we employed is relatively noninformative, it is not surprising that the estimated posterior mean is very similar to the maximum likelihood estimate: $\widehat{\boldsymbol{\beta}} = (0.978, -5.346, 1.142)^\top$. *Marginal likelihood* computation for the logit model is discussed in Example 14.3. ☞ 498

6.2 GIBBS SAMPLER

The **Gibbs sampler** can be viewed as a particular instance of the Metropolis–Hastings algorithm for generating n-dimensional random vectors [31]. Due to its importance it is presented separately. The distinguishing feature of the Gibbs sampler is that the underlying Markov chain is constructed from a sequence of conditional distributions, in either a deterministic or random fashion.

Suppose that we wish to sample a random vector $\mathbf{X} = (X_1, \ldots, X_n)$ according to a target pdf $f(\mathbf{x})$. Let $f(x_i\,|\,x_1, \ldots, x_{i-1}, x_{i+1}, \ldots, x_n)$ represent the conditional pdf of the i-th component, X_i, given the other components $x_1, \ldots, x_{i-1}, x_{i+1}, \ldots, x_n$. Here we use the Bayesian notation introduced in Appendix B.3. ☞ 672

Algorithm 6.2 (Gibbs Sampler) *Given an initial state $\mathbf{X}_0$, iterate the following steps for $t = 0, 1, \ldots$.*

1. *For a given $\mathbf{X}_t$, generate $\mathbf{Y} = (Y_1, \ldots, Y_n)$ as follows:*
   (a) *Draw Y_1 from the conditional pdf $f(x_1\,|\,X_{t,2}, \ldots, X_{t,n})$.*
   (b) *Draw Y_i from $f(x_i\,|\,Y_1, \ldots, Y_{i-1}, X_{t,i+1}, \ldots, X_{t,n})$, $\quad i = 2, \ldots, n-1$.*
   (c) *Draw Y_n from $f(x_n\,|\,Y_1, \ldots, Y_{n-1})$.*
2. *Let $\mathbf{X}_{t+1} = \mathbf{Y}$.*

The transition pdf is given by

$$\kappa_{1\to n}(\mathbf{y}\,|\,\mathbf{x}) = \prod_{i=1}^{n} f(y_i\,|\,y_1, \ldots, y_{i-1}, x_{i+1}, \ldots, x_n)\,, \tag{6.6}$$

where the subscript $1 \to n$ indicates that the components of vector $\mathbf{x}$ are updated in the order $1 \to 2 \to 3 \to \cdots \to n$. Note that in the Gibbs sampler *every* "proposal" $\mathbf{y}$, is accepted. The transition density of the reverse move $\mathbf{y} \to \mathbf{x}$, in which the vector $\mathbf{y}$ is updated in the order $n \to n-1 \to n-2 \to \cdots \to 1$ is

$$\kappa_{n\to 1}(\mathbf{x}\,|\,\mathbf{y}) = \prod_{i=1}^{n} f(x_i\,|\,y_1, \ldots, y_{i-1}, x_{i+1}, \ldots, x_n)\,.$$

Hammersley and Clifford [36] prove the following result. ☞ 207

Theorem 6.2.1 (Hammersley–Clifford) *Let $f(x_i)$ be the i-th marginal density of the pdf $f(\mathbf{x})$. Suppose that density $f(\mathbf{x})$ satisfies the* **positivity condition**, *that is, for every $\mathbf{y} \in \{\mathbf{x} : f(x_i) > 0,\ i = 1, \ldots, n\}$, we have $f(\mathbf{y}) > 0$. Then,*

$$f(\mathbf{y})\, \kappa_{n \to 1}(\mathbf{x} \,|\, \mathbf{y}) = f(\mathbf{x})\, \kappa_{1 \to n}(\mathbf{y} \,|\, \mathbf{x}) .$$

Proof (outline): Observe that

$$\begin{aligned}
\frac{\kappa_{1 \to n}(\mathbf{y} \,|\, \mathbf{x})}{\kappa_{n \to 1}(\mathbf{x} \,|\, \mathbf{y})} &= \prod_{i=1}^{n} \frac{f(y_i \,|\, y_1, \ldots, y_{i-1}, x_{i+1}, \ldots, x_n)}{f(x_i \,|\, y_1, \ldots, y_{i-1}, x_{i+1}, \ldots, x_n)} \\
&= \prod_{i=1}^{n} \frac{f(y_1, \ldots, y_i, x_{i+1}, \ldots, x_n)}{f(y_1, \ldots, y_{i-1}, x_i, \ldots, x_n)} \\
&= \frac{f(\mathbf{y}) \prod_{i=1}^{n-1} f(y_1, \ldots, y_i, x_{i+1}, \ldots, x_n)}{f(\mathbf{x}) \prod_{j=2}^{n} f(y_1, \ldots, y_{j-1}, x_j, \ldots, x_n)} \\
&= \frac{f(\mathbf{y}) \prod_{i=1}^{n-1} f(y_1, \ldots, y_i, x_{i+1}, \ldots, x_n)}{f(\mathbf{x}) \prod_{j=1}^{n-1} f(y_1, \ldots, y_j, x_{j+1}, \ldots, x_n)} = \frac{f(\mathbf{y})}{f(\mathbf{x})} .
\end{aligned}$$

The result follows by rearranging the last identity.

The Hammersley–Clifford condition is similar to the detailed balance condition for the Metropolis–Hastings sampler, because integrating both sides with respect
☞ 634 to $\mathbf{x}$ yields the global balance equation:

$$\int f(\mathbf{x})\, \kappa_{1 \to n}(\mathbf{y} \,|\, \mathbf{x})\, \mathrm{d}\mathbf{y} = f(\mathbf{y}) ,$$

from which we can conclude that f is the stationary pdf of the Markov chain with transition density $\kappa_{1 \to n}(\mathbf{y} \,|\, \mathbf{x})$. In addition, it can be shown [65] that the positivity assumption on f implies that the Gibbs Markov chain is irreducible and that f is its limiting pdf. In practice the positivity condition is difficult to verify. However, there are a number of weaker and more technical conditions (see [43, 65]) which ensure that the limiting pdf of the process $\{\mathbf{X}_t, t = 1, 2, \ldots\}$ generated via the Gibbs sampler is f, and that the convergence to f is geometrically fast.

Algorithm 6.2 presents a **systematic** (coordinatewise) Gibbs sampler. That is, the components of vector $\mathbf{X}$ are updated in the coordinatewise order $1 \to 2 \to \cdots \to n$. The completion of all the conditional sampling steps in the specified order is called a **cycle**. Alternative updating of the components of vector $\mathbf{X}$ are possible. In the **reversible Gibbs sampler** a single cycle consists of the coordinatewise updating

$$1 \to 2 \to \cdots \to n-1 \to n \to n-1 \to \cdots \to 2 \to 1 .$$

In the **random sweep/scan Gibbs sampler** a single cycle can either consist of one or several coordinates selected uniformly from the integers $1, \ldots, n$, or a random permutation $\pi_1 \to \pi_2 \to \cdots \to \pi_n$ of all coordinates. In all cases, except for the systematic Gibbs sampler, the resulting Markov chain $\{\mathbf{X}_t, t = 1, 2, \ldots\}$ is
☞ 635 *reversible*. In the case where a cycle consists of a single randomly selected coordinate, the random Gibbs sampler can be formally viewed as a Metropolis–Hastings sampler with transition function

$$q(\mathbf{y} \,|\, \mathbf{x}) = \frac{1}{n} f(y_i \,|\, x_1, \ldots, x_{i-1}, x_{i+1}, \ldots, x_n) = \frac{1}{n} \frac{f(\mathbf{y})}{\sum_{y_i} f(\mathbf{y})} ,$$

where $\mathbf{y} = (x_1, \ldots, x_{i-1}, y_i, x_{i+1}, \ldots, x_n)$. Since $\sum_{y_i} f(\mathbf{y})$ can also be written as $\sum_{x_i} f(\mathbf{x})$, we have

$$\frac{f(\mathbf{y})\, q(\mathbf{x} \,|\, \mathbf{y})}{f(\mathbf{x})\, q(\mathbf{y} \,|\, \mathbf{x})} = \frac{f(\mathbf{y})\, f(\mathbf{x})}{f(\mathbf{x})\, f(\mathbf{y})} = 1\,,$$

so that the acceptance probability $\alpha(\mathbf{x}, \mathbf{y})$ is 1 in this case.

■ EXAMPLE 6.3 (Zero-Inflated Poisson Model)

Gibbs sampling is one of the main computational techniques used in Bayesian analysis. In the **zero-inflated Poisson** model, the random data $X_1, \ldots, X_n$ are assumed to be of the form $X_i = R_i\, Y_i$, where the $Y_1, \ldots, Y_n \sim_{\text{iid}} \mathsf{Poi}(\lambda)$ are independent of $R_1, \ldots, R_n \sim_{\text{iid}} \mathsf{Ber}(p)$. Given an outcome $\mathbf{x} = (x_1, \ldots, x_n)$, the objective is to estimate both λ and p. A typical Bayesian data analysis gives the following hierarchical model: ☞ 673

- $p \sim \mathsf{U}(0,1)$ (prior for p),
- $(\lambda \,|\, p) \sim \mathsf{Gamma}(a, b)$ (prior for λ),
- $(r_i \,|\, p, \lambda) \sim \mathsf{Ber}(p)$ independently (from the model above),
- $(x_i \,|\, \mathbf{r}, \lambda, p) \sim \mathsf{Poi}(\lambda\, r_i)$ independently (from the model above),

where a and b are known parameters. It follows that the joint pdf of all parameters and $\mathbf{x}$ is

$$\begin{aligned} f(\mathbf{x}, \mathbf{r}, \lambda, p) &= \frac{b^a \lambda^{a-1} \mathrm{e}^{-b\lambda}}{\Gamma(a)} \prod_{i=1}^{n} \frac{\mathrm{e}^{-\lambda\, r_i} (\lambda\, r_i)^{x_i}}{x_i!}\, p^{r_i} (1-p)^{1-r_i} \\ &= \frac{b^a \lambda^{a-1} \mathrm{e}^{-b\lambda}}{\Gamma(a)}\, \mathrm{e}^{-\lambda \sum_i r_i}\, p^{\sum_i r_i}\, (1-p)^{n - \sum_i r_i}\, \lambda^{\sum_i x_i} \prod_{i=1}^{n} \frac{r_i^{x_i}}{(x_i)!}\,. \end{aligned}$$

The posterior pdf $f(\lambda, p, \mathbf{r} \,|\, \mathbf{x}) \propto f(\mathbf{x}, \mathbf{r}, \lambda, p)$ is of large dimension, which makes analytical computation using Bayes' formula intractable. Instead, the Gibbs sampler (Algorithm 6.2) provides a convenient tool for approximate sampling and exploration of the posterior. Here the conditionals of the posterior are:

- $f(\lambda \,|\, p, \mathbf{r}, \mathbf{x}) \propto \lambda^{a-1+\sum_i x_i}\, \mathrm{e}^{-\lambda(b + \sum_i r_i)}$,
- $f(p \,|\, \lambda, \mathbf{r}, \mathbf{x}) \propto p^{\sum_i r_i}\, (1-p)^{n - \sum_i r_i}$,
- $f(r_k \,|\, \lambda, p, \mathbf{x}) \propto \left(\dfrac{p\, \mathrm{e}^{-\lambda}}{1-p}\right)^{r_k} r_k^{x_k}\,.$

In other words, we have:

- $(\lambda \,|\, p, \mathbf{r}, \mathbf{x}) \sim \mathsf{Gamma}\Big(a + \sum_i x_i,\; b + \sum_i r_i\Big)$,
- $(p \,|\, \lambda, \mathbf{r}, \mathbf{x}) \sim \mathsf{Beta}\Big(1 + \sum_i r_i,\; n + 1 - \sum_i r_i\Big)$,
- $(r_k \,|\, \lambda, p, \mathbf{x}) \sim \mathsf{Ber}\left(\dfrac{p\, \mathrm{e}^{-\lambda}}{p\, \mathrm{e}^{-\lambda} + (1-p) \mathrm{I}_{\{x_k = 0\}}}\right)$.

To test the accuracy of the Gibbs sampler (Algorithm 6.2), we generate $n = 100$ random data points from the zero-inflated Poisson model using parameters $p = 0.3$

and $\lambda = 2$. To recover the parameters from the data, we choose $a = 1$ and $b = 1$ for the prior distribution of λ, and generate a (dependent) sample of size 10^5 from the posterior distribution using the Gibbs sampler. A 95% Bayesian confidence interval (credible interval) is constructed using the script below. ☞ 672 Note that MATLAB's *statistics toolbox* function gamrnd(a,b) draws from the $\mathsf{Gamma}(a, 1/b)$ distribution. The estimated Bayesian confidence intervals are $(1.33, 2.58)$ for λ and $(0.185, 0.391)$ for p. Observe that the true values lie within these intervals.

```
%zip.m
n=100; p=.3; lambda=2;
% generate ZIP random variables
data=poissrnd(lambda,n,1).*(rand(n,1)<p);
% now try to recover the ZIP parameters from the data
P=rand; % starting guess for p
lam=gamrnd(1,1); % starting guess for lambda
r=(rand(n,1)<P); % starting guess for  r
Sum_data=sum(data);
gibbs_sample=zeros(10^5,2);
% apply the Gibbs sampler
for k=1:10^5
    Sum_r=sum(r);
    lam=gamrnd(1+Sum_data,1/(1+Sum_r));
    P=betarnd(1+Sum_r,n+1-Sum_r);
    prob=exp(-lam)*P./(exp(-lam)*P+(1-P)*(data==0));
    r=(rand(n,1)<prob);
    gibbs_sample(k,:)=[P,lam];
end
% 95% probability interval for lambda
prctile(gibbs_sample(:,2),[2.5,97.5])
% 95% probability interval for p
prctile(gibbs_sample(:,1),[2.5,97.5])
```

Gibbs sampling is advantageous whenever it is easy to sample from the conditional distributions of the joint density. Note that it is not necessary to update each component of the random vector $\mathbf{X}$ individually. Instead, blocks or groups of variables can be updated simultaneously. For example, to sample from the joint pdf $f(x_1, x_2, x_3)$ we can consider the following version of the Gibbs sampler.

Algorithm 6.3 (Grouped Gibbs Sampler) *To sample from $f(x_1, x_2, x_3)$ with a given initial state $\mathbf{X}_0$, iterate the following steps for $t = 0, 1, 2, \ldots$.*

1. *For a given $\mathbf{X}_t = (X_{t,1}, X_{t,2}, X_{t,3})$, generate $\mathbf{Y} = (Y_1, Y_2, Y_3)$ as follows:*

   (a) Draw (Y_1, Y_2) from the conditional pdf $f(y_1, y_2 \,|\, X_{t,3})$.

   (b) Draw Y_3 from the conditional pdf $f(y_3 \,|\, Y_1, Y_2)$.

2. *Let $\mathbf{X}_{t+1} = \mathbf{Y}$.*

The grouped variables in Algorithm 6.3 are x_1 and x_2. Significant speed-up of the convergence of the chain can be achieved when highly correlated variables are grouped together [69]. Grouped Gibbs sampling is an essential idea in the *Swendsen–Wang* algorithm, to generate samples from the Potts model; see Example 6.12.

The essential idea of the Gibbs sampler — updating some component of the random vector while holding the other components fixed — is useful in many instances where the state variable is a random variable taking values in a general space, not just in $\mathbb{R}^n$, see [41]. The next example illustrates how this idea can be used to compute the normalizing constant of a complex multidimensional density.

■ EXAMPLE 6.4 (Chib's Method)

Suppose we wish to estimate the normalizing constant $\mathcal{Z}$ of

$$f(\mathbf{x}) = \frac{p(\mathbf{x})}{\mathcal{Z}}, \quad \mathbf{x} = (x_1, \ldots, x_n) ,$$

where $p(\mathbf{x})$ is known. Rearranging gives the identity

$$\mathcal{Z} = \frac{p(\mathbf{x})}{f(\mathbf{x})} \quad \text{for all } \mathbf{x} \in \{\mathbf{x} : f(\mathbf{x}) > 0\} .$$

In particular, the identity holds for a single point $\mathbf{x} = \mathbf{x}^*$. Thus, to compute $\mathcal{Z}$ we need to evaluate $f(\mathbf{x})$ at a point $\mathbf{x}^*$. For numerical accuracy, the point $\mathbf{x}^*$ is generally taken to be in a high-density region of f. We proceed to estimate $f(\mathbf{x}^*)$ in the following way.

By the *product rule* of probability theory, we can write $f(\mathbf{x}^*)$ as ☞ 616

$$f(\mathbf{x}^*) = f(x_1^*)\, f(x_2^* \,|\, x_1^*)\, f(x_3^* \,|\, x_1^*, x_2^*) \cdots f(x_n^* \,|\, x_1^*, \ldots, x_{n-1}^*) . \tag{6.7}$$

It is assumed that each of the conditional densities

$$f(x_i \,|\, x_1, \ldots, x_{i-1}, x_{i+1}, \ldots, x_n), \quad i = 1, \ldots, n ,$$

is known. In particular, the conditional $f(x_n^* \,|\, x_1^*, \ldots, x_{n-1}^*)$ in (6.7) is known. However, all other densities in the factorization of $f(\mathbf{x}^*)$ are generally unknown and have to be estimated. The first term on the right-hand side of (6.7), $f(x_1^*)$, can be estimated via

$$\widehat{f}_1 = \frac{1}{N} \sum_{i=1}^{N} f(x_1^* \,|\, X_2^{(i)}, \ldots, X_n^{(i)}) , \tag{6.8}$$

where $(X_1^{(i)}, \ldots, X_n^{(i)}) \sim f(\mathbf{x})$, $i = 1, \ldots, N$ are obtained from a run of the Gibbs sampler. The values of the first component x_1 are ignored. Similarly, the k-th term, $f(x_k^* \,|\, x_1^*, \ldots, x_{k-1}^*)$, can be estimated via

$$\widehat{f}_k = \frac{1}{N} \sum_{i=1}^{N} f(x_k^* \,|\, x_1^*, \ldots, x_{k-1}^*, X_{k+1}^{(i)}, \ldots, X_n^{(i)}), \quad k \leqslant n-1 , \tag{6.9}$$

where $(X_k^{(i)}, X_{k+1}^{(i)}, \ldots, X_n^{(i)}) \sim f(x_k, \ldots, x_n \,|\, x_1^*, \ldots, x_{k-1}^*)$, $i = 1, \ldots, N$ are obtained from a *different* Gibbs run in which $(x_1, \ldots, x_{k-1})$ is fixed to $(x_1^*, \ldots, x_{k-1}^*)$, the component x_k is discarded, and the Gibbs cycle runs over $(x_k, \ldots, x_n)$. Note that by assumption the last term $f_n = f(x_n^* \,|\, x_1^*, \ldots, x_{n-1}^*)$ is available and need not be estimated. Thus, an estimator of $\mathcal{Z}$ is:

$$\widehat{\mathcal{Z}} = \frac{p(\mathbf{x}^*)}{f_n \prod_{k=1}^{n-1} \widehat{f}_k} \,. \tag{6.10}$$

The above procedure was proposed in [18] and is summarized below.

Algorithm 6.4 (Chib's Method) *Given a point $\mathbf{x}^* = (x_1^*, \ldots, x_n^*)$ in a high-density region of $f(\mathbf{x})$, perform the following steps.*

1. *Generate $(X_1^{(i)}, X_2^{(i)}, \ldots, X_n^{(i)}) \sim f(\mathbf{x})$, $i = 1, \ldots, N$ using a run of the Gibbs sampler and compute $\widehat{f}_1$ as given in* (6.8).
2. *For $k = 2, \ldots, n-1$, run a Gibbs sampler in which $(x_1, \ldots, x_{k-1})$ is fixed to be $(x_1^*, \ldots, x_{k-1}^*)$ and the Gibbs cycle runs over $(x_k, \ldots, x_n)$. Given draws $(X_{k+1}^{(i)}, \ldots, X_n^{(i)})$, $i = 1, \ldots, N$, from the Gibbs sampler, compute the estimator $\widehat{f}_k$ via* (6.9).
3. *Evaluate $f_n = f(x_n^* \,|\, x_1^*, \ldots, x_{n-1}^*)$ and deliver the estimator* (6.10).

As a numerical example, consider the case where

$$p(\mathbf{x}) = \mathrm{e}^{-(x_1 + \cdots + x_5)}\, \mathrm{I}_{\{S(\mathbf{x}) > 8\}}, \quad \mathbf{x} = (x_1, \ldots, x_5), \quad \mathbf{x} \in \mathbb{R}_+^5 \,,$$

and

$$S(\mathbf{x}) = \min\{x_1 + x_4,\; x_1 + x_3 + x_5,\; x_2 + x_3 + x_4,\; x_2 + x_5\} \,.$$

☞ 348 Thus, $S(\mathbf{x})$ represents the shortest path from A to B in the bridge network in Figure 9.1, and $\mathcal{Z}$ is the probability that the shortest path exceeds 8 when the lengths of the links are $X_1 \ldots, X_5 \sim_{\text{iid}} \mathsf{Exp}(1)$. We have

$$f(x_i \,|\, \mathbf{x}_{-i}) = \mathrm{e}^{-x_i + \beta_i}, \quad x_i > \beta_i, \quad i = 1, \ldots, 5 \,,$$

where $\mathbf{x}_{-i}$ denotes the vector $\mathbf{x}$ with the i-th element removed and ($x^+ \stackrel{\text{def}}{=} \max\{0, x\}$)

$$\begin{aligned}
\beta_1 &= (8 - \min\{x_4, x_3 + x_5\})^+ \\
\beta_2 &= (8 - \min\{x_5, x_3 + x_4\})^+ \\
\beta_3 &= (8 - \min\{x_1 + x_5, x_2 + x_4\})^+ \\
\beta_4 &= (8 - \min\{x_1, x_2 + x_3\})^+ \\
\beta_5 &= (8 - \min\{x_2, x_1 + x_3\})^+ \,.
\end{aligned}$$

☞ 50 Sampling from these truncated exponential densities is straightforward: to obtain $X_i \sim f(x_i \,|\, \mathbf{x}_{-i})$ generate $U_i \sim \mathsf{U}(0,1)$ and set $X_i = \beta_i - \ln U_i$. The MATLAB code below applies Chib's method with $N = 10^4$. The constants $\{\beta_i\}$ are coded in the function `bet.m`. This setup gives a typical estimate of $\widehat{\mathcal{Z}} = 3.55 \times 10^{-6}$. The relative error can be estimated by running the algorithm independently a number of times. Keeping $\mathbf{x}^*$ fixed to be the same across all independent runs, we obtain an estimated relative error of 5% from 10 runs. Note that the choice of $\mathbf{x}^*$ can significantly affect the performance, see [18].

```
%chibs.m
clear all,clc
N=10^4;
xs=ones(1,5)*8; % x-star
for j=1:4 % number of conditional densities
    pi(j)=0;
    x=xs;
    for iter=1:N
        for k=j:5
            x(k)=bet(x,k)-log(rand);
        end
        pi(j)=pi(j)+exp(-xs(j)+bet(x,j));
    end
    pi(j)=pi(j)/N;
end
% step 3
pi(5)=exp(-(xs(5)-bet(xs,5)))
% estimator
exp(-sum(xs))/prod(pi)
```

```
function out=bet(x,idx)

switch idx
    case 1
        out=8-min([x(4),x(3)+x(5)]);
    case 2
        out=8-min([x(5),x(3)+x(4)]);
    case 3
        out=8-min([x(1)+x(5),x(2)+x(4)]);
    case 4
        out=8-min([x(1),x(2)+x(3)]);
    case 5
        out=8-min([x(2),x(1)+x(3)]);
end

out=max(out,0);
```

Chib's method is widely used to compute the marginal likelihood, which plays an important role in Bayesian model selection. The original method [18] is only applicable when the Gibbs sampler is used for simulation from the posterior, that is, when the normalizing constants of all full conditionals are known. The method has been extended to settings where the Metropolis–Hasting sampler is used instead, see [19]. For an interacting particle splitting approach to computing the marginal likelihood see Example 14.3.

☞ 673
☞ 498

6.3 SPECIALIZED SAMPLERS

While MCMC is a generic method and can be used to approximately generate random variables from virtually any target distribution, regardless of its dimensionality and complexity, potential problems with the MCMC method include:

1. The resulting samples are often highly correlated.
2. The Markov chain may fail to explore all of the modes of the target density.
3. The estimates obtained via MCMC samples often tend to have much greater variances than those obtained from independent sampling from the target distribution.

Various attempts have been made to overcome these difficulties. A number of extensions of the Metropolis–Hastings algorithm have been suggested that aim to speed up convergence to the steady-state regime and reduce the correlation among samples [43, 65]. In this section we review some of these enhancements.

6.3.1 Hit-and-Run Sampler

The **hit-and-run** sampler, pioneered by Smith [75], is among the first MCMC samplers in the category of **line samplers** [1]. As in previous sections, the objective is to sample from a target distribution $f(\mathbf{x}) = p(\mathbf{x})/\mathcal{Z}$ on $\mathscr{X} \subseteq \mathbb{R}^n$.

We first describe the original hit-and-run sampler for generating from a *uniform* distribution on a bounded open region $\mathscr{X}$ of $\mathbb{R}^n$. At each iteration, starting from a current point $\mathbf{x}$, a **direction vector d** is generated uniformly on the surface of an n-dimensional hypersphere. The intersection of the corresponding bidirectional line through $\mathbf{x}$ and the enclosing box of $\mathscr{X}$ defines a line segment $\mathscr{L}$. The next point $\mathbf{y}$ is then selected uniformly from the intersection of $\mathscr{L}$ and $\mathscr{X}$.

Figure 6.3 illustrates the hit-and-run algorithm for generating uniformly from the set $\mathscr{X}$ (the gray region) which is bounded by a rectangle. Given the point $\mathbf{x}$ in $\mathscr{X}$, the direction $\mathbf{d}$ is generated, which defines the line segment $\mathscr{L} = uv$. Then, a point $\mathbf{y}$ is chosen uniformly on $\mathscr{L} \cap \mathscr{X}$, for example, by the acceptance–rejection method, that is, generate a point uniformly on $\mathscr{L}$ and then accept this point only if it lies in $\mathscr{X}$.

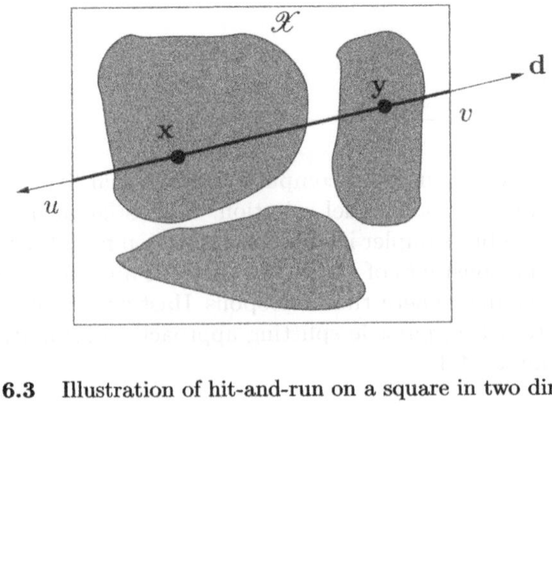

Figure 6.3 Illustration of hit-and-run on a square in two dimensions.

The uniform hit-and-run sampler asymptotically generates uniformly distributed points over *arbitrary* open regions of $\mathbb{R}^n$; see [75]. One desirable property of hit-and-run is that it can globally reach any point in the set in one step; that is, there is a positive probability of sampling any neighborhood in the set. Lovász [45] proves that hit-and-run on a convex body in n dimensions produces an approximately uniformly distributed sample in polynomial time, $\mathcal{O}(n^3)$. He notes that in practice the hit-and-run algorithm appears to offer the most rapid convergence to a uniform distribution [45, 46]. Hit-and-run is unique in that it only takes polynomial time to get out of a corner, whereas, in contrast, *ball walk* takes exponential time to get out of a corner [47].

We now describe a more general version of the hit-and-run algorithm for sampling from *any* strictly positive continuous pdf $f(\mathbf{x}) = p(\mathbf{x})/\mathcal{Z}$ on any region $\mathscr{X}$ — bounded or unbounded [16, 17]. Similar to the Metropolis–Hastings algorithm we generate a proposal move, which is then accepted or rejected with probability that depends on f. A proposal move $\mathbf{y}$ is generated by stepping away from the current point (at iteration t) $\mathbf{x}$ in the direction $\mathbf{d}$ with a step of size λ. The step size λ at iteration t is generated from a proposal density $g_t(\lambda \mid \mathbf{d}, \mathbf{x})$. The candidate point $\mathbf{y} = \mathbf{x} + \lambda\,\mathbf{d}$ is then accepted with probability

$$\alpha(\mathbf{x}, \mathbf{y}) = \min\left\{\frac{f(\mathbf{y})\, g_t\left(|\lambda| \mid -\operatorname{sgn}(\lambda)\,\mathbf{d}, \mathbf{y}\right)}{f(\mathbf{x})\, g_t\left(|\lambda| \mid \operatorname{sgn}(\lambda)\,\mathbf{d}, \mathbf{x}\right)}, 1\right\}, \tag{6.11}$$

as in the Metropolis–Hastings acceptance criterion (6.1); otherwise, the Markov chain remains in the current state $\mathbf{x}$. The condition (6.11) ensures that g_t satisfies the detailed balance condition:

$$g_t\left(\|\mathbf{x} - \mathbf{y}\| \,\middle|\, \frac{\mathbf{y} - \mathbf{x}}{\|\mathbf{x} - \mathbf{y}\|}, \mathbf{x}\right)\alpha(\mathbf{x}, \mathbf{y}) f(\mathbf{x}) = g_t\left(\|\mathbf{x} - \mathbf{y}\| \,\middle|\, \frac{\mathbf{x} - \mathbf{y}}{\|\mathbf{x} - \mathbf{y}\|}, \mathbf{y}\right)\alpha(\mathbf{y}, \mathbf{x}) f(\mathbf{y}) .$$

We refer to proposal densities that satisfy the detailed balance equations as **valid** proposals. At iteration t, let

$$\mathscr{M}_t \overset{\text{def}}{=} \{\lambda \in \mathbb{R} : \mathbf{x} + \lambda\,\mathbf{d} \in \mathscr{X}\} .$$

A valid proposal density $g_t(\lambda \mid \mathbf{d}, \mathbf{x})$ can be one of the following:

- $g_t(\lambda \mid \mathbf{d}, \mathbf{x}) = \widetilde{g}_t(\mathbf{x} + \lambda\,\mathbf{d})$, $\lambda \in \mathbb{R}$. The acceptance probability (6.11) then simplifies to

  $$\alpha(\mathbf{x}, \mathbf{y}) = \min\left\{\frac{f(\mathbf{y})\, \widetilde{g}_t(\mathbf{x})}{f(\mathbf{x})\, \widetilde{g}_t(\mathbf{y})}, 1\right\} .$$

  A common choice is

  $$g_t(\lambda \mid \mathbf{d}, \mathbf{x}) = \frac{f(\mathbf{x} + \lambda\,\mathbf{d})}{\int_{\mathscr{M}_t} f(\mathbf{x} + u\,\mathbf{d})\, du}, \quad \lambda \in \mathscr{M}_t , \tag{6.12}$$

  in which case (6.11) further simplifies to $\alpha(\mathbf{x}, \mathbf{y}) = 1$.

- $g_t(\lambda \mid \mathbf{d}, \mathbf{x}) = \widetilde{g}_t(\lambda)$, $\lambda \in \mathbb{R}$, is a symmetric ($\widetilde{g}_t(\lambda) = \widetilde{g}_t(-\lambda)$) continuous pdf that may depend only on $\mathscr{M}_t$. The acceptance probability (6.11) then simplifies to

  $$\alpha(\mathbf{x}, \mathbf{y}) = \min\{f(\mathbf{y})/f(\mathbf{x}), 1\} .$$

If $\mathscr{X}$ is unbounded it is common to choose $\widetilde{g}_t(\lambda)$ as the normal pdf with mean 0 and variance that depends on $\mathscr{M}_t$. Alternatively, if $\mathscr{X}$ is bounded a common choice is

$$\widetilde{g}_t(\lambda) = \frac{\mathrm{I}_{\{\lambda \in \mathscr{M}_t\}}}{\int_{\mathbb{R}} \mathrm{I}_{\{u \in \mathscr{M}_t\}}\, du} .$$

In summary, the hit-and-run algorithm reads as follows.

Algorithm 6.5 (Hit-and-Run)

1. *Initialize with* $\mathbf{X}_1 \in \mathscr{X}$ *and set* $t = 1$.
2. *Generate a random direction* $\mathbf{d}_t$ *according to a uniform distribution on the unit* n*-dimensional hypersphere. In other words, generate:*

$$\mathbf{d}_t = \left(\frac{Z_1}{\|\mathbf{Z}\|}, \ldots, \frac{Z_n}{\|\mathbf{Z}\|}\right)^\top, \quad Z_1, \ldots, Z_n \overset{\text{iid}}{\sim} \mathsf{N}(0,1) ,$$

where $\|\mathbf{Z}\| = \sqrt{Z_1^2 + \cdots + Z_n^2}$.

3. *Generate* λ *from a valid proposal density* $g_t(\lambda \,|\, \mathbf{d}_t, \mathbf{X}_t)$.
4. *Set* $\mathbf{Y} = \mathbf{X}_t + \lambda\, \mathbf{d}_t$, *and let:*

$$\mathbf{X}_{t+1} = \begin{cases} \mathbf{Y} & \text{with probability } \alpha(\mathbf{X}_t, \mathbf{Y}) \text{ in } (6.11) \\ \mathbf{X}_t & \text{otherwise.} \end{cases}$$

5. *If a stopping criterion is met, stop; otherwise, increment* t *and repeat from Step 2.*

Note that the proposal (6.12) gives a Gibbs-type sampling algorithm in which every candidate point is accepted and $\mathbf{X}_{t+1} \neq \mathbf{X}_t$.

■ EXAMPLE 6.5 (Truncated Multivariate Normal Generator)

A common computational problem in Bayesian data analysis involves sampling from a truncated multivariate normal pdf:

$$f(\mathbf{x}) \propto p(\mathbf{x}) = \exp\Big(-\frac{1}{2}(\mathbf{x}-\boldsymbol{\mu})^\top \Sigma^{-1} (\mathbf{x}-\boldsymbol{\mu})\Big)\, \mathrm{I}_{\{\mathbf{x} \in \mathscr{X}\}} ,$$

where $\mathscr{X} \subset \mathbb{R}^n$. For specific examples, see [6, 13, 16, 49, 50]. The hit-and-run sampler can be used to sample efficiently in such cases. With (6.12) as the proposal, the pdf of λ is:

$$g_t(\lambda \,|\, \mathbf{d}, \mathbf{x}) \propto \exp\left(-\frac{\mathbf{d}^\top \Sigma^{-1} \mathbf{d}}{2}\, \lambda^2 - \mathbf{d}^\top \Sigma^{-1} (\mathbf{x}-\boldsymbol{\mu})\, \lambda\right) \mathrm{I}_{\{\mathbf{x}+\lambda\, \mathbf{d} \in \mathscr{X}\}} ,$$

which corresponds to a truncated univariate normal distribution with mean

$$-\frac{\mathbf{d}^\top \Sigma^{-1} (\mathbf{x}-\boldsymbol{\mu})}{\mathbf{d}^\top \Sigma^{-1} \mathbf{d}}$$

and variance $(\mathbf{d}^\top\Sigma^{-1}\mathbf{d})^{-1}$, truncated to the set $\mathscr{M}_t = \{\lambda : \mathbf{x} + \lambda\,\mathbf{d} \in \mathscr{X}\}$. For example, suppose that $\mathscr{X} = \{\mathbf{x} \in \mathbb{R}^2 : \mathbf{x}^\top\mathbf{x} > 25\}$, then denoting $D = (\mathbf{d}^\top\mathbf{x})^2 - \mathbf{x}^\top\mathbf{x} + 25$, we have two cases:

- $\mathscr{M}_t \equiv \mathbb{R}$, if $D < 0$;
- $\mathscr{M}_t \equiv (-\infty, \; -\sqrt{D} - \mathbf{d}^\top\mathbf{x}] \cup [\sqrt{D} - \mathbf{d}^\top\mathbf{x}, \; \infty)$, if $D > 0$.

The following code implements this particular example and generates 10^4 points in $\mathbb{R}^2$ with $\boldsymbol{\mu} = (1/2, 1/2)^\top$ and covariance matrix with $\Sigma_{11} = \Sigma_{22} = 1$, $\Sigma_{12} = 0.9$. Figure 6.4 shows a scatter plot of the output and the boundary of $\mathscr{X}$.

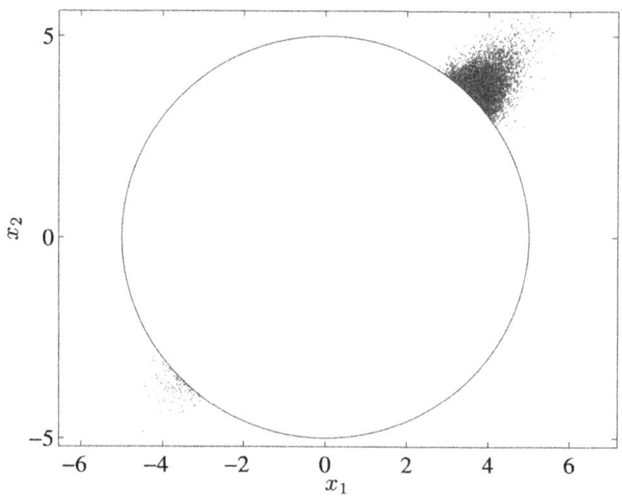

Figure 6.4 Hit-and-run sampling from a truncated bivariate normal density.

```
%Hit_and_run.m
Sig=[1,0.9;0.9,1]; Mu=[1,1]'/2; B=chol(Sig)';
x=[5,5]';                %starting point
T=10^4; data=nan(T,2);% number of points desired
for t=1:T
    d=randn(2,1); d=d/norm(d);
    D=(d'*x)^2-x'*x+25;
    z1=B\d; z2=B\(x-Mu);
    % determine mean and variance of lambda dist.
    sig=1/norm(z1); mu=-z1'*z2*sig^2;
    if D<0
        lam=mu+randn*sig;
    else
        lam=normt2(mu,sig,-sqrt(D)-d'*x,sqrt(D)-d'*x);
    end
    x=x+lam*d;
    data(t,:)=x';
```

```
end
plot(data(:,1),data(:,2),'r.'),axis equal,
hold on
ezplot('x^2+y^2-25')
```

The code above uses the function normt2.m, which draws from the $\mathsf{N}(\mu, \sigma^2)$ distribution truncated on the set $(-\infty, a] \cup [b, \infty)$, via the inverse-transform method.

```
function out=normt2(mu,sig,a,b)
pb=normcdf((b-mu)/sig);
pa=normcdf((a-mu)/sig);
if rand<pa/(pa+1-pb)
      out=mu+sig*norminv(pa*rand(size(mu)));
else
      out=mu+sig*norminv((1-pb)*rand(size(mu))+pb);
end
```

Note that while the Gibbs or Metropolis–Hastings samplers can be used for generation from truncated multivariate distributions [22, 64], the hit-and-run sampler has been observed to perform better in such settings, see Chen and Schmeiser [14, 15]. In particular, Chen and Schmeiser argue that the hit-and-run approach is especially useful in cases where the variables are highly constrained.

■ EXAMPLE 6.6 (Simulating a Stock Price Process)

☞ 526 Consider a typical stock price trajectory $S_{t_1}, S_{t_2}, \ldots, S_{t_n}$ used in option pricing (see Section 15.2):

$$S_{t_k} = S_0 \exp\left(\left\{r - \frac{\sigma^2}{2}\right\} k\,\delta + \sigma\sqrt{\delta}\sum_{i=1}^{k} X_i\right), \quad k = 1, \ldots, n\,, \tag{6.13}$$

☞ 536 where $\delta = T/n$, $t_k = k\,\delta$, and $X_1, \ldots, X_n \sim_{\text{iid}} \mathsf{N}(0, 1)$. In this example, $r, \sigma, K, \beta, S_0, n, T$ are given parameters that specify the value of a down-and-in Asian call option. We are interested in sampling from the minimum variance importance sampling density for the estimation of the value of a down-and-in Asian call option with payoff $(S_{t_n} - K)^+ \,\mathrm{I}\{\min_i S_{t_i} \leqslant \beta\}$, where β is the *barrier* of the down-and-in option; see Example 15.8. This is equivalent to sampling $\mathbf{X} = (X_1, \ldots, X_n)^\top$ from the pdf

$$f(\mathbf{x}) \propto p(\mathbf{x}) = H(\mathbf{x})\, \mathrm{e}^{-\frac{1}{2}\mathbf{x}^\top \mathbf{x}}, \quad H(\mathbf{X}) = (S_{t_n} - K)^+ \,\mathrm{I}\Big\{\min_{1 \leqslant i \leqslant n} S_{t_i} \leqslant \beta\Big\}\,,$$

and then computing the stock price trajectory $\{S_{t_k}\}$ from (6.13). In other words, under f the final stock price $S_{t_n} = S_T$ is above K and one of the past stock prices $S_{t_1}, \ldots, S_{t_{n-1}}$ is below the barrier β. With the parameters $(r, \sigma, K, \beta, S_0, n, T) = (0.07, 0.2, 1.2, 0.8, 1, 180, 180/365)$, typical realizations of the stock price process under f are depicted in Figure 6.5. The thick smooth line is the average over all $N = 10^5$ trajectories; that is, it is an estimate of $\mathbb{E}_f S_{t_i}$ for all i.

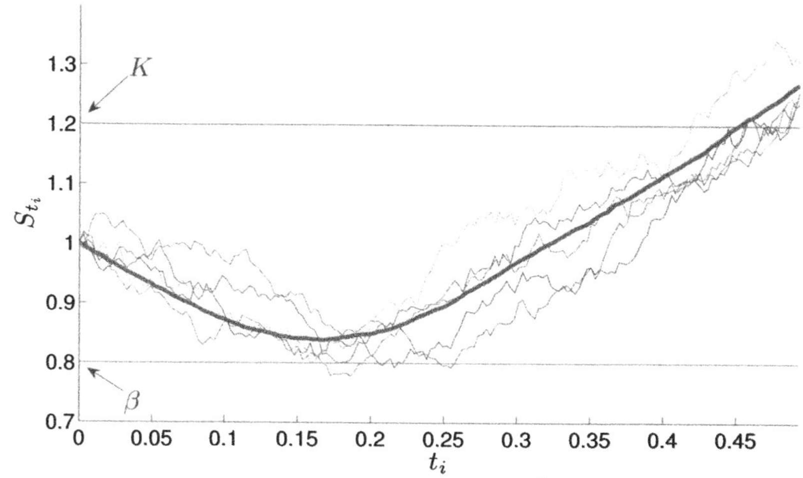

Figure 6.5 Five stock price trajectories under f. The thick smooth line shows the average of all $N = 10^5$ trajectories.

To generate the process $\{S_{t_i}\}$ or, equivalently, the vector $\mathbf{X} \sim f$, we implement Algorithm 6.5 with the proposal $g_t(\lambda \,|\, \mathbf{d}, \mathbf{x}) = \widetilde{g}_t(\mathbf{x} + \lambda\, \mathbf{d}) \propto \mathrm{e}^{-\frac{1}{2}\|\mathbf{x}+\lambda\,\mathbf{d}\|^2}$, giving the acceptance probability $\alpha(\mathbf{x}, \mathbf{y}) = \min\{H(\mathbf{y})/H(\mathbf{x}), 1\}$. Figure 6.6 shows the mean of all vectors $\mathbf{X}_1, \ldots, \mathbf{X}_N \stackrel{\text{approx.}}{\sim} f(\mathbf{x})$ as an estimate of $\mathbb{E}_f \mathbf{X}$.

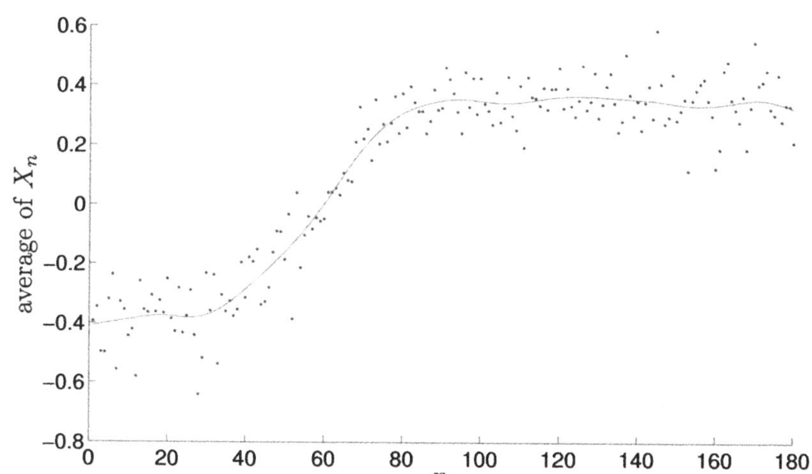

Figure 6.6 The empirical mean approximating $\mathbb{E}_f \mathbf{X}$ together with a smoothing spline showing the trend.

The code below calls the function `down_in_call.m`, which evaluates $H(\mathbf{x})$ for the given set of parameters. In addition, the code uses the function `csaps.m` from the MATLAB spline toolbox. If this function is available to the user, then the last three lines can be uncommented.

```
%down_and_in_Call.m
clear all,
r=.07; % annual interest
sig=.2; % stock volatility
K=1.2; % strike price
b=.8; % barrier
S_0=1; % initial stock price
n=180;  % number of stock price observations
T=n/365; % length of observation period (in years)
dt=T/n; % time step
N=10^5; % length of chain
x=[-ones(1,60),ones(1,n-60)]*0.4; % initial starting point
[H,path]=down_in_call(x,dt,r,sig,S_0,K,b); % evaluate H(x)

% now we simulate N sample paths of the stock process
% and compute  averages
mu=0;paths=0;
for i=1:N
    % apply hit-and-run
    d=randn(1,n); d=d/norm(d);
    lam=-d*x'+randn;
    y=x+lam*d; % make proposal
    % evaluate H(y)
    [H_new,path_new]=down_in_call(y,dt,r,sig,S_0,K,b);
    % accept or reject the proposal
    if rand<min(H_new/H,1)
        x=y;   % update
        H=H_new;
        path=path_new;
    end
    mu=mu+x/N;            % compute an estimate of E[X]
    paths=paths+path/N; % compute average stock price trajectory
    if mod(i,2*10^4)==0 % plot every 2*10^4-th step of the chain
        plot(0:dt:T,path,0:dt:T,0*path+b,0:dt:T,0*path+K)
        axis([0,T,b-0.1,K+.2]),hold all
        pause(.1)
    end
end
% plot the average price trajectory
plot(0:dt:T,paths,'r','LineWidth',3)
figure(2)
plot(mu,'k.'), hold on
% smooth the trajectory of E[X] using a spline
%pp = csaps(dt:dt:T,mu,1/(1+(dt*10)^3));
%mu_t = fnval(pp,[dt:dt:T]);
%plot(mu_t,'r') % plot the smoothed trajectory
```

```
function [H,S_t]=down_in_call(z,dt,r,sig,S_0,K,b)
% implements H(x)
y=(r-sig^2/2)*dt+sig*sqrt(dt)*z;
S_t=exp(cumsum([log(S_0),y]));
H=max(S_t(end)-K,0)*any(S_t<=b);
```

■ **EXAMPLE 6.7 (The Holmes–Diaconis–Ross Method)**

Suppose we wish to estimate the probability

$$\ell = \mathbb{P}(S(\mathbf{X}) \geqslant \gamma), \quad \mathbf{X} \sim f_0(\mathbf{x}), \quad \mathbf{x} = (x_1, \ldots, x_n)^\top,$$

where $S : \mathbb{R}^n \to \mathbb{R}$ is a given function, γ is a threshold parameter such that ℓ is a very small probability, and $f_0(\mathbf{x})$ is a *known* pdf. The **Holmes–Diaconis–Ross** method [8, 9, 23, 72] is a method for estimating ℓ using MCMC. It is related to the particle splitting method in Section 14.2. ☞ 382 ☞ 485

Let $-\infty = \gamma_0 < \gamma_1 < \cdots < \gamma_{T-1} < \gamma_T = \gamma$ be an increasing sequence of thresholds. Define the sequence of pdfs

$$f_t(\mathbf{x}) = \frac{f_0(\mathbf{x})\, \mathrm{I}_{\{S(\mathbf{x}) \geqslant \gamma_t\}}}{\ell_t}, \quad t = 0, 1, \ldots, T\,,$$

where $\ell_t = \mathbb{P}(S(\mathbf{X}) \geqslant \gamma_t)$. Define the conditional probabilities

$$c_t = \frac{\ell_t}{\ell_{t-1}} = \mathbb{P}(S(\mathbf{X}) \geqslant \gamma_t \,|\, S(\mathbf{X}) \geqslant \gamma_{t-1}), \quad \mathbf{X} \sim f_0,\ t = 1, \ldots, T\,.$$

Let $\kappa_t(\mathbf{x}\,|\,\mathbf{y})$ be the transition density of a Markov chain sampler with stationary pdf f_t. The Holmes–Diaconis–Ross algorithm is as follows.

Algorithm 6.6 (Holmes–Diaconis–Ross) *Given a parameter m (the approximate simulation effort at each level t) and a sequence $\gamma_0, \gamma_1, \ldots, \gamma_T$, set the counter $t = 2$ and execute the following steps.*

1. **Initialization.** *Compute*

$$B_1 = \min\left\{k : \sum_{i=0}^{k} G_i > m\right\},$$

where $G_0, G_1, \ldots,$ are independent $\mathsf{Geom}\,(c_1)$ *random variables. The variables $G_0, G_1, \ldots \sim_{\text{iid}} \mathsf{Geom}\,(c_1)$ can be generated by executing the following steps for $i = 0, 1, \ldots$.*

   (a) Let $G_i = 1$.

   (b) Generate $\mathbf{X}_1 \sim f_0(\mathbf{x})$.

   (c) If $S(\mathbf{X}_1) < \gamma_1$, increment $G_i = G_i + 1$ and go to Step (b); otherwise, output G_i as an outcome from $\mathsf{Geom}\,(c_1)$.

2. **MCMC sampling.** *Compute*

$$B_t = \min\left\{k : \sum_{i=0}^{k} G_i > m\right\},$$

where $G_0, G_1, \ldots,$ *are* dependent *random variables generated according to a Markov chain by setting* $\mathbf{Y} = \mathbf{X}_{t-1}$ *and then executing the following steps for* $i = 0, 1, \ldots$

(a) *Let* $G_i = 1$.

(b) *Generate* $\mathbf{X}_t \sim \kappa_{t-1}(\mathbf{x} \mid \mathbf{Y})$.

(c) *If* $S(\mathbf{X}_t) < \gamma_t$ *and* $G_i < m+1$, *increment* $G_i = G_i + 1$, *reset (overwrite)* $\mathbf{Y} = \mathbf{X}_t$, *and go to Step (b); otherwise, output* G_i *as an* approximate *outcome from* $\mathsf{Geom}(c_t)$.

3. **Stopping condition.** *If* $t = T$, *go to Step 4. If* $B_t = 0$, *set*

$$B_{t+1} = B_{t+2} = \cdots = B_T = 0$$

and go to Step 4; otherwise, reset $t = t + 1$ *and repeat from Step 2.*

4. **Final estimate.** *Let* $\widehat{c}_t = B_t/m$ *for all* t. *Output* $\mathbf{X}$ *as an approximate random variable from* $f_T(\mathbf{x})$ *and deliver the estimator*

$$\widehat{\ell} = \prod_{t=1}^{T} \widehat{c}_t \, .$$

Note that $B_1 \sim \mathsf{Bin}(m, c_1)$ (see Property 6 on Page 88) and therefore B_1/m is an unbiased estimator of c_1. However, unlike the exact sampling from $f_0(\mathbf{x})$ in Step 1, the sampling from $f_{t-1}(\mathbf{x})$ via the transition density κ_{t-1} in Step 2 is only approximate. As a result, each random variable B_t, $t = 2, 3, \ldots,$ is *approximately* distributed from $\mathsf{Bin}(m, c_t)$, $t = 2, 3, \ldots,$ so that B_t/m is an estimator of c_t.

Figure 6.7 illustrates the typical evolution of the Markov chain in the Holmes–Diaconis–Ross algorithm for the two-dimensional case ($\mathbf{X} = (X_1, X_2)$). The three level sets $\{\mathbf{x} : S(\mathbf{x}) = \gamma_t\}$, $t = 1, 2, 3$ are plotted as nested curves. Suppose that $m = 24$ and that $B_1 = 1$ with $S(\mathbf{X}_1) \geqslant \gamma_1$. We have $\widehat{c}_1 = 1/24$. To estimate c_2, starting from point $\mathbf{X}_1$ a Markov chain with transition density $\kappa_1(\mathbf{x} \mid \mathbf{y})$ (solid line) is run until S reaches γ_2 in $G_0 = 10$ steps. The first point above γ_2 is encircled. The same chain continues to run and S is above γ_2 for the second time after $G_0 + G_1 = 15$ ($G_1 = 5$) steps. Similarly, S is above γ_2 for the third time after $G_0 + G_1 + G_2 = 24$ ($G_2 = 9$) steps, and for the fourth time after $G_0 + G_1 + G_2 + G_3 = 25$ ($G_3 = 1$) steps. Since $m = 24$ we have $B_2 = 3$ and $\widehat{c}_2 = 3/24$. The 25-th state of the solid line chain is denoted by a star, because it is the initial state for a new Markov chain (dashed line) with transition density $\kappa_2(\mathbf{x} \mid \mathbf{y})$. For the dashed chain we have four points above γ_3 with $(G_0, G_1, G_2, G_3) = (9, 2, 2, 18)$, and hence $B_3 = 3$ and $\widehat{c}_3 = 3/24$. It follows that $\widehat{\ell}_3 = 9/24^3$.

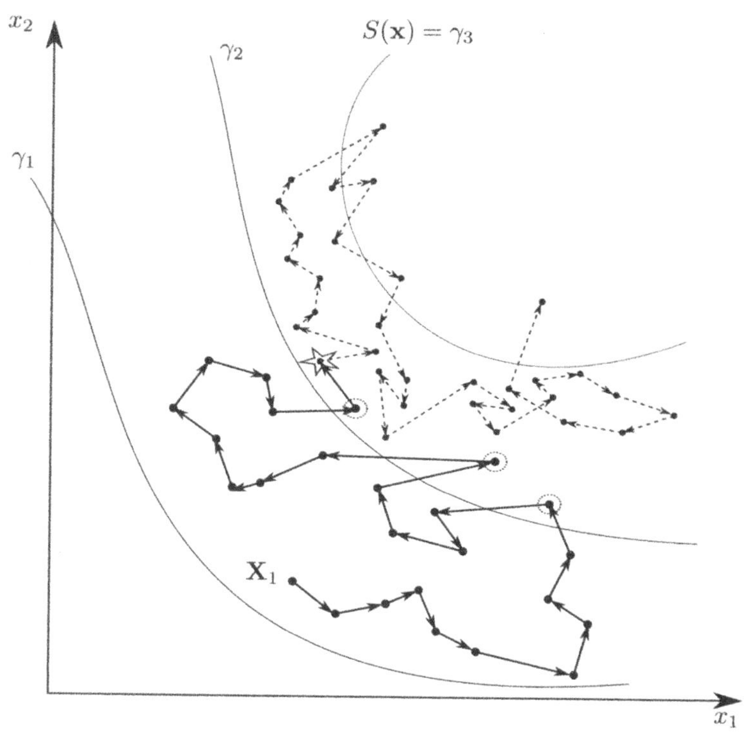

Figure 6.7 Typical evolution of the chain in the Holmes–Diaconis–Ross algorithm.

As a numerical example, we consider the shortest path problem in Example 6.4. Namely, we wish to estimate $\ell = \mathbb{P}(S(\mathbf{V}) \geqslant 8)$, $\mathbf{V} = (V_1, \ldots, V_5)$, where $V_1, \ldots, V_5 \sim_{\text{iid}} \mathsf{Exp}(1)$, and

$$S(\mathbf{v}) = S(v_1, \ldots, v_5) = \min\{v_1 + v_4,\ v_1 + v_3 + v_5,\ v_2 + v_3 + v_4,\ v_2 + v_5\} .$$

As in Example 6.4, the function $S(\mathbf{v})$ represents the shortest path from A to B in the bridge network in Figure 9.1, and ℓ is the probability that the shortest path exceeds 8. ☞ 348

The hit-and-run sampler does not apply directly to generate the Markov chain in Step 2 of the Holmes–Diaconis–Ross Algorithm 6.6. However, we can easily modify the problem in order to make the application of the hit-and-run sampler straightforward. In Example 6.5 we show that sampling from a truncated multivariate normal density is simple via the hit-and-run sampler. Observe that every integral of the form $\mathbb{E}\,\mathrm{I}_{\{S(\mathbf{V})\geqslant\gamma\}}$, where the components $V_1, \ldots, V_n$ of $\mathbf{V}$ are iid, can be written as an expectation with respect to a multivariate normal density as follows: $\mathbb{E}\,\mathrm{I}_{\{S(h(\mathbf{X}))\geqslant\gamma\}}$, where $\mathbf{X} \sim \mathsf{N}(\mathbf{0}, I)$ and h is a transformation such that $h(\mathbf{X})$ has the same distribution as $\mathbf{V}$. Thus, to simplify the implementation of the hit-and-run sampler we formulate the estimation problem so that $\ell = \mathbb{P}(S(h(X_1), \ldots, h(X_5)) \geqslant 8)$, with $X_1, \ldots, X_5 \sim_{\text{iid}} \mathsf{N}(0,1)$ and $h(x) = -\ln \Phi(x)$, where Φ is the cdf of the standard normal distribution. It follows that $f_0(\mathbf{x}) = (2\pi)^{-5/2}\mathrm{e}^{-\mathbf{x}^\top\mathbf{x}/2}$, $\mathbf{x} \in \mathbb{R}^5$, and $f_1, \ldots, f_T$ is a sequence of truncated multivariate normal pdfs. A move from $\mathbf{y}$ to $\mathbf{x}$ using the hit-and-run algorithm is as follows.

Algorithm 6.7 (Performing a Hit-and-Run Move According to $\kappa_t(\mathbf{x} \mid \mathbf{y})$) *Given* $\mathbf{y} = (y_1, \ldots, y_5)^\top$ *such that* $S(h(y_1), \ldots, h(y_5)) \geqslant \gamma_t$, *execute the steps:*

1. *Generate a random direction vector* $\mathbf{d}$ *uniformly distributed on the unit 5-dimensional sphere. Given* $\mathbf{d}$, *generate* $\Lambda \sim \mathsf{N}(-\mathbf{y}^\top \mathbf{d},\ 1)$. *Set* $\mathbf{x} = \mathbf{y} + \Lambda \mathbf{d}$.

2. *If* $S(h(x_1), \ldots, h(x_5)) \geqslant \gamma_t$, *output* $\mathbf{X} = \mathbf{x}$; *otherwise, output* $\mathbf{X} = \mathbf{y}$.

Using 30 independent runs of Algorithm 6.6 with $(\gamma_1, \ldots, \gamma_8) = (1, \ldots, 8)$ and $m = 10^4$, we obtain the estimate $\widehat{\ell} = 3.46 \times 10^{-6}$ with an estimated relative error of 3.5%. The function hdr.m below implements Algorithm 6.6 and hit_run.m implements the hit-and-run sampling step.

```
%HDR/main_hdr.m
m=10^4;N=30;gam=1:8;
ell=nan(N,1);
for i=1:N
    c=hdr(gam,m);
    ell(i)=prod(c);
end
mean(ell) % estimate
std(ell)/mean(ell)/sqrt(N) % relative error
```

```
function [c,x]=hdr(gam,m)
% gam is a vector with increasing levels
T=length(gam); c=zeros(T,1);
sum=0; B(1)=-1; % negative binomial
while sum<m+1
    x=randn(1,5); G=1;       % generate from f_0
    while S(x)<gam(1)
        G=G+1;x=randn(1,5); % generate from f_0
    end
    sum=sum+G; B(1)=B(1)+1; % binomial r.v.
end

for t=2:T
    sum=0; B(t)=-1; % negative binomial
    while sum<m+1
        x=hit_run(x,gam(t-1)); G=1;
        while (S(x)<gam(t))&&(G<m+1)
            G=G+1; x=hit_run(x,gam(t-1));
        end
        sum=sum+G; B(t)=B(t)+1; % binomial r.v.
    end
    [gam(t),B(t)/m]
    % stopping condition
    if (B(t)==0)|(t==T),  break, end
```

```
end
c=B/m; % estimate conditional probabilities
```

```
function x=hit_run(x,gam)
% hit-and-run sampler
n=length(x);
d=randn(1,n); d=d/norm(d); % sample direction
lam=-x*d'+randn;
y=x+lam*d; % make proposal
if S(y)>gam
    x=y;
end
```

```
function out=S(x)
x=-log(normcdf(x));
out=min([x(1)+x(4),x(1)+x(3)+x(5),x(2)+x(3)+x(4),x(2)+x(5)]);
```

A discrete hit-and-run version has been developed in Baumert et al. [5, 53]; see also [73]. An alternative approach for discrete problems is to use the continuous hit-and-run algorithm with a transformation that maps the continuous random variables generated by the sampler into the desired discrete random variables. More precisely, the idea is to use a transformation h such that the random variable $h(\mathbf{U})$, where $\mathbf{U} \sim \mathsf{U}(0,1)^n$, has the same distribution as the discrete random variable $\mathbf{Y}$. Examples of such transformations are given in the discussion of the *inverse-transform method*, see Section 3.1.1. ☞ 45

The hit-and-run algorithm can be embedded within an optimization framework to yield various global optimization algorithms, see [53, 71, 77, 79]. It has been applied successfully to practical problems including composite material design and shape optimization, and has been shown to have polynomial complexity, on average, for a class of quadratic programs [79]. In Section 12.3 we show how to turn any MCMC sampler into an optimization algorithm, using simulated annealing. ☞ 449

6.3.2 Shake-and-Bake Sampler

The **shake-and-bake** algorithms [7] form a class of MCMC algorithms for generating (approximately) uniform points on the boundary $\partial\mathscr{X}$ of a bounded nonempty convex polytope with non-zero volume. The polytope is defined by a system of linear inequalities $A\mathbf{x} \leqslant \mathbf{b}$, where $A = (\mathbf{a}_1, \ldots, \mathbf{a}_m)^\top$ is an $m \times n$ matrix with normalized rows ($\|\mathbf{a}_i\| = 1$ for all i) and $\mathbf{b} = (b_1, \ldots, b_m)^\top$ is a vector. It is assumed that there are no redundant rows in A, in that removing any of the rows changes the polytope. The boundary of the polytope is then

$$\partial\mathscr{X} = \bigcup_{i=1}^{m} \left\{\mathbf{x} : \mathbf{a}_i^\top \mathbf{x} = b_i,\ \mathbf{a}_j^\top \mathbf{x} < b_j \ \text{ for all } j \neq i\right\} .$$

That is, the boundary is the set of points for which exactly one of the m inequality constraints is active.

An iterative step of the shake-and-bake algorithm consists of the following procedure. Given a point $\mathbf{X}_t$ on $\partial\mathscr{X}$, generate a random *feasible* search direction vector $\mathbf{d}$. A feasible direction is one for which $\mathbf{a}_k^\top \mathbf{d} < 0$ whenever the k-th constraint is active at $\mathbf{X}_t$. The intersection point of the bidirectional line through $\mathbf{X}_t$ with the boundary $\partial\mathscr{X}$ closest to $\mathbf{X}_t$ is called the **hit point**. In other words, the hit point is the intersection point of the search direction $\mathbf{d}$ with the constraint that becomes active first. The hit point then becomes the starting point $\mathbf{X}_{t+1}$ for the next iteration.

Figure 6.8 illustrates the shake-and-bake algorithm for generating points uniformly on the boundary of a two-dimensional polygon.

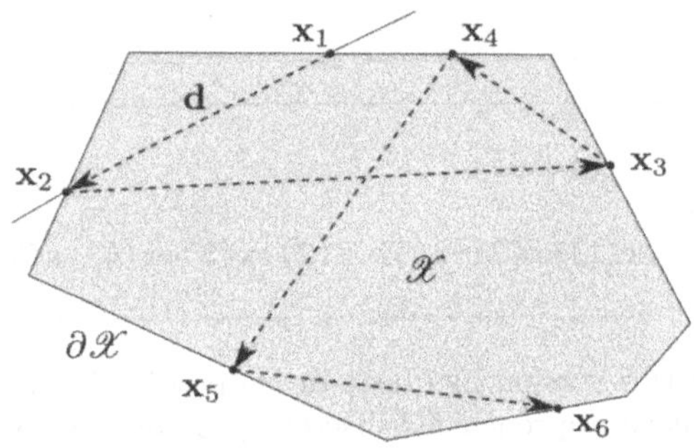

Figure 6.8 Illustration of the (running) shake-and-bake algorithm.

Given the point $\mathbf{x}_1$ on $\partial\mathscr{X}$, the direction $\mathbf{d}$ is generated, which hits $\partial\mathscr{X}$ at $\mathbf{x}_2$. Similarly, a random search direction vector through $\mathbf{x}_2$ generates $\mathbf{x}_3$ and so on. To determine the hit point $\mathbf{x}_2$ one computes all the intersection points of the line $\mathbf{x}_1 + \lambda\mathbf{d}$, $\lambda > 0$, passing through $\mathbf{x}_1$ in the direction $\mathbf{d}$ with all the hyperplanes (facets of $\mathscr{X}$) defined by the constraints $\mathbf{a}_i^\top \mathbf{x} = b_i$, $i = 1, \ldots, m$. These intersection points correspond to the values

$$\lambda_i = \frac{b_i - \mathbf{a}_i^\top \mathbf{x}_1}{\mathbf{a}_i^\top \mathbf{d}}, \quad i = 1, \ldots, m\,.$$

The hit point $\mathbf{x}_2$ is the intersection point corresponding to the *smallest* positive value in the set $\{\lambda_i\}$. In other words $\mathbf{x}_2 = \mathbf{x}_1 + \lambda^* \mathbf{d}$, where

$$\lambda^* = \min\{\lambda_i : \lambda_i > 0\}\,.$$

It remains to specify the mechanism for the generation of the search direction vector $\mathbf{d}$. Suppose that the k-th constraint is active at the current point $\mathbf{x}_1$; that is, $\mathbf{a}_k^\top \mathbf{x}_1 = b_k$ (and hence $\lambda_k = 0$). Generate a point $\mathbf{y}_1$ uniformly distributed on the
☞ 74 surface of the n-dimensional unit hypersphere centered at the origin, as illustrated in Figure 6.9.

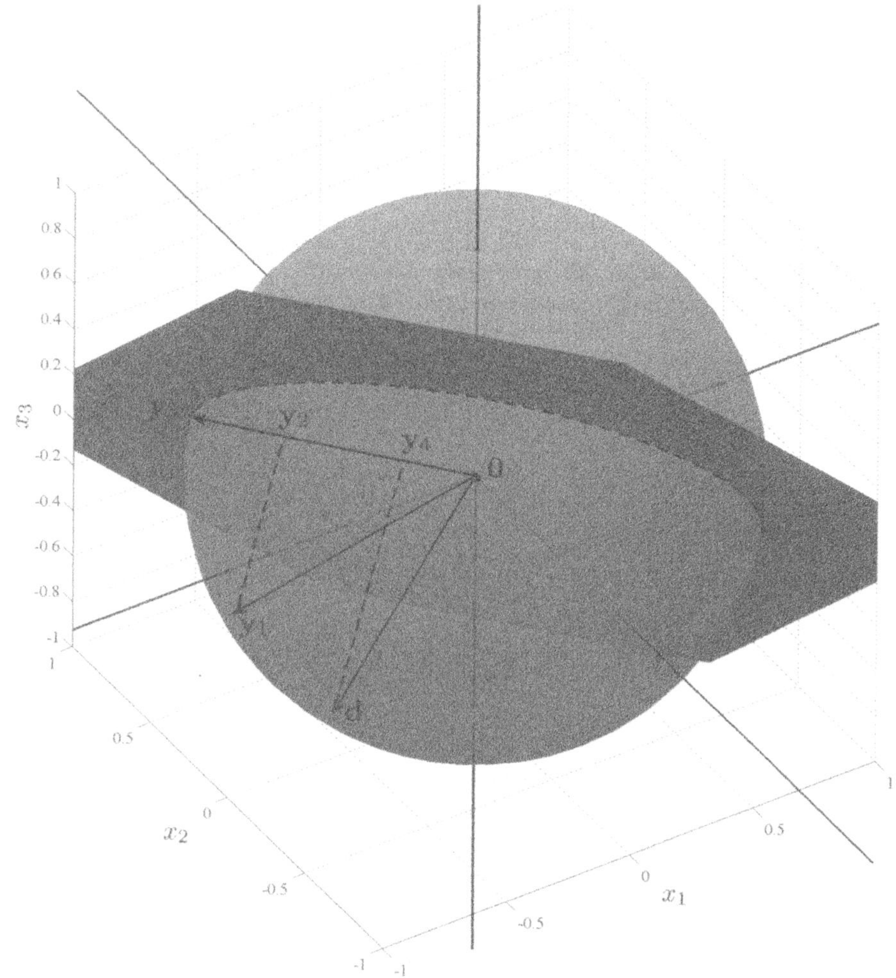

Figure 6.9 Construction of the random search direction $\mathbf{d}$. The hyperplane $\{\mathbf{x} : \mathbf{a}_k^\top \mathbf{x} = 0\}$ is shown cutting through the sphere. The lower part of the sphere that is cut by the hyperplane lies in the region for which $\{\mathbf{x} : \mathbf{a}_k^\top \mathbf{x} < 0\}$.

Compute the projection $\mathbf{y}_2$ of $\mathbf{y}_1$ onto the hyperplane $\{\mathbf{x} : \mathbf{a}_k^\top \mathbf{x} = 0\}$, that is, $\mathbf{y}_2 = \mathbf{y}_1 - (\mathbf{a}_k^\top \mathbf{y}_1)\mathbf{a}_k$. Next, set $\mathbf{y}_3 = \mathbf{y}_2 / \|\mathbf{y}_2\|$, so that $\|\mathbf{y}_3\| = 1$ and $\mathbf{y}_3$ lies at the intersection of the unit sphere and the hyperplane. Compute $\mathbf{y}_4 = R\,\mathbf{y}_3$, where R is distributed according to the radius of a uniformly chosen point in the unit ball; that is, $R^{n-1} \sim \mathsf{U}(0,1)$. Finally, the search direction $\mathbf{d}$ is defined as the unit vector whose projection on the hyperplane $\{\mathbf{x} : \mathbf{a}_k^\top \mathbf{x} = 0\}$ is given by the vector $\mathbf{y}_4$ under the condition $\mathbf{a}_k^\top \mathbf{d} < 0$. In other words the search direction is given by $\mathbf{d} = \mathbf{y}_4 - \sqrt{1 - R^2}\, \mathbf{a}_k$. This iterative procedure can be summarized as follows.

Algorithm 6.8 (Shake-and-Bake) *Given an initial state* $\mathbf{X}_0$, *iterate the following steps for* $t = 0, 1, \ldots$.

1. *Given* $\mathbf{X}_t \in \partial\mathscr{X}$, *assume that the k-th constraint is active, that is,* $\mathbf{a}_k^\top \mathbf{X}_t = b_k$.

2. *Construct a search direction vector* $\mathbf{d}$ *using the following steps.*

   (a) *Generate a point* $\mathbf{Y}_1$ *uniformly distributed on the surface of the n-dimensional unit hypersphere centered at the origin.*

   (b) *Compute* $\mathbf{Y}_2 = \mathbf{Y}_1 - (\mathbf{a}_k^\top \mathbf{Y}_1)\,\mathbf{a}_k$ *and* $\mathbf{Y}_3 = \frac{\mathbf{Y}_2}{\|\mathbf{Y}_2\|}$.

   (c) *Compute* $R = U^{1/(n-1)}$, *where* $U \sim \mathsf{U}(0,1)$, *and output*

$$\mathbf{d} = R\,\mathbf{Y}_3 - \sqrt{1-R^2}\,\mathbf{a}_k\,.$$

3. *Update to the next hit point* $\mathbf{X}_{t+1} = \mathbf{X}_t + \lambda^* \mathbf{d}$, *where*

$$\lambda^* = \min\{\lambda_i : \lambda_i > 0\}, \quad \lambda_i = \frac{b_i - \mathbf{a}_i^\top \mathbf{X}_t}{\mathbf{a}_i^\top \mathbf{d}}, \; i = 1, \ldots, m\,.$$

There are different variants of the shake-and-bake algorithm [7]. Algorithm 6.8 is referred to as the **running** shake-and-bake algorithm, because each new hit point automatically becomes the starting point for the next iteration. In an alternative version, called the **limping** shake-and-bake algorithm, a new hit point is subjected to an acceptance–rejection step, as a result of which the algorithm may start from the same point over a number of iterations. Boender et al. [7] show that of all shake-and-bake versions Algorithm 6.8 has the fastest theoretical rate of convergence as an MCMC sampler.

Similar to the hit-and-run algorithm, the shake-and-bake algorithm can be used for optimization purposes. For example, optimizing a concave function over $\mathscr{X}$ is achieved by searching for optimal values on the boundary $\partial\mathscr{X}$ [7, 62].

■ EXAMPLE 6.8 (Sampling on a Polyhedron)

Consider the polyhedron defined by the set $\{\mathbf{x} : A\mathbf{x} \leqslant \mathbf{b}\}$, where $\mathbf{b} = (0, 0, 0, \frac{2}{\sqrt{6}})^\top$ and

$$A = \begin{bmatrix} 0 & 0 & -1 \\ -1 & 0 & 0 \\ \frac{1}{\sqrt{5}} & \frac{-2}{\sqrt{5}} & 0 \\ \frac{1}{\sqrt{6}} & \frac{2}{\sqrt{6}} & \frac{1}{\sqrt{6}} \end{bmatrix}.$$

We ran Algorithm 6.8 for 10^3 iterations starting from $\mathbf{x}_1 = (1/2, 1/2, 0)^\top$. The result is displayed in Figure 6.10.

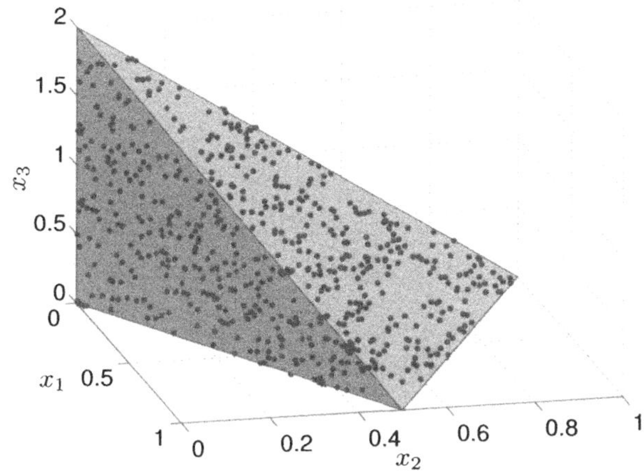

Figure 6.10 A thousand points (approximately) uniformly distributed on the surface of a polyhedron.

To verify the accuracy of the shake-and-bake algorithm, we can count the number of times each facet of the polyhedron is visited. In this simple example the number of times each constraint $\sum_j A_{ij}x_j = b_i$, $i = 1, 2, 3, 4$, is active has distribution

$$(0.1301,\ 0.2602,\ 0.2909,\ 0.3187)\ ,$$

which is easily obtained by computing the area of all triangular faces of the polyhedron. Since the empirically observed frequencies of (0.1334, 0.2543, 0.2957, 0.3166) are close to the true ones, we conclude that the algorithm converges satisfactorily. The following MATLAB code implements Algorithm 6.8.

```
%snb_polyhedron.m
A=[0,0,-1;-1,0,0;1,-2,0;1,2,1];
A(4,:)=A(4,:)/sqrt(6);
A(3,:)=A(3,:)/sqrt(5);
b=[0;0;0;2/sqrt(6)];
x=[1/2,1/2,0]'; %initial point
k=1;            %which constraint is active at x
T=10^3;         %run shake-and-bake 10^3  times
a=A(k,:); data=nan(T,length(x)); %preallocate memory
for t=1:T
    Y1=randn(1,3);
    Y1=Y1./sqrt(sum(Y1.^2));
    Y2=Y1- (a*Y1')*a;
    Y3=Y2/sqrt(sum(Y2.^2));
    R=rand^(1/2);
    d=R*Y3-sqrt(1-R^2)*a;
    lam=(b-A*x)./(A*d');
    % next compute lam_s=lambda_star and update k
```

```
    lam(k)=inf; lam(lam<0)=inf; [lam_s,k]=min(lam);
    x=x+lam_s*d'; % update to new hit point
    a=A(k,:);           % new active constraint
    data(t,:)=x; % accumulate data
    Index(t)=k;
end
% display data
plot3(data(:,1),data(:,2),data(:,3),'k.','MarkerSize',10)
```

6.3.3 Metropolis–Gibbs Hybrids

Every cycle of the Gibbs sampler (Algorithm 6.2) requires sampling from conditional pdfs of the form $f(y_i \,|\, \mathbf{x}_{-i})$, where $\mathbf{x}_{-i}$ denotes the vector $\mathbf{x}$ with the i-th element removed. In many practical problems some or all of the univariate conditional pdfs cannot be simulated easily. In such cases one can approximately sample from $f(y_i \,|\, \mathbf{x}_{-i})$ by taking a Metropolis–Hastings move (a single iteration of Algorithm 6.1) with proposal $q_i(y_i \,|\, \mathbf{x}_{-i})$ and target pdf $f(y_i \,|\, \mathbf{x}_{-i})$. Note that making multiple Metropolis–Hastings moves to better approximate the conditional $f(y_i \,|\, \mathbf{x}_{-i})$ does not necessarily lead to faster convergence toward the joint density $f(\mathbf{x})$, see [15, 32]. Some other hybrid algorithms use the Metropolis–Hastings Algorithm 6.1 with a proposal $q(\mathbf{y} \,|\, \mathbf{x})$ corresponding to a single cycle of the Gibbs sampler, that is,

$$q(\mathbf{y} \,|\, \mathbf{x}) = f(y_1 \,|\, \mathbf{x}_{-1}) \Big(\prod_{i=2}^{n-1} f(y_i \,|\, y_1, \ldots, y_{i-1}, x_{i+1}, \ldots, x_n) \Big) f(y_n \,|\, \mathbf{y}_{-n}) \,.$$

All such hybrid algorithms are valid since the stationary distribution remains unchanged [32]. Theoretical results relating to such hybrids are given in [58], where the authors find that an acceptance rate of 0.234 for the Metropolis–Hastings step is optimal in some sense.

For the case where Gibbs sampling is applied on a discrete space, Liu [42] proposes that instead of sampling directly from a given conditional pdf $f(y_i \,|\, \mathbf{x}_{-i})$, one samples from $f(y_i \,|\, \mathbf{x}_{-i})$ conditional on $y_i \neq x_i$ and then applies the Metropolis–Hastings acceptance criterion (6.1). This gives the following hybrid algorithm used as an alternative to sampling directly from the conditional pdf $f(y_i \,|\, \mathbf{x}_{-i})$.

Algorithm 6.9 (Metropolis–Gibbs Hybrid on a Discrete Space) *Given an initial state* $\mathbf{X}_0$, *iterate the following steps for* $t = 0, 1, \ldots$.

1. *Given* $\mathbf{X}_t = \mathbf{x}$, *generate* Y_i *from* $f(y_i \,|\, \mathbf{x}_{-i})$ *conditionally on* $Y_i \neq x_i$; *that is, generate* Y_i *from the pdf*

$$\frac{f(y_i \,|\, \mathbf{x}_{-i})}{1 - f(x_i \,|\, \mathbf{x}_{-i})}, \quad y_i \neq x_i \,.$$

   Set $\mathbf{Y} = (x_1, \ldots, x_{i-1}, Y_i, x_{i+1}, \ldots, x_n)$.

2. *Apply the Metropolis–Hastings acceptance criterion* (6.1) *with*

$$\alpha(\mathbf{x}, \mathbf{Y}) = \min \left\{ \frac{1 - f(x_i \,|\, \mathbf{x}_{-i})}{1 - f(Y_i \,|\, \mathbf{x}_{-i})}, 1 \right\} .$$

Liu [42] shows that the Markov chain sequence obtained via Algorithm 6.9 gives an ergodic estimator (6.4) with smaller variance than would otherwise be obtained by sampling directly from the conditional pdf $f(y_i \mid \mathbf{x}_{-i})$.

6.3.4 Multiple-Try Metropolis–Hastings

The **multiple-try Metropolis–Hastings** [44] algorithm is an extension of the Metropolis–Hastings algorithm (Algorithm 6.1) that generates M multiple proposals and then resamples the proposals in an attempt to encourage large-step transitions of the chain. One of the simplest versions of the algorithm assumes that the proposal density $q(\mathbf{y} \mid \mathbf{x})$ is symmetric and reads as follows.

Algorithm 6.10 (Multiple-Try Metropolis–Hastings) *Initialize with some* $\mathbf{X}_0$ *for which* $f(\mathbf{X}_0) > 0$. *Given the parameter* M *and the* symmetric *proposal density* $q(\mathbf{y} \mid \mathbf{x})$, *perform the following steps for each* $t = 0, 1, 2 \ldots, T$.

1. *Generate proposals* $\mathbf{Y}_1, \ldots, \mathbf{Y}_M \stackrel{\text{iid}}{\sim} q(\mathbf{y} \mid \mathbf{X}_t)$.

2. *Sample a random index* J *from the set* $\{1, \ldots, M\}$ *such that*

$$\mathbb{P}(J = j) = \frac{f(\mathbf{Y}_j)}{f(\mathbf{Y}_1) + \cdots + f(\mathbf{Y}_M)}, \quad j = 1, \ldots, M .$$

3. *Given* J, *generate proposals* $\mathbf{Z}_1, \ldots, \mathbf{Z}_{M-1} \stackrel{\text{iid}}{\sim} q(\mathbf{z} \mid \mathbf{Y}_J)$ *and set* $\mathbf{Z}_M = \mathbf{X}_t$.

4. *Generate* $U \sim \mathsf{U}(0, 1)$ *and deliver*

$$\mathbf{X}_{t+1} = \begin{cases} \mathbf{Y}_J & \text{if } U \leqslant \alpha(\mathbf{X}_t, \mathbf{Y}_J) \\ \mathbf{X}_t & \text{otherwise} , \end{cases}$$

where

$$\alpha(\mathbf{X}_t, \mathbf{Y}_J) = \min\left\{ \frac{f(\mathbf{Y}_1) + \cdots + f(\mathbf{Y}_M)}{f(\mathbf{Z}_1) + \cdots + f(\mathbf{Z}_M)}, 1 \right\} .$$

For a proof that f is the invariant distribution of the Markov chain $\{\mathbf{X}_t\}$, see [44].

Liu et al. [44] provide examples where the multiple-try Metropolis–Hastings sampler performs better than the simpler Metropolis–Hastings sampler. The multiple-try Metropolis–Hastings sampler can be generalized and enhanced using antithetic and stratified sampling ideas, see [20].

■ **EXAMPLE 6.9 (Two-Humps Density)**

Consider sampling from the bimodal pdf

$$f(\mathbf{x}; \lambda) \propto \exp\left(-\frac{x_1^2 + x_2^2 + (x_1 x_2)^2 - 2\lambda x_1 x_2}{2} \right), \quad \mathbf{x} \in \mathbb{R}^2,$$

for some parameter λ, say $\lambda = 12$. The density is depicted in the left panel of Figure 6.11. The code below implements the multiple-try Metropolis–Hastings algorithms using 10^3 steps and $M = 100$ proposal points. The result is depicted in the right panel of Figure 6.11. Compare this with the right panel of Figure 14.5, which shows that the Gibbs sampler is not suitable for this problem. ☞ 514

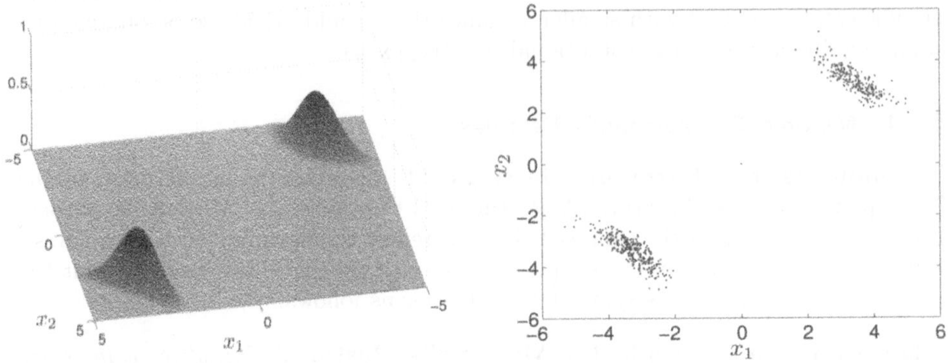

Figure 6.11 Left panel: plot of the two-humps density. Right panel: empirical distribution using 10^3 steps of the multiple-try Metropolis–Hastings algorithm with $M = 10^2$ proposal points. The initial point is the origin.

```
%multiple_try.m
T=10^3;M=100; % set up parameters
sigma=5; lam=12;
% define target pdf
f=@(x)exp(-(x(1)^2+x(2)^2+(x(1)*x(2))^2-2*lam*x(1)*x(2))/2);
X=[0,0]; % X_0
data=nan(T,2); count=0;
for t=1:T
    % step 1
    Y=repmat(X,M,1)+randn(M,2)*sigma;
    % step 2
    for i=1:M
        p(i)=f(Y(i,:));
    end
    Sum_p=sum(p);p=p/Sum_p;
    [dummy,J]=histc(rand,[0,cumsum(p)]);
    % step 3
    Z=repmat(Y(J,:),M-1,1)+randn(M-1,2)*sigma;
    Z(M,:)=X;
    % step 4
    for i=1:M
        w(i)=f(Z(i,:));
    end
    if rand<min(Sum_p/sum(w),1)
        X=Y(J,:); count=count+1;
    end
    data(t,:)=X;
end
count/T % acceptance rate estimate
plot(data(:,1),data(:,2),'.')
```

6.3.5 Auxiliary Variable Methods

Auxiliary variable methods form a general class of methods in computational statistics that exploit the fact that every density $f(\mathbf{x})$ can be viewed as a marginal density:

$$f(\mathbf{x}) = \int f(\mathbf{x}, \mathbf{y})\, \mathrm{d}\mathbf{y} ,$$

where $f(\mathbf{x}, \mathbf{y})$ is the joint density of random variables $\mathbf{X}$ and $\mathbf{Y}$. Here, $\mathbf{Y}$ is called an **auxiliary** or **latent** variable. The vector $(\mathbf{X}, \mathbf{Y})$ is said to be an **augmented** version of $\mathbf{X}$. The auxiliary variable $\mathbf{Y}$ may be a natural part of a statistical model with hidden or unobserved data. However, in general $\mathbf{Y}$ is an artificially introduced variable that has no natural interpretation within the statistical model (as given by $f(\mathbf{x})$), and is used purely as a computational device. One of the earliest auxiliary variable methods is the **expectation-maximization** (EM) method for likelihood optimization [51]. In the EM algorithm the random variables $\mathbf{X}$ are augmented by hidden or unobserved variables $\mathbf{Y}$ to obtain the so-called completed log-likelihood; see Section D.7 for a discussion. The idea of introducing auxiliary variables to make statistical inference easier is also known under the name **data augmentation**. ☞ 711

Another example of data augmentation is the composition method of Section 3.1.2.6. Namely, suppose we want to sample from the mixture pdf ☞ 53

$$f(x) = \sum_{i=1}^{K} p_i\, f_i(x) .$$

To simplify sampling from f, let Y be the discrete random variable taking values in $\{1, \ldots, K\}$ with probabilities

$$\mathbb{P}(Y = y) = p_y, \quad y = 1, \ldots, K ,$$

and consider sampling from the joint density $f(x, y) = p_y f_y(x)$: first draw Y according to $\{p_y\}$ and then sample X conditional on $Y = y$; that is, sample from $f_y(x)$. By simply ignoring Y we obtain a sample from the marginal density $f(x)$.

The auxiliary variable technique is usually specific to the setting in which it is applied and there are no general rules for a successful design of auxiliary variables [38]. The following examples illustrate the idea.

■ **EXAMPLE 6.10 (Bayesian Analysis of the Probit Model)**

The **probit** regression model is an alternative to the logit model given in Example 6.2. The only difference is that in the probit model the Bernoulli success probabilities $\{p_i\}$ take the form $p_i = \Phi\left(\mathbf{x}_i^\top \boldsymbol{\beta}\right)$, where Φ is the cdf of the standard normal distribution. Thus, the Bayesian model can be summarized as:

- Prior: $f(\boldsymbol{\beta}) \propto \exp\left(-\frac{1}{2}(\boldsymbol{\beta} - \boldsymbol{\beta}_0)^\top V_0^{-1}(\boldsymbol{\beta} - \boldsymbol{\beta}_0)\right)$, $\boldsymbol{\beta} \in \mathbb{R}^k$.
- Likelihood: $f(\mathbf{y} \,|\, \boldsymbol{\beta}) = \prod_{i=1}^{n} p_i^{y_i}(1 - p_i)^{1-y_i}$, $p_i = \Phi\left(\mathbf{x}_i^\top \boldsymbol{\beta}\right)$.

While the random walk sampler (Algorithm 6.1.2) can be easily applied here to sample from the posterior, a more efficient approach is to introduce the auxiliary vector $\mathbf{y}^* = (y_1^*, \ldots, y_n^*)^\top$ such that, given $\boldsymbol{\beta}$, $\{y_i^*\} \sim_{\text{iid}} \mathsf{N}(\mathbf{x}_i^\top \boldsymbol{\beta}, 1)$. Then, the response variables can be written as $y_i = \mathrm{I}_{\{y_i^* > 0\}}$ and

$$
\begin{aligned}
f(\boldsymbol{\beta}, \mathbf{y}^* \,|\, \mathbf{y}) &\propto f(\boldsymbol{\beta}) f(\mathbf{y}^* \,|\, \boldsymbol{\beta}) f(\mathbf{y} \,|\, \mathbf{y}^*, \boldsymbol{\beta}) \\
&\propto f(\boldsymbol{\beta}) \prod_{i=1}^{n} \mathrm{e}^{-\frac{(y_i^* - \mathbf{x}_i^\top \boldsymbol{\beta})^2}{2}} \left(\mathrm{I}_{\{y_i=1\}} \, \mathrm{I}_{\{y_i^* > 0\}} + \mathrm{I}_{\{y_i=0\}} \, \mathrm{I}_{\{y_i^* \leqslant 0\}} \right) .
\end{aligned}
$$

Thus, to sample from the posterior $f(\boldsymbol{\beta} \,|\, \mathbf{y})$ we can apply a grouped or blocked Gibbs sampler (Algorithm 6.3) to the joint density $f(\boldsymbol{\beta}, \mathbf{y}^* \,|\, \mathbf{y})$ to obtain a sample $\{(\boldsymbol{\beta}_i, \mathbf{y}_i^*)\}$ and then ignore the $\{\mathbf{y}_i^*\}$ variables. Each iteration of the grouped Gibbs sampler consists of two parts. First, sample from the conditional pdf

$$
f(\boldsymbol{\beta} \,|\, \mathbf{y}^*, \mathbf{y}) \propto f(\boldsymbol{\beta}) \prod_{i=1}^{n} \mathrm{e}^{-\frac{(y_i^* - \mathbf{x}_i^\top \boldsymbol{\beta})^2}{2}} ,
$$

which corresponds to a $\mathsf{N}(\boldsymbol{\mu}, \Sigma)$ distribution with $\Sigma = (V_0^{-1} + \mathbf{X}^\top \mathbf{X})^{-1}$ and $\boldsymbol{\mu} = \Sigma(V_0^{-1} \boldsymbol{\beta}_0 + \mathbf{X}^\top \mathbf{y}^*)$. Second, sample from $f(\mathbf{y}^* \,|\, \boldsymbol{\beta}, \mathbf{y}) = \prod_i f(y_i^* \,|\, \boldsymbol{\beta}, \mathbf{y})$ with truncated normal marginal pdfs:

$$
f(y_i^* \,|\, \boldsymbol{\beta}, \mathbf{y}) \propto \begin{cases} \mathrm{e}^{-\frac{(y_i^* - \mathbf{x}_i^\top \boldsymbol{\beta})^2}{2}} \, \mathrm{I}_{\{y_i^* > 0\}}, & \text{if } y_i = 1 \\ \mathrm{e}^{-\frac{(y_i^* - \mathbf{x}_i^\top \boldsymbol{\beta})^2}{2}} \, \mathrm{I}_{\{y_i^* \leqslant 0\}}, & \text{if } y_i = 0 . \end{cases}
$$

The following MATLAB code implements this procedure using an artificial data set. Although for $k = 3$ this is not strictly necessary, the code computes the Cholesky decomposition of the 3×3 matrix $V_0^{-1} + \mathbf{X}^\top \mathbf{X} = \Sigma^{-1}$ to avoid using the numerically inferior `inv.m` function. Note that we arbitrarily set $V_0 = 100 \, \mathrm{diag}(\mathbf{1})$, that is, V_0 is a diagonal matrix with diagonal entries equal to 100.

```
%probit_model.m
clear all,clc
n=5000; % number of data points (y_1,...,y_n)
k=3;    % number of explanatory variables
% generate artificial dataset
randn('seed', 12345);  rand('seed', 67890);
truebeta = [1 -5.5 1]';
X = [ ones(n,1) randn(n,k-1)*0.1 ];  % design matrix
Y = binornd(1,normcdf(X*truebeta));
bo=zeros(k,1); % we set Vo=100*eye(k);
I1=find(Y==1); Io=find(Y==0);
Y_star=randn(n,1);
%compute the Cholesky decomp. to avoid using inv.m
L=chol(eye(k)/100+X'*X);
T=10^4; data=nan(T,k); %allocate memory
for t=1:T
    % sample beta given Y^*
    b=L\(L'\(bo/100+X'*Y_star))+L\randn(k,1);
    % sample Y^* given beta
    M=X*b;
    Y_star(I1)=normt(M(I1),1,0,inf);
```

```
    Y_star(Io)=normt(M(Io),1,-inf,0);
    data(t,:)=b';
end
b_hat=mean(data)
Cov_hat=cov(data)
```

The code uses the normt.m function from Example 3.6. The posterior mean $\mathbb{E}[\boldsymbol{\beta} \mid \mathbf{y}]$ and covariance $\mathrm{Cov}(\boldsymbol{\beta} \mid \mathbf{y})$ are estimated as ☞ 51

$$\widehat{\mathbb{E}[\boldsymbol{\beta} \mid \mathbf{y}]} = \begin{pmatrix} 0.993 \\ -5.4186 \\ 1.143 \end{pmatrix}, \quad \widehat{\mathrm{Cov}(\boldsymbol{\beta} \mid \mathbf{y})} = \begin{pmatrix} 0.0007 & -0.0028 & 0.0007 \\ -0.0028 & 0.0643 & -0.0051 \\ 0.0007 & -0.0051 & 0.0469 \end{pmatrix}.$$

Compare this with the true value of $\boldsymbol{\beta} = (1, -5.5, 1)^\top$.

■ **EXAMPLE 6.11 (Slice Sampler)**

Suppose we wish to generate samples from the pdf

$$f(\mathbf{x}) = b \prod_{k=1}^{m} p_k(\mathbf{x}), \quad \mathbf{x} \in \mathscr{X}, \tag{6.14}$$

where b is a known or unknown constant, and the $\{p_k\}$ are known positive functions — not necessarily densities. Introduce the auxiliary variables $\mathbf{y} = (y_1, \ldots, y_m)$ such that the joint density of $\mathbf{y}$ and $\mathbf{x}$ is given by

$$f(\mathbf{x}, \mathbf{y}) \propto \prod_{k=1}^{m} \mathrm{I}_{\{0 \leqslant y_k \leqslant p_k(\mathbf{x})\}}. \tag{6.15}$$

In this way, given $\mathbf{x}$, each y_i has uniform density on the interval $[0, p_k(\mathbf{x})]$, and the marginal density $\int f(\mathbf{x}, \mathbf{y})\, d\mathbf{y}$ is equal to the target density (6.14). In addition, note that all $\{y_i\}$ are independent. The idea of the **slice sampler** [59] is to apply a grouped Gibbs sampler on the augmented space by iteratively sampling from the conditional densities $f(\mathbf{x} \mid \mathbf{y})$ and $f(\mathbf{y} \mid \mathbf{x})$.

Algorithm 6.11 (Slice Sampler) *Let* $f(\mathbf{x})$ *be of the form* (6.14). *Initialize with* $\mathbf{X}_1 \in \mathscr{X}$ *and* $t = 1$.

1. *Generate* $\mathbf{Y}$ *from the conditional density* $f(\mathbf{y} \mid \mathbf{X}_t)$; *that is, for* $k = 1, \ldots, m$ *let* $Y_k = U_k\, p_k(\mathbf{X}_t)$, *where* $U_1, \ldots, U_m \sim_{\text{iid}} \mathsf{U}(0, 1)$.

2. *Generate* $\mathbf{X}_{t+1}$ *from the conditional density* $f(\mathbf{x} \mid \mathbf{Y})$; *that is, draw* $\mathbf{X}_{t+1}$ *uniformly from the set* $\{\mathbf{x} : p_k(\mathbf{x}) \geqslant Y_k, k = 1, \ldots, m\} \cap \mathscr{X}$.

3. *Stop if a stopping criterion is met; otherwise, set* $t = t + 1$ *and repeat from Step 2.*

While sampling from $f(\mathbf{y} \mid \mathbf{x})$ is relatively easy, generating random variables from the conditional pdf $f(\mathbf{x} \mid \mathbf{y})$ can be a challenging task [59].

To illustrate the algorithm, we consider sampling from a truncated gamma density. Note that generation from the gamma distribution (Section 4.2.6) is a non- ☞ 113 trivial task. Algorithm 4.33 requires the generation of one uniform and one normal random variable and can be quite inefficient in cases where the aim is to obtain samples from the truncated gamma density on an interval $[a, b]$. The slice sampler provides an alternative (approximate) method for sampling from the truncated gamma density, which is easy to code and requires the generation of two uniform random variables. Here the target is given by:

$$f(x) \propto x^{\alpha-1}\,\mathrm{e}^{-\lambda x}, \quad \alpha > 1,\ x \in [a, b]\,,$$

and we use the slice sampler with $p_1(x) = x^{\alpha-1}$ and $p_2(x) = \mathrm{e}^{-\lambda x}$. Suppose that at iteration t, $X_t = z$, and u_1 and u_2 have been obtained in Step 1. Then, in Step 2, X_{t+1} is drawn uniformly from the set

$$\left\{x : \frac{p_1(x)}{p_1(z)} \geqslant u_1,\ \frac{p_2(x)}{p_2(z)} \geqslant u_2\right\} \cap [a, b]\,,$$

which implies that the interval from which to sample is such that

$$\max\left\{z\,u_1^{\frac{1}{\alpha-1}},\ a\right\} \leqslant x \leqslant \min\left\{z - \frac{\ln u_2}{\lambda},\ b\right\}, \quad \alpha > 1,\ \lambda > 0\,.$$

This leads to the MATLAB implementation below, which also plots a kernel density estimate using the function kde.m from Section 8.5. ☞ 319

Figure 6.12 depicts the kernel density estimate (restricted on the interval $[1, 6]$), along with the true pdf (solid line). We see that the two are in close agreement.

```
%slice_sampler.m
lam=1;alpha=2; a=1;b=6;
x=2; T=10^4; data=nan(T,1);
for t=1:T
   u=rand(1,2);
   Up=min(x-log(u(2))/lam,b);
   Lo=max(x*u(1)^(1/(alpha-1)),a);
   x=rand*(Up-Lo)+Lo;
   data(t)=x;
end
% plot density estimate and compare with true one
[bandwidth,density,xmesh]=kde(data,2^13,a,b);
plot(xmesh,density),hold on
C=gamcdf(b,alpha,1/lam)-gamcdf(a,alpha,1/lam);
plot(xmesh,gampdf(xmesh,alpha,1/lam)/C,'r')
```

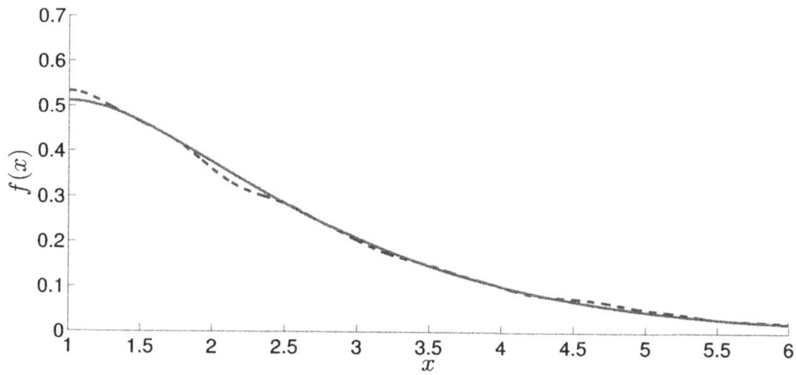

Figure 6.12 True density (solid line) and kernel density estimate (dashed line) of samples produced by the slice sampler.

For various applications and extensions of the slice sampler see [12, 56, 68]. In particular, Roberts and Rosenthal [68] propose a version of the slice sampler suitable for the frequently used class of log-concave densities. Convergence results for the slice sampler are given in [57, 67]. In particular, [67] establishes geometric ergodicity under appropriate assumptions and derives an upper bound for the total variation distance between the sampling and the target distributions. ☞ 63

■ EXAMPLE 6.12 (Potts Model)

Let $\mathbf{x} = (x_1, \ldots, x_J)$ be a vector with each $x_i \in \{1, \ldots, K\}$ for some integer $K \geqslant 2$, and let $\mathscr{X} = \{1, \ldots, K\}^J$ be the set of all such vectors. Suppose we wish to generate random vectors from the discrete multivariate **Boltzmann** pdf

$$f(\mathbf{x}) = \frac{e^{-E(\mathbf{x})}}{\mathcal{Z}}, \quad \mathbf{x} \in \mathscr{X}, \tag{6.16}$$

where $\mathcal{Z}$ is a normalization constant called the **partition function**, and E is the so-called **energy function** defined as

$$E(\mathbf{x}) = -\beta \sum_{i<j} \psi_{ij} \, I_{\{x_i = x_j\}} + \frac{\beta}{2} \sum_{i<j} \psi_{ij}, \quad \beta > 0, \tag{6.17}$$

with ψ_{ij} a given $J \times J$ symmetric matrix containing only zeros and ones. The Boltzmann pdf arises in Bayesian variable selection and spatial statistics [60]. In particular, it arises in the **Potts model** — a model in statistical mechanics for the interaction of idealized magnets located on a d-dimensional lattice [21]. In the basic two-dimensional case we have a lattice $\{1, \ldots, n\} \times \{1, \ldots, n\}$ with $J = n^2$ spatial positions (sites). Each of the n^2 sites has four nearest neighbors, with the exception of the boundary sites, which have either two or three nearest neighbors; see Figure 6.13.

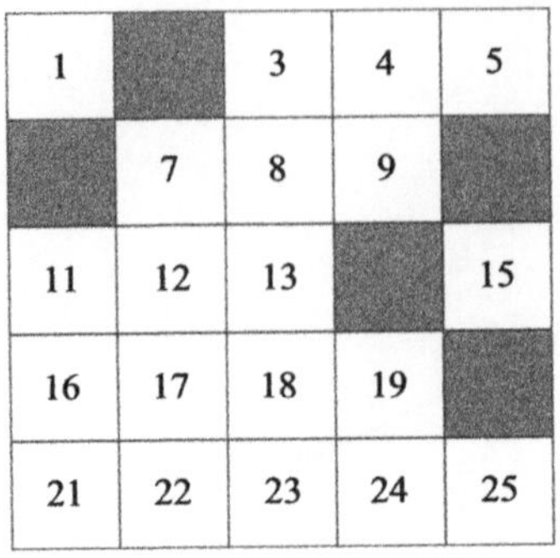

Figure 6.13 A lattice with $J = 5^2 = 25$ sites. The boundary site 1 has two neighbors (dark sites) and boundary site 15 has three neighbors.

The symmetric matrix (ψ_{ij}) is used to indicate whether sites i and j on the $n \times n$ lattice are neighbors:

$$\psi_{ij} = \begin{cases} 1 & \text{if } i \text{ and } j \text{ are neighbors,} \\ 0 & \text{otherwise.} \end{cases} \tag{6.18}$$

Let $\{1, \ldots, J\}$ with $J = n^2$ be the enumeration of all the sites. To each site we can assign a **color** x_i out of K possible colors. Then, the **configuration** of the lattice is described by the vector $\mathbf{x} = (x_1, \ldots, x_J)$. The case $K = 2$ gives the **Ising model** (see, for example, [21]), which has $2^J = 2^{n^2}$ possible configurations. In the Ising model it is common to write the configuration and the energy E in terms of variables $\mathbf{s} = (s_1, \ldots, s_{n^2})$, where $s_i = 2\,\mathrm{I}_{\{x_i=1\}} - 1$ for all i. The energy (6.17) can then be written in terms of $\mathbf{s}$ as

$$E(\mathbf{x}) = -\frac{\beta}{2} \sum_{i \leftrightarrow j} \left(\mathrm{I}_{\{x_i = x_j\}} - \frac{1}{2} \right) = -\frac{\beta}{4} \sum_{i \leftrightarrow j} s_i\, s_j \ ,$$

where $\leftrightarrow$ indicates that the summation is taken over neighboring sites (i, j).

By generating random configurations from $f(\mathbf{x})$ one can estimate quantities of interest such as the partition function $\mathcal{Z}$ and the mean value $\mathbb{E}_f E(\mathbf{X})$. Unfortunately, none of the techniques in Chapter 3 can be easily applied to sample from $f(\mathbf{x})$. Thus, one has to resort to MCMC.

We now show how to generate a sample from the target pdf $f(\mathbf{x})$ in (6.16) using the auxiliary variable method of Swendsen and Wang [26, 38, 76]. First, observe that the target density is of the form (6.14):

$$f(\mathbf{x}) \propto \prod_{i<j} \mathrm{e}^{\beta\, \psi_{ij}\, \mathrm{I}_{\{x_i = x_j\}}} .$$

Second, define the *auxiliary* random variables $Y_{ij}, 1 \leqslant i < j \leqslant J$ such that the joint pdf of $\mathbf{X}$ and $\mathbf{Y}$ is given by

$$f(\mathbf{x}, \mathbf{y}) \propto \prod_{i<j} \mathrm{I}\left\{ 0 < y_{ij} < \mathrm{e}^{\beta\, \psi_{ij}\, \mathrm{I}_{\{x_i = x_j\}}} \right\} .$$

In this way, given $\mathbf{x}$, the variables y_{ij} are independent and uniformly distributed on the interval

$$\left(0,\ e^{\beta\,\psi_{ij}\,\mathrm{I}_{\{x_i=x_j\}}}\right),$$

and the marginal density $\int f(\mathbf{x},\mathbf{y})\,\mathrm{d}\mathbf{y}$ is equal to the target pdf $f(\mathbf{x}) = e^{-E(\mathbf{x})}/\mathcal{Z}$. The idea now is to sample from the joint pdf $f(\mathbf{x},\mathbf{y})$ via the Gibbs sampler on the augmented space, by iteratively sampling from the conditional densities $f(\mathbf{x}\,|\,\mathbf{y})$ and $f(\mathbf{y}\,|\,\mathbf{x})$. Sampling from the conditional density $f(\mathbf{y}\,|\,\mathbf{x})$ is straightforward since $y_{ij} \sim \mathsf{U}(0, e^{\beta\,\psi_{ij}\,\mathrm{I}_{\{x_i=x_j\}}})$ for all i, j, independently. To sample from the conditional density $f(\mathbf{x}\,|\,\mathbf{y})$, note first that $\exp\left(\beta\,\psi_{ij}\,\mathrm{I}_{\{x_i=x_j\}}\right) \geqslant 1$ for all $i < j$. Second, observe that for each $i < j$, either $y_{ij} \in [0,1]$ or $y_{ij} \in (1, e^{\beta}]$. If $y_{ij} \leqslant 1$, then the variables X_i and X_j are not constrained and each of them has a uniform distribution over the states $1,\ldots,K$. Alternatively, if $y_{ij} > 1$, then the variables X_i and X_j are constrained to be in the same state, $X_i = X_j$, such that X_i $(= X_j)$ is uniformly distributed over $1,\ldots,K$. Thus, if all variables for which $y_{ij} > 1$, $i < j$ are gathered into clusters, then the color of the sites within each cluster is the same and uniformly distributed over all the K possible colors. The same holds for the remaining variables, which can be viewed as clusters with a single member, see Figure 6.14. This idea leads to the Swendsen–Wang algorithm.

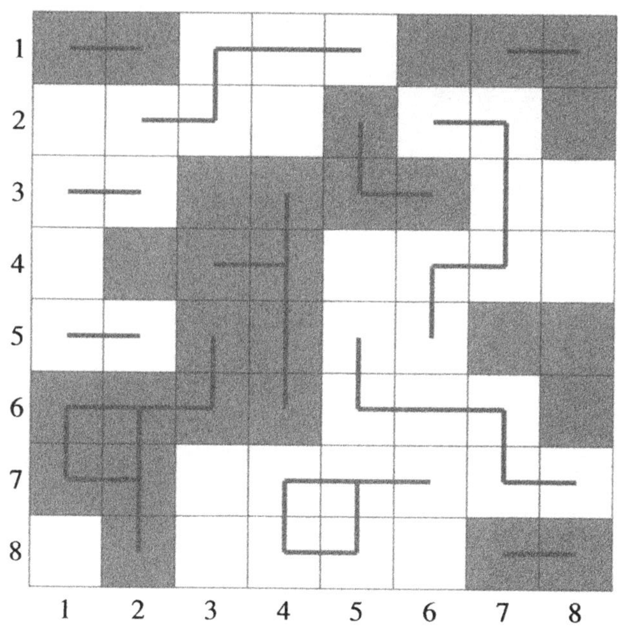

Figure 6.14 Clustering on a two-dimensional lattice with $n = 8$ and two colors ($K = 2$). Sites bonded with a line belong to a single cluster of one color. There are 17 single-site clusters.

Algorithm 6.12 (Swendsen–Wang) *For a given lattice with neighboring sites characterized by matrix* (ψ_{ij}), *initialize by generating a configuration* $\mathbf{X} = (X_1, \ldots, X_J)$ *with* X_i *uniformly distributed over* $1, \ldots, K$ *for all* i. *Then, iterate the following steps.*

1. *Given* $\mathbf{X}$, *generate* $Y_{ij} \overset{\text{iid}}{\sim} \mathsf{U}\left(0, \mathrm{e}^{\beta\, \psi_{ij}\, \mathrm{I}_{\{X_i = X_j\}}}\right)$ *for all* $1 \leqslant i < j \leqslant J$. *Set* $B_{ij} = \mathrm{I}_{\{Y_{ij} \geqslant 1\}}$ *for all* $i < j$.

2. *Given* $\{B_{ij}\}$, *generate* $\mathbf{X}$ *by clustering all the sites and choosing each cluster color independently and uniformly from the* K *possible colors.*

Note that in Step 2 we need only know the values of the Bernoulli variables $\{B_{ij}\}$. Thus, in Step 1 we can generate $B_{ij} \sim_{\text{iid}} \mathsf{Ber}(\mathrm{I}_{\{X_i = X_j\}}(1 - \mathrm{e}^{-\beta\psi_{ij}}))$, for $1 \leqslant i < j \leqslant J$ directly, instead of first generating $\{Y_{ij}\}$ and then computing $\{B_{ij}\}$.

As a numerical example consider sampling on a two-dimensional lattice with $n = 20$ and three colors ($K = 3$). We set $\beta = 0.8$. Figure 6.15 shows an outcome from f after 40 steps of the Swendsen–Wang algorithm.

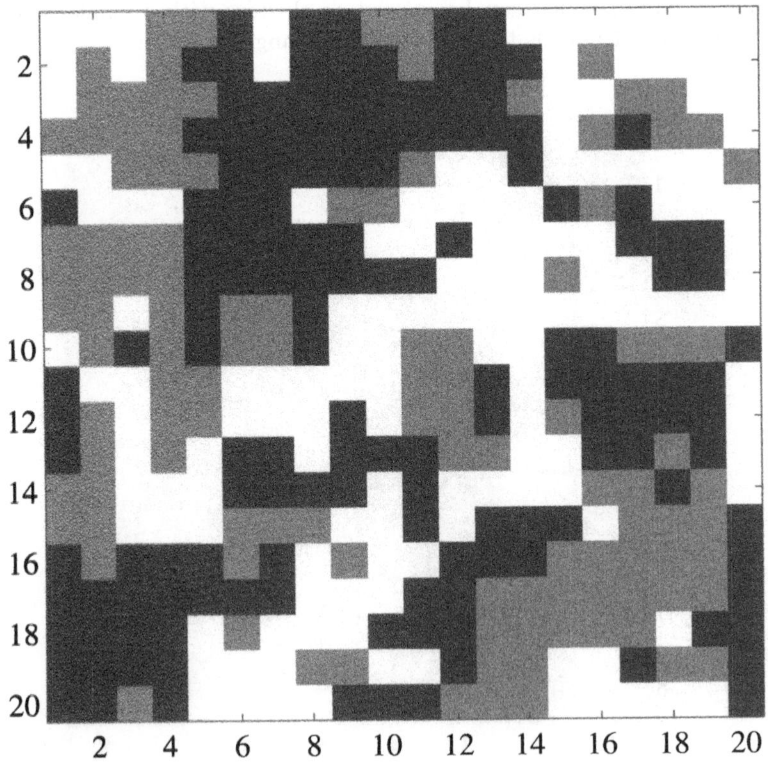

Figure 6.15 Configuration of **X** after 40 iterations of the Swendsen–Wang algorithm.

The algorithm is implemented in the following MATLAB code, which uses the functions **findneigh.m** (given in Example 5.1) and **mkclust.m** (given after the script below).

☞ 155

```
% Potts.m
set(0,'RecursionLimit',100000) % used by clust.m
n = 20; % size of square lattice
N = 40; %number of Gibbs steps
nels =n*(5*n-4); % number of ones in psi_ij
beta =.8 ;   %
a = zeros(1,nels); % preallocation of memory
b = zeros(1,nels);
c = zeros(1,nels);
k=0;
%set up sparse Psi matrix
for i=1:n
    for j=1:n
        A = findneigh(i,j,n); %index of the neighbors of site (i,j)
        number_of_neigh = size(A,1);
        for h=1:number_of_neigh  % convert (i,j) to a linear index
            a(k+h)=sub2ind([n;n],j,i);
            b(k+h)=sub2ind([n;n],A(h,2),A(h,1));
            c(k+h) = 1;
        end
        k = k+number_of_neigh;
    end
end
Psi = sparse(a,b,c,n^2,n^2); % build adjacency matrix
K=3; % number of colors;
x = ceil(rand(1,n^2)*K);  %initial state
for iter=1:N
    iter
    %step 1 of S-W
    B=sparse([],[],[],n^2,n^2); % allocate for {B_ij}
    for i=1:n^2
        neighbors_of_i = find(Psi(i,:));
        for k= neighbors_of_i
            if i < k
                B(i,k) = (rand  < (1 - exp(-beta))*(x(i)==x(k)));
                B(k,i) = B(i,k);
            end
        end
    end

    %step 2 of S-W
    [nclust,C] = mkclust(B);% given B, cluster sites
    lims = [0,find(~C)]; % tells where the zeros are
    csizes = diff(lims)-1; %cluster sizes
    for k=1:nclust
        xc = ceil(rand*K); % sample colors uniformly
        for l=(lims(k)+1):(lims(k)+csizes(k))
```

```
            x(C(l)) = xc; % assign color xc to  the remaining sites
        end
    end
    imagesc(reshape(x-1,n,n)) % plot the lattice and state of X
    colormap gray
    pause(.01)
end
```

```
function [nc,C] = mkclust(B)
%given the auxiliary variables {B_ij} organized in the symmetric
%matrix B, this function clusters the sites;
%output:
% nc - number of clusters
% C  - gives the indexes of sites belonging to each cluster,
%        separated by zeroes;
n = size(B,1);
S = 1:n;
nc = 0;
C = [];
while ~isempty(S)
    i = min(S);
    c = clust([i],i,B);
    S = setdiff(S,c);
    C = [C,c,0];
    nc = nc+1;
end
```

```
function out= clust(c,i,A)
out = unique([c,i]);
for j=setdiff(find(A(i,:)),out)
    out = sort(unique(clust(out,j,A)));
end
```

Generalizations of the Swendsen–Wang algorithm are given in [3, 4]. The authors apply a generalized Swendsen–Wang algorithm to arbitrary probability densities defined on graph partitions. The primary focus is on image analysis problems, for which the authors provide numerical results showing that their generalized Swendsen–Wang algorithm is at least two orders of magnitude faster in terms of CPU time than the Gibbs sampler. Convergence results for the Swendsen–Wang algorithm are discussed in [78], where sharp bounds for the mixing time for various values of β are provided.

6.3.6 Reversible Jump Sampler

So far, we have assumed that the dimension of the multivariate random variables $\mathbf{X}_0, \mathbf{X}_1, \ldots$ generated by an MCMC algorithm remains the same across iterations. **Reversible jump samplers** [34] are MCMC algorithms specifically designed to sample from target spaces that contain vectors of different dimensions. This often occurs in Bayesian inference when different models for a given data set are considered. For example, suppose that we are given data $\mathbf{z}$ and have a countable set of possible models $\{1, 2, 3, \ldots, M\}$ for the data. Suppose each model is characterized by a parameter vector $\boldsymbol{\beta}^{(m)} = (\beta_1, \ldots, \beta_m)$ of dimension m such that:

- The prior distribution over the set of models is given by $f(m)$, $m = 1, \ldots, M$.
- The prior distribution of $\boldsymbol{\beta}^{(m)}$, given model m, is $f(\boldsymbol{\beta}^{(m)} \mid m)$.
- The likelihood of the data, given model m with parameter $\boldsymbol{\beta}^{(m)}$, is $f(\mathbf{z} \mid \boldsymbol{\beta}^{(m)}, m)$.

In order to determine the most plausible model and draw inference for the corresponding vector of parameters, we wish to generate random variables of different dimensions from the posterior density

$$f(\boldsymbol{\beta}^{(m)}, m \mid \mathbf{z}) \propto f(\mathbf{z} \mid \boldsymbol{\beta}^{(m)}, m)\, f(\boldsymbol{\beta}^{(m)} \mid m) f(m) \,.$$

In the general setting (not necessarily Bayesian), we wish to sample from a joint density $f(\mathbf{x}, m)$, where $\dim(\mathbf{x}) = m$. The reversible jump sampler jumps between spaces of different dimensions according to a set of allowed jumps (also called moves). If we allow only jumps between vectors that differ in dimension by at most 1, then possible jumps are: $\mathbf{x}_1 \to \mathbf{x}_1'$, $\mathbf{x}_1 \to (\mathbf{x}_1', x_2')$, $(\mathbf{x}_1, x_2) \to \mathbf{x}_1'$. The reversible jump sampler may be viewed as a generalization of the Metropolis–Hastings sampler in which a move is proposed from the density $q(n, \mathbf{y} \mid \mathbf{x}) = q(n \mid \mathbf{x})\, q(\mathbf{y} \mid \mathbf{x}, n)$. Thus, given the current state $\mathbf{x}$, a new dimension n is sampled from $q(n \mid \mathbf{x})$, where typically $q(n \mid \mathbf{x}) = q(n \mid m)$; that is, q depends on the current dimension only. Given the new dimension n, a new state $\mathbf{y}$ with $\dim(\mathbf{y}) = n$ is selected according to transition function $q(\mathbf{y} \mid \mathbf{x}, n)$. This gives the following algorithm.

Algorithm 6.13 (Reversible Jump Sampler) *Given the current state* $\mathbf{X}_t$ *with* $\dim(\mathbf{X}_t) = m$, *iterate the following steps.*

1. *Generate* $n \sim q(n \mid m)$.
2. *Generate* $\mathbf{Y} \sim q(\mathbf{y} \mid \mathbf{X}_t, n)$ *with* $\dim(\mathbf{Y}) = n$.
3. *Generate* $U' \sim \mathsf{U}(0,1)$ *and deliver*

$$\mathbf{X}_{t+1} = \begin{cases} \mathbf{Y} & \text{if } U' \leqslant \alpha(\mathbf{X}_t, \mathbf{Y}) \\ \mathbf{X}_t & \text{otherwise}\,, \end{cases} \tag{6.19}$$

where

$$\alpha(\mathbf{x}, \mathbf{y}) = \min\left\{ \frac{f(\mathbf{y}, n)\, q(m \mid n)\, q(\mathbf{x} \mid \mathbf{y}, m)}{f(\mathbf{x}, m)\, q(n \mid m)\, q(\mathbf{y} \mid \mathbf{x}, n)}, 1 \right\} \,. \tag{6.20}$$

A difficulty in applying the reversible jump sampler is the selection of an appropriate transition density $q(\mathbf{y} \mid \mathbf{x}, n)$ for the generation of the transdimensional proposal $\mathbf{Y}$. After sampling a new dimension from $q(n \mid m)$, a common way to construct the transition $\mathbf{x} \to \mathbf{y}$ is to generate auxiliary variables $\mathbf{u}$ and $\mathbf{v}$ such that $(\mathbf{y}, \mathbf{v})$ matches the dimension of $(\mathbf{x}, \mathbf{u})$:

$$n + \dim(\mathbf{v}) = m + \dim(\mathbf{u}) \; .$$

Suppose that for the move $\mathbf{x} \to \mathbf{y}$ we draw $\mathbf{U}$ according to some density $g(\mathbf{u} \mid n, \mathbf{x})$ and set

$$(\mathbf{Y}, \mathbf{V}) = \phi(\mathbf{X}, \mathbf{U})$$

for some differentiable bijective transformation ϕ depending on m and n. Further, assume that the reverse move, $\mathbf{y} \to \mathbf{x}$, requires generation of $\mathbf{V} \sim g(\mathbf{v} \mid m, \mathbf{y})$. By

☞ 620

(A.33) we can formally write the joint density of $(\mathbf{Y}, \mathbf{V})$ in terms of the joint density of $(\mathbf{X}, \mathbf{U})$ (suppressing n and m in the notation):

$$g(\mathbf{v} \mid \mathbf{y})\, q(\mathbf{y}) = \frac{g(\mathbf{u} \mid \mathbf{x})\, q(\mathbf{x})}{\left| \det(J_\phi(\mathbf{x}, \mathbf{u})) \right|} \; ,$$

where $|\det(J_\phi(\mathbf{x}, \mathbf{u}))|$ is the absolute value of the determinant of the matrix of Jacobi of ϕ at $(\mathbf{x}, \mathbf{u})$. Rearranging the last equation and using $q(\mathbf{x} \mid \mathbf{y})\, q(\mathbf{y}) = q(\mathbf{y} \mid \mathbf{x})\, q(\mathbf{x})$ gives:

$$\frac{q(\mathbf{x} \mid \mathbf{y})}{q(\mathbf{y} \mid \mathbf{x})} = \frac{q(\mathbf{x})}{q(\mathbf{y})} = \frac{g(\mathbf{v} \mid \mathbf{y})}{g(\mathbf{u} \mid \mathbf{x})} \left| \det(J_\phi(\mathbf{x}, \mathbf{u})) \right| .$$

Thus, the acceptance rate (6.20) can be written as [34, 35]:

$$\alpha(\mathbf{x}, \mathbf{y}) = \min \left\{ \frac{f(\mathbf{y}, n)\, q(m \mid n)\, g(\mathbf{v} \mid m, \mathbf{y})}{f(\mathbf{x}, m)\, q(n \mid m)\, g(\mathbf{u} \mid n, \mathbf{x})} \left| \det(J_\phi(\mathbf{x}, \mathbf{u})) \right|, \, 1 \right\} .$$

The following simple example is adapted from [34] and illustrates the workings of the algorithm.

■ EXAMPLE 6.13 (Model Choice in Regression)

Suppose the data $\mathbf{z} = (z_1, \ldots, z_N)$ is the outcome of independent random variables $\{Z_i\}$ of the form

$$Z_i = \sum_{k=0}^{m-1} \beta_k \, a_i^k + \varepsilon_i, \quad \varepsilon_i \sim \mathsf{N}(0, 1), \quad i = 1, \ldots, N \; , \tag{6.21}$$

where $a_1, \ldots, a_N$ are known variables, and the parameters $m \in \{1, 2, 3\}$ and $\boldsymbol{\beta}^{(m)} = (\beta_0, \ldots, \beta_{m-1})$ are unknown. Taking uniform (that is, constant) priors for m and $\boldsymbol{\beta}$ gives the posterior density

$$f(\boldsymbol{\beta}^{(m)}, m \mid \mathbf{z}) \propto \exp\left(-\frac{1}{2} \sum_{i=1}^{N} \left(z_i - \sum_{k=0}^{m-1} \beta_k \, a_i^k \right)^2 \right) . \tag{6.22}$$

The objective is to draw from the posterior to obtain information about the parameters $\boldsymbol{\beta}^{(m)}$ and also about which model (expressed by m) is the most appropriate.

Figure 6.16 shows data $\mathbf{z}$ simulated from (6.21) with $N = 101$, $a_i = (i-1)/20, i = 1, \ldots, 101$, and $(\beta_0, \beta_1, \beta_2) = (1, 0.3, 0.15)$. From the scatter plot it is not clear if a linear model ($m = 2$) or a quadratic model ($m = 3$) is more appropriate. To assess the appropriate model denote $\mathbf{x} = \boldsymbol{\beta}^{(m)}$ and run a reversible jump sampler by selecting a dimension $n \in \{1, 2, 3\}$ at random and then proposing a move $\mathbf{x} \to \mathbf{y}$ according to the n-dimensional $\mathsf{N}(\mathbf{0}, I)$ distribution with pdf $q(\mathbf{y} \mid \mathbf{x}, n) = q(\mathbf{y} \mid n)$.

In other words, we have the following iterative procedure:

Algorithm 6.14 (Reversible Jump Sampler for Regression)

1. *Given the current state* $\mathbf{X}_t$ *with* $\dim(\mathbf{X}_t) = m$, *generate* $n \sim \mathsf{DU}(1, 3)$, *that is, choose* $n \in \{1, 2, 3\}$ *with equal probability.*

2. *Generate* $\mathbf{Y} \sim \mathsf{N}(\mathbf{0}, I)$, *where* $\dim(\mathbf{Y}) = n$.

3. *Generate* $U' \sim \mathsf{U}(0, 1)$ *and deliver*

$$\mathbf{X}_{t+1} = \begin{cases} \mathbf{Y} & \text{if } U' \leqslant \min\left\{\frac{f(\mathbf{Y}, n \mid \mathbf{z})\, q(\mathbf{X}_t \mid m)}{f(\mathbf{X}_t, m \mid \mathbf{z})\, q(\mathbf{Y} \mid n)}, 1\right\}, \\ \mathbf{X}_t & \text{otherwise}, \end{cases}$$

where $q(\mathbf{y} \mid n)$ *is the pdf of* $\mathbf{Y}$.

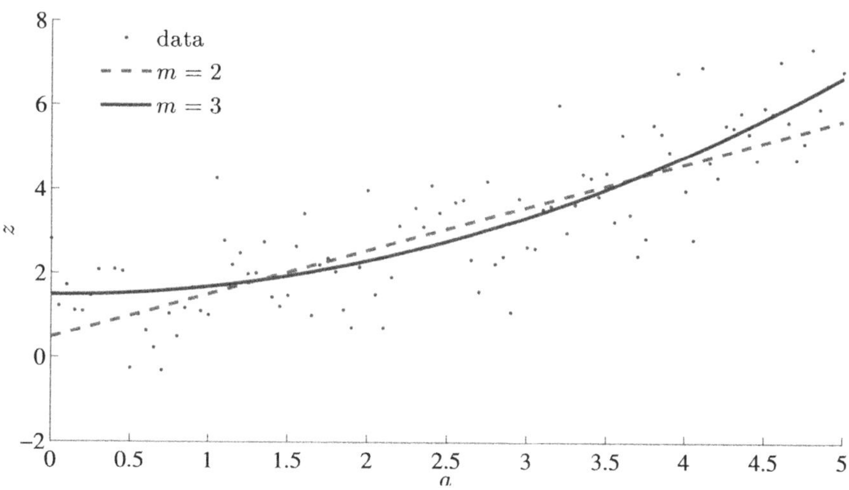

Figure 6.16 Regression data and fitted models for $m = 2$ and $m = 3$ (linear and quadratic).

The above reversible jump sampler is implemented in the code that follows. We ran the chain for $T = 10^5$ steps and produced 1008 two-dimensional vectors $\boldsymbol{\beta}^{(2)}$ and 98984 three-dimensional vectors $\boldsymbol{\beta}^{(3)}$. This gives posterior probabilities of approximately 0.0101 and 0.9898 for model $m = 2$ and $m = 3$, respectively. The posterior probability for the constant model is negligible. This indicates that the quadratic model has the best fit. The regression parameters $\boldsymbol{\beta}^{(m)}$ are estimated via the sample means of the $\{\mathbf{X}_t\}$, for $m = 2$ or 3, and are $(0.48, 1.03)$ and $(1.49, -0.01, 0.21)$. The corresponding regression curves are depicted in Figure 6.16.

```
%Reversible_jump.m
randn('state',4)
a=(0:100)'/20;
b=[1,0.3,0.15];
A=[a.^0,a,a.^2]; % coefficient matrix
z=A*b'+randn(101,1); % generate data
% posterior density
f=@(beta)exp(-0.5*norm(z-A(:,1:length(beta))*beta')^2);
% proposal
g=@(u)(exp(-0.5*norm(u)^2)/(2*pi)^(length(u)/2));
m=1;x=randn(1,m); T=10^5; % initialize
data=nan(T,1); beta0=0; beta1=beta0; beta2=beta1;
for t=1:T
    n=ceil(rand*3);
    y=randn(1,n); % make proposal
    if rand<min(f(y)/f(x)*g(x)/g(y),1)
        x=y; m=n;     % accept or reject proposal
    end
    data(t)=m;
    % determine parameters for each 'm'
    if m==1
        beta0=beta0+x;
    elseif m==2
        beta1=beta1+x;
    elseif m==3
        beta2=beta2+x;
    end
end
beta0=beta0/sum(data==1);
beta1=beta1/sum(data==2);
beta2=beta2/sum(data==3);
% plot models for each 'm'
plot(a,z,'.'), hold on
y2 = beta1(1)*A(:,1) + beta1(2)*A(:,2);
y3 = beta2(1)*A(:,1) + beta2(2)*A(:,2)  + beta2(3)*A(:,3);
plot(a,y2,'r');
plot(a,y3,'b');
figure(2), prob=[sum(data==1),sum(data==2),sum(data==3)]/T;
stem(prob), xlim([0.8,3.5]) % plot posterior model probabilities
```

More general constructions of proposals are given in [11]. There, auxiliary variables are used to make all models part of a maximal-dimensional space. The inclusion of the auxiliary variables aims to make the sampling fixed-dimensional, instead of transdimensional. Applications to time series models are given in [27].

One of the difficulties in using the reversible jump sampler is assessment of convergence. Various diagnostic tests for convergence are considered in [10, 28, 74]. Lunn et al. [48] use graphical models to help facilitate the implementation of the reversible jump sampler.

The reversible jump sampler has successfully been used in combination with *sequential Monte Carlo* methods [39, 40], where chains are run in parallel on spaces of different dimension. ☞ 482

6.4 IMPLEMENTATION ISSUES

Key issues when using MCMC are the following.

1. *Stationarity*: Determining the so-called **burn-in** period of the chain, that is, the number of points in the sequence $X_0, X_1, \ldots, X_T$ that have to be discarded before the chain is deemed to be in a stationary (steady-state) regime and the dependence on the initial point X_0 becomes irrelevant. ☞ 226

2. *Iid sampling*: Choosing the number of steps T that have to be used so that the empirical distribution of $X_0, X_1, \ldots, X_T$ (possibly after some burn-in period) provides a good approximation to the target pdf. A good approximation is one that can provide (approximately) iid random variables via *subsampling* or thinning of the chain $X_0, X_1, X_2, \ldots$, for some large enough t^*:

$$X_0, X_{t^*}, X_{2t^*}, \ldots .$$

3. *Convergence to average*: Selecting the number of steps T needed for the ergodic average (6.4) to converge to the true value $\mathbb{E}_f H(X)$.

Typically, stationarity (Point 1) is much easier to achieve and is a necessary condition for Points 2 and 3, which depend on how rapidly the chain converges to its limiting distribution (also known as the **mixing speed**). In other words, a stationary chain does not guarantee the ability to generate iid random variables from the target.

The problem of selecting the appropriate burn-in period and total number of steps T is still not resolved satisfactorily. Most approaches perform statistical hypothesis testing for the steady-state regime of the chain. Such tests are known as **convergence diagnostics**. All such tests are designed to detect nonstationary behavior of the chain. But they cannot be used to establish whether the chain is in fact stationary. In other words, if a chain passes a convergence diagnostic test, then this should be considered a necessary but not sufficient condition that the chain is actually in stationarity.

A simple and popular diagnostic is based on the covariance method described in Section 8.3.1. Fast decay of the estimated autocovariance function $\{\widehat{R}(k)\}$ is desirable and is an indication of a well-mixing chain, as illustrated in Examples 6.1 and 8.5. ☞ 309

Another commonly used convergence diagnostics is the *Gelman–Rubin* test [10, 30, 33].

Algorithm 6.15 (Gelman–Rubin Convergence Diagnostic)

1. *Generate M independent chains of length $T + D$ such that they are starting from different points in the support of the density. Discard the first D steps of the chains as burn-in and denote the remaining output of the chains by $X_{i1}, \ldots, X_{iT}$ for $i = 1, \ldots, M$.*

2. *Select a statistic, say $H_{it} = H(X_{it})$, and compute:*

$$\text{row means:} \quad \bar{H}_{i\bullet} = \frac{1}{T}\sum_{j=1}^{T} H_{ij} \,,$$

$$\text{overall mean:} \quad \bar{H} = \frac{1}{M}\sum_{i=1}^{M} \bar{H}_{i\bullet} \,,$$

$$\text{between chain variance:} \quad B = \frac{1}{M}\sum_{i=1}^{M} (\bar{H}_{i\bullet} - \bar{H})^2 \,,$$

$$\text{within chain variance:} \quad W = \frac{1}{M\,T}\sum_{j=1}^{T}\sum_{i=1}^{M} (H_{ij} - \bar{H}_{i\bullet})^2 \,,$$

$$\text{posterior variance:} \quad \widehat{\sigma^2} = \frac{T-1}{T} W + B \,.$$

3. *For a given T define the ratio $r_T = \frac{\widehat{\sigma^2}}{W}$. Monitor the sequence $r_T, r_{T+1}, \ldots,$ and run the MCMC algorithm until r_T is less than, say,* 1.1.

The idea behind the test is that asymptotically $\widehat{\sigma^2}$ and W must be equivalent, but for a finite T the quantity $\widehat{\sigma^2}$ overestimates the true variance $\text{Var}_f(H(X))$, while W underestimates it, see [29].

☞ 509
For another diagnostic based on to the Gelman–Rubin test, see Algorithm 14.7.

6.5 PERFECT SAMPLING

Suppose that we wish to generate a random variable X taking values in $\{1, \ldots, m\}$ according to a discrete density $f(i), i = 1, \ldots, m$. The MCMC method generates samples $\{X_t\}$ that are only *asymptotically* distributed according to f:

$$\lim_{t\to\infty} \mathbb{P}(X_t = i) = f(i) \,.$$

Surprisingly, it is possible to use Markov chain methodology to generate samples *exactly* distributed according to f. This methodology is called **perfect sampling**.

Propp and Wilson [63] proposed the following perfect sampling scheme. Let $\{X_t\}$ be a Markov chain with state space $\{1, \ldots, m\}$, transition matrix P, and stationary pdf f. We assume that $\{X_t\}$ has a limiting and stationary distribution, and, in particular, that it is aperiodic and irreducible. The idea of perfect sampling is to identify a finite stopping time in the past, $t = -T < 0$, so that the statistical properties of a chain started at $t = -T$ and evolved forward in time to $t = 0$ are the same as the properties of a chain started from infinitely long ago, $t = -\infty$, and

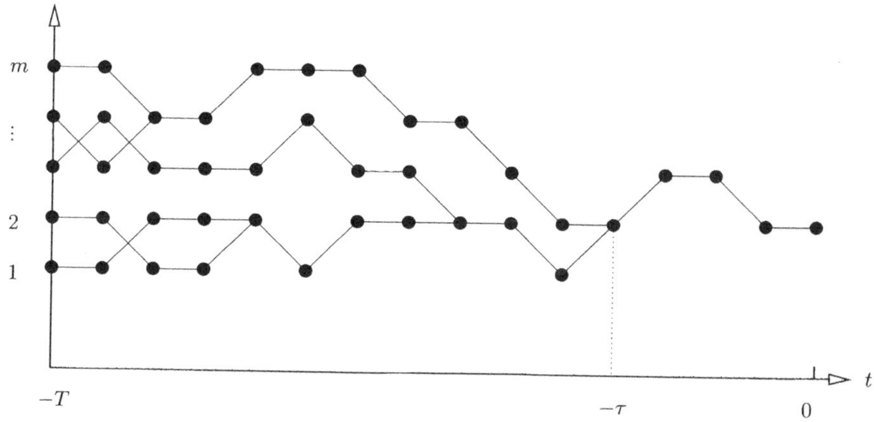

Figure 6.17 All Markov chains have coalesced at time $-\tau$.

evolved forward in time to $t = 0$. Since the latter chain is in stationarity, so will be the chain started from $t = -T$. Thus, we wish to generate a stopping time $-T$, and states $\{X_t,\ t = 0, -1, -2, \ldots, X_{-T}\}$ in such a way that X_0 has pdf f. This is done as follows.

First, given X_{-1}, we can use Algorithm 5.6 to draw X_0 from the m-point distribution corresponding to the X_{-1}-st row of P. This can, for example, be done using the inverse-transform method, which requires the generation of a random variable ☞ 45 $U_0 \sim \mathsf{U}(0,1)$. In other words, X_0 depends on X_{-1} and U_0. Similarly, X_{-1} can be generated from X_{-2} and $U_{-1} \sim \mathsf{U}(0,1)$. In general, for any negative time $-t$, the random variable X_0 depends on X_{-t} and the independent random variables $U_{-t+1}, \ldots, U_0 \sim_{\text{iid}} \mathsf{U}(0,1)$.

Second, consider starting m chains from every state $1, \ldots, m$ in the state space, using the *same* random numbers $\mathbf{U}_{-t+1} = (U_{-t+1}, \ldots, U_{-2}, U_{-1}, U_0)$ to evolve the chains. Thus, the chains are dependent, because they use common random variables. ☞ 349

Third, if two Markov chain paths coincide, or **coalesce**, at some time, then from that time onwards, both paths will be identical. The paths are said to be **coupled**. For example, in Figure 6.17 all the m paths have coalesced at time $-\tau$, and $-\tau$ can be interpreted as the time at which the initial states of the chains have been "forgotten".

Finally, Propp and Wilson show that with probability 1 there exists a finite negative time $-T$ such that all m paths will have coalesced before or at time 0. In other words, there exists, with probability 1, a stopping time $-T < 0$ such that by time 0 all m chains coupled by the common random variables $\mathbf{U}_{-T+1}$ have coalesced into a single path (are coupled).

Since a chain started from $t = -\infty$ will be in one of the m possible states at time $t = -T$, it follows that one of the m chains evolved forward in time from $-T$ is in *stationarity*. Moreover, since this stationary chain is coupled with all other chains, by time $t = 0$ all of the m coupled chains are distributed according to f at time $t = 0$.

To compute $-T$ we can work backwards from $t = 0$, by first generating U_{-1}, and checking if $-T = -1$. If this is not the case, generate U_{-2} and check if $-T = -2$, etc. This leads to the following algorithm [63], called **coupling from the past**.

Algorithm 6.16 (Coupling From the Past)

1. *Generate* $U_0 \sim \mathsf{U}(0,1)$. *Set* $\mathbf{U}_0 = U_0$. *Set* $t = -1$.

2. *Generate* m *Markov chains, starting at* t *from each of the states* $1, \ldots, m$, *using the same random vector* $\mathbf{U}_{t+1}$.

3. *Check if all chains have coalesced before or at time 0. If so, return the common value of the chains at time 0 and stop; otherwise, generate* $U_t \sim \mathsf{U}(0,1)$, *reset* $\mathbf{U}_t = (U_t, \mathbf{U}_{t+1})$ *and* $t = t - 1$, *and repeat from Step 2.*

Although coupling from the past returns an exact sample from the target f, the algorithm has practical limitations. First, the technique is difficult to use for most continuous simulation systems. Second, in many cases coupling from the past requires a large number of iterations before coalescence occurs, that is, the stopping time T may be too large. Finally, unless the sample space possesses the so-called stochastic monotonicity property [24], evolving the Markov chains from all possible states in the sample space is difficult due to large memory requirements.

Further Reading

Markov chain Monte Carlo is one of the principal tools of statistical computing and Bayesian analysis. A comprehensive discussion of MCMC techniques can be found in [65], and practical applications are discussed in [33]. For more details on the use of MCMC in Bayesian analysis we refer to [29]. For an in-depth and rigorous treatment of MCMC convergence results see [55]. An early paper on stationarity detection in Markov chains, and closely related to perfect sampling is [2].

REFERENCES

1. D. J. Aldous and J. Fill. Reversible Markov chains and random walks on graphs. Available at: `http://www.stat.berkeley.edu/~aldous/RWG/book.html`, 2002.

2. S. Asmussen, P. W. Glynn, and H. Thorisson. Stationary detection in the initial transient problem. *ACM Transactions on Modeling and Computer Simulation*, 2(2):130–157, 1992.

3. A. Barbu and S.-C. Zhu. Generalizing Swendsen–Wang to sampling arbitrary posterior probabilities. *IEEE Transactions on Pattern Analysis and Machine Intelligence*, 27(8):1239–1253, 2005.

4. A. Barbu and S.-C. Zhu. Generalizing Swendsen-Wang for image analysis. *Journal of Computational and Graphical Statistics*, 16(4):877–900, 2007.

5. S. Baumert, A. Ghate, S. Kiatsupaibul, Y. Shen, R. L. Smith, and Z. B. Zabinsky. Discrete hit-and-run for sampling points from arbitrary distributions over subsets of integer hyperrectangles. *Operations Research*, 57(3):727–739, 2009.

6. J. O. Berger and M.-H. Chen. Predicting retirement patterns: Prediction for a multinomial distribution with constrained parameter space. *Journal of the Royal Statistical Society, Series D*, 42(4):427–443, 1993.

7. C. G. E. Boender, R. J. Caron, J. F. McDonald, A. H. G. Rinnooy Kan, H. E. Romeijn, R. L. Smith, J. Telgen, and A. C. F. Vorst. Shake-and-bake algorithms for generating uniform points on the boundary of bounded polyhedra. *Operations Research*, 39(6):945–954, 1991.

8. Z. I. Botev. *The Generalized Splitting method for Combinatorial Counting and Static Rare-Event Probability Estimation.* PhD thesis, University of Queensland, available at: `http://espace.library.uq.edu.au/view/UQ:198531`, 2009.

9. Z. I. Botev and D. P. Kroese. The generalized cross entropy method, with applications to probability density estimation. *Methodology and Computing in Applied Probability*, DOI: 10.1007/s11009-009-9133-7, 2009.

10. S. P. Brooks and P. Giudici. Markov chain Monte Carlo convergence assessment via two-way analysis of variance. *Journal of Computational and Graphical Statistics*, 9(2):266–285, 2000.

11. S. P. Brooks, P. Giudici, and A. Philippe. Efficient construction of reversible jump Markov chain Monte Carlo proposal distributions. *Journal of the Royal Statistical Society, Series B*, 65(1):3–39, 2003.

12. G. Casella, K. L. Mengersen, C. P. Robert, and D. M. Titterington. Perfect samplers for mixtures of distributions. *Journal of the Royal Statistical Society, Series B*, 64(4):777–790, 2002.

13. M.-H. Chen and J. J. Deely. Bayesian analysis for a constrained linear multiple regression problem for predicting the new crop of apples. *Journal of Agricultural, Biological, and Environmental Statistics*, 1(4):467–489, 1996.

14. M.-H. Chen and B. Schmeiser. Performance of the Gibbs, hit-and-run, and Metropolis samplers. *Journal of Computational and Graphical Statistics*, 2(3):251–272, 1993.

15. M.-H. Chen and B. Schmeiser. Toward black-box sampling: A random-direction interior-point Markov chain approach. *Journal of Computational and Graphical Statistics*, 7(1):1–22, 1998.

16. M.-H. Chen and B. W. Schmeiser. General hit-and-run Monte Carlo sampling for evaluating multidimensional integrals. *Operations Research Letters*, 19(4):161–169, 1996.

17. M.-H. Chen, Q. M. Shao, and J. G. Ibrahim. *Monte Carlo Methods in Bayesian Computation.* Springer-Verlag, New York, 2000.

18. S. Chib. Marginal likelihood from the Gibbs output. *Journal of the American Statistical Association*, 90(432):1313–1321, 1995.

19. S. Chib and I. Jeliazkov. Marginal likelihood from the Metropolis-Hastings output. *Journal of the American Statistical Association*, 96(453):270–281, 2001.

20. R. V. Craiu and C. Lemieux. Acceleration of the multiple-try Metropolis algorithm using antithetic and stratified sampling. *Statistics and Computing*, 17(2):109–120, 2007.

21. N. A. C. Cressie. *Statistics for Spatial Data.* John Wiley & Sons, New York, second edition, 1993.

22. P. Damien and S. G. Walker. Sampling truncated normal, beta, and gamma densities. *Journal of Computational and Graphical Statistics*, 10(2):206–215, 1998.

23. P. Diaconis and S. Holmes. Three examples of Monte Carlo Markov chains: At the interface between statistical computing, computer science, and statistical mechanics. *Discrete Probability and Algorithms (Aldous et al. editors)*, pages 43–56, 1995.

24. P. Djuric, Y. Huang, and T. Ghirmai. Perfect sampling: a review and applications to signal processing. *IEEE Transactions on Signal Processing*, 50(2):345–356, 2002.

25. A. J. Dobson and A. G. Barnett. *Introduction to Generalized Linear Models*. Chapman & Hall, Boca Raton, FL, third edition, 2008.

26. R. G. Edwards and A. D. Sokal. Generalization of the Fortuin-Kasteleyn-Swendsen-Wang representation and Monte Carlo algorithm. *Physical Review D*, 38(6):2009–2012, 1988.

27. R. S. Ehlers and S. P. Brooks. Adaptive proposal construction for reversible jump MCMC. *Scandinavian Journal of Statistics*, 35(1):677–690, 2003.

28. Y. Fan, G. W. Peters, and S. A. Sisson. Automating and evaluating reversible jump MCMC proposal distributions. *Statistics and Computing*, 19(4):409–421, 2009.

29. A. Gelman, J. B. Carlin, H. S. Stern, and D. B. Rubin. *Bayesian Data Analysis*. Chapman & Hall, Boca Raton, FL, second edition, 2003.

30. A. Gelman and D. Rubin. Inference from iterative simulation using multiple sequences (with discussion). *Statistical Science*, 7(2):457–511, 1992.

31. S. Geman and D. Geman. Stochastic relaxation, Gibbs distribution and the Bayesian restoration of images. *IEEE Transactions on Pattern Analysis and Machine Intelligence*, 6(6):721–741, 1984.

32. W. R. Gilks, N. G. Best, and K. K. C. Tan. Adaptive rejection Metropolis sampling within Gibbs sampling. *Journal of the Royal Statistical Society, Series C*, 44(4):455–472, 1995.

33. W. R. Gilks, S. Richardson, and D. J. Spiegelhalter. *Markov Chain Monte Carlo in Practice*. Chapman & Hall, New York, 1996.

34. P. J. Green. Reversible jump Markov chain Monte Carlo computation and Bayesian model determination. *Biometrika*, 82(4):711–732, 1995.

35. P. J. Green. *Highly Structured Stochastic Systems*, volume 35, chapter : Trans-Dimensional Markov Chain Monte Carlo, pages 179–198. Oxford University Press, London, 2003.

36. J. Hammersley and M. Clifford. Markov fields on finite graphs and lattices. Available at: `http://www.statslab.cam.ac.uk/~grg/books/hammfest/hamm-cliff.pdf`. Unpublished manuscript, 1970.

37. W. K. Hastings. Monte Carlo sampling methods using Markov chains and their applications. *Biometrika*, 57(1):92–109, 1970.

38. D. M. Higdon. Auxiliary variable methods for Markov chain Monte Carlo with applications. *Journal of the American Statistical Association*, 93(442):585–595, 1998.

39. A. Jasra, A. Doucet, D. A. Stephens, and C. C. Holmes. Interacting sequential Monte Carlo samplers for trans-dimensional simulation. *Computational Statistics and Data Analysis*, 52(4):1765–1791, 2008.

40. A. Jasra, D. A. Stephens, and C. C. Holmes. Population-based reversible jump Markov chain Monte Carlo. *Biometrika*, 94(4):787–807, 2007.

41. J. M. Keith, D. P. Kroese, and D. Bryant. A generalized Markov chain sampler. *Methodology and Computing in Applied Probability*, 6(1):29–53, 2004.

42. J. S. Liu. Metropolized Gibbs sampler: an improvement. Preprint, University of Stanford, available at: `www.fas.harvard.edu/~junliu/TechRept/96folder/mgibbs.ps`, 1995.

43. J. S. Liu. *Monte Carlo Strategies in Scientific Computing*. Springer-Verlag, New York, 2001.

44. J. S. Liu, F. Liang, and W. H. Wong. The multiple-try method and local optimization in Metropolis sampling. *Journal of the American Statistical Association*, 95(449):121–134, 2000.

45. L. Lovász. Hit-and-run mixes fast. *Mathematical Programming*, 86(3):443–461, 1999.

46. L. Lovász and S. Vempala. Hit-and-run is fast and fun. Technical report, Microsoft Research, MSR-TR-2003-05, 2003.

47. L. Lovász and S. Vempala. Hit-and-run from a corner. *SIAM Journal of Computing*, 35(4):985–1005, 2006.

48. D. J. Lunn, N. Best, and J. C. Whittaker. Generic reversible jump MCMC using graphical models. *Statistics and Computing*, 19(4):395–408, 2009.

49. R. E. McCulloch, N. G. Polson, and P. E. Rossi. A Bayesian analysis of the multinomial probit model with fully identified parameters. *Journal of Econometrics*, 99(1):173–193, 2000.

50. R. E. McCulloch and P. E. Rossi. An exact likelihood analysis of the multinomial probit model. *Journal of Econometrics*, 64(1&2):207–240, 1994.

51. G. J. McLachlan and T. Krishnan. *The EM Algorithm and Extensions*. John Wiley & Sons, Hoboken, New Jersey, second edition, 2008.

52. K. Mengersen and R. Tweedie. Rates of convergence of the Hastings and Metropolis algorithms. *Annals of Statistics*, 24(1):101–121, 1996.

53. H. O. Mete, Y. Shen, Z. B. Zabinsky, S. Kiatsupaibul, and R. L. Smith. Pattern discrete and mixed hit-and-run for global optimization. *Journal of Global Optimization*, DOI: 10.1007/s10898-010-9534-8, 2010.

54. N. Metropolis, A. W. Rosenbluth, M. N. Rosenbluth, A. H. Teller, and E. Teller. Equations of state calculations by fast computing machines. *Journal of Chemical Physics*, 21(6):1087–1092, 1953.

55. S. Meyn and R. L. Tweedie. *Markov Chains and Stochastic Stability*. Cambridge University Press, London, second edition, 2009.

56. A. Mira, J. Møller, and G. O. Roberts. Perfect slice samplers. *Journal of the Royal Statistical Society, Series B*, 63(3):593–606, 2001.

57. A. Mira and L. Tierney. Efficiency and convergence properties of slice samplers. *Scandinavian Journal of Statistics*, 29(1):1–12, 2002.

58. P. Neal and G. O. Roberts. Optimal scaling for partially updating MCMC algorithms. *Annals of Applied Probability*, 16(2):475–515, 2006.

59. R. Neal. Slice sampling (with discussion). *Annals of Statistics*, 31(3):705–767, 2003.

60. D. J. Nott and P. J. Green. Bayesian variable selection and the Swendsen–Wang algorithm. *Journal of Computational and Graphical Statistics*, 13(1):141–157, 2004.

61. R. E. Pfiefer. Surface area inequalities for ellipsoids using Minkowski sums. *Geometriae Dedicata*, 28(2):171–179, 1988.

62. B. T. Polyak and E. N. Gryazina. Randomized methods based on new Monte Carlo schemes for control and optimization. *Annals of Operations Research*, DOI: 10.1007/s10479-009-0585-5, 2009.

63. J. G. Propp and D. B. Wilson. Exact sampling with coupled Markov chains and applications to statistical mechanics. *Random Structures and Algorithms*, 9(1&2):223–252, 1996.

64. C. P. Robert. Simulation of truncated normal variables. *Statistics and Computing*, 5(2):121–125, 1995.

65. C. P. Robert and G. Casella. *Monte Carlo Statistical Methods.* Springer-Verlag, New York, second edition, 2004.

66. G. O. Roberts and J. S. Rosenthal. Optimal scaling of discrete approximations to Langevin diffusions. *Journal of the Royal Statistical Society, Series B*, 60(1):255–268, 1998.

67. G. O. Roberts and J. S. Rosenthal. Convergence of slice sampler Markov chains. *Journal of the Royal Statistical Society, Series B*, 61(3):643–660, 1999.

68. G. O. Roberts and J. S. Rosenthal. The polar slice sampler. *Stochastic Models*, 18(2):257–280, 2002.

69. G. O. Roberts and S. Sahu. Updating schemes, correlation structure, blocking and parameterization for the Gibbs sampler. *Journal of the Royal Statistical Society, Series B*, 59(2):291–317, 1997.

70. G. O. Roberts and R. Tweedie. Exponential convergence for Langevin diffusions and their discrete approximations. *Bernoulli*, 2(4):341–363, 1996.

71. H. E. Romeijn and R. L. Smith. Simulated annealing for constrained global optimization. *Journal of Global Optimization*, 5(2):101–126, 1994.

72. S. M. Ross. *Simulation.* Academic Press, New York, third edition, 2002.

73. Y. Shen. *Annealing Adaptive Search with Hit-and-Run Sampling Methods for Stochastic Global Optimization Algorithms.* PhD thesis, University of Washington, 2005.

74. S. A. Sisson and Y. Fan. A distance-based diagnostic for trans-dimensional Markov chains. *Statistics and Computing*, 17(4):357–367, 2007.

75. R. L. Smith. Efficient Monte Carlo procedures for generating points uniformly distributed over bounded regions. *Operations Research*, 32(6):1296–1308, 1984.

76. R. H. Swendsen and J.-S. Wang. Nonuniversal critical dynamics in Monte Carlo simulations. *Physical Review Letters*, 58(2):86–88, 1987.

77. W. Wang, A. Ghate, and Z. B. Zabinsky. Adaptive parameterized improved hit-and-run for global optimization. *Optimization Methods and Software*, 24(4&5):569–594, 2009.

78. L. Yun. *Mixing Time of the Swendsen–Wang Dynamics on the Complete Graph and Trees.* PhD thesis, University of California, Berkeley, 2009.

79. Z. B. Zabinsky, R. L. Smith, J. F. McDonald, H. E. Romeijn, and D. E. Kaufman. Improving hit-and-run for global optimization. *Journal of Global Optimization*, 3(2):171–192, 1993.

CHAPTER 7

DISCRETE EVENT SIMULATION

Monte Carlo simulation generally involves simple algorithms for computing numerical quantities (expectations, probabilities) that are formulated in terms of basic random objects (random vectors, stochastic processes). In this respect Monte Carlo simulation differs from large-scale simulation modeling, in which the objective is to understand the workings of a real-life system by imitating it as well as possible on a computer. Applications are found in telecommunications, production scheduling, traffic control, reliability and maintenance, military planning, and inventory control, to mention but a few.

Despite their different levels of complexity and focus (computational versus modeling), Monte Carlo and simulation methods are much the same, as they both involve computer implementations of random experiments. As a result, most of the concepts and statistical techniques in this book can be applied in a general simulation setting. On the other hand, large-scale simulation requires a higher level of modeling and programming structure than Monte Carlo. The purpose of this chapter is to provide a brief introduction to the most common aspects of computer simulation and modeling, in particular with regard to discrete event systems.

7.1 SIMULATION MODELS

Simulation models aim to imitate the behavior of real-life systems. By a **system** we mean a collection of interacting entities or **objects** forming a complex whole. As a concrete example, consider a post office. Here, arriving customers queue up at one of several counters where they wait to be served. After having been served, the customers either leave the post office or reenter other queues to complete additional transactions. The *objects* of this system are the customers, the servers, and the queues. These objects are characterized by certain numerical *attributes*, such as the numbers of available waiting places in the queues, the customer types (high/low priority), the service times and arrival rates of customers, and so on. The attributes and the interaction between the objects determine the behavior of the system over time.

A first step in trying to better understand the workings of a real-life system is to make a *mathematical model* for the system, which summarizes the essential parts of the system in mathematical language, involving variables, parameters, formulas, probability distributions, relations, diagrams, etc. In order to be useful, a model must necessarily incorporate elements of realism and simplicity — two ideals that are usually in conflict. On the one hand, the model should serve as a reasonably close approximation to the real system and incorporate most of the important aspects of the real system. On the other hand, the model must not be overly complex so as to preclude its understanding and manipulation. When the model is relatively simple, it may be possible to study it *analytically*; that is, closed-form expressions may be available that describe the behavior of certain aspects of the system. For more complex systems, analytical approaches are usually much more difficult or impossible to realize. Instead the system is often analyzed *numerically* through *computer simulation*, with the assumption that the simulated system will bear enough resemblance to the real system to draw valid conclusions about the latter. The situation is schematized in Figure 7.1.

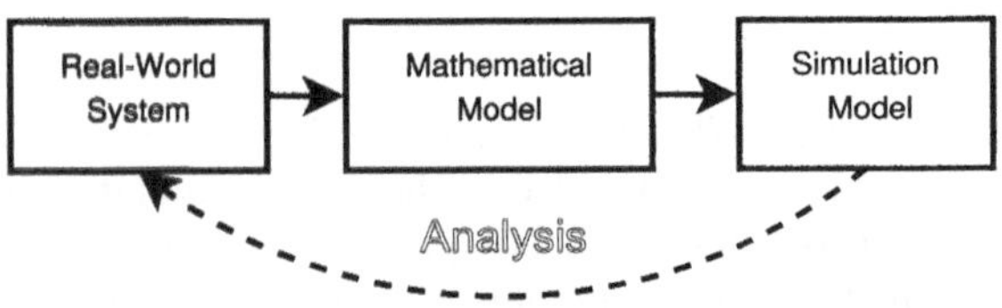

Figure 7.1 Simulation modeling and analysis.

Going back to our post office example, a standard mathematical model for such service systems is that of a **queueing system** — an example is depicted in Figure 7.2. In this model, customers arrive at random times, wait in queues, are processed by servers within random periods of time, and randomly move between queues, just as in the real-life system. However, unlike real life, the arrival and service processes are described by precise probability laws. Likewise, the routing protocol (which queue to go to next) and service discipline (the way in which clients are served, for example, first-in-first-out) follow specific rules.

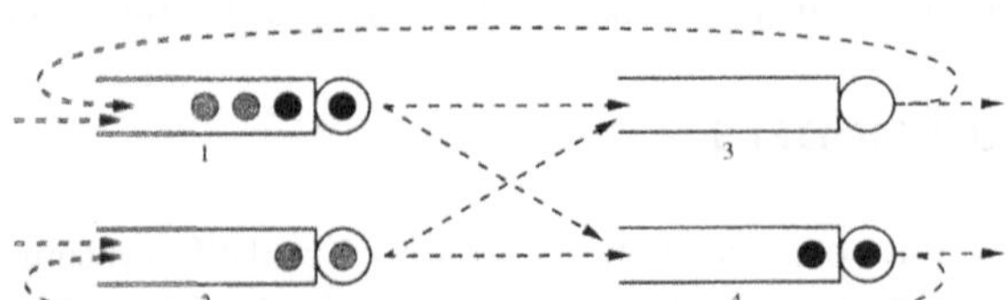

Figure 7.2 A queueing system with different types of customers.

It is possible to analyze various aspects of a queueing system analytically, under certain simplifying conditions; indeed, much research has gone into this area — see, for example, [1, 9, 18]. However, many performance measures of queueing systems, even of very simple ones, can only be derived via simulation.

A queueing system is a typical example of a **discrete event system**. These are systems where the system's "state" is changed at discrete points in time through the occurrence of certain *events*. For example, in the queueing system the state could be the numbers of customers in various queues, which changes whenever an arrival or departure event takes place. In practice such systems are *stochastic* and *dynamic* in nature, as they typically involve random variables and evolve over time. Discrete event systems can be readily simulated on a computer by specifying precisely *when* events occur and *how* they affect the system state, the details of which form the basis for the remainder of this chapter.

7.2 DISCRETE EVENT SYSTEMS

Discrete event systems model the behavior of a wide variety of systems in engineering and operations research. Applications can be found, for example, in production scheduling, reliability, traffic and transportation, inventory control, manufacturing, defence, finance, telecommunications, and computer systems. Two main ingredients in a discrete event simulation study are:

- **System state:** The collection of variables/attributes needed to describe the system at a particular time, relative to the object of study. In general, the collection of states, $\{\mathbf{X}_t, t \geqslant 0\}$ say, forms a stochastic process taking values in some state space.

- **Event:** An instantaneous occurrence that may change the state of the system. Each event is characterized by:

  - **Event time**: The time at which the event occurs.
  - **Event type**: An identifier that determines how the event affects the system state at the event time and thereafter.

Discrete event systems are observed only at the event times. Between event times the system is allowed to change only in a predictable (that is, deterministic) way. Because of their dynamic nature, discrete event systems require a time-keeping mechanism called a **simulation clock** to advance the simulation time from one event to the next. To keep track of events, the simulation maintains a list of all pending events. This list is called the **event list**, and its task is to maintain all pending events in *chronological* order; that is, events are ordered by their time of occurrence. In particular, the most imminent event is always located at the head of the event list.

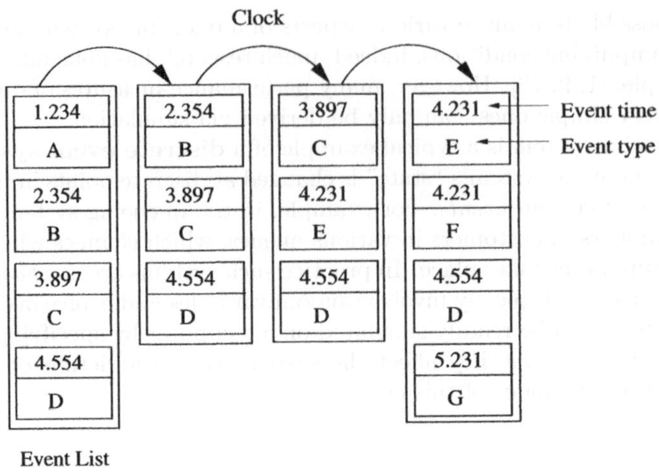

Figure 7.3 The advancement of the simulation clock and event list.

The situation is illustrated in Figure 7.3. The simulation starts by loading the initial events into the event list (chronologically ordered), in this case four events. Next, the most imminent event is unloaded from the event list for execution, and the simulation clock is advanced to its occurrence time, 1.234. After this event is processed and removed, the clock is advanced to the next event, which occurs at time 2.354. In the course of executing a "current" event, based on its type, the state of the system is updated, and future events are possibly generated and loaded into (or deleted from) the event list. In the above example, the third event — of type C, occurring at time 3.897 — schedules a new event of type E at time 4.231. Subsequently, the event of type E schedules an event of type F to occur at the same time 4.231, and an event of type G at the later time of 5.231.

There are two fundamental approaches to implementing a discrete event simulation program:

- **Event-oriented approach**: Here separate subroutines are specified for every type of event. The task of each event subroutine is to update the system state and to schedule new events in the event list. The principal role of the main program is to progress through the event list and call the corresponding subroutines at each of the event times.

- **Process-oriented approach**: Events can often be grouped into **processes** — sequences of related events. For example, in a reliability system, the events and actions associated with the failure and repair of a specific machine form a relevant process. Intuitively, a process can be viewed as a repeating program that can be interrupted and (re)activated at certain times, and in turn can manipulate other processes and data structures. In the process-oriented setting there is again a simulation clock and a list that keeps track of which event happens when, but this list contains *processes* rather than individual events. The process at the top of the list is the one that is currently active.

Our focus will be on event-oriented simulation, which is generally easier to implement in a general-purpose programming language. In addition, event-oriented programs tend to be faster than process-oriented ones. However, process-oriented simulation programs tend to be simpler conceptually, and lend themselves to easy implementation in object-oriented programming languages, allowing one to build large-scale simulation projects. There are many freely available object-oriented simulation environments nowadays, such as SSJ, J-Sim, and C++Sim, all of which have been inspired by the pioneering simulation language SIMULA [5].

7.3 EVENT-ORIENTED APPROACH

Figure 7.4 gives a flowchart of a general event-oriented simulation program. The program starts by initializing the state variables, statistical counters, and system parameters. This includes setting up the event list. The initialization could be carried out by a separate subroutine. The simulation clock mechanism is implemented as a loop consisting of three steps: (1) determine the current event time and type from the head of the event list, (2) invoke a separate event routine corresponding to this (current) event, (3) gather statistical data and advance the event list to the next event, which now becomes the current event. These steps are repeated until some stopping criterion is met — for example, when the event time exceeds some set time T. Finally, the statistical information gathered during the simulation run is reported and the simulation stops.

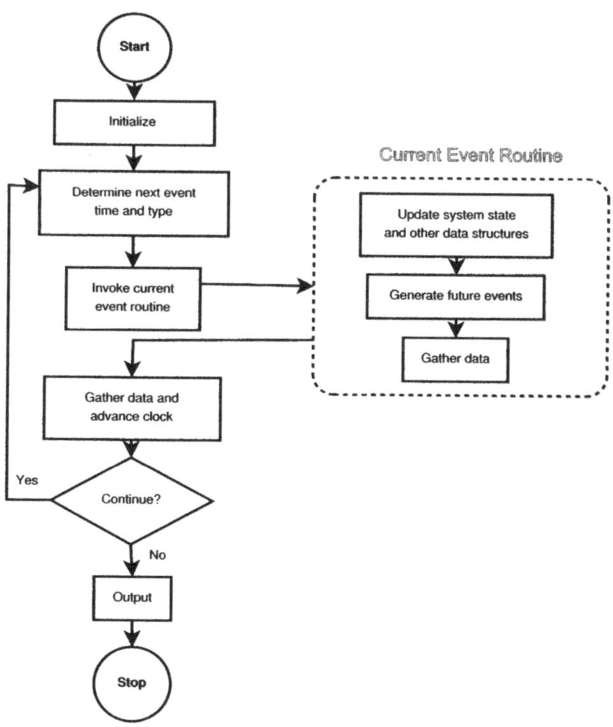

Figure 7.4 Event-oriented program flow chart.

The main modeling and programming effort lies in the specification of the event routines. For each event one needs to specify how the system state and other data structures (queues, counters, etc.) are affected by its occurrence. In addition each event can trigger other events, which need to be scheduled in the event list. Finally, statistical data may be gathered during (or usually at the end of) an event routine.

To elucidate the interactions between events, it is sometimes useful to construct an **event graph** [17]. Here the vertices in the graph correspond to events and an arc from an event A to an event B indicates that the occurrence of event A "triggers" event B; that is, schedules an occurrence of B in the event list. Figure 7.5 shows a possible event graph for the queueing system in Figure 7.2. There are six events, corresponding to arrivals at queue 1 and 2, and departures from queue 1, 2, 3, and 4. A departure at queue 4 (D4), for example, can trigger another departure at the same queue, if upon departure there are still people in the waiting line at queue 4 — one of these will move to the server and be served. In addition, D4 can trigger a departure at queue 2, if the customer that currently departs returns to queue 2 and at the same time there are no customers in the waiting line at queue 2. Dashed arcs indicate that the events (A1 and A2) have to be scheduled at the initialization phase.

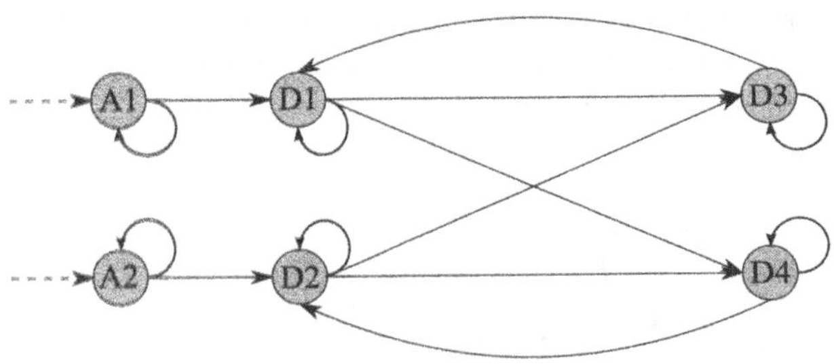

Figure 7.5 An event graph.

The event list can be implemented in different ways. It can be a simple linear array, or a more complicated data structure [11, Chapter 2]. Often the event list is constructed as a **doubly linked list**, where each event record contains pointers to the previous and the next record in the list, as illustrated in Figure 7.6. The insertion and deletion of event records is easy in such lists. However, searching a linked lists of size n in a sequential manner takes $\mathcal{O}(n)$ operations, which is inefficient for large n. In such cases it may be better to implement the event list as a binary tree or indexed list.

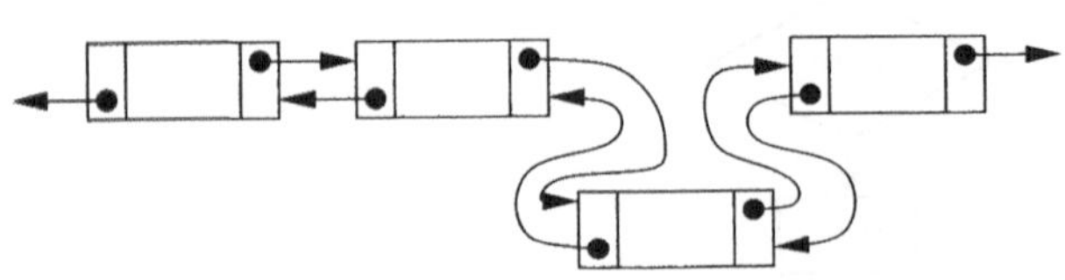

Figure 7.6 Insertion of an event record in a doubly linked event list.

The general steps in setting up an event-oriented simulation study are as follows.

1. Construct a mathematical model for the system.
2. Identify the system state, depending on which aspect of the system is being investigated.
3. Set up the appropriate data structures: queues, lists, variables, statistical counters, etc.
4. Identify the possible events and their interactions, for example using an event graph.
5. Set up an appropriate event list and main simulation routine.
6. Implement subroutines for each of the possible events. Draw flow charts for each of the subroutines if necessary.
7. Test and debug. This is a crucial and often time-consuming step. To check for programming errors, verify that the system behaves properly by stepping through the program and/or by plotting realizations of the state process. Compare the simulation results with known special cases.
8. Perform a statistical analysis of the data.
9. Report the results.

■ **EXAMPLE 7.1** **($GI/G/1$ Queueing System)**

In the classical $GI/G/1$ queueing system customers arrive to a single queue according to a renewal process with interarrival times distributed according to some cdf F. Customers are served by a single server in the order in which they arrive. If a customer arrives when the server is busy serving another customer, the arriving customer joins the end of the waiting line. The service times of the customers are independent of each other and of the interarrival times, and have a common distribution function G. The special case where the interarrival and service times are exponentially distributed corresponds to the $M/M/1$ queue discussed in Examples A.8 and A.9 on Pages 630–631. A quantity of interest is the steady-state number of customers in the system. We wish to assess this via simulation. ☞ 630

Let the system *state* at time t correspond to the number of customers in the system at that time, denoted by X_t. Note that the stochastic process $\{X_t\}$ is a regenerative process; the regeneration times can be chosen to be the arrival times of customers that find the system empty. We assume $X_0 = 0$; that is, the system is initially empty.

There are two types of *events*: arrival events and departure events. Suppose t is an event time. If the event is an arrival event, the state (the number of customers in the system) is increased by 1; in the case of a departure event the state is decreased by 1. An arrival event also triggers the next arrival event at time $t + A$, where $A \sim F$. In addition, if the system is empty at time t (that is, at the arrival event), then the arrival triggers a departure event at $t + D$, where $D \sim G$. Each departure event at time t triggers another departure event at time $t + D$, where $D \sim G$, provided that there is at least one customer next in line to be served.

Since our programming environment is MATLAB, we simply take the event list to be a 3×2 matrix, where each row corresponds to an event-time and event-type ($1 =$ arrival, $2 =$ departure). Note that at any time no more than three events are scheduled: the current event, and at most two future events (either one arrival or an arrival and a departure). After each time steps of the clock, the current event record is emptied (set to (∞, ∞)) and the event list is resorted, so that its first element corresponds to the next current event.

The following MATLAB code implements the $GI/G/1$ simulation model, using $\mathsf{Exp}(\lambda)$ interarrival and $\mathsf{Exp}(\mu)$ service times — these can be readily replaced with arbitrary distributions, if required. During the simulation, the integral $I_t = \int_0^t X_s \mathrm{d}s$ (represented by the variable `tot`) is updated after each step of the clock. The simulation is stopped at the first event time, say T_1, greater than or equal to $T = 10000$. The output of the simulation is I_{T_1}/T_1, which for large T approximates the steady-state number of customers in the system (see also Section 8.3.3). ☞ 313

In the $M/M/1$ case with $\lambda < \mu$ we may verify the correctness of the program by comparing its output with the theoretical value

$$\lim_{T \to \infty} \frac{1}{T} \int_0^T X_t \, \mathrm{d}t = \frac{\varrho}{1 - \varrho} ,$$

where $\varrho = \lambda/\mu < 1$ is the traffic intensity; see Example A.11. ☞ 639

```
%MM1/main.m
mu = 3; lambda = 2;
rho = lambda/mu;
T = 10000;
x = 0; xold = 0; % initialize current state and previous state
ev_list = inf*ones(3,2); % initialize event list
t = 0; told = 0;  % initialize current and previous event time
tot = 0;  %
ev_list(1,:) = [- log(rand)/lambda, 1]; %schedule the first arrival
N_ev = 1;                       % number of scheduled events
while t < T
    t = ev_list(1,1);
    ev_type = ev_list(1,2);
    switch ev_type
        case 1
            arrival
        case 2
            departure
    end
    ev_list(1,:) = [inf,inf];
    N_ev = N_ev - 1;
    ev_list = sortrows(ev_list,1); % sort event list
    tot =tot + xold*(t - told);
    xold = x; told =t;
end
res = tot/t
exact = rho/(1-rho)
```

```
%MM1/arrival.m
N_ev = N_ev + 1;
ev_list(N_ev,:) = [t - log(rand)/lambda, 1]; %schedule new arrival
if x == 0  % if queue is empty
    N_ev = N_ev + 1;
    ev_list(N_ev,:) = [t - log(rand)/mu, 2]; % schedule departure
end
x = x+1;
```

```
%MM1/departure.m
x = x-1; % go out of queue
if x ~= 0
    N_ev = N_ev + 1;
    ev_list(N_ev,:) = [t - log(rand)/mu, 2]; % schedule departure
end
```

7.4 MORE EXAMPLES OF DISCRETE EVENT SIMULATION

7.4.1 Inventory System

The **(s, S) policy** is a classic control policy for inventory systems of the following form. Demand for a certain commodity occurs over time according to a renewal process with an interarrival cdf F. The size of the demand is distributed according to a cdf G and is independent of the arrival process. When a demand occurs it is either filled or back-ordered (to be satisfied by delayed deliveries). The **net inventory** (on-hand inventory minus back orders) at time t is denoted by X_t, and the **inventory position** (net inventory plus on-order inventory) by Y_t. The (s, S) policy prescribes that at any time t when a demand of size D is received that would reduce the inventory position to less than s (that is, $Y_{t-} - D < s$, where Y_{t-} denotes the inventory position just before t), an order of size $S - (Y_{t-} - D)$ is placed, which brings the inventory position immediately back to S. Otherwise, no action is taken. The order arrives R time units after it is placed, where the **lead time** R is distributed according to some cdf H that is assumed to be independent of the demand process. Both inventory processes are illustrated in Figure 7.7 for $s = 5$, $S = 10$, and a lead time of 1.

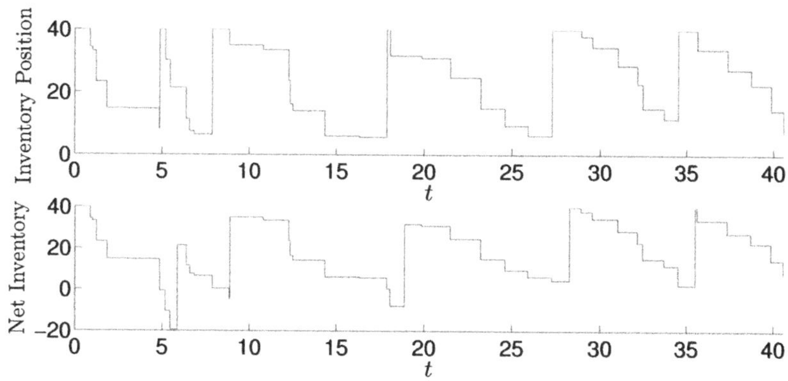

Figure 7.7 Sample paths for the two inventory processes.

Under the back-order policy and the above assumptions, both the inventory position process $\{Y_t\}$ and the net inventory process $\{X_t\}$ are regenerative. In particular, each process regenerates when it is raised to S. For example, each time an order is placed, the inventory position process regenerates.

A natural state for the system at time t is here the vector (X_t, Y_t). There are two types of event, which are depicted in the event graph in Figure 7.8. A demand arrival event at some time t and of size D triggers another demand arrival event, at time $t + A$ of size D_1, where $A \sim F$ and $D_1 \sim G$, independently. If $Y_{t-} - D \geqslant s$, the state (X_t, Y_t) is changed to $(X_t - D, Y_t - D)$. However, if $Y_{t-} - D < s$, then an order of size $S - (Y_{t-} - D)$ is scheduled for time $t + T$, where $T \sim H$; and the state (X_t, Y_t) is changed to (X_t, S). An order event of size O at some time t does not trigger any additional events, but changes the state (X_t, Y_t) to $(X_t + O, Y_t)$. At initialization one demand arrival event needs to be scheduled.

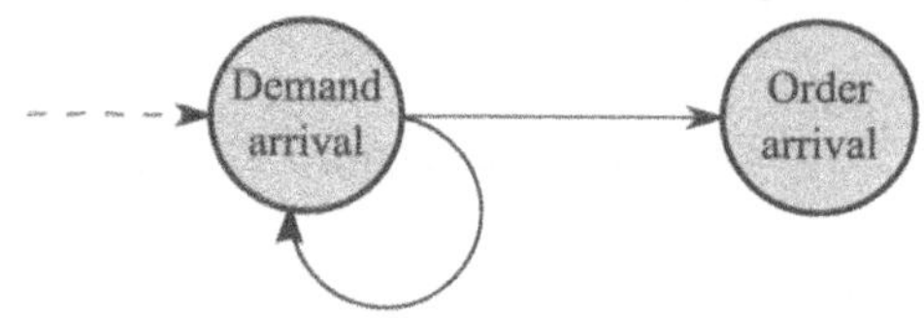

Figure 7.8 The event graph for the inventory system.

The MATLAB programs that follow implement an event-oriented simulation algorithm for the above system. The event list is here a $k \times 3$ matrix, where k is initially 3 but can increase during the simulation. Note that there is always one demand event scheduled, but there could be several order events scheduled simultaneously. The first two columns correspond to the event time and type (1 = demand, 2 = order). The third column contains the demand size or order size. During the course of the simulation the program keeps track of the total amount of time the net inventory position is negative, and the number of orders placed. At the end the long-run average cost per unit of time is given, computed as

$$C = c_1 S + c_2 f_{\text{neg}} + c_3 f_{\text{ord}},$$

where S is the upper limit in the (s, S) policy, f_{neg} is the fraction of time where the net inventory process is negative, and f_{ord} is the frequency of orders (per unit of time). A possible aim of the simulation is to select the cost-optimal values of s and S, for given values of c_1, c_2, and c_3 (see also Example 12.6). ☞ 458

In the following example code, the demand size, interarrival, and lead time distributions are taken to be $\mathsf{U}(0, 10)$, $\mathsf{Exp}(1/5)$, and $\mathsf{U}(5, 10)$, respectively, but can be easily modified to any specific distribution by changing the corresponding MATLAB functions. The cost parameters are taken as $c_1 = 5$, $c_2 = 500$, and $c_3 = 100$.

The subroutine `plottrace.m` can be used to plot realizations of the inventory processes (uncomment the corresponding lines in the code).

```
%Inventory/main.m
s = 10; S = 40;
T = 200000;
x = S; y = S; %x net inventory. y inventory position
xold = x;
t = 0; told=0;
% xx = x; yy=y; tt=0;  %uncomment for plotting
ev_list = inf*ones(3,3); %entries: (time, type, order-size)
totneg = 0; num_ord = 0;
ev_list(1,:) = [interarrival,1,demandsize]; %schedule first demand
N_ev = 1;                    % number of events scheduled
while t < T
    t = ev_list(1,1);
    ev_type = ev_list(1,2);
    ev_par = ev_list(1,3);
    switch ev_type
        case 1
            demand;
        case 2
            order;
    end
    % tt=[tt,t]; xx=[xx,x]; yy=[yy,y]; %uncomment for plotting
    N_ev = N_ev - 1;
    ev_list(1,:) = [inf,inf,inf];
    ev_list = sortrows(ev_list,1); % sort event list

    totneg = totneg + (xold < 0)*(t - told);
    xold = x; told = t;
end
frac_neg = totneg/t
freq_ord = num_ord/t
c1 = 5; c2 = 500; c3 = 100;
cost = c1*S + c2*frac_neg + c3*freq_ord  %cost per unit of time
% plottrace %uncomment for plotting
```

```
%Inventory/demand.m
demsize = ev_par;
x = x - demsize;
y = y - demsize;
if (y <s)   % if net inventory is under s limit
    N_ev = N_ev + 1;
    ev_list(N_ev,:) = [t + leadtime,2,S-y]; % schedule order
    % size of order is S - x
    y = S;
end
N_ev = N_ev + 1;
ev_list(N_ev,:) = [t + interarrival,1,demandsize]; %schedule demand
```

```
%Inventory/order.m
ord = ev_par;
num_ord = num_ord + 1;
x = x + ord;
```

```
%Inventory/interarrival.m
function out = interarrival;
out = -log(rand)*5;
```

```
%Inventory/demandsize.m
function out = demandsize
out = 10*rand;
```

```
%Inventory/leadtime.m
function out=leadtime;
out = 5 + rand*5;
```

```
%Inventory/plottrace.m
figure(1),subplot(2,1,1),
for i =1:length(yy)-1,
  line([tt(i),tt(i+1)],[yy(i),yy(i)]);
  line([tt(i+1),tt(i+1)],[yy(i),yy(i+1)]);
end

aa=axis;
axis([0,tt(end),aa(3),aa(4)]),xlabel('t'),ylabel('Inventory Pos.');

subplot(2,1,2),
for i =1:length(xx)-1,
  line([tt(i),tt(i+1)],[xx(i),xx(i)]);
  line([tt(i+1),tt(i+1)],[xx(i),xx(i+1)]);
end

aa=axis;
axis([0,tt(end),aa(3),aa(4)]),xlabel('t'),ylabel('Net Inventory')
```

7.4.2 Tandem Queue

The purpose of this example is to show how the basic $GI/G/1$ model discussed in Example 7.1 can be readily modified to handle more complicated queueing systems. Consider the model depicted in Figure 7.9, where customers who leave the first $GI/G/1$ queue enter a second queue, where they are served by a different server, or take a place in a waiting line if that server is busy. The service times in the second queue have a fixed cdf H. All service and interarrival times are independent.

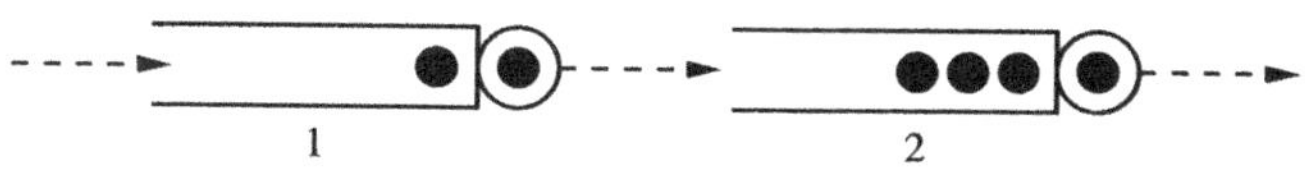

Figure 7.9 A tandem queueing system.

The state of the system at time t could be the numbers of customers, X_t and Y_t, in the first and second queue, respectively, where we regard a customer who is being served as part of the queue. Figure 7.10 depicts a typical realization of the queue length processes $\{X_t, t \geqslant 0\}$ and $\{Y_t, t \geqslant 0\}$, obtained via discrete event simulation, where the interarrival times and the service times are all $\mathsf{U}(0,1)$ distributed.

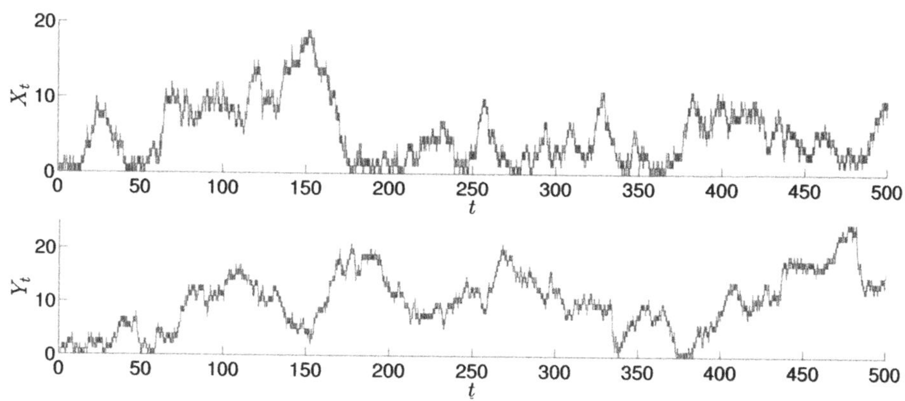

Figure 7.10 Realizations of the queue length processes $\{X_t, t \geqslant 0\}$ and $\{Y_t, t \geqslant 0\}$.

The system state (X_t, Y_t) only changes when one of the following events occur: an arrival to the first queue, a departure from the first queue, and a departure from the second queue. A typical sequence of events is given in Figure 7.11. Arrivals to the second queue do not require separate events, as they coincide with departures from the first queue.

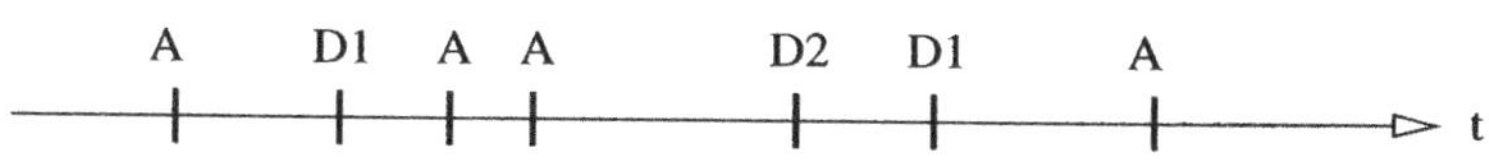

Figure 7.11 A sequence of discrete events: A = arrival, D1 = departure from the first queue, D2 = departure from the second queue.

The following MATLAB programs implement the tandem system with $\mathsf{U}(0,1)$ interarrival and service times. The main difference with the implementation for the

$GI/G/1$ queue in Example 7.1 is the additional event subroutine departure2.m. Another difference is that during the simulation the total time the second server is busy is updated (stored in **s2busy**). The output of the program is the average occupancy **s2busy/t** for server 2 and the total number of events occurring during the simulation period. Finally, when the appropriate lines are uncommented, the program plots a realization of the state processes, as in Figure 7.10.

```
%TandemQ/main.m
T = 5000;
totevents = 0;
x = 0; y = 0; yold =0;
s2busy = 0; % total time server 2 is busy
% xx=x; yy=y;
tt=0;
ev_list = inf*ones(4,2);
t = 0; told = 0;
ev_list(1,:) = [rand, 1]; %schedule the first arrival
N_ev = 1;                 % number of events
while t < T
    totevents = totevents+1;
    t = ev_list(1,1);
    %tt=[tt,t];
    ev_type = ev_list(1,2);
    switch ev_type
        case 1
            arrival
        case 2
            departure1
        case 3
            departure2
    end
    ev_list(1,:) = [inf,inf];
    N_ev = N_ev - 1;
    ev_list = sortrows(ev_list,1); % sort event list
    % xx=[xx,x]; yy=[yy,y];
    s2busy = s2busy + (yold > 0)*(t - told);
    yold = y; told=t;
end
% plottraces;
res = s2busy/t
totevents
```

```
%TandemQ/arrival.m
N_ev = N_ev + 1;
ev_list(N_ev,:) = [t + rand, 1]; %schedule new arrival
if x == 0  % if queue is empty
   N_ev = N_ev + 1;
```

```
    ev_list(N_ev,:) = [t + rand, 2]; % schedule departure at queue 1
end
x = x+1;
```

```
%TandemQ/departure1.m
x = x-1; % go out of first queue
if x ~= 0
    N_ev = N_ev + 1;
    ev_list(N_ev,:) = [t + rand, 2]; % schedule departure at queue 1
end
if y == 0
    N_ev = N_ev + 1;
    ev_list(N_ev,:) = [t + rand, 3]; % schedule departure at queue 2
end
y = y + 1; % go in second queue
```

```
%TandemQ/departure2.m
y = y-1; % go out of second queue
if y ~= 0
  N_ev = N_ev + 1;
  ev_list(N_ev,:) = [t + rand, 3]; % schedule departure at queue 2
end
```

```
%TandemQ/plottraces.m
figure,subplot(2,1,1),
for i =1:length(xx)-1,
    line([tt(i),tt(i+1)],[xx(i),xx(i)]);
    line([tt(i+1),tt(i+1)],[xx(i),xx(i+1)]);
end
aa=axis;
axis([0,tt(end),aa(3),aa(4)]),xlabel('t'),ylabel('Queue 1');

subplot(2,1,2),
for i =1:length(yy)-1,
    line([tt(i),tt(i+1)],[yy(i),yy(i)]);
    line([tt(i+1),tt(i+1)],[yy(i+1),yy(i)]);
end
aa=axis;
axis([0,tt(end),aa(3),aa(4)]),xlabel('t'),ylabel('Queue 2'),
```

7.4.3 Repairman Problem

Consider a repair system consisting of m repairmen and $n \geqslant m$ nonidentical machines, as depicted in Figure 7.12 (where $m = 4$ and $n = 8$).

When a machine breaks down it is immediately repaired by one of the available repairmen; if none are available the machine is placed into a waiting room to be served in first-come-first-served order once a repairman becomes available. As soon as a machine is repaired (and is then as good as new), the repairman takes the first machine from the waiting room, if there are any; otherwise, the repairman remains idle until the next machine failure occurs. Each machine is assumed to have a fixed lifetime distribution and repair time distribution. The lifetimes and repair times are assumed to be independent of each other.

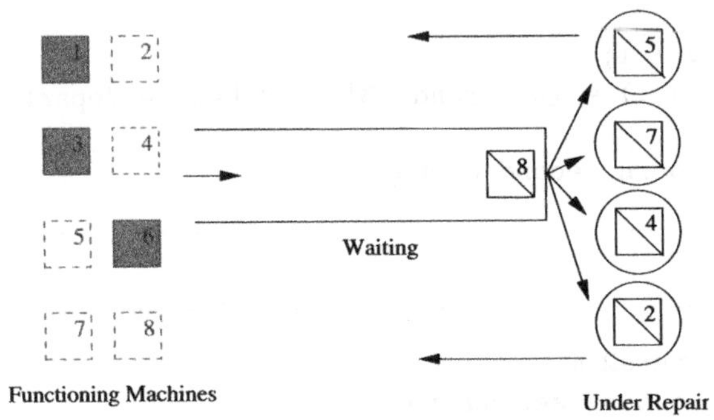

Figure 7.12 A repair system.

Two performance measures of interest are the utilization of the machines (how many machines are active over the long run) and the utilization of the repairmen (how many are busy over the long run). These quantities need to be assessed in general via simulation. Figure 7.13 depicts a typical realization of the number of busy repairmen and working machines for the MATLAB implementation below.

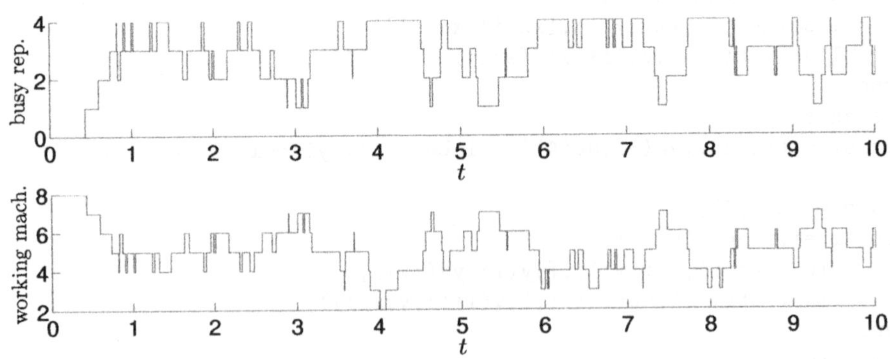

Figure 7.13 A realization of the processes describing the number of busy repairmen and functioning machines.

The *state* of the system at time t is characterized by the configuration of machines that are operational, under repair, and waiting to be repaired. The order of machines in the waiting queue is essential, as the machines are not necessarily identical. Therefore, the state must include an ordered list (queue) Q_t of waiting machine numbers. Other state variables are the number of available repairmen R_t and the number of failed machines F_t.

There are two types of events for this system: failure events and repair events. Each event triggers the execution of the corresponding failure or repair procedure. Because each machine can have different failure and repair distributions, the event procedures are machine dependent.

In the program below the event list is simply implemented as a 9×3 array where the columns correspond to the event time, event type (1=failure, 2=repair) and machine number, respectively. Note that at each time no more than $n + 1$ events are scheduled, including the current one. The waiting queue is implemented as a dynamic array.

Upon failure of a machine, a repair needs to be scheduled at a time equal to the current time plus the required repair time for the machine only if there is a repairman available to carry out the repairs. Otherwise, the machine is placed in the waiting queue. The number of failed machines is always increased by 1.

Upon repair, the number of failed machines is decreased by 1. The machine that is just repaired is scheduled for a failure, after the lifetime of the machine. If the "failed" queue is not empty the repairman takes the next machine from the queue and schedules a corresponding repair event. Otherwise, the number of available repairmen is increased by 1.

In the MATLAB program below the lifetime of machine i is assumed to be $\mathsf{Weib}(i, 1)$ distributed, $i = 1, \ldots, 8$ and the repair times are $\mathsf{U}(0, 1)$ distributed. Realizations of $\{m - R_t\}$ and $\{n - F_t\}$ can be obtained by uncommenting the appropriate lines in the main program.

```
%Repairman/main.m
global alpha
T = 1000;
mach_num = 0; repairq = [];
nrep = 4; nmach = 8; %number of repairmen and machines
alpha = [1,2,3,4,5,6,7,8];
r = nrep; %the number of repairmen available
f = 0; %the number of machines failed
%rr= r; ff=f; tt=0; %save the history (needed for plotting)
tot_util_rep = 0;
tot_util_mach = 0;
rold = r; fold = f; told = 0;

ev_list = inf*ones(nmach+1, 3); %event time, type, machine number
t = 0;
for i=1:nmach
    ev_list(i,:) = [lifetime(i), 1,i]; %schedule the failures
end
ev_list = sortrows(ev_list,1); % sort event list
```

```
N_ev = nmach;
while t < T
    t = ev_list(1,1);
    %tt=[tt,t];
    ev_type = ev_list(1,2);
    mach_num = ev_list(1,3);
    switch ev_type
        case 1
            failure
        case 2
            repair
    end
    N_ev = N_ev - 1;
    ev_list(1,:) = [inf,inf,inf];
    ev_list = sortrows(ev_list,1); % sort event list
    tot_util_rep = tot_util_rep + (nrep - rold)*(t - told);
    tot_util_mach = tot_util_mach + (nmach - fold)*(t - told);
    rold=r; told=t; fold =f;
%   rr=[rr,r];ff=[ff,f];
end
fprintf('repair util. = %g, machine util. = %g\n', ...
         tot_util_rep/t, tot_util_mach/t);
%plottrace
```

```
%Repairman/failure.m
if (r > 0) %repairman available
    N_ev = N_ev + 1;
    %schedule repair
    ev_list(N_ev,:) = [t + repairtime(mach_num), 2,mach_num];
    r = r -1;
else
    repairq = [repairq,mach_num];
end
f = f+ 1;
```

```
%Repairman/repair.m
f = f - 1; % one less failed
sq = size(repairq,2);
if (sq > 0) %still one in the queue
    N_ev = N_ev + 1;
    %schedule next repair
    ev_list(N_ev,:) = [t + repairtime(mach_num), 2,repairq(1)];
    repairq = repairq(2:sq); %remove machine
else
    r = r+1;
end
```

```
N_ev = N_ev + 1;
%schedule failure of current machine
ev_list(N_ev,:) = [t + lifetime(mach_num), 1,mach_num];
```

```
%Repairman/plottrace.m
figure(1)
subplot(2,1,1),
for i =1:length(rr)-1,
    line([tt(i),tt(i+1)],[(nrep - rr(i)),(nrep -rr(i))]);
    line([tt(i+1),tt(i+1)],[(nrep -rr(i)),(nrep -rr(i+1))]);
end
aa=axis;
axis([0,tt(end),aa(3),aa(4)]),xlabel('t'),ylabel('busy rep.');
subplot(2,1,2),
for i =1:length(ff)-1,
    line([tt(i),tt(i+1)],[(nmach - ff(i)),(nmach - ff(i))]);
    line([tt(i+1),tt(i+1)],[(nmach - ff(i)),(nmach - ff(i+1))]);
end
aa=axis;
axis([0,tt(end),aa(3),aa(4)]),xlabel('t'),ylabel('working mach.');
```

```
%Repairman/lifetime.m
function out = lifetime(i)
global alpha
out = (-log(rand))^(1/alpha(i));
```

```
%Repairman/repairtime.m
function out = repairtime(i)
out = rand;
```

Further Reading

One of the first books on Monte Carlo simulation is Hammersley and Handscomb [7]. Kalos and Whitlock [8] is another classical reference, and Ross [15] is a good modern starting point. The event- and process-oriented approaches to discrete event simulation are elegantly explained in Mitrani [14]. Among the great variety of discrete-event simulation books, all focusing on different aspects of the modeling and simulation process, we mention [2, 4, 6, 12, 16]. Additional introductory texts are [3] and [13]. The choice of computer language in which to implement a simulation program is very subjective. The simple models discussed in this chapter can be implemented in any standard computer language, even standard MATLAB, although it does not provide easy event list manipulation. Commercial simulation environments such as ARENA/SIMAN and SIMSCRIPT II.5 make the implementation of larger models much easier — the book [10] introduces the reader to simulation with ARENA. Alternatively, various free SIMULA-like Java packages exist that offer fast implementation of event- and process-oriented simulation programs. Examples are SSJ, developed by Simard and L'Ecuyer: `www.iro.umontreal.ca/~simardr/ssj/`, DSOL developed by the Technical University Delft: `sk-3.tbm.tudelft.nl/simulation/`, and J-SIM: `www.j-sim.zcu.cz/`.

REFERENCES

1. S. Asmussen. *Applied Probability and Queues.* John Wiley & Sons, New York, 1987.
2. J. S. Banks, J. S. Carson II, B. L. Nelson, and D. M. Nicol. *Discrete-Event System Simulation.* Prentice-Hall, Englewood Cliffs, NJ, fifth edition, 2009.
3. C. G. Cassandras and S. Lafortune. *Introduction to Discrete Event Systems.* Springer-Verlag, New York, second edition, 2007.
4. J. R. Clymer. *System Analysis Using Simulation and Markov Models.* Prentice Hall, Englewood Cliffs, NJ, 1990.
5. O.-J. Dahl and K. Nygaard. SIMULA: an ALGOL-based simulation language. *Communications of the ACM*, 9(9):671–678, 1966.
6. G. S. Fishman. *Discrete Event Simulation: Modeling, Programming, and Analysis.* Springer-Verlag, New York, 2001.
7. J. M. Hammersley and D. C. Handscomb. *Monte Carlo Methods.* John Wiley & Sons, New York, 1964.
8. M. H. Kalos and P. A. Whitlock. *Monte Carlo Methods, Volume I: Basics.* John Wiley & Sons, New York, 1986.
9. F. P. Kelly. *Reversibility and Stochastic Networks.* John Wiley & Sons, New York, 1979.
10. W. Kelton, R. Sadowski, and N. Swets. *Simulation with Arena.* McGraw-Hill, New York, fifth edition, 2009.
11. D. E. Knuth. *The Art of Computer Programming*, volume 1: Fundamental Algorithms. Addison-Wesley, Reading, MA, third edition, 1997.
12. A. M. Law and W. D. Kelton. *Simulation Modeling and Analysis.* McGraw-Hill, New York, third edition, 2000.
13. L. M. Leemis and S. K. Park. *Discrete-Event Simulation: A First Course.* Prentice-Hall, Englewood Cliffs, NJ, 2006.
14. I. Mitrani. *Simulation Techniques for Discrete Event Systems.* Cambridge University Press, Cambridge, 1982.
15. S. M. Ross. *Simulation.* Academic Press, New York, third edition, 2002.
16. R. Y. Rubinstein and B. Melamed. *Modern Simulation and Modeling.* John Wiley & Sons, New York, 1998.
17. L. Schruben. Simulation modeling with event graphs. *Communications of the ACM*, 26(11):957–963, 1983.
18. R. W. Wolff. *Stochastic Modeling and the Theory of Queues.* Prentice-Hall, Englewood Cliffs, NJ, 1989.

CHAPTER 8

STATISTICAL ANALYSIS OF SIMULATION DATA

This chapter describes various statistical techniques that can be used for analyzing random data produced by Monte Carlo simulation experiments. Further background on (mathematical) statistics can be found in Appendix B. ☞ 653

8.1 SIMULATION DATA

Similar to real-world data, computer-simulated data may take many different forms. In contrast to real-world data, however, simulation data are strictly reproducible, because the simulation model is completely known. Moreover, the amount of data that can be gathered from the model is limited only by the maximum amount of time one wishes to impose on the simulation runs and the amount of storage space on a computer.

The data produced by a simulation experiment can be viewed as outcomes of random variables, random vectors, time series, or stochastic processes. In general the objective of the simulation is to draw conclusions about various characteristics of these random objects, such as their expectations, correlations, and distributions.

Before embarking on a mathematical analysis of the data it may be worthwhile to detect patterns in the data through visualization and summarization. We mention a few useful standard techniques. In many cases the underlying assumption is that the data, say $X_1, \ldots, X_N$, form an **iid sample** from some distribution; that is, $X_1, \ldots, X_N$ are independent and identically distributed according to some known or unknown distribution.

8.1.1 Data Visualization

Let $X_1, \ldots, X_N$ be an iid sample from some distribution. The following graphical techniques are often used to visualize the data; various other graphical techniques may be found in, for example, [23].

1. *Scatter plot*: For a d-dimensional scatter plot ($d = 1, 2$, or 3), plot the data as points in $\mathbb{R}^d$.

2. *Histogram*: Partition the real line into a finite number of intervals or *classes*. Count how many data points fall in each class. For each class plot a rectangle whose width corresponds to the length of the class and whose area (or sometimes height) corresponds to the number (or alternatively, to the frequency) of points in that class.

3. *Empirical cdf*: The empirical cdf is an estimate of the true cdf of the data. It is a nondecreasing step function which jumps up by an amount of $1/N$ at each of the data points. More information on the empirical cdf is given in Section 8.4.

4. *Density plot*: A density plot provides an estimate of the true pdf of the data. More information on the density plots and kernel density estimation is given in Section 8.5.

■ EXAMPLE 8.1 (Visualizing Gamma Data)

Suppose $X_1, \ldots, X_N, Y_1, \ldots, Y_N \stackrel{\text{iid}}{\sim} \mathsf{Exp}(1)$. Let $Z_i = X_i + Y_i$, $i = 1, \ldots, N$. Note that $Z_i \sim \mathsf{Gamma}(2, 1)$. The following MATLAB program provides four graphical views of the data for $N = 1000$, as displayed in Figure 8.1.

```
%stateda.m
N = 10^3;
x = -log(rand(1,N)); %the data
y = -log(rand(1,N)); %the data
z = x + y;
subplot(2,2,1), hist(z,20);
[f,xi] = ksdensity(z);    %matlab's kernel density function
subplot(2,2,2), plot(xi,f);
subplot(2,2,3), ecdf(z); %empirical cdf
subplot(2,2,4), scatter(x,z,'.');
```

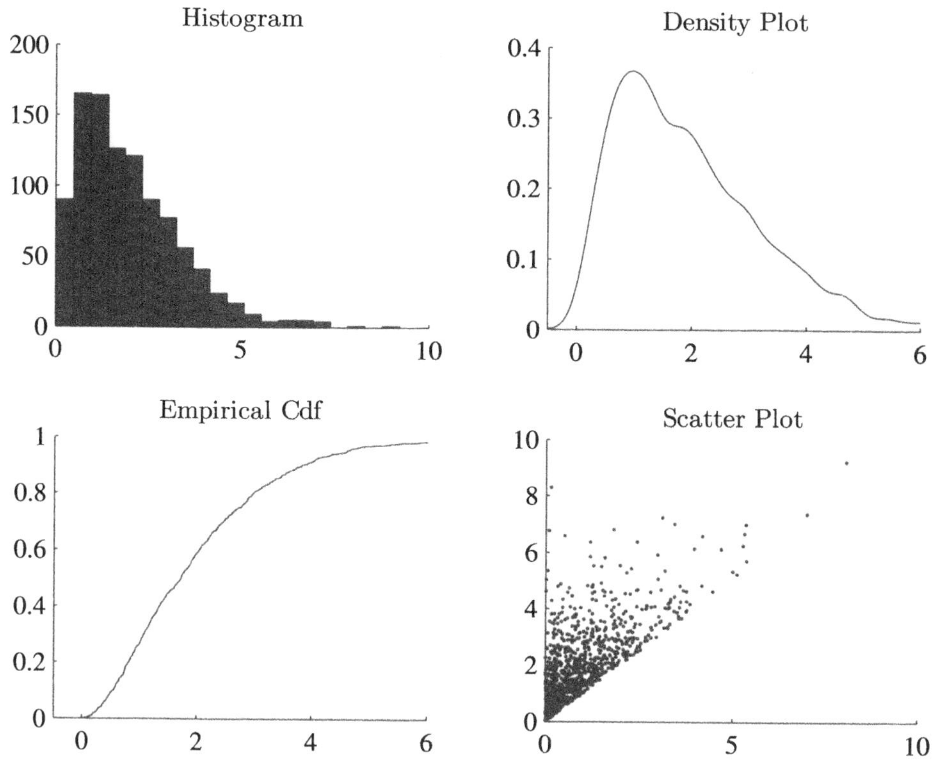

Figure 8.1 Graphical displays of data.

8.1.2 Data Summarization

Let $X_1, \ldots, X_N$ be an iid sample from some distribution, and denote the ordered data by $X_{(1)} \leqslant \cdots \leqslant X_{(N)}$. It is often useful to summarize various characteristics of the sample as given in the following list.

1. *Centrality characteristics*:

   (a) The **sample mean** gives the average of the data:

$$\bar{X} = \frac{1}{N} \sum_{i=1}^{N} X_i \, .$$

   (b) The **sample median** is

$$\widetilde{X} = \begin{cases} X_{((N+1)/2)} & \text{if } N \text{ is odd} \\ (X_{(N/2)} + X_{(N/2+1)})/2 & \text{if } N \text{ is even.} \end{cases}$$

2. *Dispersion characteristics*:

   (a) The **sample variance** is

$$S^2 = \frac{1}{N-1}\sum_{i=1}^{N}(X_i - \bar{X})^2 = \frac{1}{N-1}\left(\sum_{i=1}^{N} X_i^2 - N\bar{X}^2\right).$$

   (b) The **sample standard deviation** is the square root of the sample variance: $S = \sqrt{S^2}$.

   (c) The **range** of the data is $X_{(N)} - X_{(1)}$.

   (d) The **sample *k*-th moment** is $\frac{1}{N}\sum_{i=1}^{N} X_i^k$.

   (e) The **sample *k*-th central moment** is $\frac{1}{N}\sum_{i=1}^{N}(X_i - \bar{X})^k$.

   (f) The **sample γ-quantile** or $\gamma \times 100$ **percentile** of $X_1, \ldots, X_N$ is $X_{(\lceil \gamma N \rceil)}$.

3. *Dependency characteristics:* Let $(X_1, Y_1), \ldots, (X_N, Y_N)$ be an iid sample from a bivariate distribution.

   (a) The **sample covariance** is

$$\frac{1}{N-1}\sum_{i=1}^{N}(X_i - \bar{X})(Y_i - \bar{Y}) .$$

   (b) The **sample correlation coefficient** is

$$\frac{\sum_{i=1}^{N}(X_i - \bar{X})(Y_i - \bar{Y})}{\sqrt{\sum_{i=1}^{N}(X_i - \bar{X})^2}\,\sqrt{\sum_{i=1}^{N}(Y_i - \bar{Y})^2}} .$$

■ **EXAMPLE 8.2 (Summarizing Gamma Data)**

Suppose, as in Example 8.1, that $X_1, \ldots, X_N, Y_1, \ldots, Y_N \stackrel{\text{iid}}{\sim} \mathsf{Exp}(1)$, so that $\{Z_i\}$ with $Z_i = X_i + Y_i$, $i = 1, \ldots, N$, forms an iid sample from $\mathsf{Gamma}(2, 1)$. The following MATLAB program contains the most common functions for data summarization.

```
%stateda2.m
N = 10^3;
x = -log(rand(1,N));
y = -log(rand(1,N));
z = x + y;
meanz = mean(z);
medz = median(z);
stdz = std(z);
varz = var(z);
maxz = max(z);
```

```
minz= min(z);
q1 = quantile(z,0.25); % first quartile
q3 = quantile(z,0.75); % third quartile
display([meanz,stdz, varz])
display([minz, q1, medz, q3,  maxz])
covyz = cov(y,z)
```

Typical output from this program is:

```
1.9945  1.4136  1.9983 %sample mean, standard deviation, variance of Z

0.0379  0.9702  1.6426   2.7192  10.7030 %min, q1, med, q3, max

0.8987  0.9215   %covariance matrix of Y and Z
0.9215  1.9983
```

In many cases the data summarization quantities can be calculated dynamically; that is, as soon as new data become available the quantity is updated. An advantage of this *online* updating is that only one sample point needs to be stored in memory rather than the entire sample. The following algorithm illustrates how this is achieved for the sample mean and sample variance.

Algorithm 8.1 (Online Calculation of the Sample Mean and Variance)

1. *Initialize* $a = X_1$, $b = X_1^2$, *and* $t = 1$.
2. *Set* $a = a + X_{t+1}$ *and* $b = b + X_{t+1}^2$.
3. *Let*

$$\bar{X}_{t+1} = \frac{a}{t+1} \tag{8.1}$$

and

$$S_{t+1}^2 = \frac{b - (t+1)\,\bar{X}_{t+1}^2}{t} . \tag{8.2}$$

4. *If* $t + 1 < N$, *set* $t = t + 1$ *and repeat from Step 2; otherwise, stop.*

8.2 ESTIMATION OF PERFORMANCE MEASURES FOR INDEPENDENT DATA

Suppose the data $Y_1, \ldots, Y_N$ from a simulation experiment are independent and identically distributed according to some known or unknown discrete or continuous pdf f. Often such data are obtained by executing N independent runs of the simulation, producing output Y_i for the i-th run. Suppose the aim of the simulation is to estimate the performance measure $\ell = \mathbb{E}Y$, with $Y \sim f$. Assuming $|\ell| < \infty$, an unbiased estimator for ℓ is the sample mean of the $\{Y_i\}$; that is,

$$\bar{Y} = \frac{1}{N} \sum_{i=1}^{N} Y_i . \tag{8.3}$$

Provided that the variance of Y, say σ^2, is finite, $\bar{Y}$ approximately has a $\mathsf{N}(\ell, \sigma^2/N)$ distribution for large N (an immediate consequence of the central limit theorem). ☞ 625 If σ^2 is unknown, it can be estimated without bias via the sample variance of the $\{Y_i\}$:

$$S^2 = \frac{1}{N-1} \sum_{i=1}^{N} (Y_i - \bar{Y})^2 , \tag{8.4}$$

which (by the law of large numbers) tends to σ^2 as $N \to \infty$. This leads to an approximate $1 - \alpha$ **confidence interval** for ℓ:

$$\left(\bar{Y} - z_{1-\alpha/2} \frac{S}{\sqrt{N}}, \quad \bar{Y} + z_{1-\alpha/2} \frac{S}{\sqrt{N}} \right) , \tag{8.5}$$

where z_γ denotes the γ-quantile of the $\mathsf{N}(0,1)$ distribution.

Instead of specifying the confidence interval, the following measures of accuracy are often reported in simulation studies:

1. *Width of confidence interval*: $2\, z_{1-\alpha/2} S/\sqrt{N}$.
2. *Half-width of confidence interval*: $z_{1-\alpha/2} S/\sqrt{N}$.
3. *Relative width of confidence interval*: $2\, z_{1-\alpha/2} S/(\bar{Y}\sqrt{N})$.
4. *Estimated standard error*: $S/\sqrt{N}$. This is an estimator for the true **standard error**, that is, the standard deviation of the estimator $\bar{Y}$, which is $\sigma/\sqrt{N}$.
5. *Estimated relative error*: $S/(\bar{Y}\sqrt{N})$. This is an estimator for the true **relative error** of the estimator $\bar{Y}$:

$$\frac{\sqrt{\mathrm{Var}(\bar{Y})}}{\mathbb{E}\bar{Y}} = \frac{\sigma}{\ell\sqrt{N}} . \tag{8.6}$$

The basic estimation procedure for independent data is summarized below. The procedure is sometimes referred to as **crude Monte Carlo** (CMC). Chapter 9 discusses various estimation techniques that can potentially improve on the accuracy of CMC. ☞ 348

Algorithm 8.2 (Crude Monte Carlo for Independent Data)

1. *Generate* $Y_1, \ldots, Y_N \overset{\text{iid}}{\sim} f$ *(for example, from independent simulation runs).*
2. *Calculate the point estimate* $\bar{Y}$ *and confidence interval* (8.5) *of* $\ell = \mathbb{E}Y$.

It is often the case that the output Y is a function of some underlying random vector or stochastic process; that is,

$$Y = H(\mathbf{X}) ,$$

where H is a real-valued performance function and $\mathbf{X}$ is a random vector or process. Provided that independent copies of Y can be generated in finite time, Algorithm 8.2 can be used for both **static** simulation models (in which case $\mathbf{X}$ is a random vector) and **dynamic** models (in which case $\mathbf{X}$ represents a time-dependent stochastic process).

■ EXAMPLE 8.3 (Monte Carlo Integration)

In **Monte Carlo integration**, simulation is used to evaluate integrals (see also Section 2.1). Consider, for example, the integral ☞ 25

$$\ell = \int_{-\infty}^{\infty}\int_{-\infty}^{\infty}\int_{-\infty}^{\infty} \sqrt{|x_1 + x_2 + x_3|}\, \mathrm{e}^{-(x_1^2+x_2^2+x_3^2)/2}\, \mathrm{d}x_1\, \mathrm{d}x_2\, \mathrm{d}x_3 \, .$$

Defining $Y = |X_1 + X_2 + X_3|^{1/2}(2\pi)^{3/2}$, with $X_1, X_2, X_3 \overset{\text{iid}}{\sim} \mathsf{N}(0,1)$, we can write $\ell = \mathbb{E}Y$. The following MATLAB program provides an estimate and 95% confidence interval for ℓ. Note that here $z_{1-\alpha/2} = z_{0.975} \approx 1.96$. Typical output is $\bar{Y} = 17.04$ with confidence interval $(17.026, 17.054)$, using a sample size of $N = 10^6$. From the fractional moments of the normal distribution (Property 7 on Page 123) we can compute the exact value: $\ell \approx 17.0418$.

```
% mcint.m
c = (2*pi)^(3/2);
H = @(x) c*sqrt(abs(sum(x,2)));
N = 10^6; alpha = 0.05;
x = randn(N,3); y = H(x);
mY = mean(y); sY = std(y);
RE = sY/mY/sqrt(N);
z = icdf('norm',1-alpha/2,0,1);
fprintf('Estimate = %g, CI = (%g, %g)\n', ...
                    mY, mY*(1- z*RE), mY*(1 + z*RE))
```

The generalization of Algorithm 8.2 to vector-valued simulation output $\mathbf{Y} = (Y_1, \ldots, Y_n) \sim f$ with performance vector $\boldsymbol{\ell} = \mathbb{E}\mathbf{Y}$ is as follows, under the assumption that each element of the expectation vector $\mathbf{Y}$ and covariance matrix Σ is finite.

Algorithm 8.3 (Estimation for Independent Vector-Valued Data)

1. *Generate* $\mathbf{Y}_1, \ldots, \mathbf{Y}_N \overset{\text{iid}}{\sim} f$ *(for example, from independent simulation runs).*
2. *Calculate the point estimate* $\bar{\mathbf{Y}} = (\bar{Y}_1, \ldots, \bar{Y}_n)^\top$, *where* $\bar{Y}_j = N^{-1}\sum_{i=1}^{N} Y_{ij}$ *and* Y_{ij} *is the* j*-th component of* $\mathbf{Y}_i$, $i = 1, \ldots, N$, $j = 1, \ldots, n$.
3. *An approximate* $1 - \alpha$ *confidence region for* $\boldsymbol{\ell}$ *is*

$$\mathscr{C} = \left\{ \mathbf{y} \in \mathbb{R}^n : (\bar{\mathbf{Y}} - \mathbf{y})^\top (\widehat{\Sigma})^{-1} (\bar{\mathbf{Y}} - \mathbf{y}) \leqslant \frac{\chi^2_{n;1-\alpha}}{N} \right\}, \tag{8.7}$$

where $\widehat{\Sigma}$ *is the sample covariance matrix,*

$$\widehat{\Sigma} = \frac{1}{N-1}\sum_{i=1}^{N} (\mathbf{Y}_i - \bar{\mathbf{Y}})(\mathbf{Y}_i - \bar{\mathbf{Y}})^\top ,$$

and $\chi^2_{n;1-\alpha}$ *is the* $(1-\alpha)$*-quantile of the* χ^2_n *distribution.*

The confidence region $\mathscr{C}$ forms the interior (including the surface) of an ellipsoid. The interpretation is that the true vector $\boldsymbol{\ell}$ is contained in the *random* confidence region with probability approximately $1-\alpha$. This follows from the fact that for large N and $\Sigma = BB^\top$, the random vector $\mathbf{Z} = \sqrt{N}\, B^{-1}(\bar{\mathbf{Y}} - \boldsymbol{\ell})$ approximately has a $\mathsf{N}(\mathbf{0}, I)$ distribution, so that

$$\mathbb{P}(\boldsymbol{\ell} \in \mathscr{C}) \approx \mathbb{P}\left((\bar{\mathbf{Y}} - \boldsymbol{\ell})^\top \Sigma^{-1} (\bar{\mathbf{Y}} - \boldsymbol{\ell}) \leqslant \frac{\chi^2_{n;1-\alpha}}{N}\right) = \mathbb{P}\left(\mathbf{Z}^\top \mathbf{Z} \leqslant \chi^2_{n;1-\alpha}\right) = 1 - \alpha\,.$$

For plotting purposes it is useful to write the surface of the confidence ellipsoid as

$$\left\{ \bar{\mathbf{Y}} + \sqrt{\frac{\chi^2_{n;1-\alpha}}{N}}\; B\,\mathbf{z} \;:\; \|\mathbf{z}\| = 1 \right\},$$

which is an affine transformation of the unit hypersphere in $\mathbb{R}^n$.

8.2.1 Delta Method

☞ 625 The **delta method** can be viewed as a generalization of the central limit theorem. Let $\mathbf{Z}_1, \mathbf{Z}_2, \ldots$ be a sequence of random vectors such that

$$\sqrt{N}\,(\mathbf{Z}_N - \boldsymbol{\mu}) \xrightarrow{d} \mathbf{K} \sim \mathsf{N}(\mathbf{0}, \Sigma)\,.$$

An example is where $\mathbf{Z}_N$ is the sample mean of iid random vectors $\mathbf{X}_1, \ldots, \mathbf{X}_N$, each with expectation vector $\boldsymbol{\mu}$ and covariance matrix Σ. The delta method states that a similar convergence result holds for a *function* of $\mathbf{Z}_N$. Specifically,

$$\sqrt{N}\,(\mathbf{g}(\mathbf{Z}_N) - \mathbf{g}(\boldsymbol{\mu})) \xrightarrow{d} \mathbf{R} \sim \mathsf{N}(\mathbf{0}, J\Sigma J^\top)\,, \tag{8.8}$$

☞ 710 where $J = J_{\mathbf{g}}(\boldsymbol{\mu}) = (\partial g_i(\boldsymbol{\mu})/\partial x_j)$ is the Jacobi matrix of $\mathbf{g}$ evaluated at $\boldsymbol{\mu}$. Note that the delta method assumes that $\mathbf{g}$ is differentiable at $\boldsymbol{\mu}$, and hence the Jacobi matrix exists. The key step in the proof is the construction of the first-order Taylor expansion of $\mathbf{g}$ around $\boldsymbol{\mu}$, which is

$$\mathbf{g}(\mathbf{Z}_N) = \mathbf{g}(\boldsymbol{\mu}) + J_{\mathbf{g}}(\boldsymbol{\mu})(\mathbf{Z}_N - \boldsymbol{\mu}) + \mathcal{O}(\|\mathbf{Z}_N - \boldsymbol{\mu}\|^2)\,.$$

As $N \to \infty$, the remainder term goes to 0 because $\mathbf{Z}_N \xrightarrow{\text{a.s.}} \boldsymbol{\mu}$. Hence, the left-hand side of (8.8) is approximately $\sqrt{N} J\,(\mathbf{Z}_N - \boldsymbol{\mu}) = J\,\sqrt{N}(\mathbf{Z}_N - \boldsymbol{\mu})$. For $N \to \infty$ this converges in distribution to a random vector $J\mathbf{K}$, where $\mathbf{K} \sim \mathsf{N}(\mathbf{0}, \Sigma)$. Thus, $\mathbf{R} = J\mathbf{K} \sim \mathsf{N}(\mathbf{0}, J\Sigma\,J^\top)$.

■ **EXAMPLE 8.4 (Ratio Estimator)**

Let $(X_1, Y_1), \ldots, (X_N; Y_N)$ be iid copies of a random vector (X, Y) with mean vector (μ_X, μ_Y) and covariance matrix Σ. Suppose we wish to estimate $\ell = \mu_X/\mu_Y$. A straightforward estimator is the so-called **ratio estimator** $\bar{X}/\bar{Y}$, the asymptotic distribution of which can be derived with the delta method. Let $\mathbf{Z}_N$ be the vector $(\bar{X}, \bar{Y})$ and $g(x, y) = x/y$. Then,

$$J_g(x, y) = \left(\frac{\partial g(x,y)}{\partial x}, \; \frac{\partial g(x,y)}{\partial y}\right) = \left(\frac{1}{y}, \; \frac{-x}{y^2}\right).$$

Let $J = J_g(\mu_X, \mu_Y)$. It follows from (8.8) that $\bar{X}/\bar{Y} = g(\bar{X}, \bar{Y})$ approximately has a $\mathsf{N}(\mu_X/\mu_Y, \frac{1}{N}\sigma^2)$ distribution, with **asymptotic variance**

$$\sigma^2 = J\Sigma J^\top = \left(\frac{1}{\mu_Y}, \quad \frac{-\mu_X}{\mu_Y^2}\right) \begin{pmatrix} \mathrm{Var}(X) & \mathrm{Cov}(X,Y) \\ \mathrm{Cov}(X,Y) & \mathrm{Var}(Y) \end{pmatrix} \begin{pmatrix} \frac{1}{\mu_Y} \\ \frac{-\mu_X}{\mu_Y^2} \end{pmatrix}$$
$$= \frac{\mathrm{Var}(X) - 2\,\ell\,\mathrm{Cov}(X,Y) + \ell^2\,\mathrm{Var}(Y)}{\mu_Y^2} \,. \tag{8.9}$$

Note that σ^2 can be easily estimated from the iid sample.

8.3 ESTIMATION OF STEADY-STATE PERFORMANCE MEASURES

Suppose that the data from a simulation experiment are described by a collection $\{X_t, t \geqslant 0\}$ of dependent random variables such that X_t converges in distribution to some random variable X as $t \to \infty$. Such data could, for example, be the result of a single run of an MCMC sampler, where X is distributed according to the limiting distribution of the Markov chain. To estimate the **steady-state** or **equilibrium** performance measure $\mathbb{E}X$ we discuss the following methods (see also [21]).

☞ 623
☞ 225

- Covariance method.
- Batch means method.
- Regenerative method.

In order to cancel the effects of time dependence and the initial distribution, it is common practice to discard the data that is collected during the **burn-in period**; that is, the period of time in which the distribution of X_t significantly deviates from that of X. However, it is not always clear how long the burn-in period should be — but see [1] for a case where this can be determined. For regenerative processes the regenerative method, discussed in Section 8.3.3, avoids the burn-in issue altogether.

8.3.1 Covariance Method

Suppose that $\{X_1, X_2, \ldots, X_N\}$ is a *stationary* process, so that no burn-in is required. An unbiased estimator for the steady-state expected value $\ell = \mathbb{E}X = \mathbb{E}X_t$ (assumed to be finite) is the sample average $\bar{X} = N^{-1}\sum_{t=1}^N X_t$. The variance of $\bar{X}$ is given by (see Section A.4.3)

☞ 631
☞ 617

$$\mathrm{Var}(\bar{X}) = \frac{1}{N^2}\left(\sum_{t=1}^N \mathrm{Var}(X_t) + 2\sum_{s=1}^{N-1}\sum_{t=s+1}^N \mathrm{Cov}(X_s, X_t)\right) . \tag{8.10}$$

Since $\{X_t\}$ is stationary, we have $\mathrm{Cov}(X_s, X_t) = \mathbb{E}[X_s X_t] - \ell^2 = R(t-s)$, where R is the **(auto)covariance function** of the stationary process. Note that $R(0) = \mathrm{Var}(X_t)$. As a consequence, we can write (8.10) as

$$N\,\mathrm{Var}(\bar{X}) = R(0) + 2\sum_{t=1}^{N-1}\left(1 - \frac{t}{N}\right) R(t) \,. \tag{8.11}$$

In many applications $R(t)$ decreases rapidly with t, so that only the first few terms in the sum (8.11) are relevant. These autocovariances, say $R(0), \ldots, R(K)$, can be estimated via their (unbiased) sample averages:

$$\widehat{R}(k) = \frac{1}{N-k-1} \sum_{t=1}^{N-k} (X_t - \bar{X}\,)(X_{t+k} - \bar{X}\,), \quad k = 0, 1, \ldots, K\,.$$

Thus, for large N the variance of $\bar{X}$ can be estimated as $\widetilde{S}^2/N$, where

$$\widetilde{S}^2 = \widehat{R}(0) + 2\sum_{t=1}^{K} \widehat{R}(t)\,.$$

To obtain confidence intervals, one again uses the central limit theorem; that is, the fact that the cdf of $\sqrt{N}(\bar{X} - \ell)$ converges to the cdf of the normal distribution with expectation 0 and asymptotic variance $\sigma^2 = \lim_{N\to\infty} N\,\mathrm{Var}(\bar{X}\,)$. Using $\widetilde{S}^2$ as an estimator for σ^2, we find that an approximate $1-\alpha$ confidence interval for ℓ is given by

$$\left(\bar{X} - z_{1-\alpha/2}\frac{\widetilde{S}}{\sqrt{N}},\quad \bar{X} + z_{1-\alpha/2}\frac{\widetilde{S}}{\sqrt{N}}\right)\,. \tag{8.12}$$

■ EXAMPLE 8.5 (Random Walk Sampler)

☞ 230 Consider the random walk sampler with target distribution $\mathsf{N}(10, 1)$ and proposal $Y \sim \mathsf{N}(x, 0.04)$. The acceptance probability is therefore

$$\alpha(x, y) = \min\left\{\exp\left(-\frac{(y-10)^2 - (x-10)^2}{2}\right), 1\right\}.$$

Suppose the objective is to estimate $\ell = \mathbb{E}X$, where X is distributed according to the steady-state distribution of the Markov chain (that is, $\mathsf{N}(10, 1)$). The following MATLAB script implements the random walk sampler and provides a 95% confidence interval for ℓ using the covariance method. The sample size is $N = 10^5$. To ensure stationarity, a burn-in period of 300 is used. The maximal lag is also set to $K = 300$. Figure 8.2 shows that the autocovariance function decays sufficiently fast to make it negligible after approximately 200 lags.

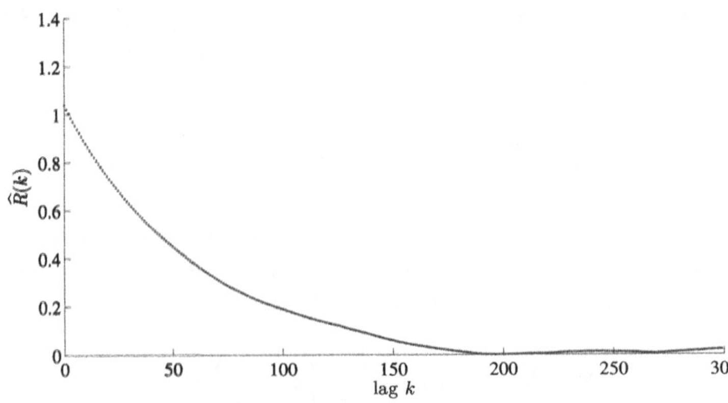

Figure 8.2 Estimated autocovariance function for the data from the random walk sampler.

A typical outcome for the confidence interval is (9.91, 10.04), which contains the true value $\ell = 10$. The asymptotic variance of the sample average for these dependent data is about a factor of 100 larger than the sample variance for independent data from the $\mathsf{N}(10, 1)$ distribution, leading to a confidence interval that is 10 times wider than would be obtained if the data were independent.

```
%covmethod.m
N=10^5; %sample size, including burn-in
tstat = 300; %burn-in period and maximal lag
sample = zeros(N,1);
sigma = 0.2; %standard deviation of proposal
X = randn*sigma; %generate an initializing point
for k=1:N
    Y = X + randn*sigma; %generate the proposal move
    if rand<min(exp(-.5*(Y-10)^2+.5*(X-10)^2), 1) %acceptance step
        X = Y; %update X
    end
    sample(k) = X; %store sample
end
K = tstat; x = sample(tstat:N);
[R,lags] = xcov(x,K,'unbiased'); %calculate covariance function
plot(lags(K+1:2*K),R(K+1:2*K),'.') %plot covariance function
S2 = R(K+1) + 2*sum(R(K+2:2*K)); %asymptotic variance
ell = mean(x);
RE = sqrt(S2)/ell/sqrt(numel(x));
fprintf('ell = %g ; CI = ( %g , %g ) \n',...
    ell,ell*(1-1.96*RE),ell*(1+1.96*RE))
```

8.3.2 Batch Means Method

In the **batch means** method, the first K values of the data $X_1, \ldots, X_M$ are discarded, corresponding to the burn-in period. The remaining $M - K$ values are divided into N batches, each of size $T = (M - K)/N$ (we assume that T is an integer). Let $X_{t,i}$ denote the t-th observation from the i-th batch. If the burn-in period is large enough, each $X_{t,i}$ has approximately the same distribution, corresponding to the steady-state variable X. Let Y_i be the sample mean of the i-th batch:

$$Y_i = \frac{1}{T} \sum_{t=1}^{T} X_{t,i}, \quad i = 1, \ldots, N .$$

The sample mean of the $\{X_{K+1}, \ldots, X_M\} = \{X_{1,1}, \ldots, X_{T,N}\}$ gives an estimator of $\ell = \mathbb{E}X$. This estimator coincides with the sample mean of the batch means $Y_1, \ldots, Y_N$:

$$\widehat{\ell} = \frac{1}{M - K} \sum_{t=K+1}^{M} X_t = \frac{1}{N} \sum_{i=1}^{N} Y_i = \bar{Y} . \tag{8.13}$$

By choosing the batch size T large enough, one can ensure *approximate independence* of the batch means $Y_1, \ldots, Y_N$, so that approximate confidence intervals can be constructed as described in Section 8.2 for independent data.

The procedure is illustrated in Figure 8.3 and summarized in Algorithm 8.4.

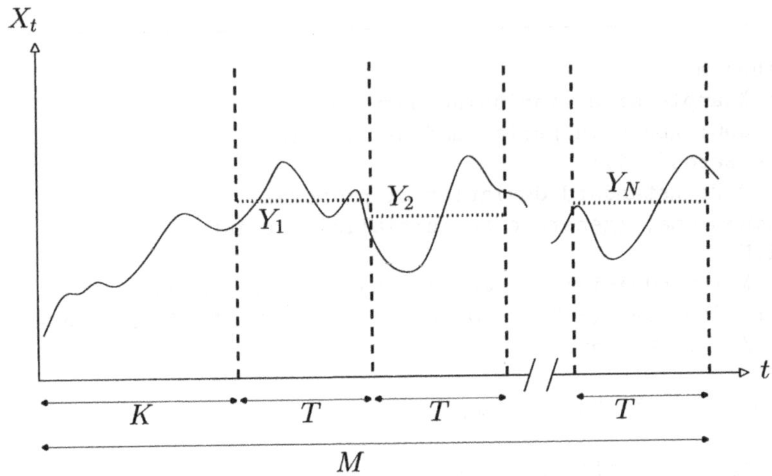

Figure 8.3 Illustration of the batch means procedure.

Algorithm 8.4 (Batch Means Method)

1. *Perform a single simulation run of length M, and delete K observations corresponding to the burn-in period.*
2. *Divide the remaining $M - K$ observations into N batches, each of length $T = (M - K)/N$.*
3. *Calculate the point estimator and the confidence interval for ℓ from* (8.13) *and* (8.5)*, respectively, where S is the sample standard deviation of $Y_1, \ldots, Y_N$.*

■ **EXAMPLE 8.6 (Random Walk on the Positive Integers)**

☞ 86 Let $I_0, I_1, \ldots \overset{\text{iid}}{\sim} \mathsf{Ber}(p)$ be a Bernoulli process for some $0 < p < 1/2$. Consider the random walk $\{X_n, n = 0, 1, 2, \ldots\}$ on $\mathbb{N}$ defined by

$$X_{n+1} = \max\{X_n + 2I_n - 1, 0\}, \quad n = 0, 1, 2 \ldots ,$$

with some $X_0 \in \mathbb{N}$. Thus, from any state $x > 0$ the random walk jumps to $x + 1$ with probability p and to $x - 1$ with probability $q = 1 - p$. From state $x = 0$ the random walk jumps to state 1 with probability p and remains in 0 with probability ☞ 633 q. By Markov chain theory (see Section A.10.2) X_n converges in distribution to $X \sim \mathsf{Geom}_0(1 - p/q)$. Suppose the objective is to estimate $\ell = \mathbb{E}X = p/(q - p)$. The following MATLAB program implements the random walk with $p = 0.25$ starting from $X_0 = 0$, and calculates a 95% confidence interval for $\ell = 0.5$ using the batch means method. A typical 95% confidence interval is $(0.480, 0.507)$.

```
%batchmeans.m
clear all
p=0.25;q=1-p;M=100000;
K = 100; %throw away
B = 300; %number of batches
N = M-K; %remaining samples
T = (M-K)/B; x = zeros(1,M);
for i=2:M
    x(i)=max(0,x(i-1)+2*(rand<p)-1);
end
y=zeros(1,B); %the batch means
for k=1:B
    y(k) = mean(x(K+1 + (k-1)*T : K + k*T));
end
ell = mean(y);
RE = std(y)/ell/sqrt(B);
true_ell = p/(q-p);
fprintf('true_ell=%g; ell=%g; CI=(%g,%g) \n',...
    true_ell, ell, ell*(1-1.96*RE), ell*(1+1.96*RE))
```

Remark 8.3.1 (Replication–Deletion) In the **replication–deletion method** N independent runs are carried out rather than a single simulation run as in the batch means method. From each replication one deletes K initial observations corresponding to the burn-in period and then calculates the point estimator and the confidence interval for ℓ via (8.13) and (8.5), respectively, exactly as in the batch means approach. Note that the confidence interval obtained with the replication–deletion method is unbiased, whereas the one obtained from the batch means method is slightly biased. However, the replication–deletion method requires deletion from *each* replication, as compared to *a single* deletion in batch means method. For this reason the replication–deletion method is not as popular as the batch means method. For more details on the replication–deletion method see [21].

8.3.3 Regenerative Method

For a discrete- or continuous-time *regenerative* process $\{X_t\}$ the existence of a limiting distribution is guaranteed under very mild conditions, see the regeneration Theorem A.9.1. Moreover, by the same theorem the behavior of the limiting distribution depends only on the behavior of the process during a typical cycle. In particular, consider a regenerative process with regeneration times $T_0, T_1, T_2, \ldots$ and cycle lengths $\tau_i = T_i - T_{i-1}$, $i = 1, 2, \ldots$. Define the **reward** in cycle i as ☞ 630

$$R_i = \sum_{t=T_{i-1}}^{T_i - 1} X_t \quad \text{or} \quad R_i = \int_{T_{i-1}}^{T_i} X_t \, \mathrm{d}t \,, \tag{8.14}$$

depending on whether $\{X_t\}$ is a discrete-time or continuous-time process. We assume for simplicity that $T_0 = 0$ and that in the discrete case the cycle lengths are not always a multiple of some integer greater than 1. Let $\tau = T_1$ be the length of the first regeneration cycle and $R = R_1$ the first reward. Then, under the conditions of Theorem A.9.1, $X_t \xrightarrow{d} X$ for some random variable X, and

$$\ell = \mathbb{E}X = \frac{\mathbb{E}R}{\mathbb{E}\tau} . \tag{8.15}$$

To estimate ℓ via simulation, note that $(R_1, \tau_1), (R_2, \tau_2), \ldots$ is a sequence of iid random vectors. The steady-state performance measure ℓ can thus be estimated via the ratio estimator

$$\widehat{\ell} = \frac{\bar{R}}{\bar{\tau}} = \frac{\sum_{i=1}^N R_i}{\sum_{i=1}^N \tau_i} . \tag{8.16}$$

The estimator $\widehat{\ell}$ is biased; that is, $\mathbb{E}\widehat{\ell} \neq \ell$. However, $\widehat{\ell}$ is **strongly consistent**; that is, it converges to ℓ with probability 1 as $N \to \infty$. This follows directly from the fact that, by the law of large numbers, $\widehat{R}$ and $\widehat{\tau}$ converge almost surely to $\mathbb{E}R$ and $\mathbb{E}\tau$, respectively.

The main advantage of the regenerative simulation method is that no burn-in period is required. A disadvantage is that the method is not as generally applicable as, for example, the batch-means method. In particular, it may be difficult to identify the regeneration points, and the regenerative cycles could be very long.

Let S_R^2, S_τ^2, and $S_{R,\tau}$ be, respectively, the sample variance of R, the sample variance of τ, and the sample covariance of R and τ, based on the data $\{(R_i, \tau_i)\}$. Then, by (8.9), an approximate $1 - \alpha$ confidence interval for ℓ is of the form

$$\left(\widehat{\ell} - z_{1-\alpha/2}\frac{S}{\sqrt{N}}, \quad \widehat{\ell} + z_{1-\alpha/2}\frac{S}{\sqrt{N}}\right) , \tag{8.17}$$

where

$$S^2 = \frac{S_R^2 - 2\widehat{\ell}\, S_{R,\tau} + \widehat{\ell}^{\,2} S_\tau^2}{\bar{\tau}^2} \tag{8.18}$$

is an estimator of the asymptotic variance of $\widehat{\ell}$; that is, of $\sigma^2 = \lim_{N\to\infty} N\,\mathrm{Var}(\widehat{\ell})$. An algorithm for constructing an approximate $1-\alpha$ confidence interval for ℓ is thus as follows.

Algorithm 8.5 (Regenerative Simulation Method)

1. *Simulate N regenerative cycles of the process $\{X_t\}$.*
2. *Compute the sequence $\{(R_i, \tau_i), i = 1, \ldots, N\}$.*
3. *Calculate the point estimator $\widehat{\ell}$ and the confidence interval of ℓ from* (8.16) *and* (8.17), *respectively.*

Note that one could also use two independent simulations of length N — one for estimating $\mathbb{E}R$ and the other for estimating $\mathbb{E}\tau$. In that case $S^2 = (S_R^2 + \widehat{\ell}^{\,2} S_\tau^2)/\bar{\tau}^2$.

Remark 8.3.2 (Steady-State and Long-Run Average Performance) If the reward in each cycle is of the form (8.14), then $\ell = \mathbb{E}X$ can be viewed as both the expected *steady-state* performance and the *long-run average* performance.

This last interpretation is valid even if the reward in each cycle is not of the form (8.14), as long as the $\{(\tau_i, R_i)\}$ are iid. In that case,

$$\ell = \lim_{t \to \infty} \frac{\sum_{i=0}^{N_t - 1} R_i}{t} = \frac{\mathbb{E}R}{\mathbb{E}\tau}, \tag{8.19}$$

where N_t is the number of regenerative cycles in $[0, t]$; see, for example, [28].

■ EXAMPLE 8.7 (Random Walk via Regenerative Simulation)

Consider the random walk $\{X_n, n = 0, 1, \ldots\}$ of Example 8.6. The MATLAB program below implements the random walk starting from $X_0 = 0$ with $p = 0.25$ and calculates a 95% confidence interval for $\ell = 0.5$ using the regenerative method. The regeneration times are taken to be the times when the process hits 0. A typical 95% confidence interval is $(0.454, 0.488)$. Figure 8.4 shows a typical outcome of the random walk for $n = 0, \ldots, 59$.

```
%regenmeth.m
clear all, p=0.25;q=1-p;
N=10000;
z = zeros(1,N);R = zeros(1,N);tau = zeros(1,N);
Rsum=0;regcount = 0;lastregtime = 1;
for i=2:N
    if rand<p
        z(i)=z(i-1)+1;
    elseif z(i-1) %if z is not zero
        z(i)=z(i-1)-1;
    end
     Rsum = Rsum + z(i);
    if z(i)==0 %regeneration detected
        regcount = regcount + 1;
        R(regcount) = Rsum;
        tau(regcount) = i - lastregtime;
        Rsum = 0;
        lastregtime = i;
    end
end
stairs(0:59,z(1:60)),hold on, plot(0:59,z(1:60),'.')
ell = mean(R)/mean(tau)
C = cov(R,tau);
s = sqrt(C(1,1) - 2*ell*C(1,2) + ell^2*C(2,2))
RE = s/mean(tau)/sqrt(N)
fprintf('ell %g ; 0.95 CI ( %g , %g ) \n', ...
             ell,ell*(1-1.96*RE),ell*(1+1.96*RE))
```

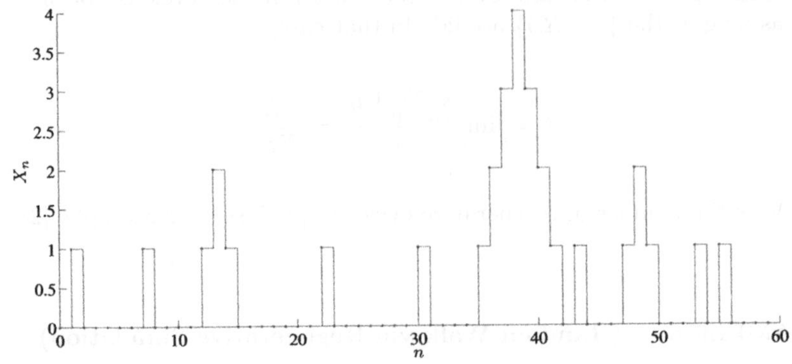

Figure 8.4 The random walk on the positive integers as a regenerative process. The regeneration times are taken to be the times when the process hits 0.

8.4 EMPIRICAL CDF

An important instrument in analyzing iid samples is the **empirical cdf**, defined as the function

$$F_N(x) = \frac{1}{N}\sum_{i=1}^{N} \mathrm{I}_{\{x_i \leqslant x\}} = \frac{|\{i : x_i \leqslant x\}|}{N}, \quad x \in \mathbb{R}\,, \tag{8.20}$$

for given data $x_1, \ldots, x_N$. The empirical cdf is a nondecreasing step function which jumps up by an amount of $1/N$ at each of the data points $\{x_i\}$. Note that F_N is right-continuous and bounded between 0 and 1. In other words, F_N is a genuine cdf. It is precisely the cdf of the random variable that takes the values $x_1, \ldots, x_N$ with probabilities $1/N, \ldots, 1/N$ (if all the observations are distinct). In Figure 8.5 the empirical cdfs are shown of iid samples of size 10 (left) and 100 (right) from the $\mathsf{Exp}(1)$ distribution. The true cdf is plotted as well.

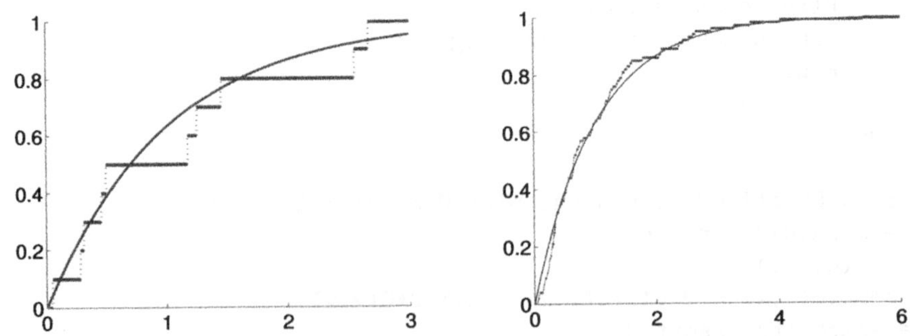

Figure 8.5 The empirical cdfs for a sample of size 10 (left) and 100 (right) from the $\mathsf{Exp}(1)$ distribution, together with the true cdf.

If instead of deterministic $\{x_i\}$, *random* X_i are taken in (8.20), then $F_N(x)$ becomes random as well. To distinguish between the deterministic and the random case, we denote the random empirical cdf by $\widehat{F}_N(x)$.

Transforming the data $X_1, \ldots, X_N$ to $U_1 = F(X_1), \ldots, U_N = F(X_N)$ gives an iid $\mathsf{U}(0,1)$ sequence of random variables, assuming that F is continuous and strictly increasing. The empirical cdf of the $\{U_i\}$, denoted by $\widehat{G}_N(u)$, is called the **reduced empirical cdf**. Letting x and u be related via $x = F^{-1}(u)$ and $u = F(x)$, we have

$$\begin{aligned}\widehat{F}_N(x) - F(x) &= \frac{1}{N}\sum_{i=1}^{N} \mathrm{I}_{\{X_i \leqslant x\}} - F(x) \\ &= \frac{1}{N}\sum_{i=1}^{N} \mathrm{I}_{\{U_i \leqslant u\}} - u = \widehat{G}_N(u) - u\,.\end{aligned}$$

The maximum distance between the empirical and the true cdf,

$$D_N = \sup_{x \in \mathbb{R}} |\widehat{F}_N(x) - F(x)| = \sup_{0 \leqslant u \leqslant 1} |\widehat{G}_N(u) - u|\,, \tag{8.21}$$

is called the **Kolmogorov statistic** of the data. Note that the distribution of D_N does not depend on F. Properties of the empirical cdf include [33]:

1. *Order statistics:* If $x_{(1)} < x_{(2)} < \cdots < x_{(N)}$ denote the (distinct) ordered samples, then
$$F_N(x_{(i)}) = \frac{i}{N}\,. \tag{8.22}$$

2. *Binomial distribution:* $N\,\widehat{F}_N(x) \sim \mathsf{Bin}(N, F(x))$ and $N\,\widehat{G}_N(u) \sim \mathsf{Bin}(N, u)$.

3. *Glivenko–Cantelli:* $D_N \xrightarrow{\text{a.s.}} 0$. Consequently, $\widehat{F}_N(x) \xrightarrow{\text{a.s.}} F(x)$, uniformly in x.

4. *Central limit theorem:* $\sqrt{N}(\widehat{F}_N(x) - F(x)) \xrightarrow{d} Z \sim \mathsf{N}(0, F(x)(1 - F(x)))$ as $N \to \infty$.

5. *Conditional Poisson:* The probability distribution of the reduced empirical cdf $\{\widehat{G}_N(u), 0 \leqslant u \leqslant 1\}$, viewed as a stochastic process on $[0,1]$, is the same as the conditional distribution of a Poisson process $\{M_u, 0 \leqslant u \leqslant 1\}$ with rate $1/N$ *given* that $M_1 = N$. ☞ 170

6. *Brownian bridge:* The stochastic process $\{\sqrt{N}(\widehat{G}_N(u) - u),\ 0 \leqslant u \leqslant 1\}$ converges in distribution to a Brownian bridge process on $[0,1]$ (see, for example, [31]). ☞ 193

7. *Kolmogorov distribution:* If F is continuous, then
$$\lim_{N\to\infty} \mathbb{P}(\sqrt{N}\,D_N \leqslant x) = \sum_{k=-\infty}^{\infty} (-1)^k \mathrm{e}^{-2(kx)^2}, \quad x > 0\,. \tag{8.23}$$

8. *Confidence interval:* An approximate $1 - \alpha$ confidence interval for $F(x)$ is
$$\left(F_N(x) - z_{1-\alpha/2}\sqrt{\tfrac{F_N(x)(1-F_N(x))}{N}},\quad F_N(x) + z_{1-\alpha/2}\sqrt{\tfrac{F_N(x)(1-F_N(x))}{N}}\right)\,.$$

Equivalently, an approximate $1-\alpha$ confidence interval for $F(x_{(i)})$ is

$$\left(\frac{i}{N} - z_{1-\alpha/2}\sqrt{\frac{i(1-i/N)}{N^2}},\quad \frac{i}{N} + z_{1-\alpha/2}\sqrt{\frac{i(1-i/N)}{N^2}}\right).$$

■ **EXAMPLE 8.8 (Confidence Bounds for the Cdf)**

In Figure 8.6 the upper and lower 90% confidence curves for the cdf F are depicted based on an iid sample of size $N = 50$ from the $\mathsf{Exp}(1)$ distribution. The true cdf is plotted as well.

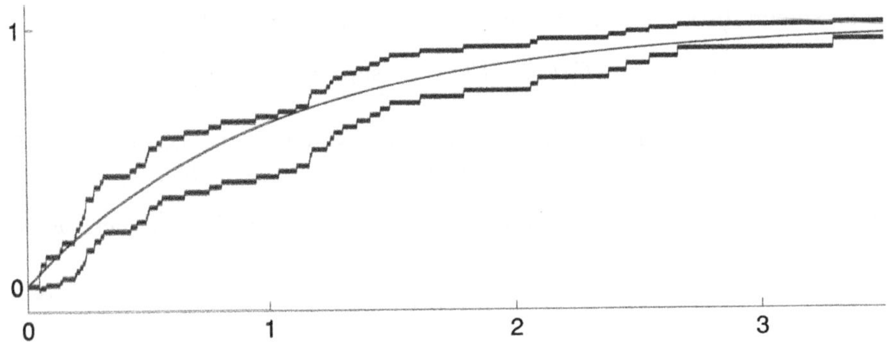

Figure 8.6 90% confidence curves for the cdf of the $\mathsf{Exp}(1)$ distribution based on 50 iid samples.

The following MATLAB code is used.

```
%empcdfr.m
clear all,clc
rand('state',123);
N = 50; %sample size
x = sort(-log(rand(1,N))); %generate and sort sample
x=[0,x]; % append 0 for plotting purposes
z =(0:N)/N;
zl = z - 1.65*sqrt(z.*(1-z)/N); % lower curve
zu = z + 1.65*sqrt(z.*(1-z)/N); % upper curve
axes('FontSize',16),hold on
for i=1:N %plot the confidence bounds
    line([x(i),x(i+1)],[zl(i),zl(i)],'LineWidth',3);
    line([x(i+1),x(i+1)],[zl(i),zl(i+1)]);
    line([x(i),x(i+1)],[zu(i),zu(i)],'LineWidth',3);
    line([x(i+1),x(i+1)],[zu(i),zu(i+1)]);
end
t = 0:0.01:max(x);plot(t,1-exp(-t))
```

8.5 KERNEL DENSITY ESTIMATION

A popular approach for estimating a probability density from (simulated) data is **kernel density estimation** [30, 35, 38]. Kernel density estimates are used in many applications, including:

- structural assessment of a distribution (multimodality, skewness, etc.) [30, 34],
- summarization of Bayesian posterior pdfs, classification and discriminant analysis [35],
- smoothed bootstrap sampling and particle filtering [15].

Let $X_1, \ldots, X_N$ be independent realizations from an unknown continuous probability density function f on some space $\mathscr{X} \subset \mathbb{R}$. A univariate **kernel density estimator** of f is defined as a function

$$\widehat{f}(x; h) = \frac{1}{Nh} \sum_{i=1}^{N} \kappa\left(\frac{x - X_i}{h}\right), \quad x \in \mathbb{R},$$

where κ is a symmetric ($\kappa(-x) = \kappa(x)$) pdf on $\mathbb{R}$, called the **kernel function**, and h is a positive parameter, called the **bandwidth**.

In a **Gaussian kernel density estimator** the kernel is standard normal; therefore, for a given scale parameter $t = h^2$ the Gaussian kernel density estimator is of the form

$$\widehat{f}(x; t) = \frac{1}{N} \sum_{i=1}^{N} \varphi(x, X_i; t), \quad x \in \mathbb{R}, \tag{8.24}$$

with

$$\varphi(x, X_i; t) = \frac{1}{\sqrt{2\pi t}} \, \mathrm{e}^{-\frac{(x - X_i)^2}{2t}} .$$

The asymptotic properties of the kernel density estimate depend crucially on the choice of the bandwidth parameter. The choice of the kernel function is much less important [38, Page 31]. Nevertheless, a preference for smoothness of the kernel density estimate for all sample sizes makes the Gaussian kernel density estimator the most popular choice.

A well-studied criterion for the quality of the estimate $\widehat{f}$ is the **mean integrated squared error** (MISE):

$$\mathrm{MISE}(t) = \mathbb{E}_f \int [\widehat{f}(x; t) - f(x)]^2 \, \mathrm{d}x ,$$

which can be decomposed into integrated squared bias and integrated variance components:

$$\mathrm{MISE}(t) = \int \Big(\underbrace{\mathbb{E}_f[\widehat{f}(x; t)] - f(x)}_{\text{pointwise bias of } \widehat{f}}\Big)^2 \mathrm{d}x + \int \underbrace{\mathrm{Var}_f(\widehat{f}(x; t))}_{\text{pointwise variance of } \widehat{f}} \, \mathrm{d}x .$$

Here the expectation and variance operators apply to the iid sample $\{X_1, \ldots, X_N\}$. An alternative error criterion is the **expected L^1 error**,

$$\mathbb{E}_f \int |\widehat{f}(x; t) - f(x)| \, \mathrm{d}x ,$$

which is scale invariant and has generally better theoretical properties [14], but is not as computationally tractable as the MISE.

For the Gaussian kernel density estimator (8.24) a first-order asymptotic approximation of the MISE (under certain regularity assumptions), is given by

$$\frac{1}{4}\,t^2\,\|f''\|^2 + \frac{1}{2N\sqrt{\pi t}}\,, \tag{8.25}$$

where $\|f''\|^2 = \int (f''(x))^2\,\mathrm{d}x$. The asymptotically optimal value of t is the minimizer

$$t^* = \left(\frac{1}{2N\sqrt{\pi}\,\|f''\|^2}\right)^{2/5}, \tag{8.26}$$

giving the optimal asymptotic rate of decay of the MISE (for a proof see [38]):

$$\mathrm{MISE}(t^*) = N^{-4/5}\,\frac{5\|f''\|^{2/5}}{4^{7/5}\pi^{2/5}} + o(N^{-4/5}), \quad N \to \infty\,. \tag{8.27}$$

In order to compute the optimal t^* for the Gaussian kernel density estimator (8.24) one needs to estimate the functional $\|f''\|^2$. The **Gaussian rule of thumb** is to assume that f is the density of the $\mathsf{N}(\widehat{\mu}, \widehat{\sigma}^2)$ distribution, where $\widehat{\mu}$ and $\widehat{\sigma}^2$ are the sample mean and variance of the data, respectively [34]. In this case $\|f''\|^2 = \widehat{\sigma}^{-5}\pi^{-1/2}3/8$ and the Gaussian rule of thumb becomes:

$$t_{\mathrm{rot}} = \left(\frac{4\,\widehat{\sigma}^5}{3\,N}\right)^{2/5} \approx 1.12\,\widehat{\sigma}^2\,N^{-2/5}\,.$$

If the data has outliers and is far from normally distributed, then a more robust rule of thumb is the following choice [35]:

$$t_{\mathrm{Rot}} = \left(\frac{4S^5}{3N}\right)^{2/5},$$

where $S = \min\left\{\widehat{\sigma}, \frac{R}{1.34}\right\}$ is a robust measure of the spread of the data. Here R is the **interquartile range** of the data defined as follows:

$$R = \underbrace{X_{(\lceil 0.75N \rceil)}}_{75\%\ \text{quantile}} - \underbrace{X_{(\lceil 0.25N \rceil)}}_{25\%\ \text{quantile}}\,.$$

■ **EXAMPLE 8.9 (Bimodal Density Estimation)**

Consider a Gaussian kernel density estimator for the estimation of the density of the skewed bimodal mixture distribution

$$\frac{3}{4}\mathsf{N}(0,1) + \frac{1}{4}\mathsf{N}\left(\frac{3}{2}, \frac{1}{3^2}\right),$$

☞ 53 by which we mean that an outcome is drawn from the $\mathsf{N}(0,1)$ distribution with probability 3/4, and similarly for the second component. We sample 10^3 points from this density and use $\sqrt{t_{\mathrm{Rot}}}$ as an estimate of the bandwidth. The left panel of Figure 8.7 shows that the density estimation using t_{Rot} is satisfactory for this problem. However, the estimation of the density of the *separated* bimodal mixture

$$\frac{1}{2}\mathsf{N}\left(-2, \frac{1}{4}\right) + \frac{1}{2}\mathsf{N}\left(2, \frac{1}{4}\right),$$

based on 10^3 points, is not so successful, as is seen from the right panel of Figure 8.7. The reason for this poor performance is that while $\min\left(\widehat{\sigma}, \frac{R}{1.34}\right)$ accurately captures the spread of the data in the skewed bimodal density, it overestimates the spread in the separated bimodal case. Thus, although the Gaussian rule of thumb is easy to implement and computationally fast, one must exercise caution when using it.

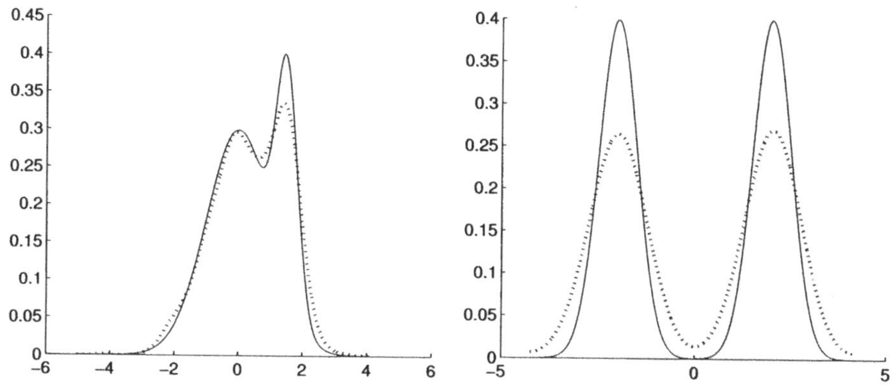

Figure 8.7 Density estimation using the robust Gaussian rule of thumb t_{Rot}. The dotted curve is the estimate and the solid line is the true density. The right panel shows that the Gaussian rule of thumb can oversmooth.

To address the shortcomings of the Gaussian rule of thumb, two conceptually different automatic data-driven bandwidth selection methods have been proposed. The first one is based on a classical performance criterion — **least squares cross validation** (LSCV) — and the second one is commonly referred to as the **plug-in bandwidth selection** method.

8.5.1 Least Squares Cross Validation

In LSCV the optimal bandwidth is defined as the global minimizer of an unbiased estimate of the **integrated squared error** (ISE), given by

$$\mathrm{ISE}(t) = \int [\widehat{f}(x;t) - f(x)]^2 \, \mathrm{d}x \, .$$

Minimization of the ISE(t) is equivalent to minimization of

$$\int [\widehat{f}(x;t)]^2 \, \mathrm{d}x - 2\,\mathbb{E}_f \widehat{f}(X;t) \, .$$

Note that unlike the MISE, the ISE is a random variable depending on the particular data. The term $\mathbb{E}_f \widehat{f}(X;t)$ can be estimated without bias via the **cross-validation estimator**:

$$\frac{1}{N} \sum_{i=1}^{N} \widehat{f}_{-i}(X_i;t) = \frac{1}{N(N-1)} \sum_{i=1}^{N} \sum_{j \neq i} \varphi(X_i, X_j; t) \, .$$

Here $\widehat{f}_{-i}$ is the Gaussian kernel density estimator based on all data points except X_i. Using the property $\int_{\mathbb{R}} \varphi(x, y; t)\, \varphi(x, z; t)\, \mathrm{d}x = \varphi(y, z; 2t)$ of the Gaussian kernel, the term $\int [\widehat{f}(x; t)]^2\, \mathrm{d}x$ can be written as:

$$\int [\widehat{f}(x;t)]^2\, \mathrm{d}x = \frac{1}{N^2} \sum_{i=1}^{N} \sum_{j=1}^{N} \varphi(X_i, X_j; 2t)\ .$$

Thus, the LSCV bandwidth $\sqrt{t_{\mathrm{LS}}}$ is formally defined as follows:

$$t_{\mathrm{LS}} = \operatorname*{argmin}_{t>0} g(t)\ ,$$

where

$$g(t) = \frac{1}{N^2} \sum_{i=1}^{N} \sum_{j=1}^{N} \varphi(X_i, X_j; 2t) - 2\frac{1}{N(N-1)} \sum_{i=1}^{N} \sum_{j \neq i} \varphi(X_i, X_j; t)\ .$$

Practical implementation of the LSCV procedure requires that we find a computational device to reduce the cost of evaluating $g(t)$. Direct evaluation of $g(t)$ for a given value of t requires $\mathcal{O}(N^2)$ evaluations of the Gaussian kernel. This computational cost can be substantially reduced by exploiting the fact that the Gaussian kernel density estimator is the solution of the PDE

$$\frac{\partial}{\partial t}\widehat{f}(x;t) = \frac{1}{2}\frac{\partial^2}{\partial x^2}\widehat{f}(x;t),\ t > 0\ , \tag{8.28}$$

with $x \in \mathbb{R}$, $\lim_{x\to\pm\infty} \widehat{f}(x;t) = 0$ and initial condition $\widehat{f}(x;0) = \Delta(x)$, where $\Delta(x) = \frac{1}{N}\sum_{i=1}^{N} \delta_{X_i}(x)$ is the **empirical density** of the data $X_1, \ldots, X_N$, and $\delta_{X_i}(x)$ is the Dirac measure at X_i [7, 9, 12]. In other words, instead of computing the Gaussian kernel density estimator $\widehat{f}(x;t)$ directly, it can be obtained by evolving the solution of the PDE (8.28) up to time t. The key observation is that (8.28) can be solved on a finite domain efficiently using an FFT-related transform. The procedure is as follows. First, without loss of generality we can consider the data to be on the unit interval $[0, 1]$. If the raw data is not on the unit interval, then it can be rescaled so that the main body of the data is in the interior of $[0, 1]$ and away from the boundaries at 0 and 1. Second, the data is binned into n bins, where n is a power of 2 — as required by the fastest implementations of the FFT. Third, note that away from the boundaries the Gaussian kernel $\varphi(x, X_i; t)$ is approximated by
☞ 714 the *theta function* [6, 8]:

$$\theta(x, X_i; t) = \sum_{k=-\infty}^{\infty} \varphi(x, 2k + X_i; t) + \varphi(x, 2k - X_i; t), \quad x \in (0,1)\ .$$

In fact, for a fixed X_i the theta function satisfies (8.28) on the interval $(0, 1)$, and the Gaussian kernel $\varphi(x, X_i; t)$ satisfies (8.28) on the domain $x \in \mathbb{R}$. The difference between the theta function and the Gaussian kernel becomes negligible as t becomes smaller in the sense that

$$\lim_{t\downarrow 0} \frac{\theta(x, X_i; t)}{\varphi(x, X_i; t)} = 1, \quad x \in (0,1)\ .$$

Application of the method of separation of variables for solving (8.28) on the interval $(0, 1)$ shows that the theta function has the following Fourier series expansion:

$$\theta(x, X_i; t) = \sum_{k=-\infty}^{\infty} \mathrm{e}^{-k^2\pi^2 t/2} \cos(k\pi x)\cos(k\pi X_i), \quad x \in (0,1)$$
$$= 1 + 2\sum_{k=1}^{\infty} \mathrm{e}^{-k^2\pi^2 t/2} \cos(k\pi x)\cos(k\pi X_i) .$$

Thus, on the interval $(0, 1)$ we can approximate the Gaussian kernel density estimator (8.24) using the truncated Fourier series expansion:

$$\widehat{f}(x;t) \approx \sum_{k=0}^{n-1} a_k \,\mathrm{e}^{-k^2\pi^2 t/2}\cos(k\pi x), \quad n \gg 1 , \tag{8.29}$$

where the coefficients $\{a_k\}_{k=0}^{n-1}$ are given by the cosine transform of the empirical data:

$$a_0 = 1, \quad a_k = \frac{2}{N}\sum_{i=1}^{N}\cos(k\pi X_i), \quad k = 1, 2, 3, \ldots . \tag{8.30}$$

Next, given the binned data over the grid of size n, we compute the coefficients $\{a_k\}_{k=0}^{n-1}$ using the fast cosine transform — an FFT-related transform, see Appendix D.5. Finally, for a given t, the values of $\widehat{f}(\cdot;t)$ on the grid are efficiently ☞ 708 computed using the inverse fast cosine transform with input the set of *smoothed* coefficients $\{a_k \,\mathrm{e}^{-k^2\pi^2 t/2}\}_{k=0}^{n-1}$.

Hence, using the theta kernel we can evaluate $\widehat{f}(\cdot;t)$ on a uniform grid of size n in $\mathcal{O}(n \ln n)$ operations. This idea is summarized in the following algorithm.

Algorithm 8.6 (Fast Evaluation of $\widehat{f}(\cdot;t)$ Using (8.29)**)** *Given iid random data $X_1, \ldots, X_N$, parameters $n(= 2^m$ for an integer $m)$ and t, execute the following steps.*

1. *Define a uniform grid of size n:*

$$y_k = \frac{k}{n}, \quad k = 0, \ldots, n .$$

Bin the data to compute

$$\widehat{f}_k = \frac{\#\{X_i : X_i \in (y_k, y_{k+1})\}}{N}, \quad k = 0, \ldots, n-1 .$$

Thus, we have that $\widehat{f}_k \approx \widehat{f}(y;0)$, $y \in (y_k, y_{k+1})$.

2. *Compute the fast cosine transform $\{\widehat{f}_k\}_{k=0}^{n-1}$ of the data and let $\{a_k\}_{k=0}^{n-1}$ be the coefficients of the fast cosine transform. We have thus efficiently computed the first n coefficients in* (8.30)*. These coefficients are computed only once and stored in memory.*

3. *Let $a_k(t) = a_k \,\mathrm{e}^{-k^2\pi^2 t/2}$ for $k = 0, \ldots, n-1$. Compute the inverse fast cosine transform of $\{a_k(t)\}_{k=0}^{n-1}$, so that we obtain the (approximate) values of $\widehat{f}(\cdot;t)$ over the uniform grid via* (8.29)*.*

If evaluation of $\widehat{f}(\cdot;t)$ is required for a different value of $t > 0$, we execute only Step 3 in the above algorithm.

The following MATLAB code implements the LSCV method using the fast cosine transform to evaluate $g(t)$. The standard MATLAB function `fminbnd.m` is used to find the minimum of $g(t)$.

```
function [bandwidth,f,y_k]=LSCV(X,n,MIN,MAX)

X=X(:); %make data a column vector
% set up the grid over which the density estimate is computed;
R=MAX-MIN; dy=R/n; y_k=MIN+[0:dy:R]; N=length(X);
%bin the data uniformly using the grid define above;
[f_k,bins]=histc(X,y_k);f_k=f_k/N;
f_k(end)=[]; y_k(1)=[]; bins(bins==n+1)=[];
a=dct1d(f_k); % discrete cosine transform of initial data
% now compute the optimal bandwidth^2 using the LSCV
t_LS= fminbnd(@g,0,.01);
% defines the LSCV function g(t) that is to be minimized
    function out=g(t)
        a_t=a.*exp(-[0:n-1]'.^2*pi^2*t/2);
        f=idct1d(a_t)/dy;
        int_f2_dy=(f'*f)*dy;
        out=int_f2_dy-2/(N-1)*sum(f(bins))+2/(N-1)/sqrt(2*pi*t*R^2);
    end

a_t=a.*exp(-[0:n-1]'.^2*pi^2*t_LS/2); % smoothed coefficients
f=idct1d(a_t)/dy;     % take the IFCT of the data
bandwidth=t_LS*R^2; % adjust the bandwidth after the rescaling

end
```

We use the following subroutine which computes the discrete cosine transform of the column vector data; see Section D.5. ☞ 708

```
function data=dct1d(data)
% computes the discrete cosine transform of a  vector
data=data(:); % make data a column vector
nrows=length(data);
% Compute weights to multiply DFT coefficients
weight = [1;2*(exp(-i*(1:nrows-1)*pi/(2*nrows))).'];
% Reorder the elements of the columns of x
data = [ data(1:2:end); data(end:-2:2) ];
% Multiply FFT by weights:
data= real(weight.* fft(data));
```

We use the following subroutine for the inverse cosine transform; see Section D.5.

```
function out = idct1d(data)
% computes the inverse discrete cosine transform of a vector
data=data(:); % make data a column vector
nrows=length(data);
% Compute weights
weights = exp(i*(0:nrows-1)*pi/(2*nrows)).';
data = real(ifft(weights.*data));
% Reorder elements
out = zeros(nrows,1);
out(1:2:nrows) = data(1:nrows/2);
out(2:2:nrows) = data(nrows:-1:nrows/2+1);
```

■ EXAMPLE 8.10 (Bimodal Density Estimation Revisited)

Figure 8.8 shows the performance of the LSCV method on the bimodal densities from Example 8.9. Notice that the estimate for the separated bimodal density (right panel) is much better than the estimate obtained using the Gaussian rule of thumb.

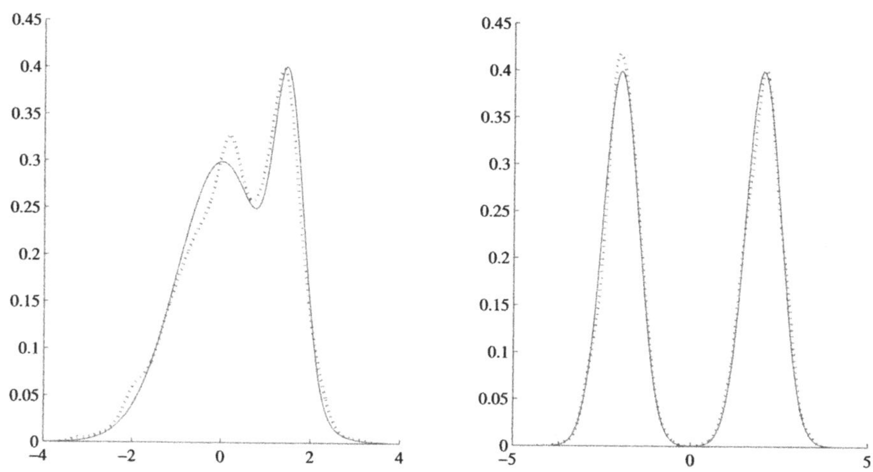

Figure 8.8 Density estimation using the LSCV method. The dotted curve is the estimate.

The bandwidth estimator computed via the LSCV method has been shown to be highly variable [27] compared to alternatives such as the plug-in estimator described in the next section. In addition, occasionally a minimizer of $g(t)$ for $t > 0$ does not exist. However, unlike the plug-in methods, the LSCV method makes no smoothness assumptions about the target density. For a more detailed discussion of the advantages and disadvantages of the LSCV method, see [22].

8.5.2 Plug-in Bandwidth Selection

An alternative to the LSCV method is the **Sheather–Jones plug-in** method [9, 32]. Recall from (8.26) that if we knew $\|f''\|^2$, then the asymptotically MISE-optimal bandwidth for (8.24) is known. One possible estimator for $\|f''\|^2$ is motivated by the identity $\|f^{(j)}\|^2 = (-1)^j \mathbb{E}_f[f^{(2j)}(X)]$, $j \geqslant 1$, where $f^{(j)}$ denotes the j-th derivative of f. The idea is to estimate $\|f^{(2)}\|^2 = \|f''\|^2$ via the estimator ($j = 2$):

$$\begin{aligned}\widehat{\|f^{(j)}\|^2} &\stackrel{\text{def}}{=} \|\widehat{f}^{(j)}(\cdot;t_j)\|^2 \\ &= \frac{1}{N^2}\sum_{k=1}^{N}\sum_{m=1}^{N}\int_{\mathbb{R}} \varphi^{(j)}(x, X_k; t_j)\,\varphi^{(j)}(x, X_m; t_j)\,\mathrm{d}x \\ &= \frac{(-1)^j}{N^2}\sum_{k=1}^{N}\sum_{m=1}^{N} \varphi^{(2j)}(X_k, X_m; 2\,t_j)\,,\end{aligned} \tag{8.31}$$

where $\widehat{f}$ is the Gaussian kernel density estimator with bandwidth $\sqrt{t_j}$, different from $\sqrt{t^*}$. Thus, we need to select a value for t_j in order to estimate $\|f^{(j)}\|^2$. A choice for t_j with good practical performance is (see [9])

$$\widehat{t}_j = \left(\frac{1 + \frac{1}{2^{j+1/2}}}{3}\; \frac{1 \times 3 \times 5 \times \cdots \times (2j-1)}{N\sqrt{\pi/2}\;\|\widehat{f^{(j+1)}}\|^2}\right)^{\frac{2}{3+2j}}. \tag{8.32}$$

Computation of $\|\widehat{f^{(j+1)}}\|^2$ requires knowledge of $\widehat{t}_{j+1}$, which in its turn requires the estimate $\widehat{t}_{j+2}$ and so on, as seen from formulas (8.31) and (8.32). We are faced with the problem of computing the infinite sequence $\{\widehat{t}_{j+k}, k \geqslant 1\}$. However, given $\widehat{t}_{l+1}$ for some integer l we can compute all $\{\widehat{t}_j,\ 1 \leqslant j \leqslant l\}$ recursively, and then estimate t^* from (8.26). This motivates the **l-stage direct plug-in bandwidth selector** [32, 38], defined as follows.

Algorithm 8.7 (l-stage Direct Plug-in Bandwidth Selector) *Given an integer $l \geqslant 3$, execute the following steps.*

1. *Compute $\|\widehat{f^{(l+2)}}\|^2$ by assuming that f is the normal pdf with mean and variance estimated from the iid sample $X_1, \ldots, X_N$. Set $j = l + 1$.*

2. *Using $\|\widehat{f^{(j+1)}}\|^2$ compute $\widehat{t}_j$ via (8.32).*

3. *Using $\widehat{t}_j$ compute $\|\widehat{f^{(j)}}\|^2$ via (8.31).*

4. *If $j > 2$, reset $j = j - 1$ and repeat from Step 2; otherwise, go to Step 5.*

5. *Use $\widehat{t}_2$ to compute $\|\widehat{f^{(2)}}\|^2$ and deliver $\widehat{t^*}$ from (8.26).*

The l-stage direct plug-in bandwidth selector thus involves the estimation of $\|\widehat{f^{(j)}}\|^2, 2 \leqslant j \leqslant l + 1$ via the plug-in estimator (8.31). To improve the method we describe the procedure in a more abstract way as follows. Denote the functional dependence of $\widehat{t}_j$ on $\widehat{t}_{j+1}$ in formula (8.32) as

$$\widehat{t}_j = \gamma_j(\widehat{t}_{j+1})\,.$$

It is then clear that $\widehat{t}_j = \gamma_j(\gamma_{j+1}(\widehat{t}_{j+2})) = \gamma_j(\gamma_{j+1}(\gamma_{j+2}(\widehat{t}_{j+3}))) = \cdots$. For simplicity of notation we define the composition

$$\gamma^{[k]}(t) = \underbrace{\gamma_1(\cdots\gamma_{k-1}(\gamma_k(t))\cdots)}_{k \text{ times}}, \quad k \geqslant 1 .$$

Inspection of formulas (8.32) and (8.26) shows that the estimate of t^* satisfies

$$\widehat{t^*} = \xi\, \widehat{t}_1 = \xi\, \gamma^{[1]}(\widehat{t}_2) = \xi\, \gamma^{[2]}(\widehat{t}_3) = \cdots = \xi\, \gamma^{[l]}(\widehat{t}_{1+l}), \quad \xi = \left(\frac{6\sqrt{2}-3}{7}\right)^{2/5} \approx 0.90 .$$

Then, for a given integer $l > 0$, the l-stage direct plug-in bandwidth selector consists of computing

$$\widehat{t^*} = \xi\, \gamma^{[l]}(\widehat{t}_{l+1}) ,$$

where $\widehat{t}_{l+1}$ is computed via (8.32) by assuming that f in $\|\widehat{f}^{(l+2)}\|^2$ is a normal density with mean and variance estimated from the data. The weakest point of this procedure is that we assume that the true f is a Gaussian density in order to compute $\|\widehat{f}^{(l+2)}\|^2$. This assumption can lead to arbitrarily bad estimates of t^*, when the true distribution is far from being normal. One of the steps in the Sheather–Jones plug-in method [32, 38] uses the l-stage direct plug-in bandwidth selector and this affects its performance adversely.

Instead, we use the **improved plug-in** method, which gives a (typically unique) solution of the nonlinear equation:

$$t = \xi\, \gamma^{[l]}(t) \quad \text{for some large enough } l\,, \tag{8.33}$$

as the (near) optimal bandwidth estimate. The improved plug-in method dispenses with any normality assumptions on f and as a consequence is more reliable and accurate than most alternative plug-in methods [3, 9].

We now give the implementation details of the improved plug-in method. Similar to the LSCV implementation in Algorithm 8.6 we assume the data to be on the interval $[0, 1]$ and use the theta function approximation in (8.29). The data is binned on a grid of size n and the coefficients $\{a_k\}_{k=0}^{n-1}$ in (8.29) are computed using the fast cosine transform. Then, for a given t,

$$\|\widehat{f}^{(j)}\|^2 \approx \frac{\pi^{2j}}{2} \sum_{k=1}^{n-1} k^{2j} a_k^2\, \mathrm{e}^{-k^2\pi^2 t} .$$

Thus, computation of the right-hand side of (8.33) is an $\mathcal{O}(n)$ operation. The following MATLAB code uses `fzero.m` to compute the root of (8.33) and uses the fast cosine transform and inverse fast cosine transform routines from Sections 8.5.1 and D.5. In addition, we use $l = 7$. Our simulation experience is that larger values of l do not change the value of the root of (8.33) in any significant way. ☞ 708

```
function [bandwidth,density,xmesh]=kde(data,n,MIN,MAX)
% set up the grid over which the density estimate is computed;
data=data(:);
R=MAX-MIN; dx=R/(n-1); xmesh=MIN+[0:dx:R]; N=length(data);
%bin the data uniformly using the grid define above;
initial_data=histc(data,xmesh)/N;
a=dct1d(initial_data); % discrete cosine transform of initial data
% now compute the optimal bandwidth^2 using the referenced method
I=[1:n-1]'.^2;
% use  fzero to solve the equation t=zeta*gamma^[l](t)
t_star=fzero(@(t)fixed_point(t,N,I,a(2:end).^2),[0,.01]);
% smooth the discrete cosine transform of initial data using t_star
a_t=a.*exp(-[0:n-1]'.^2*pi^2*t_star/2);
% now apply the inverse discrete cosine transform
density=idct1d(a_t)/dx;
% take the rescaling of the data into account
bandwidth=sqrt(t_star)*R;
```

```
function  out=fixed_point(t,N,I,a2)
% this implements the function t-zeta*gamma^[l](t)
l=7;
f=1/2*pi^(2*l)*sum(I.^l.*a2.*exp(-I*pi^2*t));
for s=l-1:-1:2
   K0=prod([1:2:2*s-1])/sqrt(pi/2);  K1=(1+(1/2)^(s+1/2))/3;
   time=(K1*K0/N/f)^(2/(3+2*s));
   f=1/2*pi^(2*s)*sum(I.^s.*a2.*exp(-I*pi^2*time));
end
out=t-(2*N*sqrt(pi)*f)^(-2/5);
```

■ EXAMPLE 8.11 (Comparison with LSCV)

Typically, the improved plug-in method is more accurate than the LSCV method and the Sheather–Jones method [9]. For example, consider estimating the **asymmetric double claw density** described by:

$$\frac{46}{100}\sum_{k=0}^{1}\mathsf{N}\left(2k-1,\left(\frac{2}{3}\right)^2\right)+\frac{1}{300}\sum_{k=1}^{3}\mathsf{N}\left(-\frac{k}{2},\frac{1}{100^2}\right)+\frac{7}{300}\sum_{k=1}^{3}\mathsf{N}\left(\frac{k}{2},\left(\frac{7}{100}\right)^2\right),$$

using N iid samples. Figure 8.9 shows the typical performance of the improved plug-in and LSCV methods with $N = 10^5$. Note that the LSCV method does not detect the three smaller claws in the left mode of the density very well. To make a more rigorous assessment of accuracy we use the error criterion,

$$\text{Ratio} = \frac{\|\widehat{f}(\cdot\,;\widehat{t^*}) - f\|^2}{\|\widehat{f}(\cdot\,;t_{\mathrm{LS}}) - f\|^2},$$

that is, the ratio of the integrated squared error of the plug-in estimator to the integrated squared error of the LSCV estimator. The second row of Table 8.1 reports the average Ratio over 10 independent simulations for different sample sizes. We can see that the performance of the LSCV method is inferior for large sample sizes.

Table 8.1 Comparison between the LSCV and improved plug-in estimators for the asymmetric double claw density.

N	10^4	10^5	10^6	10^7
Ratio	1.01	0.37	0.55	0.0083

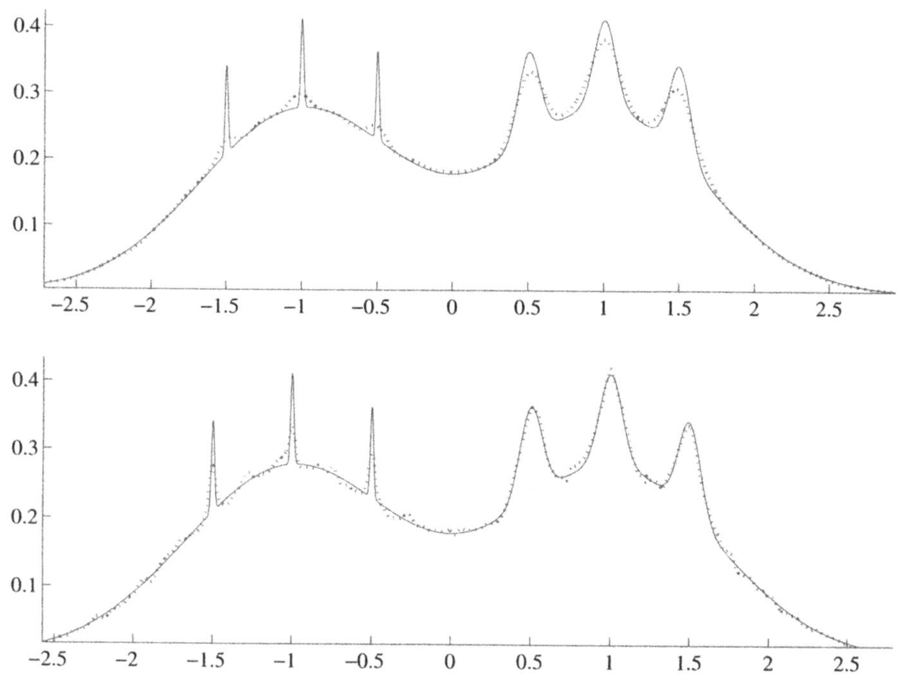

Figure 8.9 Estimation of the asymmetric double claw density using the LSCV method (top panel) and the improved plug-in method (bottom panel). The dashed curve is the estimate.

In addition to the computational advantages of using the approximation (8.29), the theta function can handle boundary effects when the support of the data is known [9]. Figure 8.10 shows the kernel density estimate based on $N = 10^2$ samples from the $\mathsf{Beta}(1,2)$ distribution using the Gaussian ($\kappa \equiv \varphi$) and the theta ($\kappa \equiv \theta$) functions as kernels with the same bandwidth of 0.0520. The theta kernel performs much better at the boundaries.

For a comprehensive simulation study of the improved plug-in method see [9], where the simple partial differential equation (8.28) can be generalized using the

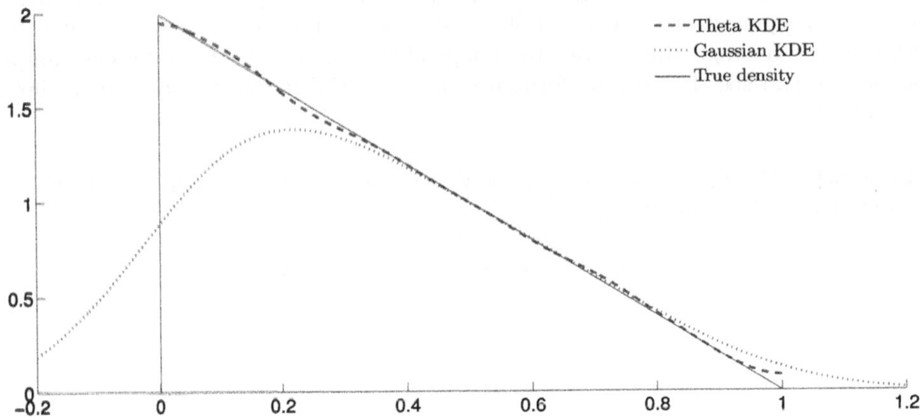

Figure 8.10 Using the theta function as the kernel ($\kappa(x) = \theta(x)$) results in better performance near the boundary of the support: $x = 0$ and $x = 1$.

linear diffusion PDE

$$\frac{\partial}{\partial t} g(x;t) = L g(x;t), \quad x \in \mathscr{X},\ t > 0\,. \tag{8.34}$$

Here:

- the linear differential operator L is of the form $\frac{1}{2}\frac{\mathrm{d}}{\mathrm{d}x}\left(a(x)\frac{\mathrm{d}}{\mathrm{d}x}\left(\frac{\cdot}{p(x)}\right)\right)$;
- $a(x)$ can be any positive function on $\mathscr{X}$ and $p(x)$ is a pilot density estimate of the true pdf f;
- both $a(x)$ and $p(x)$ have bounded second derivatives;
- the initial condition is $g(x,0) = \Delta(x)$, where $\Delta(x)$ is the empirical density.

The advantages of this more general model for density estimation stem from the fact that (8.34) defines a diffusion process with limiting and stationary pdf p. In addition, the diffusion model (8.34) subsumes many of the so-called *variable location-scale kernel density estimators* [2, 20, 29, 36], and it performs substantially better than the simpler Gaussian kernel density estimator [3].

Remark 8.5.1 (Multivariate Kernel Density Estimation) In general, bandwidth selection for *multivariate kernel density estimation* is significantly more difficult than its univariate counterpart [9, 19, 26, 37]. One difficulty is that the support of the density in high dimensions can have an arbitrarily complex geometry. For example, we can consider multivariate kernel density estimation over arbitrary connected domains that satisfy certain regularity conditions, and which require the solution of a two-dimensional diffusion PDE over this complex domain. For more details on this approach see [9].

8.6 RESAMPLING AND THE BOOTSTRAP METHOD

The idea behind resampling is very simple: an iid sample $\mathbf{x} = (x_1, \ldots, x_n)$ from some unknown cdf F represents our best knowledge about F if no further assumptions on F are made. Consequently, if further gathering of data is expensive or impossible, one can still **resample** the $\{x_i\}$ by drawing from the empirical cdf F_n. In this way one can approximately repeat the experiment that gave the original data many times. This is useful if one wants to assess the distributional properties of some statistic $T = T(\mathbf{X})$ of the data $\mathbf{X} = (X_1, \ldots, X_n)$, such as the variance, bias, or mean square error.

The **bootstrap method** is a formalization of the resampling idea. Suppose we wish to estimate a performance measure $\ell = \mathbb{E}_F h(T)$ of T, where h is a real-valued function. It is assumed that T does not depend on the order of the $\{X_i\}$. To estimate ℓ via Monte Carlo one could draw independent replications $\mathbf{X}_1, \ldots, \mathbf{X}_N$ of $\mathbf{X}$ and use the sample average $\widehat{\ell}_F = N^{-1} \sum_{i=1}^{N} h(T_i)$, where $T_i = T(\mathbf{X}_i)$, $i = 1, \ldots, N$, as an estimator for ℓ. However, it may be too time-consuming, or simply not feasible, to obtain such replications. An alternative is to resample the original data. That is, given an outcome $\mathbf{x} = (x_1, \ldots, x_n)$ of $\mathbf{X}$, draw an iid sample $\mathbf{X}^* = (X_1^*, \ldots, X_n^*)$ from the empirical cdf F_n, and repeat this N times to obtain conditionally independent vectors $\mathbf{X}_1^*, \ldots, \mathbf{X}_N^*$ with the same distribution as $\mathbf{X}^*$. For large n, F_n is close to F and consequently $\mathbb{E}_F h(T)$ can be approximated well by $\mathbb{E}_{F_n} h(T)$. The latter is usually difficult to evaluate, but can be simply estimated via Monte Carlo simulation as

$$\frac{1}{N} \sum_{i=1}^{N} h(T_i^*) \,, \tag{8.35}$$

where $T_i^* = T(\mathbf{X}_i^*)$, $i = 1, \ldots, N$. Table 8.2 gives various commonly used bootstrap estimators. The bootstrap procedure is summarized as follows.

Algorithm 8.8 (Bootstrap Method) *Let* $\mathbf{X} = (X_1, \ldots, X_n)$ *be the original data. Set* $i = 1$.

1. *Draw* $U_1, \ldots, U_n \overset{\text{iid}}{\sim} \mathsf{U}(0,1)$*, set* $I_j = \lceil n U_j \rceil$ *and* $X_j^* = X_{I_j}$, $j = 1, \ldots, n$.

2. *Set* $\mathbf{X}_i^* = (X_1^*, \ldots, X_n^*)$ *and* $T_i^* = T(\mathbf{X}_i^*)$.

3. *If* $i < N$*, set* $i = i + 1$ *and go to Step 1; otherwise, go to Step 4.*

4. *Return* (8.35) *as the bootstrap estimate of* ℓ.

Estimators $\overline{T}^*$ and $(S^*)^2$ are simply the sample mean and sample variance of the $\{T_i^*\}$. When T is an estimator of some unknown parameter θ, the bootstrap estimators of its bias and mean square error are obtained by replacing θ with T. The first bootstrap confidence interval in Table 8.2 corresponds to the usual approximate confidence interval (8.5) for the $\{T_i\}$. This is called the **normal method**. The second confidence interval consists of the $\alpha/2$ and $1 - \alpha/2$ sample quantiles of the $\{T_i^*\}$. This is called the **percentile method** [17].

Table 8.2 Common bootstrap estimators.

Quantity to estimate	Estimator
$\mathbb{E}T$	$\overline{T}^* = \frac{1}{N}\sum_{i=1}^{N} T_i^*$
$\mathrm{Var}(T)$	$(S^*)^2 = \frac{1}{N-1}\sum_{i=1}^{N}(T_i^* - \overline{T}^*)^2$
Mean square error: $\mathbb{E}(T-\theta)^2$	$\frac{1}{N}\sum_{i=1}^{N}(T_i^* - T)^2$
Bias: $\mathbb{E}T - \theta$	$\overline{T}^* - T$
$1-\alpha$ confidence interval for $\mathbb{E}T$ (normal)	$\overline{T}^* \pm z_{1-\alpha/2}\, S^*/\sqrt{N}$
$1-\alpha$ confidence interval for $\mathbb{E}T$ (percentile)	$\left(T^*_{(\lfloor N\alpha/2 \rfloor)},\, T^*_{(\lceil N(1-\alpha/2)\rceil)}\right)$

■ **EXAMPLE 8.12 (Bootstrapping the Ratio Estimator)**

Let the data $(X_1, Y_1), \ldots, (X_n, Y_n)$ be independent copies of some random vector (X, Y). Suppose we wish to estimate the ratio of expectations μ_X/μ_Y via the ratio estimator $T = \bar{X}/\bar{Y}$; see also Example 8.4. The bootstrap method provides an easy way to derive confidence intervals for μ_X/μ_Y without the need for an asymptotic analysis.

For concreteness, consider $n = 100$ iid copies of (X, Y) with $X \sim \mathsf{N}(10, 25)$ and $Y = XZ$, where $Z \sim \mathsf{U}(0,1)$ is independent of X. Hence, $\mu_X/\mu_Y = 10/5 = 2$. We have plotted in Figure 8.11 a kernel density estimate of the bootstrap values $T_1^*, \ldots, T_N^*$ (solid line) and the pdf of the $\mathsf{N}(\bar{X}/\bar{Y}, S^2/N)$ distribution (dashed line) for each of three data sets (three iid samples of size n), where S^2 is the estimate of the asymptotic variance of the ratio estimator as in (8.18).

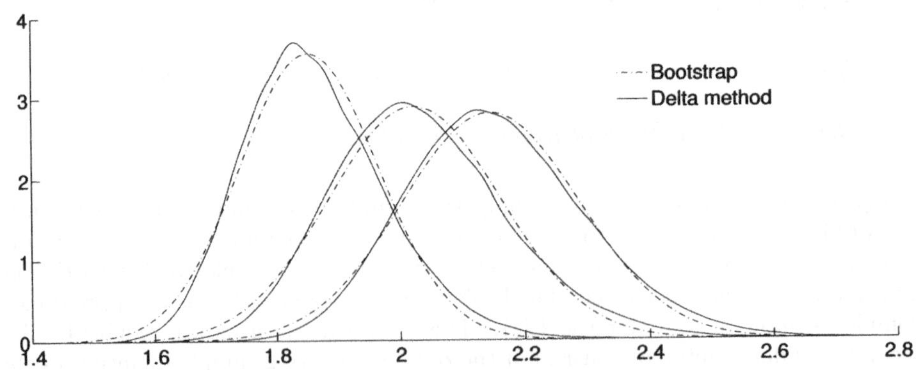

Figure 8.11 The $\alpha/2$ and $1-\alpha/2$ quantiles of each pair of pdfs correspond to $1-\alpha$ confidence intervals for the delta and bootstrap methods.

The $\alpha/2$ and $1 - \alpha/2$ quantiles of these pdfs correspond to $1 - \alpha$ confidence intervals. The figure illustrates that a bootstrap confidence interval (percentile method) is close to the corresponding approximate confidence interval obtained via the delta method. The following MATLAB code has been used. It uses the `kde.m` function from Section 8.5.2.

```
%resampratio.m
n = 100; %size of data
N = 50000; %resample size
T = zeros(1,N);
hold on
for count=1:3  %generate three sets of data
    xorg = 10 + 5*randn(1,n); %original x data
    yorg = rand(1,n).*xorg;    %original y data
    x = zeros(1,n);
    y = zeros(1,n);
    T = zeros(1,N); %bootstrap values for ratio estimator
    for i=1:N
        ind = ceil(n*rand(1,n)); % draw random indices
        x = xorg(ind);  % resampled y data
        y = yorg(ind);  % resampled x data
        T(i) = mean(x)/mean(y);
    end
    Torg =  mean(xorg)/mean(yorg);
    cv = cov(xorg,yorg);
    S2 = Torg^2*(var(xorg)/mean(xorg)^2 ...
       + var(yorg)/mean(yorg)^2 - 2*cv(1,2)/mean(xorg)/mean(yorg));
    tt = [Torg-4*sqrt(S2/n):0.01: Torg+4*sqrt(S2/n)];
    z = normpdf(tt,Torg,sqrt(S2/n));
    plot(tt,z,'r-.')
    [bandwidth,density,xmesh]=kde(T,2^14,min(tt),max(tt));
    plot(xmesh,density)
end
hold off
```

8.7 GOODNESS OF FIT

Goodness of fit procedures can be used to assess how well simulation data fit a specified statistical model. This is particularly important for the construction of good random number generators, where the output should resemble a sequence of iid U(0,1) random variables, see Section 1.5.2. Goodness of fit procedures are also ☞ 14 frequently encountered in Monte Carlo investigations of asymptotic convergence results; for example, how well the distribution of a finite sum of iid random variables is approximated by a normal distribution.

Goodness of fit approaches fall roughly into three categories: (1) graphical procedures, (2) statistical tests based on the empirical cdf, and (3) statistical tests based on binning of the data.

8.7.1 Graphical Procedures

Suppose $x_1, \ldots, x_n$ are the output data of a simulation. The $\{x_i\}$ could for example be the output of an MCMC sampler, in which case the data are dependent. We describe a number of methods that can be used to verify graphically whether the data represents (possibly dependent) samples from some pdf f or cdf F.

A straightforward graphical approach is to compare a kernel density estimate of the data to the proposed pdf f, as in Section 8.5, or to compare the empirical cdf to F, as in Section 8.4. An alternative is to use a **probability plot**. This refers to a general procedure where the data are plotted in such a way that the points should lie approximately on a straight line if the data come from the hypothesized pdf f or cdf F.

The two most common examples of probability plots are the **p-p plot** (p stands for percentile) and the **q-q plot** (q stands for quantile).

In the p-p plot each $F(x_i)$ is plotted against the corresponding $F_n(x_i)$ of the empirical cdf. In particular, using ordered observations, $F(x_{(i)})$ is plotted against i/n, which is the same as plotting the reduced empirical cdf. In the q-q plot each $x_{(i)}$ is plotted against the corresponding $F^{-1}(i/(n+1))$. This gives the following algorithms.

Algorithm 8.9 (One-Sample p-p Plot With Known Theorized Cdf) *Let $x_1, \ldots, x_n$ be the data and F the theorized cdf.*

1. *Order the data: $x_{(1)} \leqslant \cdots \leqslant x_{(n)}$.*
2. *For $i = 1, \ldots, n$ plot i/n (y-axis) against $F(x_{(i)})$ (x-axis).*
3. *Examine if the points lie approximately on a straight line with slope 1.*

Algorithm 8.10 (One-Sample q-q Plot With Known Theorized Cdf) *Let $x_1, \ldots, x_n$ be the data and F the theorized cdf.*

1. *Order the data: $x_{(1)} \leqslant \cdots \leqslant x_{(n)}$.*
2. *For $i = 1, \ldots, n$ plot $F^{-1}(i/(n+1))$ (y-axis) against $x_{(i)}$ (x-axis).*
3. *Examine if the points lie approximately on a straight line with slope 1.*

A q-q plot can also be used to test whether two samples $x_1, \ldots, x_m$ and $y_1, \ldots, y_n$ share a common cdf F by replacing the unknown $F^{-1}(i/(m+1))$ with the sample $i/(m+1)$-th quantile of the $\{y_i\}$; that is, the order statistic $y_{(\lceil ni/(m+1)\rceil)}$.

Algorithm 8.11 (Two-Sample q-q Plot) *Let $x_1, \ldots, x_m$ and $y_1, \ldots, y_n$ be the data.*

1. *Order the data: $x_{(1)} \leqslant \cdots \leqslant x_{(m)}$ and $y_{(1)} \leqslant \cdots \leqslant y_{(n)}$.*
2. *For $i = 1, \ldots, m$ plot $y_{(\lceil ni/(m+1)\rceil)}$ (y-axis) against $x_{(i)}$ (x-axis).*
3. *Examine if the points lie approximately on a straight line with slope 1.*

A q-q plot can also be used to check whether the data $x_1, \ldots, x_n$ could have come from a *family* of distributions. The most common case is to verify whether the data come from a location–scale family of cdfs $\{F(x; \mu, \sigma)\}$ with some base cdf $F(\cdot; 0, 1) = \mathring{F}(\cdot)$. Thus,

$$F(x; \mu, \sigma) = \mathring{F}\left(\frac{x - \mu}{\sigma}\right),$$

so that

$$x = \mu + \sigma \mathring{F}^{-1}(F(x;\mu,\sigma)) .$$

By replacing $F(x;\mu,\sigma)$ with the empirical cdf estimate $F_n(x) = F_n(x;\mu,\sigma)$, we see that the points $\{(x_i, \mathring{F}^{-1}(F_n(x_i)))\}$ will lie approximately on a straight line if the data are distributed according to *some* member of this location–scale family.

Algorithm 8.12 (One-Sample q-q Plot for a Location–Scale Family) *Let $x_1, \ldots, x_n$ be the data and $\mathring{F}$ the base cdf of the location–scale family.*

1. *Order the data $x_{(1)} \leqslant \cdots \leqslant x_{(n)}$.*
2. *For $i = 1, \ldots, n$ plot $\mathring{F}^{-1}(i/(n+1))$ (y-axis) against $x_{(i)}$ (x-axis).*
3. *Examine if the points lie approximately on a straight line.*

Useful additional information from a q-q or p-p plot can be gained by carrying out a regression analysis on the plotted points. In particular, for Algorithm 8.12 the estimated slope and intercept of the regression line give the scale and location parameters of the location–scale family, respectively. If the hypothesized family of distributions is not location–scale it may still be possible to employ a q-q plot by transforming the data such that the transformed data do have a location–scale form. An example is the $\mathsf{Weib}(\alpha, \lambda)$ distribution. Namely, if $X \sim \mathsf{Weib}(\alpha, \lambda)$, then $-\ln X \sim \mathsf{Gumbel}(\ln \lambda, 1/\alpha)$, which is a location–scale distribution; see Property 3 on Page 138.

Finally, the estimated correlation coefficient provides a measure for the goodness of the fit.

■ EXAMPLE 8.13 (Normal q-q Plots for the Gamma Distribution)

Consider four iid samples from a $\mathsf{Gamma}(k, 1)$ distribution with $k = 1, 10, 10, 100$ and sample sizes $n = 100, 100, 1000, 10000$, respectively. Figure 8.12 shows the q-q plots obtained via Algorithm 8.12 with the normal distribution as the theorized location–scale family.

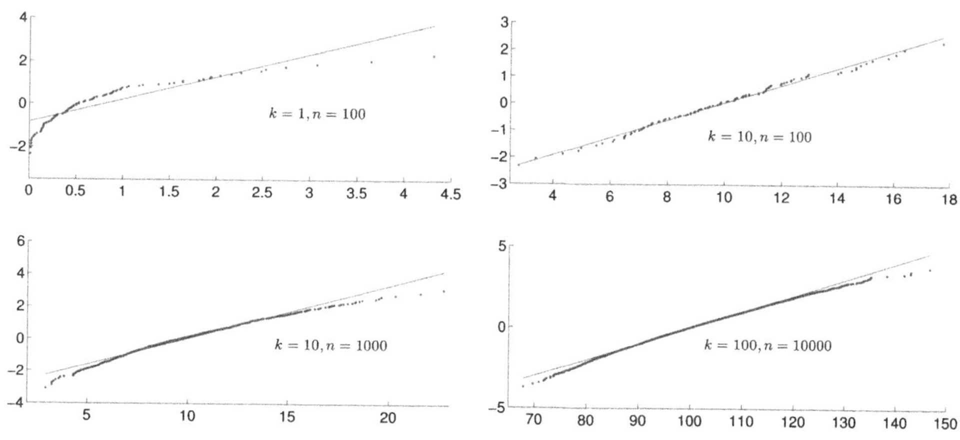

Figure 8.12 Normal q-q plots for an iid sample of size n from the $\mathsf{Gamma}(k, 1)$ distribution for various values of k. The x-axis shows the sorted data $x_{(i)}$, $i = 1, \ldots, n$ and the y-axis shows the corresponding theorized normal inverse cdf value $\Phi^{-1}(i/(n+1))$.

By the central limit theorem the distribution of $\mathsf{Gamma}(k, 1)$ is approximately normal for large k. The points in the normal q-q plot for $k = 10$ and $n = 100$ lie approximately on a straight line, indicating that the sample is difficult to distinguish from normal data, as opposed to the q-q plot for the $\mathsf{Gamma}(1, 1) \equiv \mathsf{Exp}(1)$ distribution. However, by increasing the sample size to $n = 1000$ a clear deviation from normality can be observed. A similar deviation can be observed even for the case where $k = 100$, when using a large sample size $n = 10000$. The following MATLAB code is used.

```
%qqplotex.m
nk = [100,1;
      100, 10;
      1000, 10;
      10000,100];
hold on
for count = 1:4;
    n= nk(count,1);
    k = nk(count,2);
    x = sum(-log(rand(k,n)),1);  %generate the data
    x = sort(x); %sort it
    i = 1:n;
    y = icdf('normal',i/(n+1),0,1); %compute the inverse cdf
    subplot(2,2,count)
    plot(x,y,'.')
    hold on
    p = polyfit(x,y,1)    %find the regression parameters
    f = polyval(p,x);     %the values of the regression line
    plot(x,f)
end
hold off
```

8.7.2 Kolmogorov–Smirnov Test

Let $x_1, \ldots, x_N$ be an iid sample from some continuous distribution. The Kolmogorov statistic D_N in (8.21), or more commonly its scaled version $K_N = \sqrt{N} D_N$, is often used as a test statistic to assess whether the true sampling cdf is equal to some given F or not. The null hypothesis is rejected for large values of the K_N. The exact distribution of K_N under the null hypothesis is given in the following theorem; see [16] and [25].

Theorem 8.7.1 (Distribution of the Kolmogorov Statistic) *For a given $u \in (0, 1)$, let $p = 2\lceil Nu \rceil - 1$ and $\delta = \lceil Nu \rceil - Nu$. Define H as the $p \times p$ matrix*

$$H = T - C - R + E\,,$$

where T is the Toeplitz matrix

$$T = \begin{pmatrix} 1 & 1 & 0 & \cdots & \cdots & \cdots & 0 \\ \frac{1}{2!} & 1 & 1 & 0 & \cdots & \cdots & 0 \\ \frac{1}{3!} & \frac{1}{2!} & 1 & 1 & 0 & \cdots & 0 \\ \vdots & \ddots & \ddots & \ddots & \ddots & \ddots & \vdots \\ \frac{1}{p!} & \frac{1}{(p-1)!} & \cdots & \cdots & \cdots & \frac{1}{2!} & 1 \end{pmatrix},$$

where C, R, and E are zero matrices except that the $(i,1)$-th element of C is $\delta^i/i!\,, i = 1,\ldots,p$, the (p,j)-th element of R is $\delta^{p-j+1}/(p-j+1)!, j = 1,2,\ldots,p$, and the $(p,1)$-th element of E is $\left(\max\{0, 2\delta - 1\}\right)^p/p!$. Then, denoting $\widetilde{H} = H^N$ as the N-th power of matrix H,

$$\mathbb{P}(K_N \leqslant \sqrt{N}u) = \mathbb{P}(D_N \leqslant u) = \frac{N!}{N^N}\,\widetilde{H}(\lceil Nu \rceil\,, \lceil Nu \rceil)\,, \quad u \in (0,1)\,, \tag{8.36}$$

where $\widetilde{H}(\lceil Nu \rceil\,, \lceil Nu \rceil)$ is the $\lceil Nu \rceil$-th diagonal element of the N-th power of H. Moreover,

$$\mathbb{P}(K \leqslant y) \stackrel{\text{def}}{=} \lim_{N\to\infty} \mathbb{P}(K_N \leqslant y) = \sum_{k=-\infty}^{\infty} (-1)^k \mathrm{e}^{-2(ky)^2}, \quad y > 0\,. \tag{8.37}$$

The limit result (8.37) follows from the fact that under the null hypothesis the stochastic process $\{\sqrt{N}(\widehat{G}_N(u) - u), u \in [0,1]\}$, where $\widehat{G}_N$ is the reduced empirical cdf, converges in distribution to a Brownian bridge process, say $\{Y_u, u \in [0,1]\}$; see, for example, [31]. Therefore, by the continuity theorem (see Property 3 on Page 624), K_N converges in distribution to the random variable $K = \sup_{0 \leqslant u \leqslant 1} |Y_u|$, the distribution of which is given in (8.23).

Algorithm 8.13 (Kolmogorov–Smirnov Test)

1. *Order the data $x_{(1)} < \cdots < x_{(N)}$.*
2. *Calculate $d_N = \max_{i=1,\ldots,N} \max\left\{\left|x_{(i)} - \frac{i}{N}\right|, \left|x_{(i)} - \frac{i-1}{N}\right|\right\}$.*
3. *Set $k_N = \sqrt{N} d_N$.*
4. *Calculate the p-value $\mathbb{P}(K_N \geqslant k_N)$ via (8.36), or, for large N, the asymptotic p-value $\mathbb{P}(K \geqslant k_N)$ via (8.37).*

■ EXAMPLE 8.14 (Kolmogorov–Smirnov Test)

Suppose the data are $N = 10^3$ independent copies of the random variable $X = Y + \varepsilon$, where $Y \sim \mathsf{Logistic}(0,1)$ and $\varepsilon \sim \mathsf{N}(0,1)$ are independent. Hence, X would be $\mathsf{Logistic}(0,1)$ distributed without the Gaussian noise term. The following MATLAB program generates the sample (it uses the $\mathsf{Logistic}(0,1)$ generator of Algorithm 4.45) ☞ 120 and applies the Kolmogorov–Smirnov test to assess the hypothesis H_0 that the sample is from the $\mathsf{Logistic}(0, 1)$ distribution. Figure 8.13 shows the reduced empirical cdf $\widehat{G}_N(u)$ and a line with slope one. The maximum distance between $\widehat{G}_N(u)$ and u is about 0.0620 in this case. Hence, the Kolmogorov–Smirnov statistic K_N takes the value $k_N = \sqrt{N}\,0.0620 \approx 1.9601$. The asymptotic p-value $\mathbb{P}_{H_0}(K_N \geqslant k_N)$ is 9.2×10^{-4} (rounded). The exact p-value is 8.6×10^{-4}. Thus, there is reasonable to strong evidence from the data to suggest that the true distribution is not $\mathsf{Logistic}(0, 1)$.

```
%kolsmir.m
N = 1000; %sample size
U=rand(1,N); x = log(U./(1-U))+randn(1,N); %generate sample
x = sort(1./(1+exp(-x)));
i=1:N;
dn_up = max(abs(x-i/N));dn_down = max(abs(x-(i-1)/N));
dn = max(dn_up, dn_down);
kn = sqrt(N)*dn; %KS statistic
k = -20:1:20;
a = (-1).^k.*exp(-2*(k.*kn).^2); %calculate KS probabilities
p = 1 - sum(a) % return the p-value
%or use matlab Statistics toolbox function kstest
[h,p,ksstat,cv] = kstest(x,[x',x'])
% plot the reduced empirical cdf
stairs([0,x],[0,i/N],'r'), hold on, line([0,1],[0,1])
```

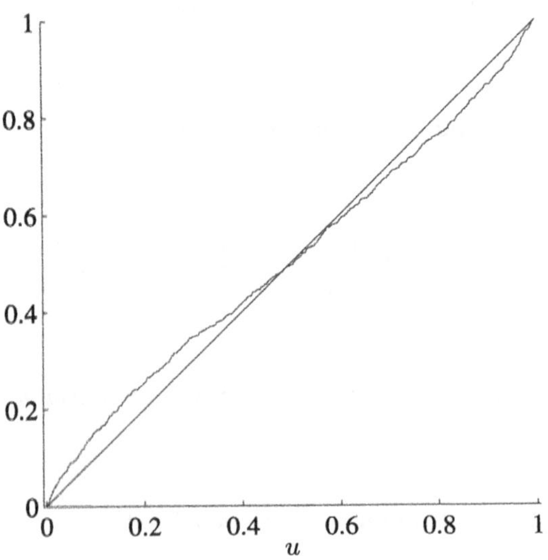

Figure 8.13 The reduced empirical cdf $\widehat{G}_N(u)$ and the line with slope one.

8.7.3 Anderson–Darling Test

The **Anderson–Darling test** is used as an alternative to the Kolmogorov–Smirnov test for testing if an iid sample $x_1, \ldots, x_N$ could have come from a given cdf F. The test statistic here is

$$\begin{aligned} A_N &= N \int_{-\infty}^{\infty} \frac{(F_N(x) - F(x))^2}{F(x)(1 - F(x))} \, \mathrm{d}F(x) = N \int_0^1 \frac{(F_N(F^{-1}(u)) - u)^2}{u(1-u)} \, \mathrm{d}u \\ &= N \sum_{i=0}^{N} \int_{y_i}^{y_{i+1}} \frac{(i/N - u)^2}{u(1-u)} \, \mathrm{d}u = -N - \frac{1}{N} \sum_{i=1}^{N} (2i - 1) \ln \left(y_i \, (1 - y_{N+1-i})\right), \end{aligned}$$

where $y_i = F(x_{(i)})$, $i = 1, \ldots, N$, and $y_0 = 0$ and $y_{N+1} = 1$. The null hypothesis is rejected for large values of A_N.

If $\widehat{G}_N$ is the random reduced empirical cdf (that is, $\widehat{G}_N(u) = \widehat{F}_N(F^{-1}(u))$), then the stochastic process $\{\sqrt{N}(\widehat{G}_N(u) - u), u \in [0,1]\}$ converges in distribution to a Brownian bridge process $\{X_u, u \in [0,1]\}$ as $N \to \infty$. Consequently, by the continuity theorem (Property 3 on Page 624) A_N converges in distribution to the random variable

$$A_\infty = \int_0^1 \frac{X_u^2}{u(1-u)} \, \mathrm{d}u \, .$$

The characteristic function of A_∞ is given in [4] as

$$\prod_{k=1}^{\infty} \left(\frac{k(k+1)}{k(k+1) - 2\mathrm{i}t} \right)^{1/2} \quad \text{and as} \quad \left(\frac{-2\pi \mathrm{i}t}{\cos\left(\frac{\pi}{2}\sqrt{1 + 8\mathrm{i}t}\right)} \right)^{1/2} .$$

The first expression shows that $A_\infty = \sum_{j=1}^{\infty} Z_j$, where the $\{Z_j\}$ are independent with $Z_j \sim \mathsf{Gamma}(1/2, j(j+1)/2)$, $j = 1, 2, \ldots$. From the characteristic function the following expression for the cdf can be found [4]:

$$\mathbb{P}(A_\infty \leqslant x) = \frac{\sqrt{2\pi}}{x} \sum_{k=0}^{\infty} \binom{-\frac{1}{2}}{k} (4k+1) \mathrm{e}^{-(4k+1)^2 \frac{\pi^2}{8x}} \int_0^\infty \mathrm{e}^{\frac{x}{8(1+w^2)} - w^2 (4k+1)^2 \frac{\pi^2}{8x}} \, \mathrm{d}w \, .$$

Numerical values for the cdf can be obtained by numerical integration of the above expression or by numerical inverse Fourier transformation of the characteristic function. Some care needs to be taken regarding the branches of the complex square root function. Marsaglia and Marsaglia [24] provide an approximating function $a(x)$ of the cdf:

$$\mathbb{P}(A_\infty \leqslant x) \approx a(x) = \begin{cases} a_1(x) & \text{if } 0 < x < 2 \\ a_2(x) & \text{if } x \geqslant 2 \, , \end{cases} \tag{8.38}$$

with

$$\begin{aligned} a_1(x) &= x^{-1/2} \mathrm{e}^{-b_1/x} (b_2 + (b_3 - (b_4 - (b_5 - (b_6 - b_7\, x)x)x)x)x) \\ a_2(x) &= \exp\left(- \mathrm{e}^{c_1 - (c_2 - (c_3 - (c_4 - (c_5 - c_6\, x)x)x)x)x} \right) , \end{aligned}$$

where the coefficients are given in Table 8.3. The approximation via $a(x)$ gives an absolute error within ± 0.0005, but the relative error increases rapidly as x increases.

Table 8.3 Coefficients used for the approximating functions.

k	b_k	c_k	d_k	m_k
1	1.2337141	1.0776	0.00022633	130.2137
2	2.00012	2.30695	6.54034	745.2337
3	0.247105	0.43424	14.6538	1705.091
4	0.0649821	0.082433	14.458	1950.646
5	0.0347962	0.008056	8.259	1116.360
6	0.0116720	0.0003146	1.91864	255.7844
7	0.00168691	—	—	—

For finite N the cdf of A_N is difficult to compute and depends on the cdf F (note that the distribution of the Kolmogorov–Smirnov statistic K_N does *not* depend on F). Numerical values are generally obtained via Monte Carlo simulation. An important special case is where F is the cdf of the $\mathsf{U}(0,1)$ distribution; that is, $F(x) = x$, $0 \leqslant x \leqslant 1$. Based on a large simulation study, Marsaglia and Marsaglia [24] propose the following approximation for finite N for the $\mathsf{U}(0,1)$ case:

$$\mathbb{P}(A_N \leqslant x) \approx a(x) + e(N, a(x)) ,$$

where $a(x)$ is defined in (8.38) and the correction function $e(N, y)$ is given by

$$e(N,y) = \begin{cases} \frac{1}{10^5}\left(\frac{370}{N^3} + \frac{78}{N^2} + \frac{6}{N}\right) g_1\left(\frac{y}{c(N)}\right) & \text{if } y < c(N) \\ \frac{1}{10^5}\left(\frac{1365}{N^2} + \frac{4213}{N}\right) g_2\left(\frac{y-c(N)}{0.8-c(N)}\right) & \text{if } c(N) \leqslant y < 0.8 \\ \frac{g_3(y)}{N} & \text{if } 0.8 \leqslant y \,, \end{cases}$$

with (see coefficients in Table 8.3)

$$\begin{aligned} c(N) &= 0.01265 + 0.1757/N \\ g_1(y) &= \sqrt{y}\,(1-y)(49y - 102) \\ g_2(y) &= -d_1 + (d_2 - (d_3 - (d_4 - (d_5 - d_6\, y)y)y)y)y \\ g_3(y) &= -m_1 + (m_2 - (m_3 - (m_4 - (m_5 - m_6\, y)y)y)y)y \,. \end{aligned}$$

8.7.4 χ^2 Tests

In a **χ^2 goodness of fit** test, the model for the data is

$$(X_1, \ldots, X_k) \sim \mathsf{Mnom}(N, p_1, \ldots, p_k) \,.$$

Each X_i can be interpreted as the total number of balls in the i-th urn, where N balls are thrown independently into urns $1, \ldots, k$ with probabilities $p_1, \ldots, p_k$; see
☞ 141
Section 4.3.2. The aim is to verify whether the $\{p_i\}$ are equal to specified values $\{\pi_i\}$. The test statistic is generally of the form

$$T = \sum_{i=1}^{k} \frac{(O_i - E_i)^2}{E_i} ,$$

where O_i is the *observed* number of observations (balls) in class (urn) i and E_i is the *expected* number of observations in class i. The distribution of T under the null hypothesis is asymptotically χ^2, and is based on the following theorems (proofs can be found, for example, in [31]).

Theorem 8.7.2 (Goodness of Fit With Known Parameters) *Suppose that* $(X_1, \ldots, X_k) \sim \mathsf{Mnom}(N, p_1, \ldots, p_k)$. *Then, for large* N

$$\sum_{i=1}^{k} \frac{(X_i - N\,p_i)^2}{N\,p_i} \overset{\text{approx.}}{\sim} \chi^2_{k-1} \,.$$

Theorem 8.7.3 (Goodness of Fit With Unknown Parameters) *Suppose that* $(X_1, \ldots, X_k) \sim \mathsf{Mnom}(N, p_1, \ldots, p_k)$, *where each* $p_i = p_i(\boldsymbol{\theta})$ *depends on an unknown parameter vector* $\boldsymbol{\theta} = (\theta_1, \ldots, \theta_r)$. *Let* $\widehat{\boldsymbol{\theta}}$ *be the MLE of* $\boldsymbol{\theta}$ *and* $\widehat{p}_i = p_i(\widehat{\boldsymbol{\theta}})$ *the MLE of* $p_i(\boldsymbol{\theta})$. *Then, for large* N ☞ 667

$$\sum_{i=1}^{k} \frac{(X_i - N\,\widehat{p}_i)^2}{N\,\widehat{p}_i} \overset{\text{approx.}}{\sim} \chi^2_{k-1-r} \,.$$

As a rule of thumb, one can use the approximations above provided that Np_i (or $N\widehat{p}_i$) is greater than 5, for all i.

We discuss two types of χ^2 goodness of fit test that frequently appear in Monte Carlo studies.

8.7.4.1 Goodness of Fit Test With Known Parameters Let $(X_1, \ldots, X_k) \sim \mathsf{Mnom}(N, p_1, \ldots, p_k)$. The objective is to test $H_0 : p_1 = \pi_1, \ldots, p_k = \pi_k$ against the alternative hypothesis that H_0 is not true, by using the test statistic

$$T = \sum_{i=1}^{k} \frac{(X_i - N\,\pi_i)^2}{N\,\pi_i} \,.$$

By Theorem 8.7.2, under H_0 and for large N, the statistic T has approximately a χ^2_{k-1} distribution. H_0 is rejected for large values of T. For an outcome t of T we therefore have the p-value

$$\mathbb{P}_{H_0}(T \geqslant t) \approx 1 - \chi^2_{k-1;t} \,,$$

where $\chi^2_{k-1;t}$ is the t-quantile of the χ^2_{k-1} distribution.

■ EXAMPLE 8.15 (χ^2 Testing for Equidistribution)

A fundamental test for random number generators is the *equidistribution* test, where the theoretical distribution is tested against the observed output; see Section 1.5.2.1. ☞ 17
Such a test can be implemented as a χ^2 goodness of fit test. As an example, suppose we generate N samples and count how many of these fall in each interval $((i-1)/k, i/k)$, $i = 1, \ldots, k$ for some integer $k > 1$. The following MATLAB program generates the data via the default uniform random number generator and performs a χ^2 test for the case $N = 10^6$ and $k = 50$. The outcome of T is in this case $t = 34.03$, which gives a p-value of 0.95. Hence, this t is likely to occur under the null hypothesis that the data are genuinely $\mathsf{U}(0,1)$ distributed, and therefore there

is no reason to reject the null hypothesis. Figure 8.14 corroborates this finding graphically.

```
%chi2eq.m
clear all, rand('state',1)
N = 10^6; k = 50; p = ones(1,k)/k; %true probabilities
u = rand(1,N); x = zeros(1,k);
for i=1:k
    x(i) = sum( (i-1)/k < u & u< i/k ); %observed count
end
t = sum((x - N*p).^2./(N*p)); %test statistic
pval = 1 - cdf('chi2',t,k-1) % p-value
bar(x,0.5)
```

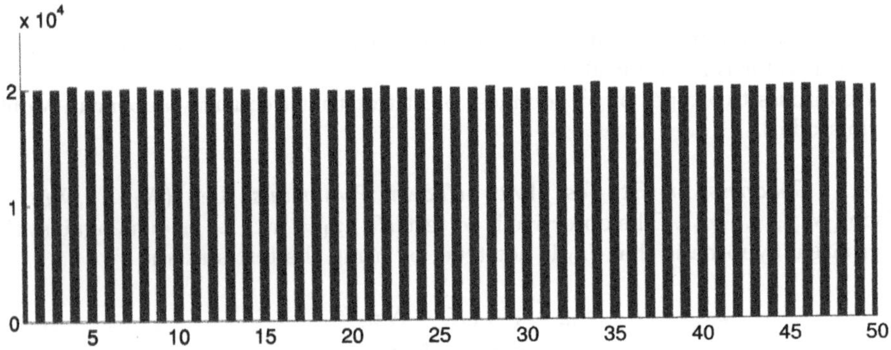

Figure 8.14 Counts of uniform numbers in the intervals $((i-1)/50, i/50)$, $i = 1, \ldots, 50$. The total sample size is $N = 10^6$.

8.7.4.2 Goodness of Fit Test With Unknown Parameters Let $(X_1, \ldots, X_k) \sim \mathsf{Mnom}(N, p_1, \ldots, p_k)$. The objective is to test $H_0 : p_1 = \pi_1(\boldsymbol{\theta}), \ldots, p_k = \pi_k(\boldsymbol{\theta})$, where the $\{\pi_i\}$ are known functions but $\boldsymbol{\theta} = (\theta_1, \ldots, \theta_r)$ is an *unknown* parameter vector, against the alternative hypothesis that H_0 is not true. The test statistic is

$$T = \sum_{i=1}^{k} \frac{(X_i - N\,\widehat{\pi}_i)^2}{N\,\widehat{\pi}_i} ,$$

where $\widehat{\pi}_i = \pi_i(\widehat{\boldsymbol{\theta}})$ is the MLE of $\pi_i(\boldsymbol{\theta})$, $i = 1, \ldots, k$. By Theorem 8.7.3, under H_0 and for large N, T approximately has a χ^2_{k-1-r} distribution. Therefore, an outcome t of T gives the approximate p-value

$$\mathbb{P}_{H_0}(T \geqslant t) \approx 1 - \chi^2_{k-1-r;t} ,$$

where $\chi^2_{k-1-r;t}$ is the t-quantile of the χ^2_{k-1-r} distribution.

■ EXAMPLE 8.16 (Contingency Tables)

Contingency tables are used to test the independence of data. Consider a random vector (U, V) taking values in $\{1, \ldots, r\} \times \{1, \ldots, c\}$. We wish to test for *independence* of U and V. To this end we take N iid copies $(U_1, V_1), \ldots, (U_N, V_N)$ of (U, V). Let $X_{ij} = \sum_{k=1}^{N} \mathrm{I}_{\{U_k = i, V_k = j\}}$ be the total number of observations in "urn" (i, j). Then $(X_{11}, \ldots, X_{rc}) \sim \mathsf{Mnom}(N, p_{11}, \ldots, p_{rc})$, with $p_{ij} = \mathbb{P}(U = i, V = j)$. The random variables U and V are independent if and only if

$$H_0 : p_{ij} = p_i \, q_j \quad \text{for all } i, j ,$$

holds for some (unknown) marginal probabilities $p_1 = \mathbb{P}(U = 1), \ldots, p_r = \mathbb{P}(U = r)$ and $q_1 = \mathbb{P}(V = 1), \ldots, q_c = \mathbb{P}(V = c)$. We can test H_0 against its negation by using the test statistic

$$T = \sum_{i=1}^{r} \sum_{j=1}^{c} \frac{(X_{ij} - N \, \widehat{p}_i \, \widehat{q}_j)^2}{N \, \widehat{p}_i \, \widehat{q}_j} ,$$

where $\widehat{p}_i = \sum_{j=1}^{c} X_{ij}/N$ and $\widehat{q}_j = \sum_{i=1}^{r} X_{ij}/N$ are the MLEs of p_i and q_j, $i = 1, \ldots, r$, $j = 1, \ldots, c$. By Theorem 8.7.3, the test statistic T approximately has a $\chi^2_{(r-1)(c-1)}$ distribution under H_0. Note that there are $r\,c$ classes, and the number of parameters that need to be estimated is $(r - 1) + (c - 1)$. H_0 is rejected for large values of T.

Further Reading

The regenerative method in a simulation context is introduced and developed by Crane and Iglehart [13]. A more complete treatment of regenerative processes is given in [5]. Fishman [18] treats the statistical analysis of simulation data in great detail. Efron and Tibshirani [17] cover the bootstrap method. For a book dedicated to graphical methods for data analysis see Chambers et al. [11]. Useful references on statistical inference and data analysis are Casella and Berger [10], Maindonald and Braun [23], and Law and Kelton [21].

REFERENCES

1. J. Abate and W. Whitt. Transient behavior of regulated Brownian motion. *Advances in Applied Probability*, 19(3):560–631, 1987.

2. I. S. Abramson. On bandwidth variation in kernel estimates—a square root law. *Annals of Statistics*, 10(4):1217–1223, 1982.

3. N. Agarwal and N. R. Aluru. A data-driven stochastic collocation approach for uncertainty quantification in MEMS. *International Journal for Numerical Methods in Engineering*, 2010. DOI: 10.1002/nme.2844.

4. T. W. Anderson and D. A. Darling. Asymptotic theory of certain "goodness of fit" criteria based on stochastic processes. *The Annals of Mathematical Statistics*, 23(2):193–212, 1952.

5. S. Asmussen. *Applied Probability and Queues.* John Wiley & Sons, New York, 1987.

6. R. Bellman. *A Brief Introduction to Theta Functions.* Holt, Rinehart and Winston, New York, 1961.

7. Z. I. Botev. A novel nonparametric density estimator. Technical report, The University of Queensland, 2006. Available from `http://espace.library.uq.edu.au/view/UQ:12535`.

8. Z. I. Botev. Nonparametric density estimation via diffusion mixing. Technical report, The University of Queensland, 2007. Available from `http://espace.library.uq.edu.au/view/UQ:120006`.

9. Z. I. Botev, J. F. Grotowski, and D. P. Kroese. Kernel density estimation via diffusion. *The Annals of Statistics*, 38(5):2916–2957, 2010.

10. G. Casella and R. L. Berger. *Statistical Inference.* Duxbury Press, Pacific Grove, second edition, 2001.

11. J. M. Chambers, W. S. Cleveland, B. Kleiner, and P. A. Tukey. *Graphical Methods for Data Analysis.* Wadsworth, Boston, 1983.

12. P. Chaudhuri and J. S. Marron. Scale space view of curve estimation. *The Annals of Statistics*, 28(2):408–428, 2000.

13. M. A. Crane and D. L. Iglehart. Simulating stable stochastic systems, II: Markov chains. *Journal of the Association for Computing Machinery*, 21(1):114–123, 1974.

14. L. Devroye and L. Györfi. *Nonparametric Density Estimation: The L_1 View.* John Wiley & Sons, New York, 1985.

15. A. Doucet, N. de Freitas, and N. Gordon. *Sequential Monte Carlo Methods in Practice.* Springer-Verlag, New York, 2001.

16. J. Durbin. *Distribution Theory for Tests Based on the Sample Distribution Function.* Society for Industrial & Applied Mathematics, Philadelphia, 1972.

17. B. Efron and R. Tibshirani. *An Introduction to the Bootstrap.* Chapman & Hall, New York, 1994.

18. G. S. Fishman. *Discrete Event Simulation: Modeling, Programming, and Analysis.* Springer-Verlag, New York, 2001.

19. M. L. Hazelton and J. C. Marshall. Linear boundary kernels for bivariate density estimation. *Statistics and Probability Letters*, 79(8):999–1003, 2009.

20. M. C. Jones, I. J. McKay, and T. C. Hu. Variable location and scale kernel density estimation. *Annals of the Institute of Statistical Mathematics*, 46(3):521–535, 1994.

21. A. M. Law and W. D. Kelton. *Simulation Modeling and Analysis.* McGraw-Hill, New York, third edition, 2000.

22. C. R. Loader. Bandwidth selection: Classical or plug-in. *The Annals of Statistics*, 27(2):415–438, 1999.

23. J. Maindonald and J. Braun. *Data Analysis and Graphics Using R: An Example-Based Approach.* Cambridge University Press, Cambridge, second edition, 2007.

24. G. Marsaglia and J. C. W. Marsaglia. Evaluating the Anderson-Darling distribution. *Journal of Statistical Software*, 9(2), 2004.

25. G. Marsaglia, W. W. Tsang, and J. Wang. Evaluating Kolmogorov's distribution. *Journal of Statistical Software*, 8(18), 2003.

26. J. C. Marshall and M. L. Hazelton. Boundary kernels for adaptive density estimators on regions with irregular boundaries. *Journal of Multivariate Analysis*, 101(4):949–963, 2010.

27. B. U. Park and J. S. Marron. Comparison of data-driven bandwidth selectors. *Journal of the American Statistical Association*, 85(409):66–72, 1990.

28. R. Y. Rubinstein and D. P. Kroese. *Simulation and the Monte Carlo Method.* John Wiley & Sons, New York, second edition, 2007.

29. M. Samiuddin and G. M. El-Sayyad. On nonparametric kernel density estimates. *Biometrika*, 77(4):865–874, 1990.

30. D. W. Scott. *Multivariate Density Estimation: Theory, Practice and Visualization.* John Wiley & Sons, New York, 1992.

31. P. K. Sen and J. M. Singer. *Large Sample Methods in Statistics.* Chapman & Hall, New York, 1993.

32. S. J. Sheather and M. C. Jones. A reliable data-based bandwidth selection method for kernel density estimation. *Journal of the Royal Statistical Society, Series B*, 53(3):683–690, 1991.

33. G. R. Shorack and J. A. Wellner. *Empirical Processes With Applications to Statistics.* SIAM, Philadelphia, 2009.

34. B. W. Silverman. *Density Estimation for Statistics and Data Analysis.* Chapman & Hall, New York, 1986.

35. J. S. Simonoff. *Smoothing Methods in Statistics.* Springer-Verlag, New York, 1996.

36. G. R. Terrell and D. W. Scott. Variable kernel density estimation. *The Annals of Statistics*, 20(3):1236–1265, 1992.

37. M. P. Wand and M. C. Jones. Multivariate plug-in bandwidth selection. *Computational Statistics*, 9:97–117, 1994.

38. M. P. Wand and M. C. Jones. *Kernel Smoothing.* Chapman & Hall, London, 1995.

CHAPTER 9

VARIANCE REDUCTION

The estimation of performance measures in Monte Carlo simulation can be made more efficient by utilizing known information about the simulation model. The more that is known about the behavior of the system, the greater the amount of variance reduction that can be achieved. The main variance reduction techniques discussed in this chapter are:

1. Antithetic random variables.
2. Control variables.
3. Conditional Monte Carlo.
4. Stratification.
5. Latin hypercube sampling.
6. Importance sampling.
7. Quasi Monte Carlo. ☞ 25

For application of variance reduction techniques in rare-event simulation see Chapter 10. In particular, Section 10.6 and Chapter 14 both present a *splitting* approach to rare-event simulation. ☞ 381 ☞ 409 ☞ 481

9.1 VARIANCE REDUCTION EXAMPLE

Each of the variance reduction methods is illustrated using the following estimation problem concerning a bridge network. The problem is sufficiently complicated to warrant Monte Carlo simulation, while easy enough to implement, so that the workings of each technique can be concisely illustrated within the same context.

■ EXAMPLE 9.1 (Bridge Network)

Consider the undirected graph in Figure 9.1, depicting a **bridge network**.

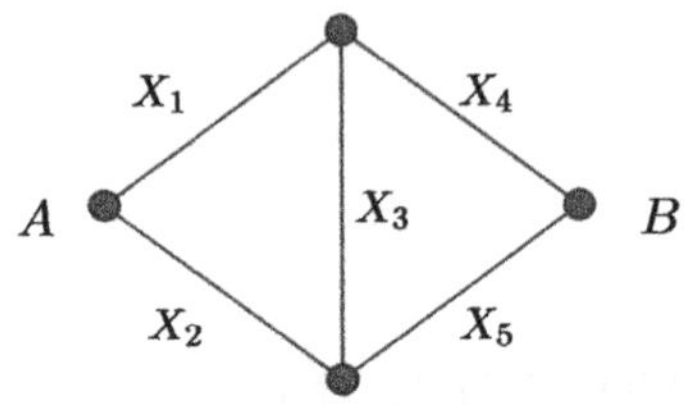

Figure 9.1 What is the expected length of the shortest path from A to B?

Suppose we wish to estimate the expected length ℓ of the shortest path between nodes (vertices) A and B, where the lengths of the links (edges) are random variables $X_1, \ldots, X_5$. We have $\ell = \mathbb{E}H(\mathbf{X})$, where

$$H(\mathbf{X}) = \min\{X_1 + X_4,\ X_1 + X_3 + X_5,\ X_2 + X_3 + X_4,\ X_2 + X_5\}\ . \tag{9.1}$$

Note that $H(\mathbf{x})$ is nondecreasing in each component of the vector $\mathbf{x}$. Suppose the lengths $\{X_i\}$ are independent and $X_i \sim \mathsf{U}(0, a_i)$, $i = 1, \ldots, 5$ with $(a_1, \ldots, a_5) = (1, 2, 3, 1, 2)$. Writing $X_i = a_i\, U_i$, $i = 1, \ldots, 5$ with $\{U_i\} \sim_{\text{iid}} \mathsf{U}(0, 1)$, we can restate the problem as the estimation of

$$\ell = \mathbb{E}h(\mathbf{U})\ , \tag{9.2}$$

where $\mathbf{U} = (U_1, \ldots, U_5)$ and $h(\mathbf{U}) = H(a_1 U_1, \ldots, a_5 U_5)$. The exact value can be determined by conditioning (see Section 9.4) and is given by

$$\ell = \frac{1339}{1440} = 0.9298611111\ldots\ .$$

☞ 306 *Crude Monte Carlo* (CMC) proceeds by generating $\mathbf{U}_1, \ldots, \mathbf{U}_N \overset{\text{iid}}{\sim} \mathsf{U}(0, 1)^5$ and returning

$$\widehat{\ell} = \frac{1}{N} \sum_{k=1}^{N} h(\mathbf{U}_k)$$

as an estimate for ℓ.

The following MATLAB program implements the CMC simulation. For a sample size of $N = 10^4$ a typical estimate is $\widehat{\ell} = 0.930$ with an estimated relative error of 0.43%.

```
%bridgeCMC.m
N = 10^4;
U = rand(N,5);
y = h(U);
est = mean(y)
percRE = std(y)/sqrt(N)/est*100
```

```
function out=h(u)
a=[1,2,3,1,2]; N = size(u,1);
X = u.*repmat(a,N,1);
Path_1=X(:,1)+X(:,4);
Path_2=X(:,1)+X(:,3)+X(:,5);
Path_3=X(:,2)+X(:,3)+X(:,4);
Path_4=X(:,2)+X(:,5);
out=min([Path_1,Path_2,Path_3,Path_4],[],2);
```

9.2 ANTITHETIC RANDOM VARIABLES

A pair of real-valued random variables (Y, Y^*) is called an **antithetic pair** if Y and Y^* have the same distribution and are *negatively correlated*. The main application of antithetic random variables in Monte Carlo estimation is based on the following theorem; see, for example, [18].

Theorem 9.2.1 (Antithetic Estimator) *Let N be an even number and let $(Y_1, Y_1^*), \ldots, (Y_{N/2}, Y_{N/2}^*)$ be independent antithetic pairs of random variables, where each Y_k and Y_k^* is distributed as Y. The* **antithetic estimator**

$$\widehat{\ell}^{(a)} = \frac{1}{N} \sum_{k=1}^{N/2} \{Y_k + Y_k^*\}, \tag{9.3}$$

is an unbiased estimator of $\ell = \mathbb{E}Y$, with variance

$$\begin{aligned}\operatorname{Var}(\widehat{\ell}^{(a)}) &= \frac{N/2}{N^2}\left(\operatorname{Var}(Y) + \operatorname{Var}(Y^*) + 2\operatorname{Cov}(Y, Y^*)\right) \\ &= \left(\operatorname{Var}(Y) + \operatorname{Cov}(Y, Y^*)\right)/N \\ &= \frac{\operatorname{Var}(Y)}{N}\left(1 + \varrho_{Y,Y^*}\right),\end{aligned}$$

where ϱ_{Y,Y^*} is the correlation between Y and Y^*.

Note that (9.3) is simply the sample mean of the independent random variables $\{(Y_k + Y_k^*)/2\}$. Since the variance of the CMC estimator $\widehat{\ell} = N^{-1}\sum_{k=1}^{N} Y_i$ is $\operatorname{Var}(Y)/N$, the above theorem shows that the use of antithetic variables leads to a smaller variance of the estimator by a factor of $1 + \varrho_{Y,Y^*}$. The amount of reduction depends crucially on the amount of negative correlation between the antithetic variables.

In general, the output of a simulation run is of the form $Y = h(\mathbf{U})$, where h is a real-valued function and $\mathbf{U} = (U_1, U_2, \ldots)$ is a random vector of iid $\mathsf{U}(0,1)$ random variables. Suppose that $\mathbf{U}^*$ is another vector of iid $\mathsf{U}(0,1)$ random variables which is dependent on $\mathbf{U}$ and for which Y and $Y^* = h(\mathbf{U}^*)$ are negatively correlated. Then (Y, Y^*) is an antithetic pair. In particular, if h is a monotone function in each of its components, then the choice $\mathbf{U}^* = \mathbf{1} - \mathbf{U}$, where $\mathbf{1}$ is the vector of 1s, yields an antithetic pair.

☞ 306 An alternative to the CMC Algorithm 8.2 for estimating $\ell = \mathbb{E}Y = \mathbb{E}h(\mathbf{U})$ is thus as follows.

Algorithm 9.1 (Antithetic Estimation for Monotone h)

1. *Generate* $Y_1 = h(\mathbf{U}_1), \ldots, Y_{N/2} = h(\mathbf{U}_{N/2})$ *from independent simulation runs.*

2. *Let* $Y_1^* = h(\mathbf{1} - \mathbf{U}_1), \ldots, Y_{N/2}^* = h(\mathbf{1} - \mathbf{U}_{N/2})$.

3. *Compute the sample covariance matrix corresponding to the pairs* $\{(Y_k, Y_k^*)\}$:

$$C = \begin{pmatrix} \frac{1}{N/2-1}\sum_{k=1}^{N/2}(Y_k - \bar{Y})^2 & \frac{1}{N/2-1}\sum_{k=1}^{N/2}(Y_k - \bar{Y})(Y_k^* - \bar{Y}^*) \\ \frac{1}{N/2-1}\sum_{k=1}^{N/2}(Y_k - \bar{Y})(Y_k^* - \bar{Y}^*) & \frac{1}{N/2-1}\sum_{k=1}^{N/2}(Y_k^* - \bar{Y}^*)^2 \end{pmatrix}.$$

4. *Estimate* ℓ *via the antithetic estimator* $\widehat{\ell}^{(a)}$ *in* (9.3) *and determine an approximate* $1-\alpha$ *confidence interval as*

$$\left(\widehat{\ell}^{(a)} - z_{1-\alpha/2} SE, \quad \widehat{\ell}^{(a)} + z_{1-\alpha/2} SE\right),$$

where SE *is the estimated standard error:*

$$SE = \sqrt{\frac{C_{1,1} + C_{2,2} + 2C_{1,2}}{2N}},$$

and z_γ *denotes the* γ*-quantile of the* $\mathsf{N}(0,1)$ *distribution.*

For each of the $N/2$ runs in Step 2 one does not necessarily have to store the complete sequence $\mathbf{U} = (U_1, U_2, \ldots)$ of random numbers in memory, but simply ☞ 2 save the random seeds for each sequence.

■ **EXAMPLE 9.2 (Antithetic Estimation for the Bridge Network)**

The following MATLAB program implements an antithetic estimator of the expected length of the shortest path ℓ in Example 9.1. A typical estimate using $N = 10^4$ samples is $\widehat{\ell}^{(a)} = 0.929$ with an estimated relative error of 0.2%. Figure 9.2 illustrates that the correlation between $h(\mathbf{U})$ and $h(\mathbf{1} - \mathbf{U})$ is relatively high in this case. The correlation coefficient is around -0.77, which means a more than fourfold reduction in simulation effort when compared to CMC. Function `h.m` in the code that follows is the same as in Example 9.1.

```
%compare_CMC_and_ARV.m
N=10^4;
U=rand(N/2,5);  % get uniform random variables
y = h(U); ya = h(1-U);
ell=(mean(y) + mean(ya))/2;
C=cov(y,ya);
var_h = sum(sum(C))/(2*N);
corr = C(1,2)/sqrt(C(1,1)*C(2,2));
fprintf('ell= %g,  RE = %g,  corr = %g\n',ell,sqrt(var_h)/ell, corr)
plot(y,ya,'.')
U = rand(N,5);
yb = h(U);
var_hb = var(yb)/N;
ReB = sqrt(var_hb)/ell
```

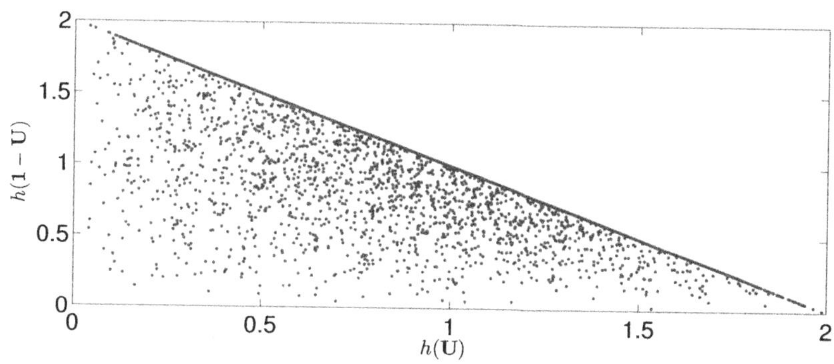

Figure 9.2 Scatter plot of $N = 10^4$ antithetic pairs (Y, Y^*) for the bridge network.

Remark 9.2.1 (Normal Antithetic Random Variables) Antithetic pairs can also be based on distributions other than the uniform. For example, suppose that $Y = H(\mathbf{Z})$, where $\mathbf{Z} = (Z_1, Z_2, \ldots)$ is a vector of iid standard normal random variables. By the inverse-transform method we can write $Y = h(\mathbf{U})$, with $h(\mathbf{u}) = H(\Phi^{-1}(u_1), \Phi^{-1}(u_2), \ldots)$, where Φ is the cdf of the $\mathsf{N}(0,1)$ distribution. Taking $\mathbf{U}^* = \mathbf{1} - \mathbf{U}$ gives $\mathbf{Z}^* = (\Phi^{-1}(U_1^*), \Phi^{-1}(U_2^*), \ldots) = -\mathbf{Z}$, so that (Y, Y^*) with $Y^* = H(-\mathbf{Z})$ forms an antithetic pair provided that Y and Y^* are negatively correlated, which is the case if H is a monotone function in each of its components. ☞ 45

9.3 CONTROL VARIABLES

Let Y be the output of a simulation run. A random variable $\widetilde{Y}$, obtained from the same simulation run, is called a **control variable** for Y if Y and $\widetilde{Y}$ are correlated (negatively or positively) and the expectation of $\widetilde{Y}$ is known. The use of control variables for variance reduction is based on the following observation.

Theorem 9.3.1 (Control Variable Estimation) *Let $Y_1,\ldots,Y_N$ be the output of N independent simulation runs, and let $\widetilde{Y}_1,\ldots,\widetilde{Y}_N$ be the corresponding control variables, with $\mathbb{E}\widetilde{Y}_k = \widetilde{\ell}$ known. Let $\varrho_{Y,\widetilde{Y}}$ be the correlation coefficient between each Y_k and $\widetilde{Y}_k$. For each $\alpha \in \mathbb{R}$ the (linear) estimator*

$$\widehat{\ell}^{(c)} = \frac{1}{N}\sum_{k=1}^{N}\left[Y_k - \alpha\left(\widetilde{Y}_k - \widetilde{\ell}\right)\right] \tag{9.4}$$

is an unbiased estimator for $\ell = \mathbb{E}Y$. The minimal variance of $\widehat{\ell}^{(c)}$ is

$$\mathrm{Var}(\widehat{\ell}^{(c)}) = \frac{1}{N}(1-\varrho^2_{Y,\widetilde{Y}})\mathrm{Var}(Y) \tag{9.5}$$

which is obtained for $\alpha = \mathrm{Cov}(Y,\widetilde{Y})/\mathrm{Var}(\widetilde{Y})$.

Usually the optimal α in (9.5) is unknown, but it can be easily estimated from the sample covariance matrix of the $\{(Y_k,\widetilde{Y}_k)\}$. This leads to the following algorithm.

Algorithm 9.2 (Control Variable Estimation)

1. *From N independent simulation runs generate $Y_1,\ldots,Y_N$ and the control variables $\widetilde{Y}_1,\ldots,\widetilde{Y}_N$.*
2. *Compute the sample covariance matrix of $\{(Y_k,\widetilde{Y}_k)\}$:*

$$C = \begin{pmatrix} \frac{1}{N-1}\sum_{k=1}^{N}(Y_k-\bar{Y})^2 & \frac{1}{N-1}\sum_{k=1}^{N}(Y_k-\bar{Y})(\widetilde{Y}_k-\bar{\widetilde{Y}}) \\ \frac{1}{N-1}\sum_{k=1}^{N}(Y_k-\bar{Y})(\widetilde{Y}_k-\bar{\widetilde{Y}}) & \frac{1}{N-1}\sum_{k=1}^{N}(\widetilde{Y}_k-\bar{\widetilde{Y}})^2 \end{pmatrix}.$$

3. *Estimate ℓ via the control variable estimator $\widehat{\ell}^{(c)}$ in (9.4) with $\alpha = C_{1,2}/C_{2,2}$ and determine an approximate $1-\alpha$ confidence interval as*

$$\left(\widehat{\ell}^{(c)} - z_{1-\alpha/2}SE,\ \ \widehat{\ell}^{(c)} + z_{1-\alpha/2}SE\right),$$

where z_γ denotes the γ-quantile of the $\mathsf{N}(0,1)$ distribution and SE is the estimated standard error:

$$SE = \sqrt{\frac{1}{N}\left(1-\frac{C^2_{1,2}}{C_{1,1}C_{2,2}}\right)C_{1,1}}\ .$$

■ **EXAMPLE 9.3 (Control Variable Estimation for the Bridge Network)**

Consider again the stochastic shortest path estimation problem for the bridge network in Example 9.1. As a control variable we can use, for example,

$$\widetilde{Y} = \min\{X_1+X_4, X_2+X_5\}\ .$$

This is particularly convenient for the current parameters $(1,2,3,1,2)$, as with high probability the shortest path will have a length equal to $\widetilde{Y}$; indeed, it will most likely have length $X_1 + X_4$, so that the latter would also be useful as a control variable. With a little calculation, the expectation of $\widetilde{Y}$ can be found to be $\mathbb{E}\widetilde{Y} = 15/16 = 0.9375$. Figure 9.3 shows the high correlation between the length of the shortest path $Y = H(\mathbf{X})$ defined in (9.1) and $\widetilde{Y}$. The corresponding correlation coefficient is around 0.98, which shows that a fiftyfold variance reduction in simulation effort is achieved compared with CMC estimation. The MATLAB program below implements the control variable estimator, using a sample size of $N = 10^4$. A typical estimate is $\widehat{\ell}^{(c)} = 0.92986$ with an estimated relative error of 0.05%. Function h.m in the code below is the same as in Example 9.1.

```
%bridgeCV.m
N=10^4;
u = rand(N,5);
Y = h(u);
Yc = hc(u);
plot(Y,Yc,'.')
C = cov(Y,Yc);
cor = C(1,2)/sqrt(C(1,1)*C(2,2))
alpha = C(1,2)/C(2,2);
yc = 15/16;
est = mean(Y - alpha*(Yc - yc))
RE = sqrt((1 - cor^2)*C(1,1)/N)/est
```

```
function out=hc(u)
a=[1,2,3,1,2];
N = size(u,1);
X = u.*repmat(a,N,1);
Path_1=X(:,1)+X(:,4);
Path_4=X(:,2)+X(:,5);
out=min([Path_1,Path_4],[],2);
```

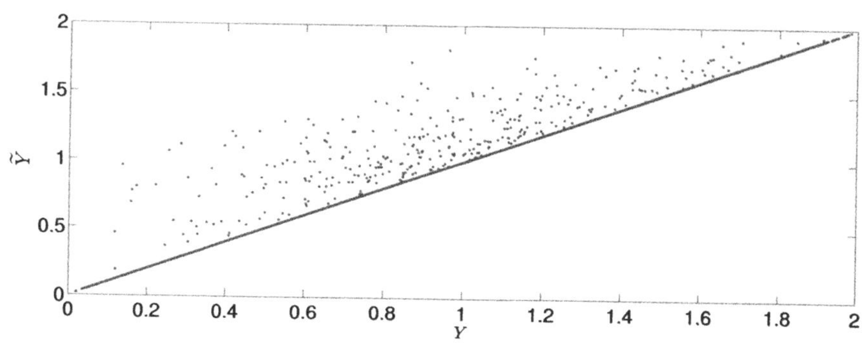

Figure 9.3 Scatter plot of $N = 10^4$ pairs $(Y, \widetilde{Y})$ for the output Y and control variable $\widetilde{Y}$ of the stochastic shortest path problem.

Remark 9.3.1 (Multiple Control Variables) Algorithm 9.2 can be extended straightforwardly to the case where more than one control variable is used for each output Y. Specifically, let $\widetilde{\mathbf{Y}} = (\widetilde{Y}_1, \ldots, \widetilde{Y}_m)^\top$ be a (column) vector of m control variables with known mean vector $\widetilde{\boldsymbol{\ell}} = \mathbb{E}\widetilde{\mathbf{Y}} = (\widetilde{\ell}_1, \ldots, \widetilde{\ell}_m)^\top$, where $\widetilde{\ell}_i = \mathbb{E}\widetilde{Y}_i$. Then, the control vector estimator of the optimal $\ell = \mathbb{E}Y$ based on independent random variables $Y_1, \ldots, Y_N$ with control vectors $\widetilde{\mathbf{Y}}_1 = (\widetilde{Y}_{11}, \ldots, \widetilde{Y}_{1m})^\top, \ldots, \widetilde{\mathbf{Y}}_N = (\widetilde{Y}_{N1}, \ldots, \widetilde{Y}_{Nm})^\top$ is given by

$$\widehat{\ell}^{(c)} = \frac{1}{N} \sum_{k=1}^{N} \left[Y_k - \boldsymbol{\alpha}^\top \left(\widetilde{\mathbf{Y}}_k - \widetilde{\boldsymbol{\ell}} \right) \right],$$

where $\boldsymbol{\alpha}$ is an estimator of the optimal vector $\boldsymbol{\alpha}^* = \Sigma_{\widetilde{\mathbf{Y}}}^{-1} \boldsymbol{\sigma}_{Y,\widetilde{\mathbf{Y}}}$. Here $\Sigma_{\widetilde{\mathbf{Y}}}$ is the $m \times m$ covariance matrix of $\widetilde{\mathbf{Y}}$, and $\boldsymbol{\sigma}_{Y,\widetilde{\mathbf{Y}}}$ is the $m \times 1$ vector whose i-th component is the covariance of Y and $\widetilde{Y}_i$, $i = 1, \ldots, m$. The variance of $\widehat{\ell}^{(c)}$ for $\boldsymbol{\alpha} = \boldsymbol{\alpha}^*$ is

$$\mathrm{Var}(\widehat{\ell}^{(c)}) = \frac{1}{N}(1 - R^2_{Y,\widetilde{\mathbf{Y}}})\mathrm{Var}(Y), \qquad (9.6)$$

where

$$R^2_{Y,\widetilde{\mathbf{Y}}} = (\boldsymbol{\sigma}_{Y,\widetilde{\mathbf{Y}}})^\top \Sigma_{\widetilde{\mathbf{Y}}}^{-1} \boldsymbol{\sigma}_{Y,\widetilde{\mathbf{Y}}} / \mathrm{Var}(Y)$$

is the square of the **multiple correlation coefficient** of Y and $\widetilde{\mathbf{Y}}$. Again, the larger $R^2_{Y,\widetilde{\mathbf{Y}}}$ is, the greater the variance reduction.

9.4 CONDITIONAL MONTE CARLO

Variance reduction using **conditional Monte Carlo** is based on the following result.

Theorem 9.4.1 (Conditional Variance) *Let Y be a random variable and $\mathbf{Z}$ a random vector. Then*

$$\mathrm{Var}(Y) = \mathbb{E}\,\mathrm{Var}(Y \mid \mathbf{Z}) + \mathrm{Var}(\,\mathbb{E}[Y \mid \mathbf{Z}]\,), \qquad (9.7)$$

and hence $\mathrm{Var}(\mathbb{E}[Y \mid \mathbf{Z}]) \leqslant \mathrm{Var}(Y)$.

Suppose that the aim is to estimate $\ell = \mathbb{E}Y$, where Y is the output from a Monte Carlo experiment, and that one can find a random variable (or vector), $\mathbf{Z} \sim g$, such that the conditional expectation $\mathbb{E}[Y \mid \mathbf{Z} = \mathbf{z}]$ can be computed analytically. By the tower property (A.28), ☞ 618

$$\ell = \mathbb{E}Y = \mathbb{E}\,\mathbb{E}[Y \mid \mathbf{Z}], \qquad (9.8)$$

it follows that $\mathbb{E}[Y \mid \mathbf{Z}]$ is an unbiased estimator of ℓ. Moreover, by Theorem 9.4.1 the variance of $\mathbb{E}[Y \mid \mathbf{Z}]$ is always smaller than or equal to the variance of Y. The conditional Monte Carlo idea is sometimes referred to as **Rao–Blackwellization**.

Algorithm 9.3 (Conditional Monte Carlo)

1. *Generate a sample* $\mathbf{Z}_1, \ldots, \mathbf{Z}_N \stackrel{\text{iid}}{\sim} g$.
2. *Calculate* $\mathbb{E}[Y \mid \mathbf{Z}_k],\ k = 1, \ldots, N$ *analytically.*
3. *Estimate* $\ell = \mathbb{E}Y$ *by*

$$\widehat{\ell}_c = \frac{1}{N} \sum_{k=1}^{N} \mathbb{E}[Y \mid \mathbf{Z}_k] \tag{9.9}$$

and determine an approximate $1 - \alpha$ *confidence interval as*

$$\left(\widehat{\ell}_c - z_{1-\alpha/2} \frac{S}{\sqrt{N}},\ \ \widehat{\ell}_c + z_{1-\alpha/2} \frac{S}{\sqrt{N}} \right),$$

where S *is the sample standard deviation of the* $\{\mathbb{E}[Y \mid \mathbf{Z}_k]\}$ *and* z_γ *denotes the* γ*-quantile of the* $\mathsf{N}(0,1)$ *distribution.*

■ EXAMPLE 9.4 (Conditional Monte Carlo for the Bridge Network)

We return to Example 9.1. Let $Z_1 = \min\{X_4, X_3 + X_5\}$, $Z_2 = \min\{X_5, X_3 + X_4\}$, $Y_1 = X_1 + Z_1$, $Y_2 = X_2 + Z_2$, and $\mathbf{Z} = (Z_1, Z_2)$. Then, $Y = H(\mathbf{X})$ can be written as

$$Y = \min\{Y_1, Y_2\},$$

where conditional upon $\mathbf{Z} = \mathbf{z}$, (Y_1, Y_2) is uniformly distributed on the rectangle $\mathscr{R}_{\mathbf{z}} = [z_1, z_1 + 1] \times [z_2, z_2 + 2]$. The conditional expectation of Y given $\mathbf{Z} = \mathbf{z}$ can be evaluated exactly, and is given by

$$\mathbb{E}[Y \mid \mathbf{Z} = \mathbf{z}] = \begin{cases} \frac{1}{2} + z_1 & \text{if } \mathbf{z} \in \mathscr{A}_0, \\ \frac{5}{12} + \frac{3z_1}{4} - \frac{z_1^2}{4} - \frac{z_1^3}{12} + \frac{z_2}{4} + \frac{z_1 z_2}{2} + \frac{z_1^2 z_2}{4} - \frac{z_2^2}{4} - \frac{z_1 z_2^2}{4} + \frac{z_2^3}{12} & \text{if } \mathbf{z} \in \mathscr{A}_1, \\ \frac{1}{12}(5 - 3z_1^2 + 3z_2 - 3z_2^2 + z_1(9 + 6z_2)) & \text{if } \mathbf{z} \in \mathscr{A}_2, \end{cases}$$

where

$$\begin{aligned} \mathscr{A}_0 &= \{\mathbf{z} : 0 \leqslant z_1 \leqslant 1,\ z_1 + 1 \leqslant z_2 \leqslant 2\}, \\ \mathscr{A}_1 &= \{\mathbf{z} : 0 \leqslant z_1 \leqslant 1,\ z_1 \leqslant z_2 \leqslant z_1 + 1\}, \\ \mathscr{A}_2 &= \{\mathbf{z} : 0 \leqslant z_1 \leqslant 1,\ 0 \leqslant z_2 \leqslant z_1\}. \end{aligned}$$

For example, if $\mathbf{z} \in \mathscr{A}_1$, then the domain $\mathscr{R}_{\mathbf{z}}$ of (Y_1, Y_2) intersects the line $y_1 = y_2$ at $y_1 = z_2$ and $y_1 = z_1 + 1$, so that

$$\begin{aligned} \mathbb{E}[Y \mid \mathbf{Z} = \mathbf{z}] = & \int_{z_1}^{z_2} \int_{z_2}^{z_2+2} y_1 \frac{1}{2} \, \mathrm{d}y_2 \, \mathrm{d}y_1 + \int_{z_2}^{z_1+1} \int_{y_1}^{z_2+2} y_1 \frac{1}{2} \, \mathrm{d}y_2 \, \mathrm{d}y_1 \\ & + \int_{z_2}^{z_1+1} \int_{z_2}^{y_1} y_2 \frac{1}{2} \, \mathrm{d}y_2 \, \mathrm{d}y_1 . \end{aligned}$$

The following MATLAB program gives an implementation of the corresponding conditional Monte Carlo estimator. A typical outcome for sample size $N = 10^4$ is

$\widehat{\ell}_c = 0.9282$ with an estimated relative error of 0.29%, compared with 0.43% for CMC, indicating more than a twofold reduction in simulation effort. Interestingly, the joint pdf of $\mathbf{Z}$ on $[0,1] \times [0,2]$ can, with considerable effort, be determined analytically, so that $\ell = \mathbb{E}Y$ can be evaluated exactly. This leads to the exact value given in the introduction:

$$\mathbb{E}Y = \frac{1339}{1440} = 0.9298611111\ldots .$$

```
%bridgeCondMC.m
N = 10^4;
S = zeros(N,1);
for i = 1:N
    u = rand(1,5);
    Z = Zcond(u);
    if Z(2)> Z(1) + 1
        S(i) = 1/2 + Z(1);
    elseif (Z(2) > Z(1))
        S(i) = 5/12 + (3*Z(1))/4 - Z(1)^2/4 - Z(1)^3/12 + Z(2)/4 ...
            + (Z(1)*Z(2))/2 + (Z(1)^2*Z(2))/4  ...
            -  Z(2)^2/4 - (Z(1)*Z(2)^2)/4 + Z(2)^3/12;
    else
        S(i) = (5 - 3*Z(1)^2 + 3*Z(2) - 3*Z(2)^2  ...
            + Z(1)*(9 + 6*Z(2)))/12;
    end
end
est = mean(S)
RE = std(S)/sqrt(N)/est
```

```
function Z=Zcond(u)
a=[1,2,3,1,2];
X = u*diag(a);
Z = [min([X(:,4), X(:,3) + X(:,5)],[],2),...
    min([X(:,5), X(:,3) + X(:,4)],[],2)];
```

9.5 STRATIFIED SAMPLING

Stratified sampling is closely related to both the composition method of Section 3.1.2.6 and the conditional Monte Carlo method discussed in the previous section. Let Y be the simulation output. The objective is to estimate $\ell = \mathbb{E}Y$. ☞ 53

Suppose that Y can be generated via the composition method. Thus, we assume that there exists a random variable Z taking values in $\{1,\ldots,m\}$, say, with known probabilities $\{p_i,\ i = 1,\ldots,m\}$. We further assume that it is easy to sample from the conditional distribution of Y given Z. The events $\{Z = i\}$, $i = 1,\ldots,m$, partition the sample space Ω into disjoint **strata**; hence the name **stratification**. Using the tower property (A.28), we can write ☞ 618

$$\ell = \mathbb{E}\,\mathbb{E}[Y \mid Z] = \sum_{i=1}^{m} p_i\,\mathbb{E}[Y \mid Z = i]\,. \tag{9.10}$$

This representation suggests that we can estimate ℓ via the following **stratified sampling estimator**:

$$\widehat{\ell}^{(s)} = \sum_{i=1}^{m} p_i \frac{1}{N_i} \sum_{j=1}^{N_i} Y_{ij} \stackrel{\text{def}}{=} \sum_{i=1}^{m} p_i \bar{Y}_{i\bullet} , \tag{9.11}$$

where Y_{ij} is the j-th of N_i independent observations from the conditional distribution of Y given $Z = i$, $i = 1, \ldots, m$. How the strata should be chosen depends very much on the problem at hand. From (9.11), the variance of the stratified sampling estimator is given by

$$\text{Var}(\widehat{\ell}^{(s)}) = \sum_{i=1}^{m} \frac{p_i^2\, \sigma_i^2}{N_i} ,$$

where $\sigma_i^2 = \text{Var}(Y \mid Z = i)$ is the variance of Y within the i-th stratum, $i = 1, \ldots, m$.

For any given choice of the strata one can select the sample sizes $\{N_i\}$ in an optimal manner, as specified in the next theorem. A simple proof based on Lagrange multipliers can be found in [10].

Theorem 9.5.1 (Optimal Allocation of Sample Sizes) *The allocation of sample sizes (rounded to integers)*

$$N_i = N \frac{p_i\, \sigma_i}{\sum_{k=1}^{m} p_k\, \sigma_k}, \quad i = 1, \ldots, m , \tag{9.12}$$

where $\sigma_i^2 = \text{Var}(Y \mid Z = i)$ *provides the smallest variance for* $\widehat{\ell}^{(s)}$ *over all choices of* $N_1, \ldots, N_m$ *for which* $\sum_i N_i = N$. *The minimum value for the variance is*

$$\frac{1}{N} \left(\sum_{i=1}^{m} p_i\, \sigma_i \right)^2 . \tag{9.13}$$

An obvious difficulty in applying Theorem 9.5.1 is that the standard deviations $\{\sigma_i\}$ are usually unknown. In practice, one can estimate the $\{\sigma_i\}$ from a pilot run and then proceed to estimate the optimal sample sizes from (9.12).

Algorithm 9.4 (Stratified Sampling)

1. *Choose the* m *strata and the sample sizes* $N_i, i = 1, \ldots, m$ — *the latter determined from* (9.12) *via a pilot run, for example.*

2. *For each stratum* $i = 1, \ldots, m$ *draw* $Y_{i1}, \ldots, Y_{iN_i}$ *independently from the conditional distribution of* Y *given* $Z = i$, *and let* $\bar{Y}_{1\bullet}, \ldots, \bar{Y}_{m\bullet}$ *be the sample means.*

3. *Estimate* ℓ *via* (9.11), *and calculate an approximate* $1 - \alpha$ *confidence interval as*

$$\left(\widehat{\ell}^{(s)} - z_{1-\alpha/2} \sqrt{\sum_{i=1}^{m} \frac{p_i^2\, \widehat{\sigma_i^2}}{N_i}} , \quad \widehat{\ell}^{(s)} + z_{1-\alpha/2} \sqrt{\sum_{i=1}^{m} \frac{p_i^2\, \widehat{\sigma_i^2}}{N_i}} \right) ,$$

where z_γ *denotes the* γ*-quantile of the* $\mathsf{N}(0, 1)$ *distribution and* $\widehat{\sigma_i^2} = \sum_{j=1}^{N_i} (Y_{ij} - \bar{Y}_{i\bullet})^2 / (N_i - 1)$ *is the sample variance of* $\{Y_{ij}, j = 1, \ldots, N_i\}$.

If the sample size for the i-th stratum is chosen to be *proportional* to p_i, that is, $N_i = p_i N$ for some overall sample size N, then

$$\mathrm{Var}(\widehat{\ell}^{(s)}) = \sum_{i=1}^{m} \frac{p_i^2 \sigma_i^2}{N_i} = \frac{1}{N}\, \mathbb{E}\, \mathrm{Var}(Y \mid Z) \leqslant \frac{1}{N} \mathrm{Var}(Y) , \tag{9.14}$$

so that the stratified estimator in this case (and hence under the optimal choice (9.12)) has a variance at least as small as the variance of the CMC estimator. This is called **proportional stratified sampling**. **Systematic sampling** [3] is proportional stratified sampling with equal weights; that is, $p_i = 1/m$ and $N_i = N/m = n$. The estimator (9.11) then reduces to

$$\widehat{\ell}^{(s)} = \frac{1}{N} \sum_{i=1}^{m} \sum_{j=1}^{n} Y_{ij} = \frac{1}{n} \sum_{j=1}^{n} \left(\frac{1}{m} \sum_{i=1}^{m} Y_{ij} \right) \stackrel{\text{def}}{=} \frac{1}{n} \sum_{j=1}^{n} \bar{Y}_{\bullet j} . \tag{9.15}$$

Note that the $\{\bar{Y}_{\bullet j}\}$ are iid random variables, so that the standard error of $\widehat{\ell}^{(s)}$ can simply be estimated as $S/\sqrt{n}$, where S is the sample standard deviation of $\{\bar{Y}_{\bullet j}\}$.

Systematic sampling is especially useful when dealing directly with uniform random variables. Specifically, let $Y = h(U)$, where $U \sim \mathsf{U}(0,1)$, and define $Z = \lceil mU \rceil$ for some fixed $m \in \{1, 2, \ldots\}$. The events $\{Z = i\} = \{(i-1)/m \leqslant U < i/m\}$, $i = 1, \ldots, m$ divide the sample space into m equiprobable strata. Sampling Y conditional on $(i-1)/m \leqslant U < i/m$ is immediate. The systematic sampling procedure for the d-dimensional case is summarized in the following algorithm.

Algorithm 9.5 (Systematic Sampling for the d-Dimensional Hypercube) *Let $\ell = \mathbb{E}h(\mathbf{U})$, where $\mathbf{U} \sim \mathsf{U}(0,1)^d$. Suppose that the k-th component of the hypercube $(0,1)^d$ is divided up into K_k equal-length intervals, $k = 1, \ldots, d$, so that $(0,1)^d$ is divided into $m = \prod_{k=1}^{d} K_k$ hyperrectangles (ignoring the boundaries)*

$$\prod_{k=1}^{d} \left(\frac{i_k}{K_k}, \frac{i_k + 1}{K_k} \right), \quad (i_1, \ldots, i_d) \in \mathscr{W} ,$$

where $\mathscr{W} = \{(i_1, \ldots, i_d) : i_k \in \{0, 1 \ldots, K_k - 1\},\ k \in \{1, \ldots, d\}\}$.

1. *For each* $\mathbf{i} = (i_1, \ldots, i_d) \in \mathscr{W}$ *generate* $\mathbf{V}_1, \ldots, \mathbf{V}_n \stackrel{\text{iid}}{\sim} \mathsf{U}(0,1)^d$ *and evaluate*

$$Y_{\mathbf{i}j} = h\left(\frac{i_1 + V_{j1}}{K_1}, \ldots, \frac{i_d + V_{jd}}{K_d} \right), \quad j = 1, \ldots, n .$$

   Let

$$\bar{Y}_{\bullet j} = \frac{1}{m} \sum_{\mathbf{i} \in \mathscr{W}} Y_{\mathbf{i}j} .$$

2. *Estimate ℓ via (9.15), which is the sample mean of $\bar{Y}_{\bullet 1}, \ldots, \bar{Y}_{\bullet n}$, and determine an approximate $1 - \alpha$ confidence interval as*

$$\left(\widehat{\ell}^{(s)} - z_{1-\alpha/2} \frac{S}{\sqrt{n}}, \ \widehat{\ell}^{(s)} + z_{1-\alpha/2} \frac{S}{\sqrt{n}} \right),$$

   where z_γ denotes the γ-quantile of the $\mathsf{N}(0,1)$ distribution and S is the sample standard deviation of $\bar{Y}_{\bullet 1}, \ldots, \bar{Y}_{\bullet n}$.

Figure 9.4 shows a typical outcome for the $d = 2$-dimensional case with $K = 5$ classes per dimension, so that the total number of strata is $m = K^d = 25$. The total sample size is $N = 150$, so that each stratum has exactly $n = N/m = 6$ samples.

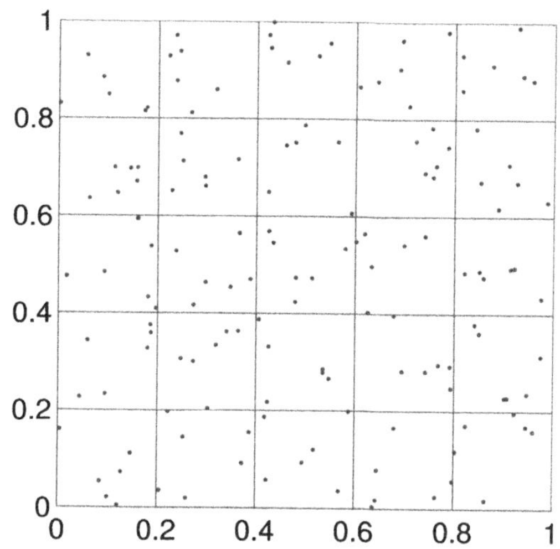

Figure 9.4 Systematic sampling on the unit square. Both dimensions are divided into $K = 5$ classes, giving a total of $m = 25$ strata. Each stratum has $n = 6$ samples for a total sample size of $N = 150$.

■ EXAMPLE 9.5 (Systematic Sampling for the Bridge Network)

We return again to Example 9.1. The MATLAB program below implements Algorithm 9.5 where each of the $d = 5$ dimensions is divided into $K = 4$ classes, giving a total of $m = 4^5 = 1024$ strata. For a total sample size of around $N = 10^4$ (the code below uses $N = 10240$) a typical estimate is $\widehat{\ell}^{(s)} = 0.9301$ with an estimated relative error of 0.13%. This means a tenfold variance reduction in comparison with CMC. Function `h.m` in the code below is the same as in Example 9.1.

```
%stratbridge.m
K = 4;
m = K^5; %number of strata
N = 10^4; %total number of samples
n = ceil(N/m); %number of samples per stratum
est = zeros(n,1);
R=(1:m)';
W=zeros(m,5);
W(:,1)=mod(R,K);
for i=2:5
    W(:,i)=(mod(R,K^i)-mod(R,K^(i-1)))./(K^(i-1));
end
```

```
for j=1:n
    V=(W+rand(m,5))./K;
    est(j)=mean(h(V));
end
mest = mean(est)
percRE = std(est)/sqrt(n)/mest*100
```

9.6 LATIN HYPERCUBE SAMPLING

The main drawback of stratification for high-dimensional estimation problems is that the number of strata grows exponentially in the number of classes. For example, if systematic sampling is applied to the d-dimensional hypercube and each coordinate is divided into K classes, then the number of strata is $m = K^d$, which is only practical for small K and d, say $K = 1, \ldots, 5$ and $d = 1, \ldots, 10$. For higher-dimensional problems a useful remedy is to apply **latin hypercube sampling** instead. The idea is to sample on the d-dimensional hypercube in such a way that only the marginal distributions are stratified. Figure 9.5 provides an illustration. In contrast to the full stratification in Figure 9.4 not all cells have the same number of samples. Instead, both the x and y coordinates are stratified in $K = 5$ classes, with 30 samples per class.

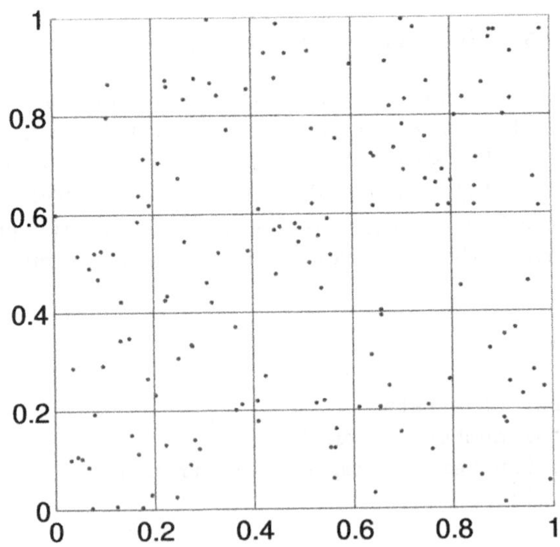

Figure 9.5 Latin hypercube sampling in dimension $d = 2$ with $K = 5$ strata per dimension and $N = 150$ samples. Each one-dimensional stratum has $n = N/K = 30$ samples.

Algorithm 9.6 (Latin Hypercube Sampling) *Starting with $i = 1$, execute the following steps:*

1. *Generate* $\mathbf{U}_1, \ldots, \mathbf{U}_K \stackrel{\text{iid}}{\sim} \mathsf{U}(0,1)^d$.
2. *Generate K independent uniform permutations,* $\mathbf{\Pi}_1, \ldots, \mathbf{\Pi}_K$, *of* $(1, \ldots, K)$.
3. *Set*
$$\mathbf{V}_k = \frac{\mathbf{\Pi}_k + \mathbf{1} - \mathbf{U}_k}{K}, \quad k = 1, \ldots, K .$$
Let
$$Y_i = \frac{1}{K} \sum_{k=1}^{K} h(\mathbf{V}_k) .$$
4. *If $i = n$, then go to Step 5; otherwise, set $i = i + 1$ and go to Step 1.*
5. *Estimate ℓ as* $\widehat{\ell}^{(h)} = \frac{1}{n} \sum_{i=1}^{n} Y_i$ *and determine an approximate* $1-\alpha$ *confidence interval as*
$$\left(\widehat{\ell}^{(h)} - z_{1-\alpha/2}\frac{S}{\sqrt{n}},\ \widehat{\ell}^{(h)} + z_{1-\alpha/2}\frac{S}{\sqrt{n}}\right),$$
where z_γ denotes the γ-quantile of the $\mathsf{N}(0,1)$ *distribution and S is the sample standard deviation of* $Y_1, \ldots, Y_n$.

■ EXAMPLE 9.6 (Latin Hypercube Sampling for the Bridge Network)

Consider again Example 9.1. The MATLAB program below implements a latin hypercube sampling scheme with $K = 50$ classes for each of the $d = 5$ dimensions. At each of the $n = 2000$ iterations, 50 points are generated in the five-dimensional hypercube, giving a total of $N = 10^4$ of such points. A typical estimate is $\widehat{\ell}^{(h)} = 0.9287$ with an estimated relative error of 0.16%, which is comparable with the 0.13% of the full stratification in Example 9.5. Function `h.m` in the code below is the same as in Example 9.1.

```
%lhcsbridge.m
d = 5;
K = 50;
N = 10^4;
n = N/K;
est = zeros(n,1);
for i = 1:n
    U = rand(K,d);
    [x,p] = sort(rand(K,d));
    V = (p + 1 - U)/K;
    est(i) = mean(h(V));
end
mean(est)
percRE = std(est)/sqrt(n)/mean(est)*100
```

9.7 IMPORTANCE SAMPLING

One of the most important variance reduction techniques is **importance sampling**. This technique is especially useful for the estimation of rare-event probabilities (see Chapter 10). The standard setting is the estimation of a quantity

$$\ell = \mathbb{E}_f H(\mathbf{X}) = \int H(\mathbf{x})\, f(\mathbf{x})\, \mathrm{d}\mathbf{x}\,, \tag{9.16}$$

where H is a real-valued function and f the probability density of a random vector $\mathbf{X}$, called the **nominal pdf**. The subscript f is added to the expectation operator to indicate that it is taken with respect to the density f.

Let g be another probability density such that $H\,f$ is **dominated** by g. That is, $g(\mathbf{x}) = 0 \Rightarrow H(\mathbf{x})\, f(\mathbf{x}) = 0$. Using the density g we can represent ℓ as

$$\ell = \int H(\mathbf{x})\, \frac{f(\mathbf{x})}{g(\mathbf{x})}\, g(\mathbf{x})\, \mathrm{d}\mathbf{x} = \mathbb{E}_g H(\mathbf{X})\, \frac{f(\mathbf{X})}{g(\mathbf{X})}\,. \tag{9.17}$$

Consequently, if $\mathbf{X}_1, \ldots, \mathbf{X}_N \sim_{\text{iid}} g$, then

$$\widehat{\ell} = \frac{1}{N} \sum_{k=1}^{N} H(\mathbf{X}_k)\, \frac{f(\mathbf{X}_k)}{g(\mathbf{X}_k)} \tag{9.18}$$

is an unbiased estimator of ℓ. This estimator is called the **importance sampling estimator** and g is called the importance sampling density. The ratio of densities,

$$W(\mathbf{x}) = \frac{f(\mathbf{x})}{g(\mathbf{x})}\,, \tag{9.19}$$

is called the **likelihood ratio** — with a slight abuse of nomenclature, as the likelihood is usually seen in statistics as a function of the parameters (see Section B.2). ☞ 664

Algorithm 9.7 (Importance Sampling Estimation)

1. *Select an importance sampling density g that dominates Hf.*

2. *Generate $\mathbf{X}_1, \ldots, \mathbf{X}_N \stackrel{\text{iid}}{\sim} g$ and let $Y_i = H(\mathbf{X}_i) f(\mathbf{X}_i)/g(\mathbf{X}_i)$, $i = 1, \ldots, N$.*

3. *Estimate ℓ via $\widehat{\ell} = \bar{Y}$ and determine an approximate $1-\alpha$ confidence interval as*

$$\left(\widehat{\ell} - z_{1-\alpha/2}\frac{S}{\sqrt{N}},\ \widehat{\ell} + z_{1-\alpha/2}\frac{S}{\sqrt{N}}\right),$$

where z_γ denotes the γ-quantile of the $\mathsf{N}(0,1)$ *distribution and S is the sample standard deviation of $Y_1, \ldots, Y_N$.*

■ **EXAMPLE 9.7 (Importance Sampling for the Bridge Network)**

The expected length of the shortest path in Example 9.1 can be written as (see (9.2))

$$\ell = \mathbb{E}h(\mathbf{U}) = \int h(\mathbf{u})\,\mathrm{d}\mathbf{u}\,,$$

where $\mathbf{U} = (U_1, \ldots, U_5)$ and $U_1, \ldots, U_5 \overset{\text{iid}}{\sim} \mathsf{U}(0,1)$. The nominal pdf is thus $f(\mathbf{u}) = 1, \mathbf{u} \in (0,1)^5$. Suppose the importance sampling pdf is of the form

$$g(\mathbf{u}) = \prod_{i=1}^{5} \nu_i\, u_i^{\nu_i - 1}\,,$$

which means that under g the components of $\mathbf{U}$ are again independent and $U_i \sim \mathsf{Beta}(\nu_i, 1)$ for some $\nu_i > 0$, $i = 1, \ldots, 5$. For the nominal (uniform) distribution we have $\nu_i = 1$, $i = 1, \ldots, 5$. Generating $\mathbf{U}$ under g is easily carried out via the inverse-transform method — see Algorithm 4.18. A good choice of $\{\nu_i\}$ is of course crucial. The MATLAB program below implements the importance sampling estimation of ℓ using, for example, $(\nu_1, \ldots, \nu_5) = (1.3, 1.1, 1, 1.3, 1.1)$. For a sample size of $N = 10^4$ a typical estimate is $\widehat{\ell} = 0.9295$ with an estimated relative error of 0.24%, which gives about a fourfold reduction in simulation effort compared with CMC estimation, despite the fact that the $\{\nu_i\}$ are all quite close to their nominal value 1. ☞ 105

```
%bridgeIS.m
N = 10^4;
nu0 = [1.3, 1.1, 1, 1.3, 1.1];
nu = repmat(nu0,N,1);
U = rand(N,5).^(1./nu);
W = prod(1./(nu.*U.^(nu - 1)),2);
y = h(U).*W;
est = mean(y)
percRE = std(y)/sqrt(N)/est*100
```

The main difficulty in importance sampling is how to choose the importance sampling distribution. A poor choice of g may seriously compromise the estimate and the confidence intervals. The following sections provide some guidance on choosing a good importance sampling distribution.

9.7.1 Minimum-Variance Density

The optimal choice g^* for the importance sampling density minimizes the variance of $\widehat{\ell}$, and is therefore the solution to the functional minimization program

$$\min_g \mathrm{Var}_g\left(H(\mathbf{X})\,\frac{f(\mathbf{X})}{g(\mathbf{X})}\right). \tag{9.20}$$

It is not difficult to show (see for example [10]) that

$$g^*(\mathbf{x}) = \frac{|H(\mathbf{x})|\,f(\mathbf{x})}{\int |H(\mathbf{x})|\,f(\mathbf{x})\,\mathrm{d}\mathbf{x}}\,. \tag{9.21}$$

In particular, if $H(\mathbf{x}) \geqslant 0$ or $H(\mathbf{x}) \leqslant 0$ then

$$g^*(\mathbf{x}) = \frac{H(\mathbf{x})\, f(\mathbf{x})}{\ell}\,, \tag{9.22}$$

in which case $\text{Var}_{g^*}(\widehat{\ell}) = \text{Var}_{g^*}(H(\mathbf{X})W(\mathbf{X})) = \text{Var}_{g^*}(\ell) = 0$, so that the estimator $\widehat{\ell}$ is *constant* under g^*. An obvious difficulty is that the evaluation of the optimal importance sampling density g^* is usually not possible. For example, $g^*(\mathbf{x})$ in (9.22) depends on the unknown quantity ℓ. Nevertheless, a good importance sampling density g should be "close" to the minimum variance density g^*.

One of the main considerations for choosing a good importance sampling pdf is that the estimator (9.18) should have finite variance. This is equivalent to the requirement that

$$\mathbb{E}_g H^2(\mathbf{X}) \frac{f^2(\mathbf{X})}{g^2(\mathbf{X})} = \mathbb{E}_f H^2(\mathbf{X}) \frac{f(\mathbf{X})}{g(\mathbf{X})} < \infty\,. \tag{9.23}$$

This suggests that g should not have lighter tails than f, and that, preferably, the likelihood ratio, f/g, should be bounded.

9.7.2 Variance Minimization Method

When the pdf f belongs to some parametric family of distributions, it is often convenient to choose the importance sampling distribution from the *same* family. In particular, suppose that $f(\cdot;\boldsymbol{\theta})$ belongs to the family

$$\{f(\cdot;\boldsymbol{\eta}),\ \boldsymbol{\eta} \in \Theta\}\,.$$

Then, the problem of finding an optimal importance sampling density in this class reduces to the following *parametric* minimization problem

$$\min_{\boldsymbol{\eta}\in\Theta} \text{Var}_{\boldsymbol{\eta}}\left(H(\mathbf{X})\, W(\mathbf{X};\boldsymbol{\theta},\boldsymbol{\eta})\right)\,, \tag{9.24}$$

where $W(\mathbf{X};\boldsymbol{\theta},\boldsymbol{\eta}) = f(\mathbf{X};\boldsymbol{\theta})/f(\mathbf{X};\boldsymbol{\eta})$. We call $\boldsymbol{\theta}$ the **nominal parameter** and $\boldsymbol{\eta}$ the **reference parameter vector** or **tilting vector**. Since under any $f(\cdot;\boldsymbol{\eta})$ the expectation of $H(\mathbf{X})\, W(\mathbf{X};\boldsymbol{\theta},\boldsymbol{\eta})$ is ℓ, the optimal solution of (9.24) coincides with that of

$$\min_{\boldsymbol{\eta}\in\Theta} V(\boldsymbol{\eta})\,, \tag{9.25}$$

where

$$V(\boldsymbol{\eta}) = \mathbb{E}_{\boldsymbol{\eta}} H^2(\mathbf{X})\, W^2(\mathbf{X};\boldsymbol{\theta},\boldsymbol{\eta}) = \mathbb{E}_{\boldsymbol{\theta}} H^2(\mathbf{X})\, W(\mathbf{X};\boldsymbol{\theta},\boldsymbol{\eta})\,. \tag{9.26}$$

We call either of the equivalent problems (9.24) and (9.25) the **variance minimization** (VM) problem; and we call the parameter vector ${}_*\boldsymbol{\eta}$ that minimizes the programs (9.24) and (9.25) the **VM-optimal reference parameter vector**. The VM problem can be viewed as a stochastic optimization problem, and can be approximately solved via Monte Carlo simulation by considering the sample average
☞ 446 version of (9.25) and (9.26):

$$\min_{\boldsymbol{\eta}\in\Theta} \widehat{V}(\boldsymbol{\eta})\,, \tag{9.27}$$

where

$$\widehat{V}(\boldsymbol{\eta}) = \frac{1}{N}\sum_{k=1}^{N} H^2(\mathbf{X}_k)\, W(\mathbf{X}_k;\boldsymbol{\theta},\boldsymbol{\eta}) \,, \tag{9.28}$$

and $\mathbf{X}_1,\ldots,\mathbf{X}_N \sim_{\text{iid}} f(\cdot;\boldsymbol{\theta})$. This problem can be solved via standard numerical optimization techniques. This gives the following modification of Algorithm 9.7.

Algorithm 9.8 (Variance Minimization Method)

1. *Select a parameterized family of importance sampling densities* $\{f(\cdot;\boldsymbol{\eta})\}$.
2. *Generate a pilot sample* $\mathbf{X}_1,\ldots,\mathbf{X}_N \stackrel{\text{iid}}{\sim} f(\cdot;\boldsymbol{\theta})$, *and determine the solution* ${}_*\widehat{\boldsymbol{\eta}}$ *to the variance minimization problem* (9.27).
3. *Generate* $\mathbf{X}_1,\ldots,\mathbf{X}_{N_1} \stackrel{\text{iid}}{\sim} f(\cdot;\,{}_*\widehat{\boldsymbol{\eta}})$ *and let* $Y_i = H(\mathbf{X}_i)f(\mathbf{X}_i;\boldsymbol{\theta})/f(\mathbf{X}_i;\,{}_*\widehat{\boldsymbol{\eta}}), i = 1,\ldots,N_1$.
4. *Estimate* ℓ *via* $\widehat{\ell} = \bar{Y}$ *and determine an approximate* $1-\alpha$ *confidence interval as*

$$\left(\widehat{\ell} - z_{1-\alpha/2}\frac{S}{\sqrt{N_1}},\ \widehat{\ell} + z_{1-\alpha/2}\frac{S}{\sqrt{N_1}}\right),$$

where z_γ *denotes the* γ*-quantile of the* $\mathsf{N}(0,1)$ *distribution and* S *is the sample standard deviation of* $Y_1,\ldots,Y_{N_1}$.

■ **EXAMPLE 9.8 (Variance Minimization for the Bridge Network)**

Consider the importance sampling approach for the bridge network in Example 9.7. There, the importance sampling distribution is the joint distribution of independent $\mathsf{Beta}(\nu_i, 1)$ random variables, for $i = 1,\ldots,5$. Hence, the reference parameter is $\boldsymbol{\nu} = (\nu_1,\ldots,\nu_5)$.

The following MATLAB program determines the optimal reference parameter vector ${}_*\widehat{\boldsymbol{\nu}}$ via the VM method using a pilot run of size $N = 10^3$ and the standard MATLAB minimization routine `fminsearch`. A typical value for ${}_*\widehat{\boldsymbol{\nu}}$ is $(1.262, 1.083, 1.016, 1.238, 1.067)$, which is similar to the one used in Example 9.7; the relative error is thus around 0.24%.

```
%vmceopt.m
N = 10^3;
U = rand(N,5);
[nu0,minv] =fminsearch(@(nu)f_var(nu,U,N),ones(1,5))
N1 = 10^4;
nu = repmat(nu0,N1,1);
U = rand(N1,5).^(1./nu);
w = prod(1./(nu.*U.^(nu - 1)),2);
y = h(U).*w;
est = mean(y)
percRE = std(y)/sqrt(N1)/est*100
```

```
function out = f_var(nu,U,N)
nu1 = repmat(nu,N,1);
W = prod(1./(nu1.*U.^(nu1 - 1)),2);
y = H(U);
out = W'*y.^2;
```

9.7.3 Cross-Entropy Method

An alternative approach to the VM method for choosing an "optimal" importance sampling distribution is based on the Kullback–Leibler cross-entropy distance, or simply **cross-entropy** (CE) distance. The CE distance between two continuous pdfs g and h is given by

$$\begin{aligned} \mathcal{D}(g,h) &= \mathbb{E}_g \ln \frac{g(\mathbf{X})}{h(\mathbf{X})} = \int g(\mathbf{x}) \ln \frac{g(\mathbf{x})}{h(\mathbf{x})}\, \mathrm{d}\mathbf{x} \\ &= \int g(\mathbf{x}) \ln g(\mathbf{x})\, \mathrm{d}\mathbf{x} - \int g(\mathbf{x}) \ln h(\mathbf{x})\, \mathrm{d}\mathbf{x}\,. \end{aligned} \tag{9.29}$$

For discrete pdfs replace the integrals with the corresponding sums. Observe that, ☞ 614 by Jensen's inequality, $\mathcal{D}(g,h) \geqslant 0$, with equality if and only if $g = h$. The CE distance is sometimes called the Kullback–Leibler *divergence*, because it is not symmetric, that is, $\mathcal{D}(g,h) \neq \mathcal{D}(h,g)$ for $g \not\equiv h$.

The idea of the CE method is to choose the importance sampling density, h say, such that the CE distance between the optimal importance sampling density g^* in (9.21) and h, is minimal. We call this the **CE-optimal pdf**. This pdf solves the *functional* optimization program $\min_h \mathcal{D}(g^*, h)$. If we optimize over all densities h, then it is immediate that the CE-optimal pdf coincides with the VM-optimal pdf g^*.

As with the VM approach in (9.24) and (9.25), we shall restrict ourselves to a parametric family of densities $\{f(\cdot;\boldsymbol{\eta}), \boldsymbol{\eta} \in \Theta\}$ that contains the nominal density $f(\cdot;\boldsymbol{\theta})$. Moreover, without any loss of generality, we only consider positive functions H. The CE method now aims to solve the *parametric* optimization problem

$$\min_{\boldsymbol{\eta}\in\Theta} \mathcal{D}\left(g^*, f(\cdot;\boldsymbol{\eta})\right)\,. \tag{9.30}$$

The optimal solution coincides with that of

$$\max_{\boldsymbol{\eta}\in\Theta} D(\boldsymbol{\eta})\,, \tag{9.31}$$

where

$$D(\boldsymbol{\eta}) = \mathbb{E}_{\boldsymbol{\theta}} H(\mathbf{X}) \ln f(\mathbf{X};\boldsymbol{\eta})\,. \tag{9.32}$$

Similar to the VM program (9.25), we call either of the equivalent programs (9.30) and (9.31) the **CE program**; and we call the parameter vector $\boldsymbol{\eta}^*$ that minimizes the program (9.30) and (9.31) the **CE-optimal reference parameter**.

☞ 446 Similar to (9.27) we can estimate $\boldsymbol{\eta}^*$ via the stochastic counterpart method as the solution of the stochastic program

$$\max_{\boldsymbol{\eta}} \widehat{D}(\boldsymbol{\eta}) = \max_{\boldsymbol{\eta}} \frac{1}{N} \sum_{k=1}^{N} H(\mathbf{X}_k) \ln f(\mathbf{X}_k; \boldsymbol{\eta}) , \tag{9.33}$$

where $\mathbf{X}_1, \ldots, \mathbf{X}_N \sim_{\text{iid}} f(\cdot; \boldsymbol{\theta})$.

In typical applications the function $\widehat{D}$ in (9.33) is convex and differentiable with respect to $\boldsymbol{\eta}$ (see [19]). In such cases the solution of (9.33) may be obtained by solving (with respect to $\boldsymbol{\eta}$) the following system of equations:

$$\frac{1}{N} \sum_{k=1}^{N} H(\mathbf{X}_k) \nabla \ln f(\mathbf{X}_k; \boldsymbol{\eta}) = \mathbf{0} , \tag{9.34}$$

where the gradient is with respect to $\boldsymbol{\eta}$. Various numerical and theoretical studies [17] have shown that the solutions to the VM and CE programs are qualitatively similar. The main advantage of the CE approach over the VM approach is that the solution to (9.33) (or (9.34)) can often be found *analytically*, as specified in the following theorem. A proof can be found in [17, Pages 69–70].

Theorem 9.7.1 (Exponential Families) *If the importance sampling density is of the form*

$$f(\mathbf{x}; \boldsymbol{\eta}) = \prod_{i=1}^{n} f_i(x_i; \eta_i) ,$$

where each $\{f_i(x_i; \eta_i), \eta_i \in \Theta_i\}$ forms a 1-parameter exponential family parameterized by the mean, then the solution to the CE program (9.33) is $\widehat{\boldsymbol{\eta}}^* = (\widehat{\eta}_1^*, \ldots, \widehat{\eta}_n^*)$, with ☞ 701

$$\widehat{\eta}_i^* = \frac{\sum_{k=1}^{N} H(\mathbf{X}_k) X_{ki}}{\sum_{k=1}^{N} H(\mathbf{X}_k)} , \quad i = 1, \ldots, n , \tag{9.35}$$

where X_{ki} is the i-th coordinate of $\mathbf{X}_k$.

For rare-event simulation the random variable $H(\mathbf{X})$ often takes the form of an indicator $\mathrm{I}_{\{S(\mathbf{X}) \geqslant \gamma\}}$. If the event $\{S(\mathbf{X}) \geqslant \gamma\}$ is rare under $f(\cdot; \boldsymbol{\theta})$, then with high probability the numerator and denominator in (9.35) are both zero, so that the CE-optimal parameter cannot be estimated in this way. Section 10.5 discusses how this can be remedied by using a multilevel approach or by sampling directly from the zero-variance importance sampling pdf g^*. ☞ 404

■ EXAMPLE 9.9 (CE Method for the Bridge Network)

In Example 9.8 the VM-optimal reference parameter is obtained by numerical minimization. We can use the CE method instead by applying (9.35) after suitable reparameterization. Note that for each i, $\mathsf{Beta}(\nu_i, 1)$ forms an exponential family, and that the corresponding expectation is $\eta_i = \nu_i/(1 + \nu_i)$. It follows that the assignment $\nu_i = \eta_i/(1 - \eta_i)$ reparameterizes the family in terms of the mean η_i.

The first four lines of the following MATLAB program implement the CE method for estimating the CE-optimal reference parameter. A typical outcome is $\widehat{\boldsymbol{\eta}} = (0.560, 0.529, 0.500, 0.571, 0.518)$, so that $\widehat{\boldsymbol{\nu}} = (1.272, 1.122, 1.000, 1.329, 1.075)$, which gives comparable results to the VM-optimal parameter vector. The corresponding relative error is estimated as 0.25%.

```
%bridgeCE.m
N = 10^3;
U = rand(N,5);
y = repmat(h(U),1,5);
v = sum(y.*U)./sum(y)
N1 = 10^4;
nu = repmat(v./(1-v),N1,1);
U = rand(N1,5).^(1./nu);
w = prod(1./(nu.*U.^(nu - 1)),2);
y = h(U).*w;
est = mean(y)
percRE = std(y)/sqrt(N1)/est*100
```

9.7.4 Weighted Importance Sampling

Algorithm 9.7 can be modified slightly by allowing the likelihood ratio (9.19) to be known *up to a constant*; that is, $W(\mathbf{X}) = f(\mathbf{X})/g(\mathbf{X}) = c\,w(\mathbf{X})$ for some known function $w(\cdot)$, but possibly unknown constant c. Since $\mathbb{E}_g W(\mathbf{X}) = 1$, we can write $\ell = \mathbb{E}_g H(\mathbf{X})\, W(\mathbf{X})$ as

$$\ell = \frac{\mathbb{E}_g H(\mathbf{X})\, W(\mathbf{X})}{\mathbb{E}_g W(\mathbf{X})} \,.$$

This suggests as an alternative to the standard importance sampling estimator (9.18) the following **weighted sample estimator**:

$$\widehat{\ell}_w = \frac{\sum_{k=1}^N H(\mathbf{X}_k)\, w_k}{\sum_{k=1}^N w_k} \,. \tag{9.36}$$

Here the $\{w_k\}$, with $w_k = w(\mathbf{X}_k)$, are interpreted as *weights* of the random sample $\{\mathbf{X}_k\}$, and the population $\{(\mathbf{X}_k, w_k)\}$ is called a **weighted sample** from $g(\mathbf{x})$. ☞ 308 Since the estimator is a ratio estimator (see Example 8.4), the weighted sample estimator (9.36) introduces some bias, which tends to 0 as N increases. Loosely speaking, we may view the weighted sample $\{(\mathbf{X}_k, w_k)\}$ as a representation of $f(\mathbf{x})$, in the sense that $\ell = \mathbb{E}_f H(\mathbf{X}) \approx \widehat{\ell}_w$ for any function H.

■ **EXAMPLE 9.10 (Weighted Sample Estimator for the Bridge Network)**

For the bridge example it is tempting to use the zero-variance pdf $g^*(\mathbf{u}) = \frac{1}{\ell}\, h(\mathbf{u})$ as the importance sampling pdf, in conjunction with (9.36). Note that $h(\mathbf{u}) \leqslant 2$ for all $\mathbf{u}$. Sampling from g^* can therefore be done via acceptance–rejection: sample $\mathbf{U} \sim \mathsf{U}(0,1)^5$ and $V \sim \mathsf{U}(0,2)$ independently, and accept $\mathbf{U}$ if $V < h(\mathbf{U})$. Since $w(\mathbf{u}) = 1/h(\mathbf{u})$, the weighted sample estimator becomes

$$\widehat{\ell}_w = \frac{1}{\frac{1}{N}\sum_{k=1}^N 1/h(\mathbf{U}_k)} \,,$$

where $\mathbf{U}_1, \ldots, \mathbf{U}_N \stackrel{\text{iid}}{\sim} g^*$. By the delta method, $\sqrt{N}(\widehat{\ell}_w - \ell)$ converges in distribution to a $\mathsf{N}(0, \sigma^2 \ell^4)$ distribution, where σ^2 is the variance of $1/h(\mathbf{U})$, which can be readily estimated from the simulation. The MATLAB program below implements the weighted sample estimator. Although the program produces unbiased estimates, the relative error is around 0.65%, which is *worse* than the one for CMC. ☞ 308

```
%wsbridge.m
clear all
N = 10^4;w   = zeros(N,1);
for k=1:N
    cont = true;
    while cont
        R = rand(1,5);v = rand*2;y = h(R);
        if v < y
            w(k) = 1/y;cont=false;
        end
    end
end
est = 1/mean(w)
percRE = std(w)*est/sqrt(N)*100
```

9.7.5 Sequential Importance Sampling

Sequential importance sampling (SIS), also called **dynamic importance sampling**, is simply importance sampling carried out in a sequential manner. To explain the SIS procedure, consider the expected performance $\ell = \mathbb{E}_f H(\mathbf{X})$ and its importance sampling estimator

$$\widehat{\ell} = \frac{1}{N} \sum_{k=1}^{N} H(\mathbf{X}_k) \frac{f(\mathbf{X}_k)}{g(\mathbf{X}_k)}, \quad \mathbf{X}_1, \ldots, \mathbf{X}_N \stackrel{\text{iid}}{\sim} g, \tag{9.37}$$

where g is the importance sampling pdf. Suppose that (a) $\mathbf{X}$ can be written as a vector $\mathbf{X} = (X_1, \ldots, X_n)$, where each of the X_i may be multidimensional; and (b) it is easy to sample from $g(\mathbf{x})$ sequentially. Specifically, suppose that $g(\mathbf{x})$ is of the form

$$g(\mathbf{x}) = g_1(x_1)\, g_2(x_2 \,|\, x_1) \cdots g_n(x_n \,|\, x_1, \ldots, x_{n-1}), \tag{9.38}$$

where it is easy to generate X_1 from the density $g_1(x_1)$, and conditional on $X_1 = x_1$ the second component from the density $g_2(x_2 \,|\, x_1)$, and so on, until one obtains a single random vector $\mathbf{X}$ from $g(\mathbf{x})$. Repeating this independently N times, one obtains an iid sample $\mathbf{X}_1, \ldots, \mathbf{X}_N$ from $g(\mathbf{x})$ and estimates ℓ according to (9.37).

To further simplify the notation we abbreviate $(x_1, \ldots, x_t)$ to $\mathbf{x}_{1:t}$ for all t. In particular, $\mathbf{x}_{1:n} = \mathbf{x}$. Typically, t can be viewed as a (discrete) time parameter and $\mathbf{x}_{1:t}$ as a path or trajectory. By the product rule of probability (A.21), the target pdf $f(\mathbf{x})$ can also be written sequentially as ☞ 616

$$f(\mathbf{x}) = f(x_1)\, f(x_2 \,|\, x_1) \cdots f(x_n \,|\, \mathbf{x}_{1:n-1}), \tag{9.39}$$

☞ 672 where we use a Bayesian notational convention (see Section B.3) for notational convenience. From (9.38) and (9.39) it follows that we can write the likelihood ratio in product form as

$$W(\mathbf{x}) = \frac{f(x_1)\, f(x_2 \,|\, x_1) \cdots f(x_n \,|\, \mathbf{x}_{1:n-1})}{g_1(x_1)\, g_2(x_2 \,|\, x_1) \cdots g_n(x_n \,|\, \mathbf{x}_{1:n-1})} . \tag{9.40}$$

If $W_t(\mathbf{x}_{1:t})$ denotes the likelihood ratio up to time t, we can write it recursively as

$$W_t(\mathbf{x}_{1:t}) = u_t\, W_{t-1}(\mathbf{x}_{1:t-1}), \quad t = 1, \ldots, n , \tag{9.41}$$

with initial weight $W_0(\mathbf{x}_{1:0}) = 1$ and **incremental weights** $u_1 = f(x_1)/g_1(x_1)$ and

$$u_t = \frac{f(x_t \,|\, \mathbf{x}_{1:t-1})}{g_t(x_t \,|\, \mathbf{x}_{1:t-1})} = \frac{f(\mathbf{x}_{1:t})}{f(\mathbf{x}_{1:t-1})\, g_t(x_t \,|\, \mathbf{x}_{1:t-1})} , \quad t = 2, \ldots, n . \tag{9.42}$$

In order to update the likelihood recursively as in (9.42) one needs to know the marginal pdf $f(\mathbf{x}_{1:t})$ for each t. This may not be simple when f does not have a Markov structure, as it requires integrating $f(\mathbf{x})$ over all $x_{t+1}, \ldots, x_n$. Instead, one can introduce a sequence of *auxiliary* pdfs $f_1, f_2, \ldots, f_n$ that are easily evaluated, and such that each $f_t(\mathbf{x}_{1:t})$ is a good approximation to $f(\mathbf{x}_{1:t})$. The terminating pdf f_n must be equal to the original f. Since

$$f(\mathbf{x}) = \frac{f_1(x_1)}{1} \frac{f_2(\mathbf{x}_{1:2})}{f_1(x_1)} \cdots \frac{f_n(\mathbf{x}_{1:n})}{f_{n-1}(\mathbf{x}_{1:n-1})} , \tag{9.43}$$

as a generalization of (9.42) we have the incremental updating weight

$$u_t = \frac{f_t(\mathbf{x}_{1:t})}{f_{t-1}(\mathbf{x}_{1:t-1})\, g_t(x_t \,|\, \mathbf{x}_{1:t-1})} , \tag{9.44}$$

for $t = 1, \ldots, n$, where we put $f_0(\mathbf{x}_{1:0}) = 1$. Note that the incremental weights u_t only need to be defined *up to a constant*, say c_t, for each t. In this case the likelihood ratio $W(\mathbf{x})$ is known up to a constant as well, say $W(\mathbf{x}) = C\, w(\mathbf{x})$, where $1/C = \mathbb{E}_g w(\mathbf{X})$ can be estimated via the corresponding sample mean. In other words, when the normalization constant is unknown, one can still estimate ℓ using the weighted sample estimator (9.36) rather than the importance sampling estimator (9.18).

Summarizing, the SIS method can be written as follows.

Algorithm 9.9 (Sequential Importance Sampling)

1. *For each* $t = 1, \ldots, n$, *sample* X_t *from* $g_t(x_t \,|\, \mathbf{x}_{1:t-1})$.

2. *Compute* $w_t = u_t\, w_{t-1}$, *where* $w_0 = 1$ *and*

$$u_t = \frac{f_t(\mathbf{X}_{1:t})}{f_{t-1}(\mathbf{X}_{1:t-1})\, g_t(X_t \,|\, \mathbf{X}_{1:t-1})}, \quad t = 1, \ldots, n . \tag{9.45}$$

3. *Repeat the steps above* N *times and estimate* ℓ *via* $\widehat{\ell}$ *in* (9.18) *or* $\widehat{\ell}_w$ *in* (9.36).

☞ 627 Applications of sequential importance sampling frequently involve random stopping times. Many importance sampling results concerning stopping times are direct generalizations of similar results for fixed times, as exemplified by the following theorem. Proofs can be found, for example, in [18, Pages 143 and 225].

Theorem 9.7.2 (Importance Sampling With a Stopping Time) *Let τ be a stopping time with respect to the stochastic process $\{X_t, t = 1, 2, \ldots\}$. Let $\mathbb{P}$ and $\widetilde{\mathbb{P}}$ be two measures under which $\mathbf{X}_{1:t} = (X_1, \ldots, X_t)$ has pdf $f_t(\mathbf{x}_{1:t})$ and $g_t(\mathbf{x}_{1:t})$, respectively, for $t = 1, 2, \ldots$. Then, for each sequence of real-valued functions H_t of $\mathbf{x}_{1:t}$, $t = 1, 2, \ldots$,*

$$\mathbb{E} \sum_{t=1}^{\tau} H_t(\mathbf{X}_{1:t}) = \widetilde{\mathbb{E}} \sum_{t=1}^{\tau} H_t(\mathbf{X}_{1:t})\, W_t \;, \tag{9.46}$$

and

$$\mathbb{E} H_\tau(\mathbf{X}_{1:\tau}) = \widetilde{\mathbb{E}} H_\tau(\mathbf{X}_{1:\tau})\, W_\tau \;, \tag{9.47}$$

where $W_t = f_t(\mathbf{X}_{1:t})/g_t(\mathbf{X}_{1:t})$ is the likelihood ratio of $\mathbf{X}_{1:t}$.

■ EXAMPLE 9.11 (Random Walk on the Integers)

Consider a random walk process $\{S_t, t = 0, 1, \ldots\}$ on the integers: $S_t = S_{t-1} + X_t$, where the $\{X_t\}$ are independent $\mathbb{P}(X_t = 1) = p$ and $\mathbb{P}(X_t = -1) = q = 1 - p$ for all $t = 1, 2, \ldots$. Suppose that $p < q$, so that the walk has a drift toward $-\infty$. Our goal is to estimate the rare-event probability ℓ of reaching state K before state 0, starting from state $0 < k \ll K$, where K is a large number. Let $\widetilde{\mathbb{P}}$ be the probability measure under which the $\{X_t\}$ are again independent, but now with $\widetilde{\mathbb{P}}(X_t = 1) = \widetilde{p}$ and $\widetilde{\mathbb{P}}(X_t = -1) = \widetilde{q} = 1 - \widetilde{p}$ for all $t = 1, 2, \ldots$. Define τ as the first time that either 0 or K is reached. As τ is a stopping time for $\{S_t\}$ we have by (9.47) that

$$\ell = \mathbb{E}\, \mathrm{I}_{\{S_\tau = K\}} = \widetilde{\mathbb{E}}\, \mathrm{I}_{\{S_\tau = K\}} W_\tau \;,$$

where $W_t, t = 1, 2, \ldots$ can be computed sequentially as $W_t = W_{t-1}\, u_t$ with

$$u_t = \begin{cases} p/\widetilde{p} & \text{if } x_t = 1 \;, \\ q/\widetilde{q} & \text{if } x_t = -1 \;, \end{cases}$$

where $W_0 = 1$. Consider the exponential family of pdfs $\{f(x;\theta),\, \theta \in \mathbb{R}\}$ defined by

$$f(x;\theta) = \mathrm{e}^{\theta x - \zeta(\theta)} f_0(x), \quad x \in \{-1, 1\} \;,$$

where $f_0(1) = p$ and $f_0(-1) = q$ (corresponding to the nominal pdf of X_t) and $\zeta(\theta) = \ln(p\mathrm{e}^{\theta} + q\mathrm{e}^{-\theta})$. Note that $\widetilde{p}$ can be related to θ via $\widetilde{p} = p\mathrm{e}^{\theta}/(p\mathrm{e}^{\theta} + q\mathrm{e}^{-\theta})$. The family can be reparameterized by the mean $v = \zeta'(\theta) = (p\mathrm{e}^{\theta} - q\mathrm{e}^{-\theta})/(p\mathrm{e}^{\theta} + q\mathrm{e}^{-\theta})$. The CE-optimal parameter v^* for estimating ℓ can be derived similarly to Theorem 9.7.1 and is given by (see [4]):

$$v^* = \frac{\mathbb{E}\, \mathrm{I}_{\{S_\tau = K\}} \sum_{i=1}^{\tau} X_i}{\mathbb{E} \tau\, \mathrm{I}_{\{S_\tau = K\}}} = \frac{(K-k)\, \mathbb{P}(S_\tau = K)}{\mathbb{E} \tau\, \mathrm{I}_{\{S_\tau = K\}}} = \frac{K-k}{\mathbb{E}[\tau \,|\, S_\tau = K]} \,.$$

Stern [21] shows that

$$\mathbb{E}\left[\tau \,|\, S_\tau = K\right] = \frac{1}{(p-q)(1-r^k)} \left[(K-k)(r^k+1) + 2K\left(\frac{r^k - r^K}{r^K - 1}\right)\right],$$

where $r = q/p$. Thus, the CE-optimal tilting parameter is

$$v^* = \frac{(K-k)(p-q)(1-r^k)}{(K-k)(r^k+1) + 2K\left(\frac{r^k - r^K}{r^K - 1}\right)} .$$

The likelihood ratio of $\mathbf{X}_{1:t} = (X_1, \ldots, X_t)$ is given by

$$W_t = \prod_{i=1}^{t} \frac{f_0(x_i)}{f(x_i;\theta)} = \mathrm{e}^{-\theta \sum_{i=1}^{t} x_i + t\,\zeta(\theta)} ,$$

where θ is related to v via $\theta = \frac{1}{2}\ln\left((1+v)q/((1-v)p)\right)$. It follows that under CE-optimal tilting,

$$\mathrm{I}_{\{S_\tau = K\}}\, W_\tau = \mathrm{I}_{\{S_\tau = K\}} \mathrm{e}^{-\theta^*(K-k) + \tau\zeta(\theta^*)} .$$

In the MATLAB code below the CE-optimal importance sampling procedure for estimating ℓ is carried out for the case $k = 10$, $K = 30$, and $p = 0.3$. The actual probability is given by

$$\ell = \frac{r^k - 1}{r^K - 1} \approx 4.3689140 \times 10^{-8} .$$

A typical estimate using $N = 10^4$ samples is 4.3685×10^{-8}, with an estimated relative error of 1.7×10^{-4}. Through experimentation we observed that the relative error is severely underestimated if N is too small, for example if $N = 1000$ for this case.

```
%gamble_CE_A.m
N = 10^4; %Run Size
results = zeros(N,1);
k = 10; K = 30; %Initial value and absorbing barrier
p = .3; q = 1 - p; r = q/p; %Actual probabilities

%Tilt the distribution using CE
v = ((K - k)*(p - q)*(1 - r^k)) / ((K-k)*(r^k + 1) ...
     + 2 * K * ((r^k - r^K) / (r^K - 1)))
theta = .5*(log((1+v)*q)-log((1-v)*p));
p_tilde = (p * exp(theta)) / (p * exp(theta) + q * exp(-theta));
q_tilde = 1 - p_tilde;

for i = 1:N
    t = 0;
    sum = k;
    while (sum ~= K) && (sum ~= 0)
        t = t+1;
        U = rand;
        sum = sum + (2*(U < p_tilde) - 1);
    end
    results(i) = exp(-theta * (K - k) + t*(log(p*exp(theta) ...
```

```
                  + q*exp(-theta))))*(sum == K);
end

ell = (r^k - 1) / (r^K - 1) %Actual Probability
ell_hat = mean(results) %Estimated Probability
RE = std(results) / sqrt(N) / ell_hat %Estimated Relative Error
```

9.7.6 Response Surface Estimation via Importance Sampling

Let the performance measure of a simulation be of the form

$$\ell(\boldsymbol{\theta}) = \mathbb{E}_{\boldsymbol{\theta}} H(\mathbf{X}) ,$$

where $\mathbf{X} \sim f(\mathbf{x}; \boldsymbol{\theta})$ depends on a parameter $\boldsymbol{\theta} \in \Theta$. Importance sampling makes it possible to gain information on a subset of the **response surface** $\{\ell(\boldsymbol{\theta}), \boldsymbol{\theta} \in \Theta\}$ from a single simulation run [2]. The idea is to write $\ell(\boldsymbol{\theta})$ as

$$\ell(\boldsymbol{\theta}) = \mathbb{E}_{\boldsymbol{\theta}_0} H(\mathbf{X})\, W(\mathbf{X}; \boldsymbol{\theta}, \boldsymbol{\theta}_0) , \tag{9.48}$$

where $\boldsymbol{\theta}_0 \in \Theta$ is such that $f(\mathbf{x}; \boldsymbol{\theta}_0) = 0$ dominates $f(\mathbf{x}; \boldsymbol{\theta}) H(\mathbf{x})$. As usual, $W(\mathbf{X}; \boldsymbol{\theta}, \boldsymbol{\theta}_0) = f(\mathbf{X}; \boldsymbol{\theta}) / f(\mathbf{X}; \boldsymbol{\theta}_0)$ is the likelihood ratio. The corresponding estimator is

$$\widehat{\ell}(\boldsymbol{\theta}; \boldsymbol{\theta}_0) = \frac{1}{N} \sum_{k=1}^{N} H(\mathbf{X}_k)\, W(\mathbf{X}_k; \boldsymbol{\theta}, \boldsymbol{\theta}_0) , \tag{9.49}$$

where $\mathbf{X}_1, \ldots, \mathbf{X}_N \stackrel{\text{iid}}{\sim} f(\mathbf{x}; \boldsymbol{\theta}_0)$. The variance of the estimator $\widehat{\ell}(\boldsymbol{\theta}; \boldsymbol{\theta}_0)$ under the importance sampling density $f(\mathbf{x}; \boldsymbol{\theta}_0)$ can be estimated by the sample variance of $\{H(\mathbf{X}_k) W(\mathbf{X}_k; \boldsymbol{\theta}, \boldsymbol{\theta}_0)\}$. The procedure is summarized in the following algorithm.

Algorithm 9.10 (Response Surface Estimation)

1. *Generate* $\mathbf{X}_1, \ldots, \mathbf{X}_N \stackrel{\text{iid}}{\sim} f(\mathbf{x}; \boldsymbol{\theta}_0)$.

2. *Estimate* $\ell(\boldsymbol{\theta})$ *via* (9.49) *and determine an approximate* $1 - \alpha$ *confidence interval as*

$$\left(\widehat{\ell} - z_{1-\alpha/2} \frac{S}{\sqrt{N}},\ \widehat{\ell} + z_{1-\alpha/2} \frac{S}{\sqrt{N}} \right) ,$$

where z_γ *denotes the* γ*-quantile of the* $\mathsf{N}(0,1)$ *distribution and* S *is the sample standard deviation of* $\{H(\mathbf{X}_k) W(\mathbf{X}_k; \boldsymbol{\theta}, \boldsymbol{\theta}_0)\}$.

Two *advantages* of the above procedure are:

1. Only a single simulation run (under $\boldsymbol{\theta}_0$) is needed to estimate the performance for many different values of $\boldsymbol{\theta}$.

2. The estimated response surface $\widehat{\ell}(\boldsymbol{\theta}; \boldsymbol{\theta}_0)$ as a function of $\boldsymbol{\theta}$ is typically piecewise differentiable, allowing easy estimation of the gradient of $\ell(\boldsymbol{\theta})$ through differentiation of $\widehat{\ell}(\boldsymbol{\theta}; \boldsymbol{\theta}_0)$. This is the idea behind the score function method for gradient estimation; see Section 11.4. ☞ 428

The main *disadvantage* is that although $\widehat{\ell}(\boldsymbol{\theta};\boldsymbol{\theta}_0)$ is an unbiased estimator of $\ell(\boldsymbol{\theta})$, its variance can be very large (or even infinite) depending on $(\boldsymbol{\theta},\boldsymbol{\theta}_0)$. This is typically the case when under $\boldsymbol{\theta}_0$ the distribution has thinner tails than under $\boldsymbol{\theta}$, which leads to a blowing up of the likelihood ratio. In such cases the estimator typically *underestimates* the true value and the standard error, so that the confidence intervals are unreliable (too small). Therefore, one should not expect to be able to reliably estimate the *whole* response surface from a sample, but only a subset thereof, sometimes called the *trust region*; see also Section C.2.2.6 and Section 11.4.1. A discussion on the dangers of importance sampling may be found, for example, in [18, Pages 209–211].

☞ 692
☞ 430

■ **EXAMPLE 9.12 (Response Surface for the Bridge Network)**

We return to Example 9.1. Let the lengths $X_1,\ldots,X_5$ of the links be independent and uniformly distributed on $(0,\theta),(0,2),(0,3),(0,1)$, and $(0,2)$, respectively. Hence, the only change in the setting of Example 9.1 is that the first component has a $\mathsf{U}(0,\theta)$ distribution, rather than a $\mathsf{U}(0,1)$ distribution. Denote the expected length of the shortest path by $\ell(\theta)$. Suppose that N iid copies of $\mathbf{X}=(X_1,\ldots,X_5)$ are available for the case where $\theta=\theta_0=3$. Then $\ell(\theta)$ can be estimated via

$$\widehat{\ell}(\theta;\theta_0)=\frac{1}{N}\sum_{k=1}^{N}H(\mathbf{X}_k)\frac{\mathrm{I}_{\{0<X_{k1}<\theta\}}/\theta}{\mathrm{I}_{\{0<X_{k1}<\theta_0\}}/\theta_0}\,.$$

Note that for $\theta>\theta_0$ the importance sampling pdf $f(\mathbf{x};\theta_0)$ does not dominate $H(\mathbf{x})f(\mathbf{x};\theta)$, so $\ell(\theta)$ can only be estimated via importance sampling for $\theta<\theta_0$. Figure 9.6 depicts a typical estimate for the response curve for the case $\theta_0=3$ using $N=10^4$ samples.

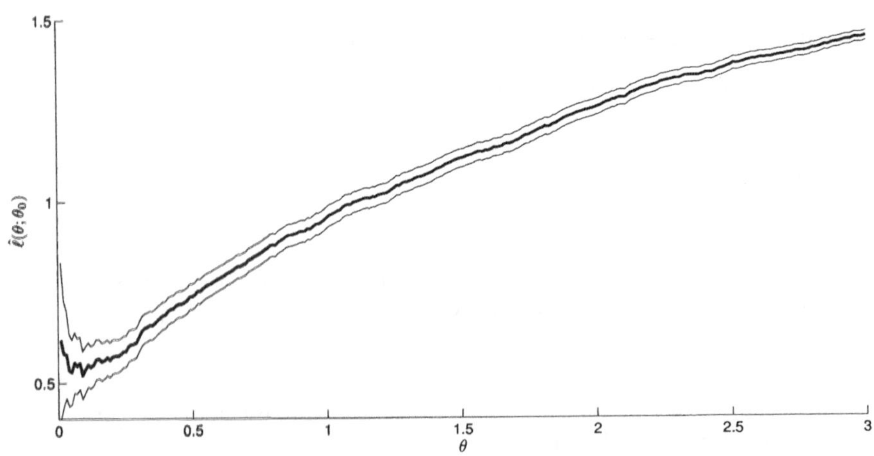

Figure 9.6 Response surface estimates with 95% confidence bounds for the uniform case.

The simulation is implemented via the following MATLAB program.

```
%responsesurfis.m
N = 10000; theta0 = 3;
a = [theta0,2,3,1,2]; u = rand(N,5);
X = u.*repmat(a,N,1); W = zeros(N,1);
y = H1(X); theta = 0:0.01:theta0;
num = numel(theta);
ell = zeros(1,num); ellL = zeros(1,num);
ellU = zeros(1,num); stell = zeros(1,num);
for i=1:num
    th = theta(i);
    W = theta0/th*(X(:,1)< th);
    ell(i) = mean(H1(X).*W);
    stell(i) = std(H1(X).*W);
    ellL(i)= ell(i) - stell(i)/sqrt(N)*1.95;
    ellU(i)= ell(i) + stell(i)/sqrt(N)*1.95;
end
plot(theta,ell, theta, ellL, theta, ellU)
```

```
function out=H1(X)
Path_1=X(:,1)+X(:,4);
Path_2=X(:,1)+X(:,3)+X(:,5);
Path_3=X(:,2)+X(:,3)+X(:,4);
Path_4=X(:,2)+X(:,5);
out=min([Path_1,Path_2,Path_3,Path_4],[],2);
```

Next, suppose that X_1 is simulated under an $\mathsf{Exp}(1/\theta_0)$ distribution instead, and that we wish to estimate how $\ell(\theta)$ behaves under general $\theta > 0$. The estimator is now

$$\widehat{\ell}(\theta;\theta_0) = \frac{1}{N}\sum_{k=1}^{N} H(\mathbf{X}_k)\frac{\mathrm{e}^{-X_{k1}/\theta}/\theta}{\mathrm{e}^{-X_{k1}/\theta_0}/\theta_0} = \frac{\theta_0}{\theta\, N}\sum_{k=1}^{N} H(\mathbf{X}_k)\,\mathrm{e}^{X_{k1}(1/\theta_0-1/\theta)} \,. \tag{9.50}$$

A typical estimate for the response curve for the case $N = 10^4$ and $\theta_0 = 3$ is depicted in Figure 9.7. The MATLAB code for this exponential case is available on the Handbook website.

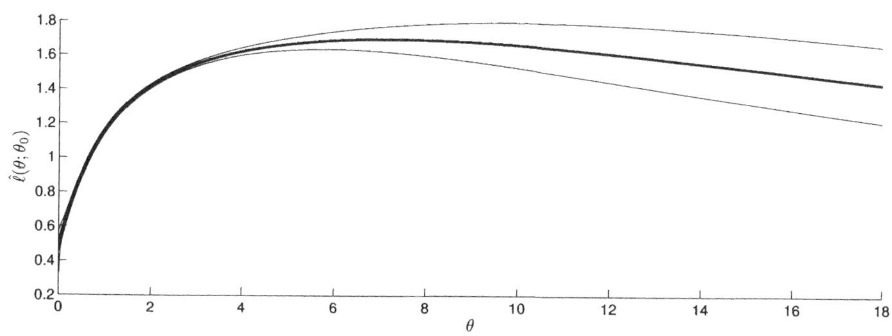

Figure 9.7 Response surface estimates with 95% confidence bounds for the exponential case.

Two significant differences with the uniform case are that (1) the estimated response curve is a smooth function of θ, and (2) it is possible to obtain estimates for all $\theta > 0$. However, as noted above, when θ is too large both the estimate and the confidence interval are unreliable. This is illustrated by the fact that the true response function must be monotone increasing in θ, whereas the estimate is decreasing from about $\theta > 7$ onward. It is not difficult to show, see [18, Page 210], that the variance of the estimator is infinite for $\theta > 2\theta_0 = 6$. Thus, it is not recommended to estimate $\ell(\theta)$ in this way for θ larger than 5, say. Note that for $\theta > \theta_0$ the importance sampling pdf $f(\mathbf{x};\theta_0)$ has thinner tails (decays quicker) in the x_1 variable than $f(\mathbf{x};\theta)$.

Finally, we mention that it is not difficult in this particular case to estimate the *entire* response function using only a single simulation run without importance sampling. The idea is to write

$$\ell(\theta) = \mathbb{E}h(\mathbf{U};\theta) ,$$

where, in the uniform case, $h(\mathbf{U};\theta) = H(\theta U_1, 2U_2, 3U_3, 2U_4, U_5)$ and $U_1,\ldots,U_5 \overset{\text{iid}}{\sim} \mathsf{U}(0,1)$, as in Example 9.1. Thus $\ell(\theta)$ is simply estimated as

$$\frac{1}{N}\sum_{k=1}^{N} h(\mathbf{U}_k;\theta) , \qquad \theta > 0 ,$$

from a single iid sample $\mathbf{U}_1,\ldots,\mathbf{U}_N \overset{\text{iid}}{\sim} \mathsf{U}(0,1)^5$.

9.8 QUASI MONTE CARLO

Quasi Monte Carlo provides a powerful way to estimate d-dimensional integrals of the form

$$\ell = \int h(\mathbf{u})\,d\mathbf{u}$$

by means of the sample average

$$\frac{1}{N}\sum_{\mathbf{u}\in\mathcal{P}_N} h(\mathbf{u}) ,$$

☞ 25 where $\mathcal{P}_N$ is a set of N quasirandom points, as is explained in Chapter 2. Error estimates can be obtained by randomizing the point set via a *random shift* (Section 2.7) and producing K independent copies of the estimator (2.10). Significant variance reduction can be obtained in this way; see, for example, [12]. The general procedure is summarized as follows.

Algorithm 9.11 (Quasi Monte Carlo Estimation)

1. *Generate a quasi Monte Carlo point set* $\mathcal{P}_N = \{\mathbf{u}_j,\ j = 1,\ldots,N\}$.

2. *Generate independent random vectors* $\mathbf{Z}_1,\ldots,\mathbf{Z}_K \overset{\text{iid}}{\sim} \mathsf{U}(0,1)^d$.

3. *Form the shifted point sets* $\mathcal{P}_N^{(i)} = (\mathcal{P}_N + \mathbf{Z}_i) \mod 1,\ i = 1,\ldots,K$.

4. Calculate

$$\widehat{\ell}_i = \frac{1}{N} \sum_{\mathbf{u} \in \mathcal{P}_N^{(i)}} h(\mathbf{u}), \quad i = 1, \ldots, K\,.$$

5. Estimate ℓ *as* $\widehat{\ell} = \frac{1}{K}\sum_{k=1}^{K} \widehat{\ell}_k$ *and determine an approximate* $1-\alpha$ *confidence interval as*

$$\left(\widehat{\ell} - z_{1-\alpha/2}\frac{S}{\sqrt{K}},\ \widehat{\ell} + z_{1-\alpha/2}\frac{S}{\sqrt{K}}\right), \tag{9.51}$$

where z_γ *denotes the* γ*-quantile of the* $\mathsf{N}(0,1)$ *distribution and* S *is the sample standard deviation of* $\widehat{\ell}_1, \ldots, \widehat{\ell}_K$.

■ EXAMPLE 9.13 (Quasi Monte Carlo for the Bridge Network)

As the expectation ℓ in (9.2) involves a relatively low-dimensional problem ($d = 5$), quasi Monte Carlo integration is expected to work well here. The MATLAB program below implements Algorithm 9.11 using a Faure point set, constructed via the MATLAB function `faure.m` defined on Page 33. The number of replications K is chosen to be 20, in order to give reasonable estimates for the relative error. The size of each of the 20 shifted point sets is $N = 500$, so that the total number of function evaluations of $h(\mathbf{u})$ is 10^4. A typical estimate is $\widehat{\ell} = 0.9308$ with an estimated relative error (that is, $S/(\sqrt{K}\widehat{\ell})$ with S as in (9.51)) of 0.072% which, for this particular problem, is better than those that were obtained by the other variance reduction methods in this chapter.

```
%brigeQMC_faure.m
K = 20;
N = 10^4/K;
F = faure(5,5,N-1);
for i=1:K
    U(:,:,i) = mod(F + repmat(rand(1,5),N,1), 1);
end
for i=1:K
    y(i) = mean(h(U(1:N,:,i)));
end
ell = mean(y);          %estimate
se = std(y)/sqrt(K); %standard error
fprintf('ell=%g, percRE = %g  \n',ell, 100*se/ell);
```

To further demonstrate the accuracy of the quasi Monte Carlo program, a typical outcome for $N = 50000$ and $K = 20$ is $\widehat{\ell} = 0.929862$, with an estimated relative error of 0.0027%. Recall that the true value is 0.929861111....

Finally, Figure 9.8 depicts, for different (quasi)random point sets, the convergence behavior of the estimator $\widehat{\ell}$ in Algorithm 9.11 using $K = 40$ repetitions and a point set of size N ranging from 8 to 10^5. All Monte Carlo methods are repeated $K = 40$ times. It is clearly seen that CMC has a significantly larger standard error (that is, $S/\sqrt{K}$ with S as in (9.51)) over the whole range of N. Moreover, CMC decreases at a slower rate than the Sobol', Faure, and Korobov point sets. The latter three are comparable in performance for this example. We used an extensible Korobov point set with parameter $a = 14471$.

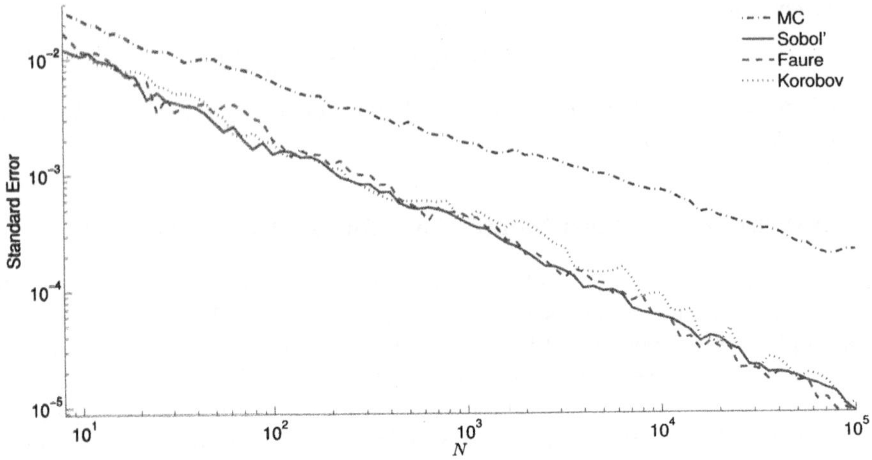

Figure 9.8 Standard error of the estimator $\widehat{\ell}$ with $K = 40$ versus the number of points N for different (quasi)random point sets.

In conclusion, in Figure 9.9 we summarize our *subjective* simulation experience of the different variance reduction techniques, as a guide for the practitioner. The figure indicates both the difficulty of implementation and the potential for improvement over CMC for each of the techniques. For completeness we include the splitting method from Chapter 14. ☞ 481

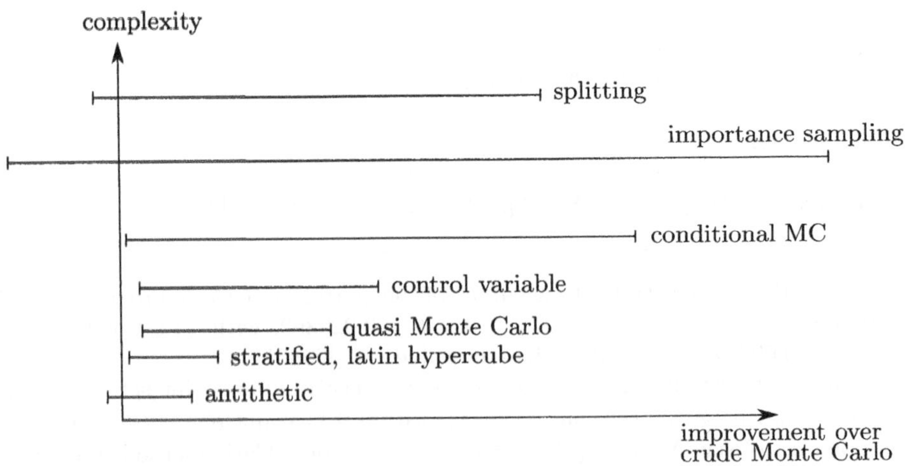

Figure 9.9 A guide for selecting a variance reduction technique.

Further Reading

The fundamental paper on variance reduction techniques is Kahn and Marshal [7]. There are many good Monte Carlo textbooks with chapters on variance reduction techniques. Among them are [1, 5, 6, 8, 9, 11, 13, 14, 15, 16, 20].

REFERENCES

1. S. Asmussen and P. W. Glynn. *Stochastic Simulation: Algorithms and Analysis.* Springer-Verlag, New York, 2007.
2. S. Asmussen and R. Y. Rubinstein. Response surface estimation and sensitivity analysis via efficient change of measure. *Stochastic Models*, 9(3):313–339, 1993.
3. W. G. Cochran. *Sampling Techniques.* John Wiley & Sons, New York, third edition, 1977.
4. P. T. de Boer, D. P. Kroese, and R. Y. Rubinstein. A fast cross-entropy method for estimating buffer overflows in queueing networks. *Management Science*, 50(7):883–895, 2004.
5. G. S. Fishman. *Monte Carlo: Concepts, Algorithms and Applications.* Springer-Verlag, New York, 1996.
6. P. Glasserman. *Monte Carlo Methods in Financial Engineering.* Springer-Verlag, New York, 2004.
7. M. Kahn and A. W. Marshall. Methods of reducing sample size in Monte Carlo computations. *Journal of the Operations Research Society of America*, 1(5):263–278, 1953.
8. J. P. C. Kleijnen. *Statistical Techniques in Simulation, Part 1.* Marcel Dekker, New York, 1974.
9. J. P. C. Kleijnen. Analysis of simulation with common random numbers: A note on Heikes et al. (1976). *Simuletter*, 11(2):7–13, 1979.
10. D. P. Kroese, T. Taimre, Z. I. Botev, and R. Y. Rubinstein. *Solutions Manual to Accompany: Simulation and the Monte Carlo Method, Second Edition.* John Wiley & Sons, New York, 2007.
11. A. M. Law and W. D. Kelton. *Simulation Modeling and Analysis.* McGraw-Hill, New York, third edition, 2000.
12. P. L'Ecuyer and C. Lemieux. Variance reduction via lattice rules. *Management Science*, 46(9):1214–1235, 2000.
13. J. S. Liu. *Monte Carlo Strategies in Scientific Computing.* Springer-Verlag, New York, 2001.
14. D. L. McLeish. *Monte Carlo Simulation and Finance.* John Wiley & Sons, New York, 2005.
15. C. P. Robert and G. Casella. *Monte Carlo Statistical Methods.* Springer-Verlag, New York, second edition, 2004.
16. S. M. Ross. *Simulation.* Academic Press, New York, third edition, 2002.
17. R. Y. Rubinstein and D. P. Kroese. *The Cross-Entropy Method: A Unified Approach to Combinatorial Optimization, Monte-Carlo Simulation and Machine Learning.* Springer-Verlag, New York, 2004.

18. R. Y. Rubinstein and D. P. Kroese. *Simulation and the Monte Carlo Method.* John Wiley & Sons, New York, second edition, 2007.

19. R. Y. Rubinstein and A. Shapiro. *Discrete Event Systems: Sensitivity Analysis and Stochastic Optimization via the Score Function Method.* John Wiley & Sons, New York, 1993.

20. I. M. Sobol'. *A Primer for the Monte Carlo Method.* CRC Press, Boca Raton, FL, 1994.

21. F. Stern. Conditional expectation of the duration in the classical ruin problem. *Mathematics Magazine,* 48(4):200–203, 1975.

CHAPTER 10

RARE-EVENT SIMULATION

In this chapter we describe algorithms for the efficient estimation of rare-event probabilities. We start by defining the notion of efficiency in the context of rare-event simulation, and then consider the following algorithms that are efficient in a particular rare-event setting.

1. Importance sampling for light tails — this covers the estimation of *stopping-time* and *overflow* probabilities using an *exponential change of measure*;

2. *Conditional Monte Carlo* for the estimation of probabilities arising from compound sums of heavy-tailed random variables;

3. *State-dependent importance sampling* for rare-event overflow probabilities;

4. General importance sampling — such as the *cross-entropy method* — with applications to financial risk modeling;

5. *Splitting methods* for estimation of *hitting* probabilities of Markov processes.

General variance reduction techniques leading to improved efficiency of estimators are discussed in Chapter 9. More details on the cross-entropy method can be found in Chapter 13. Splitting methods in non-Markovian settings and the closely related *sequential Monte Carlo* framework are discussed in Chapter 14.

☞ 347
☞ 463
☞ 481

10.1 EFFICIENCY OF ESTIMATORS

Suppose $\widehat{\ell}$ is an unbiased estimator of $\ell = \mathbb{P}(A) = \mathbb{E}\, I_A$, where ℓ is small, say, 10^{-4} or less. Event A is called a **rare event** and ℓ is called a **rare-event probability**. Often A is of the form $A = \{S(\mathbf{X}) \geqslant \gamma\}$ for some function $S : \mathbb{R}^n \to \mathbb{R}$, vector $\mathbf{X} = (X_1, \ldots, X_n)^\top$, and **level** or **threshold** parameter γ. If $Z_1, \ldots, Z_N$ are independent replications of some random variable Z for which $\mathbb{E}Z = \ell$, then

$$\widehat{\ell} = \frac{1}{N}\sum_{i=1}^{N} Z_i \,, \tag{10.1}$$

is an unbiased estimator of ℓ. In particular, if $Z = I_A$, then (10.1) is the crude
☞ 305 Monte Carlo (CMC) estimator of $\mathbb{P}(A)$; see Section 8.2.

The accuracy of the estimator $\widehat{\ell}$ can be measured in terms of its relative error:

$$\frac{\sqrt{\mathrm{Var}(\widehat{\ell})}}{\ell} = \frac{\sigma}{\ell\sqrt{N}} \,,$$

where $\sigma = \sqrt{\mathrm{Var}(Z)}$. This can be estimated from the iid sample $Z_1, \ldots, Z_N$ via

$$\frac{\widehat{\sigma}}{\widehat{\ell}\sqrt{N}} \,, \quad \text{where} \quad \widehat{\sigma} = \sqrt{\frac{1}{N}\sum_{i=1}^{N}(Z_i - \widehat{\ell})^2} \,.$$

In the rare-event context, ℓ typically depends on a *rarity* parameter; for example, $\ell = \ell(\gamma) = \mathbb{P}(S(\mathbf{X}) \geqslant \gamma)$, such that $\ell(\gamma) \to 0$ as $\gamma \to \infty$. It is of interest to study the asymptotic properties of $\widehat{\ell}(\gamma)$ as a function of γ.

For a random variable $Z = Z(\gamma)$ such that $\mathbb{E}Z(\gamma) = \ell(\gamma) \to 0$ as $\gamma \to \infty$, we have the following **second-order efficiency measures**; see, for example, [6, 10, 43, 47].

1. The estimator (10.1) is said to have **(asymptotically) vanishing relative error** if
$$\limsup_{\gamma \to \infty} \frac{\mathrm{Var}(Z(\gamma))}{\ell^2(\gamma)} = 0 \,.$$

2. The estimator (10.1) is said to have **bounded relative error** if
$$\limsup_{\gamma \to \infty} \frac{\mathrm{Var}(Z(\gamma))}{\ell^2(\gamma)} \leqslant K < \infty \,,$$
where K is a constant that does not depend on γ.

3. The estimator (10.1) is said to be **logarithmically efficient** if
$$\limsup_{\gamma \to \infty} \frac{\mathrm{Var}(Z(\gamma))}{\ell^{2-\varepsilon}(\gamma)} = 0 \quad \text{for all } \varepsilon > 0 \,,$$
or equivalently if
$$\liminf_{\gamma \to \infty} \left| \frac{\ln \mathrm{Var}(Z(\gamma))}{\ln \ell^2(\gamma)} \right| \geqslant 1 \,.$$

Arranged from the strongest to the weakest condition we have:

vanishing relative error $\Rightarrow$ bounded relative error $\Rightarrow$ logarithmic efficiency.

In most cases of interest it is much more difficult to find an estimator with bounded relative error than it is to find a logarithmically efficient estimator. Estimators with vanishing relative error are given in [15, 42, 45]. In practice, to verify any of the second-order efficiency measures it is convenient to work with the second moment $\mathbb{E}Z^2(\gamma)$ instead of the variance.

Remark 10.1.1 (Simulation Time) The efficiency measures above do not take into account the simulation time, say τ, to generate one random variable Z. One can account for it by replacing $\text{Var}(Z(\gamma))$ with $\tau\,\text{Var}(Z(\gamma))$ in the definition of any of the second-order efficiency measures. This gives the **work normalized squared relative error** or **relative time variance product** ☞ 656

$$\frac{\tau\,\text{Var}(Z(\gamma))}{\ell^2(\gamma)}\,.$$

A potential problem with the second-order efficiency measures is that they do not convey any information about the accuracy of the estimator $\widehat{\sigma}^2$ of the true variance σ^2. As a result, the quality of the central limit approximation $\widehat{\ell} \overset{\text{approx.}}{\sim} \mathsf{N}(\ell, \widehat{\sigma}^2/N)$ may be difficult to quantify and any inference based on this approximation can be challenged. To address this issue one has to consider higher moments of $Z(\gamma)$. This motivates the following more general **higher-order efficiency measures** [15].

1. The estimator (10.1) is said to have **vanishing relative centered moment of order k** if for $k \in [1, \infty)$
$$\limsup_{\gamma\to\infty} \frac{\mathbb{E}|Z(\gamma)-\ell(\gamma)|^k}{\ell^k(\gamma)} = 0\,.$$
2. The estimator (10.1) is said to have **bounded relative moment of order k** if for $k \in [1, \infty)$
$$\limsup_{\gamma\to\infty} \frac{\mathbb{E}Z^k(\gamma)}{\ell^k(\gamma)} \leqslant K < \infty\,,$$
where K is a constant independent of γ.
3. The estimator (10.1) is said to be **logarithmically efficient of order k** if
$$\limsup_{\gamma\to\infty} \frac{\ln \mathbb{E}Z^k(\gamma)}{k \ln \ell(\gamma)} = 1\,.$$
4. The estimator (10.1) is said to have a **bounded normal approximation** if
$$\limsup_{\gamma\to\infty} \frac{\mathbb{E}|Z(\gamma)-\ell(\gamma)|^3}{(\text{Var}(Z))^{3/2}} < \infty\,. \tag{10.2}$$

Note that the bounded normal approximation property is not equivalent to bounded relative moment of order 3. The motivation for introducing the concept of

☞ 625 bounded normal approximation is derived from the *Berry–Esséen* theorem, which implies that under (10.2) the normal approximation to the cdf of the standardized estimator $\widehat{\ell}$ is accurate up to order $\mathcal{O}(N^{-1/2})$, uniformly in γ; see also [63].

Properties of the higher-order efficiency measures include [15]:

1. *Vanishing relative centered moment of order* k implies *vanishing relative centered moment of order* m for $k \geqslant m \geqslant 1$.

2. *Vanishing relative centered moment of order* k (for $k > 1$) is equivalent to *bounded relative moment of order* k with upper bound 1, that is,

$$\limsup_{\gamma\to\infty} \frac{\mathbb{E}|Z(\gamma) - \ell(\gamma)|^k}{\ell(\gamma)^k} = 0 \quad \Leftrightarrow \quad \limsup_{\gamma\to\infty} \frac{\mathbb{E}Z^k(\gamma)}{\ell^k(\gamma)} = 1 .$$

3. *Bounded relative moment of order* 2 is equivalent to *bounded relative error.*

4. *Bounded relative moment of order* k implies *bounded relative moment of order* m for $k \geqslant m \geqslant 1$.

5. If $Z(\gamma)$ has *bounded relative moment of order* mk, then $Z^m(\gamma)$ has *bounded relative moment of order* k.

6. If $c_1 \leqslant \mathrm{Var}(Z)/\ell^2 \leqslant c_2$ for some constants c_1 and c_2, then *bounded relative moment of order* $2k$ for Z implies *bounded relative moment of order* k for $\widehat{\sigma}^2$ for any $k \geqslant 1$.

7. If estimator (10.1) has *vanishing relative centered moment of order* $(1+\varepsilon)$ for some $\varepsilon > 0$, then the pdf of Z converges to the zero-variance importance sampling density for the estimation of $\ell(\gamma)$.

■ **EXAMPLE 10.1 (The CMC Estimator is not Logarithmically Efficient)**

Consider the estimator (10.1) with $Z = Z(\gamma) = \mathrm{I}_{\{S(\mathbf{X}) \geqslant \gamma\}}$. We have

$$\mathbb{E}Z^2 = \mathbb{E}Z = \ell ,$$

so that

$$\lim_{\gamma\to\infty} \frac{\ln \mathbb{E}Z^2}{\ln \ell^2} = \frac{\ln \ell}{\ln \ell^2} = \frac{1}{2} < 1 .$$

Hence, the CMC estimator is not logarithmically efficient. Indeed, for small ℓ the relative error, κ say, of the CMC estimator satisfies

$$\kappa = \frac{\sqrt{\mathrm{Var}(\widehat{\ell})}}{\mathbb{E}\widehat{\ell}} = \sqrt{\frac{1-\ell}{N\,\ell}} \approx \sqrt{\frac{1}{N\,\ell}} . \tag{10.3}$$

If, for example, $\ell = 10^{-6}$, then in order to estimate ℓ accurately with relative error of 1%, we need to choose a sample size

$$N \approx \frac{1}{\kappa^2 \ell} = 10^{10} .$$

This shows that alternative estimators must be found to estimate $\ell(\gamma)$ for large γ.

■ **EXAMPLE 10.2 (An Estimator With Bounded Relative Error)**

Suppose we wish to estimate via simulation the probability $\ell = \mathbb{P}(X \geqslant \gamma)$, where $X \sim \mathsf{Exp}(u^{-1})$, with pdf f. We thus know the exact $\ell = \mathrm{e}^{-\gamma/u}$. Suppose further that γ is large compared to u, so that ℓ is a rare-event probability. We wish to use importance sampling with a Cauchy importance sampling density $g(x;\mu,\tau) = \frac{\tau}{\pi}\left(\tau^2 + (x-\mu)^2\right)^{-1}$, where the location and scale parameters μ and τ are chosen to give minimal variance for the corresponding importance sampling estimator (10.1) with $Z = f(X)\mathrm{I}_{\{X \geqslant \gamma\}}/g(X;\mu,\tau)$. This means minimizing

$$\mathbb{E}_g Z^2 = \mathbb{E}_f \frac{f(X)\,\mathrm{I}_{\{X\geqslant\gamma\}}}{g(X;\mu,\tau)} \tag{10.4}$$

with respect to the parameters μ and τ. Now (10.4) gives

$$\int_\gamma^\infty \frac{1}{\tau}\mathrm{e}^{-2x/u}\left(\tau^2 + (x-\mu)^2\right)\mathrm{d}x = \frac{u\,\mathrm{e}^{-2\gamma/u}}{4\tau}\left(2\tau^2 + (\gamma-\mu)^2 + (u+\gamma-\mu)^2\right),$$

which has minima at $(\mu,\tau) = (\gamma + u/2, \pm u/2)$. We take the solution with $\tau > 0$, giving $({}_*\mu,\ {}_*\tau) = (\gamma + u/2, u/2)$ as the parameter pair that minimizes the variance of $\widehat{\ell}$. The corresponding minimum value is

$$\mathbb{E}_g Z^2 = \int_\gamma^\infty u^{-2}\mathrm{e}^{-2x/u}\frac{2\pi}{u}\left(\frac{u^2}{4} + \left(x - \left(\gamma + \frac{u}{2}\right)\right)^2\right)\mathrm{d}x = \frac{\pi}{2}\,\ell^2,$$

and the relative error of (10.1) is given by

$$\frac{\sqrt{(\frac{\pi}{2}\ell^2 - \ell^2)/N}}{\ell} = \sqrt{\frac{\pi-2}{2N}}.$$

The last expression does not depend on γ, and hence the estimator has bounded relative error.

10.2 IMPORTANCE SAMPLING METHODS FOR LIGHT TAILS

In this section we describe some of the classical algorithms for estimating rare-event probabilities using importance sampling. These algorithms apply to a restricted class of problems in which the rare-event probability decays exponentially or faster, and are shown to have desirable efficiency properties within that class. In more complex settings the knowhow obtained from such restricted and simple settings is often used to reduce the variance significantly, but usually without achieving any of the efficiency criteria of the previous section. We first introduce the following notation and assumptions.

- Under probability measure $\mathbb{P}$, let $\mathbf{X} = (X_1, \ldots, X_n)^\top$ be a vector of iid *light-tailed* random variables, each with cdf F, pdf f, and finite expectation $\mathbb{E}X_i = \mathbb{E}X = \mu$. ☞ 703
- Denote $S_n = S(\mathbf{X}) = \sum_{i=1}^n X_i$.
- Define an exponential family of pdfs $\{f_\theta, \theta \in \Theta\}$ via ☞ 701

$$f_\theta(x) = \frac{e^{\theta x}}{M(\theta)} f(x) = e^{\theta x - \ln M(\theta)} f(x),$$

where $M(\theta) = \int e^{\theta x} f(x) dx, \ \theta \in \Theta$ is the *moment generating function* of X, and Θ is the set of θ for which $M(\theta)$ is well-defined and finite. The pdf f_θ is said to be derived from f via an **exponential twist** θ. Note that the moment generating function is convex, with $M(0) = 1$ and $M'(0) = \mu$.

- Let $\mathbb{P}_\theta$ denote the probability measure under which $X_1, \ldots, X_n \sim_{\text{iid}} f_\theta$. The corresponding likelihood ratio is given by ☞ 362

$$W(\mathbf{X}; \theta) = \frac{f(X_1) \cdots f(X_n)}{f_\theta(X_1) \cdots f_\theta(X_n)} = e^{-S_n \theta + n\zeta(\theta)},$$

where $\zeta(\theta) = \ln M(\theta)$ is the **cumulant function** of X.

- Let μ_θ be the changed increment mean (drift):

$$\mu_\theta = \mathbb{E}_\theta X = \mathbb{E} X \frac{e^{\theta X}}{M(\theta)} = \frac{M'(\theta)}{M(\theta)} = \zeta'(\theta).$$

10.2.1 Estimation of Stopping Time Probabilities

One of the earliest rare-event simulation algorithms is Siegmund's algorithm [61] for the estimation of

$$\ell = \mathbb{P}(\tau < \infty), \quad \text{where } \tau = \inf\{n : S_n \geqslant \gamma\}. \tag{10.5}$$

It is assumed that $\mu < 0$, so that ℓ becomes small as $\gamma \to \infty$. We also assume that there exists a ${}_*\theta > 0$ such that $M({}_*\theta) = 1$. Because $M(\theta)$ is a convex function with $M(0) = 1$ and $M'(0) = \mu < 0$, such a ${}_*\theta$ exists in the light-tailed case if, for example, $\lim_{\theta \to \theta_{\max}} M(\theta) = \infty$ for some $0 < \theta_{\max} \leqslant \infty$.

Algorithm 10.1 (Siegmund's Algorithm) *To estimate $\ell = \mathbb{P}(\tau < \infty)$ from simulation, execute the following steps.*

1. *Compute the root ${}_*\theta > 0$ of $M(\theta) = 1$ (or equivalently $\zeta(\theta) = 0$).*

2. *Set $S_0 = 0$. Until $S_n \geqslant \gamma$, set*

$$S_{n+1} = S_n + X_{n+1}, \quad n = 0, 1, 2, \ldots,$$

where $X_1, X_2, \ldots \sim_{\text{iid}} f_{{}_*\theta}$. Let τ be the smallest n for which $S_n \geqslant \gamma$.

3. *Set $Z = W(\mathbf{X}; {}_*\theta) = e^{-S_\tau \, {}_*\theta}$.*

4. *Repeat Steps 2 and 3 to obtain N independent replications $Z_1, \ldots, Z_N$ of Z and return their sample mean as an estimator of ℓ.*

Note that under the importance sampling pdf $f_{{}_*\theta}$ the drift $\mu_{{}_*\theta}$ is positive and $\mathbb{P}_{{}_*\theta}(\tau < \infty) = 1$.

The representation $\ell = \mathbb{E} e^{-S_\tau \, {}_*\theta}$ has various implications. For example, define the **overshoot** at time τ to be $\xi(\gamma) = S_\tau - \gamma$, so that

$$\ell = e^{-\gamma \, {}_*\theta} \, \mathbb{E} e^{-\xi(\gamma) \, {}_*\theta}.$$

Note that the "saw-tooth" process $\{\xi(\gamma), \gamma \geqslant 0\}$ is *regenerative*. As a result, $\mathbb{E}\mathrm{e}^{-\xi(\gamma)\,{}_*\theta} \to C$ for some constant $0 < C < \infty$ as $\gamma \to \infty$ — under the assumptions of Theorem A.9.1. This leads to the celebrated **Cramér–Lundberg approximation**: ☞ 631

$$\mathbb{P}(\tau < \infty) \approx C\mathrm{e}^{-\gamma\,{}_*\theta} \quad \text{as } \gamma \to \infty\,.$$

Another implication is the following efficiency result [7, Page 165].

Theorem 10.2.1 (Efficiency of Siegmund's Algorithm) *Siegmund's estimator has bounded relative error. Moreover, Siegmund's estimator is the only logarithmically efficient estimator among all importance sampling estimators whose change of measure is of the form $X_1, \ldots, X_n \sim_{\text{iid}} g$ for some importance sampling pdf g.*

Siegmund's algorithm has applications in the computation of ruin probabilities and probabilities arising in queueing theory [4].

■ **EXAMPLE 10.3 (Random Walk)**

As a numerical example, consider the case where $X \sim \mathsf{N}(\mu, 1)$, $\mu < 0$. In this case $M(\theta) = \mathrm{e}^{\mu\theta+\theta^2/2}$, and the nontrivial solution of $M(\theta) = 1$ is ${}_*\theta = -2\mu$. Hence, $f_{{}_*\theta}(x) = \mathrm{e}^{-2\mu x} f(x) \propto \mathrm{e}^{-(x+\mu)^2/2}$, which shows that $f_{{}_*\theta}$ is the pdf of the $\mathsf{N}(-\mu, 1)$ distribution. The MATLAB code below implements Algorithm 10.1 with $\mu = -1$ and $\gamma = 13$. A typical outcome is $\widehat{\ell} = 1.6367 \times 10^{-12}$ with relative error of 0.1%.

```
%siegmund.m
mu=-1;N=10^6;gamma=13;W=nan(N,1);
for i=1:N
    S=0;
    while S<gamma
        X=-mu+randn;
        S=S+X;
    end
    W(i)=exp(2*mu*S);
end
ell_hat=mean(W),RE=std(W)/mean(W)/sqrt(N)
```

■ **EXAMPLE 10.4 (Waiting Time in a $GI/G/1$ Queue)**

In this example we consider the $GI/G/1$ single server queue model with interarrival times $A_1, A_2, \ldots \sim_{\text{iid}} F_A$ and service times $B_1, B_2, \ldots, \sim_{\text{iid}} F_B$. Let Y_n denote the waiting time of the n-th customer (excluding the service time). Then, the process $\{Y_n, n = 1, 2, \ldots\}$ satisfies the **Lindley recursion** ☞ 287

$$Y_n = \max\{Y_{n-1} + B_n - A_n,\, 0\}\,, \quad n = 1, 2, \ldots\,, \tag{10.6}$$

with $Y_0 = 0$. If the expected service time $\mathbb{E}\,B$ is less than the expected interarrival time $\mathbb{E}\,A$, then (under mild conditions; see Theorem A.9.1) Y_n converges in distri- ☞ 631

bution to a steady-state waiting time Y as $n \to \infty$. Suppose we wish to estimate the rare-event probability $\mathbb{P}(Y \geqslant \gamma)$ for large γ. Standard steady-state simulation
☞ 312 techniques such as the batch-means and regenerative methods are inefficient for this problem, as event $\{Y_n \geqslant \gamma\}$ will rarely occur during a simulation run. Instead, one can use importance sampling as follows. Define $S_n = B_n - A_n$, $n = 1, 2, \ldots$ and $S_0 = 0$. Then $\{Y_n\}$ may be viewed as the **reflection** at 0 of the random walk $\{S_n\}$; that is,

$$Y_n = S_n - \min_{0 \leqslant i \leqslant n} S_i = \max_{0 \leqslant i \leqslant n} \{S_n - S_{n-i}\} .$$

Because $\{(S_n - S_{n-i}), i = 0, 1, \ldots, n\}$ has the same distribution as $\{S_i, i = 0, 1, \ldots, n\}$, Y_n has the same distribution as $\max_{0 \leqslant i \leqslant n} S_i$. Consequently, $\mathbb{P}(Y \geqslant \gamma) = \mathbb{P}(\max\{S_0, S_1, \ldots\} \geqslant \gamma) = \mathbb{P}(\tau < \infty)$, where $\tau = \inf\{n : S_n \geqslant \gamma\}$. The probability $\ell = \mathbb{P}(\tau < \infty)$ can be efficiently estimated via Siegmund's algorithm using an exponential twist ${}_*\theta$ that is the positive root of

$$M(\theta) = \mathbb{E}\,\mathrm{e}^{\theta(B-A)} = M_B(\theta) M_A(-\theta) ,$$

where M_A and M_B are the moment generating functions of the interarrival and service times, respectively.

For the $M/M/1$ case, where $A \sim \mathsf{Exp}(\lambda)$ and $B \sim \mathsf{Exp}(\mu)$ we have $M(\theta) = \frac{\mu}{\mu-\theta}\frac{\lambda}{\lambda+\theta}$, so that ${}_*\theta = \mu - \lambda$. Thus, if we denote by $M_\theta(t)$ the moment generating function of the increment X under f_θ, then

$$M_{{}_*\theta}(t) = \mathbb{E}_{{}_*\theta}\,\mathrm{e}^{tX} = \mathbb{E}\,\mathrm{e}^{(\mu-\lambda+t)X} = \frac{\lambda}{\lambda - t}\frac{\mu}{\mu + t} ,$$

which shows that under $f_{{}_*\theta}$ the increment X is the difference of two independent exponential random variables with rates λ and μ, respectively. In other words, under $f_{{}_*\theta}$ the interarrival and service times are *interchanged.* The code below implements Algorithm 10.1. With $\gamma = 13, \mu = 2$, and $\lambda = 1/2$, we obtain $\ell = 8.501 \times 10^{-10}$ with an estimated relative error of 0.3%. Note that for this case the exact value is known (see, for example, [3]):

$$\ell(\gamma) = \frac{\lambda}{\mu}\mathrm{e}^{-(\mu-\lambda)\gamma} \approx 8.496 \times 10^{-10}.$$

```
%waitGG1.m
N=10^5;gamma=13;W=nan(N,1);mu=2; lam=1/2;
for i=1:N
    S=0;
    while S<gamma
        X=-log(rand)/lam+log(rand)/mu;
        S=S+X;
    end
    W(i)=exp(-(mu-lam)*S);
end
mean(W),std(W)/mean(W)/sqrt(N)
```

10.2.2 Estimation of Overflow Probabilities

Consider the efficient estimation of $\ell_n = \mathbb{P}(S_n \geqslant nb)$ for large n and fixed $b > \mu$. Note that here the rarity parameter is n and $\mu = \mathbb{E}X$ is not restricted to be negative.

Algorithm 10.2 (Estimation of $\mathbb{P}(S_n \geqslant nb)$) *To estimate $\ell_n = \mathbb{P}(S_n \geqslant nb)$ from simulation, execute the following steps.*

1. *Compute the root θ^* of $\zeta'(\theta) = \mathbb{E}_\theta X = b$.*
2. *Generate $X_1, \ldots, X_n \sim_{\text{iid}} f_{\theta^*}$ and compute the likelihood ratio $W_n = \mathrm{e}^{-\theta^* S_n + n\zeta(\theta^*)}$.*
3. *Use N independent replications of S_n and W_n to compute the estimator:*

$$\widehat{\ell}_n = \frac{1}{N}\sum_{k=1}^{N} W_n^{(k)}\, \mathrm{I}_{\{S_n^{(k)} \geqslant nb\}} \,. \tag{10.7}$$

The efficiency of the algorithm is given by the following result [17, 44, 55].

Theorem 10.2.2 (Efficiency of the Overflow Probability Estimator) *Of all importance sampling estimators whose change of measure is of the form $X_1, \ldots, X_n \sim_{\text{iid}} g$ for some g, the estimator with $g = f_{\theta^*}$ in Algorithm 10.2 is the only one that is logarithmically efficient. Subject to some smoothness conditions on $M(\theta)$ the first-order asymptotic behavior of ℓ_n is*

$$\ell_n = \frac{\mathrm{e}^{-n(\theta^* b - \zeta(\theta^*))}}{\theta^* \, |\zeta''(\theta^*)| \sqrt{2\pi n}} (1 + o(1)) \quad \text{as } n \to \infty \,. \tag{10.8}$$

Regarding the higher-order efficiency properties, we have that the estimator (10.7) is logarithmically efficient of order k for any $k \geqslant 2$ with

$$\lim_{n\to\infty} \frac{\ln \mathbb{E}Z^k}{kn} = \theta^* b - \zeta(\theta^*).$$

Under additional assumptions, the state-dependent importance sampling in Section 10.4 can provide an estimator with bounded relative moment of order k, see [15].

■ **EXAMPLE 10.5 (Neyman–Pearson Test)**

Let $Y \sim \mathsf{Laplace}(-1, 1)$; that is, the pdf of Y is given by $f_Y(y) = \frac{1}{2}\mathrm{e}^{-|y+1|}$. Consider the transformation $X = g(Y)$ with $g(y) = -1$ for $y \leqslant -1$, $g(y) = y$ for $-1 < y < 1$, and $g(y) = 1$ for $y \geqslant 1$. Then X has pdf

$$f(x) = \frac{1}{2}\mathrm{I}_{\{x=-1\}} + \frac{1}{2}\mathrm{e}^{-x-1}\mathrm{I}_{\{-1<x<1\}} + \frac{1}{2\,\mathrm{e}^2}\mathrm{I}_{\{x=1\}} \,,$$

with respect to the sum of the Lebesgue measure and the Dirac measure at -1 and 1. The cdf corresponding to f is depicted in the left panel of Figure 10.1. We are interested in computing $\ell_n = \mathbb{P}(S_n \geqslant 0) = \mathbb{P}(X_1 + \cdots + X_n \geqslant 0)$. This problem arises in the computation of the error rates for a Neyman–Pearson test using a log-likelihood ratio [17].

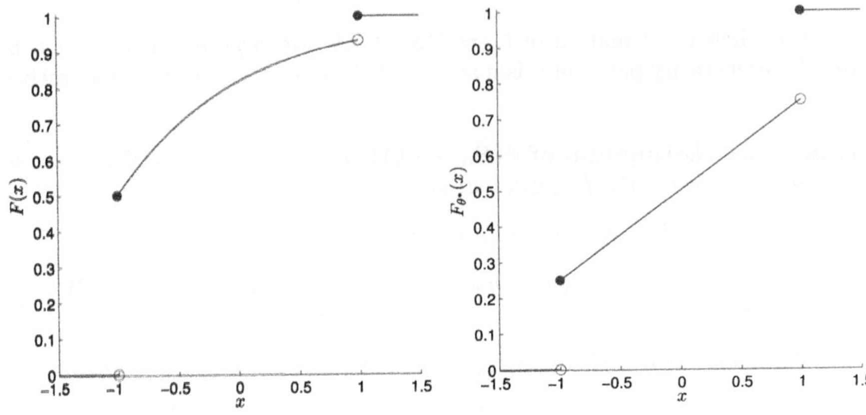

Figure 10.1 Cumulative distribution functions $F(x)$ (left panel) and $F_{\theta^*}(x)$ (right panel), where $F_{\theta^*}(x)$ is the cdf corresponding to the optimal importance sampling pdf $f_{\theta^*}(x)$.

The moment generating function of X is

$$M(\theta) = \begin{cases} \frac{e^{\theta-2}+e^{-\theta}}{2} + \frac{e^{\theta-2}-e^{-\theta}}{2(\theta-1)}, & \theta \neq 1 \\ 2e^{-1}, & \theta = 1 \end{cases},$$

with $M'(1) = 0$. Therefore, $\theta^* = 1$ and the corresponding importance sampling pdf is

$$f_{\theta^*}(x) = \frac{e^{x+1}}{2} f(x) = \frac{1}{4}.$$

The right panel of Figure 10.1 shows the cdf corresponding to f_{θ^*}. The likelihood ratio is $W_n = e^{-S_n-n+n\ln 2} = 2^n e^{-S_n-n}$. For $n = 16$ the code below gives $\widehat{\ell}_n = 8.22 \times 10^{-4}$ with an estimated relative error of 0.6%.

```
%NeyPea.m
N=10^5; n=16;W=nan(N,1);
for i=1:N
    S=0;
    for j=1:n
        U=rand;
        if U<1/4
            x=-1;
        elseif U<1/2
            x=1;
        else
            x=2*rand-1;
        end
        S=S+x;
    end
    W(i)=exp(-S+n*(log(2)-1))*(S>=0);
end
ell=mean(W), std(W)/sqrt(N)/mean(W)
```

10.2.3 Estimation For Compound Poisson Sums

In this section we consider the problem of estimating

$$\ell(\gamma) = \mathbb{P}(S_R \geqslant \gamma) = \mathbb{P}(X_1 + \cdots + X_R \geqslant \gamma) ,$$

where $R \sim \mathsf{Poi}(\lambda)$ is independent of $X_1, X_2, \ldots \sim_{\text{iid}} f$. The random variable S_R is called a **compound Poisson sum**. We further assume that the $\{X_i\}$ are positive and that f is light-tailed and satisfies either one of the following conditions:

- f decays like the density of a $\mathsf{Gamma}(\alpha, \lambda)$ distribution: ☞ 113

  $$f(x) = c\, x^{\alpha-1} \mathrm{e}^{-\lambda x}(1 + o(1)) \quad \text{as } x \to \infty ,$$

  where $c > 0$ is a constant. Examples include the pdfs of the exponential, phase-type, and inverse Gaussian (or Wald) distributions. ☞ 135

- $\int_0^\infty f^\alpha(x)\, \mathrm{d}x < \infty$ for any $\alpha \in (1, 2)$ and f can be written as:

  $$f(x) = q(x)\, \mathrm{e}^{-h(x)} ,$$

  where $0 < q(x) < \infty$ and $h(x)$ is ultimately convex in the right tail of f, that is, h is convex on $[a, b)$ for $a < b = \sup\{x : f(x) > 0\}$. Examples of such pdfs f include densities with finite support or the pdf of the $\mathsf{Weib}(\alpha, \lambda)$ distribution for $\alpha > 1$. ☞ 137

Let $M(t)$ be the moment generating function of X and ζ the cumulant function of S_R. By conditioning on R we find $\zeta(t) = \ln \mathbb{E}\mathrm{e}^{tS_R} = \ln \mathrm{e}^{\lambda(M(t)-1)} = \lambda(M(t) - 1)$. Under an exponential twist of S_R, with parameter θ, we have for the cumulant function of S_R:

$$\begin{aligned}\zeta_\theta(t) &= \ln \mathbb{E}_\theta\, \mathrm{e}^{tS_R} = \ln \mathbb{E}\, \mathrm{e}^{tS_R} \mathrm{e}^{\theta S_R - \zeta(\theta)} = \zeta(t+\theta) - \zeta(\theta) \\ &= \lambda(M(t+\theta) - M(\theta)) \end{aligned} \tag{10.9}$$

and, in particular,

$$\mathbb{E}_\theta S_R = \zeta'_\theta(0) = \lambda M'(\theta) . \tag{10.10}$$

By writing (10.9) as $\lambda_\theta(M_\theta(t) - 1)$, where $\lambda_\theta = \lambda\, M(\theta)$ and $M_\theta(t) = M(t+\theta)/M(\theta)$, we see that to obtain realizations of S_R under θ, we may simulate compound Poisson sums with rate λ_θ and increments X with moment generating function M_θ. It remains to find a good twisting parameter θ. One straightforward choice is to take θ such that (10.10) matches γ. This leads to the following algorithm.

Algorithm 10.3 (Estimation of $\mathbb{P}(S_R \geqslant \gamma)$ via Exponential Twisting)

1. *Compute the root θ^* of the equation $\lambda M'(\theta) = \gamma$, where $M(\theta)$ is the moment generating function of X.*

2. *Simulate $R \sim \mathsf{Poi}(\lambda M(\theta^*))$.*

3. *Generate R iid random variables $X_1, \ldots, X_R$, with moment generating function*

   $$M_{\theta^*}(t) = \frac{M(t + \theta^*)}{M(\theta^*)} .$$

   Set $S_R = X_1 + \cdots + X_R$.

4. *Generate N independent replications of S_R and deliver the importance sampling estimator*

$$\widehat{\ell}(\gamma) = \frac{1}{N}\sum_{k=1}^{N} \mathrm{I}_{\{S_R^{(k)} \geqslant \gamma\}} \exp\Big(-\theta^* S_R^{(k)} + \lambda(M(\theta^*)-1)\Big) . \tag{10.11}$$

Regarding the efficiency of the estimator, we have the following result [44].

Theorem 10.2.3 (Logarithmic Efficiency) *The importance sampling estimator* (10.11) *is logarithmically efficient and*

$$\ell(\gamma) = \mathbb{P}(S_R \geqslant \gamma) = \frac{\exp\Big(-\theta^*\gamma + \lambda\,(M(\theta^*)-1)\Big)}{\theta^*\sqrt{2\pi\lambda M''(\theta^*)}}\,(1+o(1)) \quad \text{as } \gamma \to \infty\,.$$

Simulation of compound Poisson sums frequently arises in insurance problems [3, 24, 29].

■ **EXAMPLE 10.6 (Estimating the Risk of Insurer Default)**

Suppose that on average $\lambda = 300$ major insurance claims per year are processed and paid by an insurance company with a core capital of 10^9 dollars. The number of claims per year is assumed to be $R \sim \mathsf{Poi}(\lambda)$. The sizes of the claims $X_1, \ldots, X_R$, measured in millions of dollars, are assumed to be iid outcomes from the $\mathsf{Gamma}(2,1)$ distribution (with pdf $f(x) = x\,\mathrm{e}^{-x}, x \geqslant 0$). We are interested in the probability that the size of all claims in a given year exceeds the core capital (that is, $X_1 + \cdots + X_R \geqslant \gamma = 10^3$). Here the moment generating function of each X_i is $M(t) = (1-t)^{-2}$, $t < 1$ and hence $\theta^* = 1 - (2\lambda/\gamma)^{1/3}$. It follows that

$$M_\theta(t) = \left(\frac{1-\theta^*}{1-\theta^*-t}\right)^2 ,$$

which is the moment generating function of the $\mathsf{Gamma}(2, 1-\theta^*)$ distribution. For the importance sampler we thus simulate the number of claims from the $\mathsf{Poi}\,(\lambda_{\theta^*})$ distribution with $\lambda_{\theta^*} = (\gamma^2\lambda/4)^{1/3}$, and the size of each claim X from the $\mathsf{Gamma}(2, (2\lambda/\gamma)^{1/3})$ distribution. We apply Algorithm 10.3 with $N = 10^4$ using the code below, and obtain 3.06×10^{-17} with an estimated relative error of 3%.

```
%compsum.m
clear all
N=10^4; W=nan(N,1); gamma=10^3;lambda=300;
theta_star=1-(2*lambda/gamma)^(1/3);
for k=1:N
    R=poissrnd((gamma^2*lambda/4)^(1/3));
    S=sum(gamrnd(2,1/(1-theta_star),1,R));
    if S>gamma
        W(k)=exp(-theta_star*S+lambda*((1-theta_star)^(-2)-1));
    else
        W(k)=0;
```

```
    end
end
mean(W), std(W)/sqrt(N)/mean(W)
```

10.3 CONDITIONING METHODS FOR HEAVY TAILS

Consider the problem of estimating

$$\ell(\gamma) = \mathbb{P}(S_n \geqslant \gamma) = \mathbb{P}(X_1 + \cdots + X_n \geqslant \gamma) , \tag{10.12}$$

where the $\{X_i\}$ are iid with cdf F from the subexponential class of distributions (see Section D.2). We are interested in the case where γ is large, making $\ell(\gamma)$ small. ☞ 703 Typically the methods for light-tailed problems do not perform well in heavy-tailed cases [6]. For example, methods based on the moment generating function are not applicable because the moment generating function is not defined on the positive real axis for heavy-tailed distributions. Instead, one of the most successful approaches is based on a conditioning idea that exploits the subexponential property:

$$\lim_{\gamma \to \infty} \frac{\mathbb{P}(S_n \geqslant \gamma)}{n\,\mathbb{P}(X_1 \geqslant \gamma)} = 1 . \tag{10.13}$$

The subexponential property essentially states that most of the time the rare event happens because a *single* variable exceeds the threshold. This is in contrast with the light-tailed case where the rare-event occurs primarily when most or all of the variables take on a large value. The following algorithm given in [8] illustrates the idea and is based on the identity

$$\ell(\gamma) = n\,\mathbb{P}(S_n \geqslant \gamma, X_n = \max_j X_j) = n\,\mathbb{E}\overline{F}\left(\left(\gamma - \sum_{j=1}^{n-1} X_j\right) \vee \max_{j \neq n} X_j\right) ,$$

where $\overline{F}(x) = 1 - F(x)$ and $a \vee b = \max\{a, b\}$.

Algorithm 10.4 (Conditional Estimator for $\mathbb{P}(S_n \geqslant \gamma)$)

1. *Generate* $X_1, \ldots, X_n \overset{\text{iid}}{\sim} F$.

2. *Compute*

$$Y = n\overline{F}\left(\left(\gamma - \sum_{j=1}^{n-1} X_j\right) \vee \max_{j \neq n} X_j\right) .$$

3. *Using N independent replications of Y, deliver the unbiased estimator*

$$\widehat{\ell}(\gamma) = \frac{1}{N} \sum_{k=1}^{N} Y_k . \tag{10.14}$$

The estimator (10.14) has the following efficiency properties [8, 42].

Theorem 10.3.1 (Efficiencies for the Heavy-Tailed Case)

1. *If F is* **regularly varying**, *then the estimator* (10.14) *has vanishing relative error. More generally,* (10.14) *can be shown [42] to have vanishing relative error if F is long-tailed and satisfies the condition:*

$$\sup_x \frac{\overline{F}(tx)}{\overline{F}(x)} < \infty, \quad t \in (0,1)\ .$$

2. *In the* **Pareto(α, λ)** *case, the estimator* (10.14) *has bounded relative error. In fact, bounded relative error holds for the more general estimator* (10.18), *where all $\{X_i\}$ are independent , but not necessarily identical Pareto random variables.*

3. *In the* **LogN(0, 1)** *case, the estimator* (10.14) *has vanishing relative error; see [42].*

4. *In the* **Weib(α, 1)** *case with* $0 < \alpha < \ln(3/2)/\ln(3) \approx 0.369$, *the estimator* (10.14) *has vanishing relative error; see [42].*

5. *If F is* **Weibull-like**, *that is,*

$$\overline{F}(x) = \frac{c\,x^{1+\lambda-\alpha}\mathrm{e}^{-x^\alpha}}{\alpha}\,(1+o(1)) \quad \textit{as } x \to \infty\ ,$$

and $0 \leqslant \alpha < \ln(3/2)/\ln(2) \approx 0.585$, *then the estimator* (10.14) *is logarithmically efficient; see [8].*

Although the above theorem considers only special cases of heavy-tailed distributions, the relevance to applications of such modeling assumptions has been cogently argued in [1, 28, 62].

10.3.1 Estimation for Compound Sums

Now consider the case $S_R = X_1 + \cdots + X_R$, where R is an integer-valued random variable with $\mathbb{E}R^2 < \infty$ and pdf f_R. As usual we assume that the sequence $X_1, X_2, \ldots$ is independent of R. In this case a good estimator of $\ell(\gamma) = \mathbb{P}(S_R \geqslant \gamma)$ combines both conditioning and control variable ideas [8]. ☞ 352

Algorithm 10.5 (Control Variable Estimator for $\mathbb{P}(S_R \geqslant \gamma)$)

1. *Generate $R \sim f_R$ and $X_1, \ldots, X_R \overset{\text{iid}}{\sim} F$, independently.*

2. *Compute*

$$Y = R\overline{F}\Big(\Big(\gamma - \sum_{j=1}^{R-1} X_j\Big) \vee \max_{j \neq R} X_j\Big) - (R - \mathbb{E}R)\,\overline{F}(\gamma)\ ,$$

where it is assumed that we have a closed-form formula for $\mathbb{E}R$.

3. Repeat Steps 1 and 2 to obtain N independent replications of Y. Deliver the unbiased estimator:

$$\widehat{\ell}(\gamma) = \frac{1}{N}\sum_{k=1}^{N} Y_k \,. \tag{10.15}$$

The efficiency properties of (10.15) are summarized in the following result [42].

Theorem 10.3.2 (Efficiencies for Random R)

1. *If F is* **regularly varying** *with index $\alpha > 0$ and*

$$\mathbb{E}R^{2(\alpha+1+c)} < \infty \quad \textit{for some } c > 0 \,,$$

then the estimator (10.14) *has vanishing relative error.*

2. *In the* **LogN(0, 1)** *case with*

$$\mathbb{E}\exp\left((\ln R)^{2+c}\right) < \infty \quad \textit{for some } c > 0 \,,$$

the estimator (10.14) *has vanishing relative error.*

3. *In the* **Weib(α, 1)** *case with $0 < \alpha < \ln(3/2)/\ln(3) \approx 0.369$ and $R \leqslant c$ for some $c > 0$, the estimator* (10.14) *has vanishing relative error.*

Asmussen and Kroese [8] report that in practical simulations the conditional estimator (10.15) performs much better than estimators based on importance sampling [5, 47].

The problem of computing the probability $\mathbb{P}(S_R \geqslant \gamma)$ arises frequently in queueing theory, telecommunications, and insurance risk [1, 56]. The following example illustrates a typical application in queueing theory.

■ EXAMPLE 10.7 (Pollaczek–Khinchin Formula)

In Example 10.4 an importance sampling procedure is given for estimating the tail distribution of the steady-state waiting time in a $GI/G/1$ queue, based on an exponential change of measure. However, the procedure only works for light-tailed distributions. In the case where the interarrival times are exponential — in which case the queueing system is said to be of $M/G/1$ type — a different estimation method can be used which also applies to heavy-tailed service time distributions. This method is based on the **Pollaczek–Khinchin** formula [3]:

> The steady-state waiting time Y of an $M/G/1$ queue with arrival rate λ and service time $B \sim G$ is distributed as the random sum $S_R = X_1 + \cdots + X_R$, where $R \sim \mathsf{Geom}_0(1 - \lambda\mathbb{E}B)$ and each X has pdf $f(x) = \mathbb{P}(B \geqslant x)/\mathbb{E}B$ corresponding to the steady-state service time.

Consider, for example, the case where the service time tail distribution is given by $\mathbb{P}(B \geqslant x) = \alpha(1+x)^{-(\alpha+1)}, x \geqslant 0$ for some $\alpha > 0$. Hence, the expected service time is $\mathbb{E}B = 1$ and the steady-state service time is $X \sim \mathsf{Pareto}(\alpha, 1)$. We wish to estimate $\ell(\gamma) = \mathbb{P}(Y \geqslant \gamma)$ for large γ and assess how close the estimate is to the

approximation $\mathbb{E}R\,\mathbb{P}(X \geqslant \gamma)$ suggested by (10.13). The following MATLAB code implements Algorithm 10.5, where γ is chosen such that $\frac{\lambda}{1-\lambda}\mathbb{P}(X \geqslant \gamma) = 10^{-11}$, $\alpha = 1/2$, and $\varrho = \lambda = 3/4$. Note that $\ell(\gamma) = \varrho\,\mathbb{P}(X_1 + \cdots + X_R \geqslant \gamma)$, where
☞ 91 $R \sim \mathsf{Geom}_0(1-\varrho)$. Using a sample size of $N = 10^3$ we obtain the estimate $\widehat{\ell}(\gamma) = 10^{-11} + 2.4 \times 10^{-25}$, with an estimated relative error of 10^{-15}.

```
%polkinex.m
rho=0.75; alpha=0.5;
gamma=((1-rho)/rho*10^(-11))^(-1/alpha)-1;
bar_F=@(x)(1+x).^(-alpha);
N=10^3; Y=nan(N,1);
for i=1:N
    R=1;
    while rand<rho
        R=R+1;
    end;
    if R==1
        val=gamma;
    else
        X=rand(1,R-1).^(-1/alpha)-1;
        S=sum(X);M=max(X); val=max(M,gamma-S);
    end
    % control variable estimator
    Y(i)=R*bar_F(val)+(1/(1-rho)-R)*bar_F(gamma);
end
format long
ell=mean(Y)*rho
RE = std(Y)/sqrt(N)/ell
```

10.3.2 Sum of Nonidentically Distributed Random Variables

Consider again the estimation of (10.12), but this time the $X_1, X_2, \ldots$ have a *different* subexponential cdfs $F_1, F_2, \ldots$. In other words, $X_1, X_2, \ldots$ is a sequence of **independent nonidentically distributed** random variables. In this case we can use the following algorithm [57], motivated by the identity

$$\mathbb{P}(S_n \geqslant \gamma) = \sum_{i=1}^{n} p(i)\frac{\mathbb{P}(S_n \geqslant \gamma, X_i = \max_j X_j)}{p(i)} = \mathbb{E}\left[\frac{\mathrm{I}\{S_n \geqslant \gamma, X_J = \max_j X_j\}}{p(J)}\right],$$

with $\overline{F}_i(x) \stackrel{\text{def}}{=} 1 - F_i(x)$ and

$$p(j) = \mathbb{P}(J = j) = \frac{\mathbb{P}(X_j \geqslant \gamma)}{\sum_{i=1}^{n}\mathbb{P}(X_i \geqslant \gamma)} = \frac{\overline{F}_j(\gamma)}{\sum_{i=1}^{n}\overline{F}_i(\gamma)}, \quad j = 1, \ldots, n. \tag{10.16}$$

Algorithm 10.6 (Sum of Independent Nonidentical Random Variables)

1. *Simulate a discrete random variable* J *with discrete pdf* $p(j)$ *given in* (10.16).

2. *Generate* $X_1, X_2, \ldots, X_{J-1}, X_{J+1}, \ldots, X_n$ *with corresponding distributions* $F_1, F_2, \ldots, F_{J-1}, F_{J+1}, \ldots, F_n$ *and compute*

$$Y = \frac{1}{p(J)}\,\overline{F}_J\Big(\Big(\gamma - \sum_{i \neq J} X_i\Big) \vee \max_{i \neq J} X_i\Big) .$$

3. *Using* N *independent replications of* Y, *deliver the unbiased estimator:*

$$\widehat{\ell}(\gamma) = \frac{1}{N}\sum_{k=1}^{N} Y_k . \tag{10.17}$$

The choice of pdf in (10.16) is close to the minimum variance pdf $p^*(i) = \mathbb{P}(X_i = \max_j X_j \mid S_n \geqslant \gamma)$. Regarding the efficiency of the estimator, we have the following result [57].

Theorem 10.3.3 (Bounded Relative Error) *The estimator* (10.17) *in Algorithm 10.6 has bounded relative error when all* $\{X_i\}$ *are independent and have either log-normal or regularly varying distributions.*

Under certain conditions it is possible to drop the independence assumption [57, Page 50]. An alternative to the estimator in Algorithm 10.6 is [23]:

$$\widehat{\ell}(\gamma) = \frac{1}{N}\sum_{k=1}^{N}\sum_{i=1}^{n}\overline{F}_i\Big(\Big(\gamma - \sum_{j \neq i} X_j^{(k)}\Big) \vee \max_{j \neq i} X_j^{(k)}\Big), \tag{10.18}$$

where $X_j^{(k)}$, $k = 1, \ldots, N$ are independent random variables with cdfs F_j, $j = 1, \ldots, n$. This estimator has bounded relative error when the $\{X_j^{(k)}\}$ have Pareto distributions, and is used for estimating the probability of large portfolio losses, where the monetary values of the losses follow a Student's t copula model. ☞ 69

Another estimator has been proposed by Juneja [45], based on the identity

$$\mathbb{P}(S_n \geqslant \gamma) = \mathbb{P}\big(\max_j X_j \geqslant \gamma\big) + \mathbb{P}\big(S_n \geqslant \gamma,\ \max_j X_j < \gamma\big) .$$

When the $\{X_i\}$ have a regularly varying distribution, then

$$\mathbb{P}(S_n \geqslant \gamma) = \mathbb{P}\big(\max_j X_j \geqslant \gamma\big)\ (1 + o(1)) \quad \text{as } \gamma \to \infty ,$$

and the so-called **residual probability** $\mathbb{P}(S_n \geqslant \gamma,\ \max_j X_j < \gamma)$ becomes asymptotically negligible. Typically, $\mathbb{P}(\max_j X_j \geqslant \gamma)$ can be evaluated exactly and all that is required is to estimate the residual probability efficiently. To this end, one can use a combined conditional and importance sampling estimator; see [45]. Note that the residual probability itself can be similarly decomposed into a leading term and a residual probability. This recursive idea has been successfully applied in network reliability estimation [18].

■ **EXAMPLE 10.8 (Sum of Log-Normals)**

Consider the case where $n = 10$ and each X_i is drawn independently from $\mathsf{LogN}(i - 10, i^2), i = 1, \ldots, n$. This problem arises in the computation of portfolio measures such as value-at-risk [9]. For $\gamma = 4 \times 10^4$ the following code, which implements Algorithm 10.6, gives an estimate of 4.6312×10^{-4} with an estimated relative error of 7×10^{-5}.

```
%sumlognor.m
clear all,clc
n=10; gamma=4*10^4; N=10^5; param=1:n;
p = 1-logncdf(gamma,param-10,sqrt(param)); p=p/sum(p);
J = randsample(n,N,'true',p); %sample index J
% generate r.v.'s X_1,X_2,...
X = lognrnd(param(ones(N,1),:)-10,sqrt(param(ones(N,1),:)));
% set the J-th entry in each row to 0.
X((J'-1)*N+(1:N))=0;
% compute estimator
Y=1-logncdf( max( [gamma-sum(X,2),X],[],2 ),J-10,sqrt(J) );
Y=Y'./p(J);
mean(Y), std(Y)/sqrt(N)/mean(Y)
```

10.4 STATE-DEPENDENT IMPORTANCE SAMPLING

In this section we review an importance sampling approach that applies to dynamic processes. The main feature of the approach is that the importance sampling change of measure depends on the progress of the underlying process in reaching the rare-event set. In other words, we have *state-dependent* importance sampling. The general framework for applying importance sampling to dynamic processes, such as Markov chains, Markov jump processes, and generalized semi-Markov processes, is given by Glynn and Iglehart [39]. The authors also discuss the application of importance sampling to steady-state simulation and to the case where the simulation horizon depends on a stopping time.

Here we review a state-dependent importance sampling procedure that yields estimators with bounded relative error for a broad class of entrance probability problems. We use the following framework.

☞ 162

- Let $\{Z_n, n = 0, 1, \ldots\}$ be a time-homogeneous Markov chain with state space $\mathscr{E}$ and transition density $p(y \mid x)$.
- Let $\mathbb{P}^z$ denote the probability measure under which $\{Z_n\}$ starts at $Z_0 = z$.
- Let $\mathscr{A} \subseteq \mathscr{R} \subseteq \mathscr{E}$; see Figure 10.2.
- Let $\tau = \inf\{n \geqslant 0 : Z_n \in \mathscr{R}\}$ be the first time that the process $\{Z_n\}$ enters the region $\mathscr{R}$.

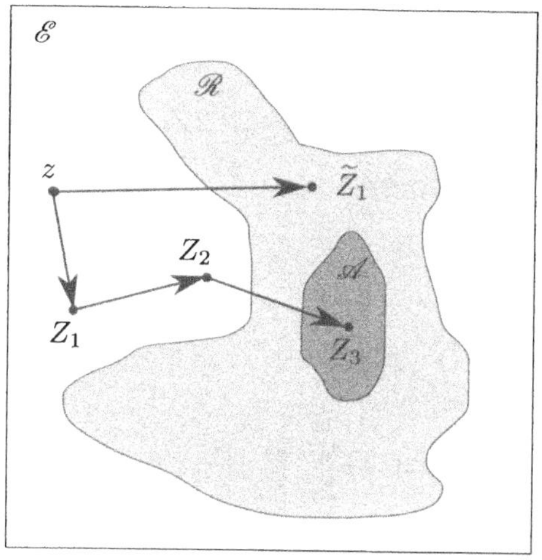

Figure 10.2 Two realizations of the Markov chain $\{Z_n, n = 0, 1, \ldots\}$, starting from $Z_0 = z$. The chain evolves until the process enters the set $\mathscr{R}$.

We are interested in estimating the probability that, starting from $z \notin \mathscr{R}$, the process enters the region $\mathscr{R}$ through the set $\mathscr{A}$; that is, the probability

$$h(z) = \mathbb{P}^z(Z_\tau \in \mathscr{A}, \tau < \infty) = \sum_{t=0}^{\infty} \mathbb{P}^z(Z_t \in \mathscr{A}, \tau = t) . \tag{10.19}$$

Many of the rare-event problems discussed in the previous sections can be formulated via an entrance probability of the form (10.19).

■ EXAMPLE 10.9 (Entrance Probabilities for Stopping Times)

Consider the stopping-time probability from Section 10.2.1:

$$\ell(\gamma) = \mathbb{P}(\inf\{n : S_n \geqslant \gamma\} < \infty), \quad S_n = X_1 + \cdots + X_n, \quad S_0 = 0 ,$$

where $X_1, X_2, \ldots \sim_{\text{iid}} f$. In this case $\{Z_n,\ n = 0, 1, \ldots\} = \{S_n,\ n = 0, 1, \ldots\}$, $Z_0 = 0$, and $\mathscr{R} = \mathscr{A} = (\gamma, \infty)$, so that $\tau = \inf\{n : Z_n \geqslant \gamma\}$, $\{Z_\tau \in \mathscr{A}\} = \{\tau < \infty\}$, and

$$\ell(\gamma) = \mathbb{P}^0(Z_\tau \in \mathscr{A}, \tau < \infty)$$

is of the form (10.19).

■ EXAMPLE 10.10 (Overflow Probabilities as an Entrance Problem)

Consider the overflow probability $\ell_n = \mathbb{P}(S_n \geqslant nb)$ in Section 10.2.2. Define the Markov chain $\{Z_k,\ k = 0, 1, \ldots\} = \{(k, S_k),\ k = 0, 1, \ldots\}$, with $Z_0 = z = (0, 0)$, and let $\mathscr{A} = \mathscr{R} = \{(n, s) : s \geqslant nb\}$. Then, $\tau = \inf\{t : S_t \geqslant nb, t = n\}$, which is either n, if $S_n \geqslant nb$, or ∞ otherwise. It follows that

$$\ell_n = \mathbb{P}(S_n \geqslant nb) = \mathbb{P}^{(0,0)}(Z_\tau \in \mathscr{A}, \tau < \infty) ,$$

which is of the form (10.19).

Returning to the general problem of estimating (10.19), observe that by the Markov property we have for all $t \geqslant 1$

$$\mathbb{P}^z(Z_t \in \mathscr{A}, \tau = t \mid Z_1 = z_1) = \mathbb{P}^{z_1}(Z_{t-1} \in \mathscr{A}, \tau = t - 1) .$$

By conditioning on Z_1 we thus have

$$\begin{aligned} h(z) &= \sum_{t=1}^{\infty} \mathbb{P}^z(Z_t \in \mathscr{A}, \tau = t) \\ &= \sum_{t=1}^{\infty} \sum_{z_1 \in \mathscr{E}} \mathbb{P}^z(Z_t \in \mathscr{A}, \tau = t \mid Z_1 = z_1)\, p(z_1 \mid z) \\ &= \sum_{z_1 \in \mathscr{E}} p(z_1 \mid z) \sum_{t=1}^{\infty} \mathbb{P}^{z_1}(Z_{t-1} \in \mathscr{A}, \tau = t - 1) \\ &= \sum_{z_1 \in \mathscr{E}} p(z_1 \mid z)\, h(z_1) , \end{aligned}$$

which shows that

$$q(y \mid x) = \frac{h(y)}{h(x)}\, p(y \mid x), \quad y \in \mathscr{E},\ x \in \mathscr{R}^c , \tag{10.20}$$

defines a proper transition density on $\mathscr{E}$. Let $\mathbb{Q}^z$ be the probability measure under which $\{Z_n\}$ is a Markov process on $\mathscr{E}$ starting at z with transition density $q(y \mid x)$. Then, it is not difficult to see [7] that under $\mathbb{Q}^z$:

1. $\mathbb{Q}^z(Z_1 = z_1, \ldots, Z_t = z_t) = \frac{h(z_t)}{h(z)}\, \mathbb{P}^z(Z_1 = z_1, \ldots, Z_t = z_t)$;
2. $\mathbb{Q}^z(Z_\tau \in \mathscr{A}, \tau < \infty) = 1$;
3. $\mathbb{Q}^z(Z_1 = z_1, \ldots, Z_\tau = z_\tau) = \mathbb{P}^z(Z_1 = z_1, \ldots, Z_\tau = z_\tau \mid Z_\tau \in \mathscr{A}, \tau < \infty)$.

In other words, the distribution of $\{Z_n\}$ under $\mathbb{Q}^z$ is the same as the distribution under $\mathbb{P}^z$ given the rare event $\{Z_\tau \in \mathscr{A}, \tau < \infty\}$. Therefore, the measure $\mathbb{Q}^z$ is the zero-variance measure for simulation of the rare event $\{Z_\tau \in \mathscr{A}, \tau < \infty\}$.

In practice one cannot simulate under $\mathbb{Q}^z$, because $h(z)$ is unknown and hence the transition density q is not available. The situation is similar to the problem of simulating from the zero-variance pdf g^* in the cross-entropy method (see Section 10.5). Suppose, however, that one has a good approximation to h, say $\widetilde{h}$. Define the transition density

$$\widetilde{q}(z \mid y) = p(z \mid y)\, \frac{\widetilde{h}(z)}{c(y)}, \quad \text{where } c(y) = \int p(z \mid y)\, \widetilde{h}(z)\, \mathrm{d}z . \tag{10.21}$$

Then, estimation of $h(z)$ can be accomplished using the following algorithm [39].

Algorithm 10.7 (State-Dependent Importance Sampling) *Given the approximation $\widetilde{h}(z)$, execute the following steps.*

1. *Simulate a trajectory $Z_1, Z_2 \ldots,$ of the Markov chain $\{Z_n\}$ until the stopping time $\tau = \inf\{t : Z_t \in \mathscr{R}\}$. The simulation is carried out under $\widetilde{\mathbb{Q}}^z$ corresponding to the transition density $\widetilde{q}$, so that the joint density of the path is (setting $z_0 = z$):*

$$\widetilde{q}(z_1 \mid z_0)\, \widetilde{q}(z_2 \mid z_1) \cdots \widetilde{q}(z_\tau \mid z_{\tau-1}) .$$

2. The corresponding likelihood ratio is

$$W(z_1,\ldots,z_T) = \frac{p(z_1\,|\,z_0)\cdots p(z_T\,|\,z_{T-1})}{\widetilde{q}(z_1\,|\,z_0)\cdots \widetilde{q}(z_T\,|\,z_{T-1})} = \prod_{t=1}^{T} \frac{c(z_{t-1})}{\widetilde{h}(z_t)} \,.$$

3. By repeating Steps 1 and 2 above, generate N instances of $Y = W(Z_1,\ldots,Z_T)\,\mathrm{I}_{\{Z_T\in\mathscr{A}\}}$ and output the estimator $\widehat{h}(z) = \frac{1}{N}\sum_{i=1}^{N} Y_i$.

To apply Algorithm 10.7 successfully, a number of issues need to be resolved [16]:

1. The choice of the transition density $\widetilde{q}$ is crucial. Typically, a candidate for $\widetilde{q}$ is suggested by considering an asymptotic approximation $\widetilde{h}(z)$ of $h(z)$.

2. It is important to be able to efficiently calculate the normalizing constant $c(y)$ in (10.21). In practice it has to be evaluated numerically. Blanchet and Liu [16] propose approximating $c(y)$ via path sampling in cases where deterministic integration is not practical.

3. The simplicity of sampling from the transition density $\widetilde{q}$ is also an important factor in the overall efficiency of Algorithm 10.7. Typically one uses the acceptance–rejection algorithm as in [14].

State-dependent importance sampling strategies are typically more efficient and reliable than their state-independent counterparts. For example, Blanchet and Glynn [14] show that if estimation of the stopping-time probability (10.5) with heavy-tailed increments is required, then their state-dependent importance sampling algorithm yields an estimator with vanishing relative error. In addition, Bassamboo et al. [11] show that there is no state-independent importance sampling algorithm that will give an efficient estimator in the regularly varying case. Note that state-dependent importance sampling algorithms are usually much more difficult to implement than their state-independent counterparts.

■ EXAMPLE 10.11 (State-Dependent and State-Independent Sampling)

Consider again the problem of estimating the overflow probability $\ell_n = \mathbb{P}(S_n \geqslant nb\,|\,S_0 = 0)$ of the random walk process $\{S_n, n = 0,1,2,\ldots\}$ with increments $X_1, X_2, \ldots \sim_{\text{iid}} f$. Define $\ell(k,s) = \mathbb{P}(S_k \geqslant s\,|\,S_0 = 0)$, so that $\ell_n = \ell(n, nb)$. Using the framework of Example 10.10 we have

$$\begin{aligned} h\big((k,s_k)\big) &= \mathbb{P}(s_k + X_{k+1} + \cdots + X_n \geqslant nb\,|\,S_k = s_k) \\ &= \mathbb{P}(S_{n-k} \geqslant nb - s_k\,|\,S_0 = 0) \\ &= \ell(n-k, nb - s_k)\,, \end{aligned}$$

and

$$\begin{aligned} h\big((k+1, s_k + x_{k+1})\big) &= \mathbb{P}(s_k + x_{k+1} + X_{k+2} + \cdots + X_n \geqslant nb\,|\,S_{k+1} = s_{k+1}) \\ &= \mathbb{P}(S_{n-k-1} \geqslant nb - s_k - x_{k+1}\,|\,S_0 = 0) \\ &= \ell(n-k-1, nb - s_k - x_{k+1})\,. \end{aligned}$$

The zero-variance transition density (10.20) can thus be written as

$$q((k+1, s_k + x_{k+1}) \,|\, (k, s_k)) = f(x_{k+1}) \frac{\ell(n-k-1, nb - s_k - x_{k+1})}{\ell(n-k, nb - s_k)} \; .$$

There are a number of possible approximations to $\ell(\,\cdot\,,\,\cdot\,)$ and hence to $q(\cdot\,|\,\cdot)$. For example, using the asymptotic approximation (10.8), we have that

$$\ell(n-k-1, nb - s_k - x_{k+1}) \approx c\,\mathrm{e}^{\theta^* x_{k+1}} \; ,$$

where c is a constant independent of x_{k+1} and θ^* is the root of $\zeta'(\theta) = \mathbb{E}_\theta X = b$. This gives the approximating transition density (10.21) as:

$$\widetilde{q}((k+1, s_k + x_{k+1}) \,|\, (k, s_k)) = f_{\theta^*}(x_{k+1}) = f(x_{k+1})\, \mathrm{e}^{\theta^* x_{k+1} - \zeta(\theta^*)} \; . \qquad (10.22)$$

Notice that (10.22) does not depend on s_k, and thus under this change of measure we generate the increments $X_1, X_2, \ldots, X_n$ as iid samples from $f_{\theta^*}(x_{k+1})$. We recognize this sampling procedure as Algorithm 10.2, which is *state-independent* importance sampling. The left panel of Figure 10.3 illustrates that the increments are chosen independent of the current state. The angle formed between each vector started at $z_k = (k, s_k)$ and the n-axis is $\tan^{-1}(b)$.

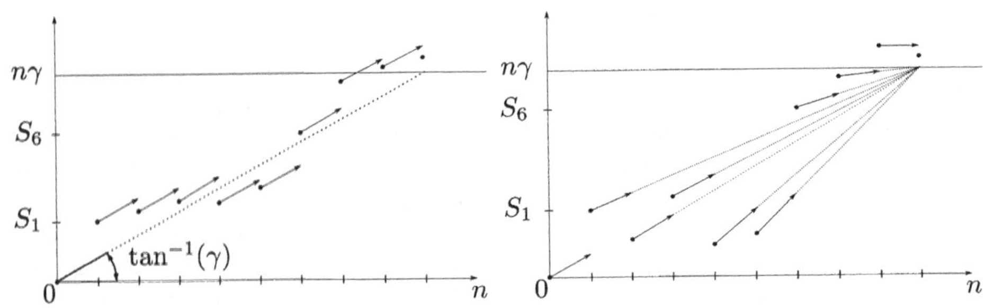

Figure 10.3 Comparison between the state-independent (left panel) and state-dependent (right panel) importance sampling. Each point on the graphs is a realization of S_k for a given k.

An alternative approximation to q takes into account the progress of the random walk $\{S_k\}$ in reaching the level nb:

$$\widetilde{q}((k+1, s_k + x_{k+1}) \,|\, (k, s_k)) = f(x_{k+1})\, \mathrm{e}^{\theta^*_{k+1} x_{k+1} - \zeta(\theta^*_{k+1})} \; , \qquad (10.23)$$

where each θ^*_{k+1}, $k = 0, \ldots, n-2$, solves the equation

$$\zeta'(\theta_{k+1}) = \frac{nb - s_k}{n - k} \; ,$$

and the final X_n is sampled conditional on $X_n \geqslant nb - s_{n-1}$. Unlike (10.22), the transition density (10.23) depends on $z_k = (k, s_k)$ and yields a state-dependent importance sampling algorithm, in which at each step k, the twisting parameter θ^*_{k+1} depends on the progress of the underlying process at step k in attaining the rare-event set. Figure 10.3 shows the difference in applying the state-dependent transition density (10.23) versus the state-independent transition density (10.22). Each vector originating from $z_k = (k, s_k)$ points to (n, nb), and hence has slope $\mathbb{E}_{\theta^*_{k+1}} X_{k+1}$. The resulting estimator has bounded relative error [13]. In contrast, Siegmund's Algorithm 10.1 only gives a logarithmically efficient estimator. More importantly, Blanchet et al. [15] and L'Ecuyer et al. [49] show that the state-dependent scheme gives an estimator with *bounded relative moment of order k*, see Page 383.

As a concrete example, consider the case where each increment of the random walk has a Laplace(0, 1) distribution. Then, $\zeta(\theta) = -\ln(1-\theta^2)$ for $|\theta| < 1$, and the likelihood ratio for a given path is

$$W = \mathrm{e}^{-\sum_{k=0}^{n-2} \theta^*_{k+1} x_{k+1} + \zeta(\theta^*_{k+1})}\, \mathbb{P}(X_n \geqslant nb - s_{n-1}) ,$$

where θ^*_{k+1} is the root of $2\theta/(1-\theta^2) = \frac{nb - s_k}{n-k}$ for $|\theta| < 1$. Sampling from the transition kernel $\widetilde{q}(z_{k+1} \mid z_k)$ is equivalent to generating $X_{k+1} = BY_1 - (1-B)Y_2$, where $B \sim \mathsf{Ber}((1+\theta^*_{k+1})/2)$, $Y_1 \sim \mathsf{Exp}(1-\theta^*_{k+1})$, and $Y_2 \sim \mathsf{Exp}(1+\theta^*_{k+1})$. The following code implements the algorithm for $n = 10$ and $b = 4$. For this particular example θ^*_{k+1} solves a quadratic equation with a unique root in the interval $(-1, 1)$. The code uses the function fzero.m to compute the unique θ^*_{k+1}. We obtained $\widehat{\ell}_n = 1.12 \times 10^{-11}$ with an estimated relative error of 3%.

```
%state_dependent_IS_Laplace.m
N=1000; S(1)=0; gamma=4; n=10; % set-up the parameters
for i=1:N
    for k=1:n-1
        % find root for twisting
        theta(k)=fzero(@(t)(  2*t/(1-t^2)...
        -(n*gamma-S(k))/(n+1-k)   ),[-0.999,.999]);
        % sample from twisted density
        if rand<(1+theta(k))/2
            x(k)=-log(rand)/(1-theta(k));
        else
            x(k)=log(rand)/(1+theta(k));
        end
        S(k+1)=S(k)+x(k); % increment the walk
    end
    % compute the likelihood ratio of the path
    W(i)=exp(-x*theta'-sum(log(1-theta.^2)))*...
    (1-cdf_laplace(n*gamma-S(end)));
end
ell=mean(W)
RE=std(W)/ell/sqrt(N)
```

10.5 CROSS-ENTROPY METHOD FOR RARE-EVENT SIMULATION

In this section we review the *cross-entropy* (CE) method in the context of rare-event
☞ 463 simulation. We refer to Chapter 13 for a more detailed account of the CE method.
☞ 366 See also Section 9.7.3 for a description of the CE method in the context of variance reduction.

While the methods presented in the previous sections apply to a narrow range of problems for which analytical approximations exist, the CE method is more broadly applicable for estimating rare-event probabilities of the form $\ell = \mathbb{P}(S(\mathbf{X}) \geqslant \gamma)$ for a performance function $S : \mathbb{R}^n \to \mathbb{R}$ and threshold γ. Here $\mathbf{X}$ is assumed to have a general probability density $f(\mathbf{x}; \mathbf{u})$, which is parameterized by a vector $\mathbf{u}$.

The main idea of the CE method is to use an importance sampling density from the same parametric family as the nominal pdf $f(\mathbf{x}; \mathbf{u})$, say $f(\mathbf{x}; \widehat{\mathbf{v}}^*)$. The parameter $\widehat{\mathbf{v}}^*$ is chosen as the estimated maximizer of the CE program (see (13.6) with $H(\mathbf{x}) = \mathrm{I}_{\{S(\mathbf{X}) \geqslant \gamma\}}$):

$$\mathbf{v}^* = \underset{\mathbf{v}}{\operatorname{argmax}} \int \mathrm{I}_{\{S(\mathbf{x}) \geqslant \gamma\}}\, f(\mathbf{x}; \mathbf{u}) \ln f(\mathbf{x}; \mathbf{v})\, \mathrm{d}\mathbf{x} = \underset{\mathbf{v}}{\operatorname{argmax}}\, \mathbb{E}_{g^*} \ln f(\mathbf{X}; \mathbf{v}) \;, \quad (10.24)$$

where $g^*(\mathbf{x}) = f(\mathbf{x})\, \mathrm{I}_{\{S(\mathbf{X}) \geqslant \gamma\}} / \ell$ is the zero-variance importance sampling pdf, which is simply the conditional pdf of $\mathbf{X} \sim f$ given $S(\mathbf{X}) \geqslant \gamma$; see (13.4).

☞ 465 In Algorithm 13.1, a general procedure is described for estimating the optimal CE parameter $\mathbf{v}^*$ using a *multilevel approach.* However, as observed in [59], such an approach may not always be necessary or desirable. We next describe how to estimate $\mathbf{v}^*$ directly from g^* without a multilevel approach. Suppose one can easily sample (approximately) from g^*. For example, we may use any of the Markov chain samplers described in Chapter 6 to simulate from g^*. Let $\mathbf{X}_1, \ldots, \mathbf{X}_N \overset{\text{approx.}}{\sim} g^*$, then we may estimate $\mathbf{v}^*$ via $\widehat{\mathbf{v}}^* = \operatorname{argmax}_{\mathbf{v}} \sum_{k=1}^{N} \ln f(\mathbf{X}_k; \mathbf{v})$. Thus, the CE
☞ 667 program reduces to a standard maximum likelihood optimization problem. Once $\widehat{\mathbf{v}}^*$ is computed we use the importance sampling estimator

$$\widehat{\ell} = \frac{1}{N_1} \sum_{k=1}^{N_1} \frac{f(\mathbf{X}_k; \mathbf{u})}{f(\mathbf{X}_k; \widehat{\mathbf{v}}^*)}\, \mathrm{I}_{\{S(\mathbf{X}_k) \geqslant \gamma\}}, \quad \mathbf{X}_1, \ldots, \mathbf{X}_{N_1} \overset{\text{iid}}{\sim} f(\cdot; \widehat{\mathbf{v}}^*) \quad (10.25)$$

to estimate ℓ. This motivates the following CE algorithm.

Algorithm 10.8 (CE Method for Rare-Event Probability Estimation)
Given the parameters N and N_1, execute the following steps.

1. *Run a Markov chain sampler to generate*

$$\mathbf{X}_1, \ldots, \mathbf{X}_N \overset{\text{approx.}}{\sim} g^*,$$

where $g^*(\mathbf{x}) \propto f(\mathbf{x})\, \mathrm{I}_{\{S(\mathbf{x}) \geqslant \gamma\}}$.

2. *Compute $\widehat{\mathbf{v}}^*$ by solving the maximum likelihood optimization problem*

$$\widehat{\mathbf{v}}^* = \underset{\mathbf{v}}{\operatorname{argmax}} \sum_{k=1}^{N} \ln f(\mathbf{X}_k; \mathbf{v}) \,.$$

3. *Given $\widehat{\mathbf{v}}^*$, deliver the importance sampling estimator* (10.25) *of the rare-event probability $\ell = \mathbb{P}(S(\mathbf{X}) \geqslant \gamma)$.*

■ **EXAMPLE 10.12 (Bridge Network Revisited)**

Consider the bridge network in Example 9.1, which is used to illustrate the various variance reduction techniques in Chapter 9. Suppose that instead of estimating the expected length of the shortest path, $\mathbb{E}\, H(\mathbf{X})$, we are interested in estimating the probability that the shortest path exceeds a given threshold γ. Thus, here we wish to estimate $\ell = \mathbb{P}(H(\mathbf{X}) \geqslant \gamma)$, where the lengths $\{X_i\}$ are independent and $X_i \sim \mathsf{U}(0, a_i)$, $i = 1, \ldots, 5$ with $(a_1, \ldots, a_5)$ being fixed parameters. Writing $X_i = a_i U_i$, $i = 1, \ldots, 5$ with $\{U_i\} \sim_{\text{iid}} \mathsf{U}(0,1)$, we can write $\ell = \mathbb{P}(h(\mathbf{U}) \geqslant \gamma)$, where $\mathbf{U} = (U_1, \ldots, U_5)$ and $h(\mathbf{U}) = H(a_1 U_1, \ldots, a_5 U_5)$, as in (9.2). ☞ 348

We use the same parametric family of importance sampling densities as in Example 9.7. In other words, the importance sampling pdf is ☞ 363

$$g(\mathbf{u}) = \prod_{i=1}^{5} g_i(u_i) \, ,$$

where each g_i is the pdf of the $\mathsf{Beta}(\nu_i, 1)$ distribution: $g_i(u) = \nu_i \, u^{\nu_i - 1}$, $u \in [0,1]$. For Step 1 of Algorithm 10.8 we use the following Gibbs sampler to simulate from $g^*(\mathbf{u}) = \mathrm{I}_{\{h(\mathbf{u}) \geqslant \gamma\}}/\ell$. ☞ 233

Algorithm 10.9 (Gibbs Sampling From g^*) *Given an initial state* $\mathbf{U}$ *such that* $h(\mathbf{U}) \geqslant \gamma$, *iterate the following steps* N *times.*

1. *Sample a random index* I *uniformly from the integers* $1, \ldots, 5$.
2. *Let* $\mathbf{U}^* = (U_1, \ldots, U_{I-1}, 0, U_{I+1}, \ldots, U_5)$; *that is,* $\mathbf{U}^*$ *is the same as* $\mathbf{U}$ *except that the* I*-th position is set to 0.*
3. *Sample* U_I *uniformly on* $(0,1)$ *such that* $U_I \geqslant \gamma - h(\mathbf{U}^*)$.
4. *Reset* $\mathbf{U} = (U_1, \ldots, U_{I-1}, U_I, U_{I+1}, \ldots, U_5)$.

Let the output of the Gibbs sampler be $\mathbf{U}_i = (U_{i1}, \ldots, U_{i5})$, $i = 1, \ldots, N$. Then, the solution of the maximum likelihood optimization problem in Step 2 of Algorithm 10.8 is

$$\widehat{v}_j^* = \frac{N}{-\sum_{i=1}^{N} \ln U_{ij}}, \qquad j = 1, \ldots, 5 \, .$$

Putting all of the above together results in the MATLAB code below. With $\gamma = 1.99$ and using the function `h.m` on Page 349, we obtain $\widehat{\ell} = 2.38 \times 10^{-5}$ with an estimated relative error of 0.6%. In this case the estimate for the optimal $\mathbf{v}^*$ is $\widehat{\mathbf{v}}^* = (295.4,\ 2.66,\ 1.26,\ 300.8,\ 2.51)$.

```
%CE_example_gibbs.m
n=5;gamma=1.99;
N=10^5;          % length of Markov chain
u=ones(1,n); % starting value for Gibbs sampling
v=0;
% run the Gibbs sampler
for i=1:N
    I=ceil(rand*n);u(I)=0;
```

```
    lower=max(gamma-h(u),0);
    u(I)=lower+(1-lower)*rand;
    v=log(u)+v;  % compute the mean of the log(u)
end
% compute the Maximum Likelihood estimate
v=-N./v;

% apply Importance sampling
N1=10^5;
nu = repmat(v,N1,1);
U = rand(N1,5).^(1./nu);
I =h(U)>gamma;
w=zeros(size(I)); % ensure w=0 where h(U)<gamma
w(I) = prod(1./(nu(I,:).*U(I,:).^(nu(I,:) - 1)),2);
% deliver the final estimator
est = mean(w)
percRE = std(w)/sqrt(N1)/est*100
```

In the next example we apply the CE method to the problem of estimating the probability of large portfolio losses in a popular financial model [40, 53]. For an alternative approach using an exponential change of measure; see [38].

■ EXAMPLE 10.13 (Credit Risk in a Normal Copula Model)

Consider a portfolio of loans consisting of n obligors each of whom has probability of default $p_i = \mathbb{P}(X_i \geqslant x_i)$, $i = 1, \ldots, n$ for some continuous latent variables $X_1, \ldots, X_n$. In other words, the i-th obligor defaults if and only if the latent variable X_i exceeds the threshold x_i. The threshold x_i is sometimes called a **default boundary**. The total loss incurred from defaults can be written as

$$L(\mathbf{X}) = \sum_{i=1}^{n} c_i \, \mathrm{I}_{\{X_i \geqslant x_i\}} , \quad \mathbf{X} = (X_1, \ldots, X_n)^\top ,$$

where each c_i is the monetary loss incurred from the default of the i-th obligor. We wish to estimate the probability of a large loss: $\ell = \mathbb{P}(L(\mathbf{X}) \geqslant \gamma)$, $\gamma \in (0, \sum_i c_i)$, where $\mathbf{X}$ is specified by the linear factor model

$$X_i = a_i \, \breve{Z}_i + \sum_{j=1}^{m} a_{ij} \, Z_j, \quad i = 1, \ldots, n ,$$

☞ 68 with $Z_1, \ldots, Z_m, \breve{Z}_1, \ldots, \breve{Z}_n \sim_{\text{iid}} \mathsf{N}(0,1)$ and $a_i^2 + \sum_{j=1}^{m} a_{ij}^2 = 1$, so that $X_i \sim \mathsf{N}(0,1)$, $i = 1, \ldots, n$. This is the *normal copula model* described in [40, 53]. For the extended case of *Student's t-copula model* see [12, 22].

Let $f(\mathbf{z}) \propto \mathrm{e}^{-\mathbf{z}^\top \mathbf{z}/2}$ be the joint density of $\mathbf{Z} = (Z_1, \ldots, Z_m, \breve{Z}_1, \ldots, \breve{Z}_n)^\top$. Then $\mathbf{X} = B\mathbf{Z}$ for some $n \times (m+n)$ matrix B depending on the $\{a_{ij}\}$. Thus ℓ can be written as $\ell = \mathbb{P}(S(\mathbf{Z}) \geqslant \gamma)$, with $S(\mathbf{Z}) = L(B\mathbf{Z})$, which can be estimated via importance sampling on $\mathbf{Z}$. Note that the minimum variance importance sampling density $g^*(\mathbf{z}) = f(\mathbf{z}) \, \mathrm{I}_{\{S(\mathbf{z}) \geqslant \gamma\}}/\ell$ is a truncated multivariate normal density. We ☞ 242 use the hit-and-run sampler to generate from g^*.

Algorithm 10.10 (Hit-and-Run) *Initialize* $t = 1$. *Given the* $(m+n) \times 1$ *vector* $\mathbf{Y}_t$ *such that* $S(\mathbf{Y}_t) \geqslant \gamma$, *iterate the following steps* N *times.*

1. *Generate* $\mathbf{d} = \left(\frac{Z_1}{\|\mathbf{Z}\|}, \ldots, \frac{Z_{m+n}}{\|\mathbf{Z}\|}\right)^\top$, $\quad Z_1, \ldots, Z_{n+m} \overset{\text{iid}}{\sim} \mathsf{N}(0,1)$.

2. *Generate* $\Lambda \sim \mathsf{N}(-\mathbf{d}^\top \mathbf{Y}_t, 1)$.

3. *If* $S(\mathbf{Y}_t + \Lambda \mathbf{d}) \geqslant \gamma$, *set* $\mathbf{Y}_{t+1} = \mathbf{Y}_t + \Lambda \mathbf{d}$; *otherwise, set* $\mathbf{Y}_{t+1} = \mathbf{Y}_t$. *Set* $t = t + 1$.

We apply the CE method to seek the optimal change of measure in the family of multivariate normal densities with mean $\boldsymbol{\mu} = (\mu_1, \ldots, \mu_{m+n})$ and covariance matrix with diagonal $\boldsymbol{\sigma}^2 = (\sigma_1^2, \ldots, \sigma_{m+n}^2)$ and zeros off the diagonal. In other words, the importance sampling density is

$$f(\mathbf{z}; \widehat{\mathbf{v}}^*) \propto \prod_{i=1}^{m+n} \mathrm{e}^{-\frac{(z_i - \widehat{\mu}_i)^2}{2\widehat{\sigma}_i^2}},$$

where (see Step 2 of Algorithm 10.8) the parameters $\widehat{\boldsymbol{\mu}}$ and $\widehat{\boldsymbol{\sigma}}^2$ are simply the sample mean and sample variance of the data, $\mathbf{Y}_1, \ldots, \mathbf{Y}_N$, generated by the hit-and-run sampler.

As a particular numerical example, consider the $m = 21$ factor model with $n = 10^3$ obligors given in [38]:

- The probabilities of default are given by $p_k = 0.01(1 + \sin(16\pi k/n))$, $k = 1, \ldots, n$.

- The monetary loss c_k increases linearly from 1 to 100 as k increases from 1 to 10^3.

- The factors $\{a_{ij}\}$ are the entries of the following $10^3 \times 21$ matrix A with block structure:

$$A = \begin{pmatrix} \mathbf{r} & \begin{bmatrix} \mathbf{f} & & \\ & \ddots & \\ & & \mathbf{f} \end{bmatrix} & \begin{matrix} G \\ \vdots \\ G \end{matrix} \end{pmatrix}, \quad \text{with} \quad G = \begin{pmatrix} \mathbf{g} & & \\ & \ddots & \\ & & \mathbf{g} \end{pmatrix},$$

  where $\mathbf{r}$ is a column vector of 1000 entries, all equal to 0.8; $\mathbf{f}$ is a column vector of 100 entries, all equal to 0.4; G is a 100×10 matrix with $\mathbf{g}$ a column vector of 10 entries, all equal to 0.4.

- The (positive) factors $\{a_i\}$ are computed from $a_i = \sqrt{1 - \sum_{j=1}^m a_{ij}^2}$.

- Set $\gamma = 4 \times 10^4$, and use sample sizes $N = 10^5$ and $N_1 = 10^5$.

The code below implements this example with the following additional modification. After running the hit-and-run sampler and computing the parameters $\widehat{\boldsymbol{\mu}}$ and $\widehat{\boldsymbol{\sigma}}^2$ we decide which components of $\mathbf{Z}$ are important, that is, which components require a change of measure. If $\Phi(\widehat{\mu}_i/\widehat{\sigma}_i) > 0.95$, where Φ is the cdf of the standard normal density, then at the 5% level the i-th component is important. Otherwise, if $\Phi(\widehat{\mu}_i/\widehat{\sigma}_i) < 0.95$, then the i-th component is simulated under the original probability measure. This is essentially the **screening** idea of Lieber, Rubinstein, and Elmakis [54, 60], in which importance sampling is applied only to a subset of the components $\mathbf{z}$ with the goal of making the likelihood ratio of the importance sampling estimator more stable.

Using this approach it turns out that only the first component of $\widehat{\boldsymbol{\mu}}$ is important and $\widehat{\mu}_1 \approx 3.9109$. The other $m+n-1$ components did not warrant a change of measure.

```
%loss_probab.m
clear all
clc
% set up parameters
n=10^3; gamma=4*10^4;m=21;
k=1:n;
p=0.01*(1+sin(16*pi*k/n));
x=norminv(1-p);
c=1:99/(n-1):100;
% set up matrix for factor design
R=ones(n,1)*0.8;
G=zeros(100,10);
for j=1:10
    G(1+(j-1)*10:10*j,j)=0.4;
end
FF=zeros(1000,10);
for j=1:10
    FF(1+(j-1)*100:100*j,j)=0.4;
end
A=[R,FF,repmat(G,10,1)];
a=sqrt(1-sum(A.^2,2));
% finish setting up matrix for factor design

% run the Hit-and-Run sampler
Y=randn(1,m+n)+4; % find a starting point
mu=0;mu2=0;N=10^5;
for i=1:N
    d=randn(1,m+n); d=d/norm(d);
    lam=-d*Y'+randn;
    Y_new=Y+lam*d; % make proposal
    if score(Y_new(1:m),Y_new(m+1:m+n),x,c,A,a)>gamma
        Y=Y_new;  % accept or reject proposal
    end
    mu=mu+Y/N; % estimate CE parameters
    mu2=mu2+Y.^2/N;
end
sig=sqrt(mu2-mu.^2);
stem(mu(1:m))
% identify the important changes of measure
p_value=1-normcdf(abs(mu)./sig);
```

```
idx=find(p_value<0.05);

%now apply importance sampling
N1=10^5;
W=zeros(N1,1); % set up likelihood ratio
for i=1:N1
    Y=randn(1,m+n); Y(idx)=Y(idx).*sig(idx)+mu(idx);
    if score(Y(1:m),Y(m+1:m+n),x,c,A,a)>gamma
        W(i)=prod(sig(idx))*...
          exp(sum((Y(idx)-mu(idx)).^2/2./sig(idx).^2-Y(idx).^2/2));
    end
end
ell=mean(W)
std(W)/sqrt(N1)/ell
Reduction_factor=(ell*(1-ell))/var(W)
```

```
function S=score(Z,Zb,x,c,A,a)
% implements the portfolio loss function
X=A*Z'+a.*Zb';
S=c*(X>x');
```

With the above code we obtain a typical estimate of $\widehat{\ell} = 7.43 \times 10^{-5}$ with an estimated relative error of 0.8%. The variance reduction over CMC is about a factor of 10^3, which is comparable with the performance of the exponential twisting used by Glasserman and Li [38]. Note that while Glasserman and Li derive the optimal exponential change of measure from theoretical arguments, here we have learned the optimal importance sampling by means of MCMC simulation.

10.6 SPLITTING METHOD

One of the first Monte Carlo techniques for rare-event probability estimation is the **splitting method**, proposed by Kahn and Harris [48] and later by Hammersley and Handscomb [41]. In the splitting technique, sample paths of a Markov process are split into multiple copies at various stages of the simulation, with the objective of generating more occurrences of the rare event. The method uses a decomposition of the state space into nested subsets so that the rare event is represented as the intersection of a nested sequence of events. Then, the probability of the rare event is the product of conditional probabilities, each of which can be estimated much more accurately than the rare-event probability itself.

A basic description of the classical splitting method is as follows. Consider a Markov process $\{\mathbf{X}_u,\ u \geqslant 0\}$ with state space $\mathscr{X} \subseteq \mathbb{R}^n$, and let S be a real-valued function on $\mathscr{X}$, referred to as the **importance function**. Assume for definiteness that $S(\mathbf{X}_0) = 0$. For any *threshold* or *level* $\gamma > 0$, let U_γ denote the first time that the process $\{S(\mathbf{X}_u),\ u \geqslant 0\}$ hits the set $[\gamma, \infty)$, and let U_0 denote the first time after 0 that the same process hits the set $(-\infty, 0]$. We assume that U_γ and

U_0 are well-defined finite stopping times with respect to the history of $\{\mathbf{X}_u\}$. One is then interested in the probability, ℓ, of the event $E_\gamma = \{U_\gamma < U_0\}$; that is, the probability that $\{S(\mathbf{X}_u)\}$ up-crosses level γ before it down-crosses level 0. Note that ℓ depends on the distribution of $\mathbf{X}_0$.

The splitting method [31, 34] is based on the observation that if $\gamma_2 > \gamma_1$, then $E_{\gamma_2} \subset E_{\gamma_1}$. Therefore, we have that $\ell = c_1 c_2$, with $c_1 = \mathbb{P}(E_{\gamma_1})$ and
☞ 616
$c_2 = \mathbb{P}(E_{\gamma_2} \mid E_{\gamma_1})$ by the product rule of probability. In many cases, estimation of $c_1 c_2$ by estimating c_1 and c_2 separately is more efficient than the direct crude Monte Carlo (CMC) estimation of ℓ. Moreover, the same argument may be used when the interval $[0, \gamma]$ is subdivided into *multiple* subintervals $[\gamma_0, \gamma_1), [\gamma_1, \gamma_2), \ldots,$ $[\gamma_{T-1}, \gamma_T]$, where $0 = \gamma_0 < \gamma_1 < \cdots < \gamma_T = \gamma$. Again, let E_{γ_t} denote the event that the process $\{S(\mathbf{X}_u)\}$ reaches level γ_t before down-crossing level 0. Since $E_{\gamma_0} \supseteq E_{\gamma_1} \supseteq \cdots \supseteq E_{\gamma_T}$ is a nested sequence of events, denoting $c_t = \mathbb{P}(E_{\gamma_t} \mid E_{\gamma_{t-1}})$, we have $\ell = \prod_{t=1}^T c_t$.

The estimation of each c_t is performed in the following way. At stage $t = 1$ we run $s_1 N_0$ (a fixed number) independent copies of $\{\mathbf{X}_u\}$ and evolve the corresponding process $\{S(\mathbf{X}_u)\}$. Each copy of $\{\mathbf{X}_u\}$ is evolved until $\{S(\mathbf{X}_u)\}$ either hits the set $(-\infty, 0]$ or up-crosses the level γ_1; that is, each copy is evolved for a time period equal to $\min\{U_{\gamma_1}, U_0\}$. The number s_1 is an integer referred to as the **splitting factor** at stage $t = 1$. Define I_j^1 to be the indicator that the j-th copy of $\{S(\mathbf{X}_u)\}$ hits the set $[\gamma_1, \infty)$ before $(-\infty, 0]$, $j = 1, \ldots, s_1 N_0$, and let N_1 be the total number of copies that up-cross γ_1; that is,

$$N_1 = \sum_{j=1}^{s_1 N_0} I_j^1 \,.$$

An unbiased estimate for c_1 is $\widehat{c}_1 = \frac{N_1}{s_1 N_0}$. For every realization of $\{S(\mathbf{X}_u)\}$ which up-crosses γ_1, we store the corresponding state $\mathbf{X}_u$ at the time u of crossing in memory. Such a state is referred to as the **entrance state** [30]. In the next stage, when $t = 2$, we start s_2 new independent copies of the chain $\{\mathbf{X}_u\}$ from each of the N_1 entrance states, giving a total of $s_2 N_1$ new chains. Again, if we let I_j^2 indicate whether the j-th copy of $\{S(\mathbf{X}_u)\}$ hits the set $[\gamma_2, \infty)$ before $(-\infty, 0]$ (with the process $\{\mathbf{X}_u\}$ starting from an entrance state at level γ_1), then $\widehat{c}_2 = \frac{N_2}{s_2 N_1}$, where $N_2 = \sum_{j=1}^{s_2 N_1} I_j^2$, is an estimate of c_2. This process is repeated for each subsequent $t = 3, \ldots, T$, such that $s_t N_{t-1}$ is the **simulation effort** at stage t, and N_t is the number of entrance states at stage t. The indicators $\{I_j^t\}$ at stage t are usually dependent, and hence the success probabilities $\{\mathbb{P}(I_j^t = 1)\}$ depend on the entrance state from which a copy of the chain $\{\mathbf{X}_u\}$ is started. It is well known [7, 34] that despite this dependence, the estimator

$$\widehat{\ell} = \prod_{t=1}^{T} \widehat{c}_t = \frac{N_T}{N_0} \prod_{t=1}^{T} s_t^{-1} \tag{10.26}$$

is unbiased.

The idea of the splitting method is illustrated in Figure 10.4, where three level sets $\{\mathbf{x} : S(\mathbf{x}) = \gamma_t\}$, $t = 0, 1, 2$ are plotted. Here three independent paths of the process $\{S(\mathbf{X}_u)\}$ are started from level $\gamma_0 = 0$. Two of these paths *die out* by down-crossing level 0 and one of the paths up-crosses level γ_1. Three new independent copies of the chain are started from the entrance state at level γ_1 (encircled

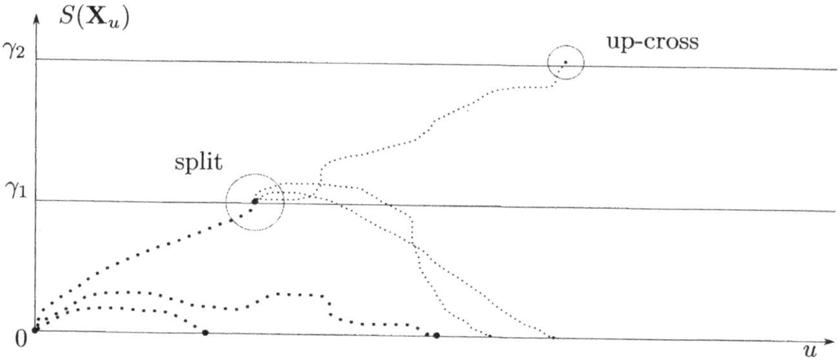

Figure 10.4 Typical evolution of the splitting of $\{S(\mathbf{X}_u)\}$.

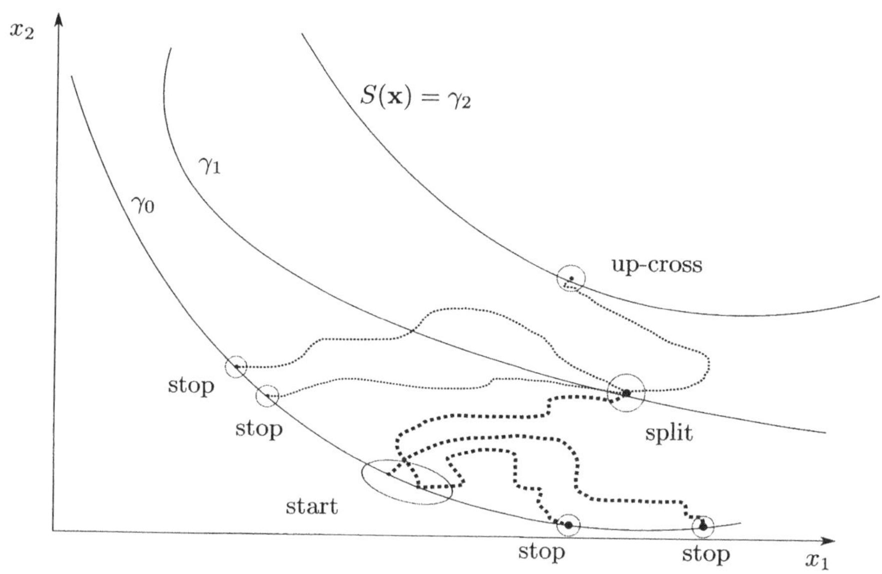

Figure 10.5 Typical evolution of the splitting algorithm for a two-dimensional Markov process $\{(X_u^{(1)}, X_u^{(2)}), u \geqslant 0\}$.

on the graph), two of these copies down-cross 0, but one copy up-crosses level γ_2. Figure 10.5 shows a typical realization of a two-dimensional Markov process $\{(X_u^{(1)}, X_u^{(2)}), u \geqslant 0\}$ that corresponds to the scenario described on Figure 10.4.

Of great importance is the choice of the importance function S, which determines how the sample space is partitioned into nested subsets. The problem of selecting an efficient importance function is similar to the problem of selecting an optimal change of measure for importance sampling. Hence, it is not surprising that it is an unresolved problem in general [32, 58]. Dean and Dupuis [25] suggest an efficient design of the importance function derived from the solution of a nonlinear partial differential equation, for cases where large deviation approximations are available.

For a given importance function S, the efficiency of the splitting method depends crucially on the number of levels T, the choice of the intermediate levels $\gamma_1, \ldots, \gamma_{T-1}$, and the splitting factors $s_1, \ldots, s_T$. Ideally one would select the levels so that the conditional probabilities $\{c_t\}$ are not too small and easily estimated via CMC. Assuming that the cost of running the Markov process is independent of t, the total simulation effort is a random variable with expected value

$$\begin{aligned}\sum_{t=1}^{T} s_t \mathbb{E} N_{t-1} &= N_0 \sum_{t=1}^{T} s_t\, \ell(\gamma_{t-1}) \prod_{j=1}^{t-1} s_j = N_0 \sum_{t=1}^{T} s_t \prod_{j=1}^{t-1} c_j\, s_j \\ &= N_0 \sum_{t=1}^{T} \frac{1}{c_t} \prod_{j=1}^{t} c_j\, s_j \,.\end{aligned} \tag{10.27}$$

An inappropriate choice of the splitting factors may lead to an exponential growth of the simulation effort. For example, if $c_j s_j = a > 1$ for all j, then the simulation effort (10.27) grows exponentially in T. This is referred to in the splitting literature as an **explosion** [35]. On the other hand, if $c_j s_j = a < 1$ for all j, then $\mathbb{E}N_T = N_0\, a^T$ decays exponentially, and with high probability N_T and $\widehat{\ell}$ will be 0, making the algorithm inefficient. Thus, it is desirable that $c_j s_j = 1$ for all j; that is, the splitting is at the **critical value** [35]. In practice, one obtains rough estimates $\{\varrho_j\}$ of $\{c_j\}$ via a pilot run and then initializes $s_t = \varrho_t^{-1}$ paths from each entrance state $j = 1, \ldots, N_t$, at every stage t. In the case where $1/\varrho_j$ is not an integer, one can generate a Bernoulli random variable with success probability $\varrho_j^{-1} - \lfloor \varrho_j^{-1} \rfloor$ and then add it to $\lfloor \varrho_j^{-1} \rfloor$ to obtain a random integer-valued splitting factor S_j with expected value $1/\varrho_j$, see [35]. This version of the splitting algorithm is called the fixed splitting implementation, because at every stage t one generates a fixed expected number of copies ϱ_t^{-1} from each entrance state. An alternative to the *fixed splitting* implementation is the **fixed effort** implementation, where the simulation effort is fixed to N at each stage, instead of fixing the number of copies. The estimator is then

$$\widehat{\ell}_{\mathrm{FE}} = \prod_{t=1}^{T} \frac{N_t}{N} \,.$$

The fixed effort implementation prevents explosions in the number of total Markov chain copies, but has the disadvantage that it is more difficult to analyze the variance of $\widehat{\ell}_{\mathrm{FE}}$; see [31].

Other implementations of the splitting method include **fixed success** splitting [33], where the number of entrance states at each stage is maintained to be a predetermined number of trajectories; and **fixed probability of success** splitting [20, 30], where the conditional probabilities $\{c_t\}$ are (approximately) equal under the proposed simulation strategy. We now summarize the fixed effort splitting [31].

Algorithm 10.11 (Fixed Effort Splitting) *Set the counter $t = 1$. Given the importance function S and the levels $\gamma_1, \ldots, \gamma_T$, execute the following steps.*

1. **Initialization.** *Generate N copies of the Markov process $\{\mathbf{X}_u\}$ where each copy is run until $\{S(\mathbf{X}_u)\}$ either hits the set $(-\infty, 0]$ or up-crosses the level γ_1. If I_j^1 is the indicator that the j-th copy of $\{S(\mathbf{X}_u)\}$ hits the set $[\gamma_1, \infty)$ before $(-\infty, 0]$, then $N_1 = \sum_{j=1}^{N} I_j^1$ is the total number of copies that up-cross γ_1. Store the entrance state for each of the N_1 copies in memory.*

2. **Bootstrap Resampling.** *Resample the* $N_t(\leqslant N)$ *entrance states uniformly to obtain a new population of* N *entrance states (with possibly repeated values).*

3. **Markov Chain Evolution.** *Start* N *independent copies of the Markov process* $\{\mathbf{X}_u\}$ *from each of the* N *bootstrapped entrance states from Step 2. Run each copy until* $\{S(\mathbf{X}_u)\}$ *either hits the set* $(-\infty, 0]$ *or up-crosses the level* γ_t. *If* I_j^t *is the indicator that the* j*-th copy of* $\{S(\mathbf{X}_u)\}$ *hits the set* $[\gamma_t, \infty)$ *before* $(-\infty, 0]$, *then* $N_t = \sum_{j=1}^{N} I_j^t$ *is the total number of copies that up-cross* γ_t. *Store the entrance states for these* N_t *copies in memory.*

4. **Stopping Condition.** *If* $t = T$ *go to Step 5. If* $N_t = 0$, *set* $N_{t+1} = N_{t+2} = \cdots = N_T = 0$ *and go to Step 5; otherwise, set* $t = t + 1$ *and repeat from Step 2.*

5. **Final Estimator.** *Deliver the estimator* $\widehat{\ell}_{\mathrm{FE}} = \prod_{t=1}^{T} N_t/N$.

An analysis of the fixed effort splitting method in an idealized asymptotic setting [31] suggests that good values for the splitting parameters are $T \approx -\ln(\ell)/2$ with thresholds $\{\gamma_t\}$ chosen such that $c_1 = \cdots = c_T \approx \mathrm{e}^{-2}$. Such a choice will (approximately) give a variance of $\mathrm{e}^2\ell^2(\ln \ell)^2/4$ per single simulation trial. For more elaborate results in this direction, including central limit results, see [21, 35] and the summary in [58].

■ **EXAMPLE 10.14 (Hitting Probability)**

Suppose the position of the particle in the plane is described by the process $\{(X_t, Y_t), t \geqslant 0\}$, where $\{X_t\}$ and $\{Y_t\}$ are independent copies of the Ornstein–Uhlenbeck SDE (see [7]): ☞ 199

$$\mathrm{d}Z_t = -Z_t\,\mathrm{d}t + \mathrm{d}W_t, \quad Z_0 = 1\,.$$

We are interested in computing the probability that the particle hits the quarter circle,

$$\{(x, y) :\ x > 0, y > 0,\ x^2 + y^2 = 25\}\,,$$

before the x- or y-axis; see Figure 10.6.

To apply the splitting method, we first need to specify an importance function. A natural, but not necessarily optimal, choice is

$$S(x, y) = \begin{cases} \sqrt{x^2 + y^2} & \text{if } x > 0 \text{ and } y > 0, \\ 0, & \text{if } x \leqslant 0 \text{ or } y \leqslant 0. \end{cases}$$

The levels

$$0 = \gamma_0 < \gamma_1 < \gamma_2 < \cdots < \gamma_T = 5$$

can then be interpreted as the increasing radii of a nested set of quarter circles. In other words, we aim to hit a circle with smaller radius before attempting to hit a circle with larger radius. These intermediate circles serve as stepping stones toward the rare-event set.

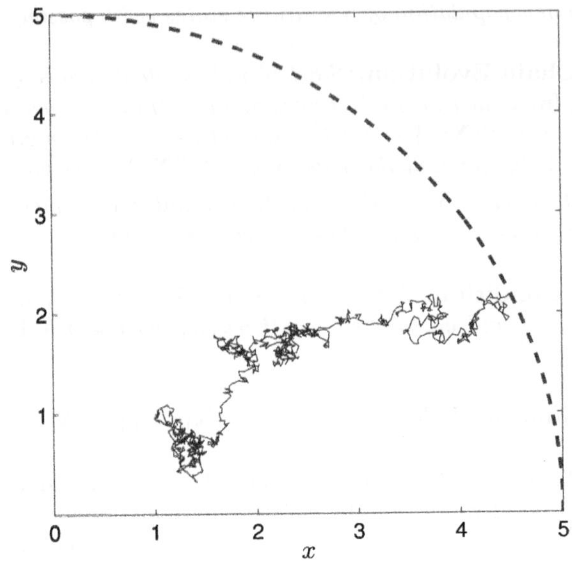

Figure 10.6 A realization of the Ornstein–Uhlenbeck process hitting the quarter circle before the x- or y-axis.

The code below uses fixed effort splitting with $N = 10^4$ and estimates the relative error from 10 independent runs. We use the levels $(\gamma_1, \gamma_2, \gamma_3, \gamma_4, \gamma_5, \gamma_6) = (3,\ 3.5,\ 4,\ 4.5,\ 4.7,\ 5)$. The function ou_split.m implements Algorithm 5.25 with step size of $h = 0.01$ and determines if the Ornstein–Uhlenbeck process has hit a quarter circle with radius γ_t. We obtained the estimate 5.6026×10^{-10} with an estimated relative error of 0.049. Figure 10.6 shows a path of the process conditional on hitting the quarter circle before the axis.

```
%OU_process_splitting.m
clear all,clc
gam=[3:0.5:4.5,4.7,5]; %splitting levels
N=10^4;
for iter=1:10 % repeat 10 independent times to estimate RE
    x_ini=1;y_ini=1;
    data=repmat([x_ini,y_ini],N,1);
    for t=1:length(gam)
        %resample the paths
        data=data(ceil(rand(N,1)*size(data,1)),:);
        elite=[];
        for i=1:N
            [indicator,x,y]=ou_split(gam(t),data(i,1),data(i,2));
            if indicator
                elite=[elite;x(end),y(end)];%store successful hits
            end
```

```
        end
        t
        c(t)=size(elite,1)/N; % conditional probability estimate
        data=elite;
    end
    ell(iter)=prod(c);
end
mean(ell)
std(ell)/sqrt(10)/mean(ell)
```

```
function [indicator,x,y,tau]=ou_split(gam,x_ini,y_ini)
% implements the exact sampling of an OU process
h=0.001; % step size
% OU updating formula for exact simulation
f=@(x,z)(exp(-h)*x+sqrt((1-exp(-2*h))/2)*z);
x(1)=x_ini; y(1)=y_ini;
tau_axis=inf; tau_circ=inf;
for i=2:10^7 % choose an arbitrarily large loop to ensure hitting
    x(i)=f(x(i-1),randn);
    y(i)=f(y(i-1),randn);
    if (x(i-1)*x(i)<0)|(y(i-1)*y(i)<0)
        tau_axis=h*i; % determine axis hitting time
    end
    if (x(i-1)^2+y(i-1)^2<gam^2)&(x(i)^2+y(i)^2>gam^2)
        tau_circ=h*i; % determine circle hitting time
    end
    tau=min(tau_circ,tau_axis);
    if isinf(tau)==0
        break
    end
end
indicator=tau_circ<tau_axis; %did we hit the circle first?
```

Further Reading

A good starting point on rare-event simulation is [7, Chapter VI], and a recent volume on many of the topics and applications of rare-event simulation is [58]. See [59] for the cross-entropy method with applications to rare-event simulation. Glasserman and Kou [37] give the first example where exponential twisting does not yield asymptotically optimal estimators. Dupuis, Sezer, and Wang [26] provide the first asymptotically optimal importance sampling for total population overflow in tandem networks; this is further generalized to general Jackson networks in Dupuis and Wang [27]. Anantharam et al. [2] show the connection between the overflow probabilities of a queueing process and its time-reversed process; see also Juneja and Nicola [46].

For an overview of splitting for rare event simulation, see [30, 36]. Various applications of the splitting method include: particle transmission [48], queueing systems [31, 32, 64, 65, 66], and reliability [19, 58]. For theoretical results about the optimal selection of the splitting levels, see [20]. A variant of the splitting method that uses quasi Monte Carlo estimators is given in [51]. Various strategies for the truncation and splitting of the Markov chain are described in [50, 52].

REFERENCES

1. R. J. Adler, R. E. Feldman, and M. S. Taqqu. *A Practical Guide to Heavy Tails: Statistical Techniques and Applications.* Birkhäuser, New York, 1998.
2. V. Anantharam, P. Heidelberger, and P. Tsoucas. Analysis of rare events in continuous time Markov chains via time reversal and fluid approximation. Technical Report RC 16280, IBM, Yorktown Heights, New York, 1990.
3. S. Asmussen. *Applied Probability and Queues.* Springer-Verlag, New York, second edition, 2003.
4. S. Asmussen and H. Albrecher. *Ruin Probabilities.* World Scientific Publishing, River Edge, NJ, 2010.
5. S. Asmussen and K. Binswanger. Simulation of ruin probabilities for subexponential claims. *ASTIN Bulletin*, 27(2):297–318, 1997.
6. S. Asmussen, K. Binswanger, and B. Højgaard. Rare events simulation for heavy-tailed distributions. *Bernoulli*, 6(2):303–322, 2000.
7. S. Asmussen and P. W. Glynn. *Stochastic Simulation: Algorithms and Analysis.* Springer-Verlag, New York, 2007.
8. S. Asmussen and D. P. Kroese. Improved algorithm for rare event simulation with heavy tails. *Advances in Applied Probability*, 38(2):545–558, 2006.
9. S. Asmussen and L. Rojas-Nandayapa. Asymptotics of sums of lognormal random variables with Gaussian copula. *Statistics & Probability Letters*, 78(16):2709–2714, 2008.
10. S. Asmussen and R. Y. Rubinstein. Steady state rare events simulation in queuing models and its complexity properties. In J. H. Dshalalow, editor, *Advances in Queuing: Theory, Methods, and Open Problems*, pages 429–462. CRC Press, Boca Raton, FL, 1995.
11. A. Bassamboo, S. Juneja, and A. Zeevi. On the efficiency loss of state-independent importance sampling in the presence of heavy-tails. *Operations Research Letters*, 35(2):251–260, 2007.
12. A. Bassamboo, S. Juneja, and A. Zeevi. Portfolio credit risk with extremal dependence: Asymptotic analysis and efficient simulation. *Operations Research*, 56(3):593–606, 2008.
13. J. Blanchet and P. W. Glynn. Strongly efficient estimators for light-tailed sums. *Proceedings of the 1st International Conference on Performance Evaluation Methodologies and Tools (ValueTools), Pisa*, 180 of ACM International Conference Proceedings Series:article 18, 2006.
14. J. Blanchet and P. W. Glynn. Efficient rare-event simulation for the maximum of heavy-tailed random walks. *The Annals of Applied Probability*, 18(4):1351–1378, 2008.

15. J. Blanchet, P. W. Glynn, P. L'Ecuyer, W. Sandmann, and B. Tuffin. Asymptotic robustness of estimators in rare-event simulation. *Proceedings of the 2007 INFORMS Simulation Society Research Workshop, Fontainebleau, France*, 2007.

16. J. Blanchet and J. C. Liu. Path-sampling for state-dependent importance sampling. *Proceedings of the 2007 Winter Simulation Conference, Washington, DC*, pages 380–388, 2007.

17. J. A. Bucklew, P. Ney, and J. S. Sadowsky. Monte Carlo simulation and large deviations theory for uniformly recurrent Markov chains. *Journal of Applied Probability*, 27(1):44–59, 1990.

18. H. Cancela and M. El Khadiri. The recursive variance-reduction simulation algorithm for network reliability evaluation. *IEEE Transactions on Reliability*, 52(2):207–212, 2003.

19. H. Cancela, L. Murray, and G. Rubino. Splitting in source-terminal network reliability estimation. *Proceedings of the 7th International Workshop on Rare Event Simulation (RESIM 2008, France)*, 2008.

20. F. Cérou and A. Guyader. Adaptive multilevel splitting for rare event analysis. *Stochastic Analysis and Applications*, 25(2):417–443, 2007.

21. F. Cérou, P. Del Moral, F. Le Gland, and P. Lezaud. Limit theorems for the multilevel splitting algorithm in the simulation of rare events. *Proceedings of the 2005 Winter Simulation Conference, Orlando*, pages 682–691, 2005.

22. J. C. C. Chan and D. P. Kroese. Efficient estimation of large portfolio loss probabilities in t-copula models. *European Journal of Operational Research*, 205(2):361–367, 2010.

23. J. C. C. Chan and D. P. Kroese. Rare-event probability estimation with conditional Monte Carlo. *Annals of Operations Research*, 2010. DOI:10.1007/s10479-009-0539-y.

24. M. G. Cruz. *Modeling, Measuring and Hedging Operation Risk*. John Wiley & Sons, Chichester, 2003.

25. T. Dean and P. Dupuis. Splitting for rare event simulation: A large deviations approach to design and analysis. *Stochastic Processes and Their Applications*, 119(2):562–587, 2008.

26. P. Dupuis, A. D. Sezer, and H. Wang. Dynamic importance sampling for queueing networks. *Annals of Applied Probability*, 17(4):1306–1346, 2007.

27. P. Dupuis and H. Wang. Importance sampling for Jackson networks. *Queueing Systems*, 62(1-2):113–157, 2009.

28. P. Embrechts, C. Klüppelberg, and T. Mikosch. *Modelling Extremal Events for Insurance and Finance*. Springer-Verlag, Berlin, 1997.

29. A. Frachot, O. Moudoulaud, and T. Roncalli. Loss distribution approach in practice. In M. K. Ong, editor, *The Basel Handbook: A Guide for Financial Practitioners*. Risk Books, London, 2004.

30. M. J. J. Garvels. *The Splitting Method in Rare Event Simulation*. PhD thesis, University of Twente, 2000.

31. M. J. J. Garvels and D. P. Kroese. A comparison of RESTART implementations. In *Proceedings of the 1998 Winter Simulation Conference*, pages 601–609, Washington, DC, 1998.

32. M. J. J. Garvels, D. P. Kroese, and J. C. W. van Ommeren. On the importance function in splitting simulation. *European Transactions on Telecommunications*, 13(4):363–371, 2002.

33. F. Le Gland and N. Oudjane. A sequential algorithm that keeps the particle system alive. In H. Blom and J. Lygeros, editors, *Stochastic Hybrid Systems: Theory*

and Safety Critical Applications, Lecture Notes in Control and Information Sciences, volume 337, pages 351–389. Springer-Verlag, New York, 2006.

34. P. Glasserman, P. Heidelberger, P. Shahabuddin, and T. Zajic. A look at multilevel splitting. In H. Niederreiter, editor, *Monte Carlo and Quasi Monte Carlo Methods 1996, Lecture Notes in Statistics*, volume 127, pages 99–108. Springer-Verlag, New York, 1996.
35. P. Glasserman, P. Heidelberger, P. Shahabuddin, and T. Zajic. A large deviations perspective on the efficiency of multilevel splitting. *IEEE Transactions on Automatic Control*, 43(12):1666–1679, 1998.
36. P. Glasserman, P. Heidelberger, P. Shahabuddin, and T. Zajic. Multilevel splitting for estimating rare event probabilities. *Operations Research*, 47(4):585–600, 1999.
37. P. Glasserman and S.-G. Kou. Limits of first passage times to rare sets in regenerative processes. *Annals of Applied Probability*, 5(2):424–445, 1995.
38. P. Glasserman and J. Li. Importance sampling for portfolio credit risk. *Management Science*, 51(11):1643–1656, 2005.
39. P. W. Glynn and D. L. Iglehart. Importance sampling for stochastic simulations. *Management Science*, 35(11):1367–1392, 1989.
40. G. Gupton, C. Finger, and M. Bhatia. Creditmetrics technical document. Technical report, J. P. Morgan & Co., New York, 1997.
41. J. M. Hammersley and D. C. Handscomb. *Monte Carlo Methods*. Methuen, London, 1964.
42. J. Hartinger and D. Kortschak. On the efficiency of the Asmussen–Kroese-estimator and its application to stop-loss transforms. *Blätter der DGVFM*, 30(2):363–377, 2009.
43. P. Heidelberger. Fast simulation of rare events in queuing and reliability models. *ACM Transactions on Modeling and Computer Simulation*, 5(1):43–85, 1995.
44. J. L. Jensen. *Saddlepoint Approximations*. Clarendon Press, Oxford, 1995.
45. S. Juneja. Estimating tail probabilities of heavy tailed distributions with asymptotically zero relative error. *Queueing Systems*, 57(2-3):115–127, 2007.
46. S. Juneja and V. Nicola. Efficient simulation of buffer overflow probabilities in Jackson networks with feedback. *ACM Transactions on Modeling and Computer Simulation*, 15(4):281–315, 2005.
47. S. Juneja and P. Shahabuddin. Simulating heavy-tailed processes using delayed hazard rate twisting. *ACM Transactions on Modeling and Computer Simulation*, 12(2):94–118, 2002.
48. H. Kahn and T. E. Harris. *Estimation of Particle Transmission by Random Sampling*. National Bureau of Standards Applied Mathematics Series, 1951.
49. P. L'Ecuyer, J. H. Blanchet, B. Tuffin, and P. W. Glynn. Asymptotic robustness of estimators in rare-event simulation. *ACM Transactions on Modeling and Computer Simulation*, 20(1):Article 6, 2010.
50. P. L'Ecuyer, V. Demers, and B. Tuffin. Splitting for rare-event simulation. *Proceedings of the 2006 Winter Simulation Conference*, pages 137–148, 2006.
51. P. L'Ecuyer, V. Demers, and B. Tuffin. Rare events, splitting, and quasi-Monte Carlo. *ACM Transactions on Modeling and Computer Simulation*, 17(2):1–44, 2007.
52. P. L'Ecuyer and B. Tuffin. Splitting and weight windows to control the likelihood ratio in importance sampling. In L. Lenzini and R. Cruz, editors, *Proceedings of the 1st International Conference on Performance Evaluation Methodologies and Tools (ValueTools), Pisa*, 2006. Article 21.

53. D. Li. On default correlations: A copula function approach. *Journal of Fixed Income*, 9(4):43–54, 2000.

54. D. Lieber, R. Y. Rubinstein, and D. Elmakis. Quick estimation of rare events in stochastic networks. *IEEE Transactions on Reliability*, 46(2):254–265, 1997.

55. V. V. Petrov. On the probabilities of large deviations for sums of independent random variables. *Theory of Probability and Its Applications*, 10(2):287–298, 1965.

56. S. I. Resnick. Heavy tail modeling and telegraphic data. *Annals of Statistics*, 25(5):1805–1869, 1997.

57. L. R. Rojas-Nandayapa. *Risk Probabilities: Asymptotics and Simulation*. PhD thesis, University of Aarhus, 2008.

58. G. Rubino and B. Tuffin, editors. *Rare Event Simulation Using Monte Carlo Methods*. John Wiley & Sons, Chichester, 2009.

59. R. Y. Rubinstein and D. P. Kroese. *The Cross-Entropy Method: A Unified Approach to Combinatorial Optimization, Monte-Carlo Simulation and Machine Learning*. Springer-Verlag, New York, 2004.

60. R. Y. Rubinstein and D. P. Kroese. *Simulation and the Monte Carlo Method*. John Wiley & Sons, New York, second edition, 2007.

61. D. Siegmund. Importance sampling in the Monte Carlo study of sequential tests. *The Annals of Statistics*, 4(4):673–684, 1976.

62. K. Sigman. Appendix: A primer on heavy-tailed distributions. *Queueing Systems*, 33(1-3):261–275, 1999. DOI: 10.1023/A:1019180230133.

63. B. Tuffin. Bounded normal approximation in simulations of highly reliable Markovian systems. *Journal of Applied Probability*, 36(4):974–986, 1999.

64. M. Villén-Altamirano and J. Villén-Altamirano. RESTART: A method for accelerating rare event simulations. In J. W. Cohen and C. D. Pack, editors, *Proceedings of the 13th International Teletraffic Congress, Queueing, Performance and Control in ATM*, pages 71–76, 1991.

65. M. Villén-Altamirano and J. Villén-Altamirano. RESTART: A straightforward method for fast simulation of rare events. In J. D. Tew, S. Manivannan, D. A. Sadowski, and A. F. Seila, editors, *Proceedings of the 1994 Winter Simulation Conference*, pages 282–289, 1994.

66. M. Villén-Altamirano and J. Villén-Altamirano. About the efficiency of RESTART. In *Proceedings of the RESIM'99 Workshop*, pages 99–128. University of Twente, The Netherlands, 1999.

CHAPTER 11

ESTIMATION OF DERIVATIVES

In this chapter we discuss four methods for gradient estimation: the *finite difference method*, *infinitesimal perturbation analysis*, the *likelihood ratio* or *score function method*, and *weak derivatives*. In addition, we discuss gradient estimation for regenerative processes. The efficient estimation of derivatives is important in sensitivity analysis of simulation output and in stochastic or noisy optimization. Details on noisy optimization are given in Chapter 12. ☞ 441

11.1 GRADIENT ESTIMATION

It is often the case that the performance measure ℓ from a Monte Carlo simulation can be viewed as a function of various parameters used in the simulation. These parameters can pertain to the distributions used in the simulation and to the mechanism under which the simulation is carried out. A typical setting is where ℓ is the expected output of a random variable Y whose value is dependent on a simulation parameter vector $\boldsymbol{\theta}$, such that

$$\ell(\boldsymbol{\theta}) = \mathbb{E}Y = \mathbb{E}_{\boldsymbol{\theta}_2} H(\mathbf{X};\boldsymbol{\theta}_1) = \int H(\mathbf{x};\boldsymbol{\theta}_1) f(\mathbf{x};\boldsymbol{\theta}_2)\,\mathrm{d}\mathbf{x}\ , \tag{11.1}$$

where $\boldsymbol{\theta} = (\boldsymbol{\theta}_1, \boldsymbol{\theta}_2)$, $H(\cdot;\boldsymbol{\theta}_1)$ is a sample performance function, and $f(\cdot;\boldsymbol{\theta}_2)$ is a pdf (for the discrete case replace the integral with a sum). In this context we refer to the parameter $\boldsymbol{\theta}_1$ as a **structural** parameter and $\boldsymbol{\theta}_2$ as a **distributional** parameter. An estimation problem formulated with only distributional parameters can often be transformed into one with only structural parameters, and vice versa; see Remark 11.1.1. It should be noted, however, that not all performance measures are of the form (11.1). For example, $\ell(\boldsymbol{\theta})$ could be the median or a quantile of $Y \sim f(\mathbf{y};\boldsymbol{\theta})$.

In addition to estimating $\ell(\boldsymbol{\theta})$ it is often relevant to estimate various *derivatives* (gradients, Hessians, etc.) of $\ell(\boldsymbol{\theta})$. Two main applications are:

1. *Sensitivity analysis*: The gradient $\nabla\ell(\boldsymbol{\theta})$ indicates how sensitive the output $\ell(\boldsymbol{\theta})$ is to small changes in the input parameters $\boldsymbol{\theta}$, and can thus be used to identify its most significant components.

2. *Stochastic optimization and root finding*: Gradient estimation is closely related to optimization through the root-finding problem $\nabla\ell(\boldsymbol{\theta}) = \mathbf{0}$, as any solution of the latter is a stationary point of ℓ and hence a candidate for a local maximum or minimum. Estimating the gradient via simulation can therefore be used to approximately determine the optimal solution(s), leading
☞ 441 to gradient-based *noisy optimization* algorithms; see Sections 12.1–12.2.

Central to the discussion of gradient estimation is the interchange between differentiation and integration. The following theorem provides sufficient conditions. See also Asmussen and Glynn [2, Chapter VII].

Theorem 11.1.1 (Interchanging Differentiation and Integration) *Let the function $g(\mathbf{x};\boldsymbol{\theta})$ be differentiable at $\boldsymbol{\theta}_0 \in \mathbb{R}^k$. Denote the corresponding gradient by $\nabla_{\boldsymbol{\theta}}\, g(\mathbf{x};\boldsymbol{\theta}_0)$. We assume that as a function of $\mathbf{x}$ this gradient is integrable. If there exists a neighborhood Θ of $\boldsymbol{\theta}_0$ and an integrable function $M(\mathbf{x};\boldsymbol{\theta}_0)$ such that for all $\boldsymbol{\theta} \in \Theta$*

$$\frac{|g(\mathbf{x};\boldsymbol{\theta}) - g(\mathbf{x};\boldsymbol{\theta}_0)|}{\|\boldsymbol{\theta}-\boldsymbol{\theta}_0\|} \leqslant M(\mathbf{x};\boldsymbol{\theta}_0)\,, \tag{11.2}$$

then

$$\nabla_{\boldsymbol{\theta}} \int g(\mathbf{x};\boldsymbol{\theta})\,\mathrm{d}\mathbf{x}\,\Big|_{\boldsymbol{\theta}=\boldsymbol{\theta}_0} = \int \nabla_{\boldsymbol{\theta}}\, g(\mathbf{x};\boldsymbol{\theta}_0)\,\mathrm{d}\mathbf{x}\,. \tag{11.3}$$

Proof: Let

$$\psi(\mathbf{x};\boldsymbol{\theta},\boldsymbol{\theta}_0) = \frac{g(\mathbf{x};\boldsymbol{\theta}) - g(\mathbf{x};\boldsymbol{\theta}_0) - (\boldsymbol{\theta}-\boldsymbol{\theta}_0)^\top \nabla_{\boldsymbol{\theta}} g(\mathbf{x};\boldsymbol{\theta}_0)}{\|\boldsymbol{\theta}-\boldsymbol{\theta}_0\|}\,.$$

Condition (11.2) implies that $|\psi(\mathbf{x};\boldsymbol{\theta},\boldsymbol{\theta}_0)| \leqslant M(\mathbf{x};\boldsymbol{\theta}_0) + \|\nabla_{\boldsymbol{\theta}} g(\mathbf{x};\boldsymbol{\theta}_0)\|$ for all $\boldsymbol{\theta} \in \Theta$. Moreover, by the existence of the gradient at $\boldsymbol{\theta}_0$, we have that $\psi(\mathbf{x};\boldsymbol{\theta},\boldsymbol{\theta}_0) \to 0$ as
☞ 615 $\boldsymbol{\theta} \to \boldsymbol{\theta}_0$. Therefore, by the dominated convergence theorem, $\int \psi(\mathbf{x};\boldsymbol{\theta},\boldsymbol{\theta}_0)\,\mathrm{d}\mathbf{x} \to 0$ as $\boldsymbol{\theta} \to \boldsymbol{\theta}_0$, which shows that (11.3) must hold.

An important special case where differentiation and integration can be inter-
☞ 701 changed arises in the theory of (natural) exponential families as summarized by the following theorem. See, for example, [13, Section 2.7] and [3, Pages 32–34].

Theorem 11.1.2 (Interchange in Exponential Families) *For any function ϕ for which*

$$\int \phi(\mathbf{x})\, \mathrm{e}^{\boldsymbol{\eta}^\top \mathbf{t}(\mathbf{x})}\, \mathrm{d}\mathbf{x} < \infty ,$$

the integral as a function of $\boldsymbol{\eta}$ has partial derivatives of all orders, for all $\boldsymbol{\eta}$ in the interior of the natural parameter space. Moreover, these derivatives can be obtained by differentiating under the integral sign. That is, ☞ 701

$$\nabla_{\boldsymbol{\eta}} \int \phi(\mathbf{x})\, \mathrm{e}^{\boldsymbol{\eta}^\top \mathbf{t}(\mathbf{x})}\, \mathrm{d}\mathbf{x} = \int \phi(\mathbf{x})\, \mathbf{t}(\mathbf{x})\, \mathrm{e}^{\boldsymbol{\eta}^\top \mathbf{t}(\mathbf{x})}\, \mathrm{d}\mathbf{x} .$$

Remark 11.1.1 (Distributional and Structural Parameters) In many cases it is possible to switch from structural to distributional parameters and vice versa by making appropriate transformations. We discuss two common situations for the case where $\mathbf{x} = x$ is scalar (generalizations to the multidimensional case are straightforward).

1. *Push-out method:* Suppose the estimation problem involves only structural parameters; for example,

$$\ell(\boldsymbol{\theta}) = \int H(x;\boldsymbol{\theta})\, f(x)\, \mathrm{d}x . \tag{11.4}$$

The **push-out method** [17] involves a (problem-dependent) change of variable $y = a(x;\boldsymbol{\theta})$ such that the transformed integral has the form

$$\ell(\boldsymbol{\theta}) = \int L(y)\, g(y;\boldsymbol{\theta})\, \mathrm{d}y . \tag{11.5}$$

In other words, the parameter $\boldsymbol{\theta}$ is "pushed-out" into the pdf g.

As a simple example consider the estimation of $\ell(\theta) = \mathbb{E}\exp(-X^\theta)$, $\theta > 0$, where $X \sim f(x)$ is a positive random variable. This is of the form (11.4) with $H(x;\theta) = \exp(-x^\theta)$. Defining $y = x^\theta$, $L(y) = \exp(-y)$, and

$$g(y;\theta) = f\big(y^{\frac{1}{\theta}}\big)\frac{1}{\theta}\, y^{\frac{1}{\theta}-1} = f(x)\,\frac{1}{\theta}\, x^{1-\theta} ,$$

the structural problem (11.4) is transformed into a distributional one (11.5).

2. *Inverse-transform method:* The inverse-transform method for random variable generation can be used to convert distributional estimation problems into structural ones. In particular, suppose $\ell(\boldsymbol{\theta}) = \mathbb{E}_{\boldsymbol{\theta}} L(Y)$ is of the form (11.5), and $G(y;\boldsymbol{\theta})$ is the cdf of $Y \sim g(y;\boldsymbol{\theta})$. By the inverse-transform method we can write $Y = G^{-1}(X;\boldsymbol{\theta})$, where $X \sim \mathsf{U}(0,1)$. If we now define $H(X;\boldsymbol{\theta}) = L(G^{-1}(X;\boldsymbol{\theta}))$, then $\ell(\boldsymbol{\theta}) = \mathbb{E}H(X;\boldsymbol{\theta})$ is of the form (11.4) with $f(x) = \mathrm{I}_{\{0<x<1\}}$. ☞ 45 ☞ 45

11.2 FINITE DIFFERENCE METHOD

Let the performance measure $\ell(\boldsymbol{\theta})$ depend on a parameter $\boldsymbol{\theta} \in \mathbb{R}^d$. Suppose $\widehat{\ell}(\boldsymbol{\theta})$ is an estimator of $\ell(\boldsymbol{\theta})$ obtained via simulation. A straightforward estimator of the

i-th component of $\nabla \ell(\boldsymbol{\theta})$, that is, $\partial \ell(\boldsymbol{\theta})/\partial \theta_i$, is the **forward difference estimator**

$$\frac{\widehat{\ell}(\boldsymbol{\theta} + \mathbf{e}_i\,\delta) - \widehat{\ell}(\boldsymbol{\theta})}{\delta} ,$$

where $\mathbf{e}_i$ denotes the i-th unit vector in $\mathbb{R}^d$ and $\delta > 0$. An alternative is the **central difference estimator**

$$\frac{\widehat{\ell}(\boldsymbol{\theta} + \mathbf{e}_i\,\delta/2) - \widehat{\ell}(\boldsymbol{\theta} - \mathbf{e}_i\,\delta/2)}{\delta} .$$

In general, both estimators are biased. The bias of the forward difference estimator is of the order $\mathcal{O}(\delta)$, whereas the bias of the central difference estimator is of the order $\mathcal{O}(\delta^2)$ (see, for example, [2, Page 209]), so that the latter estimator is generally preferred. However, the forward difference estimator requires the evaluation of only $d+1$ points per estimate, while the central difference estimator requires evaluation of $2d$ points per estimate.

A good choice of δ depends on various factors. It should be small enough to reduce the bias, but large enough to keep the variance of the estimator small. The choice of δ is usually determined via a trial run in which the variance of the estimator is assessed.

It is important to implement the finite difference method using **common random variables** (CRVs). The idea is similar to that of antithetic random variables and is as follows. As both terms in the difference estimator are produced via a simulation algorithm, they can be viewed as functions of a stream of independent uniform random variables. The important point to notice is that both terms *need not be independent.* In fact (considering only the central difference estimator), if we denote $Z_1 = \widehat{\ell}(\boldsymbol{\theta} - \mathbf{e}_i\,\delta/2)$ and $Z_2 = \widehat{\ell}(\boldsymbol{\theta} + \mathbf{e}_i\,\delta/2)$, then

☞ 349

$$\mathrm{Var}(Z_2 - Z_1) = \mathrm{Var}(Z_1) + \mathrm{Var}(Z_2) - 2\,\mathrm{Cov}(Z_1, Z_2) , \tag{11.6}$$

so that the variance of the estimator $(Z_2 - Z_1)/\delta$ can be reduced (relative to the independent case) by an amount $2\,\mathrm{Cov}(Z_1, Z_2)/\delta^2$, provided that Z_1 and Z_2 are positively correlated. This can be achieved in practice by taking the *same* random numbers in the simulation procedure to generate Z_1 and Z_2. Because δ is typically small, the correlation between Z_1 and Z_2 is typically close to 1, so that a large variance reduction can be achieved relative to the case where Z_1 and Z_2 are independent.

For the case where $\ell(\boldsymbol{\theta}) = \mathbb{E}Y = \mathbb{E}H(\mathbf{X};\boldsymbol{\theta}) = \mathbb{E}h(\mathbf{U};\boldsymbol{\theta})$, with $\mathbf{U} \sim \mathsf{U}(0,1)^d$, this leads to the following algorithm.

Algorithm 11.1 (Central Difference Estimation With CRVs)

1. *Generate* $\mathbf{U}_1, \ldots, \mathbf{U}_N \overset{\text{iid}}{\sim} \mathsf{U}(0,1)^d$.
2. *Let* $L_k = h(\mathbf{U}_k; \boldsymbol{\theta} - \mathbf{e}_i\,\delta/2)$ *and* $R_k = h(\mathbf{U}_k; \boldsymbol{\theta} + \mathbf{e}_i\,\delta/2)$, $k = 1, \ldots, N$.
3. *Compute the sample covariance matrix corresponding to the pairs* $\{(L_k, R_k)\}$*:*

$$C = \begin{pmatrix} \frac{1}{N-1}\sum_{k=1}^N (L_k - \bar{L})^2 & \frac{1}{N-1}\sum_{k=1}^N (L_k - \bar{L})(R_k - \bar{R}) \\ \frac{1}{N-1}\sum_{k=1}^N (L_k - \bar{L})(R_k - \bar{R}) & \frac{1}{N-1}\sum_{k=1}^N (R_k - \bar{R})^2 \end{pmatrix} .$$

4. Estimate $\partial \ell(\boldsymbol{\theta})/\partial \theta_i$ via the central difference estimator

$$\frac{\bar{R}-\bar{L}}{\delta}$$

with an estimated standard error of

$$SE = \frac{1}{\delta}\sqrt{\frac{C_{1,1}+C_{2,2}-2\,C_{1,2}}{N}} \,.$$

■ **EXAMPLE 11.1 (Bridge Network via the Finite Difference Method)**

Consider the bridge network in Example 9.1 on Page 348. There, the performance measure ℓ is the expected length of the shortest path between the two end nodes and is of the form $\ell(\mathbf{a}) = h(\mathbf{U}; \mathbf{a})$, where $\mathbf{U} \sim \mathsf{U}(0,1)^5$ and $\mathbf{a}$ is a parameter vector. We wish to estimate the gradient of $\ell(\mathbf{a})$ at $\mathbf{a} = (1,2,3,1,2)^\top$. The central difference estimator is implemented in the MATLAB program below. A typical estimate for the gradient based on $N = 10^6$ samples is

$$\widehat{\nabla \ell}(\mathbf{a}) = \begin{pmatrix} 0.3977 \pm 0.0003 \\ 0.0316 \pm 0.0001 \\ 0.00257 \pm 0.00002 \\ 0.3981 \pm 0.0003 \\ 0.0316 \pm 0.0001 \end{pmatrix},$$

where the notation $x \pm \varepsilon$ indicates an estimate of x with an estimated standard error of ε. The above estimate suggests that the expected length of the shortest path is most sensitive to changes in the lengths of components 1 and 4, as is to be expected for these parameter values, because the shortest path is highly likely to consist of these two edges. Component 3 contributes very little to the shortest path and its gradient is close to 0.

```
%fd_bridge.m
N = 10^6;
a = [1,2,3,1,2];
delta = 10^-3;
u = rand(N,5);
for comp=1:5
  de = zeros(1,5);
  de(comp) = delta;
  L = h1(u,a - de/2);
  R = h1(u,a + de/2);
  c = cov(L,R);
  se = sqrt((c(1,1) + c(2,2) - 2*c(1,2))/N)/delta;
  gr = (mean(R) - mean(L))/delta;
  fprintf('%g pm %3.1e\n', gr, se);
end
```

```
function out=h1(u,a)
N = size(u,1);
X = u.*repmat(a,N,1);
Path_1=X(:,1)+X(:,4);
Path_2=X(:,1)+X(:,3)+X(:,5);
Path_3=X(:,2)+X(:,3)+X(:,4);
Path_4=X(:,2)+X(:,5);
out=min([Path_1,Path_2,Path_3,Path_4],[],2);
```

11.3 INFINITESIMAL PERTURBATION ANALYSIS

Infinitesimal perturbation analysis (IPA) concerns the estimation of the gradient of a performance measure $\ell(\boldsymbol{\theta})$ of the form (11.1) with only *structural* parameters. In particular, the objective is to estimate $\nabla\ell(\boldsymbol{\theta}) = \nabla_{\boldsymbol{\theta}}\mathbb{E}H(\mathbf{X};\boldsymbol{\theta})$ for some function $H(\mathbf{x};\boldsymbol{\theta})$ and $\mathbf{X} \sim f(\mathbf{x})$ through an interchange of the gradient and the expectation operator; that is,

$$\nabla_{\boldsymbol{\theta}}\mathbb{E}H(\mathbf{X};\boldsymbol{\theta}) = \mathbb{E}\nabla_{\boldsymbol{\theta}}H(\mathbf{X};\boldsymbol{\theta}) . \tag{11.7}$$

Such an interchange is allowed under certain regularity conditions on H, see Theorem 11.1.1. If (11.7) holds, then the gradient can be estimated via crude Monte Carlo as

$$\widehat{\nabla\ell}(\boldsymbol{\theta}) = \frac{1}{N}\sum_{k=1}^{N}\nabla_{\boldsymbol{\theta}}H(\mathbf{X}_k;\boldsymbol{\theta}) , \tag{11.8}$$

where $\mathbf{X}_1,\ldots,\mathbf{X}_N \sim_{\text{iid}} f$. In contrast to the finite difference method, the IPA estimator is unbiased. Moreover, because the procedure is basically a crude Monte Carlo method, its rate of convergence is $\mathcal{O}(1/\sqrt{N})$; see also [4, 8, 20]. The IPA procedure is summarized in the following algorithm.

Algorithm 11.2 (IPA Estimation)

1. *Generate* $\mathbf{X}_1,\ldots,\mathbf{X}_N \stackrel{\text{iid}}{\sim} f$.
2. *Evaluate* $\nabla_{\boldsymbol{\theta}}H(\mathbf{X}_k;\boldsymbol{\theta})$, $k = 1,\ldots,N$ *and estimate the gradient of* $\ell(\boldsymbol{\theta})$ *via* (11.8). *Determine an approximate* $1-\alpha$ *confidence interval as*

$$\left(\widehat{\nabla\ell}(\boldsymbol{\theta}) - z_{1-\alpha/2}\, S/\sqrt{N},\ \widehat{\nabla\ell}(\boldsymbol{\theta}) + z_{1-\alpha/2}\, S/\sqrt{N}\right),$$

where S *is the sample standard deviation of* $\{\nabla_{\boldsymbol{\theta}}H(\mathbf{X}_k;\boldsymbol{\theta})\}$ *and* z_γ *denotes the* γ*-quantile of the* $\mathsf{N}(0,1)$ *distribution.*

■ EXAMPLE 11.2 (Bridge Network via IPA)

We consider the same derivative estimation problem as in Example 11.1, but deal with it via IPA. Denote the four possible paths in the bridge network by

$$\mathcal{P}_1 = \{1,4\}, \qquad \mathcal{P}_2 = \{1,3,5\}, \qquad \mathcal{P}_3 = \{2,3,4\}, \qquad \mathcal{P}_4 = \{2,5\} .$$

Then we can write

$$h(\mathbf{U};\mathbf{a}) = \min_{k=1,\ldots,4}\sum_{i\in\mathcal{P}_k} a_i\, U_i . \tag{11.9}$$

Let $K \in \{1,2,3,4\}$ be the (random) index of the minimum-length path; hence, $h(\mathbf{U};\mathbf{a}) = \sum_{i \in \mathcal{P}_K} a_i\, U_i$. The partial derivatives of $h(\mathbf{U};\mathbf{a})$ now follow immediately from (11.9):

$$\frac{\partial h(\mathbf{U};\mathbf{a})}{\partial a_i} = \begin{cases} U_i & \text{if } K \in \mathcal{A}_i, \\ 0 & \text{otherwise,} \end{cases}$$

where $\mathcal{A}_i$ is the set of indices of all paths that contain component i; that is,

$$\mathcal{A}_1 = \{1,2\}, \quad \mathcal{A}_2 = \{3,4\}, \quad \mathcal{A}_3 = \{2,3\}, \quad \mathcal{A}_4 = \{1,3\}, \quad \mathcal{A}_5 = \{2,4\}\,.$$

The IPA procedure is implemented in the MATLAB program below. A typical estimate for the gradient at $\mathbf{a} = (1,2,3,1,2)^\top$ is

$$\widehat{\nabla \ell}(\mathbf{a}) = \begin{pmatrix} 0.3980 \pm 0.0003 \\ 0.0316 \pm 0.0001 \\ 0.00255 \pm 0.00002 \\ 0.3979 \pm 0.0003 \\ 0.0316 \pm 0.0001 \end{pmatrix},$$

where the same notation is used as in Example 11.1. We see that the accuracy is similar to that of the central difference method with common random numbers. However, the IPA estimate is unbiased.

```
%ipabridge.m
N = 10^6;
a = [1,2,3,1,2];
A = [1,2;3,4;2,3;1,3;2,4];
u = rand(N,5);
for comp=1:5
    dh = zeros(N,1);
    [y,K] = HK(u,a);
    ind = find(K == A(comp,1) | K==A(comp,2));
    dh(ind) = u(ind,comp);
    gr = mean(dh);
    se = std(dh)/sqrt(N);
    fprintf('%g pm %3.1e\n', gr, se);
end
```

```
function [y,K]=HK(u,a)
N = size(u,1);
X = u.*repmat(a,N,1);
Path_1=X(:,1)+X(:,4);
Path_2=X(:,1)+X(:,3)+X(:,5);
Path_3=X(:,2)+X(:,3)+X(:,4);
Path_4=X(:,2)+X(:,5);
[y,K] =min([Path_1,Path_2,Path_3,Path_4],[],2);
```

11.4 SCORE FUNCTION METHOD

In the **score function method**, also called the **likelihood ratio method**, the performance function $\ell(\boldsymbol{\theta})$ is assumed to be of the form (11.1) with only *distributional* parameters. In particular, the objective is to estimate (in the continuous case) the gradient of

$$\ell(\boldsymbol{\theta}) = \mathbb{E}_{\boldsymbol{\theta}} H(\mathbf{X}) = \int H(\mathbf{x})\, f(\mathbf{x};\boldsymbol{\theta})\, \mathrm{d}\mathbf{x}$$

for some function H and pdf f (for the discrete case replace the integral with a sum). As with the IPA method the key is to interchange the gradient and the integral; that is,

$$\nabla_{\boldsymbol{\theta}} \int H(\mathbf{X})\, f(\mathbf{x};\boldsymbol{\theta})\, \mathrm{d}\mathbf{x} = \int H(\mathbf{X})\, \nabla_{\boldsymbol{\theta}} f(\mathbf{x};\boldsymbol{\theta})\, \mathrm{d}\mathbf{x}\,, \tag{11.10}$$

which is allowed under quite general conditions (see Theorem 11.1.1). Note that the right-hand side of (11.10) can be written as

$$\begin{aligned}\int H(\mathbf{X})\, \nabla_{\boldsymbol{\theta}} f(\mathbf{x};\boldsymbol{\theta})\, \mathrm{d}\mathbf{x} &= \int H(\mathbf{X})\, \frac{\nabla_{\boldsymbol{\theta}} f(\mathbf{x};\boldsymbol{\theta})}{f(\mathbf{x};\boldsymbol{\theta})} f(\mathbf{x};\boldsymbol{\theta})\, \mathrm{d}\mathbf{x} \\ &= \int H(\mathbf{X})\, [\nabla_{\boldsymbol{\theta}} \ln f(\mathbf{x};\boldsymbol{\theta})]\, f(\mathbf{x};\boldsymbol{\theta})\, \mathrm{d}\mathbf{x} \\ &= \mathbb{E}_{\boldsymbol{\theta}} H(\mathbf{X})\, \mathcal{S}(\boldsymbol{\theta};\mathbf{X})\,,\end{aligned}$$

☞ 664 where $\mathcal{S}(\boldsymbol{\theta};\mathbf{x}) = \nabla_{\boldsymbol{\theta}} \ln f(\mathbf{x};\boldsymbol{\theta})$ is the *score function* of f; see Section B.2. Hence, if (11.10) holds, the gradient can be estimated via crude Monte Carlo as

$$\widehat{\nabla \ell}(\boldsymbol{\theta}) = \frac{1}{N} \sum_{k=1}^{N} H(\mathbf{X}_k)\, \mathcal{S}(\boldsymbol{\theta};\mathbf{X}_k)\,, \tag{11.11}$$

where $\mathbf{X}_1, \ldots, \mathbf{X}_N \sim_{\text{iid}} f$. The score function estimator is unbiased, and, being the sample mean of iid random variables, achieves $\mathcal{O}(1/\sqrt{N})$ convergence.

Algorithm 11.3 (Gradient Estimation via the Score Function Method)

1. *Generate* $\mathbf{X}_1, \ldots, \mathbf{X}_N \overset{\text{iid}}{\sim} f(\cdot;\boldsymbol{\theta})$.
2. *Evaluate the scores* $\mathcal{S}(\boldsymbol{\theta};\mathbf{X}_k)$, $k = 1, \ldots, N$ *and estimate the gradient of* $\ell(\boldsymbol{\theta})$ *via* (11.11). *Determine an approximate* $1-\alpha$ *confidence interval as*

$$\left(\widehat{\nabla \ell}(\boldsymbol{\theta}) - z_{1-\alpha/2}\, \widehat{\sigma}/\sqrt{N},\ \widehat{\nabla \ell}(\boldsymbol{\theta}) + z_{1-\alpha/2}\, \widehat{\sigma}/\sqrt{N}\right),$$

where $\widehat{\sigma}$ *is the sample standard deviation of* $\{H(\mathbf{X}_k)\mathcal{S}(\boldsymbol{\theta};\mathbf{X}_k)\}$ *and* z_γ *denotes the* γ*-quantile of the* $\mathsf{N}(0,1)$ *distribution.*

Remark 11.4.1 (Higher-Order Derivatives) Higher-order derivatives of ℓ can be estimated in a similar fashion. Specifically, the r-th order derivative is given by

$$\nabla^r \ell(\boldsymbol{\theta}) = \mathbb{E}_{\boldsymbol{\theta}} \left[H(\mathbf{X})\, \mathcal{S}^{(r)}(\boldsymbol{\theta};\mathbf{X})\right], \tag{11.12}$$

where

$$\mathcal{S}^{(r)}(\boldsymbol{\theta};\mathbf{x}) = \frac{\nabla^r_{\boldsymbol{\theta}} f(\mathbf{x};\boldsymbol{\theta})}{f(\mathbf{x};\boldsymbol{\theta})} \tag{11.13}$$

is the **r-th order score function**, $r = 0, 1, 2, \ldots$. In particular, $\mathcal{S}^{(0)}(\boldsymbol{\theta};\mathbf{x}) = 1$ (by definition), $\mathcal{S}^{(1)}(\boldsymbol{\theta};\mathbf{x}) = \mathcal{S}(\boldsymbol{\theta};\mathbf{x}) = \nabla_{\boldsymbol{\theta}} \ln f(\mathbf{x};\boldsymbol{\theta})$, and $\mathcal{S}^{(2)}(\boldsymbol{\theta};\mathbf{x})$ can be represented as

$$\begin{aligned} \mathcal{S}^{(2)}(\boldsymbol{\theta};\mathbf{x}) &= \nabla_{\boldsymbol{\theta}}\mathcal{S}(\boldsymbol{\theta};\mathbf{x}) + \mathcal{S}(\boldsymbol{\theta};\mathbf{x})\,\mathcal{S}(\boldsymbol{\theta};\mathbf{x})^{\top} \\ &= \nabla^2_{\boldsymbol{\theta}} \ln f(\mathbf{x};\boldsymbol{\theta}) + \nabla_{\boldsymbol{\theta}} \ln f(\mathbf{x};\boldsymbol{\theta})\,[\nabla_{\boldsymbol{\theta}} \ln f(\mathbf{x};\boldsymbol{\theta})]^{\top} . \end{aligned} \tag{11.14}$$

The higher-order $\nabla^r \ell(\boldsymbol{\theta})$, $r = 0, 1, \ldots$, can be estimated via simulation as

$$\widehat{\nabla^r \ell}(\boldsymbol{\theta}) = \frac{1}{N}\sum_{k=1}^{N} H(\mathbf{X}_k)\,\mathcal{S}^{(r)}(\boldsymbol{\theta};\mathbf{X}_k) . \tag{11.15}$$

It follows that the function $\ell(\boldsymbol{\theta})$, and *all* the sensitivities $\nabla^r \ell(\boldsymbol{\theta})$ can be estimated from a single simulation, because in (11.12) all of them are expressed as expectations with respect to the same pdf, $f(\mathbf{x};\boldsymbol{\theta})$.

■ **EXAMPLE 11.3 (Bridge Network via the Score Function Method)**

Consider again the derivative estimation problem in Examples 11.1 and 11.2. As in Example 9.1 on Page 348 we can write

$$\ell(\mathbf{a}) = \int H(\mathbf{x}) f(\mathbf{x};\mathbf{a})\,\mathrm{d}\mathbf{x} ,$$

with $H(\mathbf{x}) = \min\{x_1 + x_4,\ x_1 + x_3 + x_5,\ x_2 + x_3 + x_4,\ x_2 + x_5\}$ and

$$f(\mathbf{x};\mathbf{a}) = \prod_{i=1}^{5} \frac{\mathrm{I}_{\{0<x_i<a_i\}}}{a_i} . \tag{11.16}$$

This is a typical example where the interchange of gradient and integral is *not* appropriate, because of the discontinuities at $a_1, \ldots, a_5$. However, the situation can easily be fixed by including a continuity correction. Taking the derivative with respect to a_1 gives,

$$\begin{aligned} \frac{\partial}{\partial a_1}\ell(\mathbf{a}) &= \frac{\partial}{\partial a_1}\int_0^{a_1}\left(\int_0^{a_2}\cdots\int_0^{a_5} H(\mathbf{x})\frac{1}{a_1\cdots a_5}\,\mathrm{d}x_2\cdots\mathrm{d}x_5\right)\mathrm{d}x_1 \\ &= \int_0^{a_2}\cdots\int_0^{a_5} H(a_1, x_2, \ldots, x_5)\frac{1}{a_1\cdots a_5}\,\mathrm{d}x_2\cdots\mathrm{d}x_5 \\ &\quad - \frac{1}{a_1}\int H(\mathbf{x}) f(\mathbf{x};\mathbf{a})\,\mathrm{d}\mathbf{x} \\ &= \frac{1}{a_1}\left(\mathbb{E}H(\mathbf{X}^*) - \mathbb{E}H(\mathbf{X})\right) , \end{aligned} \tag{11.17}$$

where $\mathbf{X} \sim f(\mathbf{x};\mathbf{a})$ and $\mathbf{X}^* \sim f(\mathbf{x};\mathbf{a}\,|\,x_1 = a_1)$. Both $\mathbb{E}H(\mathbf{X}^*)$ and $\mathbb{E}H(\mathbf{X})$ or $\mathbb{E}[H(\mathbf{X}^*) - H(\mathbf{X})]$ can easily be estimated via Monte Carlo. The other partial derivatives follow by symmetry.

The following MATLAB program implements the procedure. The results are similar to those of the IPA and finite difference methods.

```
%sfbridge.m
N = 10^6;
a = [1,2,3,1,2];
u = rand(N,5);
for comp=1:5
    X = u.*repmat(a,N,1);
    hx = H(X);
    X(:,comp) = a(comp);
    hxs = H(X);
    R = (-hx + hxs)/a(comp);
    gr = mean(R);
    se = std(R)/sqrt(N);
    fprintf('%g pm %3.1e\n',  gr, se);
end
```

```
function out=H(X)
Path_1=X(:,1)+X(:,4);
Path_2=X(:,1)+X(:,3)+X(:,5);
Path_3=X(:,2)+X(:,3)+X(:,4);
Path_4=X(:,2)+X(:,5);

out=min([Path_1,Path_2,Path_3,Path_4],[],2);
```

11.4.1 Score Function Method With Importance Sampling

By combining the score function method with importance sampling one can estimate the derivatives $\nabla^r \ell(\boldsymbol{\theta}) = \mathbb{E}_{\boldsymbol{\theta}}[H(\mathbf{X})\, \mathcal{S}^{(r)}(\boldsymbol{\theta};\mathbf{X})]$ simultaneously for several values of $\boldsymbol{\theta} \in \Theta$, using a single simulation run. The idea is a generalization of
☞ 373 the arguments in Section 9.7.6. In particular, let $g(\mathbf{x})$ be an importance sampling density. Then $\nabla^r \ell(\boldsymbol{\theta})$ can be written as

$$\nabla^r \ell(\boldsymbol{\theta}) = \mathbb{E}_g[H(\mathbf{X})\, \mathcal{S}^{(r)}(\boldsymbol{\theta};\mathbf{X})\, W(\mathbf{X};\boldsymbol{\theta})] \,, \tag{11.18}$$

where

$$W(\mathbf{x};\boldsymbol{\theta}) = \frac{f(\mathbf{x};\boldsymbol{\theta})}{g(\mathbf{x})} \tag{11.19}$$

is the likelihood ratio of $f(\mathbf{x};\boldsymbol{\theta})$ and $g(\mathbf{x})$. The importance sampling estimator of $\nabla^r \ell(\boldsymbol{\theta})$ can be written as

$$\widehat{\nabla^r \ell}(\boldsymbol{\theta}) = \frac{1}{N}\sum_{k=1}^{N} H(\mathbf{X}_k)\, \mathcal{S}^{(r)}(\boldsymbol{\theta};\mathbf{X}_k)\, W(\mathbf{X}_k;\boldsymbol{\theta}) \,, \tag{11.20}$$

where $\mathbf{X}_1, \ldots, \mathbf{X}_N \sim_{\text{iid}} g$. Note that $\widehat{\nabla^r \ell}(\boldsymbol{\theta})$ is an unbiased estimator of $\nabla^r \ell(\boldsymbol{\theta})$ for *all* $\boldsymbol{\theta}$. This means that by varying $\boldsymbol{\theta}$ and keeping g fixed we can, in principle, estimate the whole *response surface* $\{\nabla^r \ell(\boldsymbol{\theta}), \boldsymbol{\theta} \in \Theta\}$ without bias from a single simulation.

Often the importance sampling distribution is chosen in the *same* class of distributions as the original one. That is, $g(\mathbf{x}) = f(\mathbf{x}; \boldsymbol{\theta}_0)$, for some $\boldsymbol{\theta}_0 \in \Theta$. If we denote the importance sampling estimator of $\ell(\boldsymbol{\theta})$ for a given $\boldsymbol{\theta}_0$ by $\widehat{\ell}(\boldsymbol{\theta}; \boldsymbol{\theta}_0)$, that is,

$$\widehat{\ell}(\boldsymbol{\theta}; \boldsymbol{\theta}_0) = \frac{1}{N} \sum_{k=1}^{N} H(\mathbf{X}_k)\, W(\mathbf{X}_k; \boldsymbol{\theta}; \boldsymbol{\theta}_0) , \tag{11.21}$$

with $W(\mathbf{x}; \boldsymbol{\theta}, \boldsymbol{\theta}_0) = f(\mathbf{x}; \boldsymbol{\theta})/f(\mathbf{x}; \boldsymbol{\theta}_0)$, and the estimators in (11.20) by $\widehat{\nabla^r \ell}(\boldsymbol{\theta}; \boldsymbol{\theta}_0)$, then

$$\widehat{\nabla^r \ell}(\boldsymbol{\theta}; \boldsymbol{\theta}_0) = \nabla^r_{\boldsymbol{\theta}}\, \widehat{\ell}(\boldsymbol{\theta}; \boldsymbol{\theta}_0) = \frac{1}{N} \sum_{k=1}^{N} H(\mathbf{X}_k)\, \mathcal{S}^{(r)}(\boldsymbol{\theta}; \mathbf{X}_k)\, W(\mathbf{X}_k; \boldsymbol{\theta}; \boldsymbol{\theta}_0) . \tag{11.22}$$

Thus, the estimators of the sensitivities are simply the sensitivities of the estimators.

For a given importance sampling pdf $f(\mathbf{x}; \boldsymbol{\theta}_0)$, the algorithm for estimating the sensitivities $\nabla^r \ell(\boldsymbol{\theta})$, $r = 0, 1, \ldots$, for multiple values of $\boldsymbol{\theta}$ from a single simulation run, is as follows.

Algorithm 11.4 (Gradient Estimation via the Score Function Method)

1. *Generate a sample* $\mathbf{X}_1, \ldots, \mathbf{X}_N \overset{\text{iid}}{\sim} f(\cdot; \boldsymbol{\theta}_0)$.

2. *Calculate the sample performance* $H(\mathbf{X}_k)$ *and the scores* $\mathcal{S}^{(r)}(\boldsymbol{\theta}; \mathbf{X}_k)$, $k = 1, \ldots, N$, *for the desired parameter* $\boldsymbol{\theta}$.

3. *Calculate* $\nabla^r_{\boldsymbol{\theta}}\, \widehat{\ell}(\boldsymbol{\theta}; \boldsymbol{\theta}_0)$ *according to* (11.22).

Confidence regions for $\nabla^r \ell(\boldsymbol{\theta})$ can be obtained by standard statistical techniques. In particular (see, for example, [18] and Section A.8), $N^{1/2}\left[\nabla^r_{\boldsymbol{\theta}}\, \widehat{\ell}(\boldsymbol{\theta}; \boldsymbol{\theta}_0) - \nabla^r \ell(\boldsymbol{\theta})\right]$ converges in distribution to a multivariate normal random vector with mean zero and covariance matrix

$$\mathrm{Cov}_{\boldsymbol{\theta}_0}(H\, \mathcal{S}^{(r)}\, W) = \mathbb{E}_{\boldsymbol{\theta}_0}\left[H^2\, W^2\, \mathcal{S}^{(r)} \mathcal{S}^{(r)\,\top}\right] - [\nabla^r \ell(\boldsymbol{\theta})][\nabla^r \ell(\boldsymbol{\theta})]^\top , \tag{11.23}$$

using the abbreviations $H = H(\mathbf{X})$, $\mathcal{S}^{(r)} = \mathcal{S}^{(r)}(\boldsymbol{\theta}; \mathbf{X})$ and $W = W(\mathbf{X}; \boldsymbol{\theta}, \boldsymbol{\theta}_0)$.

■ **EXAMPLE 11.4 (Gradient Estimation for the Bridge Network)**

We return to (9.50) in Example 9.12, which gives the estimator for the expected length of the minimal path in the bridge network as a function of θ. Here $X_1 \sim \mathsf{Exp}(1/\theta)$ and the simulation is carried out under $\theta = \theta_0 = 3$. We wish to estimate the derivative $\nabla \ell(\theta; \theta_0)$ for all θ in the neighborhood of θ_0. This is achieved by ☞ 375 differentiating $\widehat{\ell}(\theta;\theta_0)$, giving

$$\begin{aligned}\widehat{\nabla\ell}(\theta;\theta_0) = \nabla\widehat{\ell}(\theta;\theta_0) &= \nabla\frac{\theta_0}{\theta N}\sum_{k=1}^{N} H(\mathbf{X}_k)\,\mathrm{e}^{X_{k1}(1/\theta_0-1/\theta)} \\ &= \frac{\theta_0}{\theta^3 N}\sum_{k=1}^{N} H(\mathbf{X}_k)\,(X_{k1}-\theta)\,\mathrm{e}^{X_{k1}(1/\theta_0-1/\theta)} \\ &= \frac{1}{N}\sum_{k=1}^{N} H(\mathbf{X}_k)\,W(\mathbf{X}_k;\theta;\theta_0)\,\mathcal{S}(\theta;X_{k1})\,,\end{aligned}$$

with $\mathcal{S}(\theta;x) = (x-\theta)/\theta^2$ being the score function corresponding to the $\mathsf{Exp}(1/\theta)$ distribution; see also Table B.11. The estimate of the derivative curve is given in Figure 11.1, which is the pathwise derivative of the estimated response curve in Figure 9.7.

☞ 666

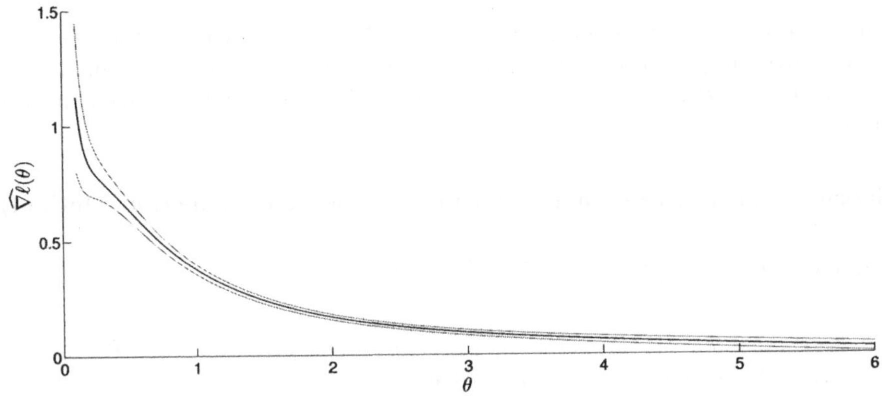

Figure 11.1 Estimates of the gradient of $\ell(\theta)$ with 95% confidence bounds.

The following MATLAB program differs from the one used for the estimation of the response surface in Example 9.12 only in the score function calculations.

```
%gradresponsesurfis.m
N = 10000;
theta0 = 3;
a = [theta0,2,3,1,2];
u = rand(N,5);
X = u.*repmat(a,N,1);
X(:,1) = -log(u(:,1))*theta0;
W = zeros(N,1);
Sc = zeros(N,1);
HX = H(X);
theta = 0.1:0.01:theta0*2;
num = numel(theta);
gradell = zeros(1,num);
```

```
gradellL = zeros(1,num);
gradellU = zeros(1,num);
stgradell = zeros(1,num);
for i=1:num
    th = theta(i);
    Sc = (-th + X(:,1))/th^2;
    W = (exp(-(X(:,1)/th))/th)./(exp(-(X(:,1)/theta0))/theta0);
    HWS = H(X).*W.*Sc;
    gradell(i) = mean(HWS);
    stgradell(i) = std(HWS);
    gradellL(i)= gradell(i) - stgradell(i)/sqrt(N)*1.95;
    gradellU(i)= gradell(i) + stgradell(i)/sqrt(N)*1.95;
end
plot(theta,gradell, theta, gradellL, theta, gradellU)
```

11.5 WEAK DERIVATIVES

Let $F(\cdot;\theta)$ be a cdf depending on a real-valued parameter θ. If there exist a constant $c(\theta)$ and cdfs $F_1(\cdot;\theta)$ and $F_2(\cdot;\theta)$ such that for every continuous bounded function H the following holds:

$$\frac{\mathrm{d}}{\mathrm{d}\theta}\int H(\mathbf{x})\,\mathrm{d}F(\mathbf{x};\theta) = c(\theta)\left(\int H(\mathbf{x})\,\mathrm{d}F_1(\mathbf{x};\theta) - \int H(\mathbf{x})\,\mathrm{d}F_2(\mathbf{x};\theta)\right), \quad (11.24)$$

then the triple $(c(\theta), F_1(\cdot;\theta), F_2(\cdot;\theta))$ is said to be a **weak derivative** for F with respect to θ. The weak derivate concept is closely related to that of the score function method [7, 14].

For any distribution it is possible to find a weak derivative triple. In general, the weak derivative triple is not unique and the most convenient representation is taken. It is possible that F is the cdf of a continuous distribution, while F_1 and/or F_2 are discrete.

Table 11.1 lists some weak derivatives for common univariate distributions. Here DSM denotes the **double-sided Maxwell distribution**, with pdf

$$f(x;\mu,\theta) = \frac{\mathrm{e}^{-\frac{(x-\mu)^2}{2\theta^2}}(x-\mu)^2}{\sqrt{2\pi}\,\theta^3}, \quad x \in \mathbb{R},$$

which defines a location–scale family of distributions with base pdf $\mathring{f}(x) = f(x;0,1) = \mathrm{e}^{-\frac{x^2}{2}}x^2/\sqrt{2\pi}$, $x \in \mathbb{R}$. Generating a random variable $X \sim \mathring{f}$ can be achieved by generating $Y \sim \mathsf{Gamma}(3/2, 1/2)$ and $B \sim \mathsf{Ber}(1/2)$ independently, and returning $X = (2B-1)\sqrt{Y}$. As a consequence, $\mu + \theta X \sim f(\cdot;\mu,\theta)$.

Table 11.1 Weak derivatives for common distributions.

Distribution	$c(\theta)$	$X \sim F_1$	$X \sim F_2$
$\mathsf{Bin}(n, \theta)$	n	$1 + \mathsf{Bin}(n-1, \theta)$	$\mathsf{Bin}(n-1, \theta)$
$\mathsf{Geom}(\theta)$	$1/\theta$	$\mathsf{Geom}(\theta)$	$\mathsf{NegBin}(2, \theta)$
$\mathsf{Poi}(\theta)$	1	$1 + \mathsf{Poi}(\theta)$	$\mathsf{Poi}(\theta)$
$\mathsf{N}(\theta, \sigma^2)$	$\frac{1}{\sigma\sqrt{2\pi}}$	$\theta + \mathsf{Weib}(2, \frac{1}{2\sigma^2})$	$\theta - \mathsf{Weib}(2, \frac{1}{2\sigma^2})$
$\mathsf{N}(\mu, \theta^2)$	$1/\theta$	$\mathsf{DSM}(\mu, \theta)$	$\mathsf{N}(\mu, \theta^2)$
$\mathsf{Gamma}(\alpha, \theta)$	α/θ	$\mathsf{Gamma}(\alpha+1, \theta)$	$\mathsf{Gamma}(\alpha, \theta)$
$\mathsf{U}(0, \theta)$	$1/\theta$	θ	$\mathsf{U}(0, \theta)$
$\mathsf{Weib}(\alpha, \theta)$	$1/\theta$	$\mathsf{Weib}(\alpha, \theta)$	$\mathsf{Gamma}(2, \theta)^{1/\alpha}$

The estimation of the gradient of $\ell(\theta) = \mathbb{E}_\theta H(\mathbf{X}) = \int H(\mathbf{x})\, \mathrm{d}F(\mathbf{x};\theta)$ is summarized in the next algorithm.

Algorithm 11.5 (Weak Derivative Estimation) *Let $(c(\theta), F_1(\cdot;\theta), F_2(\cdot;\theta))$ be a weak derivative triple for $F(\cdot;\theta)$ and let H be a continuous bounded function.*

1. *Generate $\mathbf{X}_1, \ldots, \mathbf{X}_N \overset{\text{iid}}{\sim} F_1$ and $\mathbf{Y}_1, \ldots, \mathbf{Y}_N \overset{\text{iid}}{\sim} F_2$, possibly dependently.*
2. *Estimate the derivative of $\ell(\theta)$ via*

$$\widehat{\nabla \ell}(\theta) = c(\theta) \frac{1}{N} \sum_{k=1}^{N} \left(H(\mathbf{X}_k) - H(\mathbf{Y}_k)\right) .$$

Determine an approximate $1-\alpha$ confidence interval as

$$\left(\widehat{\nabla \ell}(\theta) - z_{1-\alpha/2}\,\widehat{\sigma}/\sqrt{N},\ \widehat{\nabla \ell}(\theta) + z_{1-\alpha/2}\,\widehat{\sigma}/\sqrt{N}\right),$$

where $\widehat{\sigma}$ is the sample standard deviation of $\{H(\mathbf{X}_k) - H(\mathbf{Y}_k)\}$ and z_γ denotes the γ-quantile of the $\mathsf{N}(0,1)$ distribution.

Taking the random samples from F_1 and F_2 *dependently*, for example via common random numbers, can give considerable variance reduction, as in the finite difference method.

■ EXAMPLE 11.5 (Weak Derivative for the Bridge Network)

Consider Example 11.3 where the gradient of $\ell(\mathbf{a})$ is estimated via the score function method using a continuity correction. The main result (11.17) can be derived directly using weak derivatives. Namely, from Table 11.1 a weak derivative triple corresponding to $X_1 \sim \mathsf{U}(0, a_1)$ is given by $(1/a_1, a_1, \mathsf{U}(0, a_1))$, and as a consequence a weak derivative triple corresponding to $\mathbf{X}$ is of the form $(1/a_1, F_1, F_2)$, where F_2 has pdf $f(\mathbf{x}; \mathbf{a})$ as given in (11.16) and F_1 has pdf $f(\mathbf{x}; \mathbf{a} \mid x_1 = a_1)$, exactly as (11.17). The estimation procedure is therefore identical to the one given in Example 11.3.

11.6 SENSITIVITY ANALYSIS FOR REGENERATIVE PROCESSES

Although most of the sensitivity results in this chapter have been formulated in terms of *static* systems, where $\mathbf{X}$ is a fixed-dimensional random vector, much of the theory carries through to *time-dependent* processes $\{X_t\}$. In particular, let $X_1, X_2, \ldots$ be an input sequence of (possibly multidimensional) random variables driving an output process $\{H_t, t = 0, 1, 2, \ldots\}$. More precisely, $H_t = H_t(\mathbf{X}_t)$, for some performance function H_t, where the vector $\mathbf{X}_t = (X_1, X_2, \ldots, X_t)$ represents the history of the input process up to time t. Let the pdf of $\mathbf{X}_t$ be given by $f_t(\mathbf{x}_t; \boldsymbol{\theta})$, which depends on some parameter vector $\boldsymbol{\theta}$. Assume that $\{H_t\}$ is a *regenerative* process with a cycle length of finite expectation. A typical example is an ergodic Markov chain. The expected steady-state performance, $\ell(\boldsymbol{\theta})$, can be written as ☞ 630

$$\ell(\boldsymbol{\theta}) = \frac{\ell_R(\boldsymbol{\theta})}{\ell_\tau(\boldsymbol{\theta})} = \frac{\mathbb{E}_{\boldsymbol{\theta}} R}{\mathbb{E}_{\boldsymbol{\theta}} \tau} = \frac{\mathbb{E}_{\boldsymbol{\theta}} \sum_{t=1}^{\tau} H_t}{\mathbb{E}_{\boldsymbol{\theta}} \tau} = \frac{\mathbb{E}_g \sum_{t=1}^{\tau} H_t \, W_t}{\mathbb{E}_g \sum_{t=1}^{\tau} W_t} , \tag{11.25}$$

where R is the reward during a cycle, τ is the cycle length, and g is an importance sampling pdf, which defines the likelihood ratios $\{W_t\}$; see Section 8.3.3 and Theorem 9.7.2. Using the fact that $W_t = W_t(\mathbf{X}_t; \boldsymbol{\theta})$ is a function of $\boldsymbol{\theta}$, but $H_t = H_t(\mathbf{X}_t)$ is not, by direct differentiation of (11.25) we have ☞ 313 ☞ 371

$$\nabla \ell(\boldsymbol{\theta}) = \frac{\nabla \ell_R(\boldsymbol{\theta})}{\ell_\tau(\boldsymbol{\theta})} - \frac{\ell_R(\boldsymbol{\theta}) \nabla \ell_\tau(\boldsymbol{\theta})}{\ell_\tau^2(\boldsymbol{\theta})}$$

$$= \frac{\mathbb{E}_g \sum_{t=1}^{\tau} H_t \, \nabla W_t}{\mathbb{E}_g \sum_{t=1}^{\tau} W_t} - \frac{\mathbb{E}_g \sum_{t=1}^{\tau} H_t \, W_t}{\mathbb{E}_g \sum_{t=1}^{\tau} W_t} \, \frac{\mathbb{E}_g \sum_{t=1}^{\tau} \nabla W_t}{\mathbb{E}_g \sum_{t=1}^{\tau} W_t} , \tag{11.26}$$

assuming that the expectations and gradients can be interchanged. Observe that $\nabla W_t = W_t \, \mathcal{S}_t$, where $\mathcal{S}_t$ is the score function $\nabla_{\boldsymbol{\theta}} \ln f_t(\mathbf{x}_t; \boldsymbol{\theta})$. Using (11.25) and (11.26), one can estimate the performance $\ell(\boldsymbol{\theta})$ and gradient $\nabla \ell(\boldsymbol{\theta})$ as follows:

$$\widehat{\ell}(\boldsymbol{\theta}) = \frac{\sum_{i=1}^{N} \sum_{t=1}^{\tau_i} H_{ti} \, W_{ti}}{\sum_{i=1}^{N} \sum_{t=1}^{\tau_i} W_{ti}} \tag{11.27}$$

and, with $\widehat{\nabla \ell}(\boldsymbol{\theta}) = \nabla \widehat{\ell}(\boldsymbol{\theta})$,

$$\widehat{\nabla \ell}(\boldsymbol{\theta}) = \frac{\sum_{i=1}^{N} \sum_{t=1}^{\tau_i} H_{ti} W_{ti} \mathcal{S}_{ti}}{\sum_{i=1}^{N} \sum_{t=1}^{\tau_i} W_{ti}} - \frac{\sum_{i=1}^{N} \sum_{t=1}^{\tau_i} H_{ti} W_{ti}}{\sum_{i=1}^{N} \sum_{t=1}^{\tau_i} W_{ti}} \, \frac{\sum_{i=1}^{N} \sum_{t=1}^{\tau_i} W_{ti} \mathcal{S}_{ti}}{\sum_{i=1}^{N} \sum_{t=1}^{\tau_i} W_{ti}} . \tag{11.28}$$

Here, $\mathcal{S}_{ti}$ is the value of the score function at the t-th step in the i-th regenerative cycle, and similarly for H_{ti} and W_{ti}. Note that, similar to the static case, only a single simulation run (under g) is required to estimate the performance $\ell(\boldsymbol{\theta})$ and the gradient $\nabla \ell(\boldsymbol{\theta})$ for different values of $\boldsymbol{\theta}$.

Algorithm 11.6 (Gradient Estimation for Regenerative Processes)

1. *Generate the process* $\{X_t\}$ *under* g *until* N *regenerative cycles have been obtained.*

2. *Simultaneously generate the output processes* $\{H_t\}$ *and* $\{\nabla W_t\} = \{W_t \, \mathcal{S}_t\}$.

3. *Calculate* $\widehat{\nabla \ell}(\boldsymbol{\theta})$ *from* (11.28).

■ EXAMPLE 11.6 (Expected Waiting Time in a $GI/G/1$ Queue)

The waiting time process in a $GI/G/1$ queue is driven by sequences of interarrival times $A_1, A_2, \ldots$ and service times $B_1, B_2, \ldots$ via the Lindley equation

$$H_t = \max\{H_{t-1} + B_t - A_t,\ 0\}, \quad t = 1, 2, \ldots, \tag{11.29}$$

☞ 387 with $H_0 = 0$; see Example 10.4. Writing $X_t = (A_t, B_t)$, the $\{X_t, t = 1, 2, \ldots\}$ are independent and identically distributed. The process $\{H_t, t = 0, 1, \ldots\}$ is a regenerative process, which regenerates at each time t where $H_t = 0$. Let $\tau > 0$ denote the first such time and let H denote the steady-state waiting time. We wish to estimate the steady-state performance

$$\ell = \mathbb{E}H = \frac{\mathbb{E}\sum_{t=1}^{\tau} H_t}{\mathbb{E}\tau} .$$

Consider, for instance, the case where $A \sim \mathsf{Exp}(\lambda)$ and $B \sim \mathsf{Exp}(\mu)$, independently. Thus, H is the steady-state waiting time in the $M/M/1$ queue, and $\mathbb{E}H = \lambda/(\mu(\mu - \lambda))$, for $\mu > \lambda$; see for example [6]. Suppose we carry out the simulation using the service rate $\widetilde{\mu}$ and wish to estimate $\ell(\mu) = \mathbb{E}H$ for different values of μ using the same simulation run. Let $(A_1, B_1), \ldots, (A_\tau, B_\tau)$ denote the interarrival and service times in the first cycle. Then, for the first cycle we have

$$W_t = W_{t-1}\,\frac{\mu\,\mathrm{e}^{-\mu B_t}}{\widetilde{\mu}\,\mathrm{e}^{-\widetilde{\mu} B_t}}, \quad t = 1, 2, \ldots, \tau \quad (W_0 = 1) ,$$

$$\mathcal{S}_t = \mathcal{S}_{t-1} + \frac{1}{\mu} - B_t, \quad t = 1, 2, \ldots, \tau \quad (\mathcal{S}_0 = 0) ,$$

and H_t is as given in (11.29). From these the sums $\sum_{t=1}^{\tau} H_t W_t$, $\sum_{t=1}^{\tau} W_t$, $\sum_{t=1}^{\tau} W_t \mathcal{S}_t$, and $\sum_{t=1}^{\tau} H_t W_t \mathcal{S}_t$ can be computed sequentially. Repeating this for the subsequent cycles, one can estimate $\ell(\mu)$ and $\nabla\ell(\mu)$ from (11.27) and (11.28), respectively. Figure 11.2 displays the estimates and true values of $\nabla\ell(\mu)$ for $1.5 \leqslant \mu \leqslant 5.5$, using a single simulation run of $N = 10^5$ cycles. The simulation is carried out under the service rate $\widetilde{\mu} = 2$ and arrival rate $\lambda = 1$.

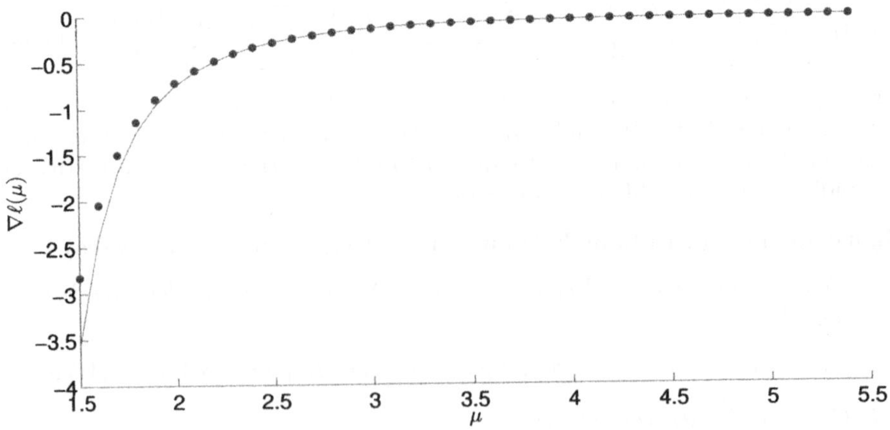

Figure 11.2 True (solid line) and estimated (points) values for the derivative of the expected steady-state waiting time as a function of μ.

We see that the gradient is estimated accurately for $\mu \geqslant 2$, but for $\mu < 2$ the quality of the estimates rapidly deteriorates. Indeed, the estimation should not be extended much below $\mu = 1.5$ as the importance sampling will result in unreliable estimates. The following MATLAB program is used.

```
%wtgrad.m
N = 10^5;
K = 40;
for i=1:K
    mu(i) = 1.5 + (i-1)*0.1;
end
mutil = 2; %simulate under this
lam = 1;
Rmat = zeros(N,K);
taumat = zeros(N,K);
gradWmat = zeros(N,K);
gradWmatHmat = zeros(N,K);
W = zeros(1,K);
Scor = zeros(1,K);
tau = zeros(1,K);
R = zeros(1,K);
for i = 1:N
    B = -log(rand)/mutil;
    W = mu.*exp(-mu.*B)/(mutil*exp(-mutil*B));
    A = -log(rand)/lam;
    H = max( B - A, 0);
    Scor = 1./mu - B;
    tau = W; %sum of W's
    R = H*W; %sum of H*W's
    gradW = Scor.*W; %sum of W*S
    gradWH = Scor.*W*H; %sum of W*S*H
    while (H > 0)
        B = -log(rand)/mutil;
        W = W.*mu.*exp(-mu.*B)/(mutil*exp(-mutil*B));
        A = -log(rand)/lam;
        H = max( H +  B - A, 0);
        Scor = Scor + 1./mu - B;
        tau = tau+W;
        R = R + H*W;
        gradW = gradW + Scor.*W; %sum of W*S's
        gradWH = gradWH + Scor.*W*H; %sum of W*S*H
    end
    taumat(i,:) = tau;
    Rmat(i,:) = R;
    gradWmat(i,:) = gradW;
    gradWmatHmat(i,:) = gradWH;
end
ell = zeros(1,K);
```

```
eltrue = lam.*(lam - 2*mu)./(lam - mu).^2./mu.^2;
for k=1:K
    ell(k) = mean(gradWmatHmat(:,k))/mean(taumat(:,k)) ...
           -(mean(Rmat(:,k))/mean(taumat(:,k))) ...
           *(mean(gradWmat(:,k))/mean(taumat(:,k)));
end
clf
hold on
plot(mu,ell,'.');
plot(mu,eltrue,'r');
hold off
```

Further Reading

The four methods for gradient estimation presented here have a long history. Overviews of the techniques can be found, for example, in [4, 11, 18, 19]. Whereas the finite difference methods simply extend ideas present in the numerical analysis literature, infinitesimal perturbation analysis was first presented in [9], and later rediscovered in [5] and [15] under the *likelihood ratio* name (see also [12] for a discussion on convergence rates). The score function method is older yet, appearing in [1] and [16]. An introduction to weak derivatives can be found in [14]. A unified view of the various gradient estimation techniques can be found in [10]. The paper also suggests hybrids between the two methods.

REFERENCES

1. V. M Aleksandrov, V. I. Sysoyev, and V. V. Shemeneva. Stochastic optimization. *Engineering Cybernetics*, 5(1):11–16, 1968.
2. S. Asmussen and P. W. Glynn. *Stochastic Simulation: Algorithms and Analysis.* Springer-Verlag, New York, 2007.
3. D. L. Brown. *Fundamentals of Statistical Exponential Families.* Institute of Mathematical Statistics, Hayward, CA, 1986.
4. P. Glasserman. *Gradient Estimation via Perturbation Analysis.* Kluwer, Norwell, MA, 1991.
5. P. W. Glynn. Likelihood ratio gradient estimation for stochastic systems. *Communications of the ACM*, 33(10):75–84, 1990.
6. D. Gross and C. M. Harris. *Fundamentals of Queueing Theory.* John Wiley & Sons, New York, second edition, 1985.
7. S. G. Henderson and B. L. Nelson, editors. *Handbooks in Operations Research and Management Science*, volume 13, chapter 19: Stochastic Gradient Estimation. North Holland, Amsterdam, 2006.
8. Y.-C. Ho and X.-R. Cao. *Perturbation Analysis of Discrete Event Dynamic Systems.* Kluwer, Norwell, MA, 1991.

9. Y.-C. Ho, M. A. Eyler, and T. T. Chien. A gradient technique for general buffer storage design in a serial production line. *International Journal on Production Research*, 17(6):557–580, 1979.

10. P. L'Ecuyer. A unified view of the IPA, SF, and LR gradient estimation techniques. *Management Science*, 36(11):1364–1383, 1990.

11. P. L'Ecuyer. An overview of derivative estimation. In B. L. Nelson, W. D. Kelton, and G. M. Clark, editors, *Proceedings of the 1991 Winter Simulation Conference*, pages 207–217, Piscataway, NJ, December 1991.

12. P. L'Ecuyer and G. Perron. On the convergence rates of IPA and FDC derivative estimators. *Operations Research*, 42(4):643–656, 1994.

13. E. L. Lehmann and J. P. Romano. *Testing Statistical Hypotheses*. Springer-Verlag, New York, third edition, 2008.

14. G. Ch. Pflug. *Optimization of Stochastic Models*. Kluwer, Boston, MA, 1996.

15. M. I. Reiman and A. Weiss. Sensitivity analysis for simulations via likelihood ratios. *Operations Research*, 37(5):830–844, 1989.

16. R. Y. Rubinstein. *Some Problems in Monte Carlo Optimization*. PhD thesis, University of Riga, Latvia, 1969. In Russian.

17. R. Y. Rubinstein. Sensitivity analysis of discrete event systems by the push out method. *Annals of Operations Research*, 39(1-4):229–250, 1993.

18. R. Y. Rubinstein and A. Shapiro. *Discrete Event Systems: Sensitivity Analysis and Stochastic Optimization via the Score Function Method*. John Wiley & Sons, New York, 1993.

19. J. C. Spall. *Introduction to Stochastic Search and Optimization: Estimation, Simulation, and Control*. John Wiley & Sons, New York, 2003.

20. R. Suri and M. A. Zazanis. Perturbation analysis gives strongly consistent sensitivity estimates for the M/G/1 queue. *Management Science*, 34(1):39–64, 1988.

CHAPTER 12

RANDOMIZED OPTIMIZATION

In this chapter we discuss optimization methods that have randomness as a core ingredient. Such randomized algorithms can be useful for solving optimization problems with many local optima and complicated constraints, possibly involving a mix of continuous and discrete variables. Randomized algorithms are also used to solve *noisy* optimization problems, in which the objective function is unknown and has to be obtained via Monte Carlo simulation.

We consider randomized optimization methods for both noisy and deterministic problems, including *stochastic approximation*, the *stochastic counterpart method*, *simulated annealing*, *evolutionary algorithms*, and the *cross-entropy method*.

We refer to Appendix C for background on deterministic optimization, and to Chapter 11 for gradient estimation techniques. Chapter 13 describes the cross-entropy method in more detail. ☞ 677 ☞ 421 ☞ 463

Throughout this chapter we use the letter S to denote the objective function.

12.1 STOCHASTIC APPROXIMATION

Suppose we have a minimization problem on $\mathscr{X} \subseteq \mathbb{R}^n$ of the form

$$\min_{\mathbf{x} \in \mathscr{X}} S(\mathbf{x}), \tag{12.1}$$

where S is an unknown function of the form $\mathbb{E}\widetilde{S}(\mathbf{x}, \boldsymbol{\xi})$, with $\boldsymbol{\xi}$ a random vector and $\widetilde{S}$ a known function. A typical example is where $S(\mathbf{x})$ is the (usually unknown) expected performance measure from a Monte Carlo simulation. Such a problem is said to be a **noisy** optimization problem, as typically only realizations of $\widetilde{S}(\mathbf{x}, \boldsymbol{\xi})$ can be observed.

Because the gradient ∇S is unknown, one cannot directly apply classical optimization methods. The **stochastic approximation method** mimics simple ☞ 690 gradient descent (see Section C.2.2.4) by replacing a deterministic gradient with a random approximation. More generally, one can approximate a subgradient instead of the gradient. It is assumed that an estimate of the gradient of S is available at any point $\mathbf{x} \in \mathscr{X}$. We denote such an estimate by $\widehat{\nabla S}(\mathbf{x})$. There are several established ways of obtaining $\widehat{\nabla S}(\mathbf{x})$. These include the finite difference method, infinitesimal perturbation analysis, the score function method, and the method of ☞ 421 weak derivatives — see Chapter 11, where S is replaced by ℓ and $\mathbf{x}$ by $\boldsymbol{\theta}$.

In direct analogy to gradient descent methods, the stochastic approximation method produces a sequence of iterates, starting with some $\mathbf{x}_1 \in \mathscr{X}$, via

$$\mathbf{x}_{t+1} = \Pi_{\mathscr{X}}\left(\mathbf{x}_t - \beta_t\, \widehat{\nabla S}(\mathbf{x}_t)\right), \tag{12.2}$$

where $\beta_1, \beta_2, \ldots$ is a sequence of strictly positive step sizes and $\Pi_{\mathscr{X}}$ is a projection operator that takes a point in $\mathbb{R}^n$ and returns a closest (typically in Euclidean distance) point in $\mathscr{X}$, ensuring that iterates remain feasible. That is, for any $\mathbf{y} \in \mathbb{R}^n$, $\Pi_{\mathscr{X}}(\mathbf{y}) \in \operatorname{argmin}_{\mathbf{z} \in \mathscr{X}} \|\mathbf{z} - \mathbf{y}\|$. Naturally, if $\mathscr{X} = \mathbb{R}^n$, then $\Pi_{\mathscr{X}}(\mathbf{y}) = \mathbf{y}$. A generic stochastic approximation algorithm is as follows.

Algorithm 12.1 (Stochastic Approximation)

1. *Initialize* $\mathbf{x}_1 \in \mathscr{X}$. *Set* $t = 1$.
2. *Obtain an estimated gradient* $\widehat{\nabla S}(\mathbf{x}_t)$ *of* S *at* $\mathbf{x}_t$.
3. *Determine a step size* β_t.
4. *Set* $\mathbf{x}_{t+1} = \Pi_{\mathscr{X}}\left(\mathbf{x}_t - \beta_t\, \widehat{\nabla S}(\mathbf{x}_t)\right)$.
5. *If a stopping criterion is met, stop; otherwise, set* $t = t + 1$ *and repeat from Step 2.*

There are many theorems on the convergence of stochastic approximation algorithms; see, for example, [14]. In particular, for an arbitrary deterministic positive sequence $\beta_1, \beta_2, \ldots$ such that

$$\sum_{t=1}^{\infty} \beta_t = \infty, \quad \sum_{t=1}^{\infty} \beta_t^2 < \infty,$$

the random sequence $\mathbf{x}_1, \mathbf{x}_2, \ldots$ converges in the mean square sense to the minimizer $\mathbf{x}^*$ of $S(\mathbf{x})$ under certain regularity conditions (see, for example, [15] for details).

We now present one of the simplest convergence theorems, drawn from Section 5.9 of [25].

Theorem 12.1.1 (Convergence of Stochastic Approximation) *Suppose the following conditions are satisfied:*

1. *The feasible set* $\mathscr{X} \subset \mathbb{R}^n$ *is convex, nonempty, closed, and bounded.*
2. $\Pi_{\mathscr{X}}$ *is the Euclidean projection operator.*
3. *The objective function* S *is well defined, finite valued, continuous, differentiable, and strictly convex in* $\mathscr{X}$ *with parameter* $\beta > 0$. *That is, there exists a* $\beta > 0$ *such that*

$$(\mathbf{y}-\mathbf{x})^\top(\nabla S(\mathbf{y}) - \nabla S(\mathbf{x})) \geqslant \beta\,\|\mathbf{y}-\mathbf{x}\|^2 \quad \textit{for all } \mathbf{x}, \mathbf{y} \in \mathscr{X} .$$

4. *The error in the stochastic gradient vector* $\widehat{\nabla S}(\mathbf{x})$ *possesses a bounded second moment. That is, for some* $K > 0$,

$$\mathbb{E}\,\|\widehat{\nabla S}(\mathbf{x})\|^2 \leqslant K^2 < \infty \quad \textit{for all } \mathbf{x} \in \mathscr{X} .$$

Then, if $\beta_t = c/t$ *for* $c > 1/(2\beta)$,

$$\mathbb{E}\,\|\mathbf{x}_t - \mathbf{x}^*\|^2 \leqslant \frac{Q(c)}{t}, \quad t = 1, 2, \ldots ,$$

where

$$Q(c) = \max\{c^2 K^2\,(2c\beta - 1)^{-1},\ \|\mathbf{x}_1 - \mathbf{x}^*\|^2\},$$

with minimal Q *attained by choosing* $c = 1/\beta$. *In other words, the expected error in terms of Euclidean distance of the iterates is of order* $\mathcal{O}(t^{-1/2})$.

Moreover, if $\mathbf{x}^*$ *is an interior point of* $\mathscr{X}$ *and if there is some constant* $L > 0$ *such that*

$$\|\nabla S(\mathbf{y}) - \nabla S(\mathbf{x})\| \leqslant L\|\mathbf{y}-\mathbf{x}\| \quad \textit{for all } \mathbf{x}, \mathbf{y} \in \mathscr{X} ,$$

(that is, $\nabla S(\mathbf{x})$ *is uniformly Lipschitz continuous in* $\mathscr{X}$*), then*

$$\mathbb{E}\,|S(\mathbf{x}_t) - S(\mathbf{x}^*)| \leqslant \frac{L\,Q(c)}{2t}, \quad t = 1, 2, \ldots .$$

In other words, the expected error in terms of Euclidean distance of the objective function values is of order $\mathcal{O}(t^{-1})$.

An attractive feature of the stochastic approximation method is its simplicity and ease of implementation in those cases where the projection $\Pi_{\mathscr{X}}$ can be easily computed. For example with **box-constraints**, where $\mathscr{X} = [a_1, b_1] \times \cdots \times [a_n, b_n]$, any component x_k of $\mathbf{x}$ is projected to a_k if $x_k < a_k$ and to b_k if $x_k > b_k$, and otherwise remains unchanged.

A weak point of the method is the ambiguity in choosing the step size sequence $\beta_1, \beta_2, \ldots$. Small step sizes lead to slow convergence and large step sizes may result in "zigzagging" behavior of the iterates. A commonly used choice is $\beta_t = c/t$ for some constant c, as suggested by Theorem 12.1.1. The practical performance of the algorithm using this step size rule depends crucially on c. This naturally leads to the idea of adaptively tuning this constant to the problem at hand.

If $\beta_t/\beta_{t+1} = 1 + o(\beta_t)$, as is the case when $\beta_t = 1/t^\gamma$ with $\gamma \in (0, 1)$, then the averaged iterate sequence defined by $\bar{\mathbf{x}}_t = \frac{1}{t}\sum_{k=1}^{t} \mathbf{x}_k$ tends to give better results than $\{\mathbf{x}_t\}$ itself; see [14, Chapter 11]. This is known as **Polyak averaging** or **iterate averaging**. One advantage here is that the algorithm will take larger step sizes than the $1/t$ case, speeding up the location of a solution.

When $\widehat{\nabla S}(\mathbf{x}_t)$ is an *unbiased* estimator of $\nabla S(\mathbf{x}_t)$ in (12.2) the stochastic approximation Algorithm 12.1 is referred to as the **Robbins–Monro** algorithm. When finite differences are used to estimate $\widehat{\nabla S}(\mathbf{x}_t)$ the resulting algorithm is known as
☞ 423 the **Kiefer–Wolfowitz** algorithm. As noted in Section 11.2, the gradient estimate usually has a bias that depends on the length of the interval corresponding to the central or forward difference estimators.

In high dimensions the **random directions** procedure can be used instead of the usual Kiefer–Wolfowitz algorithm, reducing the number of function evaluations per gradient estimate to two. This can be achieved as follows. Let $\mathbf{D}_1, \mathbf{D}_2, \ldots$ be a sequence of random direction vectors in $\mathbb{R}^n$, typically satisfying (though these are not strictly required [14]) the following conditions.

- The vectors are iid and symmetrically distributed with respect to each of the coordinate axes, with $\mathbb{E}\boldsymbol{D}_t\boldsymbol{D}_t^\top = I$ and $\|\boldsymbol{D}_t\|^2 = n$.

For example, one can take each $\boldsymbol{D}_t$ distributed uniformly on the sphere of radius $\sqrt{n}$, or each $\boldsymbol{D}_t$ distributed uniformly on $\{-1,1\}^n$ (that is, each component takes the values ± 1 with probability 1/2). However, the random directions method can exhibit poor behavior if the number of iterations is not sufficiently large (see [14]).

■ EXAMPLE 12.1 (Noisy Optimization by Stochastic Approximation)

We illustrate the stochastic approximation procedure via a simple problem of the form (12.1), with

$$\widetilde{S}(\mathbf{x}, \boldsymbol{\xi}) = \|\boldsymbol{\xi} - \mathbf{x}\|^2, \qquad \boldsymbol{\xi} \sim \mathsf{N}(\boldsymbol{\mu}, I) .$$

The function $S(\mathbf{x}) = \mathbb{E}\widetilde{S}(\mathbf{x}, \boldsymbol{\xi})$ has its minimum at $\mathbf{x}^* = \boldsymbol{\mu}$, with $S(\mathbf{x}^*) = n$. For this example we have $\nabla S(\mathbf{x}) = 2(\mathbf{x} - \boldsymbol{\mu})$. An unbiased (Robbins–Monro) estimator is

$$\widehat{\nabla S}(\mathbf{x})_{\mathrm{RM}} = \frac{1}{N}\sum_{k=1}^{N} 2(\mathbf{x} - \boldsymbol{\xi}_k) ,$$

where $\boldsymbol{\xi}_1, \ldots, \boldsymbol{\xi}_N \sim_{\mathrm{iid}} \mathsf{N}(\boldsymbol{\mu}, I)$. A central difference (Kiefer–Wolfowitz) estimator, with difference interval δ, is

$$(\widehat{\nabla S}(\mathbf{x})_{\mathrm{KW}})_i = \frac{1}{N}\sum_{k=1}^{N} \frac{\widetilde{S}(\mathbf{x} + \mathbf{e}_i\,\delta/2, \boldsymbol{\xi}_k) - \widetilde{S}(\mathbf{x} - \mathbf{e}_i\,\delta/2, \boldsymbol{\zeta}_k)}{\delta}, \quad i = 1, \ldots, n ,$$

☞ 423 where $\boldsymbol{\xi}_1, \boldsymbol{\zeta}_1, \ldots, \boldsymbol{\xi}_N, \boldsymbol{\zeta}_N \sim_{\mathrm{iid}} \mathsf{N}(\boldsymbol{\mu}, I)$. As observed in Section 11.2, the variance of this estimator can be reduced significantly by using common random variables. A practical way to do this here is to take $\boldsymbol{\zeta}_k = \boldsymbol{\xi}_k$ for each $k = 1, \ldots, N$.

Figure 12.1 illustrates the typical performance of the Robbins–Monro and Kiefer–Wolfowitz algorithms for this problem. The Kiefer–Wolfowitz algorithm is run with and without using common random variables. For each method we use 10^4 iterations. The problem dimension is $n = 100$ and $\boldsymbol{\mu} = (n, n-1, \ldots, 2, 1)^\top$. Each gradient estimate is computed using $N = 10$ independent trials. The MATLAB implementation is given below.

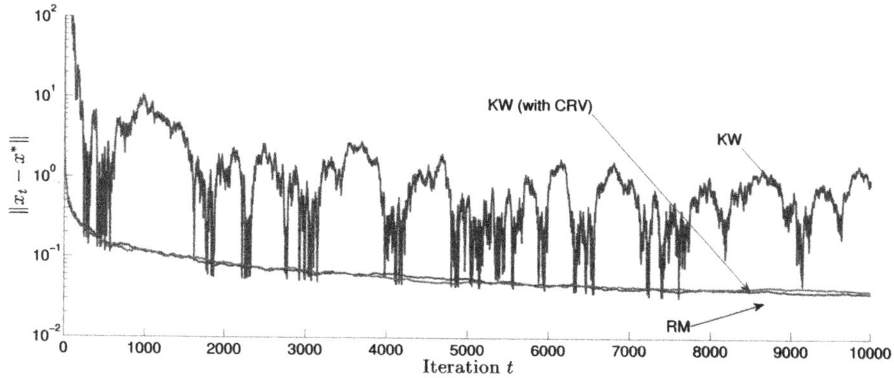

Figure 12.1 Typical convergence of Robbins–Monro and Kiefer–Wolfowitz algorithms (with and without making use of common random variables).

```
%StochApprox.m
maxits=10^4; % number of iterations
n=10^2; % dimension
N=10^1; % number of trials
mu=(n:-1:1); rmu=repmat(mu,N,1); % problem data
L=zeros(N,n); R=L; % allocate space for the central diff. estimator

c=1; % constant for the step size
delta = 1; % constant for the FD sequence
betat=@(t) c./t; % step size functions
deltat=@(t) delta/t.^(1/6); % difference interval function

xrm=10.*n.*ones(1,n); % initial Robbins-Monro iterate
xkw=10.*n.*ones(1,n); % initial Kiefer-Wolfowitz iterate
xkwCRV=10.*n.*ones(1,n); % initial Kiefer-Wolfowitz iterate w. CRV

% allocate space for the convergence history of each iterate
rmhist=zeros(1,maxits);
kwhist=zeros(1,maxits);
kwCRVhist=zeros(1,maxits);
% compute initial distance to optimal solution
rmhist(1)=sqrt(sum((xrm-mu).^2));
kwhist(1)=sqrt(sum((xkw-mu).^2));
kwCRVhist(1)=sqrt(sum((xkwCRV-mu).^2));

t=1; % iteration Counter
while (t<maxits)
    % RM gradient est.
    xi=rmu+randn(N,n);
    grm=mean(2.*(repmat(xrm,N,1)-xi),1); % unbiased est.
    % KW gradient est.
```

```
    xiL=rmu+randn(N,n);
    xiR=rmu+randn(N,n);
    xkwN=repmat(xkw,N,1);
    e1=zeros(1,n);e1(1)=deltat(t)/2;
    ekN=repmat(e1,N,1);
    for k=1:n
        L(:,k)=sum((xiL-(xkwN+ekN)).^2,2);
        R(:,k)=sum((xiR-(xkwN-ekN)).^2,2);
        ekN=circshift(ekN,[0 1]);
    end
    gkw=mean((L-R)./deltat(t),1);
    % KW gradient est. with CRV
    xiL=rmu+randn(N,n);
    xiR=xiL; % practical CRV
    xkwCRVN=repmat(xkwCRV,N,1);
    for k=1:n
        L(:,k)=sum((xiL-(xkwCRVN+ekN)).^2,2);
        R(:,k)=sum((xiR-(xkwCRVN-ekN)).^2,2);
        ekN=circshift(ekN,[0 1]);
    end
    gkwCRV=mean((L-R)./deltat(t),1);
    % Update Iterates
    xrm=xrm-betat(t).*grm;
    xkw=xkw-betat(t).*gkw;
    xkwCRV=xkwCRV-betat(t).*gkwCRV;
    % increase iteration counter and record new distance to optimum
    t=t+1;
    rmhist(t)=sqrt(sum((xrm-mu).^2));
    kwhist(t)=sqrt(sum((xkw-mu).^2));
    kwCRVhist(t)=sqrt(sum((xkwCRV-mu).^2));
end
% plot the results
tt=(1:1:(maxits));
figure,semilogy(tt,rmhist,'k-',tt,kwhist,'b-',tt,kwCRVhist,'r-',...
    'Linewidth',1.5)
```

12.2 STOCHASTIC COUNTERPART METHOD

Consider again the noisy optimization setting (12.1). The idea of the **stochastic counterpart method** (also called **sample average approximation**) is to replace the noisy optimization problem (12.1) with

$$\min_{\mathbf{x} \in \mathbb{R}^n} \widehat{S}(\mathbf{x}), \tag{12.3}$$

where

$$\widehat{S}(\mathbf{x}) = \frac{1}{N} \sum_{i=1}^{N} \widetilde{S}(\mathbf{x}, \boldsymbol{\xi}_i)$$

is a sample average estimator of $S(\mathbf{x}) = \mathbb{E}\widetilde{S}(\mathbf{x}, \boldsymbol{\xi})$ on the basis of N iid samples $\boldsymbol{\xi}_1, \ldots, \boldsymbol{\xi}_N$.

A solution $\widehat{\mathbf{x}}^*$ to this sample average version is taken to be an estimator of a solution $\mathbf{x}^*$ to the original problem (12.1). Note that (12.3) is a *deterministic* optimization problem to which any of the methods in Appendix C could apply.

■ EXAMPLE 12.2 (Stochastic Counterpart)

Consider the following parametric optimization problem that arises when applying the CE method: given a family of densities $\{f(\cdot;\mathbf{v}), \mathbf{v} \in \mathscr{V}\}$ and a target density g, locate the CE optimal parameter $\mathbf{v}^*$ that maximizes ☞ 463

$$\mathcal{D}(\mathbf{v}) \stackrel{\text{def}}{=} \mathbb{E}_g \ln f(\mathbf{Z};\mathbf{v}) = \mathbb{E}_p \left[\frac{g(\mathbf{Z})}{p(\mathbf{Z})} \ln f(\mathbf{Z};\mathbf{v}) \right],$$

where p is any pdf that *dominates* g; that is, $p(\mathbf{z}) = 0 \Rightarrow g(\mathbf{z}) = 0$. Typically this optimization problem is difficult to solve, but one can consider solving the stochastic counterpart instead, here given by

$$\max_{\mathbf{v}\in\mathscr{V}} \widehat{\mathcal{D}}(\mathbf{v}) \stackrel{\text{def}}{=} \max_{\mathbf{v}\in\mathscr{V}} \frac{1}{N} \sum_{k=1}^{N} \frac{g(\mathbf{Z}_k)}{p(\mathbf{Z}_k)} \ln f(\mathbf{Z}_k;\mathbf{v}), \tag{12.4}$$

where $\mathbf{Z}_1, \ldots, \mathbf{Z}_N \sim_{\text{iid}} p$. For various parametric families $\{f(\cdot;\mathbf{v})\}$ this proxy problem is solvable analytically, providing the key updating formulas for the CE method.

As a particular instance, suppose f is a Cauchy density, given by

$$f(z;\mathbf{v}) = \frac{1}{\pi\sigma} \frac{1}{1 + \left(\frac{z-\mu}{\sigma}\right)^2}, \quad \mathbf{v} = (\mu, \sigma),$$

and the target density is

$$g(z) \propto \left|\exp(-z^2)\cos(3\pi z)\right| \exp\left(-\frac{1}{2}\left(\frac{z-1}{\sqrt{2}}\right)^2\right).$$

The standard Cauchy density $p(z) = 1/(\pi(1+z^2))$ dominates $g(z)$.

Figure 12.2 depicts the pdfs obtained by solving 100 independent instances of the stochastic counterpart (12.4) for this problem, using approximating sample sizes of $N = 10^3$ and $N = 10^5$, respectively.

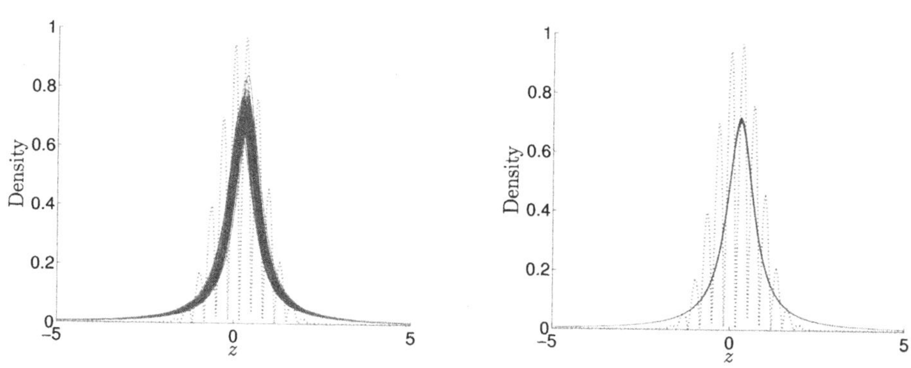

Figure 12.2 The pdfs of 100 independent solutions of the stochastic counterpart procedure using $N = 10^3$ (left) and $N = 10^5$ (right) samples. The dotted line is the target density g.

Figure 12.3 plots a typical sequence of estimates for μ and σ as a function of the sample size $N = 1, \ldots, 10^5$. The estimates of μ and σ obtained in this way strongly suggest convergence to an optimal value.

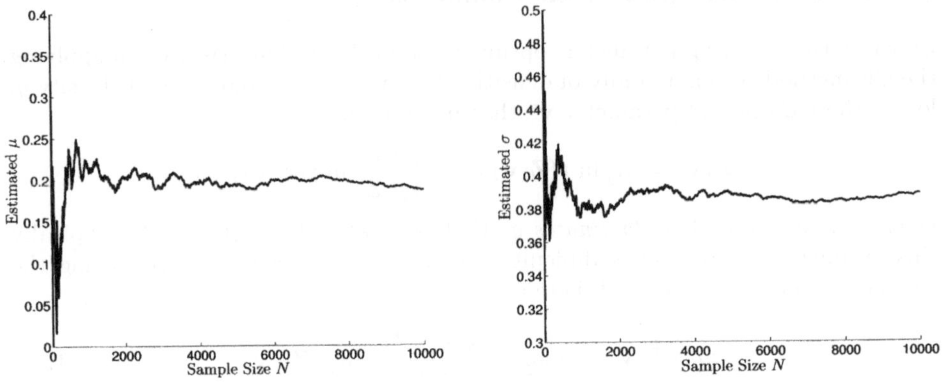

Figure 12.3 The estimates for μ and σ obtained by solving a sequence of stochastic counterpart problems for increasing N.

MATLAB code for the left panel in Figure 12.2 follows. Simply change the variable `N` to obtain the right panel. Code for Figure 12.3 is quite similar, and can be found on the Handbook website as `SCMb.m`.

```
%SCM.m
clear all
N=10^3; % Sample size
M=10^2; % Number of trials

g=@(Z) abs(exp(-Z.^2).*cos(3.*pi.*Z)).*...
    exp(-0.5.*((Z-1)./sqrt(2)).^2)./sqrt(2*pi*2);
h=@(Z,mu,sigma) (sigma>0)./(pi.*sigma.*(1+((Z-mu)./sigma).^2));
p=@(Z) 1./(pi.*(1+Z.^2)); % standard cauchy
f=@(x,Z) sum((g(Z)./p(Z)).*log(h(Z,x(1),x(2))));% St. Counterpart

approxnormg=.2330967533;% Approx. norm. const. for g (for plotting)
zz=linspace(-5,5,10^3);% Range to plot densities over
figure,hold on
for k=1:M
    Z=randn(N,1)./randn(N,1);% Z_1,...,Z_N are iid with density p
    sol=fminsearch(@(x) -f(x,Z),[1,2]);% Solve the SC
    plot(zz,h(zz,sol(1),sol(2)),'r-')
end
plot(zz,g(zz)./approxnormg,'k:','LineWidth',2) % Plot g
hold off
```

12.3 SIMULATED ANNEALING

Simulated annealing is a Markov chain Monte Carlo technique for approximately locating a global maximum of a given density $f(\mathbf{x})$. The idea is to create a sequence of points $\mathbf{X}_1, \mathbf{X}_2, \ldots$ that are approximately distributed according to pdfs $f_1(\mathbf{x}), f_2(\mathbf{x}), \ldots$ with $f_t(\mathbf{x}) \propto f(\mathbf{x})^{1/T_t}$, where $T_1, T_2, \ldots$ is a sequence of **temperatures** (known as the **annealing schedule**) that decreases to 0. If each $\mathbf{X}_t$ were sampled *exactly* from $f(\mathbf{x})^{1/T_t}$, then X_t would converge to a global maximum of $f(\mathbf{x})$ as $T_t \to 0$. However, in practice sampling is *approximate* and convergence to a global maximum is not assured. ☞ 225

A high-level simulated annealing algorithm is as follows.

Algorithm 12.2 (Simulated Annealing)

1. *Choose a starting state* $\mathbf{X}_0$ *and an initial temperature* T_0. *Set* $t = 1$.
2. *Select a temperature* $T_t \leqslant T_{t-1}$ *according to the annealing schedule.*
3. *Approximately generate a new state* $\mathbf{X}_t$ *from* $f_t(\mathbf{x}) \propto (f(\mathbf{x}))^{1/T_t}$.
4. *Set* $t = t + 1$ *and repeat from Step 2 until stopping.*

The most common application for simulated annealing is in optimization. In particular, consider the minimization problem

$$\min_{\mathbf{x} \in \mathscr{X}} S(\mathbf{x})$$

for some deterministic real-valued function $S(\mathbf{x})$. Define the **Boltzmann pdf** as

$$f(\mathbf{x}) \propto \mathrm{e}^{-S(\mathbf{x})}, \quad \mathbf{x} \in \mathscr{X} .$$

For $T > 0$ close to 0 the global maximum of $f(\mathbf{x})^{1/T} \propto \exp(-S(\mathbf{x})/T)$ is close to the global minimum of $S(\mathbf{x})$. Hence, by applying simulated annealing to the Boltzmann pdf, one can also minimize $S(\mathbf{x})$. Maximization problems can be handled in a similar way, by using a Boltzmann pdf $f(\mathbf{x}) \propto \exp(S(\mathbf{x}))$. Note that this may not define a valid pdf if the exponential terms are not normalizable.

There are many different ways to implement simulated annealing algorithms, depending on (1) the choice of Markov chain Monte Carlo sampling algorithm, (2) the length of the Markov chain between temperature updates, and (3) the annealing schedule. A popular annealing schedule is **geometric cooling**, where $T_t = \beta T_{t-1}$, $t = 1, 2, \ldots$, for a given initial temperature T_0 and a **cooling factor** $\beta \in (0, 1)$. Appropriate values for T_0 and β are problem-dependent, and this has traditionally required tuning on the part of the user. Theoretical results on adaptive tuning schemes may be found, for example, in [22, 26, 27].

The following algorithm describes a popular simulated annealing framework for minimization, which uses a random walk sampler; that is, a Metropolis–Hastings sampler with a symmetric proposal distribution. Note that the temperature is updated after a *single* step of the Markov chain. ☞ 230

Algorithm 12.3 (Simulated Annealing for Minimization)

1. *Initialize the starting state* $\mathbf{X}_0$ *and temperature* T_0*. Set* $t = 1$.
2. *Select a temperature* $T_t \leqslant T_{t-1}$ *from the annealing schedule.*
3. *Generate a candidate state* $\mathbf{Y}$ *from the symmetric proposal density* $q(\mathbf{Y} \mid \mathbf{X}_t) = q(\mathbf{X}_t \mid \mathbf{Y})$.
4. *Compute the acceptance probability*

$$\alpha(\mathbf{X}_t, \mathbf{Y}) = \min\left\{ \mathrm{e}^{-\frac{(S(\mathbf{Y})-S(\mathbf{X}_t))}{T_t}}, 1 \right\} .$$

Generate $U \sim \mathsf{U}(0,1)$ *and set*

$$\mathbf{X}_{t+1} = \begin{cases} \mathbf{Y} & \text{if } U \leqslant \alpha(\mathbf{X}_t, \mathbf{Y}) , \\ \mathbf{X}_t & \text{if } U > \alpha(\mathbf{X}_t, \mathbf{Y}) . \end{cases}$$

5. *Set* $t = t + 1$ *and repeat from Step 2 until a stopping criterion is met.*

■ **EXAMPLE 12.3 (Continuous Optimization by Simulated Annealing)**

We illustrate the simulated annealing Algorithm 12.3 by applying it to the minimization of the *trigonometric function*. For n-dimensional $\mathbf{x}$, this function is given by ☞ 697

$$S(\mathbf{x}) = 1 + \sum_{i=1}^{n} \left(8\sin^2(\eta(x_i - x_i^*)^2) + 6\sin^2(2\eta(x_i - x_i^*)^2) + \mu(x_i - x_i^*)^2 \right),$$

and has minimal value of 1 at $\mathbf{x} = \mathbf{x}^*$. In this example, we choose $n = 10$, $\mathbf{x}^* = (10, \ldots, 10)$, $\eta = 0.8$, and $\mu = 0.1$.

In order to implement the method with Metropolis–Hastings sampling, there are four ingredients we must specify: (1) an appropriate initialization of the algorithm; (2) an annealing schedule $\{T_t\}$; (3) a proposal pdf q; and (4) a stopping criterion.

For initialization, we set $T_0 = 10$ and draw $\mathbf{X}_0$ uniformly from the n-dimensional hypercube $[-50, 50]^n$. We use a geometric cooling scheme with cooling factor $\beta = 0.99999$. The (symmetric) proposal distribution (starting from $\mathbf{x}$) is taken to be $\mathsf{N}(\mathbf{x}, \sigma^2 I)$, with $\sigma = 0.75$. The algorithm is set to stop after 10^6 iterations. The MATLAB code is given below.

```
%SA.m
% Initialization
n=10; % dimension of the problem
beta=0.99999; % Factor in geometric cooling
sigma=ones(1,n).*0.75; % Variances for the proposal
N=1; %Number of steps to perform of MH
maxits=10^6; % Number of iterations
xstar=10.*ones(1,n); eta=0.8; mu=0.1;
S=@(X) 1+sum(mu.*(X-xstar).^2+6.*(sin(2.*eta.*(X-xstar).^2)).^2+...
    8.*(sin(eta.*(X-xstar).^2)).^2);
T=10; % Initial temperature
```

```
a=-50;b=50; X=(b-a).*rand(1,n)+a; %Initialize X
Sx=S(X); % Score the initial sample

t=1; % Initialize iteration counter
while (t<=maxits)
    T=beta*T; % Select New Temperature
    % Generate New State
    it=1;
    while (it<=N)
        Y=X+sigma.*randn(1,n);
        Sy=S(Y);
        alpha=min(exp(-(Sy-Sx)/T),1);
        if rand<=alpha
           X=Y; Sx=Sy;
        end
        it=it+1;
    end
    t=t+1; % Increment Iteration
end
[X,Sx,T] % Display final state, score, and temperature
```

For this example, we can sample exactly from the Boltzmann distribution via a straightforward application of *acceptance–rejection* (see Section 3.1.5). As a consequence, we can see precisely how well our approximate sampling performs with respect to the ideal case. ☞ 59

In Figure 12.4, the average performance per iteration over 10 independent trials is plotted for both the approximate Metropolis–Hastings sampling from the Boltzmann pdf used above and exact sampling from the Boltzmann pdf via acceptance–rejection. For these parameters, the approximate scheme typically fails to locate the optimal solution.

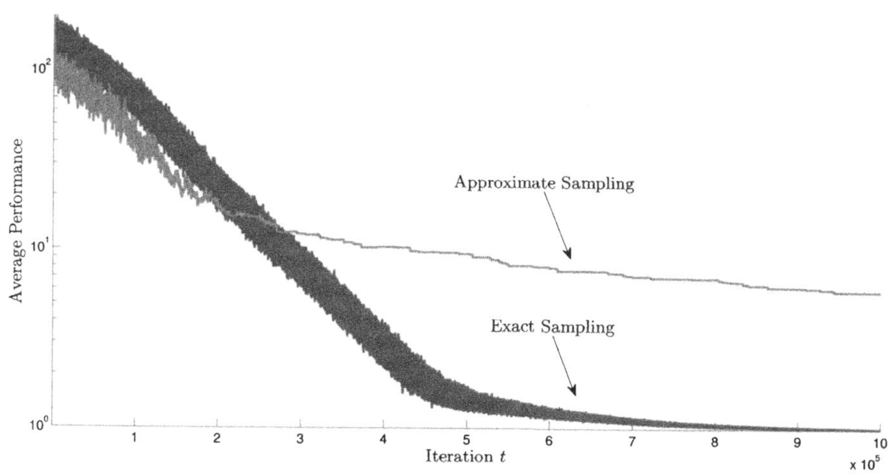

Figure 12.4 Average performance per iteration of exact and approximate sampling from the Boltzmann pdf, over 10 independent trials.

The MATLAB code used for this figure can be found on the Handbook website as SAnnealing_Multi_SA.m.

12.4 EVOLUTIONARY ALGORITHMS

Evolutionary algorithms refer to any metaheuristic framework that is inspired by the process of natural evolution. An algorithm of this type begins with a **population** $\mathscr{P}$ of **individuals** $\mathbf{x}$: objects such as points in $\mathbb{R}^n$, paths in a graph, etc. The population "evolves" from generation to generation in two stages. First, some **selection** mechanism is applied to create a new population. Second, some **alteration** mechanism is applied to the newly created population.

The objective is to create a population of individuals with superior qualities with respect to some performance measure(s) on the population. The simplest example is where $\mathscr{P}$ is a collection of n-dimensional points $\mathbf{x}$ and the goal is to minimize some objective function $S(\mathbf{x})$. In this case the **evaluation** of $\mathscr{P}$ corresponds to calculating $S(\mathbf{x})$ for all $\mathbf{x} \in \mathscr{P}$. Typically this information is used in the selection phase, for instance by only permitting the best performing 10% to be involved in creating the new generation.

The general framework for an evolutionary algorithm is summarized next.

Algorithm 12.4 (Generic Evolutionary Algorithm)

1. *Set* $t = 0$. *Initialize a population of individuals* $\mathscr{P}_t$. *Evaluate* $\mathscr{P}_t$.
2. *Select a new population* $\mathscr{P}_{t+1}$ *from* $\mathscr{P}_t$.
3. *Alter the population* $\mathscr{P}_{t+1}$.
4. *Evaluate* $\mathscr{P}_{t+1}$.
5. *If a stopping criterion is met, stop; otherwise, set* $t = t + 1$ *and repeat from Step 2.*

To appreciate the diversity of selection and alteration operations, we refer the reader to [3, 4] and [6].

There are many well-known heuristic algorithms under the umbrella of evolutionary algorithms. We will discuss three in particular: genetic algorithms, differential evolution, and estimation of distribution algorithms.

12.4.1 Genetic Algorithms

The **genetic algorithm** metaheuristic is traditionally applied to discrete optimization problems. Individuals in the population are vectors, coded to represent potential solutions to the optimization problem. Each individual is ranked according to a fitness criterion (typically just the objective function value associated with that individual). A new population is then formed as children of the previous population. This is often the result of **cross-over** and **mutation** operations applied to the fittest individuals.

Suppose that individuals in the population are n-dimensional binary vectors $\mathbf{x}$, and the goal is to minimize some objective function $S(\mathbf{x})$. A possible cross-over mechanism in this case is **one-point crossover**: given two parents $\mathbf{x}$ and $\mathbf{y}$, and a random location r between 0 and n, create a new individual $\mathbf{z} = (x_1, \ldots, x_r, y_{r+1}, \ldots, y_n)$ whose first r components are copied from the first parent and the remaining $n - r$ components from the second parent.

Determining the M "fittest" individuals could be via **tournament selection** (for example, [6, Pages 75–80]). In basic tournament selection with tournaments of size K, this involves selecting K individuals uniformly from the population, and selecting the individual with the lowest objective function value as the winner. The winner then joins the reproduction pool. This process is repeated M times, until the desired number of fittest individuals is selected.

A typical binary encoded genetic algorithm is as follows.

Algorithm 12.5 (Binary Encoded Genetic Algorithm)

1. *Set* $t = 0$. *Initialize a population of individuals* $\mathscr{P}_t = \{\mathbf{x}_1^t, \ldots, \mathbf{x}_N^t\}$ *via uniform sampling over* $\{0,1\}^n$. *Evaluate* $\mathscr{P}_t$.

2. *Construct a* **reproduction pool** $\mathscr{R}_t$ *of individuals from* $\mathscr{P}_t$ *via tournament selection.*

3. **Combine** *individuals in the reproduction pool to obtain an intermediate population* $\mathscr{C}_t$ *via one-point crossover.*

4. **Mutate** *the intermediate population* $\mathscr{C}_t$ *by flipping each component of each binary vector independently with probability* $p = 1/n$. *Denote the resulting population by* $\mathscr{S}_t$.

5. **Create** *the new generation as* $\mathscr{P}_{t+1} = \mathscr{S}_t$. *Evaluate* $\mathscr{P}_{t+1}$.

6. *If a stopping criterion is met, stop; otherwise, set* $t = t + 1$ *and repeat from Step 2.*

■ EXAMPLE 12.4 (Genetic Algorithm for the Satisfiability Problem)

We illustrate the binary encoded genetic Algorithm 12.5 by applying it to solving the *satisfiability problem* (SAT). The problem is to find a binary vector $\mathbf{x} = (x_1, x_2, \ldots, x_n) \in \{0,1\}^n$ which, given a set of m clause functions $c_j(\cdot)$, satisfies all of them. Each clause function returns $c_j(\mathbf{x}) = 1$ if $\mathbf{x}$ satisfies clause j and $c_j(\mathbf{x}) = 0$ otherwise. (See Section C.3.1 for more details.) ☞ 694

In Algorithm 12.5, we select a population and reproduction pool size of $N = 10^4$, a tournament size of $K = 2$ (binary tournament selection), and we run the algorithm for 10^3 iterations.

In Figure 12.5, the scores of the best and worst performing individuals are plotted for a typical algorithm run on the difficult problem `F34-5-23-31` from `http://www.is.titech.ac.jp/~watanabe/gensat/a2/index.html`. Note that there are 361 clauses and 81 literals, and that the algorithm in this case locates solutions that satisfy at most 359 clauses.

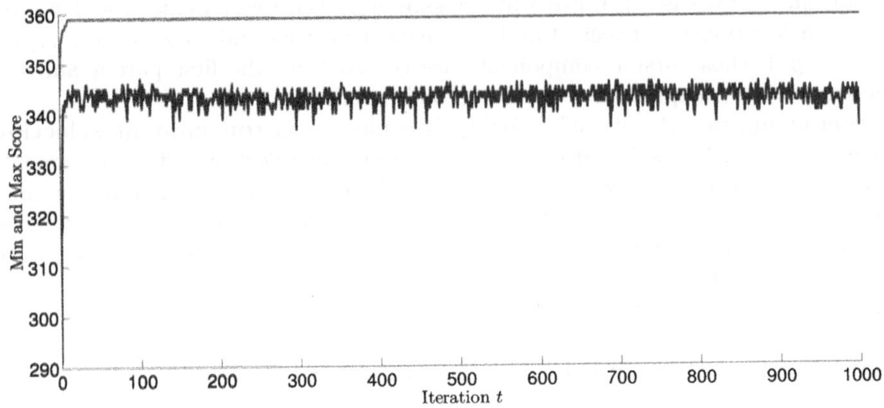

Figure 12.5 Typical best and worst performances of individuals using Algorithm 12.5 on the `F34-5-23-31` problem.

The MATLAB code used for this example can be found on the Handbook website as `GA_ex_fig.m`.

12.4.2 Differential Evolution

The method of **differential evolution** [20] is traditionally applied to continuous optimization problems. In its simplest version, on each iteration, a new population of points is constructed from the old parent population by moving each of the old points by a fixed step size in a direction determined by taking the difference of two other randomly determined points. The new population then produces a child population through crossover with the old parent population. Finally, each child only replaces its corresponding parent in the new parent population if it has a better performance.

A typical differential evolution algorithm for minimization of a function $S(\mathbf{x})$ is as follows.

Algorithm 12.6 (Differential Evolution Algorithm for Minimization)

1. *Set $t = 0$. Initialize a population of individuals $\mathscr{P}_t = \{\mathbf{x}_1^t, \ldots, \mathbf{x}_N^t\}$, say via uniform sampling over a known bounding box.*

2. *For each individual in the population, say individual $\mathbf{x}_k^t$:*

   (a) *Construct a vector $\mathbf{y}_k^{t+1} = \mathbf{x}_{R_1}^t + \alpha\,(\mathbf{x}_{R_2}^t - \mathbf{x}_{R_3}^t)$, where $R_1 \neq R_2 \neq R_3$ are three integers uniformly sampled from the set $\{1, 2, \ldots, k-1, k+1, \ldots, N\}$.*

   (b) *Apply* **binary crossover** *between $\mathbf{y}_k^{t+1}$ and $\mathbf{x}_k^t$ to obtain a trial vector $\widetilde{\mathbf{x}}_k^{t+1}$; that is,*

   $$\widetilde{\mathbf{x}}_k^{t+1} = (U_1\, y_{k,1}^{t+1} + (1 - U_1)\, x_{k,1}^t, \ldots, U_n\, y_{k,n}^{t+1} + (1 - U_n)\, x_{k,n}^t)\,,$$

   where $U_1, \ldots, U_d \sim_{\text{iid}} \mathsf{Ber}(p)$. Additionally, select a random index I, uniformly distributed on $\{1, \ldots, n\}$ and set $\widetilde{x}_{k,I}^{t+1} = y_{k,I}^{t+1}$.

(c) *If* $S(\widetilde{\mathbf{x}}_k^{t+1}) \leqslant S(\mathbf{x}_k^t)$, *set* $\mathbf{x}_k^{t+1} = \widetilde{\mathbf{x}}_k^{t+1}$; *otherwise, retain the old individual via* $\mathbf{x}_k^{t+1} = \mathbf{x}_k^t$.

3. *If a stopping criterion is met, stop; otherwise, set* $t = t + 1$ *and repeat from Step 2.*

The **scaling factor** α and the **crossover factor** p are algorithm parameters. Typical values to try are $\alpha = 0.8$ and $p = 0.9$, with a suggested population size of $N = 10\,n$.

■ EXAMPLE 12.5 (Differential Evolution)

We illustrate differential evolution by applying it to minimize the 50-dimensional Rosenbrock function (see Section C.4.1.3), given by ☞ 696

$$S(\mathbf{x}) = \sum_{i=1}^{49} \left(100\,(x_{i+1} - x_i^2)^2 + (x_i - 1)^2\right) , \tag{12.5}$$

which has minimal value of $S(\mathbf{x}) = 0$ at $\mathbf{x} = (1, \ldots, 1)$.

The population size is $N = 50$, the scaling factor is $\alpha = 0.8$, and the crossover probability is $p = 0.9$. The population is initialized by sampling uniformly on $[-50, 50]^{50}$ and is stopped after 5×10^4 iterations.

Figure 12.6 shows the typical progress of differential evolution on this problem. The best and worst performance function values per iteration are depicted. A MATLAB implementation follows.

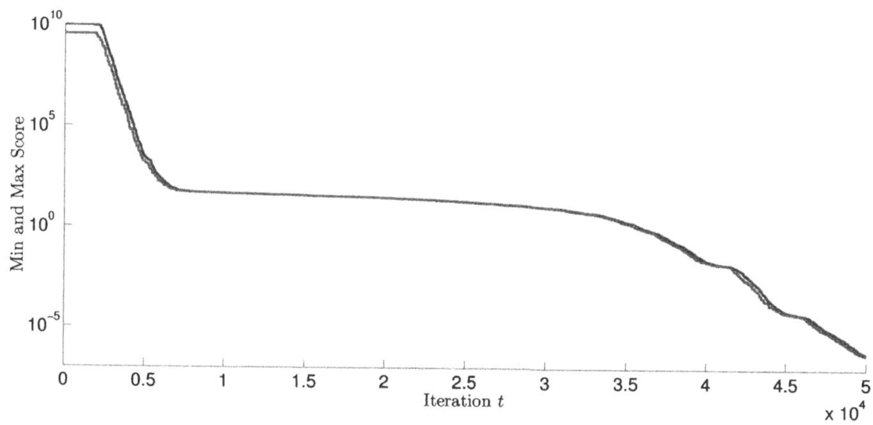

Figure 12.6 Best and worst performance function values on a typical run of a differential evolution algorithm on the 50-dimensional Rosenbrock function.

```
%DE_ex.m
M=50; % Population Size
n=50; % Dimension of the problem
F=0.8; % Scaling factor
CR=0.9; % Crossover factor
```

```
maxits=5*10^4; % Maximum number of iterations
Smaxhist=NaN.*ones(1,maxits); Sminhist=NaN.*ones(1,maxits);
% Rosenbrock Function
S=@(X)sum(100.*(X(:,2:n)-X(:,1:(n-1)).^2).^2+(X(:,1:(n-1))-1).^2,2);
a=-50; b=50; X=(b-a).*rand(M,n)+a; % Initialize population
t=1; % Iteration Counter
while (t<maxits)
    SX=S(X); [SX,idx]=sort(SX,1,'ascend'); % Score and Sort
    Smaxhist(t)=SX(M); Sminhist(t)=SX(1); % Update histories
    % Construct the new generation
    for i=1:M
        % Mutation
        r=[1:i-1,i+1:M];
        r=r(randperm(M-1));
        V=X(r(1),:)+F.*(X(r(2),:)-X(r(3),:));
        % Binomial Crossover
        U=X(i,:);
        idxr=1+floor(rand(1).*n);
        for j=1:n
            if (rand(1)<=CR)||(j==idxr)
                U(j)=V(j);
            end
        end
        if S(U)<=S(X(i,:))
            X(i,:)=U;
        end
    end
    t=t+1;
end
SX=S(X); [SX,idx]=sort(SX,1,'ascend'); % Score and Sort
Smaxhist(t)=SX(M); Sminhist(t)=SX(1); % Update histories
% Display worst & best score, and best performing sample
[SX(M),SX(1),X(idx(1),:)]
% Plot the results
figure, plot((1:1:t),Smaxhist,'k-',(1:1:t),Sminhist,'r-')
```

12.4.3 Estimation of Distribution Algorithms

The **estimation of distribution** [16] heuristic differs from the genetic and differential evolution algorithms in that successive populations are not directly constructed from previous ones. Instead of directly manipulating the individuals in the population, one uses the population to establish a probability distribution from which a subsequent generation of candidates is drawn. Typically, these replace (or are merged with) individuals in the existing population, according to their performance. A generic estimation of distribution algorithm looks as follows.

Algorithm 12.7 (Generic Estimation of Distribution Algorithm)

1. *Set* $t = 0$. *Initialize a population of individuals* $\mathscr{P}_t$. *Evaluate* $\mathscr{P}_t$.
2. *Select an intermediate population* $\mathscr{R}_t$ *from* $\mathscr{P}_t$.
3. *Estimate a distribution* F_{t+1} *from* $\mathscr{R}_t$.
4. *Sample the new population* $\mathscr{P}_{t+1}$ *according to* F_{t+1}.
5. *Evaluate* $\mathscr{P}_{t+1}$.
6. *If a stopping criterion is met, stop; otherwise, set* $t = t + 1$ *and repeat from Step 2.*

This has many similarities with the CE method discussed in the next section. Indeed, if one selects the best $\varrho \times 100\%$ (with $\varrho \in (0, 1]$) of individuals on the basis of the performance function, then, for example, the *univariate marginal distribution algorithm* of [16] is identical to the standard CE algorithm for optimization, with $\mathscr{X} = \{0, 1\}^n$ and a sampling distribution formed of independent Bernoulli components.

12.5 CROSS-ENTROPY METHOD FOR OPTIMIZATION

The **cross-entropy** (CE) method can be used for both deterministic and noisy optimization. Consider first the deterministic minimization problem

$$\min_{x \in \mathscr{X}} S(\mathbf{x}) ,$$

where S is some real-valued performance function on a set $\mathscr{X}$. The basic idea of the CE method for optimization is to define a parametric family of probability densities $\{f(\cdot; \mathbf{v}), \mathbf{v} \in \mathscr{V}\}$ on the state space $\mathscr{X}$, and to iteratively update the parameter $\mathbf{v}$ so that $f(\cdot; \mathbf{v})$ places more mass closer to solutions than on the previous iteration. In practice, this results in an algorithm with two basic phases:

- *Sampling*: Samples $\mathbf{X}_1, \ldots, \mathbf{X}_N$ are drawn independently according to $f(\cdot; \mathbf{v})$. The objective function S is evaluated at these points.
- *Updating*: A new parameter $\widehat{\mathbf{v}}$ is selected on the basis of those $\mathbf{X}_i$ for which $S(\mathbf{X}_i) \leqslant \widehat{\gamma}$ for some level $\widehat{\gamma}$. These $\{\mathbf{X}_i\}$ form the **elite sample** set, $\mathscr{E}$.

At each iteration the level parameter $\widehat{\gamma}$ is chosen as the worst performance (for minimization: the largest) of the best performing N^{e} samples, and the parameter $\mathbf{v}$ is updated as

$$\widehat{\mathbf{v}} = \underset{\mathbf{v} \in \mathscr{V}}{\operatorname{argmax}} \sum_{\mathbf{X} \in \mathscr{E}} \ln f(\mathbf{X}; \mathbf{v}) . \tag{12.6}$$

This updating formula is the result of minimizing the Kullback–Leibler or CE distance between the conditional density of $\mathbf{X} \sim f(\mathbf{x}; \mathbf{v})$ given $S(\mathbf{X}) \leqslant \widehat{\gamma}$, and $f(\mathbf{x}; \widehat{\mathbf{v}})$; see Chapter 13 for more details. Note that (12.6) yields the maximum likelihood estimator of $\mathbf{v}$ based on the elite samples. Hence, for many specific families of distributions, including exponential families, explicit solutions can be found. An ☞ 463 ☞ 667 important example is where $\mathbf{X} \sim \mathsf{N}(\boldsymbol{\mu}, \text{diag}(\boldsymbol{\sigma}^2))$, where the mean vector $\boldsymbol{\mu}$ and the vector of variances $\boldsymbol{\sigma}^2$ are updated via the sample mean and sample variance of the elite samples. This is known as **normal updating**.

A generic CE algorithm for minimization is as follows.

Algorithm 12.8 (CE Algorithm for Minimization)

1. *Choose an initial parameter vector* $\widehat{\mathbf{v}}_0$. *Let* $N^{\mathrm{e}} = \lceil \varrho N \rceil$ *be the number of elite samples. Set* $t = 1$.

2. *Generate* $\mathbf{X}_1, \ldots, \mathbf{X}_N \sim_{\text{iid}} f(\cdot; \widehat{\mathbf{v}}_{t-1})$. *Calculate the performances* $S(\mathbf{X}_i)$ *for all* i *and order them from smallest to largest:* $S_{(1)} \leqslant \ldots \leqslant S_{(N)}$. *Let* $\widehat{\gamma}_t$ *be the sample* ϱ*-quantile of performances; that is,* $\widehat{\gamma}_t = S_{(N^{\mathrm{e}})}$.

3. *Use the* same *sample* $\mathbf{X}_1, \ldots, \mathbf{X}_N$ *and set*

$$\widehat{\mathbf{v}}_t = \underset{\mathbf{v}}{\operatorname{argmax}} \sum_{k=1}^{N} \mathrm{I}_{\{S(\mathbf{X}_k) \leqslant \widehat{\gamma}_t\}} \ln f(\mathbf{X}_k; \mathbf{v}) . \tag{12.7}$$

4. *If some stopping criterion is met, stop; otherwise, set* $t = t + 1$ *and return to Step 2.*

Algorithm 12.8 can easily be modified to minimize *noisy* functions $S(\mathbf{x}) = \mathbb{E}\widetilde{S}(\mathbf{x}, \boldsymbol{\xi})$, as defined in (12.1). The only change required in the algorithm is that every function value $S(\mathbf{x})$ be replaced by its estimate $\widehat{S}(\mathbf{x})$.

In practice, to include **smoothing** of the parameter vectors, given a vector of *smoothing parameters* $\boldsymbol{\alpha}$, replace Step 3 of Algorithm 12.8 by

3'. *Use the* same *sample* $\mathbf{X}_1, \ldots, \mathbf{X}_N$ *and set*

$$\widetilde{\mathbf{v}}_t = \underset{\mathbf{v}}{\operatorname{argmax}} \sum_{k=1}^{N} \mathrm{I}_{\{S(\mathbf{X}_k) \leqslant \widehat{\gamma}_t\}} \ln f(\mathbf{X}_k; \mathbf{v}) , \tag{12.8}$$

updating $\widehat{\mathbf{v}}_t$ *as*

$$\widehat{\mathbf{v}}_t = \text{diag}(\boldsymbol{\alpha})\, \widetilde{\mathbf{v}}_t + \text{diag}(\mathbf{1} - \boldsymbol{\alpha})\, \widehat{\mathbf{v}}_{t-1} .$$

■ **EXAMPLE 12.6 ((r, R) Policy Optimization)**

☞ 289 Consider the (r, R) inventory system of Section 7.4.1 (we use the notation (r, R) instead of (s, S) to avoid confusion with the objective function S). There, the long-run average cost per unit of time to run the system is given as

$$S(r, R) = c_1 R + c_2 f_{\text{neg}} + c_3 f_{\text{ord}} ,$$

where r and R are the lower and upper limit in the (r, R) policy, f_{neg} is the fraction of time where the net inventory process is negative, and f_{ord} is the frequency of orders (per unit of time).

For given values of c_1, c_2, and c_3, we wish to find the policy constants (r, R) that minimize S. This gives the minimization problem

$$\begin{aligned} &\underset{r,R}{\operatorname{argmin}} \quad S(r,R) \\ &\text{such that} \quad r \geqslant 0, R \geqslant r\,. \end{aligned}$$

This is a *noisy* problem when f_{neg} and f_{ord} cannot be computed analytically but are estimates from realizations of the inventory process for a given policy (r, R). To do this, the system is simulated until a time horizon T, yielding estimates $\widehat{f}_{\text{neg}}$ and $\widehat{f}_{\text{ord}}$ which are then used to form an estimated cost $\widehat{S}$.

For definiteness, take the constants $c_1 = 5$, $c_2 = 500$, and $c_3 = 100$, and a time horizon of $T = 1000$ days. The interarrival, demand size, and lead time distributions are taken as $\mathsf{Exp}(1/5)$, $\mathsf{U}(0,10)$, and $\mathsf{U}(5,10)$, respectively.

To solve this problem via CE, we must specify a parametric family of sampling distributions. For simplicity, we sample two-dimensional Gaussian random vectors with independent components, $\mathbf{X} = (X_1, X_2)$. We associate realizations of X_1 and X_2 with r and R in the (r, R) policy, respectively. Policies with $r < 0$, $R < 0$, or $R < r$, incur a penalty of $+\infty$.

In the MATLAB implementation below we use a sample size of $N = 100$ and an elite proportion of $\varrho = 0.1$, giving $N^e = \lceil N\varrho \rceil = 10$ elite samples in each iteration. The algorithm is stopped once the largest standard deviation of the Gaussian sampling distribution drops below $\varepsilon = 10^{-4}$. The initial parameters for the Gaussian distribution are $\mu_0 = (100, 100)^\top$ and $\Sigma_0 = \text{diag}(100^2, 100^2)$.

A typical estimated optimal policy is $(r, R) = (15.56, 19.42)$, giving an estimated average cost of 149.6, with corresponding negative inventory fraction $\widehat{f}_{\text{neg}} = 0.0779$ and order frequency $\widehat{f}_{\text{ord}} = 0.1356$. This indeed has lower cost than a reasonable guess of say $(r, R) = (10, 40)$, for which we estimate a cost of 231.9, with $\widehat{f}_{\text{neg}} = 0.0578$ and $\widehat{f}_{\text{ord}} = 0.0302$. The function `f.m` used below is simply adapted from Section 7.4.1 and is available on the Handbook website. ☞ 289

```
%opt_policy.m
epsilon=10^(-4);
N=100; rho=0.1; alpha=1; beta=.5; Ne=ceil(N.*rho); %alg. parameters
mu=[100,100]; sig=[100,100]; muhist=mu;sighist=sig; %initialize v
while max(sig)>epsilon
    x=repmat(mu,N,1)+repmat(sig,N,1).*randn(N,2);
    for k=1:N
      if (x(k,1)>=0)&(x(k,2)>=x(k,1))
        S(k)=f(x(k,1),x(k,2)); %score if policy is feasible
      else
        S(k)=inf; % otherwise apply a penalty
      end
    end
    [S,I]=sort(S); % sort performances
    mu=alpha.*mean(x(I(1:Ne),:))+(1-alpha).*mu; %update means
    sig=beta.*std(x(I(1:Ne),:),1,1)+(1-beta).*sig; %upd. std devs
    muhist=[muhist;mu];sighist=[sighist;sig];
    [mu, sig, S(1),S(Ne)] % Display param. vect. & best and worst
end
```

12.6 OTHER RANDOMIZED OPTIMIZATION TECHNIQUES

Of the many other randomized optimization techniques we mention the following.

Response surface methods: The idea is to construct a model of the objective function (for example, by using regression techniques) from carefully selected sample–score pairs, say via spatial experimental design (see, for example, [10]).

Particle swarm optimization: This is a population- or agent-based optimization heuristic, inspired by social behavior such as the flocking of birds. Denote the population on iteration t by $\mathbf{X}_1^t, \ldots, \mathbf{X}_N^t$. Let $\mathbf{X}_k^{*t}$ be the best individual from $\mathbf{X}_k^1, \ldots, \mathbf{X}_k^t$. Let G be the globally best index, so that $\mathbf{X}_G^{*t}$ is the best solution of any individual up to time t.

The positions are updated in a simple way. First, *velocities* are determined via

$$\mathbf{V}_k^t = \mathbf{V}_k^{t-1} + c_1 U_1 \left(\mathbf{X}_k^{*(t-1)} - \mathbf{X}_k^{t-1}\right) + c_2 U_2 \left(\mathbf{X}_G^{*(t-1)} - \mathbf{X}_k^{t-1}\right),$$

where $c_1, c_2 > 0$ are constants (typically set to $c_1 = c_2 = 2$) and $U_1, U_2 \sim_{\text{iid}} \mathsf{U}(0,1)$. Second, the positions are updated via

$$\mathbf{X}_k^t = \mathbf{X}_k^{t-1} + \mathbf{V}_k^t .$$

See [11] for an early description of this heuristic.

Tabu search: The idea of this metaheuristic is to keep track of recent moves through the search space, building up a "short term memory", and then using that information to construct a list of admissible candidate moves that the algorithm may take on the next iteration. The admissible moves are usually a restriction of all possible moves by forbidding (or rendering taboo) certain attributes of those moves. This encourages global exploration of the search space. See [9] for a tutorial.

In addition, Monte Carlo ideas can be incorporated into existing techniques, for example using quasi Monte Carlo in optimization [28] or the Monte Carlo versions of
☞ 711 the expectation–maximization algorithm (for example, [29], where the expectation step is replaced by its stochastic counterpart).

Further Reading

For overviews of simulation-based optimization, see [2, 10]. Early work on stochastic approximation appears in [12, 21]. Convergence proofs can be found in [5, 14]; see also [17]. Research toward more robust algorithms is ongoing [19]. For more details on the stochastic counterpart method, we refer to [8, 23]. For simulated annealing, see [1, 13], and [27] for work on adaptive cooling schedules. A wide range of evolutionary algorithms can be found in [3, 4, 6, 18]. For differential evolution in particular, we refer to [20], and see [16] for the estimation of distribution algorithm. The monograph [24] contains more details and applications of the cross-entropy method. A closely related approach is the probability collectives theory [30, 31]. Finally, we mention *harmony search*, a relatively recent population-based metaheuristic that uses a musical metaphor [7].

REFERENCES

1. E. Aarts and J. Korst. *Simulated Annealing and Boltzmann Machines: A Stochastic Approach to Combinatorial Optimization and Neural Computers.* John Wiley & Sons, New York, 1989.
2. S. Andradóttir. A review of simulation optimization techniques. In D. J. Medeiros, E. F. Watson, J. S. Carson, and M. S. Manivannan, editors, *Proceedings of the 1998 Winter Simulation Conference, Washington, DC*, pages 151–158, 1998.
3. T. Bäck, D. B. Fogel, and Z. Michalewicz, editors. *Evolutionary Computation 1: Basic Algorithms and Operators.* Institute of Physics Publishing, Bristol, 2000.
4. T. Bäck, D. B. Fogel, and Z. Michalewicz, editors. *Evolutionary Computation 2: Advanced Algorithms and Operators.* Institute of Physics Publishing, Bristol, 2000.
5. H.-F. Chen. *Stochastic Approximation and Its Applications.* Kluwer, Dordrecht, 2002.
6. D. Dumitrescu, B. Lazzerini, L. C. Jain, and A. Dumitrescu, editors. *Evolutionary Computation.* CRC Press, Boca Raton, FL, 2000.
7. Z. W. Geem, J. H. Kim, and G. V. Loganathan. A new heuristic optimization algorithm: Harmony search. *Simulation*, 76(2):60–68, 2001.
8. C. J. Geyer and E. A. Thompson. Annealing Markov chain Monte-Carlo with applications to ancestral inference. *Journal of the American Statistical Association*, 90(431):909–920, 1995.
9. F. Glover. Tabu search: A tutorial. *Interfaces*, 20(4):74–94, 1990.
10. A. Gosavi. *Simulation-Based Optimization: Parametric Optimization Techniques and Reinforcement Learning.* Kluwer, Boston, 2003.
11. J. Kennedy and R. Eberhart. Particle swarm optimization. In *Proceedings of the 1995 IEEE International Conference on Neural Networks*, volume 4, pages 1942–1948, 1995.
12. J. Kiefer and J. Wolfowitz. Stochastic estimation of the maximum of a regression function. *The Annals of Mathematical Statistics*, 23(3):462–466, 1952.
13. S. Kirkpatrick, C. D. Gelatt, and M. P. Vecchi. Optimization by simulated annealing. *Science*, 220(4598):671–680, 1983.
14. H. J. Kushner and G. G. Yin. *Stochastic Approximation and Recursive Algorithms and Applications.* Springer-Verlag, New York, second edition, 2003.
15. T. L. Lai. Stochastic approximation. *The Annals of Statistics*, 31(2):391–406, 2003.
16. P. Larrañaga and J. A. Lozano, editors. *Estimation of Distribution Algorithms: A New Tool for Evolutionary Computation.* Kluwer, Boston, 2002.
17. P. L'Ecuyer and G. Yin. Budget-dependent convergence rate for stochastic approximation. *SIAM Journal on Optimization*, 8(1):217–247, 1989.
18. Z. Michalewicz. *Genetic Algorithms + Data Structures = Evolution Programs.* Springer-Verlag, Berlin, third edition, 1996.
19. A. Nemirovski, A. Juditsky, G. Lan, and A. Shapiro. Robust stochastic approximation approach to stochastic programming. *SIAM Journal on Optimization*, 19(4):1574–1609, 2009.
20. K. V. Price, R. M. Storn, and J. A. Lampinen. *Differential Evolution: A Practical Approach to Global Optimization.* Springer-Verlag, Berlin, 2005.
21. H. Robbins and S. Monro. A stochastic approximation method. *The Annals of Mathematical Statistics*, 22(3):400–407, 1951.

22. S. Rubenthaler, T. Rydén, and M. Wiktorsson. Fast simulated annealing in $\mathbb{R}^d$ with an application to maximum likelihood estimation in state-space models. *Stochastic Processes and Their Applications*, 119(6):1912–1931, 2009.

23. R. Y. Rubinstein. *Some Problems in Monte Carlo Optimization*. PhD thesis, University of Riga, Latvia, 1969. In Russian.

24. R. Y. Rubinstein and D. P. Kroese. *The Cross-Entropy Method: A Unified Approach to Combinatorial Optimization, Monte-Carlo Simulation, and Machine Learning*. Springer-Verlag, New York, 2004.

25. A. Shapiro, D. Dentcheva, and A. Ruszczyński. *Lectures on Stochastic Programming: Modeling and Theory*. SIAM, Philadelphia, 2009.

26. Y. Shen. *Annealing Adaptive Search with Hit-and-Run Sampling Methods for Stochastic Global Optimization Algorithms*. PhD thesis, University of Washington, 2005.

27. Y. Shen, S. Kiatsupaibul, Z. B. Zabinsky, and R. L. Smith. An analytically derived cooling schedule for simulated annealing. *Journal of Global Optimization*, 38(2):333–365, 2007.

28. Y. Wang and K.-T. Fang. Number theoretic method in applied statistics. *Chinese Annals of Mathematics, Series B*, 11(1):51–65, 1990.

29. G. C. G. Wei and M. A. Tanner. A Monte Carlo implementation of the EM algorithm and the poor man's data augmentation algorithms. *Journal of the American Statistical Association*, 85(411):699–704, 1990.

30. D. H. Wolpert. Finding bounded rational equilibria part I: Iterative focusing. In T. Vincent, editor, *Proceedings of the Eleventh International Symposium on Dynamic Games and Applications*, Tucson, Arizona, 2004.

31. D. H. Wolpert. Finding bounded rational equilibria part II: Alternative Lagrangians and uncountable move spaces. In T. Vincent, editor, *Proceedings of the Eleventh International Symposium on Dynamic Games and Applications*, Tucson, Arizona, 2004.

CHAPTER 13

CROSS-ENTROPY METHOD

The cross-entropy methodology provides a systematic way to design simple and efficient simulation procedures. This chapter describes the method for:

1. *Importance sampling* (see also Section 9.7.3); ☞ 366

2. *Rare-event simulation* (see also Section 10.5); ☞ 404

3. *Optimization*, with examples of *discrete*, *continuous*, and *noisy* problems (see also Section 12.5). ☞ 457

13.1 CROSS-ENTROPY METHOD

The **cross-entropy (CE) method** is a generic Monte Carlo technique for solving complicated estimation and optimization problems. The approach was introduced by Rubinstein in [42, 43], extending his earlier work on variance minimization methods for rare-event probability estimation [41].

The CE method can be applied to two types of problems:

1. **Estimation**: Estimate $\ell = \mathbb{E}H(\mathbf{X})$, where $\mathbf{X}$ is a random variable or vector taking values in some set $\mathscr{X}$ and H is a function on $\mathscr{X}$. An important special case is the estimation of a probability $\ell = \mathbb{P}(S(\mathbf{X}) \geqslant \gamma)$, where S is another function on $\mathscr{X}$.

Handbook of Monte Carlo Methods. By D. P. Kroese, T. Taimre, Z. I. Botev

2. **Optimization**: Maximize or minimize $S(\mathbf{x})$ over all $\mathbf{x} \in \mathscr{X}$, where S is an objective function on $\mathscr{X}$. S can be either a known or a noisy function. In the latter case the objective function needs to be estimated, for example, via simulation.

In the estimation setting of Section 13.2, the CE method can be viewed as an adaptive importance sampling procedure that uses the CE or Kullback–Leibler divergence as a measure of closeness between two sampling distributions. In the optimization setting of Section 13.3, the optimization problem is first translated into a rare-event estimation problem and then the CE method for estimation is used as an adaptive algorithm to locate the optimum.

13.2 CROSS-ENTROPY METHOD FOR ESTIMATION

Consider the estimation of

$$\ell = \mathbb{E}_f H(\mathbf{X}) = \int H(\mathbf{x})\, f(\mathbf{x})\, \mathrm{d}\mathbf{x}\,, \tag{13.1}$$

where H is a sample performance function and f is the probability density of the random vector $\mathbf{X}$. For notational convenience it is assumed that $\mathbf{X}$ is continuous. If $\mathbf{X}$ is discrete, simply replace the integral in (13.1) by a sum. Let g be another probability density such that $g(\mathbf{x}) = 0$ implies that $H(\mathbf{x})\, f(\mathbf{x}) = 0$ for all $\mathbf{x}$. Then, we can represent ℓ as

$$\ell = \int H(\mathbf{x})\, \frac{f(\mathbf{x})}{g(\mathbf{x})}\, g(\mathbf{x})\, \mathrm{d}\mathbf{x} = \mathbb{E}_g H(\mathbf{X})\, \frac{f(\mathbf{X})}{g(\mathbf{X})}\,. \tag{13.2}$$

Consequently, if $\mathbf{X}_1, \ldots, \mathbf{X}_N \sim_{\text{iid}} g$, then

$$\widehat{\ell} = \frac{1}{N} \sum_{k=1}^{N} H(\mathbf{X}_k)\, \frac{f(\mathbf{X}_k)}{g(\mathbf{X}_k)} \tag{13.3}$$

☞ 362 is an unbiased *importance sampling* estimator of ℓ. The optimal (minimum variance) importance sampling probability density is given by

$$g^*(\mathbf{x}) \propto |H(\mathbf{x})|\, f(\mathbf{x})\,, \tag{13.4}$$

(see, for example, [47, Page 132]), whose normalization constant is unknown. The idea of the CE method is to choose the importance sampling density g in a specified class of densities $\mathscr{G}$ such that the *Kullback–Leibler divergence* (see (9.29) on Page 366) between the optimal importance sampling density g^* and g is minimal. That is, find a $g \in \mathscr{G}$ that minimizes

$$\mathcal{D}(g^*, g) = \mathbb{E}_{g^*}\left[\ln \frac{g^*(\mathbf{X})}{g(\mathbf{X})}\right]\,. \tag{13.5}$$

In most cases of interest the sample performance function H is nonnegative, and the nominal probability density f is parameterized by a finite-dimensional vector $\mathbf{u}$; that is, $f(\mathbf{x}) = f(\mathbf{x}; \mathbf{u})$. It is then customary to choose the importance sampling probability density g in the *same* family of probability densities; thus,

$g(\mathbf{x}) = f(\mathbf{x}; \mathbf{v})$ for some **reference parameter v**. The CE minimization procedure then reduces to finding an optimal reference parameter vector, $\mathbf{v}^*$ say, by cross-entropy minimization:

$$\mathbf{v}^* = \underset{\mathbf{v}}{\operatorname{argmax}} \int H(\mathbf{x}) f(\mathbf{x}; \mathbf{u}) \ln f(\mathbf{x}; \mathbf{v}) \, \mathrm{d}\mathbf{x}\,, \tag{13.6}$$

which in turn can be estimated via simulation by solving with respect to $\mathbf{v}$ the stochastic counterpart program ☞ 446

$$\max_{\mathbf{v}} \frac{1}{N} \sum_{k=1}^{N} H(\mathbf{X}_k) \frac{f(\mathbf{X}_k; \mathbf{u})}{f(\mathbf{X}_k; \mathbf{w})} \ln f(\mathbf{X}_k; \mathbf{v})\,, \tag{13.7}$$

where $\mathbf{X}_1, \ldots, \mathbf{X}_N \sim_{\text{iid}} f(\cdot; \mathbf{w})$, for any reference parameter $\mathbf{w}$. The maximization (13.7) can often be solved *analytically*, in particular when the class of sampling distributions forms an exponential family; see, for example, [47, Pages 319–320]. Analytical formulas for the solution to (13.7) can be found whenever explicit expressions for the maximum likelihood estimators of the parameters can be found, see for example [15, Page 36]. ☞ 667

Often ℓ in (13.1) is of the form $\mathbb{P}(S(\mathbf{X}) \geqslant \gamma)$ for some performance function S and level γ, in which case $H(\mathbf{x})$ takes the form of an indicator function: $H(\mathbf{x}) = \mathrm{I}_{\{S(\mathbf{x}) \geqslant \gamma\}}$, so that (13.7) becomes

$$\max_{\mathbf{v}} \frac{1}{N} \sum_{\mathbf{X}_k \in \mathscr{E}} \frac{f(\mathbf{X}_k; \mathbf{u})}{f(\mathbf{X}_k; \mathbf{w})} \ln f(\mathbf{X}_k; \mathbf{v})\,, \tag{13.8}$$

and $\mathscr{E}$ is the **elite** set of samples: those $\mathbf{X}_k$ for which $S(\mathbf{X}_k) \geqslant \gamma$.

A complication in solving (13.8) occurs when ℓ is a rare-event probability; that is, a very small probability (say less than 10^{-4}). Then, for a moderate sample size N most or all of the values $H(\mathbf{X}_k)$ in (13.8) are zero, and the maximization problem becomes useless. To overcome this difficulty, the following *multilevel* CE procedure is used (see, for example, [47, Page 238]).

Algorithm 13.1 (Multilevel CE Algorithm for Estimating $\mathbb{P}(S(\mathbf{X}) \geqslant \gamma)$)

1. *Define* $\widehat{\mathbf{v}}_0 = \mathbf{u}$. *Let* $N^{\mathrm{e}} = \lceil \varrho N \rceil$. *Set* $t = 1$ *(iteration counter).*

2. *Generate* $\mathbf{X}_1, \ldots, \mathbf{X}_N \sim_{\text{iid}} f(\cdot; \widehat{\mathbf{v}}_{t-1})$. *Calculate the performances* $S(\mathbf{X}_i)$ *for all* i *and order them from smallest to largest:* $S_{(1)} \leqslant \ldots \leqslant S_{(N)}$. *Let* $\widehat{\gamma}_t$ *be the sample* $(1-\varrho)$*-quantile of performances; that is,* $\widehat{\gamma}_t = S_{(N-N^{\mathrm{e}}+1)}$. *If* $\widehat{\gamma}_t > \gamma$, *reset* $\widehat{\gamma}_t$ *to* γ.

3. *Use the* same *sample* $\mathbf{X}_1, \ldots, \mathbf{X}_N$ *to solve the stochastic program* (13.8) *with* $\mathbf{w} = \widehat{\mathbf{v}}_{t-1}$. *Denote the solution by* $\widehat{\mathbf{v}}_t$.

4. *If* $\widehat{\gamma}_t < \gamma$, *set the counter* $t = t + 1$ *and reiterate from Step 2; otherwise, proceed with Step 5.*

5. *Let* T *be the final iteration counter. Generate* $\mathbf{X}_1, \ldots, \mathbf{X}_{N_1} \sim_{\text{iid}} f(\cdot; \widehat{\mathbf{v}}_T)$ *and estimate* ℓ *via importance sampling as in* (13.3).

The algorithm requires specification of the family of sampling probability densities $\{f(\cdot;\mathbf{v}), \mathbf{v} \in \mathscr{V}\}$, the sample sizes N and N_1, and the rarity parameter ϱ (typically between 0.01 and 0.1). Typical values for the sample sizes are $N = 10^3$ and $N_1 = 10^5$. Under certain technical conditions the deterministic version of Algorithm 13.1 is guaranteed to terminate (reach level γ) provided that ϱ is chosen small enough; see [46, Section 3.5].

■ **EXAMPLE 13.1 (Rare-Event Probability Estimation)**

Suppose the problem is to estimate $\ell = \mathbb{P}(\min\{X_1,\ldots,X_n\} \geqslant \gamma)$, where $X_k \sim \mathsf{Beta}(1, u_k/(1-u_k))$, $k = 1,\ldots,n$ independently. Note that this parameterization ensures that $\mathbb{E}X_k = u_k$, and that we do not assume that the $\{u_k\}$ are the same. However, if $u_1 = \cdots = u_n = 1/2$, then we have $X_1,\ldots,X_n \sim_{\text{iid}} \mathsf{U}(0,1)$. In that case $\ell = (1-\gamma)^n$. In particular, if we take $n = 5$, $\gamma = 0.999$, and $u_k = 1/2$ for all k, then $\ell = 10^{-15}$.

For these particular problem parameters, typical output of Algorithm 13.1 using a rarity parameter of $\varrho = 0.1$, and sample sizes of $N = 10^3$ and $N_1 = 10^6$, is given below.

Table 13.1 Typical convergence of parameter vectors with multilevel CE.

t	$\widehat{\gamma}_t$			$\widehat{v}_t$		
0	-	0.5	0.5	0.5	0.5	0.5
1	0.60117	0.79938	0.80341	0.79699	0.79992	0.80048
2	0.88164	0.93913	0.94094	0.94190	0.94138	0.94065
3	0.96869	0.98423	0.98429	0.98446	0.98383	0.98432
4	0.99184	0.99586	0.99588	0.99590	0.99601	0.99590
5	0.99791	0.99896	0.99895	0.99893	0.99897	0.99896
6	0.999	0.99950	0.99949	0.99949	0.99950	0.99950

This gives a final estimate of $\widehat{\ell} = 1.0035 \times 10^{-15}$ with an estimated relative error of 0.003.

In this example, we can calculate the exact CE optimal parameters from (13.6). With $u_1 = \cdots = u_n = 1/2$, and due to independence, each component of the optimal parameter vector is solved from

$$v^* = \underset{v\in(0,1)}{\operatorname{argmax}} \int_\gamma^1 \ln\left(\left(\frac{v-1}{v}\right) x^{(v/(v-1)-1)}\right) \mathrm{d}x\,.$$

The solution is given by

$$v^* = \frac{1-\gamma}{2(1-\gamma)+\gamma\ln\gamma}\,.$$

With $\gamma = 0.999$, this gives $v^* = 0.99950$ to five significant digits, which is as found via the multilevel algorithm in Table 13.1. MATLAB code for this example follows.

```
%CEest.m
f='minbeta'; % performance function name
gam=0.999; % Desired level parameter
n=5;  % dimension of problem
N=10^5; % sample size
rho=0.01;
N1=10^6; % Final estimate sample size
N_el=round(N*rho);  % elite sample size
u=0.5.*ones(1,n); % Nominal reference parameter in Beta(1,u/(1-u))
v=u; gt=-inf; % Initialize v and gamma
maxits=10^5; % Fail-safe stopping after maxits exceeded
it=0; tic
while (gt<gam)&(it<maxits)
  it=it+1;
  % Generate and score X's
  X=rand(N,n).^(1./repmat(v./(1-v),N,1)); % Beta(1,v/(1-v))
  S=feval(f,X); [U,I]=sort(S);  % Score & Sort
  % Update Gamma_t
  gt=U(N-N_el+1);
  if gt>gam, gt=gam; N_el=N-find(U>=gam,1)+1; end
  Y=X(I(N-N_el+1:N),:);
  % Calculate likelihood ratios and update v
  W=prod(repmat((u./(1-u))./(v./(1-v)),N_el,1).*...
      Y.^(repmat((u./(1-u))-(v./(1-v)),N_el,1)),2);
  v=sum(repmat(W,1,n).*Y)./sum(repmat(W,1,n));
  [gt,v] % Display gamma and v
end
% Final estimation step
X1=rand(N1,n).^(1./repmat(v./(1-v),N1,1));
S1=feval(f,X1);
W1=prod(repmat((u./(1-u))./(v./(1-v)),N1,1).*...
    X1.^(repmat((u./(1-u))-(v./(1-v)),N1,1)),2);
H1=(S1>=gam);
ell=mean(W1.*H1);
re=sqrt((mean((W1.*H1).^2)/(ell^2))-1)/sqrt(N1);
% Display final results
time=toc; disp(time), disp(v), disp(ell), disp(re)
ell_true=(1-gam)^n;disp(ell_true) % Display true quantity
```

```
function out=minbeta(X)
out=min(X,[],2);
```

This multilevel approach is really a proxy for the ideal situation in which one can sample directly from g^*, and determine the CE optimal parameters as the maximum likelihood estimates for the given parametric family — see Section 10.5 ☞ 404 and the corresponding Algorithm 10.8, which uses MCMC to obtain an approximate sample from g^*.

13.3 CROSS-ENTROPY METHOD FOR OPTIMIZATION

Let S be a real-valued performance function on $\mathscr{X}$. Suppose we wish to find the maximum of S over $\mathscr{X}$, and the corresponding state $\mathbf{x}^*$ at which this maximum is attained (assuming for simplicity that there is only one such state). Denoting the maximum by γ^*, we thus have

$$S(\mathbf{x}^*) = \gamma^* = \max_{\mathbf{x} \in \mathscr{X}} S(\mathbf{x}) \, . \tag{13.9}$$

☞ 677 This setting includes many types of optimization problems: discrete (combinatorial), continuous, mixed, and constrained problems. If one is interested in minimizing rather than maximizing S, one can simply maximize $-S$.

Now associate with the above problem the estimation of the probability $\ell = \mathbb{P}(S(\mathbf{X}) \geqslant \gamma)$, where $\mathbf{X}$ has some pdf $f(\mathbf{x}; \mathbf{u})$ on $\mathscr{X}$ (for example corresponding to the uniform distribution on $\mathscr{X}$) and γ is some level. If γ is chosen close to the unknown γ^*, then ℓ is typically a rare-event probability, and the CE approach of Section 13.2 can be used to find an importance sampling distribution close to the theoretically optimal importance sampling density, which concentrates all its mass at the point $\mathbf{x}^*$. Sampling from such a distribution thus produces optimal or near-optimal states. A main difference with the CE method for rare-event simulation is that, in the optimization setting, the final level $\gamma = \gamma^*$ is not known in advance. The CE method for optimization produces a sequence of levels $\{\widehat{\gamma}_t\}$ and reference parameters $\{\widehat{\mathbf{v}}_t\}$ such that the former tends to the optimal γ^* and the latter to the optimal reference vector $\mathbf{v}^*$ corresponding to the point mass at $\mathbf{x}^*$; see, for example, [47, Page 251].

Given the class of sampling probability densities $\{f(\cdot; \mathbf{v}), \mathbf{v} \in \mathscr{V}\}$, the sample size N, and the rarity parameter ϱ, the CE algorithm for optimization is as follows.

Algorithm 13.2 (CE Algorithm for Optimization)

1. *Choose an initial parameter vector* $\widehat{\mathbf{v}}_0$. *Let* $N^{\mathrm{e}} = \lceil \varrho N \rceil$. *Set the level counter to* $t = 1$.

2. *Generate* $\mathbf{X}_1, \ldots, \mathbf{X}_N \sim_{\text{iid}} f(\cdot; \widehat{\mathbf{v}}_{t-1})$. *Calculate the performances* $S(\mathbf{X}_i)$ *for all* i *and order them from smallest to largest:* $S_{(1)} \leqslant \ldots \leqslant S_{(N)}$. *Let* $\widehat{\gamma}_t$ *be the sample* $(1 - \varrho)$*-quantile of performances; that is,* $\widehat{\gamma}_t = S_{(N - N^{\mathrm{e}} + 1)}$.

3. *Use the* same *sample* $\mathbf{X}_1, \ldots, \mathbf{X}_N$ *to determine the elite set of samples* $\mathscr{E}_t = \{\mathbf{X}_k \, : \, S(X_k) \geqslant \widehat{\gamma}_t\}$ *and solve the stochastic program*

$$\max_{\mathbf{v}} \sum_{\mathbf{X}_k \in \mathscr{E}_t} \ln f(\mathbf{X}_k; \mathbf{v}) \, . \tag{13.10}$$

   Denote the solution by $\widehat{\mathbf{v}}_t$.

4. *If some stopping criterion is met, stop; otherwise, set* $t = t + 1$, *and return to Step 2.*

Any CE algorithm for optimization thus involves the following two main iterative phases:

1. **Generate** an iid sample of objects in the search space $\mathscr{X}$ (trajectories, vectors, etc.) according to a specified probability distribution.
2. **Update** the parameters of that distribution, based on the N^{e} best performing samples (the elite samples), using cross-entropy minimization.

There are two key differences between Algorithm 13.1 and Algorithm 13.2: (1) Step 5 is missing for optimization, and (2) the likelihood ratio term $f(\mathbf{X}_k; \mathbf{u})/f(\mathbf{X}_k; \widehat{\mathbf{v}}_{t-1})$ in (13.7) is missing in (13.10).

Often a smoothed updating rule is used, in which the parameter vector $\widehat{\mathbf{v}}_t$ is taken as

$$\widehat{\mathbf{v}}_t = \mathrm{diag}(\boldsymbol{\alpha})\, \widetilde{\mathbf{v}}_t + \mathrm{diag}(\mathbf{1} - \boldsymbol{\alpha})\, \widehat{\mathbf{v}}_{t-1} \,, \tag{13.11}$$

where $\widetilde{\mathbf{v}}_t$ is the solution to (13.10) and $\boldsymbol{\alpha}$ is a vector of **smoothing parameters**, with each component in $[0, 1]$. Many other modifications can be found in [27, 46, 47] and in the list of references. When there are two or more optimal solutions, the CE algorithm typically "fluctuates" between the solutions before focusing on one of the solutions. The effect that smoothing has on convergence is discussed in detail in [13]. In particular, it is shown that with appropriate smoothing the CE method converges and finds the optimal solution with probability arbitrarily close to 1. Necessary and sufficient conditions for this are also given. Other convergence results can be found in [33].

13.3.1 Combinatorial Optimization

When the state space $\mathscr{X}$ is finite, the optimization problem (13.9) is often referred to as a **discrete** or **combinatorial optimization** problem. For example, $\mathscr{X}$ could be the space of combinatorial objects such as binary vectors, trees, paths through graphs, etc. To apply the CE method, one first needs to specify a convenient parameterized random mechanism to generate objects in $\mathscr{X}$. For example, when $\mathscr{X}$ is the set of binary vectors of length n, an easy generation mechanism is to draw each component independently from a Bernoulli distribution; that is, $\mathbf{X} = (X_1, \ldots, X_n)$, where $X_i \sim \mathsf{Ber}(p_i)$, $i = 1, \ldots, n$, independently. Given an elite sample set $\mathscr{E}$ of size N^{e}, the updating formula is then [15, Page 56]

$$\widehat{p}_i = \frac{\sum_{\mathbf{X} \in \mathscr{E}} X_i}{N^{\mathrm{e}}}, \quad i = 1, \ldots, n \,. \tag{13.12}$$

A possible stopping rule for combinatorial optimization problems is to stop when the overall best objective value does not change over a number of iterations. Alternatively, one could stop when the sampling distribution has "degenerated" enough. In particular, in the Bernoulli case (13.12) one could stop when all $\{p_i\}$ are less than some small distance $\varepsilon > 0$ away from either 0 or 1.

■ **EXAMPLE 13.2 (Satisfiability Problem)**

We illustrate the CE optimization Algorithm 13.2 by applying it to the satisfiability problem (SAT) considered in Example 12.4. (See Section C.3.1 for more details on the SAT problem.) ☞ 453 ☞ 694

We take our sampling pdf g to be of the form

$$g(\mathbf{x}) = \prod_{i=1}^{n} p_i^{x_i} (1 - p_i)^{1 - x_i} .$$

In this case, the i-th component of $\mathbf{x}$ is generated according to the $\mathsf{Ber}(p_i)$ distribution, independently of all other components.

The Bernoulli probabilities are updated using (13.12), where the set of elite samples $\mathscr{E}$ on iteration t is the proportion ϱ of best performers — in this case, the proportion ϱ of samples that satisfy the most clauses.

If we write $\widehat{\mathbf{p}}_t = (\widehat{p}_{t1}, \ldots, \widehat{p}_{tn})$ for the solution from (13.12) in iteration t, then the idea is that the sequence $\widehat{\mathbf{p}}_0, \widehat{\mathbf{p}}_1, \ldots$ converges to one of the solutions to the satisfiability problem.

We run the algorithm with a sample size of $N = 10^4$ and a rarity parameter of $\varrho = 0.1$, giving $N^{\mathrm{e}} = 10^3$ elite samples per iteration. We take $\widehat{\mathbf{p}}_0 = (0.5, \ldots, 0.5)$ and use a constant smoothing parameter of $\alpha = 0.5$. Finally, our algorithm is stopped after 10^3 iterations, or if the vector $\mathbf{p}_t$ has *degenerated*: if every component p_{tk} is within $\varepsilon = 10^{-3}$ of either 0 or 1.

In Figure 13.1, the scores of the best and worst performing elite samples are plotted for the problem `F34-5-23-31`. As with the binary coded genetic algorithm in Example 12.4, the best solutions found only satisfied 359 out of 361 clauses.

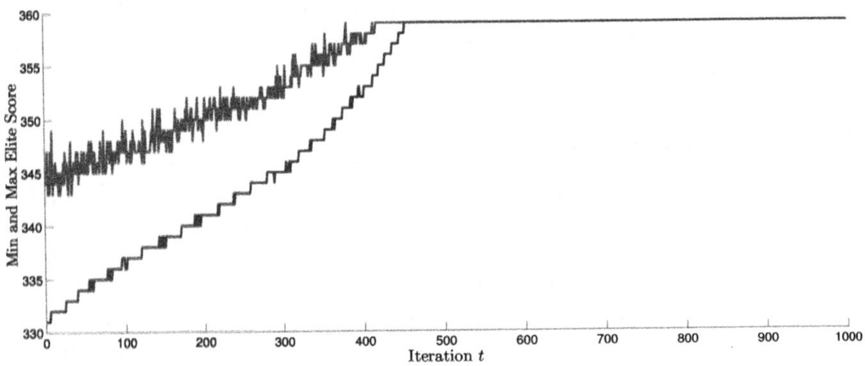

Figure 13.1 Typical best and worst elite samples using Algorithm 13.2 on the `F34-5-23-31` problem.

MATLAB code that implements the basic algorithm is given below. It is assumed there is an efficient function `sC.m` that accepts N trial vectors and computes the number of clauses that each satisfies. Our implementation of `sC.m` is available on the Handbook website.

```
%CESAT.m
%Assumes *sparse* A is loaded in the workspace
[m,n]=size(A); % Dimensions of the problem
maxits=10^3; epsilon=1e-3;
N=10^4; rho=0.1; Ne=ceil(rho*N);
alpha=0.7; % Smoothing parameter
p = 0.5*ones(1,n); it=0; best=-inf; xbest=[];
Smaxhist=zeros(1,maxits); % Allocate history memory
Sminhist=zeros(1,maxits);
while (max(min(p,1-p)) > epsilon) && (it < maxits) && (best<m)
```

```
    it = it + 1;
    x = double((rand(N,n) < repmat(p,N,1)));
    SX = sC(A,x);
    [sortSX,iX] = sortrows([x SX],n+1);
    indices=iX(N- Ne + 1:N);
    if sortSX(N,n+1)>best
      best=sortSX(N,n+1); xbest=sortSX(N,1:n);
    end
    Smaxhist(it)=sortSX(N,n+1);Sminhist(it)=sortSX(N-Ne+1,n+1);
    p=alpha.*mean(sortSX(indices,1:n))+(1-alpha).*p;
    disp([it,sortSX(N,n+1),sortSX(N-Ne+1,n+1),p])
end
disp([best xbest])
figure,plot((1:1:it),Smaxhist,'r-',(1:1:it),Sminhist,'k-')
```

13.3.2 Continuous Optimization

It is also possible to apply the CE algorithm to continuous optimization problems; in particular, when $\mathscr{X} = \mathbb{R}^n$. The sampling distribution on $\mathbb{R}^n$ can be quite arbitrary and does not need to be related to the function that is being optimized. The generation of a random vector $\mathbf{X} = (X_1, \ldots, X_n) \in \mathbb{R}^n$ is usually established by drawing the coordinates independently from some two-parameter distribution. In most applications a normal (Gaussian) distribution is employed for each component. Thus, the sampling distribution for $\mathbf{X}$ is characterized by a vector $\boldsymbol{\mu}$ of means and a vector $\boldsymbol{\sigma}$ of standard deviations. At each iteration of the CE algorithm these parameter vectors are updated simply as the vectors of sample means and sample standard deviations of the elements in the elite set; see, for example, [27]. During the course of the algorithm, the sequence of mean vectors ideally tends to the maximizer $\mathbf{x}^*$, while the vector of standard deviations tends to the zero vector. In short, one should obtain a degenerated probability density with all mass concentrated in the vicinity of the point $\mathbf{x}^*$. A possible stopping criterion is to stop when all standard deviations are smaller than some ε. This scheme is referred to as **normal updating**.

In what follows, we give examples of CE applied to unconstrained, constrained, and noisy continuous optimization problems. In each case, we employ normal updating.

■ **EXAMPLE 13.3 (Maximizing the Peaks Function)**

Suppose we wish to maximize MATLAB's peaks function, given by

$$S(\mathbf{x}) = 3\,(1-x_1)^2\,\mathrm{e}^{-x_1^2-(x_2+1)^2} - 10\left(\frac{x_1}{5} - x_1^3 - x_2^5\right)\mathrm{e}^{-x_1^2-x_2^2} - \frac{1}{3}\,\mathrm{e}^{-(x_1+1)^2-x_2^2}\,.$$

This function has three local maxima and three local minima, with global maximum at $\mathbf{x}^* \approx (-0.0093, 1.58)$ of $\gamma^* = S(\mathbf{x}^*) \approx 8.1$.

With normal updating, the choice of the initial value for $\boldsymbol{\mu}$ is not important, so we choose $\boldsymbol{\mu} = (-3,-3)$ arbitrarily. However, the initial standard deviations should be chosen large enough to ensure an initially "uniform" sampling of the region of interest, hence $\boldsymbol{\sigma} = (10, 10)$ is chosen. The CE algorithm is stopped when all standard deviations of the sampling distribution are less than some small ε, say $\varepsilon = 10^{-5}$. A typical evolution of the mean of the sampling distribution is depicted in Figure 13.2.

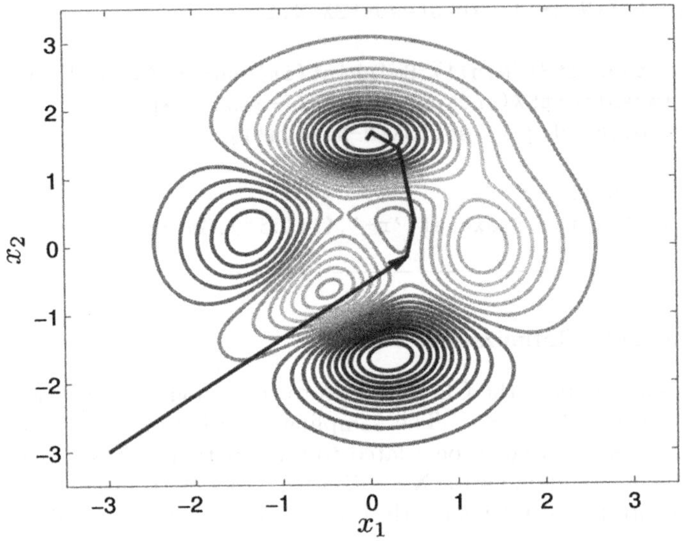

Figure 13.2 A typical evolution of the mean vector with normal updating for the peaks function.

A MATLAB implementation of CE Algorithm 13.2 is given below. The peaks function is stored in a separate file S.m.

```
%peaks\simplepeaks.m
n = 2;                                              % dimension
mu = [-3,-3]; sigma = 3*ones(1,n); N = 100; eps = 1E-5; rho=0.1;
while max(sigma) > eps
   X = randn(N,n)*diag(sigma)+ mu(ones(N,1),:);
   SX= S(X);                                        %Compute the performance
   sortSX = sortrows([X, SX],n+1);
   Elite = sortSX((1-rho)*N:N,1:n);                 % elite samples
   mu = mean(Elite,1);                              % take sample mean row-wise
   sigma = std(Elite,1);                            % take sample st.dev. row-wise
   [S(mu),mu,max(sigma)]                            % output the result
end
```

```
function out = S(X)
out =  3*(1-X(:,1)).^2.*exp(-X(:,1).^2 - (X(:,2)+1).^2) ...
 - 10*(X(:,1)/5 - X(:,1).^3 - X(:,2).^5) ...
 .*exp(-X(:,1).^2-X(:,2).^2) - 1/3*exp(-(X(:,1)+1).^2 - X(:,2).^2);
```

13.3.3 Constrained Optimization

In order to apply the CE method to constrained maximization problems, we must first put the problem in the framework of (13.9). Let $\mathscr{X}$ be a region defined by some system of inequalities:

$$G_i(\mathbf{x}) \leqslant 0, \quad i = 1, \ldots, k \; . \tag{13.13}$$

Two approaches can be adopted to solve the program (13.9) with constraints (13.13) (see also Section C.2.1). The first approach uses *acceptance–rejection*: generate a random vector $\mathbf{X}$ from, for example, a multivariate normal distribution with independent components, and accept or reject it depending on whether the sample falls in $\mathscr{X}$ or not. Alternatively, one could sample directly from a truncated distribution (for example, a truncated normal distribution) or combine such a method with acceptance-rejection. Once a fixed number of such vectors has been accepted, the parameters of the normal distribution can be updated in exactly the same way as in the unconstrained case — simply via the sample mean and standard deviation of the elite samples. A drawback of this method is that a large number of samples could be rejected before a feasible sample is found. ☞ 685

The second approach is the *penalty approach* (see Section C.2.1.1). Here, the objective function is modified to

$$\widetilde{S}(\mathbf{x}) = S(\mathbf{x}) + \sum_{i=1}^{k} H_i \, \max\{G_i(\mathbf{x}), 0\} \; , \tag{13.14}$$

where $H_i < 0$ measures the importance (cost) of the i-th penalty.

Thus, by reducing the constrained problem ((13.9) and (13.13)) to an unconstrained one ((13.9) with $\widetilde{S}$ instead of S), one can again apply Algorithm 13.2. Further details on constrained multiextremal optimization with the CE method may be found in [27].

■ **EXAMPLE 13.4 (MLE for the Dirichlet Distribution)**

Suppose that we are given data $\mathbf{x}_1, \ldots, \mathbf{x}_n \sim_{\text{iid}} \mathsf{Dirichlet}(\boldsymbol{\alpha})$, where $\boldsymbol{\alpha} = (\alpha_1, \ldots, \alpha_K)^\top$ is an unknown parameter vector satisfying $\alpha_i > 0$, $i = 1, \ldots, K$. The conditions on $\boldsymbol{\alpha}$ provide natural inequality constraints: $G_i(\boldsymbol{\alpha}) \equiv -\alpha_i \leqslant 0$, $i = 1, \ldots, K$.

We will use Algorithm 13.2 with normal updating to obtain the MLE by direct maximization of the log-likelihood for the Dirichlet distribution given the data. However, due to the constraints for a valid parameter vector $\boldsymbol{\alpha}$, we apply a penalty of $H_1 = \cdots = H_K = -\infty$ whenever a constraint is violated.

Figure 13.3 displays the distance between the mean vector of the sampling distribution and the true MLE calculated via a fixed-point technique [35] for a data size of $n = 100$ points from the Dirichlet(1, 2, 3, 4, 5) distribution. The CE parameters are a sample size of $N = 10^4$ and an elite sample size of $N^e = 10^3$. No smoothing parameter is applied to the mean vector, but a constant smoothing parameter of 0.5 is applied to each of the standard deviations.

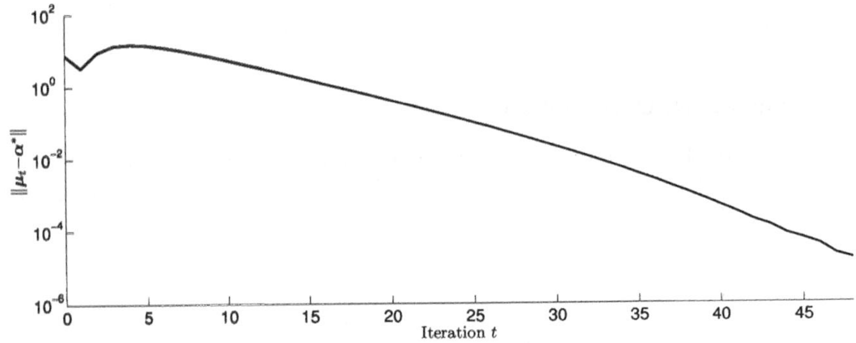

Figure 13.3 Typical evolution of the Euclidean distance between the mean vector and the MLE $\boldsymbol{\alpha}^*$.

```
%MLE\ce_dirichlet_mle_fp.m
clear all;
a=(1:1:5); n=100;
K=length(a); %K dim vector
data=dirichletrnd(a,n,K); % Generate data
epsilon=10^(-4); % For
N=10^4; rho=0.1; alphamu=1; alphasig=0.5; Ne=ceil(N.*rho);
mu=zeros(1,K); sig=ones(1,K).*10; % Initial parameters
muhist=mu;sighist=sig; % History
while max(sig)>epsilon
    x=repmat(mu,N,1)+repmat(sig,N,1).*randn(N,K); % Sample
    S=dirichlet_log_like(data,x,n,K); [S,I]=sort(S); % Score & Sort
    mu=alphamu.*mean(x(I(N-Ne+1:N),:))+(1-alphamu).*mu;
    sig=alphasig.*std(x(I(N-Ne+1:N),:),1,1)+(1-alphasig).*sig;
    muhist=[muhist;mu];sighist=[sighist;sig]; % Update History
    [mu, sig, S(end),S(N-Ne+1)]
end
% For comparison, compute the MLE using a fixed-point method
afp=dirichlet_MLE_FP(data,K);
disp([afp,dirichlet_log_like(data,afp,n,K)])
```

The function `dirichlet_log_like.m` calculates the log-likelihood of the set of trial parameters for the given data set.

```
function out=dirichlet_log_like(data,x,n,K)
out=zeros(size(x,1),1);
I=any(x<=0,2);nI=~I;
out(I)=-inf;
out(nI)=n.*(log(gamma(sum(x(nI,:),2)))-sum(log(gamma(x(nI,:))),2));
for k=1:n
    out(nI)=out(nI)+sum((x(nI,1:(K-1))-1).*...
        repmat(log(data(k,1:(K-1))),sum(nI),1),2)+(x(nI,K)-1).*...
        repmat(log(1-sum(data(k,1:(K-1)),2)),sum(nI),1);
end
```

The function dirichletrnd.m generates Dirichlet distributed realizations as in Algorithm 4.67. ☞ 141

```
function out=dirichletrnd(a,n,K)
out=zeros(n,K);
for k=1:n
  temp=zeros(1,K);
  for i=1:(K)
    temp(i)=gamrnd(a(i),1);
  end
  out(k,:)=temp./sum(temp);
end
```

The function dirichlet_MLE_FP.m computes the MLE using a fixed-point technique [35].

```
function afp=dirichlet_MLE_FP(data,K)
%Compute Dirichlet MLE via a fixed-point technique
logpdata=mean(log(data),1);
afp=ones(1,K); afpold=-inf.*afp;
while sqrt(sum((afp-afpold).^2))>10^(-12)
   afpold=afp; s=sum(afpold);
   for k=1:K
     y=(psi(s)+logpdata(k));
     if y>=-2.22
         ak=exp(y)+0.5;
     else
         ak=-1/(y-psi(1));
     end
     akold=-inf;
     while abs(ak-akold)>10^(-12)
        akold=ak; ak=akold - ((psi(akold)-y)/psi(1,akold));
     end
     afp(k)=ak;
   end
end
```

13.3.4 Noisy Optimization

Noisy (or *stochastic*) optimization problems — in which the objective function is corrupted with noise — arise in many contexts, for example, in stochastic scheduling and stochastic shortest/longest path problems, and simulation-based optimization [48]. The CE method can be easily modified to deal with noisy optimization problems. Consider the maximization problem (13.9) and assume that the performance function is noisy. In particular, suppose that $S(\mathbf{x}) = \mathbb{E}\widehat{S}(\mathbf{x})$ is *not* available, but that a sample value $\widehat{S}(\mathbf{x})$ (unbiased estimate of $\mathbb{E}\widehat{S}(\mathbf{x})$) is available, for example via simulation. The principal modification of the Algorithm 13.2 is to replace $S(\mathbf{x})$ by $\widehat{S}(\mathbf{x})$. In addition, one may need to increase the sample size in order to reduce the effect of the noise. Although various applications indicate the usefulness of the CE approach for noisy optimization (see, for example, [1, 25, 26, 36]), little is known regarding theoretical convergence results in the noisy case. Spall [51, Section 2.4] discusses various divergence results for general types of stochastic methods. A possible stopping criterion is to stop when the sampling distribution has degenerated enough. Another possibility is to stop the stochastic process when the sequence of levels $\{\widehat{\gamma}_t\}$ has reached stationarity; see for example [46, Page 207].

■ **EXAMPLE 13.5 (Noisy Peaks Function)**

This example is a noisy version of Example 13.3, for which the performance function S has been corrupted by standard normal noise: $\widehat{S}(\mathbf{x}) = S(\mathbf{x}) + \varepsilon$, $\varepsilon \sim \mathsf{N}(0,1)$. The following MATLAB code provides a simple implementation of the CE algorithm to maximize the peaks function when the sample performance values are corrupted by noise in this way. The CE parameters and the function `S.m` are the same as in Example 13.3. Typical evolution of the mean of the sampling distribution is depicted in Figure 13.4.

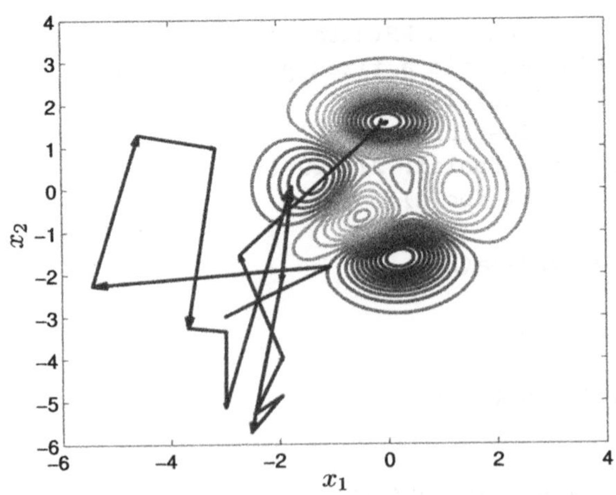

Figure 13.4 Typical evolution of the mean vector with normal updating for the noisy peaks function.

```
%peaks\simplepeaks_noisy.m
n = 2;                                 % dimension
mu = [-3,-3]; sigma = 3*ones(1,n); N = 100; eps = 1E-5; rho=0.1;
while max(sigma) > eps
   X = randn(N,n)*diag(sigma)+ mu(ones(N,1),:);
   SX= S(X);                          %Compute the performance
   SX= SX+randn(N,1);                 %Corrupt with noise
   sortSX = sortrows([X, SX],n+1);
   Elite = sortSX((1-rho)*N:N,1:n); % elite samples
   mu = mean(Elite,1);                % take sample mean row-wise
   sigma = std(Elite,1);              % take sample st.dev. row-wise
   [S(mu)+randn,mu,max(sigma)]              % output the result
end
```

Further Reading

An easy tutorial on the CE method is given in [15]. A more comprehensive treatment can be found in [46]; see also [47, Chapter 8]. The CE method home page can be found at `www.cemethod.org`.

The CE method has been successfully applied to a diverse range of estimation and optimization problems, including buffer allocation [1], queueing models of telecommunication systems [14, 16], model fitting for the truck fleet problem [3], optimal control of HIV/AIDS spread [49, 50], signal detection [30], combinatorial auctions [9], DNA sequence alignment [24, 39], scheduling and vehicle routing [4, 8, 11, 20, 23, 54], neural and reinforcement learning [31, 32, 34, 53, 55], project management [12], rare-event simulation with light- and heavy-tail distributions [2, 10, 21, 28], and clustering analysis [5, 6, 29]. Applications to combinatorial optimization problems include the *max-cut*, *traveling salesman*, and *Hamiltonian cycle* problems, see [17, 43, 44, 45]. For estimation and (noisy) optimization problems in the context of network reliability and design see [7, 22, 25, 26, 36, 37, 38, 40]. Importance sampling methods that use generalizations of the cross–entropy distance are developed in [52]. The standard CE algorithms are highly parallelizable. For details on parallel implementations of the CE method, see [18, 19].

REFERENCES

1. G. Alon, D. P. Kroese, T. Raviv, and R. Y. Rubinstein. Application of the cross-entropy method to the buffer allocation problem in a simulation-based environment. *Annals of Operations Research*, 134(1):137–151, 2005.

2. S. Asmussen, D. P. Kroese, and R. Y. Rubinstein. Heavy tails, importance sampling and cross-entropy. *Stochastic Models*, 21(1):57–76, 2005.

3. A. Belay, E. J. O'Brien, and D. P. Kroese. Truck fleet model for design and assessment of flexible pavements. *Journal of Sound and Vibration*, 311(3-5):1161–1174, 2008.

4. I. Bendavid and B. Golany. Setting gates for activities in the stochastic project scheduling problem through the cross entropy methodology. *Annals of Operations Research*, 172(1):259–276, 2009.

5. Z. I. Botev and D. P. Kroese. Global likelihood optimization via the cross-entropy method, with an application to mixture models. In R. G. Ingalls, M. D. Rossetti, J. S. Smith, and B. A. Peters, editors, *Proceedings of the 2004 Winter Simulation Conference*, pages 529–535, Washington, DC, December 2004.

6. A. Boubezoula, S. Paris, and M. Ouladsinea. Application of the cross entropy method to the GLVQ algorithm. *Pattern Recognition*, 41(10):3173–3178, 2008.

7. M. Caserta and M. Cabo-Nodar. A cross entropy based algorithm for reliability problems. *Journal of Heuristics*, 15(5):479–501, 2009.

8. M. Caserta and E. Quiñonez-Ricob. A cross entropy-Lagrangean hybrid algorithm for the multi-item capacitated lot-sizing problem with setup times. *Computers & Operations Research*, 36(2):530–548, 2009.

9. J. C. C. Chan and D. P. Kroese. Randomized methods for solving the winner determination problem in combinatorial auctions. In *Proceedings of the 2008 Winter Simulation Conference*, pages 1344–1349, 2008.

10. J. C. C. Chan and D. P. Kroese. Rare-event probability estimation with conditional Monte Carlo. *Annals of Operations Research*, 2009. DOI: 10.1007/s10479-009-0539-y.

11. K. Chepuri and T. Homem-de-Mello. Solving the vehicle routing problem with stochastic demands using the cross entropy method. *Annals of Operations Research*, 134(1):153–181, 2005.

12. I. Cohen, B. Golany, and A. Shtub. Managing stochastic finite capacity multi-project systems through the cross-entropy method. *Annals of Operations Research*, 134(1):183–199, 2005.

13. A. Costa, J. Owen, and D. P. Kroese. Convergence properties of the cross-entropy method for discrete optimization. *Operations Research Letters*, 35(5):573–580, 2007.

14. P.-T. de Boer. *Analysis and Efficient Simulation of Queueing Models of Telecommunication Systems*. PhD thesis, University of Twente, 2000.

15. P.-T. de Boer, D. P. Kroese, S. Mannor, and R. Y. Rubinstein. A tutorial on the cross-entropy method. *Annals of Operations Research*, 134(1):19–67, 2005.

16. P.-T. de Boer, D. P. Kroese, and R. Y. Rubinstein. A fast cross-entropy method for estimating buffer overflows in queueing networks. *Management Science*, 50(7):883–895, 2004.

17. A. Eshragh, J. A. Filar, and M. Haythorpe. A hybrid simulation-optimization algorithm for the Hamiltonian cycle problem. *Annals of Operations Research*, 2009. DOI: 10.1007/s10479-009-0565-9.

18. G. E. Evans. *Parallel and Sequential Monte Carlo Methods with Applications*. PhD thesis, The University of Queensland, Australia, 2009.

19. G. E. Evans, J. M. Keith, and D. P. Kroese. Parallel cross-entropy optimization. In *Proceedings of the 2007 Winter Simulation Conference*, pages 2196–2202, Washington, DC, 2007.

20. B. E. Helvik and O. Wittner. Using the cross-entropy method to guide/govern mobile agent's path finding in networks. In S. Pierre and R. Glitho, editors, *Mobile Agents for Telecommunication Applications: Third International Workshop, MATA 2001, Montreal*, pages 255–268, New York, 2001. Springer-Verlag.

21. T. Homem-de-Mello. A study on the cross-entropy method for rare event probability estimation. *INFORMS Journal on Computing*, 19(3):381–394, 2007.

22. K.-P. Hui, N. Bean, M. Kraetzl, and D. P. Kroese. The cross-entropy method for network reliability estimation. *Annals of Operations Research*, 134:101–118, 2005.

23. E. Ianovsky and J. Kreimer. An optimal routing policy for unmanned aerial vehicles (analytical and cross-entropy simulation approach). *Annals of Operations Research*, 2009. DOI: 10.1007/s10479-009-0609-1.

24. J. Keith and D. P. Kroese. Sequence alignment by rare event simulation. In *Proceedings of the 2002 Winter Simulation Conference*, pages 320–327, San Diego, 2002.

25. D. P. Kroese and K.-P. Hui. In: *Computational Intelligence in Reliability Engineering*, chapter 3: Applications of the Cross-Entropy Method in Reliability. Springer-Verlag, New York, 2006.

26. D. P. Kroese, S. Nariai, and K.-P. Hui. Network reliability optimization via the cross-entropy method. *IEEE Transactions on Reliability*, 56(2):275–287, 2007.

27. D. P. Kroese, S. Porotsky, and R. Y. Rubinstein. The cross-entropy method for continuous multi-extremal optimization. *Methodology and Computing in Applied Probability*, 8(3):383–407, 2006.

28. D. P. Kroese and R. Y. Rubinstein. The transform likelihood ratio method for rare event simulation with heavy tails. *Queueing Systems*, 46(3-4):317–351, 2004.

29. D. P. Kroese, R. Y. Rubinstein, and T. Taimre. Application of the cross-entropy method to clustering and vector quantization. *Journal of Global Optimization*, 37(1):137–157, 2007.

30. Z. Liu, A. Doucet, and S. S. Singh. The cross-entropy method for blind multiuser detection. In *IEEE International Symposium on Information Theory*, page 510, Chicago, 2004. Piscataway.

31. A. Lörincza, Z. Palotaia, and G. Szirtesb. Spike-based cross-entropy method for reconstruction. *Neurocomputing*, 71(16–18):3635–3639, 2008.

32. S. Mannor, R. Y. Rubinstein, and Y. Gat. The cross-entropy method for fast policy search. In *The 20th International Conference on Machine Learning (ICML-2003)*, pages 512–519, Washington, DC, 2003.

33. L. Margolin. On the convergence of the cross-entropy method. *Annals of Operations Research*, 134(1):201–214, 2005.

34. I. Menache, S. Mannor, and N. Shimkin. Basis function adaption in temporal difference reinforcement learning. *Annals of Operations Research*, 134(1):215–238, 2005.

35. T. P. Minka. Estimating a Dirichlet distribution, 2000. Available at `http://research.microsoft.com/en-us/um/people/minka/papers/dirichlet/`.

36. S. Nariai. *Cross-Entropy Methods in Telecommunication Systems*. PhD thesis, The University of Queensland, Australia, 2009.

37. S. Nariai and D. P. Kroese. On the design of multi-type networks via the cross-entropy method. In *Proceedings of the Fifth International Workshop on the Design of Reliable Communication Networks (DRCN)*, pages 109–114, Naples, 2005.

38. S. Nariai, D. P. Kroese, and K.-P. Hui. Designing an optimal network using the cross-entropy method. In *Intelligent Data Engineering and Automated Learning*, Lecture Notes in Computer Science, pages 228–233, New York, 2005. Springer-Verlag.

39. V. Pihur, S. Datta, and S. Datta. Weighted rank aggregation of cluster validation measures: a Monte Carlo cross-entropy approach. *Bioinformatics*, 23(13):1607–1615, 2007.

40. A. Ridder. Importance sampling simulations of Markovian reliability systems using cross-entropy. *Annals of Operations Research*, 134(1):119–136, 2005.

41. R. Y. Rubinstein. Optimization of computer simulation models with rare events. *European Journal of Operational Research*, 99(1):89–112, 1997.

42. R. Y. Rubinstein. The cross-entropy method for combinatorial and continuous optimization. *Methodology and Computing in Applied Probability*, 1(2):127–190, 1999.

43. R. Y. Rubinstein. Combinatorial optimization, cross-entropy, ants and rare events. In S. Uryasev and P. M. Pardalos, editors, *Stochastic Optimization: Algorithms and Applications*, pages 304–358, Kluwer, Dordrecht, 2001.

44. R. Y. Rubinstein. Combinatorial optimization via cross-entropy. In S. Gass and C. Harris, editors, *Encyclopedia of Operations Research and Management Sciences*, pages 102–106, Kluwer, Dordrecht, 2001.

45. R. Y. Rubinstein. The cross-entropy method and rare-events for maximal cut and bipartition problems. *ACM Transactions on Modelling and Computer Simulation*, 12(1):27–53, 2002.

46. R. Y. Rubinstein and D. P. Kroese. *The Cross-Entropy Method: A Unified Approach to Combinatorial Optimization, Monte Carlo Simulation and Machine Learning*. Springer-Verlag, New York, 2004.

47. R. Y. Rubinstein and D. P. Kroese. *Simulation and the Monte Carlo Method*. John Wiley & Sons, New York, second edition, 2007.

48. R. Y. Rubinstein and B. Melamed. *Modern Simulation and Modeling*. John Wiley & Sons, New York, 1998.

49. A. Sani. *Stochastic Modelling and Intervention of the Spread of HIV/AIDS*. PhD thesis, The University of Queensland, Australia, 2009.

50. A. Sani and D. P. Kroese. Controling the number of HIV infectives in a mobile population. *Mathematical Biosciences*, 213(2):103–112, 2008.

51. J. C. Spall. *Introduction to Stochastic Search and Optimization: Estimation, Simulation, and Control*. John Wiley & Sons, New York, 2003.

52. T. Taimre. *Advances in Cross-Entropy Methods*. PhD thesis, The University of Queensland, Australia, 2009.

53. A. Unveren and A. Acan. Multi-objective optimization with cross entropy method: Stochastic learning with clustered Pareto fronts. In *IEEE Congress on Evolutionary Computation (CEC 2007)*, pages 3065–3071, Singapore, 2007.

54. J. Wang, X. Gao, J. Shi, and Z. Li. Double unmanned aerial vehicle's path planning for scout via cross-entropy method. In *8th ACIS International Conference on Software Engineering, Artificial Intelligence, Networking, and Parallel/Distributed Computing*, volume 2, pages 632–635, Qingdao, 2007.

55. Y. Wu and C. Fyfe. Topology preserving mappings using cross entropy adaptation. In *7th WSEAS International Conference on Artificial Intelligence, Knowledge Engineering and Data Bases*, pages 176–181, Cambridge, 2008.

CHAPTER 14

PARTICLE METHODS

There is a large diversity of iterative Monte Carlo methods, known under various names in different fields of engineering and science as *evolutionary* or *population Monte Carlo*, *sequential Monte Carlo*, *branching* or *interactive particle filters*, *particle splitting*, *genealogical tree models*, and *mean field* or *Feynman–Kac particle models*. In this chapter we focus on particle splitting methods with applications in the following areas.

1. Combinatorial counting problems such as the *satisfiability (SAT) counting* problem;

2. *Bayesian marginal likelihood* estimation and simulation from posterior densities;

3. Combinatorial optimization problems such as the *binary knapsack problem*, *the traveling salesman problem*, and *the quadratic assignment problem*;

4. Simulation from complex multidimensional pdfs with applications to *network reliability* estimation.

For an account of the *classical* splitting method for estimation of hitting probabilities of Markov processes, see Section 10.6. In particular, the *fixed effort* and *fixed splitting* methods in Section 10.6 are precursors to the particle splitting approach. Sequential Monte Carlo is closely related to the sequential importance sampling method described in Section 9.7.5. For an MCMC approach to estimating marginal likelihoods and sampling from posterior densities, we refer to Section 6.2. Finally, Section 16.5 gives a detailed account on network reliability estimation via splitting.

☞ 409
☞ 369
☞ 233
☞ 567

14.1 SEQUENTIAL MONTE CARLO

Iterative or sequential Monte Carlo methods form a broad class of techniques that combine sequential importance sampling and bootstrap resampling. A large number of sequential Monte Carlo algorithms can be described in the following generic form.

☞ 369 Suppose we have the setting of sequential importance sampling in Section 9.7.5. In other words, we wish to sample from a density $f(\mathbf{x})$ using an importance sampling density

$$g(\mathbf{x}) = g_1(x_1)\, g_2(x_2 \,|\, x_1) \cdots g_n(x_n \,|\, \mathbf{x}_{1:n-1}) \;,$$

where $\mathbf{x}_{1:t} = (x_1, \ldots, x_t)$ for all t, with $\mathbf{x} = \mathbf{x}_{1:n}$. Each x_i is possibly multi-dimensional and sampling from $g_t(x_t \,|\, \mathbf{x}_{1:t-1})$ is assumed to be easy. Each element in the set $\{\mathbf{x}_{1:t}\}$ is referred to as a **particle**. Let $f_1, f_2, \ldots, f_n$ be a sequence of *auxiliary* pdfs that are easily evaluated and such that each $f_t(\mathbf{x}_{1:t})$ is a good approximation to $f(\mathbf{x}_{1:t})$, with $f_n(\mathbf{x}) = f(\mathbf{x})$. Then, the likelihood ratio $f(\mathbf{x})/g(\mathbf{x})$ can be written as

$$\frac{f_1(x_1)}{g_1(x_1)} \frac{f_2(\mathbf{x}_{1:2})}{f_1(x_1)\, g_2(x_2 \,|\, x_1)} \frac{f_3(\mathbf{x}_{1:3})}{f_2(\mathbf{x}_{1:2})\, g_3(x_3 \,|\, \mathbf{x}_{1:2})} \cdots \frac{f_n(\mathbf{x}_{1:n})}{f_{n-1}(\mathbf{x}_{1:n-1})\, g_n(x_n \,|\, \mathbf{x}_{1:n-1})} \,,$$

and can be computed recursively via $w_t = u_t\, w_{t-1}$, $t = 1, \ldots, n$, with $w_0 = 1$ and

$$u_t = \frac{f_t(\mathbf{x}_{1:t})}{f_{t-1}(\mathbf{x}_{1:t-1})\, g_t(x_t \,|\, \mathbf{x}_{1:t-1})} \,, \qquad f_0(\mathbf{x}_{1:0}) = 1 \,.$$

To sample (approximately) from $f(\mathbf{x})$ we can use the following algorithm.

Algorithm 14.1 (Sequential Monte Carlo)

1. **Initialization.** *Sample* $X_1^1, \ldots, X_1^N \overset{\text{iid}}{\sim} g_1(x_1)$ *and set* $t = 1$.

2. **Importance Sampling.** *For each* $j = 1, \ldots, N$, *sample* $Y_t^j \sim g_t(y_t \,|\, \mathbf{X}_{1:t-1}^j)$, *and compute the importance weights*

$$u_{t,j} = \frac{f_t(\mathbf{Z}_{1:t}^j)}{f_{t-1}(\mathbf{X}_{1:t-1}^j)\, g_t(Y_t^j \,|\, \mathbf{X}_{1:t-1}^j)} \,,$$

   where $\mathbf{Z}_{1:t}^j = (\mathbf{X}_{1:t-1}^j, Y_t^j)$. *Renormalize the weights so that* $\sum_{j=1}^N u_{t,j} = 1$.

3. **Selection.** *Given the population of particles* $\mathbf{Z}_{1:t}^1, \ldots, \mathbf{Z}_{1:t}^N$, *generate the new population* $\mathbf{X}_{1:t}^1, \ldots, \mathbf{X}_{1:t}^N$ *by sampling independently* N *times from the mixture pdf*

$$\sum_{j=1}^N u_{t,j}\, \mathrm{I}_{\{\mathbf{x}_{1:t} = \mathbf{Z}_{1:t}^j\}} \,.$$

4. **Stopping condition.** *If* $t = n$, *exit and output the population of particles* $\mathbf{X}_{1:t}^1, \ldots, \mathbf{X}_{1:t}^N$; *otherwise, increment* $t = t + 1$ *and repeat from Step 2.*

The selection Step 3 above corrects for the sampling bias introduced by sampling
☞ 362 from the density g rather than f — similar to the likelihood ratio in importance sampling. Note that removing the selection step from the algorithm above results
☞ 370 in the simpler SIS Algorithm 9.9.

The selection Step 3 is known under various names: **bootstrap filter**, **bootstrap resampling**, and **sample importance resampling** (see [16, 30]). Note that this step is equivalent to sampling with replacement from a multinomial distribution. To reduce the variability due to using a bootstrap filter, it is common to use stratification. The following algorithm is one of many possible ways to use stratified sampling within the bootstrap filter [22, 24]. ☞ 356

Algorithm 14.2 (Stratified Resampling) *To sample N times with replacement from the population $1, 2, \ldots, n$ according to the probabilities $p_1, p_2, \ldots, p_n$, execute the following steps.*

1. *Create $c_j = \lfloor N\, p_j \rfloor$ copies of each element j in $1, 2, \ldots, n$. Let*

$$N_r = N - \sum_{j=1}^{n} c_j$$

and store the copies within the first $N - N_r$ elements of a vector of length N as follows:

$$\mathbf{s} = (\underbrace{1, \ldots, 1}_{c_1}, \underbrace{2, \ldots, 2}_{c_2}, \cdots, \underbrace{n, \ldots, n}_{c_n}, \underbrace{0, \ldots, 0}_{N_r}) \;.$$

Note that the vector is padded with N_r zeros to ensure that its length is N. If $N_r = 0$, go to Step 3; otherwise, proceed.

2. *Sample N_r times uniformly and* without replacement *from the set of numbers $1, 2, \ldots, n$. Denote the resulting random population by $Z_1, \ldots, Z_{N_r}$. Set*

$$\mathbf{s} = (\underbrace{1, \ldots, 1, 2, \ldots, 2, \cdots, n, \ldots, n}_{N-N_r}, Z_1, \ldots, Z_{N_r}) \;.$$

3. *Randomly permute the elements of* $\mathbf{s}$ *(using, for example, Algorithm 3.31) and output* $\mathbf{s}$. ☞ 79

Note that there is no loss of generality by resampling the population $1, \ldots, n$, because the elements in $1, \ldots, n$ can be viewed as the indices of a more general population $\mathbf{X}_1, \ldots, \mathbf{X}_n$.

The following MATLAB function implements the stratification algorithm.

```
function s=stratified_resample(p,N)
% input:  [n,1] vector of resampling probabilities prob=[p1;...;pn]
%               where p(i) corresponds to the probability
%               of resampling the i-th particle;
%        scalar 'N' denoting the number of required samples;
% output: [N,1] vector 's' which gives the
%               indices (ranging from 1 to n) of the
%               particles forming the new sample;
p=p(:);n=size(p,1);
%normalize any weights so that they are proper probabilities
p=p/sum(p);
c=floor(N*p);
```

```
s=zeros(N,1);   %preallocate memory
% deterministic resampling (step 1)
cum=0; % cumulative sum of deterministic copies
for i=1:n
    det_copy=c(i);
    s(cum+1:cum+det_copy)=i*ones(det_copy,1);
    cum=cum+det_copy;
end
N_r=N-sum(c); % residual number
% start step 2
if N_r~=0 % perform residual sampling if needed
    s(N-N_r+1:N)=resample(n,N_r);
end
% destroy the structure induced by the construction of s
% by randomly reallocating the indices of s;
s=s(randperm(N)); %step 3
```

The function resample.m below implements random uniform sampling from the population $1, \ldots, n$ without replacement. The function samples k times, where $k < n$. The function is most efficient when $k \ll n$. For k close to n, a more efficient approach for sampling without replacement is given in Example 3.22.

☞ 79

```
function x=resample(n,k)
% Samples the numbers 1,...,n uniformly
% without replacement k times, where k<n;
x = zeros(1,n); S = 0;
while S < k
    x(ceil(n * rand(1,k-S))) = 1;
    S = sum(x);
end
x = find(x > 0);
% destroy sorted structure
x = x(randperm(k));
```

Figure 14.1 illustrates resampling the population $1, \ldots, 10$ one hundred times ($N = 100$) according to the probabilities

$$p_k = \frac{2k}{100}, \quad k = 1, \ldots, 10 .$$

The left (right) panel shows the result of resampling without (with) stratification. The resampling probabilities are indicated as circles and the random sample is represented as dots stacked on top of each other. Note that in the left panel, point 1 is not present in the random sample, and although $p_{10} > p_9$, point 10 is sampled fewer times than point 9. In contrast, in the right panel point 1 is sampled once (in agreement with its expected value of $Np_1 = 1$) and the outcome is consistent with the ordering $p_1 < p_2 < \cdots < p_{10}$. In summary, the outcome in the right panel is consistent with lower sampling variability.

Sequential Monte Carlo methods were initially designed for state-space models such as target tracking, but have gradually found applications in areas such as computer vision, pattern recognition, and Bayesian inference; see [10, 11].

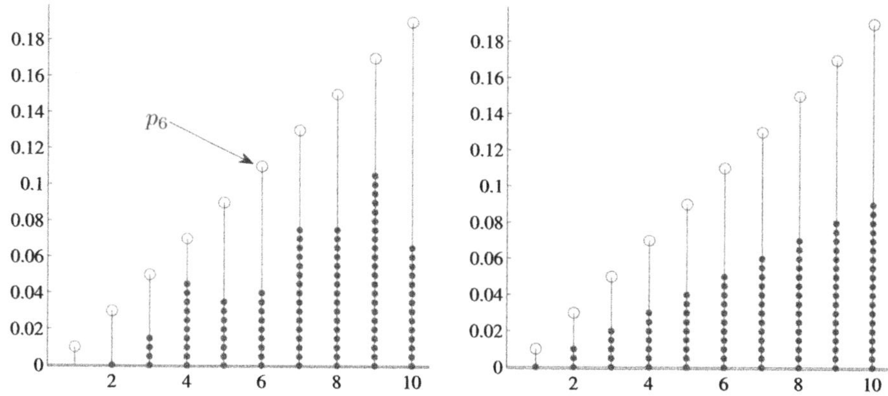

Figure 14.1 The left panel shows a result of resampling of $1, \ldots, 10$ without any stratification, and the right panel shows the result of resampling of $1, \ldots, 10$ with stratification. The resampling probabilities, $p_k = 2k/100$, $k = 1, \ldots, 10$, are indicated with circles, and the outcome of the resampling as dots stacked on top of each other.

14.2 PARTICLE SPLITTING

The remainder of this chapter deals with the **generalized splitting** (GS) algorithm [1, 2, 7], which can be viewed as a special case of the Holmes–Diaconis–Ross algorithm described in Example 6.7, and which extends the applicability of the ☞ 247 classical fixed effort splitting method [14] to both static (that is, time-independent) and non-Markovian models; see Section 10.6. ☞ 409

In addition to its links with the classical splitting of Kahn and Harris [21], the GS algorithm is also a special case of an improved sequential Monte Carlo algorithm [20, 27] with the following important difference. In sequential Monte Carlo we use bootstrap sampling (possibly stratified) to replicate the same particle in a population numerous times. This typically decreases the diversity of the Monte Carlo population. In contrast, there is no resampling step in the GS algorithm. As a result the diversity in the Monte Carlo population is improved. In addition, the unbiasedness property of the GS estimator is not a direct consequence of similar results for other sequential Monte Carlo estimators [26, 27].

The GS method involves the following general framework (see also (9.16) and ☞ 362 (13.1)). Let ℓ be the expected performance of a stochastic system of the form ☞ 464

$$\ell = \mathbb{E}\, H(\mathbf{X}) = \int H(\mathbf{x})\, f(\mathbf{x})\, \mathrm{d}\mathbf{x}\,, \quad \mathbf{X} \sim f\,, \tag{14.1}$$

where H is a real-valued function. When $\mathbf{X}$ is discrete the integral is replaced with a sum. We call f the **nominal** pdf. A special case of (14.1) is obtained when

$H(\mathbf{x}) = \mathrm{I}_{\{S(\mathbf{x}) \geqslant \gamma\}}$, where S is a given function — called the **importance function** — and γ is a **threshold** or **level** parameter such that

$$\ell \stackrel{\text{def}}{=} \ell(\gamma) = \mathbb{E}\,\mathrm{I}_{\{S(\mathbf{X}) \geqslant \gamma\}} = \mathbb{P}(S(\mathbf{X}) \geqslant \gamma), \quad \mathbf{X} \sim f, \tag{14.2}$$

☞ 381 is a rare-event probability. Another special case of (14.1) is obtained when $H(\mathbf{x}) = \mathrm{e}^{-S(\mathbf{x})/\gamma}$, which arises frequently in statistical mechanics in the estimation of the
☞ 263 partition function [29].

Using the GS algorithm, we construct unbiased estimators for rare-event probabilities of the form (14.2) and, in general, multidimensional integrals of the form (14.1). The GS method tackles these static non-Markovian problems by artificially constructing a Markov chain using, for example, Gibbs or hit-and-run moves, and then applying the splitting idea to the Markov process induced by these moves.

14.3 SPLITTING FOR STATIC RARE-EVENT PROBABILITY ESTIMATION

We first explain how one can obtain unbiased estimates of the rare-event probability (14.2). Choose the importance function S and partition the interval $(-\infty, \gamma]$ by using intermediate levels $-\infty = \gamma_0 \leqslant \gamma_1 \leqslant \ldots \leqslant \gamma_{T-1} \leqslant \gamma_T = \gamma$. Note that, unlike in classical splitting, $\gamma_0 = -\infty$. We assume that the sequence of levels is chosen such that the conditional probabilities $\mathbb{P}(S(\mathbf{X}) \geqslant \gamma_t \,|\, S(\mathbf{X}) \geqslant \gamma_{t-1}) = c_t$, $t = 1, \ldots, T$, are not rare-event probabilities, and that we have estimates $\{\varrho_t\}$ of $\{c_t\}$ available. These estimates cannot usually be determined online, but in Section 14.4 we explain how we can construct the sequence $\{(\gamma_t, \varrho_t)\}_{t=1}^{T}$ using the ADAM algorithm (see also [2]). Without loss of generality, we assume that generating random variables from the nominal pdf f is easy.

Algorithm 14.3 (GS Algorithm for Estimating $\ell = \mathbb{P}(S(\mathbf{X}) \geqslant \gamma)$) *Given a sequence $\{(\gamma_t, \varrho_t)\}_{t=1}^{T}$ and a sample size N, execute the following steps.*

1. **Initialization.** *Set $t = 1$ and $N_0 = \varrho_1 \left\lfloor \frac{N}{\varrho_1} \right\rfloor$ (which ensures that N_0/ϱ_1 is an integer). Generate*

$$\mathbf{X}_1, \ldots, \mathbf{X}_{N_0/\varrho_1} \stackrel{\text{iid}}{\sim} f$$

and denote $\mathcal{X}_0 = \{\mathbf{X}_1, \ldots, \mathbf{X}_{N_0/\varrho_1}\}$. Let $\mathcal{X}_1$ be the largest subset of elements $\mathbf{X}$ in $\mathcal{X}_0$ for which $S(\mathbf{X}) \geqslant \gamma_1$, and let N_1 be the size of $\mathcal{X}_1$. If $N_1 = 0$, go to Step 4.

2. **Markov chain sampling.** *For each $\mathbf{X}_i$ in $\mathcal{X}_t = \{\mathbf{X}_1, \ldots, \mathbf{X}_{N_t}\}$, sample independently:*

$$\mathbf{Y}_{i,j} \sim \kappa_t(\mathbf{y} \,|\, \mathbf{Y}_{i,j-1}), \quad \mathbf{Y}_{i,0} = \mathbf{X}_i, \quad j = 1, \ldots, \mathcal{S}_{ti}\,, \tag{14.3}$$

where $\mathcal{S}_{ti}$ is the splitting factor

$$\mathcal{S}_{ti} - \left\lfloor \frac{1}{\varrho_{t+1}} \right\rfloor \stackrel{\text{iid}}{\sim} \mathsf{Ber}\left(\frac{1}{\varrho_{t+1}} - \left\lfloor \frac{1}{\varrho_{t+1}} \right\rfloor \right), \quad i = 1, \ldots, N_t\,.$$

Here $\kappa_t(\mathbf{y} \,|\, \mathbf{Y}_{i,j-1})$ is a Markov transition density with stationary pdf

$$f_t(\mathbf{y}) \stackrel{\text{def}}{=} \frac{f(\mathbf{y})\, \mathrm{I}_{\{S(\mathbf{y}) \geqslant \gamma_t\}}}{\ell(\gamma_t)}\,.$$

Reset

$$\mathcal{X}_t = \left\{ \mathbf{Y}_{1,1}\,, \mathbf{Y}_{1,2}\,, \ldots, \mathbf{Y}_{1,\mathcal{S}_{t1}}\,, \cdots\cdots\cdots\,, \mathbf{Y}_{N_t,1}\,, \mathbf{Y}_{N_t,2}\,, \ldots, \mathbf{Y}_{N_t,\mathcal{S}_{tN_t}} \right\},$$

where $\mathcal{X}_t$ *contains* $|\mathcal{X}_t| = \sum_{i=1}^{N_t} \mathcal{S}_{ti}$ *elements and* $\mathbb{E}[\,|\mathcal{X}_t| \mid N_t\,] = \frac{N_t}{\varrho_{t+1}}$.

3. **Updating.** *Let* $\mathcal{X}_{t+1} = \{\mathbf{X}_1, \ldots, \mathbf{X}_{N_{t+1}}\}$ *be the largest subset of elements in* $\mathcal{X}_t$ *for which* $S(\mathbf{X}) \geqslant \gamma_{t+1}$. *Here,* N_{t+1} *is the size of* $\mathcal{X}_{t+1}$. *Increase the level counter:* $t = t + 1$.

4. **Stopping condition.** *If* $t = T$ *go to Step 5. If* $N_t = 0$, *set* $N_{t+1} = N_{t+2} = \cdots = N_T = 0$ *and go to Step 5; otherwise, repeat from Step 2.*

5. **Final estimator.** *Deliver the unbiased estimate of the rare-event probability,*

$$\widehat{\ell} = \frac{N_T}{N_0} \prod_{t=1}^{T} \varrho_t\,, \tag{14.4}$$

and the unbiased estimate of the variance of the estimator $\widehat{\ell}$:

$$\widehat{\mathrm{Var}(\widehat{\ell})} = \frac{\prod_{t=1}^{T} \varrho_t^2}{N_0(N_0 - \varrho_1)} \sum_{i=1}^{N_0/\varrho_1} \left(O_i - \frac{\varrho_1}{N_0} N_T \right)^2, \tag{14.5}$$

where O_i *denotes the number of points in* $\mathcal{X}_T$ *that share a common history with the i-th point from the initial population* $\mathcal{X}_0$ *and have their* S *value on or above level* γ *at the final iteration.*

Ideally, we would like to draw from the conditional density $f_t(\mathbf{y})$ at each stage t of the algorithm. However, this is typically impossible. Instead, in Step 2 of Algorithm 14.3 we use MCMC to approximately sample from $f_t(\mathbf{y})$. In particular, a move from $\mathbf{X}$ to $\mathbf{Y}$ (using the transition density $\kappa_t(\mathbf{y}\,|\,\mathbf{x})$) can, for example, consist of drawing $\mathbf{Y}$ from the conditional pdf

$$Y_i \sim f_t(y_i \,|\, Y_1, \ldots, Y_{i-1}, X_{i+1}, \ldots, X_n), \quad i = 1, \ldots, n\,,$$

as in the Gibbs sampling method (Section 6.2). The transition density is then ☞ 233

$$\kappa_t(\mathbf{y}\,|\,\mathbf{x}) = \prod_{i=1}^{n} f_t(y_i \,|\, y_1, \ldots y_{i-1}, x_{i+1}, \ldots, x_n)\,. \tag{14.6}$$

Alternatively, a move from $\mathbf{X}$ to $\mathbf{Y}$ may consist of a hit-and-run move (Section 6.3.1): ☞ 240

1. Generate a uniformly distributed point on the surface of the n-dimensional hypersphere:

$$\mathbf{d} = \left(\frac{Z_1}{||\mathbf{Z}||}, \ldots, \frac{Z_n}{||\mathbf{Z}||} \right)^{\top}, \quad Z_1, \ldots, Z_n \overset{\text{iid}}{\sim} \mathsf{N}(0,1)\,.$$

2. Given the current state $\mathbf{X}$, generate Λ from the density

$$\widetilde{g}(\lambda \,|\, \mathbf{X}, \mathbf{d}) = \frac{f(\mathbf{X} + \lambda\, \mathbf{d})}{\int_{-\infty}^{\infty} f(\mathbf{X} + u\, \mathbf{d})\, du}\,.$$

3. The new state of the chain is

$$\mathbf{Y} = \begin{cases} \mathbf{X} + \Lambda \mathbf{d} & \text{if} \quad S(\mathbf{X} + \Lambda \mathbf{d}) \geqslant \gamma_t\,, \\ \mathbf{X} & \text{otherwise}\,. \end{cases}$$

We illustrate Algorithm 14.3 on a typical problem of the form (14.2) with three levels ($T = 3$). Figure 14.2 depicts the GS recipe as applied to a particular two-dimensional rare-event simulation problem.

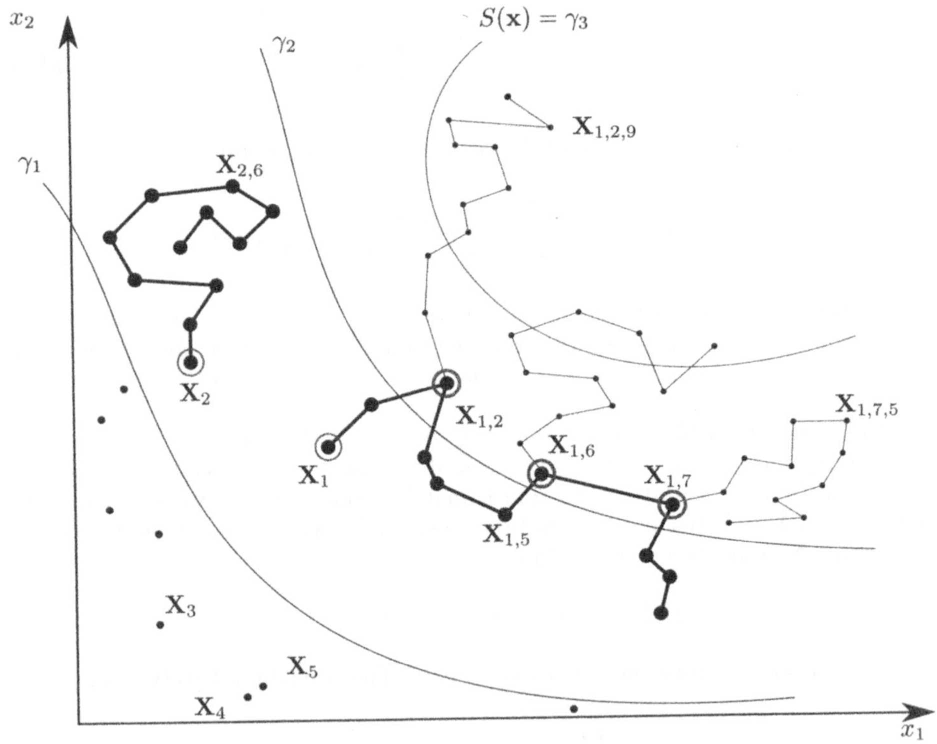

Figure 14.2 Typical evolution of the GS algorithm in a two-dimensional state space.

The three level sets $\{\mathbf{x} : S(\mathbf{x}) = \gamma_t\}$, $t = 1, 2, 3$ are plotted as nested curves and the **entrance states** for stages 1 and 2 are encircled. We assume that the $\{\gamma_t\}$ are given and that $\varrho_t = 1/10$ for all t; that is, the splitting factors are: $s_t = \varrho_t^{-1} = 10$ for all t. Initially, at stage $t = 1$, we generate $N_0/\varrho_1 = 10$ independent points from the density $f(\mathbf{x})$. We denote the points $\mathbf{X}_1, \ldots, \mathbf{X}_{10}$. Two of these points, namely $\mathbf{X}_1$ and $\mathbf{X}_2$, are such that both $S(\mathbf{X}_1)$ and $S(\mathbf{X}_2)$ are above the γ_1 threshold. Points $\mathbf{X}_1$ and $\mathbf{X}_2$ are thus the entrance states for the next stage of the algorithm, and

$N_1 = \sum_{j=1}^{10} \mathrm{I}\{S(\mathbf{X}_j) \geqslant \gamma_1\} = 2$. In stage $t = 2$ we start independent Markov chains from each of the entrance states $\mathbf{X}_1$ and $\mathbf{X}_2$. The only requirement is that each Markov chain has a stationary distribution equal to the conditional distribution of $\mathbf{X}$ given that $S(\mathbf{X}) \geqslant \gamma_1$, where $\mathbf{X} \sim f$. The length (the number of steps) of both chains is equal to $s_2 = 10$. Thus, the simulation effort for $t = 2$ is $N_1 \times s_2 = 20$. In other words, in stage $t = 2$ we generate

$$\mathbf{X}_{i,j} \sim \kappa_1(\mathbf{x} \,|\, \mathbf{X}_{i,j-1}), \quad j = 1, \ldots, 10, \;\; i = 1, 2 \,,$$

where $\mathbf{X}_{i,0} = \mathbf{X}_i$, and $\kappa_1(\cdot \,|\, \cdot)$ is a Markov transition density with stationary pdf f_1 given by ($t = 1$)

$$f_t(\mathbf{x}) = \frac{f(\mathbf{x})\, \mathrm{I}_{\{S(\mathbf{x}) \geqslant \gamma_t\}}}{\ell(\gamma_t)} \,. \tag{14.7}$$

Figure 14.2 depicts the Markov chains as branches sprouting from points $\mathbf{X}_1$ and $\mathbf{X}_2$. Note that these branches are drawn thicker than branches generated at stage $t = 3$ to distinguish the stages from each other. None of the points on the $\mathbf{X}_2$ branch have a value S above γ_2. Points $\mathbf{X}_{1,2}, \mathbf{X}_{1,6}, \mathbf{X}_{1,7}$ from the $\mathbf{X}_1$ branch (encircled) make it above the γ_2 threshold. These three points will be the entrance states for stage $t = 3$ of the algorithm. Thus,

$$N_2 = \sum_{j=1}^{10} \left(\mathrm{I}_{\{S(\mathbf{X}_{1,j}) \geqslant \gamma_2\}} + \mathrm{I}_{\{S(\mathbf{X}_{2,j}) \geqslant \gamma_2\}} \right) = 3 \,.$$

In the final stage we start three independent Markov chains with stationary density f_2 from each of the entrance states $\mathbf{X}_{1,2}, \mathbf{X}_{1,6}$, and $\mathbf{X}_{1,7}$. The length of each chain is $s_3 = 10$. Thus, the simulation effort for stage $t = 3$ is $3 \times 10 = 30$, and we generate

$$\mathbf{X}_{1,j,k} \sim \kappa_2(\mathbf{x} \,|\, \mathbf{X}_{1,j,k-1}), \quad k = 1, \ldots, 10, \; j = 2, 6, 7 \,,$$

where $\mathbf{X}_{1,j,0} = \mathbf{X}_{1,j}$, and $\kappa_2(\cdot \,|\, \cdot)$ is a Markov transition density with stationary density f_2 defined in (14.7). Figure 14.2 shows that the points that up-cross level γ_3 are $\mathbf{X}_{1,2,k}$, $k = 3, \ldots, 10$ and $\mathbf{X}_{1,6,k}$, $k = 7, 8, 10$. Thus, in the last stage $T = 3$ we have $N_T = 11$. Finally, an estimator of $\ell(\gamma_3)$ is

$$\widehat{\ell}(\gamma_3) = \frac{N_T}{N_0} \prod_{t=1}^{3} s_t^{-1},$$

and this gives the estimate 11×10^{-3}. Proposition 14.3.1 on Page 492 shows that such an estimator is unbiased.

Note that if ϱ_t^{-1} is an integer, then the length of each Markov chain started from an entrance state at stage t is deterministic, and if ϱ_t^{-1} is not an integer, then, as in the classical splitting method, the length of each Markov chain is taken to be a random integer-valued splitting factor with expected value ϱ_t^{-1}. Even though all entrance states at stage t are assigned a random splitting factor, these factors are independent and have the same expected value. In Algorithm 14.3 such random splitting factors are conveniently constructed by generating independent Bernoulli random variables with a common success probability $\varrho_t^{-1} - \lfloor \varrho_t^{-1} \rfloor$ and then adding $\lfloor \varrho_t^{-1} \rfloor$ to the Bernoulli variables.

Although Algorithm 14.3 has strong similarities with the classical splitting method (in particular, fixed splitting) in Section 10.6, there are some important differences. First, as seen from Figure 14.2, during the initialization of the GS algorithm we do not run Markov chains, but generate iid vectors from the density f in (14.2). In contrast, in the classical fixed splitting algorithm one always initializes with a population of independent Markov chains. Second, in Algorithm 14.3 the first level is always $\gamma_0 = -\infty$, while in the classical splitting algorithm γ_0 is usually finite. Finally, while in the classical fixed splitting we run the *same* Markov process throughout all stages of the algorithm, in the GS algorithm the stationary distribution of the Markov chains *changes* across the stages. More precisely, in the GS algorithm, the stationary distribution at stage t is f_{t-1}. As a result of this, *no Markov chain paths can go below level* γ_{t-1} *in stage* t of the GS algorithm. In contrast, the paths in classical fixed splitting can down-cross thresholds and even down-cross level γ_0.

☞ 409

■ EXAMPLE 14.1 (SAT Counting Problem)

☞ 694

We wish to apply the GS method to a SAT counting problem. There are many different mathematical formulations of the SAT problem [17]. Here we use a formulation that is convenient for the problems from the SATLIB project [19] with website `www.satlib.org`. Let $\mathbf{x} = (x_1, \ldots, x_n)^\top$, $\mathbf{x} \in \{0,1\}^n$ denote a **truth assignment** and let $A = (A_{ij})$ denote an $m \times n$ **clause matrix**; that is, all elements of A belong to the set $\{-1, 0, 1\}$. Define $\mathbf{b} = (b_1, \ldots, b_m)^\top$ as the vector with entries $b_i = 1 - \sum_{j=1}^{n} \mathrm{I}_{\{A_{ij}=-1\}}$. In the standard SAT problem one is interested in finding a truth assignment $\mathbf{x}$ for which $A\mathbf{x} \geqslant \mathbf{b}$. In the **SAT counting problem**, one is interested in finding the *total number* of truth assignments $\mathbf{x}$ for which $A\mathbf{x} \geqslant \mathbf{b}$. The SAT counting problem is considered more complex than the SAT problem [34, 36], and in fact the SAT counting problem is known to be a #P-complete problem. Using this formulation the SAT counting problem reduces to estimating the size of the set

$$\mathscr{X}^* = \left\{\mathbf{x} \in \{0,1\}^n : \sum_{i=1}^{m} \mathrm{I}\left\{\sum_{j=1}^{n} A_{ij} x_j \geqslant b_i\right\} \geqslant m\right\}.$$

To estimate $|\mathscr{X}^*|$ we consider the problem of estimating the probability

$$\ell = \mathbb{P}\left(S(\mathbf{X}) \geqslant m\right), \quad \{X_j\} \overset{\text{iid}}{\sim} \mathsf{Ber}(1/2), \quad S(\mathbf{x}) = \sum_{i=1}^{m} \mathrm{I}\left\{\sum_{j=1}^{n} A_{ij} x_j \geqslant b_i\right\}, \quad (14.8)$$

via Algorithm 14.3. Thus, each row of A represents a clause and $S(\mathbf{x})$ is the number of clauses that are satisfied. The size of the set is then estimated from the relation $|\mathscr{X}^*| = \ell\, 2^n$. As a numerical example, consider the `uf75-01` problem from the SATLIB website, with $m = 325$, $n = 75$. We apply Algorithm 14.3 with $N = 10^4$ and the splitting factors and levels given in Table 14.1, giving a total simulation effort of about 2.8×10^6 samples

The Markov transition density κ_t in Step 2 of Algorithm 14.3 is given by (14.6), and the stationary pdf is

$$f_t(\mathbf{x}) = \frac{1}{2^n \ell(\gamma_t)} \mathrm{I}\left\{\sum_{i=1}^{m} \mathrm{I}\left\{\sum_{j=1}^{n} A_{ij} x_j \geqslant b_i\right\} \geqslant \gamma_t\right\}, \quad \mathbf{x} \in \{0,1\}^n.$$

In other words, a move from $\mathbf{x}$ to $\mathbf{y}$ using the transition density $\kappa_t(\mathbf{y} \mid \mathbf{x})$ consists of the following Gibbs sampling procedure.

1. Given a state $\mathbf{x}$ such that $S(\mathbf{x}) \geqslant \gamma_t$, generate $Y_1 \sim f_t(y_1 \mid x_2, \ldots, x_n)$.
2. For each $k = 2, \ldots, n-1$, generate $Y_k \sim f_t(y_k \mid Y_1, \ldots, Y_{k-1}, x_{k+1}, \ldots, x_n)$.
3. Finally, generate $Y_n \sim f_t(y_n \mid Y_1, \ldots, Y_{n-1})$.

Note that one can write the conditional density of Y_k as

$$f_t(y_k \mid Y_1, \ldots, Y_{k-1}, x_{k+1}, \ldots, x_n) = \begin{cases} p_k, & y_k = 1 \\ 1 - p_k, & y_k = 0 \,, \end{cases}$$

where

$$p_k = \frac{\mathrm{I}\{s_k^+ \geqslant \gamma_t\}}{\mathrm{I}\{s_k^+ \geqslant \gamma_t\} + \mathrm{I}\{s_k^- \geqslant \gamma_t\}} \,,$$

with $s_k^- = S(Y_1, \ldots, Y_{k-1}, 0, x_{k+1}, \ldots, x_n)$, $s_k^+ = S(Y_1, \ldots, Y_{k-1}, 1, x_{k+1}, \ldots, x_n)$.

With the setup above, we obtain a typical estimate of $\widehat{|\mathscr{X}^*|} = 2.31 \times 10^3$ with an estimated relative error of 5.8%. Total enumeration of all possible truth assignments for which $A\mathbf{x} \geqslant \mathbf{b}$ would require the equivalent of a simulation effort of size $2^{75} \approx 3.7 \times 10^{22}$ and is hence impracticable. To achieve the same relative error using crude Monte Carlo would require a sample size of approximately $N = 4.8 \times 10^{21}$. Thus, we see that with a minimal amount of additional work the GS algorithm has reduced the simulation effort over CMC by a factor of approximately 10^{15}. Note that a better choice for the importance function S may allow for all conditional probabilities $c_t = \mathbb{P}(S(\mathbf{X}) \geqslant \gamma_t \mid S(\mathbf{X}) \geqslant \gamma_{t-1})$ to be approximately equal, thus giving an estimator with a smaller relative error. However the optimal choice of the importance function in splitting is still an unresolved problem [13, 25].

Table 14.1 The sequence of levels and splitting factors used in Algorithm 14.3 the SAT problem uf75-01.

t	γ_t	ϱ_t	t	γ_t	ϱ_t	t	γ_t	ϱ_t
1	285	0.4750	11	305	0.4359	21	315	0.2235
2	289	0.4996	12	306	0.4239	22	316	0.2071
3	292	0.4429	13	307	0.3990	23	317	0.1892
4	294	0.4912	14	308	0.3669	24	318	0.1722
5	296	0.4369	15	309	0.3658	25	319	0.1363
6	298	0.3829	16	310	0.3166	26	320	0.1413
7	300	0.3277	17	311	0.3072	27	321	0.1237
8	302	0.2676	18	312	0.3104	28	322	0.0953
9	303	0.4982	19	313	0.2722	29	323	0.0742
10	304	0.4505	20	314	0.2253	30	324	0.0415
						31	325	0.0115

One additional benefit of our simulation is that we can put a deterministic lower bound on $|\mathscr{X}^*|$. This can be done as follows. The population $\mathcal{X}_T$ at the final iteration of Algorithm 14.3 is approximately uniformly distributed over the set $\mathscr{X}^*$ and as a result can be used to find some of the distinct solutions of the SAT problem. We ran Algorithm 14.3 ten times with $N = 10^4$ and were able to find

Algorithm 14.4 (ADAM Algorithm) *Given the sample size N and the parameters $\varrho \in (0,1)$ and γ, execute the following steps.*

1. **Initialization.** *Set the counter $t = 1$.*

   (a) *Generate $\mathbf{X}_1, \ldots, \mathbf{X}_N \sim f$ and denote $\mathcal{X}_0 = \{\mathbf{X}_1, \ldots, \mathbf{X}_N\}$.*

   (b) *Denote $S_i = S(\mathbf{X}_i)$ for all $\mathbf{X}_i \in \mathcal{X}_{t-1}$, and let*

$$\widetilde{\gamma}_t = \underset{\gamma \in \{S_1, \ldots, S_N\}}{\operatorname{argmin}} \left\{ \frac{1}{N} \sum_{i=1}^{N} \mathrm{I}_{\{S_i \geqslant \gamma\}} \leqslant \varrho \right\} . \tag{14.10}$$

   That is, $\widetilde{\gamma}_t$ is the smallest value from among $S(\mathbf{X}_1), \ldots, S(\mathbf{X}_N)$ such that $\frac{1}{N}\sum_{i=1}^{N} \mathrm{I}\{S(\mathbf{X}_i) \geqslant \widetilde{\gamma}_t\} \leqslant \varrho$. Set $\gamma_t = \min\{\gamma, \widetilde{\gamma}_t\}$. Let $\mathcal{X}_t$ be the subset of all elements in $\mathcal{X}_{t-1}$ for which $S(\mathbf{X}) \geqslant \gamma_t$. Let $N_t = |\mathcal{X}_t|$. Then, $\varrho_t = \frac{N_t}{N}$ is an approximation of the probability $c_t = \mathbb{P}_f(S(\mathbf{X}) \geqslant \gamma_t \mid S(\mathbf{X}) \geqslant \gamma_{t-1})$, setting $\gamma_0 = -\infty$.

2. **Markov chain sampling.** *Identical to Step 2 of Algorithm 14.3, except that in (14.3) the splitting factors are generated in a different way, namely,*

$$\mathcal{S}_{ti} = \left\lfloor \frac{N}{N_t} \right\rfloor + B_i, \quad i = 1, \ldots, N_t \, .$$

   Here, the random variables $B_1, \ldots, B_{N_t}$ are $\mathsf{Ber}(1/2)$ random variables conditional on $\sum_{i=1}^{N_t} B_i = (N \bmod N_t)$. More precisely, $(B_1, \ldots, B_{N_t})$ is a binary vector with joint pdf

$$\mathbb{P}(B_1 = b_1, \ldots, B_{N_t} = b_{N_t}) = \frac{(N_t - r)!\, r!}{N_t!} \mathrm{I}_{\{b_1 + \cdots + b_{N_t} = r\}}, \quad b_i \in \{0, 1\} ,$$

   where $r = (N \bmod N_t)$. As a consequence of the generation of the splitting factors, after resetting

$$\mathcal{X}_t = \left\{ \mathbf{Y}_{1,1}, \mathbf{Y}_{1,2}, \ldots, \mathbf{Y}_{1,\mathcal{S}_{t1}}, \cdots\cdots\cdots, \mathbf{Y}_{N_t,1}, \mathbf{Y}_{N_t,2}, \ldots, \mathbf{Y}_{N_t,\mathcal{S}_{tN_t}} \right\},$$

   the set $\mathcal{X}_t$ contains exactly N elements.

3. **Updating and Estimation.** *Update the counter $t = t + 1$ and proceed as in part (b) of Step 1.*

4. **Stopping condition.** *If $\gamma_t = \gamma$, set $T = t$ and go to Step 5; otherwise, repeat from Step 2.*

5. **Final estimates.** *Deliver the estimated levels $\gamma_1, \ldots, \gamma_T$, the splitting factors $\varrho_1, \ldots, \varrho_T$, and the estimate of the rare-event probability:*

$$\widehat{\ell}_{\mathrm{ADAM}} = \prod_{t=1}^{T} \varrho_t = \frac{\prod_{t=1}^{T} N_t}{N^T} . \tag{14.11}$$

The main differences between the ADAM algorithm and the GS algorithm are the following. First, the difference in Step 2 of the ADAM algorithm is that the splitting factors are generated in a way that fixes the simulation effort at each stage to be N (see [14]). Second, as seen from (14.10), the levels $\{\gamma_t\}$ are determined online using the random populations $\{\mathcal{X}_t\}$. As a consequence of these differences, the estimator $\widehat{\ell}_{\text{ADAM}}$ is not unbiased and the algorithm does not provide a simple estimate for the variance of $\widehat{\ell}_{\text{ADAM}}$.

The ADAM algorithm can be used as a stand-alone algorithm in the sense that it can provide an estimate of ℓ without the need for any preliminary simulation. For example, in the SAT counting problem of Example 14.1 for the cost of 3.1×10^6 samples ($N = 10^5$, $\varrho = 0.5$) Algorithm 14.4 gives an estimate of $\widehat{|\mathscr{X}^*|} = 2.26 \times 10^3$ with estimated relative error of 3%. This estimate is close to the one obtained using GS with importance sampling, see Table 14.2.

14.5 ESTIMATION OF MULTIDIMENSIONAL INTEGRALS

In the previous section we show how one can estimate rare-event probabilities of the form (14.2) using either ADAM or GS. In this section we extend the applicability of these algorithms to the more general problem of estimating integrals of the form (14.1). To this end we rewrite (14.1) using the following notation:

$$\mathcal{Z} = \mathbb{E}\, H(\mathbf{Z}) = \int p(\mathbf{z})\, H(\mathbf{z})\, \mathrm{d}\mathbf{z}, \quad \mathbf{Z} \sim p\,, \tag{14.12}$$

so that now the aim is to estimate $\mathcal{Z}$. We show that the GS algorithm provides an unbiased estimate of $\mathcal{Z}$ and note that it can be easily extended to deal with estimation of sensitivities [3]. First, note that

$$\mathbb{E}\, H(\mathbf{Z}) = 2\,\mathbb{E}[H(\mathbf{Z})\, \mathrm{I}_{\{H(\mathbf{Z}) \geqslant 0\}}] - \mathbb{E}|H(\mathbf{Z})|\,,$$

so that without loss of generality we can consider estimating (14.12) for $H(\mathbf{z}) \geqslant 0$. Second, let $\widetilde{p}(\mathbf{z})$ be a proposal density from which it is easy to sample and which dominates $p(\mathbf{z})H(\mathbf{z})$ in the sense that

$$p(\mathbf{z})H(\mathbf{z}) \leqslant \mathrm{e}^{a\,\gamma + b}\, \widetilde{p}(\mathbf{z}), \qquad \text{for all } \mathbf{z} \in \mathbb{R}^n\,, \tag{14.13}$$

for some $\gamma \geqslant (\ln \mathcal{Z} - b)/a$, where $a > 0$ and $b \in \mathbb{R}$ are fixed but otherwise arbitrary constants. Note that the constant $\mathrm{e}^{a\,\gamma+b}$ is an upper bound on $\mathcal{Z}$. Typically we have $\mathrm{e}^{a\,\gamma+b} \gg \mathcal{Z}$ or $\gamma \gg (\ln \mathcal{Z} - b)/a$, and in many cases it is natural to choose $\widetilde{p}(\mathbf{z}) = p(\mathbf{z})$.

Algorithm 14.5 (Estimation of $\mathcal{Z}$) *Suppose we are given a proposal density $\widetilde{p}(\mathbf{z})$, parameters γ, a, b such that (14.13) holds, and algorithm A, where A denotes either the GS or the ADAM algorithm. Then, execute the following steps.*

1. **Estimation of ℓ.** *Use algorithm A to compute an estimate $\widehat{\ell}(\gamma)$ of*

$$\ell(\gamma) = \mathbb{E}\, \mathrm{I}_{\{S(\mathbf{X}) \geqslant \gamma\}} = \int f(\mathbf{x})\, \mathrm{I}_{\{S(\mathbf{x}) \geqslant \gamma\}}\, \mathrm{d}\mathbf{x}\,, \tag{14.14}$$

where the vector $\mathbf{x} = (\mathbf{z}, u)^\top \in \mathbb{R}^n \times (0,1)$ augments $\mathbf{z}$ with the variable $u \in (0,1)$, the value $S(\mathbf{x})$ is given by

$$S(\mathbf{x}) = \frac{1}{a} \ln \left(\frac{p(\mathbf{z})H(\mathbf{z})}{u\, \widetilde{p}(\mathbf{z})} \right) - \frac{b}{a}\,,$$

and the density $f(\mathbf{x})$ by

$$f(\mathbf{x}) = \widetilde{p}(\mathbf{z})\, \mathrm{I}_{\{0<u<1\}}, \quad \mathbf{x} \in \mathbb{R}^{n+1}\,.$$

2. **Estimation of $\mathcal{Z}$.** *Estimate $\mathcal{Z}$ in* (14.12) *by*

$$\widehat{\mathcal{Z}} = \mathrm{e}^{a\,\gamma+b}\, \widehat{\ell}(\gamma)\,.$$

The following proposition shows that the estimate $\widehat{\mathcal{Z}}$ is unbiased if A is the GS algorithm.

Proposition 14.5.1 (Relation Between ℓ and $\mathcal{Z}$) *The pdf*

$$\frac{p(\mathbf{z})H(\mathbf{z})}{\mathcal{Z}} \tag{14.15}$$

is the marginal density of $f_T(\mathbf{x}) = \frac{1}{\ell(\gamma_T)} f(\mathbf{x})\, \mathrm{I}_{\{S(\mathbf{x}) \geqslant \gamma_T\}}$ (recall that $\gamma_T = \gamma$), and $\mathcal{Z} = \mathrm{e}^{a\,\gamma+b}\ell(\gamma)$.

Proof: Note that u is an *auxiliary variable* similar to the one used in the acceptance–rejection method for random variable generation. From (14.13) it follows that (with $\mathbf{x} = (\mathbf{z}, u)^\top$ so that $x_{n+1} = u$): ☞ 59

$$\int_{\mathbb{R}} f_T(\mathbf{x})\, \mathrm{d}x_{n+1} = \int_{\mathbb{R}} \frac{\widetilde{p}(\mathbf{z})\, \mathrm{I}\{0 \leqslant u \leqslant 1\}}{\ell(\gamma)}\, \mathrm{I}\left\{\frac{1}{a} \ln\left(\frac{p(\mathbf{z})\, H(\mathbf{z})}{u\, \widetilde{p}(\mathbf{z})}\right) - \frac{b}{a} \geqslant \gamma\right\} \mathrm{d}u$$
$$= \frac{\widetilde{p}(\mathbf{z})}{\ell(\gamma)} \int_0^1 \mathrm{I}\left\{u \leqslant \frac{p(\mathbf{z})\, H(\mathbf{z})}{\mathrm{e}^{a\gamma+b}\, \widetilde{p}(\mathbf{z})}\right\} \mathrm{d}u = \frac{p(\mathbf{z})\, H(\mathbf{z})}{\ell(\gamma)\, \mathrm{e}^{a\gamma+b}}\,,$$

because $\frac{p(\mathbf{z})\, H(\mathbf{z})}{\mathrm{e}^{a\gamma+b}\, \widetilde{p}(\mathbf{z})} \leqslant 1$ for all $\mathbf{z}$ by (14.13). Since $\mathcal{Z}$ is the normalizing constant of $p(\mathbf{z})H(\mathbf{z})$, we conclude that $\mathcal{Z} = \mathrm{e}^{a\,\gamma+b}\ell(\gamma)$.

To illustrate the efficiency of Algorithm 14.5 in estimating (14.12), we consider three examples.

■ EXAMPLE 14.2 (Two Humps Function)

The following toy example is adapted from [34]. Consider the problem of estimating without bias the normalizing constant $\mathcal{Z} = \mathcal{Z}(\lambda)$ of the pdf proportional to

$$h(\mathbf{z}; \lambda) = \exp\left(-\frac{z_1^2 + z_2^2 + (z_1 z_2)^2 - 2\lambda z_1 z_2}{2}\right), \quad \mathbf{z} \in \mathbb{R}^2,$$

for some parameter $\lambda \in \mathbb{R}$, say $\lambda = 12$. The density $h(\mathbf{z}; 12)/\mathcal{Z}(12)$ is plotted in Figure 14.3.

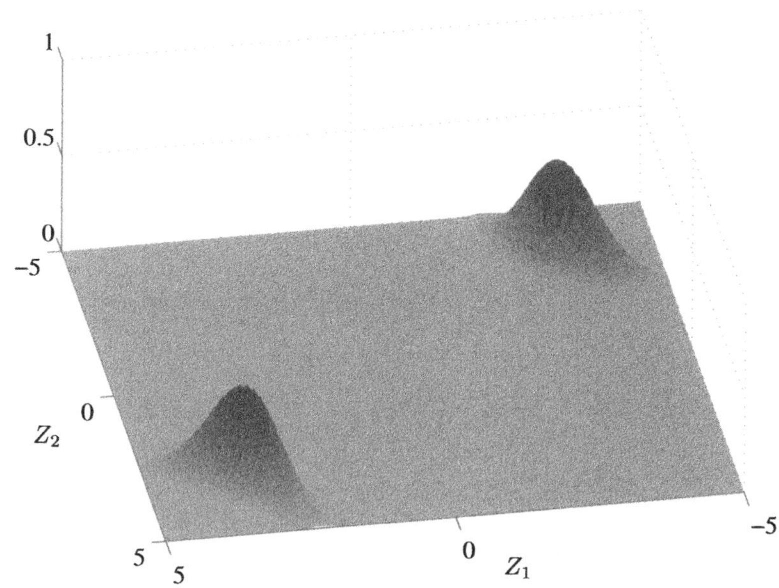

Figure 14.3 The surface of the two humps density $h(\mathbf{z}; 12)/\mathcal{Z}(12)$.

This is a problem of the form (14.12) with

$$p(\mathbf{z}) = \frac{1}{2\pi} \exp\left(-\frac{z_1^2 + z_2^2}{2}\right) \quad \text{and} \quad H(\mathbf{z}) = 2\pi \exp\left(-\frac{(z_1 z_2)^2 - 2\lambda z_1 z_2}{2}\right).$$

We apply Algorithm 14.5 with $\widetilde{p}(\mathbf{z}) = p(\mathbf{z})$, $a = \frac{1}{2}$, and $b = \frac{\lambda^2}{2} + \ln(2\pi)$. Then, (14.14) can be written as

$$\ell(\gamma) = \mathbb{P}(S(\mathbf{X}) \geqslant \gamma) = \mathbb{P}\left(-(Z_1 Z_2 - \lambda)^2 - 2\ln U \geqslant \gamma\right),$$

where the vector $\mathbf{x} = (\mathbf{z}, u)^{\top}$ is augmented such that

$$f(\mathbf{x}) = \frac{1}{2\pi} \exp\left(-\frac{z_1^2 + z_2^2}{2}\right) \mathrm{I}_{\{0<u<1\}}.$$

We take the level $\gamma = 0$, so that (14.13) holds. To apply the GS algorithm for the estimation of (14.14), we need to specify the transition pdf κ_t with stationary density

$$f_t(\mathbf{x}) = \frac{f(\mathbf{x})\, \mathrm{I}_{\{S(\mathbf{x}) \geqslant \gamma_t\}}}{\ell(\gamma_t)} = \frac{f(\mathbf{x})\, \mathrm{I}\left\{-(z_1 z_2 - \lambda)^2 - 2\ln u \geqslant \gamma_t\right\}}{\ell(\gamma_t)}.$$

A move from $\mathbf{X}$ to $\mathbf{X}^*$ using the transition density $\kappa_t(\mathbf{X}^* \mid \mathbf{X})$ consists of the following (systematic) Gibbs sampling procedure.

Algorithm 14.6 (Transition Density $\kappa_t(\mathbf{X}^* \mid \mathbf{X})$)

1. *Given a state* $\mathbf{X} = (Z_1, Z_2, U)^\top$ *for which* $S(\mathbf{X}) \geqslant \gamma_t$*, generate* $Z_1^* \sim f_t(z_1^* \mid Z_2, U)$*; that is, draw* Z_1^* *from a truncated standard normal density on the interval* (I_1, I_2)*, where* $I_1 = \min\{\frac{\lambda-\mu}{Z_2}, \frac{\lambda+\mu}{Z_2}\}$*,* $I_2 = \max\{\frac{\lambda-\mu}{Z_2}, \frac{\lambda+\mu}{Z_2}\}$*, and* $\mu = \sqrt{-\gamma_t - 2\ln U}$.
2. *Generate* $Z_2^* \sim f_t(z_2^* \mid Z_1^*, U)$*; that is, draw* Z_2^* *from a truncated standard normal density on the interval* (I_1^*, I_2^*)*, where* $I_1^* = \min\{\frac{\lambda-\mu}{Z_1^*}, \frac{\lambda+\mu}{Z_1^*}\}$*,* $I_2^* = \max\{\frac{\lambda-\mu}{Z_1^*}, \frac{\lambda+\mu}{Z_1^*}\}$*, and* $\mu = \sqrt{-\gamma_t - 2\ln U}$.
3. *Generate* $U^* \sim f_t(u^* \mid Z_1^*, Z_2^*)$*; that is, a uniform random variable* U^* *on the interval* $(0, \mu^*)$*, where*

$$\mu^* = \min\left\{1, \exp\left(-\frac{\gamma_t + (Z_1^* Z_2^* - \lambda)^2}{2}\right)\right\}.$$

Output $\mathbf{X}^* = (Z_1^*, Z_2^*, U^*)^\top$.

To estimate $\ell = \ell(0)$ we apply the GS algorithm with $N = 2000$, $\lambda = 12$, and the levels and splitting factors in Table 14.3.

Table 14.3 Levels and splitting factors used to compute the normalizing constant of $h(\mathbf{z})$.

t	1	2	3	4	5	6
γ_t	−117.91	−77.03	−44.78	−20.18	−4.40	0
ϱ_t	0.1	0.1	0.1	0.1	0.1	0.2853

We obtain a typical estimate of $\widehat{\ell} = 2.92 \times 10^{-6}$ with an estimated relative error of 5%. Hence, in Step 2 of Algorithm 14.5 we obtain $\widehat{\mathcal{Z}} = \widehat{\ell}\, \mathrm{e}^{a\,\gamma+b} = \widehat{\ell}\, 2\pi \mathrm{e}^{\lambda^2/2} = 3.41 \times 10^{26}$ with an estimated relative error of 5%. Note that Table 14.3 is computed using the ADAM algorithm with $\varrho = 0.1$, $N = 2000$, and using the same transition density κ_t. The combined simulation effort of the GS algorithm and the ADAM algorithm is about 1.2×10^5 samples. In contrast, crude Monte Carlo estimation of $\mathcal{Z}$ via the estimator $\frac{1}{M}\sum_{i=1}^{M} H(\mathbf{Z}_i)$, $\{\mathbf{Z}_i\} \sim_{\text{iid}} p(\mathbf{z})$, $M = 1.2 \times 10^5$ gives an estimate of 1.6×10^{26} with an estimated relative error of 60%.

For this two-dimensional example we can verify the simulation results using deterministic quadrature. An approximate value $\mathcal{Z}(12) \approx 3.5390 \times 10^{26}$ was obtained using the deterministic recursive Simpson's rule [12]. The constant $\mathcal{Z}$ in the next two examples, however, cannot be easily computed using an alternative method due to the high-dimensionality of the problem.

■ EXAMPLE 14.3 (Bayesian Marginal Likelihood)

☞ 231 In Example 6.2 we consider Bayesian inference on the model parameters $\boldsymbol{\beta} = (\beta_1, \ldots, \beta_k)^\top$ of the logistic model. The model is summarized as follows:

- Prior: $f(\boldsymbol{\beta}) \propto \exp\left(-\frac{1}{2\sigma^2}\|\boldsymbol{\beta} - \boldsymbol{\beta}_0\|^2\right)$, $\boldsymbol{\beta} \in \mathbb{R}^k$, with $\boldsymbol{\beta}_0$ and σ given.

- Likelihood: $f(\breve{\mathbf{y}} \mid \boldsymbol{\beta}) = \prod_{i=1}^n p_i^{\breve{y}_i}(1-p_i)^{1-\breve{y}_i}$, $p_i^{-1} = 1 + \exp(-\breve{\mathbf{x}}_i^\top \boldsymbol{\beta})$, where $\breve{\mathbf{x}}_i = (\breve{x}_{i1}, \breve{x}_{i2}, \ldots, \breve{x}_{ik})^\top$ are the explanatory variables for the i-th response and $\breve{y}_1, \ldots, \breve{y}_n$ are the binary response data.

The objective is to generate random vectors from the posterior pdf $f(\boldsymbol{\beta} \mid \breve{\mathbf{y}}) \propto f(\boldsymbol{\beta}) f(\breve{\mathbf{y}} \mid \boldsymbol{\beta})$ and to estimate the *marginal likelihood* ☞ 673

$$f(\breve{\mathbf{y}}) = \int f(\boldsymbol{\beta}) f(\breve{\mathbf{y}} \mid \boldsymbol{\beta})\, \mathrm{d}\boldsymbol{\beta}\,.$$

Estimation of the marginal likelihood is of significant interest in Bayesian model selection [8, 9, 18, 28]. Here we apply the framework of Algorithm 14.5 as follows.

- In (14.12) we take $\mathbf{z} = \boldsymbol{\beta}$, $p(\mathbf{z}) = f(\boldsymbol{\beta})$, and $H(\mathbf{z}) \equiv f(\breve{\mathbf{y}} \mid \boldsymbol{\beta})$, so that $\mathcal{Z} = f(\breve{\mathbf{y}})$.
- In (14.13) we let $\widetilde{p}(\mathbf{z}) = f(\boldsymbol{\beta})$ and choose γ (with $a = 1, b = 0$) to be a conservative estimate of the maximum of the logarithm of the likelihood:

$$\gamma \geqslant \max_{\boldsymbol{\beta}} \ln f(\breve{\mathbf{y}} \mid \boldsymbol{\beta})\,.$$

Then, the function $S(\mathbf{x})$ in Algorithm 14.5 becomes (recall $\mathbf{z} = \boldsymbol{\beta}$ and $\mathbf{x} = (\mathbf{z}, u)^\top$):

$$S(\boldsymbol{\beta}, u) = -\sum_{i=1}^n \breve{y}_i \ln\left(1 + \mathrm{e}^{-\breve{\mathbf{x}}_i^\top \boldsymbol{\beta}}\right) + (1 - \breve{y}_i)\left(\breve{\mathbf{x}}_i^\top \boldsymbol{\beta} + \ln\left(1 + \mathrm{e}^{-\breve{\mathbf{x}}_i^\top \boldsymbol{\beta}}\right)\right) - \ln u\,,$$

and the marginal likelihood estimate is $\widehat{f}(\breve{\mathbf{y}}) = \mathrm{e}^{\gamma}\,\widehat{\ell}(\gamma)$, where $\ell(\gamma)$ is given in (14.14).

As a numerical example we consider the same data used in Example 6.2, that is, $\sigma = 10$ and $\boldsymbol{\beta}_0 = \mathbf{0}$, and use the ADAM Algorithm 14.4 as a subroutine in Algorithm 14.5 with $N = 10^3$, $\varrho = 0.1$, and γ chosen to be a conservative estimate of the maximum log-likelihood value (estimated via Newton–Raphson); here we ☞ 688 take the Newton–Raphson estimate plus 5. We choose $\kappa_t(\mathbf{X}^* \mid \mathbf{X})$ as the transition pdf of the following hybrid of the hit-and-run and Gibbs sampler. ☞ 242

Algorithm 14.7 (Gibbs and Hit-and-Run Hybrid for $\kappa_t(\mathbf{X}^* \mid \mathbf{X})$)

1. *Given a state $\mathbf{X} = (\mathbf{Z}, U)^\top$ for which $S(\mathbf{X}) \geqslant \gamma_t$, generate from the conditional pdf*

$$f_t(\mathbf{x} \mid U) \propto p(\mathbf{z})\, \mathrm{I}_{\{0<U<1\}}\, \mathrm{I}_{\{S(\mathbf{x}) \geqslant \gamma_t\}}$$

using the hit-and-run algorithm. In other words, generate a vector $\mathbf{d}$ *uniformly distributed over the surface the n-dimensional unit hypersphere. Then, generate $\Lambda \sim \mathsf{N}(-\mathbf{x}^\top \mathbf{d}, 10^2)$. If $S(\mathbf{Z} + \Lambda\mathbf{d}, U) \geqslant \gamma_t$, set $\mathbf{Z}^* = \mathbf{Z} + \Lambda\mathbf{d}$; otherwise, set $\mathbf{Z}^* = \mathbf{Z}$.*

2. *Generate $U^* \sim f_t(\mathbf{x} \mid \mathbf{Z}^*)$; that is, a uniform random variable U^* on the interval $(0, \mu^*)$, where*

$$\mu^* = \min\{1,\ \exp(-\gamma_t + S(\mathbf{Z}^*, U) + \ln U)\}\,.$$

Set $\mathbf{X} = (\mathbf{Z}^*, U^*)^\top$.

3. *Repeat Steps 1 and 2 above* 10 *times and output $\mathbf{X}^* = (\mathbf{Z}^*, U^*)^\top$.*

Running the MATLAB script Bayesian_split.m below we obtain the point estimate $\ln \widehat{f}(\breve{\mathbf{y}}) = \gamma + \ln \widehat{\ell}(\gamma) = -2828.90$ with an estimated 95% confidence interval $(-2829.24,\ -2828.55)$. The script runs ADAM ten independent times and uses the output to construct a confidence interval.

We now comment on the MATLAB programs below that generate the output. The script Bayesian_split.m runs the function adam.m ten times. The function adam.m implements the ADAM algorithm, and is written in a generic form using a number of subroutines. Applying ADAM to a different problem necessitates changes only to some of its subroutines. These subroutines are as follows.

1. nominal_pdf.m — implements sampling from the nominal pdf $f(\mathbf{x})$.
2. mcmc.m — implements sampling from the transition pdf $\kappa_t(\mathbf{y} \mid \mathbf{x})$. For this particular problem it implements the hit-and-run and Gibbs hybrid described above.
3. S.m — implements the problem-specific importance function.
4. stratified_split.m — generates the splitting factors required in Step 2 of the ADAM algorithm, and is the same for all problems.
5. threshold.m — implements the computation of the level γ_t (see (14.10) in Step 1 of ADAM), and is the same for all problems.

```
% Bayesian_split.m
clear all,clc
global Y X
n=5000; % number of data points (y_1,...,y_n)
k=3;    % number of explanatory variables
% generate artificial dataset
randn('state', 12345);  rand('state', 67890);
truebeta = [1 -5.5 1]';
X = [ ones(n,1) randn(n,k-1)*0.1 ];  % design matrix
Y = binornd(1,1./(1+exp(-X*truebeta)));
bo=zeros(k,1); % we use Vo=100*eye(k);

% determine the Maximum Likelihood using Newton Raphson
err=inf; b=bo; % initial guess
while norm(err)>10^(-3)
    p=1./(1+exp(-X*b));
    g=X'*(Y-p);
    H=-X'*diag(p.^2.*(1./p - 1))*X;
    err=H\g; % compute Newton-Raphson correction
    b=b-err; % update Newton guess for MLE
end

log_H_b=S([b',1]); %logarithm of MLE
N=10^3; rho=0.1; Gamma=log_H_b+5; %conservative upper bound
% GS algorithm starts here
```

```
for k=1:10
    [el,gam,beta]=adam(N,Gamma,rho); ell(k)=el;
end
RE=std(ell)/mean(ell)/sqrt(10)
log_Z=Gamma+log(mean(ell))
%95% confidence interval
[log_Z-1.96*RE,log_Z+1.96*RE]
```

Next is the function implementing ADAM.

```
function [ell,gam,X]=adam(N,Gamma,rho)
%ADAM algorithm
%output: estimated 'ell';
%        a vector of estimated levels 'gam';
%        final population X with approximate distribution f_T
for k=1:N
    [Xp,Sp]=nominal_pdf;  %sample from the nominal pdf f(x)
    X(k,:)=Xp;       Score(k)=Sp;
end
% determine gamma as per step 1 of ADAM
gam(1)=threshold(Score,rho,Gamma);
I=Score>=gam(1);
Nt=sum(I); c(1)=Nt/N; X=X(I,:); Score=Score(I);
S_best=inf;  G=nan(1,15);

for t=2:10^3
    SPLITS=stratified_split(N,Nt); %generate splitting factors
    for i=1:Nt
        Xp=X(i,:);
        for chain_length=1:SPLITS(i)
            % apply markov transition pdf Step 2 of ADAM
            Xp=mcmc(Xp,gam(t-1));
            X(end+1,:)=Xp; Score(end+1)=S(Xp);
        end
        X(1,:)=[]; Score(1)=[];
    end
    % compute new level/threshold
    gam(t)=threshold(Score,rho,Gamma);
    I=Score>=gam(t);Nt=sum(I);
    c(t)=Nt/length(Score); X=X(I,:); Score=Score(I);

    [c(t),gam(t)] % monitor output

    if gam(t)==Gamma break, end
end
ell=prod(c);
```

Next is the function implementing the nominal pdf f.

```
function [x,Score]=nominal_pdf
% generate from the nominal pdf f(x)
x=[randn(1,3)*10,rand];
Score=S(x);
```

Next is the function that takes 10 steps according to the hybrid Gibbs and hit-and-run sampler.

```
function x=mcmc(x,gam)

k=length(x)-1;
for iter=1:10
    d=randn(1,k); d=d/norm(d); % sample direction
    lam=-x(1:k)*d'+randn*10;
    y=x(1:k)+lam*d; % make proposal

    if S([y,x(k+1)])>=gam
        x(1:k)=y;
    end
end
x(k+1)=rand*min(exp(-gam+(S(x)+log(x(k+1)))),1);
```

Next is the function implementing the importance function S.

```
function out=S(x)
global Y X
k=length(x)-1;
b=x(1:k)';
u=x(k+1);
out=-Y'*log(1+exp(-X*b))-(1-Y)'*(X*b+log(1+exp(-X*b)))-log(u);
```

Next is the function that generates the splitting factors via the conditional Bernoulli random variables in Step 2 of the ADAM algorithm.

```
function out=stratified_split(N,Nt)
r=mod(N,Nt);
if abs(r)<0.01
   r=0;
end
B=zeros(Nt,1);
B(randsample(Nt,r))=1;
out=B+floor(N/Nt);
```

Next is the function implementing the computation of each threshold γ_t.

```
function gam=threshold(S,rho,Gamma)
% determine threshold
S=sort(S);
for i=1:length(S)
   if  mean(S>=S(i))<=rho, break ,end
end
gam=min(S(i),Gamma);
```

■ EXAMPLE 14.4 (Rosenbrock Function)

The following example is adapted from [34]. Consider computing the normalizing constant of the pdf proportional to

$$h(\mathbf{z};\lambda) = \mathrm{e}^{-\lambda R(\mathbf{z})}, \quad z_i \in [-2,2],\ i = 1,\ldots,n\ ,$$

where

$$R(\mathbf{z}) = \sum_{i=1}^{n-1} \left(100(z_{i+1} - z_i^2)^2 + (z_i - 1)^2\right)$$

is the Rosenbrock function in $\mathbb{R}^n$, see Appendix C.4.1.3. Again, the problem is of the form (14.12), with $p(\mathbf{z}) = 1/4^n$, $\mathbf{z} \in [-2,2]^n$, and $H(\mathbf{z}) = 4^n h(\mathbf{z};\lambda)$. Let $\widetilde{p}(\mathbf{z}) = p(\mathbf{z})$, $a = \lambda$, and $b = \ln 4^n$. Then, (14.14) can be written as ☞ 696

$$\ell(\gamma) = \mathbb{P}\left(-\frac{\ln U}{\lambda} - R(\mathbf{Z}) \geqslant \gamma\right),$$

where

$$f(\mathbf{x}) = \frac{\prod_{i=1}^n \mathrm{I}_{\{-2\leqslant z_i \leqslant 2\}}}{4^n}\,\mathrm{I}_{\{0<u<1\}}, \quad \mathbf{x} = (\mathbf{z},u)^\top\ .$$

We take the level $\gamma = 0$, such that (14.13) is a tight bound. To estimate ℓ we apply the ADAM Algorithm 14.4 using a transition density $\kappa_t(\mathbf{x}^* \,|\, \mathbf{x})$ with stationary pdf

$$f_t(\mathbf{x}) = \frac{f(\mathbf{x})\,\mathrm{I}\left\{-\frac{\ln u}{\lambda} - R(\mathbf{z}) \geqslant \gamma_t\right\}}{\ell(\gamma_t)}\ .$$

A move from $\mathbf{X} = (\mathbf{Z},U)^\top$ to $\mathbf{X}^* = (\mathbf{Z}^*,U^*)^\top$ uses Gibbs sampling as follows. Given a state $\mathbf{X} = (\mathbf{Z},U)^\top$ such that $S(\mathbf{X}) \geqslant \gamma_t$, we generate $U^* \sim f_t(u\,|\,\mathbf{Z})$. Then, for each $j = 1,\ldots,n-1$ we generate $Z_j^* \sim f_t(z_j\,|\,U^*, Z_1^*,\ldots,Z_{j-1}^*, Z_{j+1},\ldots,Z_n)$. The distribution of each Z_j^* is uniform on the set $\{[r_1,r_2] \cup [r_3,r_4]\} \cap [-2,2]$, where $r_1 < r_2$, $r_3 < r_4$ are the real roots of a certain quartic equation [3]. Depending on the coefficients, the quartic equation has either 4 or 2 real roots (in which case $r_3 = r_1$ and $r_4 = r_2$). Finally, we generate $Z_n^* \sim f_t(z_n\,|\,U^*, Z_1^*,\ldots,Z_{n-1}^*)$. The random variable Z_n^* has uniform distribution on the set $[r_1,r_2] \cap [-2,2]$, where $r_1 < r_2$ are the roots of a certain quadratic equation.

As a numerical example, consider the case where $\lambda = 10^4$ and $n = 10$. We run the ADAM algorithm 400 independent times with $\varrho = 0.5$ and $N = 1000$, and obtain a typical estimate of $\widehat{\ell} = 9.7 \times 10^{-36}$ with an estimated relative error (using the data from the 400 runs) of 7%. Therefore, $\widehat{\mathcal{Z}} = \widehat{\ell}\, \mathrm{e}^{a\gamma+b} = \widehat{\ell}\, 4^{10} \approx 1.0 \times 10^{-29}$ with an estimated relative error of 7%. Each run of the ADAM algorithm takes about 117 iterations ($T = 117$), giving a total simulation effort of $400\,N\,T = 46.8 \times 10^6$ samples. For the same simulation effort the crude Monte Carlo estimator $\frac{1}{M}\sum_{i=1}^{M} \exp(-\lambda R(\mathbf{Z}))$ with $M = 46.8 \times 10^6$ and $\mathbf{Z}_1, \ldots, \mathbf{Z}_M \sim_{\text{iid}} \mathsf{U}(-2,2)^{10}$, gives an estimated relative error of 99.9%. To achieve a relative error of 7% using crude Monte Carlo estimation would require a simulation effort of approximately 2×10^{37} samples.

Numerical minimization of the Rosenbrock function $R(\mathbf{z})$ is commonly used as a test case for a wide range of numerical optimization routines [33]. The function $R(\mathbf{z})$ has a global minimum of 0 at $\mathbf{z} = (1, \ldots, 1)^\top$. One way in which $R(\mathbf{z})$ could be minimized is to sample approximately from the Boltzmann density $\mathrm{e}^{-\lambda R(\mathbf{z})}/\mathcal{Z}$, $\mathbf{z} \in [-2, 2]^n$ for a large value of λ. In Example 14.4, as a consequence of estimating the constant $\mathcal{Z}$ using Algorithm 14.5, we also obtain an estimate for the global minimizer of $R(\mathbf{z})$. In particular, the population $\mathcal{X}_T = \{\mathbf{X}_1, \ldots, \mathbf{X}_{N_T}\}$ at the final iteration of the ADAM algorithm is approximately distributed according to the stationary density $f_T(\mathbf{x})$. Hence, $\mathbf{Z}_i$ in $\mathbf{X}_i = (\mathbf{Z}_i, U_i)^\top$ is approximately distributed according to the marginal (Boltzmann density) $p(\mathbf{z})H(\mathbf{z})/\mathcal{Z} = \mathrm{e}^{-\lambda R(\mathbf{z})}/\mathcal{Z}$, and we can use any $\mathbf{Z}_i^*$ for which $R(\mathbf{Z}_i^*) = \min_j R(\mathbf{Z}_j)$ as an estimate for the global minimizer of $R(\mathbf{z})$. For the numerical example considered above we obtain

$$\mathbf{Z}_i^* = (1.00,\ 0.99,\ 0.99,\ 0.99,\ 1.00,\ 1.00,\ 1.00,\ 1.00,\ 1.00,\ 1.00)^\top,$$

with $R(\mathbf{Z}_i^*) \approx 5 \times 10^{-5}$. The result is close to the true minimizer $(1, \ldots, 1)^\top$. We obtain similar results for $n = 100$. Thus, Example 14.4 illustrates how one can use Algorithm 14.5 (with A set to ADAM) as an optimization algorithm. This is similar to the simulated annealing algorithm, in which the Metropolis–Hastings sampler is used to approximately sample from the Boltzmann density and minimize the function $R(\mathbf{z})$. In the next section we show how to use splitting to solve difficult combinatorial optimization problems. ☞ 449

14.6 COMBINATORIAL OPTIMIZATION VIA SPLITTING

In this section we show how we can use the ADAM algorithm (Algorithm 14.4) as an optimization heuristic. Suppose the problem is to maximize a real-valued function $S(\mathbf{x})$ on some finite set $\mathbf{x} \in \mathscr{X}$. To every problem of this kind we can associate the problem of estimating the probability

$$\ell(\gamma) = \mathbb{P}(S(\mathbf{X}) \geqslant \gamma), \quad \mathbf{X} \sim f\,,$$

where $f(\mathbf{x})$ is typically the uniform pdf on the set $\mathscr{X}$ and γ is large enough to make $\ell(\gamma)$ a rare-event probability. An important difference between the estimation and the optimization setting is that here we are not interested in obtaining an unbiased estimate for the probability $\ell(\gamma)$ itself. Rather, we only wish to sample (approximately) from the sequence of pdfs

$$f_t(\mathbf{x}) = \frac{f(\mathbf{x})\, \mathrm{I}_{\{S(\mathbf{x}) \geqslant \gamma_t\}}}{\ell(\gamma_t)}, \quad \gamma_1 < \gamma_2 < \gamma_3 < \cdots\,,$$

for as large a value of t as possible. Given this objective, we set $\gamma = \infty$ and run the ADAM algorithm to compute an ever-increasing sequence of levels $\gamma_1 < \gamma_2 < \cdots$. For an appropriate stopping condition we modify Steps 4 and 5 in Algorithm 14.4 to obtain the following optimization algorithm.

Algorithm 14.8 (ADAM for Maximization) *Steps 1, 2, and 3 are the same as in Algorithm 14.4.*

4. **Stopping Condition.** *If there is no progress in increasing γ_t over a number of iterations; that is, if $\gamma_t = \gamma_{t-1} = \cdots = \gamma_{t-d}$ for some user-specified positive integer d, set $T = t$ and go to Step 5; otherwise, repeat from Step 2.*

5. **Final estimates.** *Deliver the vector $\mathbf{X}^*$ from the set*

$$\mathcal{X}_T = \{\mathbf{X}_1, \ldots, \mathbf{X}_N\}$$

for which $S(\mathbf{X}_i)$ is maximal as an estimate for the global maximizer of (14.16).

Next, we consider a number of combinatorial optimization problems from Appendix C.3. ☞ 694

14.6.1 Knapsack Problem

A well-known difficult combinatorial optimization problem is the binary knapsack problem, specified as:

$$\begin{aligned} &\max_{\mathbf{x}} \sum_{j=1}^{n} p_j\, x_j, \quad \mathbf{x} = (x_1, \ldots, x_n) \in \{0,1\}^n\,, \\ &\text{subject to: } \sum_{j=1}^{n} w_{ij}\, x_j \leqslant c_i, \quad i = 1, \ldots, m\,. \end{aligned} \tag{14.16}$$

Here $\{p_i\}$ and $\{w_{ij}\}$ are positive weights and $\{c_i\}$ are positive cost parameters. To make (14.16) easier to handle as an estimation problem, we note that (14.16) is equivalent to $\max_{\mathbf{x} \in \{0,1\}^n} S(\mathbf{x})$, where $\mathbf{x}$ is a binary vector and

$$S(\mathbf{x}) = \widetilde{S}(\mathbf{x}) + \sum_{j=1}^{n} p_j\, x_j \stackrel{\text{def}}{=} \alpha \sum_{i=1}^{m} \mathrm{I}_{\{\sum_{j=1}^{n} w_{ij} x_j > c_i\}} + \sum_{j=1}^{n} p_j\, x_j\,,$$

with $\alpha = -\sum_{j=1}^{n} p_j$. Note that the constant α is such that if $\mathbf{x}$ satisfies all the constraints in (14.16), then $\widetilde{S}(\mathbf{x}) = 0$ and $S(\mathbf{x}) = \sum_{j=1}^{n} p_j\, x_j \geqslant 0$. Alternatively, if $\mathbf{x}$ does not satisfy all of the constraints in (14.16), then $\widetilde{S}(\mathbf{x}) \leqslant -\sum_{j=1}^{n} p_j\, x_j$ and $S(\mathbf{x}) \leqslant 0$. In other words, $\widetilde{S}$ is a penalty function. To this optimization problem ☞ 685 we can associate the problem of estimating the rare-event probability

$$\ell(\gamma) = \mathbb{P}\left(S(\mathbf{X}) \geqslant \gamma\right), \quad \gamma \in \left(0,\ \sum_{j=1}^{n} p_j\right], \quad \mathbf{X} \sim f\,,$$

where $f(\mathbf{x}) = \frac{1}{2^n}$ for all $\mathbf{x} \in \{0,1\}^n$; that is, $\mathbf{X}$ is a vector of independent Bernoulli random variables with success probability 1/2. The transition density $\kappa_t(\mathbf{y} \mid \mathbf{x})$ in Algorithm 14.8 is defined through the following Gibbs sampling procedure:

1. Given a state $\mathbf{x}$ such that $S(\mathbf{x}) \geqslant \gamma_t$, generate $Y_1 \sim f_t(y_1 \mid x_2, \ldots, x_n)$.
2. For each $k = 2, \ldots, n-1$, generate $Y_k \sim f_t(y_k \mid Y_1, \ldots, Y_{k-1}, x_{k+1}, \ldots, x_n)$.
3. Finally, generate $Y_n \sim f_t(y_n \mid Y_1, \ldots, Y_{n-1})$.

Here, the conditional density is given by

$$f_t(y_k \mid \mathbf{y}_{-k}) \propto \mathrm{I}\left\{ \widetilde{S}(\mathbf{y}) + p_k\, y_k \geqslant \gamma_t - \sum_{j \neq k} p_j\, y_j \right\},$$

where $\mathbf{y}_{-k}$ denotes the vector $\mathbf{y}$ with the k-th element removed. Sampling a random variable Y_k from such a conditional pdf can be accomplished as follows. Draw $B \sim \mathsf{Ber}(1/2)$. If $S(y_1, \ldots, y_{k-1}, B, y_{k+1}, \ldots, y_n) \geqslant \gamma_t$, then set $Y_k = B$; otherwise, set $Y_k = 1 - B$.

As a particular example, consider the knapsack problem with data, $\{p_j, w_{ij}, c_i\}$, defined in the appendix of [35]. The data is also available at: `http://people.brunel.ac.uk/~mastjjb/jeb/orlib/files/mknap2.txt` as `Sento1.dat`. The problem has 30 constraints and 60 variables. For Algorithm 14.8 we select $\varrho = 0.01$ and $N = 10^4$, and the algorithm is stopped after no progress is observed ($d = 1$). We run Algorithm 14.8 ten independent times. The algorithm always finds the optimal solution. A typical evolution of the algorithm is given in Table 14.4. The total sample size is 10^5, which is about a factor of 10^{-13} of the total effort needed for the complete enumeration of the 2^{60} possible binary vectors.

Table 14.4 A typical evolution of Algorithm 14.4 as an optimization routine for the **`Sento1.dat`** knapsack problem. The maximum value for this problem is 6704.

t	1	2	3	4	5	6	7	8
γ_t	−34758	−6428	3484	4705	5520.5	6043	6538	6704

14.6.2 Traveling Salesman Problem

☞ 695 The objective of the TSP on n nodes is to minimize the cost function

$$S(\mathbf{x}) = \sum_{i=1}^{n-1} C(x_i, x_{i+1}) + C(x_n, x_1)\,,$$

where $\mathbf{x} = (x_1, \ldots, x_n)$ is a permutation of $(1, \ldots, n)$, x_i is the i-th node to be visited in a tour represented by $\mathbf{x}$, and $C(i,j)$ is the cost from node i to node j. We now discuss how to apply the ADAM algorithm to find the optimal tour. The sequence of distributions from which we wish to sample approximately is

$$f_t(\mathbf{x}) = \frac{f(\mathbf{x})\, \mathrm{I}_{\{-S(\mathbf{x}) \geqslant \gamma_t\}}}{\ell(\gamma_t)}, \quad \mathbf{x} \in \mathscr{X}, \quad t = 1, 2, \ldots,$$

where $f(\mathbf{x})$ is the uniform density over the set $\mathscr{X}$ of all possible tours (that is, the set of all possible permutations of $(1, \ldots, n)$), and $\{\gamma_t\}$ is an increasing sequence of levels.

Table 14.5 Case studies for the symmetric TSP.

file	$S(\mathbf{x}^*)$	min	mean	max	$\bar{T}$
burma14	3323	3323	3323	3323	45.8
ulysses16	6859	6859	6859	6859	53.2
ulysses22	7013	7013	7013	7013	79.7
bayg29	1610	1610	1610	1610	113
bays29	2020	2020	2020	2020	110.7
dantzig42	699	699	699	699	188.2
eil51	426	426	427.6	430	249.1
berlin52	7542	7542	7542	7542	232.5
st70	675	675	675.8	680	370.6
eil76	538	538	543.9	547	428.6
pr76	108159	108159	108216	108304	372.3
a280	2579	2581	2594.4	2633	2026.7
ch130	6110	6110	6125.9	6172	742.8
eil101	629	643	647.6	654	620.4
gr120	6942	6956	6969.2	6991	668.4
gr137	69853	69853	69911.8	70121	781
kroA100	21282	21282	21311.2	21379	527.2
kroB100	22141	22141	22201	22330	526.7
kroC100	20749	20749	20774.4	20880	528.8
kroD100	21294	21294	21295.5	21309	529
kroE100	22068	22068	22123.1	22160	526.5
lin105	14379	14379	14385.6	14401	561.2
pr107	44303	44303	44305.3	44326	569.3
pr124	59030	59030	59048.4	59076	691.5
pr136	96772	97102	97278.2	97477	760
pr144	58537	58537	58640.9	59364	827.6
pr152	73682	73682	73833	74035	886.6
rat99	1211	1211	1213.2	1218	561.5
rd100	7910	7910	7910.8	7916	534.7
si175	21407	21013	21027.2	21051	1066.3
st70	675	676	677.6	681	369.3
swiss42	1273	1273	1273	1273	180
u159	42080	42080	42383	42509	934.4

We apply Algorithm 14.8 with the transition pdf $\kappa_t(\mathbf{y} \mid \mathbf{x})$ defined via the following conditional sampling procedure.

1. Draw a pair of indices I and J uniformly distributed over the integers $1, \ldots, n$, conditional on $I < J$. These can be generated, for example, by sampling $U_1, U_2 \sim_{\text{iid}} \mathsf{DU}(1, n)$ until $U_1 \neq U_2$, and then assigning $I = \min\{U_1, U_2\}$ and $J = \max\{U_1, U_2\}$. Let $(I, J) = (i, j)$ be the outcome.

2. Given a tour $\mathbf{x}$ such that $S(\mathbf{x}) \geqslant \gamma_t$, update $\mathbf{x}$ to $\mathbf{y}$ with probability $\mathrm{I}_{\{-S(\mathbf{y}) \geqslant \gamma_t\}}$, where the tour $\mathbf{y}$ is identical to the tour $\mathbf{x}$ with the tour segment between the x_i-th and x_j-th cities reversed. For example, if $n = 10$, $\mathbf{x} = (1, 2, 3, 4, 5, 6, 7, 8, 9, 10)$, and $(i, j) = (4, 9)$, then we update the tour $\mathbf{x}$ to the tour $\mathbf{y} = (1, 2, 3, 9, 8, 7, 6, 5, 4, 10)$ if $-S(\mathbf{y}) \geqslant \gamma_t$; otherwise, we do not update $\mathbf{x}$.

3. Repeat Steps 1 and 2 above b times.

The conditional sampling is similar to the **2-opt** heuristic commonly employed in conjunction with simulated annealing; see [34, Page 189].

In the course of the conditional sampling, the performance function is updated as follows ($j > i$):

$$S(\mathbf{y}) = S(\mathbf{x}) - C(x_{i-1}, x_i) - C(x_j, x_{j+1}) + C(x_{i-1}, x_j) + C(x_i, x_{j+1}) .$$

We now present a number of numerical experiments which demonstrate the performance of the algorithm. Table 14.5 summarizes the results from a number of benchmark problems from the TSP library: `http://www.iwr.uni-heidelberg.de/groups/comopt/software/TSPLIB95/tsp/`. The experiments are repeated ten times and the average, minimum, and maximum of $S(\mathbf{x})$ are recorded. The parameters of Algorithm 14.8 are $\varrho = 0.5$, $N = 10^2$, $d = 14$, and the 2-opt updating is repeated $b = 50\,n$ times, where n is the size of the problem. $\bar{T}$ is the average number of iterations it takes for ADAM to reach the final iteration. The second column shows the length of the optimal path: $S(\mathbf{x}^*)$.

For the first eleven test problems the algorithm finds the optimal solution in all cases out of ten trials. In some cases, the algorithm finds the optimal solution ten out of ten times. The number of iterations required to solve a problem increases with the size of the problem and is roughly equal to $\log_\varrho(n!)$, which is approximately $\frac{1}{2}\log_\varrho(2\pi n) + n\log_\varrho(n/\mathrm{e})$.

14.6.3 Quadratic Assignment Problem

The objective of the quadratic assignment problem (see Section C.3.5 in the Appendix) is to minimize the cost function ☞ 695

$$S(\mathbf{x}) = \sum_{i=1}^{n}\sum_{j=1}^{n} F_{ij}\, D(x_i, x_j) ,$$

where $\mathbf{x} = (x_1, \ldots, x_n)$ is a permutation of $(1, \ldots, n)$, and F and D are $n \times n$ symmetric matrices.

We apply Algorithm 14.8 with the transition density $\kappa_t(\mathbf{y} \,|\, \mathbf{x})$ specified via the following conditional sampling procedure.

1. Draw a pair of indices I and J uniformly distributed over the integers $1, \ldots, n$, conditional on $I \neq J$. Let $(I, J) = (i, j)$ be the outcome.

2. Given a permutation $\mathbf{x} = (x_1, \ldots, x_n)$ such that $S(\mathbf{x}) \geqslant \gamma_t$, update $\mathbf{x}$ to $\mathbf{y}$ with probability $\mathrm{I}_{\{-S(\mathbf{y}) \geqslant \gamma_t\}}$, where the vector $\mathbf{y}$ is identical to the vector $\mathbf{x}$ with the i-th and j-th elements swapped. For example, if $\mathbf{x} = (1, 2, 3, 4, 5, 6, 7, 8, 9, 10)$ and $(i, j) = (3, 7)$, then $\mathbf{y} = (1, 2, 7, 4, 5, 6, 3, 8, 9, 10)$.

3. Repeat Steps 1 and 2 above b times.

In the course of the conditional sampling, the performance function is updated via:

$$S(\mathbf{y}) = S(\mathbf{x}) + 2\sum_{k \neq i,j} (F_{kj} - F_{ki})(D(x_k, x_i) - D(x_k, x_j)) .$$

Table 14.6 Case studies for the symmetric quadratic assignment problem.

file	$S(\mathbf{x}^*)$	min	mean	max	$\bar{T}$
chr12a.dat	9552	9552	9552	9552	45.2
chr12b.dat	9742	9742	9742	9742	45.4
chr12c.dat	11156	11156	11159	11186	42.5
chr15a.dat	9896	9896	9942.8	10070	53
chr15b.dat	7990	7990	8100	8210	53.4
chr15c.dat	9504	9504	10039	10954	53.4
chr18a.dat	11098	11098	11102.4	60	64
chr18b.dat	1534	1534	1534	1534	57.3
chr20a.dat	2192	2192	2344	2406	66.9
chr20b.dat	2298	2352	2457.8	2496	64.5
chr20c.dat	14142	14142	14476.8	14812	77.7
chr22a.dat	6156	6156	6208.6	6298	81.4
chr22b.dat	6194	6194	6290.4	6362	75.3
chr25a.dat	3796	3796	4095.6	4286	90.1

Table 14.6 summarizes the results from a number of symmetric benchmark problems from the quadratic assignment problem library: `http://www.seas.upenn.edu/qaplib/inst.html`. The experiments are repeated ten times and the average, minimum, and maximum of $S(\mathbf{x})$ are recorded. The parameters of the ADAM algorithm are $\varrho = 0.5$ and $N = 10^3$. For the conditional sampling we take $b = n$, where n is the size of the problem. Again, $\bar{T}$ is the average number of iterations it takes for ADAM to reach the final iteration. The second column shows the value at the optimal solution.

We see that the algorithm finds the optimal solution in all cases except `chr20b.dat`.

14.7 MARKOV CHAIN MONTE CARLO WITH SPLITTING

In this section we consider using the GS Algorithm 14.3 as an alternative to standard MCMC sampling from multidimensional pdfs of the form

$$f_t(\mathbf{x}) = \frac{f(\mathbf{x})\, \mathrm{I}_{\{S(\mathbf{x}) \geqslant \gamma_t\}}}{\ell(\gamma_t)} . \tag{14.17}$$

Note that since (14.15) can be viewed as a marginal density of (14.17) for $t = T$, sampling from (14.15) can be achieved by sampling from (14.17). We show that the population $\mathcal{X}_T$ in the final stage of the GS algorithm can be treated as a sample from the multidimensional pdf (14.17) even in cases where standard Markov chain Monte Carlo algorithms are impractical due to poor mixing. In addition, we provide a convergence diagnostic which tests the hypothesis that the population $\mathcal{X}_T$ is drawn from the target pdf (14.17). Deciding when a Markov chain has converged is an important problem in applications of MCMC. Many methods for diagnosing convergence have been proposed, ranging from simple graphical methods to computationally intensive hypothesis tests [5, 6]. ☞ 273

For clarity of presentation we explain how to sample from (14.17) in a separate algorithm, in which the transition density is *reversible*.

Algorithm 14.9 (Splitting Sampler) *Given a sequence* $\{(\gamma_t, \varrho_t)\}_{t=1}^T$, *set* $s_t = \lceil \varrho_{t+1}^{-1} \rceil$ *for all* $t < T$ *and execute the following steps.*

1. **Initialize.** *Set the level counter* $t = 1$. *Until* $S(\mathbf{X}) \geqslant \gamma_1$, *keep generating* $\mathbf{X} \sim f$. *Let* $\mathbf{X}^1 = \mathbf{X}$ *be the output. Note that* $\mathbf{X}^1$ *has density*

$$f_1(\mathbf{x}) = f(\mathbf{x})\, \mathrm{I}_{\{S(\mathbf{x}) \geqslant \gamma_1\}}/c_1 \; .$$

2. **Markov chain sampling.** *Generate*

$$\mathbf{Y}_j \overset{\text{iid}}{\sim} \kappa_t(\mathbf{y} \,|\, \mathbf{X}^t), \quad j = 1, \ldots, s_t \; , \tag{14.18}$$

where $\kappa_t(\mathbf{y} \,|\, \mathbf{X}^t)$ *is a* **reversible** *Markov transition density with stationary pdf* $f_t(\mathbf{y})$. *Let*

$$N_{t+1} = \sum_{j=1}^{s_t} \mathrm{I}_{\{S(\mathbf{Y}_j) \geqslant \gamma_{t+1}\}} \; .$$

If $N_{t+1} = 0$, *repeat from Step 1; otherwise, continue with Step 3.*

3. **Updating.** *Let* $\mathbf{X}^{t+1}$ *be a uniformly sampled point from the set of points* $\{\mathbf{Y}_1, \ldots, \mathbf{Y}_{s_t}\}$ *for which* $S(\mathbf{X}^{t+1}) \geqslant \gamma_{t+1}$. *The pdf of* $\mathbf{X}^{t+1}$ *is thus given by the conditional density*

$$\frac{\mathrm{I}_{\{S(\mathbf{x}^{t+1}) \geqslant \gamma_{t+1}\}}\, \kappa_t(\mathbf{x}^{t+1} \,|\, \mathbf{X}^t)}{c_{t+1}(\mathbf{X}^t)} \; , \tag{14.19}$$

where $c_{t+1}(\mathbf{y}) = \int \mathrm{I}_{\{S(\mathbf{x}) \geqslant \gamma_{t+1}\}}\, \kappa_t(\mathbf{x} \,|\, \mathbf{y})\, d\mathbf{x}$ *is the probability that a move of the Markov chain starting in state* $\mathbf{y}$ *has a performance above* γ_{t+1}. *Note that an unbiased estimate of* $c_{t+1}(\mathbf{X}^t)$ *is*

$$\widehat{c}_{t+1}(\mathbf{X}^t) = \frac{N_{t+1}}{s_t} \; ,$$

so that $\mathbb{E}[\widehat{c}_{t+1}(\mathbf{X}^t) \,|\, \mathbf{X}^t] = c_{t+1}(\mathbf{X}^t)$. *Set the counter* $t = t + 1$.

4. **Final Output.** *If* $t = T$, *output* $\{\widehat{c}_{t+1}(\mathbf{X}^t)\}_{t=1}^{T-1}$ *and* $(\mathbf{X}^1, \ldots, \mathbf{X}^T)$; *otherwise, repeat from Step 2.*

A diagnostic test is based on the following result.

Proposition 14.7.1 (Convergence Diagnostic) *If for any possible outcome* $(\mathbf{x}^1, \ldots, \mathbf{x}^T)$ *of Algorithm 14.9 the sum* $\sum_{t=1}^{T-1} \ln c_{t+1}(\mathbf{x}^t)$ *is constant, then the final state* $\mathbf{X}^T$ *from Step 4 of Algorithm 14.9 has the pdf*

$$f_T(\mathbf{x}) = \frac{\mathrm{I}_{\{S(\mathbf{x}) \geqslant \gamma\}}\, f(\mathbf{x})}{\ell(\gamma)} \; .$$

In other words, if $\sum_{t=1}^{T-1} \ln c_{t+1}(\mathbf{x}^t)$ *does not depend on* $(\mathbf{x}^1, \ldots, \mathbf{x}^T)$, *then the Markov chain of Algorithm 14.9 is in stationarity.*

Proof: First, the joint pdf of $(\mathbf{X}^1, \ldots, \mathbf{X}^T)$ is:

$$\widehat{f}_T(\mathbf{x}^1, \ldots, \mathbf{x}^T) = \frac{f(\mathbf{x}^1)\, \mathrm{I}_{\{S(\mathbf{x}^1) \geqslant \gamma_1\}}}{c_1} \prod_{t=1}^{T-1} \frac{\mathrm{I}_{\{S(\mathbf{x}^{t+1}) \geqslant \gamma_{t+1}\}}\, \kappa_t(\mathbf{x}^{t+1} \,|\, \mathbf{x}^t)}{c_{t+1}(\mathbf{x}^t)} \; .$$

Using the reversibility of the transition densities $\{\kappa_t\}$, we can write the joint pdf as

$$\widehat{f}_T(\mathbf{x}^1,\ldots,\mathbf{x}^T) = \frac{f(\mathbf{x}^T)\,\mathrm{I}_{\{S(\mathbf{x}^T)\geqslant\gamma_T\}}}{c_1}\prod_{t=1}^{T-1}\frac{\kappa_t(\mathbf{x}^t\,|\,\mathbf{x}^{t+1})}{c_{t+1}(\mathbf{x}^t)}\,. \tag{14.20}$$

Ideally, we would like the joint pdf in (14.20) to be identical to the target:

$$f_T(\mathbf{x}^1,\ldots,\mathbf{x}^T) = \frac{f(\mathbf{x}^T)\,\mathrm{I}_{\{S(\mathbf{x}^T)\geqslant\gamma_T\}}}{\ell(\gamma)}\prod_{t=1}^{T-1}\kappa_t(\mathbf{x}^t\,|\,\mathbf{x}^{t+1})\,, \tag{14.21}$$

because then $\mathbf{x}^T$ has the desired marginal density $f_T(\mathbf{x})$. We can measure how close the sampling density $\widehat{f}_T(\mathbf{x}^1,\ldots,\mathbf{x}^T)$ is from the target density $f_T(\mathbf{x}^1,\ldots,\mathbf{x}^T)$ using any distance measure from Csisár's **ϕ-divergence** family of measures [4, 34]. A convenient member of Csisár's family of measures is the **χ^2 goodness of fit** divergence defined as

$$\mathcal{D}_2(p,q) = \frac{1}{2}\int\frac{(p(\mathbf{x})-q(\mathbf{x}))^2}{p(\mathbf{x})}\,\mathrm{d}\mathbf{x} = -\frac{1}{2}+\frac{1}{2}\mathbb{E}_p\frac{q^2(\mathbf{X})}{p^2(\mathbf{X})}\,,$$

for any given pair of pdfs p and q. Thus, we can measure the closeness between the sampling pdf (14.20) and the target pdf (14.21) via

$$\mathcal{D}_2(\widehat{f}_T,f_T) = -\frac{1}{2}+\frac{1}{2}\mathbb{E}_{\widehat{f}_T}\prod_{t=1}^{T-1}\frac{c_{t+1}^2(\mathbf{X}^t)}{c_{t+1}^2}\,.$$

Hence, after rearranging, we have

$$2\,\frac{\ell^2(\gamma)}{c_1^2}\,\mathcal{D}_2(\widehat{f}_T,f_T) = \mathbb{E}_{\widehat{f}_T}\prod_{t=1}^{T-1}c_{t+1}(\mathbf{X}^t) - \frac{\ell^2(\gamma)}{c_1^2} = \mathrm{Var}_{\widehat{f}_T}\left(\prod_{t=1}^{T-1}c_{t+1}(\mathbf{X}^t)\right),$$

where we have used the fact that $\mathbb{E}_{\widehat{f}_T}\prod_{t=1}^{T-1}c_{t+1}(\mathbf{X}^t) = \ell(\gamma)/c_1$. It follows that the distance between (14.20) and (14.21) is 0 if and only if $\mathrm{Var}_{\widehat{f}_T}\left(\prod_{t=1}^{T-1}c_{t+1}(\mathbf{X}^t)\right) = 0$. In other words, if $\prod_{t=1}^{T-1}c_{t+1}(\mathbf{X}^t)$ (or $\sum_{t=1}^{T-1}\ln c_{t+1}(\mathbf{X}^t)$) is a constant, then the pdfs (14.20) and (14.21) are identical and $\mathbf{X}^T$ has the desired marginal pdf. This completes the proof.

The following algorithm is an adaptation of the Gelman–Rubin test (see Remark 6.4) for testing whether the sum $\sum_{t=1}^{T-1}\ln c_{t+1}(\mathbf{X}^t)$ is a constant. ☞ 273

Algorithm 14.10 (Particle Splitting and Gelman–Rubin Diagnostic)

1. *Let* $(\mathbf{X}_1^1,\ldots,\mathbf{X}_1^T),\ldots,(\mathbf{X}_M^1,\ldots,\mathbf{X}_M^T)\sim\widehat{f}_T(\mathbf{x}^1,\ldots,\mathbf{x}^T)$ *be a population from the sampling density* (14.20) *obtained via Algorithm 14.9, and let* C *be the following* $(T-1)\times M$ *matrix of estimates*

$$C = \begin{bmatrix} \ln\widehat{c}_2(\mathbf{X}_1^1) & \ln\widehat{c}_3(\mathbf{X}_1^2) & \cdots & \ln\widehat{c}_T(\mathbf{X}_1^{T-1}) \\ \ln\widehat{c}_2(\mathbf{X}_2^1) & \ln\widehat{c}_3(\mathbf{X}_2^2) & \cdots & \ln\widehat{c}_T(\mathbf{X}_2^{T-1}) \\ \ln\widehat{c}_2(\mathbf{X}_3^1) & \ln\widehat{c}_3(\mathbf{X}_3^2) & \cdots & \ln\widehat{c}_T(\mathbf{X}_3^{T-1}) \\ \vdots & \vdots & \ddots & \vdots \\ \ln\widehat{c}_2(\mathbf{X}_M^1) & \ln\widehat{c}_3(\mathbf{X}_M^2) & \cdots & \ln\widehat{c}_T(\mathbf{X}_M^{T-1}) \end{bmatrix},$$

where the i*-th row depends on* $(\mathbf{X}_i^1,\mathbf{X}_i^2,\ldots,\mathbf{X}_i^{T-1})$.

2. Compute the following statistics:

$$\textit{row means:}\quad \bar{C}_{i\bullet} = \frac{1}{T-1}\sum_{j=1}^{T-1} C_{ij}\,,$$

$$\textit{column means:}\quad \bar{C}_{\bullet j} = \frac{1}{M}\sum_{i=1}^{M} C_{ij}\,,$$

$$\textit{overall mean:}\quad \bar{C} = \frac{1}{T-1}\sum_{j=1}^{T-1} \bar{C}_{\bullet j}\,,$$

$$\textit{row effect sum of squares:}\quad \mathrm{SS} = \sum_{i=1}^{M}(\bar{C}_{i\bullet} - \bar{C})^2\,,$$

$$\textit{within row variance:}\quad \mathrm{V} = \frac{1}{(M-1)(T-1)^2}\sum_{j=1}^{T-1}\sum_{i=1}^{M}(C_{ij} - \bar{C}_{\bullet j})^2\,.$$

3. Under the hypothesis that $\sum_{t=1}^{T-1} \ln c_{t+1}(\mathbf{X}^t)$ *is a constant, and assuming an approximately normal distribution for each of the row and column means* $\{\bar{C}_{i\bullet}, \bar{C}_{\bullet j}\}$*, the test statistic* $\mathcal{T} = \mathrm{SS}/\mathrm{V}$ *has an approximately* χ^2_{M-1} *distribution.*

If the chain in Algorithm 14.9 samples according to the target, then the sums across each row of matrix C should be roughly the same. The two-way analysis-of-variance test in Algorithm 14.10 simply tests for row effects in matrix C. Each column of C represents a different level, while each row of C represents a given factor.

Finally, we caution that most diagnostics frequently successfully detect undesirable Markov chain behavior (slow mixing or lack of stationarity), but they can never be used to demonstrate in any meaningful way that the Markov chain accurately samples from the target pdf.

■ EXAMPLE 14.5 (Comparison With Gibbs Sampling)

To illustrate the performance of the splitting sampler, we consider the problem of sampling from the pdf in Example 14.2; that is, $h(\mathbf{z}; 12)/\mathcal{Z}(12)$. We run Algorithm 14.5 with exactly the same setup as in Example 14.2, except that the three steps in Algorithm 14.6 are executed in a random order, resulting in random Gibbs sampling, as opposed to systematic Gibbs sampling. The random Gibbs sampling ensures that the transition density $\kappa_t(\mathbf{X}^* \,|\, \mathbf{X})$ is reversible. Figure 14.4 shows the empirical distribution of $\mathbf{Z}$ at levels $(\gamma_0, \gamma_1, \gamma_3, \gamma_6) = (-\infty, -117.91, -44.78, 0)$. The $\gamma_0 = -\infty$ case shows the sample from the standard Gaussian proposal $p(\mathbf{z})$, and the γ_6 case shows 2030 points approximately distributed from the target density given in Figure 14.3. Notice how the two distinct modes emerge gradually. The p-value from Step 3 of Algorithm 14.10 is 0.1, thus failing to detect transient behavior and supporting the hypothesis that the chain samples according to the target. In addition, the proportion of points in each mode at the final stage is roughly equal to 1/2, namely, 1009 points belong to the upper-right mode and 1021 points belong to the lower-left mode.

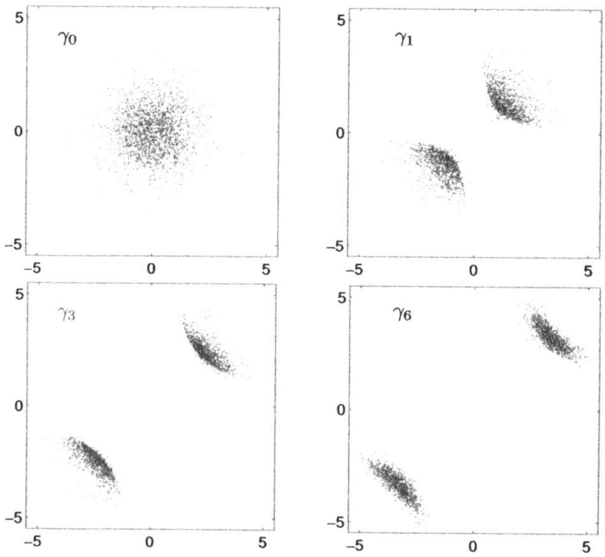

Figure 14.4 The empirical distribution of $\mathbf{Z}$ conditional on $S(\mathbf{X}) \geqslant \gamma_t$ for $t = 0, 1, 3, 6$, respectively.

Note that the standard Gibbs sampler applied to $h(\mathbf{z}; 12)/\mathcal{Z}(12)$ fails to sample from both modes. In particular, starting from $(0, 0)$ we iterate the following steps 10^9 times.

- Given (Z_1, Z_2), generate $Z_1^* \sim \mathsf{N}\left(\frac{\lambda Z_2}{1+Z_2^2}, \frac{1}{Z_2^2+1}\right)$.
- Given (Z_1^*, Z_2), generate $Z_2^* \sim \mathsf{N}\left(\frac{\lambda Z_1^*}{1+(Z_1^*)^2}, \frac{1}{(Z_1^*)^2+1}\right)$.
- Update $(Z_1, Z_2) = (Z_1^*, Z_2^*)$.

The right panel of Figure 14.5 shows that the standard Gibbs sampler results in a chain which is trapped in one of the two modes and fails to mix satisfactorily in 10^9 steps. For plotting purposes the chain of length 10^9 is thinned to 10^3 points, where we keep the 10^6-th, 2×10^6-th, 3×10^6-th, ... points from the original Markov chain sequence. Our numerical experience suggests that the performance of the Gibbs sampler is affected by the starting value of the chain. In contrast, the problem of selecting starting values for the chains is either mitigated or does not exist for the splitting sampler. Overall, the splitting sampler explores the support of the target density better than the standard Gibbs sampler.

For a comparison with other samplers such as the *equi-energy* sampler [23], we refer to [3]. The sampling problem in the last example can be resolved by means other than MCMC or splitting. For example, the acceptance–rejection method is still feasible in two dimensions. The next example, however, considers a problem for which there is no simple alternative to Markov chain sampling. ☞ 59

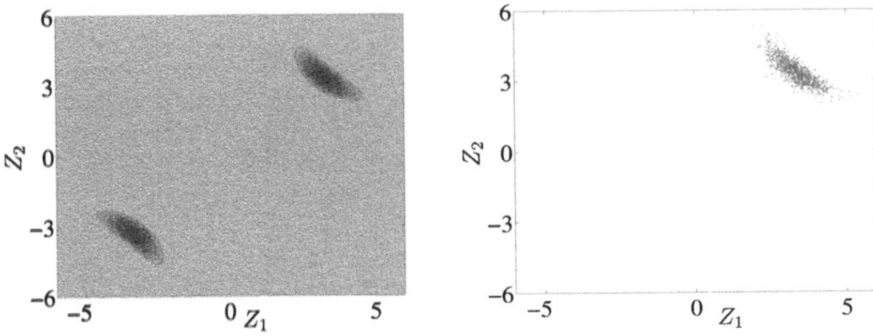

Figure 14.5 Left panel: contour plot of the two-humps density. Right panel: empirical distribution of the output of the standard Gibbs sampler.

■ EXAMPLE 14.6 (Network Simulation Using ADAM)

In Examples 16.4 and 16.5 on Pages 567 and 571 we consider estimating the probability that nodes 1 and 20 of the dodecahedron network in Figure 16.5 are not connected, given that all of the edges that connect the nodes fail with probability $q = 10^{-3}$. Here we are interested in simulating the repair-time configuration ☞ 552 $\mathbf{X} = (X_1, \ldots, X_m)^\top$ conditional on the dodecahedron network being failed (that is, there is no path of operating edges between nodes 1 and 20). In other words, we wish to simulate from the conditional density

$$\frac{f(\mathbf{x})\, \mathrm{I}_{\{S(\mathbf{x})>1\}}}{\ell}, \quad \mathbf{x} = (x_1, \ldots, x_m)^\top,$$

where $S(\mathbf{x})$ is given by (16.1), ℓ is given by (16.2), and

$$f(\mathbf{x}) = (2\pi\sigma^2)^{-m/2} \mathrm{e}^{-\mathbf{x}^\top \mathbf{x}/(2\sigma^2)}$$

with $\sigma = -1/\Phi^{-1}(q)$. Such simulation gives us insight into which edges are most likely to cause the failure of the network so that these weak edges can be strengthened to increase the reliability of the network. We consider the hit-and-run sampler ☞ 240 as a standard MCMC approach.

Algorithm 14.11 (Hit-and-Run for Reliability Network Simulation)
Given a configuration $\mathbf{x}_t$ of repair times such that $S(\mathbf{x}_t) > 1$, execute the following steps.

1. *Generate a random direction vector $\mathbf{d}$ uniformly distributed on the unit m-dimensional sphere. Given $\mathbf{d}$, generate $\Lambda \sim \mathsf{N}(-\mathbf{x}_t^\top \mathbf{d}, \sigma^2)$.*
2. *If $S(\mathbf{x}_t + \Lambda \mathbf{d}) > 1$, reset $\mathbf{x}_t = \mathbf{x}_t + \Lambda \mathbf{d}$; otherwise leave $\mathbf{x}_t$ unchanged.*
3. *Repeat Steps 1 and 2 above 100 times and output $\mathbf{X}_{t+1} = \mathbf{x}_t$ as the next state of the Markov chain.*

☞ 569 The above procedure is equivalent to Algorithm 16.8 for the case where all edges of the network fail with the same probability.

The sampler may require an inordinately large number of iterations to visit all the modes of the target density, which is needed to provide a good empirical

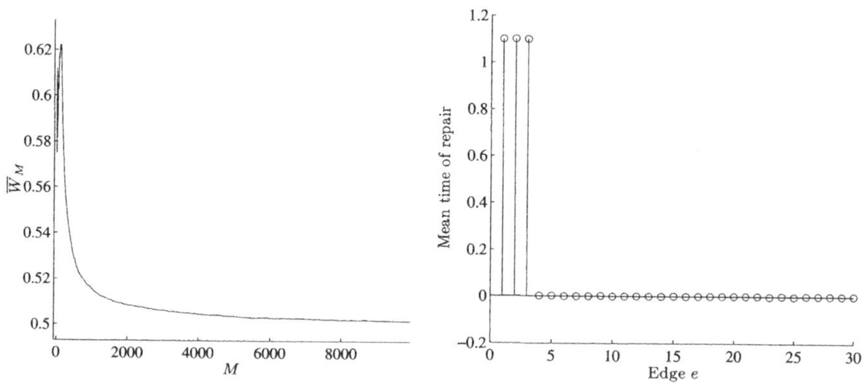

Figure 14.6 Hit-and-run sampler for the dodecahedron network started from $\mathbf{x}_1 = (1.01, \ldots, 1.01)$. The left panel shows the convergence of the empirical mean $\overline{W}_M$ to the incorrect value of 0.5. The right panel shows the mean time of repair for each edge conditional on the network being nonoperational.

approximation to the whole surface of the target density. The difficulty in exploring all the modes of the target density (also known as poor mixing behavior of the chain) translates into slow convergence of the empirical average

$$\overline{W}_M = \frac{1}{M}\sum_{t=1}^{M} W(\mathbf{X}_t) \to \mathbb{E}W(\mathbf{X}), \quad \text{as } M \to \infty\,,$$

for an arbitrary function W. To assess the convergence of the hit-and-run sampler we monitor the empirical average $\overline{W}_M$ with

$$W(\mathbf{x}) = \left(\frac{\sigma}{\widetilde{\sigma}}\right)^3 \exp\left(-\frac{x_1^2 + x_2^2 + x_3^2}{2}\left(\frac{1}{\widetilde{\sigma}^2} - \frac{1}{\sigma^2}\right)\right), \quad \widetilde{\sigma} = \frac{-1}{\Phi^{-1}(0.0926^3)}\,.$$

In this case we have

$$\mathbb{E}W(\mathbf{X}) = \frac{1}{\ell}\mathbb{P}(S(\widetilde{\mathbf{X}}) > 1), \quad \mathbf{X} \sim f,$$

where $\widetilde{X}_1, \widetilde{X}_2, \widetilde{X}_3 \sim_{\text{iid}} \mathsf{N}(0, \widetilde{\sigma}^2)$ and $\widetilde{X}_4, \ldots, \widetilde{X}_m \sim_{\text{iid}} \mathsf{N}(0, \sigma^2)$. It follows that $\widetilde{\ell} = \mathbb{P}(S(\widetilde{\mathbf{X}}) > 1)$ is the probability that the network is nonoperational given that edges 1, 2, 3 have unreliabilities 0.0926^3 and all the others 10^{-3}. We can use any of the methods in Chapter 16 to estimate the probabilities $\ell \approx 2.00 \times 10^{-9}$, $\widetilde{\ell} \approx 1.50 \times 10^{-9}$, correct to two significant figures. Thus, we expect that as $M \to \infty$,

$$\overline{W}_M \to \mathbb{E}W(\mathbf{X}) \approx 0.75\,.$$

The left panel in Figure 14.6 shows the behavior of the empirical average $\overline{W}_M$ as a function of M. We can see that the empirical average fails to converge to the correct value of 0.75 after 10^4 iterations. Although not displayed on the graph, we continued to run the hit-and-run sampler until $M = 10^6$, giving a final empirical

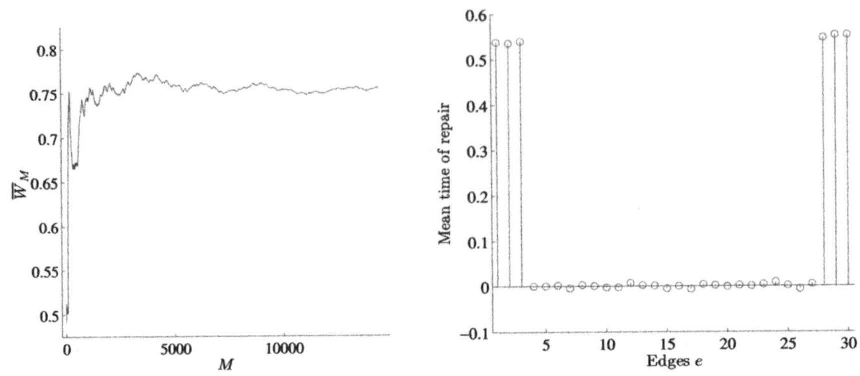

Figure 14.7 ADAM as a sampler for the dodecahedron network. The left panel shows the convergence of the empirical mean to the correct value of 0.75. The right panel shows the mean repair time for each edge conditional on the network being nonoperational.

average of approximately 0.50. Thus, even for large M the ergodic average fails to converge to the correct value. The right panel of Figure 14.6 shows the empirical average:

$$\frac{1}{M}\sum_{t=1}^{M} X_e^{(t)}, \quad \mathbf{X}_t = (X_1^{(t)} \ldots, X_m^{(t)}) \, ,$$

for each of the edges $e = 1, \ldots, m$. From the graph we can deduce that the most likely reason for the network to be nonoperational is the simultaneous failure of edges $\{1, 2, 3\}$. Note that restarting the hit-and-run sampler from different initial conditions does not improve the performance. In addition, alternative samplers such as the Gibbs algorithm perform similarly.

As alternatives to standard MCMC methods, we can use either the splitting
☞ 570 sampler (Algorithm 14.9) or the output of ADAM Algorithm 16.9. Figure 14.7 shows the performance of the ADAM Algorithm 16.9 with $N = 10^4$ and $\varrho = 0.5$, using the hit-and-run Algorithm 14.11 to sample from the transition density $\kappa_t(\mathbf{x} \,|\, \mathbf{y})$. The left panel shows the empirical average $\overline{W}_M$ as M varies from 1 to N_T, where $\mathbf{X}_1, \ldots, \mathbf{X}_{N_T}$ is the population from the last iteration of ADAM. The empirical average appears to settle around the value of 0.75, which suggests that ADAM is performing well as a sampler. In addition, the right panel shows that both modes of the target density are visited and that the most likely reason for the network to be nonoperational is that either edges $\{1, 2, 3\}$ or edges $\{28, 29, 30\}$ fail. In contrast, Figure 14.6 shows that the hit-and-run sampler gets stuck in one of the modes and fails to explore the other mode even after a large number of Markov chain iterations. The overall conclusion is that the output of ADAM provides a better approximation of the target density than the output of the hit-and-run and Gibbs samplers.

Further Reading

Given the large number of possibilities and variations in designing Monte Carlo algorithms that use a population of Markov chains or instrumental densities in an iterative way, it is difficult to describe a single generic algorithm that will apply in every setting [11]. Despite this, the main ingredients behind all such algorithms are summarized in [27] and studied in-depth in [26]. Details on the generalized splitting and ADAM methods can be found in [3]. For early references on splitting with a fixed number of particles, see [13, 14].

For additional work on using the particle splitting method to generate points uniformly within a volume, we refer to [15]. For a number of heuristics that aim to improve the performance of the particle splitting method see [31, 32]. Note, however, that these heuristics make the estimator biased and the estimation of the corresponding relative error more difficult.

REFERENCES

1. Z. I. Botev. Three examples of a practical exact Markov chain sampling. Technical report, School of Mathematics and Physics, The University of Queensland, `http://espace.library.uq.edu.au/view/UQ:130865`, 2007.
2. Z. I. Botev. An algorithm for rare-event probability estimation using the product rule of probability theory. Technical report, School of Mathematics and Physics, The University of Queensland, `http://espace.library.uq.edu.au/view/UQ:151299`, 2008.
3. Z. I. Botev. Splitting methods for efficient combinatorial counting and rare-event probability estimation. Technical report, School of Mathematics and Physics, The University of Queensland, `http://espace.library.uq.edu.au/view/UQ:178513`, 2009.
4. Z. I. Botev and D. P. Kroese. The generalized cross-entropy method, with applications to probability density estimation. *Methodology and Computing in Applied Probability*, DOI:10.1007/s11009-009-9133-7, 2009.
5. S. P. Brooks, P. Dellaportas, and G. O. Roberts. An approach to diagnosing total variation convergence of MCMC algorithms. *Journal of Computational and Graphical Statistics*, 6(3):251–265, 1997.
6. S. P. Brooks and G. O. Roberts. Convergence assessment techniques for Markov chain Monte Carlo. *Statistics and Computing*, 8(4):319–335, 1998.
7. F. Cérou, P. Del Moral, T. Furon, and A. Guyader. Rare-event simulation for a static distribution. Technical report, INRIA-00350762, 2009.
8. S. Chib. Marginal likelihood from the Gibbs output. *Journal of the American Statistical Association*, 90(432):1313–1321, 1995.
9. S. Chib and I. Jeliazkov. Marginal likelihood from the Metropolis–Hastings output. *Journal of the American Statistical Association*, 96(453):270–281, 2001.
10. N. Chopin. A sequential particle filter for static models. *Biometrika*, 89(3):539–551, 2002.
11. A. Doucet, N. de Freitas, and N. Gordon. *Sequential Monte Carlo Methods in Practice.* Springer-Verlag, New York, 2001.
12. W. Gander and W. Gautschi. Adaptive quadrature - revisited. *BIT Numerical Mathematics*, 40(1):84–101, 2000.

13. M. J. J. Garvels. *The Splitting Method in Rare Event Simulation.* PhD thesis, University of Twente, 2000.

14. M. J. J. Garvels and D. P. Kroese. A comparison of RESTART implementations. In *Proceedings of the 1998 Winter Simulation Conference*, pages 601–609, Washington, DC, 1998.

15. P. Glynn, A. Dolgin, R. Y. Rubinstein, and R. Vaisman. How to generate uniform samples on discrete sets using the splitting method. *Probability in Engineering and Information Sciences*, DOI:10.1017/S0269964810000057, 2009.

16. N. Gordon, J. Salmond, and A. Smith. A novel approach to non-linear/non-Gaussian Bayesian state estimation. *IEEE Proceedings on Radar and Signal Processing*, 140(2):107–113, 1993.

17. J. Gu, P. W. Purdom, J. Franco, and B. W. Wah. Algorithms for the satisfiability (SAT) problem: A survey. In *Satisfiability Problem: Theory and Applications*, pages 19–152. American Mathematical Society, Providence, RI, 1997.

18. C. Han and B. P. Carlin. Markov chain Monte Carlo methods for computing Bayes factors: A comparative review. *Journal of the American Statistical Association*, 96(455):1122–1132, 2001.

19. H. H. Hoos and T. Stützle. SATLIB: An online resource for research on SAT. In: SAT 2000, I. P. Gent, H. v. Maaren, T. Walsh, editors, pages 283–292. `www.satlib.org`, IOS Press, 2000.

20. A. M. Johansen, P. Del Moral, and A. Doucet. Sequential Monte Carlo samplers for rare events. In *Proceedings of the 6th International Workshop on Rare Event Simulation, Bamberg, Germany*, 2006.

21. H. Kahn and T. E. Harris. *Estimation of Particle Transmission by Random Sampling.* National Bureau of Standards Applied Mathematics Series, 1951.

22. G. Kitagawa. Monte Carlo filter and smoother for non-Gaussian non-linear state space models. *Journal of Computational and Graphical Statistics*, 5(1):1–25, 1996.

23. S. C. Kou, Q. Zhou, and W. H. Wong. Equi-energy sampler with applications in statistical inference and statistical mechanics. *The Annals of Statistics*, 34(4):1581–1619, 2006.

24. P. L'Ecuyer, V. Demeres, and B. Tuffin. Splitting for rare-event simulation. *Proceedings of the 2006 Winter Simulation Conference*, pages 137–148, 2006.

25. P. L'Ecuyer, V. Demers, and B. Tuffin. Rare events, splitting, and quasi-Monte Carlo. *ACM Transactions on Modeling and Computer Simulation*, 17(2):1–44, 2007.

26. P. Del Moral. *Feynman–Kac Formulae: Genealogical and Interacting Particle Systems with Applications.* Springer-Verlag, New York, 2004.

27. P. Del Moral, A. Doucet, and A. Jasra. Sequential Monte Carlo Samplers. *Journal of the Royal Statistical Society, Series B*, 68(3):411–436, 2006.

28. P. Del Moral, A. Doucet, and A. Jasra. Sequential Monte Carlo for Bayesian computation. In *Proceedings of the Eighth Valencia International Meeting on Bayesian Statistics*, pages 1–34, Valencia, Spain, 2007.

29. C. P. Robert and G. Casella. *Monte Carlo Statistical Methods.* Springer-Verlag, New York, second edition, 2004.

30. D. Rubin. *Multiple Imputation for Nonresponse in Surveys.* John Wiley & Sons, New York, 1987.

31. R. Y. Rubinstein. The Gibbs cloner for combinatorial optimization, counting and sampling. *Methodology and Computing in Applied Probability*, 11(2):491–549, 2009.

32. R. Y. Rubinstein. Why the classic randomized algorithms do not work and how to make them work. *Methodology and Computing in Applied Probability*, 12(1):1–50, 2010.

33. R. Y. Rubinstein and D. P. Kroese. *The Cross-Entropy Method: A Unified Approach to Combinatorial Optimization, Monte-Carlo Optimization, and Machine Learning.* Springer-Verlag, New York, 2004.

34. R. Y. Rubinstein and D. P. Kroese. *Simulation and the Monte Carlo Method.* John Wiley & Sons, New York, second edition, 2007.

35. S. Senju and Y. Toyoda. An approach to linear programming with 0-1 variables. *Management Science*, 15(4):B196–B207, 1968.

36. D. J. A. Welsh. *Complexity: Knots, Coloring and Counting.* Cambridge University Press, Cambridge, 1993.

CHAPTER 15

APPLICATIONS TO FINANCE

Monte Carlo methods are frequently encountered in financial engineering. In this chapter we highlight some of the main Monte Carlo techniques used in option pricing:

1. The *control variable method* for pricing Asian call options; ☞ 351
2. The *conditional Monte Carlo* method for pricing European call options with stochastic volatility, or with a barrier; ☞ 354
3. The *importance sampling* method for pricing barrier options; ☞ 362
4. *Infinitesimal perturbation analysis* for estimating the sensitivities of European call options; ☞ 426
5. The *score function method*, combined with importance sampling, for estimating the Greeks of barrier options. ☞ 428

We refer to Sections A.13 and 5.6 for details on stochastic differential equations and the generation of diffusion processes, respectively. ☞ 643 ☞ 183

15.1 STANDARD MODEL

A standard modeling framework for mathematical finance is the following, based on five ingredients. We assume that a probability space $(\Omega, \mathcal{F}, \mathbb{P})$ is given with a filtration $\{\mathcal{F}_t, t \geqslant 0\}$.

1. Stock price model (risky assets). Let $S_{t,1}, \ldots, S_{t,m}$ be the prices of m risky financial assets at time t — typically stocks in an equity market. The basket of assets described by the stochastic process $\{\mathbf{S}_t = (S_{t,1}, \ldots, S_{t,m})^\top, t \geqslant 0\}$ is assumed to evolve in time according to the multidimensional SDE

$$\mathrm{d}\mathbf{S}_t = \boldsymbol{\mu}_t \mathrm{d}t + \boldsymbol{\sigma}_t \,\mathrm{d}\mathbf{W}_t \,, \quad \mathbf{S}_0 = (s_1, \ldots, s_m)^\top ,$$

where $\{\mathbf{W}_t\}$ is an n-dimensional Wiener process under measure $\mathbb{P}$ and filtration $\{\mathcal{F}_t, t \geqslant 0\}$, and

$$\boldsymbol{\mu}_t = \boldsymbol{\mu}(\mathbf{S}_t, t) = \begin{pmatrix} \mu_{t,1} \\ \vdots \\ \mu_{t,m} \end{pmatrix}, \quad \boldsymbol{\sigma}_t = \boldsymbol{\sigma}(\mathbf{S}_t, t) = \begin{pmatrix} \sigma_{t,11} & \cdots & \sigma_{t,1n} \\ \vdots & \ddots & \vdots \\ \sigma_{t,m1} & \cdots & \sigma_{t,mn} \end{pmatrix}$$

are deterministic functions of $\mathbf{S}_t$ and t. The process $\{\mathbf{S}_t, t \geqslant 0\}$ is thus an m-
☞ 643 dimensional diffusion process as described in Section A.13. In addition to this we assume:

- The stocks do not pay dividends.
- There are no transaction costs.
- All securities are perfectly divisible; that is, it is possible to buy any fraction of a share.
- There are no restrictions on *short selling*; that is, it is possible to hold a negative number of units of a given share.

2. Bonds (risk-free assets). Suppose the market has at least one risk-free asset, which is modeled by the deterministic differential equation

$$B_t' = r_t B_t, \quad B_0 = 1 \,,$$

where r_t is the instantaneous interest rate at time t. Obviously, $B_t = \exp\left(\int_0^t r_s \,\mathrm{d}s\right)$. Such risk-free assets can be government bonds or interest bearing money market accounts.

3. Self-financing assumption. Under the assumptions of points 1. and 2. above, the value V_t of a portfolio containing ψ_t units of bonds and $\boldsymbol{\phi}_t = (\phi_{t,1}, \ldots, \phi_{t,m})^\top$ units of various stocks at a given time t is given by

$$V_t = B_t \,\psi_t + \sum_{i=1}^{m} \phi_{t,i} S_{t,i} = B_t \,\psi_t + \boldsymbol{\phi}_t^\top \mathbf{S}_t \,. \tag{15.1}$$

The particular value of the portfolio at time t typically depends on the units of assets $(\psi_t, \boldsymbol{\phi}_t)$ held just before time t. The stochastic process $\{(\psi_t, \boldsymbol{\phi}_t), t \geqslant 0\}$ can thus be interpreted as a **trading strategy**. More precisely, we call the process a trading strategy if it is adapted to the filtration $\{\mathcal{F}_t, t \geqslant 0\}$. By the product rule
☞ 642 for Itô processes (A.65), an infinitesimal change in the value of the portfolio is given by

$$\mathrm{d}V_t = \left(B_t \,\mathrm{d}\psi_t + \mathbf{S}_t^\top \mathrm{d}\boldsymbol{\phi}_t\right) + \psi_t \,\mathrm{d}B_t + \boldsymbol{\phi}_t^\top \mathrm{d}\mathbf{S}_t \,.$$

The portfolio is called **self-financing** if

$$B_t\,\mathrm{d}\psi_t + \mathbf{S}_t^\top \mathrm{d}\boldsymbol{\phi}_t = 0\,,$$

in which case

$$V_T = V_0 + \int_0^T \psi_t\,\mathrm{d}B_t + \int_0^T \boldsymbol{\phi}_t^\top \mathrm{d}\mathbf{S}_t\,.$$

In other words, for a given trading strategy $\{(\psi_t, \boldsymbol{\phi}_t)\}$, changes in the value of the portfolio are solely due to changes in the value of assets that are held at any given time.

4. Arbitrage-free market. A market model is said to have arbitrage opportunities if it is possible to create positive wealth from zero or negative initial wealth without incurring any risk. A market model is called **arbitrage-free** on $[0, T]$ if, in essence, for every self-financing trading strategy $\{(\psi_t, \boldsymbol{\phi}_t), 0 \leqslant t \leqslant T\}$,

$$V_0 = 0 \Rightarrow \mathbb{P}(V_T > 0) = 0\,.$$

An important result in financial engineering is that a market is arbitrage-free if there exists a probability measure $\mathbb{Q}$ such that the discounted asset price process $\{\widetilde{\mathbf{S}}_t, 0 \leqslant t \leqslant T\}$, with $\widetilde{\mathbf{S}}_t = B_t^{-1}\mathbf{S}_t$, is an $(\mathcal{F}_t, \mathbb{Q})$-martingale [16, 26]. For the standard model this arbitrage-free condition is satisfied if the system of equations

$$\boldsymbol{\sigma}_t\,\mathbf{u}_t = \boldsymbol{\mu}_t - r_t\,\mathbf{S}_t \tag{15.2}$$

has a solution $\mathbf{u}_t$, $t \in [0, T]$ for which $\mathbb{E}\exp\left(\frac{1}{2}\int_0^T \mathbf{u}_t^\top \mathbf{u}_t\,\mathrm{d}t\right) < \infty$ (*Novikov's condition*) holds. To see this, observe that, by the product rule for Itô processes, we have ☞ 642

$$\begin{aligned}
\mathrm{d}\widetilde{\mathbf{S}}_t &= B_t^{-1}(\boldsymbol{\mu}_t - r_t\,\mathbf{S}_t)\,\mathrm{d}t + B_t^{-1}\boldsymbol{\sigma}_t\,\mathrm{d}\mathbf{W}_t \\
&= B_t^{-1}\boldsymbol{\sigma}_t(\mathbf{u}_t\,\mathrm{d}t + \mathrm{d}\mathbf{W}_t) \\
&= B_t^{-1}\boldsymbol{\sigma}_t\,\mathrm{d}\mathbf{Z}_t\,,
\end{aligned}$$

where $\mathrm{d}\mathbf{Z}_t = \mathbf{u}_t\,\mathrm{d}t + \mathrm{d}\mathbf{W}_t$ defines an Itô process. Let

$$M_t = \exp\left(\int_0^t \mathbf{u}_s^\top\,\mathrm{d}\mathbf{W}_s - \frac{1}{2}\int_0^t \mathbf{u}_s^\top \mathbf{u}_s\,\mathrm{d}s\right).$$

Then, by Girsanov's theorem (see Page 642), $\{M_t, t \geqslant 0\}$ is an $(\mathcal{F}_t, \mathbb{P})$-martingale, and under the new measure $\mathbb{Q}(A) \stackrel{\text{def}}{=} \mathbb{E}[M_T \mathrm{I}_A]$ for all $A \in \mathcal{F}_T$, the process $\{\mathbf{Z}_t, 0 \leqslant t \leqslant T\}$ is a Wiener process. It follows that under $\mathbb{Q}$ the process $\{\widetilde{\mathbf{S}}_t, 0 \leqslant t \leqslant T\}$ is a martingale with respect to $\{\mathcal{F}_t, 0 \leqslant t \leqslant T\}$, which had to be shown. $\mathbb{Q}$ is called the **risk-neutral measure**. Note that the existence of a solution to (15.2) implies the existence of the risk-neutral measure $\mathbb{Q}$ and vice-versa.

The j-th component of the vector $\mathbf{u}_t$ is called the **market price of risk** associated with the j-th component of $\mathbf{W}_t$. In the scalar case $\mathbf{u}_t = u_t$ is simply called the market price of risk. This interpretation comes from the following observation. Suppose that $\boldsymbol{\mu}_t = \mu S_t$ and $\boldsymbol{\sigma}_t = \sigma S_t$, so that $\{S_t\}$ is a geometric Brownian motion process. Then, from

$$\mu = r_t + \sigma u_t$$

we can see that the excess return $\mu - r_t$ generated by a risky asset S_t is proportional to σ with constant of proportionality u_t. In this sense u_t measures the excess return (above the risk-free return) that investors demand per unit risk σ.

Let $\widetilde{V}_t = B_t^{-1} V_t = \psi_t + \boldsymbol{\phi}_t^\top B_t^{-1} \mathbf{S}_t = \psi_t + \boldsymbol{\phi}_t^\top \widetilde{\mathbf{S}}_t$ be the discounted value of the portfolio of stocks and bonds. Under the arbitrage-free and self-financing assumptions we can write

$$\mathrm{d}\widetilde{V}_t = \boldsymbol{\phi}_t^\top \mathrm{d}\widetilde{\mathbf{S}}_t = B_t^{-1} \boldsymbol{\phi}_t^\top \boldsymbol{\sigma}_t \, \mathrm{d}\mathbf{Z}_t \, .$$

It follows that under $\mathbb{Q}$ the process $\{\widetilde{V}_t, 0 \leqslant t \leqslant T\}$ is a martingale with respect to $\{\mathcal{F}_t, 0 \leqslant t \leqslant T\}$. As a consequence,

$$V_t = B_t \, \mathbb{E}_{\mathbb{Q}}[B_T^{-1} V_T \,|\, \mathcal{F}_t], \quad t \leqslant T \, . \tag{15.3}$$

The last equation gives us a formula for valuing the portfolio at time $t \leqslant T$, given that we know the stochastic behavior of the portfolio under $\mathbb{Q}$.

5. Complete market assumption. A market is **complete** if the value of every financial instrument contingent on the market assets can be replicated using the self-financing trading strategy $\{(\psi_t, \boldsymbol{\phi}_t)\}$ described above. It can be shown [5, 11, 18] that the market is complete if and only if the solution of (15.2) is *unique*.

Using the five assumptions above, we describe the mechanism for pricing a **European call option (European put option)**, which is a contract granting the holder the right, but not the obligation, to buy (sell) an amount of stock S_t at a fixed price K at some future time T. The price K is called the **strike price**. T is called the **maturity** or **expiration time**. The value of the call (put) option at maturity is referred to as the option **payoff**. The payoff at maturity for a call option is $(S_T - K)^+ = \max\{S_T - K, 0\}$ and for a put option is $(K - S_T)^+$. Suppose for concreteness that we are dealing with a call option and denote its value at time $t \leqslant T$ by $C_t = C(S_t, K, r_t, T-t)$, where $T - t$ is the time until maturity. Then, the price of the option, C_t, is given by the value V_t of the portfolio (15.1), under the condition that the portfolio replicates the payoff of the option. In other words, the price is derived from (15.3):

$$C_t = V_t = B_t \, \mathbb{E}_{\mathbb{Q}}[B_T^{-1} C_T \,|\, \mathcal{F}_t], \quad t \leqslant T \, , \tag{15.4}$$

where

$$C_T = V_T = V_0 + \int_0^T \psi_t \, \mathrm{d}B_t + \int_0^T \boldsymbol{\phi}_t^\top \mathrm{d}\mathbf{S}_t \, .$$

■ **EXAMPLE 15.1 (Black–Scholes Model)**

☞ 196 Suppose the stock price process satisfies the geometric Brownian motion SDE

$$\mathrm{d}S_t = \mu \, S_t \, \mathrm{d}t + \sigma \, S_t \, \mathrm{d}W_t \, ,$$

where $\{W_t, t \geqslant 0\}$ is a Wiener process under $\mathbb{P}$. The parameters μ and σ are called the **drift** and **volatility**, respectively. We assume that the risk-free asset (for example, government bond) yields $B_t = B_0 \, \mathrm{e}^{rt} = \mathrm{e}^{rt}$ units after time t, where r is the risk-free annual interest rate. Then, since

$$(\sigma S_t) u_t = (\mu S_t) - r S_t$$

has a unique solution, the market is complete and the market price of risk is $u_t = (\mu - r)/\sigma$. Hence, we can price any contingent claim via a self-financing trading strategy using the underlying securities. The risk-neutral (arbitrage-free) price of a European call option C_t is then derived from (15.4):

$$C_t = \mathbb{E}_{\mathbb{Q}}[\mathrm{e}^{-r(T-t)} C_T \mid \mathcal{F}_t] = \mathbb{E}_{\mathbb{Q}}[\mathrm{e}^{-r(T-t)} (S_T - K)^+ \mid S_t]$$
$$= \int_{\mathrm{e}^{-r(T-t)}K}^{\infty} (x - \mathrm{e}^{-r(T-t)}K)\, p(x)\, \mathrm{d}x$$

where $p(x)$ is the pdf of $\widetilde{S}_T = \mathrm{e}^{-rT} S_T$ conditioned on S_t under the risk-neutral measure $\mathbb{Q}$. Under $\mathbb{Q}$ the discounted process $\{\widetilde{S}_t\}$ satisfies the SDE $\mathrm{d}\widetilde{S}_t = \sigma\, \widetilde{S}_t\, \mathrm{d}Z_t$, where $\{Z_t\}$ is a Wiener process. This has strong solution $\widetilde{S}_t = S_0\, \mathrm{e}^{\sigma Z_t - \frac{1}{2}\sigma^2 t}$, $t \geqslant 0$. It follows that

$$(\widetilde{S}_T \mid S_t) \sim \mathsf{LogN}\left(\ln(S_t) - \frac{1}{2}\sigma^2 (T-t)\,,\ \sigma^2 (T-t)\right),$$

and the value of C_t can be computed analytically. The analytical expression for the price is called the **Black–Scholes** formula:

$$\mathrm{BS}(S_t, K, r, T-t, \sigma) \stackrel{\mathrm{def}}{=} S_t \Phi(a_1) - K\mathrm{e}^{-r(T-t)} \Phi(a_2), \quad 0 \leqslant t \leqslant T, \tag{15.5}$$

where $(a_1 > a_2)$

$$a_{1,2} = \frac{\ln(S_t/K) + \left(r \pm \frac{\sigma^2}{2}\right)(T-t)}{\sigma\sqrt{T-t}}. \tag{15.6}$$

It can be shown via a simple arbitrage argument [17, 18] that the value $P_t = P(S_t, K, r_t, T-t)$ of the European put option is related to the value C_t of the European call option via the **put–call parity**:

$$P_t = C_t - S_t + K\mathrm{e}^{-r(T-t)}.$$

Hence, for a European put option the Black–Scholes formula gives

$$P_t = K\mathrm{e}^{-r(T-t)} \Phi(-a_2) - S_t \Phi(-a_1).$$

The details of the option contract can vary. For example, the option may be written over a different underlying asset (stock, commodity, interest rate, or stock market index). Table 15.1 lists a variety of option contracts. The most common type of options are the European- and American-style options. The first can be exercised only at maturity and the second can be exercised on or before the expiration date. American options are more difficult to price since the possibility of exercise at any time prior to maturity requires solving an optimal stopping and stochastic optimization problem [10, 14]. Table 15.2 lists some variants of the vanilla call option (Example 15.1) and their respective payoffs.

■ **EXAMPLE 15.2 (A Compound Call Option)**

Consider an option expiring at T_1 with strike price K_1 to buy a call option expiring at $T_2 > T_1$ with strike price K_2. If we denote the price of the option expiring at T_1 (T_2) by $C_{t,1}$ ($C_{t,2}$), then

$$C_{0,1} = \mathrm{e}^{-rT_1}\, \mathbb{E}_{\mathbb{Q}}[(C_{T_1,2} - K_1)^+ \,|\, S_0] \,,$$

where

$$C_{T_1,2} = \mathrm{e}^{-r(T_2-T_1)}\, \mathbb{E}_{\mathbb{Q}}[(S_{T_2} - K_2)^+ \,|\, S_{T_1}] \,.$$

15.2 PRICING VIA MONTE CARLO SIMULATION

For most of the options in Table 15.2 there is no explicit formula such as (15.5) for the price of the option at time t. In such cases the price given by (15.4) is approximated via Monte Carlo methods. A quite general procedure for pricing European style options using Monte Carlo methods is as follows.

1. Assume we are given an SDE under probability measure $\mathbb{Q}$, which models the risk-neutral behavior of the underlying asset(s) from which the option derives its value. If the SDE is given under measure $\mathbb{P}$, then use Girsanov's theorem or otherwise to derive the evolution equation of the asset prices under the risk-neutral probability measure $\mathbb{Q}$.

2. Simulate N sample paths of the asset prices under $\mathbb{Q}$ over the relevant time horizon, say, $[0, T]$. This step usually requires a numerical scheme to approximate the solution of the SDE.

3. Evaluate the discounted payoff of each asset on each sample path, as determined by the specifics of the asset.

4. Compute a Monte Carlo estimate of the theoretical option value in (15.4) using the N discounted cash flows over the sample paths.

Note that there are two possible sources of error in approximating (15.4). The first one is associated with a discretization of the SDE via the numerical scheme in Step 2. The second source of error is the variance of the Monte Carlo estimator in Step 4. Typically, the Monte Carlo variance is much larger than the error introduced by the numerical SDE scheme. Thus, Step 4 more often than not calls for a variance reduction technique. Very often the event that the payoff of a given option is positive is a rare event. In such settings rare-event simulation techniques are invaluable.

As an illustration of this recipe, we give the following example of pricing an Asian call option.

■ **EXAMPLE 15.3 (Asian Call Option)**

Suppose we wish to price a European style Asian call option with maturity T and strike price K. The payoff of such an option at maturity is

$$C_T = (A_T - K)^+,$$

Table 15.1 Different types of option.

Option name	Description
European	May only be exercised on expiration.
American	May be exercised on any trading day on or before expiration.
Bermudan	May be exercised only on specified dates on or before expiration.
Barrier	The underlying security's price must pass a given barrier before it can be exercised.
Compound	An option on another option, which presents the holder with two separate exercise dates and decisions.
Cross/Composite	An option on some underlying asset in one currency with a strike price denominated in another currency.
Exchange	Gives the holder the right to exchange one asset for another at maturity.
Lookback	Dependent on the whole price path; the owner has the right to buy (sell) the underlying instrument at its lowest (highest) price over some preceding period.
Asian/Average	The payoff is determined by the average underlying price over some predetermined period of time.
Basket	An option on the weighted average of several underlying assets.
Rainbow	A basket option where the asset weightings depend on the final performances of the components in the basket; for example, an option on the worst-performing of several stocks.
Digital/Binary/All-or-nothing	The payoff at maturity is either some amount (units) of cash (assets) fixed in advance or nothing at all.

where A_t is the average price of the stock over the interval $[0, t]$:

$$A_t = \frac{1}{t} \int_0^t S_u \, \mathrm{d}u, \qquad t \in [0, T] \, .$$

Assume that the stock price is driven by the SDE model

$$\mathrm{d}S_t = \mu \, S_t \, \mathrm{d}t + \sigma \, S_t \, \mathrm{d}W_t \, ,$$

where $\{W_t\}$ is a Wiener process under $\mathbb{P}$. We denote the risk-free annual interest rate by r. The price of the Asian call option at time t is given by the risk-neutral

Table 15.2 A list of some common (vanilla) and not so common (exotic) call options with their payoff — all written on the underlying asset S_t.

Name	Payoff	Notes
Vanilla European	$(S_T - K)^+$	K is the strike price
Cash-or-nothing	$\mathrm{I}\{S_T > K\}$	
Down-and-out	$(S_T - K)^+ \mathrm{I}\Big\{\min_{T} S_\tau \geqslant \beta\Big\}$	β is a fixed barrier
Discretely monitored down-and-out	$(S_T - K)^+ \mathrm{I}\Big\{\min_{1 \leqslant i \leqslant n} S_{t_i} \geqslant \beta\Big\}$	
Discretely monitored down-and-in	$(S_T - K)^+ \mathrm{I}\Big\{\min_{1 \leqslant i \leqslant n} S_{t_i} \leqslant \beta\Big\}$	
Lookback	$S_T - \min_t S_t$	
Hindsight	$\left(\max_t S_t - K\right)^+$	
Asian/Average	$(A_T - K)^+$	$A_T = \dfrac{1}{T}\displaystyle\int_0^T S_t \,\mathrm{d}t$
Discretely sampled Asian	$(\widehat{A}_T - K)^+$	$\widehat{A}_T = \dfrac{1}{n+1}\displaystyle\sum_{i=0}^{n} S_{t_i}$
Discretely monitored digital down-and-in	$\mathrm{I}\{S_T > K\}\,\mathrm{I}\Big\{\min_{1 \leqslant i \leqslant n} S_{t_i} \leqslant \beta\Big\}$	

formula

$$C_t = \mathrm{e}^{-r(T-t)}\mathbb{E}_{\mathbb{Q}}[(A_T - K)^+ \mid \mathcal{F}_t], \quad t \in [0, T] .$$

We describe each of the four steps of the Monte Carlo procedure for this case.

Step 1. We must first establish the behavior of the stock price process under the risk-neutral measure $\mathbb{Q}$. As observed in Example 15.1, the discounted stock price process under $\mathbb{Q}$ satisfies

$$\mathrm{d}\widetilde{S}_t = \sigma\, \widetilde{S}_t \,\mathrm{d}Z_t \,,$$

where $\{Z_t\}$ is a Wiener process. It follows that the evolution of the undiscounted stock price under $\mathbb{Q}$ satisfies the SDE

$$\mathrm{d}S_t = r\, S_t \,\mathrm{d}t + \sigma\, S_t \,\mathrm{d}Z_t \,. \tag{15.7}$$

Thus, under the risk-neutral measure the drift is given by r — the risk-free rate of return.

Step 2. For this model there is no need to solve the SDE (15.7) numerically, since the solution is available analytically: ☞ 196

$$S_t = S_0\, \mathrm{e}^{(r-\sigma^2/2)t+\sigma Z_t} \,.$$

For large n the average stock price A_T can be approximated well via

$$\bar{S}_T = \frac{1}{n+1}\sum_{i=0}^{n} S_{t_i}, \quad t_i = \frac{iT}{n}, \ i = 0, \ldots, n\,,$$

where

$$S_{t_i} = S_0 \exp\left((r - \sigma^2/2)\, t_i + \sigma Z_{t_i}\right).$$

Typically n is chosen to be the number of trading days in the period $[0, T]$.

Step 3. The discounted payoff X of the Asian option at time $t = 0$ is

$$X = \mathrm{e}^{-rT}(\bar{S}_T - K)^+ .$$

Step 4. If $X_1, \ldots, X_N$ are N independent realizations of the discounted payoff, then the crude Monte Carlo estimator of the price of the Asian option is

$$\widehat{C}_0 = \frac{1}{N}\sum_{k=1}^{N} X_i\,.$$

The following MATLAB code implements this estimator for the set of parameters: $(r, \sigma, K, S_0, T) = (0.07, 0.2, 35, 40, 4/12)$, and $n = 88$.

```
%asian_option_CMC.m
r=.07; % annual interest
sig=0.2; % volatility
K=35; % stike price
S_0=40; % initial stock price
T=4/12; %maturity in 4 months, which is 4/12 of the year
n= 88;  % there are approx. 88 trading days in 4 months
dt=T/n; % time step
% simulate N sample paths of the stock process
N=10^4; X=nan(N,1); % number of sample path simulations
for i=1:N
    path=(r-sig^2/2)*dt+sig*sqrt(dt)*randn(1,n);
    path=cumprod([S_0,exp(path)]);
    X(i)=exp(-r*T)*max(mean(path)-K,0);
end
c_0=mean(X), Rel_error=std(X)/c_0/sqrt(N)
width=std(X)*norminv(0.975)/sqrt(N)
CI=[c_0-width,c_0+width]
```

With $N = 10^4$, we obtain a typical estimate of $\widehat{C}_0 = 5.38$ with an estimated relative error of 0.48% and a 95% confidence interval of $[5.33, 5.43]$. There is still about \$0.10 uncertainty in the price, and any mispricing on this scale could result in an arbitrage opportunity.

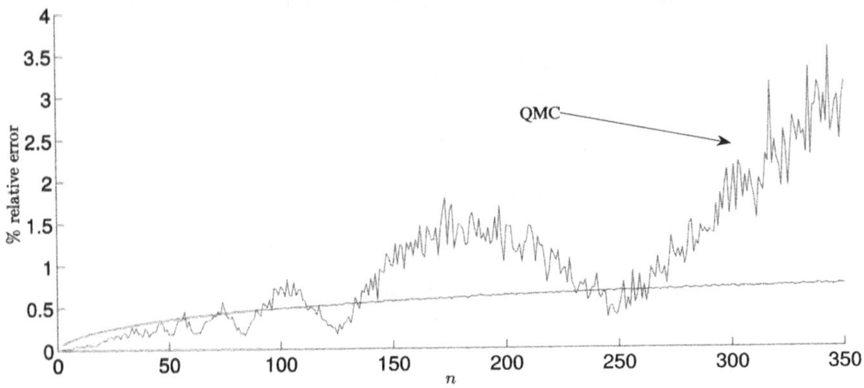

Figure 15.1 Relative error of the crude Monte Carlo estimator and the quasi Monte Carlo estimator for increasing values of the dimensionality n.

■ EXAMPLE 15.4 (Quasi Monte Carlo)

We investigate the accuracy in pricing the Asian option from Example 15.3 for various values of T (maturity time) using crude Monte Carlo and quasi Monte Carlo integration. We set $T = n/365$, where n is the number of trading days. Note that n is also the dimensionality of the integration problem. ☞ 25

For quasi Monte Carlo we use the random shift procedure described in Algorithm 9.11 as follows. Let $\mathcal{P}_N$ be the Faure quasirandom point set constructed in Algorithm 2.2 on Page 32. The number of replications M is chosen to be 40 and $N = 250$. For the crude Monte Carlo estimator we use 10^4 points so that both crude and quasi Monte Carlo use the same number of points. Using the script below we obtained Figure 15.1, which shows the estimated percentage relative error of the crude and quasi Monte Carlo estimator. From the figure we can conclude that quasi Monte Carlo outperforms crude Monte Carlo for $n < 100$. In other words, quasi Monte Carlo is more effective for lower-dimensional integration problems, but is eventually outperformed by crude Monte Carlo. In addition, as seen from the numerous dips and peaks in the error, the quasi Monte Carlo relative error can vary quite unpredictably. ☞ 376

We use two functions:

- `H.m` implements the estimator $\widehat{\ell}_i = \frac{1}{N}\sum_{\mathbf{u}\in\mathcal{P}_N^{(i)}} h(\mathbf{u})$ in Step 4 of Algorithm 9.11;
- `asian_option_QMC.m` implements the crude and quasi Monte Carlo estimator similar to `bridgeQMC.m` in Example 9.13.

```
%QMC_run_script.m
clear all,clc,trading_days=2:350;err=[];
for n=trading_days
    err=[err;asian_option_QMC(n)];
    n
```

```
end
plot(trading_days,err(:,1)), hold on
plot(trading_days,err(:,2),'r')
```

```
function out=asian_option_QMC(n)
% 'n' is the number of trading days;
% output of function is the relative error
% of CMC and QMC for a given 'n';

% option details; make T a function of n
T=n/365; %maturity in n days, which is n/365 of the year
r=.07; % annual interest
sig=0.2; % volatility
K=35; % stike price
S_0=40; % initial stock price
dt=T/n; % time step
% set up the quasi-random numbers
M = 40;
N = 10^4/M;
F = faure(n,n,N-1);
% QMC estimation
for i=1:M
    U = mod(F + repmat(rand(1,n),N,1), 1);
    y(i) = h(U,N,n,T,r,sig,K,S_0,dt);
end
ell = mean(y); %estimate
QMC_RE = std(y)/sqrt(M)/ell; % rel. error
% CMC estimation with N*M points
X=nan(N*M,1); % number of sample path simulations
for i=1:N*M
    X(i)=h(rand(1,n),1,n,T,r,sig,K,S_0,dt);
end
c_0=mean(X);
CMC_RE=std(X)/c_0/sqrt(N*M);
out=[CMC_RE, QMC_RE ]*100;
```

```
function c_0=h(U,N,n,T,r,sig,K,S_0,dt)
X=nan(N,1); % number of sample path simulations
for i=1:N
    path=(r-sig^2/2)*dt+sig*sqrt(dt)*norminv(U(i,:));
    path=cumprod([S_0,exp(path)]);
    X(i)=exp(-r*T)*max(mean(path)-K,0);
end
c_0=mean(X);
```

For a more detailed comparison of quasi and crude Monte Carlo see [7, 14]. To improve the performance of the quasi Monte Carlo method, it is common to generate the sample paths of the option using a Brownian bridge [14]. An efficient quasi Monte Carlo algorithm for pricing Asian, lookback, and barrier (that is, path-dependent) options under the more complicated variance gamma model is presented in [2].

■ EXAMPLE 15.5 (Asian Call Pricing With a Control Variable)

While quasi Monte Carlo can be effective in reducing the variance by a factor of 4 or 5, we can obtain a significantly more accurate estimate of the price of the Asian
☞ 352 option (Example 15.3) using a control variable variance reduction technique.

Let G_T be the geometric average of the stock price:

$$G_T = \exp\left(\int_0^T \ln S_t \,\mathrm{d}t\right).$$

For large n, G_T can be approximated via the geometric mean

$$\widehat{G}_t = \left(\prod_{i=0}^{n} S_{t_i}\right)^{1/(n+1)}, \quad \text{where } S_{t_i} = S_0 \exp\left(\left(r - \frac{\sigma^2}{2}\right) t_i + \sigma Z_{t_i}\right),$$

with negligible error (see, for example, [8]). Since $\widehat{G}_T$ is expected to be closely correlated to $\bar{S}_T$, and since the expectation of G_T can be computed (see below), the following control variable is suggested:

$$\widetilde{X} = \mathrm{e}^{-rT}(\widehat{G}_T - K)^+ \, .$$

The expected value $\mathbb{E}_{\mathbb{Q}}(G_T - K)^+$ is computed as follows. First, using the property that $\mathrm{Cov}(Z_s, Z_t) = \min\{s, t\}$, it can be shown that

$$\ln \widehat{G}_T \sim \mathsf{N}\left(\ln(S_0) + \frac{T}{2}\left(r - \frac{\sigma^2}{2}\right), \frac{\sigma^2}{(n+1)^2} \sum_{i=0}^{n}\sum_{j=0}^{n} \min\{t_i, t_j\}\right).$$

Since

$$\lim_{n\to\infty} \frac{1}{(n+1)^2} \sum_{i=0}^{n}\sum_{j=0}^{n} \min\{t_i, t_j\} = \frac{1}{T^2}\int_0^T \int_0^T \min\{u, v\} \,\mathrm{d}u \,\mathrm{d}v = \frac{T}{3},$$

by considering the limiting distribution of $\widehat{G}_T$ as $n \to \infty$, we arrive at

$$\ln G_T \sim \mathsf{N}\left(\ln(S_0) + \frac{T}{2}\left(r - \frac{\sigma^2}{2}\right), \frac{\sigma^2 T}{3}\right).$$

Hence, straightforward integral manipulations similar to the ones used to derive (15.5) lead to

$$\mathrm{e}^{-rT}\mathbb{E}_{\mathbb{Q}}(G_T - K)^+ = \mathrm{e}^{-\frac{(6r+\sigma^2)T}{12}} S_0 \Phi(a_1) - K\mathrm{e}^{-rT}\Phi(a_2) \, ,$$

where $(a_1 > a_2)$

$$a_{1,2} = \frac{\ln(S_0/K) + \frac{1}{2}\left(r - \frac{\sigma^2}{6} \pm \frac{\sigma^2}{3}\right)T}{\sigma\sqrt{T/3}} \, .$$

The following MATLAB code implements the control variable approach using $\widetilde{X}$.

```
% asian_option_Control_variable.m
r=.07; % annual interest
sig=0.2; % volatility
K=35; % stike price
S_0=40; % initial stock price
T=4/12; % maturity in 4 months, which is 4/12 of the year
n= 88;  % there are approx. 88 trading days in 4 months
dt=T/n; % time step
% Simulate N sample paths of the stock process
N=10^4; X=nan(N,1); tX=X; % number of sample path simulations
for i=1:N
    path=(r-sig^2/2)*dt+sig*sqrt(dt)*randn(1,n);
    path=cumprod([S_0,exp(path)]);
    X(i)=exp(-r*T)*max(mean(path)-K,0);
    tX(i)=exp(-r*T)*max(prod(path)^(1/(n+1))-K,0);
end
c_0=mean(X)
width=std(X)*norminv(0.975)/sqrt(N);
CI=[c_0-width, c_0+width]

% compute expectation of control variable
a1=(log(S_0/K)+(r-sig^2/6+sig^2/3)*T/2)/sig/sqrt(T/3);
a2=(log(S_0/K)+(r-sig^2/6-sig^2/3)*T/2)/sig/sqrt(T/3);
geo_call=exp(-(6*r+sig^2)*T/12)*S_0*normcdf(a1)-...
         K*exp(-r*T)*normcdf(a2);

Cov=cov([X,tX]);
alpha=Cov(1,2)/Cov(1,1); % optimal linear control

ell_c=c_0-alpha*mean(tX-geo_call)
width=1.96*sqrt((1-Cov(1,2)^2/Cov(1,1)/Cov(2,2))/N*Cov(1,1));
CI=[ell_c-width,ell_c+width]
```

For the same simulation effort as in Example 15.3 ($N = 10^4$) we obtain a typical estimate of $\widehat{C}_0 = 5.356$ with 95% confidence interval $[5.355, 5.357]$. Thus, if prices on the stock exchange are quoted to within \$ 0.01, then the control variable method provides a satisfactory estimate in this case. The variance reduction over the crude Monte Carlo estimator is roughly a factor of 10^3. Note that since the Asian option uses the whole stock price history, it is less sensitive to price manipulation than the vanilla European call option in Example 15.1, whose payoff depends only on the final stock price S_T.

Early references on the control variable approach for pricing Asian options include [6] and [23]. The analytical pricing using the geometric average can be used to provide bounds on the price of the arithmetic Asian option, see [27, 29]. For a discussion of the simulation and discretization error in pricing continuous arithmetic Asian options, see [1]. For pricing of discretely monitored Asian options when the underlying asset evolves according to a Lévy process, see [13].

■ **EXAMPLE 15.6 (Conditional Estimator I)**

Consider pricing a European call option with stochastic volatility. Suppose that the underlying stock price under the risk-neutral probability measure $\mathbb{Q}$ evolves according to

$$\mathrm{d}S_t = r\,S_t\,\mathrm{d}t + V_t\,S_t\,\mathrm{d}W_{t,1},$$

where the volatility is not deterministic ($V_t \neq \sigma$) but is a **mean-reverting process**, given by the SDE

$$\mathrm{d}V_t = \alpha(\sigma - V_t)\,\mathrm{d}t + \sigma_0\,V_t\,\mathrm{d}W_{t,2}\;.$$

Here σ is the long-term mean volatility, and α is the rate at which reversion to the long-term mean occurs. Note that $\{W_{t,1}\}$ and $\{W_{t,2}\}$ are assumed to be independent processes.

The price of the option thus depends both on the stochastic behavior of the stock price and the volatility. There is no simple closed-form formula for the price of this option. However, conditional on a realization $\{V_t, 0 \leqslant t \leqslant T\}$, the evolution of the stock price is available analytically:

$$S_T = S_0 \exp\left(rT - \frac{1}{2}\int_0^T V_s^2\,\mathrm{d}s + \int_0^T V_s\,\mathrm{d}W_{s,1}\right).$$

Therefore, if $\widetilde{S}_T = \mathrm{e}^{-rT} S_T$ is the discounted stock price at maturity, we have conditional on $\{V_t\}$

$$\widetilde{S}_T \sim \mathsf{LogN}\left(\ln(S_0) - \frac{1}{2}\bar{\sigma}^2 T,\; \bar{\sigma}^2 T\right),$$

where $\bar{\sigma}^2 = \frac{1}{T}\int_0^T V_s^2\,\mathrm{d}s$ is the average squared volatility. Thus, using the derivation in Example 15.1 we can compute the conditional expectation

$$\mathrm{e}^{-rT}\mathbb{E}_{\mathbb{Q}}[(S_T - K)^+ \mid S_0, V_t, 0 \leqslant t \leqslant T] = \mathrm{BS}\,(S_0, K, r, T, \bar{\sigma}),$$

where $\mathrm{BS}(S_0, K, r, T, \sigma)$ is the option price given by the Black–Scholes formula (15.5) with initial price S_0, strike price K, interest rate r, maturity T, and volatility σ. The code below implements the conditional estimator

$$X = \mathrm{BS}\,(S_0, K, r, T, \bar{\sigma})\;.$$

☞ 185 We use the Euler scheme to approximate the solution of the mean-reverting process. The code requires the function `BS.m`, which implements (15.5). We obtain a typical 95% confidence interval of [0.219, 0.221]. For the same simulation effort of $N = 10^3$ we obtained the much wider interval [0.209, 0.246] using crude Monte Carlo. An early reference using this conditioning idea is [17].

```
% EU_Call_conditional_est.m
r=0.05; sig_0=1; % volatility
alpha=10;  sig=1; % long term volatility
K=0.85; % strike price
S_0=1; V_0=0; % initial stock and volatility
T=90/365; % maturity in 90 days
n= 90;  % use 90 trading days
dt=T/n; % time step

% now we simulate N sample paths of the stock process
N=10^3; X=nan(N,1);
for k=1:N
    V=nan(1,n); V(1)=V_0;
    for i=1:n-1
        V(i+1)=V(i)+dt*alpha*(sig-V(i))+sqrt(dt)*sig_0*V(i)*randn;
    end
    sigma_bar=sqrt(sum(V.^2)*dt/T);
    X(k)=BS(S_0,K,r,T,sigma_bar); %conditional estimator
end
c_0=mean(X), Rel_error=std(X)/c_0/sqrt(N)
width=std(X)*norminv(0.975)/sqrt(N)
CI=[c_0-width,c_0+width] %95% confidence interval
```

■ **EXAMPLE 15.7 (Conditional Estimator II)**

As a further illustration of the conditional Monte Carlo method, consider pricing a **down-and-in** call option with a discretely monitored barrier [7]. This option is also known as a discretely monitored knock-in call; see Table 15.2. The stock price evolution under the risk-neutral probability measure $\mathbb{Q}$ is a geometric Brownian motion with drift r and volatility σ. The monitoring instants are $0 = t_0 < t_1 < \cdots < t_n = T$ and the option price is

$$\mathrm{e}^{-rT}\mathbb{E}_{\mathbb{Q}}\left[(S_T - K)^+\,\mathrm{I}_{\{\tau_\beta \leqslant T\}}\right],$$

where β is the barrier and $\tau_\beta = \min_i\{t_i : S_{t_i} \leqslant \beta\}$ is the first monitoring time at which the barrier is breached. There is no closed formula for the price of the option, but by conditioning on $\mathcal{H} = \{S_{t_0}, S_{t_1}, \ldots, S_{\tau_\beta}\}$, we have

$$\begin{aligned}
\mathrm{e}^{-rT}\mathbb{E}_{\mathbb{Q}}\left[(S_T - K)^+\mathrm{I}_{\{\tau_\beta \leqslant T\}}\right] &= \mathrm{e}^{-rT}\mathbb{E}_{\mathbb{Q}}\left[\mathbb{E}_{\mathbb{Q}}\left[(S_T - K)^+\,\mathrm{I}_{\{\tau_\beta \leqslant T\}}\,|\,\mathcal{H}\right]\right] \\
&= \mathrm{e}^{-rT}\mathbb{E}_{\mathbb{Q}}\left[\mathbb{E}_{\mathbb{Q}}[(S_T - K)^+\,|\,\mathcal{H}]\,\mathrm{I}_{\{\tau_\beta \leqslant T\}}\right] \\
&= \mathbb{E}_{\mathbb{Q}}\left[\mathrm{BS}(S_{\tau_\beta}, K, r, T - \tau_\beta, \sigma)\,\mathrm{I}_{\{\tau_\beta \leqslant T\}}\right],
\end{aligned}$$

where $\mathrm{BS}(\cdot)$ refers to the Black–Scholes formula (15.5). Thus, we can simulate N copies of

$$X = \mathrm{BS}(S_{\tau_\beta}, K, r, T - \tau_\beta, \sigma)\,\mathrm{I}_{\{\tau_\beta \leqslant T\}}$$

and take their average as an estimator of the option price. In other words, the conditional estimator is equivalent to simulating until either the barrier is crossed or the option expires. If the barrier is breached, then X is the Black–Scholes price with initial price S_{τ_β} and maturity $T - \tau_\beta$, otherwise $X = 0$.

With $N = 10^4$ we obtain a typical 95% confidence interval of $[0.355, 0.358]$, which represents an approximately hundredfold variance reduction over crude Monte Carlo. Note that the value of the down-and-in call option increases when the volatility σ increases, because large market volatility makes the event of a positive payoff more likely.

```
%down_and_in_Call_conditional_est.m
r=.07; % annual interest
sig=2; %stock volatility
K=1.1; % strike price
b=.9; % barrier
S_0=1; % initial stock price
n=180;  % number of stock price observations
T=n/365; % length of observation period (in years)
dt=T/n; % time step
% simulate N sample paths of the stock process
N=10^5; X=zeros(N,1); % number of sample path simulations
for i=1:N
    path=(r-sig^2/2)*dt+sig*sqrt(dt)*randn(1,n);
    path=cumprod([S_0,exp(path)]);
    %index of first breach of barrier
    index=find(path(1:end-1)<=b,1,'First');
    if ~isempty(index)
        tau=dt*(index-1); S_tau=path(index);
        X(i)=BS(S_tau,K,r,T-tau,sig);
    end
end
c_0=mean(X), Rel_error=std(X)/c_0/sqrt(N)
width=std(X)*norminv(0.975)/sqrt(N)
CI=[c_0-width,c_0+width]
```

For this problem it is possible to design a better but more complicated conditional estimator called the *filtered* Monte Carlo estimator, see [7].

■ EXAMPLE 15.8 (Importance Sampling)

Consider again the down-and-in call option with a discretely monitored barrier in Example 15.7. The payoff at maturity can be written as

$$H(\mathbf{Z}) = (S_{t_n} - K)^+ \, \mathrm{I}\Big\{ \min_{1 \leqslant i \leqslant n} S_{t_i} \leqslant \beta \Big\},$$

where (under the risk-neutral measure $\mathbb{Q}$)

$$S_{t_k} = S_0 \exp\left(\left\{r - \frac{\sigma^2}{2}\right\} k\,\delta + \sigma\sqrt{\delta}\sum_{i=1}^{k} Z_i\right), \tag{15.8}$$

where $\mathbf{Z} = (Z_1, \ldots, Z_n)^\top \sim \mathsf{N}(\mathbf{0}, I)$, $\delta = T/n$, and $t_k = k\,\delta$ for $k = 1, 2, \ldots, n$. Note that the event of a positive payoff can easily be rare and hence the computation of the option price is amenable to rare-event probability estimation methods such as importance sampling. To select a good importance sampling density, we use the cross-entropy (CE) method. First, we can write $\mathrm{e}^{rT} C_0 = \mathbb{E}_{\mathbb{Q}} H(\mathbf{Z})$. Thus, the minimum variance pdf for the estimation of $\mathbb{E}_{\mathbb{Q}} H(\mathbf{Z})$ is ☞ 362 ☞ 463

$$f(\mathbf{z}) = \frac{\varphi(\mathbf{z}) H(\mathbf{z})}{\mathbb{E}_{\mathbb{Q}} H(\mathbf{Z})},$$

where $\varphi(\mathbf{z})$ denotes the pdf of the standard normal random vector $\mathbf{Z}$. If we look for an optimal change of measure in the family $\mathsf{N}(\boldsymbol{\mu}, I)$, $\boldsymbol{\mu} \in \mathbb{R}^n$, then we obtain the CE optimal parameters

$$\mu_i = \frac{\mathbb{E}_{\mathbb{Q}} H(\mathbf{Z})\, Z_i}{\mathbb{E}_{\mathbb{Q}} H(\mathbf{Z})} = \mathbb{E}_f Z_i, \quad i = 1, \ldots, n.$$

To estimate $\boldsymbol{\mu}$ we simulate approximately from f using MCMC, in particular using the hit-and-run Algorithm 6.5. This problem is solved in Example 6.6, where the estimated $\boldsymbol{\mu}$ is depicted on Figure 6.6. ☞ 242

In the following code we use the same parameters for the option as in Example 6.6, and assume that the estimated parameter $\boldsymbol{\mu}$ obtained from the execution of the code in Example 6.6 is loaded into the workspace.

We obtain a typical estimate of 1.07×10^{-7} with an estimated relative error of 6%. Thus, an option over a million units of shares is worth approximately \$ 0.1. For this example using the conditional Monte Carlo estimator in Example 15.7 gives a comparable relative error.

```
%down_and_in_Call_Importance_Sampling.m
N=10^5; X=zeros(N,1); W=X; %preallocate memory
for i=1:N
    z=mu+randn(1,n); % importance sampling
    [H,path]=down_in_call(z,dt,r,sig,S_0,K,b);
    W(i)=exp( -.5*( z*z'-sum((z-mu).^2) )); % likelihood ratio
  X(i)=W(i)*exp(-r*T)*H;
end

mean(X)
std(X)/mean(X)/sqrt(N)
```

Importance sampling where the drift and/or the volatility of the underlying SDE model is changed using methods other than the CE method are investigated in [14, 15].

15.3 SENSITIVITIES

Suppose a financial institution or bank sells a call option $C_t = C(S_t, K, r_t, T-t)$ to one of its customers. The bank is exposed to the risk that the underlying stock price S_T will be significantly above the strike price K and will have to pay the customer the amount $S_T - K$. Ideally, the bank would like to offset or minimize its exposure to the unpredictable price fluctuations of the stock price process $\{S_t\}$, and profit from a secure stream of fees and commissions generated from providing call options as a financial product to the market participants. The minimization of this risk exposure is called **hedging**.

Recall from the standard model that there exists a trading strategy $\{(\psi_t, \phi_t), t \geqslant 0\}$ so that a portfolio containing ψ_t units of bonds and ϕ_t units of risky assets will replicate the payoff of an option contract with price C_t. Loosely speaking, to hedge against its potential call option liabilities, the bank can use part of the revenue from selling the option to create and hold a portfolio worth V_t that will (approximately) match the value of C_t at any time. In other words, the bank's net liability will be $V_t - C_t \approx 0$. The reason that exact matching is impossible is that the price process $\{S_t\}$ is changing (almost) continuously, but the bank is not able to trade continuously (using $\{(\psi_t, \phi_t), t \geqslant 0\}$) in order to adjust the replicating portfolio. Instead, the bank readjusts the portfolio at small time steps using a discrete time trading strategy $\{(\psi_{t_j}, \phi_{t_j}), j = 0, 1, 2, \ldots\}$, and avoids large discrepancies between the portfolio and the option by using local approximations for the behavior of C_t.
☞ 710 A possible local approximation is based on Taylor's theorem. For example, suppose the bank can trade at time instants $\{\ldots, t-\delta, t, t+\delta, \ldots\}$, and wishes to set up a portfolio worth $V_t = \psi_t + \phi_t S_t$ that (approximately) matches the value C_t of the option at the next trading instant $t + \delta$ (effectively hedging against its option liability in the small time interval $[t, t+\delta]$). Then, matching the values $V_t = C_t$ and the derivatives $\frac{\partial V_t}{\partial S_t} = \frac{\partial C_t}{\partial S_t}$, and solving for (ψ_t, ϕ_t) creates the portfolio with $\phi_t = \frac{\partial C_t}{\partial S_t}$ units of the underlying stock and $\psi_t = C_t - \frac{\partial C_t}{\partial S_t} S_t$ units of bonds. Depending on the quality of the Taylor approximation, such a portfolio may closely replicate the value of $C_{t+\delta}$.

■ **EXAMPLE 15.9 (Hedging With the Black–Scholes Formula)**

Suppose the price of the call option C_t is given by the Black–Scholes formula (15.5). Let a_1 and a_2 be defined in (15.6), and φ and Φ be the pdf and cdf of the standard normal distribution, respectively. Then, by the chain and product rules for differentiation we have

$$\frac{\partial C_t}{\partial S_t} = \Phi(a_1) + S_t \Phi'(a_1)\frac{\partial a_1}{\partial S_t} - K\mathrm{e}^{-r(T-t)}\Phi'(a_2)\frac{\partial a_2}{\partial S_t} = \Phi(a_1)\,,$$

where the last equality follows from:

$$\begin{aligned} S_t\Phi'(a_1)\frac{\partial a_1}{\partial S_t} - K\mathrm{e}^{-r(T-t)}\Phi'(a_2)\frac{\partial a_2}{\partial S_t} &= \frac{1}{S_t\sigma\sqrt{T-t}}\Big(S_t\Phi'(a_1) - K\mathrm{e}^{-r(T-t)}\Phi'(a_2)\Big) \\ &= \frac{\varphi(a_1)}{S_t\sigma\sqrt{T-t}}\Big(S_t - K\mathrm{e}^{a_1\sigma\sqrt{T-t}-\sigma^2\frac{T-t}{2}-r(T-t)}\Big) \\ &= \frac{\varphi(a_1)}{S_t\sigma\sqrt{T-t}}\Big(S_t - K\mathrm{e}^{\ln(\frac{S_t}{K})}\Big) = 0\,. \end{aligned}$$

Thus, the bank will have to hold $\phi_t = \Phi(a_1)$ units of stock to hedge against its option liabilities.

In general, the price of an option depends not only on the initial underlying asset price, but also the volatility, the remaining time until expiry, and the interest rate. For various hedging strategies it is of interest to compute the *sensitivity* of the option price to each of these parameters. Table 15.3 describes the nomenclature used for the various partial derivatives and a short financial interpretation of the mathematical formula. These sensitivities are named after Greek letters and are collectively referred to as **Greeks**. The Greeks for the Black–Scholes model in Example 15.1 can be calculated analytically and are given in Table 15.4. ☞ 421

Table 15.3 The Greeks of a call option.

Name	Formula	Description
Delta	$\Delta_t = \frac{\partial C_t}{\partial S_t}$	Measures the sensitivity of the option's price to the underlying stock price.
Gamma	$\Gamma_t = \frac{\partial^2 C_t}{\partial S_t^2}$	Measures the sensitivity of the option's Delta to the underlying stock price.
Theta	$\Theta_t = -\frac{\partial C_t}{\partial t}$	Measures the sensitivity of the option's price to the time remaining until expiry.
Vega	$\nu_t = \frac{\partial C_t}{\partial \sigma_t}$	Measures the sensitivity of the option's price to the volatility.
Rho	$\varrho_t = \frac{\partial C_t}{\partial r_t}$	Measures the sensitivity of the option's price to the interest rate.

Table 15.4 The Greeks for a call and put option under the Black–Scholes model in Example 15.1.

Name	Call	Put
Delta	$\Phi(a_1)$	$\Phi(a_1) - 1$
Gamma	$\frac{\varphi(a_1)}{S_t\, \sigma\, \sqrt{T-t}}$	Same as Gamma of call
Theta	$S_t\, \varphi(a_1) \frac{\sigma}{2\sqrt{T-t}} + rK\mathrm{e}^{-r(T-t)}\Phi(a_2)$	$S_t\varphi(a_1) \frac{\sigma}{2\sqrt{T-t}} - rK\mathrm{e}^{-r(T-t)}\Phi(-a_2)$
Vega	$S_t\, \varphi(a_1)\sqrt{T-t}$	Same as Vega of call
Rho	$(T-t)K\mathrm{e}^{-r(T-t)}\Phi(a_2)$	$-(T-t)K\mathrm{e}^{-r(T-t)}\Phi(-a_2)$

For models other than the Black–Scholes model, accurate computation of the sensitivities presents a challenge, which is addressed (to some extent) by Monte

Carlo techniques. Next, we describe two of the most common techniques. For a detailed discussion see [14].

15.3.1 Pathwise Derivative Estimation

☞ 426 In the financial engineering literature the infinitesimal perturbation analysis (IPA) discussed in Section 11.3 is usually referred to as **pathwise derivative estimation**. Suppose we wish to estimate $\ell'(\theta) = \frac{\mathrm{d}\ell}{\mathrm{d}\theta}(\theta)$, where

$$\ell(\theta) = \mathbb{E}_f H(\mathbf{X};\theta) = \int H(\mathbf{X};\theta) f(\mathbf{x})\, \mathrm{d}\mathbf{x}\,, \quad \theta \in \Theta\,, \tag{15.9}$$

for some open set Θ. In a financial setting H could be the payoff of an option and $f(\mathbf{x})$ a multivariate Gaussian pdf that is used to approximate the underlying Wiener process. Here the parameter θ is a structural sensitivity parameter (Section
☞ 421 11.1), and therefore we can consider the pathwise derivative estimator

$$\widehat{\ell'}(\theta) = \frac{1}{N}\sum_{k=1}^{N} H'(\mathbf{X}_k;\theta), \quad \mathbf{X}_1,\ldots,\mathbf{X}_N \stackrel{\text{iid}}{\sim} f\,,$$

where it is assumed that for a random $\mathbf{X}$ the derivative

$$H'(\mathbf{X};\theta) = \frac{\mathrm{d}}{\mathrm{d}\theta} H(\mathbf{X};\theta)$$

exists almost surely for all $\theta \in \Theta$. The multidimensional version of this estimator is given in (11.8). If the payoff of the option is smooth enough so that the expectation operator and the derivative operator can be interchanged (see (11.7)), that is,

$$\frac{\mathrm{d}}{\mathrm{d}\theta}\mathbb{E}_f H(\mathbf{X};\theta) = \mathbb{E}_f \frac{\mathrm{d}}{\mathrm{d}\theta} H(\mathbf{X};\theta)\,,$$

then the pathwise gradient estimator is consistent and unbiased. A sufficient con-
☞ 422 dition for this is given in Theorem 11.1.1, which essentially requires the uniform integrability of $(H(\mathbf{X};\theta+\varepsilon) - H(\mathbf{X};\theta))/\varepsilon$, $\varepsilon > 0$. Alternative sufficient conditions are given in [14, Page 393] and [7]. For most practical situations a simple rule of thumb that guarantees the consistency and unbiasedness of the estimator is that the option payoff is continuous in the parameter of interest. Such smoothness conditions are frequently quite restrictive. For example, the payoff of an all-or-nothing call option (see Table 15.2) is not continuous. Thus, the pathwise derivative estimates are not broadly applicable. However, whenever they are, these estimates are generally the most accurate among possible alternatives. An alternative method that applies more broadly is given in the next section.

■ **EXAMPLE 15.10 (Pathwise Derivative Estimation)**

Consider estimating the Delta of the Asian call option in Example 15.3 with payoff $(\bar{S}_T - K)^+$. The pathwise derivative of the payoff is

$$\frac{\mathrm{d}C_T}{\mathrm{d}S_0} = \frac{\mathrm{d}}{\mathrm{d}\bar{S}_T}(\bar{S}_T - K)^+ \frac{\mathrm{d}\bar{S}_T}{\mathrm{d}S_0} = \mathrm{I}_{\{\bar{S}_T > K\}} \frac{\bar{S}_T}{S_0}\,,$$

where we have used the fact that $\frac{\mathrm{d}S_{t_i}}{\mathrm{d}S_0} = S_{t_i}/S_0$. Similarly, the pathwise derivative for estimating the Vega is

$$\frac{\mathrm{d}C_T}{\mathrm{d}\sigma} = \mathrm{I}_{\{\bar{S}_T > K\}} \frac{\mathrm{d}\bar{S}_T}{\mathrm{d}\sigma} = \frac{\mathrm{I}_{\{\bar{S}_T > K\}}}{n+1} \sum_{k=0}^{n} S_{t_k} \left(-\sigma t_k + \sqrt{\delta} \sum_{i=1}^{k} Z_i \right),$$

where S_{t_k} is defined in (15.8). For convenience of implementation in the code that follows, we use the identity

$$-\sigma t_k + \sqrt{\delta} \sum_{i=1}^{k} Z_i = \frac{\ln(S_{t_k}/S_0) - (r + \sigma^2/2)t_k}{\sigma} .$$

With a simulation effort of $N = 10^6$ and using the same parameters as in Example 15.3 we obtain typical 95% confidence intervals of $[0.9758, 0.9763]$ and $[0.38, 0.43]$ for the Delta and Vega, respectively. Note that the confidence interval for the Vega is much wider than the confidence interval for the Delta. Intuitively this is due to the fact that only the distribution of S_{t_1} depends directly on S_0, and, given S_{t_1}, subsequent values of the stock price are independent of S_0. However, every $S_{t_k}, k \geqslant 1$ depends on σ. Thus, the estimator for Vega is much more variable. For a survey article which compares many methods for evaluating sensitivities of Asian options see [8].

```
% sensitivities_pathwise_Crude_Monte_Carlo.m
r=.07; % annual interest
sig=0.2; % volatility
K=35; % strike price
S_0=40; % initial stock price
T=4/12; %maturity in 4 months, which is 4/12 of the year
n= 88;  % there are approx. 88 trading days in 4 months
dt=T/n; % time step
% Simulate N sample paths of the stock process
N=10^6; Delta=nan(N,1);Vega=Delta;
for i=1:N
    path=(r-sig^2/2)*dt+sig*sqrt(dt)*randn(1,n);
    path=cumprod([S_0,exp(path)]);
    A=mean(path); % average path/stock price
    Delta(i)=exp(-r*T)*(A>=K)*A/S_0;
    Vega(i)=exp(-r*T)*(A>=K)*...
     sum( path.*(log(path/S_0)-(r+.5*sig^2)*[0:dt:T]))/(n+1)/sig;
end
D=mean(Delta); Rel_err=std(Delta)/sqrt(N)
width=std(Delta)*norminv(0.975)/sqrt(N);
CI_D=[D-width, D+width]
V=mean(Vega); Rel_err=std(Vega)/sqrt(N)
width=std(Vega)*norminv(0.975)/sqrt(N);
CI_V=[V-width, V+width]
```

15.3.2 Score Function Method

☞ 423 As pointed out in Remark 11.1.1, it is often possible to switch from a structural interpretation of the parameter θ to a distributional interpretation, and vice versa. In other words, we can frequently rewrite (15.9) as

$$\ell(\theta) = \int H(\mathbf{x}) f(\mathbf{x};\theta)\,\mathrm{d}\mathbf{x}\,,$$

and therefore if the interchange of the differential and expectation operators in (11.10) is justified, then

$$\ell'(\theta) = \int f(\mathbf{x};\theta) H(\mathbf{x}) \frac{\frac{\mathrm{d}f}{\mathrm{d}\theta}(\mathbf{x};\theta)}{f(\mathbf{x};\theta)} \mathrm{d}\mathbf{x} = \mathbb{E}_\theta H(\mathbf{X})\, \mathcal{S}(\theta;\mathbf{X})\,.$$

Sufficient conditions on the pdf f that ensure the validity of the interchange are given by Theorem 11.1.1. In contrast to the pathwise derivative estimation setting, where restrictions on the payoff function H are imposed, here the payoff function is free from any restrictions, and all smoothness requirements are imposed on the pdf f. Thus, the score function method is much more broadly applicable than the pathwise derivative estimation approach. For cases where the score function method is not applicable, see [14, Page 407]. Assuming that the interchange in (11.10) is justified, we can then use the crude Monte Carlo estimator

$$\ell'(\theta) = \frac{1}{N}\sum_{k=1}^{N} H(\mathbf{X}_k)\, \mathcal{S}(\theta;\mathbf{X}_k)\,, \quad \mathbf{X}_1,\ldots,\mathbf{X}_N \stackrel{\text{iid}}{\sim} f(\cdot;\theta)\,.$$

However, this estimator is usually not efficient. For example, for barrier options the payoff H could be 0 almost always. In Example 15.11 we therefore use importance sampling to improve the efficiency. For a general discussion of the score function
☞ 431 method combined with importance sampling, see Section 11.4.1 and Algorithm 11.4.

In summary, the advantages of the score function method are:

- It is more broadly applicable than the pathwise approach, because it does not impose any smoothness conditions on the payoff.

- Once the score $\mathcal{S}(\theta;\mathbf{X}_k)$ is computed, the specific form of the payoff H is irrelevant and we could use the same score to compute the same sensitivity for options with different payoffs.

Disadvantages of the score function method include:

- It requires explicit knowledge of the pdf of the stock price path.

- It typically gives estimators with larger variance than the variance of the pathwise derivative estimators. More importantly, the variance of the score function estimator may be infinite.

■ **EXAMPLE 15.11 (Score Function Method With Importance Sampling)**

Consider pricing the call option with discretely monitored barrier in Example 15.8 using importance sampling. Abbreviating S_{t_i} to S_i for notational convenience, we can write the value of the option as

$$\mathrm{e}^{-rT}\mathbb{E}_f H(\mathbf{Z}) = \mathrm{e}^{-rT}\mathbb{E}_p\big[(S_n - K)^+ \, \mathrm{I}_{\{\min_i S_i \leqslant \beta\}}\big] ,$$

where the joint density of the stock price path $(S_1, \ldots, S_n)$ can be factorized as

$$p(s_1, \ldots, s_n) = p(s_1 \,|\, s_0)\, p(s_2 \,|\, s_1) \cdots p(s_n \,|\, s_{n-1}) .$$

Here

$$p(s_i \,|\, s_{i-1}) = \frac{1}{s_i\, \sigma \sqrt{\delta}}\, \varphi\left(\frac{\ln(s_i/s_{i-1}) - (r - \sigma^2/2)\delta}{\sigma\sqrt{\delta}}\right), \quad i = 1, \ldots, n ,$$

where φ is the pdf of the standard normal distribution. In other words, $p(s_i \,|\, s_{i-1})$ is the pdf of the

$$\mathsf{LogN}\left(\ln(s_{i-1}) + \left(r - \frac{\sigma^2}{2}\right)\delta,\ \delta\,\sigma^2\right)$$

distribution and $\mathbf{Z} = (Z_1, \ldots, Z_n)$ is related to $(S_1, \ldots, S_n)$ via (15.8). Then, the score for computing the Delta of the option is

$$\frac{\partial}{\partial s_0} \ln p(s_1, \ldots, s_n) = \sum_{i=1}^{n} \frac{\partial}{\partial s_0} \ln p(s_i \,|\, s_{i-1}) = \frac{z_1}{s_0\, \sigma \sqrt{\delta}} ,$$

and the score for computing the Vega of the option is

$$\frac{\partial}{\partial \sigma} \ln p(s_1, \ldots, s_n) = \sum_{i=1}^{n} \frac{\partial}{\partial \sigma} \ln p(s_i \,|\, s_{i-1}) = \sum_{i=1}^{n} \frac{z_i^2 - 1}{\sigma} - z_i \sqrt{\delta} .$$

If we abbreviate the score for the desired sensitivity (Delta, Vega, or other sensitivities) to $\mathcal{S}(\mathbf{z})$, then the crude Monte Carlo estimator of the desired sensitivity is

$$\frac{\mathrm{e}^{-rT}}{N} \sum_{i=1}^{N} H(\mathbf{Z}_i)\, \mathcal{S}(\mathbf{Z}_i), \quad \mathbf{Z}_1, \ldots, \mathbf{Z}_N \stackrel{\text{iid}}{\sim} f(\mathbf{z}) ,$$

which is not an efficient estimator for the payoff of barrier options. Therefore, similar to Example 15.8 we use the hit-and-run Algorithm 6.5 to estimate the parameter $\boldsymbol{\mu}$ of the optimal importance sampling density in the family of multivariate normal distributions $\mathsf{N}(\boldsymbol{\mu}, I)$. In other words, we estimate the mean $\boldsymbol{\mu}$ of the pdf

$$f(\mathbf{z}) \propto \mathrm{e}^{-\mathbf{z}^\top \mathbf{z}/2} H(\mathbf{z})\, |\mathcal{S}(\mathbf{z})| ,$$

by $\widehat{\boldsymbol{\mu}}$ and then use $\mathsf{N}(\widehat{\boldsymbol{\mu}}, I)$ as the (near) optimal importance sampling density. The code below implements this importance sampling scheme for a given set of option parameters. The hit-and-run sampler is run for 10^5 iterations and the importance sampling estimator is based on an iid sample of size 10^6. We obtain a typical estimate of the Delta of -4.20×10^{-6} with an estimated relative error of 6.5%.

The code can be modified to obtain an estimate of the Vega. A typical estimate is 1.28×10^{-5} with an estimated relative error of 4%.

```
% sensitivities_IS_with_CE.m
r=.07; % annual interest
sig=.2; %stock volatility
K=1.2; % strike price
b=.8; % barrier
S_0=1; % initial stock price
n=180;  % number of stock price observations
T=n/365; % length of observation period (in years)
dt=T/n; % time step
N=10^5; % length of chain
x=[-ones(1,60),ones(1,n-60)]*0.4; % initial starting point
[HS,path]=payoff_times_score(x,dt,r,sig,S_0,K,b);%evaluate H(x)*S(x)

% now we simulate N sample paths of the stock process
% and compute  averages
mu=0;paths=0;
for i=1:N
    % apply hit-and-run
    d=randn(1,n); d=d/norm(d);
    lam=-d*x'+randn;
    y=x+lam*d; % make proposal
    % evaluate H(y)
    [HS_new,path_new]=payoff_times_score(y,dt,r,sig,S_0,K,b);
    % accept or reject the proposal
    if rand<min(abs(HS_new/HS),1)
        x=y;    % update
        HS=HS_new;
        path=path_new;
    end
    mu=mu+x/N;              % compute an estimate of E[X]
    paths=paths+path/N; % compute average stock price trajectory
    if mod(i,2*10^4)==0  % plot every 10^3-th step of the chain
        plot(0:dt:T,path,0:dt:T,0*path+b,0:dt:T,0*path+K)
        axis([0,T,b-0.1,K+.2]),hold all
        pause(.1)
    end
end
plot(0:dt:T,paths,'r','LineWidth',3) %plot average price trajectory
figure(2)
plot(mu,'k.'), hold on
% smooth the trajectory of E[X] using a spline
pp = csaps(dt:dt:T,mu,1/(1+(dt*10)^3));
mu_t = fnval(pp,[dt:dt:T]);
plot(mu_t,'r') %plot the smoothed trajectory
```

```
% the importance sampling estimator starts here
N=10^6; X=zeros(N,1); W=X; % number of sample path simulations
for i=1:N
    z=mu+randn(1,n);
    [HS,path]=payoff_times_score(z,dt,r,sig,S_0,K,b);
    W(i)=exp(-.5*( z*z'-sum((z-mu).^2)  )); % likelihood ratio
    X(i)=W(i)*exp(-r*T)*HS;
end
mean(X)
std(X)/mean(X)/sqrt(N)
```

The hit-and-run implementation uses the following function.

```
function [HS,S_t]=payoff_times_score(z,dt,r,sig,S_0,K,b)
% implements H(x)*S(x)
y=(r-sig^2/2)*dt+sig*sqrt(dt)*z;
S_t=exp(cumsum([log(S_0),y]));
%sensitivity_factor=z(1)/(S_0*sig*sqrt(dt)); % for Vega
sensitivity_factor=sum((z.^2-1)/sig-z*sqrt(dt)); % for Delta
HS=max(S_t(end)-K,0)*sensitivity_factor*any(S_t<=b);
```

■ **EXAMPLE 15.12 (Estimation of Gamma)**

For estimating higher-order derivatives it is possible to construct estimators that combine both the likelihood ratio and pathwise approaches. For example, to estimate the Gamma of a European call option with discounted payoff $\mathrm{e}^{-rT}(S_T - K)^+$ under the Black–Scholes stock price model, we may first construct the likelihood ratio estimator

$$X = \mathrm{e}^{-rT}(S_T - K)^+ \frac{Z_1}{S_0\,\sigma\sqrt{\delta}}\,,$$

and then take the pathwise derivative (recall that $\frac{\mathrm{d}S_T}{\mathrm{d}S_0} = S_T/S_0$):

$$\frac{\mathrm{d}X}{\mathrm{d}S_0} = \mathrm{e}^{-rT}\frac{Z_1}{S_0^2\,\sigma\,\sqrt{\delta}}\,\mathrm{I}_{\{S_T>K\}}\,K\,.$$

Then, the crude Monte Carlo estimator is the average of N realizations of $\frac{\mathrm{d}X}{\mathrm{d}S_0}$. Such mixed estimators are typically much more accurate than pure likelihood ratio estimators [14].

Finally, we mention that the finite difference method described in Section 11.3 ☞ 426 is also sometimes used in the estimation of the sensitivities of options, and it is the easiest method to implement. However, under similar conditions that make the pathwise derivative method inapplicable, the finite difference method yields estimators with high mean square error. Thus, using the finite difference method does not circumvent the smoothness problems of the payoff function. In general, while easy to implement, the finite difference method gives estimators that are biased and inefficient. We recommend its use under the rare circumstances when the payoff fails to be almost surely continuous in the parameter of interest and the relevant densities driving the asset prices are not explicitly known (so that the score function method is not applicable). For a detailed discussion of sensitivity estimation we refer to [8] and the references therein.

Further Reading

American style options are significantly more difficult to price than their European counterparts. For the computational challenges associated with American options we refer to [10, 14]. An undergraduate level presentation of deterministic and Monte Carlo methods in option pricing that requires minimal knowledge in stochastic calculus is [4]. Graduate level texts with practical numerical examples are [3] and [28]. These authors cover deterministic PDE methods such as finite difference and finite element methods for pricing financial derivatives. Texts that focus more on the theoretical aspects of financial mathematics with rigorous coverage of continuous-time stochastic calculus include [12] and [18]. A classic research monograph on pricing exotic options in complete and incomplete markets is [22]. Texts with an emphasis on Monte Carlo methods include [14, 19], and the handbook of mathematical finance [21]. In addition, McLeish [25] provides simple examples with MATLAB code. Financial models based on nonhomogeneous Lévy jump processes with pricing and hedging theory in incomplete markets are covered in [9, 20]. For the application of quasi Monte Carlo methods in finance, see, for example, [24].

REFERENCES

1. K. Abramowicz and O. Seleznjev. On the error of the Monte Carlo pricing method for Asian option. *Journal of Numerical and Applied Mathematics*, 96(1):1–10, 2008.
2. A. N. Avramidis and P. L'Ecuyer. Efficient Monte Carlo and quasi-Monte Carlo option pricing under the variance gamma model. *Management Science*, 52(12):1930–1944, 2006.
3. K. Back. *A Course in Derivative Securities: Introduction to Theory and Computation.* Springer-Verlag, Berlin, 2005.
4. F. E. Benth. *Option Theory With Stochastic Analysis: An Introduction to Mathematical Finance.* Springer-Verlag, Berlin, 2004.
5. N. H. Bingham and R. Kiesel. *Risk-Neutral Valuation.* Springer-Verlag, London, second edition, 2004.
6. P. Boyle. Options: A Monte Carlo approach. *Journal of Financial Economics*, 4(3):323–338, 1977.
7. P. Boyle, M. Broadie, and P. Glasserman. Monte Carlo methods for security pricing. *Journal of Economic Dynamics and Control*, 21(8 & 9):1267–1321, 1997.
8. P. Boyle and A. Potapchik. Prices and sensitivities of Asian options: A survey. *Insurance: Mathematics and Economics*, 42(1):189 – 211, 2008.
9. R. Cont and P. Tankov. *Financial Modelling With Jump Processes.* Chapman & Hall, Boca Raton, FL, 2004.

10. J. Detemple. *American-style Derivatives: Valuation and Computation.* Chapman & Hall/CRC Financial Mathematics Series. CRC Press, Boca Raton, FL, 2006.

11. D. Duffie. *Dynamic Asset Pricing Theory.* Princeton University Press, Princeton, third edition, 2001.

12. R. J. Elliot and P. E. Kopp. *Mathematics of Financial Markets.* Springer-Verlag, New York, second edition, 2005.

13. G. Fusai and A. Meucci. Pricing discretely monitored Asian options under Lévy processes. *Journal of Banking & Finance*, 32(10):2076–2088, 2008.

14. P. Glasserman. *Monte Carlo Methods in Financial Engineering.* Springer-Verlag, New York, 2004.

15. P. Glasserman, P. Heidelberger, and P. Shahabuddin. Asymptotically optimal importance sampling and stratification for path-dependent options. *Mathematical Finance*, 9(2):117 – 152, 1999.

16. J. M. Harrison and D. Kreps. Martingales and arbitrage in multiperiod securities markets. *Journal of Economic Theory*, 20(3):381–408, 1979.

17. J. C. Hull. *Options, Futures, and Other Derivatives.* Pearson/Prentice Hall, London, seventh edition, 2008.

18. P. J. Hunt and J. E. Kennedy. *Financial Derivatives in Theory and Practice.* John Wiley & Sons, Chichester, second edition, 2004.

19. P. Jäckel. *Monte Carlo Methods in Finance.* John Wiley & Sons, New York, 2002.

20. M. S. Joshi. *The Concepts and Practice and Mathematical Finance.* Cambridge University Press, Cambridge, 2003.

21. E. Jouini, J. Cvitanić, and M. Musiela. *Option Pricing, Interest Rates and Risk Management.* Cambridge University Press, Cambridge, 2001.

22. I. Karatzas and S. E. Shreve. *Methods of Mathematical Finance.* Springer-Verlag, New York, 1998.

23. A. G. Z. Kemna and A. C. F. Vorst. A pricing method for options based on average asset values. *Journal of Banking & Finance*, 14(1):113–129, 1990.

24. P. L'Ecuyer. Quasi-Monte Carlo methods with applications in finance. *Finance and Stochastics*, 13(3):307–349, 2009.

25. D. L. McLeish. *Monte Carlo Simulation and Finance.* John Wiley & Sons, New York, 2005.

26. M. Musiela and M. Rutkowski. *Martingale Methods in Financial Modelling.* Springer-Verlag, New York, second edition, 2005.

27. J. A. Nielsen and K. Sandmann. A pricing method for options based on average asset values. *Journal of Financial and Quantitative Analysis*, 38(2):449–473, 2003.

28. R. U. Seydel. *Tools for Computational Finance.* Springer-Verlag, Berlin, fourth edition, 2009.

29. M. Vanmaele, G. Deelstra, J. Liinev, J. Dhaene, and M. J. Goovaerts. Bounds for the price of discrete arithmetic Asian options. *Journal of Computational and Applied Mathematics*, 185(1):51–90, 2006.

CHAPTER 16

APPLICATIONS TO NETWORK RELIABILITY

Network reliability problems arise in many areas of engineering and computer science, such as telecommunications, transportation, energy supply systems, and computer networking. In this chapter we survey some of the most widely used algorithms, with emphasis on those that are easy to implement:

1. The *permutation Monte Carlo* method — a conditional Monte Carlo approach for estimating network unreliabilities, in cases where the edges of the network fail independently; ☞ 354

2. The *importance sampling* method, combined with MCMC sampling; ☞ 362

3. The *generalized splitting* method for estimating network unreliabilities in cases where the edge failures are dependent. ☞ 485

16.1 NETWORK RELIABILITY

Let $\mathcal{G}(V, E, K)$ be an undirected graph (network) with V being the set of n nodes (or vertices), E being the set of edges (or links), and $K \subseteq V$ being a set of **terminal nodes** such that $|K| \geqslant 2$.

Associated with each edge $e \in E$ is a Bernoulli random variable B_e such that $\{B_e = 1\}$ corresponds to the event that the edge is operational (on) and $\{B_e = 0\}$ corresponds to the event that the edge has failed (off). Then, labeling the edges

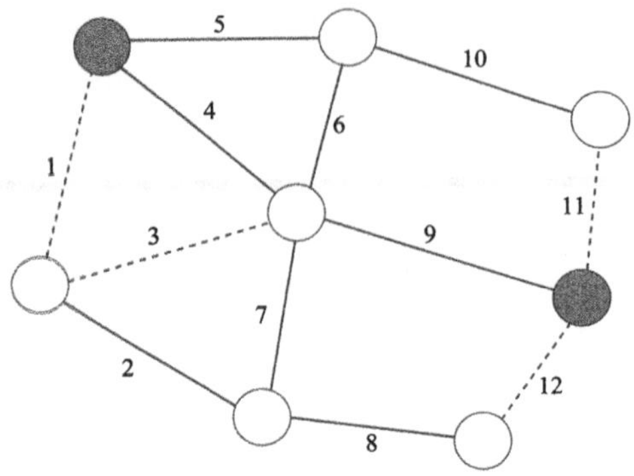

Figure 16.1 A reliability network. The network works if the two terminal nodes (filled circles) are connected by functioning links. Failed links are indicated by dashed lines.

from 1 to m, the vector $\mathbf{B} = (B_1, \ldots, B_m)$ describes one of the 2^m possible **configurations** of such a **reliability network**. Figure 16.1 shows a reliability network with two terminal nodes and $m = 12$ edges. The configuration in Figure 16.1 is $(0, 1, 0, 1, 1, 1, 1, 1, 1, 1, 0, 0)$. Denote the set of all possible configurations by $\mathscr{B}$. Let $f_{\mathbf{B}}(\mathbf{b}) = \mathbb{P}(\mathbf{B} = \mathbf{b})$ be the discrete pdf of the binary vector $\mathbf{B}$. We assume that each B_e is a Bernoulli variable with

$$\mathbb{P}(B_e = 1) = p_e = 1 - q_e, \quad e = 1, \ldots, m .$$

The **reliability** $\widetilde{\ell}$ of the network is defined as the probability that the nodes in K are connected by operational edges. Therefore, we can write

$$\widetilde{\ell} = \mathbb{E} H(\mathbf{B}) = \sum_{\mathbf{b} \in \mathscr{B}} f_{\mathbf{B}}(\mathbf{b}) \, H(\mathbf{b}) ,$$

where $H(\mathbf{b}) = 1$ if the nodes in K are connected and $H(\mathbf{b}) = 0$, otherwise. The function $H(\mathbf{b})$ is usually referred to as the **structure function** of the network (see, for example, [4]), and $\ell = 1 - \widetilde{\ell}$ is called the **unreliability** of the network, which is typically a rare-event probability. Two special cases of interest include:

1. *Two-terminal network reliability*: Here $|K| = 2$ and the two nodes of interest are called the **source** and **sink**.

2. *All-terminal network reliability*: Here $K = V$; that is, we wish to estimate the probability that all nodes in the corresponding graph are connected.

Since ℓ is typically very small, the crude Monte Carlo estimator

$$\widehat{\ell} = 1 - \frac{1}{N} \sum_{e=1}^{N} H(\mathbf{B}_e), \quad \mathbf{B}_1, \ldots, \mathbf{B}_N \stackrel{\text{iid}}{\sim} f_{\mathbf{B}}(\mathbf{b}) ,$$

which has a relative error of $\sqrt{(1 - \ell)/(N\ell)}$, is impractical. The next section considers an alternative formulation of the problem, in which the static simulation model above is reformulated as a dynamic simulation model.

16.2 EVOLUTION MODEL FOR A STATIC NETWORK

A different formulation of the network reliability estimation problem views the static network as a snapshot of a dynamic one at a particular point in time. This approach is adopted in the **evolution model** pioneered by Elperin et al. [14, 15] and Lomonosov [29]. The dynamic reformulation of the static model leads to a conditional Monte Carlo estimator for small unreliabilities that uses theory from Markov processes. Here we present an extended version of the idea showing that the evolution approach can be viewed as a data augmentation technique. This extension is useful when implementing the variance reduction techniques in subsequent sections.

Let $\mathbf{X} = (X_1, \ldots, X_m)$ be a vector of continuous latent variables with joint density $f(\mathbf{x})$ on $\mathscr{X} \subseteq \mathbb{R}^m$ such that each component B_e in $\mathbf{B}$ can be expressed in terms of these latent variables via

$$B_e = \mathrm{I}_{\{X_e \leqslant 1\}}, \quad e = 1, \ldots, m \,.$$

Hence, we have

$$\mathbb{P}(X_e > 1) = \mathbb{P}(B_e = 0) = 1 - p_e = q_e, \quad e = 1, \ldots, m \,,$$

and

$$\ell = 1 - \mathbb{E}_{f_{\mathbf{B}}} H(\mathbf{B}) = 1 - \mathbb{E}_f H(\widetilde{\mathbf{B}}(\mathbf{X}))\,,$$

where $\widetilde{\mathbf{B}}(\cdot)$ is a function from $\mathscr{X}$ to $\mathscr{B}$. If the $\{B_e\}$ are independent, then $f(\mathbf{x}) = \prod_{e=1}^m f_e(x_e)$ for some pdfs $\{f_e\}$ satisfying $\int_1^\infty f_e(x)\,\mathrm{d}x = q_e$, $e = 1, \ldots, m$.

It is useful to interpret each continuous variable X_e as a random **time of repair** of the e-th edge, so that at some time t_0 in the past (possibly $-\infty$ or 0) the e-th edge is in a failed state and becomes operational at time X_e. Thus, $\{X_e > 1\}$ is interpreted as the event that the time of repair of the e-th edge exceeds 1 and becomes operational at a time later than 1.

Let $S(\mathbf{X})$ denote the random time at which the network becomes operational, given the vector of times of repair $\mathbf{X}$. In the *two-terminal* case the network becomes operational if there is an (edge-)path between the source and sink; that is, a sequence of nonidentical edges $(v_0, v_1), (v_1, v_2), \ldots, (v_{k-1}, v_k)$ where v_0 denotes the source and v_k the sink. For example, in Figure 16.1 two possible paths are $\mathcal{P}_1 = (1, 2, 7, 4, 5, 10, 11)$ and $\mathcal{P}_2 = (5, 10, 11)$. Here we use a labeling of the edges rather than identifying the edges via vertex pairs. A path is said to be a **minimal path** if it does not contain another (shorter) path. For example, $\mathcal{P}_2$ is a minimal path, but $\mathcal{P}_1$ is not. A **cut** in a graph is a set of edges such that, if deleted from the graph, the terminal nodes become disconnected. For example, the set $\{1, 4, 5\}$ of edges on Figure 16.1 is a cut. Its removal disconnects the terminal nodes. A **minimal cut** is a cut that does not contain another (smaller) cut. Using these concepts we can formally write the time of repair of the system, $S(\mathbf{X})$, as

$$S(\mathbf{X}) = \min_{\mathcal{P} \in \mathscr{P}} \max_{e \in \mathcal{P}} X_e = \max_{\mathcal{C} \in \mathscr{C}} \min_{e \in \mathcal{C}} X_e\,, \tag{16.1}$$

where

- $\mathscr{P} = \{\mathcal{P}_j\}$ is the set of all minimal paths between the source and sink; each $\mathcal{P}_j$, $j = 1, \ldots, |\mathscr{P}|$, represents a sequence of edges connecting the source and the sink;

- $\mathscr{C} = \{\mathcal{C}_j\}$ is the set of all minimal cuts in the graph; each $\mathcal{C}_j$, $j = 1, \ldots, |\mathscr{C}|$, is a set of nodes describing a minimal cut on the graph.

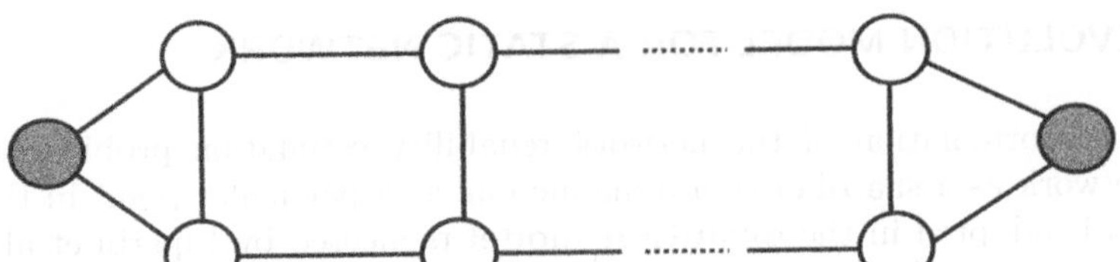

Figure 16.2 Network with ladder topology. The source and sink are filled circles.

In the case of *all terminal* reliability, paths are replaced by spanning trees [31, 32].

In terms of the vector $\mathbf{X}$ of times of repair, the unreliability of the system can be written as

$$\ell = \mathbb{P}(S(\mathbf{X}) > 1), \quad \mathbf{X} \sim f(\mathbf{x}). \tag{16.2}$$

In other words, if we imagine a dynamic network whose state changes over time because its edges are being repaired, then the state of the static network is a snapshot of the dynamic one at time 1.

The number of possible minimal paths or minimal cuts can grow exponentially with the size of the graph [32]. For example, for the *ladder* topology in Figure 16.2 with m edges we have $2^{(m+1)/3}$ possible paths connecting the source and the sink. Note that there are $(m+1)^2/9$ minimal cuts. For a complete graph with n nodes, the number of spanning trees connecting all nodes is n^{n-2} and there are $2^{n-1}-1$ minimal cuts. This large number of cuts and paths makes the direct evaluation of $S(\mathbf{X})$ via (16.1) impractical. Instead, the value $S(\mathbf{X})$ can be computed using the following simple algorithm, which uses a *depth first search* [25].

Algorithm 16.1 (Computation of $S(\mathbf{X})$) *Given a vector $\mathbf{X} = (X_1, \ldots, X_m)$ of times of repair and the network $\mathcal{G}(V, E, K)$, set $b = 1$ and execute the following steps.*

1. *Let $\boldsymbol{\pi} = (\pi_1, \ldots, \pi_m)$ be the permutation of the edges $1, \ldots, m$ such that*

$$X_{\pi_1} < X_{\pi_2} < \cdots < X_{\pi_m}.$$

2. *Consider the network $\mathcal{G}$ in which the edges $\pi_1, \ldots, \pi_b$ are working and the rest, $\pi_{b+1}, \ldots, \pi_m$, are failed.*

3. *Use depth first search to check if the network is operational. If the network is operational stop and go to Step 4; otherwise, increment $b = b + 1$ and repeat from Step 2.*

4. *Output $S(\mathbf{X}) = X_{\pi_b}$ as the time at which the network becomes operational.*

We call the b for which $S(\mathbf{X}) = X_{\pi_b}$ the **critical number** for vector $\mathbf{X}$. Note that b is a function of the permutation $\boldsymbol{\pi}$. If we implement the depth first search using an adjacency matrix, see [25], and update the adjacency matrix every time a new link is added, the complexity is $\mathcal{O}(n^2 + m)$. The MATLAB function that follows implements the algorithm for the case where the network is operational if node 1 is connected to node n. The topology of the graph is passed to the function via the global variable `GRAPH`, which is a structure array with field names describing all relevant information about a graph object. In particular, the structure field name E is accessed via the command E=GRAPH.E and contains an $m \times 2$ matrix E that indicates which edges of the graph are operational. For example, the graph in Figure 16.3 on Page 558 is described by the 18×2 matrix

$$E = \begin{pmatrix} 1 & 1 & 1 & 2 & 2 & 3 & 4 & 5 & 6 & 7 & 7 & 7 & 8 & 9 & 10 & 11 & 12 & 13 \\ 2 & 3 & 4 & 5 & 6 & 7 & 7 & 8 & 8 & 9 & 10 & 11 & 9 & 13 & 14 & 12 & 14 & 14 \end{pmatrix}^{\top} . \quad (16.3)$$

```
function [Sx,b,perm]=S(X)
% Computes the (possibly random) time at which a network becomes
% operational given the (random) times of repair X
% Inputs:  X     - a configuration of times of repair
%         GRAPH - structure containing the edges of the graph;
%                 defined as global variable
% Outputs: Sx   - the time at which the network becomes operational
%          b - the critical number for configuration X;
%          perm     - the permutation induced by sorting X;

global GRAPH
E=GRAPH.E;
n=max(E(:)); %number of nodes
A=zeros(n);  %incidence matrix
[x_sorted,perm]=sort(X); % find permutation pi
b=0; % critical number

for i=perm(:)'
    b=b+1;
    e=E(i,:); % which edge is up
    A(e,e)=1; % indicate that the nodes of 'e' are connected
    % find with which other nodes e(1) and e(2) are connected
    y=A(e(1),:) | A(e(2),:);
    A(y,y)=1; % indicate the complete connectivity of the nodes
    struc_func=A(1,n); % if nodes 1 and n are connected, set H(X)=1
    % struc_func=all(A(1,:)); use this for all-terminal rlbty
    % if structure function is one, exit
     if struc_func, break,  end
end
Sx=x_sorted(b);
```

The following script illustrates the use of the function S.m above.

```
% using_Sx.m
clear all,clc
global  GRAPH % create global object GRAPH
  E=[1,1,1,2,2,3,4,5,6,7,7,7,8,9,10,11,12,13;
      2,3,4,5,6,7,7,8,8,9,10,11,9,13,14,12,14,14]';
```

```
GRAPH.E=E; % create structure field name 'E' to store matrix E
x=rand(1,18); % generate some random times
Sx=S(x) % call function S(x)
```

An alternative to Algorithm 16.1 for the two-terminal network reliability case is to use Dijkstra's well-known shortest path algorithm [12]. Dijkstra's algorithm computes the shortest path connecting nodes 1 and n; that is, $\min_{\mathcal{P}\in\mathscr{P}} \sum_{e\in\mathcal{P}} X_e$. Since

$$(X_1^\alpha + \cdots + X_k^\alpha)^{1/\alpha} \approx \max\{X_1, \ldots, X_k\}$$

for α large enough, we can use Dijkstra's algorithm to approximate

$$S(\mathbf{X}) \approx \left(\min_{\mathcal{P}\in\mathscr{P}} \sum_{e\in\mathcal{P}} X_e^\alpha \right)^{1/\alpha} .$$

We found the approximation to be satisfactory for $\alpha = 100$. The advantage of using Dijkstra's algorithm is that the complexity of computing $S(\mathbf{X})$ becomes $\mathcal{O}(m \ln n)$, which can give significant speed-up for large graphs. For graphs with nodes of approximately $n = 100$, our experience is that Algorithm 16.1 runs faster in MATLAB.

16.3 CONDITIONAL MONTE CARLO

In this section we consider a popular method for the efficient estimation of the network unreliability for the case where the edges of the network fail independently. ☞ 354 The **permutation Monte Carlo** method [14] is a conditional Monte Carlo approach specially designed for this setting. The idea is as follows. Consider the network $\mathcal{G}(V, E, K)$ in which each edge e has an exponential repair time X_e with repair rate $\lambda(e) = -\ln(1 - p_e)$. Thus, here we have

$$f(\mathbf{x}) = \mathrm{e}^{-\sum_{e\in E} \lambda(e) x_e} \prod_{e\in E} \lambda(e), \quad \mathbf{x} \in \mathbb{R}_+^m , \tag{16.4}$$

and

$$\mathbb{P}(X_e > 1) = \mathrm{e}^{-\lambda(e)} = 1 - p_e = q_e .$$

At time $t_0 = 0$ all edges are failed. Recall that the state of edge e is expressed as $B_e = \mathrm{I}_{\{X_e \leqslant 1\}}$. Let us extend this definition by defining

$$B_e(t) = \mathrm{I}_{\{X_e \leqslant t\}} ,$$

which corresponds to the state of the edge at time t. Then, the vector $\mathbf{B}(t) = (B_1(t), \ldots, B_m(t))$, defines the configuration of the network at time t. Since the ☞ 635 repair times are exponentially distributed, $\{\mathbf{B}(t), t \geqslant 0\}$ is a Markov jump process with state space $\{0,1\}^m$, starting at state $(0, 0, \ldots, 0)$. This process is called the **construction process** of the network. The state of the static network is a snapshot of the state of the stochastic process $\{H(\mathbf{B}(t)), t \geqslant 0\}$ at time $t = 1$.

Let $\mathbf{\Pi} = (\Pi_1, \ldots, \Pi_m)$ denote the permutation of $1, \ldots, m$ defined by the sorted repair times:

$$X_{\Pi_1} < X_{\Pi_2} < \cdots < X_{\Pi_m} .$$

Thus, $\mathbf{\Pi}$ denotes the *order* in which the edges become operational. Let $A_1, A_1 + A_2, \ldots, A_1 + \cdots + A_m$ denote the times at which those edges are constructed. Thus, the $\{A_i\}$ are the sojourn times of the Markov jump process $\{\mathbf{B}(t)\}$. The random vector $\mathbf{\Pi}$ takes values in the space of permutations of E. For a given permutation $\boldsymbol{\pi} = (\pi_1, \pi_2, \ldots, \pi_m)$ define:

$$\begin{aligned} E_1 &= E, \\ E_i &= E_{i-1} \setminus \{\pi_{i-1}\},\ 2 \leqslant i \leqslant m\ , \\ \lambda(E_i) &= \sum_{e \in E_i} \lambda(e)\ , \end{aligned} \tag{16.5}$$

and let

$$b(\boldsymbol{\pi}) = \underset{i}{\operatorname{argmin}}\{i :\ H(E_{i+1}) = 1\} \tag{16.6}$$

be the critical number for $\boldsymbol{\pi}$; that is, the smallest number of edges required to make the system operational in the order defined by $\boldsymbol{\pi}$. Here $H(E_i)$ is defined as $H(\mathbf{B})$, with $B_e = 0$ for all $e \in E_i$ and $B_e = 1$ for all $e \notin E_i$. The critical number $b(\boldsymbol{\pi})$ can also be defined via the latent continuous random variables (times of repair $\mathbf{X}$):

$$X_{b(\boldsymbol{\pi})} = S(\mathbf{X})\ .$$

From the path properties of Markov jump processes and the properties of the exponential distribution (see Sections A.11 and 4.2.3), it follows that ☞ 635 ☞ 108

$$\mathbb{P}(\mathbf{\Pi} = \boldsymbol{\pi}) = \prod_{j=1}^{m} \frac{\lambda(\pi_j)}{\lambda(E_j)}\ . \tag{16.7}$$

Moreover, conditional on $\{\mathbf{\Pi} = \boldsymbol{\pi}\}$, the sojourn times $A_1, \ldots, A_m$ are independent, and each A_i is exponentially distributed with parameter $\lambda(E_i)$, $i = 1, \ldots, m$. By conditioning on $\mathbf{\Pi}$ we have

$$\ell = \sum_{\boldsymbol{\pi}} \mathbb{P}(\mathbf{\Pi} = \boldsymbol{\pi})\, \mathbb{P}(S(\mathbf{X}) > 1 \,|\, \mathbf{\Pi} = \boldsymbol{\pi})\ . \tag{16.8}$$

Using the definitions of A_i and $b(\boldsymbol{\pi})$, we can write the last probability in terms of convolutions of exponential distribution functions. Namely, for any $t \geqslant 0$ we have

$$\mathbb{P}(S(\mathbf{X}) > t \,|\, \mathbf{\Pi} = \boldsymbol{\pi}) = \mathbb{P}(A_1 + \cdots + A_{b(\boldsymbol{\pi})} > t \,|\, \mathbf{\Pi} = \boldsymbol{\pi})\ . \tag{16.9}$$

Let

$$G_t(\boldsymbol{\pi}) = \mathbb{P}(S(\mathbf{X}) > t \,|\, \mathbf{\Pi} = \boldsymbol{\pi})\ , \tag{16.10}$$

as given in (16.9). Equation (16.8) can then be rewritten as

$$\ell = \mathbb{E}\, G_1(\mathbf{\Pi})\ . \tag{16.11}$$

This suggests the following estimator. Let $\mathbf{\Pi}_1, \ldots, \mathbf{\Pi}_N$ be iid random permutations, each distributed according to (16.7). Then,

$$\widehat{\ell}_{\text{PMC}} = \frac{1}{N} \sum_{i=1}^{N} G_1(\mathbf{\Pi}_i) \tag{16.12}$$

is an unbiased estimator for ℓ. For this estimator to be of practical use, we need to be able to compute the conditional probability in (16.9) exactly. More generally, we wish to compute a probability of the form

$$\mathbb{P}(A_1 + \cdots + A_b > t) ,$$

where $A_i \sim \mathsf{Exp}(\nu_i)$, $i = 1, \ldots, b$, independently, all $\{\nu_i\}$ are distinct, and without loss of generality $\{\nu_i\}$ are ordered in descending order: $\nu_1 > \nu_2 > \cdots > \nu_b$. This
☞ 127 probability is related to the generalized Erlang distribution and under the above conditions it can be expressed as a linear combination of exponential functions:

$$\mathbb{P}(A_{b-k+1} + \cdots + A_b > t) = \sum_{j=1}^{k} \omega_{k,j} \exp(-\nu_{b-j+1}\, t), \quad k = 1, \ldots, b,$$

where $\sum_{j=1}^{k} \omega_{k,j} = 1$. The coefficients $\{\omega_{i,j}\}$ can be stored in a $b \times b$ matrix and can be computed using the algorithm below in $\mathcal{O}(b^2)$ time, see [1, 14].

Algorithm 16.2 (Exact Computation of $\mathbb{P}(A_1 + \cdots + A_b > t)$)

1. *Set* $\omega_{1,1} = 1$. *For* $k = 1, \ldots, b-1$ *and* $j = 1, \ldots, k$, *sequentially compute:*

$$\omega_{k+1,j} = \omega_{k,j} \frac{\nu_{b-k}}{\nu_{b-k} - \nu_{b-j+1}}, \quad \omega_{k+1,k+1} = 1 - \sum_{j=1}^{k} \omega_{k+1,j} .$$

2. *Output*

$$\sum_{j=1}^{b} \omega_{b,j} \exp(-\nu_{b-j+1}\, t)$$

as the exact value of $\mathbb{P}(A_1 + \cdots + A_b > t)$, *with* $A_i \sim \mathsf{Exp}(\nu_i)$, $i = 1, \ldots, b$, *independently, and* $\nu_1 > \nu_2 > \cdots > \nu_b$.

The MATLAB code below implements the algorithm. For alternative methods for evaluating the convolution, see [17].

```
function ell=convolution(t,nu)
% computes P(A_1+...+A_b>t) exactly,
% where A_i ~ Exp(nu(i)) independently;
% nu has to be decreasing (sorted) sequence
b=length(nu); % parameters of the waiting times
w=zeros(b,b); % b is critical number
w(1,1)=1;
for k=1:b-1
    for j=1:k
    w(k+1,j)=w(k,j)*nu(b-k)/(nu(b-k)-nu(b-j+1));
    w(k+1,k+1)=1-sum(w(k+1,1:k));
    end
end
ell=w(b,:)*exp(-nu(end:-1:1)'*t); % probability
```

We are now ready to state the permutation Monte Carlo algorithm.

Algorithm 16.3 (Permutation Monte Carlo)

1. *Draw independent repair times:*

$$X_e \sim \mathsf{Exp}(\lambda(e)), \quad e = 1, \ldots, m\,,$$

where $\lambda(e) = -\ln q_e$, *and let the permutation* $\mathbf{\Pi} = (\Pi_1, \ldots, \Pi_m)$ *be defined by the indices of the sorted repair times:*

$$X_{\Pi_1} < X_{\Pi_2} < \cdots < X_{\Pi_m}\,.$$

2. *Compute the random time* $S(\mathbf{X})$ *at which the network becomes operational using Algorithm 16.1.*

3. *Determine the critical number* $b = b(\mathbf{\Pi})$; *that is, find the index* b *for which*

$$S(\mathbf{X}) = X_{\Pi_b}\,.$$

4. *Compute the rates* $\lambda(E_1) > \lambda(E_2) > \cdots > \lambda(E_b)$ *via* (16.5).

5. *Compute the probability*

$$G_1(\mathbf{\Pi}) = \mathbb{P}(A_1 + \cdots + A_b > 1)\,,$$

where $A_i \sim \mathsf{Exp}(\lambda(E_i))$, $i = 1, \ldots, b$, *independently, using Algorithm 16.2.*

6. *Repeat Steps 1–5 independently* N *times and deliver* (16.12) *as the estimator for* ℓ.

Elperin et al. [14] show that the relative error of the estimator (16.12) is uniformly bounded for all values of the edge reliabilities. However, for each generated configuration the computation of the conditional probability (16.10) for $t = 1$ is of the order $\mathcal{O}(n^2)$ using the currently best implementation, see [14, 33].

■ **EXAMPLE 16.1 (Two-Terminal Reliability)**

Consider the network in Figure 16.3 with all edges failing independently with the same probability q. We are interested in estimating the probability that nodes 1 and 14 fail to be connected. Table 16.1 shows the estimated unreliabilities for various values of q.

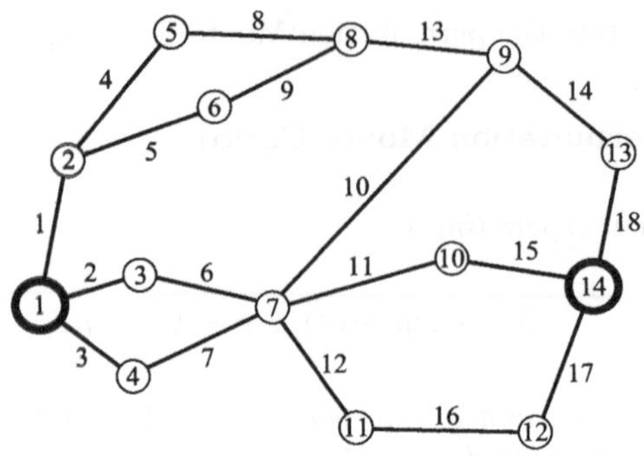

Figure 16.3 A reliability network with two terminal nodes (bold circles). Both the edges and the vertices are labeled. The network works if the terminal nodes can be connected via operating edges. The network unreliability is $\ell = 24q^3 + o(q^3)$ when all edges have the same unreliability q.

Table 16.1 Network unreliability for various edge unreliabilities q.

q	$\widehat{\ell}$	RE %	q	$\widehat{\ell}$	RE %
10^{-1}	1.88×10^{-2}	0.78	10^{-6}	1.98×10^{-17}	2.00
10^{-2}	1.96×10^{-5}	1.79	10^{-7}	1.99×10^{-20}	1.99
10^{-3}	1.97×10^{-8}	1.98	10^{-8}	2.02×10^{-23}	1.98
10^{-4}	2.04×10^{-11}	1.96	10^{-9}	2.00×10^{-26}	1.99
10^{-5}	1.96×10^{-14}	2.01	10^{-10}	1.95×10^{-29}	2.01

The table is obtained using the MATLAB code below with $N = 10^5$. Note that the matrix E given in (16.3) is assumed to be loaded into the workspace. The last column of the table provides the estimated relative error. Note that the relative error does not deteriorate as the unreliability decreases. This is in line with the results of Elperin et al. [14].

```
% hetero_PMC.m
global GRAPH
GRAPH.E=E;
tab=[];
for iter=1:10
    m=size(E,1);      % number of edges
    n=max(E(:));      % number of nodes
    p=ones(m,1)*(1-0.1^iter);
    lam=-log(1-p'); % repair time rates
    L=sum(lam);
    N=10^5; ell=nan(N,1);
    for i=1:N
        x=-log(rand(1,m))./lam; % sample repair times
        [Sx,crit,perm]=S(x);     % compute S(X)
```

```
        % compute rates for convolution
        LAM_perm=L-cumsum([0,lam(perm(1:crit-1))]);
        % compute probability given configuration
        ell(i)=convolution(1,LAM_perm);
    end
    tab=[tab;mean(ell), std(ell)/mean(ell)/sqrt(length(ell))]
end
```

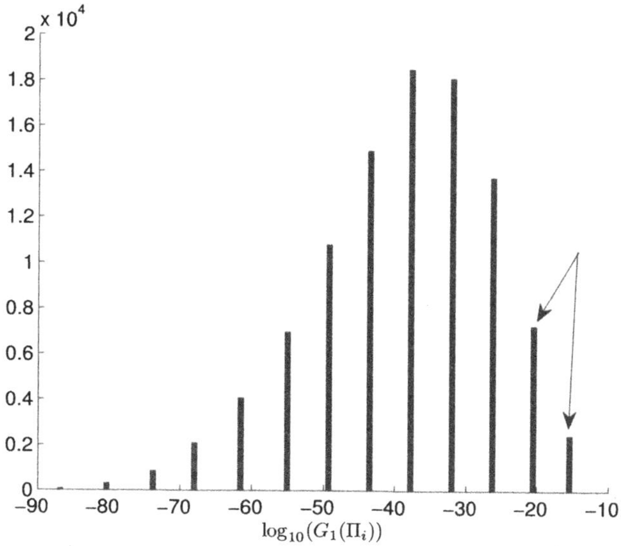

Figure 16.4 The empirical distribution of the conditional probabilities $\{G_1(\mathbf{\Pi}_i)\}$ used to compute the estimator (16.12). The main contribution to the estimator comes from the relatively few configurations in the right tail, indicated by arrows.

Figure 16.4 shows a histogram of the empirical distribution of the conditional probabilities $\{G_1(\mathbf{\Pi}_i)\}$ on a logarithmic scale for the case $q = 10^{-6}$. Since ℓ is of the order 10^{-17}, the main contribution to the estimator (16.12) comes from the relatively few configurations corresponding to the right tail of the distribution (indicated by arrows). In other words, most of the randomly generated repair time vectors correspond to a conditional probability $G_1(\mathbf{\Pi})$ close to zero. Thus, a major part of the effort to compute the probability $G_1(\mathbf{\Pi})$ in Step 5 of Algorithm 16.3 is wasted, because it does not contribute much to the final estimator. In the next section we will describe a possible remedy for this problem.

In general, the total computational effort of running Algorithm 16.3 consists of the effort of evaluating $S(\mathbf{x})$ and the effort of computing the conditional probability $G_1(\boldsymbol{\pi})$ for the corresponding permutation $\boldsymbol{\pi}$. In the special case where all edges have the *same* reliability, the function G_1 depends only on the critical number $b(\boldsymbol{\pi})$ and not on the particular permutation $\boldsymbol{\pi}$. As a consequence, G_1 takes values in a set of size at most m, see Figure 16.4. We can precompute all the m possible

values of G_1 and store them in a table, indexed by the value of b, to reduce the computational overhead. For every random configuration we only compute the critical number and find the corresponding probability G_1 from the table. Further
☞ 463 variance reduction is possible if one uses the CE method to improve the estimator (16.12). For details on the combined CE-PMC estimator, see [22, 27].

Next, we describe a modification of the permutation Monte Carlo algorithm that reduces the average cost of computing the conditional probabilities $\{G_1(\boldsymbol{\pi}_i)\}$.

16.3.1 Leap–Evolve Algorithm

Recall from Figure 16.4 and Example 16.1 that a large fraction of the conditional probabilities $\{G_1(\boldsymbol{\Pi}_i)\}$ are close to zero and may not contribute much to the estimator (16.12). This observation is valid even if the edges fail with different probabilities. To alleviate this problem, Lomonosov [29] suggested a modification of permutation Monte Carlo called the **leap–evolve** algorithm, which can be described as follows. Let $\gamma \in [0, 1]$. The configuration of the network at time γ can be described by

$$_*\mathbf{X} = (X_{\Pi_1}, X_{\Pi_2}, \ldots, X_{\Pi_p}) \,,$$

where

$$X_{\Pi_1} < \cdots < X_{\Pi_{p-1}} < X_{\Pi_p} < \gamma < X_{\Pi_{p+1}} < \cdots < X_{\Pi_m} \,.$$

Thus, $_*\mathbf{X}$ is the vector of repair times of the edges that are operational by time γ. Denote the repair times of the failed edges at time γ by

$$\mathbf{X}_* = (X_{\Pi_{p+1}}, \ldots, X_{\Pi_m}) \,.$$

By conditioning on $_*\mathbf{X}$ and the permutation $\boldsymbol{\Pi}_* \stackrel{\text{def}}{=} (\Pi_{p+1}, \ldots, \Pi_m)$ (the ordering of the future repair times), we find that

$$\begin{aligned}
\ell &= \mathbb{P}(S(\mathbf{X}) > 1) = \mathbb{P}(S(\mathbf{X}) > 1, S(\mathbf{X}) > \gamma) \\
&= \mathbb{E}\, \mathrm{I}_{\{S(\mathbf{X})>1\}} \mathrm{I}_{\{S(\mathbf{X})>\gamma\}} \\
&= \mathbb{E}\, \mathbb{E}[\mathrm{I}_{\{S(\mathbf{X})>1\}}\, \mathrm{I}_{\{S(\mathbf{X})>\gamma\}} \,|\, {}_*\mathbf{X}, \boldsymbol{\Pi}_*] \\
&= \mathbb{E}\, \mathrm{I}_{\{S(\mathbf{X})>\gamma\}} \mathbb{E}[\mathrm{I}_{\{S(\mathbf{X})>1\}} \,|\, {}_*\mathbf{X}, \boldsymbol{\Pi}_*] \\
&= \mathbb{E}\, \mathrm{I}_{\{S(\mathbf{X})>\gamma\}} G_{1-\gamma}(\boldsymbol{\Pi}_*) \,.
\end{aligned}$$

Here,

$$G_{1-\gamma}(\boldsymbol{\Pi}_*) = \mathbb{P}(A_{p+1} + A_{p+2} + \cdots + A_{b(\boldsymbol{\pi})} > 1 - \gamma) \,,$$

where $A_{p+1}, A_{p+2}, \ldots, A_{b(\boldsymbol{\pi})}$ are independent exponential random variables with corresponding rates

$$\sum_{i=p+1}^{m} \lambda(\pi_i), \; \sum_{i=p+2}^{m} \lambda(\pi_i) \,, \ldots, \; \sum_{i=b(\boldsymbol{\pi})}^{m} \lambda(\pi_i) \,.$$

Note that if $\gamma = 0$, then $\mathbb{E}\mathrm{I}_{\{S(\mathbf{X})>\gamma\}} G_{1-\gamma}(\boldsymbol{\Pi}_*)$ reduces to (16.11). The idea of the leap–evolve scheme is to reduce the number of evaluations of the function $G_{1-\gamma}(\boldsymbol{\Pi}_*)$ using a judicious choice of γ so that $\mathrm{I}_{\{S(\mathbf{X})>\gamma\}}$ is an indicator of the importance of the repair time vector $\mathbf{X}$.

Algorithm 16.4 (Leap–Evolve Estimator) *For a given γ, execute the steps.*

1. *Generate the repair-time vector $\mathbf{X} \sim f(\mathbf{x})$, where f is given by (16.4). Let $\mathbf{\Pi} = \boldsymbol{\pi}$ be the permutation corresponding to $\mathbf{X}$.*

2. *Compute $S(\mathbf{X})$ using Algorithm 16.1 or otherwise.*

3. *Compute the critical number $b = b(\boldsymbol{\pi})$ such that $S(\mathbf{X}) = X_{\pi_b}$.*

4. *If $S(\mathbf{X}) < \gamma$, set $Z_i = 0$; otherwise, compute $\boldsymbol{\pi}_* = (\pi_{p+1}, \ldots, \pi_m)$, where $\gamma < X_{\pi_{p+1}} < \cdots < X_{\pi_m}$ and use Algorithm 16.2 to compute $Z_i = G_{1-\gamma}(\boldsymbol{\pi}_*)$.*

5. *Repeat Steps 1–4 N times and deliver the unbiased estimator $\widehat{\ell}_{\text{leap}} = \frac{1}{N}\sum_{i=1}^{N} Z_i$.*

The improvement of the leap–evolve scheme over the simpler permutation Monte Carlo method depends very much on the choice of γ and the ability to generate repair time configurations $\mathbf{X}$ conditional on $\{S(\mathbf{X}) > \gamma\}$. There is no known optimal way to select γ.

■ EXAMPLE 16.2 (Leap–Evolve)

Consider the network in Figure 16.3 with the same unreliability for all edges: $q = 10^{-6}$. We arbitrarily set $\gamma = 0.1$ and use the MATLAB code below to estimate the probability that nodes 1 and 14 are not connected. The matrix given by (16.3) is assumed to be loaded into the workspace. Using $N = 2 \times 10^5$ we obtained an estimated relative error of 2.4%. This is comparable to the original permutation Monte Carlo implementation, but in the leap–evolve version $G_{1-\gamma}$ was evaluated only 66000 times reducing the total CPU time by a factor of 0.6.

```
% leap_evolve.m
global GRAPH
GRAPH.E=E;
m=size(E,1);      % number of edges
n=max(E(:));      % number of nodes
p=ones(m,1)*(1-0.1^6);
lam=-log(1-p'); % repair time rates
N=2*10^5;
ell=nan(N,1);
gamma=0.1;
for i=1:N
    x=-log(rand(1,m))./lam; % sample repair times
    [Sx,crit,perm]=S(x);    % compute S(X)
    if Sx<gamma
        ell(i)=0;
    else
        idx=find(sort(x)>=gamma,1,'first'); %compute p
        L=sum(lam(perm(idx:end))); % compute rates for conv.
        LAM_perm=L-cumsum([0,lam(perm(idx:crit-1))]);
        ell(i)=convolution(1-gamma,LAM_perm);
```

```
    end
end
mean(ell), std(ell)/mean(ell)/sqrt(length(ell))
```

In addition to the leap–evolve scheme, other extensions of the permutation Monte Carlo algorithm include the **merge process** [14] and the **tree cut and merge** algorithm [21], both of which perform much better than standard permutation Monte Carlo. A MATLAB implementation of the merge process is given in `merge_process.m` and `merge.m` on the Handbook website.

16.4 IMPORTANCE SAMPLING FOR NETWORK RELIABILITY

In this section we consider two importance sampling methods for network reliability estimation.

16.4.1 Importance Sampling Using Bounds

The importance sampling approach in this section requires some prior knowledge of the network in the form of bounds on $S(\mathbf{x})$ for its efficient implementation. The cuts and paths of a graph provide a convenient method for constructing such bounds.

Let $\mathscr{C}^* \subseteq \mathscr{C}$ and $\mathscr{P}^* \subseteq \mathscr{P}$ be a collection of minimal cuts and minimal paths of the graph $\mathcal{G}$, respectively. Then, we can use the cut and path sets to provide the following lower and upper bounds on $S(\mathbf{X})$ (see [8, 18, 33]):

$$S_L(\mathbf{X}) \stackrel{\text{def}}{=} \max_{C \in \mathscr{C}^*} \min_{e \in C} X_e \leqslant S(\mathbf{X}) \leqslant \min_{\mathcal{P} \in \mathscr{P}^*} \max_{e \in \mathcal{P}} X_e \stackrel{\text{def}}{=} S_U(\mathbf{X}) . \tag{16.13}$$

Typically, it is assumed that the paths and cuts are **edge-disjoint**; that is, a given edge can belong to no more than one path or cut in $\mathscr{C}^*$ and $\mathscr{P}^*$. Such subsets can be constructed in polynomial time [18, 31].

The method proposed by Cancela et al. [8] and inspired by Fishman's approach [18] uses a set of disjoint paths to construct an importance sampling density for a homogeneous network. The approach has been shown to be quite efficient in combination with quasi Monte Carlo, see [8].

Let $\mathscr{P}^*$ be an edge-disjoint set of paths and define

$$\ell_U = \mathbb{P}(S_U(\mathbf{X}) > 1) = \mathbb{P}\left(\min_{\mathcal{P} \in \mathscr{P}^*} \max_{e \in \mathcal{P}} X_e > 1\right) = \prod_{\mathcal{P} \in \mathscr{P}^*} \left(1 - \prod_{e \in \mathcal{P}} p_e\right).$$

Here ℓ_U is readily computable. Now suppose we can sample from the importance sampling density

$$f_U(\mathbf{x}) = \frac{f(\mathbf{x})\, \mathrm{I}_{\{S_U(\mathbf{X})>1\}}}{\ell_U} .$$

Notice that f_U is the conditional density of $\mathbf{X}$ given that at least one edge in every path in $\mathscr{P}^*$ is not operational at time 1. We can use the following importance sampling estimator to estimate ℓ:

$$\widehat{\ell}_{\text{path}} = \frac{1}{N} \sum_{i=1}^{N} \frac{f(\mathbf{X}_i)}{f_U(\mathbf{X}_i)} \mathrm{I}_{\{S(\mathbf{X}_i)>1\}} = \frac{\ell_U}{N} \sum_{i=1}^{N} \mathrm{I}_{\{S(\mathbf{X}_i)>1\}} , \tag{16.14}$$

where $\mathbf{X}_1, \ldots, \mathbf{X}_N \sim_{\text{iid}} f_U$. To sample from f_U we use the following subroutine from [8].

Algorithm 16.5 (Sampling From f_U)

1. *Take a path $\mathcal{P} = (e_1, \ldots, e_k)$ in $\mathscr{P}^*$, and compute*

$$a_i = \frac{(1 - p_{e_i}) \prod_{j=1}^{i-1} p_{e_j}}{1 - \prod_{j=1}^{k} p_{e_j}}, \quad i = 1, \ldots, k\,.$$

2. *Select an index $I \in \{1, \ldots, k\}$ with probability $\mathbb{P}(I = i) = a_i$.*

3. *For each $i = 1, \ldots, I - 1$, generate an exponential random variable Y_i with parameter $\lambda(e_i) = -\ln(1 - p_{e_i})$, conditional on $Y_i \leqslant 1$.*

4. *Generate Y_I from $\mathsf{Exp}(\lambda(e_I))$, conditional on $Y_I > 1$; that is, set $Y_I = 1 + Z$, where $Z \sim \mathsf{Exp}(\lambda(e_I))$.*

5. *For each $i = I + 1, \ldots, k$, generate $Y_i \sim \mathsf{Exp}(\lambda(e_i))$ independently.*

6. *Assign $X_{e_j} = Y_j$, $j = 1, \ldots, k$. Remove path $\mathcal{P}$ from the set of paths $\mathscr{P}^*$. If $\mathscr{P}^*$ has no more paths, continue with Step 7; otherwise, return to Step 1.*

7. *The repair time of any edge e that does not belong to any of the paths in $\mathscr{P}^*$ is sampled independently from the corresponding marginal of $f(\mathbf{x})$; that is, $X_e \sim \mathsf{Exp}(\lambda(e))$.*

8. *Output $\mathbf{X} = (X_1, \ldots, X_m)$ as a sample from $f_U(\mathbf{x})$.*

The following MATLAB code implements the algorithm.

```
function Y=path_sampling(p_bar)
%samples the repair times  that belong to the paths
k=length(p_bar);
% step 2
P=cumprod(p_bar);
p_star=(1-p_bar).*[1,P(1:end-1)];
p_star=p_star/(1-P(end));
% step 3
[dummy,idx]=histc(rand,[0,cumsum(p_star)]);
% step 4
lam=-log(1-p_bar); %exponential rates
for i=1:idx-1
    Y(i)=expt(lam(i),0,1);
end
% step 5
Y(idx)=1-log(rand)/lam(idx);
% step 6
for i=idx+1:k
    Y(i)= -log(rand)/lam(i);
end
```

```
function x=f_bar(p,paths)
% implements sampling from f_bar
% p is a vector with the corresponding reliabilities
% 'paths' is a cell array of edges describing the paths;

%step 7 and 8
m=length(p);x=zeros(1,m); p=p(:)';
for j=1:size(paths,2)
    P=paths{j}; % paths have to be disjoint!
    x(P)=path_sampling(p(P));
end
idx=find(x==0); % find the edges that are not on the paths
% sample edges not belonging to the paths
x(idx)=-log(rand(1,length(idx)))./(-log(1-p(idx)));
```

■ **EXAMPLE 16.3 (Dodecahedron Network)**

Consider the dodecahedron network in Figure 16.5. Suppose that all the edges have the same unreliability of $q = 0.001$. Let $\mathscr{P}^* = \{\mathcal{P}_1, \mathcal{P}_2, \mathcal{P}_3\}$ consist of the three paths with edges

$$\mathcal{P}_1 = (1,\ 4,\ 11,\ 20,\ 28)\ ,$$
$$\mathcal{P}_2 = (3,\ 8,\ 17,\ 26,\ 30)\ ,$$
$$\mathcal{P}_3 = (2,\ 6,\ 14,\ 23,\ 29)\ .$$

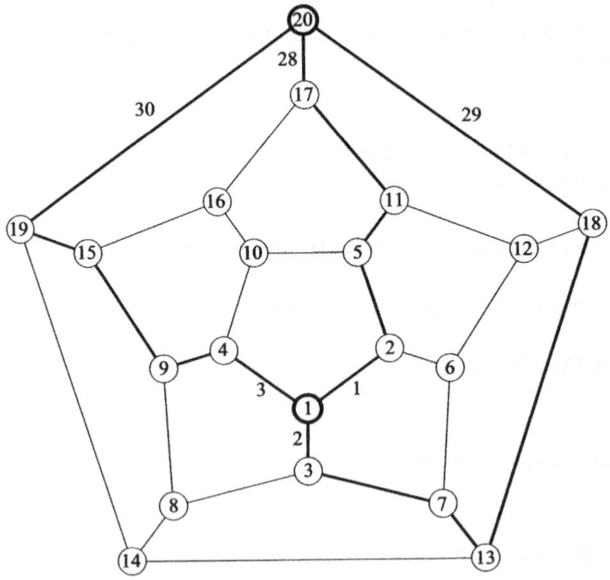

Figure 16.5 A dodecahedron network with 20 nodes and 30 links. Edges 1, 2, 3 and 28, 29, 30 are indicated.

The edges belonging to these three paths are depicted in Figure 16.5 with thicker lines. Note that $\mathcal{P}_1$ is a path on the set of nodes $1 \to 2 \to 5 \to 11 \to 17 \to 20$, path $\mathcal{P}_2$ connects the nodes $1 \to 4 \to 9 \to 15 \to 19 \to 20$, and path $\mathcal{P}_3$ connects the nodes $1 \to 3 \to 7 \to 13 \to 18 \to 20$.

Using the code below with $N = 10^5$ the estimator (16.14) gives a typical estimate of 1.95×10^{-9} with an estimated relative error of 2.5%. Noting that the corresponding crude Monte Carlo estimator has variance $\ell(1-\ell)/N$, the variance reduction factor compared with crude Monte Carlo is approximately 5×10^6.

```
% IS_bounds.m
% the matrix E is assumed to be loaded into the workspace
global  GRAPH
GRAPH.E=E;
m=size(E,1);     % number of edges
p=ones(m,1)*(1-0.1^3);
% list the edges comprising the paths
paths{1}=[1      4     11     20     28];
paths{2}=[3      8     17     26     30];
paths{3}=[2      6     14     23     29];
ell_U=1; % compute normalizing constant
for i=1:size(paths,2)
    P=paths{i};
    ell_U=ell_U*(1-prod(p(P)));
end
% compute IS estimator
N=10^5; ell=nan(N,1);
for i=1:N
    x=f_bar(p,paths);
    ell(i)=(S(x)>1)*ell_U;
end
[mean(ell), std(ell)/mean(ell)/sqrt(N)]
```

16.4.2 Importance Sampling With Conditional Monte Carlo

In this section we consider an importance sampling method that can be applied to quite general rare-event simulation problems. We first present the method in a general setting and then explain how to apply it to reliability problems.

We wish to estimate the normalizing constant ℓ of the pdf

$$\pi(\mathbf{x}) \stackrel{\text{def}}{=} \frac{f(\mathbf{x})\, \mathrm{I}_{\{S(\mathbf{x})>1\}}}{\ell}, \quad \mathbf{x} = (x_1, \ldots, x_m)^\top,$$

where $f(\mathbf{x})$ and $S(\mathbf{x})$ are given functions that can be evaluated. Assume that the conditional densities

$$\pi(x_e \,|\, \mathbf{x}_{-e}) \stackrel{\text{def}}{=} \pi(x_e \,|\, x_1, \ldots, x_{e-1}, x_{e+1}, \ldots, x_m), \quad e = 1, \ldots, m,$$

are available in closed form, and that one can easily sample from them. Using the conditional densities $\{\pi(x_e \,|\, \mathbf{x}_{-e})\}$ we can approximate the marginal densities of

☞ 225 $\pi(\mathbf{x})$ as follows. Use a Markov chain Monte Carlo sampler to obtain the approximate sample:

$$\mathbf{X}^{(1)}, \ldots, \mathbf{X}^{(M)} \overset{\text{approx.}}{\sim} \pi(\mathbf{x}) .$$

Then, an estimator of each marginal pdf $\pi(x_e) = \int \pi(\mathbf{x}) \, \mathrm{d}\mathbf{x}_{-e}$ is

$$\widehat{\pi}(x_e) = \frac{1}{M} \sum_{k=1}^{M} \pi(x_e \,|\, \mathbf{X}_{-e}^{(k)}), \quad e = 1, \ldots, m .$$

It is clear that sampling from the pdf $\widehat{\pi}(x_e)$ is straightforward using the composition method, and that we can evaluate $\widehat{\pi}(x_e)$ at any point. Define an importance sampling density of the product form ☞ 53

$$g(\mathbf{x}) \stackrel{\text{def}}{=} \prod_{e=1}^{m} \widehat{\pi}(x_e) = \prod_{e=1}^{m} \frac{1}{M} \sum_{k=1}^{M} \pi(x_e \,|\, \mathbf{X}_{-e}^{(k)}) .$$

Then, an unbiased estimator of ℓ is

$$\widehat{\ell} = \frac{1}{N} \sum_{j=1}^{N} \frac{f(\mathbf{Y}_j) \, \mathrm{I}_{\{S(\mathbf{Y}_j) > 1\}}}{g(\mathbf{Y}_j)}, \quad \mathbf{Y}_1, \ldots, \mathbf{Y}_N \overset{\text{iid}}{\sim} g(\mathbf{y}) .$$

Care must be taken to ensure that $g(\mathbf{x})$ dominates $\pi(\mathbf{x})$ (in the sense that $g(\mathbf{x}) = 0 \Rightarrow \pi(\mathbf{x}) = 0$ for all $\mathbf{x}$).

To apply this approach to estimating the unreliability ℓ in (16.2) for a homogeneous network (that is, each edge has the same unreliability q), we let $f(\mathbf{x})$ be the pdf of the $\mathsf{N}(\mathbf{0}, \sigma^2 I)$ distribution with $\sigma = -1/\Phi^{-1}(q)$ (where Φ^{-1} is the inverse of the cdf of the standard normal distribution), so that each $X_e \sim \mathsf{N}(0, \sigma^2)$ and $\mathbb{P}(X_e > 1) = q$ for all e. The conditional pdfs of $\pi(\mathbf{x})$ are ($e = 1, \ldots, m$):

$$\pi(x_e \,|\, \mathbf{X}_{-e}) = \frac{1}{\sqrt{2\pi}\sigma} \, \mathrm{e}^{-\frac{x_e^2}{2\sigma^2}} \left(\mathrm{I}_{\{S_e > 1\}} + \frac{1 - \mathrm{I}_{\{S_e > 1\}}}{q} \, \mathrm{I}_{\{x_e > 1\}} \right),$$

where $\{S_e > 1\}$ is the event that $S(X_1, \ldots, X_{e-1}, 0, X_{e+1}, \ldots, X_m) > 1$. Therefore, the estimated marginal densities become

$$\widehat{\pi}(x_e) = \frac{1}{\sqrt{2\pi}\sigma} \, \mathrm{e}^{-\frac{x_e^2}{2\sigma^2}} \left(\widehat{w}_e + \frac{1 - \widehat{w}_e}{q} \, \mathrm{I}_{\{x_e > 1\}} \right), \tag{16.15}$$

where

$$\widehat{w}_e = \frac{1}{M} \sum_{k=1}^{M} \mathrm{I}_{\{S_e^{(k)} > 1\}} ,$$

and $\{S_e^{(k)} > 1\}$ is the event that $S(X_1^{(k)}, \ldots, X_{e-1}^{(k)}, 0, X_{e+1}^{(k)}, \ldots, X_m^{(k)}) > 1$. In fact, we can write the *exact* marginal pdf of edge e as (16.15) with $\widehat{w}_e$ replaced by w_e, where w_e is the probability that the network is not operational for any time of repair X_e. In other words, the marginal density of each time of repair X_e is a mixture density of two Gaussian densities — one of which is truncated. Random variable generation from a truncated Gaussian density is explained in Example 3.7. ☞ 51

It follows that the likelihood ratio is given by

$$\frac{f(\mathbf{y})}{g(\mathbf{y})} = \prod_{e=1}^{m} \left(\widehat{w}_e + \frac{1 - \widehat{w}_e}{q} \mathrm{I}_{\{y_e > 1\}} \right)^{-1} .$$

This motivates the following algorithm.

Algorithm 16.6 (Importance Sampling Using Conditional Densities)

1. *Use an MCMC algorithm or otherwise to generate*

$$\mathbf{X}^{(1)}, \ldots, \mathbf{X}^{(M)} \overset{\text{approx.}}{\sim} \pi(\mathbf{x}) .$$

2. *Use the sample from Step 1 to construct the importance sampling pdf* $g(\mathbf{x}) = \prod_{e=1}^{m} \widehat{\pi}(x_e)$*, where* $\widehat{\pi}(x_e)$ *is defined in* (16.15).

3. *Generate* $\mathbf{Y}_1, \ldots, \mathbf{Y}_N \overset{\text{iid}}{\sim} g(\mathbf{y})$ *and deliver the estimator*

$$\widehat{\ell} = \frac{1}{N} \sum_{j=1}^{N} \mathrm{I}_{\{S(\mathbf{Y}_j) > 1\}} \prod_{e=1}^{m} \left(\widehat{w}_e + \frac{1 - \widehat{w}_e}{q} \mathrm{I}_{\{Y_{je} > 1\}} \right)^{-1} , \qquad (16.16)$$

where Y_{je} *is the e-th component of* $\mathbf{Y}_j$.

■ EXAMPLE 16.4 (Dodecahedron Network Revisited)

As a numerical example, consider the dodecahedron network in Figure 16.5 with all edges having unreliability $q = 0.001$. Example 14.6 on Page 514 explains how we can generate an approximate sample from $\pi(\mathbf{x})$ using MCMC and the splitting method. Using the sample generated with the ADAM method we obtain $\widehat{w}_e = 0.412$ ☞ 493 for $e = 1, 2, 3$, $\widehat{w}_e = 0.59$ for $e = 28, 29, 30$, and $w_e = 1$ for all other edges. With an iid sample of size $N = 10^6$ from $g(\mathbf{y})$ the estimator (16.16) gives a typical estimate of $\widehat{\ell} = 2.01 \times 10^{-9}$ with an estimated relative error of 0.76%. MATLAB code for this example can be found on the Handbook website as `relbty_marginals.m`.

16.5 SPLITTING METHOD

In this section we show how to apply the *generalized splitting* (GS) algorithm of Section 14.2 to estimate the reliabilities of a given network. The method has the ☞ 485 advantage that it easily handles cases of dependent edge failures. While the methods described in Sections 16.4.1 and 16.3 are typically more efficient, they are applicable only when the edges fail independently.

Before describing the GS algorithm for reliability estimation, we introduce the following latent variable structure for the estimation of the network unreliability ℓ in (16.2). Let $\mathbf{X} = (X_1, \ldots, X_m)^\top \sim f(\mathbf{x})$, where f is the pdf of the $\mathsf{N}(\mathbf{0}, \Sigma)$ distribution. Here the diagonal elements of Σ are

$$\Sigma_{ee} = \sigma_e^2 = \mathrm{Var}(X_e) = \frac{1}{(\Phi^{-1}(q_e))^2}, \quad e = 1, \ldots, m ,$$

where Φ^{-1} is the inverse of the cdf of the standard normal distribution. It follows that $\mathbb{P}(X_e > 1) = q_e$.

A frequently made assumption in network reliability models is that the edges fail independently of each other [33]; that is, the off-diagonal elements of Σ are all zero and so $X_1, \ldots, X_m$ are *independent*. A simple model that incorporates failure *dependencies* is to assume a full covariance matrix Σ. Then, for a given Σ, the Bernoulli variables $B_e = \mathrm{I}_{\{X_e \leqslant 1\}}$, $e = 1, \ldots, m$ have covariance structure

$$\mathrm{Var}(B_e) = p_e(1 - p_e), \quad e = 1, \ldots, m \,,$$
$$\mathrm{Cov}(B_i, B_j) = \Phi_2\left(\frac{1}{\sigma_i}, \frac{1}{\sigma_j}; \frac{\Sigma_{ij}}{\sigma_i \sigma_j}\right) - p_i\, p_j \,, \quad i \neq j \,,$$

where $\Phi_2(x, y; \varrho)$ is the cdf of the bivariate normal density evaluated at (x, y) with zero mean, unit variance in both components, and correlation coefficient ϱ. Thus,
☞ 68 the distribution of the latent variables $\mathbf{X}$ represents a *Gaussian copula* model for the Bernoulli vector $\mathbf{B}$. Conversely, if the covariance structure of the Bernoulli variables $\mathbf{B}$ is given, then, under suitable conditions [2, 24], one can determine the corresponding matrix Σ.

Having defined the latent variable structure of the evolution model, we now explain how to apply the GS method. Using a pilot run, suppose that we obtain the levels

$$-\infty = \gamma_0 < \gamma_1 < \cdots < \gamma_{T-1} < \gamma_T = 1 \,,$$

and the estimates $\{\varrho_t\}$ of the conditional probabilities

$$\varrho_t \approx \mathbb{P}_f(S(\mathbf{X}) \geqslant \gamma_t \,|\, S(\mathbf{X}) \geqslant \gamma_{t-1}) \,, \quad t = 1, \ldots, T \,.$$

Thus $\{\varrho_t^{-1}\}$ are *splitting factors*. Algorithm 16.9 is used to construct the sequence
☞ 493 $\{(\gamma_t, \varrho_t)\}_{t=1}^{T}$ using a pilot run of the algorithm below (see also Algorithm 14.4). The GS algorithm for network reliability estimation then reads as follows.

Algorithm 16.7 (Network Reliability Estimation) *Given a sequence of levels and splitting factors* $\{(\gamma_t, \varrho_t)\}_{t=1}^{T}$ *and a sample size* N, *set the counter* $t = 1$ *and execute the following steps.*

1. **Initialization.** *Let* $N_0 = \varrho_1 \left\lfloor \frac{N}{\varrho_1} \right\rfloor$. *Generate independent time of repair vectors*
$$\mathbf{X}_1, \ldots, \mathbf{X}_{N_0/\varrho_1} \overset{\text{iid}}{\sim} \mathsf{N}(\mathbf{0}, \Sigma) \,.$$
Denote $\mathcal{X}_0 = \{\mathbf{X}_1, \ldots, \mathbf{X}_{N_0/\varrho_1}\}$. *Let* $\mathcal{X}_1 = \{\mathbf{X}_1, \ldots, \mathbf{X}_{N_1}\}$ *be the largest subset of vectors in* $\mathcal{X}_0$ *for which* $S(\mathbf{X}) \geqslant \gamma_1$ *(use Algorithm 16.1 to evaluate* S*). Thus,* N_1 *is the number of vectors for which the network is nonoperational at time* γ_1.

2. **Markov chain sampling.** *For each* $\mathbf{X}_i$ *in* $\mathcal{X}_t = \{\mathbf{X}_1, \ldots, \mathbf{X}_{N_t}\}$ *sample new vectors*
$$\mathbf{Y}_{i,j} \sim \kappa_t(\mathbf{y} \,|\, \mathbf{Y}_{i,j-1}), \quad \mathbf{Y}_{i,0} = \mathbf{X}_i, \qquad j = 1, \ldots, \mathcal{S}_{ti} \,, \tag{16.17}$$
where $\mathcal{S}_{ti}$ *is a random splitting factor*
$$\mathcal{S}_{ti} - \left\lfloor \frac{1}{\varrho_{t+1}} \right\rfloor \overset{\text{iid}}{\sim} \mathsf{Ber}\left(\frac{1}{\varrho_{t+1}} - \left\lfloor \frac{1}{\varrho_{t+1}} \right\rfloor\right), \quad i = 1, \ldots, N_t \,,$$

and $\kappa_t(\mathbf{y} \mid \mathbf{Y}_{i,j-1})$ is a Markov transition density with initial state $\mathbf{Y}_{i,j-1}$ and stationary pdf

$$f_t(\mathbf{y}) \propto \mathrm{I}_{\{S(\mathbf{y})>\gamma_t\}} \, \mathrm{e}^{-\frac{1}{2}\mathbf{y}^\top \Sigma^{-1} \mathbf{y}}.$$

Reset

$$\mathcal{X}_t \equiv \left\{ \mathbf{Y}_{1,1}, \mathbf{Y}_{1,2}, \ldots, \mathbf{Y}_{1,\mathcal{S}_{t1}}, \cdots\cdots\cdots, \mathbf{Y}_{N_t,1}, \mathbf{Y}_{N_t,2}, \ldots, \mathbf{Y}_{N_t,\mathcal{S}_{tN_t}} \right\}.$$

3. **Updating.** *Let $\mathcal{X}_{t+1} = \{\mathbf{X}_1, \ldots, \mathbf{X}_{N_{t+1}}\}$ be the largest subset of vectors in $\mathcal{X}_t$ for which the network is nonoperational at time γ_{t+1}, that is, $S(\mathbf{X}) \geqslant \gamma_{t+1}$. Here, N_{t+1} is the random size of $\mathcal{X}_{t+1}$. Set the counter $t = t + 1$.*

4. **Stopping condition.** *If $t = T$ go to Step 5. If $N_t = 0$, set $N_{t+1} = N_{t+2} = \cdots = N_T = 0$ and go to Step 5; otherwise, repeat from Step 2.*

5. **Final estimator.** *Deliver the unbiased estimate of the unreliability*

$$\widehat{\ell} = \frac{N_T}{N_0} \prod_{t=1}^{T} \varrho_t .$$

and the unbiased estimate of the variance

$$\widehat{\mathrm{Var}(\widehat{\ell})} = \frac{\prod_{t=1}^{T} \varrho_t^2}{N_0(N_0 - \varrho_1)} \sum_{i=1}^{N_0/\varrho_1} \left(O_i - \frac{\varrho_1}{N_0} N_T \right)^2, \tag{16.18}$$

where O_i denotes the number of points in $\mathcal{X}_T$ that share a common history with the i-th point from the initial population $\mathcal{X}_0$ and have their S value above the level $\gamma_T = 1$ at the final stage.

To completely specify Algorithm 16.7, we need to define the transition density κ_t. The following algorithm defines a transition density with stationary pdf f_t via the hit-and-run sampler. ☞ 240

Algorithm 16.8 (Transition Density $\kappa_t(\mathbf{y} \mid \mathbf{x})$) *Given a vector $\mathbf{x}$ of times of repair such that $S(\mathbf{x}) \geqslant \gamma_t$, execute the following steps.*

1. *Generate a random direction vector $\mathbf{d}$ uniformly distributed on the unit m-dimensional sphere:*

$$\mathbf{d} = \left(\frac{Z_1}{\|\mathbf{Z}\|}, \ldots, \frac{Z_m}{\|\mathbf{Z}\|} \right)^\top, \quad \mathbf{Z} \sim \mathsf{N}(\mathbf{0}, I).$$

2. *Generate a random scale factor*

$$\Lambda \sim \mathsf{N}\left(-\frac{\mathbf{x}^\top \Sigma^{-1} \mathbf{d}}{\mathbf{d}^\top \Sigma^{-1} \mathbf{d}}, \frac{1}{\mathbf{d}^\top \Sigma^{-1} \mathbf{d}} \right).$$

3. *If $S(\mathbf{x} + \Lambda \mathbf{d}) \geqslant \gamma_t$, set $\mathbf{x} = \mathbf{x} + \Lambda \mathbf{d}$; otherwise, leave $\mathbf{x}$ unchanged.*

4. *Repeat Steps 1, 2, and 3 above, say, 100 times and output $\mathbf{Y} = \mathbf{x}$ as the next state of the Markov chain.*

☞ 706 To speed up Step 2 above, the Cholesky factor of Σ can be precomputed and stored in memory to facilitate the computation of $\Sigma^{-1}\mathbf{d}$ via backward and forward substitution. For a diagonal covariance matrix $\Sigma = \sigma^2 I$, Step 2 reduces to $\Lambda \sim$ ☞ 514 $\mathsf{N}(-\mathbf{x}^\top \mathbf{d}, \sigma^2)$; see also Algorithm 14.11.

The pilot algorithm used to compute the levels $\{\gamma_t\}$ and splitting factors $\{\varrho_t^{-1}\}$ is given as follows.

Algorithm 16.9 (ADAM for Network Reliability) *Given the sample size N and the rarity parameter $\varrho \in (0,1)$, execute the following steps.*

1. **Initialization.** *Set the counter $t = 1$.*

   (a) *Generate independent time of repair vectors $\mathbf{X}_1, \ldots, \mathbf{X}_N \stackrel{\text{iid}}{\sim} \mathsf{N}(\mathbf{0}, \Sigma)$ and denote $\mathcal{X}_0 = \{\mathbf{X}_1, \ldots, \mathbf{X}_N\}$.*

   (b) *Denote $S_i = S(\mathbf{X}_i)$ for all $\mathbf{X}_i \in \mathcal{X}_{t-1}$, and let*

$$\widetilde{\gamma}_t = \underset{\gamma \in \{S_1,\ldots,S_N\}}{\operatorname{argmin}} \left\{ \frac{1}{N} \sum_{i=1}^{N} \mathrm{I}_{\{S(\mathbf{X}_i) > \gamma\}} \leqslant \varrho \right\} . \tag{16.19}$$

   That is, $\widetilde{\gamma}_t$ is the smallest value from among $S(\mathbf{X}_1), \ldots, S(\mathbf{X}_N)$ such that $\frac{1}{N}\sum_{i=1}^{N} \mathrm{I}_{\{S(\mathbf{X}_i) > \widetilde{\gamma}_t\}} \leqslant \varrho$. Set $\gamma_t = \min\{1, \widetilde{\gamma}_t\}$. Let $\mathcal{X}_t = \{\mathbf{X}_1, \ldots, \mathbf{X}_{N_t}\}$ be the vectors in $\mathcal{X}_{t-1}$ for which the network is nonoperational at time γ_t (that is, $S(\mathbf{X}) > \gamma_t$). Then, $\varrho_t = \frac{N_t}{N}$ is an approximation of the probability $c_t = \mathbb{P}_f(S(\mathbf{X}) \geqslant \gamma_t \mid S(\mathbf{X}) \geqslant \gamma_{t-1})$, setting $\gamma_0 = -\infty$.

2. **Markov chain sampling.** *Identical to Step 2 of Algorithm 16.7, except that in (16.17) the splitting factors are generated in a different way, namely,*

$$\mathcal{S}_{ti} = \left\lfloor \frac{N}{N_t} \right\rfloor + B_i, \quad i = 1, \ldots, N_t .$$

   Here, each $B_1, \ldots, B_{N_t}$ are $\mathsf{Ber}(1/2)$ random variables with joint pdf

$$\mathbb{P}(B_1 = b_1, \ldots, B_{N_t} = b_{N_t}) = \frac{(N_t - r)!\, r!}{N_t!} \mathrm{I}_{\{b_1 + \cdots + b_{N_t} = r\}} ,$$

   where $r = (N \bmod N_t)$. As a consequence of the generation of the splitting factors, after resetting

$$\mathcal{X}_t = \left\{ \mathbf{Y}_{1,1}, \mathbf{Y}_{1,2}, \ldots, \mathbf{Y}_{1,\mathcal{S}_{t1}}, \cdots\cdots\cdots, \mathbf{Y}_{N_t,1}, \mathbf{Y}_{N_t,2}, \ldots, \mathbf{Y}_{N_t,\mathcal{S}_{tN_t}} \right\},$$

   the set $\mathcal{X}_t$ contains exactly N elements.

3. **Updating and estimation.** *Update the counter $t = t + 1$ and proceed as in part (b) of Step 1.*

4. **Stopping condition.** *If $\gamma_t = \gamma$, set $T = t$ and go to Step 5; otherwise, repeat from Step 2.*

5. **Final estimates.** *Deliver the estimated levels $\gamma_1, \ldots, \gamma_T$ and the splitting factors $\varrho_1^{-1}, \ldots, \varrho_T^{-1}$.*

■ **EXAMPLE 16.5 (Dodecahedron Network)**

Consider again the dodecahedron network in Figure 16.5 with common edge unreliability $q = 10^{-3}$. Then, we have

$$\{X_i\} \overset{\text{iid}}{\sim} \mathsf{N}(0, \sigma^2), \quad \sigma = -1/\Phi^{-1}(q),$$

and $\Sigma = \sigma^2 I$. Using the levels and splitting factors in Table 16.2 with $N = 5 \times 10^3$ we obtain a typical estimate of $\widehat{\ell} = 1.97 \times 10^{-9}$ with an estimated relative error of 4%.

Table 16.2 The sequence $\{(\gamma_t, \varrho_t)\}$ used for the setup of Algorithm 16.7. The sequence is obtained using Algorithm 16.9 with $N = 10^3$ and $\varrho = 0.1$.

t	1	2	3	4	5	6	7	8	9
ϱ_t	0.1	0.1	0.1	0.1	0.1	0.1	0.1	0.1	0.2
γ_t	0.20	0.33	0.46	0.58	0.68	0.77	0.86	0.94	1

The most computationally intensive part of Algorithm 16.7 is the evaluation of $S(\mathbf{x})$ for a given vector. All other elements of the algorithm take negligible computational time. Thus, we can measure the simulation effort by counting the number of times we need to evaluate $S(\mathbf{x})$. In this example, the total simulation effort is equivalent to approximately 4×10^7 evaluations of the function $S(\mathbf{x})$.

Figure 16.6 shows the estimates of $\mathbb{E}_{f_T} X_e$, $e = 1, \ldots, m$ obtained using the empirical mean of the vectors in $\mathcal{X}_T$, where

$$f_T(\mathbf{x}) = \frac{f(\mathbf{x})\, \mathrm{I}_{\{S(\mathbf{x})>1\}}}{\ell},$$

is the conditional density of $\mathbf{x}$ given that the network is nonoperational at time 1.

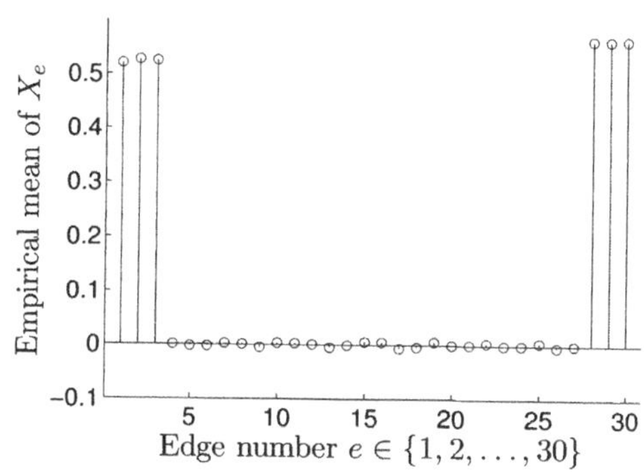

Figure 16.6 The estimated mean of the vectors $\mathbb{E}_{f_T} X_e$, $e = 1, \ldots, m$ under the minimum variance density $f_T(\mathbf{x}) \propto f(\mathbf{x})\, \mathrm{I}_{\{S(\mathbf{x})>1\}}$.

Note that the mean times of repair of edges 1, 2, 3 and 28, 29, 30 are much larger than their values under the nominal $\mathsf{N}(\mathbf{0}, \sigma^2 I)$ distribution, while the mean times of repair of all other edges remain close to 0. We call all edges with such "tilted" mean times of repair the **critical edges**. Thus, in this case the edges 1, 2, 3, 28, 29, 30 are all critical edges. Figure 16.7 shows the estimated correlation between the edges in the final population $\mathcal{X}_T$:

$$\text{Corr}(X_i, X_j) = \frac{\text{Cov}(X_i, X_j)}{\sqrt{\text{Var}(X_i)\text{Var}(X_j)}}, \quad \mathbf{X} \sim f_T(\mathbf{x}) .$$

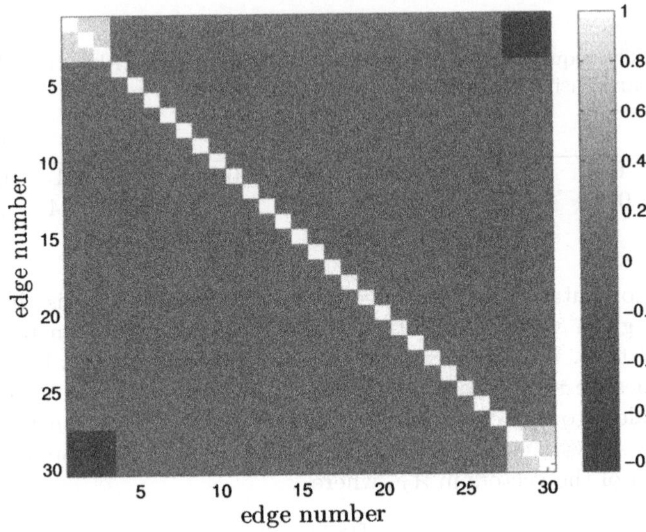

Figure 16.7 Correlation matrix estimated from the vectors at the final stage of the algorithm. All such vectors satisfy $S(\mathbf{x}) > 1$.

Edges $\{1, 2, 3\}$ are strongly positively correlated (similarly for $\{28, 29, 30\}$) and the two sets of edges $\{1, 2, 3\}$ and $\{28, 29, 30\}$ are strongly negatively correlated. In other words, failures of the network occur primarily because either edge sets $\{1, 2, 3\}$ or $\{28, 29, 30\}$ fail, but not both.

■ EXAMPLE 16.6 (Edge Dependency)

Suppose the unreliability of each edge in the dodecahedron network in Figure 16.5 is $q = 10^{-6}$. Assume that the edge failures are dependent in such a way that all times of repair are positively correlated with covariance matrix $\Sigma = \sigma^2\widetilde{\Sigma}$, where $\sigma = -1/\Phi^{-1}(q)$ and $\widetilde{\Sigma}$ is a matrix with 1s along the diagonal and all other entries equal to 0.5.

Using the levels and splitting factors in Table 16.3 with $N = 15000$ we obtain a typical estimate of $\widehat{\ell} = 6.74 \times 10^{-10}$ with an estimated relative error of 2.8%. The total simulation effort is equivalent to approximately 1.3×10^8 evaluations of the function $S(\mathbf{x})$. If the edges were to fail independently (that is $\widetilde{\Sigma} = I$), then the unreliability would be of the order 10^{-18}.

Table 16.3 The sequence $\{(\gamma_t, \varrho_t)\}$ used for Algorithm 16.7. The sequence is obtained using Algorithm 16.9 with $N = 10^3$ and $\varrho = 0.1$.

t	1	2	3	4	5	6	7	8	9	10
ϱ_t	0.1	0.1	0.1	0.1	0.1	0.1	0.1	0.1	0.1	0.8
γ_t	0.22	0.40	0.51	0.61	0.70	0.78	0.86	0.92	0.99	1

16.5.1 Acceleration Using Bounds

Typically, the main cost in using Algorithm 16.7 lies in the evaluation of the performance $S(\mathbf{X})$. It is possible to avoid evaluating $S(\mathbf{X})$ most of the time using bounds. Note that in Steps 1 and 3 of Algorithm 16.7 we are not interested in the exact value of $S(\mathbf{X})$ per se, but rather in the value of the binary random variable $B = \mathrm{I}_{\{S(\mathbf{X})>\gamma_t\}}$, that is, in testing the condition $S(\mathbf{X}) > \gamma_t$. The same is true for Step 3 in Algorithm 16.8. The computation of the indicator B can be substantially accelerated using the bounds on S given in (16.13).

This idea is similar to the *squeezing* technique used to accelerate the acceptance–rejection method in Section 3.1.5 and is summarized in the following subroutine of Algorithm 16.7. ☞ 59

Algorithm 16.10 (Evaluating $\mathrm{I}_{\{S(\mathbf{X})>\gamma_t\}}$ via Squeezing)

1. *If $S_L(\mathbf{X}) > \gamma_t$, output $B = 1$ and exit; otherwise, go to Step 2.*
2. *If $S_U(\mathbf{X}) \leqslant \gamma_t$, output $B = 0$ and exit; otherwise, go to Step 3.*
3. *Evaluate $S(\mathbf{X})$ using Algorithm 16.1 and output $B = \mathrm{I}_{\{S(\mathbf{X})>\gamma_t\}}$.*

Depending on the choice of bounds, application of Algorithm 16.10 can result in a substantial reduction in the number of evaluations of S. In addition, if S_L and S_U can be computed easily there will be a substantial speed-up of Algorithm 16.7.

A good heuristic for constructing suitable sets $\mathscr{C}^*$ and $\mathscr{P}^*$ is to require that the cuts (paths) in $\mathscr{C}^*$ ($\mathscr{P}^*$) contain the critical edges identified using the pilot Algorithm 16.9. For example, from Figure 16.6 we can deduce that the cut and path sets should contain edges 1, 2, 3, 28, 29, 30. We need not assume that the paths and cuts are edge-disjoint, as is usually assumed in the literature [18]. In other words, a given edge can belong to more than one path or cut.

■ **EXAMPLE 16.7 (Dodecahedron Network Using Bounds)**

We repeat the simulation in Example 16.5 using Algorithm 16.10 as a subroutine whenever evaluation of the indicator $\mathrm{I}_{\{S(\mathbf{X})>\gamma_t\}}$ is needed. The total number of evaluations of $\{S(\mathbf{X}) > \gamma_t\}$ is still 4×10^7, but the total number of evaluations of the function $S(\mathbf{x})$ is 4×10^6, which is 10% of the original number of samples (4×10^7). The rest of the evaluations of the indicator are helped by the upper (36%) and lower (46%) bounds.

Repeating the simulation in Example 16.6 using Algorithm 16.10 results in 7×10^7 evaluations of $S(\mathbf{X})$, which is about 50% of the simulation effort required without squeezing.

Further Reading

Exact calculation of network reliability is a difficult #P-complete problem [3, 10]. Although approximation [6] and bounding [4, 5, 11, 16, 23] techniques are available, their accuracy and scope are very much dependent on the properties of the network (such as size and topology). Currently, for large networks simulation techniques appear to be the only viable approach to approximating the unreliability. However, in highly reliable networks, such as modern communication networks, the probability of network failure is a rare-event probability in a static simulation setting. Thus, methods such as crude Monte Carlo are not practical. Various variance reduction techniques have been developed to produce better estimates: dagger sampling [28], control variables [18], sequential sampling [13], conditional Monte Carlo [14], conditional Monte Carlo using combinatorial analysis [19, 30], importance sampling [8, 27], conditional Monte Carlo combined with approximate zero-variance importance sampling [9], and (one of the most efficient and specialized approaches) recursive variance reduction [7], see also [26] for a similar proposal. While the recursive variance reduction method appears to be the most efficient method, it is currently applicable only when the edges fail independently. For a comprehensive survey of algorithms and historical developments, see [33]. For a monograph dedicated to novel evolution models for network reliability estimation, we refer to [20].

REFERENCES

1. S. V. Amari. Closed-form expressions for distribution of sum of exponential random variables. *IEEE Transactions on Reliability*, 46(4):519–522, 1997.

2. A. N. Avramidis, N. Channouf, and P. L'Ecuyer. Efficient correlation matching for fitting discrete multivariate distributions with arbitrary marginals and normal-copula dependence. *INFORMS Journal on Computing*, 21(1):88–106, 2009.

3. M. O. Ball and J. S. Provan. Bounds on the reliability polynomial for shellable independence systems. *SIAM Journal on Algebraic and Discrete Methods*, 3(2):166–181, 1982.

4. R. Barlow and F. Proschan. *Statistical Theory of Reliability and Life Testing*. Holt, Rinehart & Wilson, New York, 1975.

5. R. E. Barlow and A. W. Marshall. Bounds for distribution with monotone hazard rate, I and II. *Annals of Mathematical Statistics*, 35(3):1234–1274, 1964.

6. Y. Burtin and B. Pittel. Asymptotic estimates of the reliability of a complex system. *Engineering Cybernetics*, 10(3):445–451, 1972.

7. H. Cancela and M. El Khadiri. The recursive variance-reduction simulation algorithm for network reliability evaluation. *IEEE Transactions on Reliability*, 52(2):207–212, 2003.

8. H. Cancela, P. L'Ecuyer, M. Lee, G. Rubino, and B. Tuffin. Analysis and improvements of path-based methods for Monte Carlo reliability evaluation of static models. In J. Faulin, A. A. Juan, S. S. Martorell Alsina, and J. E. Ramirez-Marquez, editors, *Simulation Methods for Reliability and Availability of Complex Systems*. Springer-Verlag, New York, 2009.

9. H. Cancela, P. L'Ecuyer, G. Rubino, and B. Tuffin. Combination of conditional Monte Carlo and approximate zero-variance importance sampling for network reliability es-

timation. In B. Johansson, S. Jain, J. Montoya-Torres, J. Hugan, and E. Yücesan, editors, *Proceedings of the 2010 Winter Simulation Conference*, 2010.

10. C. J. Colbourn. *The Combinatorics of Network Reliability.* Oxford University Press, Oxford, 1987.

11. C. J. Colbourn, L. D. Nel, T. B. Boffey, and D. F. Yates. Network reliability and the probabilistic estimation of damage from fire spread. *Annals of Operations Research*, 50(1):173–185, 1994.

12. E. W. Dijkstra. A note on two problems in connexion with graphs. *Numerische Mathematik*, 1(1):269–271, 1959.

13. M. Easton and C. Wong. Sequential destruction method for Monte Carlo evaluation of system reliability. *IEEE Transactions on Reliability*, 29(1):27–32, 1980.

14. T. Elperin, I. B. Gertsbakh, and M. Lomonosov. Estimation of network reliability using graph evolution models. *IEEE Transactions on Reliability*, 40(5):572–581, 1991.

15. T. Elperin, I. B. Gertsbakh, and M. Lomonosov. An evolution model for Monte Carlo estimation of equilibrium network renewal parameters. *Probability in the Engineering and Informational Sciences*, 6(4):457–469, 1992.

16. J. D. Esary, F. Proschan, and D. W. Walkup. Association of random variables, with applications. *Annals of Mathematical Statistics*, 38(5):1466–1473, 1967.

17. S. Favaro and S. G. Walker. On the distribution of sums of independent exponential random variables via Wilks' integral representation. *Acta Applicandae Mathematicae*, 109(3):1035–1042, 2010.

18. G. Fishman. A Monte Carlo sampling plan for estimating network reliability. *Operations Research*, 34(4):581–594, 1980.

19. I. Gertsbakh and Y. Shpungin. Network reliability importance measures: Combinatorics and Monte Carlo based computations. *WSEAS Transactions on Computers*, 7(4):216–227, 2008.

20. I. B. Gertsbakh and Y. Shpungin. *Models of Network Reliability: Analysis, Combinatorics and Monte Carlo.* Taylor & Francis Group, Boca Raton, FL, 2010.

21. K.-P. Hui, N. Bean, M. Kraetzl, and D. P. Kroese. The tree cut and merge algorithm for estimation of network reliability. *Probability in the Engineering and Informational Sciences*, 17(1):24–45, 2003.

22. K.-P. Hui, N. Bean, M. Kraetzl, and D. P. Kroese. The cross-entropy method for network reliability estimation. *Annals of Operations Research*, 134(1):101–118, 2005.

23. C.-C. Jane and Y. W. Liah. A dynamic bounding algorithm for approximating multistate two-terminal reliability. *European Journal of Operational Research*, 205(3):625–637, 2010.

24. H. Joe. *Multivariate Models and Dependence Concepts.* Chapman & Hall, London, 1997.

25. D. E. Knuth. *The Art of Computer Programming Volume 1.* Addison-Wesley, Boston, third edition, 1997.

26. A. Konak. Combining network reductions and simulation to estimate network reliability. In S. G. Henderson, B. Biller, M.-H. Hsieh, J. Shortle, J. D. Tew, and R. R. Barton, editors, *Proceedings of the 2007 Winter Simulation Conference*, pages 2301–2305, 2007.

27. D. P. Kroese and K.-P. Hui. Applications of the cross-entropy method in reliability. In G. Levitin, editor, *Computational Intelligence in Reliability Engineering*, volume 40, pages 37–82. Springer-Verlag, Berlin, 2007.

28. H. Kumamoto, K. Tanaka, K. Inoue, and E. J. Henley. Dagger sampling Monte Carlo for system unavailability evaluation. *IEEE Transactions on Reliability*, 29(2):376–380, 1980.

29. M. Lomonosov. On Monte Carlo estimates in network reliability. *Probability in the Engineering and Informational Sciences*, 8(2):245–265, 1994.

30. M. Lomonosov and Y. Shpungin. Combinatorics and reliability Monte Carlo. *Random Structures and Algorithms*, 14(4):329–343, 1999.

31. E. Manzi, M. Labbe, G. Latouche, and F. Maffioli. Fishman's sampling plan for computing network reliability. *IEEE Transactions on Reliability*, 50(1):41–46, 2001.

32. G. Rubino. Network reliability evaluation. In J. Walrand, K. Bagchi, and G. W. Zobrist, editors, *Network Performance Modeling and Simulation*, chapter 11. Blackwell Scientific Publications, Amsterdam, 1998.

33. G. Rubino and B. Tuffin (editors). *Rare Event Simulation*. John Wiley & Sons, New York, 2009.

CHAPTER 17

APPLICATIONS TO DIFFERENTIAL EQUATIONS

The first Monte Carlo algorithm designed for a computer by John von Neumann in 1947 involved solving a Boltzmann differential equation via simulation of an associated stochastic process. In this chapter we highlight various connections between Monte Carlo simulation and differential equations. In particular, in Section 17.1 we explore the relation between stochastic and partial differential equations. In Section 17.2 we look at simulating transport processes and their connections to certain partial differential equations. Finally, in Section 17.3 we examine the scaling of Markov jump processes, giving a functional law of large numbers that relates the process to the solution of a system of ordinary differential equations. In addition, a functional central limit theorem approximation is given in the form of an approximating diffusion process.

We refer to Chapter 5 for stochastic process generation and to Sections A.9–A.13 ☞ 153
for background on Markov processes, including diffusion processes. ☞ 626

17.1 CONNECTIONS BETWEEN STOCHASTIC AND PARTIAL DIFFERENTIAL EQUATIONS

A fundamental link between differential equations and stochastic processes is found in the theory of Markov processes through the Kolmogorov backward equations (giving rise to Feynman–Kac type formulas) and Kolmogorov forward equations. Feynman–Kac type formulas enable one to write the solution of linear second-order partial differential equations of parabolic (including elliptic) type as an expectation of a function of a Markov process. This representation lends itself directly to Monte Carlo techniques. The links forged this way then enable us to:

- Evaluate expected values of functionals of SDE solutions by solving an associated partial differential equation.
- Solve certain partial differential equations by computing expectations of functionals of an associated SDE.

The Monte Carlo viewpoint becomes increasingly attractive as the problem dimension increases, if the solution is only required at a small number of points, or if the differential operator is degenerate [21].

We consider partial differential equations associated with the diffusion processes
☞ 645 described by the multidimensional SDE

$$\mathrm{d}\mathbf{X}_t = \mathbf{a}(\mathbf{X}_t, t)\,\mathrm{d}t + B(\mathbf{X}_t, t)\,\mathrm{d}\mathbf{W}_t\,, \tag{17.1}$$

with $d \times 1$ drift vector $\mathbf{a}(\mathbf{x}, t)$ and $d \times m$ dispersion matrix $B(\mathbf{x}, t)$, so that $C(\mathbf{x}, t) = B(\mathbf{x}, t)B(\mathbf{x}, t)^\top$ is the diffusion matrix.

Let $\mathbb{P}^{\mathbf{x},t}$ denote the probability measure under which the diffusion process $\{\mathbf{X}_s\}$ starts from state $\mathbf{x}$ at time t; that is, $\mathbb{P}^{\mathbf{x},t}(\mathbf{X}_t = \mathbf{x}) = 1$. The corresponding expectation is denoted by $\mathbb{E}^{\mathbf{x},t}$. When the initial time is 0, we write $\mathbb{P}^{\mathbf{x}} = \mathbb{P}^{\mathbf{x},0}$ and $\mathbb{E}^{\mathbf{x}} = \mathbb{E}^{\mathbf{x},0}$.

For each $t \in [0, T]$, let

$$L_t = \sum_{i=1}^{d} a_i(\mathbf{x}, t)\frac{\partial}{\partial x_i} + \frac{1}{2}\sum_{i=1}^{d}\sum_{j=1}^{d} C_{ij}(\mathbf{x}, t)\frac{\partial^2}{\partial x_i\,\partial x_j} \tag{17.2}$$

☞ 709 be the associated differential operator acting on functions $u(\mathbf{x}, t) \in \mathcal{C}^2 \times \mathcal{C}^1$. This operator is extended by the *infinitesimal generator* of the Markov process described by the SDE (17.1), defined as

$$A_t u(\mathbf{x}) = \lim_{s \downarrow 0} \frac{\mathbb{E}^{\mathbf{x},t} u(\mathbf{X}_{t+s}) - u(\mathbf{x})}{s}\,,$$

acting on functions u for which the limit above exists.

If the coefficients $\mathbf{a}$ and B do not depend on t, the corresponding SDE (and diffusion process) is time-homogeneous:

$$\mathrm{d}\mathbf{X}_t = \mathbf{a}(\mathbf{X}_t)\,\mathrm{d}t + B(\mathbf{X}_t)\,\mathrm{d}\mathbf{W}_t\,, \tag{17.3}$$

and the differential operator is

$$L = \sum_{i=1}^{d} a_i(\mathbf{x})\frac{\partial}{\partial x_i} + \frac{1}{2}\sum_{i=1}^{d}\sum_{j=1}^{d} C_{ij}(\mathbf{x})\frac{\partial^2}{\partial x_i\,\partial x_j}\,. \tag{17.4}$$

Throughout the rest of this section, we make the following assumptions.

Assumptions 17.1.1 (Continuity and Growth)

1. *The diffusion matrix C is* **uniformly elliptic***; that is, there exists a constant $\delta > 0$ such that*

$$\boldsymbol{\xi}^{\top} C \boldsymbol{\xi} \geqslant \delta \, \|\boldsymbol{\xi}\|^2$$

holds for all $\boldsymbol{\xi} \in \mathbb{R}^d$ and every point $\mathbf{x}$ or $(\mathbf{x}, t)$ for which the matrix C is defined. Note that, of necessity, C is positive semidefinite.

2. *The SDE* (17.1) *has a unique strong solution — see Theorem A.13.2.* ☞ 646

3. *Functions k and g appear as problem data in what follows. We assume that they are* **uniformly Hölder continuous** *for some exponent α. More precisely, given a Euclidean space $\mathscr{X} \subseteq \mathbb{R}^d$ and a function $f : \mathscr{X} \to \mathbb{R}$, f is said to be uniformly Hölder continuous in $\mathscr{X}$ with exponent $\alpha \in [0, 1]$, if there is some constant M so that*

$$\sup_{\mathbf{x},\mathbf{y} \in \mathscr{X}} \frac{|f(\mathbf{y}) - f(\mathbf{x})|}{\|\mathbf{y} - \mathbf{x}\|^{\alpha}} \leqslant M \, .$$

If $\alpha = 1$, uniform Hölder continuity is referred to as **uniform Lipschitz continuity** *in $\mathscr{X}$. If $\alpha = 0$, uniform Hölder continuity is simply boundedness of the function f in $\mathscr{X}$.*

4. *The SDE coefficient functions $\{a_i\}$ and $\{B_{ij}\}$ are uniformly Lipschitz continuous in their arguments. More precisely, if the coefficient functions are time dependent and defined in $\mathscr{X} \times [0, T]$, then they are uniformly Lipschitz continuous in $\mathscr{X} \times [0, T]$. Similarly, if the coefficient functions are independent of time and defined in $\mathscr{X}$, then they are uniformly Lipschitz continuous in $\mathscr{X}$.*

17.1.1 Boundary Value Problems

Suppose that $\mathscr{X}$ is a bounded domain (that is, a connected open subset) of $\mathbb{R}^d$ with boundary set $\partial\mathscr{X}$. Let $\overline{\mathscr{X}}$ denote the *closure* of $\mathscr{X}$. A PDE problem involving the differential operator L in (17.4) is said to be a **boundary value problem** or a **Dirichlet problem** if the objective is to find a function u on $\overline{\mathscr{X}}$ that solves

$$\begin{aligned} (L - k(\mathbf{x}))\, u(\mathbf{x}) &= -g(\mathbf{x}) \quad \text{in } \mathscr{X} \, , \\ u(\mathbf{x}) &= f(\mathbf{x}) \quad \text{on } \partial\mathscr{X} \, , \end{aligned} \tag{17.5}$$

for given functions $k : \overline{\mathscr{X}} \to [0, \infty)$, $g : \overline{\mathscr{X}} \to \mathbb{R}$, and $f : \partial\mathscr{X} \to \mathbb{R}$.

A unique solution u to the Dirichlet problem exists that is of class $\mathcal{C}^2$ in $\mathscr{X}$ and $\mathcal{C}^1$ on $\partial\mathscr{X}$ if, in addition to Assumptions 17.1.1, the following smoothness conditions for the boundary are satisfied.

Assumptions 17.1.2 (Boundary Smoothness)

1. **Continuous boundary function***: The function f is continuous on $\partial\mathscr{X}$.*

2. **Exterior sphere property***: For every point $\mathbf{y} \in \partial\mathscr{X}$, there exists a ball $\mathscr{B}(\mathbf{y})$ such that $\overline{\mathscr{B}}(\mathbf{y}) \cap \mathscr{X} = \emptyset$ and $\overline{\mathscr{B}}(\mathbf{y}) \cap \partial\mathscr{X} = \{\mathbf{y}\}$.*

Under Assumptions 17.1.1 and 17.1.2, the solution u is given by the stochastic representation contained in the following theorem.

Theorem 17.1.1 (Stochastic Representation for the Dirichlet Problem) *Let $\tau = \inf_{t\geqslant 0}\{\mathbf{X}_t \notin \mathscr{X}\}$ be the first time that the process $\{\mathbf{X}_t, t \geqslant 0\}$ in (17.3) (with associated operator L in (17.4)) exits $\mathscr{X}$. Under Assumptions 17.1.1 and 17.1.2, $\mathbb{E}^{\mathbf{x}}\tau < \infty$ for all $\mathbf{x} \in \mathscr{X}$. Moreover, there exists a unique function $u : \overline{\mathscr{X}} \to \mathbb{R}$ that is $\mathcal{C}^2$ in $\mathscr{X}$ and $\mathcal{C}^1$ on $\partial\mathscr{X}$ that satisfies the Dirichlet problem (17.5) admitting the stochastic representation*

$$u(\mathbf{x}) = \mathbb{E}^{\mathbf{x}}\left[f(\mathbf{X}_\tau)\,K(\tau) + \int_0^\tau g(\mathbf{X}_t)\,K(t)\,\mathrm{d}t\right], \quad \mathbf{x} \in \overline{\mathscr{X}}, \tag{17.6}$$

where

$$K(t) = \exp\left(-\int_0^t k(\mathbf{X}_s)\,\mathrm{d}s\right).$$

A sketch of the proof is as follows [13, Pages 253, 276, 365, and 393]. Denote by $\{\mathbf{X}_t\}$ the process that solves (17.3). Assume there exists a solution u to (17.5).
☞ 642 Apply Itô's formula (A.64) to the function $u(\mathbf{X}_s)\,K(s)$, yielding

$$\begin{aligned} u(\mathbf{X}_s)\,K(s) &= u(\mathbf{X}_0) + \int_0^s K(r)\,(L - k(\mathbf{X}_r))\,u(\mathbf{X}_r)\,\mathrm{d}r \\ &+ \sum_{i=1}^{d}\sum_{k=1}^{m}\int_0^s B_{ik}(\mathbf{X}_r, r)\frac{\partial}{\partial x_i}(u(\mathbf{X}_r)\,K(r))\,\mathrm{d}W_r^{(k)}. \end{aligned} \tag{17.7}$$

Let $\mathscr{X}_1 \subseteq \mathscr{X}_2 \subseteq \ldots$ be an increasing sequence of open sets satisfying $\overline{\mathscr{X}_n} \subset \mathscr{X}$ and $\cup_{n=1}^{\infty}\mathscr{X}_n = \mathscr{X}$. Associate first exit times with each set, defined by $\tau_n = \inf\{t \geqslant 0 : \mathbf{X}_t \notin \mathscr{X}_n\}$. This sequence of stopping times satisfies $\lim_{n\to\infty}\tau_n = \tau$.

For each $n \geqslant 1$ and $t \in [0, \min\{\tau_n, s\}]$, equation (17.5) is satisfied, meaning $(L - k(\mathbf{X}_t))u(\mathbf{X}_t) = -g(\mathbf{X}_t)$, and so each process $Y^{(n)} = \{Y_s^{(n)}\}$ defined by

$$Y_s^{(n)} = u(\mathbf{X}_{\min\{\tau_n,s\}})\,K(\min\{\tau_n, s\}) + \int_0^{\min\{\tau_n,s\}} g(\mathbf{X}_r)\,K(r)\,\mathrm{d}r, \quad 0 \leqslant s < \infty,$$

is a $\mathbb{P}^{\mathbf{x}}$ martingale for $\mathbf{x} \in \mathscr{X}$.

Under Assumptions 17.1.1 and 17.1.2 each of these processes is bounded and $\mathbb{E}^{\mathbf{x}}\tau < \infty$. Hence, the process $Y = \lim_n Y^{(n)}$, with

$$Y_s = u(\mathbf{X}_{\min\{\tau,s\}})\,K(\min\{\tau, s\}) + \int_0^{\min\{\tau,s\}} g(\mathbf{X}_r)\,K(r)\,\mathrm{d}r, \quad 0 \leqslant s < \infty,$$

is a uniformly integrable $\mathbb{P}^{\mathbf{x}}$ martingale. Taking expectations of (17.7) and using the fact that Y is a martingale, so that $\mathbb{E}Y_0 = \mathbb{E}Y_\tau$, we see that

$$u(\mathbf{x}) = \mathbb{E}^{\mathbf{x}}\left[f(\mathbf{X}_\tau)\,K(\tau) + \int_0^\tau g(\mathbf{X}_r)\,K(r)\,\mathrm{d}r\right].$$

☞ 181 The canonical Dirichlet problem is that for **Laplace's heat equation**, namely

$$\begin{aligned} \frac{1}{2}\Delta u(\mathbf{x}) &= 0 \quad \text{in } \mathscr{X}, \\ u(\mathbf{x}) &= f(\mathbf{x}) \quad \text{on } \partial\mathscr{X}, \end{aligned} \tag{17.8}$$

where Δ is the Laplace operator $\Delta = \sum_{i=1}^{d} \sum_{j=1}^{d} \frac{\partial^2}{\partial x_i \partial x_j}$. Physically, the solution u to this problem describes the equilibrium temperature obtained by keeping the temperature on the boundary of a homogeneous material set to f and allowing heat flow until a steady-state temperature distribution is reached.

The associated SDE has no drift, and is simply given by

$$\mathrm{d}\mathbf{X}_t = \mathrm{d}\mathbf{W}_t\,,$$

and the solution u has the simple stochastic representation

$$u(\mathbf{x}) = \mathbb{E}^{\mathbf{x}} f(\mathbf{X}_\tau)\,.$$

■ EXAMPLE 17.1 (Solving a Dirichlet Problem via Simulation)

Suppose our Dirichlet problem is specified as follows.

- The region $\mathscr{X}$ is defined by $\mathscr{X} = \{(x, y) \,:\, x^2 + y^2 < 5^2,\, x^4 + y^4 > 2.5^4\}$.
- We have functions $f(x, y) = 0$, $k(x, y) = 0$, $g(x, y) = \sin(x)\cos(y)$, and

$$\mathbf{a}(x, y) = \begin{pmatrix} 1 \\ 1 \end{pmatrix} \quad \text{and} \quad C = \begin{pmatrix} 1.04 & 0.6 \\ 0.6 & 4.04 \end{pmatrix}.$$

Thus the Dirichlet problem we wish to solve is given by

$$\begin{aligned} 0.52\,\frac{\partial^2 u(x,y)}{\partial x^2} + 0.6\,\frac{\partial^2 u(x,y)}{\partial x \partial y} + 2.02\,\frac{\partial^2 u(x,y)}{\partial y^2} & \\ + \frac{\partial u(x,y)}{\partial x} + \frac{\partial u(x,y)}{\partial y} &= -\sin(x)\cos(y) \quad \text{in } \mathscr{X}\,, \\ u(x,y) &= 0 \quad \text{on } \partial\mathscr{X}\,. \end{aligned}$$

The condition of uniform ellipticity is equivalent here to requiring that there exists some $\delta > 0$ such that the matrix

$$C_\delta = \begin{pmatrix} 1.04 - \delta & 0.6 \\ 0.6 & 4.04 - \delta \end{pmatrix}$$

is positive semidefinite. This is the case when, for example, $\delta = 0.04$. The functions C_{ik}, a_i, and k are all constant, and so are trivially Hölder continuous in $\mathscr{X}$. The function g is bounded in $\mathscr{X}$ (and indeed $\mathbb{R}^2$), so is also Hölder continuous. The function f is constant, satisfying the continuous boundary condition. Finally, we can find a ball of constant diameter to roll along any portion of the boundary without it "getting stuck" (that is, touching more than one point on $\partial\mathscr{X}$), so the boundary $\partial\mathscr{X}$ satisfies the exterior sphere property.

Thus there exists a unique solution $u(x, y)$ to this Dirichlet problem, and it admits the stochastic representation

$$u(x, y) = \mathbb{E}^{(x,y)} \int_0^\tau \sin(X_t)\cos(Y_t)\,\mathrm{d}t\,, \quad (x, y) \in \overline{\mathscr{X}}\,, \tag{17.9}$$

where $\mathbf{Z}_t = (X_t, Y_t)^\top$ is a diffusion process evolving according to

$$\mathrm{d}\mathbf{Z}_t = \begin{pmatrix}1\\1\end{pmatrix}\mathrm{d}t + \begin{pmatrix}1 & 0.2\\0.2 & 2\end{pmatrix}\mathrm{d}\mathbf{W}_t\,.$$

The dispersion matrix is chosen to be a matrix square root of the diffusion matrix C. For any point $(x, y) \in \overline{\mathscr{X}}$, we can obtain an approximate estimate of $u(x, y)$ via

$$\widehat{u}(x,y) = \frac{1}{N}\sum_{i=1}^{N}\sum_{j=1}^{\kappa^{(i)}} h\sin(\widehat{X}_{jh}^{(i)})\cos(\widehat{Y}_{jh}^{(i)})\,,$$

where each $\{(\widehat{X}_{jh}^{(i)}, \widehat{Y}_{jh}^{(i)})^\top, j = 1, 2, \ldots\}$ is a realization of the multidimensional Euler approximation for stochastic differential equations with constant step size h (see Section 5.6), each $\widehat{\tau}^{(i)}$ is taken to be the first time that the i-th such realization leaves $\mathscr{X}$, and each $\kappa^{(i)} = \mathrm{argmax}_j\{(j-1)h < \widehat{\tau}^{(i)}\}$ is the number of steps that realization i takes to leave $\mathscr{X}$ for the first time. The integral (17.9) is approximated by a step function, with jumps occurring at the time discretization, and values straightforwardly obtained as the integrand evaluated at the Euler approximation.

☞ 183

Figure 17.1 depicts an estimated solution on a 100×100 regular mesh over $[-5, 5]^2$, with the solution at each point of the mesh estimated as the average of $N = 1000$ realizations of the Euler approximation using a constant step size of $h = 10^{-3}$.

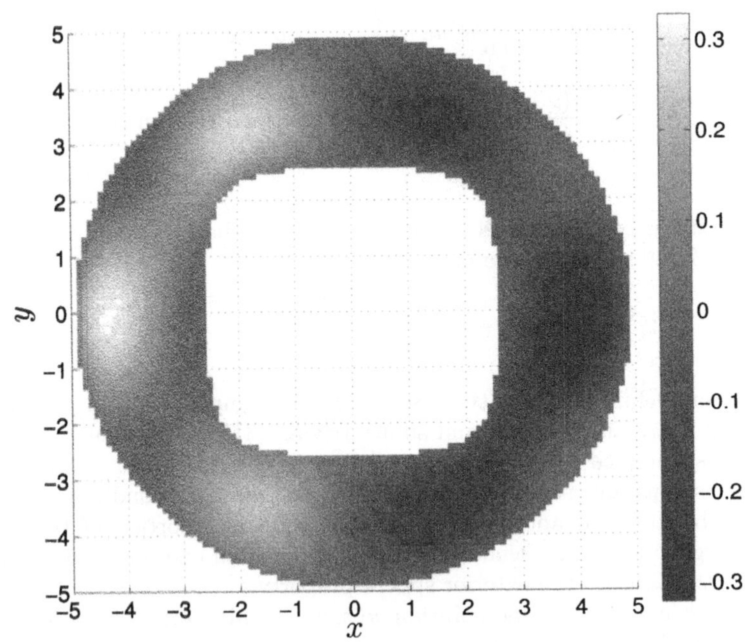

Figure 17.1 Estimated solution $u(x, y)$ to a Dirichlet problem.

The MATLAB code is given below.

```
% dirichlet_ex_script.m
% A Dirichlet example in 2D.
% Assumes k=0.

h=10.^(-3); % Timestep
nx=100; ny=100; ne=10^3;
xs=linspace(-5,5,nx);
ys=linspace(-5,5,ny);

u=zeros(nx,ny);

b=[1;1]; sig=[1,.2;.2,2];
bh=b.*h; sigh=sqrt(h).*sig;

for i=1:nx % For each x grid-point
  for j=1:ny % For each y grid-point
    x0=[xs(i);ys(j)];
    if ((x0(1)^2+x0(2)^2)>=5^2)||(x0(1)^4+x0(2)^4<=2.5^4)
      u(i,j)=NaN;
    else
      for k=1:ne % For each realization of the process

        notstopped=true;
        t=0; X=x0; intg=0;
        while (notstopped)
          t=t+h;

          X=X+bh+sigh*randn(2,1); % Euler Approx.

          % Check the Boundary condition
          if ((X(1)^2+X(2)^2)>=5^2)||(X(1)^4+X(2)^4<=2.5^4)
            notstopped=false;
          end

          intg=intg+h*sin(X(1))*cos(X(2)); % Riemann approx.
        end
        u(i,j)=u(i,j)+intg;
      end
      u(i,j)=u(i,j)/ne;
    end
  end
end

figure,surf(xs,ys,u')
```

Remark 17.1.1 (Errors Arising in Practice) In Example 17.1, we approximately estimated u by

- replacing the expectation by a Monte Carlo estimate of N realizations of the SDE process;
- approximately simulating the SDE process via Euler's method;
- evaluating the stochastic integral approximately via a step function.

Each of these steps produces approximation or estimation error. The Monte Carlo error can be reduced by applying variance reduction techniques to the original process [21, 28, 29] and [15, Chapter 16]. The approximation error due to using Euler's method can be mitigated by using a smaller step size or an approximation scheme of a higher convergence order [15]. Finally, we can use more sophisticated ways to approximate the stochastic integral, see [15, Pages 529–539] for details.

17.1.2 Terminal Value Problems

A PDE problem involving the differential operator L_t in (17.2) is said to be a **terminal value problem** or a **Cauchy problem** if it is of the form

$$\begin{aligned} \left(L_t + \frac{\partial}{\partial t} - k(\mathbf{x},t)\right) u_t(\mathbf{x}) &= -g(\mathbf{x},t) \quad \text{in } \mathbb{R}^d \times [0,T)\,, \\ u_t(\mathbf{x}) &= f(\mathbf{x}) \quad \text{on } \mathbb{R}^d \times \{T\}\,, \end{aligned} \tag{17.10}$$

where $T > 0$ is a fixed terminal time, and $k : \mathbb{R}^d \times [0,T] \to [0,\infty)$, $g : \mathbb{R}^d \times [0,T] \to \mathbb{R}$, and $f : \mathbb{R}^d \to \mathbb{R}$ are continuous functions.

A unique solution u_t to the Cauchy problem exists if, in addition to Assumptions 17.1.1, the following conditions are satisfied [13, Page 268]:

Assumptions 17.1.3 (Boundedness and Growth)

1. **Boundedness**: *The functions $C_{ik}(\mathbf{x},t)$, $a_i(\mathbf{x},t)$, and $k(\mathbf{x},t)$ are bounded in $\mathbb{R}^d \times [0,T]$.*
2. **Polynomial growth**: *The functions $f(\mathbf{x})$ and $g(\mathbf{x},t)$ satisfy $|f(\mathbf{x})| \leqslant M(1+\|\mathbf{x}\|^{2\lambda})$ for all $\mathbf{x} \in \mathbb{R}^d$ and $|g(\mathbf{x},t)| \leqslant M(1+\|\mathbf{x}\|^{2\lambda})$ for all $(\mathbf{x},t) \in \mathbb{R}^d \times [0,T]$ for some constants $M > 0$ and $\lambda \geqslant 1$.*

Moreover, u_t is given by the stochastic representation contained in the following theorem.

Theorem 17.1.2 (Feynman–Kac Representation for the Cauchy Problem) *Under Assumptions 17.1.1 and 17.1.3, there is a unique function $u_t(\mathbf{x}) : \mathbb{R}^d \times [0,T] \to \mathbb{R}$ that has continuous first partial derivatives in t, continuous second partial derivatives in $\mathbf{x}$, and satisfies the Cauchy problem (17.10). Moreover, $u_t(\mathbf{x})$ admits the stochastic representation*

$$u_t(\mathbf{x}) = \mathbb{E}^{\mathbf{x},t}\left[f(\mathbf{X}_T)\,K(t,T) + \int_t^T g(\mathbf{X}_s,s)K(t,s)\,\mathrm{d}s\right]$$

in $\mathbb{R}^d \times [0,T]$, *where*

$$K(t,s) = \exp\left(-\int_t^s k(\mathbf{X}_r, r)\,\mathrm{d}r\right).$$

There are alternative sufficient conditions for which a solution to the Cauchy problem exists, is unique, and has the above stochastic representation — see [13] and the references therein.

A sketch proof is as follows [13]. Denote by $\{\mathbf{X}_t\}$ the process that solves (17.1). Assume there exists a solution u_t to (17.10). For fixed t, identify $\mathbf{Y}_s \equiv \mathbf{X}_{t+s}$, and apply Itô's formula to the function $u_{t+s}(\mathbf{Y}_s)\exp\left(-\int_0^s k_{t+r}(\mathbf{Y}_r, t+r)\mathrm{d}r\right) \equiv u_{t+s}(\mathbf{X}_{t+s})\,K(t,t+s)$, $s \geqslant 0$, yielding

$$\begin{aligned}
u_{t+s}(\mathbf{X}_{t+s})\,K(t,t+s) &= u_t(\mathbf{X}_t) \\
&+ \int_0^s K(t,t+r)\left(L_{t+r} + \frac{\partial}{\partial t} - k_{t+r}(\mathbf{X}_{t+r})\right) u_{t+r}(\mathbf{X}_{t+r})\,\mathrm{d}r \\
&+ \sum_{i=1}^d \sum_{k=1}^m \int_0^s B_{ik}(\mathbf{X}_{t+r}, t+r)\frac{\partial}{\partial x_i}(u_{t+r}(\mathbf{X}_{t+r})\,K(t,t+r))\,\mathrm{d}W_{t+r}^{(k)}.
\end{aligned} \tag{17.11}$$

For each $s \in [0, T-t]$, equation (17.10) is satisfied, and so each process

$$u_{t+s}(\mathbf{X}_{t+s})\,K(t,t+s) + \int_0^s g(\mathbf{X}_{t+r}, t+r)\,K(t,t+r)\,\mathrm{d}r\,, \quad 0 \leqslant s < T-t$$

is a $\mathbb{P}^{\mathbf{x},t}$ martingale under Assumptions 17.1.1 and 17.1.3. Taking expectations of (17.11) together with the terminal condition, we see that

$$u_t(\mathbf{x}) = \mathbb{E}^{\mathbf{x},t}\left[f(\mathbf{X}_T)\,K(t,T) + \int_t^T g(\mathbf{X}_r, r)\,K(t,r)\,\mathrm{d}r\right].$$

Remark 17.1.2 (Initial Value Problem) For fixed $T > 0$, a solution $v_t(\mathbf{x}) : \mathbb{R}^d \times (0,T] \to \mathbb{R}$ to the **initial value problem**

$$\left(L_t - \frac{\partial}{\partial t} - k(\mathbf{x},t)\right) v_t(\mathbf{x}) = -g(\mathbf{x},t) \quad \text{in } \mathbb{R}^d \times (0,T]\,,$$

$$v_t(\mathbf{x}) = f(\mathbf{x}) \quad \text{on } \mathbb{R}^d \times \{0\}\,,$$

is obtained simply by setting $v_t(\mathbf{x}) = u_{T-t}(\mathbf{x})$, where u_s is as given in Theorem 17.1.2. The only difference is that the process $\{\mathbf{X}_t\}$ is the solution of the SDE with time-reversed coefficients given by

$$\mathrm{d}\mathbf{X}_t = \mathbf{a}(\mathbf{X}_t, T-t)\,\mathrm{d}t + B(\mathbf{X}_t, T-t)\,\mathrm{d}\mathbf{W}_t\,, \quad 0 \leqslant t \leqslant T\,,$$

rather than of the SDE given in (17.1).

17.1.3 Terminal–Boundary Problems

We now consider problems with both terminal and boundary conditions. As in Section 17.1.1, suppose that we have a bounded domain $\mathscr{X}$ of $\mathbb{R}^d$ with boundary set $\partial\mathscr{X}$. Let $T > 0$ be an arbitrary fixed terminal time.

A PDE problem involving the differential operator L_t in (17.2) is said to be a **terminal–boundary value problem** if it is of the form

$$\begin{aligned}\left(L_t + \frac{\partial}{\partial t} - k(\mathbf{x},t)\right) u_t(\mathbf{x}) &= -g(\mathbf{x},t) \quad \text{in } \mathscr{X} \times [0,T)\,, \\ u_t(\mathbf{x}) &= f(\mathbf{x}) \quad \text{on } \mathscr{X} \times \{T\}\,, \\ u_t(\mathbf{x}) &= h(\mathbf{x},t) \quad \text{on } \partial\mathscr{X} \times [0,T)\,,\end{aligned} \tag{17.12}$$

with functions $k : \overline{\mathscr{X} \times [0,T)} \to [0,\infty)$, $g : \overline{\mathscr{X} \times [0,T)} \to \mathbb{R}$, $f : \overline{\mathscr{X}} \to \mathbb{R}$, and $h : \partial\mathscr{X} \times [0,T) \to \mathbb{R}$.

A unique solution u_t exists if, in addition to Assumptions 17.1.1, the following conditions are satisfied:

Assumptions 17.1.4 (Smoothness and Continuity)

1. **Smooth boundary**: *The boundary $\partial\mathscr{X}$ is $\mathcal{C}^2$.*

2. **Continuous and consistent data**: *The function f is continuous on $\overline{\mathscr{X}}$, the function h is continuous on $\overline{\partial\mathscr{X} \times [0,T)}$, and $h(\mathbf{x},T) = f(\mathbf{x})$ on $\partial\mathscr{X}$.*

Theorem 17.1.3 (Representation for Terminal–Boundary Value Problem) *Under Assumptions 17.1.1 and 17.1.4, there is a unique solution $u_t(\mathbf{x}) : \mathscr{X} \times [0,T) \to \mathbb{R}$ satisfying the terminal–boundary value problem (17.12) and admitting the stochastic representation*

$$u_t(\mathbf{x}) = \mathbb{E}^{\mathbf{x},t}\left[\left(h(\mathbf{X}_\tau,\tau)\,\mathrm{I}_{\{\tau<T\}} + f(\mathbf{X}_T)\,\mathrm{I}_{\{\tau=T\}}\right)K(t,\tau) + \int_t^T g(\mathbf{X}_s,s)\,K(t,s)\,\mathrm{d}s\right]\,,$$

where

$$K(t,s) = \exp\left(-\int_t^s k(\mathbf{X}_r,r)\,\mathrm{d}r\right)\,,$$

and $\tau = \inf\{r \in [t,T) : \mathbf{X}_r \notin \mathscr{X}\}$ if it exists, and $\tau = T$ otherwise.

The proof is a combination of those for the Dirichlet and the Cauchy problems.

■ EXAMPLE 17.2 (Solving a Terminal–Boundary Value Problem)

Extending Example 17.1, suppose that we wish to prescribe a terminal condition and augment the boundary condition with time, by setting $f(x,y) = 0$ and $h(x,y,t) = \sin((T-t)xy)$, leaving $k(x,y,t) = 0$, but augmenting $g(x,y,t) = \mathrm{e}^{T-t}\sin(x)\cos(y)$. This gives a terminal–boundary problem written out as

$$\begin{aligned}&\frac{\partial u_t(x,y)}{\partial t} + 0.52\,\frac{\partial^2 u_t(x,y)}{\partial x^2} + 0.6\,\frac{\partial^2 u_t(x,y)}{\partial x \partial y} + 2.02\,\frac{\partial^2 u_t(x,y)}{\partial y^2} \\ &\qquad + \frac{\partial u_t(x,y)}{\partial x} + \frac{\partial u_t(x,y)}{\partial y} = -\mathrm{e}^{T-t}\sin(x)\cos(y) \quad \text{in } \mathscr{X} \times [0,T)\,, \\ &\qquad\qquad u_t(x,y) = 0 \quad \text{on } \mathscr{X} \times \{T\}\,, \\ &\qquad\qquad u_t(x,y) = \sin((T-t)xy) \quad \text{on } \partial\mathscr{X} \times [0,T)\,.\end{aligned}$$

We can employ the same approximation ideas as in Example 17.1 to obtain solutions to this equation. Figure 17.2 depicts the estimated solution at four time instants, $t = 0, 1/3, 2/3, 1$, with terminal time $T = 1$.

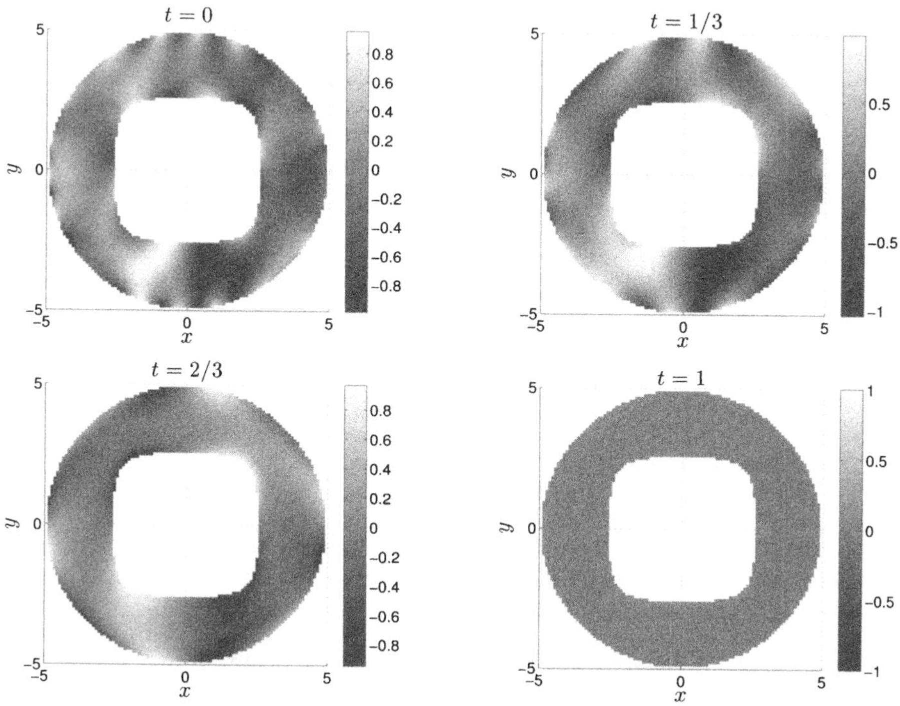

Figure 17.2 Estimated solution $u_t(x, y)$ to a terminal–boundary type problem at times $t = 0$, $t = 1/3$, $t = 2/3$, and $t = T = 1$.

The MATLAB code for this example is a simple extension of that given in Example 17.1. The corresponding program `termbound_ex_script.m` can be found on the Handbook website.

17.2 TRANSPORT PROCESSES AND EQUATIONS

Transport processes are Markov processes that model the position and velocity of objects that undergo collisions. The position component has continuous sample paths with *no* diffusion component, but the velocity component experiences "shocks". Between shocks, the process evolves according to a system of ordinary differential equations, and when a shock occurs, it possibly changes the direction of the process in an abrupt way, but maintains its spatial continuity. Figure 17.3 illustrates this generic behavior, showing the position portion of a possible transport process on the plane.

More formally, a transport process is a Markov process $\{\mathbf{Z}_t, t \geqslant 0\}$ of the form $\mathbf{Z}_s = (\mathbf{X}_s, \mathbf{V}_s)$ (interpreted as a column vector), with position component $\mathbf{X}_s \in \mathbb{R}^d$ and velocity component $\mathbf{V}_s \in \mathbb{R}^k$, that can be constructed as follows [19]:

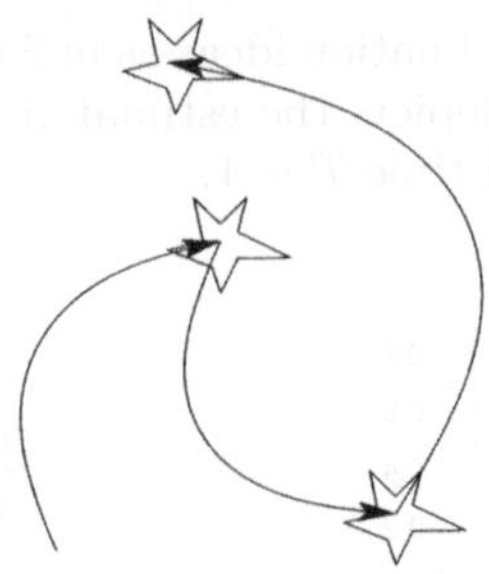

Figure 17.3 The position portion of a transport process on the plane remains continuous while experiencing shocks.

- Between shocks, the process satisfies an ordinary differential equation

$$\frac{\mathrm{d}\mathbf{Z}_t}{\mathrm{d}t} = \mathbf{a}(\mathbf{Z}_t).$$

- The times of shocks $T_1, T_2, \ldots$ occur according to a homogeneous Poisson process with rate $\overline{\lambda} = \sup_{\mathbf{z}} \lambda(\mathbf{z})$, where $\lambda(\mathbf{z}) \geqslant 0$ is a given rate function. ☞ 170

- At the time of a shock T, the process jumps from state $\mathbf{z} = (\mathbf{x}, \mathbf{v})$ to state $\mathbf{Z}' = (\mathbf{x}, \mathbf{V}')$ with probability $\lambda(\mathbf{z})/\overline{\lambda}$, where $\mathbf{V}' \sim \pi(\mathbf{v}' \,|\, \mathbf{z})$, and π is a time-homogeneous transition density for the velocities.

This construction ensures that $\{\mathbf{Z}_t\}$ is a time-homogeneous Markov process with infinitesimal generator L acting on bounded $\mathcal{C}^1$ functions $u : \mathbb{R}^{d+k} \to \mathbb{R}$ via

$$Lu(\mathbf{z}) = \underbrace{\mathbf{a}(\mathbf{z})^\top \nabla_{\mathbf{z}} u(\mathbf{z})}_{\text{pure drift part}} + \underbrace{\int \lambda(\mathbf{z})(u(\mathbf{z}') - u(\mathbf{z}))\, \pi(\mathbf{v}' \,|\, \mathbf{z})\, \mathrm{d}\mathbf{v}'}_{\text{pure shock part}},$$

where $\mathbf{z}' = (\mathbf{x}, \mathbf{v}')$.

Consider the initial value problem

$$\begin{aligned} \left(L - \frac{\partial}{\partial t} - k(\mathbf{z})\right) u_t(\mathbf{z}) &= -g(\mathbf{z}) \quad \text{in } \mathbb{R}^{d+k} \times (0, \infty), \\ u_t(\mathbf{z}) &= f(\mathbf{z}) \quad \text{on } \mathbb{R}^{d+k} \times \{0\}, \end{aligned} \tag{17.13}$$

where k, f, and g are bounded $\mathcal{C}^1$ functions from $\mathbb{R}^{d+k}$ to $\mathbb{R}$.

The following theorem gives a stochastic representation for the solution u.

Theorem 17.2.1 (Feynman–Kac for the General Transport Process) *If λ and $\pi(\cdot \,|\, \mathbf{z})$ are bounded $\mathcal{C}^1$ functions, then the unique $\mathcal{C}^1$ function that solves* (17.13) *has the following stochastic representation:*

$$u_t(\mathbf{z}) = \mathbb{E}^{\mathbf{z}} \left[f(\mathbf{Z}_t)\, K(t) + \int_0^t g(\mathbf{Z}_s)\, K(s)\, \mathrm{d}s \right],$$

where

$$K(t) = \exp\left(- \int_0^t k(\mathbf{Z}_r)\, \mathrm{d}r \right).$$

The adjoint operator of L, which we will denote by L^*, acts on u via

$$L^*u(\mathbf{z}) = \nabla_{\mathbf{z}} \cdot (\mathbf{a}(\mathbf{z})u(\mathbf{z})) + \int (\lambda(\mathbf{z}')u(\mathbf{z}') - \lambda(\mathbf{z})u(\mathbf{z}))\, \pi(\mathbf{v} \,|\, \mathbf{z}')\, \mathrm{d}\mathbf{v}',$$

where $\mathbf{z}' = (\mathbf{x}, \mathbf{v}')$, and the notation $\nabla_{\mathbf{z}} \cdot \boldsymbol{f}$ means $\sum_i \frac{\partial f}{\partial z_i}$.

Let $u_t(\mathbf{z})$ be the solution of the initial value problem

$$\begin{aligned} \left(L^* - \frac{\partial}{\partial t} - k(\mathbf{z})\right) u_t(\mathbf{z}) &= -g(\mathbf{z}, t) \quad \text{in } \mathbb{R}^{d+k} \times (0, \infty), \\ u_t(\mathbf{z}) &= f(\mathbf{z}) \quad \text{on } \mathbb{R}^{d+k} \times \{0\}. \end{aligned} \tag{17.14}$$

where k, $f \geqslant 0$, and $g \geqslant 0$ are bounded $\mathcal{C}^1$ functions, such that f and g can be normalized to be pdfs at each time t. Denote the normalizing constants by $\int f(\mathbf{z})\, \mathrm{d}\mathbf{z} = \alpha < \infty$, and $\int g(\mathbf{z}, t)\, \mathrm{d}\mathbf{z} = \beta(t) < \infty$ for $t \geqslant 0$. Define normalized versions as $\overline{f}(\mathbf{z}) = f(\mathbf{z})/\alpha$ and $\overline{g}(\mathbf{z}, t) = g(\mathbf{z}, t)/\beta(t)$, so that $\overline{f}$ and $\overline{g}$ are both pdfs for each instant of time. Then there is a generalization of the Kolmogorov forward (or Fokker–Planck) equation for this process, given in the following theorem.

Theorem 17.2.2 (Fokker–Planck for the General Transport Process) *For all bounded $\mathcal{C}^1$ functions $\phi : \mathbb{R}^{d+k} \to \mathbb{R}$, we have*

$$\int u_t(\mathbf{z})\, \phi(\mathbf{z})\, \mathrm{d}\mathbf{z} = \alpha\, \mathbb{E}^0_{\overline{f}} \left[\phi(\mathbf{Z}_t)\, K(0, t)\right] + \int_0^t \beta(s)\, \mathbb{E}^s_{\overline{g}(\cdot, s)} \left[\phi(\mathbf{Z}_t)\, K(s, t)\right] \mathrm{d}s,$$

where the notation $\mathbb{E}^s_h[\,\cdot\,]$ is shorthand for $\mathbb{E}[\,\cdot \,|\, \mathbf{Z}_s]$ with $\mathbf{Z}_s \sim h$, and

$$K(s, t) = \exp\left(-\int_s^t k(\mathbf{Z}_r)\, \mathrm{d}r\right).$$

17.2.1 Application to Transport Equations

An important application of the Feynman–Kac and Fokker–Planck representations for the general transport process is to solving *transport equations* and the more general *Vlasov's equation*. These are equations that describe how the population density of physical particles evolves when that population is behaving according to a transport process. If we denote the population density of physical particles by $u_t(\mathbf{z}) : \mathbb{R}^{2d} \times [0, \infty) \to [0, \infty)$, we do not necessarily have that $u_t(\mathbf{z})$ is a pdf for each t. This is due to the possible creation and annihilation of particles.

More precisely, the **transport equations** can be written as the initial value problem

$$\begin{aligned} \left(Q - \mathbf{v}^\top \nabla_{\mathbf{x}} - \frac{\partial}{\partial t} - k(\mathbf{z})\right) u_t(\mathbf{z}) &= -g(\mathbf{z}, t) \quad \text{in } \mathbb{R}^{2d} \times (0, \infty) \\ u_t(\mathbf{z}) &= f(\mathbf{z}) \quad \text{on } \mathbb{R}^{2d} \times \{0\}, \end{aligned}$$

where $\mathbf{z} = (\mathbf{x}, \mathbf{v})$ with $\mathbf{x}, \mathbf{v} \in \mathbb{R}^d$, $f \geqslant 0$ is a bounded function corresponding to an initial "distribution" of particles, $g(\cdot, t) \geqslant 0$ is a bounded $\mathcal{C}^1$ function for each t corresponding to a "source" of particles, $k \geqslant 0$ is a bounded function that "removes" particles, and Q is an integral operator with respect to the velocity component $\mathbf{v}$

that describes how the velocity of a particle changes after experiencing a "shock". The operator Q depends on the position component $\mathbf{x}$ and acts on functions u via

$$Qu(\mathbf{z}) = \int q(\mathbf{v}' \,|\, \mathbf{z})\, u(\mathbf{z}')\, \mathrm{d}\mathbf{v}' ,$$

where $\mathbf{z}' = (\mathbf{x}, \mathbf{v}')$ and $q \geqslant 0$ is a bounded transition density for the velocities $\mathbf{v}$.

More generally, **Vlasov's equation** applies when particles can also undergo acceleration:

$$\begin{aligned}\left(Q - \mathbf{v}^\top \nabla_{\mathbf{x}} - \nabla_{\mathbf{v}} \cdot \mathbf{a}_2(\mathbf{z}) - \frac{\partial}{\partial t} - k(\mathbf{z})\right) u_t(\mathbf{z}) &= -g(\mathbf{z}, t) \quad \text{in } \mathbb{R}^{2d} \times (0, \infty) \\ u_t(\mathbf{z}) &= f(\mathbf{z}) \quad \text{on } \mathbb{R}^{2d} \times \{0\} ,\end{aligned} \tag{17.15}$$

where $\mathbf{a}_2$ corresponds to an acceleration term that acts on the velocity component.

In the next two sections, we give the Feynman–Kac and Fokker–Planck representations for Vlasov's equation, which can be seen as special cases of the representations for general transport processes. The first, based on the backward equation, suggests the preferred Monte Carlo algorithm for evaluating the solution at one (or a small number of) point(s). The second, based on the forward equation, suggests the preferred Monte Carlo algorithm for evaluating the solution over a larger region.

17.2.1.1 Backward Equation Representation We can write Vlasov's equation in the same form as (17.13), that is, in the form

$$\begin{aligned}\left(L - \frac{\partial}{\partial t} - \widetilde{k}(\mathbf{z})\right) u_t(\mathbf{z}) &= -g(\mathbf{z}, t) \quad \text{in } \mathbb{R}^{2d} \times (0, \infty) \\ u_t(\mathbf{z}) &= f(\mathbf{z}) \quad \text{on } \mathbb{R}^{2d} \times \{0\} .\end{aligned} \tag{17.16}$$

This is done by relating the terms that appear in (17.15) to those in (17.16) as

$$\mathbf{a}(\mathbf{z}) = \begin{pmatrix} -\mathbf{v} \\ -\mathbf{a}_2(\mathbf{z}) \end{pmatrix} ,$$

$$\lambda(\mathbf{z}) = \int q(\mathbf{v}' \,|\, \mathbf{z})\, \mathrm{d}\mathbf{v}' ,$$

$$\pi(\mathbf{v}' \,|\, \mathbf{z}) = \frac{q(\mathbf{v}' \,|\, \mathbf{z})}{\lambda(\mathbf{z})} ,$$

and

$$\widetilde{k}(\mathbf{z}) = k(\mathbf{z}) - \lambda(\mathbf{z}) + \nabla_{\mathbf{v}} \cdot \mathbf{a}_2(\mathbf{z}) .$$

This results in an operator L acting on functions u via

$$Lu(\mathbf{z}) = -\mathbf{v}^\top \nabla_{\mathbf{x}} u(\mathbf{z}) - \mathbf{a}_2^\top \nabla_{\mathbf{v}} u(\mathbf{z}) + \int \lambda(\mathbf{z})(u(\mathbf{z}') - u(\mathbf{z}))\, \pi(\mathbf{v}' \,|\, \mathbf{z})\, \mathrm{d}\mathbf{v}' ,$$

where $\mathbf{z}' = (\mathbf{x}, \mathbf{v}')$.

This also implies that between shocks the associated transport process $\{\mathbf{Z}_t\} = \{(\mathbf{X}_t, \mathbf{V}_t)\}$ satisfies the system of ordinary differential equations

$$\begin{aligned}\frac{\mathrm{d}\mathbf{X}_t}{\mathrm{d}t} &= -\mathbf{V}_t , \\ \frac{\mathrm{d}\mathbf{V}_t}{\mathrm{d}t} &= -\mathbf{a}_2(\mathbf{Z}_t) .\end{aligned} \tag{17.17}$$

Further, by Theorem 17.2.1, the solution u_t of (17.16) has the stochastic representation

$$u_t(\mathbf{z}) = \mathbb{E}^{\mathbf{z}}\left[f(\mathbf{Z}_t)\,\widetilde{K}(t) + \int_0^t g(\mathbf{Z}_s, s)\,\widetilde{K}(s)\,\mathrm{d}s\right],$$

with

$$\widetilde{K}(t) = \exp\left(-\int_0^t \widetilde{k}(\mathbf{Z}_r)\,\mathrm{d}r\right). \tag{17.18}$$

We can approximate $u_t(\mathbf{z})$ for a fixed tuple $(\mathbf{z}, t)$ as follows. First, simulate N independent copies of $\{\mathbf{Z}_t\}$ via the following algorithm.

Algorithm 17.1 (Backward Equation Viewpoint)

1. *Set* $\mathbf{Z}_0 = (\mathbf{x}, \mathbf{v})$.
2. *Given some* $\overline{\lambda} = \sup_{\mathbf{z}} \lambda(\mathbf{z})$, *simulate a Poisson process with intensity* $\overline{\lambda}$, *denoting times of shocks as* $T_1, T_2, \ldots$.
3. *Between shocks,* $\mathbf{Z}_t$ *solves the system of ordinary differential equations* (17.17).
4. *At the time of a shock* T, *the process moves from state* $\mathbf{z} = (\mathbf{x}, \mathbf{v})$ *to state* $\mathbf{Z}' = (\mathbf{x}, \mathbf{V}')$ *with probability* $\lambda(\mathbf{z})/\overline{\lambda}$, *where* $\mathbf{V}' \sim \pi(\mathbf{v}' \,|\, \mathbf{z})$.
5. $\widetilde{K}(t)$ *in* (17.18) *is solved from the ordinary differential equation*

$$\frac{\mathrm{d}w(t)}{\mathrm{d}t} = -\widetilde{k}(\mathbf{Z}_t)\,w(t),$$

with initial condition $w(0) = 1/N$, *where* $w(t) = \widetilde{K}(t)/N$ *gives the relation between* w *and* $\widetilde{K}$.

Next, approximate $u_t(\mathbf{z})$ via

$$\widehat{u}_t(\mathbf{z}) = \sum_{k=1}^{N} f(\mathbf{Z}_t^{(k)})\,w^{(k)}(t) + \int_0^t g(s, \mathbf{Z}_s^{(k)})\,w^{(k)}(s)\,\mathrm{d}s\,.$$

17.2.1.2 Forward Equation Representation We can also write Vlasov's equation as a Kolmogorov forward (or Fokker–Planck) type equation, of the same form as (17.14), that is,

$$\begin{aligned}\left(L^* - \frac{\partial}{\partial t} - \widetilde{k}(\mathbf{z})\right) u_t(\mathbf{z}) &= -g(\mathbf{z}, t) \quad \text{in } \mathbb{R}^{2d} \times (0, \infty)\\ u_t(\mathbf{z}) &= f(\mathbf{z}) \quad \text{on } \mathbb{R}^{2d} \times \{0\}\,.\end{aligned} \tag{17.19}$$

This is done by relating the terms that appear in (17.15) with those in (17.19) as

$$\mathbf{a}(\mathbf{z}) = \begin{pmatrix} -\mathbf{v} \\ -\mathbf{a}_2(\mathbf{z}) \end{pmatrix},$$

$$\lambda(\mathbf{z}) = \int q(\mathbf{v} \,|\, \mathbf{z}')\,\mathrm{d}\mathbf{v}',$$

$$\pi(\mathbf{v} \,|\, \mathbf{z}') = \frac{q(\mathbf{v}' \,|\, \mathbf{z})}{\lambda(\mathbf{z})},$$

and

$$\widetilde{k}(\mathbf{z}) = k(\mathbf{z}) - \lambda(\mathbf{z}),$$

where $\mathbf{z}' = (\mathbf{x}, \mathbf{v}')$. This results in an operator L^* acting on functions u via

$$L^* u_t(\mathbf{z}) = -\mathbf{v}^\top \nabla_{\mathbf{x}} u_t(\mathbf{z}) - \nabla_{\mathbf{v}} \cdot (\mathbf{a}_2(\mathbf{z})\, u_t(\mathbf{z})) + \int (\lambda(\mathbf{z}')\, u_t(\mathbf{z}') - \lambda(\mathbf{z})\, u_t(\mathbf{z}))\, \pi(\mathbf{v} \,|\, \mathbf{z}')\, \mathrm{d}\mathbf{v}'.$$

This also implies that between jumps the associated transport process $\{\mathbf{Z}_t\} = \{(\mathbf{X}_t, \mathbf{V}_t)\}$ satisfies the system of ordinary differential equations

$$\begin{aligned} \frac{\mathrm{d}\mathbf{X}_t}{\mathrm{d}t} &= \mathbf{V}_t, \\ \frac{\mathrm{d}\mathbf{V}_t}{\mathrm{d}t} &= \mathbf{a}_2(\mathbf{Z}_t). \end{aligned} \tag{17.20}$$

Note that the signs in (17.17) and (17.20) differ.

We will now assume that $g = 0$. However, the cases $g \geqslant 0$ and $g \neq 0$ are possible too (see [19, Pages 58–59] for details). Further, we will assume that f is a positive function. Denote the normalized function as $\overline{f} = f/\alpha$, where $\alpha = \int f(\mathbf{z})\, \mathrm{d}\mathbf{z}$. In this regime, from Theorem 17.2.2, for every continuous bounded function $\phi : \mathbb{R}^{2d} \to \mathbb{R}$, we have

$$\int u_t(\mathbf{z})\, \phi(\mathbf{z})\, \mathrm{d}\mathbf{z} = \alpha\, \mathbb{E}^0_{\overline{f}} \left[\phi(\mathbf{Z}_t)\, \widetilde{K}(t) \right]$$

where $\mathbb{E}^s_h[\,\cdot\,]$ is shorthand for $\mathbb{E}[\,\cdot\,|\,\mathbf{Z}_s]$ with $\mathbf{Z}_s \sim h$, and

$$\widetilde{K}(t) = \exp\left(- \int_0^t \widetilde{k}(\mathbf{Z}_r)\, \mathrm{d}r \right). \tag{17.21}$$

We can approximate $u_t(\mathbf{z})$ via a discrete approximation on N points. First, generate N realizations of the associated transport process $\{\mathbf{Z}_t\}$ over the time interval $[0, t]$ via the following algorithm.

Algorithm 17.2 (Forward Equation Viewpoint)

1. *Generate an initial state* $\mathbf{Z}_0 \sim \overline{f}$.
2. *Given some* $\overline{\lambda} = \sup_{\mathbf{z}} \lambda(\mathbf{z})$, *simulate a Poisson process with intensity* $\overline{\lambda}$, *denoting times of shocks as* $T_1, T_2, \ldots$.
3. *Between shocks,* $\mathbf{Z}_s$ *solves the system of ordinary differential equations* (17.20).
4. *At the time of a shock* T, *the process moves from state* $\mathbf{z} = (\mathbf{x}, \mathbf{v})$ *to state* $\mathbf{Z}' = (\mathbf{x}, \mathbf{V}')$ *with probability* $\lambda(\mathbf{z})/\overline{\lambda}$, *where* $\mathbf{V}' \sim \pi(\mathbf{v} \,|\, \mathbf{x}, \mathbf{v}')$.
5. $\widetilde{K}(t)$ *in* (17.21) *is solved from the ordinary differential equation*

$$\frac{\mathrm{d}w(t)}{\mathrm{d}t} = -\widetilde{k}(\mathbf{Z}_t)\, w(t),$$

with initial condition $w(0) = 1/N$, *where* $w(t) = \widetilde{K}(t)/N$ *gives the relation between* w *and* $\widetilde{K}$.

Next, given a set of N realizations of $\{\mathbf{Z}_t\}$, approximate $u_s(\mathbf{z})$ for any $s \in [0, t]$ and $\mathbf{z}$ via

$$\widehat{u}_s(\mathbf{z}) = \alpha \sum_{i=1}^{N} w^{(i)}(s)\, \mathrm{I}_{\{\mathbf{Z}_s^{(i)} = \mathbf{z}\}} \,.$$

In particular, this means that we approximate, for any function H,

$$\int H(\mathbf{z})\, u_t(\mathbf{z})\, \mathrm{d}\mathbf{z} \tag{17.22}$$

by

$$\alpha \sum_{i=1}^{N} w^{(i)}(t)\, H(\mathbf{Z}_t^{(i)}) \,.$$

17.2.2 Boltzmann Equation

Boltzmann-type equations describe the evolution of a population of particles that travel in straight line segments between collisions, and undergo pairwise collisions. They form a particular type of transport equation, where the actual shocks are interpreted as collisions between particles.

Following [19, Chapter 4], we will consider only one type of particle (without internal energy). Denote the population (concentration) at time $t \geqslant 0$ with position $\mathbf{x} \in \mathbb{R}^3$ and velocity $\mathbf{v} \in \mathbb{R}^3$ by $u_t(\mathbf{z})$, where $\mathbf{z} = (\mathbf{x}, \mathbf{v})$. The **Boltzmann equation** can be written as

$$\left(Q - \mathbf{v}^\top \nabla_{\mathbf{x}} - \frac{\partial}{\partial t}\right) u_t(\mathbf{z}) = 0 \,,$$

where Q is an operator describing the behavior of collisions. It acts only on the velocity component, and has the form (suppressing $\mathbf{x}$)

$$Qu_t(\mathbf{v}) = \int_{\mathbb{R}^3} \int_{\mathscr{S}^2} q\left(\mathbf{v} - \mathbf{v}_1, \boldsymbol{\xi}^\top \frac{(\mathbf{v} - \mathbf{v}_1)}{\|\mathbf{v} - \mathbf{v}_1\|}\right) (u_t(\mathbf{v}')u_t(\mathbf{v}_1') - u_t(\mathbf{v})u_t(\mathbf{v}_1))\, \mathrm{d}\boldsymbol{\xi}\, \mathrm{d}\mathbf{v}_1 \,,$$

where $\mathscr{S}^2 = \{\boldsymbol{\xi} \in \mathbb{R}^3 : \|\boldsymbol{\xi}\| = 1\}$ is the surface of the unit ball in $\mathbb{R}^3$, and the new velocities of the particles, $\mathbf{v}'$ and $\mathbf{v}_1'$, have the following properties:

1. **Conservation of momentum**: $\mathbf{v}' + \mathbf{v}_1' = \mathbf{v} + \mathbf{v}_1$.
2. **Conservation of kinetic energy**: $\|\mathbf{v}'\|^2 + \|\mathbf{v}_1'\|^2 = \|\mathbf{v}\|^2 + \|\mathbf{v}_1\|^2$.

The new, postcollision velocities can be written in terms of the original, precollision velocities $\mathbf{v}$ and $\mathbf{v}_1$, together with the collision direction (represented by) $\boldsymbol{\xi}$ as:

$$\begin{aligned}
\mathbf{v}' &= \frac{1}{2}(\mathbf{v} + \mathbf{v}_1) + \frac{1}{2}\boldsymbol{\xi}\, \|\mathbf{v} - \mathbf{v}_1\| \,, \\
\mathbf{v}_1' &= \frac{1}{2}(\mathbf{v} + \mathbf{v}_1) - \frac{1}{2}\boldsymbol{\xi}\, \|\mathbf{v} - \mathbf{v}_1\| \,.
\end{aligned}$$

Further, $q(\mathbf{w}, \mu)$ is a function on $\mathbb{R}^3 \times [-1, 1]$ that depends only on the relative velocity between the two particles undergoing collision $\|\mathbf{w}\| = \|\mathbf{v} - \mathbf{v}_1\|$, and the cosine of the angle between the relative pre and postcollision velocities $\mu = \cos(\theta) = \frac{(\mathbf{v}-\mathbf{v}_1)^\top(\mathbf{v}'-\mathbf{v}_1')}{\|\mathbf{v}-\mathbf{v}_1\|\|\mathbf{v}'-\mathbf{v}_1'\|}$. Typical cases are:

- **Hard spheres:** $q(\mathbf{w}, \mu) = C\,\|\mathbf{w}\|$ for some constant C.
- **Variable hard spheres:** $q(\mathbf{w}, \mu) = C\,\|\mathbf{w}\|^{\alpha}$ for some constants C and $\alpha > 0$.

The function $\nu(\mathbf{w}, \mu) = q(\mathbf{w}, \mu)/\|\mathbf{w}\|$ is known as the **microscopic cross-section.**

In the **spatially homogeneous** framework, that is, when $\nabla_{\mathbf{x}} u_t(\mathbf{z}) = \mathbf{0}$, we can ignore the spatial component, and simply write $u_t(\mathbf{v})$. The Boltzmann equation then becomes

$$\left(Q - \frac{\partial}{\partial t}\right) u_t(\mathbf{v}) = 0\,.$$

Suppose we have an initial condition $u_0(\mathbf{v}) \geqslant 0$ that is normalized to 1 so that it is a pdf over $\mathbf{v} \in E \subseteq \mathbb{R}^3$, and that $q(\mathbf{w}, \mu)$ does not depend on μ. We wish to determine $u_t(\mathbf{v})$ over the interval $[0, T]$ for some fixed $T > 0$. We can proceed by considering a related process and equation that will allow us to approximate the solution to the original problem.

For fixed integer $N > 0$, define a Markov process $\mathbf{V}^N(t) = (\mathbf{V}_1^N(t), \ldots, \mathbf{V}_N^N(t))$, known as the **Bird collision process**, as follows.

1. Fix the state space as $E^{(N)}$, the N-th **symmetric power** of the velocity space E, so that two elements in $E^{(N)}$ are equivalent if one can be obtained from the other by simply permuting the component indices.

2. Set the initial distribution such that each component $\mathbf{V}_k^N(0)$ is distributed according to $u_0(\mathbf{v})$ for $k = 1, \ldots, N$.

3. The times between collisions are exponential, with parameter

$$\frac{1}{N} \sum_{i=1}^{N} \sum_{j=1}^{N} q(\mathbf{V}_i^N - \mathbf{V}_j^N)\,,$$

where we have suppressed the time parameter.

4. Given that a collision has occurred, a pair of particles (i, j) is chosen to collide with probability

$$\frac{q(\mathbf{V}_i^N - \mathbf{V}_j^N)}{\sum_{i=1}^{N} \sum_{j=1}^{N} q(\mathbf{V}_i^N - \mathbf{V}_j^N)}\,,$$

where we have suppressed the time parameter.

5. Once a pair has been chosen to collide, the process moves from state $(\mathbf{v}_1, \ldots, \mathbf{v}_N)$ to state $(\mathbf{v}_1, \ldots, \mathbf{v}_i', \ldots, \mathbf{v}_j', \ldots, \mathbf{v}_N)$, where the two new velocities for particles i and j are given by

$$\mathbf{v}_i' = \frac{1}{2}(\mathbf{v}_i + \mathbf{v}_j) + \frac{1}{2}\,\boldsymbol{\Xi}\,\|\mathbf{v}_i - \mathbf{v}_j\|\,,$$
$$\mathbf{v}_j' = \frac{1}{2}(\mathbf{v}_i + \mathbf{v}_j) - \frac{1}{2}\,\boldsymbol{\Xi}\,\|\mathbf{v}_i - \mathbf{v}_j\|\,,$$

where $\boldsymbol{\Xi}$ is drawn uniformly from $\mathscr{S}^2$, the surface of the unit ball in $\mathbb{R}^3$. See Section 3.3.3 for algorithms to generate $\boldsymbol{\Xi}$.

☞ 74

The density $h_N(\cdot, t) : E^{(N)} \to \mathbb{R}$ for this process satisfies the Kolmogorov forward PDE

$$\frac{\partial}{\partial t} h_N(\mathbf{v}, t) = \frac{1}{N} \sum_{i=1}^{N} \sum_{j \neq i}^{N} \int_{\mathscr{S}^2} q(\mathbf{v}_i - \mathbf{v}_j) \left(h_N(\mathbf{v}'_{ij}, t) - h_N(\mathbf{v}, t) \right) \mathrm{d}\boldsymbol{\xi} ,$$

$$h_N(\mathbf{v}, 0) = \prod_{k=1}^{N} u_0(\mathbf{v}_k) ,$$

where $\mathbf{v}'_{ij} = (\mathbf{v}_1, \ldots, \mathbf{v}'_i, \ldots, \mathbf{v}'_j, \ldots, \mathbf{v}_N)$ is a vector that matches $\mathbf{v}$ in all but components i and j, which are related to $\mathbf{v}_i$ and $\mathbf{v}_j$ exactly as described in Step 5 for the process.

It turns out [19, Page 98] that, for all t, the marginal density

$$p_N(\mathbf{v}, t) = \int \cdots \int h_N(\mathbf{v}, \mathbf{v}_2, \ldots, \mathbf{v}_N, t) \, \mathrm{d}\mathbf{v}_2 \ldots \mathrm{d}\mathbf{v}_N \to u_t(\mathbf{v}) \quad \text{as } N \to \infty ,$$

where u_t is the solution to the original (spatially homogeneous) Boltzmann problem. Furthermore, we can approximate the density $u_t(\mathbf{v})$ via Monte Carlo by constructing a realization of $\{\mathbf{V}^N(t)\}$ at time t and determining an empirical approximation:

$$\widehat{u}_t(\mathbf{v}) = \frac{1}{N} \sum_{k=1}^{N} \mathrm{I}_{\{\mathbf{V}_k^N(t) = \mathbf{v}\}} .$$

By the law of large numbers, we have $\widehat{u}_t(\mathbf{v}) \to u_t(\mathbf{v})$ as $N \to \infty$ [19, Page 98]. In particular, this means that we can approximate, for any function H,

$$\int H(\mathbf{v}) \, u_t(\mathbf{v}) \, \mathrm{d}\mathbf{v}$$

by the crude Monte Carlo estimator

$$\frac{1}{N} \sum_{k=1}^{N} H(\mathbf{V}_k^N(t)) .$$

■ EXAMPLE 17.3 (Boltzmann Equation)

For concreteness, suppose that the initial distribution is

$$u_0(\mathbf{v}) = \exp\left(-(v_1 + v_2 + v_3)\right) , \quad v_1, v_2, v_3 \geqslant 0 ,$$

meaning that each of the three velocity components can be drawn independently from the $\mathsf{Exp}(1)$ distribution. Suppose further that we adopt the hard sphere collision model, so that $q(\mathbf{w}) = C\|\mathbf{w}\|$, where we will set $C = 1$ for simplicity. Figure 17.4 depicts an estimated curve $H(\mathbf{v}) = v_1^3 + v_2^3 + v_3^3$ over the time interval $[0, 1]$ on the basis of $N = 1000$ realizations of the associated Bird collision process.

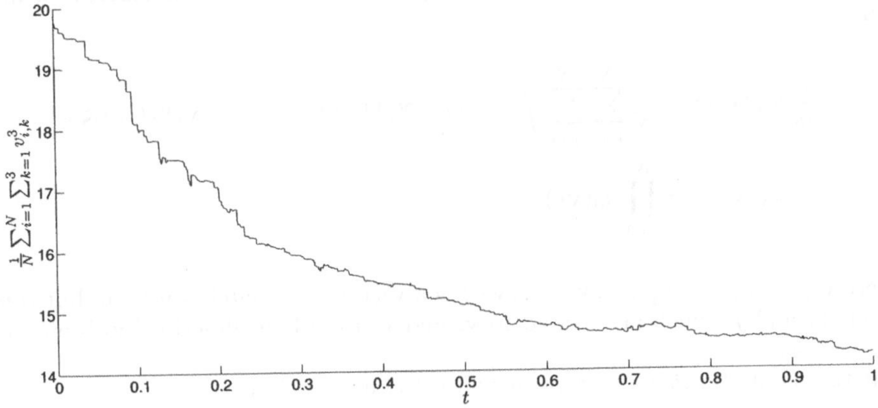

Figure 17.4 Estimated $H(\mathbf{v}) = v_1^3 + v_2^3 + v_3^3$ for $t \in [0, 1]$.

The following MATLAB code was used.

```
function Boltzmann_ex
% Simple Boltzmann equation example
% Spatially homogeneous regime
N=10^3; d=3;% Number of particles; Dimension
T=10^0; % Final time
v=-log(rand(N,d));
rho=1;
w=rho.*ones(N,1)./N;
C=1; alpha=1; % Collision constants
pairs=nchoosek((1:1:N),2);
t=0;
hh=mean(sum(v.^3,2));
tt=t;
while t<=T
  qij=C.*((sum((v(pairs(:,1),:)-v(pairs(:,2),:)).^2,2)).^(alpha/2));
  sumqij=sum(qij);
  lambda=rho*sumqij/N;
  tau=-log(rand(1))/lambda;
  t=t+tau % Time of the new collision
  if t>T
    t=T; break;
  end
  p=[0;cumsum(qij./sumqij)];
  r=rand(1);
  pidx=find(p>=r,1)-1; % Index of the colliding pair
  % Uniformly distributed on the Unit sphere in \R^3
  sigma=randn(1,3); sigma=sigma./sqrt(sum(sigma.^2));
  a=(v(pairs(pidx,1),:)+v(pairs(pidx,2),:))./2;
  b=(sum((v(pairs(pidx,1),:)-v(pairs(pidx,2),:)).^2)^(0.5))/2;
```

```
  vprime=a+sigma.*b;
  v1prime=a-sigma.*b;
  v(pairs(pidx,1),:)=vprime;
  v(pairs(pidx,2),:)=v1prime;
  tt=[tt,t];
  hh=[hh,mean(sum(v.^3,2))];
end
figure,plot(tt,hh,'k-')
```

17.3 CONNECTIONS TO ODES THROUGH SCALING

A further link between stochastic processes and deterministic differential equations arises through the scaling of certain Markov processes in space or time. On one scale, the scaled process can lose all stochasticity, and instead satisfy a system of ODEs. This is similar to the standard law of large numbers, and is a first-order approximation. If scaled deviations of the stochastic process from its first-order approximation are considered, a second-order result analogous to the central limit theorem can be obtained. In this section, we consider Markov jump processes, whose transition rates are of a form that is amenable to such scaling.

Suppose that we have a family $\{\boldsymbol{J}_N(\cdot)\}$ of Markov jump process models (see Section 5.3) indexed by $N > 0$, where $\boldsymbol{J}_N(\cdot)$ takes values in $\mathscr{Y}_N \subseteq \mathbb{Z}^d$, and has transition rate matrix $Q_N = (q_N(\mathbf{u}, \mathbf{v}),\ \mathbf{u}, \mathbf{v} \in \mathscr{Y}_N)$. ☞ 166

The process $\boldsymbol{J}_N$ could for example describe:

1. The number of each type of molecule present in a chemical reaction.

2. The number of individuals in a population that are in each phase of the spread of a disease. For example, an individual may be infective, susceptible, or immune.

3. The number of customers present at each queue in a Markovian queuing network.

The index N may be thought of as describing the size of the process. For example, N could be associated with the volume of the solvent in which a chemical reaction is taking place, or the total number of individuals present in a population.

The transition rate matrix Q_N and the state space $\mathscr{Y}_N$ form the model description — characterizing the rates at which molecules react in a chemical reaction process, for example.

If the family of processes is such that all of the transition rates depend on the current state $\mathbf{u}$ only through its "density" $\mathbf{u}/N$, at least in an asymptotic sense, then under certain conditions we may describe the large N behavior of the "density process" $\{\mathbf{X}_N(t)\} = \{\boldsymbol{J}_N(t)/N, t \geqslant 0\}$ through a set of ordinary differential equations.

More precisely, suppose that there is a subset $\mathscr{E}$ of $\mathbb{R}^d$ and a family $\{f_N, N > 0\}$ of continuous functions, with $f_N : \mathscr{E} \times \mathbb{Z}^d \to \mathbb{R}$, such that

$$q_N(\mathbf{u}, \mathbf{u} + \mathbf{v}) = N f_N\left(\frac{\mathbf{u}}{N}, \mathbf{v}\right), \quad \mathbf{v} \neq \mathbf{0}.$$

If there exists a function $\mathbf{F} : \mathscr{E} \to \mathbb{R}^d$ such that $\{\mathbf{F}_N\}$, given by $\mathbf{F}_N(\mathbf{x}) = \sum_{\mathbf{v}} \mathbf{v} f_N(\mathbf{x}, \mathbf{v})$, $\mathbf{x} \in \mathscr{E}$, converges pointwise to $\mathbf{F}$ on $\mathscr{E}$, then the family of Markov jump processes is said to be **asymptotically density dependent**. If f_N (and hence $\mathbf{F}_N$) is the same for all N, then the family is called **density dependent**.

The following functional law of large numbers establishes convergence of the family of scaled **density processes** $\{\mathbf{X}_N(\cdot)\}$ to the unique deterministic trajectory that corresponds to the solution of an appropriate system of ordinary differential equations. For a proof, see [6, Chapter 11] and [16, 24].

Theorem 17.3.1 (Functional Law of Large Numbers) *Suppose that $f_N(\cdot, \mathbf{v})$ is bounded for each $\mathbf{v}$ and N, that $\mathbf{F}$ is Lipschitz continuous on $\mathscr{E}$ and that $\{\mathbf{F}_N\}$ converges uniformly to $\mathbf{F}$ on $\mathscr{E}$. Then, if $\lim_{N\to\infty} \mathbf{X}_N(0) = \mathbf{x}_0$, the density process $\mathbf{X}_N(\cdot)$ converges uniformly in probability on $[0, t]$ to the unique deterministic trajectory $\mathbf{x}(\cdot)$ satisfying $\mathbf{x}(0) = \mathbf{x}_0$ and*

$$\frac{\mathrm{d}}{\mathrm{d}s}\mathbf{x}(s) = \mathbf{F}(\mathbf{x}(s)), \quad \mathbf{x}(s) \in \mathscr{E}, \; s \in [0, t]. \tag{17.23}$$

This enables us to associate the microscopic behavior of our stochastic system with its macroscopic behavior in a precise way. We illustrate this with an example of a chemical reaction.

■ EXAMPLE 17.4 (Chemical Reaction)

Suppose we have the following chemical reaction occurring inside a volume V [10]:

$$X \underset{c_2}{\overset{c_1}{\rightleftharpoons}} Y, \quad 2X \underset{c_4}{\overset{c_3}{\rightleftharpoons}} Z, \quad W + X \underset{c_6}{\overset{c_5}{\rightleftharpoons}} 2X,$$

where $c_1, \ldots, c_6$ are volume independent **reaction rates** (see, for example, [27]). This stochastic chemical reaction can be viewed as a four-dimensional Markov jump process with system state at time t represented by the vector $\boldsymbol{J}_V(t) = (w_V(t), x_V(t), y_V(t), z_V(t))^\top$. Following [18], the transition rates are:

$$\begin{aligned}
q_V((w, x, y, z), (w, x-1, y+1, z)) &= c_1\, x \\
q_V((w, x, y, z), (w, x+1, y-1, z)) &= c_2\, y \\
q_V((w, x, y, z), (w, x-2, y, z+1)) &= c_3\, x(x-1)/(2V) \\
q_V((w, x, y, z), (w, x+2, y, z-1)) &= c_4\, z \\
q_V((w, x, y, z), (w-1, x+1, y, z)) &= c_5\, w\, x/V \\
q_V((w, x, y, z), (w+1, x-1, y, z)) &= c_6\, x(x-1)/(2V).
\end{aligned}$$

The process is asymptotically density dependent, with

$$\mathbf{F}(\mathbf{x}) = \begin{pmatrix} c_6 \frac{x_2^2}{2} - c_5 x_1 x_2 \\ c_2 x_3 + 2c_4 x_4 + c_5 x_1 x_2 - c_1 x_2 - c_3 x_2^2 - c_6 \frac{x_2^2}{2} \\ c_1 x_2 - c_2 x_3 \\ c_3 \frac{x_2^2}{2} - c_4 x_4 \end{pmatrix}.$$

If the density process $\{\mathbf{X}_V(t)\} = \{\boldsymbol{J}_V(t)/V, t \geqslant 0\}$ has initial value satisfying $\lim_{V\to\infty} \mathbf{X}_V(0) = \mathbf{x}_0$, then the deterministic limit on $[0, t]$ is the solution to the corresponding system of ODEs in (17.23), with initial condition $\mathbf{x}(0) = \mathbf{x}_0$.

The Markov jump process $\{\boldsymbol{J}_V(t)\}$ is straightforward to simulate (see Section 5.3), and the system of ordinary differential equations can be solved using standard numerical techniques. Figure 17.5 shows the convergence of the first component of the process $w_V(t)$ to the deterministic limit described here, where we have chosen rates $c_1 = c_2 = c_3 = c_4 = c_5 = 1$ and $c_6 = 10$, and initial concentrations of $w_V(0) = x_V(0) = y_V(0) = z_V(0) = 100V$. ☞ 166

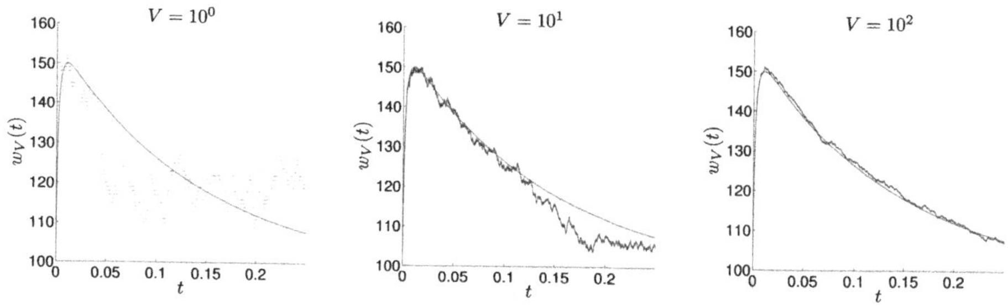

Figure 17.5 Realizations of the first component of the process $\boldsymbol{J}_V$, together with its deterministic limit over the time interval $[0, 0.25]$. This illustrates convergence of the scaled process to a deterministic limit as the volume V increases.

We also observe the Markov jump process reaching steady-state behavior over time, in correspondence with the deterministic system. This is illustrated in Figure 17.6.

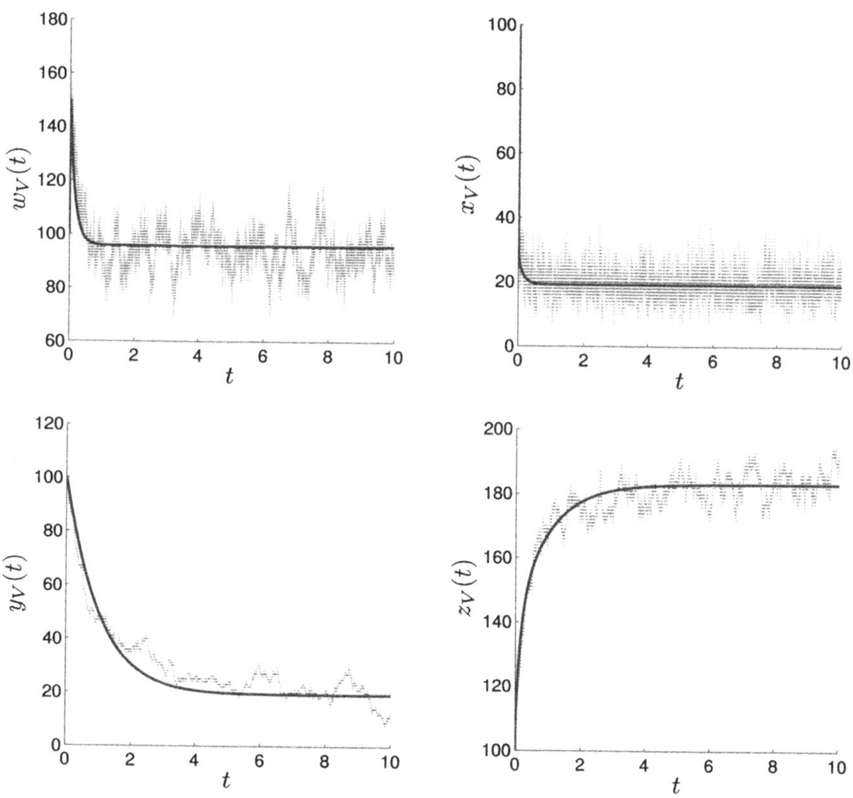

Figure 17.6 Realizations of all the components of the process, together with their deterministic limits, over the time interval $[0, 10]$, with volume $V = 10^0$.

The MATLAB code used is given below. The ODE function **gillespie_ode.m** is a direct transcription of the given system.

```
% chemical_ex.m
c = [1,1,1,1,1,10]; Vs=[10^0,10^1,10^2]; nV=length(Vs);
w0 = 100; x0=100; y0=100; z0=100;
% Deterministic Limit
t0=0; t1=0.25;
options = odeset('RelTol',1e-4);
[T,Y] = ode45(@(t,y) gillespie_ode(t,y,c),[t0 t1], ...
              [w0,x0,y0,z0],options);
for k=1:nV
    V=Vs(k); % Volume (scaling) parameter
    w=w0*V; x=x0*V; y=y0*V; z=z0*V;
    ww=[];xx=[];yy=[];zz=[];tt=[];
    t = 0;
    while t<t1
        ww=[ww,w]; xx=[xx,x]; yy=[yy,y]; zz=[zz,z]; tt=[tt,t];
        v = [x, y, x*(x-1)/(2*V), z, w*x/V, x*(x-1)/(2*V)];
        a = c.*v;
        lam = sum(a);
        p = a/sum(a);
        t = t -log(rand)/lam;
        r = min(find(cumsum(p)>=rand));
        switch r
            case 1
                x = x-1; y = y+1;
            case 2
                x = x+1; y = y-1;
            case 3
                x = x-2; z = z+1;
            case 4
                x = x+2; z = z-1;
            case 5
                x = x+1; w = w-1;
            case 6
                x = x-1; w = w+1;
        end
    end
end
```

```
function dx = gillespie_ode(t,x,c)
dx=zeros(4,1);
dx(1) = c(6)*(x(2)^2)/2 - c(5)*x(1)*x(2);
dx(2) = c(2)*x(3) + 2*c(4)*x(4) + c(5)*x(1)*x(2) -...
        c(1)*x(2) - c(3)*(x(2)^2)-c(6)*(x(2)^2)/2;
dx(3) = c(1)*x(2) - c(2)*x(3);
dx(4) = c(3)*(x(2)^2)/2 - c(4)*x(4);
```

Assuming that additional second-order conditions are satisfied, there is a functional central limit theorem that also applies. Loosely speaking, it states that for large N the scaled process around the deterministic trajectory behaves like a *Gaussian diffusion* process (see Section 5.1). For a proof, see [6, Chapter 11] and [17, 24]. ☞ 154

Theorem 17.3.2 (Functional Central Limit Theorem) *Suppose that $f_N(\cdot, \mathbf{v})$ is bounded for each $\mathbf{v}$ and N, that $\mathbf{F}$ is Lipschitz continuous and has uniformly continuous first partial derivatives on $\mathscr{E}$, and that*

$$\lim_{N\to\infty} \sup_{\mathbf{x}\in\mathscr{E}} \sqrt{N}\, \|\mathbf{F}_N(\mathbf{x}) - \mathbf{F}(\mathbf{x})\| = 0\,.$$

Suppose also that the sequence $\{G_N\}$, where G_N is a $d \times d$ matrix with entries

$$G_N(i,j) = \sum_{\mathbf{v}} v_i\, v_j\, f_N(\mathbf{x}, \mathbf{v}), \quad \mathbf{x} \in \mathscr{E}\,,$$

converges uniformly to G, where G is uniformly continuous on $\mathscr{E}$. Let $\mathbf{x}_0 \in \mathscr{E}$ and let $\mathbf{x}(\cdot)$ be the unique trajectory satisfying $\mathbf{x}(0) = \mathbf{x}_0$ and (17.23). Then, if

$$\lim_{N\to\infty} \sqrt{N}\, (\mathbf{X}_N(0) - \mathbf{x}_0) = \mathbf{z}\,,$$

the family of processes $\{\mathbf{Z}_N(\cdot)\}$, defined by

$$\mathbf{Z}_N(s) = \sqrt{N}\, (\mathbf{X}_N(s) - \mathbf{x}(s))\,, \qquad 0 \leqslant s \leqslant t\,,$$

converges in distribution on $[0, t]$ to a Gaussian diffusion $\mathbf{Z}(\cdot)$ with initial value $\mathbf{Z}(0) = \mathbf{z}$ and with mean vector $\boldsymbol{\mu}_s = \mathbb{E}\mathbf{Z}(s) = M_s\mathbf{z}$, where $M_s = \exp(\int_0^s B_u \mathrm{d}u)$ and $B_s = J_{\mathbf{F}}(\mathbf{x}(s))$. Here, $J_{\mathbf{F}}$ is the Jacobi matrix of $\mathbf{F}$. The covariance matrix of $\mathbf{Z}(\cdot)$ is

$$\Sigma_s = M_s \left(\int_0^s M_u^{-1} G(\mathbf{x}(u)) (M_u^{-1})^\top \mathrm{d}u \right) M_s^\top\,.$$

If the initial point is an equilibrium point (that is, $\mathbf{x}_0 = \mathbf{x}^*$, where $\mathbf{x}^*$ satisfies $\mathbf{F}(\mathbf{x}^*) = \mathbf{0}$), then, under the conditions of Theorem 17.3.2, we have that the process $\mathbf{Z}_N(\cdot)$ converges to an *Ornstein–Uhlenbeck* process. A precise statement follows [24, 25].

Theorem 17.3.3 (Central Limit Theorem at an Equilibrium Point) *If $\mathbf{x}^*$ satisfies $\mathbf{F}(\mathbf{x}^*) = \mathbf{0}$ then, under the conditions of Theorem 17.3.2, the family $\{\mathbf{Z}_N(\cdot)\}$, defined by*

$$\mathbf{Z}_N(s) = \sqrt{N}\, (\mathbf{X}_N(s) - \mathbf{x}^*), \quad 0 \leqslant s \leqslant t\,,$$

converges in distribution on $[0, t]$ to an Ornstein–Uhlenbeck process $\mathbf{Z}(\cdot)$ with initial value $\mathbf{Z}(0) = \mathbf{z}$, local drift matrix $B = J_{\mathbf{F}}(\mathbf{x}^*)$, and local covariance matrix $G = G(\mathbf{x}^*)$. In particular, $\mathbf{Z}(s)$ is normally distributed with mean vector $\boldsymbol{\mu}_s = \exp(Bs)\mathbf{z}$ and covariance matrix

$$\Sigma_s = \int_0^s \exp(Bu)\, G \exp(B^\top u)\, \mathrm{d}u = \Sigma - \exp(Bs)\, \Sigma \exp(B^\top s)\,,$$

where the stationary covariance matrix Σ satisfies $B\Sigma + \Sigma B^\top + G = 0$.

Further Reading

When discussing connections between stochastic and partial differential equations in Section 17.1, we relied mainly on [13, 22], with additional reference to [7, 8, 9, 14]. There are stochastic representations for other problems, such as those with Neumann-type boundary conditions, see for example, [7] for details. For variance reduction techniques, see [20], and [15, Chapter 16] for an overview (see also [21, 28, 29]). For the general transport and Boltzmann equations of Section 17.2, see [19, 23]. Booth [5] discusses the different viewpoint that transport equation practitioners have toward variance reduction (contrasted to the standard practice in the simulation literature). The diffusion approximation of Markov jump processes in Section 17.3 follows [24, 25, 26], which summarize, extend, and apply the work in [1, 2, 3, 4, 16, 17, 18]. The chemical example is based on [10]. Finally, [11] and [12] provide historical information on some of the earliest applications of the Monte Carlo method to neutron transport.

REFERENCES

1. A. D. Barbour. On a functional central limit theorem for Markov population processes. *Advances in Applied Probability*, 6(1):21–39, 1974.
2. A. D. Barbour. Quasi-stationary distributions in Markov population processes. *Advances in Applied Probability*, 8(2):296–314, 1976.
3. A. D. Barbour. Density-dependent Markov population processes. In W. Jäger, H. Rost, and P. Tautu, editors, *Biological Growth and Spread*, volume 38 of *Lecture Notes in Biomathematics*, pages 36–49. Springer-Verlag, Berlin, 1980.
4. A. D. Barbour. Equilibrium distributions Markov population processes. *Advances in Applied Probability*, 12(3):591–614, 1980.
5. T. Booth. Particle transport applications. In G. Rubino and B. Tuffin, editors, *Rare Event Simulation Using Monte Carlo Methods*, pages 215–242. John Wiley & Sons, Chichester, 2009.
6. S. N. Ethier and T. G. Kurtz. *Markov Processes: Characterization and Convergence.* John Wiley & Sons, New York, 1986.
7. M. I. Freidlin. *Functional Integration and Partial Differential Equations.* Princeton University Press, Princeton, New Jersey, 1985.
8. A. Friedman. *Stochastic Differential Equations and Applications: Volume 1.* Academic Press, New York, 1975.
9. A. Friedman. *Stochastic Differential Equations and Applications: Volume 2.* Academic Press, New York, 1976.
10. D. T. Gillespie. A general method for numerically simulating the stochastic time evolution of coupled chemical reactions. *Journal of Computational Physics*, 22(4):403–434, 1976.
11. J. S. Hendricks. A Monte Carlo code for particle transport. *Los Alamos Science*, 22:30–43, 1994.
12. C. C. Hurd. A note on early Monte Carlo computations and scientific meetings. *Annals of the History of Computing*, 7(2):141–155, 1985.
13. I. Karatzas and S. E. Shreve. *Brownian Motion and Stochastic Calculus.* Springer-Verlag, New York, second edition, 1991.

14. F. C. Klebaner. *Introduction to Stochastic Calculus with Applications.* Imperial College Press, London, second edition, 2005.

15. P. E. Kloeden and E. Platen. *Numerical Solution of Stochastic Differential Equations.* Springer-Verlag, Berlin, 1999. Corrected third printing.

16. T. G. Kurtz. Solutions of ordinary differential equations as limits of pure jump Markov processes. *Journal of Applied Probability*, 7(1):49–58, 1970.

17. T. G. Kurtz. Limit theorems for sequences of jump Markov processes approximating ordinary differential processes. *Journal of Applied Probability*, 8(2):344–356, 1971.

18. T. G. Kurtz. The relationship between stochastic and deterministic models in chemical reactions. *The Journal of Chemical Physics*, 57(7):2976–2978, 1972.

19. B. Lapeyre, É. Pardoux, and R. Sentis. *Introduction to Monte-Carlo Methods for Transport and Diffusion Equations.* Oxford University Press, Oxford, 2003.

20. G. N. Milstein. *Numerical Integration of Stochastic Differential Equations.* Kluwer Academic Publishers, Dordrecht, 1995.

21. N. J. Newton. Variance reduction for simulated diffusions. *SIAM Journal on Applied Mathematics*, 54(6):1780–1805, 1994.

22. B. Øksendal. *Stochastic Differential Equations: An Introduction with Applications.* Springer-Verlag, Berlin, fifth edition, 1998.

23. M. A. Pinsky. *Lectures on Random Evolution.* World Scientific, Singapore, 1991.

24. P. K. Pollett. On a model for interference between searching insect parasites. *Journal of the Australian Mathematical Society, Series B*, 32(2):133–150, 1990.

25. P. K. Pollett. Diffusion approximations for a circuit switching network with random alternative routing. *Australian Telecommunication Research*, 25(2):45–51, 1991.

26. P. K. Pollett. Diffusion approximations for ecological models. In F. Ghassemi, editor, *Proceedings of the International Congress on Modelling and Simulation*, volume 2, pages 843–848, Canberra, Australia, 2001. Modelling and Simulation Society of Australia and New Zealand.

27. P. K. Pollett and A. Vassallo. Diffusion approximations for some simple chemical reaction schemes. *Advances in Applied Probability*, 24(4):875–893, 1992.

28. W. Wagner. Monte Carlo evaluation of functionals of solutions of stochastic differential equations. Variance reduction and numerical examples. *Stochastic Analysis and Applications*, 6(4):447–468, 1988.

29. W. Wagner. Unbiased Monte Carlo estimators for functionals of weak solutions of stochastic differential equations. *Stochastics and Stochastic Reports*, 28(1):1–20, 1989.

APPENDIX A
PROBABILITY AND STOCHASTIC PROCESSES

The purpose of this chapter is to review some fundamental concepts in probability and stochastic processes, and to familiarize the reader with the notation in this book.

A.1 RANDOM EXPERIMENTS AND PROBABILITY SPACES

The basic notion in probability theory is that of a **random experiment**: an experiment whose outcome cannot be determined in advance. Mathematically, a random experiment is modeled via a **probability space** $(\Omega, \mathcal{H}, \mathbb{P})$, where:

- Ω is the set of all possible outcomes of the experiment, called the **sample space**.
- $\mathcal{H}$ is the collection of all subsets of Ω to which a probability can be assigned; such subsets are called **events**. The collection $\mathcal{H}$ is assumed to contain Ω itself, be closed under complements ($A \in \mathcal{H} \Rightarrow A^c \in \mathcal{H}$), and be closed under countable unions ($A_1, A_2, \ldots \in \mathcal{H} \Rightarrow \cup_i A_i \in \mathcal{H}$). Such a collection is called a **σ-algebra**.
- $\mathbb{P}$ is a **probability measure**, which assigns to each event A a number $\mathbb{P}(A)$ between 0 and 1, indicating the likelihood that the outcome of the random experiment lies in A.

Any probability measure $\mathbb{P}$ must satisfy the following **Kolmogorov axioms**:

1. $\mathbb{P}(A) \geqslant 0$ for all $A \in \mathcal{H}$.
2. $\mathbb{P}(\Omega) = 1$.
3. For any sequence $A_1, A_2, \ldots$ of disjoint (that is, nonoverlapping) events,

$$\mathbb{P}\Big(\bigcup_i A_i\Big) = \sum_i \mathbb{P}(A_i)\,. \tag{A.1}$$

The axioms ensure that the probability of any event lies between 0 and 1. An event that happens with probability 1 is called an **almost sure** (a.s.) event. The requirement (A.1) is often referred to as the **sum rule** of probability. It simply states that if an event can happen in a number of different but not simultaneous ways, the probability of that event is the sum of the probabilities of the comprising events.

■ EXAMPLE A.1 (Discrete Sample Space)

In many applications the sample space is **countable**, that is, $\Omega = \{a_1, a_2, \ldots\}$. In this case the easiest way to specify a probability measure $\mathbb{P}$ is to first assign a probability p_i to each **elementary event** $\{a_i\}$, with $\sum_i p_i = 1$, and then to define

$$\mathbb{P}(A) = \sum_{i:a_i \in A} p_i \quad \text{for all } A \subseteq \Omega\,.$$

Here the collection of events $\mathcal{H}$ can be taken to be equal to the collection of *all* subsets of Ω. The triple $(\Omega, \mathcal{H}, \mathbb{P})$ is called a **discrete probability space**.

This idea is graphically represented in Figure A.1. Each element a_i, represented by a black dot, is assigned a probability weight p_i, indicated by the size of the dot. The probability of the event A is simply the sum of the weights of all the outcomes in A.

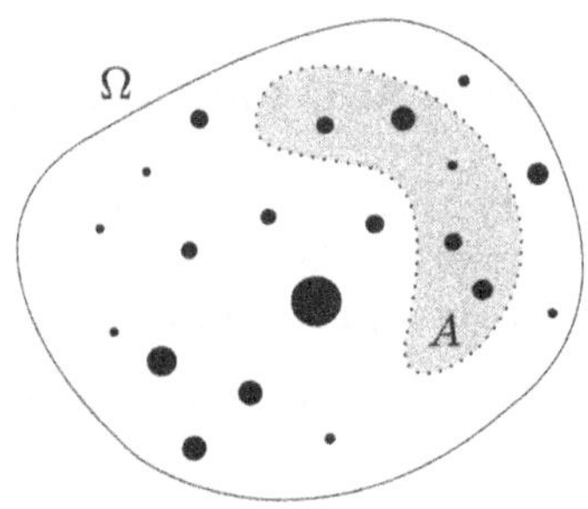

Figure A.1 A discrete sample space.

Remark A.1.1 (Equilikely Principle) A special case of a discrete probability space occurs when a random experiment has finitely many and *equally likely* outcomes. In this case the probability measure is given by

$$\mathbb{P}(A) = \frac{|A|}{|\Omega|}\,, \tag{A.2}$$

where $|A|$ denotes the number of outcomes in A and $|\Omega|$ the total number of outcomes. Thus, the calculation of probabilities reduces to counting. This is called the **equilikely principle**.

A.1.1 Properties of a Probability Measure

The following properties of a probability measure follow directly from the Kolmogorov axioms. Proofs can be found, for example, in [5, 25].

1. *Complement*: $\mathbb{P}(A^c) = 1 - \mathbb{P}(A)$.
2. *Monotonicity*: $A \subseteq B \Rightarrow \mathbb{P}(A) \leqslant \mathbb{P}(B)$.
3. *Sum rule*: $\{A_i\}$ disjoint $\Rightarrow \mathbb{P}(\cup_i A_i) = \sum_i \mathbb{P}(A_i)$.
4. *Inclusion–exclusion*:
$$\mathbb{P}(\cup_i A_i) = \sum_i \mathbb{P}(A_i) - \sum_{i<j} \mathbb{P}(A_i \cap A_j) + \sum_{i<j<k} \mathbb{P}(A_i \cap A_j \cap A_k) - \cdots .$$
In particular, $\mathbb{P}(A \cup B) = \mathbb{P}(A) + \mathbb{P}(B) - \mathbb{P}(A \cap B)$.
5. *Continuity from below*: Let $A_1, A_2, \ldots$ be an increasing sequence of events, that is, $A_1 \subseteq A_2 \subseteq \cdots \subseteq A$, with $A = \cup_n A_n$. Then, the sequence $\mathbb{P}(A_1), \mathbb{P}(A_2), \ldots$ increases monotonely to $\mathbb{P}(A)$.
6. *Continuity from above*: Let $A_1, A_2, \ldots$ be a decreasing sequence of events, that is, $A_1 \supseteq A_2 \supseteq \cdots \supseteq A$, with $A = \cap_n A_n$. Then, the sequence $\mathbb{P}(A_1), \mathbb{P}(A_2), \ldots$ decreases monotonely to $\mathbb{P}(A)$.
7. *Boole's inequality*: $\mathbb{P}(\cup_i A_i) \leqslant \sum_i \mathbb{P}(A_i)$.
8. *Borel–Cantelli*: Let $A_1, A_2, \ldots$ be a sequence of events, and let $\limsup A_n = \cap_m \cup_{n \geqslant m} A_n$ denote the event that infinitely many A_n occur. Then,
$$\sum_n \mathbb{P}(A_n) < \infty \quad \Rightarrow \quad \mathbb{P}(\limsup A_n) = 0 .$$
Under the additional assumption that the $\{A_i\}$ are *pairwise independent*, ☞ 616
$$\sum_n \mathbb{P}(A_n) = \infty \quad \Rightarrow \quad \mathbb{P}(\limsup A_n) = 1 .$$

A.2 RANDOM VARIABLES AND PROBABILITY DISTRIBUTIONS

It is often convenient to describe a random experiment via **random variables**, representing numerical measurements of the experiment. Random variables are usually denoted by capital letters from the last part of the alphabet. A vector $\mathbf{X} = (X_1, \ldots, X_n)$ of random variables is called a **random vector**. A collection of random variables $\{X_t, t \in \mathscr{T}\}$, where $\mathscr{T}$ is any index set, is called a **stochastic process**. The set of possible values for X_t (assuming this is independent of t)

is called the **state space** of the process. Stochastic processes are discussed in
☞ 153 Sections A.9–A.13. Chapter 5 is devoted to random process generation.

From a mathematical point of view, a random variable X taking values in some set E is a function $X : \Omega \to E$ such that

$$\{X \in B\} \stackrel{\text{def}}{=} \{\omega \in \Omega : X(\omega) \in B\} \in \mathcal{H} \quad \text{for all } B \in \mathcal{E} \,,$$

where $\mathcal{E}$ is a σ-algebra on E. The pair $(E, \mathcal{E})$ is called a **measurable space**. If not otherwise specified we assume that X is a **numerical** random variable; that is, $E = \mathbb{R}$. It is sometimes useful to have E as the extended real line $\overline{\mathbb{R}} = \mathbb{R} \cup \{\pm\infty\}$. In either case, $\mathcal{E}$ is the corresponding Borel σ-algebra. The **Borel σ-algebra** is the smallest σ-algebra on $\mathbb{R}$ or $\overline{\mathbb{R}}$ that contains all intervals (or, equivalently, all open sets). Elements of this σ-algebra are called **Borel sets** — for example, a countable union of intervals is a Borel set. The **Lebesgue measure** m is the unique measure on the Borel σ-algebra such that $m([a, b]) = b - a$. Similar definitions hold for n-dimensional Euclidean spaces, replacing intervals by rectangles, etc.

Define

$$P_X(B) = \mathbb{P}(X \in B), \quad B \in \mathcal{E} \,.$$

Then, P_X is a probability measure on $(E, \mathcal{E})$. It is called the **distribution** of X. The probability distribution P_X of a numerical random variable X is completely determined by its **cumulative distribution function** (cdf), defined by

$$F(x) = P_X([-\infty, x]) = \mathbb{P}(X \leqslant x), \quad x \in \overline{\mathbb{R}} \,.$$

The following properties of a cdf F are a direct consequence of the Kolmogorov axioms. For proofs see, for example, [25].

1. *Right-continuous*: $\lim_{h \downarrow 0} F(x + h) = F(x)$.

2. *Increasing*: $x \leqslant y \Rightarrow F(x) \leqslant F(y)$.

3. *Bounded*: $0 \leqslant F(x) \leqslant 1$.

Conversely, to each function F satisfying the above properties corresponds exactly one distribution P_X; see, for example, [5, Theorem 2.2.2]. Figure A.2 shows a generic cdf.

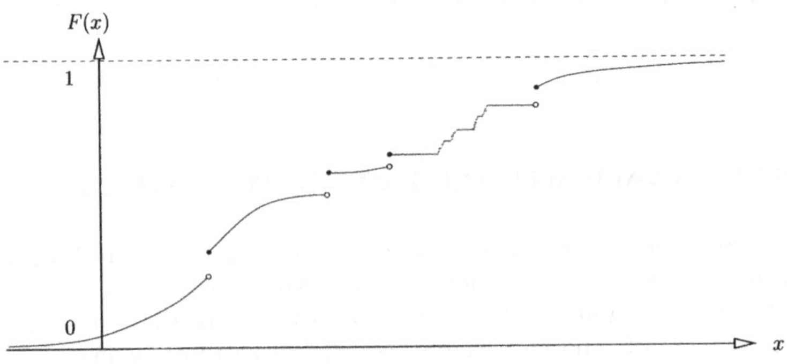

Figure A.2 A cumulative distribution function (cdf).

A cdf F_d is called **discrete** if there exist numbers $x_1, x_2, \ldots$ and probabilities $0 < f(x_i) \leqslant 1$ summing up to 1, such that for all x

$$F_d(x) = \sum_{x_i \leqslant x} f(x_i) . \tag{A.3}$$

Such a cdf is piecewise constant and has jumps of sizes $f(x_1), f(x_2), \ldots$ at points $x_1, x_2, \ldots$, respectively.

A cdf F_c is called **absolutely continuous** if there exists a positive function f such that for all x

$$F_c(x) = \int_{-\infty}^{x} f(u)\,du . \tag{A.4}$$

Note that such an F_c is differentiable (and hence continuous) with derivative f. However, in general the derivative F_c' of a continuous cdf F_c does not necessarily satisfy (A.4). A typical example of a continuous cdf whose derivative is 0 almost everywhere — and hence violates (A.4) — is the **Cantor function**, depicted in Figure A.3. Such continuous cdfs are said to be **singular**. Most distributions used in practice are either discrete or absolutely continuous, or a mixture thereof.

■ **EXAMPLE A.2 (Cantor Function)**

The Cantor function is constructed in the following way. Let $F(1) = 1$. Divide the interval $[0, 1)$ into three equal parts: $[0, \frac{1}{3})$, $[\frac{1}{3}, \frac{2}{3})$, and $[\frac{2}{3}, 1)$. Define $F(x) = \frac{1}{2}$ for $x \in [\frac{1}{3}, \frac{2}{3})$. Next, divide $[0, \frac{1}{3})$ into three subintervals $[0, \frac{1}{9}), [\frac{1}{9}, \frac{2}{9})$, and $[\frac{2}{9}, \frac{3}{9})$ and divide $[\frac{2}{3}, 1)$ into $[\frac{6}{9}, \frac{7}{9}), [\frac{7}{9}, \frac{8}{9})$, and $[\frac{8}{9}, 1)$. Let F have the value $\frac{1}{4}$ on $[\frac{1}{9}, \frac{2}{9})$ and $\frac{3}{4}$ on $[\frac{7}{9}, \frac{8}{9})$. Now divide each of the four remaining subintervals again into three parts. Assign the values $\frac{1}{8}, \frac{3}{8}, \frac{5}{8}$, and $\frac{7}{8}$ to the middle intervals, and continue this process indefinitely. This cdf is continuous, but its derivative is 0 almost everywhere.

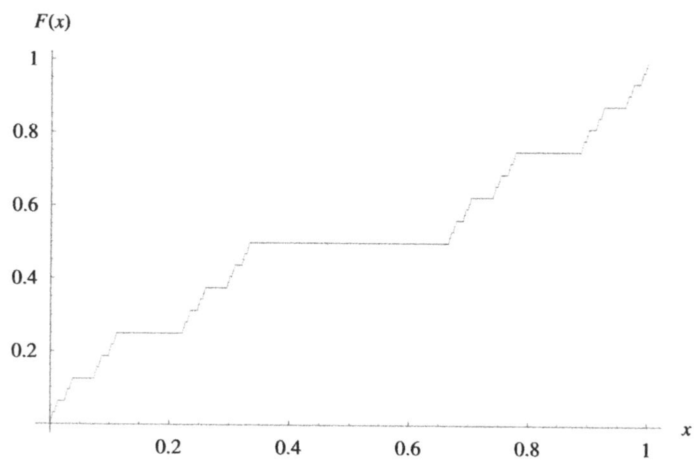

Figure A.3 The Cantor function is a continuous singular cdf.

It can be shown (see, for example, [5, Chapter 1]) that every cdf F can be written as the unique convex combination, or **mixture**, of a discrete, an absolutely continuous, and a continuous singular cdf:

$$F(x) = \alpha_1 F_d(x) + \alpha_2 F_c(x) + \alpha_3 F_s(x), \quad \text{where} \quad \alpha_1 + \alpha_2 + \alpha_3 = 1,$$

and $\alpha_k \geqslant 0$ for $k = 1, 2, 3$.

A.2.1 Probability Density

A probability distribution on some measurable space $(E, \mathcal{E})$ is often of the form

$$P_X(B) = \int_B f(x)\, \mathrm{d}m(x), \quad B \in \mathcal{E},$$

where m is some measure on $(E, \mathcal{E})$. We say that P_X has a **probability density function** (pdf), or simply **density**, f with respect to m.

EXAMPLE A.3 (Discrete Distribution)

Suppose a random variable X has a discrete cdf, as in (A.3). Thus, X takes values in some finite or countable set of points $E = \{x_1, x_2, \ldots\}$, with $\mathbb{P}(X = x_i) = f(x_i) > 0$, $i = 1, 2, \ldots$. Define $f(x) = 0$ for all $x \notin E$. Let $\mathcal{E}$ denote the collection of all subsets of E and let m be the **counting measure** on $(E, \mathcal{E})$, that is, $m(B)$ is the number of points in E that lie in the set B. Then, we see that the distribution P_X of X satisfies

$$P_X(B) = \mathbb{P}(X \in B) = \sum_{x_i \in B} f(x_i) = \int_B f(x)\, \mathrm{d}m(x) \quad \text{for all } B \subseteq E. \qquad \text{(A.5)}$$

In other words, X has a density f with respect to the counting measure m. Such a random variable is called **discrete** and P_X is called a **discrete distribution**. Such a distribution is thus completely specified by its (discrete) pdf, and probabilities can be evaluated via summation, as in (A.5). This is illustrated in Figure A.4.
☞ 85 Many specific discrete distributions are given in Chapter 4.

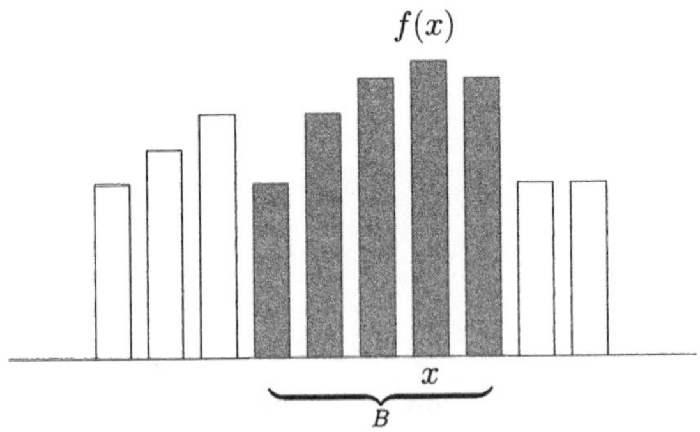

Figure A.4 Discrete probability density function (pdf). The shaded area corresponds to the probability $\mathbb{P}(X \in B)$.

■ **EXAMPLE A.4 (Absolutely Continuous Distribution)**

Suppose a random variable X has an absolutely continuous cdf, as in (A.4). Then, the distribution P_X of X satisfies

$$P_X(B) = \mathbb{P}(X \in B) = \int_B f(x)\,\mathrm{d}x = \int_B f(x)\,\mathrm{d}m(x) \tag{A.6}$$

for all Borel sets B, where m is the Lebesgue measure. The distribution P_X is said to be **absolutely continuous** with respect to the Lebesgue measure, and f is the corresponding pdf. As a consequence, such a distribution is completely specified by its pdf, and probabilities can be evaluated via integration. This is illustrated in Figure A.5. Many specific absolutely continuous distributions are given in Chapter 4. ☞ 85

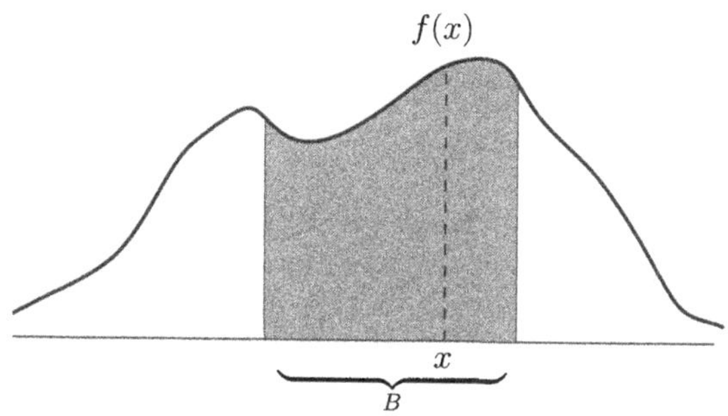

Figure A.5 Absolutely continuous probability density function (pdf). The shaded area corresponds to the probability $\mathbb{P}(X \in B)$.

We can view $f(x)$ as the probability "density" at $X = x$, in the sense that, for small h,

$$\mathbb{P}(x \leqslant X \leqslant x + h) = \int_x^{x+h} f(u)\,\mathrm{d}u \approx h\,f(x)\,.$$

Remark A.2.1 (Probability Density and Probability Mass) It is important to note that we deliberately use the *same* name, "pdf", and symbol, f, in both the discrete and the absolutely continuous case, rather than distinguish between a probability mass function (pmf) and probability density function (pdf). The reason is that from a measure-theoretic point of view the pdf plays exactly the same role in the discrete and absolutely continuous cases. The only difference is the measure m. We use the notation $X \sim \mathsf{Dist}$, $X \sim f$, and $X \sim F$ to indicate that X has distribution Dist, pdf f, and cdf F.

A.2.2 Joint Distributions

Distributions for random vectors and stochastic processes can be specified in much the same way as for random variables. In particular, the distribution of a random vector $\mathbf{X} = (X_1, \ldots, X_n)$ is completely determined by specifying the **joint cdf** F, defined by

$$F(x_1, \ldots, x_n) = \mathbb{P}(X_1 \leqslant x_1, \ldots, X_n \leqslant x_n), \quad x_i \in \mathbb{R},\ i = 1, \ldots, n\,.$$

Similarly, the distribution of a stochastic process $\{X_t, t \in \mathscr{T}\}$, with $\mathscr{T} \subseteq \mathbb{R}$, is completely determined by its **finite-dimensional distributions**; that is, the distributions of the random vectors $(X_{t_1}, \ldots, X_{t_n})$ for any choice of n and $t_1, \ldots, t_n$.

By analogy to the one-dimensional case, a random vector $\mathbf{X}$ taking values in $\mathbb{R}^n$ is said to have a pdf f with respect to some measure m, if

$$\mathbb{P}(\mathbf{X} \in B) = \int_B f(\mathbf{x})\, \mathrm{d}m(\mathbf{x}) , \tag{A.7}$$

for all n-dimensional Borel sets B. The important cases are when m is either the counting measure or the Lebesgue measure.

The **marginal pdfs** can be recovered from the joint pdf by "integrating out the other variables". For example, for a random vector (X, Y) with pdf f with respect to the Lebesgue measure on $\mathbb{R}^2$, the pdf f_X of X is given by

$$f_X(x) = \int f(x,y)\, \mathrm{d}y .$$

Remark A.2.2 (Multivariate Singular Distributions) Continuous singular distributions are much more likely to be encountered in the multidimensional setting. For example, if a numerical random vector takes values exclusively in a lower dimensional subset, then the distribution has a derivative of 0 almost everywhere with respect to the Lebesgue measure and so is singular.

A.3 EXPECTATION AND VARIANCE

It is often useful to consider various numerical characteristics of a random variable or its distribution. For example, two such quantities are the expectation and variance. The first measures the mean value of the distribution; the second measures the spread or dispersion of the distribution. The intuitive definition of the expectation of a discrete random variable X is that it is the average of the possible values that X can take, weighted by the corresponding probabilities; that is,

$$\mathbb{E}X = \sum_x x\, \mathbb{P}(X = x) .$$

Similarly, for an absolutely continuous random variable X the expectation is given by

$$\mathbb{E}X = \int x\, f(x)\, \mathrm{d}x .$$

Both definitions are part of a more general framework, in which the **expectation** of X is defined as the abstract integral

$$\mathbb{E}X = \int X\, \mathrm{d}\mathbb{P} , \tag{A.8}$$

which is defined in four steps (see, for example, [5, Chapter 3] and [12]):

1. If X is an **indicator function** of some event A, that is,

$$X = \mathrm{I}_A \stackrel{\text{def}}{=} \begin{cases} 1 & \text{if } \omega \in A \\ 0 & \text{otherwise ,} \end{cases} \tag{A.9}$$

   then

$$\mathbb{E}X \stackrel{\text{def}}{=} \mathbb{P}(A) \,. \tag{A.10}$$

2. If X is positive and **simple**, that is, $X = \sum_{i=1}^{n} a_i \, \mathrm{I}_{A_i}$ for some positive (possibly infinite) numbers $a_1, \ldots, a_n$ and events $A_1, \ldots, A_n$, then

$$\mathbb{E}X \stackrel{\text{def}}{=} \sum_{i=1}^{n} a_i \, \mathbb{P}(A_i) \,, \tag{A.11}$$

   with $\infty \times 0$ and $0 \times \infty$ defined to be 0.

3. If X is a positive random variable, then

$$\mathbb{E}X \stackrel{\text{def}}{=} \lim_{n \to \infty} \mathbb{E}X_n \,, \tag{A.12}$$

   where $X_1, X_2, \ldots$ is any sequence of simple random variables that increases almost surely to X. We write ☞ 623

$$X_n \stackrel{a.s.}{\nearrow} X \,.$$

   It can be shown [12] that such a sequence exists, and that the limit in (A.12) exists (possibly infinity) and does not depend on the increasing sequence of simple random variables.

4. Finally, for a general (not necessarily positive) random variable X, write $X = X^+ - X^-$, where $X^+ = \max\{X, 0\}$ and $X^- = \max\{-X, 0\}$ are the positive and negative parts of X (note that both are $\geqslant 0$), and define

$$\mathbb{E}X \stackrel{\text{def}}{=} \mathbb{E}X^+ - \mathbb{E}X^- \,,$$

   provided that at least one of the right-hand-side terms is finite ($\infty - \infty$ is not well-defined).

Random variables X for which $\mathbb{E}|X| < \infty$ (and hence the expectation is finite) are called **integrable**. It is not difficult to show [3, Page 216] that a random variable X is integrable if and only if

$$\lim_{c \to \infty} \mathbb{E}|X| \, \mathrm{I}_{\{|X| > c\}} = 0 \,.$$

A random variable X is said to be **square integrable** if $\mathbb{E}X^2 < \infty$. A sequence of random variables $X_1, X_2, \ldots$ is said to be **integrable** if

$$\sup_n \mathbb{E}|X_n| < \infty \,.$$

A sequence of random variables $X_1, X_2, \ldots$ is said to be **uniformly integrable**, if

$$\lim_{c \to \infty} \sup_n \mathbb{E}|X_n| \, \mathrm{I}_{\{|X_n| > c\}} = 0 \,.$$

In particular, $X_1, X_2, \ldots$ must be integrable. Moreover, if for some $\varepsilon > 0$

$$\sup_n \mathbb{E}|X_n|^{1+\varepsilon} < \infty ,$$

then the sequence $X_1, X_2, \ldots$ is uniformly integrable. Another sufficient condition for uniform integrability [2, Page 32] is the existence of an integrable random variable Y such that $\mathbb{P}(|X_n| \geqslant x) \leqslant \mathbb{P}(|Y| \geqslant x)$ for all x and n. For continuous-time stochastic processes $\{X_t, t \geqslant 0\}$, integrability and uniform integrability are defined in the same way, replacing the discrete n with a continuous t.

For the purpose of calculating expectations, the following theorem is indispensable.

Theorem A.3.1 (Expected Value) Let X be a random variable with distribution P_X and cdf F, and let g be a numerical function, then (provided that the integral exists)

$$\mathbb{E}\, g(X) = \int g(X)\, \mathrm{d}\mathbb{P} = \int_{\mathbb{R}} g(x)\, \mathrm{d}P_X(x) \stackrel{\text{def}}{=} \int_{-\infty}^{\infty} g(x)\, \mathrm{d}F(x) . \qquad \text{(A.13)}$$

The last integral in (A.13) is called a **Lebesgue–Stieltjes** integral. In most cases of practical interest this integral can be determined via elementary summation or Riemann integration, in which it can be viewed as a Riemann–Stieltjes integral [3, Page 228]. In particular, when X is discrete with pdf f, (A.13) reduces to

$$\mathbb{E}\, g(X) = \sum_x g(x) f(x) , \qquad \text{(A.14)}$$

and in the absolutely continuous case (A.13) becomes

$$\mathbb{E}\, g(X) = \int_{-\infty}^{\infty} g(x)\, f(x)\, \mathrm{d}x . \qquad \text{(A.15)}$$

Theorem A.3.1 can be readily generalized to random vectors. In particular, if $\boldsymbol{X} = (X_1, \ldots, X_n)$ is a random vector with (n-dimensional) cdf F, and g a numerical function on $\mathbb{R}^n$, then

$$\mathbb{E}\, g(\mathbf{X}) = \int_{\mathbb{R}^n} g(\mathbf{x})\, \mathrm{d}F(\mathbf{x}) . \qquad \text{(A.16)}$$

A.3.1 Properties of the Expectation

☞ 623 Below, $X, X_1, X_2, \ldots$, and Y are random variables, and $\mathbf{X}$ is a random vector. We write $X_n \xrightarrow{\text{a.s.}} X$ to indicate that the sequence $X_1, X_2, \ldots$ converges almost surely to X. Note that Properties 1, 2, 4, and 7 below follow directly from the definition of the expectation. Proofs of the other properties can be found, for example, in [3, Chapter 3]. See also Section A.8.

1. *Positivity*: For positive random variables the expectation always exists (possibly $+\infty$).

2. *Linearity*: $\mathbb{E}(aX + bY) = a\,\mathbb{E}X + b\,\mathbb{E}Y$ for $a, b \in \mathbb{R}$.

3. *Monotonicity*: If $X \geqslant Y$, then $\mathbb{E}X \geqslant \mathbb{E}Y$.

4. *Indicator*: If I_A is the indicator of the event A, then $\mathbb{E}\, I_A = \mathbb{P}(A)$.

5. *Jensen's inequality*: Let $\mathscr{X}$ be a convex subset of $\mathbb{R}^n$ and $h : \mathscr{X} \to \mathbb{R}$ be a convex measurable function. Let $\mathbf{X}$ be a random vector taking values in $\mathscr{X}$, such that $\mathbb{E}\mathbf{X} = (\mathbb{E}X_1, \ldots, \mathbb{E}X_n)$ is finite. Then, $\mathbb{E}h(\mathbf{X})$ exists and ☞ 679

$$\mathbb{E}h(\mathbf{X}) \geqslant h(\mathbb{E}\mathbf{X}) .$$

6. *Fatou's lemma*: If $X_n \geqslant 0$, then

$$\mathbb{E} \liminf_{n\to\infty} X_n \leqslant \liminf_{n\to\infty} \mathbb{E}X_n .$$

7. *Monotone convergence theorem*: Suppose $\mathbb{E}X_n$ exists for some n, then

$$X_n \overset{a.s.}{\nearrow} X \quad \Rightarrow \quad \mathbb{E}X_n \nearrow \mathbb{E}X .$$

8. *Dominated convergence theorem*: Suppose $|X_n| \leqslant Y$ for all n, where $\mathbb{E}Y < \infty$. Then,

$$X_n \xrightarrow{\text{a.s.}} X \quad \Rightarrow \quad \mathbb{E}X_n \to \mathbb{E}X .$$

A.3.2 Variance

The **variance** of a random variable X, denoted by $\text{Var}(X)$ (or sometimes σ^2) is defined by

$$\text{Var}(X) = \mathbb{E}(X - \mathbb{E}X)^2 = \mathbb{E}X^2 - (\mathbb{E}X)^2 .$$

The square root of the variance is called the **standard deviation**.

In general, the mean and the variance do not give enough information to completely specify the distribution of a random variable. However, they may provide useful bounds. We give three such bounds. A proof of Kolmogorov's inequality may be found in [5, Page 116].

1. *Markov's inequality*: For any positive random variable X with expectation μ,

$$\mathbb{P}(X \geqslant x) \leqslant \frac{\mu}{x}, \quad x \geqslant 0 . \tag{A.17}$$

2. *Chebyshev's inequality*: Let X be a random variable with finite expectation and variance, μ and σ^2, respectively. Then,

$$\mathbb{P}(|X - \mu| \geqslant x) \leqslant \frac{\sigma^2}{x^2}, \quad x \geqslant 0 . \tag{A.18}$$

3. *Kolmogorov's inequality*: Let $X_1, X_2, \ldots$ be a sequence of independent random variables. Let $S_1, S_2, \ldots$ be the sequence of partial sums, defined by $S_n = X_1 + \cdots + X_n$ and assumed to have finite expectations and variances, $\{\mu_n\}$ and $\{\sigma_n^2\}$, respectively. Then,

$$\mathbb{P}\left(\max_{1\leqslant i\leqslant n} |S_i - \mu_i| \geqslant x\right) \leqslant \frac{\sigma_n^2}{x^2}, \quad x \geqslant 0 . \tag{A.19}$$

A.4 CONDITIONING AND INDEPENDENCE

Conditional probabilities and conditional distributions are used to model additional information on a random experiment. Independence is used to model *lack* of such information.

A.4.1 Conditional Probability

Suppose some event $B \subseteq \Omega$ occurs. Given this fact, event A will occur if and only if $A \cap B$ occurs, and the relative chance of A occurring is therefore $\mathbb{P}(A \cap B)/\mathbb{P}(B)$, provided $\mathbb{P}(B) > 0$. This leads to the definition of the **conditional probability** of A given B:

$$\mathbb{P}(A \mid B) = \frac{\mathbb{P}(A \cap B)}{\mathbb{P}(B)}, \quad \text{if } \mathbb{P}(B) > 0 . \tag{A.20}$$

The above definition breaks down if $\mathbb{P}(B) = 0$. Such conditional probabilities must be treated with more care [3].

Three important consequences of the definition of conditional probability are:

1. *Product rule*: For any sequence of events $A_1, A_2, \ldots, A_n$,

$$\mathbb{P}(A_1 \cdots A_n) = \mathbb{P}(A_1)\,\mathbb{P}(A_2 \mid A_1)\,\mathbb{P}(A_3 \mid A_1 A_2) \cdots \mathbb{P}(A_n \mid A_1 \cdots A_{n-1}) , \tag{A.21}$$

   using the abbreviation $A_1 A_2 \cdots A_k = A_1 \cap A_2 \cap \cdots \cap A_k$.

2. *Law of total probability*: If $\{B_i\}$ forms a **partition** of Ω (that is, $B_i \cap B_j = \emptyset, i \neq j$ and $\cup_i B_i = \Omega$), then for any event A

$$\mathbb{P}(A) = \sum_i \mathbb{P}(A \mid B_i)\,\mathbb{P}(B_i) . \tag{A.22}$$

3. *Bayes' rule*: Let $\{B_i\}$ form a partition of Ω. Then, for any event A with $\mathbb{P}(A) > 0$,

$$\mathbb{P}(B_j \mid A) = \frac{\mathbb{P}(A \mid B_j)\,\mathbb{P}(B_j)}{\sum_i \mathbb{P}(A \mid B_i)\,\mathbb{P}(B_i)} . \tag{A.23}$$

A.4.2 Independence

Two events A and B are said to be **independent** if the knowledge that B has occurred does not change the probability that A occurs. That is, A, B independent $\Leftrightarrow \mathbb{P}(A \mid B) = \mathbb{P}(A)$. Since $\mathbb{P}(A \mid B)\,\mathbb{P}(B) = \mathbb{P}(A \cap B)$, an alternative definition of independence is

$$A,\ B \text{ independent} \Leftrightarrow \mathbb{P}(A \cap B) = \mathbb{P}(A)\,\mathbb{P}(B) .$$

This definition covers the case where $\mathbb{P}(B) = 0$ and can be extended to arbitrarily many events: events $A_1, A_2, \ldots$ are said to be **independent** if for any k and any choice of distinct indices $i_1, \ldots, i_k$,

$$\mathbb{P}(A_{i_1} \cap A_{i_2} \cap \cdots \cap A_{i_k}) = \mathbb{P}(A_{i_1})\,\mathbb{P}(A_{i_2}) \cdots \mathbb{P}(A_{i_k}) .$$

The $\{A_i\}$ are said to be **pairwise independent** if every two events are independent.

The concept of independence can also be formulated for *random variables.* Random variables $X_1, X_2, \ldots$ are said to be **independent** if the events $\{X_{i_1} \in B_1\}, \ldots, \{X_{i_n} \in B_n\}$ are independent for all finite choices of n, distinct indices $i_1, \ldots, i_n$, and Borel sets $B_1, \ldots, B_n$.

An important characterization of independent random variables is the following (for a proof, see [25], for example).

Theorem A.4.1 (Product of Marginal Pdfs) *Random variables $X_1, \ldots, X_n$ with marginal pdfs $f_{X_1}, \ldots, f_{X_n}$ and joint pdf f are independent if and only if*

$$f(x_1, \ldots, x_n) = f_{X_1}(x_1) \cdots f_{X_n}(x_n) \quad \textit{for all } x_1, \ldots, x_n \,. \tag{A.24}$$

Many probabilistic models involve random variables $X_1, X_2, \ldots$ that are **independent and identically distributed**, abbreviated as **iid**. We will use this abbreviation throughout this book.

A.4.3 Covariance

The **covariance** of two random variables X and Y with expectations μ_X and μ_Y, respectively, is defined as

$$\mathrm{Cov}(X, Y) = \mathbb{E}[(X - \mu_X)(Y - \mu_Y)] \,.$$

This is a measure for the amount of linear dependency between the variables. Let $\sigma_X^2 = \mathrm{Var}(X)$ and $\sigma_Y^2 = \mathrm{Var}(Y)$. A scaled version of the covariance is given by the **correlation coefficient**,

$$\varrho(X, Y) = \frac{\mathrm{Cov}(X, Y)}{\sigma_X \, \sigma_Y} \,.$$

Below we use the notation $\mu_X = \mathbb{E}X$ and $\sigma_X^2 = \mathrm{Var}(X)$. The following properties follow directly from the definitions of variance and covariance.

1. $\mathrm{Var}(X) = \mathbb{E}X^2 - \mu_X^2$.
2. $\mathrm{Var}(aX + b) = a^2 \, \sigma_X^2$.
3. $\mathrm{Cov}(X, Y) = \mathbb{E}[XY] - \mu_X \, \mu_Y$.
4. $\mathrm{Cov}(X, Y) = \mathrm{Cov}(Y, X)$.
5. $-\sigma_X \sigma_Y \leqslant \mathrm{Cov}(X, Y) \leqslant \sigma_X \sigma_Y$.
6. $\mathrm{Cov}(aX + bY, Z) = a \, \mathrm{Cov}(X, Z) + b \, \mathrm{Cov}(Y, Z)$.
7. $\mathrm{Cov}(X, X) = \sigma_X^2$.
8. $\mathrm{Var}(X + Y) = \sigma_X^2 + \sigma_Y^2 + 2 \, \mathrm{Cov}(X, Y)$.
9. If X and Y are independent, then $\mathrm{Cov}(X, Y) = 0$.

As a consequence of Properties 2 and 8 we have that for any sequence of *independent* random variables $X_1, \ldots, X_n$ with variances $\sigma_1^2, \ldots, \sigma_n^2$,

$$\mathrm{Var}(a_1 X_1 + a_2 X_2 + \cdots + a_n X_n) = a_1^2\,\sigma_1^2 + a_2^2\,\sigma_2^2 + \cdots + a_n^2\,\sigma_n^2 \,, \tag{A.25}$$

for any choice of constants $a_1, \ldots, a_n$.

For random vectors, such as $\mathbf{X} = (X_1, \ldots, X_n)$, it is convenient to write the expectations and covariances in vector notation. It will usually be clear from the context whether we interpret $\mathbf{X}$ as a *row* or a *column vector*. In some cases, for example, with matrix multiplication, we make the distinction explicit. For a random (column) vector $\mathbf{X}$ we define its **expectation vector** as the vector of expectations

$$\boldsymbol{\mu} = (\mu_1, \ldots, \mu_n)^\top = (\mathbb{E}X_1, \ldots, \mathbb{E}X_n)^\top \,.$$

The **covariance matrix** Σ is defined as the matrix whose (i,j)-th element is

$$\mathrm{Cov}(X_i, X_j) = \mathbb{E}[(X_i - \mu_i)(X_j - \mu_j)] \,.$$

If we define the expectation of a vector (matrix) to be the vector (matrix) of expectations, then we can compactly write

$$\boldsymbol{\mu} = \mathbb{E}\mathbf{X}$$

and

$$\Sigma = \mathbb{E}[(\mathbf{X} - \boldsymbol{\mu})(\mathbf{X} - \boldsymbol{\mu})^\top] \,.$$

A.4.4 Conditional Density and Expectation

Suppose X and Y are both discrete or both absolutely continuous, with joint pdf f, and suppose $f_X(x) > 0$. Then, the **conditional pdf** of Y given $X = x$ is given by

$$f_{Y|X}(y\,|\,x) = \frac{f(x,y)}{f_X(x)} \quad \text{for all } y \,. \tag{A.26}$$

In the discrete case the formula is a direct translation of (A.20), with $f_{Y|X}(y\,|\,x) = \mathbb{P}(Y = y\,|\,X = x)$. In the absolutely continuous case a similar interpretation, in terms of densities, can be used (see, for example, [25, Page 221]). The corresponding distribution is called the **conditional distribution** of Y given $X = x$, and the corresponding **conditional expectation** is

$$\mathbb{E}[Y\,|\,X = x] = \begin{cases} \sum_y y\, f_{Y|X}(y\,|\,x) & \text{discrete case,} \\ \int y\, f_{Y|X}(y\,|\,x)\,\mathrm{d}y & \text{absolutely continuous case.} \end{cases} \tag{A.27}$$

Note that $\mathbb{E}[Y\,|\,X = x]$ is a function of x. The corresponding random variable is written as $\mathbb{E}[Y\,|\,X]$. A similar formalism can be used when conditioning on a sequence of random variables $X_1, \ldots, X_n$ or on a σ-algebra; see, for example, [5, Chapter 9]. The conditional expectation has similar properties to the ordinary expectation in Section A.3.1. Other useful properties (see, for example, [28]) are:

1. *Tower property*: If $\mathbb{E}Y$ exists, then

$$\mathbb{E}\,\mathbb{E}[Y\,|\,X] = \mathbb{E}Y \,. \tag{A.28}$$

2. *Taking out what is known*: If $\mathbb{E}Y$ exists, then

$$\mathbb{E}[XY \mid X] = X\mathbb{E}Y .$$

3. *Orthogonal projection*: If Y is square integrable, then $\mathbb{E}[Y|X]$ is the function $h(X)$ that minimizes $\mathbb{E}(Y - h(X))^2$.

A.5 L^P SPACE

Let $(\Omega, \mathcal{H}, \mathbb{P})$ be a probability space and X a numerical random variable. For $p \in [1, \infty)$ define

$$\|X\|_p = (\mathbb{E}|X|^p)^{\frac{1}{p}}$$

and let

$$\|X\|_\infty = \inf\{x : \mathbb{P}(|X| \leqslant x) = 1\} .$$

For each $p \in [1, \infty]$ we denote by L^p the collection of all numerical random variables X for which $\|X\|_p < \infty$. In particular L^1 is comprised of all integrable random variables and L^2 is comprised of all square integrable random variables.

The following properties of L^p spaces can be found, for example, in [26, Chapter 3].

1. *Positivity*: $\|X\|_p \geqslant 0$, and $\|X\|_p = 0 \Leftrightarrow X = 0$ (a.s.).
2. *Multiplication with a constant*: $\|c\,X\|_p = |c|\,\|X\|_p$.
3. *Minkowski's (triangle) inequality*: $\|X + Y\|_p \leqslant \|X\|_p + \|Y\|_p$.
4. *Hölder's inequality*: For $p, q, r \in [1, \infty]$ with $\frac{1}{p} + \frac{1}{q} = \frac{1}{r}$,

$$\|XY\|_r \leqslant \|X\|_p\,\|Y\|_q . \tag{A.29}$$

The particular case with $p = q = 2$ and $r = 1$ is called **Schwarz's inequality**.

5. *Monotonicity*: If $1 \leqslant p < q \leqslant \infty$, then $\|X\|_p \leqslant \|X\|_q$.

The space L^p is a *linear space*. The first three properties above identify $\|\cdot\|_p$ as a norm on this space, provided that random variables that are almost surely equal are identified as one and the same. Of particular importance is L^2, which is in fact a *Hilbert space*, with inner product

$$\langle X, Y\rangle = \mathbb{E}[XY] .$$

We denote the L^2 norm simply by $\|\cdot\|$, suppressing the subscript.

For random variables in L^2 the concepts of variance and covariance have a geometric interpretation. Namely, if X and Y are zero-mean random variables (their expectation is 0), then

$$\mathrm{Var}(X) = \|X\|^2 \quad \text{and} \quad \mathrm{Cov}(X, Y) = \langle X, Y\rangle .$$

Another important use of L^2 spaces is in conditioning. Let X and Y be random variables. Define $\mathcal{K}$ to be the space of functions of X that are square integrable.

If the moment generating function is finite in an open interval containing 0, then the integer moments $\{\mathbb{E}X^k\}$ exist, are finite, and uniquely determine the distribution of X. Moreover, in that case the following properties hold (see, for example, [5]):

1. *Moment property*: $\mathbb{E}X^k = M^{(k)}(0)$, $k \geqslant 1$.

2. *Taylor's theorem*:

$$M(t) = \sum_{k=0}^{\infty} \frac{\mathbb{E}X^k}{k!} t^k .$$

Remark A.7.1 (Infinite Moment Generating Function) If the moment generating function is not finite in any open interval containing 0, then the sequence of integer moments, even if they are all finite, is not sufficient to uniquely characterize the distribution of a random variable; see, for example, [13].

A.7.3 Characteristic Function

The most general transform concept is that of the characteristic function. Every random variable has a characteristic function. It is closely related to the classical Fourier transform of a function and has superior analytical properties to the moment generating function.

The **characteristic function** of a random variable X with cdf F, is the function $\phi : \mathbb{R} \to \mathbb{C}$, defined by

$$\phi(t) = \mathbb{E}\,\mathrm{e}^{\mathrm{i}tX} = \int_{-\infty}^{\infty} \mathrm{e}^{\mathrm{i}tx}\,\mathrm{d}F(x), \qquad t \in \mathbb{R} ,$$

or, equivalently,

$$\phi(t) = \mathbb{E}\cos(tX) + \mathrm{i}\,\mathbb{E}\sin(tX), \qquad t \in \mathbb{R} .$$

Note that $\phi(0) = 1$ and $|\phi(t)| \leqslant 1$. Some other properties are (for proofs see [5], for example):

1. *Moment property*: If $\mathbb{E}|X|^n < \infty$, then, for $k = 1, 2, \ldots, n$, $\phi^{(k)}$ is finite and continuous on $\mathbb{R}$, with

$$\phi^{(k)}(t) = \mathrm{i}^k\,\mathbb{E}\left[X^k\,\mathrm{e}^{\mathrm{i}tX}\right], \qquad t \in \mathbb{R} ,$$

and so, in particular, $\mathbb{E}X^k = (-\mathrm{i})^k \phi^{(k)}(0)$.

2. *Taylor's theorem*: If $\mathbb{E}|X|^n < \infty$, then, in a neighborhood of 0,

$$\phi(t) = \sum_{k=0}^{n} \frac{\mathbb{E}X^k}{k!} (\mathrm{i}t)^k + o(t^n) .$$

3. *Continuity*: Let $F_1, F_2, \ldots$ be a sequence of cdfs, with characteristic functions $\phi_1, \phi_2, \ldots$. If $\phi_n(t) \to \phi(t)$, pointwise, and $\phi(t)$ is continuous at $t = 0$, then there exists a cdf F such that F_n *converges weakly* (see Page 623) to F, and ϕ is its characteristic function.

4. $\phi(t)$ is uniformly continuous on $\mathbb{R}$.

5. $\phi_{(-X)}(t) = \overline{\phi_X(t)}$, $t \in \mathbb{R}$, from which it follows that a random variable is *symmetric* around 0 (that is, X and $-X$ are identically distributed) if and only if its characteristic function is *real-valued.*

A.8 LIMIT THEOREMS

Let $(\Omega, \mathcal{H}, \mathbb{P})$ be a probability space, and let $X_1, X_2, \ldots, X$ be random variables taking values in a metric space E with distance ϱ and equipped with a σ-algebra $\mathcal{E}$. A typical example is $E = \mathbb{R}^n$ with ϱ the Euclidean distance $\varrho(\mathbf{x}, \mathbf{y}) = \|\mathbf{x} - \mathbf{y}\|$; see also [2]. Recall that for numerical random variables $E = \mathbb{R}$ and $\varrho(x, y) = |x - y|$.

A.8.1 Modes of Convergence

We have the following definitions of the different modes of convergence of random variables.

1. *Almost sure convergence*: The sequence of numerical random variables $X_1, X_2, \ldots$ is said to converge **almost surely** to a numerical random variable X, denoted $X_n \xrightarrow{\text{a.s.}} X$, if
$$\mathbb{P}\left(\lim_{n\to\infty} X_n = X\right) = 1\,.$$

2. *Convergence in L^p-norm*: The sequence of numerical random variables $X_1, X_2, \ldots$ is said to converge **in L^p-norm** to a numerical random variable X, denoted $X_n \xrightarrow{L^p} X$, if
$$\lim_{n\to\infty} \mathbb{E}|X_n - X|^p = 0\,,$$
or, equivalently, if $\lim_{n\to\infty} \|X_n - X\|_p = 0$, where $\|\cdot\|_p$ denotes the L^p norm. Convergence in L^2-norm is often called **mean square convergence.** ☞ 619

3. *Convergence in probability*: The sequence $X_1, X_2, \ldots$ is said to converge **in probability** to X, denoted $X_n \xrightarrow{\mathbb{P}} X$, if
$$\lim_{n\to\infty} \mathbb{P}(\varrho(X_n, X) < \varepsilon) = 1 \quad \text{for all } \varepsilon > 0\,.$$

4. *Convergence in distribution*: Let P_{X_n} be the distribution of X_n and P_X the distribution of X. The sequence $X_1, X_2, \ldots$ is said to converge **in distribution** to X, denoted $X_n \xrightarrow{d} X$, if the distribution P_{X_n} **converges weakly** to P_X, that is,
$$\lim_{n\to\infty} P_{X_n}(A) = P_X(A)$$
for all sets $A \in \mathcal{E}$ such that $P_X(\partial A) = 0$, where $\partial A \in \mathcal{E}$ is the boundary of A. An equivalent definition is that
$$\lim_{n\to\infty} \mathbb{E}h(X_n) = \mathbb{E}h(X)$$

for all bounded continuous functions $h : E \to \mathbb{R}$.

5. *Complete convergence*: A sequence of random variables $X_1, X_2, \ldots$ is said to converge **completely** to X, denoted $X_n \xrightarrow{\text{cpl.}} X$, if

$$\sum_n \mathbb{P}(\varrho(X_n, X) > \varepsilon) < \infty \quad \text{for all } \varepsilon > 0\,.$$

The most general relationships among the various modes of convergence for numerical random variables are shown on the following diagram. Proofs can be found in [2] and [3]. See also [14].

$$\begin{array}{ccccc}
\boxed{X_n \xrightarrow{\text{cpl.}} X} & \Rightarrow & \boxed{X_n \xrightarrow{\text{a.s.}} X} & & \\
 & & \Downarrow & & \\
 & & \boxed{X_n \xrightarrow{\mathbb{P}} X} & \Rightarrow & \boxed{X_n \xrightarrow{d} X} \\
 & & \Uparrow & & \\
\boxed{X_n \xrightarrow{L^p} X} & \overset{p \geqslant q}{\Rightarrow} & \boxed{X_n \xrightarrow{L^q} X} & &
\end{array}$$

A.8.2 Converse Results on Modes of Convergence

1. *Convergence in distribution to a constant* [2, Page 24]: Let c be a constant element of E. Then,

$$X_n \xrightarrow{d} c \quad \Rightarrow \quad X_n \xrightarrow{\mathbb{P}} c\,.$$

2. *Convergence in probability combined with uniform integrability* [28, Page 131]:
☞ 613
Suppose the numerical random variables $\{X_n\}$ are uniformly integrable. Then, for $p \geqslant 1$,

$$X_n \xrightarrow{\mathbb{P}} X \quad \Rightarrow \quad X_n \xrightarrow{L^p} X\,.$$

This includes the case where $|X_n| \leqslant Y$ for all n with $\mathbb{E}Y < \infty$ (dominated convergence).

3. *Continuity theorem* [2, Page 30]: Let $h : E \to E'$ be a measurable function, with $(E', \mathcal{E}')$ a measurable space and E' equipped with metric ϱ'. Let $D_h \in \mathcal{E}$ be the set of discontinuities of h. If $\mathbb{P}(X \in D_h) = 0$ (in particular, when h is continuous), then

$$X_n \xrightarrow{d} X \quad \Rightarrow \quad h(X_n) \xrightarrow{d} h(X)\,.$$

A special case is **Slutsky's theorem**: if $E = \mathbb{R}^2$ and $E' = \mathbb{R}$, then we have $(X_n, Y_n) \xrightarrow{d} (X, c)$, where $c \in \mathbb{R}$ is a constant, implies $h(X_n, Y_n) \xrightarrow{d} h(X, c)$ for all continuous functions $h : \mathbb{R}^2 \to \mathbb{R}$.

4. *Finite expectation of infinite series*: Let $X_n \geqslant 0$. If the infinite series $\sum_n X_n$ has finite expectation, then $X_n \xrightarrow{\text{a.s.}} 0$.

5. *Skorohod representation* [11, Page 271]: If $X_n \xrightarrow{d} X$ with corresponding distributions P_{X_n} and P_X, then there exist random variables $\widetilde{X}_1, \widetilde{X}_2, \ldots, \widetilde{X}$

in $(E, \mathcal{E})$ with distributions P_{X_n} for each n and P_X, respectively, such that $\widetilde{X}_n \xrightarrow{\text{a.s.}} \widetilde{X}$.

6. *Monotone convergence*: Suppose $\mathbb{E}X_n$ exists for some n. Then, for any $p \geqslant 1$,

$$X_n \overset{a.s.}{\nearrow} X \quad \Rightarrow \quad X_n \overset{L^p}{\nearrow} X \,.$$

A.8.3 Law of Large Numbers and Central Limit Theorem

We briefly discuss two of the main results in probability: the law of large numbers and the central limit theorem. Both are associated with sums of independent random variables. For details, see, for example, [3, Pages 85, 357, and 385].

Let $X_1, X_2, \ldots$ be iid random variables with expectation μ. The law of large numbers states that the sample average $(X_1 + \cdots + X_n)/n$ is close to μ for large n.

Theorem A.8.1 (Strong Law of Large Numbers) *Let $X_1, \ldots, X_n$ be iid with expectation μ. Then,*

$$\frac{X_1 + \cdots + X_n}{n} \xrightarrow{\text{a.s.}} \mu \quad \text{as } n \to \infty \,.$$

The central limit theorem describes the limiting distribution of the sum $S_n = X_1 + \cdots + X_n$. Loosely, it states that the random sum S_n has a distribution that is approximately normal (Gaussian) when n is large. The more precise statement is given next.

Theorem A.8.2 (Central Limit Theorem) *Let $X_1, \ldots, X_n$ be iid with expectation μ and variance $\sigma^2 < \infty$. Then,*

$$\frac{S_n - n\mu}{\sigma\sqrt{n}} \xrightarrow{d} Y \sim \mathsf{N}(0,1) \quad \text{as } n \to \infty \,.$$

In other words, for large n the random sum S_n has a distribution that is approximately normal with expectation $n\mu$ and variance $n\sigma^2$. Under the extra condition that $\mathbb{E}|X - \mu|^3 < \infty$, precise error bounds can be found on the standardized cdf of S_n. Below, Φ is the cdf of $Y \sim \mathsf{N}(0,1)$.

Theorem A.8.3 (Berry–Esséen) *Let $X_1, \ldots, X_n$ be iid with expectation μ and variance $\sigma^2 < \infty$. Then, for all n,*

$$\sup_x \left| \mathbb{P}\left(\frac{S_n - n\mu}{\sigma\sqrt{n}} \leqslant x \right) - \Phi(x) \right| \leqslant K \frac{\mathbb{E}|X_1 - \mu|^3}{\sqrt{n}\,\sigma^3} \,,$$

for some constant $K > 0$ that does not depend on n or the distribution of X_1.

For a proof, see, for example, [5, Page 224]. The smallest constant K found to date is $K = 0.7056$, see [27].

Theorem A.8.4 (Multivariate Central Limit Theorem) *Let $\mathbf{X}_1, \ldots, \mathbf{X}_n$ be iid random vectors with expectation vector $\boldsymbol{\mu}$ and finite covariance matrix Σ. Define $\mathbf{S}_n = \mathbf{X}_1 + \cdots + \mathbf{X}_n$. Then,*

$$\frac{\mathbf{S}_n - n\boldsymbol{\mu}}{\sqrt{n}} \xrightarrow{d} \mathbf{Y} \sim \mathsf{N}(\mathbf{0}, \Sigma) \quad \text{as } n \to \infty \,.$$

A.9 STOCHASTIC PROCESSES

A **stochastic process** or **random process** is a collection of random variables $\{X_t, t \in \mathscr{T}\}$ on a probability space $(\Omega, \mathcal{H}, \mathbb{P})$, where $\mathscr{T}$ is any index set. The set E of possible values for X_t (assuming this is independent of t) is called the **state space** of the process. The index set $\mathscr{T}$ is often taken to be a countable or continuous subset of $\mathbb{R}$, and so a stochastic process is often thought of as a random variable evolving through time, with X_t representing the state of the process at time t.

The distribution of a stochastic process $X = \{X_t, t \in \mathscr{T}\}$, with $\mathscr{T} \subseteq \mathbb{R}$, is completely determined by its finite-dimensional distributions; that is, the distributions of the random vectors $(X_{t_1}, \ldots, X_{t_n})$ for any choice of n and $t_1, \ldots, t_n$. However, the finite-dimensional distributions do not completely determine the sample path behavior of a stochastic process; see, for example, [3, Page 308]. Hence, questions of continuity and differentiability cannot be answered by examining the finite-dimensional distributions alone. Processes that share the same finite-dimensional distributions are called **versions** of each other. If, in addition, the processes share the same probability space, then they are called **modifications** of each other. For a consistent system of finite-dimensional distributions it is always possible to choose a version of the stochastic process that (almost surely) has separable paths [3, Pages 526–527]. A path $\{x_t, t \in \mathscr{T}\}$ is said to be **separable** if there exists a countable dense subset $\mathscr{D}$ of $\mathscr{T}$, such that for each $t \in \mathscr{T}$ there exists a sequence $t_1, t_2, \dots \in \mathscr{D}$ with $t_n \to t$ and $x_{t_n} \to x_t$. The sample path behavior of a separable process is determined by its finite-dimensional distributions. We will assume henceforth that we are dealing with the separable versions of stochastic processes.

■ EXAMPLE A.5 (Bernoulli Process)

A basic example of a stochastic process is any collection $\{X_1, X_2, \ldots\}$ of iid random variables. When $X_t \sim_{\text{iid}} \mathsf{Ber}(p)$ for $t = 1, 2, \ldots$ the process is called a **Bernoulli process**. Here the state space is $E = \{0, 1\}$ and the index set is $\mathscr{T} = \{1, 2, \ldots\}$. The process models the random experiment where a biased coin is tossed indefinitely. The beginning of a typical **sample path** of the process for $p = 0.5$ is given in Figure A.6.

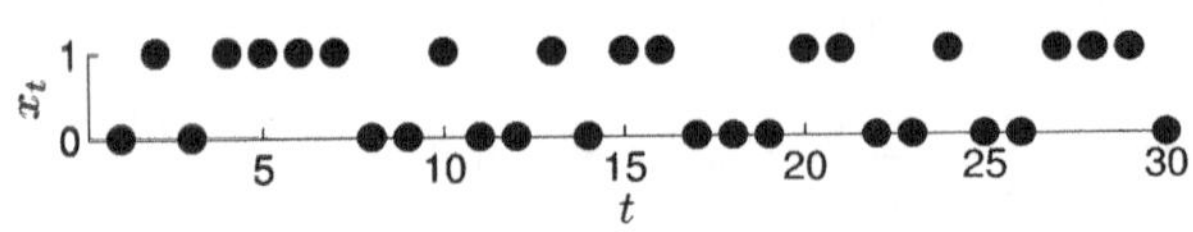

Figure A.6 A typical sample path for a Bernoulli process with $p = 0.5$.

The description and study of real-valued stochastic processes that evolve over time are facilitated by the following notions. In all cases $\mathscr{T}$ is assumed to be one of $\mathbb{N}, \mathbb{Z}, \mathbb{R}_+$, or $\mathbb{R}$.

A collection $\{\mathcal{H}_t\} = \{\mathcal{H}_t, t \in \mathscr{T}\}$ of σ-algebras of events, with the property that $\mathcal{H}_t \subseteq \mathcal{H}_{t+s}$ for any $s \geqslant 0$ and $t \in \mathscr{T}$, is called a **filtration** or **history**. A filtration is called **right-continuous** if $\mathcal{H}_t = \mathcal{H}_{t+} \stackrel{\text{def}}{=} \cap_{s>t} \mathcal{H}_s$ for all t. A filtration can be thought of as an increasing flow of information about some random phenomenon. A stochastic process $\{X_t, t \in \mathscr{T}\}$ is called **adapted** to a filtration $\{\mathcal{H}_t\}$ if X_t is $\{\mathcal{H}_t\}$-measurable, for every $t \in \mathscr{T}$; that is, $X_t \in \mathcal{H}_t$ for all t. Intuitively, $\mathcal{H}_s, s \leqslant t$ contains the complete history of the process up to time t. A random variable $\tau \in \mathscr{T}$ is said to be a **stopping time** with respect to $\{\mathcal{H}_t\}$ if for each $t \in \mathscr{T}$ the event $\{\tau \leqslant t\}$ lies in $\mathcal{H}_t$. Intuitively, τ is a stopping time if one can decide if it has occurred by time t on the basis of the information (contained in $\mathcal{H}_t$) up until time t.

■ **EXAMPLE A.6 (Bernoulli Process Continued)**

A rich variety of stochastic processes can be derived from a Bernoulli process $\{X_t, t = 1, 2, \ldots\}$. For example, define $S_0 = 0$ and $S_t = S_{t-1} + X_t$, $t = 1, 2, \ldots$. Process $\{S_t\}$ is an example of a **random walk** process. Let $\mathcal{H}_t$ be the history of the Bernoulli process up until time t. Note that $\{S_t\}$ is adapted to the filtration $\{\mathcal{H}_t\}$, because all information regarding $S_1, \ldots, S_t$ can be obtained from $X_1, \ldots, X_t$ and vice versa. Let τ_n be the first time that $\{S_t\}$ crosses level n, that is, $\tau_n = \inf\{t : S_t \geqslant n\}$. Then, τ_n is a stopping time with respect to $\{\mathcal{H}_t\}$, because the occurrence of $\{\tau_n \leqslant t\}$ can be decided upon using information about $X_1, \ldots, X_t$ only.

A.9.1 Gaussian Property

A real-valued stochastic process $\{X_t, t \in \mathscr{T}\}$ is said to be **Gaussian** if all its finite-dimensional distributions are Gaussian (normal); that is, if the vector $(X_{t_1}, \ldots, X_{t_n})$ is multidimensional Gaussian for any choice of n and $t_1, \ldots, t_n \in \mathscr{T}$, or equivalently, if any linear combination $\sum_{i=1}^{n} b_i X_{t_i}$ has a Gaussian distribution. Gaussian processes can thus be thought of as generalizations of Gaussian random vectors. ☞ 143

The probability distribution of a Gaussian process is determined completely by its **expectation function**

$$\mu_t = \mathbb{E} X_t, \quad t \in \mathscr{T}$$

and **covariance function**

$$\sigma_{s,t} = \mathrm{Cov}(X_s, X_t), \quad s, t \in \mathscr{T} .$$

A zero-mean Gaussian process is one for which $\mu_t = 0$ for all t. The generation of Gaussian processes is discussed in Section 5.1. ☞ 154

■ **EXAMPLE A.7 (Wiener Process)**

The **Wiener process** $\{W_t, t \geqslant 0\}$ can be defined as a zero-mean Gaussian process with covariance function

$$\sigma_{s,t} = \min\{s, t\} , \quad s, t \geqslant 0 .$$

It forms the basis of a great variety of other stochastic processes; see Chapter 5 and Sections A.12–A.13. The Wiener process has many interesting properties and characterizations, which are further discussed in Section 5.5. ☞ 177

A.9.2 Markov Property

A stochastic process $\{X_t, t \in \mathscr{T}\}$ on $(\Omega, \mathcal{H}, \mathbb{P})$, with index set $\mathscr{T} \subseteq \mathbb{R}$ and state space E (equipped with a σ-algebra $\mathcal{E}$), is said to be a **Markov process** if for every $s \geqslant 0$, $t \in \mathscr{T}$, and $A \in \mathcal{E}$ it satisfies the **Markov property**:

$$\mathbb{P}(X_{t+s} \in A \,|\, \mathcal{H}_t) = \mathbb{P}(X_{t+s} \in A \,|\, X_t)\,, \tag{A.34}$$

where $\mathcal{H}_t$ is the history of the process up until time t. The Markov process is said to be **time-homogeneous** if the conditional probability $P_s(x, A) = \mathbb{P}(X_{t+s} \in A \,|\, X_t = x)$ does not depend on t for any fixed s. The function P_s is called the (s-step) **transition kernel** of the Markov process. When (A.34) holds for any stopping time τ instead of a fixed t, then $\{X_t\}$ is said to have the **strong Markov property**.

The Markov property can be expressed as

$$(X_{t+s} \,|\, X_u, u \leqslant t) \quad \sim \quad (X_{t+s} \,|\, X_t)\,, \tag{A.35}$$

which emphasizes that the conditional future distributions of the sample path given the entire sample path history are the same as those given only the present state. In other words, for a Markov process the conditional distribution of the "future" variable X_{t+s} given the entire past of the process $\{X_u, u \leqslant t\}$ is the same as the conditional distribution of X_{t+s} given only the "present" X_t.

We assume from now on that the Markov process is time-homogeneous, unless otherwise specified. Markov processes come in many different varieties, depending on the choice of index set $\mathscr{T}$ and state space E. In most cases of practical interest $\mathscr{T} = \mathbb{N}$ or $\mathbb{R}_+$ and $E \subseteq \mathbb{R}^n$. In addition, in many cases P_t is of the form

$$P_t(x, A) = \int_{y \in A} p_t(x, y)\, \mathrm{d}y \quad \text{or} \quad P_t(x, A) = \sum_{y \in A} p_t(x, y)\,, \tag{A.36}$$

in the continuous and discrete case, respectively. Here, $p_t(x, y)$ is the **transition kernel density**. In this case the finite-dimensional distributions of the Markov process (and hence the distribution of the entire process) are determined by the family of transition kernels $\{P_t, t \geqslant 0\}$ and the distribution of X_0 — the **initial distribution** of the Markov process. Namely, by the product rule (A.21) and the Markov property the joint probability density f of any random vector $(X_0, X_{t_1}, \ldots, X_{t_n})$ satisfies

$$f(x_0, x_1, \ldots, x_n) = f_{X_0}(x_0)\, p_{t_1}(x_0, x_1)\, p_{t_2 - t_1}(x_1, x_2) \cdots p_{t_n - t_{n-1}}(x_{n-1}, x_n)\,,$$

where f_{X_0} is the density of X_0. The kernel P_t can be viewed as a linear operator $f \mapsto P_t f$ acting on suitable functions f, such that

$$P_t f(x) \stackrel{\text{def}}{=} \mathbb{E}^x f(X_t) = \int P_t(x, \mathrm{d}y) f(y)\,.$$

Here $\mathbb{E}^x$ denotes the expectation operator under which the process starts in x at time 0. An important property of $\{P_t, t \geqslant 0\}$ is the **semigroup** property:

$$P_{s+t} = P_s\, P_t \quad \text{for all } s, t \geqslant 0\,. \tag{A.37}$$

These are the **Chapman–Kolmogorov** equations.

Sections A.10 and A.11 discuss discrete-time Markov processes, often called **Markov chains**, and continuous-time Markov processes in greater detail.

A.9.3 Martingale Property

A **martingale** is a real-valued stochastic process $X = \{X_t, t \in \mathscr{T}\}$, with $\mathscr{T} \subseteq \mathbb{R}$, such that:

1. X is adapted to a filtration $\{\mathcal{H}_t\}$.

2. $\mathbb{E}|X_t| < \infty$ for all $t \in \mathscr{T}$.

3. For any $s \leqslant t \in \mathscr{T}$,

$$\mathbb{E}[X_t \,|\, \mathcal{H}_s] = X_s\,, \quad \text{a.s.} \tag{A.38}$$

The state X_t of the process can be interpreted as the fortune at time t of a gambler playing a game. In this context a martingale can be thought of as a "fair game", in the sense that the gambler's fortune in the future is expected to be the same as the gambler's current fortune, given all the past and present information on the game. In some cases it is important to stress the filtration $\{\mathcal{H}_t\}$ and probability measure $\mathbb{P}$ under which the above martingale conditions hold.

A process X is called a **submartingale** if (A.38) holds with "=" replaced by "$\geqslant$". An $\boldsymbol{L^p}$**-(sub)martingale** is a (sub)martingale for which $\mathbb{E}|X_t|^p < \infty$ for all t. One usually distinguishes between **discrete-time** ($\mathscr{T} = \mathbb{N}$ or $\mathbb{Z}$) and **continuous-time** ($\mathscr{T} = \mathbb{R}_+$ or $\mathbb{R}$) martingales. The properties of continuous-time martingales are similar to those of the discrete-time equivalents, but often additional regularity conditions are required. We list a number of properties of martingales. For proofs, see [7].

1. *Sample path regularity*: Let $X = \{X_t, t \geqslant 0\}$ be a submartingale such that $t \mapsto \mathbb{E}X_t$ is continuous. Then, X has a modification that has right-continuous and left-limited paths (this is automatically so if X is a martingale).

2. *Maximum bound*: Let $\{X_t, t = 0, 1, 2, \ldots\}$ be an L^p-martingale for some $p \geqslant 1$. Then,

$$\mathbb{P}\Big(\max_{0 \leqslant t \leqslant n} |X_t| \geqslant x\Big) \leqslant \frac{\mathbb{E}|X_n|^p}{x^p}, \quad x \geqslant 0\,.$$

3. *Convergence*: Let the process $X = \{X_t, t = 0, 1, 2, \ldots\}$ be a (sub)martingale. If $\sup_n \mathbb{E}X_n^+ < \infty$, where $x^+ = \max\{x, 0\}$, then X converges almost surely to an integrable random variable X_∞.

4. *Optional sampling*: Let $X = \{X_t, t \geqslant 0\}$ be a (sub)martingale and $\tau_1, \tau_2, \ldots$ be a sequence of stopping times such that $\tau_i \leqslant K_i$ for some deterministic sequence $K_1, K_2, \ldots < \infty$. Then, $\{X_{\tau_i}, i = 1, 2, \ldots\}$ is a (sub)martingale with respect to filtration $\{\mathcal{H}_{\tau_i}\}$.

5. *Optional stopping*: Let the process $X = \{X_t, t \geqslant 0\}$ be a martingale and τ a finite stopping time. If X is uniformly integrable, then $X_\tau = \mathbb{E}[X_\infty \,|\, \mathcal{H}_\tau]$ and $\mathbb{E}X_\tau = \mathbb{E}X_0$.

6. *Criterion for martingale:* Let $\{X_t, t \geqslant 0\}$ be a process such that $\mathbb{E}|X_\tau| < \infty$ and $\mathbb{E}X_\tau = \mathbb{E}X_0$ for every bounded stopping time $\tau \leqslant K < \infty$. Then, X is a martingale.

7. *Submartingale implying martingale:* Let $X = \{X_t, t \geqslant 0\}$ be a submartingale on $t \in [0, T]$. If $\mathbb{E}X_T = \mathbb{E}X_0$, then X is a martingale on $t \in [0, T]$.

8. *Martingale representation:* Let $\{X_t, 0 \leqslant t \leqslant T\}$ be a square-integrable martingale. Then there exists a unique process $\{\phi_t\}$ adapted to $\{\mathcal{H}_t\}$ such that:

   (a) $\mathbb{E} \int_0^T \phi_t^2 \, \mathrm{d}t < \infty$;

   (b) $X_t = X_0 + \int_0^t \phi_s \, \mathrm{d}W_s$, $t \in [0, T]$, where $\{W_t, t \geqslant 0\}$ is a Wiener process adapted to $\{\mathcal{H}_t\}$.

A.9.4 Regenerative Property

A real-valued stochastic process $X = \{X_t, t \geqslant 0\}$ is said to be **regenerative** if there exist times $T_0 \leqslant T_1 < T_2 < T_3 < \ldots$ of the form $T_n = A_1 + \cdots + A_n$, $n = 1, 2, \ldots$, where the $\{A_i\}$ are iid, such that conditional on $X_s, s \leqslant T_n$ the process $\{X_{T_n+t}, t \geqslant 0\}$ has the same distribution as $\{X_{T_0+t}, t \geqslant 0\}$. In other words, a regenerative process "regenerates" itself at times $T_0, T_1, \ldots$. That is, given the history of the process up to time T_n, the process after T_n behaves probabilistically as if it has started afresh. The $\{T_n\}$ are called **regeneration times**. When $T_0 = 0$, the process is called **pure**; otherwise, it is called **delayed**.

■ **EXAMPLE A.8** ($M/M/1$ **Queue**)

The $M/M/1$ **queueing system** describes a service facility where customers arrive at certain random times and are served by a single server. Arriving customers who find the server busy wait in the queue. Customers are served in the order in which they arrive. The interarrival times are iid exponential random variables with rates λ, and the service times of customers are iid exponential random variables with rates μ. Finally, the service times are independent of the interarrival times. Assume that at $T_0 = 0$ the system is empty and that mean service time is smaller than the mean interarrival time. Let X_t be the number of customers in the system at time t. Then $\{X_t, t \geqslant 0\}$ is a regenerative process. Namely, let T_1 be the first time the system becomes empty again after a service completion. The probabilistic behavior of the process $\{X_t\}$ from T_1 onwards is exactly the same as from $t = 0$ onwards, even if we knew the complete history up to time T_1. Let T_2 be the next time the system becomes empty after a service completion, and so on. Figure A.7 shows a realization of this process.

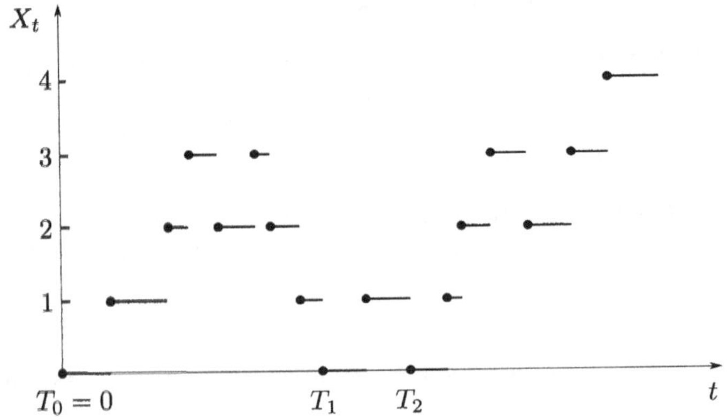

Figure A.7 The number of customers in an $M/M/1$ queue as a function of time.

The process $\{T_n\}$ of renewal times forms a so-called **renewal process**; the corresponding $\{A_n\}$ are called **cycle lengths**. The following main property of regenerative processes is derived from the properties of renewal processes; see, for example, [4, Chapter 9]. A random variable A is said to have a **lattice** distribution if A takes values in the lattice $\{a + b\,n, n \in \mathbb{Z}\}$ for some values of a and b ($b \neq 0$); b is called the **period**.

Theorem A.9.1 (Regeneration Theorem) *Let $\{X_t\}$ be a continuous-time regenerative process with right-continuous paths and nonlattice distribution of the cycle length with expectation $\mu = \mathbb{E}A_1 < \infty$. Then, X_t converges in distribution to a random variable X, such that for all f*

$$\mathbb{E}f(X) = \frac{1}{\mu}\mathbb{E}\int_{T_0}^{T_1} f(X_s)\,\mathrm{d}s\,, \tag{A.39}$$

provided that the expectation exists.

Let $\{X_t\}$ be a discrete-time regenerative process with cycle length distribution of period $b = 1$ and expectation $\mu = \mathbb{E}A_1 < \infty$. Then, X_n converges in distribution to a random variable X, such that for all f

$$\mathbb{E}f(X) = \frac{1}{\mu}\mathbb{E}\sum_{k=T_0}^{T_1-1} f(X_k)\,, \tag{A.40}$$

provided that the expectation exists.

In other words, if G_t denotes the cdf of X_t ($G_t(x) = \mathbb{P}(X_t \leqslant x)$), then under the mild conditions above, there *exists* a continuous cdf G such that $\lim_{t\to\infty} G_t(x) = G(x)$ for all x.

Often G_t is difficult to calculate, but G is usually much easier to find, via equation (A.39) or (A.40). Moreover, with the existence of G guaranteed, we can now give a precise meaning to the behavior of the stochastic process "in the stationary situation" or "in equilibrium".

■ **EXAMPLE A.9 ($M/M/1$ Queue Continued)**

As in Example A.8, let X_t denote the number of customers in an $M/M/1$ queueing system at time t. When the arrival rate is smaller than the service rate, $\{X_t\}$ is a regenerative process, and hence X_t converges in distribution to a random variable X that can be interpreted as the number of customers in the system "in equilibrium" or far into the future. Similarly, the expected steady-state number of customers in the stationary situation simply refers to the expectation of X.

A.9.5 Stationarity and Reversibility

A stochastic process $\{X_t, t \in \mathscr{T}\}$ is said to be **strongly stationary** if the distributions of the random vectors $(X_{t_1}, \ldots, X_{t_n})$ and $(X_{t_1+s}, \ldots, X_{t_n+s})$ are the same for any choice of n and $s, t_1, \ldots, t_n \in \mathscr{T}$.

A stochastic process $\{X_t, t \in \mathscr{T}\}$ is said to be **weakly stationary** if both the expectation function $\{\mathbb{E}X_t\}$ and covariance function $\{\mathrm{Cov}(X_t, X_{t+s})\}$ do not depend on t. The function $R(s) = \mathrm{Cov}(X_t, X_{t+s})$ is then called the **autocovariance function**.

In other words, the distribution of a strongly stationary process is invariant under time shifts (or space shifts in cases where $\mathscr{T}$ is a spatial index set). For weakly stationary processes the covariance function is invariant under time shifts. A strongly stationary process is weakly stationary whenever its mean and covariance function exist. In particular, this is the case when $\mathbb{E}X_t^2 < \infty$, $t \in \mathscr{T}$. However, a weakly stationary process is not necessarily strongly stationary. A notable exception to this are Gaussian processes (see Section A.9.1), as their finite-dimensional distributions depend only on the corresponding means and covariances.

A strongly stationary stochastic process $\{X_t\}$ with index set $\mathbb{Z}$ or $\mathbb{R}$ is said to be **reversible** if, for any positive integer n and for all $t_1, \ldots, t_n$, the vector $(X_{t_1}, \ldots, X_{t_n})$ has the same distribution as $(X_{-t_1}, \ldots, X_{-t_n})$. One way to visualize reversible processes is to imagine that we have taken a video of the stochastic process which we may run in forward and reverse time. If we cannot detect whether the video is running forward or backward, the process is reversible.

A.10 MARKOV CHAINS

A Markov process (see Section A.9.2) with a countable index set $\mathscr{T}$ is called a **Markov chain**. Below, we assume that the index set is either $\mathbb{N}$ or $\mathbb{Z}$ and that the chain is time-homogeneous. Generating realizations of a Markov chain is discussed in Section 5.2. ☞ 162

Recall from Section A.9.2 that the transition kernel $P_t(x, A)$ of a general time-homogeneous Markov process gives the probability that starting from x the chain ends up in set A after t discrete time steps. Of particular importance for Markov chains is the one-step transition kernel P_1. If the state space E is countable, say $E = \mathbb{N}$, we can write its (discrete) density as

$$p(x, y) = P_1(x, \{y\}) = \mathbb{P}(X_{t+1} = y \mid X_t = x), \quad x, y \in E, \quad t \in \mathbb{N}. \tag{A.41}$$

We can arrange these one-step transition probabilities in a one-step **transition matrix** P with (x, y)-th entry given by $p(x, y)$. Similarly, P_t is represented by the t-step transition matrix with (x, y)-th element $p_t(x, y) = P_t(x, \{y\})$. Note that the elements of P_t in every row are nonnegative and sum up to unity. Such a matrix is called a **stochastic** matrix. If additionally every column sums to unity, then the matrix is called **doubly stochastic**.

By the Chapman–Kolmogorov equations (A.37), the t-step transition matrix is in fact equal to the t-th power of P; that is, $P_t = P^t$. It follows that if $\boldsymbol{\pi}_t = (\mathbb{P}(X_t = k), k \in E)$ is the row vector representing the probability distribution of X_t, then

$$\boldsymbol{\pi}_t = \boldsymbol{\pi}_0 \, P^t \quad \text{for all } t = 0, 1, \ldots, \tag{A.42}$$

where P^0 is the identity matrix.

When E is nondenumerable, for example $E = \mathbb{R}$, and P_t has a density p_t as in (A.36), the one-step transition matrix is replaced by the one-step transition density $p(x, y) = p_1(x, y)$. The Chapman–Kolmogorov equations for the transition densities become

$$p_{t+s}(x, y) = \int_E p_s(x, z) \, p_t(z, y) \, \mathrm{d}z, \quad s, t \in \mathbb{N}, \quad x, y \in E. \tag{A.43}$$

A convenient way to describe a discrete-state Markov chain X is through its **transition graph**. States are indicated by the nodes of the graph (without the weight labels), and a strictly positive (> 0) transition probability $p(x, y)$ from state x to y is indicated by an arrow from x to y with weight $p(x, y)$. An example of a transition graph is given in Figure A.8.

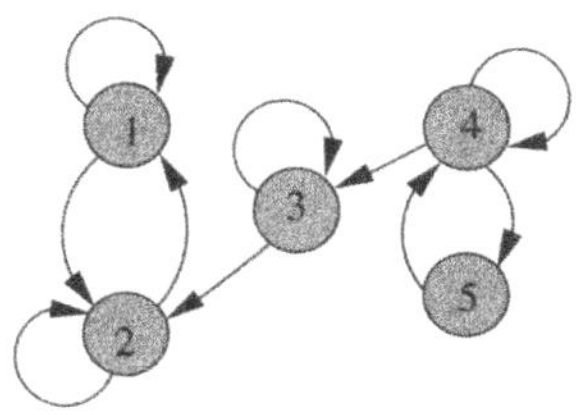

Figure A.8 A transition graph of a discrete-state Markov chain.

A.10.1 Classification of States

Let $X = \{X_t, t = 0, 1, \dots\}$ be a time-homogeneous Markov chain with state space E. Let x and y be arbitrary states in E. Let T denote the time that the chain first visits state y, or first returns to y if it started there; and let N_y denote the total number of visits to y from time 0 onwards. We write $\mathbb{P}^y(A)$ for $\mathbb{P}(A \mid X_0 = y)$ for any event A. We denote the corresponding expectation operator by $\mathbb{E}^y$. The states of a Markov chain are typically classified as follows.

1. A state y is called a **recurrent** state if $\mathbb{P}^y(T < \infty) = 1$; otherwise, it is called **transient**. A recurrent state y is called **positive recurrent** if $\mathbb{E}^y T < \infty$; otherwise, it is called **null-recurrent**.

2. A state y is said to be **periodic with period** $\boldsymbol{\delta}$, if $\delta \geqslant 2$ is the largest integer for which $\mathbb{P}^y(T = n\delta, \text{ for some } n \geqslant 1) = 1$. If $\delta = 1$, the state is said to be **aperiodic**.

3. If $p_t(x, y) > 0$ for some $t \geqslant 0$, then x is said to **lead to** y — written as $x \to y$. If $x \to y$ and $y \to x$, then x and y are said to **communicate** — written as $x \leftrightarrow y$. A set of states $C \subseteq E$ is called a **communicating class** if, for any pair $x, y \in C$, $x \leftrightarrow y$, and further that for every $x \in C$ there is no $y \in E \setminus C$ such that $x \leftrightarrow y$. If E is the only communicating class, the Markov chain is said to be **irreducible**.

4. A set of states $A \subseteq E$ such that $\sum_{y \in A} p(x, y) = 1$ for all $x \in A$ is called a **closed** set. A state x is called an **absorbing** state if $\{x\}$ is closed.

Recurrence and transience are class properties; that is, the elements in each communicating class are either all recurrent or all transient. Figure A.8 shows the transition graph of a Markov chain with three communicating classes.

A.10.2 Limiting Behavior

The limiting or steady-state behavior of Markov chains as $t \to \infty$ is of considerable interest and importance, and is often simpler to describe and analyze than the transient behavior of the chain for fixed t.

For simplicity, assume that the state space E is countable. Then, a Markov chain $\{X_t\}$ is a discrete-time regenerative process, where possible renewal times are the times when the process returns to a specific state. Irreducibility and aperiodicity ensure, via Theorem A.9.1, that

$$\lim_{t\to\infty} p_t(x,y) = \pi(y) \,, \tag{A.44}$$

for some $\pi(y) \in [0,1]$. Moreover, $\pi(y) > 0$ if y is positive recurrent and $\pi(y) = 0$ otherwise. The intuitive reason behind this result is that the process "forgets" where it was initially if it goes on long enough. Thus, provided that $\pi(y) \geqslant 0$ and $\sum_y \pi(y) = 1$, the numbers $\{\pi(y), y \in E\}$ form the **limiting distribution** of the Markov chain. Note that these conditions are not always satisfied. For example, they are clearly not satisfied if the Markov chain is transient, and they may not be satisfied even if the chain is recurrent (namely when the states are null-recurrent). When $E = \{0,1,2,\ldots\}$, then the limiting distribution is usually identified with the row vector $\boldsymbol{\pi} = (\pi_0, \pi_1, \ldots)$. The following is proved, for example, in [4].

Theorem A.10.1 (Limiting Distribution) *For an irreducible aperiodic Markov chain with transition matrix P, if the limiting distribution $\boldsymbol{\pi}$ exists, then $\boldsymbol{\pi}$ is uniquely determined by the solution of the constrained system of equations*

$$\boldsymbol{\pi} = \boldsymbol{\pi} P, \quad \sum_{y\in E} \pi_y = 1, \quad \pi_y \geqslant 0 \quad \textit{for all } y \in E \,. \tag{A.45}$$

In fact, the solution of (A.45) *will automatically be strictly positive* ($\pi_y > 0$). *Conversely, if there exists a row vector $\boldsymbol{\pi}$ satisfying* (A.45), *then $\boldsymbol{\pi}$ is the limiting distribution of the Markov chain. In addition, $\pi_y > 0$ for all y, and all states are positive recurrent.*

Let X be a Markov chain with limiting distribution $\boldsymbol{\pi}$. Suppose $\boldsymbol{\pi}_0 = \boldsymbol{\pi}$. Then, combining (A.42) and (A.45), we have $\boldsymbol{\pi}_t = \boldsymbol{\pi}$. Thus, if the initial distribution of the Markov chain is equal to the limiting distribution, then the distribution of X_t is the same for all t and is given by this limiting distribution. For any Markov chain, any $\boldsymbol{\pi}$ which satisfies (A.45) is called a **stationary distribution**, because using $\boldsymbol{\pi}$ as an initial distribution renders the Markov chain a stationary process.

Noting that $\sum_y p(x,y) = 1$, we can rewrite (A.45) as the system of equations

$$\sum_y \pi(x)\, p(x,y) = \sum_y \pi(y)\, p(y,x) \quad \text{for all } x \in E \,. \tag{A.46}$$

These are called the **global balance equations**. We can interpret (A.45) as the statement that the "probability flux" out of x is balanced with the probability flux into x. An important generalization, which follows directly from (A.46), states that the same balancing of probability fluxes holds for an arbitrary set A. That is, for every set $A \subseteq E$ of states we have

$$\sum_{x\in A}\sum_{y\notin A} \pi(x)\, p(x,y) = \sum_{x\in A}\sum_{y\notin A} \pi(y)\, p(y,x) \,. \tag{A.47}$$

A.10.3 Reversibility

A good way to think of the global balance equations (A.46) is that they balance the probability flux out of each state x with the probability flux into state x. For *reversible* (see Section A.9.5) Markov chains a much stronger form of balance equations holds, where the probability flux from state x to state y is balanced with that from state y to state x. The following theorem is proved in [17, 20].

Theorem A.10.2 (Reversible Markov Chain) *A stationary Markov chain is reversible if and only if there exists a collection of positive numbers $\{\pi(x),\ x \in E\}$, summing to unity that satisfy the* **detailed (or local) balance equations**

$$\pi(x)\, p(x, y) = \pi(y)\, p(y, x)\ , \quad x, y \in E\ . \tag{A.48}$$

Whenever there exists such a collection $\{\pi(x)\}$, it is the stationary distribution of the process.

The following gives a simple criterion for reversibility based on the transition probabilities. A proof can be found in [17, Page 21].

Theorem A.10.3 (Kolmogorov's Criterion) *A stationary Markov chain is reversible if and only if its transition probabilities satisfy*

$$p(x_1, x_2)\, p(x_2, x_3) \cdots p(x_{n-1}, x_n)\, p(x_n, x_1) = p(x_1, x_n)\, p(x_n, x_{n-1}) \cdots p(x_2, x_1) \tag{A.49}$$

for all finite loops of states $x_1, \ldots, x_n, x_1$.

The idea is quite intuitive: if the process in forward time is more likely to traverse a certain closed loop in one direction than in the opposite direction, then in backward time it will exhibit the opposite behavior, and hence we have a criterion for detecting the direction of time. If such "looping" behavior does not occur, the process must be reversible.

A.11 MARKOV JUMP PROCESSES

A **Markov jump process** is a Markov process (see Section A.9.2) with a continuous index set and a discrete (that is, countable) state space E. Generating realizations of a Markov jump process is discussed in Section 5.3. For simplicity ☞ 166 we assume that the Markov jump process is time-homogeneous and that the index set is either $\mathbb{R}$ or $\mathbb{R}_+$. Let $p_t(x, y) = P_t(x, \{y\}) = \mathbb{P}(X_t = y \,|\, X_0 = x)$ denote the **transition probability** from x to y in $t \geqslant 0$ time units. Similar to a Markov chain with a discrete state space, we can arrange the transition probabilities into a matrix $(p_t(x, y))$. With a slight abuse of notation we will also write this matrix as P_t. We will call the family $\{P_t, t \geqslant 0\}$, or P_t viewed as a function t, a **transition function**. It is said to be **standard** if $\lim_{t \downarrow 0} P_t = I$ (the identity matrix) and **honest** if $P_t \mathbf{1} = \mathbf{1}$ for all t, where $\mathbf{1}$ is a column vector of ones. We will consider only standard transition functions.

The analogue of the one-step transition matrix for Markov chains is the ***Q*-matrix** defined as

$$Q = P_0' = \lim_{t \downarrow 0} \frac{P_t - I}{t}\ . \tag{A.50}$$

The (x, y)-th entry $(x \neq y)$ of Q, denoted $q(x, y)$, is called the **transition rate** from x to y. The x-th diagonal entry, $q(x, x)$, is written as $-q_x$. It can be shown [1] that

(a) $0 \leqslant q(x, y) \leqslant \infty$, $x \neq y$,

(b) $\sum_{y \neq x} q(x, y) \leqslant q_x$.

A state x is said to be **stable** if $q_x < \infty$; and **instantaneous** if $q_x = \infty$. If $q_x = 0$ the state x is called **absorbing**.

A Markov jump process is usually defined by specifying a matrix Q that satisfies the properties (a) and (b) above. Such a matrix is again called a Q-matrix. It is said to be **stable** if all the states are stable, **uniformly bounded** if $\sup_x q_x < \infty$, and **conservative** if $Q\mathbf{1} = \mathbf{0}$. Finally, Q is called **regular** if it is conservative and

$$Q\mathbf{z} = \lambda \mathbf{z}, \quad -1 \leqslant z_i \leqslant 1 \text{ for all } i \,,$$

has the unique trivial solution $\mathbf{z} = \mathbf{0}$ for all $\lambda > 0$. The following theorem is proved in [1].

Theorem A.11.1 (Sample Path Behavior) *For each stable and conservative Q-matrix there exists a Markov jump process X whose paths are right-continuous step functions up to a certain random time T_∞. Moreover, the sample path behavior up to T_∞ can be described as follows:*

1. *Given its past, the probability that X jumps from its current state x to state y is $K(x, y) = q(x, y)/q_x$.*

2. *The amount of time that X spends in state y has an $\mathsf{Exp}(q_y)$ distribution, independent of the past history.*

A typical sample path of X is sketched in Figure A.9. The process jumps at times $T_1, T_2, \ldots$ to states $Y_1, Y_2, \ldots$, staying an exponentially distributed length of time in each state.

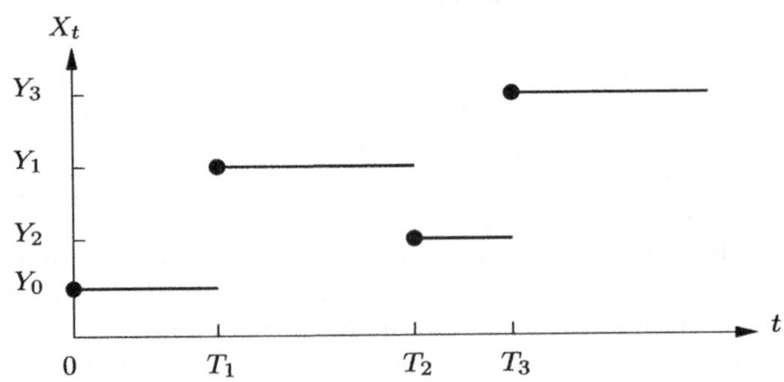

Figure A.9 A sample path of a Markov jump process $\{X_t, t \geqslant 0\}$.

The first statement of Theorem A.11.1 implies that the process $\{Y_n, n \in \mathbb{N}\}$ is in fact a time-homogeneous Markov chain, with one-step transition matrix $K = (K(x, y))$. This Markov chain is called the **embedded Markov chain** or the **jump chain**.

A convenient way to describe a Markov jump process is through its **transition rate graph** (see, for example, Figure A.10). This is similar to a transition graph for Markov chains. The states are represented by the nodes of the graph, and a transition rate from state x to y is indicated by an arrow from x to y with weight $q(x, y)$.

Classification concepts such as irreducibility, communication, recurrence, and transience are defined in the same way as for a Markov chain; see Section A.10.1. Note, however, that there is no concept of periodicity for Markov jump processes.

■ **EXAMPLE A.10 (Birth and Death Process)**

A **birth and death process** is a Markov jump process with a transition rate graph of the form given in Figure A.10. Imagine that X_t represents the total number of individuals in a population at time t. Jumps to the right correspond to "births", and jumps to the left to "deaths". The **birth rates** $\{b_i\}$ and the **death rates** $\{d_i\}$ may differ from state to state. Many applications of Markov chains involve processes of this kind.

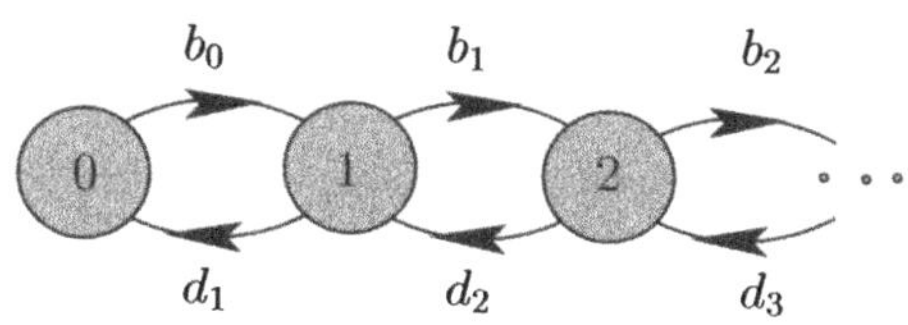

Figure A.10 The transition rate graph of a birth and death process.

Note that the process jumps from one state to the next according to a Markov chain with transition probabilities $K_{0,1} = 1$, $K_{i,i+1} = b_i/(b_i + d_i)$ and $K_{i,i-1} = d_i/(b_i + d_i)$, $i = 1, 2, \ldots$. Moreover, it spends an $\mathsf{Exp}(b_0)$ amount of time in state 0 and an $\mathsf{Exp}(b_i + d_i)$ amount of time in state $i \neq 0$.

Theorem A.11.2 (Kolmogorov Equations) *Any transition function P_t with conservative Q-matrix Q satisfies the* **Kolmogorov backward equations***:*

$$P_t' = Q\,P_t, \quad t \geqslant 0\,. \tag{A.51}$$

This is easy to see when P_t and Q are finite-dimensional, as, by the Chapman–Kolmogorov equations (A.37), $\lim_{h\downarrow 0}(P_{t+h} - P_t)/h = \lim_{h\downarrow 0}(P_h - I)/h\,P_t = Q\,P_t$. In a similar way, finite-dimensional transition functions satisfy the **Kolmogorov forward equations**:

$$P_t' = P_t\,Q, \quad t \geqslant 0\,. \tag{A.52}$$

The proof for infinite-dimensional transition functions is not as straightforward and requires certain regularity conditions on Q — for example, Q being conservative, as in Theorem A.11.2. Indeed, for some transition functions the forward equations may not hold at all. However, a converse result to the above theorem is as follows [1, Page 70].

Theorem A.11.3 (Minimal Transition Function) *For any stable Q-matrix Q there exists a transition function P_t^M that is the solution to both the backward and forward equations and is minimal in the sense that $P_t^M \leqslant P_t$ for any other solution P_t of either the backward or forward equation. If P_t^M is honest, it is the unique solution to the backward and forward equations.*

The Markov jump process with P_t^M as its transition function is called the **minimal** Q**-process** and corresponds to the Markov jump process X in Theorem A.11.1.

For a Markov jump process we usually only have knowledge of the Q-matrix Q, and so directly verifying whether or not P_t^M is honest may not be easy or even possible. However, it is often possible to determine the honesty of P_t^M indirectly via inspection of Q, as is seen from the following theorem [1].

Theorem A.11.4 (Regular Q-matrix) *If Q is regular then the minimal solution to the Kolmogorov backward equations is honest, and is therefore the unique solution to the forward and backward equation. In particular, this is the case when Q is conservative and uniformly bounded.*

In most applications the Markov jump process is defined by a conservative uniformly bounded Q-matrix (in particular, when the Q-matrix is of finite dimensions). The transition matrix (function) is then the unique solution to the Kolmogorov differential equations, and can be written in matrix-exponential form as

$$P_t = \mathrm{e}^{Qt} = \sum_{k=0}^{\infty} \frac{t^k Q^k}{k!} .$$

A.11.1 Limiting Behavior

The limiting behavior of Markov jump processes is akin to that of the Markov chains discussed in Section A.10.2.

Theorem A.11.5 (Limiting Distribution) *Let $\{X_t, t \geqslant 0\}$ be an irreducible Markov jump process with regular Q-matrix Q. Then, irrespective of x,*

$$\lim_{t\to\infty} \mathbb{P}(X_t = y \mid X_0 = x) = \pi(y) , \tag{A.53}$$

for some number $\pi(y) \geqslant 0$. Moreover, the row vector $\boldsymbol{\pi} = \{\pi(y)\}$ is the solution to

$$\boldsymbol{\pi} Q = \mathbf{0}, \quad \sum_{y \in E} \pi(y) = 1 , \tag{A.54}$$

provided such a solution exists, in which case all states are positive recurrent. If such a solution does not exist, then $\boldsymbol{\pi} = \mathbf{0}$.

As in the Markov chain case, $\boldsymbol{\pi}$ defines the **limiting distribution** of X. Any solution $\boldsymbol{\pi}$ of (A.54) with $\sum_y \pi(y) = 1$ is called a **stationary distribution**, because taking it as the initial distribution of the Markov jump process renders the process stationary. Equations (A.54) are, as in the Markov chain case, called the **global balance equations**, and can be written as

$$\sum_{y \neq x} \pi(x)\, q(x,y) = \sum_{y \neq x} \pi(y)\, q(y,x) \quad \text{for all } x \in E , \tag{A.55}$$

balancing the "probability flux" out of x with that into x. The global balance equations are readily generalized to (A.47), replacing the transition probabilities with transition rates. More importantly, if the process is *reversible* then the stationary distribution can be found from the **detailed balance equations**:

$$\pi(x)\, q(x,y) = \pi(y)\, q(y,x)\ , \quad x, y \in E\ . \tag{A.56}$$

Reversibility can be easily verified by checking that "looping" does not occur, that is, via Kolmogorov's criterion (A.49), replacing the probabilities p with rates q. The criterion in this case is thus given by

$$q(x_1,x_2)\, q(x_2,x_3) \cdots q(x_{n-1},x_n)\, q(x_n,x_1) = q(x_1,x_n)\, q(x_n,x_{n-1}) \cdots q(x_2,x_1)$$

for all finite loops of states $x_1, \ldots, x_n, x_1$.

■ **EXAMPLE A.11 ($M/M/1$ Queue Continued)**

Let X_t denote the number of customers in an $M/M/1$ queueing system at time $t \geqslant 0$; see Examples A.8 and A.9. The process $\{X_t, t \geqslant 0\}$ is an irreducible birth and death process with birth rates λ and death rates μ. The system of equations (A.54) has a unique solution

$$\pi(y) = (1-\varrho)\varrho^y, \quad y = 0, 1, 2, \ldots\ , \tag{A.57}$$

where $\varrho = \lambda/\mu$, if and only if $\varrho < 1$. For $\lambda < \mu$ all the states are therefore positive recurrent. Note that any birth and death process is reversible. As a consequence (A.57) can be found directly from the local balance equations

$$\pi(y)\, \lambda = \pi(y+1)\, \mu, \quad y = 0, 1, \ldots\ .$$

Theorem A.9.1 shows that for $\varrho < 1$ the steady-state expected number of customers in the system is $\mathbb{E}X = \varrho/(1-\varrho)$.

A.12 ITÔ INTEGRAL AND ITÔ PROCESSES

An important class of stochastic processes — that of **Itô processes** — is constructed from the Wiener process via the notion of the Itô integral. The Wiener process is discussed in more detail in Section 5.5, but here we only consider its role in Itô integration. The Itô integral provides the mathematical justification of integrals of the form ☞ 177

$$\int_0^T F_t \,\mathrm{d}W_t\ ,$$

where $W = \{W_t\}$ is a Wiener process and $F = \{F_t\}$ is a stochastic process. In its simplest form the Itô integral is defined for processes F that are **predictable** [18] with respect to the history of W and satisfy

$$\mathbb{E}\int_0^T F_s^2 \,\mathrm{d}s < \infty\ . \tag{A.58}$$

We will denote this class of integrands by $\mathscr{H}_T$. A sufficient condition for predictability is that the process is left-continuous and adapted — so, F_t may depend on $\{W_s, s \leqslant t\}$ but not on $\{W_s, s > t\}$. Let $t \leqslant T$ and $F \in \mathscr{H}_T$. The **Itô integral** of F with respect to W over $[0, t]$ is defined as

$$\int_0^t F_s \, \mathrm{d}W_s \stackrel{\text{def}}{=} \lim_{n\to\infty} \sum_{k=0}^{n-1} F_{t_k}(W_{t_{k+1}} - W_{t_k}), \quad 0 = t_0 < \cdots < t_n = t\,, \tag{A.59}$$

where $\lim_{n\to\infty} \max_k\{t_{k+1} - t_k\} = 0$, and the convergence is in the mean square sense (see Section A.8.1).

Remark A.12.1 (Stochastic Integral) The Itô integral is an example of a **stochastic integral**. The general theory of stochastic integration [21, 23] allows $\{W_t\}$ to be replaced by **semimartingales** — processes that can be decomposed as the sum of a (local) martingale and a process of finite variation — and the integrand process $\{F_t\}$ by predictable processes that satisfy weaker conditions than (A.58). In particular, it can be shown that the limit (A.59) still exists, but in probability rather than in the mean square sense, if (A.58) is replaced by

$$\int_0^T F_s^2 \, \mathrm{d}s < \infty \quad \text{a.s.} \tag{A.60}$$

An **Itô process** is any stochastic process $\{X_t, 0 \leqslant t \leqslant T\}$ that can be written in the form

$$X_t = X_0 + \int_0^t \mu_s \, \mathrm{d}s + \int_0^t \sigma_s \, \mathrm{d}W_s, \quad 0 \leqslant t \leqslant T\,,$$

where $\{\mu_t\}$ is adapted, with $\int_0^T |\mu_t| \, \mathrm{d}t < \infty$ and $\{\sigma_t\} \in \mathscr{H}_T$. The above integral equation is usually written in the shorthand differential form

$$\mathrm{d}X_t = \mu_t \, \mathrm{d}t + \sigma_t \, \mathrm{d}W_t\,. \tag{A.61}$$

Note that the coefficients μ_t and σ_t may depend on the whole path $\{W_s, s \leqslant t\}$. An ***m*-dimensional Itô process** $\{\mathbf{X}_t\} = \{(X_{t,1}, \ldots, X_{t,m})^\top\}$ driven by an n-dimensional Wiener process $\{\mathbf{W}_t\} = \{(W_{t,1}, \ldots, W_{t,n})^\top\}$ can be defined analogously via the differential expression

$$\mathrm{d}X_{t,i} = \mu_{t,i} \, \mathrm{d}t + \sum_{j=1}^{n} \sigma_{t,ij} \, \mathrm{d}W_{t,j}, \quad i = 1, \ldots, m\,,$$

written in matrix–vector notation as

$$\mathrm{d}\mathbf{X}_t = \boldsymbol{\mu}_t \, \mathrm{d}t + \boldsymbol{\sigma}_t \, \mathrm{d}\mathbf{W}_t\,, \tag{A.62}$$

where

$$\boldsymbol{\mu}_t = \begin{pmatrix} \mu_{t,1} \\ \vdots \\ \mu_{t,m} \end{pmatrix} \quad \text{and} \quad \boldsymbol{\sigma}_t = \begin{pmatrix} \sigma_{t,11} & \cdots & \sigma_{t,1n} \\ \vdots & \ddots & \vdots \\ \sigma_{t,m1} & \cdots & \sigma_{t,mn} \end{pmatrix}.$$

An Itô process is an example of a semimartingale. As a special case of the general theory of stochastic integration with respect to such processes one may define integration with respect to Itô processes. In particular, (see, for example, [18]) if $X = \{X_t\}$ is an Itô process and $F = \{F_t\} \in \mathscr{H}_T$, then the stochastic integral of F with respect to X is defined as:

$$\int_0^t F_s \mathrm{d}X_s \stackrel{\text{def}}{=} \int_0^t F_s\, \mu_s\, \mathrm{d}s + \int_0^t F_s\, \sigma_s\, \mathrm{d}W_s, \quad 0 \leqslant t \leqslant T\,.$$

Let $X = \{X_t\}$ and $Y = \{Y_t\}$ be two processes adapted to the same filtration. Then,

$$[X,Y]_t \stackrel{\text{def}}{=} \lim_{n\to\infty} \sum_{k=0}^{n-1} (X_{t_{k+1}} - X_{t_k})(Y_{t_{k+1}} - Y_{t_k})\,,$$

where $0 = t_0 < \cdots < t_n = t$ and $\lim_{n\to\infty} \max_k\{t_{k+1} - t_k\} = 0$, is called the **covariation** between the processes X and Y. The special case $[X,X]_t$, denoted $[X]_t$, is called the **quadratic variation** of X.

Below we list a number of properties of Itô integrals and Itô processes. Proofs may be found in [23], for example.

1. *Isometry property:* If $F, G \in \mathscr{H}_T$, then for any $0 \leqslant t \leqslant T$,

$$\mathbb{E} \int_0^t F_s\, \mathrm{d}W_s \int_0^t G_s\, \mathrm{d}W_s = \mathbb{E} \int_0^t F_s\, G_s\, \mathrm{d}s\,.$$

2. *Martingale property:* If $F \in \mathscr{H}_T$, then the Itô process defined by

$$Y_t = \int_0^t F_s\, \mathrm{d}W_s, \quad 0 \leqslant t \leqslant T\,,$$

is a square-integrable martingale.

3. *Quadratic variation and covariation:* Let $\mathrm{d}X_t = \mu_t\, \mathrm{d}t + \sigma_t\, \mathrm{d}W_t$ and $\mathrm{d}Y_t = \nu_t\, \mathrm{d}t + \varrho_t\, \mathrm{d}W_t$ define two Itô processes with respect to the *same* Wiener process $\{W_t\}$. Then,

$$[X,Y]_t = \int_0^t \sigma_s\, \varrho_s\, \mathrm{d}s\,.$$

In shorthand differential form the covariation and the quadratic variation are

$$\mathrm{d}[X,Y]_t = \sigma_t\, \varrho_t\, \mathrm{d}t \quad \text{and} \quad \mathrm{d}[X]_t = \sigma_t^2\, \mathrm{d}t\,, \quad \text{respectively.}$$

4. *Covariance for multivariate Itô process:* Let $\{\mathbf{X}_t\}$ be an m-dimensional Itô process. Then using the formal rules (see [18]) $(\mathrm{d}t)^2 = \mathrm{d}t\, \mathrm{d}W_{t,i} = 0$ and $\mathrm{d}W_{t,i}\, \mathrm{d}W_{t,j} = \delta_{ij}\, \mathrm{d}t$, where $\delta_{ij} = 1$ if $i = j$ and 0 otherwise, we can write:

$$\mathrm{d}[X_{\cdot,i}, X_{\cdot,j}]_t = \mathrm{d}X_{t,i}\, \mathrm{d}X_{t,j} = \sum_{k=1}^n \sigma_{t,ik}\, \sigma_{t,jk}\, \mathrm{d}t\,, \quad i,j \in \{1,\ldots,m\}\,.$$

5. *Itô's lemma*: Let $\mathrm{d}X_t = \mu_t\, \mathrm{d}t + \sigma_t\, \mathrm{d}W_t$ define an Itô process and let $f(x) : \mathbb{R} \to \mathbb{R}$ be twice continuously differentiable with first and second derivatives f' and f'', respectively. Then,

$$f(X_t) = f(X_0) + \int_0^t f'(X_s)\, \mathrm{d}X_s + \frac{1}{2} \int_0^t f''(X_s)\, \sigma_s^2\, \mathrm{d}s \tag{A.63}$$

or, in differential form:

$$\mathrm{d}f(X_t) = f'(X_t)\,\mathrm{d}X_t + \frac{1}{2} f''(X_t)\,\sigma_t^2\,\mathrm{d}t\,.$$

Compare this with the corresponding **chain rule** of ordinary calculus: $\mathrm{d}f(x(t)) = f'(x(t))\,\mathrm{d}x(t)$.

6. *Itô's lemma in* $\mathbb{R}^m$: Let $\{\mathbf{X}_t\}$ be an m-dimensional Itô process, and $f : \mathbb{R}^m \to \mathbb{R}$ be twice continuously differentiable in all variables, then

$$\mathrm{d}f(\mathbf{X}_t) = \sum_{i=1}^{m} \partial_i f(\mathbf{X}_t)\,\mathrm{d}X_{t,i} + \frac{1}{2}\sum_{i=1}^{m}\sum_{j=1}^{m} \partial_{ij} f(\mathbf{X}_t)\,\mathrm{d}[X_{\cdot,i}, X_{\cdot,j}]_t\,. \tag{A.64}$$

A special case is the **product rule** for Itô processes:

$$\mathrm{d}(X_t Y_t) = Y_t\,\mathrm{d}X_t + X_t\,\mathrm{d}Y_t + \mathrm{d}[X,Y]_t\,. \tag{A.65}$$

The corresponding integral form is the Itô **integration by parts** formula. Another special case is $\mathbf{X}_t = (X_t, t)^\top$, where t ($\geqslant 0$) is deterministic and the process $\{X_t\}$ is governed by $\mathrm{d}X_t = \mu_t\,\mathrm{d}t + \sigma_t\,\mathrm{d}W_t$. Then,

$$\mathrm{d}f(X_t, t) = \left(\frac{\partial f}{\partial t}(\mathbf{X}_t) + \mu_t \frac{\partial f}{\partial x}(\mathbf{X}_t) + \frac{\sigma_t^2}{2}\frac{\partial^2 f}{\partial x^2}(\mathbf{X}_t)\right)\mathrm{d}t + \sigma_t \frac{\partial f}{\partial x}(\mathbf{X}_t)\,\mathrm{d}W_t\,. \tag{A.66}$$

7. *Gaussian process for deterministic integrands*: If $f(t,s)$ is a nonrandom function with $\int_0^t f^2(t,s)\,\mathrm{d}s < \infty$ for any $0 \leqslant t \leqslant T$, then the Itô integral

$$Y_t = \int_0^t f(t,s)\,\mathrm{d}W_s \tag{A.67}$$

defines a Gaussian process $\{Y_t, 0 \leqslant t \leqslant T\}$ with mean zero and covariance function

$$\mathrm{Cov}(Y_s, Y_t) = \int_0^{\min\{s,t\}} f(t,u) f(s,u)\,\mathrm{d}u\,.$$

Note that, unless $f(t,s) = f(s)$, $\{Y_t\}$ need not be a martingale. If $f(t,s) = f(s)$, then by the integration by parts formula we also have

$$Y_t = W_t f(t) - \int_0^t W_s\,\mathrm{d}f(s)\,.$$

8. *Time-change*: Let $\{W_t\}$ be a Wiener process, and define $Z_t = W_{C_t}$, $t \geqslant 0$, for some given deterministic function $C_t = \int_0^t f^2(s)\,\mathrm{d}s < \infty$ for all $t \leqslant T$. Then, the stochastic process $\{Z_t, 0 \leqslant t \leqslant T\}$ has the same distribution as the Itô integral process $\{Y_t\}$ defined in (A.67) with $f(t,s) = f(s)$.

9. *Girsanov's theorem*: Let $\mathrm{d}\mathbf{X}_t = \boldsymbol{\mu}_t\,\mathrm{d}t + \mathrm{d}\mathbf{W}_t$ define a multidimensional Itô process under probability measure $\mathbb{P}$ with respect to a filtration $\mathcal{F} =$

$\{\mathcal{F}_t, t \geqslant 0\}$. Assume that $\{\boldsymbol{\mu}_t, t \geqslant 0\}$ satisfies **Novikov's condition**: $\mathbb{E}\exp(\frac{1}{2}\int_0^t \boldsymbol{\mu}_s^\top \boldsymbol{\mu}_s \,\mathrm{d}s) < \infty$. For each $t \geqslant 0$ define

$$M_t = \exp\left(\int_0^t \boldsymbol{\mu}_s^\top \,\mathrm{d}\mathbf{W}_s - \frac{1}{2}\int_0^t \boldsymbol{\mu}_s^\top \boldsymbol{\mu}_s \,\mathrm{d}s\right) .$$

Then $\{M_t, t \geqslant 0\}$ is a martingale with respect to $\mathcal{F}$. For a fixed $T \geqslant 0$ let $\mathbb{P}_T$ denote the restriction of $\mathbb{P}$ to $\mathcal{F}_T$. Define a new measure $\widetilde{\mathbb{P}}_T$ by

$$\widetilde{\mathbb{P}}_T(A) = \mathbb{E}_T M_T \mathrm{I}_A\,, \quad A \in \mathcal{F}_T\,,$$

so that

$$\mathbb{P}_T(A) = \widetilde{\mathbb{E}}_T \mathrm{I}_A / M_T\,.$$

Then under $\widetilde{\mathbb{P}}_T$ the process $\{\mathbf{X}_t, 0 \leqslant t \leqslant T\}$ is a Wiener process.

Remark A.12.2 (Stratonovich Integral) Let W be a Wiener process and $X \in \mathscr{H}_T$. For any $0 \leqslant t \leqslant T$ let

$$\int_0^t X_s \circ \mathrm{d}W_s \stackrel{\text{def}}{=} \lim_{n\to\infty} \sum_{k=0}^{n-1} \frac{X_{t_k} + X_{t_{k+1}}}{2}(W_{t_{k+1}} - W_{t_k}), \quad 0 = t_0 < \cdots < t_n = t\,,$$

where $\lim_{n\to\infty} \max_k\{t_{k+1} - t_k\} = 0$ and convergence is in the mean square sense. This defines the **Stratonovich integral** of X with respect to W over $[0, t]$. This integral does not in general define a martingale, and therefore most of the properties above do not directly apply. However, the Stratonovich integral has the advantage that it formally obeys the standard calculus formulas. In particular, for a three times continuously differentiable function f, the Stratonovich integral formally satisfies the ordinary chain rule

$$\mathrm{d}f(X_t) = f'(X_t) \circ \mathrm{d}X_t\,.$$

A.13 DIFFUSION PROCESSES

Let $\{W_t\}$ be a Wiener process, and $a(x,t)$ and $b(x,t)$ be deterministic functions. A **stochastic differential equation** (SDE) for a stochastic process $\{X_t\}$ is an expression of the form

$$\mathrm{d}X_t = a(X_t, t)\,\mathrm{d}t + b(X_t, t)\,\mathrm{d}W_t\,. \tag{A.68}$$

The coefficient a is called the **drift** and b^2 (or sometimes b) the **diffusion** coefficient. When a and b do not depend on t explicitly (that is, $a(x,t) = \widetilde{a}(x)$, and $b(x,t) = \widetilde{b}(x)$), the SDE is said to be **autonomous** or **homogeneous**. When a and b are linear in x, the SDE is said to be **linear**.

Intuitively, the process $\{X_t\}$ is specified by a "noisy ODE", relating its derivative at t to a function of its present value X_t and an additional noise term. Mathematically, $\{X_t\}$ is the solution to the integral equation

$$X_t = X_0 + \int_0^t a(X_s, s)\,\mathrm{d}s + \int_0^t b(X_s, s)\,\mathrm{d}W_s\,, \tag{A.69}$$

where the last integral is defined in the Itô sense. Note that when $b \equiv 0$, we obtain an ordinary differential equation.

Remark A.13.1 (Diffusion-Type SDE) Although SDEs of the type above are by far the most common, it should be noted that there exist more general SDEs [18, 19], where, for example, a and b depend on the whole history of $\{X_s, s \leqslant t\}$ rather than only on t and X_t. The special case (A.68) is also referred to as a **diffusion-type** SDE.

A stochastic process $\{X_t\}$ is said to be a **strong solution** to the SDE (A.68) if X_t is a function of t and the underlying Wiener process $\{W_s, s \leqslant t\}$, and satisfies (A.69). It is called a **weak solution** if (A.69) holds for *some* Wiener process.

The following theorem gives conditions for existence and uniqueness of strong solutions on an interval $[0, T]$. A proof can be found, for example, in [19].

Theorem A.13.1 (Existence and Uniqueness of Strong Solutions) *Suppose the following conditions are satisfied:*

1. **Linear growth condition**: *There is a constant C such that for all $t \in [0, T]$*

$$|a(x,t)| + |b(x,t)| \leqslant C(1 + |x|) \quad \text{for all } x\,. \tag{A.70}$$

2. **Local Lipschitz continuity in x**: *For every $K > 0$ there is a constant D_K such that for all $t \in [0, T]$*

$$|a(x,t) - a(y,t)| + |b(x,t) - b(y,t)| \leqslant D_K|x - y| \quad \text{for all } x, y \in [-K, K]\,. \tag{A.71}$$

3. *X_0 is independent of $\{W_t, 0 \leqslant t \leqslant T\}$ and has finite variance.*

Then the SDE (A.68) *has a unique strong solution on* $[0, T]$. *In addition, the solution has almost surely continuous paths, is a strong Markov process, and* $\int_0^T \mathbb{E}X_s^2 \,\mathrm{d}s < \infty$.

The linear growth condition ensures that each path of the SDE does not "explode"; that is, the path does not tend to $\pm\infty$ within a finite interval of time. Note that a similar condition is required for ordinary differential equations. For example, the differential equation $\mathrm{d}x(t) = x^2(t)\,\mathrm{d}t$, $x(0) = a$ has a "local" solution $x(t) = a/(1 - at)$ on the interval $[0, 1/a)$ rather than a "global" solution on $\mathbb{R}_+$. Removing Condition 1 still gives a unique strong solution, but only up to a (random) time of explosion.

Local Lipschitz continuity ensures that solutions of SDEs can be constructed via an iterative procedure, similar to that for ordinary differential equations (Picard iteration [23]). As a measure of smoothness of a function, this condition lies between continuity and differentiability. In particular, if a and b are continuously differentiable in x, or, more generally, if their derivatives in x are uniformly bounded on $[0, T]$, then they satisfy (A.71).

Weak solutions to the SDE (A.68) exist under slightly more general conditions; see for example [23]. In particular, if for each t the functions a and b are bounded and continuous in x. Such solutions are only defined through their probability distributions, rather than pathwise via the Wiener process W.

■ **EXAMPLE A.12 (Linear SDE)**

For a linear SDE

$$\mathrm{d}X_t = (\alpha_t + \beta_t X_t)\,\mathrm{d}t + (\gamma_t + \delta_t X_t)\,\mathrm{d}W_t\ ,$$

the (strong) solution can be given explicitly as the product $X_t = U_t\, V_t$, with

$$U_t = \exp\left\{\int_0^t \left(\beta_s - \frac{1}{2}\delta_s^2\right)\mathrm{d}s + \int_0^t \delta_s\,\mathrm{d}W_s\right\},$$

$$V_t = X_0 + \int_0^t \frac{\alpha_s - \gamma_s\delta_s}{U_s}\mathrm{d}s + \int_0^t \frac{\gamma_s}{U_s}\,\mathrm{d}W_s\ .$$

In particular, if $\delta_t \equiv 0$ and $\beta_t \equiv \beta$ (constant), then

$$X_t = \mathrm{e}^{\beta t}\left(X_0 + \int_0^t \mathrm{e}^{-\beta s}\alpha_s\,\mathrm{d}s + \int_0^t \mathrm{e}^{-\beta s}\gamma_s\,\mathrm{d}W_s\right),$$

and $\{X_t\}$ is therefore a Gaussian process, provided that the distribution of X_0 is Gaussian (this includes the case where X_0 is a constant). See [19, Pages 110–113] for more details.

A solution $\{X_t\}$ to (A.68) or, more precisely to (A.69), is called a **diffusion process**, or, more specifically, an **Itô diffusion**. From Theorem A.13.1, Itô diffusions are Markov processes with continuous paths. Let $X = \{X_t\}$ be an Itô diffusion with drift and diffusion coefficients a and b^2, respectively. The meaning of these terms becomes clear when considering the infinitesimal behavior of X. In particular, by (A.69),

$$X_{t+h} - X_t = \int_t^{t+h} a(X_s, s)\,\mathrm{d}s + \int_t^{t+h} b(X_s, s)\,\mathrm{d}W_s\ .$$

Taking the conditional expectation given $X_t = x$ on both sides yields

$$\mathbb{E}[X_{t+h} - x \mid X_t = x] = a(x,t)\,h + o(h)\ ,$$

since the expectation of the second integral in the above integral equation is 0, due to the martingale property of the Itô integral. Similarly, using the isometry Property 1 on Page 641,

$$\mathrm{Var}(X_{t+h} - x \mid X_t = x) = \mathbb{E}[(X_{t+h} - x - a(x,t)\,h)^2 \mid X_t = x] + o(h) = b^2(x,t)\,h + o(h)\ .$$

In other words, given that the process is at position x at time t, the displacement of X in the next $h \ll 1$ time units has expectation $a(x,t)h$ and variance $b^2(x,t)h$.

Remark A.13.2 (Boundary Behavior) We have considered only diffusions on the whole real line. Diffusions on a half-line or intervals are also possible. For such processes the behavior at the boundary needs to be specified, in addition to the behavior in the interior of the domain described by the SDE. See, for example [8, 16].

The analogue of (A.68) in $\mathbb{R}^m$ is given by the **multidimensional SDE**

$$\mathrm{d}\mathbf{X}_t = \mathbf{a}(\mathbf{X}_t, t)\,\mathrm{d}t + B(\mathbf{X}_t, t)\,\mathrm{d}\mathbf{W}_t\ , \tag{A.72}$$

where $\{\mathbf{W}_t\}$ is an n-dimensional Wiener process, $\mathbf{a}(\mathbf{x}, t)$ is an m-dimensional vector (the drift) and $B(\mathbf{x}, t)$ an $m \times n$ matrix, for each $\mathbf{x} \in \mathbb{R}^m$ and $t \in \mathbb{R}$. The $m \times m$ matrix $C = BB^\top$ is called the **diffusion matrix**.

As with the one-dimensional case, existence and uniqueness of strong solutions to multidimensional SDEs relies on certain Lipschitz and linear growth conditions. In particular, we have the following multidimensional version of Theorem A.13.1 (see [18, Page 173]):

Theorem A.13.2 (Strong Solutions of Multidimensional SDEs) *Suppose the following conditions are satisfied, where for a matrix A, $\|A\| \stackrel{\text{def}}{=} \sqrt{\text{tr}(AA^\top)}$:*

1. **Linear growth condition**: *There is a constant C such that for all $t \in [0, T]$*

$$\|\mathbf{a}(\mathbf{x}, t)\| + \|B(\mathbf{x}, t)\| \leqslant C(1 + \|\mathbf{x}\|) \quad \text{for all } \mathbf{x} . \tag{A.73}$$

2. **Local Lipschitz continuity in x**: *For every $K > 0$ there is a constant D_K such that for all $t \in [0, T]$*

$$\|\mathbf{a}(\mathbf{x}, t) - \mathbf{a}(\mathbf{y}, t)\| + \|B(\mathbf{x}, t) - B(\mathbf{y}, t)\| \leqslant D_K \|\mathbf{x} - \mathbf{y}\| \quad \text{for all } \|\mathbf{x}\|, \|\mathbf{y}\| \leqslant K . \tag{A.74}$$

3. *$\mathbf{X}_0$ is independent of $\{\mathbf{W}_t, 0 \leqslant t \leqslant T\}$ and $\mathbb{E}\|\mathbf{X}_0\|^2 < \infty$.*

Then the SDE (A.68) *has a unique strong solution on* $[0, T]$.

A.13.1 Kolmogorov Equations

For autonomous SDEs, that is, those of the form

$$\mathrm{d}X_t = a(X_t)\,\mathrm{d}t + b(X_t)\,\mathrm{d}W_t , \tag{A.75}$$

the corresponding diffusion process is a *time-homogeneous* Markov process. The corresponding transition kernel P_t can be found from the Kolmogorov backward (and under more restrictive conditions) from the Kolmogorov forward equations. To see this, let L be the linear elliptic differential operator

$$Lf(x) = a(x)f'(x) + \frac{1}{2}b^2(x)f''(x) \tag{A.76}$$

acting on all twice continuously differentiable functions on compact sets. Then, by Itô's formula,

$$M_t \stackrel{\text{def}}{=} f(X_t) - f(X_0) - \int_0^t Lf(X_s)\,\mathrm{d}s$$

defines a martingale on $[0, T]$. In particular, denoting by $\mathbb{E}^x$ the expectation operator under which the process starts at x, we have

$$\mathbb{E}^x f(X_t) = f(x) + \int_0^t \mathbb{E}^x Lf(X_s)\,\mathrm{d}s , \tag{A.77}$$

where the interchange of expectation and integral is allowed by Fubini's theorem. It follows that

$$Lf(x) = \lim_{t \downarrow 0} \frac{\mathbb{E}^x f(X_t) - f(x)}{t} . \tag{A.78}$$

The limit in (A.78) also defines the **infinitesimal generator** of the Markov process. The domain of the infinitesimal generator consists of all bounded measurable functions for which the limit exists — this includes the domain of L, hence the infinitesimal generator extends L. Let P_t be the transition kernel of the Markov process and define the operator P_t by

$$P_t f(x) = \int P_t(x, \mathrm{d}y) f(y) = \mathbb{E}^x f(X_t) .$$

Then, by (A.77), we obtain the **Kolmogorov forward equations**:

$$P_t' f = P_t L f . \tag{A.79}$$

Moreover, by the Chapman–Kolmogorov equations we have $P_{t+s} f(x) = P_s P_t f(x) = \mathbb{E}^x P_t f(X_s)$, and therefore

$$\frac{1}{s} \left\{ P_{t+s} f(x) - P_t f(x) \right\} = \frac{1}{s} \left\{ \mathbb{E}^x P_t f(X_s) - P_t f(x) \right\} .$$

Letting $s \downarrow 0$, we obtain the **Kolmogorov backward equations**:

$$P_t' f = L P_t f . \tag{A.80}$$

If P_t has a transition density p_t, then we can write the last equation as

$$\frac{\mathrm{d}}{\mathrm{d}t} \int p_t(x,y) f(y) \,\mathrm{d}y = \int L p_t(x,y) f(y) \,\mathrm{d}y ,$$

so that $p_t(x,y)$ for fixed y satisfies the Kolmogorov backward equations:

$$\begin{aligned} \frac{\partial}{\partial t} p_t(x,y) &= L p_t(x,y) \\ &= a(x) \frac{\partial}{\partial x} p_t(x,y) + \frac{1}{2} b^2(x) \frac{\partial^2}{\partial x^2} p_t(x,y) . \end{aligned} \tag{A.81}$$

Similarly, (A.79) can be written as $\frac{\mathrm{d}}{\mathrm{d}t} \int p_t(x,y) f(y) \,\mathrm{d}y = \int p_t(x,y) L f(y) \,\mathrm{d}y = \int f(y) L^* p_t(x,y) \,\mathrm{d}y$, where L^* (acting here on y) is the adjoint operator of L defined by $\int g(y) L h(y) \,\mathrm{d}y = \int h(y) L^* g(y) \,\mathrm{d}y$. Hence, for fixed x the density $p_t(x,y)$ satisfies the Kolmogorov forward equations, also called the **Fokker–Planck equations**:

$$\begin{aligned} \frac{\partial}{\partial t} p_t(x,y) &= L^* p_t(x,y) \\ &= -\frac{\partial}{\partial y} \left(a(y)\, p_t(x,y) \right) + \frac{1}{2} \frac{\partial^2}{\partial y^2} \left(b^2(y)\, p_t(x,y) \right) . \end{aligned} \tag{A.82}$$

Sufficient conditions on $a(x)$ and $b(x)$ such that $p_t(x,y)$ exists and is the unique solution to the forward and backward equations are that $a(x)$ and $b(x)$ have partial derivatives up to order two, which are bounded and satisfy a Lipschitz condition; see also [18]. This illustrates the important connection between partial differential equations of the form $u_t' = L u_t$ and diffusion processes. Indeed, given an elliptic operator L, the pdf of the corresponding diffusion process gives the fundamental solution (Green's function) of the partial differential equation — see also Chapter 17. ☞ 577

Remark A.13.3 (Operators for Multidimensional SDEs) For multidimensional SDEs of the form (A.72) the infinitesimal generator extends the operator

$$Lf(\mathbf{x}) = \sum_{i=1}^{m} a_i(\mathbf{x}) \frac{\partial}{\partial x_i} f(\mathbf{x}) + \frac{1}{2} \sum_{i=1}^{m} \sum_{j=1}^{m} C_{ij}(\mathbf{x}) \frac{\partial^2}{\partial x_i \, \partial x_j} f(\mathbf{x}) ,$$

where $\{a_i\}$ are the components of $\mathbf{a}$ and $\{C_{ij}\}$ the components of $C = BB^\top$.

A.13.2 Stationary Distribution

Consider again the diffusion governed by the autonomous SDE (A.75). Suppose that

$$\pi(y) = \int p_t(x, y)\, \pi(x)\, \mathrm{d}x ,$$

where p_t is the transition density of the diffusion. Then $\pi(x)$ is called a **stationary** or **invariant density** of the diffusion (A.75). If the initial state X_0 has density $\pi(x)$, then $\{X_t, t \geqslant 0\}$ is a stationary process.

Theorem A.13.3 (Stationary Distribution) *If the stationary density of* (A.75) *exists and is twice continuously differentiable, then it solves the ODE*

$$L^*\pi = 0 \quad \Leftrightarrow \quad \frac{1}{2} \frac{\mathrm{d}^2}{\mathrm{d}y^2} \left(b^2(y)\, \pi(y)\right) - \frac{\mathrm{d}}{\mathrm{d}y} \left(a(y)\, \pi(y)\right) = 0 ,$$

where L^* is the adjoint operator in (A.82). *The stationary density that solves the ODE is of the form*

$$\pi(x) = \frac{c}{b^2(x)} \exp\left(\int_{x_0}^{x} \frac{2\, a(y)}{b^2(y)} \mathrm{d}y \right) ,$$

where x_0 is an arbitrary constant and c is a constant such that $\int \pi(y)\, \mathrm{d}y = 1$.

For a rigorous discussion of the conditions for existence of π, see [22].

Loosely speaking, if the diffusion process is in the stationary regime, its distribution does not change in time. Hence, the transition density p_t is independent of time and the partial derivative with respect to t in (A.82) is zero, giving the equation $L^*\pi = 0$ satisfied by the stationary density.

A.13.3 Feynman–Kac Formula

The Feynman–Kac formula establishes an important relationship between stochastic processes and linear parabolic PDEs. The result can be used to approximate the solution of a PDE via Monte Carlo methods. Alternatively, conditional expec-
☞ 577 tations of a diffusion process can be computed by solving a PDE, see Chapter 17.

For each $t \geqslant 0$ let L_t be the linear elliptic differential operator

$$L_t u(x,t) = a(x,t) \frac{\partial}{\partial x} u(x,t) + \frac{1}{2}\, b^2(x,t) \frac{\partial^2}{\partial x^2} u(x,t)$$

acting on all twice continuously differentiable functions on compact sets.

Theorem A.13.4 (Feynman–Kac Formula) *Let $k(x,t)$ and $f(x)$ be bounded functions and let the process $\{X_t, 0 \leqslant t \leqslant T\}$ evolve according to (A.68). Assume that the solution to the PDE*

$$\left(L_t + \frac{\partial}{\partial t} - k(x,t)\right) u(x,t) = 0, \quad x \in \mathbb{R},\ t \in [0,T]\,,$$

with final condition $u(x,T) = f(x)$ exists. Then, the solution is unique and given by

$$u(x,t) = \mathbb{E}\left[\mathrm{e}^{-\int_t^T k(X_s,s)\,\mathrm{d}s} f(X_T) \,\middle|\, X_t = x\right], \quad t \in [0,T]\,.$$

We explain why the formula is plausible (for a detailed treatment see [10, 22]). Define the process $\{Y_t\}$ via $Y_t = \mathrm{e}^{-\int_0^t k(X_s,s)\,\mathrm{d}s}$. Then, applying the Itô formula (A.66) we have

$$\mathrm{d}(Y_t\, u(X_t,t)) = Y_t\left(\left(L_t + \frac{\partial}{\partial t} - k(X_t,t)\right) u(X_t,t)\,\mathrm{d}t + b(X_t,t)\,\frac{\partial u}{\partial x}(X_t,t)\,\mathrm{d}W_t\right)\,.$$

Since u is the solution to the PDE, the drift term is 0. Since Y_t is bounded by assumption, it can be shown [10] that existence and uniqueness of the solution of the PDE implies that $\int_0^T \mathbb{E}\,|Y_t\, u(X_t,t)| < \infty$. Therefore, the process

$$Y_t\, u(X_t,t) = \int_0^t Y_s\, b(X_s,s)\,\frac{\partial u}{\partial x}(X_s,s)\,\mathrm{d}W_s$$

is a martingale. Using the Markov property of the SDE (see Theorem A.13.1) and the final condition we obtain

$$Y_t\, u(X_t,t) = \mathbb{E}[\,Y_T\, u(X_T,T)\,|\, X_s,\ 0 \leqslant s \leqslant t] = \mathbb{E}[\,Y_T\, f(X_T)\,|\, X_t]\,,$$

which after rearrangement yields the desired result. For multidimensional analogues of the Feynman–Kac formula see Chapter 17.

A.13.4 Exit Times

Diffusion processes are often studied through their exit times from an interval. Below we assume that $\{X_t\}$ is a homogeneous diffusion process defined by the SDE (A.75) and satisfying the existence and uniqueness conditions of Theorem A.13.1.

Let $[l,r]$ (with $l < r$) be an arbitrary interval, and let τ_l and τ_r be the first times that the process hits l and r, respectively. Let $\tau = \min\{\tau_l, \tau_r\} = \tau_l \wedge \tau_r$ be the first **exit time** from the interval $[l,r]$.

The following results may, for example, be found in [18]. Central in the proof is the fact that, by Itô's lemma, the process $\{M_t\}$ defined by

$$M_t = f(X_{t\wedge\tau}) - \int_0^{t\wedge\tau} Lf(X_u)\,\mathrm{d}u \quad \text{is a martingale.}$$

Theorem A.13.5 (Exit Times) *Let the diffusion coefficient $b(x)$ be a strictly positive and continuous function on $[l,r]$, and let f be any twice continuously differentiable function. Then the following holds:*

1. *The function s given by $s(x) = \mathbb{E}^x \tau$ satisfies the differential equation*

$$Ls = -1, \quad \textit{with} \quad s(l) = 0, \quad s(r) = 0\ ,$$

where operator L is given in (A.76).

2. *Any nonconstant positive solution of $Lh = 0$ is of the form*

$$h(x; x_0, y_0) = \int_{x_0}^{x} \exp\left(-\int_{y_0}^{y} \frac{2a(u)}{b^2(u)}\,\mathrm{d}u\right) \mathrm{d}y\ ,$$

for some arbitrary constants x_0, y_0. These are called **harmonic** *functions for L.*

3. *For any such harmonic function,*

$$\mathbb{P}^x(\tau_l < \tau_r) = \frac{h(r) - h(x)}{h(r) - h(l)}\ .$$

Further Reading

An easy introduction to probability theory with many examples can be found in [25]. More detailed textbooks include [11] and [28]. Classical references on probability theory are [5] and [9]. A good non-measure-theoretic introduction to stochastic processes is [24]. A detailed treatment of Markov processes can be found in [7], and a handy text on Markov processes with countable state spaces is [1]. An accessible measure-theoretic introduction to probability theory, including stochastic processes, can be found in [3]. For many examples in probability theory and stochastic processes, see Feller's two volumes [8, 9]. Other good references for stochastic processes are [4, 15, 16], and the classic [6].

REFERENCES

1. W. J. Anderson. *Continuous-Time Markov Chains: An Applications-Oriented Approach.* Springer-Verlag, New York, 1991.
2. P. Billingsley. *Convergence of Probability Measures.* John Wiley & Sons, New York, 1968.
3. P. Billingsley. *Probability and Measure.* John Wiley & Sons, New York, third edition, 1995.
4. E. Çinlar. *Introduction to Stochastic Processes.* Prentice Hall, Englewood Cliffs, NJ, 1975.
5. K. L. Chung. *A Course in Probability Theory.* Academic Press, New York, second edition, 1974.
6. J. L. Doob. *Stochastic Processes.* John Wiley & Sons, New York, 1953.
7. S. N. Ethier and T. G. Kurtz. *Markov Processes: Characterization and Convergence.* John Wiley & Sons, New York, 1986.

8. W. Feller. *An Introduction to Probability Theory and Its Applications*, volume II. John Wiley & Sons, New York, 1966.

9. W. Feller. *An Introduction to Probability Theory and Its Applications*, volume I. John Wiley & Sons, New York, second edition, 1970.

10. D. Freedman. *Brownian Motion and Diffusion.* Springer-Verlag, New York, 1971.

11. G. R. Grimmett and D. R. Stirzaker. *Probability and Random Processes.* Oxford University Press, Oxford, third edition, 2001.

12. P. R. Halmos. *Measure Theory.* Springer-Verlag, New York, second edition, 1978.

13. C. C. Heyde. On a property of the lognormal distribution. *Journal of the Royal Statistical Society, Series B*, 25(2):392–393, 1963.

14. P. L. Hsu and H. Robbins. Complete convergence and the law of large numbers. *Proceedings of the National Academy of Sciences, U.S.A.*, 33(2):25–31, 1947.

15. S. Karlin and H. M. Taylor. *A First Course in Stochastic Processes.* Academic Press, New York, second edition, 1975.

16. S. Karlin and H. M. Taylor. *A Second Course in Stochastic Processes.* Academic Press, New York, 1981.

17. F. P. Kelly. *Reversibility and Stochastic Networks.* John Wiley & Sons, New York, 1979.

18. F. C. Klebaner. *Introduction to Stochastic Calculus with Applications.* Imperial College Press, London, second edition, 2005.

19. P. E. Kloeden and E. Platen. *Numerical Solution of Stochastic Differential Equations.* Springer-Verlag, Berlin, 1999. Corrected third printing.

20. J. R. Norris. *Markov Chains.* Cambridge University Press, Cambridge, 1997.

21. B. Øksendal. *Stochastic Differential Equations.* Springer-Verlag, Berlin, fifth edition, 2003.

22. R. G. Pinsky. *Positive Harmonic Functions and Diffusion.* Cambridge University Press, Cambridge, 1995.

23. P. E. Protter. *Stochastic Integration and Differential Equations.* Springer-Verlag, Heidelberg, second edition, 2005.

24. S. M. Ross. *Stochastic Processes.* John Wiley & Sons, New York, second edition, 1996.

25. S. M. Ross. *A First Course in Probability.* Prentice Hall, Englewood Cliffs, NJ, seventh edition, 2005.

26. W. Rudin. *Real and Complex Analysis.* McGraw-Hill, New York, third edition, 1987.

27. I. G. Shevtsova. Sharpening of the upper bound of the absolute constant in the Berry–Esséen inequality. *Theory of Probability and Its Applications*, 51(3):549–553, 2007.

28. D. Williams. *Probability with Martingales.* Cambridge University Press, Cambridge, 1991.

APPENDIX B

ELEMENTS OF MATHEMATICAL STATISTICS

B.1 STATISTICAL INFERENCE

Statistics deals with the gathering, summarization, analysis, and interpretation of data. The two main branches of statistics are:

1. *Classical statistics*: Here the data object $\mathbf{x}$ is viewed as the outcome of a random object $\mathbf{X}$ described by a probabilistic model — usually the model is specified up to a (multidimensional) parameter; that is, $\mathbf{X} \sim f(\cdot;\boldsymbol{\theta})$ for some $\boldsymbol{\theta}$. The statistical inference is then purely concerned with the model and in particular with the parameter $\boldsymbol{\theta}$. For example, on the basis of the data one may wish to

   (a) estimate the parameter,

   (b) perform statistical tests on the parameter, or

   (c) validate the model.

2. *Bayesian statistics*: In this approach the model parameter $\boldsymbol{\theta}$ is itself random: $\boldsymbol{\theta} \sim f(\boldsymbol{\theta})$. Bayes' formula $f(\boldsymbol{\theta} \,|\, \mathbf{x}) \propto f(\mathbf{x} \,|\, \boldsymbol{\theta}) f(\boldsymbol{\theta})$ is used to update the distribution of the parameter based on the observed data $\mathbf{x}$.

Mathematical statistics uses probability theory and other branches of mathematics to study data from a purely mathematical standpoint.

Handbook of Monte Carlo Methods. By D. P. Kroese, T. Taimre, Z. I. Botev

B.1.1 Classical Models

Let $\mathbf{x}$ represent the observed data, viewed as the outcome of the random data $\mathbf{X}$. For example, $\mathbf{X}$ could be a random vector $(X_1, \ldots, X_n)^\top$. A real- or vector-valued function of the data is called a **statistic**. For example, if $\mathbf{X} = (X_1, \ldots, X_n)^\top$, then the sample mean $T = T(\mathbf{X}) = (X_1 + \cdots + X_n)/n$ is one such statistic. It is customary to use the *same* letter for both the function T and the random variable $T(\mathbf{X})$. We write T for statistics taking values in $\mathbb{R}$ and $\mathbf{T}$ for statistics taking values in $\mathbb{R}^d$ for some $d \geqslant 2$. It is important that a statistic be *computable*; that is, it cannot depend on any unknown parameters.

We summarize some classical models for data.

B.1.1.1 Iid Sample The data $X_1, \ldots, X_n$ are independent and identically distributed:

$$X_1, \ldots, X_n \overset{\text{iid}}{\sim} \mathsf{Dist}\ ,$$

according to some known or unknown distribution Dist. Often the sampling distribution is specified up to an unknown parameter $\boldsymbol{\theta}$, with $\boldsymbol{\theta} \in \Theta$. An iid sample is often called a **random sample** in the statistics literature. Note that the word "sample" can refer to both a collection of random variables and to a single random variable. It should be clear from the context which meaning is being used.

A standard model for data is:

$$X_1, \ldots, X_n \overset{\text{iid}}{\sim} \mathsf{N}(\mu, \sigma^2)\ ,$$

in which case $\boldsymbol{\theta} = (\mu, \sigma^2)$ and $\Theta = \mathbb{R} \times \mathbb{R}_+$.

B.1.1.2 Analysis of Variance In a one-way analysis of variance the objective is to compare the means $\mu_1, \ldots, \mu_k$ of k independent groups (or **levels**) of normal **responses**, all responses having the same variance σ^2. Specifically, denoting the i-th response at level j by X_{ij}, $i = 1, \ldots, n_j$, $j = 1, \ldots, k$, where n_j is the sample size of the j-th group, the model is

$$X_{ij} = \mu_j + \varepsilon_{ij}, \quad i = 1, \ldots, n_j, \quad j = 1, \ldots, k\ , \quad \{\varepsilon_{ij}\} \overset{\text{iid}}{\sim} \mathsf{N}(0, \sigma^2)\ ,$$

or, equivalently,

$$X_{ij} \sim \mathsf{N}(\mu_j, \sigma^2), \quad i = 1, \ldots, n_j, \quad j = 1, \ldots, k\ , \text{ independently}\ .$$

B.1.1.3 Regression Regression models are used to describe functional relationships between **explanatory** variables and **response** variables. In a **linear regression** model, the relationship is linear. Defining Y_i as the i-th response variable and x_i as the fixed (that is, deterministic) i-th explanatory variable, a standard model is

$$Y_i = \beta_0 + \beta_1 x_i + \varepsilon_i, \quad i = 1, \ldots, n, \quad \{\varepsilon_i\} \overset{\text{iid}}{\sim} \mathsf{N}(0, \sigma^2) \tag{B.1}$$

for certain *unknown* parameters β_0, β_1, and σ^2. The line

$$y = \beta_0 + \beta_1 x \tag{B.2}$$

is called the **regression line**. By replacing it with a general curve $y = g(x; \boldsymbol{\theta})$ one obtains a general regression model. For example, $y = \beta_0 + \beta_1 x + \beta_2 x^2$ gives a **quadratic regression** model, and $y = \mathbf{x}^\top \boldsymbol{\beta}$, where $\mathbf{x}$ is a multidimensional explanatory variable and $\boldsymbol{\beta}$ a parameter vector, gives a **multiple linear regression** model.

B.1.1.4 Linear Model A data vector $\mathbf{Y} = (Y_1, \ldots, Y_n)^\top$ is said to satisfy a **linear model** if

$$\mathbf{Y} = A\boldsymbol{\beta} + \boldsymbol{\varepsilon}, \quad \boldsymbol{\varepsilon} \sim \mathsf{N}(\mathbf{0}, \sigma^2 I) \tag{B.3}$$

for some $n \times k$ matrix A (the **design matrix**), a k-dimensional vector of **parameters** $\boldsymbol{\beta} = (\beta_1, \ldots, \beta_k)^\top$, and a vector $\boldsymbol{\varepsilon} = (\varepsilon_1, \ldots, \varepsilon_n)^\top$ of iid $\mathsf{N}(0, \sigma^2)$-distributed **error** terms. The analysis of variance and regression models are special cases.

For an outcome $\mathbf{y}$, the **least squares method** can be used to fit the model to the data. In particular, the optimal $\widehat{\boldsymbol{\beta}}$ is chosen such that the Euclidean distance between $\mathbf{y}$ and $A\widehat{\boldsymbol{\beta}}$ is minimal. Equivalently, $\widehat{\boldsymbol{\beta}}$ is the solution to

$$\nabla_{\boldsymbol{\beta}} \|\mathbf{y} - A\boldsymbol{\beta}\|^2 = A^\top(\mathbf{y} - A\boldsymbol{\beta}) = \mathbf{0}.$$

These linear set of equations are called the **normal equations**. Therefore, if $A^\top A$ is *invertible* (A can always be chosen such that this is the case), then

$$\widehat{\boldsymbol{\beta}} = (A^\top A)^{-1} A^\top \mathbf{y}.$$

In practice, we never compute the inverse $(A^\top A)^{-1}$, but compute $\widehat{\boldsymbol{\beta}}$ from the normal equations using, for example, Gaussian elimination. Geometrically, $\widehat{\boldsymbol{\beta}}$ is the projection of $\mathbf{y}$ onto the subspace spanned by the columns of A. Moreover, it is not difficult to show that $\widehat{\boldsymbol{\beta}}$ is precisely the maximum likelihood estimate of $\boldsymbol{\beta}$ (see Section B.2.1).

B.1.2 Sufficient Statistics

A **sufficient** statistic for a parameter (vector) $\boldsymbol{\theta}$ is a statistic that captures all the information about $\boldsymbol{\theta}$ contained in the data. This means that we can *summarize* the data via a sufficient statistic, sometimes giving a tremendous reduction in data.

If $\mathbf{T}(\mathbf{X})$ is a sufficient statistic for $\boldsymbol{\theta}$, then any inference about $\boldsymbol{\theta}$ depends on the sample $\mathbf{X} = (X_1, \ldots, X_n)^\top$ only through the value $\mathbf{T}(\mathbf{X})$. More precisely, a statistic $\mathbf{T}(\mathbf{X})$ is called a **sufficient statistic** for $\boldsymbol{\theta}$ if the conditional distribution of $\mathbf{X}$ given $\mathbf{T}(\mathbf{X})$ does not depend on $\boldsymbol{\theta}$. The workhorse for establishing sufficiency is the following theorem. A proof can be found, for example, in [4].

Theorem B.1.1 (Factorization Theorem) *Let $f(\mathbf{x};\boldsymbol{\theta})$ denote the joint pdf of the data $\mathbf{X} = (X_1, \ldots, X_n)^\top$. A statistic $\mathbf{T}(\mathbf{X})$ is sufficient for $\boldsymbol{\theta}$ if and only if there exist functions $g(\mathbf{t}, \boldsymbol{\theta})$ and $h(\mathbf{x})$ such that for all data points $\mathbf{x}$ and all parameter points $\boldsymbol{\theta}$,*

$$f(\mathbf{x};\boldsymbol{\theta}) = g(\mathbf{T}(\mathbf{x}), \boldsymbol{\theta})\, h(\mathbf{x}). \tag{B.4}$$

■ **EXAMPLE B.1 (Sufficient Statistics for Exponential Families)**

Sufficiency is particularly easy to establish for exponential families. Suppose that $X_1, \ldots, X_n$ is an iid sample from the exponential family with pdf ☞ 701

$$\mathring{f}(x;\boldsymbol{\theta}) = c(\boldsymbol{\theta})\, \mathrm{e}^{\sum_{i=1}^m \eta_i(\boldsymbol{\theta})\, t_i(x)}\, \mathring{h}(x),$$

where $\{\eta_i\}$ are linearly independent. The pdf of $\mathbf{X} = (X_1, \ldots, X_n)^\top$ is therefore

$$f(\mathbf{x};\boldsymbol{\theta}) = \underbrace{c(\boldsymbol{\theta})^n\, \mathrm{e}^{\sum_{i=1}^m \eta_i(\boldsymbol{\theta})\, \sum_{k=1}^n t_i(x_k)}}_{g(\mathbf{T}(\mathbf{x}),\boldsymbol{\theta})} \underbrace{\prod_{k=1}^n \mathring{h}(x_k)}_{h(\mathbf{x})}.$$

A direct consequence of the factorization theorem is that

$$\mathbf{T}(\mathbf{X}) = \left(\sum_{k=1}^{n} t_1(X_k), \ldots, \sum_{k=1}^{n} t_m(X_k)\right)$$

is a sufficient statistic for $\boldsymbol{\theta}$.

■ **EXAMPLE B.2 (Sufficient Statistics for the Normal Distribution)**

As a particular instance of Example B.1, consider the $\mathsf{N}(\mu, \sigma^2)$ case. Thus, $\boldsymbol{\theta} =$
☞ 702 (μ, σ^2), and from Table D.1 it follows that a sufficient statistic for $\boldsymbol{\theta}$ is $\mathbf{T} = (T_1, T_2)$, with $T_1 = \sum_{k=1}^n X_k$ and $T_2 = \sum_{k=1}^n X_k^2$. This means that for the standard data model, the data can be summarized via only T_1 and T_2.

Moreover, it is not difficult to see that any 1-to-1 function of a sufficient statistic again yields a sufficient statistic. Hence, the sample mean $\widetilde{T}_1 = \bar{X}$ and the sample variance

$$\widetilde{T}_2 = \frac{1}{n-1} \sum_{k=1}^{n} (X_k - \bar{X})^2 = \frac{1}{n-1} \left(\sum_{k=1}^{n} X_k^2 - n\bar{X}^2 \right)$$

also form a pair of sufficient statistics, because the mapping

$$\widetilde{T}_1 = \frac{T_1}{n} \quad \text{and} \quad \widetilde{T}_2 = \frac{1}{n-1} \left(T_2 - T_1^2/n\right)$$

is invertible.

B.1.3 Estimation

Suppose the distribution of the data $\mathbf{X}$ is completely specified up to an unknown parameter vector $\boldsymbol{\theta}$. The aim is to estimate $\boldsymbol{\theta}$ on the basis of the observed data $\mathbf{x}$ only. (An alternative could be to estimate $\boldsymbol{\eta} = \mathbf{g}(\boldsymbol{\theta})$ for some vector-valued function $\mathbf{g}$.) Specifically, the goal is to find an **estimator** $\mathbf{T} = \mathbf{T}(\mathbf{X})$ that is close to the unknown $\boldsymbol{\theta}$. The corresponding outcome $\mathbf{t} = \mathbf{T}(\mathbf{x})$ is the **estimate** of $\boldsymbol{\theta}$. The **bias** of an estimator $\mathbf{T}$ of $\boldsymbol{\theta}$ is defined as $\mathbf{T} - \boldsymbol{\theta}$. An estimator $\mathbf{T}$ of $\boldsymbol{\theta}$ is said to be **unbiased** if $\mathbb{E}_{\boldsymbol{\theta}}\mathbf{T} = \boldsymbol{\theta}$. We often write $\widehat{\boldsymbol{\theta}}$ for both an estimator and estimate of $\boldsymbol{\theta}$. The **mean square error** (MSE) of a real-valued estimator T is defined as

$$\text{MSE} = \mathbb{E}_\theta (T - \theta)^2 \ .$$

An estimator T_1 is said to be more **efficient** than an estimator T_2 if the MSE of T_1 is smaller than the MSE of T_2. The MSE can be written as the sum

$$\text{MSE} = (\mathbb{E}_\theta T - \theta)^2 + \text{Var}_\theta(T) \ .$$

The first term measures the unbiasedness and the second is the variance of the estimator. In particular, for an *unbiased* estimator the MSE of an estimator is simply equal to its variance.

For simulation purposes it is often important to include the *running time* of the estimator in efficiency comparisons. One way to compare two unbiased estimators
☞ 383 T_1 and T_2 is to compare their **relative time variance products**,

$$\frac{\tau_i \operatorname{Var}(T_i)}{(\mathbb{E}T_i)^2}, \quad i = 1, 2, \tag{B.5}$$

where τ_1 and τ_2 are the times required to calculate the estimators T_1 and T_2, respectively. In this scheme, T_1 is considered more efficient than T_2 if its relative time variance product is smaller.

Two systematic approaches for constructing sound estimators are:

- the maximum likelihood method; see Section B.2.1,
- the method of moments, discussed next.

B.1.3.1 Method of Moments Suppose $x_1, \ldots, x_n$ are outcomes from an iid sample $X_1, \ldots, X_n \sim_{\text{iid}} f(x; \boldsymbol{\theta})$, where $\boldsymbol{\theta} = (\theta_1, \ldots, \theta_k)$ is unknown. The moments of the sampling distribution can be easily estimated. Namely, if $X \sim f(x; \boldsymbol{\theta})$, then the r-th moment of X, that is, $\mu_r(\boldsymbol{\theta}) = \mathbb{E}_{\boldsymbol{\theta}} X^r$ (assuming it exists), can be estimated through the **sample r-th moment**

$$m_r = \frac{1}{n} \sum_{i=1}^{n} x_i^r .$$

The **method of moments** procedure involves choosing the estimate $\widehat{\boldsymbol{\theta}}$ of $\boldsymbol{\theta}$ such that each of the first k sample and true moments are matched:

$$m_r = \mu_r(\widehat{\boldsymbol{\theta}}), \quad r = 1, 2, \ldots, k .$$

In general, this set of equations is nonlinear, and so its solution often has to be found numerically.

■ EXAMPLE B.3 (Sample Mean and Sample Variance)

Suppose the data is given by $\mathbf{X} = (X_1, \ldots, X_n)^\top$, where the $\{X_i\}$ form an iid sample from a general distribution with mean μ and variance $\sigma^2 < \infty$. Matching the first two moments gives the set of equations

$$\frac{1}{n} \sum_{i=1}^{n} x_i = \mu ,$$

$$\frac{1}{n} \sum_{i=1}^{n} x_i^2 = \mu^2 + \sigma^2 .$$

The method of moments estimates for μ and σ^2 are therefore the **sample mean**

$$\widehat{\mu} = \bar{x} = \frac{1}{n} \sum_{i=1}^{n} x_i , \tag{B.6}$$

and

$$\widehat{\sigma^2} = \frac{1}{n} \sum_{i=1}^{n} x_i^2 - (\bar{x})^2 = \frac{1}{n} \sum_{i=1}^{n} (x_i - \bar{x})^2 . \tag{B.7}$$

The corresponding estimator for μ, $\bar{X}$, is unbiased. However, the estimator for σ^2 is biased: $\mathbb{E}\widehat{\sigma^2} = \sigma^2(n-1)/n$. An unbiased estimator is the **sample variance**

$$S^2 = \widehat{\sigma^2}\frac{n}{n-1} = \frac{1}{n-1}\sum_{i=1}^{n}(X_i - \bar{X})^2 .$$

The square root of the sample variance $S = \sqrt{S^2}$ is called the **sample standard deviation.**

B.1.3.2 Confidence Interval An essential part in any estimation procedure is to provide an assessment of the *accuracy* of the estimate. Indeed, without information on its accuracy the estimate itself would be meaningless. Confidence intervals (sometimes called **interval estimates**) provide a precise way of describing the uncertainty in the estimate.

Let $X_1, \ldots, X_n$ be random variables with a joint distribution depending on a parameter $\theta \in \Theta$. Let $T_1 < T_2$ be statistics (thus, $T_i = T_i(X_1, \ldots, X_n)$, $i = 1, 2$ are functions of the data, but not of θ).

1. The random interval (T_1, T_2) is called a **stochastic confidence interval** for θ with confidence $1 - \alpha$ if

$$\mathbb{P}_\theta(T_1 < \theta < T_2) \geqslant 1 - \alpha \quad \text{for all } \theta \in \Theta . \tag{B.8}$$

2. If t_1 and t_2 are the observed values of T_1 and T_2, then the interval (t_1, t_2) is called the **(numerical) confidence interval** for θ with confidence $1 - \alpha$ for every $\theta \in \Theta$.

3. If (B.8) only holds approximately, the interval is called an **approximate confidence interval**.

4. The probability $\mathbb{P}_\theta(T_1 < \theta < T_2)$ is called the **coverage probability**. For a $1 - \alpha$ confidence interval, it must be at least $1 - \alpha$.

For multidimensional parameters $\boldsymbol{\theta} \in \mathbb{R}^d$ the stochastic confidence interval is replaced with a stochastic **confidence region** $\mathscr{C} \subset \mathbb{R}^d$ such that $\mathbb{P}_{\boldsymbol{\theta}}(\boldsymbol{\theta} \in \mathscr{C}) \geqslant 1 - \alpha$ for all $\boldsymbol{\theta}$.

The systematic construction of (approximate) confidence intervals often involves *likelihood methods*, see Section B.2. Another approach is to use the *bootstrap* ☞ 331 *method*, see Section 8.6. The analogue of a confidence interval in Bayesian analysis is called a **credible interval**; see Section B.3.

■ EXAMPLE B.4 (Approximate Confidence Interval for the Mean)

Let $X_1, X_2, \ldots, X_n$ be an iid sample from a distribution with mean μ and variance $\sigma^2 < \infty$ (both assumed to be unknown). By the central limit theorem and the law of large numbers,

$$T = \frac{\bar{X} - \mu}{S/\sqrt{n}} \overset{\text{approx.}}{\sim} \mathsf{N}(0, 1) ,$$

for large n, where S is the sample standard deviation. Rearranging the approximate equality $\mathbb{P}(|T| \leqslant z_{1-\alpha/2}) \approx 1 - \alpha$, where $z_{1-\alpha/2}$ is the $1 - \alpha/2$ quantile of the standard normal distribution, yields

$$\mathbb{P}\left(\bar{X} - z_{1-\alpha/2}\frac{S}{\sqrt{n}} \leqslant \mu \leqslant \bar{X} + z_{1-\alpha/2}\frac{S}{\sqrt{n}}\right) \approx 1 - \alpha\,,$$

so that

$$\left(\bar{X} - z_{1-\alpha/2}\frac{S}{\sqrt{n}},\ \bar{X} + z_{1-\alpha/2}\frac{S}{\sqrt{n}}\right), \text{ abbreviated as } \bar{X} \pm z_{1-\alpha/2}\frac{S}{\sqrt{n}}\,, \tag{B.9}$$

is an approximate stochastic $1 - \alpha$ confidence interval for μ.

Since (B.9) is an asymptotic result only, care should be taken when applying it to cases where the sample size is small or moderate and the sampling distribution is heavily skewed. For one- and two-sample normal (Gaussian) data Table B.1 provides *exact* confidence intervals for various parameters. The model for the two-sample data is $X_1, \ldots, X_m \sim_{\text{iid}} \mathsf{N}(\mu_X, \sigma_X^2)$ and $Y_1, \ldots, Y_n \sim_{\text{iid}} \mathsf{N}(\mu_Y, \sigma_Y^2)$, where $X_1, \ldots, X_m, Y_1, \ldots, Y_n$ are independent. All parameters are assumed to be unknown.

Table B.1 Exact confidence intervals for normal data with unknown mean and variance.

Parameter	Exact $1 - \alpha$ confidence interval	Condition
μ_X	$\bar{X} \pm t_{m-1;1-\alpha/2}\dfrac{S_X}{\sqrt{m}}$	
σ_X^2	$\left(\dfrac{(m-1)S_X^2}{\chi^2_{m-1;1-\alpha/2}},\ \dfrac{(m-1)S_X^2}{\chi^2_{m-1;\alpha/2}}\right)$	
$\mu_X - \mu_Y$	$\bar{X} - \bar{Y} \pm t_{m+n-2;1-\alpha/2}\, S_p\sqrt{\dfrac{1}{m} + \dfrac{1}{n}}$	$\sigma_X^2 = \sigma_Y^2$
σ_X^2/σ_Y^2	$\left(F_{n-1,m-1;\alpha/2}\,\dfrac{S_X^2}{S_Y^2},\ F_{n-1,m-1;1-\alpha/2}\,\dfrac{S_X^2}{S_Y^2}\right)$	

Here $S_p^2 = \frac{\sum_{i=1}^{m}(X_i - \bar{X})^2 + \sum_{j=1}^{n}(Y_j - \bar{Y})^2}{m+n-2}$ is the **pooled sample variance**, $t_{n;\gamma}$ is the γ quantile of the t_n distribution, and $F_{m,n;\gamma}$ is the γ quantile of the $\mathsf{F}(m,n)$ distribution.

For one- and two-sample data from the binomial distribution, described by the model $X \sim \mathsf{Bin}(m, p_X)$ and $Y \sim \mathsf{Bin}(n, p_Y)$ independently, approximate $(1 - \alpha)$ confidence intervals for p_X and $p_X - p_Y$ are given in Table B.2. We use the notation $\widehat{p}_X = X/m$ and $\widehat{p}_Y = Y/n$.

Table B.2 Approximate confidence intervals for binomial data.

Parameter	Approximate $1-\alpha$ confidence interval
p_X	$\widehat{p}_X \pm z_{1-\alpha/2}\sqrt{\dfrac{\widehat{p}_X(1-\widehat{p}_X)}{m}}$
$p_X - p_Y$	$\widehat{p}_X - \widehat{p}_Y \pm z_{1-\alpha/2}\sqrt{\dfrac{\widehat{p}_X(1-\widehat{p}_X)}{m} + \dfrac{\widehat{p}_Y(1-\widehat{p}_Y)}{n}}$

Finally, Table B.3 gives exact confidence intervals for various parameters of the linear regression model (B.1).

Table B.3 Exact confidence intervals for normal regression data.

Parameter	Exact $1-\alpha$ confidence interval
β_0	$\widehat{\beta}_0 \pm t_{n-2;1-\alpha/2}\,\widetilde{S}\sqrt{\dfrac{\sum_{i=1}^n x_i^2}{nS_{xx}}}$
β_1	$\widehat{\beta}_1 \pm t_{n-2;1-\alpha/2}\,\widetilde{S}\sqrt{\dfrac{1}{S_{xx}}}$
$\beta_0 + \beta_1 x$	$\widehat{Y} \pm t_{n-2;1-\alpha/2}\,\widetilde{S}\sqrt{\dfrac{1}{n} + \dfrac{(x-\bar{x})^2}{S_{xx}}}$
σ^2	$\left(\dfrac{(n-2)\widetilde{S}^2}{\chi^2_{n-2;1-\alpha/2}}, \dfrac{(n-2)\widetilde{S}^2}{\chi^2_{n-2;\alpha/2}}\right)$

Here $\widehat{\beta}_1 = \frac{S_{xY}}{S_{xx}}$, $\widehat{\beta}_0 = \bar{Y} - \widehat{\beta}_1\bar{x}$, $\widetilde{S}^2 = \frac{1}{n-2}\sum_{i=1}^n (Y_i - \widehat{\beta}_0 - \widehat{\beta}_1 x_i)^2$, $S_{xx} = \sum_{i=1}^n (x_i - \bar{x})^2$, and $S_{xY} = \sum_{i=1}^n (x_i - \bar{x})(Y_i - \bar{Y})$.

B.1.4 Hypothesis Testing

Suppose the model for the data $\mathbf{X}$ is described by a family of probability distributions that depend on a parameter $\boldsymbol{\theta} \in \Theta$. The aim of **hypothesis testing** is to decide, on the basis of the observed data $\mathbf{x}$, which of two competing hypotheses, $H_0 : \boldsymbol{\theta} \in \Theta_0$ (the **null hypothesis**) and $H_1 : \boldsymbol{\theta} \in \Theta_1$ (the **alternative hypothesis**), holds true.

In classical statistics the null hypothesis and alternative hypothesis do not play equivalent roles. H_0 contains the "status quo" statement, and is only rejected if the observed data are very unlikely to have happened under H_0.

The decision whether to accept or reject H_0 is dependent on the outcome of a **test statistic** $\mathbf{T} = \mathbf{T}(\mathbf{X})$. For simplicity, we discuss only the one-dimensional case $\mathbf{T} \equiv T$. Two (related) types of decision rules are generally used:

1. **Decision rule 1**: *Reject H_0 if T falls in the critical region.*

Here the **critical region** is any appropriately chosen region in $\mathbb{R}$. In practice a critical region is one of the following:

- *left one-sided:* $(-\infty, c]$,
- *right one-sided:* $[c, \infty)$,
- *two-sided:* $(-\infty, c_1] \cup [c_2, \infty)$.

For example, for a right one-sided test, H_0 is rejected if the outcome of the test statistic is too large. The endpoints c, c_1, and c_2 of the critical regions are called **critical values**.

2. **Decision rule 2**: *Reject H_0 if the p-value is smaller than some p_0.*

   The **p-value** is the probability that under H_0 the (random) test statistic takes a value as extreme as or more extreme than the one observed. In particular, if t is the observed outcome of the test statistic T, then

   - *left one-sided test:* $p = \mathbb{P}_{H_0}(T \leqslant t)$,
   - *right one-sided:* $p = \mathbb{P}_{H_0}(T \geqslant t)$,
   - *two-sided:* $p = \min\{2\mathbb{P}_{H_0}(T \leqslant t),\ 2\mathbb{P}_{H_0}(T \geqslant t)\}$.

   The smaller the p-value, the greater the strength of the evidence against H_0 provided by the data. As a rule of thumb:

   $p < 0.10$ suggestive evidence,
   $p < 0.05$ reasonable evidence,
   $p < 0.01$ strong evidence.

Whether the first or the second decision rule is used, one can make two types of errors, as depicted in Table B.4.

Table B.4 Type I and II errors in hypothesis testing.

	True statement	
Decision	*H_0 is true*	*H_1 is true*
Accept H_0	Correct	Type II Error
Reject H_0	Type I Error	Correct

The **power** of the test at $\theta \in \Theta_1$ is defined as the probability that H_0 is rejected (correctly). That is,

$$\text{Power}(\theta) = \mathbb{P}_\theta(T \in \text{Critical Region}) = 1 - \mathbb{P}_\theta(\text{Type II Error})\ .$$

The function $\theta \mapsto \text{Power}(\theta)$, with $\theta \in \Theta_1$ is called the **power curve**.

The choice of the test statistic and the corresponding critical region involves a multiobjective optimization criterion, whereby both the probabilities of a type I and type II error should, ideally, be chosen as small as possible. Unfortunately, these probabilities compete with each other. For example, if the critical region is made larger (smaller), the probability of a type II error is reduced (increased), but at the same time the probability of a type I error is increased (reduced).

Since the type I error is considered more serious, Neyman and Pearson [8] suggested the following approach: choose the critical region such that the probability of a type II error is as small as possible, while keeping the probability of a type I error below a predetermined small **significance level** α.

Remark B.1.1 (Equivalence of Decision Rules) Note that decision rule 1 and 2 are equivalent in the following sense:

Reject H_0 if T falls in the critical region, at significance level α.

$\Leftrightarrow$

Reject H_0 if the p-value is $\leqslant$ significance level α.

In other words, the p-value of the test is the smallest level of significance that would lead to the rejection of H_0.

In general, a statistical test involves the following steps:

1. Formulate an appropriate statistical model for the data.

2. Give the null and alternative hypotheses.

3. Determine the test statistic.

4. Determine the distribution of the test statistic under H_0.

5. Calculate the outcome of the test statistic.

6. Calculate the p-value *or* calculate the critical region, given a preselected significance level α.

7. Accept or reject H_0.

The actual choice of an appropriate test statistic is akin to selecting a good estimator for the unknown parameter $\boldsymbol{\theta}$. The test statistic should summarize the information about $\boldsymbol{\theta}$ and make it possible to distinguish between the alternative hypotheses. The likelihood ratio test provides a systematic approach to constructing powerful test statistics; see Section B.2.3.

We conclude with a number of standard tests involving normal and binomial data. Below, z_γ denotes the γ quantile of the $\mathsf{N}(0,1)$ distribution. The γ quantiles of the χ^2_n, t_n, and $\mathsf{F}(m,n)$ distributions are denoted by $\chi^2_{n;\gamma}$, $t_{n;\gamma}$, and $F_{m,n;\gamma}$, respectively. Details may be found in [1], for example.

Table B.5 Normal distribution, one sample: testing μ.

Model:	$X_1,\ldots,X_n \overset{\text{iid}}{\sim} \mathsf{N}(\mu,\sigma^2)$	
H_0:	$\mu = \mu_0$	
Test statistic:	$T = \frac{\bar{X}-\mu_0}{S/\sqrt{n}}$	
Null distribution:	$T \sim \mathsf{t}_{n-1}$	
Reject H_0 if:	$T \geqslant t_{n-1;1-\alpha}$	$H_1 : \mu > \mu_0$
	$T \leqslant -t_{n-1;1-\alpha}$	$H_1 : \mu < \mu_0$
	$T \leqslant -t_{n-1;1-\alpha/2}$ or $T \geqslant t_{n-1;1-\alpha/2}$	$H_1 : \mu \neq \mu_0$

Table B.6 Normal distribution, one sample: testing σ^2.

Model:	$X_1, \ldots, X_n \overset{\text{iid}}{\sim} \mathsf{N}(\mu, \sigma^2)$	
H_0:	$\sigma^2 = \sigma_0^2$	
Test statistic:	$T = S^2(n-1)/\sigma_0^2$	
Null distribution:	$T \sim \chi^2_{n-1}$	
Reject H_0 if:	$T \geqslant \chi^2_{n-1;1-\alpha}$	$H_1 : \sigma^2 > \sigma_0^2$
	$T \leqslant \chi^2_{n-1;\alpha}$	$H_1 : \sigma^2 < \sigma_0^2$
	$T \geqslant \chi^2_{n-1;1-\alpha/2}$ or $T \leqslant \chi^2_{n-1;\alpha/2}$	$H_1 : \sigma^2 \neq \sigma_0^2$

Table B.7 Normal distribution, two samples: testing $\mu_X - \mu_Y$.

Model:	$X_1, \ldots, X_m \overset{\text{iid}}{\sim} \mathsf{N}(\mu_X, \sigma^2)$, $Y_1, \ldots, Y_n \overset{\text{iid}}{\sim} \mathsf{N}(\mu_Y, \sigma^2)$ $X_1, \ldots, X_m, Y_1, \ldots, Y_n$ are independent	
H_0:	$\mu_X = \mu_Y$	
Test statistic:	$T = \frac{\bar{X} - \bar{Y}}{S_p\sqrt{\frac{1}{m} + \frac{1}{n}}}$	
Null distribution:	$T \sim \mathsf{t}_{n+m-2}$	
Reject H_0 if:	$T \geqslant t_{n+m-2;1-\alpha}$	$H_1 : \mu_X > \mu_Y$
	$T \leqslant -t_{n+m-2;1-\alpha}$	$H_1 : \mu_X < \mu_Y$
	$T \leqslant -t_{n+m-2;1-\alpha/2}$ or $T \geqslant t_{n+m-2;1-\alpha/2}$	$H_1 : \mu_X \neq \mu_Y$

Here $S_p^2 = \frac{\sum_{i=1}^m (X_i - \bar{X})^2 + \sum_{j=1}^n (Y_j - \bar{Y})^2}{m+n-2}$ is the *pooled* sample variance. Note that in Table B.7 the variances of the two samples are assumed to be equal. If $\{X_i\}$ and $\{Y_i\}$ are assumed to have different variances and the sample sizes are large, then one can use the test statistic

$$T = \frac{\bar{X} - \bar{Y}}{\sqrt{\frac{S_X^2}{m} + \frac{S_Y^2}{n}}} \, ,$$

which under H_0 approximately has a $\mathsf{N}(0, 1)$ distribution. An alternative approach is to use Welch's t-test [9].

Table B.8 Normal distribution, two samples: testing σ_X^2/σ_Y^2.

Model:	$X_1, \ldots, X_m \overset{\text{iid}}{\sim} \mathsf{N}(\mu_X, \sigma_X^2)$, $Y_1, \ldots, Y_n \overset{\text{iid}}{\sim} \mathsf{N}(\mu_Y, \sigma_Y^2)$ $X_1, \ldots, X_m, Y_1, \ldots, Y_n$ are independent	
H_0:	$\sigma_X^2 = \sigma_Y^2$	
Test statistic:	$T = S_X^2/S_Y^2$	
Null distribution:	$T \sim \mathsf{F}(m-1, n-1)$	
Reject H_0 if:	$T \geqslant F_{m-1,n-1;1-\alpha}$	$H_1 : \sigma_X^2 > \sigma_Y^2$
	$T \leqslant F_{m-1,n-1;\alpha}$	$H_1 : \sigma_X^2 < \sigma_Y^2$
	$T \geqslant F_{m-1,n-1;1-\alpha/2}$ or $T \leqslant F_{m-1,n-1;\alpha/2}$	$H_1 : \sigma_X^2 \neq \sigma_Y^2$

Table B.9 Binomial distribution, one sample: testing p.

Model:	$X \sim \mathsf{Bin}(n, p)$	
H_0:	$p = p_0$	
Test statistic:	$T = X$	
Null distribution:	$\mathsf{Bin}(n, p_0)$	
Reject H_0 if:	$X \geqslant c$, where c is the smallest integer such that $\mathbb{P}_{H_0}(X \geqslant c) \leqslant \alpha$	$H_1 : p > p_0$
	$X \leqslant c$, where c is the largest integer such that $\mathbb{P}_{H_0}(X \leqslant c) \leqslant \alpha$	$H_1 : p < p_0$
	$X \leqslant c_1$ or $X \geqslant c_2$, where c_1 is the largest integer such that $\mathbb{P}_{H_0}(X \leqslant c_1) \leqslant \alpha/2$ and c_2 is the smallest integer such that $\mathbb{P}_{H_0}(X \geqslant c_2) \leqslant \alpha/2$	$H_1 : p \neq p_0$

For large n an alternative is to use the test statistic

$$Z = \frac{X - np_0}{\sqrt{np_0(1 - p_0)}} ,$$

which under H_0 approximately has a $\mathsf{N}(0, 1)$ distribution. The null hypothesis is then rejected if $Z \geqslant z_{1-\alpha}$ for $H_1 : p > p_0$, $Z \leqslant -z_{1-\alpha}$ for $H_1 : p < p_0$, and $[Z \leqslant -z_{1-\alpha/2}$ or $Z \geqslant z_{1-\alpha/2}]$ for $H_1 : p \neq p_0$.

Table B.10 Binomial distribution, two samples: testing $p_X - p_Y$.

Model:	$X \sim \mathsf{Bin}(m, p_X)$ and $Y \sim \mathsf{Bin}(n, p_Y)$ independent	
H_0:	$p_X = p_Y$	
Test statistic:	$T = \frac{\widehat{p}_X - \widehat{p}_Y}{\sqrt{\widehat{p}(1-\widehat{p})\left(\frac{1}{m}+\frac{1}{n}\right)}}$	
Null distribution:	$\mathsf{N}(0, 1)$ (approx.)	
Reject H_0 if:	$Z \geqslant z_{1-\alpha}$	$H_1 : p_X > p_Y$
	$Z \leqslant -z_{1-\alpha}$	$H_1 : p_X < p_Y$
	$Z \geqslant z_{1-\alpha/2}$ or $Z \leqslant -z_{1-\alpha/2}$	$H_1 : p_X \neq p_Y$

Here $\widehat{p}_X = X/m$, $\widehat{p}_Y = Y/n$, and $\widehat{p} = (X + Y)/(m + n)$.

B.2 LIKELIHOOD

The concept of *likelihood* is central in statistics. It describes in a precise way the information about model parameters that is contained in the observed data.

Let $\mathbf{X} = (X_1, \ldots, X_n)^\top$ be a random vector that is distributed according to a pdf $f(\mathbf{x}; \boldsymbol{\theta})$ (discrete or continuous) with parameter vector $\boldsymbol{\theta} = (\theta_1, \ldots, \theta_d)^\top \in \Theta$. Let $\mathbf{x}$ be an outcome of $\mathbf{X}$. The function $\mathcal{L}(\boldsymbol{\theta}; \mathbf{x}) = f(\mathbf{x}; \boldsymbol{\theta})$, $\boldsymbol{\theta} \in \Theta$, is called the **likelihood function** of $\boldsymbol{\theta}$, based on $\mathbf{x}$. The (natural) logarithm of the likelihood function is called the **log-likelihood function** and is denoted by l. The gradient of the log-likelihood function l is called the **score function**, and is denoted by $\mathcal{S}$. Hence,

$$\mathcal{S}(\boldsymbol{\theta};\mathbf{x}) = \begin{pmatrix} \frac{\partial l(\boldsymbol{\theta};\mathbf{x})}{\partial \theta_1} \\ \frac{\partial l(\boldsymbol{\theta};\mathbf{x})}{\partial \theta_2} \\ \vdots \\ \frac{\partial l(\boldsymbol{\theta};\mathbf{x})}{\partial \theta_d} \end{pmatrix} = \nabla_{\boldsymbol{\theta}} \ln \mathcal{L}(\boldsymbol{\theta};\mathbf{x}) = \frac{\nabla_{\boldsymbol{\theta}} f(\mathbf{x};\boldsymbol{\theta})}{f(\mathbf{x};\boldsymbol{\theta})} . \tag{B.10}$$

If θ is one-dimensional, the score function is thus defined as

$$\mathcal{S}(\theta;\mathbf{x}) = \frac{\mathrm{d}}{\mathrm{d}\theta} l(\theta;\mathbf{x}) = \frac{\mathrm{d}}{\mathrm{d}\theta} \ln \mathcal{L}(\theta;\mathbf{x}) = \frac{\frac{\mathrm{d}}{\mathrm{d}\theta} f(\mathbf{x};\theta)}{f(\mathbf{x};\theta)} .$$

The *random vector* $\mathcal{S}(\boldsymbol{\theta}) = \mathcal{S}(\boldsymbol{\theta};\mathbf{X})$ with $\mathbf{X} \sim f(\cdot;\boldsymbol{\theta})$ is called the **efficient score** or simply **score**. The covariance matrix $\mathcal{I}(\boldsymbol{\theta})$ of the score $\mathcal{S}(\boldsymbol{\theta})$ is called the **Fisher information matrix**. Note that $\mathcal{L}$ is a function of $\boldsymbol{\theta}$ for fixed $\mathbf{x}$, whereas $f(\mathbf{x};\boldsymbol{\theta})$ is viewed as a function of $\mathbf{x}$ for fixed $\boldsymbol{\theta}$. Similarly, l and $\mathcal{S}$ and $\mathcal{I}$ are functions of $\boldsymbol{\theta}$. The expectation of the score $\mathcal{S}(\boldsymbol{\theta})$ is equal to the zero vector:

$$\begin{aligned} \mathbb{E}_{\boldsymbol{\theta}} \mathcal{S}(\boldsymbol{\theta}) &= \int \frac{\nabla_{\boldsymbol{\theta}} f(\mathbf{x};\boldsymbol{\theta})}{f(\mathbf{x};\boldsymbol{\theta})} f(\mathbf{x};\boldsymbol{\theta})\,\mathrm{d}\mathbf{x} \\ &= \int \nabla_{\boldsymbol{\theta}} f(\mathbf{x};\boldsymbol{\theta})\,\mathrm{d}\mathbf{x} = \nabla_{\boldsymbol{\theta}} \int f(\mathbf{x};\boldsymbol{\theta})\,\mathrm{d}\mathbf{x} = \nabla_{\boldsymbol{\theta}} 1 = \mathbf{0} , \end{aligned}$$

provided the interchange of differentiation and integration is justified. In particular, this is allowed for natural exponential families; see [6]. We will assume henceforth that $\mathbb{E}_{\boldsymbol{\theta}} \mathcal{S}(\boldsymbol{\theta}) = \mathbf{0}$.

Table B.11 displays the score functions $\mathcal{S}(\boldsymbol{\theta};x)$ calculated from (B.10) for some commonly used distributions. In this table ψ refers to the digamma function. ☞ 716

The concepts of likelihood and score are particularly useful in the case where $X_1,\ldots,X_n$ form an iid sample from some pdf $\mathring{f}$; that is, $X_1,\ldots,X_n \sim_{\text{iid}} \mathring{f}(\cdot;\boldsymbol{\theta})$. In that case, the likelihood of $\boldsymbol{\theta}$ given the data $\mathbf{x} = (x_1,\ldots,x_n)^\top$ is the product

$$\mathcal{L}(\boldsymbol{\theta};\mathbf{x}) = \prod_{i=1}^{n} \mathring{f}(x_i;\boldsymbol{\theta}) . \tag{B.11}$$

Consequently, the log-likelihood is the sum $l(\boldsymbol{\theta};\mathbf{x}) = \sum_{i=1}^{n} \ln \mathring{f}(x_i;\boldsymbol{\theta})$, and the score is

$$\mathcal{S}(\boldsymbol{\theta};\mathbf{X}) = \sum_{i=1}^{n} \mathring{\mathcal{S}}(\boldsymbol{\theta};X_i) , \tag{B.12}$$

where $\mathring{\mathcal{S}}(\boldsymbol{\theta};x)$ is the score function corresponding to $\mathring{f}(x;\boldsymbol{\theta})$. It follows that the information matrix satisfies

$$\mathcal{I}(\boldsymbol{\theta}) = n\,\mathring{\mathcal{I}}(\boldsymbol{\theta}) ,$$

where $\mathring{\mathcal{I}}(\boldsymbol{\theta})$ is the information matrix corresponding to $\mathring{f}$.

Note that the random vectors $\{\mathring{\mathcal{S}}(\boldsymbol{\theta};X_i)\}$ are independent and identically distributed with mean vector $\mathbf{0}$ and covariance matrix $\mathring{\mathcal{I}}(\boldsymbol{\theta})$. The law of large numbers and the central limit theorem now lead directly to two important properties of the ☞ 625

Table B.11 Score functions for commonly used distributions.

Distribution	$\boldsymbol{\theta}$	$\mathcal{S}(\boldsymbol{\theta}; x)$
$\mathsf{Exp}(\lambda)$	λ	$\lambda^{-1} - x$
$\mathsf{Gamma}(\alpha, \lambda)$	(α, λ)	$\left(\ln(\lambda x) - \psi(\alpha),\ \alpha\lambda^{-1} - x\right)^\top$
$\mathsf{N}(\mu, \sigma^2)$	(μ, σ)	$\left(\sigma^{-2}(x-\mu),\ -\sigma^{-1} + \sigma^{-3}(x-\mu)^2\right)^\top$
$\mathsf{Weib}(\alpha, \lambda)$	(α, λ)	$\left(\alpha^{-1} + \ln(\lambda x)[1 - (\lambda x)^\alpha],\ \frac{\alpha}{\lambda}[1 - (\lambda x)^\alpha]\right)^\top$
$\mathsf{Bin}(n, p)$	p	$\dfrac{x - np}{p(1-p)}$
$\mathsf{Poi}(\lambda)$	λ	$\dfrac{x}{\lambda} - 1$
$\mathsf{Geom}(p)$	p	$\dfrac{1 - p\,x}{p(1-p)}$

score of an iid sample.

1. *Law of large numbers*: As $n \to \infty$,

$$\frac{1}{n}\mathcal{S}(\boldsymbol{\theta}; \mathbf{X}) \to \mathbb{E}_{\boldsymbol{\theta}}\mathring{\mathcal{S}}(\boldsymbol{\theta}; X) = \mathbf{0}\,, \tag{B.13}$$

since the expected score is the zero vector.

2. *Central limit theorem*: For large n

$$\mathcal{S}(\boldsymbol{\theta}; \mathbf{X}) \overset{\text{approx.}}{\sim} \mathsf{N}(\mathbf{0}, n\,\mathring{\mathcal{I}}(\boldsymbol{\theta}))\,. \tag{B.14}$$

■ EXAMPLE B.5 (Bernoulli Random Sample)

Let $X_1, \ldots, X_n \sim_{\text{iid}} \mathsf{Ber}(p)$. Then, for a given observation $\mathbf{x} = (x_1, \ldots, x_n)^\top$, the likelihood of p is given by

$$\mathcal{L}(p; \mathbf{x}) = \prod_{i=1}^{n} p^{x_i}(1-p)^{1-x_i} = p^x(1-p)^{n-x}, \quad 0 < p < 1\,, \tag{B.15}$$

where $x = x_1 + \cdots + x_n$. The log-likelihood is $l(p) = x \ln p + (n - x)\ln(1-p)$. Through differentiation with respect to p, we find the score function:

$$\mathcal{S}(p; x) = \frac{x}{p} - \frac{n-x}{1-p} = \frac{x}{p(1-p)} - \frac{n}{1-p}\,. \tag{B.16}$$

The corresponding score $\mathcal{S}(p)$ is obtained by replacing x with $X \sim \mathsf{Bin}(n, p)$. The expectation of $\mathcal{S}(p)$ is 0 and its variance (the information matrix/number) is

$$\mathcal{I}(p) = \frac{\mathrm{Var}(X)}{p^2(1-p)^2} = \frac{n}{p(1-p)}\,.$$

Hence, for large n, $\mathcal{S}(p)$ approximately has a $\mathsf{N}(0, n/(p(1-p)))$ distribution.

Other properties of the likelihood and score include (for proofs see, for example, [1]):

1. *Natural exponential family:* For an exponential family in canonical form ☞ 701

$$f(\mathbf{x};\boldsymbol{\eta}) = \mathrm{e}^{\boldsymbol{\eta}^\top \mathbf{t}(\mathbf{x}) - A(\boldsymbol{\eta})} h(\mathbf{x}) , \tag{B.17}$$

with A as in (D.3), the log-likelihood function is $l(\boldsymbol{\eta};\mathbf{x}) = \boldsymbol{\eta}^\top \mathbf{t}(\mathbf{x}) - A(\boldsymbol{\eta}) + \ln h(\mathbf{x})$, so that the score function becomes

$$\mathcal{S}(\boldsymbol{\eta};\mathbf{x}) = \mathbf{t}(\mathbf{x}) - \nabla A(\boldsymbol{\eta}) \ = \mathbf{t}(\mathbf{x}) - \mathbb{E}_{\boldsymbol{\eta}} \mathbf{t}(\mathbf{X}) . \tag{B.18}$$

It follows that the information matrix is the covariance matrix of $\mathbf{t}(\mathbf{X})$:

$$\mathcal{I}(\boldsymbol{\eta}) = \mathrm{Cov}(\mathbf{t}(\mathbf{X})) = \nabla^2 A(\boldsymbol{\eta}) . \tag{B.19}$$

2. *Information matrix*: An alternative expression for the information matrix is

$$\mathcal{I}(\boldsymbol{\theta}) = -\mathbb{E}_{\boldsymbol{\theta}} H(\boldsymbol{\theta};\mathbf{X}) , \tag{B.20}$$

where $H(\boldsymbol{\theta};\mathbf{X})$ is the Hessian of $l(\boldsymbol{\theta};\mathbf{X})$; that is, the (random) matrix

$$H(\boldsymbol{\theta};\mathbf{X}) = \left(\frac{\partial^2 \ln f(\mathbf{X};\boldsymbol{\theta})}{\partial\theta_i \partial\theta_j}\right) = \left(\frac{\partial^2 l(\boldsymbol{\theta};\mathbf{X})}{\partial\theta_i \partial\theta_j}\right) = \left(\frac{\partial \mathcal{S}_i(\boldsymbol{\theta};\mathbf{X})}{\partial\theta_j}\right) ,$$

where $\mathcal{S}_i$ denotes the i-th component of the score. This alternative expression is valid under mild conditions (which are satisfied for exponential families) that allow the interchange of the order of integration and differentiation [6].

3. *Cramér–Rao*: Let $\mathbf{X} \sim f(\mathbf{x};\boldsymbol{\theta})$. The variance of any unbiased estimator $Z = Z(\mathbf{X})$ of $g(\boldsymbol{\theta})$, where g is a $\mathcal{C}^1$ function, is bounded from below by

$$\mathrm{Var}(Z) \geqslant (\nabla g(\boldsymbol{\theta}))^\top \, \mathcal{I}^{-1}(\boldsymbol{\theta}) \, \nabla g(\boldsymbol{\theta}) . \tag{B.21}$$

4. *Location–scale families*: For location–scale families $\{f(x;\mu,\sigma)\}$ the Fisher information does not depend on μ. In particular, for location families it is constant.

B.2.1 Likelihood Methods for Estimation

Let $\mathbf{x}$ be the observed data from the model $\mathbf{X} \sim f(\mathbf{x};\boldsymbol{\theta})$, yielding the likelihood function $\mathcal{L}(\boldsymbol{\theta};\mathbf{x}) = f(\mathbf{x};\boldsymbol{\theta})$. The **maximum likelihood estimate** (MLE) of $\boldsymbol{\theta}$ is a vector $\widehat{\boldsymbol{\theta}}$ such that $\mathcal{L}(\widehat{\boldsymbol{\theta}};\mathbf{x}) \geqslant \mathcal{L}(\boldsymbol{\theta};\mathbf{x})$ for all $\boldsymbol{\theta}$ in the parameter space Θ. The corresponding random variable (a function of $\mathbf{X}$), also denoted $\widehat{\boldsymbol{\theta}}$, is called the **maximum likelihood estimator** (also abbreviated as MLE).

Since the natural logarithm is an increasing function, maximization of $\mathcal{L}(\boldsymbol{\theta};\mathbf{x})$ is equivalent to maximization of the log-likelihood $l(\boldsymbol{\theta};\mathbf{x})$. This is often easier, especially when $\mathbf{X}$ is an iid sample from some sampling distribution.

If $l(\boldsymbol{\theta};\mathbf{x})$ is a differentiable function with respect to $\boldsymbol{\theta}$ and the maximum is attained in the *interior* of Θ, *and there exists a unique maximum*, then the MLE of $\boldsymbol{\theta}$ can be found by differentiating $l(\boldsymbol{\theta};\mathbf{x})$ with respect to $\boldsymbol{\theta}$ — more precisely, by solving

$$\nabla_{\boldsymbol{\theta}}\, l(\boldsymbol{\theta};\mathbf{x}) = \mathbf{0}\,.$$

In other words, the MLE is obtained by finding a root of the score; that is, by solving

$$\mathcal{S}(\boldsymbol{\theta};\mathbf{x}) = \mathbf{0}\,. \tag{B.22}$$

Properties of the maximum likelihood estimator include (see, for example, [6, Page 444]):

1. *Consistency*: The maximum likelihood estimator $\widehat{\boldsymbol{\theta}}$ is **consistent**. That is, with probability tending to 1 as $n \to \infty$ the likelihood equation has a solution $\widehat{\boldsymbol{\theta}}$ such that for all $\varepsilon > 0$
$$\mathbb{P}(\|\widehat{\boldsymbol{\theta}} - \boldsymbol{\theta}\| > \varepsilon) \to 0\,.$$

2. *Asymptotic normality*: Suppose that $\widehat{\boldsymbol{\theta}}_1, \widehat{\boldsymbol{\theta}}_2, \ldots$ is a sequence of consistent maximum likelihood estimators for $\boldsymbol{\theta}$. Then, $\sqrt{n}\,(\widehat{\boldsymbol{\theta}}_n - \boldsymbol{\theta})$ converges in distribution to a $\mathsf{N}(\mathbf{0}, \mathring{\mathcal{I}}^{-1}(\boldsymbol{\theta}))$-distributed random vector as $n \to \infty$. In other words
$$\widehat{\boldsymbol{\theta}}_n \overset{\text{approx.}}{\sim} \mathsf{N}(\boldsymbol{\theta}, \mathring{\mathcal{I}}^{-1}(\boldsymbol{\theta})/n)\,.$$

3. *Invariance*: Let $\widehat{\boldsymbol{\theta}}$ be the MLE of $\boldsymbol{\theta}$. Then, for any function $\mathbf{g}$ the MLE of $\mathbf{g}(\boldsymbol{\theta})$ is $\mathbf{g}(\widehat{\boldsymbol{\theta}})$.

Note that Property 1 only says that there *exists* a sequence of MLEs $\widehat{\boldsymbol{\theta}}_1, \widehat{\boldsymbol{\theta}}_2, \ldots$ that converge (in probability) to the true $\boldsymbol{\theta}$. When there are multiple local maxima, a particular sequence $\widehat{\boldsymbol{\theta}}_1, \widehat{\boldsymbol{\theta}}_2, \ldots$ may in fact converge to a local maximum.

Theorem B.2.1 (Exponential Families) *For natural exponential families of the form* (B.17) *the MLE is found by solving*

$$\mathbf{t}(\mathbf{x}) - \nabla A(\boldsymbol{\eta}) = \mathbf{t}(\mathbf{x}) - \mathbb{E}_{\boldsymbol{\eta}}\,\mathbf{t}(\mathbf{X}) = \mathbf{0}\,. \tag{B.23}$$

That is, $\boldsymbol{\eta}$ is chosen such that the observed and expected values of $\mathbf{t}(\mathbf{X})$ *are matched.*

■ EXAMPLE B.6 (MLE for the Gamma Distribution)

We wish to estimate both parameters of a $\mathsf{Gamma}(\alpha, \lambda)$ distribution, based on an iid sample $\mathbf{x} = (x_1, \ldots, x_n)$. The corresponding pdf is

$$\mathring{f}(x;\alpha,\lambda) = \frac{\lambda^\alpha x^{\alpha-1} \mathrm{e}^{-\lambda x}}{\Gamma(\alpha)}, \quad x \geqslant 0\,,$$

☞ 702 which is of the form (B.17) with $\mathbf{t}(x) = (x, \ln x)^\top$, $\boldsymbol{\eta} = (-\lambda,\ \alpha - 1)^\top$ and $A(\boldsymbol{\eta}) = -(\eta_2+1)\ln(-\eta_1) + \ln\Gamma(\eta_2+1)$; see also Table D.1. Consequently, the score function is given by

$$\mathring{\mathcal{S}}(\boldsymbol{\eta};x) = \mathbf{t}(x) - \nabla A(\boldsymbol{\eta}) = \begin{pmatrix} x + (\eta_2+1)/\eta_1 \\ \ln x + \ln(-\eta_1) - \psi(\eta_2+1) \end{pmatrix},$$

where ψ is the digamma function. The information matrix is ☞ 716

$$\mathring{\mathcal{I}}(\boldsymbol{\eta}) = \nabla^2 A(\boldsymbol{\eta}) = \begin{pmatrix} (\eta_2+1)/\eta_1^2 & -1/\eta_1 \\ -1/\eta_1 & \psi'(\eta_2+1) \end{pmatrix} .$$

The score function corresponding to the iid sample is therefore

$$\mathcal{S}(\boldsymbol{\eta};\mathbf{x}) = \sum_{i=1}^n \mathring{\mathcal{S}}(\boldsymbol{\eta};x_i) = \begin{pmatrix} \sum_{i=1}^n x_i + n(\eta_2+1)/\eta_1 \\ \sum_{i=1}^n \ln x_i + n(\ln(-\eta_1) - \psi(\eta_2+1)) \end{pmatrix}$$

and the information matrix is $\mathcal{I}(\boldsymbol{\eta}) = n\,\mathring{\mathcal{I}}(\boldsymbol{\eta})$. Estimates of the parameters are found by numerically solving $\mathcal{S}(\boldsymbol{\eta};\mathbf{x}) = \mathbf{0}$ (continued in Example B.8).

B.2.1.1 Score Intervals The score function can also be used to construct confidence intervals. We consider only the one-dimensional case; that is, $\theta \in \mathbb{R}$. Let $\mathbf{X} = (X_1,\ldots,X_n)^\top$ be an iid sample from some sampling distribution $\mathring{f}$. Because of the normal approximation (B.14), the statistic $\mathcal{S}(\theta;\mathbf{X})/\sqrt{\mathcal{I}(\theta)}$ is approximately standard normal, and hence

$$\left\{\theta : -z_{1-\alpha/2} \leqslant \frac{\mathcal{S}(\theta;\mathbf{X})}{\sqrt{n\,\mathring{\mathcal{I}}(\theta)}} \leqslant z_{1-\alpha/2}\right\}$$

is an approximate $1-\alpha$ **confidence set** (not necessarily an interval).

■ **EXAMPLE B.7 (Score Interval for a Bernoulli Random Sample)**

Let $\mathbf{X}$ be an iid sample from $\mathsf{Ber}(p)$. The information matrix is $\mathcal{I}(p) = n/(p(1-p))$ and the score is $\mathcal{S}(p;\mathbf{X}) = n(\bar{X}-p)/(p(1-p))$; see (B.16). So the confidence set becomes

$$\left\{p \;:\; -z_{1-\alpha/2} \leqslant \frac{n(\bar{X}-p)}{p(1-p)} \times \sqrt{\frac{p(1-p)}{n}} \leqslant z_{1-\alpha/2}\right\}$$
$$= \left\{p \;:\; -z_{1-\alpha/2} \leqslant \frac{\bar{X}-p}{\sqrt{p(1-p)/n}} \leqslant z_{1-\alpha/2}\right\}$$

By solving the quadratic equation $(\bar{X}-p)^2 = a^2\,p(1-p)/n$ with respect to p, this confidence set can be written as the interval $\{T_1 \leqslant p \leqslant T_2\}$ with

$$T_{1,2} = \frac{a^2 + 2n\bar{X} \mp a\sqrt{a^2 - 4n(\bar{X}-1)\bar{X}}}{2\,(a^2+n)} ,$$

where $a = z_{1-\alpha/2}$. This **score interval** has much better coverage behavior than the confidence interval in Table B.2, over the complete range of p.

B.2.2 Numerical Methods for Likelihood Maximization

It is frequently not possible to find the MLE $\widehat{\boldsymbol{\theta}}$ in an explicit form. In that case one needs to solve the equation $\mathcal{S}(\boldsymbol{\theta}) = \mathcal{S}(\boldsymbol{\theta};\mathbf{x}) = \mathbf{0}$ numerically via a root-finding

procedure. A well-known method is the *Newton–Raphson* procedure (see also Sec-
☞ 688 tion C.2.2.1). Starting from a guess $\boldsymbol{\theta}$, a "better" guess is obtained by approximating the score via a linear function. More precisely, suppose that $\boldsymbol{\theta}$ is the initial guess for the root. If the latter is reasonably close to $\widehat{\boldsymbol{\theta}}$, a first-order Taylor approximation around $\boldsymbol{\theta}$ gives

$$\mathcal{S}(\widehat{\boldsymbol{\theta}}) \approx \mathcal{S}(\boldsymbol{\theta}) + \nabla\mathcal{S}(\boldsymbol{\theta})(\widehat{\boldsymbol{\theta}} - \boldsymbol{\theta}) = \mathcal{S}(\boldsymbol{\theta}) + H(\boldsymbol{\theta})(\widehat{\boldsymbol{\theta}} - \boldsymbol{\theta}) ,$$

where $H(\boldsymbol{\theta}) = H(\boldsymbol{\theta}; \mathbf{x})$ is the Hessian of the log-likelihood, that is, the matrix of second-order partial derivatives of l. Since $\mathcal{S}(\widehat{\boldsymbol{\theta}}) = \mathbf{0}$, we have $\widehat{\boldsymbol{\theta}} \approx \boldsymbol{\theta} - H^{-1}(\boldsymbol{\theta})\,\mathcal{S}(\boldsymbol{\theta})$. This suggests the following iterative scheme.

Algorithm B.1 (Newton–Raphson Scheme for the MLE)

1. *Start with an initial guess* $\boldsymbol{\theta}_0$. *Set* $t = 0$.
2. *Set*
$$\boldsymbol{\theta}_{t+1} = \boldsymbol{\theta}_t - H^{-1}(\boldsymbol{\theta}_t)\,\mathcal{S}(\boldsymbol{\theta}_t) . \tag{B.24}$$
3. *If* $\mathcal{S}(\boldsymbol{\theta}_{t+1}) < \varepsilon$ *for some small* $\varepsilon > 0$, *then return* $\boldsymbol{\theta}_{t+1}$ *as the MLE; otherwise, set* $t = t + 1$ *and go to Step 2.*

To implement the Newton–Raphson scheme, it is often crucial to come up with a good starting value for the parameter vector. One natural way to obtain a good guess is to match the sample and theoretical moments via the method of moments; see Section B.1.3.1.

Notice that $H(\boldsymbol{\theta}) = H(\boldsymbol{\theta}; \mathbf{x})$ depends on the parameter $\boldsymbol{\theta}$ and data $\mathbf{x}$, and may be quite complicated. However, the expectation of $H(\boldsymbol{\theta}; \mathbf{X})$ under $\boldsymbol{\theta}$ is simply the negative of the information matrix $\mathcal{I}(\boldsymbol{\theta})$, which does not depend on the data. This suggests the alternative to (B.24):

$$\boldsymbol{\theta}_{t+1} = \boldsymbol{\theta}_t + \mathcal{I}^{-1}(\boldsymbol{\theta}_t)\,\mathcal{S}(\boldsymbol{\theta}_t) , \tag{B.25}$$

which may be easier to implement if the information matrix is readily available.

■ EXAMPLE B.8 (MLE for the Gamma Distribution)

We continue Example B.6 to find MLEs for the parameters of the $\mathsf{Gamma}(\alpha, \lambda)$ distribution. The initial guess is obtained by matching the expectation and variance to the sample mean and sample variance, $\bar{x} = \sum_{i=1}^n x_i/n$ and $s^2 = \sum_{i=1}^n (x_i - \bar{x})^2/(n-1)$, respectively. Since for $X \sim \mathsf{Gamma}(\alpha, \lambda)$, $\mathbb{E}X = \alpha/\lambda$ and $\mathrm{Var}(X) = \alpha/\lambda^2$, this leads to the initial guess $\boldsymbol{\eta}_0 = (-\lambda_0, \alpha_0 - 1)^\top$, where $\alpha_0 = \frac{\bar{x}^2}{s^2}$ and $\lambda_0 = \frac{\bar{x}}{s^2}$. The following MATLAB program implements the Newton–Raphson scheme to find the MLE for $\alpha = 3$ and $\lambda = 0.05$.

```
%gammMLE.m
n = 100;
alpha = 3; lambda = 0.05;
x = gamrnd(alpha,1/lambda,1,n);
sumlogx = sum(log(x)); sumx = sum(x);
alp = mean(x)^2/var(x); lam = mean(x)/var(x); % initial guess
```

```
eta = [-lam;alp - 1]; S = Inf;
while  sum(abs(S) > 10^(-5)) > 0
    S = [sumx + n*(eta(2) + 1)/eta(1); ...
          sumlogx + n*(log(-eta(1)) - psi(eta(2) + 1))];
    I = n * [ (eta(2)+1)/eta(1)^2, -1/eta(1); ...
         -1/eta(1) , psi(1,eta(2)+1)];
    eta = eta + I\S;
end
fprintf('lam_hat = %g , alpha_hat = %g \n',-eta(1),1+eta(2))
```

B.2.3 Likelihood Methods for Hypothesis Testing

Let $X_1, \ldots, X_n$ be an iid sample from a distribution with unknown parameter $\boldsymbol{\theta} \in \Theta$. Write $\mathbf{X}$ for the corresponding random vector, and denote the likelihood by $\mathcal{L}(\boldsymbol{\theta}; \mathbf{x})$. Suppose Θ_0 and Θ_1 are two nonoverlapping subsets of Θ, such that $\Theta_0 \cup \Theta_1 = \Theta$.

The **likelihood ratio statistic** is defined as

$$\Lambda = \frac{\max_{\boldsymbol{\theta} \in \Theta_0} \mathcal{L}(\boldsymbol{\theta}; \mathbf{X})}{\max_{\boldsymbol{\theta} \in \Theta} \mathcal{L}(\boldsymbol{\theta}; \mathbf{X})} = \frac{\mathcal{L}(\widehat{\boldsymbol{\theta}}_0; \mathbf{X})}{\mathcal{L}(\widehat{\boldsymbol{\theta}}; \mathbf{X})} ,$$

where $\widehat{\boldsymbol{\theta}}$ is the maximum likelihood estimator of $\boldsymbol{\theta}$ and $\widehat{\boldsymbol{\theta}}_0$ the maximum likelihood estimator of $\boldsymbol{\theta}$ over Θ_0 only.

The likelihood ratio statistic Λ can be used as a test statistic for testing the hypotheses

$$H_0: \quad \boldsymbol{\theta} \in \Theta_0 ,$$
$$H_1: \quad \boldsymbol{\theta} \in \Theta_1 .$$

The critical region is $(0, \lambda^*]$; that is, reject H_0 if Λ is smaller than some critical value λ^*. To determine λ^* one needs to know the distribution of Λ under H_0. In general this is a difficult task, but it is sometimes possible to derive the distribution of a function of Λ under H_0. This is then taken as the test statistic. The critical region follows by inspection.

■ EXAMPLE B.9 (Likelihood Ratio Method for Gaussian Data)

Suppose $X_1, \ldots, X_n \overset{\text{iid}}{\sim} \mathsf{N}(\mu, \sigma^2)$, with μ and σ^2 unknown. We wish to test

$$H_0: \quad \mu = \mu_0 ,$$
$$H_1: \quad \mu \neq \mu_0 .$$

The likelihood function is given by

$$\mathcal{L}(\mu, \sigma^2; \mathbf{X}) = \left(\frac{1}{2\pi\sigma^2}\right)^{n/2} \exp\left(-\frac{1}{2}\sum_{i=1}^{n} \frac{(X_i - \mu)^2}{\sigma^2}\right) .$$

Maximizing $\mathcal{L}$ over Θ gives the maximum likelihood estimate $(\widehat{\mu}, \widehat{\sigma^2})$ given in (B.6) and (B.7). Maximizing $\mathcal{L}$ over $\Theta_0 = \{(\mu_0, \sigma^2), \sigma^2 > 0\}$ gives the estimate $(\mu_0, \widetilde{\sigma^2})$, with

$$\widetilde{\sigma^2} = \frac{1}{n}\sum_{i=1}^{n}(X_i - \mu_0)^2 \; .$$

Hence,

$$\Lambda \;=\; \frac{\mathcal{L}(\mu_0, \widetilde{\sigma^2}; \mathbf{X})}{\mathcal{L}(\widehat{\mu}, \widehat{\sigma^2}; \mathbf{X})} = \left(\frac{\sum_{i=1}^{n}(X_i - \bar{X})^2}{\sum_{i=1}^{n}(X_i - \mu_0)^2}\right)^{n/2} = \left(1 + \frac{1}{n-1}T^2\right)^{-n/2} ,$$

where $T = \frac{\bar{X} - \mu_0}{S/\sqrt{n}}$ and S is the sample standard deviation. Rejecting H_0 for small values of Λ is equivalent to rejecting H_0 for large values of $|T|$. Moreover, under H_0, T has a t_{n-1} distribution. Thus, the likelihood ratio method yields the test in Table B.5.

The asymptotic distribution of the likelihood ratio statistic under H_0 can be derived in several cases, in particular when Θ_0 consists of only one point $\boldsymbol{\theta}_0$. Under H_0 the log-likelihood function satisfies

$$-2\ln\Lambda = -2(l(\boldsymbol{\theta}_0) - l(\widehat{\boldsymbol{\theta}})) \; .$$

☞ 711 A second-order Taylor expansion at $\boldsymbol{\theta}_0$ around $\widehat{\boldsymbol{\theta}}$ gives

$$l(\boldsymbol{\theta}_0) = l(\widehat{\boldsymbol{\theta}}) + (\nabla l(\widehat{\boldsymbol{\theta}}))^\top(\widehat{\boldsymbol{\theta}} - \boldsymbol{\theta}_0) + \frac{1}{2}(\widehat{\boldsymbol{\theta}} - \boldsymbol{\theta}_0)^\top \nabla^2 l(\widehat{\boldsymbol{\theta}})(\widehat{\boldsymbol{\theta}} - \boldsymbol{\theta}_0) + \mathcal{O}(\|\widehat{\boldsymbol{\theta}} - \boldsymbol{\theta}_0\|^3) \; .$$

Because $\nabla l(\widehat{\boldsymbol{\theta}}) = \mathbf{0}$ and $\nabla^2 l(\widehat{\boldsymbol{\theta}}) \approx -\mathcal{I}(\boldsymbol{\theta}_0)$, where $\mathcal{I}$ is the information matrix, we have

$$-2\ln\Lambda \approx (\widehat{\boldsymbol{\theta}} - \boldsymbol{\theta}_0)^\top \mathcal{I}(\boldsymbol{\theta}_0)(\widehat{\boldsymbol{\theta}} - \boldsymbol{\theta}_0) \; .$$

By the central limit theorem $\widehat{\boldsymbol{\theta}} - \boldsymbol{\theta}_0$ has approximately a $\mathsf{N}(\mathbf{0}, \mathcal{I}^{-1}(\boldsymbol{\theta}_0))$ distribution under H_0. Thus, for a large sample size, we have that $-2\ln\Lambda$ is approximately distributed as $\mathbf{X}^\top\mathcal{I}(\boldsymbol{\theta}_0)\mathbf{X}$ with $\mathbf{X} \sim \mathsf{N}(\mathbf{0}, \mathcal{I}^{-1}(\boldsymbol{\theta}_0))$, which has a χ^2_k distribution, where k is the dimension of $\boldsymbol{\theta}$. This gives the following theorem; see also [1].

Theorem B.2.2 (Asymptotic Distribution of Likelihood Ratio Statistic) *For a k-dimensional parameter space, if the null hypothesis has only one value $H_0 : \boldsymbol{\theta} = \boldsymbol{\theta}_0$ and the alternative hypothesis is $H_1 : \boldsymbol{\theta} \neq \boldsymbol{\theta}_0$, then under some mild regularity conditions (which are satisfied for exponential families):*

$$-2\ln\Lambda \overset{\text{approx.}}{\sim} \chi^2_k \quad \textit{for large } n \; .$$

B.3 BAYESIAN STATISTICS

Bayesian statistics is a branch of statistics that is centered around Bayes' formula (A.23). The type of statistical reasoning here is somewhat different from that in classical statistics. In particular, model parameters are usually treated as random rather than fixed quantities and Bayesian statistics uses different notational conventions from those in classical statistics. The two main differences in notation are:

1. Pdfs and conditional pdfs always use the *same letter* f (sometimes p is used instead of f). That is, instead of writing $f_X(x)$ and $f_{X \mid Y}(x \mid y)$ for the pdf of X and the conditional pdf of X given Y, one simply writes $f(x)$ and $f(x \mid y)$. If Y is a different random variable, its pdf (at y) is thus denoted by $f(y)$. This particular style of notation is typical in Bayesian analysis and can be of great descriptive value, despite its apparent ambiguity. We will use this notation whenever we work in a Bayesian setting.

2. One does not usually indicate random variables by capital letters and their outcomes by lower case letters. It is assumed that it is clear from the context whether a variable x or θ should be interpreted as a number or a random variable.

In Bayesian statistics the data $\mathbf{x}$ is modeled via a conditional pdf $f(\mathbf{x} \mid \boldsymbol{\theta})$, called the **likelihood**, that depends on a random parameter $\boldsymbol{\theta}$ taking values in some set Θ. The **a priori** information about $\boldsymbol{\theta}$ (that is, the knowledge about $\boldsymbol{\theta}$ without using any information from the data) is summarized by the pdf of $\boldsymbol{\theta}$, which is called the **prior** pdf. Additional knowledge about $\boldsymbol{\theta}$ obtained from the observed data $\mathbf{x}$ is given by the conditional pdf $f(\boldsymbol{\theta} \mid \mathbf{x})$, called the **posterior** pdf. The posterior and prior pdfs are related via Bayes' formula (replace the integral with a sum in the discrete case):

$$f(\boldsymbol{\theta} \mid \mathbf{x}) = \frac{f(\mathbf{x} \mid \boldsymbol{\theta})\, f(\boldsymbol{\theta})}{\int f(\mathbf{x} \mid \boldsymbol{\theta})\, f(\boldsymbol{\theta})\, \mathrm{d}\boldsymbol{\theta}} \propto f(\mathbf{x} \mid \boldsymbol{\theta})\, f(\boldsymbol{\theta})\,. \tag{B.26}$$

The denominator in (B.26),

$$f(\mathbf{x}) = \int f(\mathbf{x} \mid \boldsymbol{\theta})\, f(\boldsymbol{\theta})\, \mathrm{d}\boldsymbol{\theta}\,,$$

is often called the **marginal likelihood** and is usually difficult to compute. A **Bayesian model** specifies the prior pdf and likelihood. Once the model is given, all inference is based on the posterior pdf in (B.26). For example, a vector for which the posterior pdf is maximal yields a point estimate for $\boldsymbol{\theta}$, called the **maximum a posteriori** estimate. Another estimate is obtained by taking the expected value of $\boldsymbol{\theta}$ under the posterior pdf. A Bayesian $1-\alpha$ confidence region, or **credible region**, is any subset $\mathscr{C} \subset \Theta$, such that

$$\mathbb{P}(\boldsymbol{\theta} \in \mathscr{C} \mid \mathbf{x}) = \int_{\boldsymbol{\theta} \in \mathscr{C}} f(\boldsymbol{\theta} \mid \mathbf{x})\, \mathrm{d}\boldsymbol{\theta} \geqslant 1 - \alpha\,. \tag{B.27}$$

Bayesian models are often constructed in a *hierarchical* way. For example, a three-parameter **hierarchical model** could be specified as follows:

$$\begin{aligned} a &\sim f(a)\,, \\ (b \mid a) &\sim f(b \mid a)\,, \\ (c \mid a, b) &\sim f(c \mid a, b)\,, \\ (\mathbf{x} \mid a, b, c) &\sim f(\mathbf{x} \mid a, b, c)\,. \end{aligned} \tag{B.28}$$

In other words, first specify the prior pdf of a, then given a specify the pdf of b, etc., until finally the likelihood as a function of all the parameters is given. This procedure allows for a straightforward evaluation of the joint pdf as the product of the conditional pdfs:

$$f(\mathbf{x}, a, b, c) = f(\mathbf{x} \,|\, a, b, c)\, f(c \,|\, a, b)\, f(b \,|\, a) f(a)\,.$$

To find the posterior $f(a, b, c \,|\, \mathbf{x})$, simply view $f(\mathbf{x}, a, b, c)$ as a function of a, b, and c for fixed $\mathbf{x}$. To find the marginal posterior pdfs, $f(a \,|\, \mathbf{x}), f(b \,|\, \mathbf{x})$, and $f(a \,|\, \mathbf{x})$, one needs to *integrate out* the other parameters. For example,

$$f(c \,|\, \mathbf{x}) = \iint f(a, b, c \,|\, \mathbf{x})\, da\, db\,.$$

■ EXAMPLE B.10 (Coin Flipping and Bayesian Learning)

Consider the random experiment where we toss a biased coin n times. Suppose that the outcomes are $x_1, \ldots, x_n$, with $x_i = 1$ if the i-th toss is heads and $x_i = 0$ otherwise, for $i = 1, \ldots, n$. A possible Bayesian model for the data is

$$p \sim \mathsf{U}(0, 1)$$
$$(x_1, \ldots, x_n \,|\, p) \overset{\text{iid}}{\sim} \mathsf{Ber}(p)\,.$$

The likelihood is therefore

$$f(\mathbf{x} \,|\, p) = \prod_{i=1}^{n} p^{x_i} (1-p)^{1-x_i} = p^s\,(1-p)^{n-s}\,,$$

where $s = x_1 + \cdots + x_n$ is the total number of heads. Since $f(p) = 1$, the posterior pdf is

$$f(p \,|\, \mathbf{x}) = c\, p^s\,(1-p)^{n-s}\,, \quad p \in [0, 1]\,,$$

which is the pdf of the $\mathsf{Beta}(s+1, n-s+1)$ distribution. The normalization constant is $c = (n+1)\binom{n}{s}$. The maximum a posteriori estimate of p is s/n, which coincides with the classical maximum likelihood estimate. The expectation of the posterior pdf is $(s+1)/(n+2)$. The graph of the pdf for $n = 100$ and $s = 1$ is given in Figure B.1. For this case a left one-sided 95% credible interval for p is $[0, 0.0461]$, where 0.0461 is the 0.95 quantile of the $\mathsf{Beta}(2, 100)$ distribution.

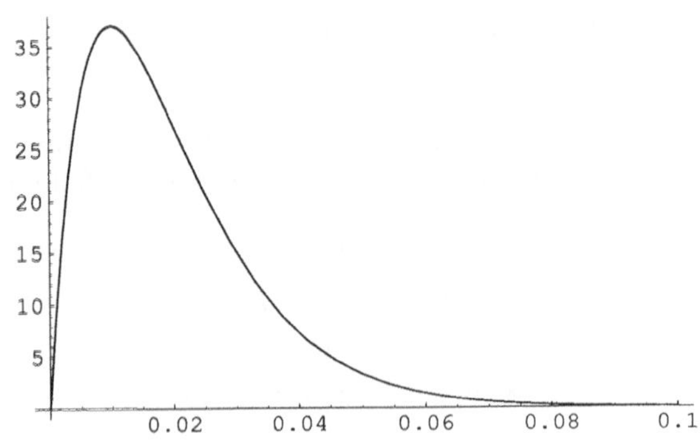

Figure B.1 Posterior pdf for p, with $n = 100$ and $s = 1$.

Evaluating or drawing from the marginal posterior distributions may not always be easy or feasible. When this is the case, one often turns to Markov chain Monte Carlo techniques; see Chapter 6. For example, in the three-parameter model (B.28) one could use the *Gibbs sampler* to sample from the posterior pdf: ☞ 225

1. Initialize a, b, c. Then iterate the following steps:
2. Draw a from $f(a \mid b, c, \mathbf{x})$.
3. Draw b from $f(b \mid a, c, \mathbf{x})$.
4. Draw c from $f(c \mid a, b, \mathbf{x})$.

After we obtain a (dependent) sample $\{(a_t, b_t, c_t)\}$ from $f(a, b, c \mid \mathbf{x})$, process only the variables of interest, for example, only the $\{c_t\}$ to obtain a dependent sample from $f(c \mid \mathbf{x})$.

B.3.1 Conjugacy

In Bayesian analysis it is convenient to have the posterior and prior pdfs in the *same* family of distributions. This property is called **conjugacy**. The advantage of conjugacy is that only the parameters of the distribution need to be updated.

Exponential families provide natural conjugate families. In particular, consider the m-dimensional exponential family ☞ 701

$$f(\mathbf{x} \mid \boldsymbol{\theta}) = c(\boldsymbol{\theta})^n \mathrm{e}^{\sum_{i=1}^m \eta_i(\boldsymbol{\theta}) \sum_{k=1}^n t_i(x_k)} \prod_{i=1}^n h(x_k) , \tag{B.29}$$

which is the joint pdf of an iid sample from an exponential family — see Example B.1. Suppose the prior pdf is chosen of the form

$$f(\boldsymbol{\theta}) \propto c(\boldsymbol{\theta})^b \mathrm{e}^{\sum_{i=1}^m \eta_i(\boldsymbol{\theta}) a_i},$$

where the proportionality constant only depends on $\mathbf{a} = (a_1, \ldots, a_m, b)$. Then, the posterior pdf becomes

$$f(\boldsymbol{\theta} \mid \mathbf{x}) \propto f(\boldsymbol{\theta}) f(\mathbf{x} \mid \boldsymbol{\theta}) \propto c(\boldsymbol{\theta})^{n+b} \mathrm{e}^{\sum_{i=1}^m \eta_i(\boldsymbol{\theta}) \left(a_i + \sum_{k=1}^n t_i(x_k)\right)},$$

where the proportionality constant does not depend on $\boldsymbol{\theta}$. Thus, $f(\boldsymbol{\theta})$ and $f(\boldsymbol{\theta} \mid \mathbf{x})$ are in the same $(m+1)$-dimensional exponential family.

■ EXAMPLE B.11 (Conjugate Prior for the Poisson Distribution)

Let $x_1, \ldots, x_n \sim_{\text{iid}} \mathsf{Poi}(\lambda)$, with sample mean $\bar{x} = (x_1 + \cdots + x_n)/n$. The joint pdf can be written in the form (B.29) as

$$f(\mathbf{x} \mid \lambda) = \mathrm{e}^{-n\lambda} \mathrm{e}^{n\bar{x} \ln \lambda} \prod_{i=1}^n \frac{1}{x_i!} ,$$

which suggests a conjugate prior of the form $f(\lambda) \propto \mathrm{e}^{-b\lambda} \mathrm{e}^{a \ln \lambda} = \mathrm{e}^{-b\lambda} \lambda^a$, corresponding to the Gamma distribution. In particular, if we take a $\mathsf{Gamma}(\alpha, \beta)$ prior for λ, that is,

$$f(\lambda) \propto \mathrm{e}^{-\beta\lambda} \lambda^{\alpha-1} ,$$

then the posterior pdf is

$$f(\lambda \mid \mathbf{x}) \propto \mathrm{e}^{-(n+\beta)\lambda} \lambda^{\alpha-1+n\bar{x}},$$

which corresponds to the $\mathsf{Gamma}(\alpha + n\bar{x}, \beta + n)$ distribution.

Further Reading

For an accessible introduction to mathematical statistics with simple applications see [5]. For more detailed overview of statistical inference, see Casella and Berger [2]. A standard reference for classical or frequentist statistical inference is [6]. An applied reference for Bayesian inference is [3]. For a survey of numerical techniques relevant to computational statistics see [7].

REFERENCES

1. P. J. Bickel and K. A. Doksum. *Mathematical Statistics*, volume I. Pearson Prentice Hall, Upper Saddle River, NJ, second edition, 2007.
2. G. Casella and R. L. Berger. *Statistical Inference.* Duxbury Press, Pacific Grove, CA, second edition, 2001.
3. A. Gelman. *Bayesian Data Analysis.* Chapman & Hall, New York, second edition, 2004.
4. R. V. Hogg and T. A. Craig. *Introduction to Mathematical Statistics.* Prentice Hall, New York, fifth edition, 1995.
5. R. J. Larsen and M. L. Marx. *An Introduction to Mathematical Statistics and Its Applications.* Prentice Hall, New York, third edition, 2001.
6. E. L. Lehmann and G. Casella. *Theory of Point Estimation.* Springer-Verlag, New York, second edition, 1998.
7. J. F. Monahan. *Numerical Methods of Statistics.* Cambridge Series in Statistical and Probabilistic Mathematics. Cambridge University Press, London, 2010.
8. J. Neyman and E. Pearson. On the problem of the most efficient tests of statistical hypotheses. *Philosophical Transactions of the Royal Society of London, Series A*, 231:289–337, 1933. DOI:10.1098/rsta.1933.0009.
9. B. L. Welch. The generalization of 'Student's' problem when several different population variances are involved. *Biometrika*, 34(1-2):28–35, 1947.

APPENDIX C

OPTIMIZATION

In this appendix we review various aspects of deterministic optimization. We refer the reader to Chapter 12 if they have a *noisy* optimization problem instead. In Section C.1 we introduce the optimization notation, discuss important classes of optimization problems, and give key theoretical results. Section C.2 deals with the practical issues that arise in optimization, together with several optimization algorithms. There is a close connection between deterministic gradient decent methods discussed in this appendix and the stochastic approximation method covered in Section 12.1. In particular, the latter can be thought of as a generalization of the same idea to a random setting. Moreover, there are many problems that can be solved without resorting to Monte Carlo techniques. In particular, it is worthwhile checking if a deterministic approach applies, as they can be much more efficient. Finally, in Section C.3, we give a small selection of test problems.

☞ 441

C.1 OPTIMIZATION THEORY

Optimization is concerned with finding minimal or maximal solutions of a real-valued **objective function** f in some set $\mathscr{X}$:

$$\min_{\mathbf{x}\in\mathscr{X}} f(\mathbf{x}) \quad \text{or} \quad \max_{\mathbf{x}\in\mathscr{X}} f(\mathbf{x}) \,. \tag{C.1}$$

Since any maximization problem can easily be converted into a minimization problem via the equivalence $\max_{\mathbf{x}} f(\mathbf{x}) \equiv -\min_{\mathbf{x}} -f(\mathbf{x})$, we focus only on minimization

Handbook of Monte Carlo Methods. By D. P. Kroese, T. Taimre, Z. I. Botev problems. We use the following terminology. A **local minimizer** of $f(\mathbf{x})$ is an element $\mathbf{x}^* \in \mathscr{X}$ such that $f(\mathbf{x}^*) \leqslant f(\mathbf{x})$ for all $\mathbf{x}$ in some neighborhood of $\mathbf{x}^*$. If $f(\mathbf{x}^*) \leqslant f(\mathbf{x})$ for all $\mathbf{x} \in \mathscr{X}$, then $\mathbf{x}^*$ is called a **global minimizer** or **global solution**. The set of global minimizers is denoted by

$$\underset{\mathbf{x}\in\mathscr{X}}{\operatorname{argmin}} f(\mathbf{x}) .$$

The function value $f(\mathbf{x}^*)$ corresponding to a local/global minimizer $\mathbf{x}^*$ is called a **local/global minimum** of $f(\mathbf{x})$.

Optimization problems may be classified by the set $\mathscr{X}$ and the objective function f. If $\mathscr{X}$ is countable, the optimization problem is called **discrete** or **combinatorial**. If instead $\mathscr{X}$ is a nondenumerable set such as $\mathbb{R}^n$ and f takes values in a nondenumerable set, then the problem is said to be **continuous**. Optimization problems that are neither discrete nor continuous are said to be **mixed**. In **functional minimization**, optimization occurs over a space of functions. Thus, the problem is again of the form (C.1), but now $\mathbf{x}$ is a *function* and $f(\mathbf{x})$ is a *functional* of $\mathbf{x}$; for example, the integral of $\|\mathbf{x}\|$ over some time interval. Many interesting problems fall into this framework, such as problems in the **calculus of variations**, in **optimal control**, and in **shape optimization**; see, for example, [21].

The search set $\mathscr{X}$ is often defined by means of **constraints**. A standard setting for constrained optimization (minimization) is the following:

$$\begin{aligned} &\min_{\mathbf{x}\in\mathscr{Y}} f(\mathbf{x}) \\ \text{subject to: } &h_i(\mathbf{x}) = 0, \quad i = 1,\ldots,m, \\ &g_i(\mathbf{x}) \leqslant 0, \quad i = 1,\ldots,k. \end{aligned} \tag{C.2}$$

Here, f is the objective function, and $\{g_i\}$ and $\{h_i\}$ are given functions so that $h_i(\mathbf{x}) = 0$ and $g_i(\mathbf{x}) \leqslant 0$ represent the **equality** and **inequality** constraints, respectively. The region $\mathscr{X} \subseteq \mathscr{Y}$ where the objective function is defined and where all the constraints are satisfied is called the **feasible region**. An optimization problem without constraints is said to be an **unconstrained** problem.

For an unconstrained continuous optimization problem, the search space $\mathscr{X}$ is often taken to be (a subset of) $\mathbb{R}^n$, and f is assumed to be a $\mathcal{C}^k$ function for sufficiently high k (typically $k = 2$ or 3 suffices); that is, its k-th order derivative is continuous. For a $\mathcal{C}^1$ function the standard approach to minimizing $f(\mathbf{x})$ is to solve the equation

$$\nabla f(\mathbf{x}) = \mathbf{0} , \tag{C.3}$$

☞ 710 where $\nabla f(\mathbf{x})$ is the **gradient** of f at $\mathbf{x}$. The solutions $\mathbf{x}^*$ to (C.3) are called **stationary points**. Stationary points can be local/global minimizers, local/global maximizers, or **saddle points** (which are neither). If, in addition, the function is $\mathcal{C}^2$, the condition

$$\mathbf{y}^\top(\nabla^2 f(\mathbf{x}^*))\,\mathbf{y} > 0 \quad \text{for all } \mathbf{y} \neq \mathbf{0} , \tag{C.4}$$

ensures that the stationary point $\mathbf{x}^*$ is a local minimizer of f. The condition (C.4)
☞ 711 states that the *Hessian* matrix of f at $\mathbf{x}^*$ is **positive definite**. We write $H \succ 0$ to indicate that a matrix H is positive definite.

In Figure C.1 we have a multiextremal objective function on $\mathscr{X} = \mathbb{R}$. There are four stationary points: two are local minimizers, one is a local maximizer, and one is neither a minimizer nor a maximizer, but a saddle point.

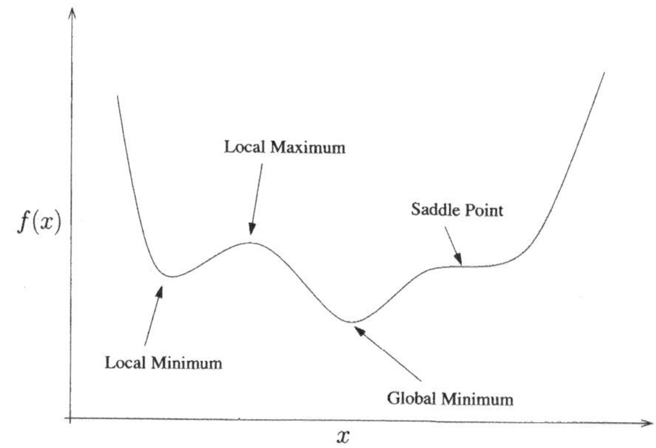

Figure C.1 A multiextremal problem in one dimension.

An important class of optimization problems is related to the notion of *convexity*. A set $\mathscr{X}$ is said to be **convex** if for all $\mathbf{x}_1, \mathbf{x}_2 \in \mathscr{X}$ it holds that $\alpha\, \mathbf{x}_1 + (1-\alpha)\, \mathbf{x}_2 \in \mathscr{X}$ for all $0 \leqslant \alpha \leqslant 1$. An optimization program of the form (C.2) is said to be a **convex programming problem** if:

1. The objective f is a **convex function**; that is, for all $\mathbf{x}, \mathbf{y} \in \mathscr{X}$ and $0 \leqslant \alpha \leqslant 1$,

$$f\big(\alpha\, \mathbf{x} + (1-\alpha)\, \mathbf{y}\big) \leqslant \alpha f(\mathbf{x}) + (1-\alpha) f(\mathbf{y})\,. \tag{C.5}$$

2. The inequality constraint functions $\{g_i\}$ are convex.

3. The equality constraint functions $\{h_i\}$ are **affine**, that is, of the form $\mathbf{a}_i^\top \mathbf{x} - b_i$. This is equivalent to both h_i and $-h_i$ being convex for all i.

A function f satisfying (C.5) with strict inequality is said to be **strictly convex**. It is said to be (strictly) **concave** if $-f$ is (strictly) convex. Assuming that $\mathscr{X}$ is an open set, convexity for $f \in \mathcal{C}^1$ is equivalent to

$$f(\mathbf{y}) \geqslant f(\mathbf{x}) + (\mathbf{y} - \mathbf{x})^\top \nabla f(\mathbf{x}) \quad \text{for all } \mathbf{x}, \mathbf{y} \in \mathscr{X}\,.$$

Moreover, for $f \in \mathcal{C}^2$ strict convexity is equivalent to the Hessian matrix being positive definite for all $\mathbf{x} \in \mathscr{X}$, and convexity is equivalent to the Hessian matrix being **positive semidefinite** for all $\mathbf{x}$; that is, $\mathbf{y}^\top \left(\nabla^2 f(\mathbf{x})\right) \mathbf{y} \geqslant 0$ for all $\mathbf{y}$ and $\mathbf{x}$. We write $H \succeq 0$ to indicate that a matrix H is positive semidefinite.

Table C.1 summarizes some commonly encountered problems, all of which are convex, with the exception of the quadratic programs with $A \not\succeq 0$.

Table C.1 Some common classes of optimization problems.

Name	$f(\mathbf{x})$	Constraints
Linear Program (LP)	$\mathbf{c}^\top\mathbf{x}$	$A\mathbf{x} = \mathbf{b}$ and $\mathbf{x} \geqslant 0$
Inequality Form LP	$\mathbf{c}^\top\mathbf{x}$	$A\mathbf{x} \leqslant \mathbf{b}$
Quadratic Program (QP)	$\frac{1}{2}\mathbf{x}^\top A\mathbf{x} + \mathbf{b}^\top\mathbf{x}$	$D\mathbf{x} \leqslant \mathbf{d},\quad E\mathbf{x} = \mathbf{e}$
Convex QP (CQP)	$\frac{1}{2}\mathbf{x}^\top A\mathbf{x} + \mathbf{b}^\top\mathbf{x}$	$D\mathbf{x} \leqslant \mathbf{d},\ E\mathbf{x} = \mathbf{e} \quad (A \succeq 0)$
Convex Quadr. Constr. QP (CQCQP)	$\frac{1}{2}\mathbf{x}^\top A\mathbf{x} + \mathbf{b}^\top\mathbf{x}$	$\mathbf{x}^\top A_i\mathbf{x} + \mathbf{b}_i^\top\mathbf{x} + c_i \leqslant 0,\quad i = 1,\ldots,m,$ $E\mathbf{x} = \mathbf{e} \quad (A, A_1, \ldots \succeq 0)$
Quadratically Constrained QP (QCQP)	$\frac{1}{2}\mathbf{x}^\top A\mathbf{x} + \mathbf{b}^\top\mathbf{x}$	$\mathbf{x}^\top A_i\mathbf{x} + \mathbf{b}_i^\top\mathbf{x} + c_i \leqslant 0,\quad i = 1,\ldots,m,$ $E\mathbf{x} = \mathbf{e}$
Second-order Cone Program (SOCP)	$\mathbf{a}^\top\mathbf{x}$	$\lVert A_i\mathbf{x}+\mathbf{b}_i\rVert \leqslant \mathbf{c}_i^\top\mathbf{x}+\mathbf{d}_i,\ i = 1, 2, \ldots, m,$ $E\mathbf{x} = \mathbf{e}$
Geometric Program (GP)	$\sum_{i=1}^{m} c_i \prod_{j=1}^{n} x_j^{a_{ij}}$	$\sum_{i=1}^{m^{(k)}} c_i^{(k)} \prod_{j=1}^{n} x_j^{a_{ij}^{(k)}} \leqslant 1,$ $d_i^{(l)} \prod_{j=1}^{n} x_j^{b_{ij}^{(l)}} = 1, \quad \mathbf{x} > \mathbf{0}$
Semidefinite Program (SDP)	$\mathbf{c}^\top\mathbf{x}$	$(A_0 + \sum_{i=1}^{n} x_i A_i) \succeq 0, \quad \{A_i\}$ symmetric
Convex Program (CVX)	$f(\mathbf{x})$ convex	$\{g_i(\mathbf{x})\}$ convex, $\{h_i(\mathbf{x})\}$ of the form $\mathbf{a}_i^\top\mathbf{x} - b_i$

A hierarchy of convex optimization problems is:

$$\mathrm{LP} \subset \mathrm{CQP} \subset \mathrm{CQCQP} \subset \mathrm{SOCP} \subset \mathrm{SDP} \subset \mathrm{CVX}.$$

Geometric programs, after a suitable transformation [4], fit into the hierarchy as follows:

$$\mathrm{LP} \subset \mathrm{GP} \subset \mathrm{CVX}.$$

Recognizing convex optimization problems or those that can be transformed to convex optimization problems can be challenging. However, once formulated as convex optimization problems, these can be efficiently solved using subgradient [19], bundle [10], and cutting-plane methods [11].

■ **EXAMPLE C.1 (Fastest Mixing Markov Chain on a Graph)**

Consider a time-homogeneous Markov chain on a graph $\mathcal{G} = (V, E)$ with one-step transition matrix P of dimension n, having non-zero (i, j)-th entry only if $(i, j) \in E$. ☞ 632

Suppose we wish to populate the non-zero entries of P in order to find the **fastest mixing** Markov chain on $\mathcal{G}$ — that is the chain whose second-largest eigenvalue modulus is smallest. This can be formulated as an SDP as follows [3, 5]:

$$\begin{aligned} &\min_{P,s} s \\ \text{such that } &-sI \preceq P - \frac{1}{n}\mathbf{1}\mathbf{1}^\top \preceq sI \\ &P\mathbf{1} = \mathbf{1} \\ &P = P^\top \\ &P_{ij} \geqslant 0\,, \quad i, j, = 1, 2, \ldots, n \\ &P_{ij} = 0\,, \quad (i, j) \notin E\,. \end{aligned}$$

This problem can then be solved via standard methods for SDPs, for example, interior-point methods, in which Newton's method is applied to a sequence of linearly constrained problems, where the inequality constraints become part of the objective function through the use of a **barrier function** (see [5, Chapter 11] for details). A subgradient method has also been developed for this problem which is preferred for large graphs (see [3] for details).

As an illustration, consider the transition graph depicted in Figure C.2.

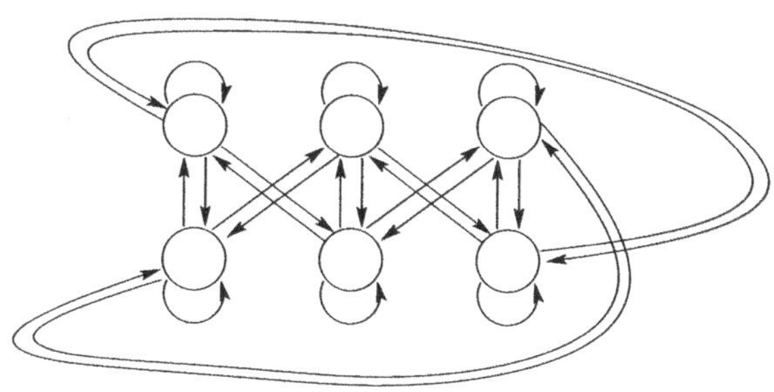

Figure C.2 Transition graph with self-loops.

In this case, the SDP is solved with a matrix

$$P^* = \begin{pmatrix} 1/3 & 0 & 0 & 2/9 & 2/9 & 2/9 \\ 0 & 1/3 & 0 & 2/9 & 2/9 & 2/9 \\ 0 & 0 & 1/3 & 2/9 & 2/9 & 2/9 \\ 2/9 & 2/9 & 2/9 & 1/3 & 0 & 0 \\ 2/9 & 2/9 & 2/9 & 0 & 1/3 & 0 \\ 2/9 & 2/9 & 2/9 & 0 & 0 & 1/3 \end{pmatrix}.$$

Note that this matrix is of the form

$$P_p = \begin{pmatrix} pI & (1-p)/3\,O \\ (1-p)/3\,O & pI \end{pmatrix},$$

where I is the 3×3 identity matrix, O is a 3×3 matrix of ones, and $p \in [0, 1]$. Indeed, by symmetry considerations, and the fact that P is a stochastic matrix, an optimal solution *must* be of the form above.

The optimal matrix, $P^* \equiv P_{1/3}$, has second largest eigenvalue modulus of 1/3. Figure C.3 plots the second largest eigenvalue modulus for matrices of the form P_p versus p.

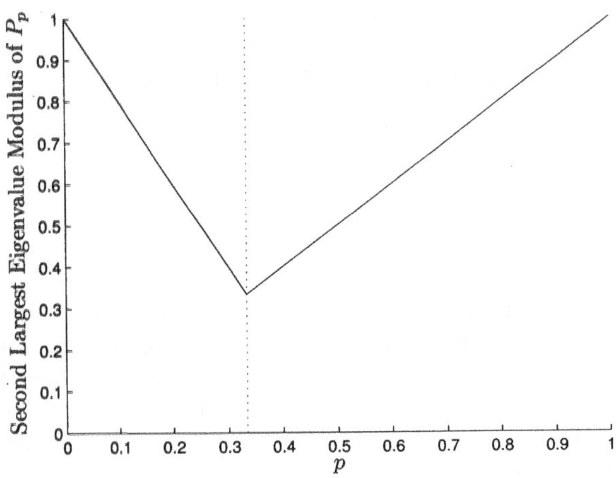

Figure C.3 The second largest eigenvalue modulus of P_p for $p \in [0, 1]$.

In some optimization problems one has many different objective functions, and one wishes to find a solution that simultaneously results in acceptable solutions over all functions. One widely used concept is that of **Pareto optimality**. Informally, a point is called Pareto optimal if one cannot improve the solution with respect to one objective function without worsening the solution of another. In a minimization setting, this can be outlined formally as follows.

- Suppose we have m functions $f_1, f_2, \ldots, f_m : \mathbb{R}^n \to \mathbb{R}$. A point $\mathbf{x}$ is said to **dominate** a point $\mathbf{y}$ if

  1. for all i, $f_i(\mathbf{x}) \leqslant f_i(\mathbf{y})$; and
  2. for at least one j, $f_j(\mathbf{x}) < f_j(\mathbf{y})$.

- A point $\mathbf{x}^* \in \mathbb{R}^n$ is called **Pareto optimal** if it is not dominated by any other point. Denote the set of Pareto optimal points by $\mathscr{X}^*$.

- The **Pareto front** is the set of the form $\{(f_1(\mathbf{x}), \ldots, f_m(\mathbf{x})) : \mathbf{x} \in \mathscr{X}^*\}$, so that any point on the Pareto front corresponds to a Pareto optimal point.

There are numerous problems that are encapsulated by the term **noisy optimization**, where the objective function f cannot be evaluated precisely but instead an estimate $\widehat{f}$ is available. This estimate can be viewed as f corrupted by some stochastic mechanism. For constrained problems, the constraint functions may only be available as estimates. The intuition here is that one may have a level of uncertainty as to whether or not a given point is feasible.

The **stochastic approximation approach** (also called the **stochastic counterpart method**) of Section 12.2 actually replaces deterministic elements of a given problem by estimates, which results in a simpler optimization problem. ☞ 446

C.1.1 Lagrangian Method

The main components of the Lagrangian method are the Lagrange multipliers and the Lagrange function. The method was developed by Lagrange in 1797 for the optimization problem (C.2) with only equality constraints. In 1951 Kuhn and Tucker extended Lagrange's method to inequality constraints. Given an optimization problem (C.2) containing only equality constraints $h_i(\mathbf{x}) = 0,\ i = 1,\ldots,m$, the **Lagrange function** is defined as

$$\mathcal{L}(\mathbf{x}, \boldsymbol{\beta}) = f(\mathbf{x}) + \sum_{i=1}^{m} \beta_i\, h_i(\mathbf{x}) ,$$

where the coefficients $\{\beta_i\}$ are called the **Lagrange multipliers**. A necessary condition for a point $\mathbf{x}^*$ to be a local minimizer of $f(\mathbf{x})$ subject to the equality constraints $h_i(\mathbf{x}) = 0,\ i = 1,\ldots,m$, is

$$\begin{aligned} \nabla_{\mathbf{x}}\, \mathcal{L}(\mathbf{x}^*, \boldsymbol{\beta}^*) &= \mathbf{0} , \\ \nabla_{\boldsymbol{\beta}}\, \mathcal{L}(\mathbf{x}^*, \boldsymbol{\beta}^*) &= \mathbf{0} , \end{aligned}$$

for some value $\boldsymbol{\beta}^*$. The above conditions are also sufficient if $\mathcal{L}(\mathbf{x}, \boldsymbol{\beta}^*)$ is a convex function of $\mathbf{x}$.

Given the original optimization problem (C.2), containing both the equality and inequality constraints, the **generalized Lagrange function**, or **Lagrangian**, is defined as

$$\mathcal{L}(\mathbf{x}, \boldsymbol{\alpha}, \boldsymbol{\beta}) = f(\mathbf{x}) + \sum_{i=1}^{k} \alpha_i\, g_i(\mathbf{x}) + \sum_{i=1}^{m} \beta_i\, h_i(\mathbf{x}) .$$

Theorem C.1.1 (Karush–Kuhn–Tucker Conditions) *A necessary condition for a point* $\mathbf{x}^*$ *to be a local minimizer of* $f(\mathbf{x})$ *in the optimization problem* (C.2) *is the existence of an* $\boldsymbol{\alpha}^*$ *and* $\boldsymbol{\beta}^*$ *such that*

$$\begin{aligned} \nabla_{\mathbf{x}}\, \mathcal{L}(\mathbf{x}^*, \boldsymbol{\alpha}^*, \boldsymbol{\beta}^*) &= \mathbf{0} , \\ \nabla_{\boldsymbol{\beta}}\, \mathcal{L}(\mathbf{x}^*, \boldsymbol{\alpha}^*, \boldsymbol{\beta}^*) &= \mathbf{0} , \\ g_i(\mathbf{x}^*) &\leqslant 0, \quad i = 1,\ldots,k , \\ \alpha_i^* &\geqslant 0, \quad i = 1,\ldots,k , \\ \alpha_i^*\, g_i(\mathbf{x}^*) &= 0, \quad i = 1,\ldots,k . \end{aligned}$$

For *convex* programs we have the following important results [5, 8]:

1. Every local solution $\mathbf{x}^*$ to a convex programming problem is a global solution and the set of global solutions is convex. If, in addition, the objective function is strictly convex, then any global solution is unique.

2. For a strictly convex programming problem with $\mathcal{C}^1$ objective and constraint functions, the KKT conditions are necessary and sufficient for a unique global solution.

C.1.2 Duality

The aim of duality is to provide an alternative formulation of an optimization problem which is often more computationally efficient or has some theoretical significance (see [8, Page 219]). The original problem (C.2) is referred to as the **primal** problem whereas the reformulated problem, based on Lagrange multipliers, is referred to as the **dual** problem. Duality theory is most relevant to convex optimization problems. It is well known that if the primal optimization problem is (strictly) convex then the dual problem is (strictly) concave and has a (unique) solution from which the (unique) optimal primal solution can be deduced.

The **Lagrange dual program** (also called the **Wolfe dual**) of the primal program (C.2), is:

$$\max_{\boldsymbol{\alpha},\boldsymbol{\beta}} \quad \mathcal{L}^*(\boldsymbol{\alpha},\boldsymbol{\beta})$$
$$\text{subject to:} \quad \boldsymbol{\alpha} \geqslant \mathbf{0}\,,$$

where $\mathcal{L}^*$ is the **Lagrange dual function:**

$$\mathcal{L}^*(\boldsymbol{\alpha},\boldsymbol{\beta}) = \inf_{\mathbf{x}\in\mathscr{X}} \mathcal{L}(\mathbf{x},\boldsymbol{\alpha},\boldsymbol{\beta})\,. \tag{C.6}$$

It is not difficult to see that if f^* is the minimal value of the primal problem, then $\mathcal{L}^*(\boldsymbol{\alpha},\boldsymbol{\beta}) \leqslant f^*$ for any $\boldsymbol{\alpha} \geqslant \mathbf{0}$ and any $\boldsymbol{\beta}$. This property is called **weak duality**. The Lagrangian dual program thus determines the best lower bound on f^*. If d^* is the optimal value for the dual problem then $d^* \leqslant f^*$. The difference $f^* - d^*$ is called the **duality gap**.

The duality gap is extremely useful for providing lower bounds for the solutions of primal problems that may be impossible to solve directly. It is important to note that for linearly constrained problems, if the primal is infeasible (does not have a solution satisfying the constraints), then the dual is either infeasible or unbounded. Conversely, if the dual is infeasible then the primal has no solution. Of crucial importance is the **strong duality** theorem, which states that for convex programs (C.2) with linear constrained functions h_i and g_i the duality gap is zero, and any $\mathbf{x}^*$ and $(\boldsymbol{\alpha}^*,\boldsymbol{\beta}^*)$ satisfying the KKT conditions are (global) solutions to the primal and dual programs, respectively. In particular, this holds for linear and convex quadratic programs (note that not all quadratic programs are convex).

For a convex primal program with $\mathcal{C}^1$ objective and constraint functions, the Lagrangian dual function (C.6) can be obtained by simply setting the gradient (with respect to $\mathbf{x}$) of the Lagrangian $\mathcal{L}(\mathbf{x},\boldsymbol{\alpha},\boldsymbol{\beta})$ to zero. One can further simplify the dual program by substituting into the Lagrangian the relations between the variables thus obtained.

Further, for a convex primal problem, if there is a **strictly feasible** point $\widetilde{\mathbf{x}}$ (that is, a feasible point satisfying all of the inequality constraints with strict inequality), then the duality gap is zero, and strong duality holds. This is known as **Slater's condition** [5, Page 226].

The Lagrange dual problem is an important example of a **saddle-point problem** or **minimax** problem. In such problems the aim is to locate a point $(\mathbf{x}^*, \mathbf{y}^*) \in \mathscr{X} \times \mathscr{Y}$ that satisfies

$$\sup_{\mathbf{y}\in\mathscr{Y}} \inf_{\mathbf{x}\in\mathscr{X}} f(\mathbf{x},\mathbf{y}) = \inf_{\mathbf{x}\in\mathscr{X}} f(\mathbf{x},\mathbf{y}^*) = f(\mathbf{x}^*,\mathbf{y}^*) = \sup_{\mathbf{y}\in\mathscr{Y}} f(\mathbf{x}^*,\mathbf{y}) = \inf_{\mathbf{x}\in\mathscr{X}} \sup_{\mathbf{y}\in\mathscr{Y}} f(\mathbf{x},\mathbf{y}) .$$

The equation

$$\sup_{\mathbf{y}\in\mathscr{Y}} \inf_{\mathbf{x}\in\mathscr{X}} f(\mathbf{x},\mathbf{y}) = \inf_{\mathbf{x}\in\mathscr{X}} \sup_{\mathbf{y}\in\mathscr{Y}} f(\mathbf{x},\mathbf{y})$$

is known as the **minimax** equality. Other problems that fall into this framework are zero-sum games in game theory; see also [6] for a number of combinatorial optimization problems that can be viewed as minimax problems.

C.2 TECHNIQUES FOR OPTIMIZATION

C.2.1 Transformation of Constrained Problems

A constrained optimization problem can sometimes be reformulated as a simpler unconstrained problem — for example, by preprocessing or transforming the variables. We discuss a number of techniques pertaining to minimization problems of the form (C.2).

C.2.1.1 Penalty Functions The overarching idea of penalty functions is to transform a constrained problem into an unconstrained problem by adding weighted constraint-violation terms to the original objective function, with the premise that the new problem has a solution that is identical or close to the original one and is easier to solve.

As an extreme example, one could replace a constrained problem with an unconstrained one with objective function

$$\widetilde{f}(\mathbf{x}) = \begin{cases} f(\mathbf{x}) & \text{if } \mathbf{x} \text{ is feasible}, \\ \infty & \text{otherwise}. \end{cases}$$

These types of transformations can change the nature of the problem — in the extreme example just given, $\widetilde{f}$ is not a differentiable function, regardless of the form of f. If $f(\mathbf{x})$ is well-defined outside the feasible region, then a simple constant-penalty function is

$$\widetilde{f}(\mathbf{x}) = \begin{cases} f(\mathbf{x}) & \text{if } \mathbf{x} \text{ is feasible}, \\ f(\mathbf{x}) + C & \text{otherwise}, \end{cases}$$

where C is some large constant. If there are only equality constraints, then

$$\widetilde{f}(\mathbf{x}) = f(\mathbf{x}) + \sum_{i=1}^{m} a_i \{h_i(\mathbf{x})\}^2$$

for some constants $a_1, \ldots, a_m > 0$ is an **exact penalty function**, in the sense that the minimizer $\widetilde{\mathbf{x}}^*$ of $\widetilde{f}$ is equal to the minimizer $\mathbf{x}^*$ of f subject to the m equality constraints $h_1, \ldots, h_m$. With the addition of inequality constraints, one could use

$$\widetilde{f}(\mathbf{x}) = f(\mathbf{x}) + \sum_{i=1}^{m} a_i \{h_i(\mathbf{x})\}^2 + \sum_{j=1}^{k} b_j \max\{g_j(\mathbf{x}), 0\}$$

for some constants $a_1, \ldots, a_m, b_1, \ldots, b_k > 0$.

Sequential penalty functions are penalty functions for which one obtains $\mathbf{x}^*$ as the limiting solution of a sequence of penalized problems. **Barrier functions** are an important example of such penalty functions. Suppose we have an inequality-only constrained problem. Then the **inverse barrier function** is of the form

$$\widetilde{f}(\mathbf{x}) = f(\mathbf{x}) - \alpha \sum_{j=1}^{k} \frac{1}{g_j(\mathbf{x})}, \quad \alpha > 0,$$

and the **logarithmic barrier function** has the form

$$\widetilde{f}(\mathbf{x}) = f(\mathbf{x}) - \alpha \sum_{j=1}^{k} \ln(-g_j(\mathbf{x})).$$

These have the property that solutions $\widetilde{\mathbf{x}}^*$ to the barrier problem tend to solutions $\mathbf{x}^*$ of the original problem as $\alpha \downarrow 0$.

Another penalty function of interest is

$$\widetilde{f}(\mathbf{x}) = \nu f(\mathbf{x}) + \sum_{i=1}^{m} |h_i(\mathbf{x})| + \sum_{j=1}^{k} \max\{g_j(\mathbf{x}), 0\},$$

where $\nu > 0$. This penalty function has the property that for appropriately chosen ν, local minimizers of $\widetilde{f}$ are local solutions to the constrained problem "to a large extent" [8].

C.2.1.2 Change of Variables Suppose in problem (C.2) the unconstrained set $\mathscr{Y}$ can be transformed to the feasible region $\mathscr{X}$ of the constrained problem via a function $\boldsymbol{\phi} : \mathbb{R}^n \to \mathbb{R}^n$ such that $\mathscr{X} = \boldsymbol{\phi}(\mathscr{Y})$. Then, (C.2) is equivalent to the minimization problem

$$\min_{\mathbf{y} \in \mathscr{Y}} f(\boldsymbol{\phi}(\mathbf{y})),$$

in the sense that a solution $\mathbf{x}^*$ of the original problem is obtained from a transformed solution $\mathbf{y}^*$ via $\mathbf{x}^* = \boldsymbol{\phi}(\mathbf{y}^*)$. Table C.2 lists some examples of possible transformations.

Table C.2 Some transformations to eliminate constraints.

Constrained	Unconstrained
$x > 0$	$\exp(y)$
$x \geqslant 0$	y^2
$a \leqslant x \leqslant b$	$a + (b - a)\sin^2(y)$

C.2.1.3 Transforming the Objective and Constraint Functions Instead of transforming the variable $\mathbf{x}$ in (C.2), one can try to transform the objective function: $\widetilde{f}(\mathbf{x}) = a(f(\mathbf{x}))$ for some real-valued function a. The same can be done for the constraint functions. In particular, if

1. $a : \mathbb{R} \to \mathbb{R}$ is a monotone increasing function,
2. $b_1, \ldots, b_m : \mathbb{R} \to \mathbb{R}$ are functions that satisfy $b_i(u) = 0 \iff u = 0$, and
3. $c_1, \ldots, c_k : \mathbb{R} \to \mathbb{R}$ satisfy $c_j(u) \leqslant 0 \iff u \leqslant 0$,

then (C.2) is equivalent to the constrained problem of the same form with objective function $\widetilde{f}(\mathbf{x}) = a(f(\mathbf{x}))$ and constraint functions $\widetilde{h}_i(\mathbf{x}) = b_i(h_i(\mathbf{x}))$ and $\widetilde{g}_j(\mathbf{x}) = c_j(g_j(\mathbf{x}))$, in the sense that any solution to the transformed problem is a solution to the original problem.

C.2.1.4 Slack Variables The inequality constraints in (C.2) can be modified by the introduction of k independent **slack variables** $y_1, \ldots, y_k$, replacing each of the k inequality constraints $g_i(\mathbf{x}) \leqslant 0$ by pairs

$$\begin{aligned} -y_i &\leqslant 0 \,, \\ g_i(\mathbf{x}) + y_i &= 0 \,. \end{aligned}$$

This gives a problem of dimension $n + k$, with $m + k$ equality constraints, and k simple inequality constraints. The resulting problem is equivalent to the original one, in the sense that any solution $(\widetilde{\mathbf{x}}^*, \widetilde{\mathbf{y}}^*)$ to the transformed problem is such that $\widetilde{\mathbf{x}}^*$ is also a solution to the original problem.

C.2.1.5 Eliminating Equality Constraints Suppose that $\mathscr{Y} = \mathbb{R}^n$ in (C.2) and that there is a function $\boldsymbol{\phi} : \mathbb{R}^l \to \mathbb{R}^n$ such that for any $\mathbf{x}$ satisfying

$$h_1(\mathbf{x}) = \cdots = h_m(\mathbf{x}) = 0 \,,$$

we have some $\mathbf{y}$ such that $\mathbf{x} = \boldsymbol{\phi}(\mathbf{y})$. This situation arises, in particular, if one can explicitly solve the m equality constraints in terms of l of the variables $\{x_i, i = 1, \ldots, n\}$. Then, we can consider a transformed problem (C.2) with $\widetilde{f}(\mathbf{y}) = f(\boldsymbol{\phi}(\mathbf{y}))$ and $\widetilde{g}_k(\mathbf{y}) = g_k(\boldsymbol{\phi}(\mathbf{y}))$. Any solution $\mathbf{y}^*$ to this problem can be transformed into a solution of the original problem by taking $\mathbf{x}^* = \boldsymbol{\phi}(\mathbf{y}^*)$.

It is also possible to replace each equality constraint $h_j(\mathbf{x}) = 0$ by a pair of inequality constraints $h_j(\mathbf{x}) \leqslant 0$ and $-h_j(\mathbf{x}) \leqslant 0$. This idea is mainly of theoretical interest as it can lead to poor practical performance.

C.2.2 Numerical Methods for Optimization and Root Finding

In order to minimize a $\mathcal{C}^1$ function $f : \mathbb{R}^n \to \mathbb{R}$ one may solve

$$\nabla f(\mathbf{x}) = \mathbf{0} \,,$$

which gives a stationary point of f. As a consequence, any technique for root-finding can be transformed into an unconstrained optimization method by attempting to locate roots of the gradient. However, as noted in Section C.1, not all stationary for any $\mathbf{u}_t \neq \mathbf{0}$. **Broyden's method** takes $\mathbf{u}_t = C_{t-1}^\top(\mathbf{x}_t - \mathbf{x}_{t-1})$. Similar to Newton's method, one usually takes a partial step size α_t at each iteration, giving

$$\mathbf{x}_{t+1} = \mathbf{x}_t - \alpha_t \, C_t \, \mathbf{f}(\mathbf{x}_t) \, ,$$

where α_t is resolved via a line search technique (see Section C.2.2.5).

In an optimization context, we simply replace $\mathbf{f}$ by ∇f, giving the iterative updating formula

$$\mathbf{x}_{t+1} = \mathbf{x}_t - \alpha_t \, C_t \nabla f(\mathbf{x}_t) \, .$$

In order to have a fast optimization method it is important that the matrix C_t is calculated quickly at each iteration t. Below, we list the most common formulas for C_t, which express C_t as a low-rank update to the previous matrix, C_{t-1}. We use the notation $\mathbf{y}_t = \mathbf{x}_t - \mathbf{x}_{t-1}$ and $\mathbf{g}_t = \nabla f(\mathbf{x}_t) - \nabla f(\mathbf{x}_{t-1})$.

1. The **Broyden–Fletcher–Goldfarb–Shanno** formula updates C_t as

$$C_t = C_{t-1} + \left(1 + \frac{\mathbf{g}_t^\top C_{t-1} \, \mathbf{g}_t}{\mathbf{y}_t^\top \, \mathbf{g}_t}\right) \frac{\mathbf{y}_t \, \mathbf{y}_t^\top}{\mathbf{y}_t^\top \, \mathbf{g}_t} - \left(\frac{\mathbf{y}_t \, \mathbf{g}_t^\top \, C_{t-1} + C_{t-1} \, \mathbf{g}_t \, \mathbf{y}_t^\top}{\mathbf{y}_t^\top \, \mathbf{g}_t}\right) .$$

   If C_{t-1} is symmetric then so is C_t, since the (rank 2) update matrix is symmetric.

2. The **(symmetric) rank one** formula is given by

$$C_t = C_{t-1} + \frac{(\mathbf{y}_t - C_{t-1} \, \mathbf{g}_t)(\mathbf{y}_t - C_{t-1} \, \mathbf{g}_t)^\top}{(\mathbf{y}_t - C_{t-1} \, \mathbf{g}_t)^\top \mathbf{g}_t} \, .$$

   This applies a rank 1 update to C_{t-1}, and if C_{t-1} is symmetric, then so is C_t. However, even if C_{t-1} is positive definite, C_t may not be. Further, the denominator $(\mathbf{y}_t - C_{t-1}\mathbf{g}_t)^\top \mathbf{g}_t$ may become small (introducing numerical difficulties) or vanish entirely.

3. The **Davidon–Fletcher–Powell** formula is given by

$$C_t = C_{t-1} + \frac{\mathbf{y}_t \, \mathbf{y}_t^\top}{\mathbf{y}_t^\top \, \mathbf{g}_t} - \frac{C_{t-1} \, \mathbf{g}_t \, \mathbf{g}_t^\top \, C_{t-1}}{\mathbf{g}_t^\top \, C_{t-1} \, \mathbf{g}_t} \, .$$

   This applies a rank 2 update to C_{t-1} to obtain C_t. If C_{t-1} is positive definite then so is C_t, whenever $\mathbf{y}_t^\top \mathbf{g}_t > 0$ (a curvature condition). Symmetry is also preserved.

In practice, the initial matrix C_0 is often set to the identity matrix I.

C.2.2.4 Projected Subgradient Method Consider the problem of minimizing a convex function f over a convex set $\mathscr{X} \subseteq \mathbb{R}^n$, where at each point $\mathbf{x}$ a subgradient of f, denoted $\mathbf{g}(\mathbf{x})$, is available. A **subgradient** $\mathbf{g}(\mathbf{x})$ is a vector that satisfies $f(\mathbf{y}) - f(\mathbf{x}) \geqslant \mathbf{g}(\mathbf{x})^\top(\mathbf{y} - \mathbf{x})$ for all $\mathbf{y} \in \mathscr{X}$.

Starting with a feasible $\mathbf{x}_1 \in \mathscr{X}$, the **projected subgradient method** produces a sequence of iterates via

$$\mathbf{x}_{t+1} = \Pi_{\mathscr{X}}(\mathbf{x}_t - \beta_t \, \mathbf{g}(\mathbf{x}_t)) \, , \qquad \text{(C.10)}$$

where $\{\beta_t\}$ is a sequence of strictly positive step sizes and $\Pi_{\mathscr{X}}$ is a **projection operator** from $\mathbb{R}^n$ to $\mathscr{X}$. Common choices for the step sizes are:

1. If the optimal objective function value f^* is known, then **Polyak's step size** (or a **dynamic step size**) is often used:

$$\beta_t = \frac{f(\mathbf{x}_t) - f^*}{\|\mathbf{g}(\mathbf{x}_t)\|^2} \,.$$

   If f^* is not known, then it is sometimes estimated on step t by $\widehat{f}_t^* = -\delta_t + \min_{s \leqslant t} f(\mathbf{x}_s)$, where $\delta_t \geqslant 0$.

2. **Constant step size**: $\beta_t = \beta$.

3. **Constant step length**: $\beta_t = \alpha/\|\mathbf{g}(\mathbf{x}_t)\|^2$. The constant step length can be viewed as a dynamically scaled constant step size, with scaling factor $\gamma_t = 1/\|\mathbf{g}(\mathbf{x}_t)\|^2$. Another popular scaling choice is $\gamma_t = 1/\max\{c, \|\mathbf{g}(\mathbf{x}_t)\|\}$, for some constant $c > 0$.

4. **Slowly decreasing step sizes**: Here $\lim_{t\to\infty} \beta_t = 0$ but at a slow enough rate such that $\sum_{t=1}^{\infty} \beta_t = \infty$.

5. **Slowly decreasing step lengths**: Here $\beta_t = \alpha_t/\|\mathbf{g}(\mathbf{x}_t)\|^2$, where $\alpha_t \geqslant 0$, $\lim_{t\to\infty} \alpha_t = 0$, and $\sum_{t=1}^{\infty} \alpha_t = \infty$.

If $f \in \mathcal{C}^1$, that is, f is differentiable, then the *only* subgradient is the (usual) gradient, and $\mathbf{g}(\mathbf{y}) = \nabla f(\mathbf{y})$. In this situation we recover the steepest descent method. When only a stochastic estimate of a subgradient $\mathbf{g}_t$ is available, we may use the stochastic approximation methods in Section 12.1. ☞ 441

C.2.2.5 Line Search Techniques A **line search problem** is a one-dimensional optimization problem of the following form. Given a function $f : \mathbb{R}^n \to \mathbb{R}$, a direction $\mathbf{d} \in \mathbb{R}^n$, and an initial point $\mathbf{x} \in \mathbb{R}^n$, solve

$$\min_{\alpha \geqslant 0} h(\alpha) \stackrel{\text{def}}{=} \min_{\alpha \geqslant 0} f(\mathbf{x} + \alpha\, \mathbf{d}) \,.$$

For many practical implementations of algorithms such as Newton's method or quasi Newton methods, an exact or approximate solution to a line search problem is desired.

A line search problem may be solved using, for example, Newton's method or bisection applied to the gradient of h. A simple scheme that does not use derivatives is **golden section search**: given a **bracket** of points $(\alpha_1, \alpha_2, \alpha_3)$ satisfying $h(\alpha_2) \leqslant \min\{h(\alpha_1), h(\alpha_3)\}$, $\alpha_1 < \alpha_2 < \alpha_3$, and the ratio condition

$$\frac{\alpha_3 - \alpha_2}{\alpha_2 - \alpha_1} = \frac{1 + \sqrt{5}}{2} \quad \text{or} \quad \frac{2}{1 + \sqrt{5}} \,,$$

a new point α_4 is chosen so that potential brackets $(\alpha_1, \alpha_2, \alpha_4)$ and $(\alpha_2, \alpha_4, \alpha_3)$ maintain the ratio condition. Whichever interval is smaller and still brackets a minimum is chosen as the next bracket of points.

When a line search problem is a subproblem of a larger optimization algorithm, for example when determining adaptive step sizes, then speed is paramount. For a

3. Compute a new center and generator via

$$\bar{\mathbf{x}}_{t+1} = \bar{\mathbf{x}}_t - \frac{1}{n+1}\frac{B_t\mathbf{g}_t}{\sqrt{\mathbf{g}_t^\top B_t\mathbf{g}_t}},$$

and

$$B_{t+1} = \frac{n^2}{n^2-1}\left(B_t - \frac{2}{n+1}\frac{(B_t\mathbf{g}_t)(B_t\mathbf{g}_t)^\top}{\mathbf{g}_t^\top B_t\mathbf{g}_t}\right),$$

defining a new ellipsoid

$$E_{t+1} = \{\mathbf{x} \in \mathbb{R}^n \,:\, (\mathbf{x} - \bar{\mathbf{x}}_{t+1})^\top B_{t+1}^{-1}(\mathbf{x} - \bar{\mathbf{x}}_{t+1}) \leqslant 1\}.$$

4. If a convergence criterion is met, stop; otherwise, set $t = t+1$ and repeat from Step 2.

C.3 SELECTED OPTIMIZATION PROBLEMS

In this section we list a number of potentially challenging discrete problems, together with a suite of commonly encountered (constrained and unconstrained) continuous test problems.

C.3.1 Satisfiability Problem

The (Boolean) satisfiability (SAT) problem plays a central role in combinatorial optimization and complexity theory. Any NP-complete problem, such as the max-cut problem, the graph-coloring problem, and the TSP can be translated *in polynomial time* into a SAT problem. We consider only the problem in *conjunctive normal form.*

In this case, the particular problem of interest is to find a binary vector $\mathbf{x} = (x_1, x_2, \ldots, x_n) \in \{0,1\}^n$ which, given a set of m true–false *clauses*, satisfies all of them, or else declare that the problem has no solution. Typically, the elements of $\mathbf{x}$ are interpreted as "true" or "false" according to whether they are 1 or 0. If we associate **clause functions** $c_1, \ldots, c_m$ with the m clauses, with the property that $c_j(\mathbf{x}) = 1$ when $\mathbf{x}$ satisfies clause j and $c_j(\mathbf{x}) = 0$ otherwise, then a SAT problem is of the form

$$\max_{\mathbf{x}\in\{0,1\}^n} \prod_{k=1}^{m} c_k(\mathbf{x}),$$

or equivalently

$$\max_{\mathbf{x}\in\{0,1\}^n} \sum_{k=1}^{m} c_k(\mathbf{x}).$$

☞ 490 For another formulation of the SAT problem using matrices, see Example 14.1.

C.3.2 Knapsack Problem

The basic problem is

$$\max_{\mathbf{x}\in\mathscr{X}} \mathbf{p}^\top\mathbf{x},$$

subject to the constraint

$$\mathbf{w}^\top \mathbf{x} \leqslant c ,$$

where both $\mathbf{p}$ and $\mathbf{w}$ are vectors of nonnegative numbers, and $\mathscr{X}$ is typically the set $\{0,1\}^n$, in which case this problem is called the **binary knapsack** problem.

The interpretation is that we have a collection of n items, with each item k having associated usefulness measured by p_k, and weight by w_k. The goal is to maximize the usefulness of items packed in the knapsack, subject to the condition that the total weight does not exceed c.

C.3.3 Max-Cut Problem

The max-cut problem can be formulated as follows. Given a weighted graph $\mathcal{G} = (V, E)$ with node set $V = \{1, 2, \ldots, n\}$ and edge set E, partition the nodes of the graph into two subsets V_1 and V_2 such that the sum of the weights (costs) of the edges between the two subsets are maximized. We assume that the costs are nonnegative (possibly 0), and are stored in a cost matrix C with each entry $C(i,j)$ as the cost (weight) of link (i,j).

More precisely, the problem is to solve

$$\max_{V_1, V_2} \sum_{(i,j) \in V_1 \times V_2} (C(i,j) + C(j,i)) .$$

C.3.4 Traveling Salesman Problem

The traveling salesman problem (TSP) can be formulated as follows. Consider a weighted graph $\mathcal{G}$ with n nodes, labeled $1, 2, \ldots, n$. The nodes represent cities and the edges represent the roads between the cities. Each edge from i to j has weight or cost $C(i,j)$, representing the length of the road. The problem is to find the shortest **tour** that visits all the cities exactly once with the exception of the starting city, which is also the terminating city. The problem is determined by the cost matrix C, where $C(i,j) = C(j,i)$ (possibly infinity).

More precisely, we wish to solve

$$\min_{\mathbf{x} \in \Pi} C(x_n, x_1) + \sum_{i=1}^{n-1} C(x_i, x_{i+1}) ,$$

where Π is the set of all permutations of $1, 2, \ldots, n$.

C.3.5 Quadratic Assignment Problem

The quadratic assignment problem has various applications such as computer chip design, optimal resource allocation, and scheduling. In the context of optimal allocation, the objective is to find an assignment of a set of n facilities to a set of n locations such that the total cost of the assignment is minimized. More precisely, the problem is to minimize the cost function

$$\min_{\mathbf{x} \in \Pi} \sum_{i=1}^{n} \sum_{j=1}^{n} F_{ij}\, D(x_i, x_j) ,$$

where Π is the set of all permutations of $1, 2, \ldots, n$ and F is an $n \times n$ matrix such that F_{ij} represents the **flow** of materials from facility i to facility j. The matrix D is such that $D(i,j)$ represents the **distance** between location i and location j. In the *symmetric* quadratic assignment problem both F and D are symmetric matrices.

C.3.6 Clustering Problem

The clustering problem reads as follows: given a dataset of points in a d-dimensional Euclidean space, partition the data into K **clusters** such that some empirical **loss function** is minimized. A typical loss function is the sum of the squared Euclidean distances between the points and their respective cluster centers. Let us denote the points as $\mathbf{y}_1, \ldots, \mathbf{y}_N$, the clusters as $\mathcal{C}_1, \ldots, \mathcal{C}_K$, and the corresponding **cluster centroids** as $\mathbf{c}_1, \ldots, \mathbf{c}_K$, determined from the nonempty clusters as

$$\mathbf{c}_i = \frac{1}{|\mathcal{C}_i|} \sum_{\mathbf{y} \in \mathcal{C}_i} \mathbf{y} \, .$$

Then the problem is

$$\min_{\mathcal{C}_1, \ldots, \mathcal{C}_K} \sum_{k=1}^{K} \sum_{\mathbf{y} \in \mathcal{C}_k} \|\mathbf{y} - \mathbf{c}_k\| \, .$$

C.4 CONTINUOUS PROBLEMS

C.4.1 Unconstrained Problems

C.4.1.1 Paviani's Function

$$f(\mathbf{x}) = \sum_{i=1}^{n} \left(\ln^2(x_i - 2) + \ln^2(10 - x_i) \right) - \left(\prod_{i=1}^{n} x_i \right)^{0.2} \, . \tag{C.11}$$

C.4.1.2 Rastrigin's Function

$$f(\mathbf{x}) = 10n + \sum_{i=1}^{n} (x_i^2 - 10\cos(2\pi x_i)) \, . \tag{C.12}$$

This has minimal value of $f(\mathbf{x}) = 0$ at $\mathbf{x} = (0, \ldots, 0)$.

C.4.1.3 Rosenbrock Function, Second de Jong's Function

$$f(\mathbf{x}) = \sum_{i=1}^{n-1} \left(100\,(x_{i+1} - x_i^2)^2 + (x_i - 1)^2\right) \, . \tag{C.13}$$

This has minimal value of $f(\mathbf{x}) = 0$ at $\mathbf{x} = (1, \ldots, 1)$.

C.4.1.4 Sphere Model, First de Jong's Function

$$f(\mathbf{x}) = \sum_{i=1}^{n} x_i^2 \, . \tag{C.14}$$

This has minimal value of $f(\mathbf{x}) = 0$ at $\mathbf{x} = (0, \ldots, 0)$.

C.4.1.5 Third de Jong's Function

$$f(\mathbf{x}) = \sum_{i=1}^{n} \lfloor x_i \rfloor \ . \tag{C.15}$$

This has minimal value of $f(\mathbf{x}) = 0$ at $\mathbf{x} = (0, \dots, 0)$.

C.4.1.6 Trigonometric Function

$$f(\mathbf{x}) = 1 + \sum_{i=1}^{n} 8\sin^2(\eta(x_i - x_i^*)^2) + 6\sin^2(2\eta(x_i - x_i^*)^2) + \mu(x_i - x_i^*)^2 \ . \tag{C.16}$$

This has minimal value of $f(\mathbf{x}) = 1$ at $\mathbf{x} = \mathbf{x}^*$.

C.4.2 Constrained Problems

C.4.2.1 Ackley's Function

$$f(\mathbf{x}) = \sum_{i=1}^{n-1} \left(20 + \mathrm{e} - 20\,\mathrm{e}^{-0.2\sqrt{0.5(x_{i+1}^2 + x_i^2)}} - \mathrm{e}^{0.5(\cos(2\pi x_{i+1}) + \cos(2\pi x_i))}\right) , \tag{C.17}$$

with $-30 \leqslant x_i \leqslant 30$. This has minimal value of $f(\mathbf{x}) = 0$ at $\mathbf{x} = (0, \dots, 0)$.

C.4.2.2 Ackley's Stretch V Sine Wave Function

$$f(\mathbf{x}) = \sum_{i=1}^{n-1} (x_{i+1}^2 + x_i^2)^{0.25} (\sin^2(50(x_{i+1}^2 + x_i^2)^{0.1}) + 1) \ , \quad -10 \leqslant x_i \leqslant 10 \ . \tag{C.18}$$

This has minimal value of $f(\mathbf{x}) = 0$ at $\mathbf{x} = (0, \dots, 0)$.

C.4.2.3 Egg Holder

$$f(\mathbf{x}) = \sum_{i=1}^{n-1} -(x_{i+1} + 47)\sin\left(\sqrt{\left|x_{i+1} + \frac{x_i}{2} + 47\right|}\right) - x_i \sin\left(\sqrt{|x_i - (x_{i+1} + 47)|}\right)$$
$$- 512 \leqslant x_i \leqslant 512 \ . \tag{C.19}$$

C.4.2.4 Griewangk's Function

$$f(\mathbf{x}) = -\prod_{i=1}^{n} \cos\left(\frac{x_i}{\sqrt{i}}\right) + \sum_{i=1}^{n} \frac{x_i^2}{4000} + 1 \ , \quad -600 \leqslant x_i \leqslant 600 \ . \tag{C.20}$$

This has minimal value of $f(\mathbf{x}) = 0$ at $\mathbf{x} = (0, \dots, 0)$.

C.4.2.5 Keane's Function

$$f(\mathbf{x}) = \left| \frac{\sum_{i=1}^{n} \cos^4(x_i) - 2\prod_{i=1}^{n} \cos^2(x_i)}{\sqrt{\sum_{i=1}^{n} i x_i^2}} \right| , \tag{C.21}$$

with constraints: $\prod_{i=1}^{n} x_i \geqslant 0.75$, $\sum_{i=1}^{n} x_i \leqslant 7.5\,n$, and $0 \leqslant x_i \leqslant 10$.

C.4.2.6 Master's Cosine Wave Function

$$f(\mathbf{x}) = -\sum_{i=1}^{n-1} e^{-\frac{1}{8}(x_{i+1}^2+0.5x_ix_{i+1}+x_i^2)} \cos\left(4\sqrt{x_{i+1}^2 + 0.5x_ix_{i+1} + x_i^2}\right), \quad \text{(C.22)}$$

with $-5 \leqslant x_i \leqslant 5$.

C.4.2.7 Michalewicz's Function

$$f(\mathbf{x}) = \sum_{i=1}^{n-1} \left(\sin(x_{i+1}) \sin^{20}\left(\frac{2x_{i+1}^2}{\pi}\right) + \sin(x_i) \sin^{20}\left(\frac{x_{i+1}^2}{\pi}\right)\right), \quad 0 \leqslant x_i \leqslant \pi. \quad \text{(C.23)}$$

C.4.2.8 Pathological Test Function

$$f(\mathbf{x}) = \sum_{i=1}^{n-1} \left(\frac{\sin^2\left(\sqrt{x_{i+1}^2 + 100x_i^2}\right) - 0.5}{0.001(x_{i+1}^2 - 2x_{i+1}x_i + x_i^2)^2 + 1.0} + 0.5\right), \quad -100 \leqslant x_i \leqslant 100. \quad \text{(C.24)}$$

C.4.2.9 Rana's Function

$$f(\mathbf{x}) = \sum_{i=1}^{n-1} \Big((x_{i+1} + 1) \cos\left(\sqrt{|x_{i+1} - x_i + 1|}\right) \sin\left(\sqrt{|x_{i+1} + x_i + 1|}\right)$$
$$+ x_i \cos\left(\sqrt{|x_{i+1} + x_i + 1|}\right) \sin\left(\sqrt{|x_{i+1} - x_i + 1|}\right)\Big), -500 \leqslant x_i \leqslant 500. \quad \text{(C.25)}$$

C.4.2.10 Schwefel's Function

$$f(\mathbf{x}) = \sum_{i=1}^{n} -x_i \sin(\sqrt{|x_i|}), \quad -512 \leqslant x_i \leqslant 512. \quad \text{(C.26)}$$

Within the constraints, this has minimal value of $f(\mathbf{x}) \approx -418.9829n$ at $\mathbf{x} \approx (420.9687, \ldots, 420.9687)$.

C.4.2.11 Sine Envelope Sine Wave Function

$$f(\mathbf{x}) = -\sum_{i=1}^{n-1} \left(\frac{\sin^2\left(\sqrt{x_{i+1}^2 + x_i^2} - 0.5\right)}{(0.001(x_{i+1}^2 + x_i^2) + 1)^2} + 0.5\right), \quad -100 \leqslant x_i \leqslant 100. \quad \text{(C.27)}$$

Further Reading

An introductory volume to the basic ideas and algorithms in optimization is [17]. A good general overview can be found in [1] and [12], with Fletcher as a classic reference [8]. For trust region algorithms, we refer to [7]. For convex optimization, as well as for practical optimization considerations, we refer to the comprehensive book of Boyd and Vandenberghe [5]. For an introduction on semidefinite programming, see [20], and see [4] for a tutorial on geometric programming. Finally, there are many techniques that we have not discussed. Among them are simplex methods (see, for example, [1, Chapter 11], or [2] for an analysis of the algorithm); interior point methods (see, for example, [1, Chapter 12]); the Nelder–Mead simplex method (see, for example, [16] for a clear description); and branch-and-bound algorithms [15].

REFERENCES

1. A. Antoniou and W.-S. Lu. *Practical Optimization: Algorithms and Engineering Applications.* Springer-Verlag, New York, 2007.
2. K. H. Borgwardt. *The Simplex Method: A Probabilistic Analysis.* Springer-Verlag, Berlin, 1987.
3. S. Boyd, P. Diaconis, and L. Xiao. Fastest mixing Markov chain on a graph. *SIAM Review*, 46(4):667–689, 2004.
4. S. Boyd, S.-J. Kim, L. Vandenberghe, and A. Hassibi. A tutorial on geometric programming. *Optimization and Engineering*, 8(1):67–127, 2007.
5. S. Boyd and L. Vandenberghe. *Convex Optimization.* Cambridge University Press, Cambridge, 2004.
6. F. Cao, D.-Z. Du, B. Gao, P.-J. Wan, and P. M. Pardalos. Minimax problems in combinatorial optimization. In D.-Z. Du and P. M. Pardalos, editors, *Minimax and Applications*, pages 269–292. Kluwer, Dordrecht, 1995.
7. A. R. Conn, N. I. M. Gould, and Ph. L. Toint. *Trust-Region Methods.* SIAM, Philadelphia, 2000.
8. R. Fletcher. *Practical Methods of Optimization.* John Wiley & Sons, New York, 1987.
9. G. H. Golub and C. F. Van Loan. *Matrix Computations.* Johns Hopkins University Press, Baltimore, third edition, 1996.
10. J.-B. Hiriart-Urruty and C. Lemarèchal. *Fundamentals of Convex Analysis.* Springer-Verlag, New York, 2001.
11. J. E. Kelley, Jr. The cutting-plane method for solving convex programs. *Journal of the Society for Industrial and Applied Mathematics*, 8(4):703–712, 1960.
12. H. Th. Jongen, K. Meer, and E. Triesch. *Optimization Theory.* Kluwer, Boston, 2004.
13. L. G. Khachiyan. A polynomial algorithm in linear programming. *Soviet Mathematics Doklady*, 20:191–194, 1979.
14. L. G. Khachiyan. Polynomial algorithms in linear programming. *USSR Computational Mathematics and Mathematical Physics*, 20:53–72, 1980.
15. E. L. Lawler and D. E. Wood. Branch-and-bound methods: A survey. *Operations Research*, 14(4):699–719, 1966.
16. K. I. M. McKinnon. Convergence of the Nelder-Mead simplex method to a nonstationary point. *SIAM Journal on Optimization*, 9(1):148–158, 1998.
17. A. Neumaier. *Introduction to Numerical Analysis.* Cambridge University Press, Cambridge, 2001.
18. N. Z. Shor. Cut-off method with space extension in convex programming problems. *Cybernetics*, 13:94–96, 1977.

19. N. Z. Shor. *Minimization Methods for Non-differentiable Functions.* Springer-Verlag, Berlin, 1985.

20. L. Vandenberghe and S. Boyd. Semidefinite programming. *SIAM Review*, 38(1):49–95, 1996.

21. Y. M. Wan. *Introduction to the Calculus of Variations and Its Applications.* Chapman & Hall, New York, 1995.

22. D. B. Yudin and A. S. Nemirovsky. Informational complexity and efficient methods for solving complex extremal problems. *Matekon*, 13:25–45, 1977.

APPENDIX D
MISCELLANY

D.1 EXPONENTIAL FAMILIES

Let $\mathbf{X} = (X_1, \ldots, X_n)^\top$ be an n-dimensional random vector with pdf $f(\cdot; \boldsymbol{\theta})$, where $\boldsymbol{\theta} = (\theta_1, \ldots, \theta_d)^\top$ is a d-dimensional parameter vector. $\mathbf{X}$ is said to belong to an m-dimensional **exponential family** if there exist $\mathbb{R}^m$-valued functions

$$\mathbf{t}(\mathbf{x}) = (t_1(\mathbf{x}), \ldots, t_m(\mathbf{x}))^\top, \quad \boldsymbol{\eta}(\boldsymbol{\theta}) = (\eta_1(\boldsymbol{\theta}), \ldots, \eta_m(\boldsymbol{\theta}))^\top, \quad h(\mathbf{x}) > 0 \,,$$

with $m \leqslant d$, and a normalizing function $c(\boldsymbol{\theta}) > 0$, such that

$$f(\mathbf{x}; \boldsymbol{\theta}) = c(\boldsymbol{\theta})\, \mathrm{e}^{\boldsymbol{\eta}(\boldsymbol{\theta})^\top \mathbf{t}(\mathbf{x})}\, h(\mathbf{x}) \,. \tag{D.1}$$

The representation of an exponential family is not unique in general. It is often convenient to reparameterize exponential families via the $\{\eta_i\}$; that is, to take the latter as the parameters rather than the $\{\theta_i\}$. The reparameterized pdf is then

$$\widetilde{f}(\mathbf{x}; \boldsymbol{\eta}) = \widetilde{c}(\boldsymbol{\eta})\, \mathrm{e}^{\boldsymbol{\eta}^\top \mathbf{t}(\mathbf{x})}\, h(\mathbf{x}) \,, \tag{D.2}$$

where $\widetilde{c}(\boldsymbol{\eta})$ is the normalization constant. Such an exponential family is said to be in **canonical form** [2, Page 52] or is said to be a **natural exponential family**. The parameter space for the **natural parameter vector** $\boldsymbol{\eta}$ is then usually chosen as large as possible; the **natural parameter space** consists of all $\boldsymbol{\eta}$ satisfying

$$\int \mathrm{e}^{\boldsymbol{\eta}^\top \mathbf{t}(\mathbf{x})}\, h(\mathbf{x})\, \mathrm{d}\mathbf{x} < \infty \,,$$

where the integral is replaced by a sum in the discrete case. It can be shown that the corresponding parameter space, E say, is *convex*. That is, if $\boldsymbol{\eta}_1$ and $\boldsymbol{\eta}_2$ are in E, then so is $\alpha\boldsymbol{\eta}_1 + (1-\alpha)\boldsymbol{\eta}_2$, for any $0 \leqslant \alpha \leqslant 1$.

Table D.1 displays the functions $c(\boldsymbol{\theta}), \eta_i(\boldsymbol{\theta}), t_i(x)$, and $h(x)$ for several commonly used univariate distributions. A dash means that the corresponding value is not used. We use x rather than $\mathbf{x}$ in the notation, because the variable x here is a scalar, not a vector. Below, B denotes the beta function; see Section D.9.1.

Table D.1 Various univariate exponential families.

Distr.	$\boldsymbol{\theta}$	$t_1(x),\ t_2(x)$	$c(\boldsymbol{\theta})$	$\eta_1(\boldsymbol{\theta}),\ \eta_2(\boldsymbol{\theta})$	$h(x)$
$\mathsf{Beta}(\alpha,\beta)$	(α,β)	$\ln x,\ \ln(1-x)$	$1/B(\alpha,\beta)$	$\alpha,\ \beta-1$	1
$\mathsf{Bin}(n,p)$	p	$x,\ -$	$(1-p)^n$	$\ln\left(\frac{p}{1-p}\right),\ -$	$\binom{n}{x}$
$\mathsf{Gamma}(\alpha,\lambda)$	(α,λ)	$x,\ \ln x$	$\frac{\lambda^\alpha}{\Gamma(\alpha)}$	$-\lambda,\ \alpha-1$	1
$\mathsf{Geom}(p)$	p	$x-1,\ -$	p	$\ln(1-p),\ -$	1
$\mathsf{N}(\mu,\sigma^2)$	(μ,σ^2)	$x,\ x^2$	$\frac{\mathrm{e}^{-\mu^2/(2\sigma^2)}}{\sigma\sqrt{2\pi}}$	$\frac{\mu}{\sigma^2},\ -\frac{1}{2\sigma^2}$	1
$\mathsf{NegBin}(r,p)$	p	$x,\ -$	p^r	$\ln(1-p),\ -$	$\frac{\Gamma(r+x)}{\Gamma(r)x!}$
$\mathsf{Poi}(\lambda)$	λ	$x,\ -$	$\mathrm{e}^{-\lambda}$	$\ln\lambda,\ -$	$\frac{1}{x!}$
$\mathsf{Wald}(\mu,\lambda)$	(λ,μ)	$x,\ 1/x$	$\mathrm{e}^{\lambda/\mu}\sqrt{\lambda}$	$-\lambda/(2\mu^2),\ -\lambda/2$	$1/\sqrt{2\pi x^3}$
$\mathsf{Weib}(\alpha,\lambda)$	(α,λ)	$\ln x,\ -x^\alpha$	$\alpha\lambda^\alpha$	$\alpha-1,\ \lambda^\alpha$	1

Examples of multivariate exponential families include:

1. *Dirichlet distribution*: The n-dimensional $\mathsf{Dirichlet}(\boldsymbol{\alpha})$ distribution, with $\boldsymbol{\alpha} = (\alpha_1,\ldots,\alpha_{n+1})^\top$ forms an $(n+1)$-dimensional exponential family: set $\boldsymbol{\theta} = \boldsymbol{\alpha}$ and $\boldsymbol{\eta}(\boldsymbol{\theta}) = \boldsymbol{\theta} - \mathbf{1}$; put $t_i(\mathbf{x}) = \ln x_i$ for $i = 1,\ldots,n$ and $t_{n+1}(\mathbf{x}) = \ln(1 - \sum_{i=1}^n x_i)$; finally, let $h(\mathbf{x}) = 1$ and
$$c(\boldsymbol{\theta}) = \frac{\Gamma\left(\sum_{i=1}^{n+1}\alpha_i\right)}{\prod_{i=1}^{n+1}\Gamma(\alpha_i)}\,.$$

2. *Multinomial distribution*: The $\mathsf{Mnom}(n,\mathbf{p})$ distribution, with parameter vector $\mathbf{p} = (p_1,\ldots,p_k)^\top$ forms a $(k-1)$-dimensional exponential family, with $\boldsymbol{\theta} = (p_1,\ldots,p_{k-1})^\top$, $t_i(\mathbf{x}) = x_i$, and $\eta_i(\boldsymbol{\theta}) = \ln(p_i/p_k)$, $i = 1,\ldots,k-1$, where $p_k = 1 - \sum_{i=1}^{k-1} p_i$. In addition, $c(\boldsymbol{\theta}) = p_k^n$ and $h(\mathbf{x}) = n!/\left(\prod_{i=1}^k x_i!\right)$, where $\mathbf{x} = (x_1,\ldots,x_{k-1})^\top$.

3. *Multivariate normal distribution*: For the n-dimensional multivariate normal case, put $\boldsymbol{\theta} = (\boldsymbol{\mu},\Sigma)$. Define $m = n(n+3)/2$ functions as follows: $t_{ij}(\mathbf{x}) =$

$x_i x_j$ for $i \leqslant j$ and $t_k(\mathbf{x}) = x_k$, $i, j, k = 1, \ldots, n$. The corresponding parameter functions are $\eta_{ij}(\boldsymbol{\theta}) = -\Sigma^{-1}_{ij}$ for $i < j$, $\eta_{ii}(\boldsymbol{\theta}) = -\frac{1}{2}\Sigma^{-1}_{ii}$, and

$$\eta_k(\boldsymbol{\theta}) = \Sigma^{-1}_{kk}\mu_k + 2\sum_{j<k}\Sigma^{-1}_{jk}\mu_j \,.$$

Finally, we have $h(\mathbf{x}) = 1$, and

$$c(\boldsymbol{\theta}) = \frac{\exp\left(-\frac{1}{2}\mathrm{tr}(\Sigma^{-1}\boldsymbol{\mu}\boldsymbol{\mu}^\top)\right)}{\sqrt{(2\pi)^n \det(\Sigma)}} \,,$$

where $\mathrm{tr}(M)$ denotes the trace of a matrix M, and $\det(M)$ the determinant of M.

For an exponential family in canonical form (D.2) the random vector $\mathbf{t}(\mathbf{X})$ is a *sufficient statistic* for $\boldsymbol{\eta}$, capturing all the information about $\boldsymbol{\eta}$ contained in the sample $\mathbf{X}$. Moreover, defining ☞ 655

$$A(\boldsymbol{\eta}) = -\ln \widetilde{c}(\boldsymbol{\eta}) = \ln \int \mathrm{e}^{\boldsymbol{\eta}^\top \mathbf{t}(\mathbf{x})}\, h(\mathbf{x})\, \mathrm{d}\mathbf{x} \,, \tag{D.3}$$

the function A provides convenient moment properties of $\mathbf{t}(\mathbf{X})$, as listed in Table D.2; see, for example, [2, Page 59].

Table D.2 Moment properties of exponential families. E is the natural parameter space.

Property		Condition
Expectation vector $\mathbb{E}\,\mathbf{t}(\mathbf{X})$	$\nabla A(\boldsymbol{\eta})$	$\boldsymbol{\eta} \in$ interior of E
Covariance matrix $\mathrm{Cov}(\mathbf{t}(\mathbf{X}))$	$\nabla^2 A(\boldsymbol{\eta})$	$\boldsymbol{\eta} \in$ interior of E
Moment generating function $\mathbb{E}\,\mathrm{e}^{\mathbf{s}^\top \mathbf{t}(\mathbf{X})}$	$\mathrm{e}^{A(\boldsymbol{\eta}+\mathbf{s})-A(\boldsymbol{\eta})}$	$\boldsymbol{\eta}$ and $\boldsymbol{\eta}+\mathbf{s} \in$ interior of E

D.2 PROPERTIES OF DISTRIBUTIONS

In this section, we list definitions of common and useful properties of random variables. The main reference is [9].

D.2.1 Tail Properties

A random variable X with cdf F is said to have a (right) **light-tailed** distribution if its moment generation function is finite for some $t > 0$. That is,

$$\mathbb{E}\,\mathrm{e}^{tX} \leqslant c < \infty \,. \tag{D.4}$$

Otherwise, that is, when $\mathbb{E}\,\mathrm{e}^{tX} = \infty$ for all $t > 0$, X is said to have a **heavy-tailed** distribution.

Note that we consider only the right tail of the distribution, as the left tail is treated analogously.

Since, for every x, $\mathbb{E}\,\mathrm{e}^{tX} \geqslant \mathbb{E}\,\mathrm{e}^{tX}\,\mathrm{I}_{\{X>x\}} \geqslant \mathrm{e}^{tx}\,\mathbb{P}(X > x)$, it follows that for any $t > 0$ and c satisfying (D.4):

$$\mathbb{P}(X > x) \leqslant c\,\mathrm{e}^{-tx} \,.$$

In other words, if X has a light tail, then $\bar{F}(x) = 1 - F(x)$ decays at an exponential rate or faster. Similarly, heavy-tailedness is equivalent to $\lim_{x\to\infty} \mathrm{e}^{tx}\bar{F}(x) = \infty$ for all $t > 0$.

Any distribution with bounded support is light-tailed. Examples of light-tailed distributions with unbounded support are:

$\mathsf{Exp}(\lambda)$	$\mathsf{Geom}(p)$	$\mathsf{Gamma}(\alpha, \lambda)$	$\mathsf{Gumbel}(\mu, \sigma)$
$\mathsf{Laplace}(\mu, \sigma)$	$\mathsf{Logistic}(\mu, \sigma)$	$\mathsf{N}(\mu, \sigma^2)$	$\mathsf{PH}(\boldsymbol{\alpha}, A)$
$\mathsf{Poi}(\lambda)$	$\mathsf{Wald}(\mu, \sigma)$	$\mathsf{Weib}(\alpha, \lambda), \alpha \geqslant 1$	

Each of the following properties imply heavy-tailedness, in the following order of generality:

Regularly varying $\Rightarrow$ Subexponential $\Rightarrow$ Long-tailed $\Rightarrow$ Heavy-tailed

1. A distribution is said to be **long-tailed** if

$$\lim_{x\to\infty} \frac{\bar{F}(x+t)}{\bar{F}(x)} = 1 \quad \text{for all } t \,.$$

2. A distribution on the interval $(0, \infty)$ is said to be **subexponential** if, with $X_1, \ldots, X_n \sim_{\text{iid}} F$,

$$\lim_{x\to\infty} \frac{\mathbb{P}(X_1 + \cdots + X_n > x)}{\mathbb{P}(X_1 > x)} = n \quad \text{for all } n \,, \qquad \text{(D.5)}$$

or equivalently

$$\lim_{x\to\infty} \frac{\mathbb{P}(X_1 + \cdots + X_n > x)}{\mathbb{P}(\max\{X_1, \ldots, X_n\} > x)} = 1 \,.$$

3. A distribution is called **regularly varying** if

$$\bar{F}(x) = \frac{L(x)}{x^{\alpha}}$$

for some $\alpha > 0$ and some function L that satisfies $L(tx)/L(x) \to 1$ as $x \to \infty$, for all $t > 0$.

Examples of regularly varying distributions are:

$\mathsf{Cauchy}(\mu, \sigma)$	$\mathsf{F}(m, n)$	$\mathsf{Fréchet}(\alpha, \mu, \sigma)$	$\mathsf{Pareto}(\alpha, \lambda)$
$\mathsf{t}_{\nu}(\mu, \sigma^2)$			

See [8] for additional properties of this class of distributions. Two families of distributions that are subexponential but not regularly varying are $\mathsf{LogN}(\mu, \sigma^2)$ and $\mathsf{Weib}(\alpha, \lambda), \alpha < 1$.

D.2.2 Stability Properties

A distribution S is called **sum-stable**, or **(weakly) stable** if the distribution of the sum of $X_1, X_2 \sim_{\text{iid}} \mathsf{S}$ is again S, up to a location and scale parameter — specifically, if for some $a > 0$ and $b \in \mathbb{R}$,

$$\frac{X_1 + X_2 - b}{a} \sim \mathsf{S}\,.$$

If $b = 0$, then S is said to be **strictly stable**. The family of all continuous sum-stable distributions is $\mathsf{Stable}(\alpha, \beta, \mu, \sigma)$. An example of a discrete stable distribution is the $\mathsf{Poi}(\lambda)$ distribution. Each of the following properties are implied by sum-stability, in the following order of generality:

Sum-stable $\Rightarrow$ Infinitely Divisible $\Rightarrow$ n-Divisible $\Rightarrow$ n-Decomposable

1. A random variable X is said to be ***n*-decomposable** if there exist n independent random variables $Y_1, \ldots, Y_n$ such that $X \sim \sum_{k=1}^{n} Y_k$.
2. A random variable X is said to be ***n*-divisible** if there exist n iid random variables $Y_1, \ldots, Y_n$ such that $X \sim \sum_{k=1}^{n} Y_k$.
3. If X is n-divisible for every n, then X is called **infinitely divisible**.

Examples of infinitely divisible distributions are:

$\mathsf{Cauchy}(\mu, \sigma)$	$\mathsf{Exp}(\lambda)$	$\mathsf{F}(m, n)$	$\mathsf{Gamma}(\alpha, \lambda)$
$\mathsf{Geom}(p)$	$\mathsf{Gumbel}(\mu, \sigma)$	$\mathsf{Laplace}(\mu, \sigma)$	$\mathsf{Logistic}(\mu, \sigma)$
$\mathsf{LogN}(\mu, \sigma^2)$	$\mathsf{NegBin}(r, p)$	$\mathsf{N}(\mu, \sigma^2)$	$\mathsf{Pareto}(\alpha, \lambda)$
$\mathsf{Poi}(\lambda)$	$\mathsf{Stable}(\alpha, \beta, \mu, \sigma)$	$\mathsf{t}_\nu(\mu, \sigma^2)$	$\mathsf{Wald}(\mu, \lambda)$
$\mathsf{Weib}(\alpha, \lambda)$, $\alpha \leqslant 1$			

Theorem D.2.1 (Lévy–Khinchin) *A random variable X has an infinitely divisible distribution on $\mathbb{R}$ if and only if its characteristic function, $\mathbb{E}\,\mathrm{e}^{\mathrm{i}sX}, s \in \mathbb{R}$, is of the form $\mathrm{e}^{\psi(s)}$, with*

$$\psi(s) = -\frac{\sigma^2 s^2}{2} + \mathrm{i}\gamma s + \int \left(\mathrm{e}^{\mathrm{i}sx} - 1 - \mathrm{i}sx\,\mathrm{I}_{\{|x| \leqslant 1\}}\right) \nu(\mathrm{d}x), \quad s \in \mathbb{R}\,,$$

for some $\sigma \geqslant 0$, $\gamma \in \mathbb{R}$, and measure $\nu(\mathrm{d}x)$ on $\mathbb{R}$ such that $\int \min\{1, x^2\}\, \nu(\mathrm{d}x) < \infty$.

A distribution S is called **max-stable** if the distribution of the maximum of $X_1, X_2 \sim_{\text{iid}} \mathsf{S}$ is again S, up to a location and scale parameter — specifically, if for some $a > 0$ and $b \in \mathbb{R}$,

$$\frac{\max\{X_1, X_2\} - b}{a} \sim \mathsf{S}\,.$$

If $b = 0$, then S is said to be **strictly max-stable**. The cdf G of a max-stable random variable satisfies the **stability postulate**: for each $t > 0$, there exist $a_t > 0$ and $b_t \in \mathbb{R}$ such that [11, Page 276]

$$G(x) = [G(a_t x + b_t)]^t\,.$$

A random variable X with cdf F is said to belong to the **domain of attraction** of (or is attracted to) a max-stable law with cdf G if there exist $a_n > 0$ and $b_n \in \mathbb{R}$ such that

$$\lim_{n\to\infty} [F(a_n x + b_n)]^n = G(x)\,.$$

Note that the cdf in the limit is the cdf of the variable $(M_n - b_n)/a_n$, where $M_n = \max\{X_1, \ldots, X_n\}$ for $n \in \mathbb{N}$.

There are only three families of continuous max-stable distributions:

1. $\mathsf{Gumbel}(\mu, \sigma)$: Examples of distributions that are attracted to a Gumbel law are:

$\mathsf{Exp}(\lambda)$	$\mathsf{Gamma}(\alpha, \lambda)$	$\mathsf{Gumbel}(\mu, \sigma)$	$\mathsf{Logistic}(\mu, \sigma)$	$\mathsf{N}(\mu, \sigma^2)$

2. $\mathsf{Fréchet}(\alpha, \mu, \sigma)$: Any regularly varying distribution with index α is attracted to a Fréchet law with the same parameter α.

3. $-\mathsf{Weib}(\alpha, \mu, \sigma)$ distributions (**reversed Weibull**): Examples of distributions that are attracted to a reversed Weibull law are the $\mathsf{U}[a, b]$ and the $\mathsf{Beta}(\alpha, \beta)$ distributions.

Note that not all distributions are attracted to one of these three laws. For example, the $\mathsf{Poi}(\lambda)$ distribution is attracted to none of these.

A distribution S is called **min-stable** if, for some $a > 0$ and $b \in \mathbb{R}$

$$\frac{\min\{X_1, X_2\} - b}{a} \sim \mathsf{S}\,, \quad X_1, X_2 \overset{\text{iid}}{\sim} \mathsf{S}\,.$$

If $b = 0$, then S is said to be **strictly min-stable**.

Again, there are only three families of continuous min-stable distributions, being the distributions of $(-X)$, where X is max-stable [7].

D.3 CHOLESKY FACTORIZATION

Any $n \times n$ covariance matrix $\Sigma = (\sigma_{ij})$ can be factorized as $\Sigma = CC^\top$, where $C = (c_{ij})$ is a lower triangular $n \times n$ matrix that is recursively defined as

$$c_{ij} = \frac{\sigma_{ij} - \sum_{k=1}^{j-1} c_{ik} c_{jk}}{\left(\sigma_{jj} - \sum_{k=1}^{j-1} c_{jk}^2\right)^{1/2}}\,, \quad \sum_{k=1}^{0} c_{ik} c_{jk} \overset{\text{def}}{=} 0, \quad 1 \leqslant j \leqslant i \leqslant n\,. \tag{D.6}$$

D.4 DISCRETE FOURIER TRANSFORM, FFT, AND CIRCULANT MATRICES

The **discrete Fourier transform** of a vector $\mathbf{x} = (x_0, \ldots, x_{N-1})^\top$ of complex numbers is the vector $\widetilde{\mathbf{x}} = (\widetilde{x}_0, \ldots, \widetilde{x}_{N-1})^\top$ defined as

$$\widetilde{x}_t = \sum_{s=0}^{N-1} \mathrm{e}^{-\frac{2\pi \mathrm{i}}{N} st}\, x_s = \sum_{s=0}^{N-1} \omega^{st}\, x_s, \quad t = 0, \ldots, N-1\,, \tag{D.7}$$

where $\omega = \exp(-2\pi i/N)$. In other words, $\widetilde{\mathbf{x}}$ is obtained from $\mathbf{x}$ via the linear transformation

$$\widetilde{\mathbf{x}} = F\mathbf{x} ,$$

where

$$F = \begin{pmatrix} 1 & 1 & 1 & \dots & 1 \\ 1 & \omega & \omega^2 & \dots & \omega^{N-1} \\ 1 & \omega^2 & \omega^4 & \dots & \omega^{2(N-1)} \\ \vdots & \vdots & \vdots & \ddots & \vdots \\ 1 & \omega^{N-1} & \omega^{2(N-1)} & \dots & \omega^{(N-1)^2} \end{pmatrix} .$$

Note that $F/\sqrt{N}$ is a unitary matrix, and hence the inverse of $F/\sqrt{N}$ is simply its complex conjugate $\overline{F}/\sqrt{N}$. It follows that $F^{-1} = \overline{F}/N$ and thus

$$x_t = \frac{1}{N} \sum_{s=0}^{N-1} \omega^{-st} \, \widetilde{x}_s, \quad t = 0, \dots, N-1 . \tag{D.8}$$

The **fast Fourier transform** (FFT) is a numerical algorithm for the fast evaluation of (D.7) and (D.8). By using a divide-and-conquer strategy, the algorithm reduces the computational complexity from $\mathcal{O}(N^2)$ (for naive evaluation of the linear transform) to $\mathcal{O}(N \ln N)$ [10].

One area where FFT techniques are useful is in computations involving circulant matrices. Let $\mathbf{c} = (c_0, c_1, \dots, c_{N-1})^\top$ be a complex-valued vector. The **circulant matrix** corresponding to $\mathbf{c}$ is the matrix $C = (c_{ij},\ i, j \in \{0, \dots, N-1\})$ with elements $c_{ij} = c_{(i-j) \bmod N}$. That is,

$$C = \begin{pmatrix} c_0 & c_{N-1} & \dots & c_2 & c_1 \\ c_1 & c_0 & c_{N-1} & & c_2 \\ \vdots & c_1 & c_0 & \ddots & \vdots \\ c_{N-2} & & \ddots & \ddots & c_{N-1} \\ c_{N-1} & c_{N-2} & \dots & c_1 & c_0 \end{pmatrix} , \tag{D.9}$$

where the columns are obtained from the vector $\mathbf{c} = (c_0, c_1, \dots, c_{N-1})^\top$ by circularly permuting the indices.

Let $\mathbf{f}_t$ be the t-th column of the discrete Fourier matrix F, $t = 0, 1, \dots, N-1$. The fundamental relation between the discrete Fourier transform and a circulant matrix C is that the eigenvalues of C are

$$\lambda_t = \mathbf{c}^\top \overline{\mathbf{f}}_t, \quad t = 0, 1, \dots, N-1 ,$$

with corresponding eigenvectors $\mathbf{f}_t$. Namely, the s-th element of $C\mathbf{f}_t$ is

$$\sum_{k=0}^{N-1} c_{(s-k) \bmod N}\, \omega^{tk} = \sum_{y=0}^{N-1} c_y\, \omega^{t(s-y)} = \underbrace{\omega^{ts}}_{s\text{-th element of } \mathbf{f}_t} \underbrace{\sum_{y=0}^{N-1} c_y\, \omega^{-ty}}_{\lambda_s} .$$

As a consequence, if we define $\boldsymbol{\lambda} = (\lambda_0, \dots, \lambda_{N-1})^\top = \bar{F}\mathbf{c}$ as the vector of eigenvalues, and $D = F\sqrt{\mathrm{diag}(\boldsymbol{\lambda}/N)}$, then $D\overline{D}^\top = F\mathrm{diag}(\boldsymbol{\lambda})\overline{F}/N = C$. Hence, D is a complex square root matrix of C.

If, moreover, C is the covariance matrix of a zero-mean Gaussian vector then realizations of this vector can be obtained in the following way. Let $D = D_1 + \mathrm{i}D_2$, where D_1 and D_2 are real-valued matrices. By definition, $C = D\overline{D}^{\top} = (D_1 + \mathrm{i}D_2)(D_1^{\top} - \mathrm{i}D_2^{\top}) = (D_1D_1^{\top} + D_2D_2^{\top}) + \mathrm{i}(D_2D_1^{\top} - D_1D_2^{\top})$. In particular, $D_1D_1^{\top} + D_2D_2^{\top} = C$. Let $\mathbf{Z} = \mathbf{Z}_1 + \mathrm{i}\mathbf{Z}_2$, where $\mathbf{Z}_1$ and $\mathbf{Z}_2$ are independent n-dimensional standard Gaussian vectors, and define $\mathbf{X} = D\mathbf{Z} = \mathbf{X}_1 + \mathrm{i}\mathbf{X}_2$, where

$$\mathbf{X}_1 = \Re(D\mathbf{Z}) = D_1\mathbf{Z}_1 - D_2\mathbf{Z}_2 \,,$$
$$\mathbf{X}_2 = \Im(D\mathbf{Z}) = D_2\mathbf{Z}_1 + D_1\mathbf{Z}_2 \,.$$

Then, $\mathbf{X}_1$ and $\mathbf{X}_2$ are dependent zero-mean Gaussian vectors, with covariance matrix C. By using the FFT one can calculate both $\boldsymbol{\lambda} = \bar{F}\mathbf{c}$ and $\mathbf{X} = F\sqrt{\mathrm{diag}(\boldsymbol{\lambda}/N)}\mathbf{Z}$ with $\mathcal{O}(N \ln N)$ complexity.

D.5 DISCRETE COSINE TRANSFORM

The discrete cosine transform is important in signal processing due to its property of compressing highly correlated data very well [10, Pages 151–153].

Let $x_0, \ldots, x_{N-1}$ be a sequence of real numbers. The one-dimensional **discrete cosine transform** is defined as the sequence

$$y_k = \alpha_k \sum_{n=0}^{N-1} x_n \cos\left(\frac{\pi(2n+1)k}{2N}\right), \quad 0 \leqslant k \leqslant N-1 \,,$$

where $\alpha_0 = 1$ and $\alpha_k = 2$, $k \geqslant 1$. The **inverse discrete cosine transform** gives back the original sequence:

$$x_n = \frac{1}{N} \sum_{k=0}^{N-1} y_k \cos\left(\frac{\pi(2n+1)k}{2N}\right), \quad 0 \leqslant n \leqslant N-1 \,.$$

Note that the discrete cosine transform is not the real part of the discrete Fourier transform. Nevertheless, the FFT can be used to compute the discrete cosine transform in $\mathcal{O}(N \ln N)$ operations as follows. For simplicity, assume that N is an even integer. For fastest performance one has to use $N = 2^m$ for some integer m.

Algorithm D.1 (Fast Cosine Transform) *To compute the discrete cosine transform of the sequence $x_0, \ldots, x_{N-1}$ execute the following steps.*

1. *Define the reordered sequence $\widetilde{x}_0, \ldots, \widetilde{x}_{N-1}$:*

$$(x_0, x_2, x_4, \ldots, x_{N-6}, x_{N-4}, x_{N-2}, x_{N-1}, x_{N-3}, x_{N-5}, \ldots, x_5, x_3, x_1),$$

so that

$$\widetilde{x}_n = x_{2n} \,,$$
$$\widetilde{x}_{N-n-1} = x_{2n+1} \,,$$

where n is such that $0 \leqslant n \leqslant \frac{N}{2} - 1$.

2. *Compute the discrete Fourier transform (recall that* $\omega = \exp(-2\pi \mathrm{i}/N)$*):*

$$z_k = \sum_{n=0}^{N-1} \widetilde{x}_n \, \omega^{kn}, \quad 0 \leqslant k \leqslant N-1\,,$$

of the sequence $\widetilde{x}_0, \ldots, \widetilde{x}_{N-1}$ *using the FFT.*

3. *Output the real part of the sequence* $\{\alpha_k \, z_k \, \mathrm{e}^{-\mathrm{i}\pi k/(2N)}\}$*:*

$$y_k = \Re\left[\alpha_k \, z_k \, \mathrm{e}^{-\mathrm{i}\pi k/(2N)}\right], \quad 0 \leqslant k \leqslant N-1$$

as the discrete cosine transform of $x_0, \ldots, x_{N-1}$.

The discrete cosine transform algorithm above is implemented in the function `dct1d.m` on Page 324. The inverse discrete cosine transform is computed using the following algorithm.

Algorithm D.2 (Fast Inverse Cosine Transform) *To compute the inverse discrete cosine transform of the sequence* $y_0, \ldots, y_{N-1}$ *execute the following steps.*

1. *Compute the inverse discrete Fourier transform*

$$z_n = \frac{1}{N} \sum_{k=0}^{N-1} \left[y_k \, \mathrm{e}^{\mathrm{i}\pi k/(2N)}\right] \omega^{-kn}, \quad 0 \leqslant n \leqslant N-1\,,$$

of the sequence $\{y_k \, \mathrm{e}^{\mathrm{i}\pi k/(2N)}\}$ *using the inverse FFT.*

2. *Output the reordered sequence* $x_1, \ldots, x_{N-1}$ *of real numbers:*

$$\begin{aligned} x_{2n} &= \Re[z_n], \\ x_{2n+1} &= \Re[z_{2(N-n-1)}], \end{aligned}$$

where $0 \leqslant n \leqslant \frac{N}{2} - 1$*, as the inverse discrete cosine transform of the sequence* $y_1, \ldots, y_{N-1}$.

The inverse discrete cosine transform algorithm above is implemented in the function `idct1d.m` on Page 324.

D.6 DIFFERENTIATION

Let $\mathscr{A}$ and $\mathscr{B}$ be sets of real numbers. The **derivative** of a function $f : \mathscr{A} \to \mathscr{B}$ at a is defined as the limit

$$f'(a) = \frac{\mathrm{d}}{\mathrm{d}x} f(a) = \lim_{x \to a} \frac{f(x) - f(a)}{x - a}\,,$$

provided the limit exists — independent of the direction from which x approaches a. If the derivative of f exists for all x in $\mathscr{A}$, then f is said to be **differentiable** on $\mathscr{A}$, with **derivative function** (or simply **derivative**) f'. The **second-order derivative** f'', or $f^{(2)}$, is the derivative of the derivative f'. Higher-order derivatives $f^{(3)}, f^{(4)}, \ldots$ are defined similarly. A function is said to be **continuously**

differentiable if its derivative is a continuous function. The collection of such functions is denoted by $\mathcal{C}^1$. Similarly, $\mathcal{C}^k$ is the collection of functions whose k-th derivative is continuous. Some standard rules for differentiation are:

1. *Sum rule:* $(f+g)' = f' + g'$.
2. *Product rule:* $(fg)' = f'g + fg'$.
3. *Quotient rule:* $\left(\frac{f}{g}\right)' = \frac{f'g - fg'}{g^2}, \quad (g \neq 0)$.
4. *Chain rule:* $(g(f(x)))' = g'(f(x))\, f'(x)$.
5. *Monomial rule:* $\frac{\mathrm{d}}{\mathrm{d}x} x^n = n\, x^{n-1}$.

One of the most useful concepts in calculus is that of the Taylor expansion of a function, which states, loosely, that each differentiable function can be described locally as a polynomial function.

Theorem D.6.1 (Taylor Expansion) *Let f have a continuous $(n+1)$-st derivative on the open interval $\mathscr{I} = (a-r, a+r)$, and let $a \in \mathscr{I}$. Then, for every $x \in \mathscr{I}$,*

$$f(x) = \sum_{k=0}^{n} \frac{f^{(k)}(a)\,(x-a)^k}{k!} + \mathcal{O}((x-a)^{n+1}) \quad as \ x \to a\,. \tag{D.10}$$

By dropping the remainder term in the right-hand side of (D.10), one obtains the n-th order **Taylor approximation** to the function f around a.

Differentiation for multivariate functions can be defined similarly to the univariate case. In a function $f(x_1, \ldots, x_n)$ of several variables, the **partial derivative** with respect to x_i, denoted $\frac{\partial f}{\partial x_i}$ or simply $\partial_i f$, is the derivative taken with respect to x_i while all other variables are held constant. Partial derivatives of partial derivatives are denoted using a similar notation. For example, the partial derivative of $\partial_i f$ with respect to x_j is denoted $\frac{\partial^2 f}{\partial x_i \partial x_j}$ or simply $\partial_{ij} f$.

The derivative of a multivariate function is defined in a similar way. In particular, let $\mathbf{f}$ be a function from $\mathbb{R}^n \to \mathbb{R}^m$ defined by

$$\begin{pmatrix} x_1 \\ x_2 \\ \vdots \\ x_n \end{pmatrix} \mapsto \begin{pmatrix} f_1(\mathbf{x}) \\ f_2(\mathbf{x}) \\ \vdots \\ f_m(\mathbf{x}) \end{pmatrix}.$$

The **derivative** of $\mathbf{f}$ at $\mathbf{x}$ is defined as the matrix of partial derivatives:

$$J_{\mathbf{f}}(\mathbf{x}) = \begin{pmatrix} \partial_1 f_1(\mathbf{x}) & \cdots & \partial_n f_1(\mathbf{x}) \\ \vdots & \ldots & \vdots \\ \partial_1 f_m(\mathbf{x}) & \cdots & \partial_n f_m(\mathbf{x}) \end{pmatrix}, \tag{D.11}$$

and is called the **Jacobi matrix** of $\mathbf{f}$ at $\mathbf{x}$. It is denoted by $J_{\mathbf{f}}(\mathbf{x})$ or also $\frac{\partial \mathbf{f}}{\partial \mathbf{x}}(\mathbf{x})$.

For a real-valued multivariate function, that is, $f : \mathbb{R}^n \to \mathbb{R}$, the **gradient** of f is the transpose of the Jacobian matrix, that is, the *column* vector

$$\nabla f(\mathbf{x}) = \begin{pmatrix} \partial_1 f(\mathbf{x}) \\ \vdots \\ \partial_n f(\mathbf{x}) \end{pmatrix}. \tag{D.12}$$

The derivative of the function $\mathbf{x} \mapsto \nabla f(\mathbf{x})$ is called the **Hessian matrix** of f, denoted $H_f(\mathbf{x})$ or $\nabla^2 f(\mathbf{x})$. In other words, the Hessian is the matrix of second derivatives:

$$\nabla^2 f(\mathbf{x}) = \begin{pmatrix} \partial_{11} f(\mathbf{x}) & \cdots & \partial_{1n} f(\mathbf{x}) \\ \vdots & \cdots & \vdots \\ \partial_{n1} f(\mathbf{x}) & \cdots & \partial_{nn} f(\mathbf{x}) \end{pmatrix} . \tag{D.13}$$

If the partial derivatives are *continuous* in a region around $\mathbf{x}$, then $\partial_{ij} f(\mathbf{x}) = \partial_{ji} f(\mathbf{x})$ and, hence, the Hessian matrix $H_f(\mathbf{x})$ is *symmetric*.

The chain rule in the multidimensional case is as follows. If $\mathbf{f} : \mathbb{R}^n \to \mathbb{R}^m$ and $\mathbf{g} : \mathbb{R}^m \to \mathbb{R}^k$, then the Jacobian matrix of the **composition** $\mathbf{g} \circ \mathbf{f}$ — which maps $\mathbf{x}$ to $\mathbf{g}(\mathbf{f}(\mathbf{x}))$ — is given by the matrix product of the Jacobian matrices of $\mathbf{g}$ and $\mathbf{f}$:

$$J_{\mathbf{g} \circ \mathbf{f}}(\mathbf{x}) = J_{\mathbf{g}}(\mathbf{f}(\mathbf{x}))\, J_{\mathbf{f}}(\mathbf{x}) .$$

Gradients and Hessian matrices feature prominently in multidimensional Taylor expansions.

Theorem D.6.2 (Multidimensional Taylor Expansions) *Let $\mathscr{X}$ be an open subset of $\mathbb{R}^n$ and let $\mathbf{a} \in \mathscr{X}$. If $f : \mathscr{X} \to \mathbb{R}$ is a $\mathcal{C}^2$ function with gradient $\nabla f(\mathbf{x})$ and Hessian matrix $H_f(\mathbf{x})$, then for every $\mathbf{x} \in \mathscr{X}$, we have*

$$f(\mathbf{x}) = f(\mathbf{a}) + [\nabla f(\mathbf{a})]^\top (\mathbf{x} - \mathbf{a}) + \mathcal{O}(\|\mathbf{x} - \mathbf{a}\|^2) \quad \text{as } \|\mathbf{x} - \mathbf{a}\| \to 0 ,$$

and

$$f(\mathbf{x}) = f(\mathbf{a}) + [\nabla f(\mathbf{a})]^\top (\mathbf{x} - \mathbf{a}) + \frac{1}{2}(\mathbf{x} - \mathbf{a})^\top H_f(\mathbf{a})\,(\mathbf{x} - \mathbf{a}) + \mathcal{O}(\|\mathbf{x} - \mathbf{a}\|^3) .$$

By dropping the $\mathcal{O}$ remainder terms, one obtains the corresponding Taylor approximations.

D.7 EXPECTATION-MAXIMIZATION (EM) ALGORITHM

The **expectation-maximization** algorithm (EM) is a general algorithm for maximization of multimodal likelihood functions, or equivalently for finding the mode of complex posterior densities. For a review of the theoretical and practical aspects of the EM algorithm we refer to McLachlan and Krishnan [13], who also describe the historical origins of the algorithm before and since Dempster et al. [6].

Suppose that, given the data $\mathbf{x} = (x_1, \ldots, x_N)$, we have the following Bayesian posterior: ☞ 672

$$f(\boldsymbol{\theta} \,|\, \mathbf{x}) = \frac{f(\mathbf{x} \,|\, \boldsymbol{\theta})\, f(\boldsymbol{\theta})}{\int f(\mathbf{x} \,|\, \boldsymbol{\theta})\, f(\boldsymbol{\theta})\, d\boldsymbol{\theta}} ,$$

where $f(\boldsymbol{\theta})$ is a given prior on the model parameters and $f(\mathbf{x} \,|\, \boldsymbol{\theta})$ is the likelihood.

We wish to compute the mode of the posterior density: $\boldsymbol{\theta}^* = \operatorname{argmax}_{\boldsymbol{\theta}} f(\boldsymbol{\theta} \,|\, \mathbf{x})$. For simplicity, we will assume that $f(\boldsymbol{\theta}) \propto 1$ and consider the more general case later. Under this assumption, finding the mode of the posterior is the same as maximizing the likelihood function:

$$\boldsymbol{\theta}^* = \operatorname*{argmax}_{\boldsymbol{\theta}} f(\mathbf{x} \,|\, \boldsymbol{\theta}) . \tag{D.14}$$

The key element of the EM algorithm is the augmentation of the data $\mathbf{x}$ with a suitable vector of *latent variables* $\mathbf{z}$ such that

$$f(\mathbf{x}\,|\,\boldsymbol{\theta}) = \int f(\mathbf{x},\mathbf{z}\,|\,\boldsymbol{\theta})\,\mathrm{d}\mathbf{z}\,.$$

The function $f(\mathbf{x},\mathbf{z}\,|\,\boldsymbol{\theta})$ is usually referred to as the **complete-data likelihood** function. The choice of the latent variables is guided by the desire to make the maximization of the complete-data likelihood: $\mathrm{argmax}_{\boldsymbol{\theta}}\, f(\mathbf{x},\mathbf{z}\,|\,\boldsymbol{\theta})$ much easier than (D.14).

Let the density of the latent variables be $g(\mathbf{z})$, then we can write:

$$\begin{aligned}
\ln f(\mathbf{x}\,|\,\boldsymbol{\theta}) &= \int g(\mathbf{z}) \ln f(\mathbf{x}\,|\,\boldsymbol{\theta})\,\mathrm{d}\mathbf{z}\\
&= \int g(\mathbf{z}) \ln\left(\frac{f(\mathbf{x},\mathbf{z}\,|\,\boldsymbol{\theta})/g(\mathbf{z})}{f(\mathbf{z}\,|\,\mathbf{x},\boldsymbol{\theta})/g(\mathbf{z})}\right)\mathrm{d}\mathbf{z}\\
&= \int g(\mathbf{z}) \ln\left(\frac{f(\mathbf{x},\mathbf{z}\,|\,\boldsymbol{\theta})}{g(\mathbf{z})}\right)\mathrm{d}\mathbf{z} - \int g(\mathbf{z}) \ln\left(\frac{f(\mathbf{z}\,|\,\mathbf{x},\boldsymbol{\theta})}{g(\mathbf{z})}\right)\mathrm{d}\mathbf{z}\\
&= \int g(\mathbf{z}) \ln\left(\frac{f(\mathbf{x},\mathbf{z}\,|\,\boldsymbol{\theta})}{g(\mathbf{z})}\right)\mathrm{d}\mathbf{z} + \mathcal{D}(g, f(\cdot\,|\,\mathbf{x},\boldsymbol{\theta}))\,, \qquad \text{(D.15)}
\end{aligned}$$

☞ 366 where $\mathcal{D}(g, f(\cdot\,|\,\mathbf{x},\boldsymbol{\theta}))$ is the Kullback–Leibler distance from the density g to $f(\cdot\,|\,\mathbf{x},\boldsymbol{\theta})$. Since $\mathcal{D} \geqslant 0$ with equality only when $g(\mathbf{z}) = f(\mathbf{z}\,|\,\mathbf{x},\boldsymbol{\theta})$, it follows that

$$\ln f(\mathbf{x}\,|\,\boldsymbol{\theta}) \geqslant \mathcal{L}(g,\boldsymbol{\theta}) \stackrel{\text{def}}{=} \int g(\mathbf{z}) \ln\left(\frac{f(\mathbf{x},\mathbf{z}\,|\,\boldsymbol{\theta})}{g(\mathbf{z})}\right)\mathrm{d}\mathbf{z}$$

for all $\boldsymbol{\theta}$ and any density g. In other words, $\mathcal{L}(g,\boldsymbol{\theta})$ is a lower bound on the likelihood that involves the complete-data likelihood. The EM algorithm then aims to increase this lower bound as much as possible as follows.

Algorithm D.3 (EM Algorithm) *Suppose we have an initial guess for the maximizer, say* $\boldsymbol{\theta}_0$. *The EM algorithm consists of iterating the following steps for* $t = 1, 2, \ldots$.

1. **Expectation Step (E-Step)**: *Given the current guess* $\boldsymbol{\theta}_{t-1}$ *maximize* $\mathcal{L}(g,\boldsymbol{\theta}_{t-1})$ *as a function of* g. *In other words, determine* g *from the functional optimization program* $\max_g \mathcal{L}(g,\boldsymbol{\theta}_{t-1})$. *From the identity* (D.15), *the exact solution is:*

   $$g_t(\mathbf{z}) \stackrel{\text{def}}{=} f(\mathbf{z}\,|\,\mathbf{x},\boldsymbol{\theta}_{t-1})\,.$$

   Compute the expectation

   $$Q_t(\boldsymbol{\theta}) \stackrel{\text{def}}{=} \mathbb{E}_{g_t} \ln f(\mathbf{x},\mathbf{Z}\,|\,\boldsymbol{\theta})\,. \qquad \text{(D.16)}$$

2. **Maximization Step (M-Step)**: *Given the current* g_t, *maximize* $\mathcal{L}(g_t,\boldsymbol{\theta})$ *as a function of* $\boldsymbol{\theta}$. *To perform this step note that, since* $\mathcal{L}(g_t,\boldsymbol{\theta}) = Q_t(\boldsymbol{\theta}) - \mathbb{E}_{g_t} \ln g_t(\mathbf{Z})$, *the maximization of* $\mathcal{L}(g_t,\boldsymbol{\theta})$ *with respect to* $\boldsymbol{\theta}$ *is equivalent to finding*

   $$\boldsymbol{\theta}_t = \underset{\boldsymbol{\theta}}{\mathrm{argmax}}\, Q_t(\boldsymbol{\theta})\,.$$

3. **Stopping Condition**: *If, for example,*

$$\left|\frac{\ln f(\mathbf{x}\,|\,\boldsymbol{\theta}_t) - \ln f(\mathbf{x}\,|\,\boldsymbol{\theta}_{t-1})}{\ln f(\mathbf{x}\,|\,\boldsymbol{\theta}_t)}\right| \leqslant \varepsilon$$

for some small tolerance ε, *terminate the algorithm.*

Note that the M-step consists of the maximization of the expected value of the logarithm of the complete-data likelihood: $Q_t(\boldsymbol{\theta})$. If $f(\mathbf{x}, \mathbf{z}\,|\,\boldsymbol{\theta})$ belongs to an exponential family, then this typically leads to a straightforward optimization problem with a unique solution.

The identity (D.15) can be used to show that the likelihood does not decrease with each iteration of the algorithm. This property is one of the strengths of the algorithm. For example, it can be used to debug computer implementations of the EM algorithm: if the likelihood is observed to decrease at any iteration, then one has detected a bug in the program.

The convergence of the sequence $\{\boldsymbol{\theta}_t\}$ to a local maximum can be guaranteed under certain continuity conditions [4, 18]. The convergence of the algorithm to a global maximum depends strongly on the starting values and in many cases an appropriate choice of starting values may not be clear. Typically, practitioners run the algorithm from different random starting points to ascertain empirically a global optimum is achieved.

The EM algorithm is closely related to the Gibbs sampler [5, 15, 16]. Both ☞ 233
algorithms exploit the idea of conditioning or artificially creating hidden or latent variables [17]. Moreover, the EM algorithm and the Gibbs sampler complement each other in the sense that, while the Gibbs sampler is frequently used to sample from a given complex posterior density, the EM algorithm is used to find the mode of the posterior density.

In cases where the prior is not proportional to 1, the EM algorithm can be easily modified so that: $\mathcal{L}(g, \boldsymbol{\theta}) = \int g(\mathbf{z}) \ln\left(f(\mathbf{x}, \mathbf{z}\,|\,\boldsymbol{\theta})/g(\mathbf{z})\right) \mathrm{d}\mathbf{z} + \ln f(\boldsymbol{\theta})$ and $Q_t(\boldsymbol{\theta}) = \mathbb{E}_{g_t} \ln f(\mathbf{x}, \mathbf{Z}\,|\,\boldsymbol{\theta}) + \ln f(\boldsymbol{\theta})$. The E-step remains the same, because it does not involve $\boldsymbol{\theta}$, and the M-step is only modified through the introduction of the term $\ln f(\boldsymbol{\theta})$.

The EM algorithm is particularly suited for fitting mixture models as shown in the next example.

■ EXAMPLE D.1 (Fitting a Gaussian Mixture Model)

Suppose we have $x_1, \ldots, x_N$, where each x_i is an independent outcome from the following Gaussian mixture model:

$$f(x\,|\,\boldsymbol{\theta}) = \sum_{r=1}^{c} \frac{w_r}{\sigma_r}\, \varphi\left(\frac{x - \mu_r}{\sigma_r}\right) .$$

Here, φ is the pdf of the $\mathsf{N}(0, 1)$ distribution, and $\boldsymbol{\theta} = (\boldsymbol{\mu}, \boldsymbol{\sigma}, \mathbf{w})$ with $\boldsymbol{\mu} = (\mu_1, \ldots, \mu_c)$, $\boldsymbol{\sigma} = (\sigma_1, \ldots, \sigma_c)$, and $\mathbf{w} = (w_1, \ldots, w_c)$. The likelihood is thus

$$f(\mathbf{x}\,|\,\boldsymbol{\theta}) = \prod_{i=1}^{N} \sum_{r=1}^{c} \frac{w_r}{\sigma_r}\, \varphi\left(\frac{x_i - \mu_r}{\sigma_r}\right).$$

Direct maximization of the likelihood can be quite costly, see, for example, [3]. To simplify the likelihood, introduce the discrete latent variables $\mathbf{z} = (z_1, \ldots, z_N) \in$

$\{1, 2, \ldots, c\}^N$ such that the complete-data likelihood can be written as

$$f(\mathbf{x}, \mathbf{z} \mid \boldsymbol{\theta}) = \prod_{i=1}^{N} \frac{w_{z_i}}{\sigma_{z_i}} \varphi\left(\frac{x_i - \mu_{z_i}}{\sigma_{z_i}}\right).$$

The variable z_i can be interpreted as an indicator of the component of the mixture model from which x_i is drawn. The density of $\mathbf{z}$ given the data is $g(\mathbf{z}) = f(\mathbf{z} \mid \mathbf{x}, \boldsymbol{\theta}) = \prod_{i=1}^{N} g_i(z_i)$, where

$$g_i(r) \stackrel{\text{def}}{=} \frac{w_r}{\sigma_r} \varphi\left(\frac{x_i - \mu_r}{\sigma_r}\right) \Big/ \sum_{k=1}^{c} \frac{w_k}{\sigma_k} \varphi\left(\frac{x_i - \mu_k}{\sigma_k}\right) \tag{D.17}$$

for $i = 1, \ldots, N$ and $r = 1, \ldots, c$. The expected complete-data likelihood in the E-step is then

$$\mathbb{E}_g \ln f(\mathbf{x}, \mathbf{Z} \mid \boldsymbol{\theta}) = \sum_{i=1}^{N} \sum_{r=1}^{c} g_i(r) \left(\ln w_r - \ln \sigma_r - \frac{(x_i - \mu_r)^2}{2\sigma_r^2}\right) + \text{constant} .$$

Therefore, in the M-step, maximization of $\mathbb{E}_g \ln f(\mathbf{x}, \mathbf{Z} \mid \boldsymbol{\theta})$ with respect to $\mathbf{w}$ (under the constraint $\sum_r w_r = 1$, $w_i \geqslant 0$ for all i), the means $\boldsymbol{\mu}$, and the variances $\boldsymbol{\sigma}$, gives the following (for $r = 1, \ldots, c$):

$$\begin{aligned} w_r &= \frac{1}{N} \sum_{i=1}^{N} g_i(r) , \\ \mu_r &= \frac{\sum_{i=1}^{N} g_i(r) x_i}{\sum_{i=1}^{N} g_i(r)} , \\ \sigma_r^2 &= \frac{\sum_{i=1}^{N} g_i(r)(x_i - \mu_r)^2}{\sum_{i=1}^{N} g_i(r)} . \end{aligned} \tag{D.18}$$

Thus, given a starting guess $\boldsymbol{\theta} = (\boldsymbol{\mu}, \boldsymbol{\sigma}, \mathbf{w})$, the EM algorithm consists of iterating the following steps until convergence:

E-Step. Given the parameters $(\boldsymbol{\mu}, \boldsymbol{\sigma}, \mathbf{w})$, compute (D.17).

M-Step. Given $g(\mathbf{z}) = \prod_i g_i(z_i)$ from the E-step, update the values for $(\boldsymbol{\mu}, \boldsymbol{\sigma}, \mathbf{w})$ using equations (D.18).

The statistics toolbox in MATLAB include the function `gmdistribution.fit`, which implements the EM algorithm for fitting Gaussian mixture models. For a comprehensive treatment of mixture models see the monographs [12, 14].

D.8 POISSON SUMMATION FORMULA

Let $f(x)$ be a continuous function on $\mathbb{R}$ such that $\int_{-\infty}^{\infty} |f(x)| \mathrm{d}x < \infty$, and let $\widetilde{f}(\omega) \stackrel{\text{def}}{=} \int_{-\infty}^{\infty} f(x) \, \mathrm{e}^{-\mathrm{i}2\pi\omega x} \mathrm{d}x$ be the *Fourier transform* of f. The **Poisson summation formula** [19] states that

$$\sum_{k=-\infty}^{\infty} f(k + x) \, \mathrm{e}^{-\mathrm{i}\pi s(2k+x)} = \sum_{k=-\infty}^{\infty} \widetilde{f}(k + s) \, \mathrm{e}^{\mathrm{i}\pi x(2k+s)} ,$$

provided that $\sum_{k=-\infty}^{\infty} f(k+x)$ converges uniformly on every finite interval and $\sum_{k=-\infty}^{\infty} |\widetilde{f}(k)| < \infty$. Special cases of the Poisson summation formula include the identities $\sum_{k=-\infty}^{\infty} f(k) = \sum_{k=-\infty}^{\infty} \widetilde{f}(k)$ and

$$\sum_{k=-\infty}^{\infty} f(k)\cos(\pi k x)\cos(\pi k y) = \frac{1}{2}\sum_{k=-\infty}^{\infty} \widetilde{f}\Big(k+\frac{x-y}{2}\Big) + \widetilde{f}\Big(k+\frac{x+y}{2}\Big), \quad x,y\in\mathbb{R}\,.$$

The last identity can be used to derive the **theta function identity**: ☞ 322

$$\begin{aligned}\theta(x,y;t) &\stackrel{\text{def}}{=} \frac{1}{\sqrt{2\pi t}}\sum_{k=-\infty}^{\infty} \mathrm{e}^{-\frac{(x+y-2k)^2}{2t}} + \mathrm{e}^{-\frac{(x-y-2k)^2}{2t}} \\ &= \sum_{k=-\infty}^{\infty} \mathrm{e}^{-k^2\pi^2 t/2}\cos(k\pi x)\cos(k\pi y), \quad t>0\,.\end{aligned}$$

D.9 SPECIAL FUNCTIONS

Here we list some special functions that make an appearance in this book (see [1] for more details).

D.9.1 Beta Function $B(\alpha,\beta)$

$$B(\alpha,\beta) = \int_0^1 t^{\alpha-1}(1-t)^{\beta-1}\mathrm{d}t = \frac{\Gamma(\alpha)\,\Gamma(\beta)}{\Gamma(\alpha+\beta)}, \quad \alpha,\beta>0\,.$$

Note that $B(\alpha,\beta)=B(\beta,\alpha)$. In MATLAB this function is implemented in `beta.m`.

D.9.2 Incomplete Beta Function $I_x(\alpha,\beta)$

$$I_x(\alpha,\beta) = \frac{1}{B(\alpha,\beta)}\int_0^x t^{\alpha-1}(1-t)^{\beta-1}\mathrm{d}t, \quad \alpha,\beta>0,\ x\in[0,1]\,.$$

Some useful identities include:

1. *Reflection property*: $I_x(\alpha,\beta) = 1 - I_{1-x}(\beta,\alpha)$.
2. *Hypergeometric function*: $I_x(\alpha,\beta) = \frac{x^\alpha}{\alpha\,B(\alpha,\beta)}\ {}_2F_1(\alpha,1-\beta;\alpha+1;x)$.

In MATLAB this function is implemented in `betainc.m`.

D.9.3 Error Function $\mathrm{erf}(x)$

$$\mathrm{erf}(x) = \frac{2}{\sqrt{\pi}}\int_0^x \mathrm{e}^{-t^2}\mathrm{d}t, \quad x\in\mathbb{R}\,.$$

Some useful relations to other special functions include:

1. *Confluent hypergeometric function*: $\mathrm{erf}(x) = \frac{2x}{\sqrt{\pi}}\ {}_1F_1(\frac{1}{2};\frac{3}{2};-x^2)$.
2. *Incomplete gamma function*: $\mathrm{erf}(x) = \mathrm{sgn}(x)\,P(\frac{1}{2},x^2)$.

In MATLAB this function is implemented in `erf.m`.

D.9.4 Digamma function $\psi(x)$

$$\psi(x) = \frac{\mathrm{d}}{\mathrm{d}x} \ln \Gamma(x) = \frac{\Gamma'(x)}{\Gamma(x)}, \quad x > 0 .$$

The digamma function has the following integral representation:

$$\psi(x) = \int_0^\infty \left(\frac{\mathrm{e}^{-t}}{t} - \frac{\mathrm{e}^{-t\,x}}{1 - \mathrm{e}^{-t}} \right) \mathrm{d}t, \quad x > 0 .$$

In MATLAB this function is implemented in `psi.m`.

D.9.5 Gamma Function $\Gamma(\alpha)$

$$\Gamma(\alpha) = \int_0^\infty \mathrm{e}^{-x} x^{\alpha-1} \, \mathrm{d}x , \quad \alpha > 0 .$$

Some useful properties of the Γ function are:

1. *Functional equation:* $\Gamma(\alpha + 1) = \alpha \, \Gamma(\alpha)$.
2. *Factorial function:* $\Gamma(n) = (n - 1)!$ for $n = 1, 2, \ldots$.
3. *Gauss multiplication formula:* $\Gamma(n\,x) = (2\pi)^{(1-n)/2} n^{n\,x - \frac{1}{2}} \prod_{k=0}^{n-1} \Gamma\left(x + \frac{k}{n}\right)$.
4. *Euler reflection formula:* $\Gamma(1 - x)\Gamma(1 + x) = \frac{\pi\,x}{\sin(\pi\,x)}$, $x \in (0, 1)$.
5. *Special values:* $\Gamma(1/2) = \sqrt{\pi}$.

In MATLAB this function is implemented in `gamma.m`.

D.9.6 Incomplete Gamma Function $P(\alpha, x)$

$$P(\alpha, x) = \frac{1}{\Gamma(\alpha)} \int_0^x \mathrm{e}^{-t} t^{\alpha-1} \mathrm{d}t, \quad \alpha > 0, \; x > 0 .$$

For positive integer values of $\alpha = n$, we have:

$$P(n, x) = 1 - \mathrm{e}^{-x} \sum_{k=0}^{n-1} \frac{x^k}{k!} .$$

In MATLAB this function is implemented in `gammainc.m`.

D.9.7 Hypergeometric Function ${}_2F_1(a, b; c; z)$

$${}_2F_1(a, b; c; z) = \frac{\Gamma(c)}{\Gamma(a)\,\Gamma(b)} \sum_{n=0}^{\infty} \frac{\Gamma(a + n)\,\Gamma(b + n)}{\Gamma(c + n)} \frac{z^n}{n!} ,$$

which converges for all $|z| < 1$ provided that c is not a negative integer. Note that the series converges for all $|z| \leqslant 1$ provided that c is not a negative integer and $\Re[c - a - b] > 0$. Some special cases are:

- ${}_2F_1(1, 1; 2; z) = -z^{-1} \ln(1 - z)$.
- ${}_2F_1(a, b; c; 1) = \frac{\Gamma(c)\Gamma(c-a-b)}{\Gamma(c-a)\Gamma(c-b)}$.
- ${}_2F_1(a, b; b; z) = (1 - z)^{-a}$.

In MATLAB this function is implemented in `hypergeom.m`.

D.9.8 Confluent Hypergeometric Function ${}_1F_1(\alpha;\gamma;x)$

$$ {}_1F_1(\alpha;\gamma;x) = \frac{\Gamma(\gamma)}{\Gamma(\alpha)\,\Gamma(\gamma-\alpha)} \int_0^1 t^{\alpha-1}(1-t)^{\gamma-\alpha-1}\,\mathrm{e}^{tx}\,\mathrm{d}t\,, \quad \gamma > \alpha > 0\,. $$

An alternative notation for ${}_1F_1(\alpha;\gamma;x)$ is $M(\alpha,\gamma,x)$. In series form, the function ${}_1F_1$ is given by

$$ {}_1F_1(\alpha;\gamma;x) = 1 + \frac{\alpha}{\gamma}\frac{x}{1!} + \frac{\alpha(\alpha+1)}{\gamma(\gamma+1)}\frac{x^2}{2!} + \cdots\,, \quad x \in \mathbb{R}\,. $$

In MATLAB this function is implemented in `hypergeom.m`.

D.9.9 Modified Bessel Function of the Second Kind $K_\nu(x)$

$$ K_\nu(x) = \int_0^\infty \mathrm{e}^{-x\cosh(t)}\cosh(\nu t)\,\mathrm{d}t, \quad x > 0\,. $$

In MATLAB this function is implemented in `besselk.m`.

REFERENCES

1. M. Abramowitz and I. A. Stegun, editors. *Handbook of Mathematical Functions with Formulas, Graphs, and Mathematical Tables.* Number 55 in National Bureau of Standards Applied Mathematics Series. United States Government Printing Office, Washington, DC, tenth edition, 1964.
2. P. J. Bickel and K. A. Doksum. *Mathematical Statistics*, volume I. Pearson Prentice Hall, Upper Saddle River, NJ, second edition, 2007.
3. Z. I. Botev and D. P. Kroese. Global likelihood optimization via the cross-entropy method, with an application to mixture models. *Proceedings of the Winter Simulation Conference, Washington, DC*, pages 529–535, 2004.
4. R. A. Boyles. On the convergence of the EM algorithm. *Journal of the Royal Statistical Society, Series B*, 45(1):47–50, 1983.
5. G. Casella and R. L. Berger. Estimation with selected binomial information or do you really believe that Dave Winfield is batting .471? *Journal of the American Statistical Association*, 89(427):1080–1090, 1994.
6. A. P. Dempster, N. M. Laird, and D. B. Rubin. Maximum likelihood from incomplete data via the EM algorithm. *Journal of the Royal Statistical Society, Series B*, 39(1):1–38, 1977.
7. P. Embrechts, C. Klüppelberg, and T. Mikosch. *Modelling Extremal Events for Insurance and Finance.* Springer-Verlag, New York, 1997.
8. P. Embrechts and N. Veraverbeke. Estimates for the probability of ruin with special emphasis on the possibility of large claims. *Insurance Mathematics and Economics*, 1(1):55–72, 1982.
9. S. Foss, D. Korshunov, and S. Zachary. *An Introduction to Heavy-tailed and Subexponential Distributions*, volume 13. Oberwolfach Preprints, ISSN 1864-7596, 2009. `http://www.mfo.de/publications/owp/2009/OWP2009_13.pdf`.
10. A. K. Jain. *Fundamentals of Digital Image Processing.* Prentice Hall, Englewood Cliffs, NJ, 1989.

11. N. L. Johnson and S. Kotz. *Distributions in Statistics: Continuous Univariate Distributions, Volume 1.* Houghton Mifflin Company, New York, 1970.

12. G. J. McLachlan and K. E. Basford. *Mixture Models: Inference and Applications to Clustering.* Marcel Dekker, New York, 1988.

13. G. J. McLachlan and T. Krishnan. *The EM Algorithm and Extensions.* John Wiley & Sons, Hoboken, NJ, second edition, 2008.

14. G. J. McLachlan and D. Peel. *Finite Mixture Models.* John Wiley & Sons, New York, 2000.

15. X.-L. Meng and D. van Dyk. The EM algorithm – an old folk-song sung to a fast new tune (with discussion). *Journal of the Royal Statistical Society, Series B*, 59(3):511–567, 1997.

16. C. P. Robert and G. Casella. *Monte Carlo Statistical Methods.* Springer-Verlag, New York, 2004.

17. R. H. Swendson and J.-S. Wang. Nonuniversal critical dynamics in Monte Carlo simulations. *Physical Review Letters*, 58(2):86–88, 1987.

18. C. F. J. Wu. On the convergence properties of the EM algorithm. *The Annals of Statistics*, 11(1):95–103, 1983.

19. A. Zygmund. *Trigonometric Series.* Cambridge University Press, Cambridge, third edition, 2003.

ACRONYMS AND ABBREVIATIONS

a.s.	Almost surely
ADAM	ADAptive Multilevel
cdf	Cumulative distribution function
CE	Cross-entropy
CMC	Crude Monte Carlo
Corr	Correlation
Cov	Covariance
CRV	Common random variables
EM	Expectation–maximization
FFT	Fast Fourier transform
GS	Generalized splitting
iid	Independent identically distributed
IPA	Infinitesimal perturbation analysis
ISE	Integrated squared error
LCG	Linear congruential generator
LSCV	Least squares cross validation
LSFR	Linear feedback shift register
KKT	Karush–Kuhn–Tucker
max-cut	Maximal cut
MCMC	Markov chain Monte Carlo
MLE	Maximum likelihood estimate (or estimator)

MISE	Mean integrated square error
MSE	Mean square error
MRG	Multiple-recursive generator
ODE	Ordinary differential equation
PDE	Partial differential equation
pdf	Probability density function
SAT	Satisfiability (problem)
SDE	Stochastic differential equation
SIS	Sequential importance sampling
TSP	Traveling salesman problem
Var	Variance
VM	Variance minimization

LIST OF SYMBOLS

$\gg$	much greater than
$\approx$	is approximately
$\propto$	is proportional to
$\oplus$	XOR (binary addition operator)
$\stackrel{\text{def}}{=}$	is defined as
$\Leftrightarrow$	is equivalent to
$\sim$	is distributed as
$\stackrel{\text{iid}}{\sim}$, $\sim_{\text{iid}}$	are independent and identically distributed as
$\stackrel{\text{approx.}}{\sim}$	is approximately distributed as
$\stackrel{\text{a.s.}}{\longrightarrow}$	converges almost surely to
$\stackrel{d}{\longrightarrow}$	converges in distribution to
$\stackrel{a.s.}{\nearrow}$	increases almost surely to
$\stackrel{L^p}{\longrightarrow}$	converges in L^p-norm to
$\stackrel{\mathbb{P}}{\longrightarrow}$	converges in probability to
$\stackrel{\text{cpl.}}{\longrightarrow}$	converges completely to
$\|\cdot\|$	Euclidean norm
∇f	gradient of f

$\nabla^2 f$	Hessian of f
$\lceil x \rceil$	smallest integer larger than x
$\lfloor x \rfloor$	largest integer smaller than x
x^+	$x^+ = \max\{x, 0\}$
$\mathbf{x}_{-i}$	vector $\mathbf{x}$ with i-th element removed
$A^\top, \mathbf{x}^\top$	transpose of matrix A or vector $\mathbf{x}$
$A \succ 0$	matrix A is positive definite
$A \succeq 0$	matrix A is positive semidefinite
$\mathrm{diag}(\mathbf{a})$	diagonal matrix with diagonal entries defined by $\mathbf{a}$
$\mathrm{tr}(A)$	trace of matrix A
$\dim(\mathbf{x})$	dimension of vector $\mathbf{x}$
$\det(A)$	determinant of matrix A

$\mathcal{B}$	Borel σ-algebra on $\mathbb{R}$
$\mathcal{B}^n$	Borel σ-algebra on $\mathbb{R}^n$
$\mathbb{C}$	set of complex numbers
d	differential symbol
$\mathbb{E}$	expectation
e	the number $2.71828\ldots$
i	the root of -1
$\Im$	imaginary part
$\mathrm{I}_A, \mathrm{I}\{A\}$	indicator function of event A
ℓ	performance measure
$\ln$	(natural) logarithm
$\mathbb{N}$	set of natural numbers $\{0, 1, \ldots\}$
φ	pdf of the standard normal distribution
Φ	cdf of the standard normal distribution
$\mathbb{P}$	probability measure
$\mathcal{O}$	big-O order symbol: $f(x) = \mathcal{O}(g(x))$ if $\lvert f(x) \rvert \leqslant \alpha g(x)$ for some constant α as $x \to a$
o	little-o order symbol: $f(x) = o(g(x))$ if $f(x)/g(x) \to 0$ as $x \to a$
$\mathbb{R}$	the real line = one-dimensional Euclidean space
$\mathbb{R}_+$	positive real line: $[0, \infty)$
$\mathbb{R}^n$	n-dimensional Euclidean space
$\Re$	real part
$S_{(i)}$	i-th order statistic
$\widehat{\boldsymbol{\theta}}$	estimate / estimator

$\boldsymbol{\theta}^*$	optimal parameter
$\mathbf{x}, \mathbf{y}$	vectors
$\mathbf{X}, \mathbf{Y}$	random vectors
$\mathscr{X}, \mathscr{Y}$	sets
$\mathbb{Z}$	set of integers $\{\ldots, -1, 0, 1, \ldots\}$
$\mathcal{C}^k$	space of functions with continuous k-th derivatives
$\mathcal{S}$	score function
$\mathcal{I}$	Fisher information matrix
$\mathcal{D}$	Kullback–Leibler divergence (cross-entropy distance)
$\mathcal{D}_2$	χ^2 divergence

LIST OF DISTRIBUTIONS

INDEX

C

D

E

F

J

K

L

M

N

O

P

Q

R

S

T

U

V

WILEY SERIES IN PROBABILITY AND STATISTICS

ESTABLISHED BY WALTER A. SHEWHART AND SAMUEL S. WILKS

The ***Wiley Series in Probability and Statistics*** is well established and authoritative. It covers many topics of current research interest in both pure and applied statistics and probability theory. Written by leading statisticians and institutions, the titles span both state-of-the-art developments in the field and classical methods.

Reflecting the wide range of current research in statistics, the series encompasses applied, methodological and theoretical statistics, ranging from applications and new techniques made possible by advances in computerized practice to rigorous treatment of theoretical approaches.

This series provides essential and invaluable reading for all statisticians, whether in academia, industry, government, or research.

† ABRAHAM and LEDOLTER · Statistical Methods for Forecasting
AGRESTI · Analysis of Ordinal Categorical Data, *Second Edition*
AGRESTI · An Introduction to Categorical Data Analysis, *Second Edition*
AGRESTI · Categorical Data Analysis, *Second Edition*
ALTMAN, GILL, and McDONALD · Numerical Issues in Statistical Computing for the Social Scientist
AMARATUNGA and CABRERA · Exploration and Analysis of DNA Microarray and Protein Array Data
ANDĚL · Mathematics of Chance
ANDERSON · An Introduction to Multivariate Statistical Analysis, *Third Edition*
* ANDERSON · The Statistical Analysis of Time Series
ANDERSON, AUQUIER, HAUCK, OAKES, VANDAELE, and WEISBERG · Statistical Methods for Comparative Studies
ANDERSON and LOYNES · The Teaching of Practical Statistics
ARMITAGE and DAVID (editors) · Advances in Biometry
ARNOLD, BALAKRISHNAN, and NAGARAJA · Records
* ARTHANARI and DODGE · Mathematical Programming in Statistics
* BAILEY · The Elements of Stochastic Processes with Applications to the Natural Sciences
BALAKRISHNAN and KOUTRAS · Runs and Scans with Applications
BALAKRISHNAN and NG · Precedence-Type Tests and Applications
BARNETT · Comparative Statistical Inference, *Third Edition*
BARNETT · Environmental Statistics
BARNETT and LEWIS · Outliers in Statistical Data, *Third Edition*
BARTOSZYNSKI and NIEWIADOMSKA-BUGAJ · Probability and Statistical Inference
BASILEVSKY · Statistical Factor Analysis and Related Methods: Theory and Applications
BASU and RIGDON · Statistical Methods for the Reliability of Repairable Systems
BATES and WATTS · Nonlinear Regression Analysis and Its Applications

*Now available in a lower priced paperback edition in the Wiley Classics Library.
†Now available in a lower priced paperback edition in the Wiley–Interscience Paperback Series.

BECHHOFER, SANTNER, and GOLDSMAN · Design and Analysis of Experiments for Statistical Selection, Screening, and Multiple Comparisons
BELSLEY · Conditioning Diagnostics: Collinearity and Weak Data in Regression
† BELSLEY, KUH, and WELSCH · Regression Diagnostics: Identifying Influential Data and Sources of Collinearity
BENDAT and PIERSOL · Random Data: Analysis and Measurement Procedures, *Fourth Edition*
BERRY, CHALONER, and GEWEKE · Bayesian Analysis in Statistics and Econometrics: Essays in Honor of Arnold Zellner
BERNARDO and SMITH · Bayesian Theory
BHAT and MILLER · Elements of Applied Stochastic Processes, *Third Edition*
BHATTACHARYA and WAYMIRE · Stochastic Processes with Applications
BILLINGSLEY · Convergence of Probability Measures, *Second Edition*
BILLINGSLEY · Probability and Measure, *Third Edition*
BIRKES and DODGE · Alternative Methods of Regression
BISGAARD and KULAHCI · Time Series Analysis and Forecasting by Example
BISWAS, DATTA, FINE, and SEGAL · Statistical Advances in the Biomedical Sciences: Clinical Trials, Epidemiology, Survival Analysis, and Bioinformatics
BLISCHKE AND MURTHY (editors) · Case Studies in Reliability and Maintenance
BLISCHKE AND MURTHY · Reliability: Modeling, Prediction, and Optimization
BLOOMFIELD · Fourier Analysis of Time Series: An Introduction, *Second Edition*
BOLLEN · Structural Equations with Latent Variables
BOLLEN and CURRAN · Latent Curve Models: A Structural Equation Perspective
BOROVKOV · Ergodicity and Stability of Stochastic Processes
BOULEAU · Numerical Methods for Stochastic Processes
BOX · Bayesian Inference in Statistical Analysis
BOX · R. A. Fisher, the Life of a Scientist
BOX and DRAPER · Response Surfaces, Mixtures, and Ridge Analyses, *Second Edition*
* BOX and DRAPER · Evolutionary Operation: A Statistical Method for Process Improvement
BOX and FRIENDS · Improving Almost Anything, *Revised Edition*
BOX, HUNTER, and HUNTER · Statistics for Experimenters: Design, Innovation, and Discovery, *Second Editon*
BOX, JENKINS, and REINSEL · Time Series Analysis: Forcasting and Control, *Fourth Edition*
BOX, LUCEÑO, and PANIAGUA-QUIÑONES · Statistical Control by Monitoring and Adjustment, *Second Edition*
BRANDIMARTE · Numerical Methods in Finance: A MATLAB-Based Introduction
† BROWN and HOLLANDER · Statistics: A Biomedical Introduction
BRUNNER, DOMHOF, and LANGER · Nonparametric Analysis of Longitudinal Data in Factorial Experiments
BUCKLEW · Large Deviation Techniques in Decision, Simulation, and Estimation
CAIROLI and DALANG · Sequential Stochastic Optimization
CASTILLO, HADI, BALAKRISHNAN, and SARABIA · Extreme Value and Related Models with Applications in Engineering and Science
CHAN · Time Series: Applications to Finance with R and S-Plus®, *Second Edition*
CHARALAMBIDES · Combinatorial Methods in Discrete Distributions
CHATTERJEE and HADI · Regression Analysis by Example, *Fourth Edition*
CHATTERJEE and HADI · Sensitivity Analysis in Linear Regression
CHERNICK · Bootstrap Methods: A Guide for Practitioners and Researchers, *Second Edition*
CHERNICK and FRIIS · Introductory Biostatistics for the Health Sciences
CHILÈS and DELFINER · Geostatistics: Modeling Spatial Uncertainty

*Now available in a lower priced paperback edition in the Wiley Classics Library.
†Now available in a lower priced paperback edition in the Wiley–Interscience Paperback Series.

CHOW and LIU · Design and Analysis of Clinical Trials: Concepts and Methodologies, *Second Edition*
CLARKE · Linear Models: The Theory and Application of Analysis of Variance
CLARKE and DISNEY · Probability and Random Processes: A First Course with Applications, *Second Edition*
* COCHRAN and COX · Experimental Designs, *Second Edition*
COLLINS and LANZA · Latent Class and Latent Transition Analysis: With Applications in the Social, Behavioral, and Health Sciences
CONGDON · Applied Bayesian Modelling
CONGDON · Bayesian Models for Categorical Data
CONGDON · Bayesian Statistical Modelling
CONOVER · Practical Nonparametric Statistics, *Third Edition*
COOK · Regression Graphics
COOK and WEISBERG · Applied Regression Including Computing and Graphics
COOK and WEISBERG · An Introduction to Regression Graphics
CORNELL · Experiments with Mixtures, Designs, Models, and the Analysis of Mixture Data, *Third Edition*
COVER and THOMAS · Elements of Information Theory
COX · A Handbook of Introductory Statistical Methods
* COX · Planning of Experiments
CRESSIE · Statistics for Spatial Data, *Revised Edition*
CSÖRGŐ and HORVÁTH · Limit Theorems in Change Point Analysis
DANIEL · Applications of Statistics to Industrial Experimentation
DANIEL · Biostatistics: A Foundation for Analysis in the Health Sciences, *Eighth Edition*
* DANIEL · Fitting Equations to Data: Computer Analysis of Multifactor Data, *Second Edition*
DASU and JOHNSON · Exploratory Data Mining and Data Cleaning
DAVID and NAGARAJA · Order Statistics, *Third Edition*
* DEGROOT, FIENBERG, and KADANE · Statistics and the Law
DEL CASTILLO · Statistical Process Adjustment for Quality Control
DeMARIS · Regression with Social Data: Modeling Continuous and Limited Response Variables
DEMIDENKO · Mixed Models: Theory and Applications
DENISON, HOLMES, MALLICK and SMITH · Bayesian Methods for Nonlinear Classification and Regression
DETTE and STUDDEN · The Theory of Canonical Moments with Applications in Statistics, Probability, and Analysis
DEY and MUKERJEE · Fractional Factorial Plans
DILLON and GOLDSTEIN · Multivariate Analysis: Methods and Applications
DODGE · Alternative Methods of Regression
* DODGE and ROMIG · Sampling Inspection Tables, *Second Edition*
* DOOB · Stochastic Processes
DOWDY, WEARDEN, and CHILKO · Statistics for Research, *Third Edition*
DRAPER and SMITH · Applied Regression Analysis, *Third Edition*
DRYDEN and MARDIA · Statistical Shape Analysis
DUDEWICZ and MISHRA · Modern Mathematical Statistics
DUNN and CLARK · Basic Statistics: A Primer for the Biomedical Sciences, *Third Edition*
DUPUIS and ELLIS · A Weak Convergence Approach to the Theory of Large Deviations
EDLER and KITSOS · Recent Advances in Quantitative Methods in Cancer and Human Health Risk Assessment
* ELANDT-JOHNSON and JOHNSON · Survival Models and Data Analysis
ENDERS · Applied Econometric Time Series

*Now available in a lower priced paperback edition in the Wiley Classics Library.
†Now available in a lower priced paperback edition in the Wiley–Interscience Paperback Series.

† ETHIER and KURTZ · Markov Processes: Characterization and Convergence
EVANS, HASTINGS, and PEACOCK · Statistical Distributions, *Third Edition*
FELLER · An Introduction to Probability Theory and Its Applications, Volume I, *Third Edition,* Revised; Volume II, *Second Edition*
FISHER and VAN BELLE · Biostatistics: A Methodology for the Health Sciences
FITZMAURICE, LAIRD, and WARE · Applied Longitudinal Analysis
* FLEISS · The Design and Analysis of Clinical Experiments
FLEISS · Statistical Methods for Rates and Proportions, *Third Edition*
† FLEMING and HARRINGTON · Counting Processes and Survival Analysis
FUJIKOSHI, ULYANOV, and SHIMIZU · Multivariate Statistics: High-Dimensional and Large-Sample Approximations
FULLER · Introduction to Statistical Time Series, *Second Edition*
† FULLER · Measurement Error Models
GALLANT · Nonlinear Statistical Models
GEISSER · Modes of Parametric Statistical Inference
GELMAN and MENG · Applied Bayesian Modeling and Causal Inference from Incomplete-Data Perspectives
GEWEKE · Contemporary Bayesian Econometrics and Statistics
GHOSH, MUKHOPADHYAY, and SEN · Sequential Estimation
GIESBRECHT and GUMPERTZ · Planning, Construction, and Statistical Analysis of Comparative Experiments
GIFI · Nonlinear Multivariate Analysis
GIVENS and HOETING · Computational Statistics
GLASSERMAN and YAO · Monotone Structure in Discrete-Event Systems
GNANADESIKAN · Methods for Statistical Data Analysis of Multivariate Observations, *Second Edition*
GOLDSTEIN and LEWIS · Assessment: Problems, Development, and Statistical Issues
GREENWOOD and NIKULIN · A Guide to Chi-Squared Testing
GROSS, SHORTLE, THOMPSON, and HARRIS · Fundamentals of Queueing Theory, *Fourth Edition*
GROSS, SHORTLE, THOMPSON, and HARRIS · Solutions Manual to Accompany Fundamentals of Queueing Theory, *Fourth Edition*
* HAHN and SHAPIRO · Statistical Models in Engineering
HAHN and MEEKER · Statistical Intervals: A Guide for Practitioners
HALD · A History of Probability and Statistics and their Applications Before 1750
HALD · A History of Mathematical Statistics from 1750 to 1930
† HAMPEL · Robust Statistics: The Approach Based on Influence Functions
HANNAN and DEISTLER · The Statistical Theory of Linear Systems
HARTUNG, KNAPP, and SINHA · Statistical Meta-Analysis with Applications
HEIBERGER · Computation for the Analysis of Designed Experiments
HEDAYAT and SINHA · Design and Inference in Finite Population Sampling
HEDEKER and GIBBONS · Longitudinal Data Analysis
HELLER · MACSYMA for Statisticians
HINKELMANN and KEMPTHORNE · Design and Analysis of Experiments, Volume 1: Introduction to Experimental Design, *Second Edition*
HINKELMANN and KEMPTHORNE · Design and Analysis of Experiments, Volume 2: Advanced Experimental Design
HOAGLIN, MOSTELLER, and TUKEY · Fundamentals of Exploratory Analysis of Variance
* HOAGLIN, MOSTELLER, and TUKEY · Exploring Data Tables, Trends and Shapes
* HOAGLIN, MOSTELLER, and TUKEY · Understanding Robust and Exploratory Data Analysis
HOCHBERG and TAMHANE · Multiple Comparison Procedures

*Now available in a lower priced paperback edition in the Wiley Classics Library.
†Now available in a lower priced paperback edition in the Wiley–Interscience Paperback Series.

HOCKING · Methods and Applications of Linear Models: Regression and the Analysis of Variance, *Second Edition*
HOEL · Introduction to Mathematical Statistics, *Fifth Edition*
HOGG and KLUGMAN · Loss Distributions
HOLLANDER and WOLFE · Nonparametric Statistical Methods, *Second Edition*
HOSMER and LEMESHOW · Applied Logistic Regression, *Second Edition*
HOSMER, LEMESHOW, and MAY · Applied Survival Analysis: Regression Modeling of Time-to-Event Data, *Second Edition*
† HUBER and RONCHETTI · Robust Statistics, *Second Edition*
HUBERTY · Applied Discriminant Analysis
HUBERTY and OLEJNIK · Applied MANOVA and Discriminant Analysis, *Second Edition*
HUNT and KENNEDY · Financial Derivatives in Theory and Practice, *Revised Edition*
HURD and MIAMEE · Periodically Correlated Random Sequences: Spectral Theory and Practice
HUSKOVA, BERAN, and DUPAC · Collected Works of Jaroslav Hajek—with Commentary
HUZURBAZAR · Flowgraph Models for Multistate Time-to-Event Data
IMAN and CONOVER · A Modern Approach to Statistics
† JACKSON · A User's Guide to Principle Components
JOHN · Statistical Methods in Engineering and Quality Assurance
JOHNSON · Multivariate Statistical Simulation
JOHNSON and BALAKRISHNAN · Advances in the Theory and Practice of Statistics: A Volume in Honor of Samuel Kotz
JOHNSON and BHATTACHARYYA · Statistics: Principles and Methods, *Fifth Edition*
JOHNSON and KOTZ · Distributions in Statistics
JOHNSON and KOTZ (editors) · Leading Personalities in Statistical Sciences: From the Seventeenth Century to the Present
JOHNSON, KOTZ, and BALAKRISHNAN · Continuous Univariate Distributions, Volume 1, *Second Edition*
JOHNSON, KOTZ, and BALAKRISHNAN · Continuous Univariate Distributions, Volume 2, *Second Edition*
JOHNSON, KOTZ, and BALAKRISHNAN · Discrete Multivariate Distributions
JOHNSON, KEMP, and KOTZ · Univariate Discrete Distributions, *Third Edition*
JUDGE, GRIFFITHS, HILL, LÜTKEPOHL, and LEE · The Theory and Practice of Econometrics, *Second Edition*
JUREČKOVÁ and SEN · Robust Statistical Procedures: Aymptotics and Interrelations
JUREK and MASON · Operator-Limit Distributions in Probability Theory
KADANE · Bayesian Methods and Ethics in a Clinical Trial Design
KADANE AND SCHUM · A Probabilistic Analysis of the Sacco and Vanzetti Evidence
KALBFLEISCH and PRENTICE · The Statistical Analysis of Failure Time Data, *Second Edition*
KARIYA and KURATA · Generalized Least Squares
KASS and VOS · Geometrical Foundations of Asymptotic Inference
† KAUFMAN and ROUSSEEUW · Finding Groups in Data: An Introduction to Cluster Analysis
KEDEM and FOKIANOS · Regression Models for Time Series Analysis
KENDALL, BARDEN, CARNE, and LE · Shape and Shape Theory
KHURI · Advanced Calculus with Applications in Statistics, *Second Edition*
KHURI, MATHEW, and SINHA · Statistical Tests for Mixed Linear Models
KLEIBER and KOTZ · Statistical Size Distributions in Economics and Actuarial Sciences
KLEMELÄ · Smoothing of Multivariate Data: Density Estimation and Visualization
KLUGMAN, PANJER, and WILLMOT · Loss Models: From Data to Decisions, *Third Edition*

*Now available in a lower priced paperback edition in the Wiley Classics Library.
†Now available in a lower priced paperback edition in the Wiley–Interscience Paperback Series.

KLUGMAN, PANJER, and WILLMOT · Solutions Manual to Accompany Loss Models: From Data to Decisions, *Third Edition*

KOTZ, BALAKRISHNAN, and JOHNSON · Continuous Multivariate Distributions, Volume 1, *Second Edition*

KOVALENKO, KUZNETZOV, and PEGG · Mathematical Theory of Reliability of Time-Dependent Systems with Practical Applications

KOWALSKI and TU · Modern Applied U-Statistics

KRISHNAMOORTHY and MATHEW · Statistical Tolerance Regions: Theory, Applications, and Computation

KROESE, TAIMRE, and BOTEV · Handbook of Monte Carlo Methods

KROONENBERG · Applied Multiway Data Analysis

KVAM and VIDAKOVIC · Nonparametric Statistics with Applications to Science and Engineering

LACHIN · Biostatistical Methods: The Assessment of Relative Risks, *Second Edition*

LAD · Operational Subjective Statistical Methods: A Mathematical, Philosophical, and Historical Introduction

LAMPERTI · Probability: A Survey of the Mathematical Theory, *Second Edition*

LANGE, RYAN, BILLARD, BRILLINGER, CONQUEST, and GREENHOUSE · Case Studies in Biometry

LARSON · Introduction to Probability Theory and Statistical Inference, *Third Edition*

LAWLESS · Statistical Models and Methods for Lifetime Data, *Second Edition*

LAWSON · Statistical Methods in Spatial Epidemiology

LE · Applied Categorical Data Analysis

LE · Applied Survival Analysis

LEE and WANG · Statistical Methods for Survival Data Analysis, *Third Edition*

LePAGE and BILLARD · Exploring the Limits of Bootstrap

LEYLAND and GOLDSTEIN (editors) · Multilevel Modelling of Health Statistics

LIAO · Statistical Group Comparison

LINDVALL · Lectures on the Coupling Method

LIN · Introductory Stochastic Analysis for Finance and Insurance

LINHART and ZUCCHINI · Model Selection

LITTLE and RUBIN · Statistical Analysis with Missing Data, *Second Edition*

LLOYD · The Statistical Analysis of Categorical Data

LOWEN and TEICH · Fractal-Based Point Processes

MAGNUS and NEUDECKER · Matrix Differential Calculus with Applications in Statistics and Econometrics, *Revised Edition*

MALLER and ZHOU · Survival Analysis with Long Term Survivors

MALLOWS · Design, Data, and Analysis by Some Friends of Cuthbert Daniel

MANN, SCHAFER, and SINGPURWALLA · Methods for Statistical Analysis of Reliability and Life Data

MANTON, WOODBURY, and TOLLEY · Statistical Applications Using Fuzzy Sets

MARCHETTE · Random Graphs for Statistical Pattern Recognition

MARDIA and JUPP · Directional Statistics

MASON, GUNST, and HESS · Statistical Design and Analysis of Experiments with Applications to Engineering and Science, *Second Edition*

McCULLOCH, SEARLE, and NEUHAUS · Generalized, Linear, and Mixed Models, *Second Edition*

McFADDEN · Management of Data in Clinical Trials, *Second Edition*

* McLACHLAN · Discriminant Analysis and Statistical Pattern Recognition

McLACHLAN, DO, and AMBROISE · Analyzing Microarray Gene Expression Data

McLACHLAN and KRISHNAN · The EM Algorithm and Extensions, *Second Edition*

McLACHLAN and PEEL · Finite Mixture Models

McNEIL · Epidemiological Research Methods

*Now available in a lower priced paperback edition in the Wiley Classics Library.

†Now available in a lower priced paperback edition in the Wiley–Interscience Paperback Series.

MEEKER and ESCOBAR · Statistical Methods for Reliability Data
MEERSCHAERT and SCHEFFLER · Limit Distributions for Sums of Independent Random Vectors: Heavy Tails in Theory and Practice
MICKEY, DUNN, and CLARK · Applied Statistics: Analysis of Variance and Regression, *Third Edition*
* MILLER · Survival Analysis, *Second Edition*
MONTGOMERY, JENNINGS, and KULAHCI · Introduction to Time Series Analysis and Forecasting
MONTGOMERY, PECK, and VINING · Introduction to Linear Regression Analysis, *Fourth Edition*
MORGENTHALER and TUKEY · Configural Polysampling: A Route to Practical Robustness
MUIRHEAD · Aspects of Multivariate Statistical Theory
MULLER and STOYAN · Comparison Methods for Stochastic Models and Risks
MURRAY · X-STAT 2.0 Statistical Experimentation, Design Data Analysis, and Nonlinear Optimization
MURTHY, XIE, and JIANG · Weibull Models
MYERS, MONTGOMERY, and ANDERSON-COOK · Response Surface Methodology: Process and Product Optimization Using Designed Experiments, *Third Edition*
MYERS, MONTGOMERY, VINING, and ROBINSON · Generalized Linear Models. With Applications in Engineering and the Sciences, *Second Edition*
† NELSON · Accelerated Testing, Statistical Models, Test Plans, and Data Analyses
† NELSON · Applied Life Data Analysis
NEWMAN · Biostatistical Methods in Epidemiology
OCHI · Applied Probability and Stochastic Processes in Engineering and Physical Sciences
OKABE, BOOTS, SUGIHARA, and CHIU · Spatial Tesselations: Concepts and Applications of Voronoi Diagrams, *Second Edition*
OLIVER and SMITH · Influence Diagrams, Belief Nets and Decision Analysis
PALTA · Quantitative Methods in Population Health: Extensions of Ordinary Regressions
PANJER · Operational Risk: Modeling and Analytics
PANKRATZ · Forecasting with Dynamic Regression Models
PANKRATZ · Forecasting with Univariate Box-Jenkins Models: Concepts and Cases
* PARZEN · Modern Probability Theory and Its Applications
PEÑA, TIAO, and TSAY · A Course in Time Series Analysis
PIANTADOSI · Clinical Trials: A Methodologic Perspective
PORT · Theoretical Probability for Applications
POURAHMADI · Foundations of Time Series Analysis and Prediction Theory
POWELL · Approximate Dynamic Programming: Solving the Curses of Dimensionality
PRESS · Bayesian Statistics: Principles, Models, and Applications
PRESS · Subjective and Objective Bayesian Statistics, *Second Edition*
PRESS and TANUR · The Subjectivity of Scientists and the Bayesian Approach
PUKELSHEIM · Optimal Experimental Design
PURI, VILAPLANA, and WERTZ · New Perspectives in Theoretical and Applied Statistics
† PUTERMAN · Markov Decision Processes: Discrete Stochastic Dynamic Programming
QIU · Image Processing and Jump Regression Analysis
* RAO · Linear Statistical Inference and Its Applications, *Second Edition*
RAUSAND and HØYLAND · System Reliability Theory: Models, Statistical Methods, and Applications, *Second Edition*
RENCHER · Linear Models in Statistics
RENCHER · Methods of Multivariate Analysis, *Second Edition*
RENCHER · Multivariate Statistical Inference with Applications

*Now available in a lower priced paperback edition in the Wiley Classics Library.
†Now available in a lower priced paperback edition in the Wiley–Interscience Paperback Series.

* RIPLEY · Spatial Statistics
* RIPLEY · Stochastic Simulation
ROBINSON · Practical Strategies for Experimenting
ROHATGI and SALEH · An Introduction to Probability and Statistics, *Second Edition*
ROLSKI, SCHMIDLI, SCHMIDT, and TEUGELS · Stochastic Processes for Insurance and Finance
ROSENBERGER and LACHIN · Randomization in Clinical Trials: Theory and Practice
ROSS · Introduction to Probability and Statistics for Engineers and Scientists
ROSSI, ALLENBY, and McCULLOCH · Bayesian Statistics and Marketing
† ROUSSEEUW and LEROY · Robust Regression and Outlier Detection
* RUBIN · Multiple Imputation for Nonresponse in Surveys
RUBINSTEIN and KROESE · Simulation and the Monte Carlo Method, *Second Edition*
RUBINSTEIN and MELAMED · Modern Simulation and Modeling
RYAN · Modern Engineering Statistics
RYAN · Modern Experimental Design
RYAN · Modern Regression Methods, *Second Edition*
RYAN · Statistical Methods for Quality Improvement, *Second Edition*
SALEH · Theory of Preliminary Test and Stein-Type Estimation with Applications
* SCHEFFE · The Analysis of Variance
SCHIMEK · Smoothing and Regression: Approaches, Computation, and Application
SCHOTT · Matrix Analysis for Statistics, *Second Edition*
SCHOUTENS · Levy Processes in Finance: Pricing Financial Derivatives
SCHUSS · Theory and Applications of Stochastic Differential Equations
SCOTT · Multivariate Density Estimation: Theory, Practice, and Visualization
† SEARLE · Linear Models for Unbalanced Data
† SEARLE · Matrix Algebra Useful for Statistics
† SEARLE, CASELLA, and McCULLOCH · Variance Components
SEARLE and WILLETT · Matrix Algebra for Applied Economics
SEBER · A Matrix Handbook For Statisticians
† SEBER · Multivariate Observations
SEBER and LEE · Linear Regression Analysis, *Second Edition*
† SEBER and WILD · Nonlinear Regression
SENNOTT · Stochastic Dynamic Programming and the Control of Queueing Systems
* SERFLING · Approximation Theorems of Mathematical Statistics
SHAFER and VOVK · Probability and Finance: It's Only a Game!
SILVAPULLE and SEN · Constrained Statistical Inference: Inequality, Order, and Shape Restrictions
SMALL and McLEISH · Hilbert Space Methods in Probability and Statistical Inference
SRIVASTAVA · Methods of Multivariate Statistics
STAPLETON · Linear Statistical Models, *Second Edition*
STAPLETON · Models for Probability and Statistical Inference: Theory and Applications
STAUDTE and SHEATHER · Robust Estimation and Testing
STOYAN, KENDALL, and MECKE · Stochastic Geometry and Its Applications, *Second Edition*
STOYAN and STOYAN · Fractals, Random Shapes and Point Fields: Methods of Geometrical Statistics
STREET and BURGESS · The Construction of Optimal Stated Choice Experiments: Theory and Methods
STYAN · The Collected Papers of T. W. Anderson: 1943–1985
SUTTON, ABRAMS, JONES, SHELDON, and SONG · Methods for Meta-Analysis in Medical Research
TAKEZAWA · Introduction to Nonparametric Regression
TAMHANE · Statistical Analysis of Designed Experiments: Theory and Applications
TANAKA · Time Series Analysis: Nonstationary and Noninvertible Distribution Theory

*Now available in a lower priced paperback edition in the Wiley Classics Library.
†Now available in a lower priced paperback edition in the Wiley–Interscience Paperback Series.

THOMPSON · Empirical Model Building
THOMPSON · Sampling, *Second Edition*
THOMPSON · Simulation: A Modeler's Approach
THOMPSON and SEBER · Adaptive Sampling
THOMPSON, WILLIAMS, and FINDLAY · Models for Investors in Real World Markets
TIAO, BISGAARD, HILL, PEÑA, and STIGLER (editors) · Box on Quality and Discovery: with Design, Control, and Robustness
TIERNEY · LISP-STAT: An Object-Oriented Environment for Statistical Computing and Dynamic Graphics
TSAY · Analysis of Financial Time Series, *Third Edition*
UPTON and FINGLETON · Spatial Data Analysis by Example, Volume II: Categorical and Directional Data
† VAN BELLE · Statistical Rules of Thumb, *Second Edition*
VAN BELLE, FISHER, HEAGERTY, and LUMLEY · Biostatistics: A Methodology for the Health Sciences, *Second Edition*
VESTRUP · The Theory of Measures and Integration
VIDAKOVIC · Statistical Modeling by Wavelets
VINOD and REAGLE · Preparing for the Worst: Incorporating Downside Risk in Stock Market Investments
WALLER and GOTWAY · Applied Spatial Statistics for Public Health Data
WEERAHANDI · Generalized Inference in Repeated Measures: Exact Methods in MANOVA and Mixed Models
WEISBERG · Applied Linear Regression, *Third Edition*
WEISBERG · Bias and Causation: Models and Judgment for Valid Comparisons
WELSH · Aspects of Statistical Inference
WESTFALL and YOUNG · Resampling-Based Multiple Testing: Examples and Methods for *p*-Value Adjustment
WHITTAKER · Graphical Models in Applied Multivariate Statistics
WINKER · Optimization Heuristics in Economics: Applications of Threshold Accepting
WONNACOTT and WONNACOTT · Econometrics, *Second Edition*
WOODING · Planning Pharmaceutical Clinical Trials: Basic Statistical Principles
WOODWORTH · Biostatistics: A Bayesian Introduction
WOOLSON and CLARKE · Statistical Methods for the Analysis of Biomedical Data, *Second Edition*
WU and HAMADA · Experiments: Planning, Analysis, and Parameter Design Optimization, *Second Edition*
WU and ZHANG · Nonparametric Regression Methods for Longitudinal Data Analysis
YANG · The Construction Theory of Denumerable Markov Processes
YOUNG, VALERO-MORA, and FRIENDLY · Visual Statistics: Seeing Data with Dynamic Interactive Graphics
ZACKS · Stage-Wise Adaptive Designs
ZELTERMAN · Discrete Distributions—Applications in the Health Sciences
* ZELLNER · An Introduction to Bayesian Inference in Econometrics
ZHOU, McCLISH, and OBUCHOWSKI · Statistical Methods in Diagnostic Medicine, *Second Edition*

*Now available in a lower priced paperback edition in the Wiley Classics Library.
†Now available in a lower priced paperback edition in the Wiley–Interscience Paperback Series.

Printed and bound by CPI Group (UK) Ltd, Croydon, CR0 4YY
16/09/2026
14963463-0006